BIOLOGY

concepts and applications

SIXTH EDITION

CECIE STARR

Christine A. Evers

Lisa Starr

THOMSON

BROOKS/COLE

Australia • Canada • Mexico • Singapore
Spain • United Kingdom • United States

PUBLISHER Jack C. Carey

VICE-PRESIDENT, EDITOR-IN-CHIEF Michelle Julet

SENIOR DEVELOPMENT EDITOR Peggy Williams

ASSOCIATE DEVELOPMENT EDITOR Suzannah Alexander

EDITORIAL ASSISTANT Chris Ziemba

COPY EDITOR Diana Starr

TECHNOLOGY PROJECT MANAGER Keli Amann

MARKETING MANAGER Ann Caven

MARKETING ASSISTANT Leyla Jowza

ADVERTISING PROJECT MANAGER Nathaniel Michelson

PROJECT MANAGER, EDITORIAL PRODUCTION Andy Marinkovich

PRINT/MEDIA BUYER Karen Hunt

PERMISSIONS EDITOR Joohee Lee

PRODUCTION SERVICE Grace Davidson & Associates

TEXT AND COVER DESIGNER Gary Head

PHOTO RESEARCHERS Myrna Engler, Terri Wright

ILLUSTRATORS Leif Buckley, Seth Gold, Gary Head, Chris Keeney, and Lisa Starr

COVER PRINTER Phoenix Color Corp

COMPOSITOR Lachina Publishing Services

PRINTER R.R. Donnelley/Willard

COVER IMAGE One of the Hawaiian honeycreepers, living representatives of a spectacular adaptive radiation in the Hawaiian Islands. Frans Lanting/Minden Pictures

Printed in the United States of America
2 3 4 5 6 7 09 08 07 06 05

Library of Congress Control Number: 2004114897
Student Edition with InfoTrac College Edition (hardcover): ISBN 0-534-46223-5
Student Edition with InfoTrac College Edition (softcover): ISBN 0-495-18835-2
Annotated Instructor's Edition: ISBN 0-534-46226-X

International Student Edition: ISBN 0-495-03198-4
(Not for sale in the United States)

For more information about our products, contact us at:
Thomson Learning Academic Resource Center

1-800-423-0563
For permission to use material from this text or product, submit a request online at http://www.thomsonrights.com.
Any additional questions about permissions can be submitted by email to thomsonrights@thomson.com.

BOOKS IN THE BROOKS/COLE BIOLOGY SERIES

Biology: The Unity and Diversity of Life, Tenth, Starr/Taggart
Engage Online for Biology: The Unity and Diversity of Life
Biology: Concepts and Applications, Sixth, Starr
Basic Concepts in Biology, Sixth, Starr
Biology Today and Tomorrow, Starr
Biology, Seventh, Solomon/Berg/Martin
Human Biology, Sixth, Starr/McMillan
Biology: A Human Emphasis, Sixth, Starr
Human Physiology, Fifth, Sherwood
Fundamentals of Physiology, Second, Sherwood
Human Physiology, Fourth, Rhoades/Pflanzer

Laboratory Manual for Biology, Fourth, Perry/Morton/Perry
Laboratory Manual for Human Biology, Morton/Perry/Perry
Photo Atlas for Biology, Perry/Morton
Photo Atlas for Anatomy and Physiology, Morton/Perry
Photo Atlas for Botany, Perry/Morton
Virtual Biology Laboratory, Beneski/Waber
Introduction to Cell and Molecular Biology, Wolfe
Molecular and Cellular Biology, Wolfe
Biotechnology: An Introduction, Second, Barnum

Introduction to Microbiology, Third, Ingraham/Ingraham
Microbiology: An Introduction, Batzing
Genetics: The Continuity of Life, Fairbanks/Anderson
Human Heredity, Seventh, Cummings
Current Perspectives in Genetics, Second, Cummings
Gene Discovery Lab, Benfey

Animal Physiology, Sherwood, Kleindorf, Yarcey
Invertebrate Zoology, Seventh, Ruppert/Fox/Barnes
Mammalogy, Fourth, Vaughan/Ryan/Czaplewski
Biology of Fishes, Second, Bond
Vertebrate Dissection, Ninth, Homberger/Walker

Plant Biology, Second, Rost/Barbour/Stocking/Murphy
Plant Physiology, Fourth, Salisbury/Ross
Introductory Botany, Berg

General Ecology, Second, Krohne
Essentials of Ecology, Third, Miller
Terrestrial Ecosystems, Second, Aber/Melillo
Living in the Environment, Fourteenth, Miller
Environmental Science, Tenth, Miller
Sustaining the Earth, Seventh, Miller
Case Studies in Environmental Science, Second, Underwood
Environmental Ethics, Third, Des Jardins
Watersheds 3—Ten Cases in Environmental Ethics, Third, Newton/Dillingham

Problem-Based Learning Activities for General Biology, Allen/Duch
The Pocket Guide to Critical Thinking, Second, Epstein

Thomson Higher Education
10 Davis Drive
Belmont, CA 94002-3098
USA

Asia (including India)
Thomson Learning
5 Shenton Way
#01-01 UIC Building
Singapore 068808

Australia/New Zealand
Thomson Learning Australia
102 Dodds Street
Southbank, Victoria 3006
Australia

Canada
Thomson Nelson
1120 Birchmount Road
Toronto, Ontario M1K 5G4

UK/Europe/Middle East/Africa
Thomson Learning
High Holborn House
50/51 Bedford Row
London WC1R 4LR
United Kingdom

CONTENTS IN BRIEF

INTRODUCTION

1 Invitation to Biology

UNIT I   PRINCIPLES OF CELLULAR LIFE

2 Life's Chemical Basis

3 Molecules of Life

4 How Cells Are Put Together

5 How Cells Work

6 Where It Starts—Photosynthesis

7 How Cells Release Chemical Energy

UNIT II   PRINCIPLES OF INHERITANCE

8 How Cells Reproduce

9 Meiosis and Sexual Reproduction

10 Observing Patterns in Inherited Traits

11 Chromosomes and Human Genetics

12 DNA Structure and Function

13 From DNA to Proteins

14 Controls Over Genes

15 Studying and Manipulating Genomes

UNIT III   PRINCIPLES OF EVOLUTION

16 Processes of Evolution

17 Evolutionary Patterns, Rates, and Trends

18 The Origin and Early Evolution of Life

UNIT IV   EVOLUTION AND BIODIVERSITY

19 Prokaryotes and Viruses

20 The Simplest Eukaryotes—Protists and Fungi

21 Plant Evolution

22 Animal Evolution—The Invertebrates

23 Animal Evolution—The Vertebrates

24 Plants and Animals—Common Challenges

UNIT V   HOW PLANTS WORK

25 Plant Tissues

26 Plant Nutrition and Transport

27 Plant Reproduction and Development

UNIT VI   HOW ANIMALS WORK

28 Animal Tissues and Organ Systems

29 Neural Control

30 Sensory Perception

31 Endocrine Control

32 How Animals Move

33 Circulation

34 Immunity

35 Respiration

36 Digestion and Human Nutrition

37 The Internal Environment

38 Animal Reproduction and Development

UNIT VII   PRINCIPLES OF ECOLOGY

39 Population Ecology

40 Community Structure and Biodiversity

41 Ecosystems

42 The Biosphere

43 Behavioral Ecology

DETAILED CONTENTS

INTRODUCTION

1 Invitation to Biology

IMPACTS, ISSUES *What Am I Doing Here?* 2

1.1 LIFE'S LEVELS OF ORGANIZATION 4
From Small to Smaller 4
From Smaller to Vast 4

1.2 OVERVIEW OF LIFE'S UNITY 6
DNA, The Basis of Inheritance 6
Energy, The Basis of Metabolism 6
Life's Responsiveness to Change 7

1.3 IF SO MUCH UNITY, WHY SO MANY SPECIES? 8

1.4 AN EVOLUTIONARY VIEW OF DIVERSITY 10

1.5 THE NATURE OF BIOLOGICAL INQUIRY 11
Observations, Hypotheses, and Tests 11
About the Word "Theory" 11

1.6 THE POWER OF EXPERIMENTAL TESTS 12
An Assumption of Cause and Effect 12
Example of an Experimental Design 12
Example of a Field Experiment 12
Bias in Reporting Results 13

1.7 THE LIMITS OF SCIENCE 14

UNIT I PRINCIPLES OF CELLULAR LIFE

2 Life's Chemical Basis

IMPACTS, ISSUES *What Are You Worth?* 18

2.1 START WITH ATOMS 20

2.2 FOCUS ON SCIENCE *Radioisotopes* 21

2.3 WHAT HAPPENS WHEN ATOM BONDS WITH ATOM? 22
Electrons and Energy Levels 22
From Atoms to Molecules 22

2.4 BONDS IN BIOLOGICAL MOLECULES 24
Ion Formation and Ionic Bonding 24
Covalent Bonding 24
Hydrogen Bonding 25

2.5 WATER'S LIFE-GIVING PROPERTIES 26
Polarity of the Water Molecule 26
Water's Temperature-Stabilizing Effects 26

Water's Solvent Properties 27
Water's Cohesion 27

2.6 ACIDS AND BASES 28
The pH Scale 28
How Do Acids and Bases Differ? 28
Salts and Water 29
Buffers Against Shifts in pH 29

3 Molecules of Life

IMPACTS, ISSUES *Science or the Supernatural?* 32

3.1 MOLECULES OF LIFE—FROM STRUCTURE TO FUNCTION 34
Carbon's Bonding Behavior 34
Functional Groups 34
How Do Cells Actually Build Organic Compounds? 35

3.2 FOCUS ON THE ENVIRONMENT *Bubble, Bubble, Toil and Trouble* 36

3.3 THE TRULY ABUNDANT CARBOHYDRATES 38
The Simple Sugars 38
Short-Chain Carbohydrates 38
Complex Carbohydrates 38

3.4 GREASY, OILY—MUST BE LIPIDS 40
Fats and Fatty Acids 40
Phospholipids 41
Waxes 41
Cholesterol and Other Sterols 41

3.5 PROTEINS—DIVERSITY IN STRUCTURE AND FUNCTION 42

3.6 WHY IS PROTEIN STRUCTURE SO IMPORTANT? 44
Just One Wrong Amino Acid . . . 44
. . . And You Get Sickle-Shaped Cells! 44
Denaturation 44

3.7 NUCLEOTIDES AND THE NUCLEIC ACIDS 46

4 How Cells Are Put Together

IMPACTS, ISSUES *Where Did Cells Come From?* 50

4.1 SO WHAT IS "A CELL"? 52
Components of All Cells 52
Why Aren't Cells Bigger? 52

4.2 FOCUS ON SCIENCE *How Do We "See" Cells?* 54
The Cell Theory 54
Some Modern Microscopes 54

4.3 ALL LIVING CELLS HAVE MEMBRANES 56

4.4 INTRODUCING PROKARYOTIC CELLS 58

4.5 INTRODUCING EUKARYOTIC CELLS 60

4.6 THE NUCLEUS 61

4.7 THE ENDOMEMBRANE SYSTEM 62
   Endoplasmic Reticulum 62
   Golgi Bodies 62
   Membranous Sacs With Diverse Functions 63

4.8 MITOCHONDRIA AND CHLOROPLASTS 64
   Mitochondria 64
   Chloroplasts 64

4.9 VISUAL SUMMARY OF EUKARYOTIC CELL
   COMPONENTS 65

4.10 THE CYTOSKELETON 66
   Moving Along With Motor Proteins 66
   Cilia, Flagella, and False Feet 67

4.11 CELL SURFACE SPECIALIZATIONS 68
   Eukaryotic Cell Walls 68
   Matrixes Between Animal Cells 69
   Cell Junctions 69

5 How Cells Work

IMPACTS, ISSUES Alcohol, Enzymes, and Your Liver 72

5.1 INPUTS AND OUTPUTS OF ENERGY 74
   The One-Way Flow of Energy 74
   Up and Down the Energy Hills 74
   ATP—The Cell's Energy Currency 75

5.2 INPUTS AND OUTPUTS OF SUBSTANCES 76
   The Nature of Metabolic Reactions 76
   Redox Reactions 76
   Types of Metabolic Pathways 77

5.3 HOW ENZYMES MAKE SUBSTANCES REACT 78

5.4 ENZYMES DON'T WORK IN A VACUUM 80
   Help From Cofactors 80
   Controls Over Enzymes 80
   Effects of Temperature, pH, and Salinity 81

5.5 DIFFUSION, MEMBRANES, AND METABOLISM 82
   What Is a Concentration Gradient? 82
   What Determines Diffusion Rates? 83
   Membrane Crossing Mechanisms 83

5.6 HOW THE MEMBRANE TRANSPORTERS WORK 84
   Passive Transport 84
   Active Transport 85

5.7 WHICH WAY WILL WATER MOVE? 86
   Movement of Water 86
   Effects of Tonicity 86
   Effects of Fluid Pressure 86

5.8 MEMBRANE TRAFFIC TO AND FROM
   THE CELL SURFACE 88
   Endocytosis and Exocytosis 88
   Membrane Cycling 89

6 Where It Starts—Photosynthesis

IMPACTS, ISSUES Pastures of the Seas 92

6.1 SUNLIGHT AS AN ENERGY SOURCE 94
   Properties of Light 94
   Pigments—The Rainbow Catchers 94

6.2 WHAT IS PHOTOSYNTHESIS, AND WHERE
   DOES IT HAPPEN? 96
   Two Stages of Reactions 96
   A Look Inside the Chloroplast 96
   Photosynthesis Changed the Biosphere 96

6.3 LIGHT-DEPENDENT REACTIONS 98
   Transducing the Absorbed Energy 98
   Making ATP and NADPH 98

6.4 FOCUS ON SCIENCE A Case of Controlled
   Energy Release 100

6.5 LIGHT-INDEPENDENT REACTIONS:
   THE SUGAR FACTORY 101

6.6 FOCUS ON THE ENVIRONMENT Different Plants,
   Different Carbon-Fixing Pathways 102

7 How Cells Release Chemical Energy

IMPACTS, ISSUES When Mitochondria Spin Their Wheels 106

7.1 OVERVIEW OF ENERGY-RELEASING
   PATHWAYS 108
   Comparison of the Main Types of
   Energy-Releasing Pathways 108
   Overview of Aerobic Respiration 108

7.2 FIRST STAGE: GLYCOLYSIS 110

7.3 SECOND STAGE OF AEROBIC RESPIRATION 112
   Acetyl–CoA Formation 112
   The Krebs Cycle 112

7.4 THIRD STAGE OF AEROBIC RESPIRATION—
THE BIG ENERGY PAYOFF 114
Electron Transfer Phosphorylation 114
Summing Up: The Energy Harvest 115

7.5 FERMENTATION PATHWAYS 116
Alcoholic Fermentation 116
Lactate Fermentation 117

7.6 ALTERNATIVE ENERGY SOURCES IN THE BODY 118
The Fate of Glucose at Mealtime and
In Between Meals 118
Energy From Fats 118
Energy From Proteins 118

7.7 FOCUS ON EVOLUTION *Perspective on Life* 120

UNIT II   PRINCIPLES OF INHERITANCE

8   How Cells Reproduce

IMPACTS, ISSUES *Henrietta's Immortal Cells* 124

8.1 OVERVIEW OF CELL DIVISION MECHANISMS 126
Mitosis, Meiosis, and the Prokaryotes 126
Key Points About Chromosome Structure 126

8.2 INTRODUCING THE CELL CYCLE 128
The Wonder of Interphase 128
Mitosis and the Chromosome Number 129

8.3 A CLOSER LOOK AT MITOSIS 130

8.4 DIVISION OF THE CYTOPLASM 132
Cleavage in Animals 132
Cell Plate Formation in Plants 133
Appreciate the Process! 133

8.5 FOCUS ON HEALTH *When Control Is Lost* 134
The Cell Cycle Revisited 134
Checkpoint Failure and Tumors 134
Characteristics of Cancer 135

9   Meiosis and Sexual Reproduction

IMPACTS, ISSUES *Why Sex?* 138

9.1 AN EVOLUTIONARY VIEW 140

9.2 OVERVIEW OF MEIOSIS 140
Think "Homologues" 140
Two Divisions, Not One 141

9.3 VISUAL TOUR OF MEIOSIS 142

9.4 HOW MEIOSIS PUTS VARIATION IN TRAITS 144
Crossing Over in Prophase I 144
Metaphase I Alignments 145

9.5 FROM GAMETES TO OFFSPRING 146
Gamete Formation in Plants 146
Gamete Formation in Animals 146
More Shufflings at Fertilization 146

10   Observing Patterns in Inherited Traits

IMPACTS, ISSUES *Menacing Mucus* 150

10.1 MENDEL'S INSIGHT INTO INHERITANCE PATTERNS 152
Mendel's Experimental Approach 152
Terms Used in Modern Genetics 153

10.2 MENDEL'S THEORY OF SEGREGATION 154
Monohybrid Cross Predictions 154
Testcrosses 155

10.3 MENDEL'S THEORY OF INDEPENDENT ASSORTMENT 156

10.4 MORE PATTERNS THAN MENDEL THOUGHT 158
ABO Blood Types—A Case of Codominance 158
Incomplete Dominance 158
Single Genes With a Wide Reach 159
When Products of Gene Pairs Interact 159

10.5 COMPLEX VARIATIONS IN TRAITS 160
Regarding the Unexpected Phenotype 160
Continuous Variation in Populations 160

10.6 GENES AND THE ENVIRONMENT 162

11   Chromosomes and Human Genetics

IMPACTS, ISSUES *Strange Genes, Tortured Minds* 166

11.1 THE CHROMOSOMAL BASIS OF INHERITANCE 168
A Rest Stop on Our Conceptual Road 168
Autosomes and Sex Chromosomes 168
Sex Determination in Humans 169

11.2 FOCUS ON SCIENCE *Karyotyping Made Easy* 170

11.3 IMPACT OF CROSSING OVER ON INHERITANCE 171

11.4 HUMAN GENETIC ANALYSIS 172

11.5 EXAMPLES OF HUMAN INHERITANCE PATTERNS 174
Autosomal Dominant Inheritance 174
Autosomal Recessive Inheritance 174
X-Linked Inheritance 175

11.6 FOCUS ON HEALTH *Too Young, Too Old* 176

11.7 ALTERED CHROMOSOMES 176
The Main Categories of Structural Change 176
Duplication 176
Inversion 176
Deletion 176
Translocation 177
Does Chromosome Structure Evolve? 177

11.8 CHANGES IN THE CHROMOSOME NUMBER *178*
Autosomal Change and Down Syndrome *178*
Changes in the Sex Chromosome Number *179*
Female Sex Chromosome Abnormalities *179*
Male Sex Chromosome Abnormalities *179*

11.9 FOCUS ON HEALTH *Prospects in Human Genetics* *180*
Bioethical Questions *180*
Some of the Options *180*
Genetic Screening *180*
Phenotypic Treatments *180*
Prenatal Diagnosis *181*
Genetic Counseling *181*
Regarding Abortion *181*
Preimplantation Diagnosis *181*

12 DNA Structure and Function

IMPACTS, ISSUES *Goodbye, Dolly* *184*

12.1 THE HUNT FOR FAME, FORTUNE, AND DNA *186*
Early and Puzzling Clues *186*
Confirmation of DNA Function *186*
Enter Watson and Crick *187*

12.2 THE DISCOVERY OF DNA'S STRUCTURE *188*
DNA's Building Blocks *188*
Patterns of Base Pairing *189*

12.3 FOCUS ON BIOETHICS *Rosalind's Story* *190*

12.4 DNA REPLICATION AND REPAIR *190*

12.5 FOCUS ON SCIENCE *Reprogramming DNA To Clone Mammals* *192*

13 From DNA to Proteins

IMPACTS, ISSUES *Ricin and Your Ribosomes* *194*

13.1 HOW IS RNA TRANSCRIBED FROM DNA? *196*
The Nature of Transcription *196*
Finishing Touches on mRNA Transcripts *197*

13.2 DECIPHERING mRNA TRANSCRIPTS *198*
The Genetic Code *198*
The Other RNAs *199*

13.3 TRANSLATING mRNA INTO PROTEIN *200*

13.4 MUTATED GENES AND THEIR PROTEIN PRODUCTS *202*
Common Mutations *202*
How Do Mutations Arise? *203*
The Proof Is in the Protein *203*

14 Controls Over Genes

IMPACTS, ISSUES *Between You and Eternity* *206*

14.1 SOME CONTROL MECHANISMS *208*

14.2 PROKARYOTIC GENE CONTROL *208*
Negative Control of the Lactose Operon *208*
Positive Control of the Lactose Operon *208*

14.3 EUKARYOTIC GENE CONTROL *210*
Same Genes, Different Cell Lineages *210*
When Controls Come Into Play *210*

14.4 EXAMPLES OF GENE CONTROLS *212*
Homeotic Genes and Body Plans *212*
X Chromosome Inactivation *212*

14.5 FOCUS ON SCIENCE *There's a Fly in My Research* *214*
Drosophila! *214*
Clues to Gene Control *214*
Genes and Patterns in Development *215*

15 Studying and Manipulating Genomes

IMPACTS, ISSUES *Golden Rice or Frankenfood?* *218*

15.1 FOCUS ON SCIENCE *Tinkering With the Molecules of Life* *220*
Emergence of Molecular Biology *220*
The Human Genome Project *221*

15.2 A MOLECULAR TOOLKIT *222*
The Scissors: Restriction Enzymes *222*
Cloning Vectors *222*
cDNA Cloning *223*

15.3 HAYSTACKS TO NEEDLES *224*
Isolating Genes *224*
PCR *225*

15.4 FOCUS ON SCIENCE *First Just Fingerprints, Now DNA Fingerprints* *226*

15.5 AUTOMATED DNA SEQUENCING *227*

15.6 PRACTICAL GENETICS *228*
  Designer Plants *228*
  Barnyard Biotech *229*

15.7 **FOCUS ON BIOETHICS** *Weighing Benefits and Risks* *230*
  Who Gets Well? *230*
  Who Gets Enhanced? *230*
  Knockout Cells and Organ Factories *231*
  Regarding "Frankenfood" *231*

15.8 BRAVE NEW WORLD *232*
  Genomics *232*
  DNA Chips *232*

UNIT III   PRINCIPLES OF EVOLUTION

16   Processes of Evolution

IMPACTS, ISSUES *Rise of the Super Rats* *236*

16.1 EARLY BELIEFS, CONFOUNDING DISCOVERIES *238*
  Questions From Biogeography *238*
  Questions From Comparative Morphology *239*
  Questions About Fossils *239*

16.2 A FLURRY OF NEW THEORIES *240*
  Squeezing New Evidence Into Old Beliefs *240*
  Voyage of the *Beagle* *240*

16.3 DARWIN'S THEORY TAKES FORM *242*
  Old Bones and Armadillos *242*
  A Key Insight—Variation in Traits *242*
  Natural Selection Defined *243*

16.4 THE NATURE OF ADAPTATION *244*
  Salt-Tolerant Tomatoes *244*
  No Polar Bears in the Desert *244*
  Adaptation to What? *245*

16.5 INDIVIDUALS DON'T EVOLVE, POPULATIONS DO *246*
  Variation in Populations *246*
  The "Gene Pool" *246*
  Stability and Change in Allele Frequencies *246*

16.6 MUTATIONS REVISITED *248*

16.7 DIRECTIONAL SELECTION *248*
  Pesticide Resistance *248*
  Antibiotic Resistance *248*
  Coat Color in Desert Mice *249*

16.8 SELECTION AGAINST OR IN FAVOR OF EXTREME PHENOTYPES *250*
  Stabilizing Selection *250*
  Disruptive Selection *251*

16.9 MAINTAINING VARIATION IN A POPULATION *252*
  Sexual Selection *252*
  Sickle-Cell Anemia—Lesser of Two Evils? *252*

16.10 GENETIC DRIFT—THE CHANCE CHANGES *254*
  Bottlenecks and the Founder Effect *254*
  Genetic Drift and Inbred Populations *255*

16.11 GENE FLOW *255*

17   Evolutionary Patterns, Rates, and Trends

IMPACTS, ISSUES *Measuring Time* *258*

17.1 FOSSILS—EVIDENCE OF ANCIENT LIFE *260*
  How Do Fossils Form? *260*
  Fossils in Sedimentary Rock Layers *261*
  Interpreting the Fossil Record *261*

17.2 **FOCUS ON SCIENCE** *Dating Pieces of the Puzzle* *262*
  Radiometric Dating *262*
  Placing Fossils in Geologic Time *262*

17.3 EVIDENCE FROM BIOGEOGRAPHY *264*
  An Outrageous Hypothesis *264*
  Drifting Continents, Changing Seas *265*

17.4 EVIDENCE FROM COMPARATIVE MORPHOLOGY *266*
  Morphological Divergence *266*
  Morphological Convergence *267*

17.5 EVIDENCE FROM PATTERNS OF DEVELOPMENT *268*

17.6 EVIDENCE FROM BIOCHEMISTRY *270*
  Protein Comparisons *270*
  Nucleic Acid Comparisons *270*
  Molecular Clocks *271*

17.7 REPRODUCTIVE ISOLATION, MAYBE NEW SPECIES *272*

17.8 THE MAIN MODEL FOR SPECIATION *274*
  The Nature of Allopatric Speciation *274*
  Allopatric Speciation on Archipelagos *274*

17.9 OTHER SPECIATION MODELS *276*
  Sympatric Speciation *276*
    Evidence From Cichlids in Africa *276*
    Polyploidy's Impact *276*
  Parapatric Speciation *277*

17.10 PATTERNS OF SPECIATION AND EXTINCTION *278*
  Branching and Unbranched Evolution *278*
  Evolutionary Trees and Rates of Change *278*
  Adaptive Radiations *278*
  Extinctions—The End of the Line *279*

17.11 HOW CAN WE ORGANIZE THE EVIDENCE? *280*
  Naming, Identifying, and Classifying Species *280*
  What's in a Name? A Cladistics View *281*

17.12 AN EVOLUTIONARY TREE OF LIFE *282*

18 The Origin and Early Evolution of Life

IMPACTS, ISSUES   *Looking for Life in All the Odd Places*   286

18.1   IN THE BEGINNING . . .   288
Conditions on the Early Earth   288
Abiotic Synthesis of Organic Compounds   289

18.2   HOW DID CELLS ORIGINATE?   290
Origin of Agents of Metabolism   290
Origin of the First Plasma Membranes   290
Origin of Self-Replicating Systems   291

18.3   THE FIRST CELLS   292

18.4   FOCUS ON SCIENCE   *Where Did Organelles Come From?*   294
Origin of the Nucleus and ER   294
Origin of Mitochondria and Chloroplasts   294
Evidence of Endosymbiosis   295

18.5   TIMELINE FOR LIFE'S ORIGIN AND EVOLUTION   296

UNIT IV   EVOLUTION AND BIODIVERSITY

19 Prokaryotes and Viruses

IMPACTS, ISSUES   *West Nile Virus Takes Off*   300

19.1   CHARACTERISTICS OF PROKARYOTIC CELLS   302
Body Plans, Shapes, and Sizes   302
Metabolic Diversity   302
Growth and Reproduction   303
Classification   303

19.2   THE BACTERIA   304
Representative Diversity   304
Regarding the "Simple" Bacteria   305

19.3   THE ARCHAEA   306
The Third Domain   306
Here, There, Everywhere   306

19.4   VIRUSES AND VIROIDS   308

19.5   FOCUS ON HEALTH   *Evolution and Infectious Diseases*   310
The Nature of Disease   310
Drug-Resistant Strains   310
Foodborne Diseases and Mad Cows   310

20 The Simplest Eukaryotes— Protists and Fungi

IMPACTS, ISSUES   *Tiny Critters, Big Impacts*   314

20.1   CHARACTERISTICS OF PROTISTS   316

20.2   THE MOST ANCIENT GROUPS   316
Flagellated Protozoans   316
Euglenoids   316
Amoeboid Protozoans   317

20.3   THE ALVEOLATES   318
Ciliated Protozoans   318
Apicomplexans   318
Dinoflagellates   319

20.4   FOCUS ON THE ENVIRONMENT   *Algal Blooms*   320

20.5   THE STRAMENOPILES   320
Oomycotes   320
Chrysophytes   321
Brown Algae   321

20.6   RED ALGAE   322

20.7   GREEN ALGAE   323

20.8   FOCUS ON THE ENVIRONMENT   *Environmental Escape Artists*   324
Consider a Green Alga   324
Consider a Slime Mold   324

20.9   CHARACTERISTICS OF FUNGI   326

20.10   FUNGAL DIVERSITY   328

20.11   FOCUS ON HEALTH   *The Unloved Few*   329

20.12   FUNGAL SYMBIONTS   330
Fungal Endophytes   330
Lichens   330
Mycorrhizae   330
As Fungi Go, So Go the Forests   331

21 Plant Evolution

IMPACTS, ISSUES   *Beginnings, and Endings*   334

21.1   EVOLUTIONARY TRENDS AMONG PLANTS   336
Roots, Stems, and Leaves   336
From Haploid to Diploid Dominance   336
Evolution of Pollen and Seeds   337

21.2 THE BRYOPHYTES—NO VASCULAR TISSUE 338

21.3 SEEDLESS VASCULAR PLANTS 340

Lycophytes 340

Horsetails 341

Ferns 341

21.4 FOCUS ON THE ENVIRONMENT Ancient Carbon Treasures 342

21.5 THE RISE OF SEED-BEARING PLANTS 343

21.6 GYMNOSPERMS—PLANTS WITH NAKED SEEDS 344

21.7 ANGIOSPERMS—THE FLOWERING PLANTS 346

21.8 A GLIMPSE INTO FLOWERING PLANT DIVERSITY 348

21.9 FOCUS ON THE ENVIRONMENT Deforestation Revisited 349

22 Animal Evolution—The Invertebrates

IMPACTS, ISSUES Old Genes, New Drugs 352

22.1 OVERVIEW OF THE ANIMAL KINGDOM 354

General Characteristics 354

Clues in Body Plans 354

22.2 ANIMAL ORIGINS 356

22.3 SPONGES—SUCCESS IN SIMPLICITY 357

22.4 CNIDARIANS—SIMPLE TISSUES, NO ORGANS 358

22.5 FLATWORMS—THE SIMPLEST ORGAN SYSTEMS 360

22.6 ANNELIDS—SEGMENTS GALORE 362

Advantages of Segmentation 362

Annelid Adaptations—A Case Study 362

22.7 THE EVOLUTIONARILY PLIABLE MOLLUSKS 364

Hiding Out, Or Not 364

A Cephalopod Need for Speed 365

22.8 ROUNDWORMS 366

22.9 ARTHROPODS—THE MOST SUCCESSFUL ANIMALS 367

22.10 A LOOK AT THE CRUSTACEANS 368

22.11 SPIDERS AND THEIR RELATIVES 369

22.12 A LOOK AT INSECT DIVERSITY 370

22.13 THE PUZZLING ECHINODERMS 372

23 Animal Evolution—The Vertebrates

IMPACTS, ISSUES Interpreting and Misinterpreting the Past 376

23.1 THE CHORDATE HERITAGE 378

Chordate Characteristics 378

Invertebrate Chordates 378

A Word of Caution 379

23.2 EVOLUTIONARY TRENDS AMONG THE VERTEBRATES 380

Early Craniates 380

Key Innovations 381

Major Vertebrate Groups 381

23.3 JAWED FISHES AND THE RISE OF TETRAPODS 382

Cartilaginous Fishes 382

"Bony Fishes" 383

23.4 AMPHIBIANS—THE FIRST TETRAPODS ON LAND 384

23.5 FOCUS ON THE ENVIRONMENT Vanishing Acts 385

23.6 THE RISE OF AMNIOTES 386

The "Reptiles" 386

The Age of Dinosaurs 386

23.7 EXISTING REPTILIAN GROUPS 388

23.8 BIRDS—THE FEATHERED ONES 390

23.9 MAMMALS 392

23.10 FROM EARLY PRIMATES TO HOMINIDS 394

Trends in Primate Evolution 394

Origins and Early Divergences 395

23.11 EMERGENCE OF EARLY HUMANS 396

23.12 EMERGENCE OF MODERN HUMANS 398

Early Big-Time Walkers 398

Where Did Modern Humans Originate? 398

24 Plants and Animals—Common Challenges

IMPACTS, ISSUES Too Hot To Handle 402

24.1 LEVELS OF STRUCTURAL ORGANIZATION 404

From Cells to Multicelled Organisms 404

Growth Versus Development 404

Structural Organization Has a History 404

The Body's Internal Environment 405

Start Thinking "Homeostasis" 405

24.2 RECURRING CHALLENGES TO SURVIVAL 406

Gas Exchange in Large Bodies 406

Internal Transport in Large Bodies 406

Maintaining the Water–Solute Balance 406

Cell-to-Cell Communication 407

On Variations in Resources and Threats 407

24.3 HOMEOSTASIS IN ANIMALS 408

Negative Feedback 408

Positive Feedback 409

24.4 DOES HOMEOSTASIS OCCUR IN PLANTS? *410*

Walling Off Threats *410*

Sand, Wind, and the Yellow Bush Lupine *410*

About Rhythmic Leaf Folding *411*

24.5 HOW CELLS RECEIVE AND RESPOND TO SIGNALS *412*

UNIT V HOW PLANTS WORK

25 Plant Tissues

IMPACTS, ISSUES *Drought Versus Civilization* *416*

25.1 OVERVIEW OF THE PLANT BODY *418*

Three Plant Tissue Systems *419*

Eudicots and Monocots—Same Tissues, Different Features *419*

25.2 TWO CATEGORIES OF PLANT TISSUES *420*

Simple Tissues *420*

Complex Tissues *421*

25.3 PRIMARY STRUCTURE OF SHOOTS *422*

Behind the Apical Meristem *422*

Inside the Stem *422*

25.4 A CLOSER LOOK AT LEAVES *424*

Leaf Similarities and Differences *424*

Leaf Fine Structure *424*

25.5 PRIMARY STRUCTURE OF ROOTS *426*

25.6 SECONDARY GROWTH—THE WOODY PLANTS *428*

26 Plant Nutrition and Transport

IMPACTS, ISSUES *Leafy Clean-Up Crews* *432*

26.1 PLANT NUTRIENTS AND AVAILABILITY IN SOIL *434*

Properties of Soil *434*

Leaching and Erosion *435*

26.2 THE ROOT OF IT ALL *436*

Specialized Absorptive Structures *436*

How Roots Control Water Uptake *437*

26.3 WATER TRANSPORT THROUGH PLANTS *438*

Transpiration Defined *438*

Cohesion–Tension Theory *438*

26.4 HOW PLANTS CONSERVE WATER *440*

The Water-Conserving Cuticle *440*

Controlled Water Loss at Stomata *440*

26.5 DISTRIBUTION OF ORGANIC COMPOUNDS IN PLANTS *442*

27 Plant Reproduction and Development

IMPACTS, ISSUES *Imperiled Sexual Partners* *446*

27.1 SEXUAL REPRODUCTION IN FLOWERING PLANTS *448*

Regarding the Flowers *448*

Coevolution With Pollinators *449*

27.2 A NEW GENERATION BEGINS *450*

Microspore and Megaspore Formation *450*

Pollination and Fertilization *450*

27.3 FROM ZYGOTES TO SEEDS IN FRUITS *452*

The Embryo Sporophyte *452*

Seed Dispersal *453*

27.4 ASEXUAL REPRODUCTION OF FLOWERING PLANTS *454*

Asexual Reproduction in Nature *454*

Induced Propagation *454*

27.5 PATTERNS OF EARLY GROWTH AND DEVELOPMENT *456*

How Do Seeds Germinate? *456*

Genetic Programs, Environmental Cues *456*

27.6 CELL–CELL COMMUNICATION IN PLANT DEVELOPMENT *458*

Major Types of Plant Hormones *458*

Other Signaling Molecules *459*

27.7 FOCUS ON THE ENVIRONMENT *Herbicides!* *460*

27.8 ABOUT THOSE TROPISMS *460*

Responding to Light *460*

Responding to Gravity *461*

Responding to Contact *461*

27.9 CONTROL OF FLORAL DEVELOPMENT *462*

How Do Plants Know When To Flower? *462*

Genes That Control Flower Formation *462*

27.10 LIFE CYCLES END, AND TURN AGAIN *464*

UNIT VI   HOW ANIMALS WORK

28   Animal Tissues and Organ Systems

IMPACTS, ISSUES   *Open or Close the Stem Cell Factories?* 468

28.1 EPITHELIAL TISSUE 470
What Is Epithelium? 470
Glandular Epithelium 470
Cell Junctions 471

28.2 CONNECTIVE TISSUES 472
Soft Connective Tissues 472
Specialized Connective Tissues 472

28.3 MUSCLE TISSUES 474

28.4 NERVOUS TISSUE 475

28.5 FOCUS ON SCIENCE *Tissue Engineering* 475

28.6 OVERVIEW OF MAJOR ORGAN SYSTEMS 476

28.7 FOCUS ON HEALTH *Human Skin—Example of an Organ System* 478
The Vitamin Connection 479
Of Suntans and Shoe-Leather Skin 479

29   Neural Control

IMPACTS, ISSUES   *In Pursuit of Ecstasy* 482

29.1 NEURONS—THE GREAT COMMUNICATORS 484
Neurons and Their Functional Zones 484
A Neuron's Plasma Membrane 485

29.2 HOW IS AN ACTION POTENTIAL PROPAGATED? 486

29.3 HOW NEURONS SEND MESSAGES TO OTHER CELLS 488
Chemical Synapses 488
Synaptic Integration 489

29.4 PATHS OF INFORMATION FLOW 490
Blocks and Cables of Neurons 490
Reflex Arcs 490

29.5 TYPES OF NERVOUS SYSTEMS 492
Regarding the Nerve Net 492
On the Importance of Having a Head 492
Evolution of the Spinal Cord and Brain 493

29.6 WHAT ARE THE MAJOR EXPRESSWAYS? 494
Peripheral Nervous System 494
Spinal Cord 495

29.7 THE VERTEBRATE BRAIN 496
The Brain's Subdivisions 496
Protection at the Blood–Brain Barrier 496
The Human Brain 497

29.8 CONSIDER THE CEREBRAL CORTEX 498
Functional Areas of the Cortex 498
Connections With the Limbic System 499
Making Memories 499

29.9 FOCUS ON HEALTH *Drugging the Brain* 500
Roots of Addiction 500
Effects of Psychoactive Drugs 500

30   Sensory Perception

IMPACTS, ISSUES   *A Whale of a Dilemma* 504

30.1 OVERVIEW OF SENSORY PATHWAYS 506

30.2 SENSING TOUCH, PRESSURE, TEMPERATURE, AND PAIN 507

30.3 SAMPLING THE CHEMICAL WORLD 508

30.4 BALANCING ACTS 509

30.5 MAKING SENSE OF SOUNDS 510
Properties of Sound 510
Evolution of Vertebrate Hearing 510

30.6 DO YOU SEE WHAT I SEE? 512
Requirements for Vision 512
The Human Eye 512

30.7 A CLOSER LOOK AT THE RETINA 514

30.8 FOCUS ON HEALTH *Visual Disorders* 515
So Near, So Far 515
Color Blindness 515
Age-Related Disorders 515

31   Endocrine Control

IMPACTS, ISSUES   *Hormones in the Balance* 518

31.1 INTRODUCING THE VERTEBRATE ENDOCRINE SYSTEM 520
Comparison of Signaling Molecules 520
Overview of the Endocrine System 520

31.2 THE NATURE OF HORMONE ACTION 522
Signal Reception, Transduction, and Response 522
Hormone Signaling Mechanisms 522

31.3 THE HYPOTHALAMUS AND PITUITARY GLAND 524
Posterior Lobe Secretions 524
Anterior Lobe Secretions 524
Abnormal Pituitary Outputs 525

31.4 FEEDBACK LOOPS IN HORMONE SECRETIONS 526
Negative Feedback and the Adrenal Cortex 526
Local Feedback and the Adrenal Medulla 527
Skewed Feedback From the Thyroid 527

31.5 DIRECT RESPONSES TO CHEMICAL CHANGE 528

Secretions From Parathyroid Glands 528

Effects of Local Signaling Molecules 528

Secretions From Pancreatic Islets 528

31.6 HORMONES AND THE ENVIRONMENT 530

Daylength and the Pineal Gland 530

Thyroid Function and Frog Deformities 530

Chemical Soups and Sperm Counts 530

Comparative Look at Invertebrates 531

32 How Animals Move

IMPACTS, ISSUES *Pumping Up Muscles* 536

32.1 WHAT IS A SKELETON? 538

32.2 ZOOMING IN ON BONES AND JOINTS 540

Bone Structure and Function 540

Bone Formation and Remodeling 540

Where Bones Meet—Skeletal Joints 541

32.3 SKELETAL–MUSCULAR SYSTEMS 542

32.4 HOW DOES SKELETAL MUSCLE CONTRACT? 544

Sliding-Filament Model for Contraction 545

32.5 ENERGY FOR CONTRACTION 546

32.6 PROPERTIES OF WHOLE MUSCLES 546

Types of Contractions 546

What Is Muscle Fatigue? 547

What Are Muscular Dystrophies? 547

Muscles, Exercise, and Aging 547

32.7 FOCUS ON HEALTH *Oh, Clostridium!* 548

33 Circulation

IMPACTS, ISSUES *And Then My Heart Stood Still* 550

33.1 THE NATURE OF BLOOD CIRCULATION 552

Evolution of Vertebrate Circulation 552

Links With the Lymphatic System 553

33.2 CHARACTERISTICS OF BLOOD 554

Functions of Blood 554

Blood Composition and Volume 554

33.3 FOCUS ON HEALTH *Blood Disorders* 556

33.4 BLOOD TYPING 556

ABO Blood Typing 557

Rh Blood Typing 557

33.5 HUMAN CARDIOVASCULAR SYSTEM 558

33.6 THE HEART IS A LONELY PUMPER 560

Heart Structure 560

How Does Cardiac Muscle Contract? 561

33.7 PRESSURE, TRANSPORT, AND FLOW DISTRIBUTION 562

Rapid Transport in Arteries 562

Distributing Blood Flow 562

Controlling Blood Pressure 563

33.8 DIFFUSION AT CAPILLARIES, THEN BACK TO THE HEART 564

Capillary Function 564

Venous Pressure 565

33.9 FOCUS ON HEALTH *Cardiovascular Disorders* 566

Good Clot, Bad Clot 566

A Silent Killer—Hypertension 566

Atherosclerosis 566

Rhythms and Arrythmias 567

Risk Factors 567

33.10 CONNECTIONS WITH THE LYMPHATIC SYSTEM 568

Lymph Vascular System 568

Lymphoid Organs and Tissues 569

34 Immunity

IMPACTS, ISSUES *The Face of AIDS* 572

34.1 OVERVIEW OF THE BODY'S DEFENSES 574

Intact Skin and Mucous Membranes 574

Nonspecific Internal Defenses 574

Evolution of Adaptive Immunity 574

34.2 THE FIRST LINE OF DEFENSE 575

34.3 SECOND LINE OF DEFENSE—NONSPECIFIC INTERNAL RESPONSES 576

Inflammatory Response 576

Antimicrobial Proteins 577

Fever 577

34.4 THIRD LINE OF DEFENSE—ADAPTIVE IMMUNITY 578

The Defenders and Their Main Targets 578

Control of Immune Responses 579

34.5 HOW LYMPHOCYTES FORM AND DO BATTLE 580

B Cells, T Cells, and Antigen Receptors 580

Where Are the Battlegrounds? 580

34.6 ANTIBODY-MEDIATED RESPONSE 582

Antibody Function 582

Classes of Immunoglobulins 582

34.7 CELL-MEDIATED RESPONSE 584

34.8 FOCUS ON SCIENCE *Immunotherapy* 585

34.9 DEFENSES ENHANCED OR COMPROMISED 586

Immunization 586

Allergies 586

Autoimmune Disorders 587

Deficient Immune Responses 587

34.10 FOCUS ON HEALTH *AIDS Revisited—Immunity Lost* 588

HIV Infection—A Titanic Struggle Begins 588

How Is HIV Transmitted? 588

What About Drugs and Vaccines? 589

35   Respiration

IMPACTS, ISSUES *Up in Smoke* 592

35.1 THE NATURE OF RESPIRATION 594

The Basis of Gas Exchange 594

Factors Influencing Gas Exchange 594

35.2 INVERTEBRATE RESPIRATION 596

35.3 FOCUS ON THE ENVIRONMENT *Gasping for Oxygen* 597

35.4 VERTEBRATE RESPIRATION 598

Gills of Fishes and Amphibians 598

Evolution of Paired Lungs 598

35.5 HUMAN RESPIRATORY SYSTEM 600

The System's Many Functions 600

From Airways Into the Lungs 601

35.6 THE RESPIRATORY CYCLE 602

Cyclic Reversals in Air Pressure Gradients 602

Living the High Life 603

35.7 GAS EXCHANGE AND TRANSPORT 604

Exchanges at the Respiratory Membrane 604

Oxygen Transport 604

Carbon Dioxide Transport 604

Balancing Air and Blood Flow Rates 605

35.8 FOCUS ON HEALTH *When the Lungs Break Down* 606

Bronchitis and Emphysema 606

Smoking's Impact 607

36   Digestion and Human Nutrition

IMPACTS, ISSUES *Hips and Hunger* 610

36.1 THE NATURE OF DIGESTIVE SYSTEMS 612

Incomplete and Complete Systems 612

Correlations With Feeding Behavior 613

36.2 THE HUMAN DIGESTIVE SYSTEM 614

Into the Mouth, Down the Tube 615

36.3 DIGESTION IN THE STOMACH AND SMALL INTESTINE 616

The Stomach 616

The Small Intestine 616

Controls Over Digestion 617

36.4 ABSORPTION FROM THE SMALL INTESTINE 618

Structure of the Intestinal Lining 618

What Are the Absorption Mechanisms? 618

36.5 WHAT HAPPENS TO ABSORBED ORGANIC COMPOUNDS? 620

36.6 THE LARGE INTESTINE 621

Colon Function 621

Colon Malfunction 621

36.7 HUMAN NUTRITIONAL REQUIREMENTS 622

A Carbohydrate–Insulin Connection 622

Good Fat, Bad Fat 623

Body-Building Proteins 623

Alternative Diets 623

36.8 VITAMINS AND MINERALS 624

36.9 FOCUS ON SCIENCE *Weighty Questions, Tantalizing Answers* 626

What Is a Good Body Weight? 626

Genes, Hormones, and Obesity 627

37   The Internal Environment

IMPACTS, ISSUES *Truth in a Test Tube* 630

37.1 MAINTAINING THE EXTRACELLULAR FLUID 632

Challenges in Water 632

Challenges on Land 632

37.2 THE HUMAN URINARY SYSTEM 634

Components of the System 634

Nephrons—Functional Units of Kidneys 634

37.3 URINE FORMATION 636

Three Related Processes 636

Reabsorption of Water and Sodium 636

37.4 FOCUS ON HEALTH *When Kidneys Break Down* 638

37.5 ACID–BASE BALANCE 638

37.6 MAINTAINING THE CORE TEMPERATURE *639*
Heat Gains and Losses *639*
Endotherm? Ectotherm? Heterotherm? *639*

37.7 TEMPERATURE REGULATION IN MAMMALS *640*
Responses to Heat Stress *640*
Responses to Cold Stress *640*

38 Animal Reproduction and Development

IMPACTS, ISSUES *Mind-Boggling Births 644*

38.1 REFLECTIONS ON SEXUAL REPRODUCTION *646*
Sexual Versus Asexual Reproduction *646*
Costs of Sexual Reproduction *646*

38.2 STAGES OF REPRODUCTION AND DEVELOPMENT *648*

38.3 EARLY MARCHING ORDERS *650*
Information in the Egg *650*
Cleavage—The Start of Multicellularity *650*
Cleavage Patterns *651*

38.4 HOW DO SPECIALIZED TISSUES AND ORGANS FORM? *652*
Cell Differentiation *652*
Morphogenesis *653*

38.5 PATTERN FORMATION *654*
Embryonic Induction *654*
A Theory of Pattern Formation *654*
Evolutionary Constraints on Development *655*

38.6 REPRODUCTIVE SYSTEM OF HUMAN MALES *656*
When Gonads Form and Become Active *656*
Structure and Function of the Reproductive System *656*
Cancers of the Prostate and Testes *657*

38.7 SPERM FORMATION *658*

38.8 REPRODUCTIVE SYSTEM OF HUMAN FEMALES *660*
Components of the System *660*
Overview of the Menstrual Cycle *660*

38.9 PREPARATIONS FOR PREGNANCY *662*
Cyclic Changes in the Ovary *662*
Cyclic Changes in the Uterus *663*

38.10 VISUAL SUMMARY OF THE MENSTRUAL CYCLE *664*

38.11 PREGNANCY HAPPENS *665*
Sexual Intercourse *665*
Fertilization *665*

38.12 FOCUS ON BIOETHICS *Control of Human Fertility 666*
The Issue *666*
Some Options *666*
Seeking or Ending Pregnancy *667*

38.13 FOCUS ON HEALTH *Sexually Transmitted Diseases 668*

38.14 FORMATION OF THE EARLY EMBRYO *670*
Cleavage and Implantation *670*
Extraembryonic Membranes *670*

38.15 EMERGENCE OF THE VERTEBRATE BODY PLAN *672*

38.16 WHY IS THE PLACENTA SO IMPORTANT? *673*

38.17 EMERGENCE OF DISTINCTLY HUMAN FEATURES *674*

38.18 FOCUS ON HEALTH *Mother as Provider, Protector, Potential Threat 676*
Nutritional Considerations *676*
Infectious Diseases *676*
Alcohol, Tobacco, and Other Drugs *677*

38.19 FROM BIRTH ONWARD *678*
Giving Birth *678*
Nourishing the Newborn *679*
Postnatal Development *679*

38.20 FOCUS ON SCIENCE *Why Do We Age and Die? 680*
Programmed Life Span Hypothesis *680*
Cumulative Assaults Hypothesis *680*

UNIT VII   PRINCIPLES OF ECOLOGY

39 Population Ecology

IMPACTS, ISSUES *The Human Touch 684*

39.1 CHARACTERISTICS OF POPULATIONS *686*

39.2 FOCUS ON SCIENCE *Elusive Heads to Count 687*

39.3 POPULATION SIZE AND EXPONENTIAL GROWTH *688*
Gains and Losses in Population Size *688*
From Zero to Exponential Growth *688*
What Is the Biotic Potential? *689*

39.4 LIMITS ON THE GROWTH OF POPULATIONS *690*

What Are the Limiting Factors? *690*

Carrying Capacity and Logistic Growth *690*

Density-Independent Factors *691*

39.5 LIFE HISTORY PATTERNS *692*

Life Tables *692*

Patterns of Survival and Reproduction *692*

39.6 FOCUS ON SCIENCE *Natural Selection and Life Histories* *694*

39.7 HUMAN POPULATION GROWTH *696*

39.8 FERTILITY RATES AND AGE STRUCTURE *698*

39.9 POPULATION GROWTH AND ECONOMIC EFFECTS *700*

Demographic Transitions *700*

A Question of Immigration Policies *700*

A Question of Resource Consumption *700*

Impacts of No Growth *701*

40 Community Structure and Biodiversity

IMPACTS, ISSUES *Fire Ants in the Pants* *704*

40.1 WHICH FACTORS SHAPE COMMUNITY STRUCTURE? *706*

The Niche *706*

Categories of Species Interactions *706*

40.2 MUTUALISM *707*

40.3 COMPETITIVE INTERACTIONS *708*

Competitive Exclusion *708*

Resource Partitioning *709*

40.4 PREDATOR–PREY INTERACTIONS *710*

Coevolution of Predators and Prey *710*

Models for Predator–Prey Interactions *710*

The Canadian Lynx and Snowshoe Hare *710*

40.5 FOCUS ON EVOLUTION *An Evolutionary Arms Race* *712*

Adaptations of Prey *712*

Adaptive Responses of Predators *713*

40.6 PARASITE–HOST INTERACTIONS *714*

Parasites and Parasitoids *714*

Uses as Biological Controls *715*

40.7 FOCUS ON EVOLUTION *Cowbird Chutzpah* *715*

40.8 FORCES CONTRIBUTING TO COMMUNITY STABILITY *716*

A Succession Model *716*

The Climax Pattern Model *717*

Cyclic, Nondirectional Changes *717*

40.9 FORCES CONTRIBUTING TO COMMUNITY INSTABILITY *718*

The Role of Keystone Species *718*

Species Introductions Tip the Balance *719*

40.10 FOCUS ON THE ENVIRONMENT *Exotic Invaders* *720*

The Plants That Ate Georgia *720*

The Alga Triumphant *720*

The Rabbits That Ate Australia *721*

40.11 BIOGEOGRAPHIC PATTERNS *722*

Mainland and Marine Patterns *722*

Island Patterns *723*

40.12 THREATS TO BIODIVERSITY *724*

On the Newly Endangered Species *724*

Habitat Losses and Fragmentation *724*

Conservation Biology *725*

40.13 SUSTAINING BIODIVERSITY *726*

Identifying Areas At Risk *726*

Economic Factors and Sustainable Development *726*

41 Ecosystems

IMPACTS, ISSUES *Bye-Bye, Blue Bayou* *730*

41.1 THE NATURE OF ECOSYSTEMS *732*

Overview of the Participants *732*

Structure of Ecosystems *733*

41.2 THE NATURE OF FOOD WEBS *734*

41.3 FOCUS ON THE ENVIRONMENT *DDT in Food Webs* *736*

41.4 STUDYING ENERGY FLOW THROUGH ECOSYSTEMS *737*

What Is Primary Productivity? *737*

Ecological Pyramids *737*

41.5 FOCUS ON SCIENCE *Energy Flow Through Silver Springs* *738*

41.6 OVERVIEW OF BIOGEOCHEMICAL CYCLES *739*

41.7 GLOBAL CYCLING OF WATER *740*

The Hydrologic Cycle *740*

The Water Crisis *740*

41.8 CARBON CYCLE *742*

41.9 FOCUS ON THE ENVIRONMENT *Greenhouse Gases, Global Warming* *744*

41.10 NITROGEN CYCLE *746*

The Cycling Processes *746*

Human Impact on the Nitrogen Cycle *747*

41.11 PHOSPHORUS CYCLE *748*

42 The Biosphere

IMPACTS, ISSUES   *Surfers, Seals, and the Sea*   752

42.1 GLOBAL AIR CIRCULATION PATTERNS   754
Climate and Temperature Zones   754
Harnessing the Sun and Wind   755

42.2 FOCUS ON THE ENVIRONMENT   *Air Circulation Patterns and Human Affairs*   756
A Fence of Wind and Ozone Thinning   756
No Wind, Lots of Pollutants, and Smog   756
Winds and Acid Rain   757

42.3 THE OCEAN, LANDFORMS, AND CLIMATES   758
Ocean Currents and Their Effects   758
Rain Shadows and Monsoons   759

42.4 REALMS OF BIODIVERSITY   760

42.5 MOISTURE-CHALLENGED BIOMES   762
Deserts, Natural and Man-Made   762
Dry Shrublands, Dry Woodlands, and Grasslands   762

42.6 MORE RAIN, MORE TREES   764
Broadleaf Forests   764
Coniferous Forests   764

42.7 BRIEF SUMMERS AND LONG, ICY WINTERS   766

42.8 DON'T FORGET THE SOILS   767

42.9 FRESHWATER PROVINCES   768
Lake Ecosystems   768
Stream Ecosystems   769

42.10 LIFE AT LAND'S END   770
Wetlands and the Intertidal Zone   770
Rocky and Sandy Coastlines   771
Coral Reefs   771

42.11 FOCUS ON THE ENVIRONMENT *Coral Bleaching*   772

42.12 THE OPEN OCEAN   772
Surprising Diversity   772
Upwelling and Downwelling   772

42.13 FOCUS ON SCIENCE   *Applying Knowledge of the Biosphere*   774

43 Behavioral Ecology

IMPACTS, ISSUES   *My Pheromones Made Me Do It*   778

43.1 BEHAVIOR'S HERITABLE BASIS   780
Genes and Behavior   780
Hormones and Behavior   780
Instinctive Behavior   781

43.2 LEARNED BEHAVIOR   782

43.3 THE ADAPTIVE VALUE OF BEHAVIOR   783

43.4 COMMUNICATION SIGNALS   784
The Nature of Communication Signals   784
Communication Displays   784
Illegitimate Signalers and Receivers   785

43.5 MATES, OFFSPRING, AND REPRODUCTIVE SUCCESS   786
Sexual Selection and Mating Behavior   786
Parental Care   787

43.6 COSTS AND BENEFITS OF SOCIAL GROUPS   788
Cooperative Predator Avoidance   788
The Selfish Herd   788
Cooperative Hunting   788
Dominance Hierarchies   789
Regarding the Costs   789

43.7 WHY SACRIFICE YOURSELF?   790
Social Insects   790
Social Mole Rats   790
Indirect Selection for Altruism   791

43.8 AN EVOLUTIONARY VIEW OF HUMAN SOCIAL BEHAVIOR   792
Human Pheromones   792
Hormones and Bonding Behavior   792
Evolutionary Questions   792

EPILOGUE   *Biological Principles and the Human Imperative*   795

Appendix I   Classification System
Appendix II   Units of Measure
Appendix III   Answers to Self-Quizzes
Appendix IV   Answers to Genetics Problems
Appendix V   Closer Look at Some Major Metabolic Pathways
Appendix VI   The Amino Acids
Appendix VII   Periodic Table of the Elements
Appendix VIII   Annotations to A Journal Article

PREFACE

What a triumph! Today, biologists offer us a sweeping story of life's unity and diversity—how living things are built, how they work, how they got that way, and where they came from. In that story are fabulous clues to human health, reproduction, our connections with everything else on Earth, even our collective survival.

With this book, we offer a coherent introduction to that story. We weave examples of problem solving and experiments through its pages to show the power of thinking critically about the natural world. We highlight core concepts, current understandings, and research trends for major fields of biological inquiry. We explain the structure and function of a broad sampling of organisms in enough detail so students can develop a working vocabulary about life's parts and processes. We selectively introduce applications that may help students sense the value of learning the core concepts.

Teachers of many millions of students already know about the effectiveness of an approach that integrates current topics with take-home lessons. They recognize, as we do, that nearly all students will stick with lively, relevant, easy-to-follow writing. Students can "get" the big picture of life—and become confident enough to think critically about the past, present, and future on their own. This is biology's greatest gift.

Make It Relevant

Biological inquiry now reaches into our lives in many direct and indirect ways. *What students learn today will have impact on how they make decisions tomorrow*—in the voting booth as well as in their personal lives.

HOW WOULD YOU VOTE? Each chapter opens with a current issue that relates to its content. For instance, protein synthesis chapter starts with how a bioweapon in the news—ricin—kills by unraveling ribosomes. The microevolution chapter starts with how a selective agent —a rodenticide—favors "super rats." Recognizing that the current generation is at ease with music videos, we created a custom *videoclip* about each introductory issue. These unique videos, in the student CD *and* instructor's Media Manager, are dynamic lecture launchers.

We ask students, *How would you vote on research or on an application related to such issues?* We return to the question in more depth at the end of the chapter. All over the country, students will vote *on-line* and access campuswide, statewide, and nationwide tallies. This interactive approach to issues reinforces the premise that individual actions can make a difference.

Make It Easy To Follow

THE BIG PICTURE As students start each chapter, they get *"the big picture"*—a mini-overview on-page and in video format. Each big-picture concept has a boxed green title. We repeat these titles at the top of appropriate text pages as ongoing reminders of how core concepts fit together.

HIGHLIGHTED KEY POINTS Students often complain that their textbooks are dry and boring. We opt for a conversational style that holds their attention, without sacrificing clarity. We encourage them to focus on the flow of content rather than hunting for bits they may be tested on. How? By highlighting the key points for them, in blue boxes at the end of each section. We keep them on target with a *running in-text summary*, and with a *section-by-section summary* at the chapter's end.

READ-ME-FIRST ART *Read-me-first diagrams* help students who are comfortable with visual learning as well as with reading. For a preview of where their reading will take them, students can first walk step by step through the art in each text section. We offer the same art in narrated, animated form on the Media Manager lecture tool and the student CD. Repeated exposure to visual material reinforces understanding, engages multiple learning styles, and allows self-paced review of challenging biological concepts.

Make It Brief, With Clear Explanations

To keep book length manageable, we were selective about which topics to include while allocating enough space to explain those topics clearly. If something is worth reading about, why reduce it to a factoid? Factoids invite mind-numbing memorization, a study habit that will not help students develop their capacity to think critically about the world and their place in it.

For instance, you can safely bet that most nonmajors simply do not want to memorize each catalytic step of crassulacean acid metabolism. They *do* want to learn about the biological basis of sex, and many female students want to know what will be going on inside them if and when they get pregnant. Good explanations can help them make their own informed decisions on many biology-related issues, including STDs, fertility drugs, prenatal diagnoses of genetic disorders, gene therapies, and abortions. Over the years, a number of student readers have written to tell us they devoured such material even when it was not assigned.

Our choices for which topics to condense, expand, or delete were not arbitrary. They reflect three decades of feedback from teachers throughout the world.

Offer Easy-to-Use Media Tools

Each chapter ends with a *Media Menu*. The menu directs students to the free CD, which starts with an issues-oriented *video* and an animation of *the big picture*. The menu also lists all *read-me-first animations*, and most of the additional animations and interactions.

Our readers have free access to an exclusive online database—*InfoTrac College Edition*, a full-text library of more than 4,000 periodicals. The Media Menu also lists sample articles, examples of *web sites* relevant to that chapter, and an expanded *How-Would-You-Vote* question.

A free *BiologyNow* CD-ROM in each book is a portal to all CD- and web-based learning tools. It is an easy, integrated way for students to do homework and assess how well they understand it. After reading a chapter, students go to http://biology.brookscole.com/starr6, enter the access code packaged with their book, and take a *diagnostic pre-test*. Based on their individual answers, a personalized learning plan directs them to text sections, art, and animations that explain questions they answer incorrectly. Students can e-mail pre-test and *post-test* results to instructors. Ideal for homework assignments, the self-grading pre-tests also can flow directly into WebCT or Blackboard gradebooks when the instructor uses the WebTutor™ ToolBox course management tool that is offered free with this book.

vMentor is a free online biology tutorial service available from 6 AM to 12 PM Monday through Saturday. For review, *interactive flashcards* define all of the book's boldface terms and have audio pronunciation guides.

InfoTrac and an annotated list of web sites make research on the pros and cons of an issue a snap. After students research the *How-Would-You-Vote* question, they can cast their vote and also see at once how others have voted. They can email information about their research into the issue, and their vote, to instructors.

MEDIA TOOLS

Media Menu End-of-chapter menu listing art on the CD-Rom, InfoTrac links, web site links, how-would-you vote question instructor, by section.

Resources Integrator All of the media for the book integrated for instructors, by section.

BiologyNow™ Diagnoses which topics students have not yet mastered and creates customized learning plans to focus their study and review.

How Would You Vote? Invitation to research and vote on a controversial question.

vMentor Free on-line help with homework, through two-way voice communication and through a computer whiteboard. Restrictions apply.*

InfoTrac College Edition Exclusive online, searchable database of periodical, plus exercises that can help guide student research.

Interactive Flashcards Definitions and audio guides to pronouncing all boldface terms in the text.

Web Sites Section-by-section annotated lists to the best biology on the Web. Internet exercises to guide research.

ACKNOWLEDGMENTS

Thanks to the advisors listed below for their ongoing impact on the book's content. John Jackson and Walt Judd both deserve special recognition for their deep commitment to excellence in education. This latest edition still reflects the influential contributions of the instructors listed on the following page, who helped shape our thinking. *Impacts/Issues* sections, custom videos, *the big picture* overviews—such features are responses to their insights from the classroom. Lisa Starr and Christine Evers are invaluable partners in research, writing, creating art, and page make-up.

Starting with Susan Badger, Thomson Learning proved why it's one of the world's foremost publishers; Sean Wakely, Michelle Julet, Kathie Head, thank you again. Keli Amann created a fine web site and managed the CD production. Peggy Williams brought tenacity, intelligence, and humor to the project. Both Andy Marinkovich and Grace Davidson took my world-class compulsivity in stride. Gary Head created functional designs and great graphics for the book; Steve Bolinger did so for the media tools. Star MacKenzie, Ann Caven, Karen Hunt, Suzannah Alexander, Diana Starr, Chris Ziemba, Myrna Engler—the list goes on. Yet no listing conveys how this team interacted to create something extraordinary. And thank you, Jack Carey, for being the first to identify the need for features, including student voting, that can further biology education.

CECIE STARR, *November 2004*

MAJOR ADVISORS AND REVIEWERS

JOHN D. JACKSON *North Hennepin Community College*
WALTER JUDD *University of Florida*

JOHN ALCOCK *Arizona State University*
CHARLOTTE BORGESON *University of Nevada*
DEBORAH C. CLARK *Middle Tennessee University*
MELANIE DeVORE *Georgia College and State University*
DANIEL FAIRBANKS *Brigham Young University*
TOM GARRISON *Orange Coast College*
DAVID GOODIN *The Scripps Research Institute*
PAUL E. HERTZ *Barnard College*
TIMOTHY JOHNSTON *Murray State University*
EUGENE KOZLOFF *University of Washington*
ELIZABETH LANDECKER–MOORE *Rowan University*
KAREN MESSLEY *Rock Valley College*
THOMAS L. ROST *University of California, Davis*
LAURALEE SHERWOOD *West Virginia University*
E. WILLIAM WISCHUSEN *Louisiana State University*
STEPHEN L. WOLFE *University of California, Davis*

CONTRIBUTORS OF INFLUENTIAL REVIEWS AND CLASS TESTS

ADAMS, DARYL   *Minnesota State University, Mankato*
ANDERSON, DENNIS   *Oklahoma City Community College*
BENDER, KRISTEN   *California State University, Long Beach*
BOGGS, LISA   *Southwestern Oklahoma State University*
BORGESON, CHARLOTTE   *University of Nevada*
BOWER, SUSAN   *Pasadena City College*
BOYD, KIMBERLY   *Cabrini College*
BRICKMAN, PEGGY   *University of Georgia*
BROWN, EVERT   *Casper College*
BRYAN, DAVID W.   *Cincinnati State College*
BURNETT, STEPHEN   *Clayton College*
BUSS, WARREN   *University of Northern Colorado*
CARTWRIGHT, PAULYN   *University of Kansas*
CASE, TED   *University of California, San Diego*
COLAVITO, MARY   *Santa Monica College*
COOK, JERRY L.   *Sam Houston State University*
DAVIS, JERRY   *University of Wisconsin, LaCrosse*
DENGLER, NANCY   *University of California, Davis*
DESAIX, JEAN   *University of North Carolina*
DIBARTOLOMEIS, SUSAN   *Millersville University of Pennsylvania*
DIEHL, FRED   *University of Virginia*
DONALD-WHITNEY, CATHY   *Collin County Community College*
DUWEL, PHILIP   *University of South Carolina, Columbia*
EAKIN, DAVID   *Eastern Kentucky University*
EBBS, STEPHEN   *Southern Illinois University*
EDLIN, GORDON   *University of Hawaii, Manoa*
ENDLER, JOHN   *University of California, Santa Barbara*
ERWIN, CINDY   *City College of San Francisco*
FOX, P. MICHAEL   *SUNY College at Brockport*
FOREMAN, KATHERINE   *Moraine Valley Community College*
GIBLIN, TARA   *Stephens College*
GILLS, RICK   *University of Wisconsin, La Crosse*
GREENE, CURTIS   *Wayne State University*
GREGG, KATHERINE   *West Virginia Wesleyan College*
HARLEY, JOHN   *Eastern Kentucky University*
HARRIS, JAMES   *Utah Valley Community College*
HELGESON, JEAN   *Collin County Community College*
HESS, WILFORD M.   *Brigham Young University*
HOUTMAN, ANNE   *Cal State, Fullerton*
HUFFMAN, DAVID   *Southwestern Texas University*
HUFFMAN, DONNA   *Calhoun Community College*
INEICHER, GEORGIA   *Hinds Community College*
JOHNSTON, TAYLOR   *Michigan State University*
JUILLERAT, FLORENCE   *Indiana University, Purdue University*
KENDRICK, BRYCE   *University of Waterloo*
HOUTMAN, ANNE   *Cal State, Fullerton*
KETELES, KRISTEN   *University of Central Arkansas*
KIRKPATRICK, LEE A.   *Glendale Community College*
KREBS, CHARLES   *University of British Columbia*
LANZA, JANET   *University of Arkansas, Little Rock*
LEICHT, BRENDA   *University of Iowa*
LOHMEIER, LYNNE   *Mississippi Gulf Coast Community College*
LORING, DAVID   *Johnson County Community College*
MACKLIN, MONICA   *Northeastern State University*
MANN, ALAN   *University of Pennsylvania*
MARTIN, KATHY   *Central Connecticut State University*
MARTIN, TERRY   *Kishwaukee College*
MASON, ROY B.   *Mount San Jacinto College*
MATTHEWS, ROBERT   *University of Georgia*
MAXWELL, JOYCE   *California State University, Northridge*
MCNABB, ANN   *Virginia Polytechnic Institute and State University*
MEIERS, SUSAN   *Western Illinois University*

MEYER, DWIGHT H.   *Queensborough Community College*
MICKLE, JAMES   *North Carolina State University*
MINOR, CHRISTINE V.   *Clemson University*
MCCLURE, JERRY   *Miami University*
MILLER, G. TYLER   *Wilmington, North Carolina*
MITCHELL, DENNIS M.   *Troy University*
MONCAYO, ABELARDO C.   *Ohio Northern University*
MOORE, IGNACIO   *Virginia Tech*
MORRISON-SHETTLER, ALLISON   *Georgia State University*
MORTON, DAVID   *Frostburg State University*
NELSON, RILEY   *Brigham Young University*
NICKLES, JON R.   *University of Alaska Anchorage*
NOLD, STEPHEN   *University of Wisconsin- Stout*
PADGETT, DONALD   *Bridgewater State College*
PENCOE, NANCY   *State University of West Georgia*
PERRY, JAMES   *University of Wisconsin, Center Fox Valley*
PITOCCHELLI, DR. JAY   *Saint Anselm College*
PLETT, HAROLD   *Fullerton College*
POLCYN, DAVID M.   *California State University, San Bernardino*
PURCELL, JERRY   *San Antonio College*
REID, BRUCE   *Kean College of New Jersey*
RENFROE, MICHAEL   *James Madison University*
REZNICK, DAVID   *California State University, Fullerton*
RICKETT, JOHN   *University of Arkansas, Little Rock*
ROHN, TROY   *Boise State University*
ROIG, MATTIE   *Broward Community College*
ROSE, GRIEG   *West Valley College*
SANDIFORD, SHAMILI A.   *College of Du Page*
SCHREIBER, FRED   *California State University, Fresno*
SELLERS, LARRY   *Louisiana Tech University*
SHONTZ, NANCY   *Grand Valley State University*
SHAPIRO, HARRIET   *San Diego State University*
SHOPPER, MARILYN   *Johnson County Community College*
SIEMENS, DAVID   *Black Hills State University*
SMITH, BRIAN   *Black Hills State University*
SMITH, JERRY   *St. Petersburg Junior College, Clearwater Campus*
STEINERT, KATHLEEN   *Bellevue Community College*
SUNDBERG, MARSHALL D.   *Emporia State University*
SUMMERS, GERALD   *University of Missouri*
SVENSSON, PETER   *West Valley College*
SWANSON, ROBERT   *North Hennepin Community College*
SWEET, SAMUEL   *University of California, Santa Barbara*
SZYMCZAK, LARRY J.   *Chicago State University*
TERHUNE, JERRY   *Jefferson Community College, University of Kentucky*
TAYLOR, JANE   *Northern Virginia Community College*
TIZARD, IAN   *Texas A&M University*
TRAYLER, BILL   *California State University at Fresno*
TROUT, RICHARD E.   *Oklahoma City Community College*
TURELL, MARSHA   *Houston Community College*
TYSER, ROBIN   *University of Wisconsin, LaCrosse*
VAJRAVELU, RANI   *University of Central Florida*
VANDERGAST, AMY   *San Diego State University*
VERHEY, STEVEN   *Central Washington University*
VICKERS, TANYA   *University of Utah*
VOGEL, THOMAS   *Western Illinois University*
WARNER, MARGARET   *Purdue University*
WEBB, JACQUELINE F.   *Villanova University*
WELCH, NICOLE TURRILL   *Middle Tennessee State University*
WELKIE, GEORGE W.   *Utah State University*
WENDEROTH, MARY PAT   *University of Washington*
WINICUR, SANDRA   *Indiana University, South Bend*
WOLFE, LORNE   *Georgia Southern University*
YONENAKA, SHANNA   *San Francisco State University*
ZAYAITZ, ANNE   *Kutztown University of Pennsylvania*

Introduction

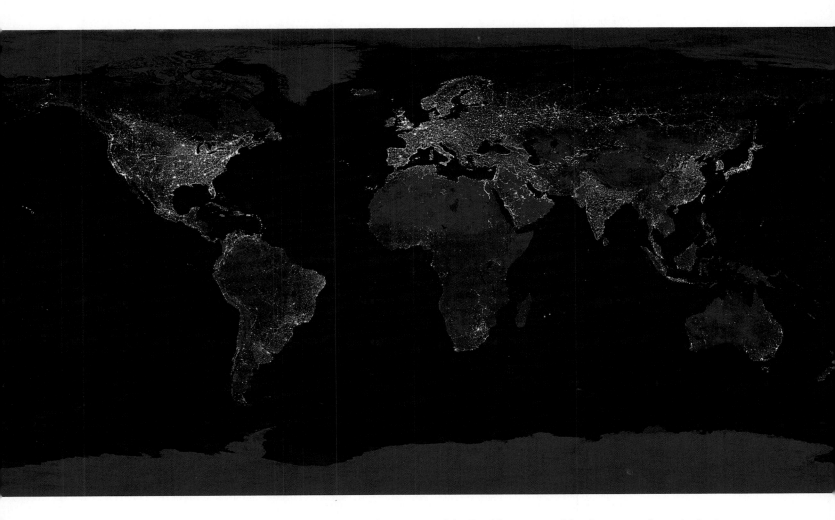

Current configurations of the Earth's oceans and land masses—the geologic stage upon which life's drama continues to unfold. This composite satellite image reveals global energy use at night by the human population. Just as biological science does, it invites you to think more deeply about the world of life—and about our impact upon it.

IMPACTS, ISSUES   *What Am I Doing Here?*

Leaf through a newspaper on any given Sunday and you may get an uneasy feeling that the world is spinning out of control. There's a lot about the Middle East, where

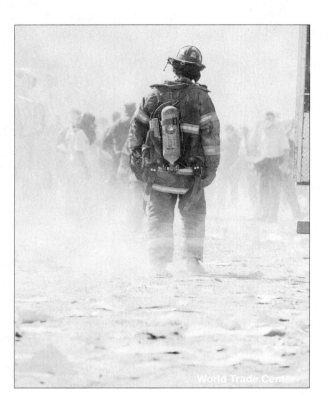

World Trade Center

great civilizations have come and gone. You won't find much on the amazing coral reefs in the surrounding seas, especially at the northern end of the Red Sea. Now the news is about oil and politics, terrorists, and war.

Think back on the 1991 Persian Gulf conflict, when thick smoke from oil fires blocked out sunlight, and black rain fell. Iraqis deliberately released about 460 million gallons of crude oil into the Gulf. Uncounted numbers of reef organisms died. So did thousands of birds.

Today Kuwaitis wonder if the oil fires caused their higher cancer rates. They join New Yorkers who are worried about developing lung problems from breathing dense, noxious dust that filled the air after the horrific terrorist attack on the World Trade Center.

Nature, too, seems to have it in for us. Cholera, the flu, and SARS pose global threats. An AIDS pandemic is unraveling the very fabric of African societies. Forests burn fiercely. Storms, droughts, and heat waves are often monstrous. Polar ice caps and once-vast glaciers are melting too rapidly, and the whole atmosphere is warming up.

It's enough to make you throw down the paper and long for the good old days, when things were so much simpler.

Of course, read up on the good old days and you'll find they weren't so good. Bioterrorists were around in 1346, when soldiers catapulted the corpses of bubonic

the big picture

Life's Underlying Unity   Life shows a hierarchy of organization, extending from the molecular level through the biosphere. Shared features at the molecular level are the basis of life's unity.

Life's Diversity   Life also shows spectacular diversity. Several million kinds of organisms already have been named, past and present, each with some traits that make it unique from all the others.

plague victims into a walled city under siege. Infected people and rats fled the city and helped fuel the Black Death, a plague that left 25 million dead in Europe. In 1918, the Spanish flu raced around the world and left somewhere between 30 and 40 million people dead. Between 1945 and 1949, about 100,000 people in the United States contracted polio, a disease that left many permanently paralyzed. In those times, too, many felt helpless in a world that seemed out of control.

What it boils down to is this: For a couple of million years, we humans and our immediate ancestors have been trying to make sense of the natural world and what we're doing in it. We observe it, come up with ideas, then test the ideas. But the more pieces of the puzzle we fit together, the bigger the puzzle gets. We now know that it is almost overwhelmingly big.

You could walk away from the challenge and simply not think. You could let others tell you what to think. Or you could choose to develop your own understanding of the puzzle.

Maybe you're interested in the pieces that affect your health, the food you eat, or your children, should you choose to reproduce. Maybe you just find life fascinating. No matter what your focus might be, you can deepen your perspective. You can learn ways to sharpen how you interpret the natural world, including human nature. This is the gift of biology, the scientific study of life.

 How Would You Vote?

The warm seas of the Middle East support some of the world's most spectacular coral reef ecosystems. Should the United States provide funding to help preserve the reefs? See the Media Menu for details, then vote online.

Explaining Unity in Diversity Evolutionary theories, especially the theory of evolution by natural selection, help us see a profound connection between life's underlying unity and its diversity.

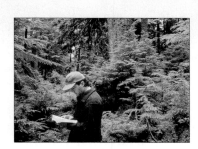

How We Know Biologists find out about life by observing, asking questions, and formulating and testing hypotheses in nature or the laboratory. They report results in ways that others can test.

1.1 Life's Levels of Organization

The world of life shows levels of organization, from the simple to the complex. Take time to see how these levels connect to get a sense of how the topics of this book are organized and where they will take you.

FROM SMALL TO SMALLER

Imagine yourself on the deck of a sailing ship, about to journey around the world. The distant horizon of a vast ocean beckons, and suddenly you sense that you are just one tiny part of the great scheme of things.

Now imagine one of your red blood cells can think. It realizes it's only a tiny part of the great scheme of your body. A string of 375 cells like itself would fit across a straightpin's head—and you have trillions of cells. One of the fat molecules at the red blood cell's surface is thinking about how small it is. A string of 1,200,000 molecules like itself would stretch across that pinhead. Now a hydrogen atom in the molecule is pondering the great scheme of a fat molecule. Figure 1.1a depicts one. It would take 53,908,355 side-by-side hydrogen atoms to stretch across the head of a pin!

FROM SMALLER TO VAST

With that single atom, you have reached the entry level of nature's great pattern of organization. Like nonliving things, all organisms are made of building blocks called atoms. At the next level are molecules. Life's unique properties emerge when certain kinds of molecules are organized into cells. These "molecules of life" are complex carbohydrates, complex fats and other lipids, proteins, DNA, and RNA (Figure 1.1b). The **cell** is the smallest unit of organization with the

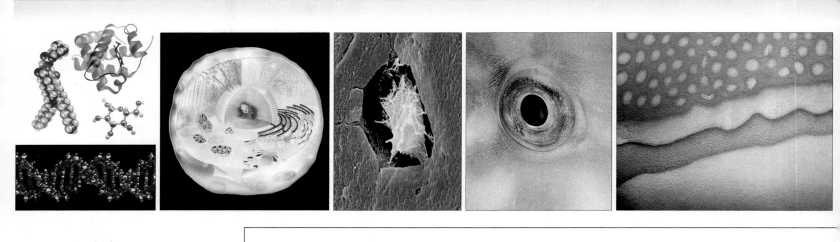

b molecule

Two or more joined atoms of the same or different elements. "Molecules of life" are complex carbohydrates, lipids, proteins, DNA, and RNA. Only living cells now make them.

c cell

Smallest unit that can live and reproduce on its own or as part of a multicelled organism. It has an outer membrane, DNA, and other components.

d tissue

Organized aggregation of cells and substances interacting in a specialized activity. Many cells (*white*) made this bone tissue from their own secretions.

e organ

Structural unit made of two or more tissues interacting in some task. A parrotfish eye is a sensory organ used in vision.

f organ system

Organs interacting physically, chemically, or both in some task. Parrotfish skin is an integumentary system with tissue layers, organs such as glands, and other parts.

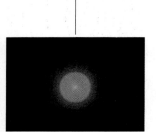

a atom

Smallest unit of an element that still retains the element's properties. Electrons, protons, and neutrons are its building blocks. This hydrogen atom's electron zips around a proton in a spherical volume of space.

Figure 1.1 Increasingly complex levels of organization in nature, extending from subatomic particles to the biosphere.

capacity to survive and reproduce on its own, given raw materials, energy inputs, information encoded in its DNA, and suitable conditions in its environment.

At the next level of organization are multicelled organisms made of specialized, interdependent cells, often organized as tissues, organs, and organ systems. A higher level of organization is the **population**, a group of single-celled or multicelled individuals of the same species occupying a specified area. A school of fish is a population (Figure 1.1*h*), as are all of the single-celled amoebas in an isolated lake.

Next comes the **community**, all populations of all species occupying one area. Its extent depends on the area specified. It might be the Red Sea, an underwater cave, or a forest in South America. It might even be a community of tiny organisms that live, reproduce, and die quickly inside the cupped petals of a flower.

The next level of organization is the **ecosystem**, or a community together with its physical and chemical environment. Finally, the **biosphere** is the highest level of life. It encompasses all regions of the Earth's crust, waters, and atmosphere in which organisms live.

This book is a journey through the globe-spanning organization of life. So take a moment to study Figure 1.1. You can use it as a road map of where each part fits in the great scheme of things.

> *Nature shows levels of organization, from the simple to the increasingly complex.*
>
> *Life's unique characteristics originate at the atomic and molecular level. They extend through cells, populations, communities, ecosystems, and the biosphere.*

Read Me First!
and watch the narrated animation on life's levels of organization

g multicelled organism

Individual made of different types of cells. Cells of most multicelled organisms, including this Red Sea parrotfish, are organized as tissues, organs, and organ systems.

h population

Group of single-celled or multicelled individuals of the same species occupying a specified area. This is a fish population in the Red Sea.

i community

All populations of all species occupying a specified area. This is part of a coral reef in the Gulf of Aqaba at the northern end of the Red Sea.

j ecosystem

A community that is interacting with its physical environment. It has inputs and outputs of energy and materials. Reef ecosystems flourish in warm, clear seawater throughout the Middle East.

k the biosphere

All regions of the Earth's waters, crust, and atmosphere that hold organisms. In the vast universe, Earth is a rare planet. Without its abundance of free-flowing water, there would be no life.

1.2 Overview of Life's Unity

"Life" isn't easy to define. It's just too big, and it's been changing for 3.9 billion years! Even so, you can frame a definition in terms of its unity and diversity. Here's the unity part: All living things grow and reproduce with the help of DNA, energy, and raw materials. They sense and respond to what is going on. But details of their traits differ among many millions of kinds of organisms. That's the diversity part—variation in traits.

DNA, THE BASIS OF INHERITANCE

You will never, ever find a rock made of nucleic acids, proteins, and complex carbohydrates and lipids. In the natural world, only living cells make these molecules. And the signature molecule of life is the nucleic acid called DNA. No chunk of granite or quartz has it.

DNA holds information for building proteins from smaller molecules, the amino acids. By analogy, if you follow suitable instructions and invest enough energy

Figure 1.2 Examples of objects built from the same materials by different assembly instructions.

in the task, you might organize a pile of a few kinds of ceramic tiles (representing amino acids) into diverse patterns (representing proteins), as in Figure 1.2.

Why are proteins so important? Many are structural materials, regulators of cell activities, and enzymes. Enzymes are the cell's main worker molecules. They build, split, and rearrange the molecules of life in ways that keep cells alive. Without enzymes, nothing much could be done with DNA's information. There would be no new organisms.

In nature, each organism inherits its DNA—and its traits—from parents. *Inheritance* means an acquisition of traits after parents transmit their DNA to offspring. Think about it. Baby storks look like storks and not like pelicans because they inherited stork DNA, which isn't exactly the same as pelican DNA.

Reproduction refers to actual mechanisms by which parents transmit DNA to offspring. For frogs, humans, trees, and other organisms, the information in DNA guides *development*—the transformation of the first cell of a new individual into a multicelled adult, typically with many different tissues and organs (Figure 1.3).

ENERGY, THE BASIS OF METABOLISM

Becoming alive and maintaining life processes requires energy—the capacity to do work. Each normal living cell has ways to obtain and convert energy from its surroundings. By a process called **metabolism**, every cell acquires and uses energy to maintain itself, grow, and make more cells.

Where does the energy come from? Nearly all of it flows from the sun into the world of life, starting with **producers**. Producers are plants and other organisms that make their own food molecules from simple raw materials. Animals and decomposers are **consumers**. They cannot make their own food; they survive by feeding on tissues of producers and other organisms.

Figure 1.3 "The insect"—actually a series of stages of development guided largely by instructions in DNA. Here, a silkworm moth, from a fertilized egg (**a**), to a larval stage called a caterpillar (**b**), to a pupal stage (**c**), to the winged form of the adult (**d,e**).

When, say, zebras browse on plants, some energy stored in plant tissues is transferred to them. Later on, energy is transferred to a lion as it devours the zebra. And it gets transferred again as decomposers go to work, acquiring energy from the remains of zebras, lions, and other organisms.

Decomposers are mostly the kinds of bacteria and fungi that break down sugars and other molecules to simpler materials. Some of the breakdown products are cycled back to producers as raw materials. Over time, energy that plants originally captured from the sun returns to the environment.

Energy happens to flow in one direction, from the environment, through producers, then consumers, and then back to the environment (Figure 1.4). These are the energy exchanges that maintain life's organization. Later on, you will see how life's interconnectedness relates to modern-day problems, including major food shortages, AIDS, cholera, acid rain, global warming, and rapid losses in biodiversity.

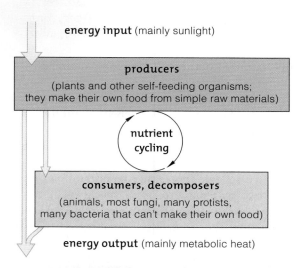

energy input (mainly sunlight)

producers
(plants and other self-feeding organisms; they make their own food from simple raw materials)

nutrient cycling

consumers, decomposers
(animals, most fungi, many protists, many bacteria that can't make their own food)

energy output (mainly metabolic heat)

Figure 1.4 The one-way flow of energy and cycling of materials in the world of life.

LIFE'S RESPONSIVENESS TO CHANGE

It's often said that only living things respond to the environment. Yet even a rock shows responsiveness, as when it yields to gravity's force and tumbles down a hill or changes its shape slowly under the repeated batterings of wind, rain, or tides.

The difference is this: Living things sense changes in their surroundings, and they make compensatory, controlled responses to them. How? With receptors. Receptors are molecules and structures that detect stimuli, which are specific kinds of energy. Different receptors respond to different stimuli. A stimulus may be sunlight energy, chemical potential energy (as when a substance is more concentrated outside a cell than inside), or the mechanical energy of a bite (Figure 1.5).

Switched-on receptors can trigger changes in cell activities. As a simple example, after you finish eating a piece of fruit, sugars leave your gut and enter your bloodstream. Think of blood and the fluid around cells as an *internal* environment, which must be kept within tolerable limits. Too much or too little sugar in blood changes that internal environment. This can cause diabetes and other medical problems. Normally, when there is too much sugar in blood, your pancreas starts secreting more insulin. Most living cells in your body have receptors for this hormone, which stimulates them to take up more sugar. When enough cells do so, the blood sugar level returns to normal.

In such ways, organisms keep the internal environment within a range that cells can tolerate. This state is called **homeostasis**, and it is one of the key defining characteristics of life.

Figure 1.5 Response to signals from pain receptors, activated by a lion cub flirting with disaster.

Organisms build proteins based on instructions in DNA, which they inherit from their parents. Organisms reproduce, grow, and stay alive by way of metabolism—ongoing energy conversions and energy transfers at the cellular level.

Organisms interact through a one-way flow of energy and a cycling of materials. Collectively, their interdependencies have global impact.

Organisms sense and respond to changing conditions in controlled ways. The responses help them maintain tolerable conditions in their internal environment.

1.3 If So Much Unity, Why So Many Species?

Although unity pervades the world of life, so does diversity. Organisms differ enormously in body form, in the functions of their body parts, and in behavior.

Superimposed on life's unity is tremendous diversity. Millions of kinds of organisms, or **species**, live on Earth. Many more lived during the past 3.9 billion years, but their lineages vanished; about 99.9 percent of all species have become extinct.

For centuries, scholars have tried to make sense of diversity. In 1735, a physician named Carolus Linnaeus devised a scheme for classifying organisms by assigning a two-part name to each species. The first part designates the *genus* (plural, genera). Each genus is one or more species grouped together on the basis of a number of traits that are unique to that group alone. The second part of the name refers to a particular species within the genus. Today, biologists attempt to sort out the relationships among species not only on the basis of observable traits, but also using evidence of descent from a common ancestor.

For instance, *Scarus gibbus* is the scientific name for the humphead parrotfish (Figure 1.1*g*). Another species in the same genus is *S. coelestinus*, the midnight parrotfish. We abbreviate a genus name once it's been spelled out in a document.

Biologists are still working out how to group the organisms. Most now favor a classification system with three domains: Bacteria, Archaea, and Eukarya (Figure 1.6). As shown in Figure 1.7, the third domain includes protists, plants, fungi, and animals.

The **archaea** and **bacteria** are single-celled. They are *prokaryotic*, meaning they do not have a nucleus (a membrane-bound sac that keeps DNA separated from the rest of the cell's interior). Prokaryotes include diverse producers or consumers. Of all groups, theirs shows the greatest metabolic diversity.

Archaea live in boiling ocean water, freezing desert rocks, sulfur-rich lakes, and other habitats as harsh as those thought to have prevailed when life originated.

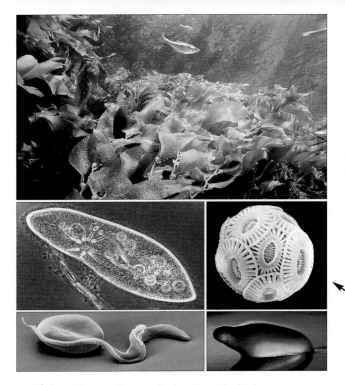

protists Diverse single-celled and multicelled eukaryotic species that range from microscopic single cells to giant seaweeds. Even this tiny sampling conveys why many biologists now believe the "protists" are many separate lineages.

archaea These prokaryotes are evolutionarily closer to the eukaryotes than to bacteria. This is a colony of methane-producing cells.

Figure 1.7 A few representatives of life's diversity.

Bacteria are sometimes called eubacteria, which means "true bacteria," to distinguish them from archaea. They are far more common than archaeans, and they live throughout the world in diverse habitats.

Plants, fungi, animals, and protists are members of the group **eukarya**, which means they have nuclei. Eukaryotes are generally larger and far more complex than the prokaryotes. The differences among protistan lineages are so great that they could be divided into several separate groups, which would result in a major reorganization of the domain.

Bacteria (EUBACTERIA)	Archaea (ARCHAEBACTERIA)	Eukarya (EUKARYOTES)

Figure 1.6 Three domains of life.

Read Me First!
*and watch
the narrated
animation on
life's diversity*

plants Generally, photosynthetic, multicelled eukaryotes, many with roots, stems, and leaves. Plants are the primary producers for ecosystems on land. Redwoods and flowering plants are examples.

fungi Single-celled and multicelled eukaryotes; mostly decomposers, also many parasites and pathogens. Without the fungal and bacterial decomposers, communities would become buried in their own wastes.

animals Multicelled eukaryotes that ingest tissues or juices of other organisms. Like this basilisk lizard, most actively move about during at least part of their life.

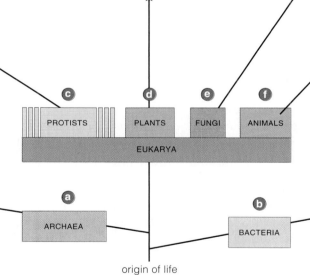

© PROTISTS ④ PLANTS ⓔ FUNGI ⓕ ANIMALS

EUKARYA

ⓐ ARCHAEA

ⓑ BACTERIA

origin of life

bacteria By far the most common prokaryotes; collectively, these single-celled species are the most metabolically diverse organisms on Earth.

Plants are multicelled, photosynthetic producers. They can make their own food by using simple raw materials and sunlight as an energy source.

Most **fungi**, such as the mushrooms sold in grocery stores, are multicelled decomposers and consumers with a distinct way of feeding. They secrete enzymes that digest food outside the fungal body, then their individual cells absorb the digested nutrients. **Animals** are multicelled consumers that ingest tissues of other organisms. Different kinds are herbivores (grazers), carnivores (meat eaters), scavengers, and parasites.

All develop by a series of embryonic stages, and they actively move about during their life.

Pulling this information together, are you getting a sense of what it means when someone says that life shows unity *and* diversity?

To make the study of life's diversity more manageable, we group organisms related by descent from a shared ancestor. We recognize three domains—archaea, bacteria, and eukarya (protists, fungi, plants, and animals).

1.4 An Evolutionary View of Diversity

How can organisms be so much alike and still show staggering diversity? A theory of evolution by natural selection explains this. For now, simply think about how it starts with a simple observation: Individuals of a population show variation in the details of their shared traits.

Your traits make you and 6.3 billion other individuals members of the human population. Traits are different aspects of an organism's form, function, or behavior. For example, humans show a range of height and hair color. All natural populations have differences among their individuals.

What causes variation in traits? **Mutations**. These are heritable changes in DNA. Some mutations lead to novel traits that make an individual better able to secure food, a mate, hiding places, and so on. We call these *adaptive* traits.

An adaptive form of a trait tends to become more common over generations, because it gives individuals a better chance to live and bear more offspring than

WILD ROCK DOVE

individuals who don't have it. When different forms of a trait are becoming more or less common, evolution is under way. To biologists, **evolution** simply means heritable change in a line of descent. Mutations, the source of new traits, provide the variation that serves as the raw material for evolution.

"Diversity" refers to variations in traits that have accumulated in lines of descent. Later chapters show the actual mechanisms that bring it about. For now, start thinking about what a great naturalist, Charles Darwin, discovered about evolution:

First, populations tend to increase in size, past the capacity of their environment to sustain them, so their members must compete for resources (food, shelter).

Second, individuals of natural populations differ from one another in the details of their shared traits. Most variation has a heritable basis.

Third, when individuals differ in their ability to survive and reproduce, the traits that help them do so tend to become more common in the population over time. This outcome is called **natural selection**.

Take a look at the pigeons in Figure 1.8. They differ in feather color, size, and other traits. Suppose pigeon breeders are looking for, say, pigeons with black, curly-tipped feathers. They allow only the pigeons with the darkest and curliest-tipped feathers to mate. In time, only pigeons with black, curly-tipped feathers make up the breeders' captive population. Lighter, less curly feathers will become less common.

Pigeon breeding is a case of *artificial* selection. One form of a trait is favored over others in an artificial environment under contrived, manipulated conditions. Darwin saw that breeding practices could be an easily understood model for *natural* selection, a favoring of some forms of a given trait over others in nature.

Just as breeders are "selective agents" promoting reproduction of particular captive pigeons, different agents operate across the range of variation in the wild. Pigeon-eating peregrine falcons are among them (Figure 1.8). Swifter or better camouflaged pigeons are more likely to avoid peregrine falcons and live long enough to reproduce, compared with not-so-swift or too-conspicuous pigeons.

Figure 1.8 Outcome of artificial selection. Just a few of the more than 300 varieties of domesticated pigeons, all descended from captive populations of wild rock doves. By contrast, peregrine falcons are one of the agents of natural selection in the wild.

Traits are variations in form, function, or behavior that arise as a result of mutations in DNA. Some traits are more adaptive than others to prevailing conditions.

Natural selection is an outcome of differences in survival and reproduction among individuals of a population that vary in one or more heritable traits. The process of evolution, or change in lines of descent, gives rise to life's diversity.

1.5 The Nature of Biological Inquiry

The preceding sections introduced some big concepts. Consider approaching this or any other collection of "facts" with a critical attitude. "Why should I accept that they have merit?" The answer requires a look at how biologists make inferences about observations, then test their inferences against actual experience.

OBSERVATIONS, HYPOTHESES, AND TESTS

To get a sense of "how to do science," you might start with practices that are common in scientific research:

1. Observe some aspect of nature, carefully check what others have found out about it, then frame a question or identify a problem related to your observation.

2. Formulate **hypotheses**, or educated guesses, about possible answers to questions or solutions to problems.

3. Using hypotheses as your guide, make a **prediction** —a statement of what you should find in the natural world if you were to go looking for it. This is often called the "if–then" process. *If* gravity does not pull objects toward the Earth, *then* it should be possible to observe apples falling up, not down, from a tree.

4. Devise ways to **test** the accuracy of predictions, as by making systematic observations, building models, and conducting experiments. **Models** are theoretical, detailed descriptions or analogies that might help us visualize something that hasn't been directly observed.

5. If your tests do not confirm a prediction, check to see what might have gone wrong. It may be that you overlooked a factor that had an impact on the results. Or maybe a hypothesis is not a good one.

6. Repeat the tests or devise new ones—the more the better, because hypotheses that withstand many tests are likely to have a higher probability of being useful.

7. Objectively analyze and report the test results as well as the conclusions you drew from them.

You might hear someone refer to these practices as "the scientific method," as if all scientists march to the drumbeat of an absolute, fixed procedure. They do not. Many observe, describe, and report on some aspect of nature, then leave the hypothesizing to others. Some scientists are lucky; they stumble onto information that they are not even looking for. Of course, it isn't always a matter of luck. Chance seems to favor a mind that has already been prepared, by education, experience, or both, to recognize what the information might mean. So it is not a single method that scientists have in common. It is a critical attitude about being shown rather than being told—that is, by accepting ideas supported by tests, and by taking a logical approach to problem solving.

ABOUT THE WORD "THEORY"

Suppose no one has disproved a hypothesis after years of rigorous tests. Suppose scientists use it to interpret more data or observations, which could involve more hypotheses. When a hypothesis meets these criteria, it may become accepted as a **scientific theory**.

You may hear people apply the word "theory" to a speculative idea, as in the expression "It's just a theory." But a scientific theory differs from speculation for this reason: *After testing the predictive power of a scientific theory many times and in many ways in the natural world, researchers have yet to find evidence that disproves it.* This is why the theory of natural selection is respected. It successfully explains diverse issues, such as how life originated, how river dams can alter ecosystems, and why antibiotics aren't working.

Maybe a well-tested theory is as close to the truth as scientists can get with known evidence. For instance, after more than a century of many thousands of tests, Darwin's theory holds, with only minor modification. We can't prove it holds under all possible conditions; that would take an infinite number of tests. As for any theory, we can only say *it has a high probability of being a good one.* Biologists do keep looking for information and devising tests that might disprove its premises.

A scientific approach to studying nature is based on asking questions, formulating hypotheses, making predictions, testing the predictions, and objectively reporting the results.

A scientific theory is a time-tested intellectual framework that is used to interpret a broad range of observations and data. Scientific theories remain open to rigorous tests, revision, and tentative acceptance or rejection.

1.6 The Power of Experimental Tests

Experiments are tests that simplify observation in nature, because conditions under which observations are made can be controlled. Well-designed experiments help you predict what you'll find in nature when a hypothesis is a good one—or won't find if it is wrong.

AN ASSUMPTION OF CAUSE AND EFFECT

A scientific experiment starts with a key premise: *Any aspect of nature has an underlying cause that can be tested by observation.* This premise is what sets science apart from faith in the supernatural ("beyond nature"). It means a scientific hypothesis must be testable in the natural world in ways that might well disprove it.

Most aspects of nature are complex, an outcome of many interacting variables. A **variable** is a specific aspect of an object or event that can differ among individuals or changes over time. Scientists simplify their observation of complex phenomena by designing experiments to test one variable at a time. They define a **control group**—a standard for comparison with one or more **experimental groups**. There are two kinds of control groups. A control group can be identical to an experimental group; except for *one variable event*, it is tested the same way as the experimental group. A control group may also differ from an experimental group in *one variable aspect*; in this case, it is tested exactly the same way as the experimental group.

EXAMPLE OF AN EXPERIMENTAL DESIGN

In 1996, the FDA approved Olestra®, a type of synthetic fat replacement made from sugar and vegetable oil, for use as a food additive. The first Olestra-containing product to reach consumers in the United States was a potato chip. Soon controversy raged. Some people complained of severe gastrointestinal distress after eating the chips. In 1998, medical researchers at Johns Hopkins University performed an experiment to test whether the new chips were indeed causing problems. Their prediction was this: *If Olestra causes intestinal problems, then people who eat products that contain Olestra will end up with gastrointestinal cramps.*

A suburban Chicago multiplex theater was chosen as the "laboratory" for this experiment. More than 1,100 people were invited to watch a movie and eat their fill of potato chips while they were there. They ranged between 13 and 88 years old. Unmarked bags each contained a family-size portion of potato chips. Some of the bags held Olestra potato chips, and the others held regular potato chips.

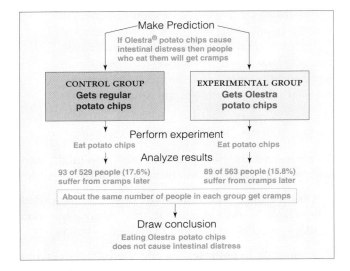

Figure 1.9 shows these steps:

Make Prediction
If Olestra® potato chips cause intestinal distress then people who eat them will get cramps

CONTROL GROUP
Gets regular potato chips

EXPERIMENTAL GROUP
Gets Olestra potato chips

Perform experiment
Eat potato chips — Eat potato chips

Analyze results
93 of 529 people (17.6%) suffer from cramps later — 89 of 563 people (15.8%) suffer from cramps later

About the same number of people in each group get cramps

Draw conclusion
Eating Olestra potato chips does not cause intestinal distress

Figure 1.9 Example of a typical sequence of steps taken in a scientific experiment.

In this experiment, both control and experimental groups consisted of a random sample of moviegoers; each group got different chips (the variable event).

Later, the researchers telephoned the moviegoers at home and tabulated reports of gastrointestinal distress. They found that 89 of 563 people (15.8 percent) who ate Olestra chips complained of stomach cramps. Of 529 people, 93 (17.6 percent) who ate the regular chips did as well. They concluded that eating Olestra potato chips—at least during one sitting—does not cause gastrointestinal distress (Figure 1.9).

EXAMPLE OF A FIELD EXPERIMENT

Consider that many toxic or unpalatable species are vividly colored, often with distinctive patterning. Predators learn to avoid individuals that display particular visual cues after eating a few of them and suffering ill consequences.

In 1879, a naturalist named Fritz Müller formulated a hypothesis about unrelated species of distasteful butterflies that show striking resemblance to one another. A visual similarity between different species that may confuse potential predators (or prey) is called **mimicry**. Müller thought such a resemblance benefits individuals of both butterfly species because they share the burden of educating predatory birds.

Durrell Kapan, an evolutionary biologist, tested the hypothesis in 2001 with a field experiment in the rain forests of Ecuador. There are two forms of *Heliconius cydno*, an unpalatable species of butterfly. One has yellow markings on its wings; the other does not.

Figure 1.10 *Heliconius* butterflies. (**a**) Two forms of *H. cydno* and (**b**) *H. eleuchia*.

(**c**) Kapan's experiment with *Heliconius* butterflies in an Ecuadoran rain forest. *H. cydno* butterflies with or without yellow markings on their wings were captured and transferred to a habitat of *H. eleuchia*, a species that also has yellow wing markings. Local predatory birds, familiar with untasty yellow *H. eleuchia*, avoided the *H. cydno* butterflies with yellow markings but ate the white ones.

a

b

Control Group
34 *H. cydno* individuals with yellow markings

Experimental Group
46 *H. cydno* individuals with white markings

Experiment
Both yellow and white forms of *H. cydno* butterflies are introduced into isolated rain forest habitat of yellow *H. eleuchia* butterflies. Numbers of individuals resighted recorded on a daily basis for two weeks.

one of the agents of selection

Results
Experimental group (*H. cydno* individuals without yellow wing markings) is selected against. 37 of the original group of 46 white butterflies disappear (80%), compared with 20 of the 34 yellow controls (58%).

c

Both resemble another unpalatable species that lives nearby, *H. eleuchia*, which also has yellow in its wings (Figure 1.10). Kapan made a prediction: Birds that had already learned not to prey on *H. eleuchia* would also avoid *H. cydno* butterflies with yellow markings.

He captured both forms of *H. cydno*. The form with no yellow markings was the experimental group, and the form with the yellow markings was the control. He released both groups into parts of the forest that held isolated populations of *H. eleuchia* butterflies. He made daily counts of how many of the transplanted butterflies survived during the next two weeks, the approximate life span of the butterflies.

Kapan found that individuals of the experimental group were less likely to survive in the new habitat (Figure 1.10c). Resident birds familiar with *H. eleuchia* butterflies most likely ate them because they did not bear the familiar visual cue—yellow markings—that signaled bad taste. The control group did better, as you can see from the test results listed in Figure 1.10c. Local birds probably had an idea of how the new butterflies would taste, and avoided them.

Kapan's test results confirmed his prediction, and it also turned out to be evidence of natural selection.

BIAS IN REPORTING RESULTS

Experimenters run a risk of interpreting data in terms of what they wish to prove or dismiss. That is why scientists prefer *quantitative* reports of experiments, with numbers or some other precise measurement. Such data give other experimenters an opportunity to confirm tests, and, perhaps more importantly, allow others to check their conclusions.

This last point gets us back to the value of thinking critically. Scientists must keep asking themselves: *Will observations or experiments show that a hypothesis is false?* They expect one another to put aside pride or bias by testing ideas in ways that may prove them wrong. Even if someone won't, others will—because science is a cooperative yet competitive community. Ideally, individuals share ideas, knowing it's as useful to expose errors as to applaud insights. They can and often do change their mind when evidence contradicts their ideas. And therein lies the strength of science.

Experiments simplify observations in nature by restricting a researcher's focus to one variable at a time.

Tests are based on the premise that any aspect of nature has one or more underlying causes. Scientific hypotheses can be tested in ways that might disprove them.

1.7 The Limits of Science

Beyond the realm of scientific inquiry, some events are unexplained. Why do we exist, for what purpose? Why do we have to die at a particular moment? Such questions lead to *subjective* answers, which come from within as an integrated outcome of all the experiences and mental connections that shape our consciousness. People differ enormously in this regard. That is why subjective answers do not readily lend themselves to scientific analysis and experiments.

This is not to say subjective answers are without value. No human society can function for long unless its individuals share a commitment to standards for making judgments, even if they are subjective. Moral, aesthetic, philosophical, and economic standards vary from one society to the next. But they all guide people in deciding what is important and good, and what is not. All attempt to give meaning to what we do.

Every so often, scientists stir up controversy when they explain something that was thought to be beyond natural explanation, or belonging to the supernatural. This is often the case when a society's moral codes are interwoven with religious interpretations of the past. Exploring a long-standing view of the natural world from a scientific perspective might be misinterpreted as questioning morality, even though the two are not the same thing.

As one example, centuries ago in Europe, Nikolaus Copernicus studied the planets and concluded that the Earth circles the sun. Today this seems obvious. Back then, it was heresy. The prevailing belief was that the Creator made the Earth—and, by extension, humans—the immovable center of the universe. One respected scholar, Galileo Galilei, studied the Copernican model of the solar system, thought it was a good one, and said so. He was forced to retract his statement, on his knees, and put the Earth back as the fixed center of things. (Word has it that he also muttered, "Even so, it *does* move.") Later, Darwin's theory of evolution also ran up against prevailing belief.

Today, as then, society has sets of standards. Those standards might be questioned when a new, natural explanation runs counter to supernatural beliefs. This doesn't mean scientists who raise questions are less moral, less lawful, less sensitive, or less caring than anyone else. It only means one more standard guides their work. Their ideas about nature must be tested in the external world, in ways that can be repeated.

The external world, not internal conviction, is the testing ground for the theories generated in science.

Summary

Section 1.1   Life shows many levels of organization. All things, living and nonliving, are made of atoms. The properties of life emerge in cells. An organism may be a single cell or multicelled. In most multicelled species, cells are organized as tissues, organs, and organ systems.

A population consists of individuals of the same species in a specified area. A community consists of all populations occupying the same area. An ecosystem is a community and its environment. The biosphere includes all regions of Earth's atmosphere, waters, and land where we find living organisms.

Section 1.2   Life shows unity. All organisms have DNA, which holds instructions for building proteins. They inherit the instructions from their parents and pass them on to offspring. All require energy and raw materials from the environment to grow and reproduce. All sense changes in the surroundings and respond to them in controlled ways (Table 1.1).

Section 1.3   Life shows tremendous diversity. Many millions of species exist; many more lived in the past. Each is unique in some aspects of its body plan, function, and behavior. We group species that are related by descent from a common ancestor. A current classification system puts species in three domains: archaea, bacteria, and eukarya. Protists, plants, fungi, and animals are eukaryotes.

Table 1.1   Summary of Life's Characteristics

Shared characteristics that reflect life's unity

1. All life forms contain "molecules of life" (complex carbohydrates, lipids, proteins, and nucleic acids).

2. Organisms consist of one or more cells.

3. Cells are constructed of the same kinds of atoms and molecules according to the same laws of energy.

4. Organisms acquire and use energy and materials to survive and reproduce.

5. Organisms sense and make controlled responses to conditions in their internal and external environments.

6. Heritable information is encoded in DNA.

7. Characteristics of individuals in a population can change over generations; the population can evolve.

Foundations for life's diversity

1. Mutations in DNA give rise to variations in traits, or details of body form, function, and behavior.

2. Traits enhancing survival and reproduction become more common in a population over generations. This process is called natural selection.

3. Diversity is the sum total of variations that accumulated in different lines of descent over the past 3.9 billion years.

Section 1.4 Mutations change DNA and give rise to new variations of heritable traits. Natural selection occurs if a variation affects survival and reproduction. A population is evolving by natural selection when an adaptive form of a trait is becoming more common.

Section 1.5 Scientific methods are varied, but all are based on a logical approach to explaining nature. Scientists observe some aspect of nature, then develop a hypothesis about what might have caused it. They use the hypothesis to make predictions that can be tested by making more observations, building models, or doing experiments.

Scientists analyze test results, draw conclusions from them, and share this information with other scientists. A hypothesis that does not hold up under repeated testing is modified or discarded. A scientific theory is a long-standing hypothesis that explains a broad range of related phenomena and has been supported by many different tests.

Section 1.6 Science cannot answer all questions. It deals only with aspects of nature that lend themselves to systematic observation, hypotheses, predictions, and experiments.

Most aspects of nature are complex, an outcome of many interacting variables. A scientific experiment allows a scientist to change one variable at a time and observe what happens. Experiments are designed so experimental groups can be compared with a control group. Scientists share their results so others can check their conclusions.

Self-Quiz
Answers in Appendix III

1. The smallest unit of life is the _____ .

2. _____ is the capacity of cells to extract energy from sources in the environment, and use it to live, grow, and reproduce.

3. _____ is a state in which the internal environment is being maintained within a tolerable range.

4. A trait is _____ if it improves an organism's ability to survive and reproduce in a given environment.

5. Differences in heritable traits arise through _____ .

6. Researchers assign all species to one of three _____ .

7. _____ secure energy from their surroundings.
   a. Producers   c. Decomposers
   b. Consumers   d. All of the above

8. DNA _____ .
   a. contains instructions for building proteins
   b. undergoes mutation
   c. is transmitted from parents to offspring
   d. all of the above

9. _____ is the acquisition of traits after parents transmit their DNA to offspring.
   a. Metabolism   c. Homeostasis
   b. Reproduction   d. Inheritance

10. A control group is _____ .
   a. a standard against which experimental groups can be compared
   b. an experiment that gives conclusive results

11. Match the terms with the most suitable descriptions.
   _____ adaptive trait
   _____ natural selection
   _____ scientific theory
   _____ hypothesis
   _____ prediction

   a. statement of what you can expect to observe in nature
   b. proposed explanation; an educated guess
   c. improves chances of surviving and reproducing
   d. related set of hypotheses that form a broadly useful, testable explanation
   e. outcome of differences in survival, reproduction among individuals of a population that differ in the details of one or more traits

Critical Thinking

1. A scientific theory about some aspect of nature rests upon inductive logic—inference of a generalized conclusion from particular instances. The assumption is that, because an outcome of some event has been observed to happen with great regularity, it will happen again. However, we can't know this for certain, because there is no way to account for all possible variables that may affect the outcome. To illustrate this point, Garvin McCain and Erwin Segal offer a parable:

Once there was a highly intelligent turkey. The turkey lived in a pen, attended by a kind, thoughtful master, and it had nothing to do but reflect upon the world's wonders and regularities. Morning always began with the sky getting light, followed by the clop, clop, clop of its master's friendly footsteps, which was followed by the appearance of delicious food. Other things varied—sometimes the morning was warm and sometimes cold—but food always followed footsteps. The sequence of events was so predictable that it eventually became the basis of the turkey's theory about the goodness of the world.

One morning, after more than a hundred confirmations of the goodness theory, the turkey listened for the clop, clop, clop, heard it, and had its head chopped off.

The turkey learned the hard way that explanations about the world only have a high or low probability of being correct. Today, some people take this uncertainty to mean that "facts are irrelevant—facts change." If that is so, should we just stop doing scientific research? Why or why not?

2. Witnesses in a court of law are asked to "swear to tell the truth, the whole truth, and nothing but the truth." What are some of the problems inherent in the question? Can you think of a better alternative?

3. Many popular magazines publish an astounding array of articles on diet, exercise, and other health-related topics. Some authors recommend a diet or dietary supplement. What kinds of evidence do you think the articles should include so that you can decide whether to accept their recommendations?

a Natalie, blindfolded, randomly plucks a jelly bean from a jar of 120 green and 280 black jelly beans. That's a ratio of 30 to 70 percent.

b The jar is hidden before she removes her blindfold. She observes a single green jelly bean in her hand and assumes the jar holds only green jelly beans.

c Still blindfolded, Natalie randomly picks 50 jelly beans from the jar and ends up with 10 green and 40 black ones.

d The larger sample leads her to assume one-fifth of the jar's jelly beans are green and four-fifths are black (a ratio of 20 to 80). Her larger sample more closely approximates the jar's green-to-black ratio. The more times Natalie repeats the sampling, the greater the chance she will come close to knowing the actual ratio.

Figure 1.11 A simple demonstration of sampling error.

4. Rarely can experimenters observe all individuals of a group. They select subsets or samples of populations, events, and other aspects of nature. They must avoid *sampling error*, which means obtaining misleading results by using subsets that aren't really representative of the whole (Figure 1.11). Test results are less likely to be distorted when a sampling is large and the test is repeated. Explain how sampling error might have affected the results of the butterfly experiment described in Section 1.6.

5. The Olestra® potato chip experiment in Section 1.6 was a *double-blind* study: Neither the subjects of the experiment nor the researchers knew which potato chips were in which bag until after all the subjects had reported. What do you think are some of the challenges for researchers performing a double-blind study?

6. In 1988 Dr. Randolph Byrd and his colleagues undertook a study of 393 patients admitted to the San Francisco General Hospital Coronary Care Unit. In the experiment, "born-again" Christian volunteers were asked to pray daily for a patient's rapid recovery and for prevention of complications and death. None of the patients knew if he or she was being prayed for or not, and none of the volunteers or patients knew each other. How each patient fared in the hospital was classified by Byrd as "good," "intermediate," or "bad." Byrd determined that the patients who had been prayed for fared a little better than those who had not. His was the first experiment to document, in a scientific fashion, statistically significant results in support of the prediction that prayer has beneficial effects on the outcome of seriously ill patients. Publication of these results engendered a storm of criticism, mostly from scientists who cited bias in Byrd's experimental design. For instance, Byrd classified the patients after the experiment had been finished. Think about how bias might play a role in interpreting medical data. Why do you think this experiment generated a dramatic response from the rest of the scientific community?

I Principles of Cellular Life

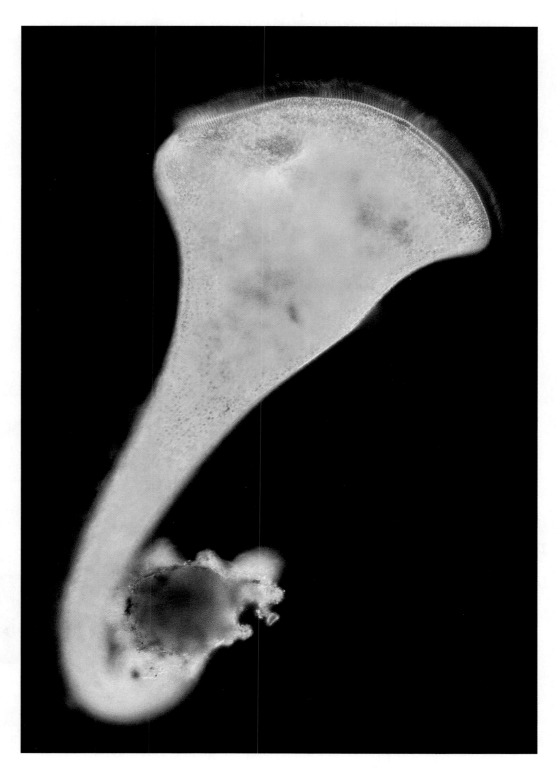

Staying alive means securing energy and raw materials from the environment. Shown here, a living cell of genus Stentor. This single-celled protist has hairlike projections around the opening to a cavity in its body, which is about two millimeters long. Its "hairs" are fused-together cilia, which beat the surrounding water. They create a current that wafts food into the cavity.

IMPACTS, ISSUES   *What Are You Worth?*

Hollywood thinks Leonardo DiCaprio is worth $20 million a picture, the Yankees think shortstop Alex Rodriguez is worth $217 million per decade, and the United States thinks the average teacher is worth $44,367 per year. Chemically, though, how much is a human body really worth (Figure 2.1a)?

Think about it. The human body is a collection of **elements**, or types of atoms. **Atoms** are fundamental substances that have mass and take up space, and cannot be broken apart by everyday means. Keep grinding up a chunk of copper, and the smallest bit you will end up with will be a lone atom of copper. Atoms are the smallest units of an element that still retain the element's properties.

Oxygen, hydrogen, carbon, and nitrogen are the most abundant elements in organisms. Next are phosphorus, potassium, sulfur, calcium, and sodium. *Trace* elements make up less than 0.01 percent of body weight. Selenium is an example.

Wait a minute! Selenium, mercury, arsenic, lead, and many other elements in the body are toxic, right? Maybe, or maybe not. As researchers

Oxygen (O)	43.00 kg	$0.021739	Chromium (Cr)	14.00 mg	0.003402
Carbon (C)	16.00 kg	6.400000	Manganese (Mg)	12.00 mg	0.001526
Hydrogen (H)	7.00 kg	0.028315	Arsenic (As)	7.00 mg	0.023576
Nitrogen (N)	1.80 kg	9.706929	Lithium (Li)	7.00 mg	0.024233
Calcium (Ca)	1.00 kg	15.500000	Cesium (Cs)	6.00 mg	0.000016
			Mercury (Hg)	6.00 mg	0.004718
Phosphorus (P)	780.00 g	68.198594	Germanium (Ge)	5.00 mg	0.130435
Potassium (K)	140.00 g	4.098737	Molybdenum (Mo)	5.00 mg	0.001260
Sulfur (S)	140.00 g	0.011623	Cobalt (Co)	3.00 mg	0.001509
Sodium (Na)	100.00 g	2.287748	Antimony (Sb)	2.00 mg	0.000243
Chlorine (Cl)	95.00 g	1.409496	Silver (Ag)	2.00 mg	0.013600
Magnesium (Mg)	19.00 g	0.444909	Niobium (Nb)	1.50 mg	0.000624
Iron (Fe)	4.20 g	0.054600	Zirconium (Zr)	1.00 mg	0.000830
Fluorine (F)	2.60 g	7.917263	Lanthanium (La)	0.80 mg	0.000566
Zinc (Zn)	2.30 g	0.088090	Gallium (Ga)	0.70 mg	0.003367
Silicon (Si)	1.00 g	0.370000	Tellurium (Te)	0.70 mg	0.000722
Rubidium (Rb)	0.68 g	1.087153	Yttrium (Y)	0.60 mg	0.005232
Strontium (Sr)	0.32 g	0.177237	Bismuth (Bi)	0.50 mg	0.000119
Bromine (Br)	0.26 g	0.012858	Thallium (Tl)	0.50 mg	0.000894
Lead (Pb)	0.12 g	0.003960	Indium (In)	0.40 mg	0.000600
			Gold (Au)	0.20 mg	0.001975
Copper (Cu)	72.00 mg	0.012961	Scandium (Sc)	0.20 mg	0.058160
Aluminum (Al)	60.00 mg	0.246804	Tantalum (Ta)	0.20 mg	0.001631
Cadmium (Cd)	50.00 mg	0.010136	Vanadium (V)	0.11 mg	0.000322
Cerium (Ce)	40.00 mg	0.043120	Thorium (Th)	0.10 mg	0.004948
Barium (Ba)	22.00 mg	0.028776	Uranium (U)	0.10 mg	0.000103
Iodine (I)	20.00 mg	0.094184			
Tin (Sn)	20.00 mg	0.005387	Samarium (Sm)	50.00 µg	0.000118
Titanium (Ti)	20.00 mg	0.010920	Beryllium (Be)	36.00 µg	0.000218
Boron (B)	18.00 mg	0.002172	Tungsten (W)	20.00 µg	0.000007
Nickel (Ni)	15.00 mg	0.031320			
Selenium (Se)	15.00 mg	0.037949			

Grand Total: $118.63

Figure 2.1 **(a)** What are you worth, chemically speaking? **(b)** Proportions of the most common elements in a human body, Earth's crust, and seawater. How are they similar? How do they differ? **a**

the big picture

Atoms and Elements

All substances are made of one or more elements. Atoms are the smallest units of matter that still retain the element's properties. They are composed of protons, neutrons, and electrons.

Why Electrons Matter

Whether an atom will interact with other atoms depends on how many electrons it has and how they are arranged. Chemical bonds unite two or more atoms.

Human		Earth's Crust		Ocean	
Oxygen	61.0%	Oxygen	46.0%	Oxygen	85.7%
Carbon	23.0	Silicon	27.0	Hydrogen	10.8
Hydrogen	10.0	Aluminum	8.2	Chlorine	2.0
Nitrogen	2.6	Iron	6.3	Sodium	1.1
Calcium	1.4	Calcium	5.0	Magnesium	0.1
Phosphorus	1.1	Magnesium	2.9	Sulfur	0.1
Potassium	0.2	Sodium	2.3	Calcium	0.04
Sulfur	0.2	Potassium	1.5	Potassium	0.03

b

decipher chemical processes peculiar to life, they are finding that many trace elements considered to be poisons actually perform essential biological functions. For instance, large doses of chromium damage nerves and cause cancer, but one form works with insulin, a hormone that helps control the glucose level in blood. A little selenium is toxic, but too little causes heart and thyroid problems. An intricate balance of the right kinds of elements keeps the body functioning properly.

One more point: Earth's crust contains the same elements as the human body, but we're not just dirt. Like all living things, the proportions and organization of our elements are unique (Figure 2.1b). And building and maintaining that organization takes tremendous input of energy (just ask any pregnant woman).

You could buy all of the elements in a 150-pound human body for about $118.63. But constructing any living thing requires a remarkably complex interplay of energy and biological molecules that is far beyond the scope of any laboratory to duplicate, at least for now.

 How Would You Vote?

Fluoride has been proven to help prevent tooth decay. But too much wrecks bones and teeth, and causes birth defects. A lot can kill you. Many communities in the United States add fluoride to their drinking water. Do you want it in yours? See the Media Menu for details, then vote online.

Atoms Bond

The molecular organization and the activities of every living thing arise from ionic, covalent, and hydrogen bonds between atoms.

No Water, No Life

Water's unique characteristics, including temperature-stabilizing effects, cohesion, and solvent properties, make life possible on Earth.

Hydrogen Ions Rule

Life is adapted to the properties of water and to changing concentrations of hydrogen ions and other substances dissolved in water.

2.1 Start With Atoms

Life's chemical properties start with protons, neutrons, and electrons. The unique character of each element actually begins with the number of protons, which is the same in all of its atoms.

Atoms, again, are the smallest units that retain the properties of an element. All atoms are made of three kinds of subatomic particles: protons, neutrons, and electrons (Figure 2.2). Each **proton** carries a positive *charge*, or a defined amount of electricity. Protons are symbolized as p^+. An atom's nucleus (core) holds one or more protons. It also holds **neutrons**, which have no charge. Zipping about the nucleus are one or more **electrons**, which carry a negative charge (e^-).

The positive charge of a proton and the negative charge of an electron balance each other. So an atom that has the same number of electrons and protons has no net electrical charge.

Each element has a unique **atomic number**, or the number of protons in the nucleus of its atoms. For example, the atomic number for hydrogen, which has one proton, is 1. For carbon, with six protons, it is 6.

Each element also has a **mass number**, equal to the total number of protons and neutrons in the atomic nucleus. For example, carbon, with six protons and six neutrons, has a mass number of 12.

Why bother with atomic and mass numbers? If you know how many electrons, protons, and neutrons the atoms of an element contain, you can predict what the

a **b** **c**

Figure 2.2 Different ways of representing atoms, using hydrogen (H) as the example. (**a**) A shell model shows the number of electrons and their relative distances from the nucleus. (**b**) Balls show relative sizes of atoms. (**c**) Electron density clouds show electron distribution around the nucleus.

chemical behavior of that element will probably be under different conditions.

Elements were being classified in terms of chemical similarities long before their subatomic particles were discovered. In 1869, Dmitry Mendeleev, known more for his extravagant hair than his discoveries (he cut it only once per year), arranged the known elements into a repeating pattern based on their chemical properties. Using gaps in this **periodic table**, Mendeleev was able to predict correctly the existence of other elements that had yet to be discovered.

The elements fall into order in the periodic table according to their atomic number (Figure 2.3). Those in the same column of the table have the same number of electrons available for interaction with other atoms. As a result, they behave in a remarkably similar way. For instance, helium, neon, radon, and other gases in the vertical column farthest to the right are called *inert* elements because none of their electrons is available for chemical interaction. Consequently, they rarely do much; they exist mostly as solitary atoms.

Not all of the elements in the periodic table occur in nature. The elements after atomic number 92 are so highly unstable that they have been produced only in very small quantities in the laboratory—sometimes no more than a single atom. They wink out of existence fast. Some elements still haven't been made.

Figure 2.3 Periodic table of the elements and Dmitry Mendeleev, who created it. Some of the symbols for elements are abbreviations for their Latin names. For instance, Pb (lead) is short for *plumbum*; the word "plumbing" is related, because ancient Romans used lead to make their water pipes.

Atoms are the smallest units of an element, or fundamental substance, that still retain the properties of that element. Ninety-two elements occur naturally on Earth.

One or more positively charged protons, negatively charged electrons, and (except for hydrogen) neutrons make up atoms.

An element's chemical properties are a direct consequence of the number of electrons it has available for interacting with other atoms.

2.2 Radioisotopes

*All elements are defined by the number of protons in their atoms—but an element's atoms can differ in their number of neutrons. We call such atoms **isotopes** of the same element. And some are radioactive.*

Henri Becquerel discovered radioactivity by accident in 1896. He put some crystals of phosphorescent uranium salts on top of an unexposed photographic plate inside a desk drawer. Between the uranium and the plate were several sheets of opaque black paper, a coin, and a metal screen. A day later, he used the film and developed it. Surprisingly, a negative image of the coin and screen appeared on it. Energy emitted by the uranium had exposed the film all around the metal. Becquerel concluded that uranium salts emit some form of "radiation" capable of going through things that light cannot penetrate. What was it?

As we now know, most elements in nature have two or more kinds of isotopes. Carbon has three, nitrogen has two, and so on. A superscript number to the left of an element's symbol is the isotope's mass number (combined number of protons and neutrons). For instance, carbon's three natural isotopes are ^{12}C (or carbon 12, the most common form, with six protons, six neutrons), ^{13}C (six protons, seven neutrons), and ^{14}C (six protons, eight neutrons).

Too many or too few neutrons in the nucleus of an atom can cause it to be unstable, or radioactive. A radioactive atom spontaneously emits energy as subatomic particles and x-rays when its nucleus disintegrates. This process, called **radioactive decay**, transforms one element into another. ^{13}C and ^{14}C are radioactive isotopes, or **radioisotopes**, of carbon. Each radioisotope decays with a particular amount of energy into a predictable, more stable product. For example, after 5,700 years, about half of the atoms in a sample of ^{14}C will have turned into ^{14}N (nitrogen) atoms. As you'll see in Chapter 17, researchers use radioactive decay to estimate the age of fossils.

Different isotopes of an element are still the same element. For the most part, carbon is carbon, regardless of how many neutrons it has. Living systems use ^{12}C the same way as ^{14}C. Knowing this, researchers or clinicians studying a certain type of molecule make **tracers**, in which a radioisotope gets substituted for a stable element in that molecule. They deliver tracers into a cell, a multicelled body, or an ecosystem. Energy from radioactive decay is like a shipping label; it helps us track the molecule of interest with instruments that detect radioactivity.

Melvin Calvin and his colleagues used a tracer, carbon dioxide gas made with ^{14}C, to discover the specific steps of photosynthesis. By steeping plants in the radioactive gas, they were able to follow the path of the radioactive carbon atoms through each reaction step in the formation of sugars and starches.

Radioisotopes also are used in medicine. *PET* (short for *Positron-Emission Tomography*) uses radioisotopes to form images of body tissues. Clinicians attach a radioisotope to glucose or another sugar. They inject this tracer into a patient, who is moved into a PET scanner (Figure 2.4*a*). Cells throughout the body absorb the tracer at different rates. The scanner then detects radiation caused by energy from the decay of the radioisotope, and that radiation is used to form an image. Such images can reveal variations and abnormalities in metabolic activity (Figure 2.4*d*).

detector ring inside PET scanner   body section inside ring

The ring intercepts emissions from the labeled molecules

Figure 2.4   (**a**) Patient moving into a PET scanner. (**b,c**) Inside, a ring of detectors intercepts radioactive emissions from labeled molecules that were injected into the patient. Computers analyze and color-code the number of emissions from each location in the scanned body region.

(**d**) Different colors in a brain scan signify differences in metabolic activity. Cells of this brain's left half absorbed and used the labeled molecules at expected rates. However, cells in the right half showed little activity. The patient was diagnosed as having a neurological disorder.

2.3 What Happens When Atom Bonds With Atom?

Atoms acquire, share, and donate electrons. Atoms of some elements do this easily; others do not. Why is this so? To come up with an answer, look to the number and arrangement of electrons in atoms.

ELECTRONS AND ENERGY LEVELS

In our world, simple physics explains the motion of an apple falling from a tree. Tiny electrons belong to a strange world where everyday physics doesn't apply. (If electrons were as big as apples, you'd be about 3.5 times taller than our solar system is wide.) Different forces bring about the motion of electrons, which can get from here to there without going in between!

We can calculate where an electron is, although not exactly. The best we can do is say that it's somewhere in a fuzzy cloud of probability density. Where it can go depends on how many other electrons are buzzing about an atom's nucleus. As it turns out, electrons can occupy orbitals, which are volumes of space around the nucleus. There are many orbitals, with different three-dimensional shapes.

An atom has about same number of electrons as protons. For most atoms, that's a lot of electrons. How are these electrons arranged, given that they repel each other? Think of an atom as a multilevel apartment building with lots of vacant rooms to rent to electrons, and a nucleus in the basement. Each "room" is one orbital, and it rents out to two electrons at most. An orbital holding one electron only has a vacancy; another electron can move in.

Each floor in that atomic apartment building corresponds to an energy level. There is only one room on the first floor (one orbital at the lowest energy level, closest to the nucleus), and it fills first. For hydrogen, the simplest atom, that room has a single electron (Figure 2.5). For helium, it has two. In other words, helium has no vacancies at the first (lowest) energy level. In larger atoms, more electrons rent second-floor rooms. If the second floor is filled, additional electrons rent third-floor rooms, and so on. *They fill orbitals at successively higher energy levels.*

The farther an electron is from the basement (the nucleus), the greater its energy. An electron in a first-floor room can't move to the second or third floor, let alone the penthouse, unless a boost of energy gets it there. Suppose it absorbs the right amount of energy from, say, sunlight, to get excited about moving up. Move it does. If nothing fills that lower room, though, the electron will quickly go back to it, emitting extra energy as it does. Later, you'll see how cells in plants and in your eyes can harness and use that energy.

FROM ATOMS TO MOLECULES

In shell models, nested "shells" correspond to energy levels. They offer us an easy way to check for electron vacancies in various atoms (Figure 2.6). Bear in mind, atoms do not look like these flat diagrams. The shells are not three-dimensional volumes of space, and they certainly don't show the electron orbitals.

Atoms that have vacancies in the outermost "shell" tend to give up, acquire, or share electrons with other atoms. This kind of electron-swapping between atoms is known as **chemical bonding** (Section 2.4). Atoms with zero vacancies rarely bond with other atoms. By contrast, the most common atoms in organisms—such as oxygen, carbon, hydrogen, nitrogen, and calcium—have vacancies in orbitals at their outermost energy level. And they do bond with other atoms.

third
energy level
(second floor)

second
energy level
(first floor)

first
energy level
(closest to the
basement)

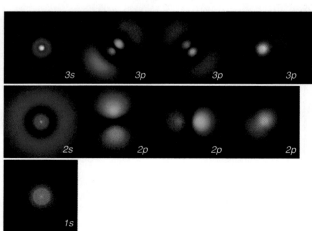

Figure 2.5 First, second, and third levels of the atomic apartment building. Each picture is a three-dimensional approximation of an electron orbital. Colors are most intense in locations where electrons are most likely to be. Orbitals farthest from the nucleus have greater energy and are more complex.

c **Third shell** shows the third set of orbitals: one *s* orbital, three *p* orbitals, and five *d* orbitals, a total of nine orbitals with room for 18 electrons. Sodium has one electron in the third shell of orbitals, and chlorine has seven. Both have vacancies, so they form chemical bonds.

SODIUM
$11p^+, 11e^-$

CHLORINE
$17p^+, 17e^-$

b **Second shell** shows the second energy level, which combines a set of one *s* orbital plus three *p* orbitals. The second shell of orbitals has room for a total of eight electrons. Carbon has six electrons, two in the first shell and four in the second shell. It has four vacancies. Oxygen has two vacancies. Both carbon and oxygen form chemical bonds.

CARBON
$6p^+, 6e^-$

OXYGEN
$8p^+, 8e^-$

a **First shell** shows the first energy level, containing a single orbital (*1s*). Hydrogen has only one electron in this orbital that can hold two. Hydrogen gives up its electron easily, becoming a chemically reactive free proton. A helium atom has two electrons in the *1s* orbital. Having no vacancies, helium does not usually form chemical bonds.

HYDROGEN
$1p^+, 1e^-$

HELIUM
$2p^+, 2e^-$

Figure 2.6 Shell model. Using this model, it is easy to see the vacancies in each atom's outer orbitals. Each circle represents all of the orbitals on one energy level. Larger circles correspond to higher energy levels. This model is highly simplified; a more realistic rendering would show the electrons as fuzzy clouds of probability density about ten thousand times bigger than the nucleus.

REACTANTS:

$12H_2O$ + $6CO_2$

WATER   CARBON DIOXIDE

24 hydrogens   6 carbons
12 oxygens   12 oxygens

sunlight energy

PRODUCTS:

$6O_2$ + $C_6H_{12}O_6$ + $6H_2O$

OXYGEN   GLUCOSE   WATER

12 oxygens   6 carbons   12 hydrogens
              12 hydrogens   6 oxygens
              6 oxygens

Figure 2.7 Chemical bookkeeping. Chemical equations are representations of reactions, or interactions between atoms and molecules. Substances entering a reaction are to the left of a reaction arrow (reactants), and products are to the right, as shown by this chemical equation for photosynthesis.

A **molecule** is simply two or more atoms of the same or different elements joined in a chemical bond. You can write a molecule's chemical composition as a formula that uses symbols for elements. A formula shows the number of each kind of atom in a molecule (Figure 2.7). Water has the chemical formula H_2O. The subscript number tells you that two hydrogen (H) atoms are present for each oxygen (O) atom.

Compounds are molecules that consist of two or more different elements in proportions that never do vary. Water is an example. All water molecules have one oxygen atom bonded to two hydrogen atoms. The ones in rain clouds, the seas, a Siberian lake, a flower's petals, your bathtub, or anywhere else always have twice as many hydrogen as oxygen atoms.

In a **mixture**, two or more molecules intermingle without chemically bonding. For instance, you can make a mixture by swirling water and sugar together. The proportions of elements in a mixture can vary.

Electrons occupy orbitals, or defined volumes of space around an atom's nucleus. Successive orbitals correspond to levels of energy, which become higher with distance from the nucleus.

One or at most two electrons can occupy any orbital. Atoms with vacancies in their highest level orbitals can interact with other atoms.

A molecule is two or more atoms joined in a chemical bond. Atoms of two or more elements are bonded together in compounds. A mixture consists of intermingled molecules.

2.4 Bonds in Biological Molecules

The distinctive properties of biological molecules start with atoms interacting at the level of electrons.

ION FORMATION AND IONIC BONDING

An electron, recall, has a negative charge equal to a proton's positive charge. When an atom has as many electrons as protons, these charges balance each other, so the atom will have a net charge of zero.

Atoms with more electrons than protons carry a net negative charge, and those with more protons than electrons carry a net positive charge. An atom that has either a positive or negative charge is known as an **ion**. Ions form when atoms gain or lose electrons.

Example: An uncharged chlorine atom has seven electrons, hence one vacancy, in the third orbital level. Chlorine tends to grab an electron from other places. That extra electron will make it a chloride ion (Cl^-), with a net negative charge. A sodium atom has a lone electron in the same orbital level, but it is easier to give that one up than to acquire seven more. If it does, it will only have second-level orbitals, and they will be full of electrons—so no vacancy. It becomes a sodium ion with a net positive charge (Na^+).

What happens when one atom gives up an electron that another accepts? The two resulting ions may stay close together, because they have opposite charges that attract each other. A close association of ions is an **ionic bond**. Figure 2.8*a* shows a crystal of table salt, or NaCl. In such crystals, ionic bonds hold the ions in an orderly arrangement.

COVALENT BONDING

In an ionic bond, one atom donates an extra electron that the other accepts. What if both atoms want an extra electron? They can *share* one of their electrons in a hybrid orbital that spans both nuclei. Each atom's vacancy becomes partly filled with a shared electron. When atoms share one or more electrons, they are joined in a **covalent bond** (Figure 2.8*b*). Such bonds are stable and are much stronger than ionic bonds.

Unlike chemical formulas, structural formulas show how atoms are physically arranged in a molecule—they reveal the bonding pattern. A single line that connects two atoms in a structural formula represents two shared electrons in one covalent bond. Molecular hydrogen, with one covalent bond, is written H—H.

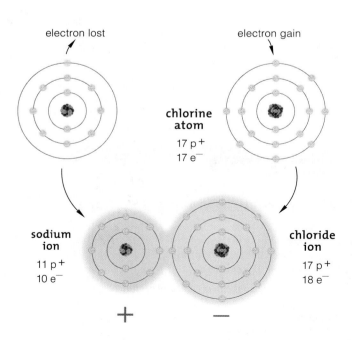

Figure 2.8 Important bonds in biological molecules.

a Ionic bonding. A sodium atom donates its extra electron to a chlorine atom.

In each crystal of table salt, or NaCl, many sodium and chloride ions stay close together because of the mutual attraction of opposite charges. Their ongoing interaction is a case of ionic bonding.

Two atoms can share two electron pairs in a *double* covalent bond. Molecular oxygen (O=O) is like this. In a *triple* covalent bond, two atoms share three pairs, as they do in molecular nitrogen (N≡N). Each time you breathe in, a stupendous number of gaseous O_2 and N_2 molecules flows into your lungs.

In a *nonpolar* covalent bond, two identical atoms share electrons equally, and the molecule shows no difference in charge between its two ends. Molecular hydrogen (H_2) has such symmetry, as do O_2 and N_2.

A *polar* covalent bond forms between atoms of different elements. One of the atoms pulls the shared electrons a little toward one end of the bond. Because the electrons spend extra time there, that part of the molecule bears a slight negative charge. The opposite end bears a slight positive charge. A water molecule (H—O—H) has two polar covalent bonds; the oxygen is negatively charged, and the hydrogens are positive.

HYDROGEN BONDING

A hydrogen atom taking part in a polar covalent bond bears a slight positive charge, so it attracts negatively charged atoms. When the negatively charged atom is bound to a different molecule or to a different part of the same molecule, the interaction between it and the hydrogen atom is called a **hydrogen bond**.

Because they are weak, hydrogen bonds form and break easily. They play crucial roles in the structure and function of biological molecules, especially with water (Section 2.5). They often form between different parts of very large molecules that have folded over on themselves, and hold them in a particular shape. They are also what holds the two nucleotide strands of large DNA molecules together. You can get a sense of these interactions from Figure 2.8c.

Ions form when atoms acquire a net charge by gaining or losing electrons. Two ions of opposite charge attract each other. They can associate in an ionic bond.

In a covalent bond, atoms share a pair of electrons. When atoms share the electrons equally, the bond is nonpolar. When the sharing is not equal, the bond is polar—slightly positive at one end, slightly negative at the other.

In a hydrogen bond, a covalently bound hydrogen atom attracts a negatively charged atom taking part in a different covalent bond.

Read Me First!
and watch the narrated animation on how atoms bond

Two hydrogen atoms, each with one proton, share two electrons in a single nonpolar covalent bond.

molecular hydrogen (H_2)
H—H

Two oxygen atoms, each with eight protons, share four electrons in a nonpolar double covalent bond.

molecular oxygen (O_2)
O=O

Oxygen has vacancies for two electrons in its highest energy level orbitals. Two hydrogen atoms can each share an electron with oxygen. The resulting two polar covalent bonds form a water molecule.

water (H_2O)
H—O—H

b Covalent bonding. Each atom becomes more stable by sharing electron pairs in hybrid orbitals.

Two molecules interacting weakly in one H bond, which can form and break easily.

hydrogen bond

water molecule   ammonia molecule

H bonds helping to hold part of two large molecules together.

Many H bonds hold DNA's two strands together along their length. Individually they are weak, but collectively stabilize DNA's large structure.

c Hydrogen bonds. Such bonds can form at a hydrogen atom that is already covalently bonded in a molecule. The atom's slight positive charge weakly attracts an atom with a slight negative charge that is already covalently bonded to something else. As shown, this can happen between one of the hydrogen atoms of a water molecule and the nitrogen atom of an ammonia molecule.

2.5 Water's Life-Giving Properties

No sprint through basic chemistry is complete unless it leads to the collection of molecules called water. Life originated in water. Organisms still live in it or they cart water around with them inside cells and tissue spaces. Many metabolic reactions use water. Cell shape and cell structure absolutely depend on it.

POLARITY OF THE WATER MOLECULE

Figure 2.9*a* shows the structure of a water molecule. Two hydrogen atoms have formed polar covalent bonds with an oxygen atom. The molecule has no net charge. Even so, the oxygen pulls the shared electrons more than the hydrogen atoms do. Thus, the molecule of water has a slightly negative "end" that's balanced out by its slightly positive "end."

A water molecule's polarity attracts other water molecules. Also, it is so attractive to sugars and other polar molecules that hydrogen bonds readily form between them. That is why polar molecules are known as **hydrophilic** (water-loving) substances.

That same polarity repels oils and other nonpolar molecules, which are **hydrophobic** (water-dreading) substances. Shake a bottle filled with water and salad oil, then set it on a table. Soon, new hydrogen bonds replace the ones that the shaking broke. The reunited water molecules push out oil molecules, which cluster as oil droplets or as an oily film at the water's surface.

The same kinds of interactions occur at the thin, oily membrane between the water inside and outside cells. Membrane organization, and life itself, starts with hydrophilic and hydrophobic interactions. You'll be reading about membrane structure in Chapter 4.

WATER'S TEMPERATURE-STABILIZING EFFECTS

Cells are mostly water, and they also release a lot of metabolic heat. Without water's hydrogen bonds, cells would cook in their own juices. How? All molecules vibrate nonstop, and they move more as they absorb heat. **Temperature** is a measure of molecular motion. Compared to most other fluids, water absorbs more heat energy before it gets measurably hotter. So water acts as a heat reservoir, and its temperature remains relatively stable. In time, increases in heat step up the motion within water molecules. Before that happens, however, much of the heat will go into disrupting hydrogen bonds between molecules.

slight negative charge
on the oxygen atom

The + and − ends balance each other; the whole molecule carries no net charge, overall.

slight positive charge
a on the hydrogen atoms

Figure 2.9 Water, a substance essential for life.

(**a**) Polarity of an individual water molecule.

(**b**) Hydrogen bonding pattern among water molecules in liquid water. Dashed lines signify hydrogen bonds, which break and reform rapidly.

(**c**) Hydrogen bonding in ice. Below 0°C, every water molecule hydrogen-bonds to four others, in a rigid three-dimensional lattice. The molecules are farther apart, or less dense, than they are in liquid water. As a result, ice floats on water.

Thanks partly to rising levels of methane and other greenhouse gases that are contributing to global warming, the Arctic ice cap is melting. At current rates, it will be gone in fifty years. So will the polar bears. Already their season for hunting seals is shorter, bears are thinner, and they are giving birth to fewer cubs.

Figure 2.10 Two spheres of hydration.

Figure 2.11 Examples of water's cohesion. (**a**) When a pebble hits liquid water and forces molecules away from the surface, the individual water molecules don't fly off every which way. They stay together in droplets. Why? Countless hydrogen bonds exert a continuous inward pull on individual molecules at the surface.

(**b**) And just how does water rise to the very top of trees? Cohesion, and evaporation from leaves, pulls it upward.

With a fairly stable water temperature, hydrogen bonds form as fast as they break. Energy inputs can increase the molecular motion so much that the bonds stay broken, and individual molecules at the water's surface escape into air. By this process, **evaporation**, heat energy converts liquid water to a gas. An energy input has overcome the attraction between molecules of water, which break free. The surface temperature of water decreases during evaporation.

Evaporative water loss helps you and some other mammals cool off when you sweat on hot, dry days. Sweat, about 99 percent water, evaporates from skin.

Below 0°C, water molecules don't move enough to break their hydrogen bonds, and they become locked in the latticelike bonding pattern of ice (Figure 2.9c). Ice is less dense than water. During winter freezes, ice sheets may form near the surface of ponds, lakes, and streams. The ice blanket "insulates" the liquid water beneath it and helps protect many fishes, frogs, and other aquatic organisms against freezing.

WATER'S SOLVENT PROPERTIES

Water is an excellent *solvent*, meaning ions and polar molecules easily dissolve in it. A dissolved substance is known as a **solute**. In general, a substance is said to be *dissolved* after water molecules cluster around ions or molecules of it and keep them dispersed in fluid.

Water molecules cluster around a solute, thereby forming a *sphere of hydration*. Spheres form around any solute in cellular fluids, tree sap, blood, the fluid in your gut, and every other fluid associated with life. Watch it happen after you pour table salt (NaCl) into a cup of water. In time, the crystals of salt separate into ions of sodium (Na^+) and chloride (Cl^-). Each Na^+ attracts the negative end of some water molecules even as Cl^- attracts the positive end of others (Figure 2.10). Spheres of hydration formed this way keep the ions dispersed in fluid.

WATER'S COHESION

Still another life-sustaining property of water is its cohesion. **Cohesion** means something is showing a capacity to resist rupturing when it is stretched, or placed under tension. You see its effect when a tossed pebble breaks the surface of a lake, a pond, or some other body of liquid water (Figure 2.11a). At or near the surface, uncountable numbers of hydrogen bonds are exerting a continuous, inward pull on individual molecules. Bonding creates a high surface tension.

Cohesion is in play inside organisms, too. Plants, for example, absorb nutrient-laden water while they grow. Very narrow columns of liquid water rise inside pipelines of vascular tissues, which extend from roots to leaves and other plant parts. On sunny days, water evaporates from leaves as molecules break free and diffuse into the air (Figure 2.11b). The cohesive force of hydrogen bonds pulls replacements into the leaf cells, in ways you'll read about in Section 26.3.

Being slightly polar, water molecules hydrogen bond to one another and to other polar (hydrophilic) substances. They tend to repel nonpolar (hydrophobic) substances.

The unique properties of liquid water make life possible. Water has cohesion, temperature-stabilizing effects, and a capacity to dissolve many substances.

2.6 Acids and Bases

Ions are dissolved in fluids inside and outside a cell, and they affect its structure and function. Among the most influential are hydrogen ions. They have far-reaching effects largely because they are chemically active and there are so many of them.

THE PH SCALE

At any instant in liquid water, some water molecules split into ions of hydrogen (H^+) and hydroxide (OH^-). These ions are the basis of the **pH scale**. The scale is a way to measure the relative amount of hydrogen ions in solutions such as seawater, blood, or sap. The greater the H^+ concentration, the lower the pH. Pure water (not rainwater or tap water) always has as many H^+ as OH^- ions. This state is neutrality, or pH 7.0 (Figure 2.12).

A one unit decrease from neutrality corresponds to a tenfold increase in H^+ concentration, and an increase by one unit corresponds to a tenfold decrease in H^+ concentration. One way to get a sense of the range is to taste baking soda (pH 9), water (pH 7), and lemon juice (pH 2).

HOW DO ACIDS AND BASES DIFFER?

Substances called **acids** *donate* hydrogen ions and **bases** *accept* hydrogen ions when dissolved in water. *Acidic* solutions, such as lemon juice, gastric fluid, and coffee, release H^+; their pH is below 7. *Basic* solutions, such as seawater, baking soda, and egg white, combine with H^+. Basic solutions (also known as alkaline solutions) have a pH above 7.

Nearly all of life's chemistry occurs near pH 7. Most of your body's internal environment (tissue fluids and blood) is between pH 7.3 and 7.5. Seawater is more basic than body fluids of the organisms living in it.

Acids and bases can be weak or strong. The weak acids, such as carbonic acid (H_2CO_3), are stingy H^+ donors. Strong acids readily give up H^+ in water. An example is the hydrochloric acid that dissociates into H^+ and Cl^- inside your stomach. The H^+ makes your gastric fluid far more acidic, which in turn activates protein-digesting enzymes.

Too much HCl can cause an *acid stomach*. Antacids taken for this condition, including milk of magnesia, release OH^- ions that combine with H^+ to reduce the pH of stomach contents.

High concentrations of strong acids or bases can disrupt ecosystems and make it impossible for cells

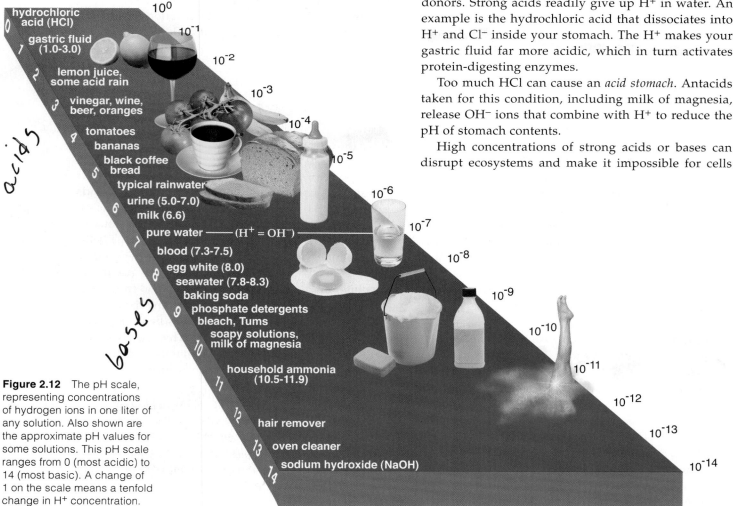

Figure 2.12 The pH scale, representing concentrations of hydrogen ions in one liter of any solution. Also shown are the approximate pH values for some solutions. This pH scale ranges from 0 (most acidic) to 14 (most basic). A change of 1 on the scale means a tenfold change in H^+ concentration.

acids

hydrochloric acid (HCl)
gastric fluid (1.0–3.0)
lemon juice, some acid rain
vinegar, wine, beer, oranges
tomatoes
bananas
black coffee
bread
typical rainwater
urine (5.0–7.0)
milk (6.6)
pure water ——— ($H^+ = OH^-$)
blood (7.3–7.5)
egg white (8.0)
seawater (7.8–8.3)
baking soda
phosphate detergents
bleach, Tums
soapy solutions, milk of magnesia
household ammonia (10.5–11.9)
hair remover
oven cleaner
sodium hydroxide (NaOH)

bases

10^0
10^{-1}
10^{-2}
10^{-3}
10^{-4}
10^{-5}
10^{-6}
10^{-7}
10^{-8}
10^{-9}
10^{-10}
10^{-11}
10^{-12}
10^{-13}
10^{-14}

Figure 2.13 Emissions of sulfur dioxide from a coal-burning power plant. Airborne pollutants such as sulfur dioxide dissolve in water vapor to form acidic solutions. They are a component of acid rain.

to survive. Read the labels on containers of ammonia, drain cleaner, and other products commonly stored in households. Many cause severe *chemical burns*. So does sulfuric acid in car batteries. Fossil fuel burning and nitrogen fertilizers release strong acids that lower the pH of rainwater (Figure 2.13). Some regions are quite sensitive to this *acid rain*. Alterations in the chemical composition of soil and water can harm organisms. We return to this topic in Section 42.2.

SALTS AND WATER

A **salt** is any compound that dissolves easily in water and releases ions *other than* H^+ and OH^-. It commonly forms when an acid interacts with a base. For example:

HCl (acid) + NaOH (base) $\rightleftharpoons$ NaCl (salt) + H_2O
HYDROCHLORIC ACID   SODIUM HYDROXIDE   SODIUM CHLORIDE

Na^+ Cl^- (ionization)

Bidirectional arrows indicate that the reaction goes in both directions. Many of the ions released when salts dissolve in fluid are important components of cellular processes. For example, ions of sodium, potassium, and calcium help nerve and muscle cells function and help plant cells take up water from soil.

BUFFERS AGAINST SHIFTS IN PH

Cells must respond fast to even slight shifts in pH, because excess H^+ or OH^- can alter the functions of biological molecules. Responses are rapid with **buffer systems**. Think of such a system as a dynamic chemical partnership between a weak acid or base and its salt. These two related chemicals work in equilibrium to counter slight shifts in pH. For example, if a small amount of a strong base enters a buffered fluid, the weak acid partner can neutralize the excess OH^- ions by donating some H^+ ions to the solution.

Most body fluids are buffered. Why? Enzymes, receptors, and all other essential biological molecules function properly only within a narrow range of pH. Deviation from the range halts cellular processes.

Carbon dioxide, a by-product of many reactions, combines with water in the blood to compose a buffer system of carbonic acid and bicarbonate ions. When blood pH rises a bit, the carbonic acid neutralizes the excess OH^- by releasing some hydrogen ions, which combine with the OH^- to form water:

$$OH^- + H_2CO_3 \longrightarrow HCO_3^- + H_2O$$
CARBONIC ACID   BICARBONATE (SALT)   WATER

When blood becomes more acidic, this salt mops up the excess H^+ and so shifts the balance of the buffer system toward the acid:

$$HCO_3^- + H^+ \longrightarrow H_2CO_3$$
BICARBONATE   CARBONIC ACID

Buffer systems can neutralize only so many excess ions. With even a slight excess above that point, the pH swings widely. When the blood pH (7.3–7.5) falls even to 7, the individual may fall into a *coma*, an often irreversible state of unconsciousness. This happens in *respiratory acidosis*. Carbon dioxide accumulates, too much carbonic acid forms, and blood pH plummets. By contrast, when the blood pH increases even to 7.8, *tetany* may occur; skeletal muscles cannot be released from contraction. In *alkalosis*, a rise in blood pH can't be reversed. Such conditions can be lethal.

> Ions dissolved in fluids on the inside and outside of cells have key roles in cell function. Acidic substances release hydrogen ions, and basic substances accept them. Salts are compounds that release ions other than H^+ and OH^-.
>
> Acid–base interactions help maintain pH, which is the H^+ concentration in a fluid. Buffer systems help control the body's acid–base balance at levels suitable for life.

(handwritten margin note: pH constant)

Summary

Introduction Chemistry helps us understand the nature of all substances that make up cells, organisms, and the Earth, its waters, and the atmosphere. Table 2.1 summarizes some key chemical terms that you will encounter throughout this book.

Table 2.1	Summary of Important Players in the Chemical Basis of Life
ATOM	Fundamental form of matter that occupies space, has mass, and cannot be broken apart by ordinary physical or chemical means.
Proton (p^+)	Positively charged particle of the atomic nucleus.
Electron (e^-)	Negatively charged particle that can occupy a volume of space (orbital) around the nucleus.
Neutron	Uncharged particle of the atomic nucleus. For a given element, the mass number is the sum of the number of protons and neutrons in the nucleus.
ELEMENT	Type of atom defined by the number of protons, which is its atomic number. Each element has unique chemical properties.
MOLECULE	Unit of matter in which two or more atoms of the same element, or different ones, are bonded together by shared electrons.
Compound	Molecule composed of two or more different elements in unvarying proportions. Water is an example.
Mixture	Intermingling of two or more elements or compounds in proportions that vary.
ISOTOPE	One of two or more forms of an element that differ in the number of neutrons in their nuclei.
Radioisotope	Unstable isotope, having an unbalanced number of protons and neutrons, that emits particles and energy.
Tracer	Molecule of a substance to which a radioisotope is attached. Together with tracking devices, it is used to identify movement or destination of the substance in a metabolic pathway, the body, or some other system.
ION	Atom in which the number of electrons differs from the number of protons; negatively or positively charged. A proton without an electron zipping around it is a hydrogen ion (H^+).
SOLUTE	Any molecule or ion dissolved in some solvent.
Hydrophilic substance	Polar molecule or molecular region that can readily dissolve in water.
Hydrophobic substance	Nonpolar molecule or molecular region that strongly resists dissolving in water.
ACID	Substance that donates H^+ when dissolved in water.
BASE	Substance that accepts H^+ when dissolved in water.
SALT	Compound that releases ions other than H^+ or OH^- when dissolved in water.

Section 2.1 All substances consist of one or more elements. Ninety-two elements are naturally occurring. An atom consists of one or more positively charged protons, negatively charged electrons, and (except for hydrogen atoms) one or more uncharged neutrons. Protons and neutrons occupy the core region, or nucleus. In elements, all of the atoms have the same number of protons.

Section 2.2 Most elements have isotopes, which are two or more forms of atoms that have the same number of protons but different numbers of neutrons. An atom is radioactive when its nucleus is unstable. All elements have one or more radioactive isotopes.

Section 2.3 Whether an atom interacts with others depends on the number and arrangement of its electrons, which occupy orbitals (volumes of space) around the atomic nucleus. When an atom has one or more vacancies in orbitals at its highest energy level, it can interact with other atoms by donating, accepting, or sharing electrons.

Section 2.4 An atom may lose or gain one or more electrons and thus become an ion, which has a positive or negative charge.

Generally, a chemical bond is a union between the electron structures of atoms.

a. In an ionic bond, a positive ion and negative ion stay together by mutual attraction of opposite charges.

b. Atoms often share one or more pairs of electrons in covalent bonds. Electron sharing is equal in nonpolar covalent bonds, and it is unequal in polar covalent bonds. Interacting atoms have no net charge overall, even though the bond can be slightly negative at one end and slightly positive at the other.

c. In a hydrogen bond, one covalently bonded atom (e.g., oxygen) that has a slight negative charge is weakly attracted to the slight positive charge of a hydrogen atom taking part in a different polar covalent bond.

Section 2.5 Polar covalent bonds join together three atoms in a water molecule (two hydrogens and one oxygen). The water molecule's polarity invites extensive hydrogen bonding between molecules in bodies of water. Such bonding is the basis of liquid water's ability to resist temperature changes (more than other fluids do), display internal cohesion, and easily dissolve polar or ionic substances. These properties make life possible.

Section 2.6 The pH of a solution indicates its hydrogen ion concentration. A typical pH range is from 0 (highest H^+ concentration, most acidic) to 14 (lowest H^+ concentration, most basic). At pH 7, or neutrality, H^+ and OH^- concentrations are equal. Acids release H^+ ions in water; bases combine with them. Buffer systems help maintain a favorable pH in internal environments. This is important because most biological processes operate only within a narrow range of pH.

Self-Quiz
Answers in Appendix III

1. Is this statement true or false: Every type of atom consists of protons, neutrons, and electrons.

2. Electrons carry a _____ charge.
   a. positive   b. negative   c. zero

3. A(n) _____ is any molecule to which a radioisotope has been attached for research or diagnostic purposes.
   a. ion   c. element
   b. isotope   d. tracer

4. Atoms share electrons unequally in a(n) _____ bond.
   a. ionic   c. polar covalent
   b. hydrogen   d. nonpolar covalent

5. In a hydrogen bond, a hydrogen atom covalently bonded to one molecule weakly interacts with a _____ part of a neighboring molecule.
   a. polar   b. nonpolar   c. hydrophobic

6. Liquid water shows _____ .
   a. polarity   d. cohesion
   b. hydrogen-bonding capacity e. b through d
   c. notable heat resistance   f. all of the above

7. Hydrogen ions (H^+) are _____ .
   a. the basis of pH values   d. dissolved in blood
   b. unbound protons   e. both a and b
   c. targets of certain buffers f. a through d

8. When dissolved in water, a(n) _____ donates H^+; however, a(n) _____ accepts H^+.

9. A(n) _____ is a dynamic chemical partnership between a weak acid and a weak base.
   a. ionic bond   c. buffer system
   b. solute   d. solvent

10. Match the terms with their most suitable description.
    ___ trace element a. atomic nucleus components
    ___ salt   b. two atoms sharing electrons
    ___ covalent   c. any polar molecule that readily
          bond   dissolves in water
    ___ hydrophilic   d. releases ions other than H^+ and
          substance OH^- when dissolved in water
    ___ protons,   e. makes up less than 0.001
          neutrons   percent of body weight

Critical Thinking

1. By weight, oxygen is the most abundant element in organisms, ocean water, and Earth's crust. Predict which element is the most abundant in the whole universe.

2. *Ozone* is a chemically active form of oxygen gas. High in Earth's atmosphere, a vast layer of it absorbs about 98 percent of the sun's harmful rays. Normal oxygen gas consists of two oxygen atoms joined in a double nonpolar covalent bond: O=O. Ozone has three covalent bonds in this arrangement: O=O—O. It is highly reactive with a variety of substances, and it gives up an oxygen atom and releases gaseous oxygen (O=O). Using what you know about chemistry, explain why you think it is so reactive.

3. Some undiluted acids are less corrosive than when diluted with a little water. In fact, lab workers are told to wipe off splashes with a towel before washing. Explain.

4. Medieval scientists and philosophers called alchemists were the predecessors of modern-day chemists. Many of them tried to transform lead (atomic number 82) into gold (atomic number 79). Explain why they never succeeded.

5. David, an inquisitive three-year-old, poked his fingers into warm water in a metal pan on the stove and didn't sense anything hot. Then he touched the pan itself and got a nasty burn. Explain why water in a metal pan heats up far more slowly than the pan itself.

6. How do many insects, and the basilisk lizard shown in Figure 1.7, walk on water?

7. Why do you think H^+ is often written as H_3O^+?

Media Menu

Student CD-ROM
  Impacts, Issues Video
    What Are You Worth?
  Big Picture Animation
    Elements, bonding patterns, and pH
  Read-Me-First Animation
    How atoms bond
  Other Animations and Interactions
    The shell model of electron distribution
    Structure of water
    How salt dissolves
    The pH scale

InfoTrac
- One-Molecule Chemistry Gets Big Reaction. *Science News*, September 2000.
- What's Water Got to Do with It? *Astronomy*, August 2001.
- Walking on Water. *Natural History*, April 2000.

Web Sites
- Web Elements: www.webelements.com/
- Chemistry Review: web.mit.edu/esgbio/www/chem/review.html
- Water Science for Schools: ga.water.usgs.gov/edu/

How Would You Vote?

Fluoride has been proven to prevent tooth decay. However, a high intake of fluoride can discolor teeth, weaken bones, and cause birth defects. Really large amounts can kill you. Many communities in the United States add fluoride to their water supply. Do you want it added to yours?

3 MOLECULES OF LIFE

About 2,000 years ago, in the mountains of Greece, the oracle of Delphi delivered prophecies from Apollo after inhaling sweet-smelling fumes that had collected in the sunken floor of her temple. Her prophecies tended to be rambling and cryptic. Why? She was babbling in a hydrocarbon-induced trance. Geologists recently found intersecting, earthquake-prone faults under the temple. When the faults slipped, methane, ethane, and ethylene

methane

ethane

ethylene

Figure 3.1 *Left:* In Greece, ruins of the Temple of Apollo, where hydrocarbon gases seep from the earth. *Above:* The oracle at Delphi, believed to dispense cryptic advice direct from Apollo to people who, like her, had no knowledge of chemistry. The invisible, hallucinogenic fumes that induced her babblings are fancifully depicted in this painting.

the big picture

No Carbon, No Life Living cells build carbohydrates, lipids, proteins, and nucleic acids from simpler organic compounds. These large molecules of life have a backbone of carbon atoms with attached functional groups that help dictate their structure and function.

Carbohydrates Carbohydrates are the most abundant biological molecules. Simple types function as quick energy sources or transportable forms of energy. Complex types function as structural materials or energy reservoirs.

seeped out from the depths. All three gases are mild narcotics. The sweet-smelling ethylene can bring on hallucinations (Figure 3.1).

Ancient Greeks thought that Apollo spoke to them through the oracle; they believed in the supernatural. Scientists looked for a natural explanation, and they found carbon compounds behind her words. Why is their explanation more compelling? It started with tested information about the structure and effects of the world's substances, and it was based on analysis of three gaseous substances drawn from the site.

All three gases are nothing more than a few carbon and hydrogen atoms; hence the name, hydrocarbons. Thanks to scientific inquiry, we now know a lot about them. For example, we know that methane was present when Earth first formed. We know it is released when volcanoes erupt, when we burn wood or peat or fossil fuels, and when termites and cattle pass gas. Methane collects in the atmosphere and in ocean depths along the continental shelves. We also know methane is one of the greenhouse gases, which you will read about in Chapter 41, and that it is a contributing factor in global warming.

In short, knowledge about lifeless substances can tell you a lot about life. It will serve you well when you turn your mind to almost any topic concerning the past, present, and future—from Greek myths, to health and disease, to forests, and to physical and chemical conditions that span the globe and affect life everywhere.

 How Would You Vote?

Undersea methane deposits might be developed as a vast supply of energy, but the environmental costs are unknown. Should we continue to move toward exploiting this resource? See the Media Menu for details, then vote online.

Lipids

Certain lipids function as energy reservoirs, others as structural components of cell membranes, as waterproofing or lubricating substances, and as signaling molecules.

Proteins

Structurally and functionally, proteins are the most diverse molecules of life. They include enzymes, structural materials, signaling molecules, and transporters.

Nucleotides and Nucleic Acids The nucleic acids DNA and RNA are made of a few kinds of nucleotide subunits. They store, retrieve, and translate genetic information that provides instructions for building proteins.

3.1 Molecules of Life—From Structure to Function

Under present-day conditions on Earth, only living cells can make complex carbohydrates and lipids, proteins, and nucleic acids. These are the molecules of life, and their structure holds clues to how each kind functions.

The molecules of life are **organic compounds**, which contain carbon and at least one hydrogen atom. They have a precise number of atoms arranged in specific ways. **Functional groups** are lone atoms or clusters of atoms that are covalently bonded to carbon atoms of organic compounds.

CARBON'S BONDING BEHAVIOR

Living things consist mainly of oxygen, hydrogen, and carbon. Most of the oxygen and hydrogen are in the form of water. Put water aside, and carbon makes up more than half of what's left.

Carbon's importance in life arises from its versatile bonding behavior. *Each carbon atom can covalently bond with as many as four other atoms.* Such bonds, in which two atoms share one, two, or three pairs of electrons, are relatively stable. They often join carbon atoms into "backbones" to which hydrogen, oxygen, and other elements are attached. The three-dimensional shapes of large organic compounds start with these bonds.

Methane, mentioned in the chapter introduction, is the simplest organic compound. It has four hydrogen atoms bonded covalently to a carbon atom (CH_4).

Figure 3.2*a* is a ball-and-stick model for glucose, an organic compound with hydrogen and oxygen bonded covalently to a backbone of six carbon atoms. Usually this backbone coils back on itself, with two carbons joined to form a ring structure (Figure 3.2*b*).

You can represent a carbon ring in different ways. A flat structural model may show the carbons but not the atoms bonded to it (Figure 3.2*c*).

We use insights into the structure of molecules to explore how cells and multicelled organisms function. For instance, virus particles can infect a cell if they dock at specific protein molecules of a cell membrane. Like Lego® blocks, membrane proteins have ridges, clefts, and charged regions that can match up with ridges, clefts, and charged regions of a protein at the surface of a virus particle.

a

b

c

Figure 3.2 Some ways of representing organic compounds. (**a**) Ball-and-stick model for glucose, linear structure. (**b**) Glucose ring structure. (**c**) Two kinds of simplified six-carbon rings.

FUNCTIONAL GROUPS

Functional groups, again, are either atoms or clusters of atoms that impart distinct properties to a molecule. The number, kind, and arrangement of functional groups influences structural and chemical properties of carbohydrates, lipids, proteins, and nucleic acids (Figure 3.3).

For example, hydrocarbons (organic molecules of only carbon and hydrogen atoms) are hydrophobic, or nonpolar. Fatty acids have chains of them, which is why lipids with fatty acid tails resist dissolving. Sugars, a class of compounds called alcohols, contain one or more polar *hydroxyl* (—OH) groups. Molecules of water quickly form hydrogen bonds with these groups; that is why sugars dissolve fast in water.

Figure 3.3 Common functional groups in the molecules of life, with examples of their chemical characteristics.

Figure 3.5 Two metabolic reactions with frequent roles in building organic compounds. (**a**) A condensation reaction, with two molecules being joined into a larger one. (**b**) Hydrolysis, a water-requiring cleavage reaction.

Figure 3.4 Observable differences in traits between female and male wood ducks (*Aix sponsa*), influenced by estrogen and testosterone. These two sex hormones have the same carbon ring structure. They differ only in the position of functional groups attached to the rings.

As another example, testosterone and estrogen are sex hormones. Both are remodeled versions of a type of lipid called cholesterol, and they differ slightly in their functional groups (Figure 3.4). That seemingly tiny difference has big consequences. Consider: Early on, a vertebrate embryo is neither male nor female; it just has a set of tubes that will slowly develop into a reproductive system. In the presence of testosterone, the tubes will become male sex organs, and male traits will develop. In the absence of testosterone, however, the tubes will become female sex organs that secrete estrogens. Those estrogens will guide the formation of distinctly female traits.

HOW DO CELLS ACTUALLY BUILD ORGANIC COMPOUNDS?

Cells build big molecules mainly from four families of small organic compounds: simple sugars, fatty acids, amino acids, and nucleotides. These compounds have two to thirty-six carbon atoms, at most. We refer to them as *monomers* when they are structural units of larger molecules. Molecules that contain repeating monomers are also known as polymers (*mono–*, one; *poly–*, many). As you will see shortly, starch can be considered a polymer of many glucose units.

How do cells actually do the construction work? At this point, just be aware that reactions by which a cell builds, rearranges, and splits apart all organic compounds require more than inputs of energy. They

also require **enzymes**: proteins that make substances react faster than they would on their own. Different enzymes mediate the following classes of reactions:

1. *Functional-group transfer*. A functional group split away from one molecule is transferred to another.

2. *Electron transfer*. An electron split away from one molecule is donated to another.

3. *Rearrangement*. One type of organic compound is converted to another by a juggling of internal bonds.

4. *Condensation*. Covalent bonding between two small molecules results in a larger molecule.

5. *Cleavage*. A molecule splits into two smaller ones.

To get a sense of these cell activities, think of what happens in a **condensation reaction**. Enzymes split an —OH group from one molecule and an H atom from another. A covalent bond between the discarded parts forms H_2O (Figure 3.5*a*). Polymers such as starch form by repeated condensation reactions.

Another example: **Hydrolysis**, a type of cleavage reaction, is like condensation in reverse (Figure 3.5*b*). Enzymes split molecules at specific groups, then attach one —OH group and a hydrogen atom derived from water to the exposed sites.

Complex carbohydrates, complex lipids, proteins, and nucleic acids are the molecules of life.

Organic compounds have diverse, three-dimensional shapes and functions that arise from their carbon backbone and with functional groups covalently bonded to it.

Enzyme-mediated reactions build the molecules of life mainly from smaller organic compounds—simple sugars, fatty acids, amino acids, and nucleotides.

3.2 Bubble, Bubble, Toil and Trouble

Why include methane, a "lifeless" hydrocarbon, in a chapter on the molecules of life? Consider: Vast methane hydrate deposits beneath the ocean floor could explode at any time, as a colossal methane belch that could actually end life as we know it.

This story started long ago, when organic remains and wastes of countless marine organisms sank to the bottom of the ocean. Over time, more and more sediments accumulated above them. Today, a few kilometers under the seafloor, this organic collection nourishes methane-producing archaea. A tremendous amount of methane, the product of metabolic activity, bubbles upward and emerges into ocean water in places called *methane seeps* (Figures 3.6 and 3.7).

In the mud near seeps, methane pressure is high. There, unrelated microorganisms associate in tight clusters. Archaea inside the clusters use methane as

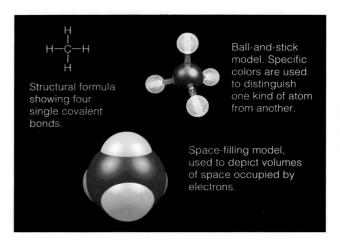

Figure 3.6 Molecular models for methane (CH_4). The ball-and-stick model is good for conveying bond angles and the distribution of a molecule's mass. The space-filling model is better for showing a molecule's surfaces.

Figure 3.7 (a) "Chimneys" of microorganisms and bubbles of methane gas almost 230 meters (750 feet) below sea level in the Black Sea.

(b) The seafloor methane cycle. Methane, formed by archaea far beneath the seafloor, seeps out through vents. Some is captured by clusters of microorganisms that release carbon dioxide and hydrogen sulfide as metabolic products. Other microbes use those products and become the basis of deep-sea food webs.

methane-eating archaea (*red*) and sulfate-eating bacteria (*green*) near seeps

methane-producing archaea

b

methane hydrates

Figure 3.8 A blob of methane hydrate on the seafloor. Notice the methane bubbles above it.

Figure 3.9 *Lystrosaurus*, about a meter long. This animal is now extinct but made it through the Permian mass extinction.

an energy source, and carbon dioxide and hydrogen are released as wastes. In a remarkable metabolic handoff, bacteria that surround them immediately use the hydrogen. During the process they convert sulfate dissolved in seawater to hydrogen sulfide.

The allied organisms accomplish a chemical feat that neither would be capable of on its own. But this doesn't account for all of the methane produced by the underground archaea. What happens to the rest? At methane seeps, high water pressure and low temperature "freezes" the bubbling methane into an icy material called *methane hydrate* (Figure 3.8).

Recently, scientists discovered vast deposits of methane hydrate around the world. They estimate that a thousand billion tons of methane are frozen on the seafloor. The deposits actually are the world's largest reserve of natural gas, but we don't have a safe, efficient way to retrieve it.

Here's the problem. The icy crystals are unstable. Methane hydrate instantly falls apart into methane gas and liquid water as soon as the temperature goes up or the pressure goes down. It doesn't take much, only a few degrees. Methane hydrate disintegration can be explosive. It can cause an irreversible chain reaction that can vaporize neighboring deposits. We have physical evidence of ancient methane hydrate explosions. Small ones pockmarked the ocean floor; immense ones caused underwater landslides that stretched from one continent to another.

The greatest of all mass extinctions occurred 250 million years ago; it marked the end of the Permian period. All but about five percent of marine species abruptly vanished. So did about 70 percent of the known plants, insects, and other species on land.

Chemical clues locked in fossils point to a huge spike in the atmospheric concentration of carbon dioxide—not just any carbon dioxide, but rather carbon dioxide that had been assembled by living things. Something caused lots of methane hydrate to disintegrate at once. In an abrupt, gargantuan burp, millions of tons of methane exploded from the seafloor. Methane-eating bacteria quickly converted most of it to carbon dioxide, which displaced most of the oxygen in the atmosphere and ocean.

Too much carbon dioxide, too little oxygen. Just imagine being instantaneously transported to the top of Mount Everest and trying to jog in the "thin air," with its lower oxygen concentration. You would pass out and die. Before the Permian's Great Dying, oxygen made up about 35 percent of the atmosphere. After the burp, its concentration plummeted to 12 percent. Nearly all of the animals on land and in the seas probably suffocated.

For a long time, scientists couldn't figure out why *Lystrosaurus* didn't become extinct along with nearly everything else. Lucky *Lystrosaurus*. Someone finally figured out that this mammal-like reptile did not suffocate because it was already adapted to the stale, oxygen-poor air of its underground burrows. As Figure 3.9 indicates, *Lystrosaurus* had a big chest cavity, big lungs, stout ribs, and a short route for gas flow inside its stubby nostrils.

The methane problem is closer than you might think. Not long ago, huge methane hydrate deposits were discovered about 96 kilometers (60 miles) off the coast of Newport, Oregon, and off the Atlantic seaboard. What is to become of small-lunged, thin-ribbed people after another methane burp?

3.3 The Truly Abundant Carbohydrates

Which biological molecules are most plentiful in nature? Carbohydrates. Most consist of carbon, hydrogen, and oxygen in a 1:2:1 ratio. Cells use different carbohydrates as structural materials and transportable or storable forms of energy. Monosaccharides, oligosaccharides, and polysaccharides are the main classes.

THE SIMPLE SUGARS

"Saccharide" is from a Greek word that means sugar. The *mono*saccharides (one sugar unit) are the simplest carbohydrates. They have at least two —OH groups bonded to their carbon backbone and one aldehyde or ketone group. Most dissolve easily in water. Common types have a backbone of five or six carbon atoms that tends to form a ring structure when dissolved.

Ribose and deoxyribose are the sugar monomers of RNA and DNA, respectively; each has five carbon atoms. Glucose has six (Figure 3.10*a*). Cells use glucose as an energy source, a structural unit, or a precursor

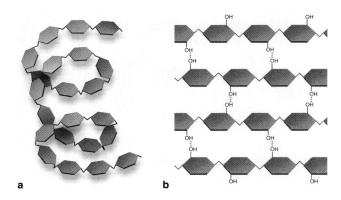

Figure 3.11 Bonding patterns for glucose units in (**a**) starch, and (**b**) cellulose. In amylose, a form of starch, a series of covalently bonded glucose units form a chain that coils. In cellulose, bonds form between glucose chains. The pattern stabilizes the chains, which can become tightly bundled.

(parent molecule) for other organic compounds, such as vitamin C (a sugar acid) and glycerol, an alcohol with three —OH groups.

SHORT-CHAIN CARBOHYDRATES

Unlike the simple sugars, an *oligo*saccharide is a short chain of covalently bonded sugar monomers. (*Oligo*– means a few.) The *di*saccharides consist of two sugar monomers. Sucrose, the most plentiful sugar in nature, contains a glucose and a fructose unit (Figure 3.10*b*). Lactose, a disaccharide in milk, has one glucose and one galactose unit. Table sugar is sucrose extracted from sugarcane and sugar beets. Many proteins and lipids have oligosaccharide side chains. Later in the book, you will come across chains with essential roles in self-recognition and immunity. You also will see how such chains are part of receptors that function in cell-to-cell communication.

COMPLEX CARBOHYDRATES

The "complex" carbohydrates, or *poly*saccharides, are straight or branched chains of many sugar monomers —often hundreds or thousands. Different kinds have one or more types of monomers. The most common polysaccharides are cellulose, starch, and glycogen. Even though all three are made of glucose, they differ a lot in their properties. Why? The answer starts with differences in covalent bonding patterns between their glucose units, which are joined together in chains.

In starch, the pattern of covalent bonding puts each glucose unit at an angle relative to the next unit in line. The chain ends up coiling like a spiral staircase

a Structure of glucose   **b** Structure of fructose

glucose   fructose

sucrose   + H₂O

Figure 3.10 (**a**,**b**) Straight-chain and ring forms of glucose and fructose. For reference purposes, carbon atoms of these simple sugars are numbered in sequence, starting at the end closest to the molecule's aldehyde or ketone group. (**c**) Condensation of two monosaccharides into a disaccharide.

a Structure of amylose, a soluble form of starch. Cells inside tree leaves briefly store excess glucose monomers as starch grains in their chloroplasts, which are tiny, membrane-bound sacs that specialize in photosynthesis.

c Glycogen. Animal cells build this polysaccharide as a storage form when the body has excess glucose. It is especially abundant in the liver and muscles of highly active animals, including fishes and people.

b Structure of cellulose. In cellulose fibers, chains of glucose units stretch side by side and hydrogen-bond at –OH groups. The many hydrogen bonds stabilize the chains in tight bundles that form long fibers. Few organisms produce enzymes that can digest this insoluble material. Cellulose is a structural component of plants and plant products, such as wood and cotton dresses.

Figure 3.12 Molecular structure of starch (**a**), cellulose (**b**), and glycogen (**c**), and their typical locations in a few organisms. All three carbohydrates consist only of glucose units.

(Figure 3.11*a*). Many —OH groups project outward from the coiled chains and make the chains accessible to certain enzymes. This is important. For example, plants briefly store their photosynthetically produced glucose in the form of starch. When free glucose is scarce, enzymes quickly hydrolyze the starch.

In cellulose, glucose chains stretch side by side and hydrogen-bond to one another, as in Figure 3.11*b*. This bonding arrangement stabilizes the chains in a tightly bundled pattern that can resist hydrolysis by most enzymes. Long fibers of cellulose are a structural part of plant cell walls (Figure 3.12*b*). Like steel rods in reinforced concrete, these fibers are tough, insoluble, and resistant to weight loads and mechanical stress, such as strong winds against stems.

In animals, glycogen is the sugar-storage equivalent of starch in plants (Figure 3.12*c*). Muscle and liver cells store a lot of it. When the sugar level in blood falls, liver cells degrade glycogen, and the released glucose enters blood. Exercise strenuously but briefly, and muscle cells tap glycogen for a burst of energy.

Figure 3.13 (*Right*) A tick's body covering is a protective cuticle reinforced with a polysaccharide called chitin (*below*). You may "hear" chitin when big spider legs clack across an aluminum oil pan on a garage floor.

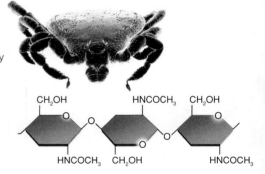

Chitin has nitrogen-containing groups attached to its glucose units. This polysaccharide derivative strengthens the external skeletons and other hard body parts of many animals, including crabs, earthworms, insects, spiders, and ticks of the sort shown in Figure 3.13. It also strengthens the cell walls of fungi.

The simple sugars (such as glucose), oligosaccharides, and polysaccharides (such as starch) are carbohydrates. Each cell requires carbohydrates as structural materials, and as storable or transportable packets of energy.

3.4 Greasy, Oily—Must Be Lipids

If something is greasy or oily to the touch, you can bet it's a lipid or has lipid parts. Cells use different lipids as energy packages, structural materials, and signaling molecules. Fats, phospholipids, and waxes have fatty acid tails. Sterols have a backbone of four carbon rings.

FATS AND FATTY ACIDS

Lipids are nonpolar hydrocarbons. Although they do not dissolve in water, they mix with other nonpolar substances, as butter does in warm cream sauce.

The lipids called **fats** have one, two, or three fatty acids attached to a glycerol molecule. A **fatty acid** has a backbone of as many as thirty-six carbon atoms, a carboxyl group at one end, and hydrogen atoms at most or all of the remaining carbons (Figure 3.14). They stretch out like flexible tails from the glycerol. *Unsaturated* fatty acids have one or more double bonds. *Saturated* types have single bonds only.

Weak interactions keep many saturated fatty acids tightly packed in animal fats. These fats are solid at room temperature. Most plant fats stay liquid at room temperature, as "vegetable oils." Their packing isn't as stable because of rigid kinks in their fatty acid tails. That's why vegetable oils flow freely.

Neutral fats such as butter, lard, and vegetable oils, are mostly **triglycerides**. Each has three fatty acid tails linked to a glycerol (Figure 3.15). Triglycerides are the most abundant lipids inside your body and its richest reservoir of energy. Gram for gram, they yield more than twice as much energy as complex carbohydrates such as starches. All vertebrates store triglycerides as droplets in fat cells that make up adipose tissue.

Layers or patches of adipose tissue insulate the body and cushion some of its parts. Like many other kinds of

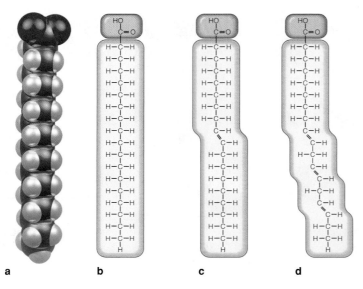

a   **b**   **c**   **d**

Figure 3.14 Three fatty acids. (**a,b**) Space-filling model and structural formula for stearic acid. The carbon backbone is fully saturated with hydrogen atoms. (**c**) Oleic acid, with a double bond in its backbone, is an unsaturated fatty acid. (**d**) Linolenic acid, also unsaturated, has three double bonds.

Figure 3.15 Condensation of (**a**) three fatty acids and one glycerol molecule into (**b**) a triglyceride. The photograph shows triglyceride-protected emperor penguins during an Antarctic blizzard.

glycerol

$+ 3H_2O$

a   three fatty acid tails   **b**   triglyceride

a

b

hydrophilic
head

two
hydrophobic
tails

c

cell membrane section

Figure 3.16 (**a**) Space-filling model, (**b**) structural formula, and (**c**) an icon for a phospholipid. This is the most common type in animal and plant cell membranes. Are its two tails saturated or unsaturated?

b

a

c

Figure 3.17 (**a**) Honeycomb: food warehouses and bee nurseries. Bees construct the compartments from their own water-repellent, waxy secretions. (**b**) Sterol backbone. (**c**) Structural formula for cholesterol, the main sterol of animal tissues. Your liver makes enough cholesterol for your body. A fat-rich diet may lead to clogged arteries.

animals, penguins of the Antarctic can keep warm in extremely cold winter months thanks to a thick layer of triglycerides beneath their skin (Figure 3.15).

PHOSPHOLIPIDS

Phospholipids have a glycerol backbone, two nonpolar fatty acid tails, and a polar head (Figure 3.16). They are a main component of cell membranes, which have two layers of lipids. The heads of one layer are dissolved in the cell's fluid interior; heads of the other layer are dissolved in the surroundings. Sandwiched between the two are the tails. You will read about membranes in Chapter 4.

WAXES

Waxes have long-chain fatty acids tightly packed and linked to long-chain alcohols or carbon rings. All have a firm consistency; all repel water. Surfaces of plants have a cuticle that contains waxes and another lipid, cutin. A plant cuticle restricts water loss and thwarts some parasites. Waxes also protect, lubricate, and lend pliability to skin and to hair. Birds secrete waxes, fats, and fatty acids that waterproof feathers. Bees use beeswax for honeycomb, which houses each new bee generation as well as honey (Figure 3.17a).

CHOLESTEROL AND OTHER STEROLS

Sterols are among the many lipids with no fatty acids. The sterols differ in the number, position, and type of their functional groups, but all have a rigid backbone of four fused-together carbon rings (Figure 3.17b).

Sterols are components of every eukaryotic cell membrane. The most common type in animal tissues is cholesterol (Figure 3.17c). Cholesterol also becomes remodeled into compounds as diverse as bile salts, steroids, and the vitamin D required for good bones and teeth. Bile salts assist in fat digestion in the small intestine. Sex hormones are vital for the formation of gametes and the development of secondary sexual traits. Such traits include the amount and distribution of hair in mammals, and feather color in birds.

Being largely hydrocarbon, lipids can intermingle with other nonpolar substances, but they resist dissolving in water.

Triglycerides, or neutral fats, have a glycerol head and three fatty acid tails. They are the major energy reservoirs. Phospholipids are a main component of cell membranes.

Sterols such as cholesterol serve as membrane components and precursors of steroid hormones and other compounds. Waxes are firm, yet pliable, components of water-repellent and lubricating substances.

3.5 Proteins—Diversity in Structure and Function

Of all large biological molecules, proteins are the most diverse. Some kinds speed reactions; others are the stuff of spider webs or feathers, bones, hair, and other body parts. Nutritious types abound in seeds and eggs. Many proteins move substances, help cells communicate, or defend against pathogens. Amazingly, cells assemble thousands of different proteins from only twenty kinds of amino acids!

An **amino acid** is a small organic compound with an amino group ($-NH_3^+$), a carboxyl group ($-COO^-$, the acid part), a hydrogen atom, and one or more atoms called its R group. In most cases, these components are attached to the same carbon atom (Figure 3.18a). Biological amino acids are shown in Appendix VI.

When a cell constructs a protein, it strings amino acids together, one after the other. Instructions coded

Read Me First!

and watch the narrated animation on peptide bond formation

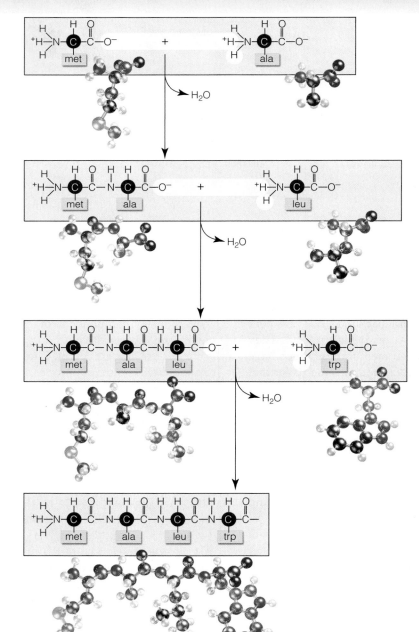

b Instructions encoded in DNA specify the order of amino acids to be joined in a polypeptide chain. The first amino acid is usually methionine (met). Alanine comes next in this one.

c In a condensation reaction, a peptide bond forms between the methionine and alanine. Leucine is next in line.

d A peptide bond forms between the alanine and the leucine (leu). Tryptophan (trp) queues up.

e The newly formed polypeptide chain. The sequence of amino acids in the growing chain is met–ala–leu–trp.

Figure 3.18 (a) Generalized formula for amino acids. The green box highlights the R group, one of the side chains that include functional groups. (b–e) Peptide bond formation during protein synthesis. Chapter 13 provides a closer look at protein synthesis.

amino group   carboxyl group

R group (20 kinds, each with distinct properties)

a

Figure 3.19 Three levels of protein structure. (**a**) Primary structure is a linear sequence of amino acids. (**b**) Many hydrogen bonds (dotted lines) along a polypeptide chain result in a helically coiled or sheetlike secondary structure. (**c**) Coils and sheets packed into stable domains represent a third structural level.

in DNA specify the order in which any of the twenty kinds of amino acids will occur in a given protein. A *peptide* bond forms as a condensation reaction joins the amino group of one amino acid and the carboxyl group of the next in line. Each **polypeptide chain** consists of three or more amino acids. The carbon backbone of this chain incorporates nitrogen atoms in this regular pattern: —N—C—C—N—C—C—.

The sequence of amino acids in a polypeptide chain is its *primary* structure. A new polypeptide chain twists, bends, loops, and folds, which is its *secondary* structure. Hydrogen bonds between R groups make some stretches of amino acids coil into a helical shape, a bit like a spiral staircase; they might make other regions form sheets or loops (Figure 3.19b). Bear in mind, the primary structure for each type of protein is unique in some respects, but similar patterns of coils, sheets, and loops do recur among them.

Much as an overly twisted rubber band coils back on itself, the coils, sheets, and loops of a protein fold up even more, into compact domains. A "domain" is a polypeptide chain or part of it that has become organized as a structurally stable unit. This third level of organization, a protein's *tertiary* structure, is what makes the protein a functional molecule. For instance, barrel-shaped domains of some proteins function as subway tunnels through membranes (Figure 3.19c).

Many proteins are two or more polypeptide chains that are bonded together or closely associated with one another. This is the fourth level of organization, or *quaternary* protein structure. Many enzymes and other proteins are globular, with multiple polypeptide chains folded into rounded shapes. Hemoglobin, described shortly, is a classic example of such a protein.

Protein structure doesn't stop here. Enzymes often attach short, linear, or branched oligosaccharides to a new polypeptide chain, making a *glyco*protein. Many glycoproteins occur at the cell surface or are secreted from cells. Lipids also get attached to many proteins. The cholesterol, triglycerides, and phospholipids that your body absorbs after a meal are transported about as components of *lipo*proteins.

Many proteins are fibrous, with polypeptide chains organized as strands or sheets. They contribute to cell shape and organization, and help cells and cell parts move about. Other proteins make up cartilage, hair, skin, and parts of muscles and brain cells.

A protein has primary structure, a sequence of amino acids covalently bonded as a polypeptide chain.

Local regions of a polypeptide chain become twisted and folded into helical coils, sheetlike arrays, and loops. These arrangements are the protein's secondary structure.

A polypeptide chain or parts of it become organized as structurally stable, compact, functional domains. Such domains are a protein's tertiary structure.

Many proteins show quaternary structure; they consist of two or more polypeptide chains.

3.6 Why Is Protein Structure So Important?

Cells are good at making proteins that are just what their DNA specifies. But sometimes a protein just turns out wrong. A different amino acid may lead to a misfolded shape that has far-reaching consequences.

JUST ONE WRONG AMINO ACID . . .

Four tightly packed polypeptides called globins make up each hemoglobin molecule. Each globin chain is folded into a pocket that cradles a **heme** group, a large organic molecule with an iron atom at its center (Figure 3.20*a*). Heme is an oxygen transporter. During its life span, each of the red blood cells in your body transports billions of oxygen molecules, all bound to the heme in globin molecules.

Globin comes in two slightly different forms, alpha and beta. Two of each form make up one hemoglobin molecule in adult humans. Glutamate is normally the sixth amino acid in the beta globin chain, but a DNA mutation sometimes puts a different amino acid—valine—in the sixth position instead (Figure 3.21*b*). Unlike glutamate, which carries an overall negative charge, valine has no net charge. As a result of that one substitution, a tiny patch of the protein changes from polar to nonpolar, which in turn causes the globin's behavior to change slightly. Hemoglobin with this mutation in its beta chain is designated HbS.

. . . AND YOU GET SICKLE-SHAPED CELLS!

Every human inherits two genes for beta globin, one from each of two parents. (Genes are units of DNA that encode heritable traits.) Cells use both genes when they make beta globin. If one is normal and the other has the valine mutation, a person makes enough normal hemoglobin and can lead a relatively normal life. But someone who inherits two mutant genes can only make the mutant hemoglobin HbS. The outcome is sickle-cell anemia, a severe genetic disorder.

As blood moves through lungs, the hemoglobin in red blood cells binds oxygen, then gives it up in body regions where oxygen levels are low. After oxygen is released, red blood cells quickly return to the lungs and pick up more. In the few moments when they have no bound oxygen, the hemoglobin molecules clump together just a bit. But HbS molecules do not form such clusters in places where oxygen levels are low. They form large, stable, rod-shaped aggregates.

Red blood cells containing these aggregates become distorted into a sickle shape (Figure 3.21*c*). The sickle cells clog tiny blood vessels called capillaries, which disrupts blood circulation. Tissues become oxygen-starved. Figure 3.21*d* lists the far-reaching effects of sickle-cell anemia on tissues and organs.

DENATURATION

The shape of a protein defines its biological activity. A globin molecule cradles heme, an enzyme speeds some reaction, a receptor transduces an energy signal. These proteins and others cannot function unless they stay coiled, folded, and packed in a precise way. Their shape depends on many hydrogen bonds and other interactions—which heat, shifts in pH, or detergents can disrupt. At such times, polypeptide chains unwind and change shape in an event called **denaturation**.

Consider albumin, a protein in the white of an egg. When you cook eggs, the heat does not disrupt the covalent bonds of albumin's primary structure. But it destroys albumin's weaker hydrogen bonds, and so the protein unfolds. When the translucent egg white turns opaque, we know albumin has been altered. For a few proteins, denaturation might be reversed if and when normal conditions return, but albumin isn't one of them. There is no way to uncook an egg.

heme   alpha globin   alpha globin

a

beta globin   beta globin

b

Figure 3.20 (**a**) Globin. This coiled polypeptide chain cradles heme, a functional group that contains an iron atom. (**b**) Hemoglobin, an oxygen-transport protein in red blood cells. This is one of the proteins with quaternary structure. It consists of four globin molecules (two alphas and two betas) held together by hydrogen bonds.

Read Me First!
and watch the narrated
animation on sickle-cell anemia

VALINE   HISTIDINE   LEUCINE   THREONINE   PROLINE   GLUTAMATE   GLUTAMATE

a Normal amino acid sequence at the start of a beta chain for hemoglobin.

VALINE   HISTIDINE   LEUCINE   THREONINE   PROLINE   VALINE   GLUTAMATE

b One amino acid substitution results in the abnormal beta chain in HbS molecules. During protein synthesis, valine was added instead of glutamate at the sixth position of the growing polypeptide chain.

c Glutamate has an overall negative charge; valine has no net charge. This difference gives rise to a water-repellent, sticky patch on HbS molecules. They stick together because of that patch, forming rod-shaped clumps that distort normally rounded red blood cells into sickle shapes. (A sickle is a farm tool that has a crescent-shaped blade.)

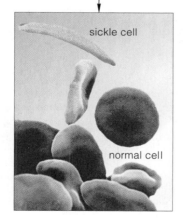

sickle cell

normal cell

Clumping of cells in bloodstream

Circulatory problems, damage to brain, lungs, heart, skeletal muscles, gut, and kidneys

Heart failure, paralysis, pneumonia, rheumatism, gut pain, kidney failure

Spleen concentrates sickle cells

Spleen enlargement

Immune system compromised

Rapid destruction of sickle cells

Anemia, causing weakness, fatigue, impaired development, heart chamber dilation

Impaired brain function, heart failure

d *Above left:* Melba Moore, celebrity spokesperson for sickle-cell anemia organizations. *Above right:* Range of symptoms for a person with two mutated genes (Hb^S) for hemoglobin's beta chain.

Figure 3.21 Sickle-cell anemia's molecular basis and its main symptoms. Sections 16.9 and 33.3 touch upon the evolutionary and ecological aspects of this genetic disorder.

So what is the big take-home lesson? Hemoglobin, hormones, enzymes, transporters—these are the kinds of proteins that help us survive. Twists and folds in their polypeptide chains form anchors, or membrane-spanning barrels, or jaws that can grip enemy agents in the body. Mutation can alter the chains enough to block or enhance an anchoring, transport, or defensive function. Sometimes the consequences are awful. Yet changes in the sequences and functional domains also give rise to variation in traits—the raw material of evolution. *Learn about the structure and function of proteins, and you are on your way to comprehending life in its richly normal and abnormal expressions.*

The structure of proteins dictates function. Mutations that alter a protein's structure can have drastic consequences on its function, and the health of organisms harboring them.

3.7 Nucleotides and the Nucleic Acids

Certain small organic compounds called nucleotides are energy carriers, enzyme helpers, and messengers. Some are the building blocks for DNA and RNA. They are, in short, central to metabolism, survival, and reproduction.

Nucleotides have one sugar, at least one phosphate group, and one nitrogen-containing base. Deoxyribose or ribose is the sugar. Both sugars have a five-carbon ring structure; ribose has an oxygen atom attached to carbon 2 of the ring and deoxyribose does not. The bases have a single or double carbon ring structure.

The nucleotide **ATP** (adenosine triphosphate) has a row of three phosphate groups attached to its sugar (Figure 3.22). ATP can readily transfer the outermost phosphate group to many other molecules and make them reactive. Such transfers are vital for metabolism.

Other nucleotides have different metabolic roles. Some are **coenzymes**, necessary for enzyme function. They move electrons and hydrogen from one reaction site to another. NAD^+ and FAD are major kinds.

Still other nucleotides act as chemical messengers within and between cells. Later in the book, you will read about one of these messengers, which is known as cAMP (cyclic adenosine monophosphate).

Certain nucleotides also function as monomers for single- and double-stranded molecules called **nucleic acids**. In such strands, a covalent bond forms between the sugar of one nucleotide and the phosphate group of the next (Figure 3.23). The nucleic acids DNA and RNA store and retrieve heritable information.

All cells start out life and maintain themselves with instructions in their double-stranded molecules of deoxyribonucleic acid, or **DNA**. This nucleic acid is made of four kinds of deoxyribonucleotides. Figure 3.23a shows their structural formulas. As you can see, the four differ only in their component base, which is adenine, guanine, thymine, or cytosine.

Figure 3.24 shows how hydrogen bonds between bases join the two strands along the length of a DNA molecule. Think of every "base pairing" as one rung of a ladder, and the two sugar–phosphate backbones as the ladder's two posts. The ladder twists and turns in a regular pattern, forming a double helical coil.

The sequence of bases in DNA encodes heritable information about all the proteins that give each new cell the potential to grow, maintain itself, and even to reproduce. Part of that sequence is unique for each species. Some parts are identical, or nearly so, among many species. We return to DNA's structure and its function in Chapter 12.

base (*blue*)

three phosphate groups

sugar (*red*)

OH   OH

Figure 3.22 Structural formula for an ATP molecule.

phosphate group

sugar (deoxyribose)

ADENINE (A) base with a double-ring structure

THYMINE (T) base with a single-ring structure

GUANINE (G) base with a double-ring structure

CYTOSINE (C) base with a single-ring structure

b

Figure 3.23 (**a**) Nucleotides of DNA. Two of the nucleotide bases, adenine and guanine, have a double-ring structure. The two others, thymine and cytosine, have a single-ring structure. (**b**) Bonding pattern between successive bases in nucleic acids.

a

hydrogen bonding
between bases

covalent
bonding
in carbon
backbone

Figure 3.24 Models for the DNA molecule.

The **RNAs** (ribonucleic acids) have four kinds of ribonucleotide monomers. Unlike DNA, most RNAs are single strands, and one base is uracil instead of thymine. One type of RNA is a messenger that carries eukaryotic DNA's protein-building instructions out of the nucleus and into the cytoplasm, where they are translated by other RNAs. Chapter 13 returns to RNA and its role in protein synthesis.

Different nucleotides serve as coenzymes, subunits of nucleic acids, energy carriers, and chemical messengers.

The nucleic acid DNA consists of two nucleotide strands joined by hydrogen bonds and twisted as a double helix. Its nucleotide sequence contains heritable instructions about how to build all of a cell's proteins.

RNA is a single-stranded nucleic acid. Different RNAs have roles in the processes by which a cell retrieves and uses genetic information in DNA to build proteins.

Summary

Section 3.1   Organic compounds consist of carbon and at least one hydrogen atom. Carbon atoms bond covalently with up to four other atoms, often forming long chains or rings. Functional groups attached to a carbon backbone contribute to an organic compound's properties. Enzyme-driven reactions synthesize carbohydrates, proteins, lipids, and nucleic acids from smaller organic subunits. Table 3.1 on the next page summarizes these compounds.

Section 3.2   Methane gas produced by archaea far below the seafloor is partially metabolized by diverse microorganisms. Millions of tons of the remaining methane has become frozen into unstable, potentially explosive methane hydrate deposits on the seafloor.

Section 3.3   Carbohydrates include simple sugars, oligosaccharides, and polysaccharides. Living cells use carbohydrates as energy sources, transportable or storage forms of energy, and structural materials.

Section 3.4   Lipids are greasy or oily compounds that tend not to dissolve in water but mix easily with nonpolar compounds, such as other lipids. Neutral fats (triglycerides), phospholipids, waxes, and sterols are lipids. Cells use lipids as major sources of energy and as structural materials.

Section 3.5   Structurally and functionally, proteins are the most diverse molecules of life. Their primary structure is a sequence of amino acids—a polypeptide chain. Such chains twist, coil, and bend into functional domains. Many proteins, including hemoglobin and most enzymes, consist of two or more chains. Certain protein aggregates form hair, muscle, connective tissue, cytoskeleton, and other materials.

Section 3.6   A protein's overall structure determines its function. Sometimes a mutation in DNA results in an amino acid substitution that can drastically alter a protein. Such changes can cause genetic diseases, including sickle-cell anemia. Weak bonds that hold a protein's shape are disrupted by temperature, pH shifts, or exposure to detergent, and usually result in the protein unfolding permanently.

Section 3.7   Nucleotides consist of sugar, phosphate, and a nitrogen-containing base. They have essential roles in metabolism, survival, and reproduction. ATP energizes many kinds of molecules by phosphate-group transfers. Other nucleotides are coenzymes or chemical messengers. DNA and RNA are nucleic acids, each composed of four kinds of nucleotide subunits.

The sequence of nucleotide bases in DNA encodes instructions for how to construct all of a cell's proteins. Different kinds of RNA molecules interact in the translation of DNA's genetic information.

Table 3.1 Summary of the Main Organic Compounds in Living Things

Category	Main Subcategories	Some Examples and Their Functions	
CARBOHYDRATES . . . contain an aldehyde or a ketone group, and one or more hydroxyl groups	**Monosaccharides** (simple sugars)	Glucose	Energy source
	Oligosaccharides (short-chain carbohydrates)	Sucrose (a disaccharide)	Most common form of sugar; the form transported through plants
	Polysaccharides (complex carbohydrates)	Starch, glycogen Cellulose	Energy storage Structural roles
LIPIDS . . . are mainly hydrocarbon; generally do not dissolve in water but do dissolve in nonpolar substances, such as other lipids	**Lipids with fatty acids** *Glycerides:* Glycerol backbone with one, two, or three fatty acid tails	Fats (e.g., butter), oils (e.g., corn oil)	Energy storage
	Phospholipids: Glycerol backbone, phosphate group, one other polar group, and (often) two fatty acids	Phosphatidylcholine	Key component of cell membranes
	Waxes: Alcohol with long-chain fatty acid tails	Waxes in cutin	Conservation of water in plants
	Lipids with no fatty acids *Sterols:* Four carbon rings; the number, position, and type of functional groups differ among sterols	Cholesterol	Component of animal cell membranes; precursor of many steroids and vitamin D
PROTEINS . . . are one or more polypeptide chains, each with as many as several thousand covalently linked amino acids	**Fibrous proteins** Long strands or sheets of polypeptide chains; often tough, water-insoluble	Keratin Collagen Enzymes	Structural component of hair, nails Structural component of bone Great increase in rates of reactions
	Globular proteins One or more polypeptide chains folded into globular shapes; many roles in cell activities	Hemoglobin Insulin Antibodies	Oxygen transport Control of glucose metabolism Tissue defense
NUCLEIC ACIDS (AND NUCLEOTIDES) . . . are chains of units (or individual units) that each consist of a five-carbon sugar, phosphate, and a nitrogen-containing base	**Adenosine phosphates**	ATP cAMP (Section 20.8)	Energy carrier Messenger in hormone regulation
	Nucleotide coenzymes	NAD^+, $NADP^+$, FAD	Transfer of electrons, protons (H^+) from one reaction site to another
	Nucleic acids Chains of thousands to millions of nucleotides	DNA, RNAs	Storage, transmission, translation of genetic information

Self-Quiz

Answers in Appendix III

1. Name the molecules of life and the families of small organic compounds from which they are built.

2. Each carbon atom can share pairs of electrons with as many as _____ other atoms.
   a. one b. two c. three d. four

3. Sugars are a class of _____ , which have one or more _____ groups.
   a. proteins; amino   c. alcohols; hydroxyl
   b. acids; phosphate   d. carbohydrates; carboxyl

4. _____ is a simple sugar (a monosaccharide).
   a. Glucose   c. Ribose   e. both a and b
   b. Sucrose   d. Chitin   f. both a and c

5. The fatty acid tails of unsaturated fats incorporate one or more _____ .
   a. single covalent bonds   b. double covalent bonds

6. Sterols are among the many lipids with no _____ .
   a. saturation   c. phosphates
   b. fatty acids   d. carbons

7. Which of the following is a class of molecules that encompasses all of the other molecules listed?
   a. triglycerides   c. waxes   e. lipids
   b. fatty acids   d. sterols   f. phospholipids

8. _____ are to proteins as _____ are to nucleic acids.
   a. Sugars; lipids   c. Amino acids; hydrogen bonds
   b. Sugars; proteins   d. Amino acids; nucleotides

9. A denatured protein has lost its _____ .
   a. hydrogen bonds   c. function
   b. shape   d. all of the above

10. Nucleotides occur in _____ .
    a. ATP   b. DNA   c. RNA   d. all are correct

11. Which of the following nucleotides is *not* found in DNA?
    a. adenine   b. uracil   c. thymine   d. guanine

12. Match the molecule with the most suitable description.
    _____ long sequence of amino acids   a. carbohydrate
    _____ a rechargeable battery in cells   b. phospholipid
    _____ glycerol, fatty acids, phosphate   c. polypeptide
    _____ two strands of nucleotides   d. DNA
    _____ one or more sugar monomers   e. ATP

Critical Thinking

1. In the following list, identify which is the carbohydrate, the fatty acid, the amino acid, and the polypeptide:

   a. $^+NH_3$—CHR—COO$^-$ c. (glycine)$_{20}$

   b. $C_6H_{12}O_6$ d. $CH_3(CH_2)_{16}COOH$

2. A clerk in a health-food store tells you that "natural" vitamin C extracts from rose hips are better than synthetic tablets of this vitamin. Given what you know about the structure of organic compounds, what would be your response? How would you design an experiment to test whether a natural and synthetic version of a vitamin differ?

3. It seems there are "good" and "bad" unsaturated fats. The double bonds of both put a bend in their fatty acid tails. But the bend in *trans* fatty acid tails keeps them aligned in the same direction along their whole length. The bend in *cis* fatty acid tails makes them zigzag (Figure 3.25).

   Some *trans* fatty acids occur naturally in beef. But most form by industrial processes that solidify vegetable oils for margarine, shortening, and the like. These substances are widely used in prepared foods (such as cookies) and in french fries and other fast-food products. *Trans* fatty acids are linked to heart attacks. Speculate on why your body might have an easier time dealing with *cis* fatty acids than *trans* fatty acids.

4. The shapes of protein domains often are clues to functions. For example, shown at left is an HLA, a type of recognition protein at the surface of vertebrate body cells. Certain cells of the immune system use HLAs to distinguish self (the body's own cells) from nonself. Each HLA has a jawlike region (*arrow*) that can bind and display nonself fragments, thus alerting the immune system to the presence of an invader or some other threat. Speculate on what may happen if a mutation caused the jawlike region to misfold.

5. Cholesterol from food or synthesized in the liver is too hydrophobic to circulate in blood; complexes of protein and lipids ferry it around. Low density lipoprotein, or *LDL*, transports cholesterol out of the liver and into cells. High density lipoprotein, or *HDL,* ferries the cholesterol that is released from dead cells back to the liver.

   High LDL levels are implicated in atherosclerosis, heart disease, and stroke. The main protein in LDL is called ApoA1, and a mutant form of it has the wrong amino acid (cysteine instead of arginine) at one location in its primary sequence. Carriers of this LDL mutation have very low levels of HDL, which is typically predictive of heart disease, but paradoxically they have no heart problems.

   Some heart patients received injections of the mutant LDL, which acted like a drain cleaner; it quickly reduced the size of cholesterol deposits in the patients' arteries.

   A few years from now, it might be possible to reverse years of damage with such treatment. However, many caution that a low-fat, low-cholesterol diet is still the best preventive measure for long-term health. Would you opt for artery-rooting treatments over a healthy diet?

cis fatty acid   *trans* fatty acid

Figure 3.25 Maybe rethink the french fries?

Media Menu

Student CD-ROM

Impacts, Issues Video
   Science or the Supernatural?
Big Picture Animation
   The chemistry of organic compounds
Read-Me-First Animation
   Peptide bond formation
   Sickle-cell anemia
Other Animations and Interactions
   Condensation and hydrolysis
   Triglyceride formation
   Structure of hemoglobin
   DNA structure

InfoTrac

- Sweet Medicines. *Scientific American*, July 2002.
- The Form Counts: Proteins, Fats, and Carbohydrates. *Consumers' Research Magazine*, August 2001.
- Sorting Fat from Fiction. *Prepared Foods*, October 2002.
- Proteins Rule. *Scientific American*, April 2002.

Web Sites

- General Chemistry Online: antoine.frostburg.edu/chem/senese/101/index.shtml
- The Molecules of Life: biop.ox.ac.uk/www/mol_of_life/index.html
- The Protein Data Bank: www.rcsb.org/pdb

How Would You Vote?

Huge methane reservoirs lie off the east coast of the United States and in arctic regions of North America. The environmental impact of disturbing the reserves is not known but might be significant. Yet using them could lessen our dependence on foreign oil. Should the government encourage use of these reserves for research and development?

IMPACTS, ISSUES   *Where Did Cells Come From?*

Do you ever think of yourself as being close to 1/1,000 of a kilometer tall? Probably not. Yet that's how we think of cells. We measure them in micrometers— in millionths of a millimeter, which is a thousandth of a meter, which is a thousandth of a kilometer. The bacteria in Figure 4.1 are a few micrometers "tall."

Somewhere in the distant past, between 3.9 billion and 2.5 billion years ago, cells no bigger than this first appeared on Earth. They were prokaryotic, meaning their DNA was exposed to the rest of their insides. They had no nucleus. At the time, the atmosphere had little free oxygen. The earliest cells probably extracted energy from their food by way of anaerobic reactions, which don't use free oxygen. Plenty of food—simple organic compounds—had already accumulated in the environment through natural geologic processes.

By 2.1 billion years ago, in tidal flats and freshwater habitats, tiny cells were slowly changing the world. Vast populations were making food by a photosynthetic pathway that released oxygen as a by-product. At first, all of that oxygen combined with iron in rocks, forming rust. When oxygen saturated all of the exposed iron deposits on Earth, free oxygen became more and more concentrated in water, then in air.

Oxygen, a reactive gas, attacks organic compounds, including the ones making up cells. The oxygen-enriched

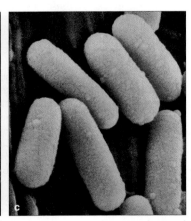

Figure 4.1 How small are cells? Shown here, bacterial cells peppering the tip of a household pin.

| 100 μm | 20 μm | 0.5 μm |

the big picture

Basic Cell Features   Nearly all cells are microscopic in size. They all start out life with a plasma membrane, a semifluid interior called cytoplasm, and an inner region of DNA. The plasma membrane helps control the flow of specific substances into and out of cells.

Cell Membrane Features   Two layers of phospholipid molecules are the structural basis of cell membranes. Proteins in this bilayer and those attached to its surfaces carry out many different membrane functions.

atmosphere became a selection pressure of global dimensions. Cell lineages that couldn't neutralize oxygen never left mud and other anaerobic habitats. In other lineages, though, mutations changed metabolic steps in ways that could neutralize oxygen, then *use* it. Aerobic respiration, an oxygen-requiring pathway, had emerged in most groups, and it proved handy in the growing competition for resources.

With so much oxygen around, the free organic compounds that were food for microorganisms became scarce. Compounds made by living cells became the sought-after sources of carbon and energy. Coinciding with this major change in available food sources, novel predators, parasites, and partners evolved. The first eukaryotic cells were among them.

The new cells ran reactions and stored things inside tiny sacs and other compartments made of membranes. One sac, the nucleus, controlled access to their DNA. Others, called mitochondria, yielded far more energy from metabolism than anaerobic reactions, enough to power more active life-styles and build larger, more complex bodies. Without such innovations in small cells, big plants and animals never would have evolved.

This chapter introduces the key defining features of prokaryotic and eukaryotic cells. It invites you to reflect on the earliest, simplest ancestors of you and all other eukaryotic forms of life. Why bother? Science is close to *creating* simple forms of life in test tubes. A bioethical line is about to be crossed.

 How Would You Vote?

Researchers are modifying prokaryotes in efforts to make the simplest form of life possible. They are creating "new" organisms by removing genes one at a time. Should this research continue? See the Media Menu for details, then vote online.

Prokaryotic Cells Compared to eukaryotic cells, prokaryotes have little internal complexity, and no nucleus. However, when taken as a group, they are the most metabolically diverse organisms. All species have prokaryotic ancestors.

Eukaryotic Cells Eukaryotic cells contain organelles. These membrane-bound compartments divide the cell interior into functional regions for specialized tasks. A major organelle, the nucleus, keeps the DNA away from cytoplasmic machinery.

4.1 So What Is "A Cell"?

Inside your body and at all of its moist surfaces, many trillions of cells live in interdependency. In northern forests, four-celled structures called pollen grains drift down from pine trees. In scummy pond water, free-living single cells called bacteria and amoebas move about. How are these cells alike, and how do they differ?

COMPONENTS OF ALL CELLS

The **cell** is the smallest unit with the properties of life: a capacity for metabolism, controlled responses to the environment, growth, and reproduction. Cells differ in size, shape, and activities, yet are alike in three respects. They start out life with a plasma membrane, a region of DNA, and cytoplasm (Figure 4.2):

1. **Plasma membrane**. A thin, outermost membrane maintains a cell as a distinct entity. It lets metabolic events proceed in controllable ways, separated from the outside environment. Yet this plasma membrane does not isolate the cell interior. It's a bit like a house with many windows and doors that don't open for just anyone. Water, oxygen, and carbon dioxide cross in and out freely. Other substances, such as nutrients and ions, get escorted across.

2. **Nucleus** or **nucleoid**. DNA occupies a membrane-bound sac (nucleus) inside the cell or, in the simplest kinds of cells, a nucleoid (part of the cytoplasm).

3. **Cytoplasm**. Cytoplasm is everything between the plasma membrane and the region of DNA. It consists of a semifluid matrix and other components, such as **ribosomes**, the structures on which proteins are built.

In structural terms, prokaryotes are the simplest cells; nothing separates their DNA from the cytoplasm. Bacteria and archaea are the only prokaryotes. All other organisms—from amoebas and trees to puffball mushrooms and elephants—are eukaryotic. Internal membranes divide the cytoplasm of eukaryotic cells into functional compartments. One compartment, the nucleus, is the key defining feature of these cells.

WHY AREN'T CELLS BIGGER?

Can any cells be observed with the unaided eye? Just a few types can. They include "yolks" of bird eggs, cells in watermelon tissues, and fish eggs. These get big because they aren't doing much, metabolically speaking, at maturity. Most metabolically active cells are too tiny to be seen by the unaided eye (Figure 4.3).

So why aren't all cells big? A physical relationship called the **surface-to-volume ratio** constrains increases in cell size. By this relationship, an object's volume increases with the cube of its diameter, but the surface area increases only with the square.

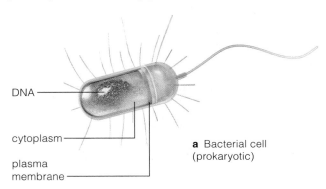

DNA

cytoplasm

plasma membrane

a Bacterial cell (prokaryotic)

DNA in nucleus

cytoplasm

plasma membrane

b Plant cell (eukaryotic)

DNA in nucleus

cytoplasm

plasma membrane

c Animal cell (eukaryotic)

Figure 4.2 Overview of the general organization of prokaryotic cells and eukaryotic cells. The three cells are not drawn to the same scale.

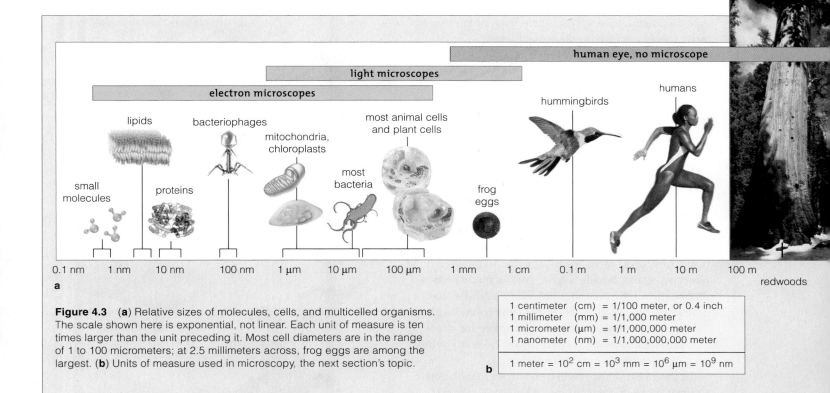

Figure 4.3 (**a**) Relative sizes of molecules, cells, and multicelled organisms. The scale shown here is exponential, not linear. Each unit of measure is ten times larger than the unit preceding it. Most cell diameters are in the range of 1 to 100 micrometers; at 2.5 millimeters across, frog eggs are among the largest. (**b**) Units of measure used in microscopy, the next section's topic.

```
1 centimeter  (cm)  = 1/100 meter, or 0.4 inch
1 millimeter  (mm)  = 1/1,000 meter
1 micrometer  (µm)  = 1/1,000,000 meter
1 nanometer   (nm)  = 1/1,000,000,000 meter
```

$$1 \text{ meter} = 10^2 \text{ cm} = 10^3 \text{ mm} = 10^6 \text{ µm} = 10^9 \text{ nm}$$

diameter (cm):	0.5	1.0	1.5
surface area (cm^2):	0.79	3.14	7.07
volume (cm^3):	0.06	0.52	1.77
surface-to-volume ratio:	13.17:1	6.04:1	3.99:1

Figure 4.4 One example of the surface-to-volume ratio. This physical relationship between increases in volume and surface area puts constraints on cell sizes and shapes.

Apply this constraint to a round cell. As Figure 4.4 shows, *if a cell expands in diameter during growth, then its volume will increase faster than its surface area does.* Suppose you induce a round cell to grow four times wider. Its volume increases 64 times (4^3). However, its surface area increases only 16 times (4^2). This means each unit of plasma membrane must service four times as much cytoplasm as before. A lot more substances have to get in and out!

If a cell's diameter is too great, the inward flow of nutrients and the outward flow of wastes just won't be fast enough to keep up with metabolic activity, and you'll end up with a dead cell.

A big, round cell also would have trouble moving materials throughout its cytoplasm. Random motions of molecules distribute materials through tiny cells. If a cell isn't tiny, you can expect it to be long or thin, or have outfoldings or infoldings that increase its surface relative to its volume. *The smaller or narrower or more frilly-surfaced the cell, the more efficiently materials cross its surface and become distributed through the interior.*

Surface-to-volume constraints also shape the body plans of multicelled species. For example, small cells attach end to end in strandlike algae, so each interacts directly with its surroundings. Cells in your muscles are as long as the muscle itself, but each one is thin enough to facilitate diffusion.

All living cells have an outermost plasma membrane, an internal region called cytoplasm, and an internal region where DNA is concentrated.

Bacteria and archaea are prokaryotic cells. Unlike eukaryotic cells, they do not have a nucleus.

As cells grow, their volume increases faster than their surface area. A surface-to-volume ratio is a physical relationship that affects metabolic activity, and thus constrains cell size. It also constrains cell shape and body plans of multicelled organisms.

4.2 How Do We "See" Cells?

Like their centuries-old forerunners, modern microscopes are our best windows on the cellular world.

THE CELL THEORY

Early in the seventeenth century, Galileo Galilei put two glass lenses inside a cylinder and peered at the patterns of an insect's eyes. He was one of the first to record a biological observation with a microscope. The study of the cellular basis of life was about to begin, first in Italy, then in France and England.

At midcentury, Robert Hooke focused a microscope on thinly sliced cork from a mature tree and saw tiny compartments (Figure 4.5). He gave them the Latin name *cellulae*, meaning small rooms—hence the origin of the biological term "cell." They actually were dead plant cell walls, which is what cork is made of, but Hooke didn't think of them as being dead because neither he nor anyone else knew cells could be alive.

Antony van Leeuwenhoek, a shopkeeper, made exceptional lenses. By the late 1600s, he was spying on sperm, protists, even a bacterium.

In the 1820s, improved lenses brought cells into sharper focus. Robert Brown, a botanist, saw an opaque spot in cells and called it a nucleus. Later, the botanist Matthias Schleiden wondered if a plant cell develops as an independent unit even though it's part of the plant. By 1839, after years of studying animal tissues, the zoologist Theodor Schwann reported that cells and cell products make up animals as well as plants—and that cells have an individual life of their own even when they are part of a multicelled species.

Rudolf Virchow, a physiologist, completed his own studies of a cell's growth and reproduction—that is, its division into daughter cells. Every cell, he decided, comes from a cell that already exists.

And so, microscopic analysis yielded three generalizations, which constitute the **cell theory**. *First*, every organism consists of one or more cells. *Second*, the cell is the smallest unit that still displays the properties of life. *Third*, the continuity of life arises directly

Figure 4.6 Generalized diagram and photograph of one kind of compound light microscope.

path of light rays (bottom to top) to eye

Ocular lens enlarges primary image formed by objective lenses.

prism that directs rays to ocular lens

Objective lenses (those closest to specimen) form the primary image. Most compound light microscopes have several.

stage (holds microscope slide in position)

Condenser lenses focus light rays through specimen.

illuminator

source of illumination (housed in the base of the microscope)

from the growth and division of single cells. Today, microscopy still supports all three insights.

SOME MODERN MICROSCOPES

Like the earlier instruments, many microscopes still use light rays to make images. Picture a series of waves moving across an ocean. Each **wavelength** is the distance from one wave's peak to the peak of the wave behind it. Light also travels in waves as it moves. In a *compound light microscope* (Figure 4.6), two or more sets of glass lenses bend light waves passing through a cell or some other specimen in ways that form an enlarged view of it.

Cells are visible under the microscope when they are thin enough for light to pass through them, but most are nearly colorless and look uniformly dense. Some colored dyes stain cells nonuniformly, and are used to make their components visible.

The best light microscopes can enlarge cells up to 2,000 times. Beyond that, cell parts appear larger but not clearer. Why? Parts that are smaller than one-half

Figure 4.5 Robert Hooke's microscope and his drawing of cell walls from cork tissue.

Figure 4.7 Generalized diagram of an electron microscope.

You can get an idea of the diameter of the lenses from this photograph of a transmission electron microscope (TEM). A beam of electrons from an electron gun moves down the microscope column and is focused by magnets.

In transmission electron microscopy, electrons pass through a thin slice of specimen, then illuminate a fluorescent screen on the monitor. The shadows cast by the specimen's internal details appear, as in Figure 4.8c.

incoming electron beam

condenser lens (focuses a beam of electrons onto specimen)

specimen

objective lens

intermediate lens

projector lens

viewing screen (or photographic film)

of a wavelength of light are too small to make light bend, so they don't show up.

Electron microscopes use magnetic lenses to bend and focus beams of electrons, which can't be focused through a glass lens (Figure 4.7). Electrons travel in wavelengths that are about 100,000 times shorter than those of visible light. That is why an electron microscope can bring into focus objects 100,000 times smaller than you can see with a light microscope.

In a *transmission* electron microscope, electrons pass through a sample and are focused into an image of the specimen's internal details (Figure 4.8c). In *scanning* electron microscopes, a beam of electrons moves back and forth across a specimen that has a thin coating of metal. The metal responds by emitting its own electrons and x-rays, which can be converted into an image of the surface. The images can have fantastic detail. Figure 4.8d shows an example.

a Light micrograph (phase-contrast process)

b Light micrograph (Nomarski process)

c Transmission electron micrograph, thin section

d Scanning electron micrograph

`10 μm`

Figure 4.8 How different microscopes reveal different aspects of the same organism—a green alga (*Scenedesmus*). All four images are at the same magnification. (**a,b**) Light micrographs. (**c**) Transmission electron micrograph. (**d**) Scanning electron micrograph. A horizontal bar below a micrograph, as in (**d**), provides a visual reference for size. One micrometer (μm) is 1/1,000,000 of 1 meter. Using the scale bar, can you estimate the length and width of a *Scenedesmus* cell?

4.3 All Living Cells Have Membranes

Cell membranes consist of a lipid bilayer in which many different kinds of proteins are embedded. The membrane is a continuous boundary layer across which the flow of substances is selectively controlled.

Think back on the phospholipids, the most abundant components of cell membranes (Section 3.4). Each has one phosphate-containing head and two fatty acid tails attached to a glycerol backbone (Figure 4.9*a*). The head is hydrophilic; it dissolves fast in water. The two tails are hydrophobic; water repels them.

When you immerse many phospholipid molecules in water, they interact with water molecules and with one another until they spontaneously cluster into a sheet or film at the water's surface. Some even line up as two layers, with their fatty acid tails sandwiched between their outward-facing hydrophilic heads. This arrangement, called a **lipid bilayer**, is the structural basis of every cell membrane (Figure 4.9*b*,*c*).

Figure 4.10 shows the **fluid mosaic model**. By this model, a cell membrane has a mixed composition—a

adhesion proteins

Adhesion proteins project outward from plasma membranes of multicelled species especially. They help cells of the same type stick together in the proper tissues.

communication proteins

Communication proteins of two adjoining cells match up, forming a direct channel between their cytoplasms. Signals flow through the channel. This channel is part of a gap junction between two heart muscle cells.

mosaic—of phospholipids, glycolipids, sterols, and proteins. Its phospholipids are diverse, with different kinds of heads and with tails that vary in length and saturation. Unsaturated fatty acids have one or more double covalent bonds in their carbon backbone, and fully saturated fatty acids have none (Section 3.4).

Also by this model, a membrane is *fluid* because of the motions and interactions of its components. Most phospholipids drift sideways, spin on their long axes, and flex their tails, so they don't bunch up as a solid layer. Most membrane phospholipids have at least one kinked (unsaturated) fatty acid tail.

Hydrogen bonds and other weak interactions help proteins associate with the phospholipids. Many of the membrane proteins span the bilayer, with hydrophilic parts extending past both of its surfaces. Others are anchored to underlying cell structures.

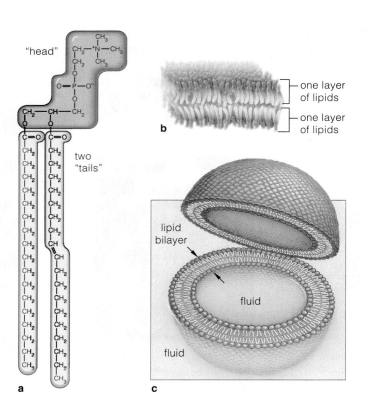

Figure 4.9 Lipid bilayer organization of cell membranes. (**a**) One of the phospholipids, the most abundant membrane components. (**b**,**c**) These lipids and others are arranged as two layers. Their hydrophobic tails are sandwiched between their hydrophilic heads in the bilayer.

Read Me First!

*and watch the
narrated animation
on cell membranes*

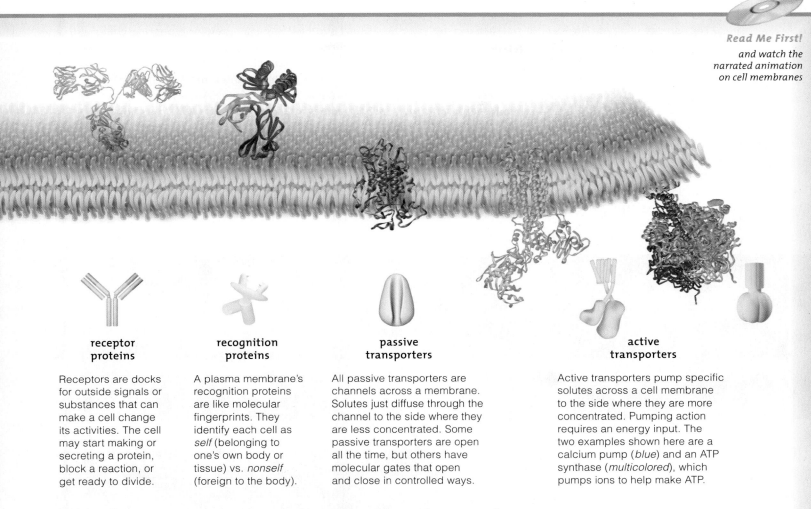

**receptor
proteins**

**recognition
proteins**

**passive
transporters**

**active
transporters**

Receptors are docks
for outside signals or
substances that can
make a cell change
its activities. The cell
may start making or
secreting a protein,
block a reaction, or
get ready to divide.

A plasma membrane's
recognition proteins
are like molecular
fingerprints. They
identify each cell as
self (belonging to
one's own body or
tissue) vs. *nonself*
(foreign to the body).

All passive transporters are
channels across a membrane.
Solutes just diffuse through the
channel to the side where they
are less concentrated. Some
passive transporters are open
all the time, but others have
molecular gates that open
and close in controlled ways.

Active transporters pump specific
solutes across a cell membrane
to the side where they are more
concentrated. Pumping action
requires an energy input. The
two examples shown here are a
calcium pump (*blue*) and an ATP
synthase (*multicolored*), which
pumps ions to help make ATP.

Figure 4.10 Part of a plasma membrane. It consists of a lipid bilayer and far more proteins
than we can show here. Later chapters use the icons below these protein ribbon models.

The lipid bilayer functions mainly as a barrier to
water-soluble substances. Proteins carry out nearly all
other membrane tasks. Many proteins are receptors
for signals. Others transport specific solutes across the
bilayer. Some transport the solutes passively; others
require an energy input. Still other proteins function
as enzymes that mediate events at the membrane.

Especially among multicelled species, the plasma
membrane bristles with diverse proteins. Recognition
proteins identify the cell as belonging to the body,
and other kinds help defend it against attacks. Special
proteins even help different cells communicate with
one another or stick together in tissues. Figure 4.10
introduces important categories of membrane proteins
and gives a brief description of their functions.

The fluid mosaic model is a good starting point for
thinking about cell membranes. But keep in mind that
membranes have different types and arrangements of
molecules. Even the two surfaces of the same bilayer
are not exactly alike. For example, carbohydrate side
chains that are attached to many proteins and lipids
project from the cell, not into it. All such differences
among plasma membranes and internal membranes
correlate with their functions.

*All cell membranes consist of two layers of lipids—mainly
phospholipids—and diverse proteins. Hydrophobic parts of
the lipids are sandwiched between hydrophilic parts, which
are dissolved in cytoplasmic fluid or in extracellular fluid.*

*All cell membranes have protein receptors, transporters,
and enzymes. The plasma membrane also incorporates
adhesion, communication, and recognition proteins.*

4.4 Introducing Prokaryotic Cells

The word prokaryote is taken to mean "before the nucleus." The name reminds us that bacteria and then archaea originated before cells with a nucleus evolved.

Prokaryotes are the smallest known cells. As a group they are the most metabolically diverse forms of life on Earth. Different kinds can exploit energy and raw materials in nearly all environments, from dry deserts to hot springs to mountain ice.

We recognize two domains of prokaryotic cells: the eubacteria, or true bacteria, and archaea (Sections 1.3 and 19.1). Cells of both groups are alike in outward appearance, size, and where they live. Even so, they differ in major ways at the molecular level. Bacteria start synthesizing each new polypeptide chain with a modified amino acid, formylmethionine. Archaea, like eukaryotes, start chains with methionine. They also have a few proteins called histones that interact with their DNA. Eukaryotic DNA has a great many histone molecules attached; bacterial DNA has none.

Most prokaryotic cells are not much wider than one micrometer; rod-shaped species are at most a few micrometers long (Figures 4.11 and 4.12). Structurally, these are the simplest cells. A semirigid or rigid wall around the plasma membrane helps impart shape to most species. Also, just under the plasma membrane,

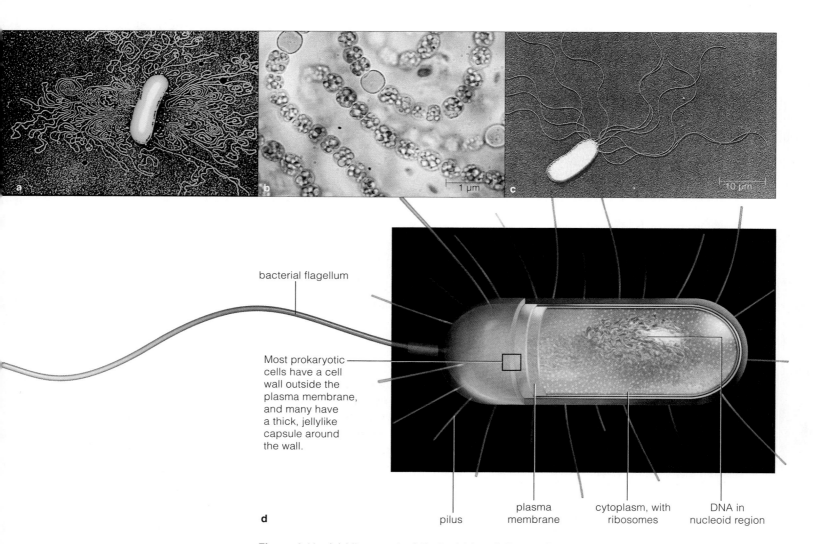

bacterial flagellum

Most prokaryotic cells have a cell wall outside the plasma membrane, and many have a thick, jellylike capsule around the wall.

d

pilus

plasma membrane

cytoplasm, with ribosomes

DNA in nucleoid region

Figure 4.11 (**a**) Micrograph of *Escherichia coli*. Researchers manipulated this bacterial cell to release its single, circular molecule of DNA. (**b**) Cells of various bacterial species are shaped like balls, rods, or corkscrews. Ball-shaped cells of *Nostoc*, a photosynthetic bacteria, stick together in a thick, jellylike sheath of their own secretions. Chapter 19 gives other examples. (**c**) Like this *Pseudomonas marginalis* cell, many species have one or more bacterial flagella that propel the cell body in fluid environments. (**d**) Generalized sketch of a typical prokaryotic body plan.

Figure 4.12 From Bitter Springs, Australia, fossilized bacteria dating back to about 850 million years ago, in Precambrian times: (**a**) a colonial form, most likely *Myxococcoides minor*, and (**b**) cells of a filamentous species (*Palaeolyngbya*).

(**c**) One of the structural adaptations seen among archaea. Many of these prokaryotic species live in extremely hostile habitats, such as the ones thought to have prevailed when life originated. Most archaea and some bacteria have a dense lattice of proteins anchored to the outer surface of their plasma membranes. The unique composition of some of these lattices may help cells withstand the insults of extreme environments, such as the near-boiling, mineral-rich water spewing from hydrothermal vents on the ocean floor.

arrays of protein filaments in the cytoplasm compose a simple internal "skeleton," a bit like the cytoskeleton of eukaryotic cells.

Sticky polysaccharides that often envelop bacterial cell walls help them attach to interesting surfaces such as river rocks, teeth, and the vagina. Many disease-causing (pathogenic) bacteria have a thick protective capsule of jellylike polysaccharides around their wall.

All cell walls are permeable to dissolved substances, which are free to move to and away from the plasma membrane. However, eukaryotic cell walls differ in their structure, as you'll see in Section 4.11.

One or more bacterial flagella often project above the cell wall. Bacterial flagella are motile but differ in structure from eukaryotic flagella (Section 4.10); they do not have an orderly, inner array of microtubules. They help cells move about in fluid habitats, including animal body fluids. Other surface projections include pili (singular, pilus). These protein filaments help many kinds of bacterial cells attach to surfaces and to one another, sometimes for transfer of genetic material.

Like eukaryotic cells, bacteria and archaea depend on their plasma membrane to selectively control the flow of substances into and out of the cytoplasm. The lipid bilayer bristles with diverse protein channels, transporters, and receptors. It incorporates built-in machinery for reactions. For example, photosynthesis proceeds at the plasma membrane in many bacterial species. Organized arrays of proteins harness light energy and convert it to chemical energy in the form of ATP, which is used to build sugars.

The cytoplasm holds many ribosomes on which polypeptide chains are built. DNA is concentrated in an irregularly shaped region of cytoplasm called the nucleoid. Prokaryotic cells inherit one molecule of DNA, in the form of a circle. We call it a bacterial chromosome. The cytoplasm of some species also holds plasmids: far smaller circles of DNA that carry just a few genes. Typically, plasmid genes confer selective advantages, such as antibiotic resistance.

One more intriguing point: In cyanobacteria, part of the plasma membrane projects into the cytoplasm, where it repeatedly folds back on itself. As it happens, pigments and other molecules of photosynthesis are embedded in the membrane, as they are in the inner membrane of chloroplasts. Were ancient cyanobacteria the forerunners of chloroplasts? Section 18.4 looks at this possibility. It is one aspect of a remarkable story about how prokaryotes gave rise to all protists, plants, fungi, and animals.

Bacteria and archaea are different groups of prokaryotic cells; their DNA is not housed inside a nucleus. Most have a permeable cell wall around their plasma membrane that structurally supports and imparts shape to the cell.

These are the simplest cells, but as a group they show the most metabolic diversity. Their metabolic activities proceed at the plasma membrane and within the cytoplasm.

4.5 Introducing Eukaryotic Cells

Figure 4.13 Transmission electron micrograph of a plant cell, cross-section. This is a photosynthetic cell from a blade of timothy grass.

Figure 4.14 Transmission electron micrograph of an animal cell, cross-section. This is a cell from a rat liver.

All cells engage in biosynthesis, dismantling tasks, and energy production, but eukaryotic cells compartmentalize these operations. Their interior is subdivided into a nucleus and other organelles having specialized functions.

Like the prokaryotes, eukaryotic cells have ribosomes in the cytoplasm. Unlike them, eukaryotic cells have an intricate internal skeleton of proteins; we call it a cytoskeleton. They also start out life with **organelles**: internal compartments such as the nucleus. *Eu–* means true; and *karyon*, meaning kernel, refers to a nucleus. Figures 4.13 and 4.14 show two eukaryotic cells.

What advantages do organelles offer? Their outer membrane encloses and sustains a microenvironment. Membrane components selectively control the types and amounts of substances entering or leaving. Their action concentrates substances for metabolic reactions, isolates toxic or disruptive ones, and exports others.

For instance, organelles called mitochondria and chloroplasts concentrate hydrogen ions in ways that lead to the formation of ATP molecules. Enzymes in lysosomes can digest large organic compounds, and they would digest the whole cell if they escaped.

In addition, just as your organ systems interact in controlled ways to keep your whole body running, specialized organelles interact in ways that keep the whole cell running. Ions and molecules move out of one organelle and into another. They move to and from the plasma membrane.

Some substances move through the cytoplasm by a series of organelles. One series functions as a *secretory* pathway. It moves new polypeptide chains from ribosomes through organelles known as ER, then through Golgi bodies, then on to the plasma membrane for release from the cell. Another series is an *endocytic* pathway; it moves substances into the cell. The substances don't travel unescorted; they are enclosed in sacs (vesicles) that have pinched off from organelle membranes or the plasma membrane. Section 4.9 is a visual summary of eukaryotic cell components.

All eukaryotic cells start out life with a nucleus and other organelles, as well as ribosomes and a cytoskeleton. Specialized cells typically incorporate additional kinds of organelles and structures.

Organelles physically separate chemical reactions, many of which are incompatible.

Organelles organize events, as when they assemble, store, or move substances along pathways to and from the plasma membrane or to specific destinations in the cytoplasm.

4.6 The Nucleus

Constructing, operating, and reproducing cells can't be done without carbohydrates, lipids, proteins, and nucleic acids. It takes a class of proteins—enzymes—to build and use these molecules. Said another way, a cell's structure and function start with proteins. And instructions for building proteins are located in DNA.

Unlike prokaryotes, eukaryotic cells have their genetic material distributed among a number of linear DNA molecules of different lengths. The term **chromosome** refers to one double-stranded DNA molecule together with the many histones and other protein molecules attached to it. Each human body cell, for instance, has forty-six chromosomes; frog cells have twenty-six. **Chromatin** is the name for the collection of DNA and proteins in any nucleus of a eukaryotic cell.

The nucleus has two functions. First, it isolates the cell's DNA from potentially damaging reactions in the cytoplasm. Second, it allows or restricts access to DNA's hereditary information through controls over receptors, transport proteins, and pores at its surface. This structural and functional separation makes it far easier to keep DNA molecules organized and also to copy them before a cell divides.

When eukaryotic cells are not dividing, you can't see individual DNA molecules. The nucleus just looks grainy in micrographs, as in Figures 4.14 and 4.15. When cells divide, they duplicate their DNA. During actual division stages, the duplicated DNA molecules become more condensed and compact, like tiny rods. At such times, the DNA no longer looks grainy; each molecule becomes visible in micrographs.

A **nuclear envelope** encloses the semifluid interior of the nucleus (nucleoplasm). It consists of two lipid bilayers studded with proteins. Many of the proteins are organized in complexes that form pores across the envelope (Figure 4.15*b*). A nucleus also contains at least one nucleolus (plural, nucleoli), a construction site where large and small subunits of ribosomes are assembled from RNA and proteins. The subunits pass through pores and enter the cytoplasm. There, large and small subunits join briefly as intact ribosomes.

The outer envelope of the nucleus keeps DNA molecules separated from the cytoplasmic machinery and thus controls access to a cell's hereditary information.

With this separation, DNA is easier to keep organized and to copy before a parent cell divides into daughter cells.

Pores across the nuclear envelope help control the passage of many substances between the nucleus and cytoplasm.

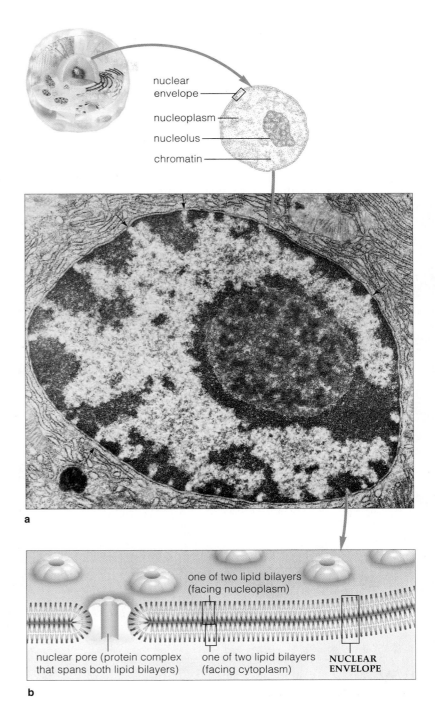

nuclear envelope
nucleoplasm
nucleolus
chromatin

a

one of two lipid bilayers (facing nucleoplasm)

nuclear pore (protein complex that spans both lipid bilayers)   one of two lipid bilayers (facing cytoplasm)   **NUCLEAR ENVELOPE**

b

Figure 4.15   (**a**) Pancreatic cell nucleus. Small arrows on the transmission electron micrograph point to pores where control systems selectively restrict or allow passage of specific substances across the nuclear envelope. (**b**) Sketch of part of the nuclear envelope. Each pore is an organized cluster of membrane proteins.

4.7 The Endomembrane System

New polypeptide chains become folded into proteins.
Some proteins are stockpiled in the cytoplasm or used
at once. Others enter flattened sacs and tubes of the
endomembrane system*: ER, Golgi bodies, and vesicles.*
All proteins that are destined for export or for insertion
into cell membranes pass through this system.

ENDOPLASMIC RETICULUM

Endoplasmic reticulum, or **ER**, is a channel that starts
at the nuclear envelope and extends through part of
the cytoplasm. Here, polypeptide chains are processed
into final proteins, and lipids are assembled. Vesicles
deliver many proteins and lipids to Golgi bodies.

Rough ER consists of flattened sacs and tubes with
ribosomes attached to their outer surface, as in Figure
4.16c. Newly forming polypeptide chains enter it or
become inserted into its membrane. They can do so
only if they contain a built-in signal (a special sequence

of fifteen to twenty amino acids). Enzymes in the
channel often modify polypeptide chains into final
form. You'll see a lot of rough ER in cells that make,
store, and secrete proteins. Example: ER-rich gland
cells in your pancreas make and secrete enzymes that
end up in your small intestine and help digest meals.

Smooth ER is ribosome-free (Figure 4.16d). It makes
lipid molecules that become part of cell membranes.
The ER also takes part in fatty acid breakdown and
degrades some toxins. Sarcoplasmic reticulum, a type
of smooth ER, functions in muscle contraction.

GOLGI BODIES

Patches of ER membrane bulge and break away as
vesicles, each with proteins inside or incorporated in
its membrane. Many vesicles fuse with **Golgi bodies**.
These organelles are folded into flattened, membrane-
bound sacs (Figure 4.16e). Golgi bodies attach sugar

nucleus
rough ER
smooth ER
Golgi body

b RNA messages
are translated into
polypeptide chains
on ribosomes. Many
chains are stockpiled
in the cytoplasm or
used at once. Others
enter the rough ER.

pore

chromatin   nucleolus   nuclear envelope   cytoplasm   ribosome   vesicle
(two lipid bilayers)

the cell nucleus   **rough ER**

a DNA instructions for making proteins are transcribed in the
nucleus and moved to the cytoplasm. RNAs are the messengers
and protein builders.

c Flattened sacs of rough ER form one continuous channel
between the nucleus and smooth ER. Polypeptide chains that
enter the channel undergo modification. They will be inserted
into organelle membranes or will be secreted from the cell.

Figure 4.16 Endomembrane system. Here, many proteins are processed, lipids are assembled, and
both products are sorted and shipped to cellular destinations or to the plasma membrane for export.

side chains to ER proteins and lipids. They also cleave some proteins. The finished products are packaged in vesicles and then shipped to lysosomes, to the plasma membrane, or to the outside of the cell.

MEMBRANOUS SACS WITH DIVERSE FUNCTIONS

Vesicles help organize metabolic activities. Different kinds bud from ER, Golgi, or plasma membranes. For instance, the **lysosomes** that bud from Golgi bodies contain enzymes that digest carbohydrates, proteins, nucleic acids, and lipids. Different vesicles transport proteins and other substances between organelles or to the outer membrane where they are expelled.

The vesicles called **peroxisomes** hold enzymes that digest fatty acids and amino acids. An important function is the breakdown of hydrogen peroxide, a toxic product of metabolism. Enzyme action converts hydrogen peroxide to water and oxygen or uses it in reactions that break down alcohol and other toxins. Drink alcohol, and peroxisomes of liver and kidney cells normally will degrade nearly half of it.

Some vesicles fuse and form large vacuoles, such as the **central vacuole** of mature plant cells. Ions, amino acids, sugars, and toxic substances accumulate in the fluid-filled interior of a central vacuole, which expands and forces the pliable cell wall to enlarge. One benefit is an increase in cell surface area.

> *Endoplasmic reticulum is a membrane-bound channel where polypeptide chains are processed and lipids are assembled. Golgi bodies further modify many of the proteins and lipids.*
>
> *Vesicles help integrate cell activities. Different kinds transport substances around the cell, and break down nutrients and toxins. In plant cells, the central vacuole functions in storage and in increasing the cell surface area.*

Read Me First!
and watch the narrated animation on the endomembrane system

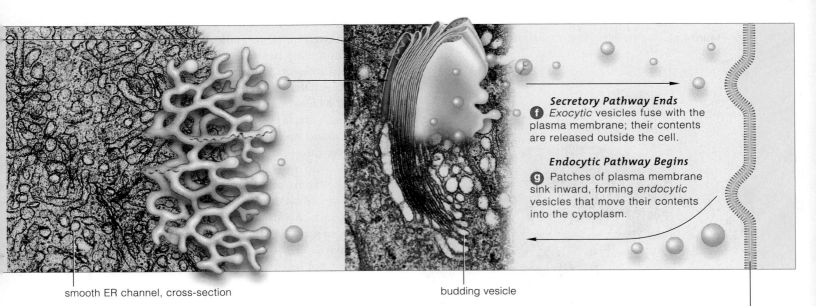

Secretory Pathway Ends
f *Exocytic* vesicles fuse with the plasma membrane; their contents are released outside the cell.

Endocytic Pathway Begins
g Patches of plasma membrane sink inward, forming *endocytic* vesicles that move their contents into the cytoplasm.

smooth ER channel, cross-section

budding vesicle

smooth ER

d Some proteins in the channel continue on to smooth ER, becoming membrane proteins or smooth ER enzymes. Some of these enzymes make lipids and inactive toxins.

Golgi body

e A Golgi body receives, processes, and repackages substances that arrive in vesicles from the ER. Different vesicles transport the substances to other parts of the cell.

plasma membrane

h Exocytic vesicles release cell products and wastes to the outside. Endocytic vesicles move nutrients, water, and other substances into the cytoplasm from outside (Section 5.5).

4.8 Mitochondria and Chloroplasts

ATP, recall, is the energy carrier that jump-starts most of the reactions in cells. So how do cells get ATP in the first place? All of them make ATP by aerobic respiration, which is completed in mitochondria. Algae and photosynthetic plant cells also contain ATP-making organelles called chloroplasts.

MITOCHONDRIA

ATP-forming reactions of aerobic respiration end in the **mitochondrion** (plural, mitochondria). Compared to prokaryotic cells, these organelles make far more ATP from the same compounds. A much-folded inner membrane divides the interior of the mitochondrion into two compartments (Figure 4.17). Hydrogen ions released from the breakdown of organic compounds accumulate in the inner compartment by operation of transport systems, of the sort described in Sections 5.6 and 6.3. As they flow back to the outer compartment, they drive the formation of ATP. Oxygen accepts the spent electrons and keeps the reactions going. Each time you breathe in, you are securing oxygen mainly for the mitochondria in your many trillions of cells.

All eukaryotic cells have one or more mitochondria. The cells that require a great deal of energy, such as those of your liver, heart, and skeletal muscles, may each contain a thousand or more.

CHLOROPLASTS

Many plant cells have plastids, which are organelles of photosynthesis, storage, or both. **Chloroplasts** are important plastids, and only photosynthetic eukaryotic cells have them. In these organelles, energy from the sun drives the formation of ATP and NADPH, which are then used in the formation of organic compounds.

A chloroplast has two outer membranes around a semifluid interior, the stroma, which bathes an inner membrane. Often this single membrane is folded back on itself as a series of stacked, flattened disks (Figure 4.18). Each stack is called a thylakoid. Embedded in the thylakoid membrane are light-trapping pigments, including chlorophylls, as well as enzymes and other proteins with roles in photosynthesis. Glucose, then sucrose, starch, and other organic compounds are built from carbon dioxide and water in the stroma.

Both mitochondria and chloroplasts are a lot like bacteria in size, structure, and biochemistry. Both have their own DNA, RNA, and ribosomes. Coincidence? Probably not, according to the theories discussed in Section 18.4.

Figure 4.17 Typical mitochondrion. This organelle specializes in producing ATP.

outer membrane
outer compartment
inner compartment
inner membrane

two outer membranes

thylakoids (inner membrane system folded into flattened disks)

Figure 4.18 Typical chloroplast, the key defining feature of photosynthetic eukaryotic cells.

Reactions that release energy from organic compounds occur at the compartmented, internal membrane of mitochondria. The reactions, which require oxygen, produce far more ATP than can be produced by any other cellular reaction.

Photosynthetic eukaryotic cells contain chloroplasts, which specialize in making sugars and other carbohydrates.

4.9 Visual Summary of Eukaryotic Cell Components

CELL WALL
Protects, structurally supports cell

CHLOROPLAST
Specializes in photosynthesis

CENTRAL VACUOLE
Increases cell surface area, stores metabolic wastes

nuclear envelope
nucleolus
DNA in nucleoplasm

NUCLEUS
Keeps DNA and its transcription into RNA away from potentially damaging reactions in cytoplasm

CYTOSKELETON
Structurally supports, imparts shape to cell; moves cell and its components

microtubules
microfilaments
intermediate filaments (not shown)

RIBOSOMES
(attached to rough ER and free in cytoplasm) Sites of protein synthesis

MITOCHONDRION
Energy powerhouse; produces many ATP by aerobic respiration

ROUGH ER
Modifies new polypeptide chains; synthesizes lipids

SMOOTH ER
Diverse roles; e.g., makes lipids, degrades fats, inactivates toxins

PLASMODESMA
Communication junction between adjoining cells

GOLGI BODY
Modifies, sorts, ships proteins and lipids for export or for insertion into cell membranes

PLASMA MEMBRANE
Selectively controls the kinds and amounts of substances moving into and out of cell; helps maintain cytoplasmic volume, composition

LYSOSOME-LIKE VESICLE
Digests, recycles materials

Figure 4.19 Typical organelles and structures of plant cells.

nuclear envelope
nucleolus
DNA in nucleoplasm

NUCLEUS
Keeps DNA and its transcription into RNA away from potentially damaging reactions in cytoplasm

CYTOSKELETON
Structurally supports, imparts shape to cell; moves cell and its components

microtubules
microfilaments
intermediate filaments

RIBOSOMES (attached to rough ER and free in cytoplasm) Sites of protein synthesis

MITOCHONDRION
Energy powerhouse; produces many ATP by aerobic respiration

ROUGH ER
Modifies new polypeptide chains; synthesizes lipids

SMOOTH ER
Diverse roles; e.g., makes lipids, degrades fats, inactivates toxins

CENTRIOLES
Special centers that produce and organize microtubules

GOLGI BODY
Modifies, sorts, ships proteins and lipids for export or for insertion into cell membranes

PLASMA MEMBRANE
Selectively controls the kinds and amounts of substances moving into and out of cell; helps maintain cytoplasmic volume, composition

LYSOSOME
Digests, recycles materials

Figure 4.20 Typical organelles and structures of animal cells.

4.10 The Cytoskeleton

Like you, all eukaryotic cells have an internal structural framework—a skeleton. Unlike your skeleton, theirs has elements that are not permanently rigid; they assemble and disassemble at different times.

tubulin subunits

In between the nucleus and plasma membrane of all eukaryotic cells is a **cytoskeleton**—an interconnected system of many protein filaments. Different parts of the system reinforce, organize, and move structures, and often the whole cell. Many parts are permanent; others form at certain times in a cell's life. Figure 4.21 shows an example from one kind of animal cell. The two major cytoskeletal elements—microtubules and microfilaments—have diverse functions. Another type, intermediate filaments, strengthens some animal cells.

Microtubules are long, hollow cylinders of tubulin subunits (Figure 4.21*a*). They organize the cell interior and form a dynamic framework that moves structures such as chromosomes to specific locations. Controls govern which microtubules grow or fall apart at any given time. Those growing in a specific direction—say, the forward end of a prowling amoeba—might get a protein cap that keeps them intact. Those at the trailing end aren't used, aren't capped, and fall apart.

Colchicine, made by the autumn crocus (*Colchicum autumnale*), is a poison. It blocks microtubule assembly, so the cells of animals that eat the plant can't divide. Western yews (*Taxus brevifolia*) make taxol, another microtubule poison. Taxol can stop the uncontrolled cell divisions that give rise to some kinds of cancers.

Microfilaments consist of two coiled-up polypeptide chains of actin monomers, as in Figure 4.21*b*. They often reinforce cell shape or cause it to change. For example, crosslinked, bundled, and gel-like arrays of microfilaments make up a reinforcing **cell cortex** that underlies the plasma membrane. Also, microfilaments anchor proteins and assist in muscle contraction. An animal cell divides as microfilaments around its midsection contract, pinching the cell in two. Although prokaryotic cells lack a cytoskeleton, some types do have microfilament-like proteins that reinforce the cell body.

The microtubules and microfilaments found in all eukaryotic cells are similar. How can they do so many different things if they are so uniform? Other proteins assist them. Among these *accessory* proteins are **motor proteins**, which move cell parts along microtubules when repeatedly energized by ATP.

Intermediate filaments are the most stable parts of some cytoskeletons (Figure 4.21*c*). They strengthen and help maintain cell structures. One type, the lamins, anchor actin and myosin of contractile units found in muscle cells. Other types anchor cells in tissues.

One or at most two kinds of intermediate filaments occur in certain animal cells. Researchers use them to identify the type of cell. *Cell typing* is a useful tool in diagnosing the tissue origin of diverse cancers.

MOVING ALONG WITH MOTOR PROTEINS

Think about the bustle at a train station during the busiest holiday season, and you get an idea of what goes on in cells. Microtubules and microfilaments are the cell's tracks. Kinesins, dyneins, myosins, and other motor proteins are the freight engines (Figure 4.22). ATP is the fuel for movement.

Some motor proteins move chromosomes. Others slide one microtubule over another, or chug along tracks inside nerve cells that extend from your spine to your toes. Many engines are organized in series, each moving some vesicle partway along the track before giving it up to the next in line. Kinesins in

a   25 nm

actin subunit

one polypeptide chain

8–12 nm

b   5–7 nm   **c**

Figure 4.21 Subunits and structure of (**a**) microtubules, (**b**) microfilaments, and (**c**) one kind of intermediate filament. The micrograph at left shows intermediate filaments (*red*) of cultured kangaroo rat cells. The *blue*-stained organelle inside each cell is the nucleus.

Figure 4.22 A motor protein, kinesin, on a microtubule. Kinesin scoots along the length of the microtubule in a hand-over-hand motion. If the microtubule is anchored near the cell's center, the kinesin moves its freight away from the center.

Figure 4.23 (**a**) Internal organization of flagella and cilia. Inside both motile structures is a 9+2 array: a ring of nine pairs of microtubules around one pair at the core. All are connected by spokes and linking elements that restrict the range of sliding. (**b**) Cilia (*gold*) on cells lining an airway that leads to human lungs.

spokes, rings of connective system

central sheath

one central pair of microtubules

plasma membrane

one of nine pairs of microtubules with dynein arms down their length

a

microtubules near base of flagellum or cilium

basal body embedded in cytoplasm

plasma membrane

b

plant cells drag chloroplasts to new positions that are more efficient for light interception as the angle of the sun changes overhead.

Different myosins can move structures along microfilaments or slide one microfilament over another. For example, muscle cells contain long fibers divided into contractile units. Each unit has side-by-side arrays of microfilaments and myosin filaments. When ATP activates it, myosin slides all microfilaments in directions that shorten each unit. When all of the units shorten, the cell itself shortens; it contracts.

CILIA, FLAGELLA, AND FALSE FEET

Besides moving internal parts, many cells move their body or extend parts of it. First, consider **flagella** (singular, flagellum) and **cilia** (singular, cilium). Both are motile structures that project from the cell surface.

Eukaryotic flagella usually are longer and not as profuse as cilia. Many eukaryotic cells swim with the help of whiplike flagella. Sperm do this. The ciliated protozoans swim by beating many cilia in synchrony. In the airways to your lungs, cilia beat nonstop; their coordinated movement sweeps out airborne bacteria and particles that otherwise might reach the lungs (Figure 4.23*b*).

Inside these motile structures is a ring of nine pairs of microtubules around a central pair. Protein spokes and links stabilize the *9+2 array*, which starts at a **centriole** (Figure 4.23*a*). This barrel-shaped structure produces and organizes microtubules, then it remains positioned below the finished array as a **basal body**.

Flagella and cilia move by a sliding mechanism. All pairs of microtubules extend the same distance into the motile structure's tip. Stubby dynein arms project from each pair in the outer ring. When ATP energizes them, the arms grab the microtubule pair in front of them, tilt in a short, downward stroke, then let go. As the bound pair slides down, its arms bind the pair in front of it, forcing it to slide down also—and so on around the ring. The microtubules can't slide too far, but each *bends* a bit. Their sliding motion is converted to a bending motion.

As a final example, some free-living cells, such as macrophages and amoebas, form **pseudopods** ("false feet"). These temporary, irregular lobes project from the cell and function in locomotion and prey capture. Pseudopods move as microfilaments elongate inside them. Motor proteins attached to the microfilaments drag the plasma membrane with them.

A cytoskeleton of protein filaments is the basis of eukaryotic cell shape, internal structure, and movement. Accessory proteins extend the range of functions for those filaments.

Microtubules move cell components. Microfilaments form flexible, linear bundles and networks that reinforce and restructure the cell surface. Intermediate filaments strengthen and maintain shapes of some animal cells.

Cell contractions and migrations, chromosome movements, and other forms of cell movements arise at organized arrays of microtubules, microfilaments, and accessory proteins.

When energized by ATP, motor proteins move in specific directions, along tracks of microtubules and microfilaments. They deliver cell components to new locations.

4.11 Cell Surface Specializations

Our survey of eukaryotic cells concludes with a look at cell walls and other specialized surface structures. Many of these architectural marvels are made primarily of cell secretions. Others are clusters of membrane proteins that connect neighboring cells, structurally and functionally.

EUKARYOTIC CELL WALLS

Single-celled eukaryotic species are directly exposed to the environment. Many have a **cell wall**, a structural component that encloses the plasma membrane. A cell wall protects and physically supports a cell. It's porous, so water and solutes easily move to and from the plasma membrane. A cell would die without these exchanges. Many single-celled protists have a wall around their plasma membranes. So do plant cells and many types of fungal cells.

For example, in the growing parts of multicelled plants, young cells secrete molecules of pectin and other glue-like polysaccharides, as well as cellulose. The cellulose molecules are laid down in the gluey matrix as ropelike strands. All of these materials are components of the plant cell's **primary wall** (Figure 4.24). The sticky primary wall cements abutting cells together. Being thin and pliable, it permits the cell to enlarge under the pressure of incoming water.

Cells that have only a thin primary wall retain the capacity to divide or change shape as they grow and develop. Many types stop enlarging when they are mature. Such cells secrete material on the primary wall's inner surface. These deposits form a lignified, rigid **secondary wall** that reinforces cell shape (Figure 4.24*d*). Secondary wall deposits are extensive and contribute more to structural support.

In woody plants, up to 25 percent of the secondary wall is made of lignin. This organic compound makes plant parts more waterproof, less susceptible to plant-attacking organisms, and stronger.

Read Me First!

and watch the narrated animation on plant cell walls

(a) Randomly oriented cellulose strands in a growing primary wall let a cell expand in all directions. Cross-oriented strands let it lengthen only.

(b) plasmodesma across primary walls of two adjoining cells

middle lamella

c middle lamella
plasma membrane
primary cell wall

d space previously filled with cytoplasm
secondary cell wall (added in layers)
primary cell wall

Figure 4.24 Plant cell walls. (**a**) Microtubules orient cellulose strands, the main construction material for plant walls. Depending on the orientations, the cell will end up round or long.

(**b,c**) Sections through three cells. Cell secretions form a middle lamella, a layer with thickened corners between the walls of adjoining cells. Many channels across adjacent walls, called plasmodesmata, directly connect the cytoplasm of plant cells.

(**d**) In many plant cells, more layers are deposited on the inside of the primary wall. They strengthen the wall and maintain its shape. When the cell dies, the stiffened walls remain. (**e**) This happens in water-conducting pipelines that thread through most plant tissues. Interconnected, stiffened walls of dead cells form the tubes.

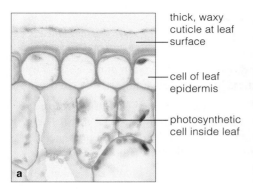

thick, waxy
cuticle at leaf
surface

cell of leaf
epidermis

photosynthetic
cell inside leaf

Figure 4.25 (**a**) Section through a plant cuticle, an outer surface layer of cell secretions. (**b**) A living cell inside bone tissue, the stuff of vertebrate skeletons.

At plant surfaces exposed to air, waxes and other cell secretions build up, forming a protective cuticle. This type of semitransparent surface covering limits water losses from aboveground parts during hot, dry days (Figure 4.25*a*).

MATRIXES BETWEEN ANIMAL CELLS

Animal cells have no cell walls. Intervening between many of them are matrixes made of cell secretions and of materials absorbed from the surroundings. For example, cartilage at the knobby ends of leg bones consists of scattered cells and protein fibers embedded in a ground substance of firm polysaccharides. Living cells also secrete the extensive, hardened matrix that we call bone tissue (Figure 4.25*b*).

CELL JUNCTIONS

Even when a wall or some other structure imprisons a cell in its own secretions, the cell still has contact with the outside world at its plasma membrane. Also, in multicelled species, membrane components extend into adjoining cells or the surrounding matrix. Among the components are **cell junctions**: molecular structures where a cell sends or receives signals or materials, or recognizes and cements itself to cells of the same type.

In plants, for instance, channels extend across the primary wall of adjacent living cells and interconnect the cytoplasm of both (Figure 4.24*b*). Each channel is a plasmodesma (plural, plasmodesmata). Substances flow quickly from cell to cell across these junctions.

In most animal tissues, three types of cell-to-cell junctions are common (Figure 4.26). *Tight* junctions link cells of most body tissues, including epithelia that line outer surfaces, internal cavities, and organs. The junctions seal abutting cells together so water-soluble substances cannot pass between them. That is why gastric fluid does not leak across the stomach lining

free surface
of epithelial
tissue (not
attached to
any other tissue)

examples
of proteins
that make
up tight
junctions

gap junctions

adhering
junction

basement
membrane

Figure 4.26 The most common types of cell junctions in animal tissues.

and damage internal tissues. *Adhering* junctions occur in skin, the heart, and in other organs subjected to stretching. *Gap* junctions link the cytoplasm of certain adjoining cells. They are open channels for a rapid flow of substances, most notably in heart muscle.

A variety of protistan, plant, and fungal cells have a porous wall that surrounds the plasma membrane.

Young plant cells have a thin primary wall pliable enough to permit expansion. Some mature cells also deposit a lignin-reinforced secondary wall that affords structural support.

Animal cells have no walls, but they and many other cells often secrete substances that help form matrixes of tissues. Junctions often occur between cells of multicelled organisms.

Summary

Section 4.1 The cell is the smallest unit that still displays the properties of life. The surface-to-volume ratio constrains size increases. All cells start out life with an outer plasma membrane, cytoplasm, and a nucleus or nucleoid area that contains DNA.

Section 4.2 Most cells are microscopically small. Different microscopes reveal cell shapes and structures.

Section 4.3 Cell membranes consist mainly of lipids and proteins. A lipid bilayer gives a membrane its fluid properties and prevents water-soluble substances from freely crossing it. Proteins embedded in the bilayer or positioned at its surfaces carry out many functions.

Section 4.4 Bacteria and archaea are prokaryotic. They have no nucleus and are the smallest, structurally simplest cells known (Table 4.1).

Sections 4.5, 4.6 Eukaryotic cells generally have diverse organelles: membranous sacs that divide the cell's interior into functional compartments (Table 4.1). Organelles physically separate chemical reactions from the rest of the cell. A nucleus isolates and protects the cell's genetic material.

Section 4.7 In the endomembrane system's ER and Golgi bodies, new polypeptide chains take on final form and lipids are assembled; both get packaged into vesicles for transport, storage, and other cell activities.

Section 4.8 The final reactions of aerobic respiration occur in mitochondria, where many ATP form. The chloroplasts of photosynthetic plant and algal cells use energy from the sun to make sugars.

Section 4.10 Eukaryotic cells have a cytoskeleton of microtubules, microfilaments, and intermediate filaments. It imparts shape and supports and moves cell parts, motile structures, and often the whole cell.

Section 4.11 Many bacterial, protistan, fungal, and plant cells have a wall around the plasma membrane. In multicelled organisms, adjoining cells form diverse structural and functional connections.

Table 4.1 Summary of Typical Components of Prokaryotic and Eukaryotic Cells

Cell Component	Function	Prokaryotic	Eukaryotic			
		Bacteria, Archaea	Protists	Fungi	Plants	Animals
Cell wall	Protection, structural support	✓*	✓*	✓	✓	None
Plasma membrane	Control of substances moving into and out of cell	✓	✓	✓	✓	✓
Nucleus	Physical separation and organization of DNA	None	✓	✓	✓	✓
DNA	Encoding of hereditary information	✓	✓	✓	✓	✓
RNA	Transcription, translation of DNA messages into polypeptide chains of specific proteins	✓	✓	✓	✓	✓
Nucleolus	Assembly of subunits of ribosomes	None	✓	✓	✓	✓
Ribosome	Protein synthesis	✓	✓	✓	✓	✓
Endoplasmic reticulum (ER)	Initial modification of many of the newly forming polypeptide chains of proteins; lipid synthesis	None	✓	✓	✓	✓
Golgi body	Final modification of proteins, lipids; sorting and packaging them for use inside cell or for export	None	✓	✓	✓	✓
Lysosome	Intracellular digestion	None	✓	✓*	✓*	✓
Mitochondrion	ATP formation	**	✓	✓	✓	✓
Photosynthetic pigments	Light–energy conversion	✓*	✓*	None	✓	None
Chloroplast	Photosynthesis; some starch storage	None	✓*	None	✓	None
Central vacuole	Increasing cell surface area; storage	None	None	✓*	✓	None
Bacterial flagellum	Locomotion through fluid surroundings	✓*	None	None	None	None
Flagellum or cilium with 9+2 microtubular array	Locomotion through or motion within fluid surroundings	None	✓*	✓*	✓*	✓
Complex cytoskeleton	Cell shape; internal organization; basis of cell movement and, in many cells, locomotion	Rudimentary***	✓*	✓*	✓*	✓

   *Known to be present in cells of at least some groups.
  **Many groups use oxygen-requiring (aerobic) pathways of ATP formation, but mitochondria are not involved.
 ***Protein filaments form a simple scaffold that helps support the cell wall in at least some species.

Figure 4.27 Cross-sections through the flagellum of a sperm cell from (**a**) a male with Kartagener syndrome and (**b**) an unaffected male. Notice the dynein arms that extend from the paired microtubules.

Self-Quiz

Answers in Appendix III

1. Cell membranes consist mainly of a _____ .
   a. carbohydrate bilayer and proteins
   b. protein bilayer and phospholipids
   c. lipid bilayer and proteins

2. Identify the components of the cells shown above.

3. Organelles _____ .
   a. are membrane-bound compartments
   b. are typical of eukaryotic cells, not prokaryotic cells
   c. separate chemical reactions in time and space
   d. All of the above are features of organelles.

4. Cells of many protists, plants, and fungi, but not animals, commonly have _____ .
   a. mitochondria   c. ribosomes
   b. a plasma membrane   d. a cell wall

5. Is this statement true or false: The plasma membrane is the outermost component of all cells. Explain your answer.

6. Unlike eukaryotic cells, prokaryotic cells _____ .
   a. lack a plasma membrane   c. have no nucleus
   b. have RNA, not DNA   d. all of the above

7. Match each cell component with its function.
   ____ mitochondrion   a. protein synthesis
   ____ chloroplast   b. initial modification of new
   ____ ribosome   polypeptide chains
   ____ rough ER   c. modification of new proteins;
   ____ Golgi body   sorting, shipping tasks
                         d. photosynthesis
                         e. formation of many ATP

Critical Thinking

1. Why is it likely that you will never meet a two-ton amoeba on a sidewalk?

2. Your professor shows you an electron micrograph of a cell with many mitochondria, Golgi bodies, and a lot of rough ER. What kinds of cellular activities would require such an abundance of the three kinds of organelles?

3. *Kartagener syndrome* is a genetic disorder caused by a mutated form of the protein dynein. Affected people have chronically irritated sinuses, and mucus builds up in the airways to their lungs. Bacteria form huge populations in the thick mucus. Their metabolic by-products and the inflammation they trigger combine to damage tissues. Males affected by the syndrome make sperm but are infertile (Figure 4.27). Some have become fathers with the help of a procedure that injects sperm cells directly into eggs. Explain how an abnormal dynein molecule could cause the observed effects.

Media Menu

Student CD-ROM

Impacts, Issues Video
   Where Did Cells Come From?
Big Picture Animation
   The unity and diversity of cells
Read-Me-First Animation
   Cell membranes
   The endomembrane system
   Plant cell walls
Other Animations and Interactions
   Surface-to-volume ratio   Light microscopy
   Lipid bilayer organization   Electron microscopy
   Eukaryotic organelles   Flagella structure

InfoTrac
   • Secrets of a Rock. *Newsweek International*, March 2002.
   • Cell Fantastyk. *Natural History*, May 2000.
   • Scientists Give Golgi Apparatus Its Own Identity. *Cancer Weekly*, December 2000.
   • Symbionts and Assassins. *Natural History*, July 2000.
   • Integral Connections. *The Scientist*, August 2001.

Web Sites
   • Cells Alive: www.cellsalive.com
   • Inside a Cell: gslc.genetics.utah.edu/units/basics/cell
   • What Is a Cell?:
     www.ncbi.nlm.nih.gov/About/primer/genetics_cell.html

How Would You Vote?

Scientists are trying to create a "minimal organism" from a living cell that has a small number of genes. They remove its genes one at a time until they have the simplest possible hereditary package that still allows survival and reproduction. Some people think it is wrong or dangerous to create "new" life forms. Do you think the research should continue?

IMPACTS, ISSUES *Alcohol, Enzymes, and Your Liver*

Consider the cells that are supposed to keep a heavy drinker alive. It makes little difference whether a drinker gulps down 12 ounces of beer, 5 ounces of wine, or 1–1/2 ounces of eighty-proof vodka. Each drink has the same amount of "alcohol" or, more precisely, ethanol.

Ethanol molecules—CH_3CH_2OH—have water-soluble and fat-soluble components, which the stomach and small intestine quickly absorb. The bloodstream moves more than 90 percent of these components to the liver, where enzymes speed their breakdown to a nontoxic form called acetate (acetic acid). However, the liver's alcohol-metabolizing enzymes can detoxify only so much in a given hour.

One of the enzymes you'll read about in this chapter is catalase, a foot soldier against toxin attacks on the body (Figure 5.1). Catalase is thought to assist alcohol dehydrogenase. When alcohol circulates in blood, these enzymes convert it to acetaldehyde. Reactions can't end there, however, because acetaldehyde is toxic at high concentrations. In healthy people at least, another kind of enzyme speeds its breakdown to nontoxic forms.

Given the liver's central role in alcohol metabolism, habitually heavy drinkers gamble with alcohol-induced liver diseases. Over time, the capacity to tolerate alcohol diminishes because there are fewer and fewer liver cells —hence fewer enzymes—for detoxification.

In *alcoholic hepatitis*, inflammation and destruction of liver tissue is widespread. Another disease, *alcoholic cirrhosis*, permanently scars the liver. In time, the liver just stops working, with devastating effects.

Figure 5.1 Ribbon model of catalase, an enzyme that helps detoxify many substances that can damage the body, such as the alcohol in beer, martinis, and other drinks.

the big picture

The One-Way Flow of Energy Energy, the capacity to do work, can be converted from one form to another but can't be created from scratch. It flows in one direction, from usable to less usable forms. Some is lost as heat with each conversion.

How Cells Use Energy Energy flows into the web of life, mainly from the sun, and flows out of it. Cells tap into the one-way flow by energy-acquiring processes, starting with photosynthesis. They convert inputs of energy to forms that keep them alive and working properly.

The liver is the largest gland in the human body, and its activity impacts everything else. You'd have a hard time digesting and absorbing food without it. Your cells would have a hard time synthesizing and taking up carbohydrates, lipids, and proteins, and staying alive.

There's more. The liver makes plasma proteins. These proteins circulate in blood and are vital for blood clotting, immunity, maintaining the fluid volume of the internal environment, and other tasks. Also, liver enzymes get rid of a lot more toxic compounds than acetaldehyde.

Binge drinking—consuming large amounts of alcohol in a brief period—is now the most serious drug problem on campuses in the United States. Consider: 44 percent of nearly 17,600 students surveyed at 140 colleges and universities are caught up in the culture of drinking. They report having five alcoholic drinks a day, on average.

Binge drinking does more than damage the liver. Put aside the related 500,000 injuries from accidents, the 70,000 cases of date rape, and the 400,000 cases of (whoops) unprotected sex among students in an average year. Binge drinking can kill before you know what hit you. Drink too much, too fast, and you can abruptly stop the beating of your heart. Think about it.

With this example, we turn to **metabolism**, the cell's capacity to acquire energy and use it to build, degrade, store, and release substances in controlled ways. At times, the activities of your cells may seem remote from your interests. But they help define who you are and what you will become, liver and all.

 How Would You Vote?

Some people have damaged their liver because they drank too much alcohol. Others have a diseased liver. There aren't enough liver donors for all the people waiting for liver transplants. Should life-style be a factor in deciding who gets a transplant? See the Media Menu for details, then vote online.

How Enzymes Work Without enzymes, substances would not react fast enough to maintain living cells, hence life itself. Controls over enzyme action also maintain life through adjustments in the concentration of substances moving across cell membranes.

Membranes and Metabolism Cells have built-in mechanisms that increase and decrease concentrations of substances across their membranes. The adjustments are essential for metabolic reactions and metabolic pathways.

5.1 Inputs and Outputs of Energy

Cells secure energy from their surroundings and use it for thousands of tasks. Energy drives metabolism—chemical work that stockpiles, builds, rearranges, and breaks down substances. It drives the mechanical work of moving cell parts, body parts, or the whole organism. It drives the electrochemical work of moving charged substances across membranes, as happens when cells make ATP.

THE ONE-WAY FLOW OF ENERGY

Energy is a capacity to do work, and you can't create it out of nothing. By the **first law of thermodynamics**, any isolated system has a finite amount of energy that cannot be added to or lost. Energy *can* be converted from one form to another. However, the total amount in the system stays the same. Motion, chemical bonds, heat, electricity, sound, nuclear forces, and gravity are examples of different forms of energy.

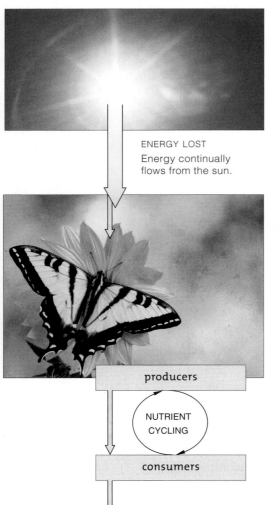

ENERGY LOST
Energy continually flows from the sun.

ENERGY GAINED
Sunlight energy reaches environments on Earth. Producers of nearly all ecosystems secure some and convert it to stored forms of energy. They and all other organisms convert stored energy to forms that can drive cellular work.

producers

NUTRIENT CYCLING

consumers

ENERGY LOST
With each conversion, there is a one-way flow of a bit of energy back to the environment. Nutrients cycle between producers and consumers.

"Entropy" is a measure of the degree of a system's disorder. By the **second law of thermodynamics**, the entropy, or disorder, of the universe always increases. Think of Egyptian pyramids—once highly organized, now crumbling, and thousands of years from now, dust. According to the second law, pyramids and all other things are on their way toward maximum entropy.

Energy is part of this big picture. It spontaneously flows toward its most disorganized form—heat. Why? Converting energy from one form to another is never 100 percent efficient. Although energy is conserved in any exchange, at least some of it dissipates as heat. It is not easy to convert heat to a different form of energy.

Can life be one glorious pocket of resistance to this depressing flow toward maximum entropy? After all, new bonds hold atoms together in orderly patterns in each new organism. Molecules get more organized and have richer stores of energy, not poorer.

Even so, the second law does apply to life on Earth. Life's main energy source is the sun, which has been losing energy ever since it formed about 5 billion years ago. Photosynthetic cells intercept light energy from the sun and convert it to chemical bond energy in sugars, starches, and other compounds. Organisms that eat plants get at the stored chemical energy by breaking and rearranging chemical bonds. With each conversion, however, a bit of energy escapes as heat. Cells don't convert that heat to other forms of energy. They simply can't use it to do work.

Overall, then, energy flows in one direction. Life can maintain its astounding organization only because it is being continually resupplied with energy that is being lost from someplace else (Figure 5.2).

UP AND DOWN THE ENERGY HILLS

Cells store and retrieve energy when they convert one molecule to another. In photosynthetic cells, sunlight energy drives ATP formation, then energy from ATP drives glucose formation (Figure 5.3*a*). Six molecules of carbon dioxide (CO_2) and six of water (H_2O) are converted to one molecule of glucose ($C_6H_{12}O_6$) and six of oxygen (O_2). Photosynthetic reactions require energy input; they are *endergonic* (meaning energy in).

Figure 5.2 A one-way flow of energy into ecosystems compensates for the one-way flow of energy out of it.

We think of glucose as a high-energy compound because it can be converted to more stable molecules for a net gain of energy. It does take an investment of energy to get the conversion reactions started, but the formation of more stable end products releases more energy than the amount invested. For example, CO_2 and H_2O are all that's left of glucose at the end of aerobic respiration. Both still have energy stored in covalent bonds, but the two products are so stable that cells cannot gain energy by converting them to something else. It's as if carbon dioxide and water are at the base of an "energy hill."

Aerobic respiration releases energy bit by bit, with many conversion steps, so cells can capture some of it efficiently. This metabolic process is like a downhill run, from high-energy glucose to low-energy carbon dioxide and water (Figure 5.3b). Such reactions, which show a net energy release, are said to be *exergonic*.

ATP—THE CELL'S ENERGY CURRENCY

All cells stay alive by *coupling* energy inputs to energy outputs, mainly with adenosine triphosphate, or **ATP**. This nucleotide consists of a five-carbon sugar (ribose), a base (adenine), and three phosphate groups (Figure 5.4a). ATP readily gives up a phosphate group to other molecules and primes them to react. Such phosphate-group transfers are known as **phosphorylations**.

ATP is the currency in a cell's economy. Cells earn it by investing in energy-releasing reactions. They spend it in energy-requiring reactions that keep them alive. We use a cartoon coin to symbolize ATP.

Because ATP is the main energy carrier for so many reactions, you might infer—correctly—that cells have ways to renew it. When ATP gives up a phosphate group, ADP (adenosine diphosphate) forms. ATP can re-form when ADP binds to inorganic phosphate (P_i) or to a phosphate group that was split from a different molecule. Regenerating ATP by this **ATP/ADP cycle** helps drive most metabolic reactions (Figure 5.4b).

Energy is the capacity to do work. It flows in one direction, from more usable to less usable forms. Heat is the least usable form of energy. Organisms maintain complex organization by being resupplied with energy from someplace else.

All organisms secure energy from outside sources. The sun is the primary source of energy for the web of life. All organisms use and store energy in chemical bonds.

ATP is the main energy carrier in all living cells. It couples energy-releasing and energy-requiring reactions. ATP primes molecules to react by transferring a phosphate group to them.

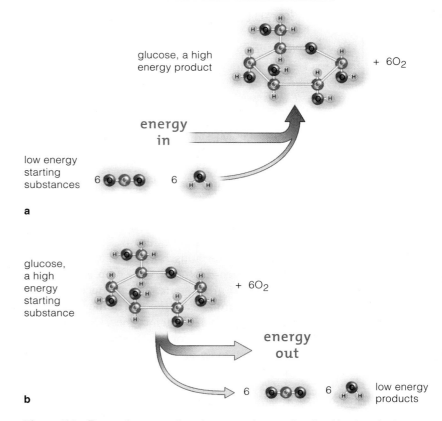

Figure 5.3 Two main categories of energy changes involved in chemical work. (**a**) Endergonic reactions, which won't run without an energy input. (**b**) Exergonic reactions, which end with a net release of usable energy.

Figure 5.4 (**a**) Ball-and-stick model for ATP, an energy carrier. (**b**) ATP couples energy-releasing reactions with energy-requiring ones. In the ATP/ADP cycle, recurring phosphate-group transfers turn ATP into ADP, and back again to ATP.

5.2 Inputs and Outputs of Substances

How cells get energy is only one aspect of metabolism. Another is the accumulation, conversion, and disposal of materials by energy-driven reactions. Most reactions are part of stepwise metabolic pathways.

THE NATURE OF METABOLIC REACTIONS

For any metabolic reaction, the starting substances are called *reactants*. Substances formed during a reaction sequence are *intermediates*, and those left at the end are the *products*. ATP and other *energy carriers* activate enzymes and other molecules by making phosphate-group transfers. *Enzymes* are catalysts: They can speed specific reactions enormously. *Cofactors* are metal ions and coenzymes such as NAD^+. They help enzymes by moving functional groups, atoms, and electrons from one reaction site in an enzyme to another. *Transport proteins* help solutes across membranes. Controls over transport proteins adjust concentrations of substances required for reactions, and so influence the timing and direction of metabolism.

Bear in mind, metabolic reactions don't always run from reactants to products. They might start out in this "forward" direction. But most also run in reverse, with products being converted back to reactants. Such reversible reactions tend to run spontaneously toward **chemical equilibrium**, when the reaction rate is about the same in both directions. For most reactions, the amounts of reactant and product molecules differ at that time (Figure 5.5). It is like a party where people drift between two rooms. The number in each room stays the same—say, thirty in one and ten in the other —even as individuals move back and forth.

Why bother to think about this? *Each cell can bring about big changes in its activities by controlling a few steps of reversible metabolic pathways.*

For instance, when your cells need a quick bit of energy, they rapidly split glucose into two pyruvate molecules. They do so by a sequence of nine enzyme-mediated steps of a pathway called glycolysis. When glucose supplies are too low, cells quickly reverse this pathway and build glucose from pyruvate and other substances. How? Six steps of the pathway happen to be reversible, and the other three are bypassed. An input of energy from ATP drives the bypass reactions in the uphill (energetically unfavorable) direction.

What if cells did not have this reverse pathway? They wouldn't be able to build glucose fast enough to compensate for episodes of starvation, when glucose supplies in blood become dangerously low.

REDOX REACTIONS

Energy flows from the environment through all living things by way of photosynthesis and other metabolic pathways. Individual cells capture free energy, store it, then release it in manageable bits. They *control* the energy they require to grow and reproduce.

Cells release energy efficiently by electron transfers, or **oxidation–reduction reactions**. In these "redox" reactions, one molecule gives up electrons (is oxidized) and another gains them (is reduced). Commonly, hydrogen atoms are released at the same time, thus becoming H^+. Being attracted to the opposite charge of the electrons, H^+ tags along with them.

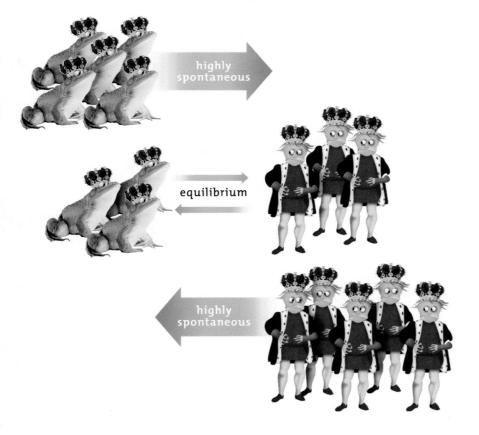

highly spontaneous

equilibrium

highly spontaneous

Figure 5.5 Chemical equilibrium. With a high concentration of reactant molecules (represented as wishful frogs), a reaction runs most strongly in the forward direction, to products (the princes). When the concentration of product molecules is high, it runs most strongly in reverse. At equilibrium, the rates of the forward and reverse reactions are the same.

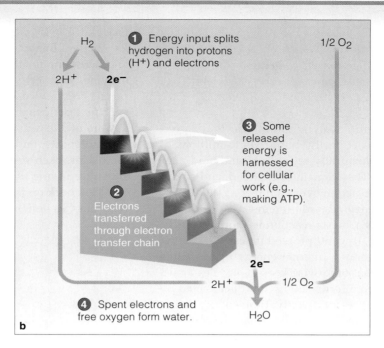

a

b

① Energy input splits hydrogen into protons (H⁺) and electrons

② Electrons transferred through electron transfer chain

③ Some released energy is harnessed for cellular work (e.g., making ATP).

④ Spent electrons and free oxygen form water.

Figure 5.6 Uncontrolled versus controlled energy release. (**a**) Free hydrogen and oxygen exposed to an electric spark react and release energy all at once. (**b**) Electron transfer chains let the same reaction proceed in small, more manageable steps that can access the released energy.

Start thinking about redox reactions, because they are central to photosynthesis and aerobic respiration. In the next two chapters, you'll see how coenzymes pick up electrons and H⁺ stripped from substrates, then deliver them to **electron transfer chains**. Such chains are membrane-bound arrays of enzymes and other molecules that accept and give up electrons in sequence. Electrons are at a higher energy level when they enter a chain than when they leave. Think of the electrons as descending a staircase and stingily losing a bit of energy at each step (Figure 5.6). For these two pathways, stepwise electron transfers concentrate H⁺ in ways that contribute to ATP formation.

TYPES OF METABOLIC PATHWAYS

We've mentioned metabolic pathways in passing, but let's now formally define them. **Metabolic pathways** are enzyme-mediated sequences of reactions in cells. Many are *biosynthetic* (or anabolic), and they require energy inputs. Examples are the assembly of glucose, starch, proteins, and other high-energy molecules from small molecules. The main biosynthetic pathway in the biosphere is photosynthesis (Figure 5.7).

Degradative (or catabolic) pathways are exergonic, overall. These reactions can break down molecules to smaller, lower energy products. Aerobic respiration releases a lot of usable energy (ATP) in the step-by-step enzymatic breakdown of glucose. It is the main degradative pathway in the biosphere (Figure 5.7).

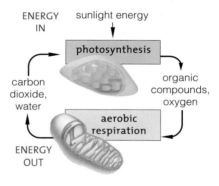

ENERGY IN — sunlight energy

photosynthesis

carbon dioxide, water

organic compounds, oxygen

aerobic respiration

ENERGY OUT

Figure 5.7 The main metabolic pathways in ecosystems. Energy input from the sun drives photosynthesis, and aerobic respiration yields a lot of usable energy. ATP forms in both pathways by way of redox reactions.

Not all metabolic pathways are linear, a straight line from the reactants to products. In cyclic pathways, the final step regenerates a reactant that is the point of entry for the reaction sequence. In branched pathways, reactants or intermediates are channeled into two or more different reaction sequences.

Metabolic pathways are orderly, enzyme-mediated reaction sequences, some biosynthetic, others degradative.

Control over a key step of a metabolic pathway can bring about rapid shifts in cell activities.

Many aspects of metabolism involve electron transfers, or oxidation–reduction reactions. Electron transfer chains are important sites of energy exchange in both photosynthesis and aerobic respiration.

5.3 How Enzymes Make Substances React

What would happen if you left a cupful of glucose out in the open? Not much. Years would pass before you would see evidence of its conversion to carbon dioxide and water. Yet that same conversion takes only a few seconds in your body. Enzymes make the difference.

Enzymes, again, are catalytic molecules; they speed rates of specific reactions by hundreds to millions of times. Enzymes chemically recognize, bind, and alter specific reactants. They remain unchanged, and so can mediate the same reaction over and over again. Except for a few RNAs, enzymes are proteins.

Regardless of whether a reaction is spontaneous or enzyme-mediated, it won't proceed unless the starting substances have enough internal energy to overcome repulsive forces that otherwise keep molecules apart. All molecules have internal energy that is affected by temperature and pressure. **Activation energy** refers to the minimum amount of internal energy that molecules must have before a reaction gets going.

Activation energy is an energy barrier—something like a hill or a brick wall (Figures 5.8 and 5.9). One way or another, that barrier must be surmounted before the reaction will proceed. Enzymes lower the barrier. How? *Compared to the surrounding environment,*

they offer a stable microenvironment that is more favorable for reaction. Enzymes are far larger than **substrates**, another name for the reactants that bind to a specific enzyme. Each enzyme has one or more **active sites**: pockets or crevices where substrates bind and where specific reactions are catalyzed (Figure 5.10).

Part of the substrate is complementary in shape, size, solubility, and charge to the active site. Because of this fit, each enzyme can recognize and bind its substrate among thousands of substances in cells.

Think back on the main types of enzyme-mediated reactions (Section 3.1). In *functional group transfers*, one

Figure 5.8 Activation energy. Reactants must have a minimum amount of internal energy before a given reaction will run to products. Sometimes they need an input of energy to get there. An enzyme enhances the reaction rate by lowering the amount of activation energy required. It makes the energy hill smaller.

Figure 5.9 A simple way to think about the energy required to get a reaction going without an enzyme (**a**) and with the help of an enzyme (**b**).

Read Me First!
and watch the narrated
animation on catalase action

a Hydrogen peroxide (H_2O_2) enters a cavity in catalase. It is the substrate for a reaction aided by an iron molecule in a heme group (*red*).

b A hydrogen of the peroxide is attracted to histidine, an amino acid projecting into the cavity. One oxygen binds the iron.

c This binding destabilizes the peroxide bond, which breaks. Water (H_2O) forms. In a later reaction, another H_2O_2 will pull the oxygen from iron, which will then be free to act again.

Figure 5.10 How catalase works. This enzyme has four polypeptide chains and four heme groups.

molecule gives up a functional group to another. In *electron transfers*, one or more electrons stripped from one molecule are donated to another. In *rearrangements*, a juggling of internal bonds converts one molecule to another. In *condensation*, two molecules are covalently bound together as a larger molecule. Finally, in *cleavage* reactions, a larger molecule splits into smaller ones.

When we talk about activation energy, *we really are talking about the energy it takes to align reactive chemical groups, briefly destabilize electric charges, and break bonds*.

Enzymes lower activation energy by restraining a reactant molecule. Binding to the active site stretches and squeezes the reactant into a certain shape, maybe next to another molecule or reactive group. This puts a substrate at its **transition state**, meaning its bonds are at the breaking point and the reaction can run easily to product.

The binding between an enzyme and its substrate is weak, and temporary (that's why the reaction does not change the enzyme). But energy is released when weak bonds form. This "binding energy" stabilizes the transition state long enough to keep the enzyme and its substrate together for the reaction.

Four mechanisms work alone or in combination to get substrates to the transition state:

Helping substrates get together. Substrate molecules rarely react at low concentrations. Binding to an active site boosts local substrate concentration by as much as ten millionfold.

Orienting substrates in positions favoring reaction. On their own, substrates collide from random directions. By contrast, weak but extensive bonds at an active site put reactive groups close together.

Shutting out water. Because of its ability to form hydrogen bonds so easily, water can interfere with the breaking and formation of bonds during reactions. Some active sites contain mostly nonpolar amino acids. The hydrophobic groups keep water away from the active site and reactions.

Inducing changes in enzyme shape. Weak interactions between the enzyme and its substrate may induce the enzyme to change its shape. By the **induced-fit model**, a substrate is not quite complementary to an active site. The enzyme bends and optimizes the fit; in doing so, it pulls the substrate to the transition state.

On their own, chemical reactions occur too slowly to sustain life. Enzymes greatly increase reaction rates by lowering the activation energy—the minimum amount of energy required to align reactive groups, destabilize electric charges, and break bonds so that products can form from reactants.

Enzymes drive their substrates to a transition state, when the reaction can most easily run to completion. This happens in the enzyme's active site.

In the active site, substrates move to the transition state by mechanisms that concentrate and orient them, that exclude water, and that induce an optimal fit with the active site.

5.4 Enzymes Don't Work In a Vacuum

Controls over enzyme function help cells respond quickly to changing conditions by triggering adjustments in metabolic reactions. Feedback mechanisms that can activate or inhibit enzymes conserve resources. Cells synthesize what conditions require—no more, no less.

HELP FROM COFACTORS

Cofactors (specific metal ions or coenzymes) help out at the active site of enzymes or taxi electrons, H^+, or functional groups to a different location. **Coenzymes** are a class of organic compounds that may or may not have a vitamin component.

One or more metal ions assist nearly a third of all known enzymes. Metal ions easily give up and accept electrons. As part of coenzymes, they help products form by shifting electron arrangements in substrates or intermediates. That is what goes on at the hemes in catalase. Heme, an organic ring structure, incorporates iron at its center. Figure 5.10 shows how iron atoms in heme coenzymes help catalase break down hydrogen peroxide to water.

Like vitamin E, catalase is one of the **antioxidants**: It helps neutralize free radicals. *Free radicals* are atoms with unpaired electrons—reactive, unbound fragments left over from reactions. As we age, we make less and less catalase, so free radicals accumulate. They attack the structure of DNA and other biological molecules.

Some coenzymes are tightly bound to an enzyme. Others, such as NAD^+ and $NADP^+$, can diffuse freely through a cell membrane or cytoplasm. Either way, coenzymes participate intimately in a reaction. Unlike enzymes, many become modified during the reaction, but they are regenerated elsewhere.

CONTROLS OVER ENZYMES

Many controls over enzymes maintain, lower, and raise the concentrations of substances. Others adjust how fast enzyme molecules are synthesized, and they activate or inhibit the ones already built.

In some cases, a molecule that acts as an activator or inhibitor reversibly binds to its own *allosteric* site, not the active site, on the enzyme (*allo–*, other; *steric*, structure). Binding alters the enzyme's shape in a way that hides or exposes the active site (Figure 5.11).

Picture a bacterial cell making tryptophan and other amino acids—the building blocks for proteins. Even when the cell has made enough proteins, tryptophan synthesis continues until its increasing concentration causes **feedback inhibition**. This means a change that results from a specific activity *shuts down the activity*.

A feedback loop starts and ends at many allosteric enzymes. In this case, unused tryptophan binds to an allosteric site on one of the enzymes in a tryptophan biosynthesis pathway. It blocks the active site, so less tryptophan is made (Figure 5.12). At times when not many tryptophan molecules are around, more active

allosteric activator
allosteric binding site vacant
enzyme active site
substrate cannot bind
active site altered, substrate can bind
a

allosteric inhibitor
allosteric binding site vacant; active site can bind substrate
active site altered, can't bind substrate
b

Figure 5.11 Allosteric control over enzyme activity. (**a**) An active site is unblocked when an activator binds to a vacant allosteric site. (**b**) An active site is blocked when an inhibitor binds to a vacant allosteric site.

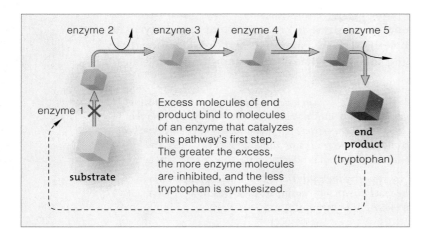

enzyme 2   enzyme 3   enzyme 4   enzyme 5

enzyme 1

Excess molecules of end product bind to molecules of an enzyme that catalyzes this pathway's first step. The greater the excess, the more enzyme molecules are inhibited, and the less tryptophan is synthesized.

end product (tryptophan)

substrate

Figure 5.12 Feedback inhibition of a metabolic pathway. Five kinds of enzymes act in sequence to convert a substrate to tryptophan.

sites remain exposed, and so the synthesis rate picks up. In such ways, feedback loops quickly adjust the concentrations of substances.

In humans and other multicelled species, enzyme controls are just amazing. They keep individual cells functioning in ways that benefit the whole body!

EFFECTS OF TEMPERATURE, PH, AND SALINITY

Temperature is a measure of molecular motion. As it rises, it boosts reaction rates both by increasing the likelihood that a substrate will bump into an enzyme and by raising a substrate molecule's internal energy. Remember, the more energy a reactant molecule has, the closer it gets to jumping that activation energy barrier and taking part in a reaction.

Above or below the range of temperature that an enzyme can tolerate, weak bonds break, and enzyme shape changes. Substrates no longer can bind to the active site, and the reaction rate falls sharply (Figure 5.13). Such declines typically occur with fevers above 44°C (112°F), which people usually cannot survive.

Enzyme action is also affected by pH (Figure 5.14). In the human body, most enzymes work best when the pH is between 6 and 8. For instance, trypsin is active in the small intestine (pH of 8 or so).

One of the notable exceptions is pepsin, a protein-digesting enzyme. Pepsin is a nonspecific protease; it chews up any proteins. It is produced in inactive form and normally becomes activated only in gastric fluid, in the stomach. Gastric fluid happens to be a highly acidic environment (pH 1–2). It's a good thing that activated pepsin is confined to the stomach. If it were to leak out (as happens with peptic ulcers), it could digest a lot of you instead of proteins in your food.

Most enzymes don't work well when the fluids in which they are dissolved are saltier or less salty than their range of tolerance. Too much or too little salt interferes with the hydrogen bonds that help hold an enzyme in its three-dimensional shape. By doing so, it inactivates the enzyme.

Many enzymes are assisted by cofactors, which are specific metal ions or coenzymes.

Enzyme action adjusts the concentrations and kinds of substances available in cells. Controls over enzymes enhance or inhibit their activity.

Enzymes work best when the cellular environment stays within limited ranges of temperature, pH, and salinity. The actual ranges differ from one type of enzyme to the next.

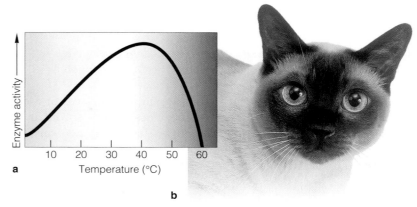

Figure 5.13 Enzymes and the environment. (**a**) How increases in temperature affect one enzyme's activity. (**b**) Temperature outside the body affects the fur color of Siamese cats. Epidermal cells that give rise to the cat's fur produce a brownish-black pigment, melanin. Tyrosinase, an enzyme in the melanin production pathway, is heat-sensitive in the Siamese. It becomes inactive in warmer parts of the cat's body, which end up with less melanin, and lighter fur. Put this cat in booties for a few weeks and its warm feet will turn white.

Figure 5.14 Enzymes and the environment. (**a**) How pH values affect three enzymes. (**b**) Cranberry plants grow best in acidic bogs. Unlike most plants, they have no nitrate reductase. This enzyme converts nitrate (NO_3) found in typical soils to metabolically useful ammonia (NH_3). Nitrogen in highly acidic soils is already in the form of ammonia (NH_4^+). (**c**) Life in wastewater from a copper mine in California. The slime streamers are microbial communities dominated by an archaean, which makes unique enzymes that help it live in this toxic, highly acidic environment.

5.5 Diffusion, Membranes, and Metabolism

What determines whether a substance will move one way or another to and from a cell, across that cell's membranes, or through the cell itself? Part of the answer has to do with something called diffusion.

Think about the water bathing the surfaces of a cell membrane. Plenty of substances are dissolved in it, but the kinds and amounts close to its two surfaces differ. The membrane itself set up the difference and is busy maintaining it. How? Each cell membrane has

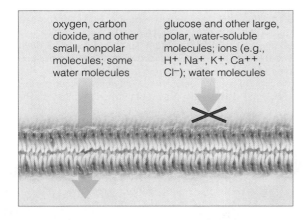

Figure 5.15 Selective permeability of cell membranes. Small, nonpolar molecules and some water molecules can cross the lipid bilayer. Ions and large, polar, water-soluble molecules and the water dissolving them cross with the help of transport proteins.

selective permeability. Its molecular structure allows some substances but not others to cross it in certain ways, at certain times.

Lipids of a membrane's bilayer are mostly nonpolar, so they let small, nonpolar molecules such as O_2 and CO_2 slip across. Water molecules are polar, but some slip through gaps that form as the hydrophobic tails of many lipids flex and bend. The bilayer itself is not permeable to ions or large, polar molecules such as glucose; these cross with the help of proteins. Water often crosses with them (Figure 5.15).

Membrane barriers and crossings are vital, because metabolism depends on the cell's capacity to increase, decrease, and maintain concentrations of molecules and ions required for reactions. They also supply cells or organelles with raw materials, get rid of wastes, and collectively maintain the cell's volume and pH.

WHAT IS A CONCENTRATION GRADIENT?

Now picture molecules or ions of some substance near a membrane. They move constantly, collide at random, and bounce off one another. When the concentration in one region is not the same as in an adjoining region, this condition is a *gradient*. A **concentration gradient** is a difference in the number per unit volume of ions or molecules of a substance between adjoining regions.

In the absence of other forces, a substance tends to move from a region where it is more concentrated to a region where it is less concentrated. At temperatures characteristic of life, *thermal energy that is inherent in molecules drives this movement*. Although the molecules are colliding and careening back and forth millions of times per second, their *net* movement is away from the place where they are most concentrated.

Diffusion is the name for the net movement of like molecules or ions down a concentration gradient. It is a factor in how substances move into, through, and out of cells. In multicelled species, it moves substances between body regions and between the body and its environment. For instance, when oxygen builds up in leaf cells, it may diffuse into air inside the leaf, then into air outside, where its concentration is lower.

Figure 5.16 Two cases of diffusion. (**a**) A drop of dye enters a bowl filled with water. Gradually, the dye molecules become evenly dispersed through the molecules of water. (**b**) The same thing happens with water molecules. Here, dye (*red*) and water (*yellow*) are added to the same bowl. Each substance will show a net movement down its own concentration gradient.

Like other substances, oxygen tends to diffuse in a direction set by its *own* concentration gradient, not by gradients of other solutes. You can see the outcome of this tendency by squeezing a drop of dye into water. The dye molecules diffuse into the region where they are not as concentrated, and water molecules move into the region where *they* are not as concentrated. Figure 5.16 shows simple examples of diffusion.

WHAT DETERMINES DIFFUSION RATES?

How fast a particular solute diffuses depends on the steepness of its concentration gradient, its size, the temperature, and electric or pressure gradients that may be present.

First, rates are high with steep gradients, because more molecules are moving out of a region of greater concentration compared to the number moving into it. Second, more heat energy in warmer regions makes molecules move faster and collide more often. Third, smaller molecules diffuse faster than large ones do.

Fourth, an electric gradient may alter the rate and direction of diffusion. An **electric gradient** is simply a difference in electric charge between adjoining regions. For example, each ion dissolved in fluids bathing a cell membrane contributes to an electric charge at one side or the other. Opposite charges attract. Therefore, the fluid with more negative charge overall exerts the greatest pull on positively charged substances, such as sodium ions. Later chapters explain how many cell activities, including ATP formation and the sending and receiving of signals in nervous systems, are based on the force of electric and concentration gradients.

Fifth, as you will see shortly, diffusion also may be affected by a **pressure gradient**. This is a difference in the exerted force per unit area in two adjoining regions.

MEMBRANE CROSSING MECHANISMS

Before getting into the actual mechanisms that move substances across membranes, study the overview in Figure 5.17. These mechanisms help supply cells and organelles with raw materials and get rid of wastes. Collectively, they help maintain the volume and pH of cells or organelles within functional ranges.

Small, nonpolar molecules such as oxygen diffuse across the membrane's lipid bilayer. Polar molecules and ions diffuse through the interior of transport proteins that span the bilayer. Passive transporters simply allow a substance to follow its concentration gradient across a membrane. The mechanism is called *passive transport*, or "facilitated" diffusion.

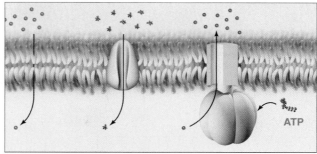

High

|
Concentration gradient across cell membrane
|

Low

| Diffusion of lipid-soluble substances across bilayer | Passive transport of water-soluble substances through channel protein; no energy input needed | Active transport through ATPase; requires energy input from ATP |

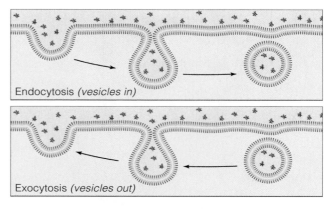

Endocytosis *(vesicles in)*

Exocytosis *(vesicles out)*

Figure 5.17 Overview of membrane crossing mechanisms.

Polar molecules cross the membrane through the interior of active transporters. The net direction of movement is against the concentration gradient, and it requires an input of energy. We call this mechanism *active transport*. Energy-activated transporters move a substance against its concentration gradient.

Other mechanisms move substances in bulk into or out of cells. *Exocytosis* involves fusion of the plasma membrane and a membrane-bound vesicle that formed inside the cytoplasm. *Endocytosis* involves an inward sinking of a patch of plasma membrane, which seals back on itself to form a vesicle inside the cytoplasm.

Diffusion is the net movement of molecules or ions of a substance into an adjoining region where they are not as concentrated.

The force of a concentration gradient can drive the directional movement of a substance across membranes. The gradient's steepness, temperature, molecular size, and electric and pressure gradients affect diffusion rates.

Cellular mechanisms increase and decrease concentration gradients across cell membranes.

5.6 How the Membrane Transporters Work

EXTRACELLULAR FLUID

passive transport protein

LIPID BILAYER

CYTOPLASM

glucose, more concentrated outside cell than inside

glucose transporter

d When the glucose binding site is again vacant, the protein resumes its original shape.

a Glucose binds to a vacant site inside the channel through the transport protein.

c Glucose becomes exposed to fluid on other side of the membrane. It detaches from the binding site and diffuses out of the channel.

b Bound glucose makes the protein change shape. Part of the channel closes behind the solute. Another part opens in front of it.

Large, polar molecules and ions can't cross a lipid bilayer; they require the help of transport proteins.

Many kinds of solutes cross a membrane by diffusing through a channel or tunnel inside transport proteins. When one solute molecule or ion enters the channel and weakly binds to the protein, the protein's shape changes. The channel closes behind the bound solute and opens in front of it, which exposes the solute to the fluid environment on the opposite side of the membrane. There, the binding site reverts to what it was before, so the solute is released.

PASSIVE TRANSPORT

In **passive transport**, a concentration gradient, electric gradient, or both drive the diffusion of a substance across a membrane, through the interior of a transport protein. The protein does not require an energy input to assist the directional movement. That is why this mechanism is also known as facilitated diffusion.

Some passive transporters are open channels, and some are channels with gates that can be opened or closed when conditions change. Others, including the glucose transporter illustrated in Figure 5.18, assist solutes across by undergoing reversible changes in their shape.

The *net* direction of movement depends on how many molecules or ions of the solute are randomly colliding with the transporters. Encounters are more frequent on the side of the membrane where the solute concentration is greatest. The solute's *net* movement tends to be toward the side of the membrane where it is less concentrated.

If nothing else were going on, passive transport would continue until concentrations on both sides of a cell membrane became equal. But other processes affect the outcome. For instance, glucose transporters help glucose from blood move into cells, which use it for biosynthesis and for quick energy. How? As fast as glucose molecules are diffusing into cells, others are being used up. By using up glucose, then, these cells maintain the gradient that favors the uptake of *more* glucose.

Figure 5.18 Passive transport. This model shows one of the glucose transporters that span the plasma membrane. Glucose crosses in both directions. The *net* movement is down its concentration gradient until its concentrations are equal on both sides of the membrane.

ACTIVE TRANSPORT

Only in a dead cell have solute concentrations become equal on both sides of membranes. Living cells never stop expending energy to pump solutes into and out of their interior. With **active transport**, energy-driven protein motors help move a specific solute across the cell membrane *against* its concentration gradient.

Only specific solutes can bind to functional groups that line the interior channel of an active transporter. When the solute enters the channel and binds to one of those groups, the transporter accepts a phosphate group from an ATP molecule. The phosphate-group transfer changes the transporter's shape in a way that releases the solute on the other side of the membrane.

Figure 5.19 focuses on a **calcium pump**. This active transporter helps keep the concentration of calcium in a cell at least a thousand times lower than outside. A different active transporter, the **sodium–potassium pump**, mediates the movement of two kinds of ions, in opposite directions. Sodium ions (Na⁺) from the cytoplasm diffuse into the open channel of the pump, where they bind to functional groups. A phosphate-group transfer by ATP prompts the pump to change shape and release the sodium ions outside the cell.

The channel through the activated pump is now open to the outside of the cell. Potassium ions (K⁺) diffuse into the pump and bind to functional groups inside. The phosphate group is released from the pump, which reverts to its original shape. When it does, the potassium ions are released to the cytoplasm.

Active transport systems help maintain membrane gradients that are essential to many processes, such as muscle contraction and nerve cell (neuron) function.

Some transport proteins are open or gated channels across cell membranes. Others change shape when solutes bind to them.

In passive transport, a solute simply diffuses through the interior of a transporter; an energy input is not necessary.

In active transport, the net diffusion of a specific solute is against its concentration gradient. The transporter must be activated by an energy input from ATP to counter the force inherent in the gradient.

higher concentration of calcium ions outside cell compared to inside

calcium pump

e The shape of the pump returns to its resting position.

a An ATP molecule binds to a calcium pump.

Pᵢ

ADP

d The shape change permits calcium to be released at opposite membrane surface. A phosphate group and ADP are released.

b Calcium enters a tunnel through the pump, binds to functional groups inside.

c The ATP transfers a phosphate group to pump. The energy input will cause pump's shape to change.

Figure 5.19 Active transport by a calcium pump. This sketch shows its channel for calcium ions. ATP transfers a phosphate group to the pump, thus providing energy that can drive the movement of calcium *against* a concentration gradient across the cell membrane.

5.7 Which Way Will Water Move?

By far, more water diffuses across cell membranes than any other substance, so the main factors that influence its directional movement deserve special attention.

MOVEMENT OF WATER

Something as trickly as a running faucet or as mighty as Niagara Falls demonstrates **bulk flow**, or the mass movement of one or more substances in response to pressure, gravity, or another external force. Bulk flow accounts for some water movement in big multicelled organisms. A beating heart generates fluid pressure that pumps blood, which is mostly water. Sap flows inside tubes in trees, and this, too, is bulk flow.

What about the movement of water into and out of cells and organelles? If the concentration of water is not equal across a cell membrane, osmosis tends to occur. **Osmosis** is the diffusion of water across a selectively permeable membrane, to a region where the water concentration is lower.

You may be asking: How can water, a liquid, be more or less "concentrated"? Its concentration actually is influenced by the concentration of *solutes* on both sides of the membrane. If you pour glucose or some other solute into a glass of water, you will increase the volume of liquid in the glass. Now the same number of water molecules will become less concentrated than they were before; they will diffuse through the larger volume of space.

Now suppose you divide the interior of a glass container with a selectively permeable membrane, one that permits the diffusion of water but not glucose (a large, polar molecule) across it. You have created a water concentration gradient. More water molecules will diffuse across the membrane, into the solution, than will diffuse back (Figure 5.20).

In cases of osmosis, "solute concentration" refers to the *total number* of molecules or ions in a specified volume of a solution. It doesn't matter whether the dissolved substance is glucose, urea, or anything else; the *type* of solute doesn't dictate water concentration.

EFFECTS OF TONICITY

Suppose you decide to test the statement that water tends to move to a region where solutes are more concentrated. You make three sacs from a membrane that water but not sucrose can cross. You fill each sac with a solution that's 2 percent sucrose, then immerse one in a liter of water. You immerse another sac in a solution that is 10 percent sucrose. And you immerse the third sac in a solution that is 2 percent sucrose.

In each experiment, tonicity dictates the extent and direction of water movement across the membrane, as Figure 5.21 shows. *Tonicity* refers to the relative solute concentrations of two fluids. When two fluids that are on opposing sides of a membrane differ in their solute concentrations, the **hypotonic solution** is the one with fewer solutes. The one having more solutes is a **hypertonic solution**. And water tends to diffuse from a hypotonic fluid to a hypertonic one. **Isotonic solutions** show no net osmotic movement.

Normally, the fluid inside your cells is isotonic with tissue fluid outside. If the fluid outside becomes far too hypotonic, too much water will diffuse into those cells and make them burst. If it gets too hypertonic, water will diffuse out, and the cells will shrivel.

Most cells have built-in mechanisms that adjust to changes in tonicity. Red blood cells don't. Figure 5.21 shows what happens to them when tonicity changes.

EFFECTS OF FLUID PRESSURE

Selective transport of solutes across the plasma membrane keeps animal cells from bursting. Cells of plants and many protists, fungi, and bacteria avoid bursting with the help of pressure on their cell walls.

water molecules   protein molecules

semipermeable membrane between two compartments

Figure 5.20 Solute concentration gradients and osmosis. A membrane divides this container. Water, but not proteins, can cross it. Pour 1 liter of water in the left compartment and 1 liter of a protein-rich solution in the right one. The proteins occupy some of the space in the right one. The net diffusion of water in this case is from left to right (large *gray* arrow).

Pressure differences as well as solute concentrations influence osmosis. Take a look at Figure 5.22. It shows how water will diffuse across a membrane between a hypotonic and a hypertonic solution until the solute concentration is the same on both sides. As you can see, the *volume* of the formerly hypertonic solution has increased (because its solutes cannot diffuse out).

Hydrostatic pressure is the force that any volume of fluid exerts against a wall, a membrane, or some other structure enclosing it. (In plants, this pressure is called *turgor*.) Hydrostatic pressure that has built up in a cell can counter the further inward diffusion of water. This **osmotic pressure** is the amount of force preventing any further increase in volume.

Think of the pliable primary wall of a young plant cell. As it matures, many vesicles start to coalesce into a large central vacuole. During cell growth, water diffuses into the vacuole and puts more fluid pressure on the cell wall. The wall expands, so the cell volume increases. Continued expansion of the wall (and of the cell) ends when enough internal fluid pressure develops to counter the water uptake.

Plant cells are vulnerable to water losses, which can occur when soil dries out or becomes too salty. Water stops diffusing in and starts diffusing out, so internal fluid pressure falls and the cytoplasm shrinks.

In later chapters, you'll see how fluid pressure has a role in the distribution of water and solutes inside the body of plants and animals.

Osmosis is a net diffusion of water between two solutions that differ in solute concentration and are separated by a selectively permeable membrane. The greater the number of molecules and ions dissolved in a given amount of water, the lower the water concentration will be.

Water tends to move osmotically to regions of greater solute concentration (from hypotonic to hypertonic solutions). There is no net diffusion between isotonic solutions.

Fluid pressure that a solution exerts against a membrane or wall influences the osmotic movement of water.

Read Me First!
and watch the narrated animation on tonicity and water movement

2% sucrose solution

1 liter of distilled water

1 liter of 10% sucrose solution

1 liter of 2% sucrose solution

Hypotonic Conditions
Water diffuses into red blood cells, which swell up

Hypertonic Conditions
Water diffuses out of the cells, which shrink

Isotonic Conditions
No net movement of water, no change in cell size or shape

Figure 5.21 Tonicity and the direction of water movement. In each of three containers, arrow widths signify the direction and the relative amounts of flow. The micrographs below each sketch show the shape of a human red blood cell that is immersed in fluids of higher, lower, or equal concentrations of solutes. The solutions inside and outside red blood cells are normally balanced. This type of cell has no way to adjust to drastic change in solute levels in its fluid surroundings.

Figure 5.22 Experiment showing an increase in fluid volume as an outcome of osmosis. A semipermeable membrane separates two compartments. Over time, the net diffusion will be the same in both directions across the membrane, but the fluid volume in the second compartment will be greater because there are more solute molecules in it.

first compartment

second compartment

hypotonic solution

hypertonic solution

membrane permeable to water but not to solutes

fluid volume rises in second compartment

5.8 Membrane Traffic To and From the Cell Surface

We leave this chapter with another look at exocytosis and endocytosis. By these mechanisms, vesicles move substances to and from the plasma membrane. Vesicles help the cell take in and expel materials in amounts that are more than transport proteins can handle.

ENDOCYTOSIS AND EXOCYTOSIS

Think back on the membrane traffic to and from a cell surface (Figure 5.17). By **exocytosis**, a vesicle moves to the surface, and the protein-studded lipid bilayer of its membrane fuses with the plasma membrane. As this exocytic vesicle is losing its identity, its contents are released to the outside (Figures 5.23 and 5.24).

There are three pathways of **endocytosis**, but all take up substances near the cell surface. A small patch of plasma membrane balloons inward and pinches off inside the cytoplasm, forming an endocytic vesicle that moves its contents to some organelle or stores them in a cytoplasmic region (Figure 5.23).

Figure 5.23 Endocytosis of cholesterol molecules.

With *receptor-mediated* endocytosis, receptors at the membrane bind to molecules of a hormone, vitamin, mineral, or another substance. A tiny pit forms in the plasma membrane beneath the receptors. The pit sinks into the cytoplasm and closes back on itself, and in this way it becomes a vesicle (Figure 5.24).

Phagocytosis ("cell eating") is a common endocytic pathway. Phagocytes such as amoebas engulf microbes, food particles, or cellular debris. In multicelled species, macrophages and some other white blood cells do this to pathogenic viruses or bacteria, cancerous body cells, and other threats.

Phagocytosis also involves receptors. Receptors that bind a target cause microfilaments to form a mesh just beneath the phagocyte's plasma membrane. When the microfilaments contract, they squeeze some cytoplasm toward the margins of the cell, forming a bulging lobe called a pseudopod (Figure 5.25). Pseudopods flow all around the target and form a vesicle. This sinks into the cell and fuses with lysosomes, the organelles of intracellular digestion. Lysosomes digest the trapped items into fragments and smaller, reusable molecules.

Bulk-phase endocytosis is not as selective. A vesicle forms around a small volume of the extracellular fluid regardless of the kinds of substances dissolved in it. This pathway continually removes patches of plasma membrane, balancing the steady additions that arrive in the form of exocytic vesicles from the cytoplasm.

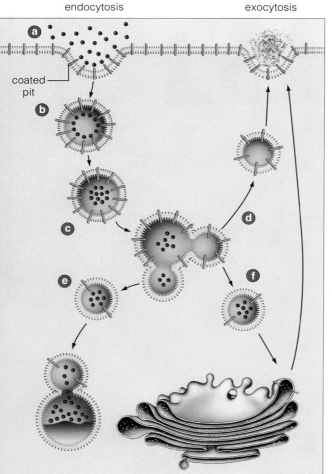

endocytosis   exocytosis

a Molecules get concentrated inside coated pits of plasma membrane.

b Endocytic vesicles form from the pits.

c Enclosed molecules are sorted and often released from receptors.

d Many sorted molecules are cycled back to the plasma membrane.

e,f Many other sorted molecules are delivered to lysosomes and stay there or are degraded. Still others are routed to spaces in the nuclear envelope and inside ER membranes, and others to Golgi bodies.

Figure 5.24 Cycling of membrane lipids and proteins. This sketch starts with receptor-mediated endocytosis. Patches of the plasma membrane form endocytic vesicles. New membrane arrives as exocytic vesicles that budded from ER membranes and Golgi bodies. The membrane initially used for endocytic vesicles will cycle receptor proteins and lipids back to the plasma membrane.

parasite   macrophage

a

b   bacterium   phagocytic vesicle

Figure 5.25 (**a**) A macrophage engulfing *Leishmania mexicana*. This parasitic protozoan causes leishmaniasis, a disease that can be fatal. Bites from infected sandflies transmit the parasite to humans.

(**b**) Phagocytosis. Lobes of an amoeba's cytoplasm surround a target. The plasma membrane of the extensions fuses to form a phagocytic vesicle. In the cytoplasm, this endocytic vesicle fuses with lysosomes, which digest its contents.

MEMBRANE CYCLING

For as long as a cell remains alive, exocytosis and endocytosis continually replace and withdraw patches of its plasma membrane, as in Figure 5.24. And they apparently do so at rates that can maintain the plasma membrane's total surface area.

As one example, neurons release neurotransmitters in bursts of exocytosis. Each neurotransmitter is a type of signaling molecule that acts on neighboring cells. An intense burst of endocytosis counterbalances each major burst of exocytosis.

Whereas transport proteins in a cell membrane deal only with ions and small molecules, exocytosis and endocytosis move large packets of materials across a plasma membrane.

By exocytosis, a cytoplasmic vesicle fuses with the plasma membrane, and its contents are released outside the cell.

By endocytosis, a small patch of the plasma membrane sinks inward and seals back on itself, forming a vesicle inside the cytoplasm. Membrane receptors often mediate this process.

Summary

Section 5.1 Cells engage in metabolism, or chemical work. They obtain and use energy to stockpile, build, rearrange, and break apart substances.

Energy in biological systems flows in one direction, from usable to less usable forms. Life maintains its complex organization by being resupplied with energy lost from someplace else. Sunlight is the ultimate energy source for the web of life.

ATP, the main energy carrier, couples reactions that release energy with reactions that require it. It primes molecules to react through phosphate-group transfers.

Section 5.2 Metabolic pathways are orderly, enzyme-mediated reaction sequences. Photosynthesis and other energy-requiring, *biosynthetic* pathways build large molecules with high energy from smaller ones. Energy-releasing, *degradative* pathways such as aerobic respiration break down large molecules to small products with lower bond energies. Table 5.1 lists the participants.

Cells increase, maintain, and lower concentrations of substances by coordinating thousands of reactions. They rapidly shift rates of metabolism by controlling a few steps of reversible pathways.

Electron transfers, or oxidation–reduction reactions, often proceed in series at cell membranes.

Section 5.3 Enzymes are catalysts; they enormously enhance rates of specific reactions and are not altered by their function. Pockets or cavities in these big molecules create favorable microenvironments for the reaction; these are the active sites.

Table 5.1 Summary of the Main Participants in Metabolic Reactions	
Reactant	Substance that enters a metabolic reaction or pathway; also called the substrate of a specific enzyme
Intermediate	Substance formed between the reactants and end products of a reaction or pathway
Product	Substance at the end of a reaction or pathway
Enzyme	A protein that greatly enhances reaction rates; a few RNAs also do this
Cofactor	Coenzyme (such as NAD^+) or metal ion; assists enzymes or taxis electrons, hydrogen, or functional groups between reaction sites
Energy carrier	Mainly ATP; couples energy-releasing reactions with energy-requiring ones
Transport protein	Protein that passively assists or actively pumps specific solutes across a cell membrane

Activation energy is the minimum internal energy that reactant molecules must have for a reaction to occur. Enzymes lower it by boosting substrate concentrations in the active site, by orienting substrates, by shutting out most or all water, and by inducing a precise fit with them.

Section 5.4   Cofactors (metal ions, coenzymes, or both) help an enzyme catalyze a reaction. Controls over enzyme action influence the kinds and amounts of substances available. Enzymes function best within a limited range of temperature, pH, and salinity.

Section 5.5   Diffusion is the movement of molecules or ions toward an adjoining region where they are less concentrated. The steepness of the concentration gradient, temperature, molecular size, and gradients in electrical charge and pressure influence diffusion rates.

Built-in cellular mechanisms work with and against gradients to move solutes across membranes.

Molecular oxygen, carbon dioxide, and other small nonpolar molecules diffuse across a membrane's lipid bilayer. Ions and large, polar molecules such as glucose cross it with the help of transport proteins. Some water moves through proteins and some through the bilayer.

Section 5.6   Many solutes cross membranes through transport proteins that act as open or gated channels or that reversibly change shape. Passive transport does not require energy input; a solute is free to follow its own concentration gradient across the membrane. Active transport requires an energy input from ATP to move a specific solute against its concentration gradient.

Section 5.7   Osmosis is the diffusion of water across a selectively permeable membrane, down the water concentration gradient. Pressure gradients can affect it.

Section 5.8   By exocytosis, a cytoplasmic vesicle fuses with the plasma membrane, and its contents are released outside. By endocytosis, a patch of plasma membrane forms a vesicle that sinks into the cytoplasm.

Self-Quiz

Answers in Appendix III

1. _____ is life's primary source of energy.
   a. Food   b. Water   c. Sunlight   d. ATP

2. Which of the following statements is *not* correct? A metabolic pathway _____ .
   a. has an orderly sequence of reaction steps
   b. is mediated by only one enzyme that starts it
   c. may be biosynthetic or degradative, overall
   d. all of the above

3. An enzyme _____ .
   a. is a protein
   b. lowers the activation energy of a reaction
   c. is destroyed by the reaction it catalyzes
   d. a and b

4. Immerse a living cell in a hypotonic solution, and water will tend to _____ .
   a. diffuse into the cell   c. show no net movement
   b. diffuse out of the cell   d. move in by endocytosis

5. _____ can readily diffuse across a lipid bilayer.
   a. Glucose   c. Carbon dioxide
   b. Oxygen   d. b and c

6. Sodium ions cross a membrane at transport proteins that receive an energy boost. This is a case of _____ .
   a. passive transport   c. facilitated diffusion
   b. active transport   d. a and c

7. Vesicle formation occurs in _____ .
   a. membrane cycling   c. endocytosis, exocytosis
   b. phagocytosis   d. all of the above

8. The rate of diffusion is affected by _____ .
   a. temperature   c. molecular size
   b. electrical gradients d. all of the above

9. Match the substance with its suitable description.
   ____ coenzyme or metal ion   a. reactant
   ____ adjusts gradients at membrane b. enzyme
   ____ substance entering a reaction c. cofactor
   ____ substance formed during   d. intermediate
        a reaction   e. product
   ____ substance at end of reaction   f. energy carrier
   ____ enhances reaction rate   g. transporter
   ____ mainly ATP   proteins

Critical Thinking

1. Cyanide, a toxic compound, binds irreversibly to an enzyme that is a component of electron transfer chains. The outcome is *cyanide poisoning*. Binding prevents the enzyme from donating electrons to a nearby acceptor molecule in the system. What effect will this have on ATP formation? From what you know of ATP's function, what effect will this have on a person's health?

2. In cells, superoxide dismutase (*below*) has a quaternary structure—it consists of two polypeptide chains. In each chain, a strandlike domain is arrayed as a barrel around a copper ion and a zinc ion (coded *red* and *blue*).

   Which part of the barrel is probably hydrophobic? Which part is hydrophilic? Do you suppose substrates bind inside or outside the barrels? Do the metal ions have a role in catalysis?

3. Catalase breaks down hydrogen peroxide, a reactive by-product of aerobic metabolism, to water and oxygen (Figure 5.10). It is a very efficient enzyme: One molecule of catalase can break down 6 million hydrogen peroxide molecules every minute. It is found in most organisms that live under aerobic conditions because hydrogen

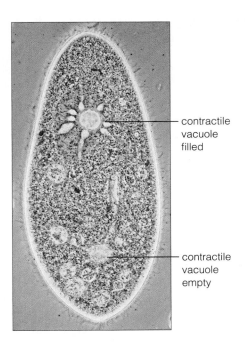

contractile
vacuole
filled

contractile
vacuole
empty

Figure 5.26 *Paramecium* contractile vacuoles.

Figure 5.27 Go ahead, name the mystery membrane mechanism.

peroxide is toxic—cells must dispose of it quickly or they risk being damaged. Peroxide is catalase's substrate; but by a neat trick, it also can inactivate other toxins, including alcohol. Can you guess what the trick is?

4. Nutritional supplements often include plant enzymes. Explain why it is not likely that plant enzymes will aid your digestion.

5. Why does applying lemon juice to sliced apples keep them from turning brown?

6. Explain why hydrogen peroxide bubbles when you dribble it on an open cut but does not bubble on skin that is unbroken.

7. Most of the cultivated fields in California are heavily irrigated. Over the years, most of the imported water has evaporated from the soil, leaving behind solutes. What problems will the altered soil cause plants?

8. Water moves osmotically into *Paramecium*, a single-celled aquatic protist. If unchecked, the influx would bloat the cell and rupture its plasma membrane, killing the cell. An energy-requiring mechanism that involves contractile vacuoles expels excess water (Figure 5.26). Water enters the vacuole's tubelike extensions and collects inside. A full vacuole contracts and squirts water out of the cell through a pore. Are *Paramecium*'s surroundings hypotonic, hypertonic, or isotonic?

9. Imagine you're a juvenile shrimp living in an *estuary*, where freshwater draining from the land mixes with saltwater from the sea. Many people own homes near a lake and want boat access to the sea. They ask their city for permission to build a canal to your estuary. If they succeed, what may happen to you?

10. Is the white blood cell shown in Figure 5.27 disposing of a worn-out red blood cell by endocytosis, phagocytosis, or both?

Media Menu

Student CD-ROM

Impacts, Issues Video
  Alcohol, Enzymes, and Your Liver
Big Picture Animation
  Energy, enzymes, and movement across membranes
Read-Me-First Animation
  Controlling energy release
  Catalase action
  Passive transport
  Active transport
  Tonicity and water movement
Other Animations and Interactions
  Activation energy interaction
  Allosteric activation
  Feedback inhibition

InfoTrac

- Harnessing the Energy. *World and I*, October 2001.
- Ion Channel Protein Contraceptive Target. *Drug Discovery & Technology News*, October 2001.
- Drug Abuse During 1970s and 1980s May Explain Doubling of Deaths from Alcoholic Liver Disease. *Hepatitis Weekly*, August 2002.

Web Sites

- National Institute on Alcohol Abuse and Alcoholism: www.niaaa.nih.gov
- Adenosine Triphosphate—ATP: www.bris.ac.uk/Depts/Chemistry/MOTM/atp/atp1.htm
- Introduction to Enzymes: www.worthington-biochem.com/introBiochem/introEnzymes.html
- Pumping Ions: www.mbl.edu/publications/LABNOTES/10.1/pumping_ions.html

How Would You Vote?

The only cure for liver failure, regardless of its cause, is a liver transplant. A shortage of livers means many potential transplant recipients die waiting. How should these organs be allocated? Should people who invited liver failure by their own abusive life-style be a lower priority for transplants than those with failure brought on by a transfusion or a genetic disorder?

IMPACTS, ISSUES   *Pastures of the Seas*

Think about the last bit of apple, celery, chicken, pizza, or any other food you ate. Where did it come from? Look past the refrigerator, the market or restaurant, or even the farm. Look to individual plants, the starting point for nearly all of your food.

Plants, and many bacteria and protists, are "self-nourishing" organisms, or **autotrophs**. They tap into an environmental energy source and use it to *make* food from simple materials. By contrast, most bacteria and protists, all fungi, and all animals are **heterotrophs**. They are not self-nourishing; they cannot make their own food. They must eat autotrophs, one another, and organic wastes. (*Hetero–* means other; in this case, "being nourished by others.")

Plants do something you'll never do. By the process of **photosynthesis**, they make food by using no more than sunlight energy, water, and carbon dioxide (CO_2). Each year, they produce 220 billion tons of sugar, enough to make 300 quadrillion sugar cubes. These photosynthetic autotrophs have been doing so for more than a billion years. That is a LOT of sugar.

Uncountable numbers of photosynthesizers also abound in the seas. A cupful of seawater may hold more than 24 million microscopically small cells of different species! Most are bacteria and protists that form "pastures of the seas"—the producers that feed most other marine organisms. Like plants, they too "bloom" in spring, when nutrients churned up from the deep support rapid population growth. Figure 6.1 is a record of an algal bloom that stretched from North Carolina to Spain.

Figure 6.1 Satellite images that convey the magnitude of photosynthetic activity during springtime in the North Atlantic Ocean. Sensors responded to concentrations of chlorophyll, which were greatest in regions coded *red*.

the big picture

Catching the Rainbow   Energy enters the world of life when chlorophyll and other photosynthetic pigments absorb energy in the sun's rays.

Overview of Photosynthesis   Photosynthesis occurs in two stages in chloroplasts. Energy from the sun is converted to chemical energy and stored in ATP and NADPH. These molecules are later used to assemble sugars from carbon dioxide and water.

$$12H_2O + 6CO_2 \xrightarrow{\text{light energy}} 6O_2 + C_6H_{12}O_6 + 6H_2O$$

water carbon dioxide   oxygen glucose water

Imagine zooming in on just one small patch of "pasture" in an Antarctic sea. There, tiny shrimplike crustaceans are rapidly eating tinier photosynthesizers, including algal cells of the sort shown in the filmstrip. Dense concentrations of such crustaceans, known as krill, are feeding other animals, including fishes, penguins, seabirds, and the immense blue whale. A single, mature whale is straining four tons of krill from the water today, as it has been doing for months. And before they themselves were eaten, the four tons of krill had munched through *1,200 tons* of the pasture!

Another point: Collectively, photosynthetic cells on land and in the seas handle staggering numbers of reactant and product molecules. By doing so, they even help shape the global climate. They also sponge up nearly half of the CO_2 we humans release each year, as by burning fossil fuels. Without them, CO_2 would accumulate faster and warm the atmosphere, which already is warming too fast.

In short, *photosynthesis is the main pathway by which energy and carbon enter the web of life.* Photosynthetic autotrophs make, use, and store organic compounds, the food for most heterotrophs. And *all* organisms release that stored energy for cellular work, mainly by aerobic respiration.

There are different types of photoautotrophs, and they perform photosynthesis in different ways. In this chapter we focus on oxygenic (oxygen-producing) photosynthesis in plants and algae.

✓ *How Would You Vote?*

Crop plants feed most of the human population. Limits on the activity of some enzymes can limit crop production. Should we genetically engineer plants to boost photosynthesis and get higher crop yields? See the Media Menu for details, then vote online.

Making ATP and NADPH   In the first stage of photosynthesis, sunlight energy becomes converted to chemical bond energy of ATP. Water molecules are broken apart, NADPH forms, and oxygen escapes into the air.

Making Sugars   The second stage is the "synthesis" part of photosynthesis. ATP delivers energy to reaction sites where sugars are built with atoms of hydrogen (delivered by NADPH), carbon, and oxygen (from carbon dioxide in the air).

6.1 Sunlight As an Energy Source

*Photosynthesis runs on a fraction of the electromagnetic spectrum, or the full range of energy radiating from the sun. Radiant energy undulates across space, something like waves crossing a sea. The horizontal distance between two successive waves is a **wavelength**.*

PROPERTIES OF LIGHT

Our story starts with energy in the sun's rays—not all of it, just light of wavelengths between 380 and 750 nanometers. These are wavelengths of visible light, the ones that drive photosynthesis.

Light is made of *photons*, which are individual packets of electromagnetic energy traveling in waves. The shorter a photon's wavelength, the higher its energy (Figure 6.2). For example, blue light has a shorter wavelength and more energy than red light:

480-nm

700-nm

blue light has more energy than red light

Photons with wavelengths shorter than violet light are energetic enough to disrupt DNA of living cells.

PIGMENTS—THE RAINBOW CATCHERS

Pigments are a class of molecules that absorb photons with particular wavelengths. Photons that a pigment cannot absorb bounce off or continue on through it; they are reflected or transmitted.

Figure 6.3 Ball-and-stick model for chlorophylls *a* and *b*, which differ by only a single functional group. In chlorophyll *b*, the group is —COO⁻, not the —CH₃ shown. The light-catching portion is the flattened ring structure—which is similar to a heme except it holds a magnesium atom instead of iron. The hydrocarbon backbone readily dissolves in the lipid bilayers of cell membranes.

Certain pigments are the molecular bridges from sunlight to photosynthesis. **Chlorophyll *a*** is the most abundant type in plants, green algae, and a number of photoautotrophic bacteria (Figure 6.3). It is the best at absorbing red and violet wavelengths. Chlorophyll *b* absorbs light at slightly different wavelengths. It is an *accessory* pigment, meaning that it enhances efficiency of photosynthesis reactions by capturing additional wavelengths. All chlorophylls reflect or transmit green wavelengths, which is why plant parts that are rich in chlorophylls appear green to us.

Accessory pigments include the **carotenoids**, which absorb blue-violet and blue-green wavelengths, and reflect red, orange, and yellow ones. Beta-carotene is a carotenoid that colors carrots and other plant parts

shortest wavelengths (most energetic) | range of most radiation reaching Earth's surface | range of heat escaping from Earth's surface | longest wavelengths (lowest energy)

gamma rays | x rays | ultraviolet radiation | near-infrared radiation | infrared radiation | microwaves | radio waves

VISIBLE LIGHT

400   450   500   550   600   650   700

Wavelengths of light (nanometers)

Figure 6.2 The electromagnetic spectrum. Energy undulates across space in waves. The distance between crests of two successive waves is a wavelength and is measured in nanometers. About 2.5 million nanometers fit in one inch. Visible light is a very small part of the spectrum, which includes all electromagnetic waves. Like many other organisms, we perceive visible light wavelengths as colors.

Figure 6.4 (**a**) Absorption spectra reveal the efficiency with which chlorophylls *a* and *b* absorb wavelengths of visible light. Peaks in the graphs reveal the wavelengths that each type of pigment absorbs best.

(**b**) Absorption spectra for beta-carotene (a carotenoid) and phycobilin. As Figure 6.6 describes, before scientists devised ways to measure absorption efficiencies, the botanist T. Engelmann figured out which colors of light were best at driving photosynthesis in a green alga.

Collectively, these and other photosynthetic pigments can capture almost the entire spectrum of visible light.

Figure 6.5 Leaf color. In spring and summer, intensely green leaves have an abundance of chlorophylls, which mask the presence of carotenoids, xanthophylls, and other accessory pigments.

In many kinds of plants, chlorophyll synthesis lags behind its breakdown in autumn, so more stable pigments show through. Cold, sunny days trigger the production of water-soluble anthocyanins in leaf cells. The anthocyanins act like a sunscreen; they protect leaves from ultraviolet radiation.

Figure 6.6 T. Englemann's study of photosynthesis in *Spirogyra*, a strandlike green alga. A long time ago, most people assumed plants withdrew raw materials for photosynthesis from soil. By 1882 a few chemists suspected that plants use light, water, and something in the air. Englemann wondered: What parts of sunlight do plants favor?

As he knew, free oxygen is released during photosynthesis. He also knew some bacteria use oxygen during aerobic respiration, as most organisms do. He hypothesized: If bacteria require oxygen, then we can expect them to gather in places where the most photosynthesis is going on. He put a water droplet containing bacterial cells on a microscope slide with the green alga *Spirogyra*. He used a crystal prism to break up a beam of sunlight and cast a spectrum of colors across the slide.

Bacteria gathered mostly where violet and red light fell on the green alga. Algal cells released more oxygen in the part illuminated by light of those colors—the very best light for photosynthesis. Compare Figure 6.4.

A crystal prism breaks up a beam of light into a spectrum of colors, which are cast across a droplet of water on a microscope slide.

orange. **Xanthophylls** are yellow, brown, purple, or blue accessory pigments; **phycobilins** are red or blue-green. Absorption spectra (singular, spectrum) give us a picture of how such photosynthetic pigments absorb different wavelengths of visible light (Figure 6.4).

The chlorophyll content in the leaves of deciduous species declines in autumn and lets the carotenoids, xanthophylls, and **anthocyanin**, a red-purple pigment, show through (Figure 6.5). Each year in New England, tourists spend a billion dollars to watch a three-week display of red, orange, and gold leaves of maples and other trees. We also can thank the deep red to purple anthocyanins for the visual appeal of many flowers and food, including blueberries, red grapes, cherries, red cabbage, and rhubarb.

Such photosynthetic pigments do not work alone. Organized arrays of them work together and harvest energy from the sun. For now, start thinking about the structure of that chlorophyll molecule in Figure 6.3—particularly the flattened ring. Here, alternating single and double covalent bonds share electrons. And these are the electrons which, when excited by inputs of energy, get photosynthesis going.

Light from the sun travels through space in waves, and wavelengths of visible light correspond to specific colors.

Chlorophyll a and diverse accessory pigments absorb specific wavelengths of visible light. They are the molecular bridge between the sun and photosynthesis.

6.2 What Is Photosynthesis and Where Does It Happen?

Sit outdoors on a warm, sunny day and you will never build your own food but you will get hot. Plants plug into the sun's energy usefully, without getting cooked.

TWO STAGES OF REACTIONS

Photosynthesis proceeds through two reaction stages. The first stage, the **light-dependent reactions**, converts light energy to chemical bond energy of ATP. Also, water molecules are split, and the coenzyme NADP+ picks up the released electrons and hydrogen. We call its reduced form NADPH. The oxygen atoms released from water molecules escape into the surroundings.

In the second stage, called the **light-independent reactions**, energy from ATP jump-starts reactions that form glucose and other carbohydrates. At these same sites, NADPH gives up electrons and hydrogen ions, which bond with carbon and oxygen to form glucose.

Here is a simple way to summarize the reactions of photosynthesis:

$$12H_2O + 6CO_2 \xrightarrow[\text{enzymes}]{\text{light energy}} 6O_2 + C_6H_{12}O_6 + 6H_2O$$

water   carbon   oxygen   glucose   water
          dioxide

A LOOK INSIDE THE CHLOROPLAST

Photosynthetic reactions differ among certain bacteria, protists, and plants. For now, focus on what goes on in **chloroplasts**, the organelles of photosynthesis in plants and algae. Each chloroplast has two outer membranes, which enclose a semifluid interior, the stroma (Figure 6.7c). Inside the stroma is the **thylakoid membrane**, a third membrane folded in ways that form a single compartment. Often the folds look like flattened channels between stacks of flattened sacs (thylakoids). The space inside the sacs is part of one continuous compartment. Sugars are built outside this compartment, in the stroma (Figure 6.8).

Embedded in all the thylakoids are **photosystems**: clusters of 200 to 300 pigments and other molecules that trap energy from the sun. Chloroplasts have two types of photosystems, called I and II (Figure 6.7d).

PHOTOSYNTHESIS CHANGED THE BIOSPHERE

Before zooming down further, to the mechanisms of photosynthesis, zoom out in your mind to the global impact of one of the steps involved. About 3.2 billion

Figure 6.7 Zooming in on sites of photosynthesis in a typical plant leaf. Two thousand chloroplasts, lined up single file, would be no wider than a dime. Think of all the chloroplasts in a corn or rice plant—each a tiny sugar-making factory—to get a sense of the magnitude of metabolic events required to feed you and every other living thing.

years ago, oxygen released by photosynthetic bacteria started accumulating in the atmosphere, which before then held little of it. As the atmosphere changed, so did the world of life. All that free oxygen favored the evolution of a novel pathway—aerobic respiration—that efficiently releases a great deal of energy from organic compounds. An oxygen-rich atmosphere was a key environmental factor in the evolution of large, active animals, which the aerobic pathway sustains. Breathing in oxygen helps keep them alive.

In the first stage of photosynthesis, sunlight energy drives ATP and NADPH formation, and oxygen is released. In chloroplasts, this stage occurs at the thylakoid membrane.

Embedded in the membrane are photosystems—clusters of pigments and other molecules—where light energy is captured.

The second stage occurs in the stroma. Energy from ATP drives the synthesis of sugars. Carbon dioxide provides carbon and oxygen atoms for the reactions. NADPH delivers the required electrons and hydrogen atoms.

The atmosphere was free of oxygen before photosynthesis evolved. Oxygen released by emerging photosynthesizers slowly accumulated. It changed the atmosphere and became a selective force in the evolution of aerobic respiration.

Figure 6.8 Overview of the two stages of photosynthesis in a chloroplast. The first stage, the light-dependent reactions, occurs at the thylakoid membrane. The second stage (light-independent reactions that produce sugars) occurs in the stroma.

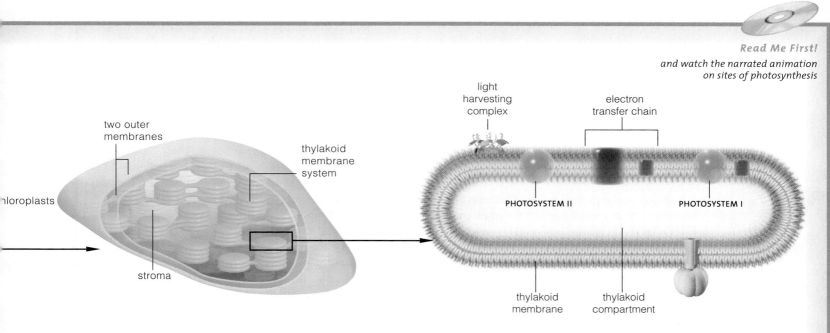

Read Me First!
and watch the narrated animation
on sites of photosynthesis

c Closer look at one chloroplast. It has two outer membranes and an inner *thylakoid* membrane in its semifluid interior (the stroma). In many cells, the inner membrane resembles stacks of flattened sacs connected by channels. The interiors of all sacs and channels interconnect, forming a single compartment.

d Components of the thylakoid membrane system that carry out the first stage of photosynthesis—the light-dependent reactions. Light-harvesting complexes capture photon energy and pass it to two types of photosystems. Electron transfer chains embedded in the membrane have roles in ATP and NADPH formation.

6.3 Light-Dependent Reactions

In the first stage of photosynthesis, the sunlight energy harvested at photosystems drives ATP formation. Water molecules are split, and their oxygen diffuses away. NADP+, a coenzyme, picks up the electrons and hydrogen, which will be used in the second stage to form sugars.

TRANSDUCING THE ABSORBED ENERGY

Suppose a photon collides with a pigment molecule that absorbs it. The photon's energy will boost one of the pigment's electrons to a higher energy level. If nothing else happens, the electron quickly will drop back to its unexcited state, losing the extra energy as it does. The energy is emitted as heat, or as another photon. Photon emissions from electrons losing extra energy are visible as fluorescent light.

In the thylakoid membrane, however, energy that excited electrons give up is kept in play. Embedded in the membrane are many photosystems. Surrounding them are hundreds of light-harvesting complexes, or circular clusterings of pigments and other proteins (Figure 6.9*a*). Pigments in light-harvesting complexes also absorb photon energy, but they don't waste it. Electrons of these pigments can hold on to energy by passing it back and forth, like a volleyball. The energy released from one cluster is passed to another, which passes it on to another, and so on until it reaches a photosystem—a reaction center.

Look back on Figure 6.3, which shows the structure of chlorophyll. Two molecules of chlorophyll *a* are at the center of every photosystem. Their flat rings face each other so closely that electrons in both rings are destabilized. When light-harvesting neighbors pass on photon energy to a photosystem, electrons come right off of that special pair of chlorophylls.

The freed electrons immediately enter an electron transfer chain positioned next to the photosystem in the thylakoid membrane. *The entry of electrons from a photosystem into an electron transfer chain is the first step in the light-dependent reactions*—in the conversion of photon energy to chemical energy for photosynthesis.

MAKING ATP AND NADPH

Let's use Figure 6.9 to track electrons that a type II photosystem gives up. **Electron transfer chains**, recall, are cell membrane components. Each is an organized array of enzymes, coenzymes, and other proteins through which electrons are transferred step-by-step. In the process of moving electrons, molecules of the chain pick up hydrogen ions (H^+) from the stroma, cart them across the thylakoid membrane, and release

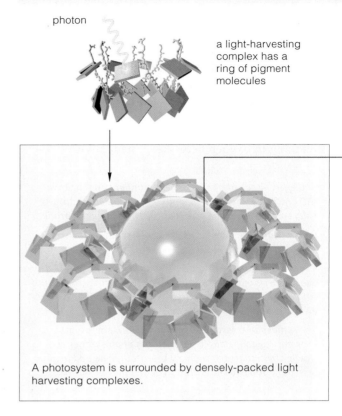

photon

a light-harvesting complex has a ring of pigment molecules

A photosystem is surrounded by densely-packed light harvesting complexes.

a In the thylakoid membrane of chloroplasts, rings of pigments can intercept photons coming from any direction. They pass captured photon energy to nearby photosystems. Each photosystem collects energy from hundreds of light-harvesting complexes surrounding it; a few are shown here.

Figure 6.9 How ATP and NADPH form during photosynthesis in chloroplasts.

them to the inner compartment. Their repeated action causes concentration and electric gradients to build up across the membrane. The combined force of those gradients attracts H^+ back toward the stroma.

The H^+ ions can only cross the membrane with the help of **ATP synthases**, as explained in Section 4.3. Ion flow through these membrane proteins causes the attachment of inorganic phosphate to a molecule of ADP in the stroma. In this way, ATP forms.

As long as electrons flow through transfer chains, the cell can keep on producing ATP. But where do the electrons come from in the first place? By the process of *photolysis*, photosystem II replaces its lost electrons by pulling them from water molecules—which then dissociate into hydrogen ions and molecular oxygen. The free oxygen diffuses out of the chloroplast, then

you are here

Read Me First!
and watch the narrated animation on photosynthetic
pathways for making ATP and NADPH

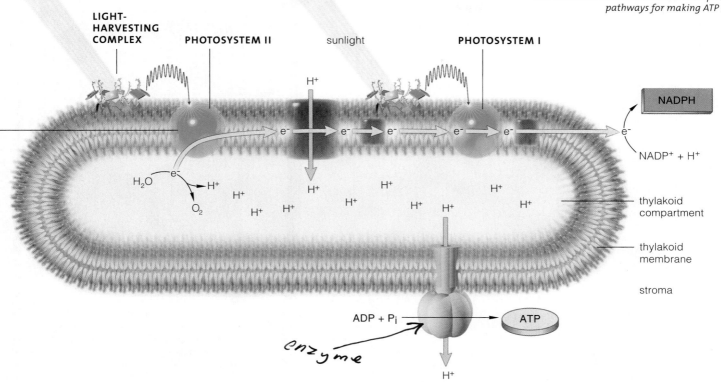

b Photon energy (*red*) causes photosystem II to lose electrons. It replaces them by pulling electrons from water molecules, which then split into oxygen and hydrogen ions (H$^+$). Oxygen leaves the cell as O$_2$.

c Electrons from photosystem II enter an electron transfer chain, which also moves H$^+$ from the stroma into the thylakoid compartment. Electrons continue on to photosystem I.

d H$^+$ concentration and electric gradients build up across the thylakoid membrane. The force of these gradients propels H$^+$ through ATP synthases, driving ATP formation.

e Photon energy (*red*) also triggers the loss of electrons from photosystem I. Through an intermediary molecule, the electrons are transferred to NADP$^+$, which also picks up H$^+$ and thereby becomes NADPH.

out of the cell and into the air. Hydrogen ions remain in the thylakoid compartment, and they contribute to the gradients that drive ATP formation.

So where do the electrons end up? After passing through the electron transfer chain, they continue on to photosystem I. There, light-harvesting complexes volley energy to a special pair of chlorophylls at the photosystem's reaction center, causing them to release electrons. An intermediary molecule transfers them to NADP$^+$, which attracts hydrogen ions at the same time. In this way, NADPH forms.

Photosystem I also runs independently in a more ancient cyclic pathway. Electrons freed from it enter an adjoining transfer chain, which moves hydrogen ions into the thylakoid compartment. As before, the resulting gradient drives ATP formation. At the end

of this chain, however, electrons are cycled back to photosystem I, and no NADPH forms.

These "noncyclic" and "cyclic" pathways operate at the same time in many photosynthetic organisms. Which one dominates at a particular time depends on metabolic demands for ATP and NADPH.

In the light-dependent reactions, sunlight energy drives the formation of ATP, NADPH, or both.

Both ATP and NADPH form by a noncyclic pathway in which electrons are pulled from water molecules, then flow through two types of photosystems, and finally to NADP$^+$. This is the photosynthetic pathway that releases free oxygen.

ATP alone forms in a cyclic pathway that starts and ends at photosystem I.

6.4 A Case of Controlled Energy Release

One of the themes threading through this book is that organisms convert one form of energy to another in highly controlled ways. The light-dependent reactions are a classic example of such conversions. Figure 6.10 walks you step-by-step through these conversions.

Read Me First!

and watch the narrated animation on energy release in photosynthesis

Cyclic Pathway of ATP Formation

Noncyclic Pathway of ATP and NADPH Formation

a Photosystem I gets a boost of photon energy from a light-harvesting cluster. It loses an electron.

b The electron passes from one molecule to another in an electron transfer chain that is embedded in the thylakoid membrane. It loses a little energy with each transfer, and ends up being reused by photosystem I.

c Molecules in the transfer chain ferry H+ across the thylakoid membrane into the inner compartment. Hydrogen ions accumulating in the compartment create an electrochemical gradient across the membrane that drives ATP synthesis, as shown in Figure 6.9.

a Photosystem II gets a boost of photon energy from a light-harvesting cluster, then loses an electron. Here, too, the electron moves through a different electron transfer chain and loses a little energy with each transfer. It ends up at photosystem I.

b Photosystem I gets a boost of photon energy from a light-harvesting complex, then loses electrons. The freed electrons, along with hydrogen ions, are used in the formation of NADPH from NADP+.

c As in the cyclic pathway, operation of the electron transfer chain puts hydrogen ions into the thylakoid compartment. In this case, hydrogens released from dissociated water molecules also enter the compartment. The H+ concentration and electric gradient across the membrane are tapped for ATP formation (Figure 6.9).

d Electrons lost from photosystem I are replaced by the electrons lost from photosystem II. Electrons lost from photosystem II are replaced by electrons from water. (Photolysis pulls water molecules apart into electrons, H+ and O_2.)

Figure 6.10 Energy transfers in the light-dependent reactions. The pair of chlorophyll *a* molecules at the center of photosystem I is designated p700. The pair in photosystem II is designated p680. The pairs respond most efficiently to wavelengths of 700 and 680 nanometers, respectively.

6.5 Light-Independent Reactions: The Sugar Factory

The chloroplast is a sugar factory, and the Calvin–Benson cycle is its machinery. These cyclic, light-independent reactions are the "synthesis" part of photosynthesis.

Sugars form in the **Calvin–Benson cycle**, which runs inside the stroma of chloroplasts (Figure 6.11). This cyclic pathway uses ATP and NADPH from the light-dependent reactions. We call them light-*independent* because they also can run in the dark, as long as ATP and NADPH are available.

ATP energy drives these sugar-building reactions, and NADPH donates hydrogen and electrons. Plants get carbon and oxygen building blocks from carbon dioxide (CO_2) in the air. Algae of aquatic habitats get them from CO_2 dissolved in water.

Rubisco, an enzyme, transfers a carbon from CO_2 to five-carbon ribulose biphosphate, or RuBP. The resulting unstable compound is the entry point for the Calvin–Benson cycle. It splits at once into two stable molecules of phosphoglycerate (PGA), each having a backbone of three carbons. The process of securing carbon from the environment by incorporating it in a stable organic compound is called **carbon fixation**.

Each PGA gets a phosphate group from ATP, and hydrogen and electrons from NADPH. For every six CO_2 fixed, twelve phosphoglyceraldehydes (PGAL) form. Ten PGAL become rearranged in a way that regenerates RuBP. The other two combine to make a six-carbon glucose with a phosphate group attached.

Most of the glucose is converted at once to sucrose or starch by other pathways that conclude the light-independent reactions. Sucrose is the main form in which carbohydrate is transported in plants; starch is the main storage form. Cells convert excess PGAL to starch, which they briefly store as starch grains in the stroma. After the sun goes down, starch is converted to sucrose for export to other cells in leaves, stems, and roots. Photosynthetic products and intermediates end up as energy sources or building blocks for all the lipids, amino acids, and other organic compounds that plants require for growth, survival, and reproduction.

Driven by ATP energy, the light-independent reactions make sugars with hydrogen and electrons from NADPH, and with carbon and oxygen from carbon dioxide.

f It takes six turns of the Calvin–Benson cycle (six carbon atoms) to make one glucose molecule.

e Ten of the PGAL get phosphate groups from ATP. In terms of energy, this primes them for an uphill run—for synthesis reactions that regenerate RuBP.

d The phosphorylated glucose enters reactions that form carbohydrate products—mainly sucrose, starch, and cellulose.

a CO_2 in air spaces inside a leaf diffuses into a photosynthetic cell. Rubisco attaches the carbon atom of CO_2 to RuBP, which starts the Calvin–Benson cycle. Each resulting intermediate splits at once into two PGAs.

b Each PGA molecule gets a phosphate group from ATP, plus hydrogen and electrons from NADPH. The resulting intermediate, PGAL, is thus primed for reaction.

c Two of the twelve PGAL molecules combine to form one molecule of glucose with an attached phosphate group.

Figure 6.11 Light-independent reactions of photosynthesis. *Brown* circles signify carbon atoms. Appendix V details the reaction steps.

6.6 Different Plants, Different Carbon-Fixing Pathways

If sunlight intensity, air temperature, rainfall, and soil composition never varied, photosynthesis might be the same in all plants. But environments differ, and so do details of photosynthesis. For example, you see such differences on hot days when water is scarce.

Take a look at the leaves in Figure 6.12. They all have a waxy cuticle that restricts water loss. Water and gases move into and out of leaves across tiny openings called **stomata** (singular, stoma). Stomata close on hot, dry days. Water and O_2 can't get out, and CO_2 can't get in. A plant's capacity to make sugars declines when its photosynthetic cells are exposed to too much O_2 and not enough CO_2.

That's why beans, sunflowers, and many other plants don't grow well in hot, dry climates unless they are steadily irrigated. We call them **C3 plants**, because the *three*-carbon PGA is the first stable intermediate of the Calvin–Benson cycle. Remember the enzyme that fixes carbon for this cycle? When oxygen builds up in leaves of C3 plants, rubisco uses oxygen—not CO_2—in an alternate reaction that yields only one molecule of PGA (Figure 6.12*a*).

In **C4 plants**, *four*-carbon oxaloacetate forms first in reactions that fix carbon twice (Figure 6.12*b*). In mesophyll cells, the C4 cycle fixes carbon no matter how much O_2 there is. This reaction delivers CO_2 directly to bundle-sheath cells, where it enters the Calvin–Benson cycle (Figure 6.12*b*). The C4 cycle keeps the CO_2 level near rubisco high enough to stop the competing reaction. C4 plants do use an extra ATP. Compared to C3 plants, though, they lose less water and make more sugar when days are dry.

CAM plants open stomata at night and fix carbon by repeated turns of a C4 cycle, then the Calvin–Benson cycle runs the next day (Figure 6.12*c*). These plants include cacti and other succulents, which have juicy, water-storing tissues and thick surface layers adapted to hot, dry climates. Some CAM plants survive prolonged droughts by closing stomata even at night. They fix CO_2 released by aerobic respiration, which supports slow growth.

In short, C3 plants, C4 plants, and CAM plants respond differently to hot, dry conditions, when their photosynthetic cells must deal with too much oxygen and not enough carbon dioxide.

The C4 cycle evolved separately in many lineages, over millions of years. Before then, CO_2 levels in air were higher, so C3 plants had the advantage in hot climates. Which cycle will be best in the future? CO_2 levels have been rising for decades and may double in fifty years. C3 plants may again have the edge.

Leaves of basswood (*Tilia americana*), a typical C3 plant.

Leaves of corn (*Zea mays*), a typical C4 plant

Beavertail cactus (*Opuntia basilaris*), one of the CAM plants

Figure 6.12 Comparison of carbon-fixing adaptations in three kinds of plants.

Read Me First!
and watch the narrated animation
on carbon-fixing adaptations

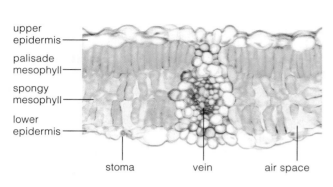

upper
epidermis
palisade
mesophyll
spongy
mesophyll
lower
epidermis

stoma   vein   air space

Basswood leaf, cross-section.

stomata closed,
no CO_2 uptake

RuBP   PGA
Calvin–
Benson
cycle

sugar

a Many C3 plants evolved in moist temperate zones. The basswood tree is one of them. On hot, dry days, it can't grow as well as C4 plants because its rubisco uses O_2 in an inefficient reaction that competes with the Calvin–Benson cycle. Not as many sugars are produced.

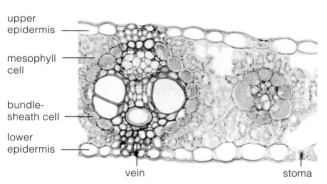

upper
epidermis

mesophyll
cell

bundle-
sheath cell

lower
epidermis

vein   stoma

Corn leaf, cross-section.

stomata closed,
no CO_2 uptake

C4   oxaloacetate
cycle   mesophyll
cell

CO_2

RuBP   PGA
Calvin–   bundle-
Benson   sheath
cycle   cell

sugar

b How C4 plants fix carbon in hot, dry weather, when there is too little CO_2 and too much O_2 inside leaves. A C4 cycle is common in grasses, corn, and other plants that evolved in the tropics. In their mesophyll cells, CO_2 gets fixed by an enzyme that ignores O_2. That reaction releases carbon dioxide in adjoining bundle sheath cells, where the Calvin–Benson cycle runs.

stoma

epidermis
mesophyll
cell

air space

Cacti have photosynthetic, fleshy stems, not leaves. This cross-section shows a stoma and mesophyll cells inside.

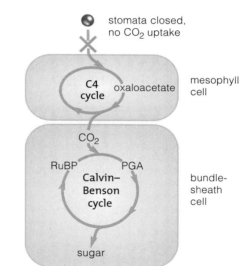

CO_2 uptake at night only

C4   runs
cycle   at
night
mesophyll
cell
Calvin–   runs
Benson   during
cycle   day

sugar

c How CAM plants fix carbon in hot, dry climates. Their stomata limit water loss by opening only at night. That is when CO_2 enters and O_2 departs. The CO_2 is fixed by a C4 cycle that runs at night. The fixed carbon enters the Calvin–Benson cycle in the same cell during daylight hours.

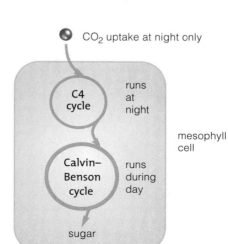

Summary

Section 6.1 Photosynthesis starts with the absorption of light energy by pigment molecules. Chlorophyll *a*, the main photosynthetic pigment, is best at absorbing violet and red wavelengths. Carotenoids and other pigments absorb characteristic wavelengths.

Section 6.2 Photosynthesis has two stages: the light-dependent and the light-independent reactions. This equation and Figure 6.13 summarize the process:

$$12H_2O + 6CO_2 \xrightarrow{\text{light energy}} 6O_2 + C_6H_{12}O_6 + 6H_2O$$

water   carbon dioxide   oxygen glucose water

In plants and algae, the light-dependent reactions occur at the thylakoid membrane, which forms a continuous compartment in the semifluid interior (stroma) of the chloroplast. Starting long ago, oxygen released by these reactions has accumulated in the atmosphere. Without it, aerobic respiration would not have evolved.

Sections 6.3, 6.4 Many clusters of pigments in the thylakoid membrane absorb photons and pass the energy to many photosystems. The light-dependent reactions start when electrons released from photosystems enter electron transfer chains. Operation of the transfer chains results in the formation of the energy carrier ATP and the reduced coenzyme NADPH.

The released electrons can move through a noncyclic or a cyclic pathway. In the noncyclic reactions, electrons lost from photosystem II enter an electron transfer chain. Electron flow through the chain causes hydrogen ions to accumulate in the thylakoid compartment. Light energy prompts photosystem I to lose electrons which, along with hydrogen ions, convert NADP$^+$ to NADPH.

Electrons lost from photosystem I are replaced by electrons from photosystem II. Photosystem II replaces its lost electrons by pulling them away from water molecules. This dissociates them into H$^+$ and O$_2$, a process called photolysis.

In the cyclic reactions, electrons from photosystem I enter a different electron transfer chain, then are recycled back to the same photosystem.

In both pathways, H$^+$ accumulation in the thylakoid compartment forms concentration and electric gradients across the thylakoid membrane. H$^+$ flows in response to the gradients, through ATP synthases. The flow causes P$_i$ to be attached to ADP in the stroma, forming ATP.

Section 6.5 Light-independent reactions, the synthesis part of photosynthesis, occur in the stroma. In C3 plants, the enzyme rubisco attaches carbon to RuBP to start the Calvin–Benson cycle. In this cyclic pathway, energy from ATP, carbon and oxygen from CO$_2$, and hydrogen and electrons from NADPH are used to make glucose, which immediately enters reactions that form the end products of photosynthesis (e.g., sucrose, cellulose, and starch).

Section 6.6 On hot, dry days, plants close stomata and conserve water, so oxygen from photosynthesis builds up in leaves. When that happens, rubisco uses oxygen instead of CO$_2$, which slows sugar production. C4 plants fix carbon twice, in two cell types. CAM plants close stomata during the day and fix carbon at night.

chloroplasts

Figure 6.13 Visual summary of photosynthesis.

Self-Quiz

Answers in Appendix III

1. Photosynthetic autotrophs use _____ from the air as a carbon source and _____ as their energy source.

2. Light-*dependent* reactions in plants occur at the _____ .
    a. thylakoid membrane   c. stroma
    b. plasma membrane   d. cytoplasm

3. In the light-*dependent* reactions, _____ .
    a. carbon dioxide is fixed   c. CO$_2$ accepts electrons
    b. ATP and NADPH form   d. sugars form

4. What accumulates inside the thylakoid compartment during the light-*dependent* reactions?
    a. glucose   c. hydrogen ions
    b. RuBP   d. carbon dioxide

5. Light-*independent* reactions proceed in the _____ .
    a. cytoplasm   b. plasma membrane   c. stroma

6. The Calvin–Benson cycle starts when _____ .
    a. light is available
    b. carbon dioxide is attached to RuBP
    c. electrons leave photosystem II

7. Match each event with its most suitable description.
    _____ ATP formation only   a. rubisco required
    _____ CO$_2$ fixation   b. ATP, NADPH required
    _____ PGAL formation   c. electrons cycled back
                                   to photosystem I

Critical Thinking

1. Imagine walking through a garden of red, white, and blue petunias. Explain each of the colors in terms of which wavelengths of light the flower is absorbing.

2. While gazing into an aquarium, you observe bubbling from an aquatic plant (Figure 6.14). What's going on?

3. In the laboratory, Krishna invites plants to take up a carbon radioisotope ($^{14}CO_2$). In which compound will the labeled carbon appear first?

4. About 200 years ago, Jan Baptista van Helmont did experiments on the nature of photosynthesis. He wanted to know where growing plants get the materials necessary for increases in size. He planted a tree seedling weighing 5 pounds in a barrel filled with 200 pounds of soil. He watered the tree regularly. Five years passed. Then van Helmont weighed the tree and the soil. The tree weighed 169 pounds, 3 ounces. The soil weighed 199 pounds, 14 ounces. Because the tree gained so much weight and the soil lost so little, he concluded the tree had gained weight by absorbing the water he had added to the barrel. Given what you know about the composition of biological molecules, why was he misguided? Knowing what you do about photosynthesis, what really happened?

5. The green alga in Figure 6.15a lives in seawater. Its main pigments absorb red light. Its accessory pigments help harvest energy in sunlit waters, and some shield it against ultraviolet radiation. Other green algae live in ponds, lakes, on rocks, even in snow. The red alga in Figure 6.15b grows on tropical reefs in clear, warm water. Its phycobilins absorb green and blue-green wavelengths that penetrate deep water. Some of its relatives live in deep, dimly lit waters; their pigments are nearly black.

   If wavelengths are such a vital source of energy, why aren't all pigments black?

   Hint: If photoautotrophs first evolved in the seas, then their pigment molecules must also have evolved in the seas. We have evidence that life arose near hydrothermal vents on the seafloor. Survival may have depended on being able to move away from weak infrared radiation (heat energy), which has been measured at vents, to keep from being boiled alive. Millions of years later, bacterial descendants were evolving near the sea surface. By one hypothesis, light-sensing machinery in deep-sea bacteria became modified for shallow-water photosynthesis.

6. Only about eight classes of pigment molecules are known, but this limited group gets around in the world. For example, animals synthesize the brownish-black melanin and some other pigments, but not carotenoids. Photoautotrophs make carotenoids, which move up through food webs, as when tiny aquatic snails graze on green algae and then flamingos eat the snails.

   Flamingos modify ingested carotenoids in plenty of ways. For instance, their cells split beta-carotene to form two molecules of vitamin A. This vitamin is the precursor of retinol, a visual pigment that transduces light into electric signals in the flamingo's eyes. Beta-carotene also gets dissolved in fat reservoirs under the skin. From there they are taken up by cells that give rise to bright pink feathers. Choose an animal and do some research into its life cycle and diet. Use your research to identify possible sources for the pigments that color its surfaces.

Figure 6.14 Leaves of *Elodea*, an aquatic plant.

Figure 6.15 (**a**) A green alga (*Codium*) from shallow coastal waters. (**b**) Red alga from a tropical reef.

Media Menu

Student CD-ROM

Impacts, Issues Video
   Pastures of the Seas
Big Picture Animation
   Harnessing light energy to build sugars
Read-Me-First Animation
   Sites of photosynthesis
   Photosynthetic pathways for making ATP and NADPH
   Energy release in photosynthesis
   The Calvin–Benson cycle
   Carbon-fixing adaptations
Other Animations and Interactions
   T. Englemann's study
   Overview of the stages of photosynthesis

InfoTrac

- Light of Our Lives. *Geographical*, January 2001.
- Sunlight at Southall Green. *Perspectives in Biology and Medicine*, Summer 2001.
- Scripps Research Gives Tiny Phytoplankton Large Role in Earth's Climate System; Study Shows Microscopic Plants Keep Planet Warm, Offers New Considerations for Iron Fertilization Efforts in Oceans. *Ascribe Higher Education News Service*, November 6, 2002.

Web Sites

- ASU Photosynthesis Center: photoscience.la.asu.edu/photosyn
- NASA Earth Observatory: earthobservatory.nasa.gov
- Calvin Nobel Prize: www.nobel.se/chemistry/laureates/1961/index.html

How Would You Vote?

The carbon-fixing enzyme rubisco is the world's most abundant protein. Plants require a lot of this enzyme because it's not that efficient. Some scientists think that using genetic engineering to modify the rubisco of plants could help increase world crop production. Would you support the use of tax dollars to fund this research?

IMPACTS, ISSUES *When Mitochondria Spin Their Wheels*

In the early 1960s, Swedish physician Rolf Luft reflected on some odd symptoms of a young patient. The woman felt weak and too hot all the time. Even on the coldest winter days, she couldn't stop sweating and her skin was flushed. She was thin in spite of a huge appetite.

Luft inferred that his patient's symptoms pointed to a metabolic disorder. Her cells seemed to be spinning their wheels. They were active, but a lot of activity was being dissipated as metabolic heat. He ordered tests designed to detect her metabolic rates. The patient's oxygen consumption was the highest ever recorded!

Microscopic examination of a tissue sample from the patient's skeletal muscles revealed mitochondria, the cell's ATP-producing powerhouses. But there were far too many of them, and they were abnormally shaped. Other studies showed that the mitochondria were engaged in aerobic respiration—yet very little ATP was forming.

The disorder, now called *Luft's syndrome*, was the first to be linked directly to a defective organelle. By analogy, someone with this mitochondrial disorder functions like a city with half of its power plants shut down. Skeletal and heart muscles, the brain, and other hard-working body parts with the highest energy demands are hurt the most.

More than a hundred other mitochondrial disorders are now known. One, a heritable genetic disease called *Friedreich's ataxia*, runs in families (Figure 7.1). Affected people develop weak muscles, loss of coordination (ataxia), and visual problems. Many die when they are young adults because of heart muscle irregularities.

Figure 7.1 Photogenic siblings with Friedreich's ataxia. Leah (*left*) began to lose balance and coordination when she was 5. She was in a wheelchair by the time she was 11. She is now diabetic, and has lost part of her hearing. Joshua (*right*) was 3 when his symptoms began. By the time he was 11, he was unable to walk. He is now legally blind.

Both young people have a heart condition called hypertrophic cardiomyopathy. Both had spinal fusion surgery. Although they have lost a large part of their fine motor skills, with the aid of adaptive equipment they continue to go to school and work in productive jobs. Leah has a part-time modeling career.

the big picture

glycolysis → aerobic respiration
glycolysis → alcoholic fermentation
glycolysis → lactate fermentation

start of glucose breakdown

three different ways of completing glucose breakdown

It All Starts With Glycolysis   All cells make ATP by breaking down organic compounds, which releases energy stored in chemical bonds. The main pathways all start in the cytoplasm, with glycolysis.

How the Aerobic Route Ends   In aerobic respiration alone, glucose breakdown is completed in mitochondria. This pathway has the greatest net energy yield from each glucose molecule.

A mutant gene and its abnormal protein product give rise to Friedreich's ataxia. The abnormal protein causes iron to accumulate inside mitochondria. Iron is required for electron transfers that drive ATP formation. But too much invites an accumulation of **free radicals**—unbound molecular fragments with the wrong number of electrons. These highly reactive fragments can attack all of the molecules of life.

Type 1 diabetes, atherosclerosis, amyotrophic lateral sclerosis (Lou Gehrig's disease), Parkinson's, Alzheimer's, and Huntington's diseases—defective mitochondria contribute to every one of these age-related problems. So when you consider mitochondria in this chapter, don't assume they are too remote from your interests. Without them, you wouldn't make enough ATP even to read about how they do it.

The preceding chapter described how plants and all other photosynthetic organisms get energy from the sun. You and all the other heterotrophs around you get some of the energy that they captured secondhand, thirdhand, and so on. Regardless of its source, energy must first be put into a form that can drive thousands of different life-sustaining reactions. That form is ATP's chemical bond energy.

You already read about the way ATP molecules form during photosynthesis. Turn now to how all organisms make ATP by tapping into the chemical bond energy of organic compounds—especially glucose.

 How Would You Vote?

Developing new drugs is costly. There's little incentive for pharmaceutical companies to target ailments, such as Friedreich's ataxia, that affect relatively few individuals. Should the government provide some funding to private companies that search for cures for diseases affecting only a small number of people? See the Media Menu for details, then vote online.

How Fermentation Routes End

In lactate and alcoholic fermentation, glucose breakdown starts *and* ends in the cytoplasm. The net energy yield is small.

What Cells Do With Food

Big meals, small meals, no meals—cells shunt carbohydrates, lipids, and proteins into breakdown pathways.

Evolutionary Connections

Aerobic respiration and photosynthesis are connected on a global scale, and that connection can be traced to the evolution of novel metabolic pathways.

7.1 Overview of Energy-Releasing Pathways

Plants make ATP during photosynthesis and use it to synthesize glucose and other carbohydrates. But all organisms, plants included, can make ATP by breaking down carbohydrates, lipids, and proteins.

Organisms stay alive by taking in energy. Plants and all other photosynthetic autotrophs get energy from the sun. Heterotrophs get energy by eating plants and one another. Regardless of its source, the energy must be in a form that can drive thousands of diverse life-sustaining reactions. Energy that becomes converted into chemical bond energy of adenosine triphosphate —ATP—serves that function.

COMPARISON OF THE MAIN TYPES OF ENERGY-RELEASING PATHWAYS

The first energy-releasing metabolic pathways were operating billions of years before Earth's oxygen-rich atmosphere evolved, so we can expect that they were *anaerobic*; the reactions did not use free oxygen. Many prokaryotes and protists still live in places where oxygen is absent or not always available. They make ATP by fermentation and other anaerobic pathways. Many eukaryotic cells still use fermentation, including skeletal muscle cells. However, the cells of nearly all species extract energy efficiently from glucose by way of **aerobic respiration**, an oxygen-dependent pathway. Each breath you take provides your actively respiring cells with a fresh supply of oxygen.

Make note of this point: *In all cells, all of the main energy-releasing pathways start with the same reactions in the cytoplasm.* During the initial reactions, **glycolysis**, enzymes cleave and rearrange a glucose molecule into two molecules of **pyruvate**, an organic compound that has a three-carbon backbone.

Once glycolysis is over, energy-releasing pathways differ. Only the aerobic pathway continues and ends in mitochondria (Figure 7.2). There, oxygen accepts and removes electrons that drove the reactions.

Only aerobic respiration delivers enough ATP to build and maintain big multicelled organisms, including redwoods and highly active animals, such as people and Canada geese.

As you examine the energy-releasing pathways in sections to follow, keep in mind that enzymes catalyze each step, and intermediates formed at one step serve as substrates for the next enzyme in the pathway.

OVERVIEW OF AEROBIC RESPIRATION

Of all energy-releasing pathways, aerobic respiration gets the most ATP for each glucose molecule. Whereas anaerobic routes have a net yield of two ATP, aerobic respiration typically yields thirty-six or more. If you were a bacterium, you would not require much ATP. Being far larger, more complex, and highly active, you depend on the aerobic pathway's high yield. When a molecule of glucose is used as the starting material, aerobic respiration can be summarized this way:

$$C_6H_{12}O_6 \ + \ 6O_2 \ \longrightarrow \ 6CO_2 \ + \ 6H_2O$$

glucose   oxygen   carbon   water
                          dioxide

Figure 7.2 Where the main energy-releasing pathways of ATP formation start and finish.

AEROBIC RESPIRATION

start (glycolysis) in cytoplasm

completed in mitochondrion

ANAEROBIC ENERGY-RELEASING PATHWAYS

start (glycolysis) in cytoplasm

completed in cytoplasm

Read Me First!
and watch the narrated animation
on aerobic respiration

cytoplasm

a glucose

2 ATP → **glycolysis** → 4 ATP

energy input to
start reactions

$e^- + H^+$

(2 ATP *net*)

2 NADH

2 pyruvate

mitochondrion

2 NADH $e^- + H^+$ → 2 CO_2

8 NADH $e^- + H^+$

b
Krebs cycle   → 4 CO_2

2 FADH$_2$ $e^- + H^+$ → 2 ATP

c

e^-

electron transfer phosphorylation

32 ATP

d

e H^+ → water

$e^- +$ oxygen

f

TYPICAL NET ENERGY YIELD: **36 ATP**

Figure 7.3 Overview of aerobic respiration. Reactions start in the cytoplasm and end in mitochondria.

(**a**) In the first stage, glycolysis, enzymes partly break down glucose to pyruvate.

(**b**) In the second stage, enzymes break down pyruvate to carbon dioxide.

(**c**) NAD$^+$ and FAD pick up the electrons and hydrogen stripped from intermediates in both stages.

(**d**) The final stage is electron transfer phosphorylation. The reduced coenzymes NADH and FADH$_2$ give up electrons to electron transfer chains. H$^+$ tags along with electrons. Electron flow through the chains sets up H$^+$ gradients, which are tapped to make ATP.

(**e**) Oxygen accepts electrons at the end of the third stage, forming water.

(**f**) From start to finish, a typical net energy yield from a glucose molecule is thirty-six ATP.

However, as you can see, the summary equation only tells us what the substances are at the start and finish of the pathway. In between are three reaction stages.

Figure 7.3 is your overview of aerobic respiration. Glycolysis, again, is the first stage. The second stage is a cyclic pathway, the **Krebs cycle**. Enzymes break down pyruvate to carbon dioxide and water, which releases many electrons and hydrogen atoms.

As you track the reactions, you'll come across two enzyme helpers, the coenzymes **NAD$^+$** (nicotinamide adenine dinucleotide) and **FAD** (flavin adenine dinucleotide). Both accept electrons and hydrogen derived from intermediates of glucose breakdown. Unbound hydrogen atoms are simply hydrogen ions (H$^+$), or naked protons. When the two coenzymes are carrying electrons and hydrogen, they are in a reduced form and may be abbreviated NADH and FADH$_2$.

Few ATP form during glycolysis or the Krebs cycle. The big energy harvest comes in the third stage, after the coenzymes give up the electrons and hydrogen to

electron transfer chains. The chains are the machinery of **electron transfer phosphorylation**. They set up H$^+$ concentration and electric gradients, which drive ATP formation at nearby membrane proteins. It is in this final stage that so many ATP molecules are produced. As it ends, oxygen inside the mitochondrion accepts the "spent" electrons from the last component of each transport system. Oxygen picks up H$^+$ at the same time and thereby forms water.

Nearly all metabolic reactions run on energy released from glucose and other organic compounds. The main energy-releasing pathways start in the cytoplasm with glycolysis, a stage of reactions that break down glucose to pyruvate.

Anaerobic pathways have a small net energy yield, typically two ATP per glucose.

Aerobic respiration, an oxygen-dependent pathway, runs to completion in mitochondria. From start (glycolysis) to finish, it typically has a net energy yield of thirty-six ATP.

7.2 The First Stage: Glycolysis

Let's track what happens to a glucose molecule in the first stage of aerobic respiration. Remember, the same steps happen in anaerobic energy-releasing pathways.

Any of several six-carbon sugars can be broken down in glycolysis. Each molecule of glucose, recall, has six carbon, twelve hydrogen, and six oxygen atoms (Section 3.3). The carbons form its backbone. During glycolysis, this one molecule is partly broken down to two molecules of pyruvate, a three-carbon compound:

$$glucose \longrightarrow \text{P}-glucose \longrightarrow 2\ pyruvate$$

The initial steps of gycolysis are *energy-requiring*, and that energy is delivered by ATP. One ATP molecule activates glucose by transferring a phosphate group to it. Then another ATP transfers a phosphate group to the intermediate that forms. Thus, it takes an energy investment of two ATP to start glycolysis (Figure 7.4a).

The second intermediate is split into one PGAL (phosphoglyceraldehyde) and one molecule with the same number of atoms arranged a bit differently. An enzyme can reversibly convert the two, and its action delivers two PGAL for the next reaction (Figure 7.4b).

In the first *energy-releasing* step of glycolysis, both PGALs are converted to intermediates that give up a phosphate group to ADP, and so ATP forms. Two later intermediates do the same thing. Thus four ATP have been formed by **substrate-level phosphorylation**. We define this metabolic event as the direct transfer of a phosphate group from the substrate of a reaction to some other molecule—in this case, to ADP.

Meanwhile, the coenzyme NAD⁺ accepts electrons and hydrogens from each PGAL, becoming NADH.

By this time, a total of four molecules of ATP have formed, but remember that two ATP were invested to get the reactions going. The *net* yield of glycolysis is two ATP and two NADH.

To summarize, glycolysis converts the bond energy of glucose to bond energy of ATP—a transportable form of energy. The electrons and hydrogen stripped from glucose and picked up by NAD⁺ will enter the next stage of reactions. And so will the end products of glycolysis—two pyruvate molecules.

> *Glycolysis is a series of reactions that partially break down glucose or other six-carbon sugars to two molecules of pyruvate. It takes two ATP to jump-start the reactions.*
>
> *Two NADH and four ATP form. However, when we subtract the two ATP required to start the reactions, the net energy yield of glycolysis is two ATP from one glucose molecule.*

glucose

GLYCOLYSIS

pyruvate

to second stage of aerobic respiration or to a different energy-releasing pathway

GLUCOSE

Figure 7.4 Glycolysis. This first stage of the main energy-releasing pathways occurs in the cytoplasm of all prokaryotic and eukaryotic cells. Glucose is the reactant in this example. Appendix V gives the structural formulas of intermediates that form during its breakdown. Two pyruvate, two NADH, and four ATP form in glycolysis. Cells invest two ATP to start the reactions however, so the *net* energy yield is two ATP.

Depending on the type of cell and environmental conditions, the pyruvate may enter the second set of reactions of the aerobic pathway, including the Krebs cycle. Or it may be used in other reactions, such as those of fermentation.

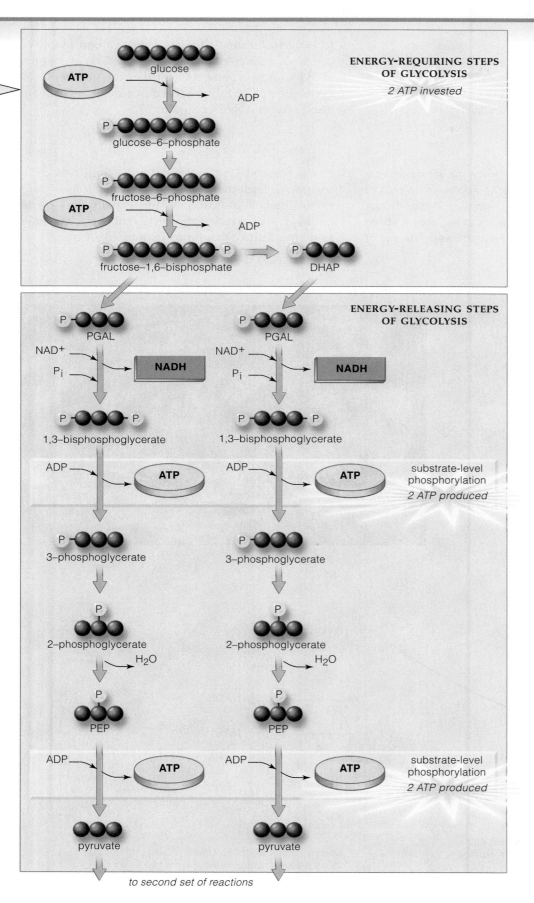

ENERGY-REQUIRING STEPS OF GLYCOLYSIS

2 ATP invested

ATP — glucose — ADP

glucose–6–phosphate

fructose–6–phosphate

ATP — ADP

fructose–1,6–bisphosphate   DHAP

ENERGY-RELEASING STEPS OF GLYCOLYSIS

PGAL

NAD⁺, Pᵢ → NADH

1,3–bisphosphoglycerate

ADP → ATP   substrate-level phosphorylation
2 ATP produced

3–phosphoglycerate

2–phosphoglycerate

H₂O

PEP

ADP → ATP   substrate-level phosphorylation
2 ATP produced

pyruvate

to second set of reactions

Track the six carbon atoms (*brown circles*) of glucose. Glycolysis requires an energy investment of two ATP:

a One ATP transfers a phosphate group to glucose, jump-starting the reactions.

b Another ATP transfers a phosphate group to an intermediate, causing it to split into two three-carbon compounds: PGAL and DHAP (dihydroxyacetone phosphate). Both have the same atoms, arranged differently. They are interconvertible, but only PGAL can continue on in glycolysis. DHAP gets converted, so two PGAL are available for the next reaction.

c Two NADH form when each PGAL gives up two electrons and a hydrogen atom to NAD⁺.

d Two intermediates each transfer a phosphate group to ADP. *Thus, two ATP have formed by direct phosphate group transfers.* The original energy investment of two ATP is now paid off.

e Two more intermediates form. Each gives up one hydrogen atom and an —OH group. These combine as water. Two molecules called PEP form by these reactions.

f Each PEP transfers a phosphate group to ADP. *Once again, two ATP have formed by substrate-level phosphorylation.*

In sum, glycolysis has a net energy yield of two ATP for each glucose molecule. Two NADH also form during the reactions, and two molecules of pyruvate are the end products.

7.3 Second Stage of Aerobic Respiration

The two pyruvate molecules formed by glycolysis can leave the cytoplasm and enter a mitochondrion. There they enter reactions that get the Krebs cycle going. Many coenzymes pick up the electrons and hydrogens released when the two pyruvates are dismantled.

to oxygen, forming CO_2. Each two-carbon fragment combines with a coenzyme (designated A) and forms acetyl–CoA, a type of cofactor that can get the Krebs cycle going. The initial breakdown of each pyruvate also yields one NADH (Figure 7.7).

ACETYL–CoA FORMATION

Start with Figure 7.5, which shows the structure of a typical mitochondrion. Figure 7.6 zooms in on part of the interior where the second-stage reactions occur. At the start of these reactions, enzyme action strips one carbon atom from each pyruvate and attaches it

THE KREBS CYCLE

The two acetyl–CoA molecules enter the Krebs cycle separately. Each transfers its two-carbon acetyl group to four-carbon oxaloacetate. Incidentally, this cyclic pathway is also called the citric acid cycle, after the first intermediate that forms (citric acid, or citrate).

Figure 7.5 *Above*, functional zones inside the mitochondrion. An inner membrane system divides this organelle's interior into an inner and an outer compartment. The second and third stages of aerobic respiration play out at this membrane.

Figure 7.6 *Right:* Overview of the number of ATP molecules and coenzymes that form in the second stage of aerobic respiration. The reactions start with two molecules of pyruvate from glycolysis. Pyruvate moves from the cytoplasm, then across the outer and inner mitochondrial membranes, into the inner compartment where the reactions take place.

Two pyruvates cross the inner mitochondrial membrane·

outer mitochondrial compartment

inner mitochondrial compartment

Krebs Cycle

2 NADH

6 NADH

2 FADH$_2$

2 ATP

Eight NADH, two FADH$_2$, and two ATP are the payoff from the complete breakdown of two pyruvates in the second-stage reactions.

6 CO_2

The six carbon atoms from two pyruvates diffuse out of the mitochondrion, then out of the cell, in six CO_2

Read Me First!
and watch the narrated animation on the Krebs cycle

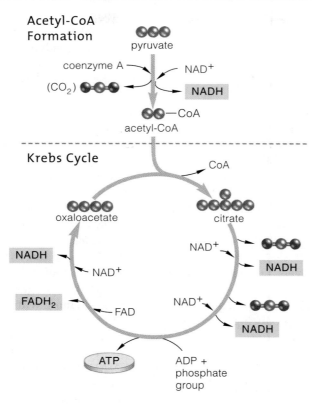

a One carbon atom is stripped from each pyruvate and is released as CO_2. The remaining fragment binds with coenzyme A, forming acetyl–CoA.

b NAD^+ picks up hydrogen and electrons, forming one NADH.

c In the first step of the Krebs cycle, acetyl–CoA transfers two carbons to oxaloacetate, forming citrate.

d In rearrangements of intermediates, another carbon atom is released as CO_2, and NADH forms as NAD^+ picks up hydrogen and electrons.

h The final steps regenerate oxaloacetate. NAD^+ picks up hydrogen and electrons, forming NADH. *At this point in the cycle, three NADH and one $FADH_2$ have formed.*

g $FADH_2$ forms as the coenzyme FAD picks up electrons and hydrogen.

e Another carbon atom is released as CO_2. Another NADH forms. *Three carbon atoms now have been released.* This balances out the three carbons that entered (in one pyruvate).

f A phosphate group is attached to ADP. At this point, one ATP has formed by substrate-level phosphorylation.

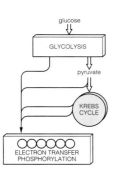

Figure 7.7 Aerobic respiration's second stage: formation of acetyl–CoA and the Krebs cycle. The reactions occur in a mitochondrion's inner compartment. It takes two turns of the cycle to break down the two pyruvates from glucose. A total of two ATP, eight NADH, two $FADH_2$, and six CO_2 molecules form. Organisms release the CO_2 from the reactions into their surroundings.

It takes two turns of the Krebs cycle to completely break down two molecules of pyruvate to CO_2 and water. Only two ATP form, which doesn't add much to the small net yield from glycolysis. However, in addition to those two NADH produced during the formation of acetyl–CoA, six more NADH and two $FADH_2$ are produced in the cycle. With their cargo of electrons and hydrogen atoms, these ten coenzymes constitute a big potential payoff for the cell.

Four more CO_2 molecules form as the Krebs cycle turns. In total, *six* carbon atoms (from two pyruvates) depart during the second stage of aerobic respiration, in six molecules of CO_2 (Figures 7.6 and 7.7). And so glucose from glycolysis has lost all of its carbons; it has become fully oxidized.

For interested students, Appendix V has a closer look at the steps of these remarkable reactions.

Aerobic respiration's second stage starts after two pyruvate molecules from glycolysis move from the cytoplasm, across the outer and inner mitochondrial membranes, and into the inner mitochondrial compartment.

Here, pyruvate is converted to acetyl–CoA, which starts the Krebs cycle. A total of two ATP and ten coenzymes (eight NADH, two $FADH_2$) form. All of pyruvate's carbons depart, in the form of carbon dioxide.

Together with two coenzymes (NADH) that formed during glycolysis, the ten from the second stage will deliver electrons and hydrogen to the third and final stage.

7.4 Third Stage of Aerobic Respiration—The Big Energy Payoff

In the aerobic pathway's third stage, energy release goes into high gear. Coenzymes from the first two stages provide the hydrogen and electrons that drive the formation of many ATP. Electron transfer chains and ATP synthases function as the machinery.

ELECTRON TRANSFER PHOSPHORYLATION

The third stage starts as coenzymes donate electrons to electron transfer chains that are located in the inner mitochondrial membrane (Figure 7.8). The flow of electrons through the chains drives the attachment of phosphate to ADP molecules. Hence the name *electron transfer phosphorylation.*

Incremental energy release, recall, is more efficient than one big burst of energy that would result in little more than a lot of unusable heat (Section 5.2). When electrons flow through transfer chains, they give up energy bit by bit, in usable parcels, to substances that can briefly store it.

The two NADH that formed in the cytoplasm (by glycolysis) can't reach the ATP-producing machinery directly. They give up their electrons and hydrogen to transport proteins, which shuttle them into the inner compartment. There, NAD+ or FAD inside pick them up. Eight NADH and two FADH$_2$ from the second stage are already inside.

When all of these coenzymes turn over electrons to transfer chains, they release hydrogen ions (H+) at the same time. Electrons passing through the chains lose a bit of energy at each step. In three parts of the chain, that energy drives the pumping of H+ into the outer compartment. There, accumulation of H+ sets up an electrochemical gradient across the inner membrane.

H+ can't diffuse across membranes. The only way it can follow the gradients, which lead back to the inner compartment, is by flowing through the interior of ATP synthases (Figures 7.8 and 7.9). H+ flow through these transport proteins drives the formation of ATP from ADP and unbound phosphate.

The last molecules in the electron transfer chains pass electrons to gaseous oxygen, which forms water after combining with H+. *Oxygen is the final acceptor of electrons stripped from glucose.* In oxygen-starved cells, electrons in the transfer chain have nowhere to go. The whole chain backs up with electrons all the way to NADPH, so no H+ gradient forms, and no ATP is made. Cells of complex organisms don't survive long without oxygen, because they can't produce enough ATP to sustain life processes.

Figure 7.8 Electron transfer phosphorylation, the third and final stage of aerobic respiration.

At the inner mitochondrial membrane, NADH and FADH$_2$ give up electrons to the transfer chains. When electrons are transferred through the chains, unbound hydrogen (H+) is shuttled across the membrane to the outer compartment:

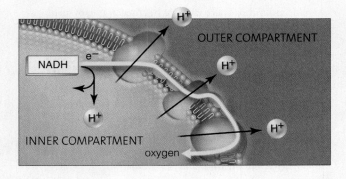

H+ concentration is now greater in the outer compartment. Concentration and electric gradients across the membrane have been set up. H+ follows these gradients through the interior of ATP synthases. The flow drives the formation of ATP from ADP and unbound phosphate (P$_i$).

Do these events sound familiar? They should. ATP forms in much the same way inside chloroplasts. H+ concentration and electric gradients across the inner thylakoid membrane drive ATP formation. In thylakoids, H+ flows in the opposite direction compared to the flow in chloroplasts.

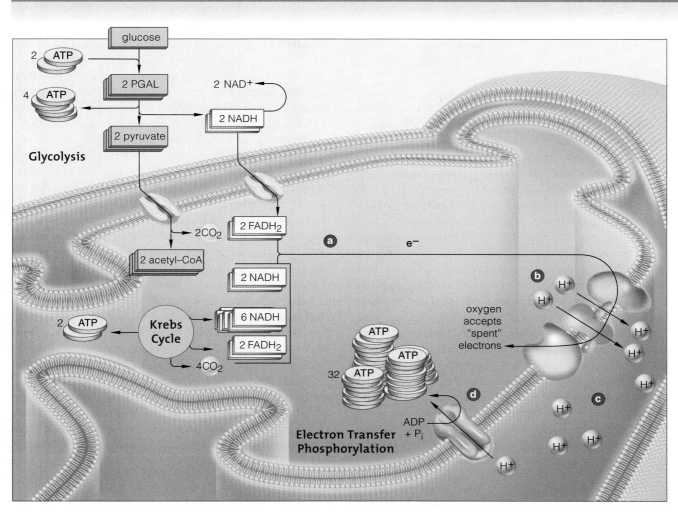

Read Me First!
and watch the narrated animation on third-stage reactions

a Electrons and hydrogen from NADH and FADH$_2$ that formed during the first and second stages enter electron transfer chains.

b As electrons are transferred through the chains, H$^+$ ions are shuttled across the inner membrane, into the outer compartment.

c H$^+$ concentration becomes greater in the outer compartment than the inner one. Chemical and electrical gradients have been established.

d Hydrogen ions follow the gradients through the interior of ATP synthases, driving ATP synthesis from ADP and phosphate (P$_i$).

Figure 7.9 Summary of the transfers of electrons and hydrogen from coenzymes involved in ATP formation in mitochondria.

SUMMING UP: THE ENERGY HARVEST

Thirty-two ATP typically form in the third stage. Add in the four produced in the first and second stages, and aerobic respiration has netted thirty-six ATP from one glucose molecule. That's a lot of ATP! Anaerobic pathways may use up eighteen glucose molecules to get the same net yield.

The actual yield varies. Shifting concentrations of reactants, intermediates, and products affect it. So does the shuttling of electrons and hydrogen from NADH that forms in the cytoplasm. Shuttling mechanisms are not the same in all cell types.

In aerobic respiration's third stage, electrons and hydrogen from coenzymes (NADH and FADH$_2$) interact with electron transfer chains in the mitochondrion's inner membrane.

Electron flow through transfer chains makes H$^+$ accumulate in the outer mitochondrial compartment. The resulting chemical and electrical gradients across the inner membrane drive the synthesis of thirty-two ATP.

Aerobic respiration typically nets thirty-six ATP molecules from each glucose molecule metabolized.

7.5 Fermentation Pathways

We turn now to the use of glucose as a substrate for fermentation pathways. These are anaerobic pathways. They don't use oxygen as the final acceptor of electrons that ultimately drive the ATP-forming machinery.

Diverse organisms are fermenters. Many are protists and bacterial species that live in marshes, bogs, mud, ocean sediments, the animal gut, canned foods, sewage treatment ponds, and other oxygen-free places. Some die when exposed to free oxygen. Bacterial species that cause botulism and many other diseases are like this. Other fermenters are indifferent to oxygen's presence. Still other kinds use oxygen, but they also can switch to fermentation when oxygen becomes scarce.

Glycolysis is the first stage of fermentation, as it is in aerobic respiration (Figure 7.4). Here, too, pyruvate and NADH form, and the net energy yield is two ATP. But fermentation reactions cannot completely degrade glucose (to carbon dioxide and water). They produce no more ATP beyond the small yield from glycolysis. *The final steps simply regenerate NAD+, the coenzyme that is essential for the breakdown reactions.* The regeneration allows glycolysis reactions to continue production of small amounts of ATP in the absence of oxygen.

Fermentation yields enough energy to sustain many single-celled anaerobic organisms. It even helps some aerobic cells when oxygen levels are stressfully low. But it isn't enough to sustain large, multicelled organisms, this being why you'll never see anaerobic elephants.

ALCOHOLIC FERMENTATION

In **alcoholic fermentation**, the three-carbon backbone of the two pyruvate molecules from glycolysis is split. The reactions result in two molecules of acetaldehyde (an intermediate having a two-carbon backbone), and two of carbon dioxide. Next, the acetaldehydes accept electrons and hydrogen from NADH, thus becoming an alcohol product called ethanol (Figure 7.10).

Some single-celled fungi called yeasts are famous for their use of this pathway. One type, *Saccharomyces cerevisiae*, makes bread dough rise. Bakers mix it with sugar, then blend both into dough. Fermenting yeast cells release carbon dioxide, and the dough expands (rises) as the gas forms bubbles in it. Oven heat forces the bubbles out of spaces they had occupied in the dough, and the alcohol product evaporates away.

Wild and cultivated strains of *Saccharomyces* are used to produce alcohol in wine. Crushed grapes are left in vats along with the yeast, which converts sugar in the juice to ethanol. Ethanol is toxic to microbes. When a fermenting brew's ethanol content nears 10 percent, yeast cells start dying and fermentation ends.

Birds get drunk when they eat too many naturally fermented berries. Landscapers avoid planting berry-producing shrubs along highways because inebriated birds fly into windshields. Also, wild turkeys, robins, and other birds get tipsy on fermenting fruit that has dropped from orchard trees.

Figure 7.10 A look at alcoholic fermentation. (**a**) Yeasts, single-celled organisms, make ATP by this anaerobic pathway.

(**b**) A vintner examining the color and clarity of one fermentation product of *Saccharomyces*. Strains of this yeast live on the sugar-rich tissues of ripe grapes.

(**c**) Carbon dioxide released from cells of *S. cerevisiae* makes bread dough rise in this bakery.

(**d**) Alcoholic fermentation. The intermediate acetaldehyde functions as the final electron acceptor. The end product of the reactions is ethanol.

Figure 7.11 Lactate fermentation. In this anaerobic pathway, the product (lactate) is the final acceptor of electrons originally stripped from glucose. The reactions have a net energy yield of two ATP (from glycolysis).

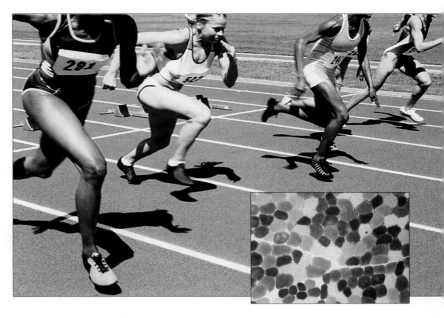

Figure 7.12 Sprinters, calling upon lactate fermentation in their muscles. The micrograph is a cross-section through part of a muscle showing three types of fibers. The lighter fibers contribute to muscle speed by producing ATP with lactate fermentation when demands for energy are high. Darker fibers use aerobic respiration and support greater levels of endurance.

LACTATE FERMENTATION

In **lactate fermentation**, NADH gives up electrons and hydrogen to two pyruvate molecules from glycolysis. The transfer converts each pyruvate to lactate, a three-carbon compound (Figure 7.11). You've probably heard of lactic acid, the non-ionized form of this compound, but lactate is by far the most common form inside living cells, which is our focus here.

Lactobacillus and some other bacteria use lactate fermentation. Their fermenting action can spoil food, yet some species have commercial uses. For instance, huge populations that break down glucose in milk give us cheeses, yogurt, buttermilk, and other dairy products. Fermenters also help in curing meats and in pickling some fruits and vegetables, such as sauerkraut. Lactate is an acid; it gives these foods a sour taste.

Lactate fermentation as well as aerobic respiration yields ATP for muscles that are partnered with bones. These skeletal muscles contain a mixture of cell types. Cells composing *slow-twitch* muscle fibers support light, steady, prolonged activity, as during marathon runs or bird migrations. Slow-twitch muscle cells make ATP only by the aerobic respiration pathway, and so they have many mitochondria. They are dark red because they contain large amounts of myoglobin, a pigment related to hemoglobin that is used to store oxygen for aerobic respiration.

By contrast, cells of pale *fast-twitch* muscle fibers have few mitochondria and no myoglobin, and use lactate fermentation to produce ATP. They are useful when demands for energy are immediate and intense, such as in weight lifting or sprints (Figure 7.12). Lactate fermentation works quickly but not for long— it does not produce enough ATP to sustain activity. That is one reason you don't see migrating chickens. Flight muscles in a chicken are the white breast meat, containing mostly fast-twitch fibers.

Short bursts of flight evolved in the ancestors of chickens, perhaps as a way of escaping predators or improving agility during territorial battles. Chickens do walk or sprint; hence the "dark meat" (slow-twitch muscle) in their thighs and legs. So what sort of breast meat can you expect to find in a migrating duck?

Section 32.5 is an overview of alternative energy pathways for muscle cells.

In fermentation pathways, an organic substance that forms during the reactions is the final acceptor of electrons originally derived from glucose.

Alcoholic fermentation and lactate fermentation both have a net energy yield of two ATP for each glucose molecule metabolized. That ATP forms during glycolysis. The remaining reactions regenerate NAD+, the coenzyme that keeps these pathways operating.

7.6 Alternative Energy Sources in the Body

So far, you've looked at what happens after glucose molecules enter an energy-releasing pathway. Now start thinking about what cells do when they have too much or too little glucose.

THE FATE OF GLUCOSE AT MEALTIME AND IN BETWEEN MEALS

What happens to glucose at mealtime? While you and all other mammals are eating, glucose and other small organic molecules are being absorbed across the gut lining, and your blood is transporting them through the body. The rising glucose concentration in blood prompts an organ, the pancreas, to secrete insulin. This hormone makes cells take up glucose faster.

Cells trap the incoming glucose by converting it to glucose–6–phosphate. This is the first intermediate of glycolysis, formed by a phosphate group transfer from ATP (Figures 7.4 and 7.13). Phosphorylated glucose can't be transported out of the cell.

When glucose intake exceeds cellular demands for energy, the body's ATP-producing machinery goes into high gear. Unless a cell is using ATP rapidly, the ATP concentration in cytoplasm rises, and glucose–6–phosphate is diverted into a biosynthesis pathway. Glucose gets stored as glycogen, one of the storage polysaccharides found in animals (Section 3.3). Liver cells and muscle cells especially favor this alternative pathway. Together, these two types of cells maintain the largest stores of glycogen in the body.

Between meals, the blood level of glucose declines. If the decline were not countered, that would be bad news for the brain, your body's glucose hog. At any time, your brain is taking up more than two-thirds of the freely circulating glucose. Why? The brain's many hundreds of millions of nerve cells (neurons) use this sugar as their preferred energy source.

The pancreas responds to glucose decline by secreting glucagon. This hormone causes liver cells to convert stored glycogen to glucose and send it back to the blood. Only liver cells do this; muscle cells won't give it up. The blood glucose level rises, and brain cells keep on functioning. Thus, *hormones control whether your cells use free glucose as an energy source or tuck it away.*

Don't let this explanation lead you to believe that your cells squirrel away huge amounts of glycogen. Glycogen makes up only 1 percent or so of the adult body's total energy reserves, the energy equivalent of two cups of cooked pasta. Unless you eat on a regular basis, you'll end up depleting your liver's small glycogen stores in less than twelve hours.

Of the total energy reserves in, say, a typical adult who eats well, 78 percent (about 10,000 kilocalories) is concentrated in body fat and 21 percent in proteins.

ENERGY FROM FATS

How does a human body access its fat reservoir? A fat molecule, recall, has a glycerol head and one, two, or three fatty acid tails. The body stores most fats as triglycerides, with three tails each. Triglycerides build up inside of the fat cells of adipose tissue. This tissue is strategically located under the skin of buttocks and other body regions.

When the blood glucose level falls, triglycerides are tapped as an energy alternative. Enzymes in fat cells cleave bonds between glycerol and fatty acids, which both enter the blood. Enzymes in the liver convert the glycerol to PGAL. And PGAL, recall, is one of the key intermediates for glycolysis (Figure 7.4). Nearly all cells of your body take up circulating fatty acids, and enzymes inside them cleave the fatty acid backbones. The fragments are converted to acetyl–CoA, which can enter the Krebs cycle.

Compared to glucose, a fatty acid tail has far more carbon-bound hydrogen atoms, so it yields more ATP. Between meals or during steady, prolonged exercise, fatty acid conversions supply about half of the ATP that muscle, liver, and kidney cells require.

What happens if you eat too many carbohydrates? Aerobic respiration, remember, converts glucose to pyruvate, then to acetyl–CoA, which enters the Krebs cycle. When too much glucose is circulating through the body, acetyl–CoA is diverted to a pathway that synthesizes fatty acids. *Too much glucose ends up as fat.*

ENERGY FROM PROTEINS

Some enzymes in your digestive system split dietary proteins into their amino acid subunits, which are then absorbed into the bloodstream. Cells use amino acids to build other proteins or nitrogen-containing compounds. Even so, if you eat more protein than your body needs, amino acids will be broken down further. Their $-NH_3^+$ group is pulled off, forming ammonia (NH_3). Depending on the types of amino acids, the leftover carbon backbones are broken down to either acetyl–CoA, pyruvate, or one of the intermediates of the Krebs cycle. Your cells can funnel any of these compounds into the Krebs cycle.

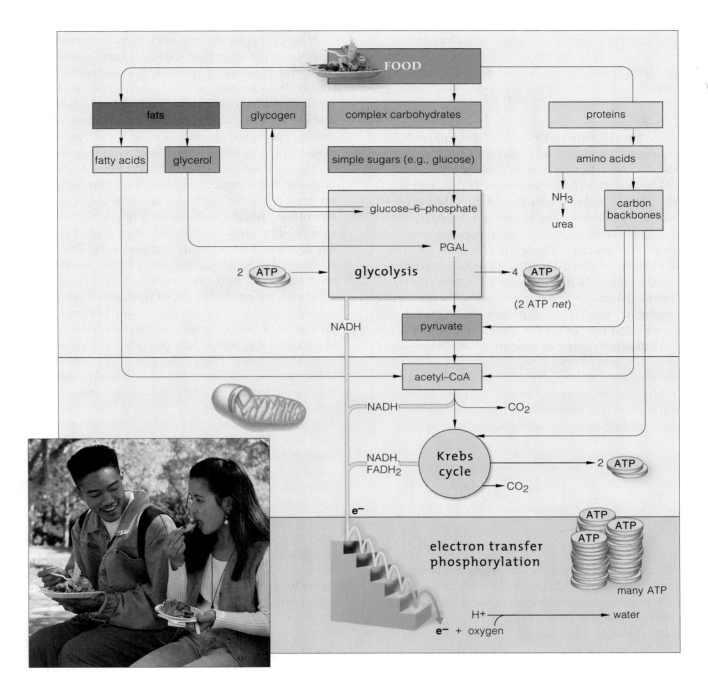

Figure 7.13 Reaction sites where a variety of organic compounds can enter the different stages of aerobic respiration. The compounds shown are alternative energy sources in humans and other mammals.

Notice how complex carbohydrates, fats, and proteins cannot enter the aerobic pathway directly. The digestive system, then individual cells, must first break apart these molecules to simpler compounds that the pathway can dismantle further.

As you can see, maintaining and accessing energy reserves is complicated business. Controlling the use of glucose is special because it is the fuel of choice for the brain. However, providing all of your cells with energy starts with the kinds of food you eat.

In humans and other mammals, the entrance of glucose or other organic compounds into an energy-releasing pathway depends on the kinds and proportions of carbohydrates, fats, and proteins in the diet.

7.7 Perspective on Life

In this unit you read about photosynthesis and aerobic respiration—the main pathways by which cells trap, store, and release energy. What you may not know is that the two pathways became linked, on a grand scale, over evolutionary time.

When life originated long ago, the atmosphere held little free oxygen. We can expect that those first cells had to make ATP by reactions similar to glycolysis, and fermentation pathways probably dominated. More than a billion years passed before the oxygen-evolving pathway of photosynthesis emerged.

Oxygen slowly accumulated in the atmosphere. Some cells now used it to accept electrons, perhaps as a chance outcome of mutated proteins in electron transfer chains. In time, some of their descendants abandoned anaerobic habitats. Among them were the forerunners of all bacteria, protists, plants, fungi, and animals that now engage in aerobic respiration.

With aerobic respiration, a great flow of carbon, hydrogen, and oxygen through metabolic pathways of living organisms came full circle. For the final products of this aerobic pathway—carbon dioxide and water—are the same materials necessary to build organic compounds in photosynthesis:

Perhaps you have difficulty seeing the connection between yourself—a highly intelligent being—and such remote-sounding events as energy flow and the cycling of carbon, hydrogen, and oxygen. Is this really the stuff of humanity?

Think back on the structure of a water molecule. Two hydrogen atoms sharing electrons with oxygen may not seem close to your daily life. Yet, through that sharing, water molecules show polarity and hydrogen-bond with one another. Their chemical behavior is a beginning for the organization of lifeless matter that leads in turn to the organization of all living things.

For now you can visualize other diverse molecules interspersed through water. Nonpolar kinds resist interaction with water; polar kinds dissolve in it. On their own, the phospholipids among them assemble into a two-layered film. Such lipid bilayers, recall, are the framework of cell membranes, hence all cells.

From the beginning, the cell has been the basic *living* unit. The essence of life is not some mysterious force; it is molecular organization and metabolic control. With a membrane to contain them, reactions *can* be controlled. With molecular mechanisms built into membranes, cells respond to energy changes and shifting concentrations of solutes in the environment. Response mechanisms operate by "telling" proteins—enzymes—when and what to build or tear down.

And it is not some mysterious force that creates proteins. DNA, the double-stranded treasurehouse of inheritance, has the chemical structure—*the chemical message*—that allows molecule to reproduce molecule, one generation after the next. In your body, DNA strands tell trillions of cells how countless molecules must be built or torn apart for their stored energy.

So yes, carbon, hydrogen, oxygen, and other atoms of organic molecules are the stuff of you, and us, and all of life. Yet it takes more than molecules to complete the picture. Life continues as long as a continuous flow of energy sustains its organization. Molecules become assembled into cells, cells into organisms, organisms into communities, and so on through the biosphere.

It takes energy inputs from the sun to maintain the levels of biological organization. And energy flows through time in one direction—from organized to less organized forms. Only as long as energy continues to flow into the great web of life can life continue in all its rich expressions.

So life is no more *and no less* than a marvelously complex system for prolonging order. Sustained with energy transfusions from the sun, life continues by its capacity for self-reproduction. With the hereditary instructions of DNA, energy and materials become organized, generation after generation. Even with the death of individuals, life elsewhere is prolonged. With each death, molecules are released and may be recycled as raw materials for new generations.

With this flow of energy and cycling of material through time, each birth is affirmation of our ongoing capacity for organization, each death a renewal.

Summary

Section 7.1 All organisms, including photosynthetic types, make ATP by the breakdown of glucose and other organic compounds. Glycolysis, the initial breakdown of one glucose to two pyruvate molecules, takes place in the cytoplasm. It is the first stage of all the main energy-releasing pathways, and it doesn't require free oxygen.

Anaerobic pathways end in the cytoplasm, and the net yield of ATP is small. An oxygen-requiring pathway called aerobic respiration continues in mitochondria, and it releases far more ATP energy from glucose.

Section 7.2 The first steps of glycolysis require an energy input of 2 ATP. Phosphate-group transfers from ATP drive the breakdown of a molecule of glucose (or another sugar) to two molecules of pyruvate, each with a three-carbon backbone. Two molecules of the coenzyme NAD^+ pick up electrons and hydrogen stripped from reaction intermediates, forming two NADH. Four ATP form during glycolysis, but the net energy yield is two ATP (because two ATP had to be invested up front).

Section 7.3 If the two pyruvates from glycolysis enter a mitochondrion, they will be fully degraded, as part of the second and third stages of aerobic respiration.

The second stage consists of acetyl–CoA formation and the Krebs cycle. Two pyruvates are converted to acetyl–CoA, and two carbon atoms depart in the form of CO_2. In two turns of the Krebs cycle (one for each pyruvate), intermediates are degraded; four more carbons escape (as CO_2). Coenzymes NAD^+ and FAD pick up electrons and hydrogen from intermediates. Two ATP form.

In total, eight NADH, two $FADH_2$, two ATP, and six CO_2 form during the aerobic pathway's second stage.

Section 7.4 The third stage of aerobic respiration proceeds at electron transfer chains and ATP synthases in the inner mitochondrial membrane. Electron transfer chains accept electrons and hydrogen from the NADH and $FADH_2$ that formed in the first two stages. Electron flow through the chains causes H^+ to accumulate in the inner mitochondrial compartment, so H^+ concentration and electric gradients build up across the membrane. H^+ flows down the gradients, through the interior of ATP synthases. This flow drives the attachment of unbound phosphate to ADP, forming many ATP.

Free oxygen picks up the electrons at the end of the transfer chains and combines with H^+, forming water.

Aerobic respiration has a typical net energy yield of thirty-six ATP for each glucose molecule metabolized.

Section 7.5 Fermentation pathways do not require oxygen, and they take place only in the cytoplasm. They use the pyruvate and ATP that formed during the first stage of reactions (glycolysis). The remaining reactions regenerate NAD^+. No more ATP forms. The net energy yield is only the two ATP that formed in glycolysis.

In alcoholic fermentation, the two pyruvates from glycolysis are converted to two acetaldehyde and two CO_2 molecules. When NADH transfers electrons and hydrogen to acetaldehyde, two ethanol molecules form and NAD^+ is regenerated.

In lactate fermentation, NAD^+ is regenerated when electrons and hydrogen are transferred from NADH to the two pyruvate molecules from glycolysis, which forms two lactate molecules as end products.

Slow-twitch and fast-twitch skeletal muscle cells support different levels of activity. Aerobic respiration and lactate fermentation occur in different cells that make up these muscles.

Section 7.6 In the human body, simple sugars from carbohydrates, glycerol and fatty acids from fats, and carbon backbones of amino acids from proteins can enter the aerobic pathway as alternative energy sources.

Section 7.7 Life's diversity, interconnections, and continuity arise from its unity at the molecular level.

Self-Quiz
Answers in Appendix III

1. Glycolysis starts and ends in the _____ .
   a. nucleus   c. plasma membrane
   b. mitochondrion   d. cytoplasm

2. Which of the following molecules does not form during glycolysis?
   a. NADH   b. pyruvate   c. $FADH_2$ d. ATP

3. Aerobic respiration is completed in the _____ .
   a. nucleus   c. plasma membrane
   b. mitochondrion   d. cytoplasm

4. In the third stage of aerobic respiration, _____ is the final acceptor of electrons from glucose.
   a. water   b. hydrogen   c. oxygen   d. NADH

5. Fill in the blanks in the diagram below.

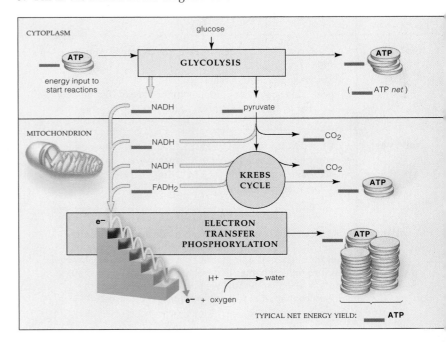

6. In alcoholic fermentation, _____ is the final acceptor of electrons stripped from glucose.
   a. oxygen
   b. pyruvate
   c. acetaldehyde
   d. sulfate

7. Fermentation pathways produce no more ATP beyond the small yield from glycolysis. The remaining reactions _____ .
   a. regenerate FAD
   b. regenerate NAD$^+$
   c. regenerate NAD
   d. regenerate FADH$_2$

8. In certain organisms and under certain conditions, _____ can be used as an energy alternative to glucose.
   a. fatty acids
   b. glycerol
   c. amino acids
   d. all of the above

9. Match the event with its most suitable description.
   ____ glycolysis
   ____ fermentation
   ____ Krebs cycle
   ____ electron transfer phosphorylation

   a. ATP, NADH, FADH$_2$, CO$_2$, and water form
   b. glucose to two pyruvates
   c. NAD$^+$ regenerated, two ATP net
   d. H$^+$ flows through ATP synthases

Media Menu

Student CD-ROM

Impacts, Issues Video
   When Mitochondria Spin Their Wheels
Big Picture Animation
   Energy-releasing pathways and links to photosynthesis
Read-Me-First Animation
   Aerobic respiration
   Glycolysis
   The Krebs cycle
   Third-stage reactions
Other Animations and Interactions
   Comparison of energy-releasing pathways
   Structure and function of a mitochondrion
   Fermentation pathways
   Alternative energy sources

InfoTrac

- My Personal Challenge. *The Exceptional Parent*, August 1998.
- Mitochondria: Cellular Energy Co.—Researchers Strive to Keep the Energy Pipeline Open in the Face of Damaging Cellular Insults. *The Scientist*, June 2002.

Web Sites

- United Mitochondrial Disease Foundation: www.umdf.org
- Friedreich's Ataxia Research Alliance: www.frda.org
- National Organization for Rare Disorders: www.raredisorders.org

How Would You Vote?

Friedreich's ataxia is devastating but relatively rare. In the United States, it affects 1 individual in 50,000. This is good news for most of us, but means that there is relatively little incentive for companies to develop treatments. Who should fund this research? Should we provide tax incentives to companies that work to find cures for rare diseases?

Critical Thinking

1. Living cells of your body absolutely do not use their nucleic acids as alternative energy sources. Suggest why.

2. Suppose you start a body-building program. You are already eating plenty of carbohydrates. Now a qualified nutritionist recommends that you start a protein-rich diet that includes protein supplements. Speculate on how extra dietary proteins will be put to use, and in which tissues.

3. Each year, Canada geese lift off in precise formation from their northern breeding grounds. They head south to spend the winter months in warmer climates. Then they make the return trip in spring. As is the case for other migratory birds, their flight muscle cells are efficient at using fatty acids as an energy source. Remember, the carbon backbone of fatty acids can be cleaved into small fragments that can be converted to acetyl–CoA for entry into the Krebs cycle.

   Suppose a lesser Canada goose from Alaska's Point Barrow has been steadily flapping along for about three thousand kilometers and is approaching Klamath Falls, Oregon. It looks down and notices a rabbit sprinting from a coyote with a taste for rabbit.

   With a stunning burst of speed, the rabbit reaches the safety of its burrow.

   Which energy-releasing pathway predominated in muscle cells in the rabbit's legs? Why was the Canada goose relying on a different pathway for most of its journey? And why wouldn't the pathway of choice in goose flight muscle cells be much good for a rabbit making a mad dash from its enemy?

4. At high altitudes, oxygen levels are low. Mountain climbers risk altitude sickness, which is characterized by shortness of breath, weakness, dizziness, and confusion.

   Curiously, early symptoms of *cyanide poisoning* are similar to altitude sickness. This highly toxic poison binds tightly to a cytochrome, the last molecule in mitochondrial electron transfer chains. When cyanide becomes bound to it, cytochrome can't transfer electrons to the next component of the chain. Explain why cytochrome shutdown might cause the same symptoms as altitude sickness.

5. ATP form in mitochondria. In warm-blooded animals, so does a lot of heat, which can be circulated in ways that help regulate body temperature. Cells of brown adipose tissue (fat) make a protein that disrupts the formation of electron transfer chains in mitochondrial membranes. H$^+$ gradients are affected, so fewer ATP form; electrons in the transfer chains give up more of their energy as heat. Because of this, some researchers are hypothesizing that brown adipose tissue may not function like white adipose tissue, which is an energy reservoir. Brown adipose tissue may function in thermogenesis, or heat production.

   Mitochondria, recall, contain their own DNA, which may have mutated independently in human populations that evolved in the Arctic and in the hot tropics. If that is so, then mitochondrial function may be adapted to climate.

   How do you suppose such a mitochondrial adaptation might affect people living where the temperature range no longer correlates with their ancestral heritage? Would you expect people whose ancestors evolved in the Arctic to be more or less likely to put on a lot of weight than those whose ancestors lived in the tropics? See *Science*, January 9, 2004: 223–226 for more information.

II Principles of Inheritance

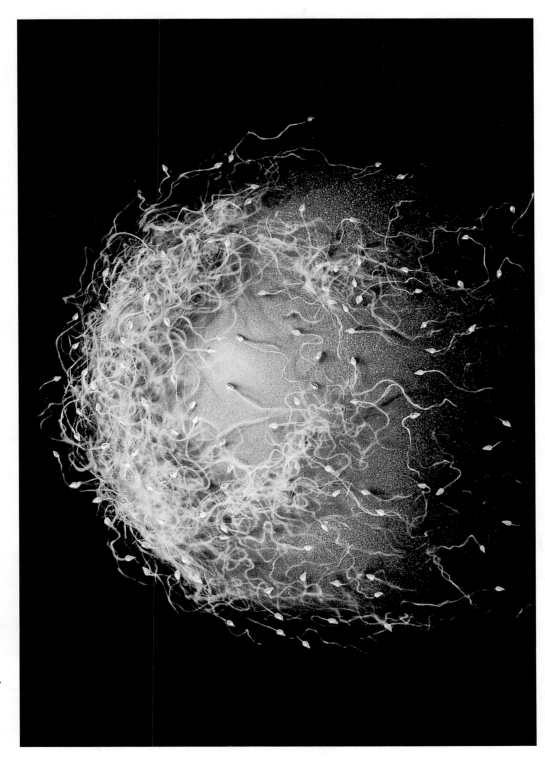

Human sperm, one of which will penetrate this mature egg and so set the stage for the development of a new individual in the image of its parents. This exquisite art is based on a scanning electron micrograph.

IMPACTS, ISSUES *Henrietta's Immortal Cells*

Each human starts out as a fertilized egg. By the time of birth, cell divisions and other processes have given rise to a body of about a trillion cells. Even in the adult, billions of cells are still dividing and replacing their damaged or worn-out predecessors.

In 1951, George and Margaret Gey of Johns Hopkins University were trying to develop a way to keep human cells dividing outside the body. An "immortal" cell lineage

could help researchers study basic life processes as well as cancer and other diseases. Using cells to study cancer would be a far better alternative than experimenting directly on patients and risking their lives.

For almost thirty years, the Geys tried to grow normal and diseased human cells. But they could not stop the cellular descendants from dying within a few weeks.

Mary Kubicek, a lab assistant, tried again and again to establish a self-perpetuating lineage of cultured human cancer cells. She was about to give up, but she prepared one last sample and named them HeLa cells. The code name signified the first two letters of the patient's first and last names.

Those HeLa cells began to divide. Four days later, there were so many cells that the researchers subdivided them into more culture tubes. The cells grew at a phenomenal rate; they divided every twenty-four hours and coated the surface of the tubes within days.

Sadly, cancer cells in the patient were dividing just as often. Six months after she had been diagnosed with cancer, malignant cells had infiltrated tissues all through her body. Two months after that, Henrietta Lacks, a young woman from Baltimore, was dead.

Figure 8.1 Dividing HeLa cells—a legacy of Henrietta Lacks, who was a casualty of cancer. Her cellular contribution to science is still helping others every day.

the big picture

What Divides, and When Eukaryotic cells reproduce by duplicating their chromosomes, getting them into genetically identical parcels by mitosis or meiosis, and dividing the parcels as well as cytoplasm among daughter cells. Prokaryotic cells divide by a different mechanism.

Mitosis Mitosis, a nuclear division mechanism, has four continuous stages: prophase, metaphase, anaphase, and telophase. During these stages, a microtubular spindle moves duplicated chromosomes so that they end up in two genetically identical nuclei.

Although Henrietta passed away, her cells lived on in the Geys' laboratory (Figure 8.1). In time, HeLa cells were shipped to research laboratories all over the world. The Geys used HeLa cells to identify precisely the viral strains that cause polio, which was rampant at the time. Tissue culture techniques developed in their laboratory were used to grow a vaccine. Other scientists used the cells to study mechanisms of cancer, viral growth, the effects of radiation, protein synthesis, and more. Some HeLa cells even traveled into space for experiments on the *Discoverer XVII* satellite. Each year, hundreds of important research projects move forward, thanks to Henrietta's immortal cells.

Henrietta was only thirty-one when runaway cell divisions killed her. Decades later, her legacy continues to help humans everywhere, through her cellular descendants that are still dividing day after day.

Understanding cell division—and, ultimately, how new individuals are put together in the image of their parents—starts with answers to three questions. *First*, what kind of information guides inheritance? *Second*, how is the information copied in a parent cell before being distributed into daughter cells? *Third*, what kinds of mechanisms actually parcel out the information to daughter cells?

We will need more than one chapter to survey the nature of cell reproduction and other mechanisms of inheritance. This chapter introduces the structures and mechanisms that cells use to reproduce.

How Would You Vote?

It is illegal to sell your organs, but you can sell your cells, including eggs, sperm, and blood cells. HeLa cells continue to be sold all over the world by cell culture firms. Should the family of Henrietta Lacks share in the profits? See the Media Menu for details, then vote online.

Cytoplasmic Division   After nuclear division, the cytoplasm divides in a way that typically puts a nucleus in each daughter cell. The cytoplasm of an animal cell is simply pinched in two. In a plant cell, a cross-wall forms in the cytoplasm and divides it.

The Cell Cycle and Cancer   The cell cycle has built-in checkpoints, or mechanisms that monitor and control the timing and rate of cell division. On rare occasions, these surveillance mechanisms fail, and cell division becomes uncontrollable. Tumor formation is the outcome.

8.1 Overview of Cell Division Mechanisms

*The continuity of life depends on **reproduction**. By this process, parents produce a new generation of cells or multicelled individuals like themselves. Cell division is the bridge between generations.*

A dividing cell faces a challenge. Each of its daughter cells must get information encoded in the parental DNA and enough cytoplasm to start up its own operation. DNA "tells" it which proteins to build. Some of the proteins are structural materials; others are enzymes that speed construction of organic compounds. If the cell does not inherit all of the required information, it will not be able to grow or function properly.

In addition, the parent cell's cytoplasm already has enzymes, organelles, and other metabolic machinery. When a daughter cell inherits what looks like a blob of cytoplasm, it really is getting start-up machinery that will keep it running until it can use information in DNA for growing on its own.

MITOSIS, MEIOSIS, AND THE PROKARYOTES

Eukaryotic cells can't just split in two, because a single nucleus holds the DNA. They do split their cytoplasm into daughter cells. But they don't do this until *after* their DNA has been copied, sorted out, and packaged by way of mitosis or meiosis.

Mitosis is a nuclear division mechanism that occurs in *somatic* cells (body cells) of multicelled eukaryotes. It is the basis of increases in body size during growth, replacements of worn-out or dead cells, and tissue repair. Many plants, animals, fungi, and single-celled protists also reproduce asexually, or make copies of themselves, by way of mitosis (Table 8.1).

Meiosis is a different nuclear division mechanism. It functions only in sexual reproduction, and it precedes the formation of gametes (such as sperm and eggs) or spores. In complex animals, gametes form from *germ* cells. As you will see in this chapter and the one that follows, meiosis and mitosis have a lot in common, but the outcomes differ.

What about prokaryotic cells—the archaea and the eubacteria? They reproduce asexually by an entirely different mechanism called prokaryotic fission. We will consider prokaryotic fission later, in Section 19.1.

KEY POINTS ABOUT CHROMOSOME STRUCTURE

Every eukaryotic cell has a characteristic number of DNA molecules, each with many attached proteins. Together, a molecule of DNA and its proteins are one **chromosome**. Chromosomes are duplicated before the cell enters nuclear division. Each chromosome and its copy stay attached to each other as **sister chromatids** until late in the nuclear division process. Think of each chromatid as one arm and leg of a sunbather stretched out on the sand (Figure 8.2).

Early in mitosis or meiosis, a chromosome coils back on itself repeatedly, to a highly condensed form, by interactions between its proteins and DNA. At high magnification, you can see the histone proteins, which look like beads on a string (Figure 8.3*d*). DNA winds

a One unduplicated chromosome

one chromatid
one chromatid
} two sister chromatids

b One chromosome (duplicated)

Figure 8.2 A simple way to visualize a eukaryotic chromosome in the unduplicated state and duplicated state. Eukaryotic cells are duplicated before mitosis or meiosis.

Table 8.1	Cell Division Mechanisms
Mechanisms	Functions
Mitosis, cytoplasmic division	In *all* multicelled eukaryotes, the basis of the following: 1. Increases in body size during growth. 2. Replacement of dead or worn-out cells. 3. Repair of damaged tissues. In single-celled and many multicelled eukaryotes, *also* the basis of asexual reproduction.
Meiosis, cytoplasmic division	In single-celled and multicelled eukaryotes, the basis of sexual reproduction; precedes gamete or spore formation (Chapter 9).
Prokaryotic fission	In bacteria and archaea only, the basis of asexual reproduction (Section 19.1).

Read Me First!

and watch the narrated animation on chromosome structural organization

centromere
(constricted region)

supercoiling of the
coiled loops of DNA

b At times when a chromosome is most condensed, the proteins associated with it interact in ways that package loops of already coiled DNA into "supercoils."

a A duplicated human chromosome in its most condnsed form.

Figure 8.3 (**a**) Scanning electron micrograph of a duplicated human chromosome in its most condensed form. (**b,c**) Interacting proteins hold loops of coiled DNA in the supercoiled array of a cylindrical fiber. (**d,e**) The most basic unit of organization is the nucleosome: part of a DNA molecule looped twice around a core of histones. The transmission electron micrographs correspond to organizational levels (**c**) and (**d**).

fiber

c At a deeper level of structural organization, the chromosomal proteins and DNA are organized as a cylindrical fiber.

twice around histone "spools." A histone–DNA spool is a **nucleosome**, a unit of structural organization.

While each duplicated chromosome is condensing, a pronounced constriction appears in a predictable location along its length. At this constriction, the **centromere**, the chromosome's sister chromatids are attached to each other (Figure 8.3). On its surface, we find kinetochores: docking sites for microtubules that will move the chromosome during nuclear division. The centromere's location is different for each type of chromosome and is one of its defining characteristics.

So what is the point of the structural organization? Tight packaging might help keep chromosomes from getting tangled up while they are moved and sorted out into parcels *during* nuclear division. Also, *between* divisions, nucleosome packaging can be loosened up in selected regions, giving enzymes access to required bits of information in the DNA.

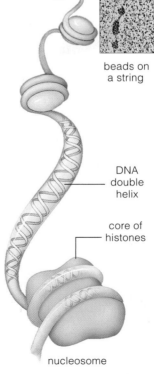

beads on
a string

d Immerse a chromosome in saltwater and it loosens to a beads-on-a-string organization What appears to be a "string" is one DNA molecule. Each "bead" is a nucleosome.

DNA
double
helix

core of
histones

When a cell divides, each daughter cell receives a required number of chromosomes and some cytoplasm. In eukaryotic cells, this involves nuclear and cytoplasmic division.

One nuclear division mechanism, mitosis, is the basis of bodily growth, cell replacements, tissue repair, and often asexual reproduction in eukaryotes.

Meiosis, another nuclear division mechanism, is the basis of sexual reproduction. It precedes gamete or spore formation.

nucleosome

e A nucleosome consists of part of a DNA molecule looped twice around a core of histone proteins.

8.2 Introducing the Cell Cycle

Let's start thinking about cell reproduction as a recurring series of events, a cycle. This isn't the same as a life cycle, which is a sequence of stages through which individuals of a species pass during their lifetime.

A **cell cycle** is a series of events from one cell division to the next (Figure 8.4). It starts when a new daughter cell forms by mitosis and cytoplasmic division; it ends when the cell divides. Mitosis, cytoplasmic division, and then interphase constitute one turn of the cycle.

THE WONDER OF INTERPHASE

During **interphase**, a cell increases in mass, roughly doubles the number of its cytoplasmic components, and duplicates its DNA. For most cells, interphase is the longest portion of the cell cycle. Biologists divide it into three stages:

G1   Interval ("*Gap*") of cell growth and functioning before the onset of DNA replication

S   Time of "*Synthesis*" (DNA replication)

G2   Second interval (*Gap*), after DNA replication when the cell prepares for division

G1, S, and G2 are code names for some events that are just amazing, considering how much DNA is stuffed in a nucleus. For example, if you could stretch out all the DNA molecules from one of your somatic cells in a single line, they would extend past the fingertips of an outstretched arm. A line of all the DNA from one salamander cell would stretch about 540 feet!

The wonder is, enzymes and other proteins in cells *selectively* access, activate, and silence information in all that DNA. They also make base-by-base copies of every DNA molecule before they divide. Most of this cellular work is completed in interphase.

G1, S, and G2 of interphase have distinct patterns of biosynthesis. Most of your cells remain in G1 while they are building proteins, carbohydrates, and lipids. Cells destined to divide enter S, when they copy their DNA and the proteins attached to it. During G2, they make the proteins that will drive mitosis.

Once S begins, DNA replication usually proceeds at about the same rate and is completed before mitosis. The rate holds for all cells of a species, so you might well wonder if the cell cycle has built-in molecular brakes. It does. Apply the brakes that are supposed to work in G1, and the cycle stalls in G1. Lift the brakes, and the cell cycle runs to completion. Said another way, *control mechanisms govern the rate of cell division.*

Imagine a car losing its brakes just as it starts down a steep mountain road. As you will read later in the chapter, that's how cancer starts. Crucial controls over division are lost, and the cell cycle can't stop turning.

The cell cycle lasts about the same amount of time for cells of the same type but varies among different types. For example, all neurons (nerve cells) in your brain remain in G1 of interphase, and usually they will not divide again. By contrast, every second, 2 to 3 million precursors of red blood cells form to replace worn-out ones circulating in your body. Early in the development of a sea urchin embryo, the number of cells doubles every two hours.

Adverse conditions often disrupt the cell cycle. When deprived of a vital nutrient, for example, the free-living cells called amoebas do not leave interphase. Even so, when any cell moves past a certain point in interphase, the cycle normally will continue regardless of the conditions outside because of built-in controls over its duration.

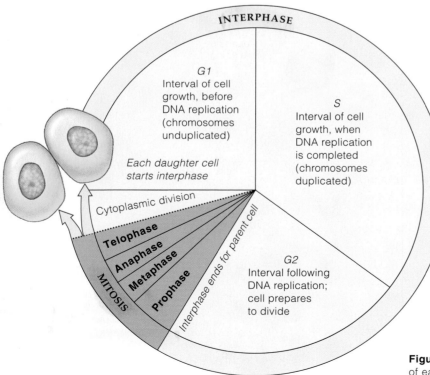

Figure 8.4 Eukaryotic cell cycle, generalized. The length of each interval differs among different cell types.

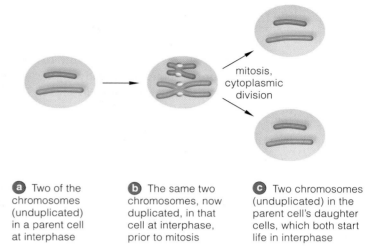

a Two of the chromosomes (unduplicated) in a parent cell at interphase

b The same two chromosomes, now duplicated, in that cell at interphase, prior to mitosis

c Two chromosomes (unduplicated) in the parent cell's daughter cells, which both start life in interphase

Figure 8.5 *Above:* One way to think about how mitosis maintains a parental chromosome number, one generation to the next.

Left: Here's an example. These are twenty-three pairs of metaphase chromosomes from a diploid cell of a human female. The last two are a pair of sex chromosomes. (In human females, such cells have an XX pairing; in males, they have an XY pairing.) When all goes well, each time a somatic cell in this female undergoes mitosis and cytoplasmic division, the daughter cells will always end up with an unduplicated set of these twenty-three pairs of chromosomes.

MITOSIS AND THE CHROMOSOME NUMBER

To know what mitosis does, you have to know that each species has a characteristic **chromosome number**, or the sum of all chromosomes in cells of a given type. Body cells of gorillas and chimpanzees have 48, pea plants have 14, and humans have 46 (Figure 8.5).

Actually, your cells have a **diploid number** (2*n*) of chromosomes; there are two of each type. Those 46 are like volumes of two sets of books numbered from 1 to 23. You have two volumes of, say, chromosome 22—*a pair of them.* Except for one sex chromosome pairing (XY), both have the same length and shape, and carry the same hereditary information about the same traits.

Think of them as two sets of books on how to build a house. Your father gave you one set. Your mother had her own ideas about wiring, plumbing, and so on. She gave you an alternate edition on the same topics, but it says slightly different things about many of them.

With mitosis, a diploid parent cell can produce two diploid daughter cells. This doesn't mean each merely gets forty-six or forty-eight or fourteen chromosomes. If only the total mattered, then one cell might get, say, two pairs of chromosome 22 and no pairs whatsoever of chromosome 9. But neither cell could function like its parent *without two of each type of chromosome.*

Mitosis has four stages: *prophase, metaphase, anaphase,* and *telophase.* All use a **bipolar mitotic spindle**. This dynamic structure is made of microtubules that grow or shrink as tubulin subunits are added or lost from their ends. The spindle forms as microtubules grow toward each other from two poles until they overlap. Some tether the duplicated chromosomes.

One chromatid of each chromosome gets attached to microtubules extending from one spindle pole, its sister gets attached to microtubules from the other pole, then they are dragged apart. A complete set of (now-unduplicated) chromosomes ends up in each half of the cell before the cytoplasm divides. That is how mitosis can maintain a parental chromosome number through turn after turn of the cell cycle (Figure 8.5).

microtubule of bipolar spindle

> *Interphase, mitosis, and cytoplasmic division constitute one turn of the cell cycle.*
>
> *During interphase, a new cell increases its mass, roughly doubles the number of its cytoplasmic components, and duplicates its chromosomes. The cycle ends after the cell undergoes mitosis and then divides its cytoplasm.*

8.3 A Closer Look at Mitosis

Let's focus on a "typical" animal cell to see how mitosis can keep the chromosome number constant, division after division, from one cell generation to the next.

By the time a cell enters **prophase**—the first stage of mitosis—its chromosomes are already duplicated, with sister chromatids joined at the centromere. They are in threadlike form, but now they start to twist and fold. By the end of prophase, they will be condensed into thick, compact, rod-shaped forms (Figure 8.6a–c).

Also before prophase, two barrel-shaped centrioles and two centrosomes started duplicating themselves next to the nucleus. A centriole, recall, gives rise to a cilium or flagellum. In animal cells, it is embedded in a centrosome, which it helps organize. A **centrosome** is a site where microtubules originate. In prophase, the duplicated centrioles move apart—as do the two centrosomes—until they are on opposite sides of the nucleus. Microtubules grow out of each centrosome. *These are the microtubules that form the bipolar spindle.*

As prophase ends, the nuclear envelope starts to break up into tiny flattened vesicles. The microtubules now interact with the chromosomes and one another. Some tether chromosomes at the docking sites called kinetochores. Others tether the chromosome arms. And still others keep on growing from centrosomes until they overlap midway between the two spindle poles. Driven by ATP energy, motor proteins (dyneins and kinesins) produce the force necessary to assemble the spindle, and to bind and move the chromosomes.

Microtubules from one pole tether one chromatid of each chromosome; microtubules from the opposite pole tether the other. They engage in a tug-of-war, growing and shrinking until they are the same length. At that point, **metaphase**, all duplicated chromosomes are aligned midway between the spindle poles. The alignment is crucial for the next stage of mitosis.

At **anaphase**, the kinetochores of sister chromatids detach from each other and take off toward opposite spindle poles. Driven by motor proteins, they move

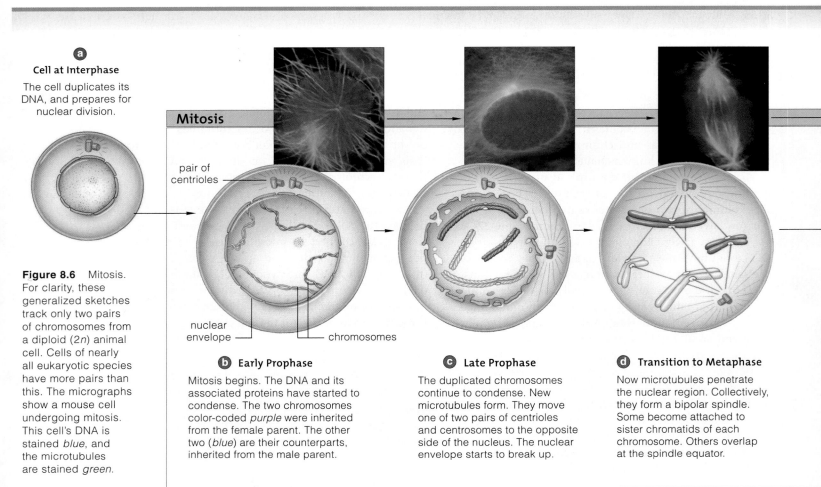

a

Cell at Interphase

The cell duplicates its DNA, and prepares for nuclear division.

Mitosis

pair of centrioles

Figure 8.6 Mitosis. For clarity, these generalized sketches track only two pairs of chromosomes from a diploid (2n) animal cell. Cells of nearly all eukaryotic species have more pairs than this. The micrographs show a mouse cell undergoing mitosis. This cell's DNA is stained *blue*, and the microtubules are stained *green*.

nuclear envelope — chromosomes

b **Early Prophase**

Mitosis begins. The DNA and its associated proteins have started to condense. The two chromosomes color-coded *purple* were inherited from the female parent. The other two (*blue*) are their counterparts, inherited from the male parent.

c **Late Prophase**

The duplicated chromosomes continue to condense. New microtubules form. They move one of two pairs of centrioles and centrosomes to the opposite side of the nucleus. The nuclear envelope starts to break up.

d **Transition to Metaphase**

Now microtubules penetrate the nuclear region. Collectively, they form a bipolar spindle. Some become attached to sister chromatids of each chromosome. Others overlap at the spindle equator.

along microtubules toward the opposite spindle poles, dragging the chromatids with them. At the same time, the microtubules are shortening at both ends even as chromatids remain attached to them. The net effect is that sister chromatids are reeled in to opposite poles.

Also at the same time, microtubules that overlap midway between the spindle poles are ratcheting past one another. Motor proteins drive their interactions, which push the two spindle poles farther apart.

Sister chromatids, recall, are genetically identical. Once they detach from each other at anaphase, each is a separate chromosome in its own right.

Telophase gets under way when one of each type of chromosome reaches a spindle pole. Each half of the cell now contains two genetically identical clusters of chromosomes. Now all the chromosomes decondense. Vesicles derived from the old nuclear envelope fuse and form patches of membrane around each cluster. Patch joins with patch until a new nuclear envelope encloses each cluster. And so two nuclei form (Figure 8.6g). In our example, the parent cell had a diploid number of chromosomes. So does each nucleus.

Once two nuclei have formed, telophase is over— and so is mitosis.

Prior to mitosis, each chromosome in a cell's nucleus is duplicated, so it consists of two sister chromatids.

In prophase, chromosomes condense to rodlike forms, and microtubules form a bipolar spindle. The nuclear envelope breaks up. Some microtubules harness the chromosomes.

At metaphase, all chromosomes are aligned midway between the spindle's poles, at its equator.

At anaphase, microtubules move sister chromatids of each chromosome apart, to opposite spindle poles.

At telophase, a new nuclear envelope forms around each of two clusters of decondensing chromosomes.

Thus mitosis forms two daughter nuclei. Each has the same chromosome number as the parent cell's nucleus.

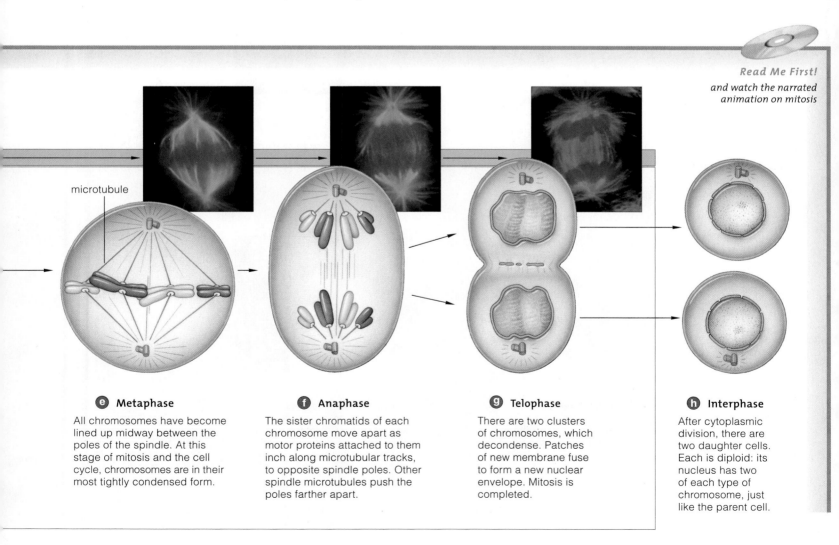

Read Me First!
and watch the narrated animation on mitosis

microtubule

e Metaphase

All chromosomes have become lined up midway between the poles of the spindle. At this stage of mitosis and the cell cycle, chromosomes are in their most tightly condensed form.

f Anaphase

The sister chromatids of each chromosome move apart as motor proteins attached to them inch along microtubular tracks, to opposite spindle poles. Other spindle microtubules push the poles farther apart.

g Telophase

There are two clusters of chromosomes, which decondense. Patches of new membrane fuse to form a new nuclear envelope. Mitosis is completed.

h Interphase

After cytoplasmic division, there are two daughter cells. Each is diploid: its nucleus has two of each type of chromosome, just like the parent cell.

8.4 Division of the Cytoplasm

The cytoplasm usually divides at some time between late anaphase and the end of telophase. The actual mechanism of cytoplasmic division—or, as it is often called, cytokinesis—differs among species.

CLEAVAGE IN ANIMALS

An animal cell divides by **cleavage**, a mechanism that pinches its cytoplasm in two. Typically, the plasma membrane starts to sink inward as a thin indentation about halfway between the cell's two poles (Figure 8.7a). This cleavage furrow is the first visible sign that

the cytoplasm in an animal cell is dividing. The furrow advances until it extends all the way around the cell. As it does, it deepens along a plane that corresponds to the equator of the former microtubular spindle.

How does this happen? In the cytoplasm just under the plasma membrane, microfilaments organized in a thin, ringlike band generate the contractile force for the cut (Figure 8.8). These cytoskeletal elements are attached to the plasma membrane. They slide past one another, as outlined in Section 4.10. As they do, they drag the plasma membrane deeper and deeper inward until the cytoplasm is partitioned. Each of the two

Read Me First!
and watch the narrated animation on cytoplasmic division

  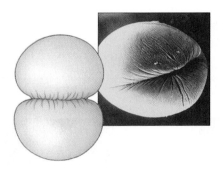

1 Mitosis is over, and the spindle is disassembling.

2 At the former spindle equator, a ring of microfilaments attached to the plasma membrane contracts.

3 As the microfilament ring shrinks in diameter, it pulls the cell surface inward.

4 Contractions continue; the cell is pinched in two.

a Animal cell division

cell plate forming

1 As mitosis ends, vesicles cluster at the spindle equator. They contain materials for a new primary cell wall.

2 Vesicle membranes fuse. The wall material is sandwiched between two new membranes that lengthen along the plane of a newly forming cell plate.

3 Cellulose is deposited inside the sandwich. In time, these deposits will form two cell walls. Others will form the middle lamella between the walls and cement them together.

4 A cell plate grows at its margins until it fuses with the parent cell plasma membrane. The primary wall of growing plant cells is still thin. New material is deposited on it.

b Plant cell division

Figure 8.7 Cytoplasmic division of an animal cell (**a**) and a plant cell (**b**).

ring of microfilaments midway between the two spindle poles, in the same plane as the spindle equator

Figure 8.8 Cleavage. Inside this animal cell, a ring of microfilaments is pinching the cytoplasm in two.

daughter cells that forms this way ends up with a nucleus, some cytoplasm, and plasma membrane.

CELL PLATE FORMATION IN PLANTS

Plant cells cannot divide the same way your cells do, because most of them have a cell wall. That wall prevents their cytoplasm from simply pinching in two.

Instead, cytoplasmic division of plant cells involves **cell plate formation**, as shown in Figure 8.7b. By this mechanism, tiny vesicles packed with wall-building materials fuse with one another and with remnants of the microtubular spindle. Together, deposits of the materials form a disklike structure called a cell plate. Deposits of cellulose accumulate at the plate. In time, they thicken enough to form a cross-wall through the cell. New plasma membrane extends across both sides of it. This wall grows until it bridges the cytoplasm and divides the parent cell in two.

APPRECIATE THE PROCESS!

Take a moment to look closely at your hands. Visualize the cells making up your palms, thumbs, and fingers. Now imagine the mitotic divisions that produced all of the cell generations that preceded them while you were developing, early on, inside your mother (Figure 8.9). And be grateful for the astonishing precision of mechanisms that led to their formation at prescribed times, in prescribed numbers, for the alternatives can be terrible indeed.

Why? Good health and survival itself depend on the proper timing and completion of cell cycle events. Some genetic disorders arise as a result of mistakes that happened during the duplication or distribution of even one chromosome. Unchecked cell divisions often destroy surrounding tissues and, ultimately, the

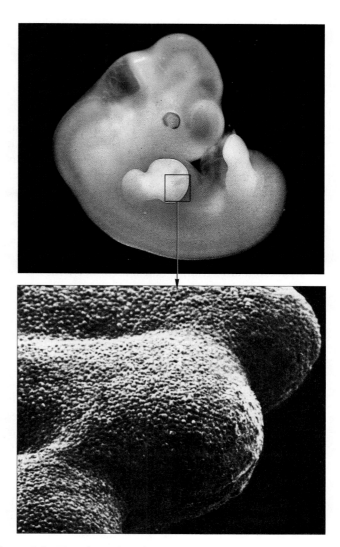

Figure 8.9 Transformation of a paddlelike structure into a human hand through mitosis, cytoplasmic divisions, and other processes of embryonic development. The scanning electron micrograph shows individual cells.

individual. Such losses can start in body cells. They can start in the germ cells that give rise to sperm and eggs, although rarely. The last section of this chapter can give you a sense of the consequences.

After mitosis, a separate mechanism cuts the cytoplasm into two daughter cells, each with a daughter nucleus.

Cleavage is a form of cytoplasmic division in animal cells. Microfilaments banded around a cell's midsection slide past one another in a way that pinches the cytoplasm in two.

Cytoplasmic division in plants often involves the formation of a cross-wall between the new plasma membranes of adjoining daughter cells.

8.5 When Control Is Lost

Growth and reproduction depend on controls over cell division. On rare occasions, something goes wrong in a somatic cell or germ cell. Cancer may be the outcome.

THE CELL CYCLE REVISITED

Millions of cells in your skin, bone marrow, gut lining, liver, and elsewhere divide and replace their worn-out, dead, and dying predecessors every second. They don't divide willy-nilly; many cellular mechanisms control cell growth, DNA replication, and division. They also control when the division machinery is put to rest.

What happens when something goes wrong? For example, if sister chromatids do not separate as they should during mitosis, one daughter cell may end up with too many chromosomes, the other with too few. Chromosomal DNA can be attacked by free radicals or peroxides, two metabolic by-products. It can be damaged by cosmic radiation, which bombards us all the time. Problems are frequent, inevitable, and must be corrected quickly.

The cell cycle has built-in checkpoints that keep errors from getting out of hand. Certain proteins—products of checkpoint genes—monitor whether the DNA gets fully replicated, whether it gets damaged, even whether nutrient concentrations are sufficient to support cell growth. The surveillance helps cells identify and correct problems.

Some checkpoint proteins make the cell cycle advance; their absence arrests it. The ones called *growth factors* invite transcription of genes that help the body grow. For instance, epidermal growth factor activates an enzyme by binding to cells in epithelial tissues; it is a signal to start mitotic cell divisions.

Other proteins inhibit cell cycle changes. Several checkpoint gene products put the brakes on mitosis when chromosomal DNA gets damaged (Figure 8.10). Some of the *kinases*, enzymes that phosphorylate other molecules, act as checkpoint gene products. When DNA is broken or incomplete, they activate other proteins in a cascade of signaling events that ultimately stop the cell cycle or induce cell death.

CHECKPOINT FAILURE AND TUMORS

Sometimes a checkpoint gene mutates so that its protein product no longer functions properly. When all checkpoint mechanisms for a particular process fail, the cell loses control over its replication cycle. Figures 8.11 through 8.13 show a few of the outcomes.

In some cases it gets stuck in mitosis and divides over and over again, with no interphase. In other cases, damaged chromosomes are replicated or cells don't die as they are supposed to, because signals calling for cell death are disabled. A growing mass of a cell's defective descendants may invade other tissues in the body, as a tumor.

In most tumor cells, at least one checkpoint protein is missing. That is why checkpoint gene products that *inhibit* mitosis are called tumor suppressors. Checkpoint genes encoding proteins that *stimulate* mitosis are called oncogenes. Mutations that affect oncogene products or the rate at which they form help transform a normal cell into a tumor cell. Mutant checkpoint genes are linked with increased risk of tumors, and sometimes they run in families.

Moles and other tumors are **neoplasms**—abnormal masses of cells that lost controls over how they grow

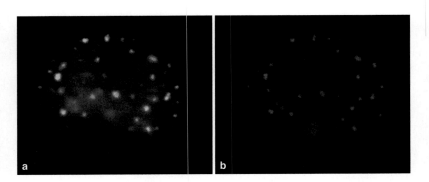

Figure 8.10 Protein products of checkpoint genes in action. DNA in the nucleus of this cell has been damaged by ionizing radiation. (**a**) *Green* spots pinpoint the location of *53BP1*, and (**b**) *red* spots pinpoint the location of *BRCA1*. Both proteins have clustered around the same chromosome breaks in a single cell nucleus. The integrated action of these proteins and others can arrest mitosis until the DNA breaks are fixed.

Figure 8.11 Scanning electron micrograph of a cervical cancer cell, the kind that killed Henrietta Lacks.

and divide. Ordinary skin moles and other *benign*, or noncancerous, neoplasms grow very slowly, and their cells retain the surface recognition proteins that are supposed to keep them in a home tissue (Figure 8.12). Unless a benign neoplasm grows too large or becomes irritating, it poses no threat to the body.

CHARACTERISTICS OF CANCER

Cancers are abnormally growing and dividing cells of a *malignant* neoplasm. They disrupt surrounding tissues, both physically and metabolically. Cancer cells are grossly disfigured. They can break loose from their home tissues. They can slip into and out of blood vessels and lymph vessels, and invade other tissues where they do not belong (Figure 8.12).

All cancer cells display four characteristics. *First*, they grow and divide abnormally. The controls on overcrowding in tissues are lost and cell populations reach abnormally high densities. The number of tiny blood vessels that service the growing cell mass also increases abnormally.

Second, the cytoplasm and plasma membrane of cancer cells become grossly altered. The membrane becomes leaky and has abnormal or lost proteins. The cytoskeleton shrinks, becomes disorganized, or both. Enzyme action shifts, as in an amplified reliance on ATP formation by glycolysis.

Third, cancer cells have a weakened capacity for adhesion. Recognition proteins are lost or altered, so they can't stay anchored in proper tissues. They break away and may establish growing colonies in distant tissues. *Metastasis* is the name for this process of abnormal cell migration and tissue invasion.

Fourth, cancer cells usually have lethal effects. Unless they are eradicated by surgery, chemotherapy, or other procedures, their uncontrollable divisions put an individual on a painful road to death.

Each year in the developed countries alone, cancers cause 15 to 20 percent of all deaths. And cancer is not just a human problem. Cancers are known to occur in most of the animal species studied to date.

Cancer is a multistep process. Researchers have already identified many of the mutant genes that contribute to it. They also are working to identify drugs that specifically target and destroy cancer cells or stop them from dividing.

HeLa cells, for instance, were used in early tests of taxol, an anticancer drug that stops spindles from disassembling. With this kind of research, we may one day have drugs that can put the brakes on cancer cells. We return to this topic in later chapters.

Read Me First!
and watch the narrated animation on cancer

benign tumor

malignant tumor

❶ Cancer cells slip out of their home tissue.

❷ The metastasizing cells become attached to the wall of a blood or lymph vessel. They secrete digestive enzymes onto it. Then they cross the wall at the breach.

❸ Cancer cells creep or tumble along inside blood vessels, then leave the bloodstream the same way they got in. They start new tumors in new tissues.

Figure 8.12 Comparison of benign and malignant tumors. Benign tumors typically are slow-growing and stay put in their home tissue. Cells of a malignant tumor can migrate abnormally through the body and establish colonies even in distant tissues.

Figure 8.13 Skin cancers. (**a**) A *basal cell carcinoma* is the most common type. This slow-growing, raised lump is typically uncolored, reddish-brown, or black.

(**b**) The second most common form is *squamous cell carcinoma*. This pink growth, firm to the touch, grows fast under the surface of skin exposed to the sun.

(**c**) *Malignant melanoma* spreads fastest. Cells form dark, encrusted lumps. They may itch like an insect bite or bleed easily.

Summary

Section 8.1 The continuity of life depends on reproduction. By this process, parents produce a new generation of individuals like themselves. Cell division is the bridge between generations. When a cell divides, its daughters each receive a required number of DNA molecules and some cytoplasm.

Mitosis and meiosis occur only in eukaryotic cells. These nuclear division mechanisms sort out a parent cell's chromosomes into daughter nuclei. A separate mechanism divides the cytoplasm. Prokaryotic cells divide by a different mechanism.

Mitosis is the basis of multicellular growth, cell replacements, and tissue repair. Many eukaryotic organisms also reproduce *asexually* by mitosis.

Meiosis, the basis of sexual reproduction, precedes the formation of gametes or spores.

A chromosome is a molecule of DNA and associated proteins. When duplicated, it consists of two sister chromatids, both attached to its centromere region by kinetochores. These are docking sites for microtubules.

Section 8.2 Each cell cycle starts when a new cell forms. It proceeds through interphase and ends when the cell reproduces by nuclear and cytoplasmic division. A cell carries out its functions in interphase. Before it divides, it increases in mass, roughly doubles the number of its cytoplasmic components, then duplicates each of its chromosomes.

Section 8.3 The sum of all chromosomes in cells of a given type is the chromosome number. Human body cells have a diploid chromosome number of 46, or two copies of 23 types of chromosome. Mitosis maintains the chromosome number, one generation to the next.

Mitosis has four continuous stages:

a. Prophase. The duplicated, threadlike chromosomes start to condense. With the help of motor proteins, new microtubules start forming a bipolar mitotic spindle. The nuclear envelope starts to break apart into tiny vesicles.

Some microtubules growing from one spindle pole tether one chromatid of each chromosome; others that are growing from the opposite pole tether its sister chromatid. Still other microtubules extending from both poles grow until they overlap at the midpoint of the newly forming spindle.

b. Metaphase. At metaphase, all chromosomes have become aligned at the spindle's midpoint.

c. Anaphase. Kinetochores detach from chromosomes, dragging the chromatids with them along microtubules, which are shortening at both ends. The microtubules that overlap ratchet past each other, pushing the spindle poles farther apart. Different motor proteins drive the movements. One of each type of parental chromosome ends up clustered together at each spindle pole.

d. Telophase. Chromosomes decondense to threadlike form. A new nuclear envelope forms around each cluster. Both nuclei have the parental chromosome number.

Fill in the blanks of the diagram below to check your understanding of the four stages of mitosis.

Section 8.4 Cytoplasmic division mechanisms differ. Animal cells undergo cleavage. A microfilament ring under the plasma membrane contracts, pinching the cytoplasm in two. In plant cells, a cross-wall forms in the cytoplasm and divides it.

Section 8.5 Checkpoint gene products are part of mechanisms that control the cell cycle. The mechanisms can stimulate or arrest the cell cycle, or even prompt cell death. Mutant checkpoint genes cause tumors to form by disrupting normal cell cycle controls.

Cancer is a multistep process involving altered cells that grow and divide abnormally. Such cells have a weakened ability to stick to one another in tissues, and sometimes they migrate to new tissues.

Self-Quiz

Answers in Appendix III

1. Mitosis and cytoplasmic division function in _____ .
   a. asexual reproduction of single-celled eukaryotes
   b. growth, tissue repair, often asexual reproduction
   c. gamete formation in prokaryotes
   d. both a and b

2. A duplicated chromosome has _____ chromatid(s).
   a. one   b. two   c. three   d. four

3. The basic unit that structurally organizes a eukaryotic chromosome is the _____ .
   a. supercoil   c. nucleosome
   b. bipolar mitotic spindle   d. microfilament

4. The chromosome number is _____ .
   a. the sum of all chromosomes in cells of a given type
   b. an identifiable feature of each species
   c. maintained by mitosis
   d. all of the above

5. A somatic cell having two of each type of chromosome has a(n) _____ chromosome number.
   a. diploid   b. haploid   c. tetraploid   d. abnormal

6. Interphase is the part of the cell cycle when _____ .
   a. a cell ceases to function
   b. a germ cell forms its spindle apparatus
   c. a cell grows and duplicates its DNA
   d. mitosis proceeds

7. After mitosis, the chromosome number of a daughter cell is _____ the parent cell's.
   a. the same as   c. rearranged compared to
   b. one-half   d. doubled compared to

8. Only _____ is not a stage of mitosis.
   a. prophase b. interphase c. metaphase d. anaphase

9. Match each stage with the events listed.
   _____ metaphase   a. sister chromatids move apart
   _____ prophase   b. chromosomes start to condense
   _____ telophase   c. daughter nuclei form
   _____ anaphase   d. all duplicated chromosomes are
                         aligned at the spindle equator

Critical Thinking

1. Figure 8.14 shows metaphase chromosomes. Name their levels of structural organization, starting with DNA molecules and histones.

2. Pacific yews (*Taxus brevifolius*) are among the slowest growing trees, which makes them vulnerable to extinction. People started stripping their bark and killing them when they heard that *taxol,* a chemical extracted from the bark, may work against breast and ovarian cancer. It takes bark from about six trees to treat one patient. Do some research and find out why taxol has potential as an anticancer drug and what has been done to protect the trees.

3. X-rays emitted from some radioisotopes damage DNA, especially in cells undergoing DNA replication. Humans exposed to high levels of x-rays face *radiation poisoning.* Hair loss and a damaged gut lining are early symptoms. Speculate why. Also speculate on why radiation exposure is used as a therapy to treat some cancers.

4. Suppose you have a way to measure the amount of DNA in a single cell during the cell cycle. You first measure the amount at the G1 phase. At what points during the remainder of the cycle would you predict changes in the amount of DNA per cell?

5. The cervix is part of the uterus, a chamber in which embryos develop. The *Pap smear* is a screening procedure that can detect *cervical cancer* in its earliest stages.

Treatments range from freezing precancerous cells or killing them with a laser beam to removal of the uterus (a hysterectomy). The treatments are more than 90 percent effective when this cancer is detected early. Survival chances plummet to less than 9 percent after it spreads.

Most cervical cancers develop slowly. Unsafe sex increases the risk. A key risk factor is infection by human

Figure 8.14 Human chromosomes at metaphase, each in the duplicated state.

papillomaviruses (HPV), which cause genital warts. Viral genes coding for the tumor-inducing proteins get inserted into the DNA of cervical cells. In one study, 91 percent of patients with cervical cancer had been infected with HPV.

Not all women request Pap smears. Many wrongly believe the procedure is costly. Many don't recognize the importance of abstinence or "safe" sex. Others simply don't want to think about whether they have cancer. Knowing what you've learned so far about the cell cycle and cancer, what would you say to a woman who falls into one or more of these groups?

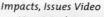

Media Menu

Student CD-ROM

Impacts, Issues Video
   Henrietta's Immortal Cells
Big Picture Animation
   Normal cell division and cancer
Read-Me-First Animation
   Chromosome structural organization
   Mitosis
   Cytoplasmic division
   Cancer
Other Animations and Interactions
   The cell cycle

InfoTrac

   • Cell Cycle Circuits Mapped. *Applied Genetics News,* October 2001.
   • HIV Protein Stops Cell Division. *Virus Weekly,* April 2002.
   • Familiar Proteins Play Unfamiliar Role In Cell Division. *Stem Cell Week,* January 2002.
   • How You Can Lower Your Cancer Risk. *Harvard Health Letter,* August 2002.

Web Sites

   • Talking Genetic Glossary: www.genome.gov/glossary.cfm
   • Mitosis World: www.bio.unc.edu/faculty/salmon/lab/mitosis
   • Animated Cell Cycle: www.cellsalive.com/cell_cycle.htm
   • The Cytokinetic Mafia Home Page: www.bio.unc.edu/faculty/salmon/lab/mafia/

How Would You Vote?

When she died, Henrietta Lacks left behind a husband and five children. The scientists who propagated the HeLa cell line never told her or her family how they were using her tissues. Today, HeLa cells are sold by cell culture firms around the world. Should her survivors get a share of the profits?

IMPACTS, ISSUES *Why Sex?*

Women and men, does and bucks, geese and ganders. Most of us take it for granted that it takes two to make offspring, and among eukaryotes it usually does—at least some of the time. Sexual reproduction combines DNA from individuals of two mating types. It started many hundreds of millions of years ago among tiny, single-celled eukaryotes, although no one knows how. An unsolved puzzle is why it happened at all. Asexual reproduction, whereby an individual makes offspring that are copies of itself, is far more efficient.

Protists and fungi routinely reproduce by mitosis. Plants and many invertebrates, including corals, sea stars, and flatworms, can form new individuals from parts that bud or break off. But almost all of these species also reproduce sexually.

For instance, some single-celled algae reproduce asexually again and again, by way of mitosis, and form huge populations exactly like themselves. Only when nitrogen is scarce do sexual cells of two types form. The fusion of two cells, one of each type, produces new individuals. The result is offspring that are not exact copies of one another or their parents.

Consider also the plant-sucking insects called aphids. In the summer nearly all are females. Each matures in less than a week and can produce as many as five new females a day from her unfertilized eggs (Figure 9.1). This process, called parthenogenesis, allows population sizes to soar rapidly. Only as autumn approaches do male aphids develop. Even then, females that manage to survive over the winter can do without the opposite

Figure 9.1 Sexual reproduction moments. (**a**) Mealybugs mating. (**b**) Poppy plant being helped by a beetle, which makes pollen deliveries for it. (**c**) Aphid giving birth. Like females of many other sexually reproducing species, this one also reproduces asexually, all by itself.

the big picture

Sexual Reproduction Sexual reproduction requires meiosis, gamete formation, and fertilization. Meiosis is a nuclear division mechanism that halves the parental chromosome number for forthcoming gametes.

Meiosis Meiosis consists of two nuclear divisions, not one. There is no interphase between divisions. It results in four haploid nuclei. When cytoplasmic division follows, each of the resulting haploid cells may function as a gamete or spore.

sex. The next spring they begin another round of female production all by themselves.

There are even a few all-female species of fishes, reptiles, and birds. No mammal is parthenogenic in nature. In 2004, however, researchers at the University of Tokyo in Japan fused two mouse eggs in a test tube to produce an embryo with all-female DNA. The embryo developed into Kaguya—the world's first fatherless mammal. She grew to adulthood, mated with a male mouse, and produced offspring of her own, as shown in the filmstrip at right.

Does this mean males could soon be unneccessary? Hardly. It wasn't easy to produce a mouse that has all-female DNA. The researchers had to turn off genes in one egg. Even then, it took more than 600 attempts before they succeeded in producing two viable embryos. Besides, the prevalance of sexually reproducing species suggests that a division into two sexes must offer selective advantages.

With this chapter, we turn to the kinds of cells that serve as the bridge between generations of organisms. Three interconnected events—meiosis, the formation of gametes, and fertilization—are the hallmarks of sexual reproduction. As you will see in many chapters throughout the book, these events occur in the life cycle of almost all eukaryotic species. Through the production of offspring with new and unique traits, they have contributed immensely to the diversity of life.

How Would You Vote?

Japanese researchers have successfully created a "fatherless" mouse that contains the genetic material from the eggs of two females. The mouse is healthy and fully fertile. Do you think researchers should be allowed to try the same process with human eggs? See the Media Menu for details, then vote online.

Gene Shufflings With Sex During meiosis, crossing over and the random alignment of chromosomes at metaphase puts different mixes of maternal and paternal genes in gametes. More mixing occurs at fertilization. Such events introduce variation in traits among offspring.

Meiosis in Life Cycles Plant and animal life cycles typically include fertilization, meiosis, and gamete formation. In plant life cycles, meiosis is followed by the formation of spores which, under favorable conditions, germinate and give rise to gamete-producing bodies.

9.1 An Evolutionary View

Asexual reproduction produces genetically identical copies of a parent. Sexual reproduction introduces variation in the details of traits among offspring.

When an orchid or aphid reproduces all by itself, what sort of offspring does it get? By the process of **asexual reproduction**, one alone produces offspring, and each offspring inherits the same number and kinds of genes as its parent. **Genes** are stretches of chromosomes—that is, of DNA molecules. The genes for each species contain all the heritable information necessary to make new individuals. Rare mutations aside, then, asexually produced individuals can only be *clones*, or genetically identical copies of the parent.

Inheritance gets far more interesting with **sexual reproduction**. The process involves meiosis, formation of gametes, and fertilization (union of the nuclei of two gametes). In most sexual reproducers, such as humans, the first cell of a new individual contains *pairs of genes* on pairs of chromosomes. Usually, one of each pair is maternal and the other paternal in origin (Figure 9.2).

If information in every pair of genes were identical down to the last detail, sexual reproduction would also produce clones. Just imagine—you, every person you know, the entire human population might be a clone, with everybody looking alike. But the two genes of a pair might *not* be identical. Why not? The molecular structure of a gene can change; it can mutate. So two genes that happen to be paired in a person's cells may "say" slightly different things about a trait. Each unique molecular form of the same gene is called an **allele**.

Such tiny differences affect thousands of traits. For example, whether your chin has a dimple depends on which pair of alleles you inherited at one chromosome location. One kind of allele at that location says "put a dimple in the chin." Another kind says "no dimple." This leads to one reason why the individuals of sexually reproducing species don't all look alike. *By sexual reproduction, offspring inherit new combinations of alleles, which lead to variations in the details of their traits.*

This chapter gets into the cellular basis of sexual reproduction. More importantly, it starts you thinking about far-reaching effects of gene shufflings at certain stages of the process. The process introduces variations in traits among offspring that are typically acted upon by agents of natural selection. Thus, *variation in traits is a foundation for evolutionary change.*

Figure 9.2
A maternal and a paternal chromosome. Any gene on one might be slightly different structurally than the same gene on the other.

Sexual reproduction introduces variation in traits by bestowing novel combinations of alleles on offspring.

9.2 Overview of Meiosis

Meiosis is a nuclear division process that divides the parental chromosome number by half in specialized reproductive cells. It is central to sexual reproduction.

THINK "HOMOLOGUES"

Think back to the preceding chapter and its focus on mitotic cell division. Unlike mitosis, **meiosis** partitions chromosomes into parcels not once but *twice* prior to cytoplasmic division. Unlike mitosis, it is the first step leading to the formation of gametes. Male and female gametes, such as sperm and eggs, fuse to form a new individual. In most multicelled eukaryotes, cells that form in specialized reproductive structures or organs give rise to gametes. Figure 9.3 shows examples of where the cellular antecedents of gametes originate.

As you know, the **chromosome number** is the sum total of chromosomes in cells of a given type. If a cell has a **diploid number** (2n), it has a *pair* of each type of chromosome, often from two parents. Except for a pairing of nonidentical sex chromosomes, each pair has the same length, shape, and assortment of genes, and they line up with each other at meiosis. We call them **homologous chromosomes** (*hom*– means alike).

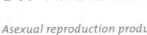

anther (where cells that give rise to male gametes originate)

ovules, inside an ovary (where cells that give rise to female gametes originate)

a Flowering plant

testis (where sperm originate)

ovary (where eggs develop)

b Human male   **c** Human female

Figure 9.3   Examples of reproductive organs, where cells that give rise to gametes originate.

Body cells of humans have 23 + 23 homologous chromosomes (Figure 9.4). So do the germ cells that give rise to human gametes. After a germ cell finishes meiosis, 23 chromosomes—one of each type—will end up in those gametes. Meiosis halves the chromosome number, so the gametes have a **haploid number** (n).

TWO DIVISIONS, NOT ONE

Meiosis is like mitosis in some ways, but the result is different. As in mitosis, a germ cell duplicates its DNA in interphase. The two DNA molecules and associated proteins stay attached at the centromere, the notably constricted region along their length. For as long as they remain attached, we call them **sister chromatids**:

one chromosome in the duplicated state

As in mitosis, the microtubules of a spindle apparatus move the chromosomes in prescribed directions.

With meiosis, however, *chromosomes go through two consecutive divisions that end with the formation of four haploid nuclei*. There is no interphase between divisions, which we call meiosis I and meiosis II:

interphase (*DNA replication before meiosis I*)	MEIOSIS I	no interphase (*no DNA replication before meiosis II*)	MEIOSIS II
	PROPHASE I		PROPHASE II
	METAPHASE I		METAPHASE II
	ANAPHASE I		ANAPHASE II
	TELOPHASE I		TELOPHASE II

In meoisis I, each duplicated chromosome aligns with its partner, *homologue to homologue*. After the two chromosomes of every pair have lined up with each other, they are moved apart:

each homologue in the cell pairs with its partner

then the partners separate

The cytoplasm typically starts to divide at some point after each homologue detaches from its partner. The two daughter cells formed this way are haploid,

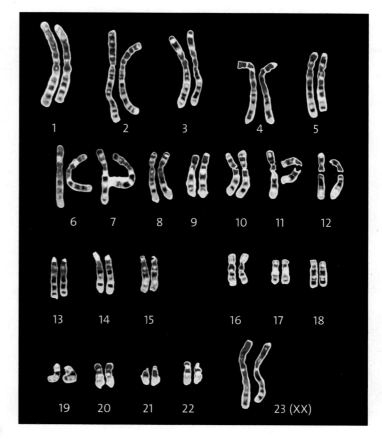

Figure 9.4 Another look at the twenty-three pairs of homologous human chromosomes. This example is from a human female, with two X chromosomes. Human males have a different pairing of sex chromosomes (XY). As in Figure 9.2, these chromosomees have been labeled with several fluorescent markers.

with *one* of each type of chromosome. Don't forget, these chromosomes are still in the duplicated state.

Next, during meiosis II, *the two sister chromatids of each chromosome are separated from each other*:

two chromosomes (unduplicated)

one chromosome (duplicated)

Each chromatid is now a separate chromosome. Next, four nuclei form, and the cytoplasm typically divides once more. The final outcome is four haploid cells. Figure 9.5, on the next two pages, offers a closer look at key events of meiosis and their consequences.

Meiosis, a nuclear division mechanism, reduces a parental cell's chromosome number by half—to a haploid number (n).

9.3 Visual Tour of Meiosis

Meiosis I

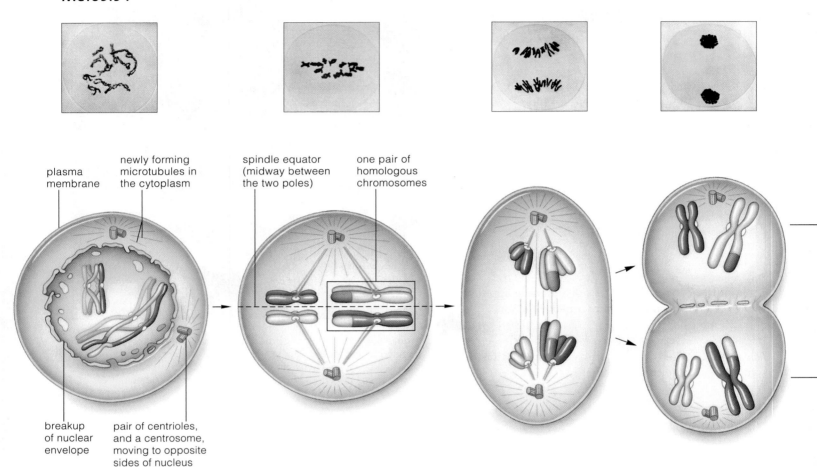

plasma membrane

newly forming microtubules in the cytoplasm

spindle equator (midway between the two poles)

one pair of homologous chromosomes

breakup of nuclear envelope

pair of centrioles, and a centrosome, moving to opposite sides of nucleus

a Prophase I

As prophase I begins, chromosomes become visible as threadlike forms. Each pairs with its homologue and usually swaps segments with it, as indicated by the breaks in color in the large chromosomes. Microtubules are forming a bipolar spindle (Section 8.3). If two pairs of centrioles are present, one pair is moved to the opposite side of the nuclear envelope, which is starting to break up.

b Metaphase I

Microtubules from one spindle pole have tethered one of each type of chromosome; microtubules from the other pole have tethered its homologue. By metaphase I, a tug-of-war between the two sets of microtubules has aligned all chromosomes midway between the poles.

c Anaphase I

Microtubules attached to each chromosome shorten and move it toward a spindle pole. Other microtubules, which extend from the poles and overlap at the spindle equator, ratchet past each other and push the two poles farther apart. Motor proteins drive the ratcheting.

d Telophase I

One of each type of chromosome has now arrived at the spindle poles. For most species, the cytoplasm divides at some point, forming two haploid cells. All chromosomes are still duplicated.

Figure 9.5 Sketches of meiosis in a generalized animal cell. This is a nuclear division mechanism. It reduces the parental chromosome number in immature reproductive cells by half, to the haploid number, for forthcoming gametes. To keep things simple, we track only two pairs of homologous chromosomes. Maternal chromosomes are shaded *purple* and paternal chromosomes *blue*.

Of the four haploid cells that form by way of meiosis and cytoplasmic divisions, one or all may develop into gametes and function in sexual reproduction. In plants, cells that form by way of meiosis may develop into spores, which take part in a stage of the life cycle that precedes gamete formation.

Read Me First!
and watch the narrated animation on meiosis

Meiosis II

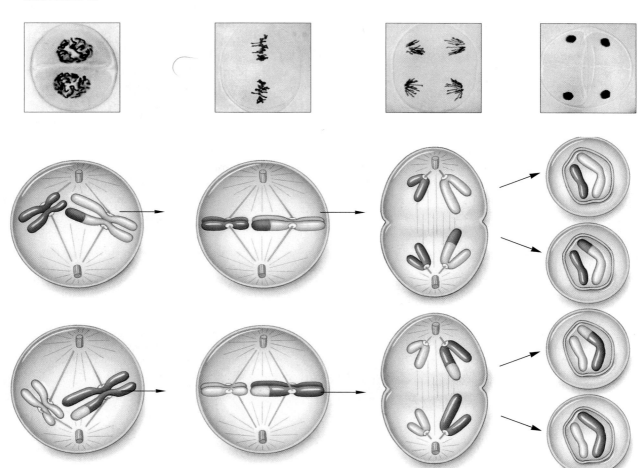

There is no DNA replication between the two nuclear divisions.

e Prophase II

A new bipolar spindle forms in each haploid cell. Microtubules have moved one member of the pair of centrioles to the opposite end of each cell. One chromatid of each chromosome becomes tethered to one spindle pole, and its sister chromatid becomes tethered to the opposite pole.

f Metaphase II

Microtubules from both spindle poles have assembled and disassembled in a tug-of-war that ended at metaphase II, when all chromosomes are positioned midway between the poles.

g Anaphase II

The attachment between sister chromatids of each chromosome breaks. Each is now a separate chromosome but is still tethered to microtubules, which move it toward a spindle pole. Other microtubules push the poles apart. A parcel of unduplicated chromosomes ends up near each pole. One of each type of chromosome is present in each parcel.

h Telophase II

In telophase II, four nuclei form as a new nuclear envelope encloses each cluster of chromosomes. After cytoplasmic division, each of the resulting daughter cells has a haploid (*n*) number of chromosomes.

9.4 How Meiosis Puts Variation in Traits

As Sections 9.2 and 9.3 make clear, the overriding function of meiosis is the reduction of a parental chromosome number by half. However, two other events that occur during meiosis have evolutionary consequences.

The preceding section mentioned in passing that pairs of homologous chromosomes swap parts of themselves during prophase I. It also showed how homologous chromosomes become aligned with their partner at metaphase I. Let's take a look at these two events, because they contribute enormously to variation in the traits of sexually reproducing species. They introduce novel combinations of alleles into the gametes that form after meiosis. Those combinations—and the way they are further mixed together at fertilization—are the start of a new generation of individuals that differ in the details of their shared traits.

CROSSING OVER IN PROPHASE I

Prophase I of meiosis is a time of much gene shuffling. Reflect on Figure 9.6a, which shows two chromosomes condensed to threadlike form. All chromosomes in a germ cell condense this way. When they do, each is drawn close to its homologue. Molecular interactions stitch homologues together point by point along their length with little space between them. The intimate, parallel orientation favors **crossing over**, a molecular interaction between a chromatid of one chromosome and a chromatid of its homologous partner. The two *nonsister* chromatids break at the same places along their length, then the two exchange corresponding segments; they swap genes.

Gene swapping would be pointless if each type of gene never varied. But remember, a gene can come in

Read Me First!

and watch the narrated animation on crossing over

a Both chromosomes shown here were duplicated during interphase, before meiosis. When prophase I is under way, sister chromatids of each chromosome are positioned so close together that they look like a single thread.

b Each chromosome becomes zippered to its homologue, so all four chromatids are tightly aligned. If the two sex chromosomes have different forms, such as X paired with Y, they still get zippered together, but only in a tiny region at their ends.

c We show the pair of chromosomes as if they already condensed only to give you an idea of what goes on. They really are in a tightly aligned, threadlike form during prophase I.

d The intimate contact encourages one crossover (and usually more) to happen at various intervals along the length of nonsister chromatids.

e Nonsister chromatids exchange segments at the crossover sites. They continue to condense into thicker, rodlike forms. By the start of metaphase I, they will be unzipped from each other.

f Crossing over breaks up old combinations of alleles and puts new ones together in the cell's pairs of homologous chromosomes.

Figure 9.6 Key events of prophase I, the first stage of meiosis. For clarity, we show only one pair of homologous chromosomes and one crossover event. Typically, more than one crossover occurs. *Blue* signifies the paternal chromosome; *purple* signifies its maternal homologue.

slightly different forms: alleles. You can bet that some number of the alleles on one chromosome will *not* be identical to their partner alleles on the homologue. Every crossover is a chance to swap slightly different versions of hereditary instructions for gene products.

We will look at the mechanism of crossing over in later chapters. For now, just remember this: *Crossing over leads to recombinations among genes of homologous chromosomes, and to variation in traits among offspring.*

METAPHASE I ALIGNMENTS

Major shufflings of intact chromosomes start during the transition from prophase I to metaphase I. Suppose this is happening right now in one of your germ cells. Crossovers have already made genetic mosaics of the chromosomes, but put this aside to simplify tracking. Just call the twenty-three chromosomes you inherited from your mother the *maternal* chromosomes, and the twenty-three inherited from your father the *paternal* chromosomes.

At metaphase I, microtubules have tethered one chromosome of each pair to one spindle pole and its homologue to the other, and all are lined up at the spindle equator (Figure 9.5b). Have they tethered all maternal chromosomes to one pole and all paternal chromosomes to the other? Maybe, but probably not. As microtubules grow outward from the poles, they latch on to the first chromosome they contact. Because the tethering is random, there is no particular pattern to the metaphase I positions of maternal and paternal chromosomes. Carry this thought one step further. After a pair of homologous chromosomes are moved apart during anaphase I, *either one* of them can end up at either spindle pole.

Think of the possibilities while tracking just three pairs of homologues. By metaphase I, these three pairs may be arranged in any one of four possible positions (Figure 9.7). This means that eight combinations (2^3) are possible for forthcoming gametes.

Cells that give rise to human gametes have twenty-three pairs of homologous chromosomes, not three. So every time a human sperm or egg forms, you can expect a total of *8,388,608* (or 2^{23}) possible combinations of maternal and paternal chromosomes!

Moreover, in each sperm or egg, many hundreds of alleles inherited from the mother might not "say" the exact same thing about hundreds of different traits as alleles inherited from the father. Are you beginning to get an idea of why such fascinating combinations of traits show up the way they do among the generations of your own family tree?

Read Me First!
and watch the narrated animation on random alignment

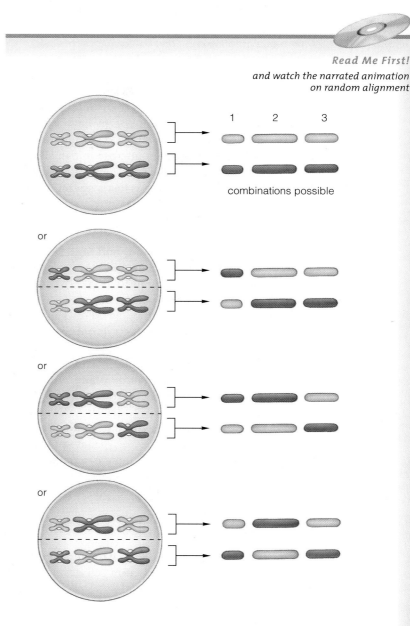

Figure 9.7 Possible outcomes for the random alignment of merely three pairs of homologous chromosomes at metaphase I. The three types of chromosomes are labeled 1, 2, and 3. With four alignments, eight combinations of maternal chromosomes (*purple*) and paternal chromosomes (*blue*) are possible in gametes.

Crossing over, an interaction between a pair of homologous chromosomes, breaks up old combinations of alleles and puts new ones together during prophase I of meiosis.

The random tethering and subsequent positioning of each pair of maternal and paternal chromosomes at metaphase I lead to different combinations of maternal and paternal traits in each new generation.

9.5 From Gametes to Offspring

What happens to the gametes that form after meiosis? Later chapters have specific examples. Here, just focus on where they fit in the life cycles of plants and animals.

Gametes are not all the same in their details. Human sperm have one tail, opossum sperm have two, and roundworm sperm have none. Crayfish sperm look like pinwheels. Most eggs are microscopic in size, yet an ostrich egg inside its shell is as big as a football. A flowering plant's male gamete is just a sperm nucleus.

GAMETE FORMATION IN PLANTS

Seasons vary for plants on land, and so fertilization must coincide with spring rains and other conditions that favor growth of the new individual. That is why life cycles of plants generally alternate between spore production and gamete production. Plant **spores** are haploid resting cells, often walled, that develop after meiosis (Figure 9.8*a*). They originate in reproductive structures of *sporophytes*, or spore-producing bodies. Pine trees, corn plants, and all other plants with roots, stems, and leaves are examples of sporophytes.

Plant spores stay dormant in dry or cold seasons. When they resume growth (germinate), they undergo mitosis and form *gametophytes*, or gamete-producing haploid bodies. For example, female gametophytes form on pine cone scales. In their tissues, gametes form by way of meiosis, as Chapter 27 explains.

GAMETE FORMATION IN ANIMALS

In animals, germ cells give rise to gametes. In a male reproductive system, a diploid germ cell develops into a large, immature cell: a primary spermatocyte. This cell enters meiosis and cytoplasmic divisions. Four haploid cells result, and they develop into spermatids (Figure 9.9). These cells undergo changes that include the formation of a tail. Each becomes a **sperm**, which is a common type of mature male gamete.

In female animals, a germ cell becomes an **oocyte**, or immature egg. Unlike sperm, an oocyte stockpiles many cytoplasmic components, and its four daughter cells differ in size and function (Figure 9.10).

As an oocyte divides after meiosis I, one daughter cell—the secondary oocyte—gets nearly all of the cytoplasm. The other cell, a first polar body, is small. Later, both of these haploid cells enter meiosis II, then cytoplasmic division. One of the secondary oocyte's daughter cells develops into a second polar body. The other receives most of the cytoplasm and develops into a gamete. A mature female gamete is called an ovum (plural, ova) or, more often, an **egg**.

And so we have one egg. The three polar bodies that formed don't function as gametes and aren't rich in nutrients or plump with cytoplasm. In time they degenerate. But the fact that they formed means the egg now has a haploid chromosome number. Also, by getting most of the cytoplasm, the egg has enough metabolic machinery to support the early divisions of the new individual.

MORE SHUFFLINGS AT FERTILIZATION

The chromosome number characteristic of the parents is restored at **fertilization**, a time when a female and male gametes unite and their haploid nuclei fuse. If meiosis did not precede it, fertilization would double

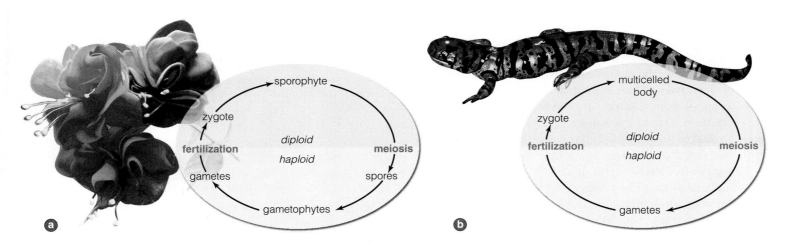

Figure 9.8 (**a**) Generalized life cycle for most plants. (**b**) Generalized life cycle for animals. The zygote is the first cell to form when the nuclei of two gametes fuse at fertilization.

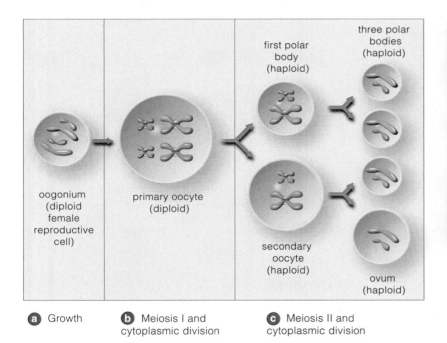

a Growth **b** Meiosis I and cytoplasmic division **c** Meiosis II and cytoplasmic division

Figure 9.9 Generalized sketch of sperm formation in animals. Figure 38.16 shows a specific example (how sperm form in human males).

Figure 9.10 Animal egg formation. Eggs are far larger than sperm and larger than the three polar bodies. The painting above, based on a scanning electron micrograph, depicts human sperm surrounding an ovum.

the chromosome number each generation. Doublings would disrupt hereditary information, usually for the worse. Why? That information is like a fine-tuned set of blueprints that must be followed exactly, page after page, to build a normal individual.

Fertilization also adds to variation among offspring. Reflect on the possibilities for humans alone. During prophase I, each human chromosome undergoes an average of two or three crossovers. Even without these crossovers, random positioning of pairs of paternal and maternal chromosomes at metaphase I results in

one of millions of possible chromosome combinations in each gamete. And of all male and female gametes that are produced, *which* two actually get together is a matter of chance. The sheer number of combinations that can exist at fertilization is staggering!

The distribution of random mixes of chromosomes into gametes, random metaphase chromosome alignments, and fertilization contribute to variation in traits of offspring.

Summary

Section 9.1 Alleles are slightly different molecular forms of the same gene, and they specify different versions of the same gene product.

Sections 9.2, 9.3 Meiosis, which consists of two nuclear divisions, is central to sexual reproduction. It halves the parental chromosome number (Figure 9.11).

Meiosis I, the first nuclear division, partitions homologous chromosomes into two clusters, both with one of each type of chromosome. All of the chromosomes were duplicated earlier, in interphase.

Prophase I: Chromosomes start condensing into rodlike forms. The nuclear envelope starts to break up. If duplicated pairs of centrioles are present, one pair moves to the opposite side of the nucleus along with a centrosome, from which new microtubules of a spindle originate. Crossing over occurs between homologues.

Metaphase I: All pairs of homologous chromosomes are positioned at the spindle equator. Microtubules have tethered the maternal or paternal chromosome of each pair to either pole, at random.

Anaphase I: Microtubules pull each chromosome away from its homologue, to opposite spindle poles.

Telophase I: Two haploid nuclei form. Cytoplasmic division typically follows.

In meiosis II, the second nuclear division, the sister chromatids of all the chromosomes are pulled away from each other and partitioned into two clusters. By the end of telophase II, four nuclei—each with a haploid chromosome number—have formed.

Section 9.4 In prophase I, *non*sister chromatids of homologous chromosomes break at corresponding sites and exchange segments. Crossing over puts new allelic combinations in chromosomes.

At metaphase I, maternal and paternal chromosomes have been randomly tethered to one spindle pole or the other, which mixes up allelic combinations even more. Alleles are randomly shuffled again when two gametes meet up at fertilization.

All three types of allele shufflings lead to variation in the details of shared traits among offspring.

Section 9.5 Meiosis, the formation of gametes, and fertilization occur in the life cycles of plants and animals. In plant sporophytes, meiosis is followed by haploid spore formation. Germinating spores give rise to gametophytes, where cells that give rise to gametes originate. In most animals, germ cells in reproductive organs give rise to sperm or eggs. Fusion of a sperm and egg nucleus at fertilization results in a zygote.

Figure 9.11 Comparison of the key features of mitosis and meiosis. We use a diploid cell with only two paternal and two maternal chromosomes. All of the chromosomes were duplicated during interphase, prior to nuclear division. Mitosis maintains the chromosome number. Meiosis halves it, to the haploid number.

Mitosis

PROPHASE	METAPHASE	ANAPHASE	TELOPHASE
	Chromosomes align at spindle equator.	Sister chromatids of chromosomes separate.	two nuclei (2*n*)

Meiosis I

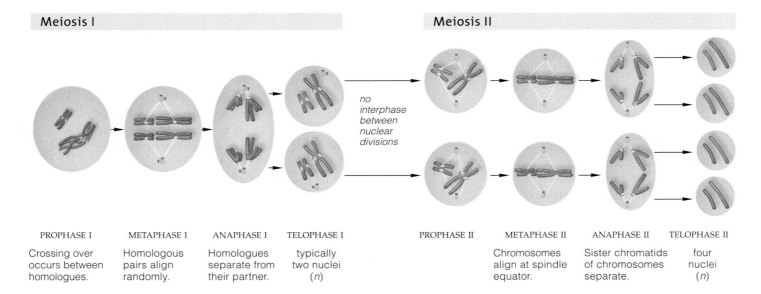

no interphase between nuclear divisions

PROPHASE I	METAPHASE I	ANAPHASE I	TELOPHASE I
Crossing over occurs between homologues.	Homologous pairs align randomly.	Homologues separate from their partner.	typically two nuclei (*n*)

Meiosis II

PROPHASE II	METAPHASE II	ANAPHASE II	TELOPHASE II
	Chromosomes align at spindle equator.	Sister chromatids of chromosomes separate.	four nuclei (*n*)

Self-Quiz

Answers in Appendix III

1. Meiosis and cytoplasmic division function in _____ .
   a. asexual reproduction of single-celled eukaryotes
   b. growth, tissue repair, often asexual reproduction
   c. sexual reproduction
   d. both b and c

2. A duplicated chromosome has _____ chromatid(s).
   a. one   b. two   c. three   d. four

3. A somatic cell having two of each type of chromosome has a(n) _____ chromosome number.
   a. diploid   b. haploid   c. tetraploid   d. abnormal

4. Sexual reproduction requires _____ .
   a. meiosis   c. gamete formation
   b. fertilization   d. all of the above

5. Generally, a pair of homologous chromosomes _____ .
   a. carry the same genes   c. interact at meiosis
   b. are the same length, shape   d. all of the above

6. Meiosis _____ the parental chromosome number.
   a. doubles   b. halves   c. maintains   d. corrupts

7. Meiosis is a division mechanism that produces _____ .
   a. two cells   c. eight cells
   b. two nuclei   d. four nuclei

8. Pairs of duplicated, homologous chromosomes end up at opposite spindle poles during _____ .
   a. prophase I   c. anaphase I
   b. prophase II   d. anaphase II

9. Sister chromatids of each duplicated chromosome end up at opposite spindle poles during _____ .
   a. prophase I   c. anaphase I
   b. prophase II   d. anaphase II

10. Match each term with its description.
    ___ chromosome   a. different molecular forms
         number   of the same gene
    ___ alleles   b. none between meiosis I, II
    ___ metaphase I   c. all chromosomes aligned
    ___ interphase   at spindle equator
                     d. all chromosomes of a given type

Critical Thinking

1. Why can we expect meiosis to give rise to genetic differences between parent cells and their daughter cells in fewer generations than mitosis?

2. As mentioned in the chapter introduction, aphids can reproduce asexually and sexually at different times of year. How might their reproductive flexibility be an adaptation to seasonal change?

3. The bdelloid rotifer lineage started at least 40 million years ago (Figure 9.12). About 360 known species of these tiny animals are found in many aquatic habitats worldwide. Speculate on why scientists were surprised to discover that all bdelloid rotifers are female.

4. Actor Viggo Mortensen inherited a gene that makes his chin dimple. Figure 9.13b shows what he might have looked like with an ordinary form of that gene. What is the name for alternative forms of the same gene?

Figure 9.12 Bdelloid rotifer.

Figure 9.13 Viggo Mortensen (**a**) with and (**b**) without a chin dimple.

Media Menu

Student CD-ROM

Impacts, Issues Video
   Why Sex?
Big Picture Animation
   Meiosis and sexual reproduction
Read-Me-First Animation
   Meiosis
   Crossing over
   Random alignment
Other Animations and Interactions
   Variation in life cycles
   Sperm formation
   Egg formation

InfoTrac

- Bdelloids: No Sex for over 40 Million Years. *Science News*, May 2000.
- Crossover Interference in Humans. *American Journal of Human Genetics*, July 2003.
- Tracking Down a Cheating Gene. *American Scientist*, March 2000.

Web Sites

- Meselson Lab: golgi.harvard.edu/meselson/research.html
- Mitosis vs. Meiosis: www.pbs.org/wgbh/nova/miracle/divide.html

How Would You Vote?

Japanese researchers have created a "fatherless" mouse from two eggs. Other scientists have coaxed unfertilized human eggs to develop into embryos. Some people object to the use of any human embryo for research purposes, and some worry about the potential to produce "fatherless" humans. Would you support a ban on this technique?

IMPACTS, ISSUES *Menacing Mucus*

Cystic fibrosis (CF) is a debilitating and ultimately fatal genetic disorder. In 1989, researchers identified the mutated gene that causes it. In 2001, the American College of Obstetricians and Gynecologists suggested that all prospective parents be screened for mutated versions of the gene. The suggestion led to the first mass screening for carriers of a genetic disorder.

CFTR, the gene's product, is a membrane transport protein. It helps chloride and water move into and out of cells that secrete mucus or sweat. More than 10 million people in the United States inherited one normal and one abnormal copy of the CFTR gene. They do have sinus problems, but no other symptoms develop. Most do not even know they carry the gene.

CF develops in anyone who inherits a mutant form of the gene from both parents. Thick, dry mucus clogs bronchial airways to their lungs and makes it hard to breathe (Figure 10.1). The mucus coating the airways is supposed to be thin enough to trap airborne particles and pathogens, so that ciliated cells lining the airways can sweep them out. Bacterial populations thrive in the thick mucus of CF patients, shown in the filmstrip.

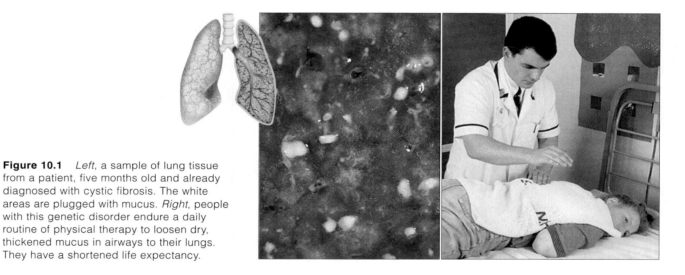

Figure 10.1 *Left*, a sample of lung tissue from a patient, five months old and already diagnosed with cystic fibrosis. The white areas are plugged with mucus. *Right*, people with this genetic disorder endure a daily routine of physical therapy to loosen dry, thickened mucus in airways to their lungs. They have a shortened life expectancy.

the big picture

An Experimental Approach Experiments with pea plants yielded the first observable evidence that parents transmit genes—units of information about heritable traits—to offspring. The experiments also revealed some underlying patterns of inheritance.

Two Theories Emerge As Mendel sensed, diploid organisms have pairs of genes and each gamete gets only one of the pairs. Also, genes on pairs of homologous chromosomes tend to be sorted out for distribution into gametes independently of gene pairs of other chromosomes.

Antibiotics help keep the pathogens under control but cannot get rid of them entirely. Also, to loosen the mucus, patients must go through daily routines of posture changes and thumps on the chest and back. Even with physiotherapy, most can expect lung failure. A double lung transplant can extend their life, but donor organs are scarce. Even if they do receive a transplant, few will live past their thirtieth birthday.

The severity of CF and the prevalence of carriers in the general population persuaded doctors to screen prospective parents—hundreds of thousands of them. By 2003, however, the law of unintended consequences took effect. Some people misunderstood the screening results. Some took unnecessary diagnostic tests to find out if their child would be normal. Confused by test results, a few may have aborted normal fetuses.

So here we are today, working our way through the genetic basis of our very lives. And where did it all start? It started in a small garden, with a monk named Gregor Mendel. By analyzing generation after generation of pea plants in experimental plots, he uncovered indirect but observable evidence of how parents bestow units of hereditary information—genes—on offspring.

This chapter starts out with the methods and some representative results of Mendel's experiments. His pioneering work remains a classic example of how a scientific approach can pry open important secrets about the natural world. To this day, it serves as the foundation for modern genetics.

How Would You Vote?

Many advances in genetics, including the ability to detect mutant genes that cause severe disorders, raise bioethical questions. Should we encourage the mass screening of prospective parents for the alleles that cause cystic fibrosis? And should we as a society encourage women to give birth only if their child will not develop severe medical problems? See the Media Menu for details, then vote online.

Beyond Mendel The traits that Mendel studied happened to follow simple dominant-to-recessive patterns of gene expression. The expression of genes for most traits is not as straightforward. Incomplete dominance and codominance are cases in point.

Less Predictable Variation Although many genes have predictable, observable effects on traits, the expression of most genes is variable. Most traits are outcomes of interactions among the products of two or more genes. Environmental factors also influence gene expression.

10.1 Mendel's Insight Into Inheritance Patterns

We turn now to recurring inheritance patterns among humans and other sexually reproducing species. You already know meiosis halves the parental chromosome number, which is restored at fertilization. Here the story picks up with some observable outcomes of these events.

More than a century ago, people wondered about the basis of inheritance. As many knew, both sperm and eggs transmit information about traits to offspring, but few suspected that the information is organized in units (genes). By the prevailing view, the father's

Figure 10.2 Gregor Mendel, the founder of modern genetics.

carpel   stamen

a Garden pea flower, cut in half. Sperm form in pollen grains, which originate in male floral parts (stamens). Eggs develop, fertilization takes place, and seeds mature in female floral parts (carpels).

b Pollen from a plant that breeds true for purple flowers is brushed onto a floral bud of a plant that breeds true for white flowers. The white flower had its stamens snipped off. This is one way to assure cross-fertilization of plants.

c Later, seeds develop inside pods of the cross-fertilized plant. An embryo within each seed develops into a mature pea plant.

d Each new plant's flower color is indirect but observable evidence that hereditary material has been transmitted from the parent plants.

Figure 10.3 Garden pea plant (*Pisum sativum*), which can self-fertilize or cross-fertilize. Experimenters can control the transfer of its hereditary material from one flower to another.

blob of information "blended" with the mother's blob at fertilization, like milk into coffee.

Carried to its logical conclusion, blending would slowly dilute a population's shared pool of hereditary information until there was only a single version of each trait. Freckled children would never pop up in a family of nonfreckled people. In time, all of the colts and fillies that are descended from a herd of white stallions and black mares would be gray. But freckles do show up, and not all horses are gray. The blending theory could scarcely explain the obvious variation in traits that people could observe with their own eyes. Even so, few disputed the theory.

"Blending" proponents dismissed Charles Darwin's theory of natural selection. According to the theory's key premise, individuals of a population vary in the details of the traits they have in common. Over the generations, variations that help an individual survive and reproduce show up among more offspring than variations that do not. Less helpful variations might persist, but among fewer individuals. They may even disappear. It is not that some versions of a trait are "blended out" of the population. Rather, *the frequency of each version of a trait among all individuals of the population may persist or change over time.*

Even before Darwin presented his theory, someone was gathering evidence that eventually would help support it. A monk, Gregor Mendel (Figure 10.2), had already guessed that sperm and eggs carry distinct "units" of information about heritable traits. After analyzing certain traits of pea plants generation after generation, he found indirect but *observable* evidence of how parents transmit genes to offspring.

MENDEL'S EXPERIMENTAL APPROACH

Mendel spent most of his adult life in Brno, a city near Vienna that is now part of the Czech Republic. Yet he was not a man of narrow interests who accidentally stumbled onto dazzling principles.

Mendel's monastery was close to European capitals that were centers of scientific inquiry. Having been raised on a farm, he was keenly aware of agricultural principles and their applications. He kept abreast of literature on breeding experiments. He also belonged to a regional agricultural society and even won awards for developing improved varieties of vegetables and fruits. Shortly after he entered the monastery, Mendel took a number of courses in mathematics, physics, and botany at the University of Vienna. Few scholars of his time showed interest in both plant breeding *and* mathematics.

a A *pair of homologous chromosomes,* each in the unduplicated state (most often, one from a male parent and its partner from a female parent)

b A *gene locus* (plural, loci), the location for a specific gene on a chromosome. *Alleles* are at corresponding loci on a pair of homologous chromosomes

c A pair of alleles may be identical or nonidentical. They are represented in the text by letters such as D or d

d Three *pairs of genes* (at three loci on this pair of homologous chromosomes); same thing as three pairs of alleles

Figure 10.4 A few genetic terms. Garden peas and other species with a diploid chromosome number have pairs of genes, on pairs of homologous chromosomes. Most genes come in slightly different molecular forms called alleles. Different alleles specify different versions of the same trait. An allele at any given location on a chromosome may or may not be identical to its partner on the homologous chromosome.

Shortly after his university training, Mendel began studying *Pisum sativum*, the garden pea plant (Figure 10.3). This plant is self-fertilizing. Its male and female gametes—call them sperm and eggs—originate in the same flower, and fertilization can occur in the same flower. A lineage of pea plants can "breed true" for certain traits. This means successive generations will be just like parents in one or more traits, as when all offspring grown from seeds of self-fertilized, white-flowered parent plants also have white flowers.

Pea plants also cross-fertilize when pollen from one plant's flower reaches another plant's flower. Mendel knew he could open the flower buds of a plant that bred true for a trait, such as white flowers, and snip out its stamens. Pollen grains, in which sperm develop, originate in stamens. Then he could brush the buds with pollen from a plant that bred true for a *different* version of the same trait—say, purple flowers.

As Mendel hypothesized, such clearly observable differences might help him track a given trait through many generations. If there were patterns to the trait's inheritance, *then those patterns might tell him something about heredity itself.*

TERMS USED IN MODERN GENETICS

In Mendel's time, no one knew about genes, meiosis, or chromosomes. As we follow his thinking, we can clarify the picture by substituting some modern terms used in inheritance studies, as stated here and in Figure 10.4:

1. **Genes** are units of information about heritable traits, transmitted from parents to offspring. Each gene has a specific location (locus) on a chromosome.

2. Cells with a **diploid** chromosome number ($2n$) have pairs of genes, on pairs of homologous chromosomes.

3. **Mutation** alters a gene's molecular structure and its message about a trait. It may cause a trait to change, as when one gene for flower color specifies white and a mutant form specifies yellow. All molecular forms of the same gene are known as **alleles**.

4. When offspring inherit a pair of *identical* alleles for a trait generation after generation, we expect them to be a true-breeding lineage. Offspring of a cross between two individuals that breed true for different forms of a trait are **hybrids**; each one has inherited *nonidentical* alleles for the trait.

5. When a pair of alleles on homologous chromosomes are identical, this is a *homozygous* condition. When the two are not identical, this is a *heterozygous* condition.

6. An allele is *dominant* when its effect on a trait masks that of any *recessive* allele paired with it. We use capital letters to signify dominant alleles and lowercase letters for recessive ones. A and a are examples.

7. Pulling this all together, a **homozygous dominant** individual has a pair of dominant alleles (AA) for the trait being studied. A **homozygous recessive** individual has a pair of recessive alleles (aa). And a **heterozygous** individual has a pair of nonidentical alleles (Aa).

8. Two terms help keep the distinction clear between genes and the traits they specify. *Genotype* refers to the particular alleles that an individual carries. *Phenotype* refers to an individual's observable traits.

9. P stands for the parents, F_1 for their first-generation offspring, and F_2 for the second-generation offspring.

Mendel hypothesized that tracking clearly observable differences in forms of a given trait might reveal patterns of inheritance.

He predicted that hereditary information is transmitted from one generation to the next as separate units (genes) and is not "diluted" at fertilization.

10.2 Mendel's Theory of Segregation

Mendel used monohybrid crosses to test his hypothesis that pea plants inherit two "units" of information for a trait, one from each parent.

Monohybrid crosses use F_1 offspring of parents that breed true for different forms of a trait ($AA \times aa = Aa$). The experiment itself is a cross between two identical F_1 heterozygotes, which are the "monohybrids" ($Aa \times Aa$).

MONOHYBRID CROSS PREDICTIONS

Mendel tracked many traits over two generations. In one set of experiments, he crossed plants that bred true for purple *or* white flowers. All F_1 offspring had purple flowers. They self-fertilized, and some of the F_2 offspring had white flowers! What was going on?

Pea plants have pairs of homologous chromosomes. Assume one plant is homozygous dominant (AA) and another is homozygous recessive (aa) for flower color. Following meiosis in both plants, each sperm or egg that forms has only one allele for flower color (Figure 10.5). Thus, when a sperm fertilizes an egg, only one outcome is possible: $A + a = Aa$.

With his background in mathematics, Mendel knew about sampling error (Chapter 1). He crossed many thousands of plants. He also counted and recorded the number of dominant and recessive forms of traits. On average, three of every four F_2 plants were dominant, and one was recessive (Figure 10.6).

The ratio hinted that fertilization is a chance event having a number of possible outcomes. Mendel knew about probability, which applies to chance events *and so could help him predict possible outcomes of his genetic crosses*. **Probability** means this: The chance that each outcome of an event will occur is proportional to the number of ways in which the event can be reached.

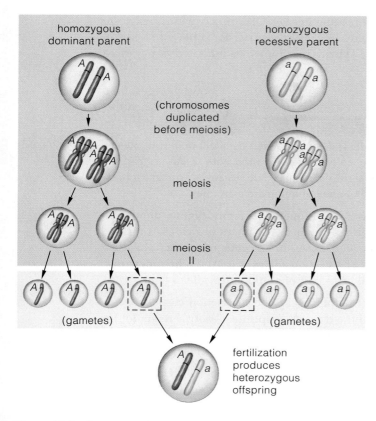

Figure 10.5 One gene of a pair segregating from the other gene in a monohybrid cross. Two parents that breed true for two versions of a trait produce only heterozygous offspring.

Figure 10.6 *Right*: Some monohybrid cross experiments with pea plants. Mendel's counts of F_2 offspring having dominant or recessive hereditary "units" (alleles). On average, the 3:1 phenotypic ratio held for traits.

Trait Studied	Dominant Form	Recessive Form	F_2 Dominant-to-Recessive Ratio
SEED SHAPE	5,474 round	1,850 wrinkled	2.96 : 1
SEED COLOR	6,022 yellow	2,001 green	3.01 : 1
POD SHAPE	882 inflated	299 wrinkled	2.95 : 1
POD COLOR	428 green	152 yellow	2.82 : 1
FLOWER COLOR	705 purple	224 white	3.15 : 1
FLOWER POSITION	651 along stem	207 at tip	3.14 : 1
STEM LENGTH	787 tall	277 dwarf	2.84 : 1

A **Punnett-square method**, explained and applied in Figure 10.7, shows the possibilities. If half of a plant's sperm or eggs are *a* and half are *A*, then we can expect four outcomes with each fertilization:

POSSIBLE EVENT	PROBABLE OUTCOME
sperm *A* meets egg *A*	1/4 *AA* offspring
sperm *A* meets egg *a*	1/4 *Aa*
sperm *a* meets egg *A*	1/4 *Aa*
sperm *a* meets egg *a*	1/4 *aa*

Each F$_2$ plant has 3 chances in 4 of inheriting at least one dominant allele (purple flowers). It has 1 chance in 4 of inheriting two recessive alleles (white flowers). That is a probable phenotypic ratio of 3:1.

Mendel's observed ratios were not *exactly* 3:1. Yet he put aside the deviations. To understand why, flip a coin several times. As we all know, a coin is as likely to end up heads as tails. But often it ends up heads, or tails, several times in a row. If you flip the coin only a few times, the observed ratio might differ a lot from the predicted ratio of 1:1. Flip it many times, and you are more likely to approach the predicted ratio.

That is why Mendel used rules of probability and counted so many offspring. He minimized sampling error deviations in the predicted results.

TESTCROSSES

Testcrosses supported Mendel's prediction. In such experimental tests, an organism shows dominance for a specified trait but its genotype is unknown, so it is crossed to a known homozygous recessive individual in a number of matings. Results may reveal whether it is homozygous dominant or heterozygous.

For example, Mendel crossed F$_1$ purple-flowered plants with true-breeding white-flowered plants. If all were homozygous dominant, then all the F$_2$ offspring would be purple flowered. If heterozygous, then only about half would. That is what happened. Half of the F$_2$ offspring had purple flowers (*Aa*) and half had white (*aa*). Go ahead and construct Punnett squares as a way to predict possible outcomes of this testcross.

Results from Mendel's monohybrid crosses became the basis of a theory of **segregation**, as stated here:

MENDEL'S THEORY OF SEGREGATION *Diploid cells have pairs of genes, on pairs of homologous chromosomes. The two genes of each pair are separated from each other during meiosis, so they end up in different gametes.*

Read Me First!
and watch the narrated animation on monohybrid crosses

a Punnett-square method

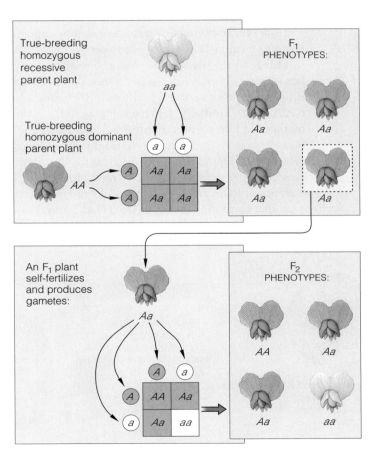

b Cross between two plants that breed true for different forms of a trait, followed by a monohybrid cross between their F$_1$ offspring

Figure 10.7 (**a**) Punnett-square method of predicting probable outcomes of genetic crosses. Circles signify gametes. *Italics* indicate dominant or recessive alleles. Possible genotypes among offspring are written in the squares. (**b**) Results from one of Mendel's monohybrid crosses. On average, the ratio of dominant-to-recessive that showed up among second-generation (F$_2$) plants was 3:1.

10.3 Mendel's Theory of Independent Assortment

In another set of experiments, Mendel used dihybrid crosses to explain how two pairs of genes assort into gametes.

*Di*hybrids are the offspring of parents that breed true for different versions of two traits. A **dihybrid cross** is an experimental intercross between F_1 dihybrids that are identically heterozygous for two pairs of genes.

Let's duplicate one of Mendel's dihybrid crosses for flower color (alleles *A* or *a*) and for height (*B* or *b*):

True-breeding parents: *AABB*   X   *aabb*

Gametes: *AB*   *AB*   *ab*   *ab*

F_1 hybrid offspring: *AaBb*

As Mendel would have predicted, F_1 offspring from this cross are all purple-flowered and tall (*AaBb*).

How will the two gene pairs assort into gametes in these F_1 plants? It depends partly on the chromosome locations of the two pairs. Assume that one pair of homologous chromosomes have the *Aa* alleles and a different pair have the *Bb* alleles. All chromosomes, recall, align midway between the spindle poles at metaphase I of meiosis (Figures 9.5 and 10.8). The one bearing the *A* or the *a* allele might be tethered to either pole. The same can happen to the chromosome bearing the *B* or *b* allele. Following meiosis, only four combinations of alleles are possible in the sperm or eggs that form: $1/4\,AB$, $1/4\,Ab$, $1/4\,aB$, and $1/4\,ab$.

①   **②**

AABB   **aabb**

homozygous dominant parent plant (purple flowers, tall stem)   homozygous recessive parent plant (white flowers, short stem)

Figure 10.9 Results from Mendel's dihybrid cross starting with parent plants that bred true for different versions of two traits: flower color and plant height. *A* and *a* signify dominant and recessive alleles for flower color. *B* and *b* signify dominant and recessive alleles for height. The Punnett square shows the F_2 combinations possible:

☐ 9/16 or 9 purple flowered, tall
☐ 3/16 or 3 purple-flowered, dwarf
☐ 3/16 or 3 white-flowered, tall
☐ 1/16 or 1 white-flowered, dwarf

Nucleus of a diploid (2*n*) reproductive cell with two pairs of homologous chromosomes

Figure 10.8 An example of independent assortment at meiosis. Either chromosome of a pair may get tethered to either spindle pole. When just two pairs are tracked, two different metaphase I alignments are possible.

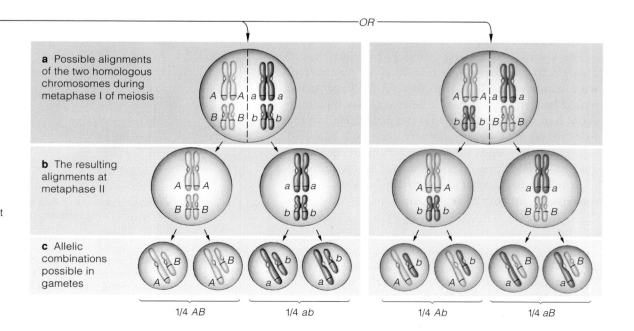

OR

a Possible alignments of the two homologous chromosomes during metaphase I of meiosis

b The resulting alignments at metaphase II

c Allelic combinations possible in gametes

1/4 *AB*   1/4 *ab*   1/4 *Ab*   1/4 *aB*

Read Me First!

and watch the narrated animation on dihybrid crosses

4 Possible outcomes of cross-fertilization of F$_1$ plants:

meiosis, gamete formation

1/4 **AB**   1/4 **A**b   1/4 a**B**   1/4 ab

meiosis, gamete formation

1/4 **AB** — 1/16 **AABB** | 1/16 **AAB**b | 1/16 **A**a**BB** | 1/16 **A**a**B**b

1/4 **A**b — 1/16 **AAB**b | 1/16 **AA**bb | 1/16 **A**a**B**b | 1/16 **A**abb

1/4 a**B** — 1/16 **A**a**BB** | 1/16 **A**a**B**b | 1/16 aa**BB** | 1/16 aa**B**b

1/4 ab — 1/16 **A**a**B**b | 1/16 **A**abb | 1/16 aa**B**b | 1/16 aabb

3 F$_1$ OUTCOME

All F$_1$ plants are *AaBb* heterozygotes (purple flowers, tall stems)

Given the alternative metaphase I alignments, many allelic combinations can result at fertilization. Simple multiplication (four sperm types × four egg types) tells us that sixteen combinations of genotypes are possible among F$_2$ offspring of a dihybrid cross (Figure 10.9).

Adding all possible phenotypes gives us a ratio of 9:3:3:1. We can expect to see 9/16 tall purple-flowered, 3/16 dwarf purple-flowered, 3/16 tall white-flowered, and 1/16 dwarf white-flowered F$_2$ plants. Results from one dihybrid cross were close to this ratio.

Mendel could only analyze numerical results from such crosses because he did not know seven pairs of homologous chromosomes carry a pea plant's "units" of inheritance. He could do no more than hypothesize that the two units for flower color were sorted out into gametes independently of the two units for height.

In time, his hypothesis became known as the theory of **independent assortment**. In modern terms, after meiosis ends, the genes on each pair of homologous chromosomes are assorted into gametes independently of how all the other pairs of homologues are sorted out. Independent assortment and hybrid intercrosses give rise to genetic variation. In a monohybrid cross for one gene pair, three genotypes are possible: *AA*, *Aa*, and *aa*. We represent this as 3^n, where *n* is the number of gene pairs. The more pairs, the more combinations are possible. If parents differ in twenty gene pairs, for instance, the number approaches 3.5 billion!

In 1866, Mendel published his idea, but apparently he was read by few and understood by no one. Today his theory of segregation still stands. However, his theory of independent assortment does not apply to *all* gene combinations, as you will see in Chapter 11.

MENDEL'S THEORY OF INDEPENDENT ASSORTMENT *As meiosis ends, genes on pairs of homologous chromosomes have been sorted out for distribution into one gamete or another, independently of gene pairs of other chromosomes.*

10.4 More Patterns Than Mendel Thought

Mendel happened to focus on traits that have clearly dominant and recessive forms. However, expression of genes for most traits is not as straightforward.

ABO BLOOD TYPES—A CASE OF CODOMINANCE

In *codominance*, a pair of nonidentical alleles affecting two phenotypes are both expressed at the same time in heterozygotes. For example, red blood cells have a type of glycolipid at the plasma membrane that helps give them their unique identity. The glycolipid comes in slightly different forms. An analytical method, *ABO blood typing*, reveals which form a person has.

An enzyme dictates the glycolipid's final structure. Humans have three alleles for this enzyme. Two, I^A and I^B, are codominant when paired. The third allele, i, is recessive; a pairing with I^A or I^B masks its effect. (If the letter for an allele is superscript, it signifies a lack of dominance.) Here we have a **multiple allele system**, the occurrence of three or more alleles for one gene locus among individuals of a population.

Each of these glycolipid molecules was assembled in the endomembrane system (Figure 4.16). First, an oligosaccharide chain was attached to a lipid, then a sugar was attached to the chain. But alleles I^A and I^B specify different versions of the enzyme that attaches the sugar. The two attach *different* sugars, which gives a glycolipid molecule a different identity: A or B.

Which alleles do you have? If you have $I^A I^A$ or $I^A i$, your blood is type A. With $I^B I^B$ or $I^B i$, it is type B. With codominant alleles $I^A I^B$, it is AB—you have both versions of the sugar-attaching enzyme. If you are homozygous recessive (*ii*), the glycolipid molecules never did get a final sugar side chain, so your blood type is not A or B. It is O (Figure 10.10).

Range of genotypes:

$I^A I^A$		$I^B I^B$	
or		or	
$I^A i$	$I^A I^B$	$I^B i$	*ii*

Blood types: A | AB | B | O

Figure 10.10 Possible allelic combinations for ABO blood typing.

homozygous parent   x   homozygous parent

All F₁ offspring heterozygous for flower color:

Cross two of the F₁ plants, and the F₂ offspring will show three phenotypes in a 1:2:1 ratio:

Figure 10.11 Incomplete dominance in heterozygous (*pink*) snapdragons, in which an allele that affects red pigment is paired with a "white" allele.

INCOMPLETE DOMINANCE

In *incomplete* dominance, one allele of a pair is not fully dominant over its partner, so the heterozygote's phenotype is *somewhere between* the two homozygotes. Cross true-breeding red and white snapdragons and their F₁ offspring will be pink-flowered. Cross two F₁ plants and you can expect to see red, white, and *pink* flowers in a certain ratio (Figure 10.11). Why the "odd" pattern? Red snapdragons have two alleles that let them make a lot of molecules of a red pigment. White snapdragons have two mutant alleles, and they are pigment-free. Pink snapdragons have one "red" allele and one "white" one. These heterozygotes make just enough pigment to color flowers pink, not red.

Two interacting gene pairs also can give rise to a phenotype that neither produces by itself. In chickens, interactions among *R* and *P* alleles specify the walnut, rose, pea, and single combs shown in Figure 10.12.

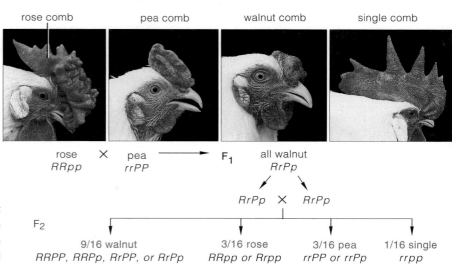

Figure 10.12 Interaction between two genes with variable effects on the comb on a chicken's head. The first cross is between a Wyandotte (rose comb) and a Brahma (pea comb). Check the outcomes by making a Punnett-square diagram.

rose comb   pea comb   walnut comb   single comb

rose × pea ⟶ F_1 all walnut
$RRpp$ $rrPP$ $RrPp$

$RrPp$ × $RrPp$

F_2

9/16 walnut | 3/16 rose | 3/16 pea | 1/16 single
RRPP, RRPp, RrPP, or RrPp | RRpp or Rrpp | rrPP or rrPp | rrpp

SINGLE GENES WITH A WIDE REACH

The alleles at one locus on a chromosome may affect two or more traits in good or bad ways. This outcome of the activity of one gene's product is **pleiotropy**. We see its effects in many genetic disorders, such as cystic fibrosis, sickle-cell anemia, and Marfan syndrome.

An autosomal dominant mutation in the gene for fibrillin causes *Marfan syndrome*. Fibrillin is a protein of connective tissues, the most abundant, widespread of all vertebrate tissues. We find many thin fibrillin strands, loose or cross-linked with the protein elastin, in the heart, blood vessels, and skin. They passively recoil after being stretched, as by the beating heart.

Altered fibrillin weakens the connective tissues in 1 of 10,000 men and women of any ethnicity. The heart, blood vessels, skin, lungs, and eyes are at risk. One of the mutations disrupts the synthesis of fibrillin 1, its secretion from cells, and its deposition. It skews the structure and function of smooth muscle cells inside the wall of the aorta, a big vessel carrying blood out of the heart. Cells infiltrate and multiply inside the wall's epithelial lining. Calcium deposits accumulate and the wall becomes inflamed. Elastic fibers split into fragments. The aorta wall, thinned and weakened, can rupture abruptly during strenuous exercise.

Until recent medical advances, Marfan syndrome killed most affected people before they were fifty years old. Flo Hyman was one of them (Figure 10.13).

WHEN PRODUCTS OF GENE PAIRS INTERACT

Traits also arise from interactions among products of two or more gene pairs. In some cases, two alleles can mask expression of another gene's alleles, and some expected phenotypes may not appear at all.

For example, several gene pairs govern fur color in Labrador retrievers. The fur appears black, yellow, or brown depending on how enzymes and other products of gene pairs synthesize melanin, a dark pigment, and deposit it in different body regions. Allele *B* (black) has a stronger effect and is dominant to *b* (brown). Alleles at another gene locus control how much melanin gets

Figure 10.13 Flo Hyman, at left, captain of the United States volleyball team that won an Olympic silver medal in 1984. Two years later, at a game in Japan, she slid to the floor and died. A dime-sized weak spot in the wall of her aorta had burst. We know at least two affected college basketball stars also died abruptly as a result of Marfan syndrome.

deposited in hair. Allele *E* permits full deposition. Two recessive alleles (*ee*) reduce it, so fur appears yellow.

Alleles at another locus (*C*) may override those two. They encode the first enzyme in a melanin-producing pathway. *CC* or *Cc* individuals do make the functional enzyme. An individual with two recessive alleles (*cc*) cannot. *Albinism*, the absence of melanin, is the result.

Some alleles are fully dominant, incompletely dominant, or codominant with a partner on the homologous chromosome.

With pleiotropy, alleles at a single locus have positive or negative impact on two or more traits.

Gene effects do not always appear together but rather appear over time. A gene's product may alter one trait, which may cause alteration in another trait, and so on.

10.5 Complex Variations in Traits

For most populations or species, individuals show rich variation for many of the same traits. Variation arises from gene mutations, cumulative gene interactions, and variations in environmental conditions.

REGARDING THE UNEXPECTED PHENOTYPE

As Mendel found out, phenotypic effects of one or two pairs of certain genes show up in predictable ratios. Two or more gene pairs also can produce phenotypes in predictable ratios. However, track some genes over the generations, and you might find that the resulting phenotypes were not what you expected.

As one example, *camptodactyly*, a rare abnormality, affects the shape and movement of fingers. Some of the people who carry a mutant allele for this heritable trait have immobile, bent fingers on both hands. Others have immobile, bent fingers on the left or right hand only. Fingers of still other people who have the mutant allele are not affected in any obvious way at all.

What causes such odd variation? Remember, most organic compounds are synthesized by a sequence of metabolic steps. *Different enzymes, each a gene product, control different steps.* Maybe one gene has mutated in a number of ways. Maybe a gene product blocks some pathway or makes it run nonstop or not long enough. Maybe poor nutrition or another variable factor in the individual's environment influences a crucial enzyme in the pathway. Such variable factors often introduce less predictable variations in the phenotypes that we otherwise associate with certain genes.

CONTINUOUS VARIATION IN POPULATIONS

Generally, individuals of a population display a range of small differences in most traits. This characteristic of natural populations, called **continuous variation**, depends mainly on how many gene products affect a given trait and on how many environmental factors impact them. The greater the number of genes

a Idealized bell-shaped curve for a population that displays continuous variation in a trait

The line of a bell-shaped curve reveals continuous variation in the population

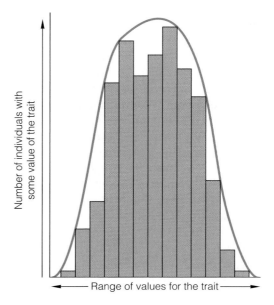

b A bell-shaped curve that corresponds to the height distribution among individual females in the far-right photograph in (**c**)

and environmental factors, the more continuous is the distribution of all versions of the trait.

Look in a mirror at your eye color. The colored part is the iris, a doughnut-shaped, pigmented structure just under the cornea. The color results from several gene products. Some products help make and distribute the light-absorbing pigment melanin, the same pigment that affects coat color in mammals. Almost black irises have dense melanin deposits, and melanin molecules absorb most of the incoming light. Deposits are not as great in brown eyes, so some unabsorbed light is reflected out. Light brown or hazel eyes have even less melanin (Figure 10.14).

Green, gray, or blue eyes do not have green, gray, or blue pigments. The iris has some melanin, but not

Figure 10.14 Examples from a range of continuous variation in human eye color. Products of different gene pairs interact in making and distributing the melanin that helps color the iris. Different combinations of alleles result in small color differences. The frequency distribution for the eye-color trait is continuous over a range from black to light blue.

Read Me First!

and watch the narrated animation on continuous variation in traits

5'3" 5'4" 5'5" 5'6" 5'7" 5'8" 5'9" 5'10" 5'11" 6'0" 6'1" 6'2" 6'3" 6'4" 6'5"
Height (feet/inches)

Figure 10.15 Continuous variation in body height, one of the traits that help characterize the human population.

(**a**) A bar graph can depict continuous variation in a population. The proportion of individuals in each category is plotted against the range of measured phenotypes. The curved line above this particular set of bars is an idealized example of the kind of bell-shaped curve that emerges for populations showing continuous variation in a trait.

(**b,c**) Jon Reiskind and Greg Pryor wanted to show the frequency distribution for height among biology students at the University of Florida. They divided students into two groups: male and female. For each group, they divided the range of possible heights, measured the students, and assigned each to the appropriate category.

4'11" 5'0" 5'1" 5'2" 5'3" 5'4" 5'5" 5'6" 5'7" 5'8" 5'9" 5'10" 5'11"
Height (feet/inches)

c Two examples of continuous variation: Biology students (males, *left*; females, *right*) organized by height.

much. Many or most of the blue wavelengths of light that do enter the eyeball are simply reflected out.

How can you describe the continuous variation of some trait in a group? Consider the students in Figure 10.15. They range from short to tall, with average heights more common than the extremes. Start out by dividing the full range of phenotypes into measurable categories—for instance, number of inches. Next, count how many students are in each category to get the relative frequencies of all phenotypes across the range of measurable values.

The chart in Figure 10.15*b* is a plot of the number of students in each height category. The shortest bars represent categories having the fewest individuals. The tallest bar signifies the category with the most. In this case, a graph line skirting the top of all the bars will be a bell-shaped curve. Such "bell curves" are typical of any trait that shows continuous variation.

Enzymes and other gene products control each step of most metabolic pathways. Mutations, interactions among genes, and environmental conditions may affect one or more steps. The outcome is variation in phenotypes.

For most traits, individuals of a population or species show continuous variation—a range of small differences.

Usually, the greater the number of genes and environmental factors that influence a trait, the more continuous the distribution of versions of that trait.

10.6 Genes and the Environment

We have mentioned, in passing, that the environment often contributes to variable gene expression among a population's individuals. Now consider a few cases.

Possibly you have noticed a Himalayan rabbit's coat color. Like a Siamese cat, this mammal has dark hair in some parts of its body and lighter hair in others. The Himalayan rabbit is homozygous for the c^h allele of the gene specifying tyrosinase. Tyrosinase is one of the enzymes involved in melanin production. The c^h allele specifies a heat-sensitive form of this enzyme. And this form is active only when the air temperature around the body is below 33°C, or 91°F.

When cells that give rise to this rabbit's hair grow under warmer conditions, they cannot make melanin, so hairs appear light. This happens in body regions that are massive enough to conserve a fair amount of metabolic heat. The ears and other slender extremities tend to lose metabolic heat faster, so they are cooler. Figure 10.16 shows one experiment that demonstrated how environmental temperatures affect this allele.

One classic experiment identified environmental effects on yarrow plants. These plants can grow from cuttings, so they are a useful experimental organism. Why? Cuttings from the same plant all have the same genotype, so experimenters can discount genes as a basis for differences that show up among them.

In this case, three yarrow cuttings were planted at three elevations. Two plants that grew at the lowest elevation and highest elevation fared best; the one at the mid-elevation grew poorly (Figure 10.17).

Icepack is strapped onto a hair-free patch.

New hair growing in patch exposed to cold is black.

Figure 10.16 Observable effect of an environmental factor that influences gene expression. A Himalayan rabbit normally has black hair only on its long ears, nose, tail, and leg regions farthest from the body mass. In one experiment, a patch of a rabbit's white coat was removed, then an icepack was secured over the hairless patch. Where the colder temperature had been maintained, the hairs that grew back were black.

Himalayan rabbits are homozygous for an allele of the gene for tyrosinase, an enzyme required to make melanin. As described in the text, this allele encodes a heat-sensitive form of the enzyme, which functions only when air temperature is below about 33°C.

a Mature cutting at high elevation (3,050 meters above sea level)

b Mature cutting at mid-elevation (1,400 meters above sea level)

c Mature cutting at low elevation (30 meters above sea level)

Figure 10.17 One experiment demonstrating the impact of environmental conditions of three different habitats on gene expression in yarrow (*Achillea millefolium*). Cuttings from the same parent plant were grown in the same soil batch but at three different elevations.

However, recall from Chapter 1 that sampling error can skew experimental results. The experimenters did the same growth experiments for *many* yarrow plants and found no consistent pattern; phenotypic variation was too great. For instance, a cutting from one plant did *best* at mid-elevation. The conclusion? For yarrow plants, at least, individuals with different genotypes react differently across a range of environments.

Similarly, plant a hydrangea in a garden and it may have pink flowers instead of the expected blue ones. Soil acidity affects the function of gene products that color hydrangea flowers.

What about humans? One of our genes codes for a transporter protein that moves serotonin across the plasma membrane of brain cells. This gene product has several effects, one of which is to counter anxiety and depression when traumatic events challenge us. For a long time, researchers have known that some people handle stress without getting too upset, while others spiral into a deep and lasting depression.

Mutation of the gene for the serotonin transporter compromises responses to stress. It is as if some of us are bicycling through life without an emotional helmet. Only when we take a fall does the phenotypic effect— depression—appear. Other genes also affect emotional states, but mutation of this one reduces our capacity to snap out of it when bad things happen.

And so we conclude this chapter, which introduces heritable and environmental factors that give rise to great variation in traits. What is the take-home lesson? Simply this: An individual's phenotype is an outcome of complex interactions among its genes, enzymes and other gene products, and the environment.

Variation in traits arises not only from gene mutations and interactions, but also in response to variations in environmental conditions that each individual faces.

Summary

Section 10.1 Genes are heritable units of information about traits. Each gene has its own locus on a particular chromosome. Different molecular forms of the same gene are known as alleles. Diploid cells have two copies of each gene, usually one inherited from each of two parents, on homologous chromosomes.

Offspring of a cross between two individuals that breed true for different forms of a trait are hybrids; each has inherited nonidentical alleles for the trait.

An individual with two dominant alleles for a trait (AA) is homozygous dominant. A homozygous recessive has two recessive alleles (aa). A heterozygote has two nonidentical alleles (Aa) for a trait. A dominant allele may mask the effect of a recessive allele partnered with it on the homologous chromosome.

Section 10.2 Results from Mendel's monohybrid crosses between F_1 offspring of true-breeding pea plants in time led to a theory of segregation: Paired genes on homologous chromosomes separate from each other at meiosis and end up in different gametes. The theory is based on a pattern of dominance and recessiveness that showed up among F_2 offspring of monohybrid crosses:

	A	a
A	AA	Aa
a	Aa	aa

AA (dominant)
Aa (dominant)
Aa (dominant)
aa (recessive)
} the expected phenotypic ratio of 3:1

Section 10.3 Mendel did dihybrid crosses between F_1 offspring of parents that bred true for two different traits. The dihybrid cross results were close to a 9:3:3:1 ratio:

9 dominant for both traits
3 dominant for A, recessive for b
3 dominant for B, recessive for a
1 recessive for both traits

His results support a theory of independent assortment: Meiosis assorts gene pairs of homologous chromosomes for forthcoming gametes independently of how gene pairs of the other chromosomes are sorted out. This is an outcome of the random alignment of all pairs of homologous chromosomes at metaphase I.

Section 10.4 Inheritance patterns are not always straightforward. Some alleles are codominant or not fully dominant. Products of gene pairs often interact in ways that influence the same trait. A single gene may have effects on two or more traits. Products of pairs of genes often interact in ways that influence the same trait. One gene may have positive or negative effects on two or more traits, a condition called pleiotropy.

Section 10.5 Mutations and interactions among gene products contribute to variation in traits among the individuals of a population. Some traits show a range of small, incremental differences—continuous variation.

Section 10.6 Environmental conditions can alter gene expression. The individuals of most populations show complex variation in traits, a combination of gene expression and exposure to environmental factors.

Self-Quiz

Answers in Appendix III

1. Alleles are _____ .
   a. different molecular forms of a gene
   b. different phenotypes
   c. self-fertilizing, true-breeding homozygotes

2. A heterozygote has a _____ for a trait being studied.
   a. pair of identical alleles
   b. pair of nonidentical alleles
   c. haploid condition, in genetic terms
   d. a and c

3. The observable traits of an organism are its _____ .
   a. phenotype   c. genotype
   b. sociobiology   d. pedigree

4. Second-generation offspring from a cross are the _____ .
   a. F_1 generation   c. hybrid generation
   b. F_2 generation   d. none of the above

5. F_1 offspring of the monohybrid cross $AA \times aa$ are _____ .
   a. all AA c. all Aa
   b. all aa d. 1/2 AA and 1/2 aa

6. Refer to Question 5. Assuming complete dominance, the F_2 generation will show a phenotypic ratio of _____ .
   a. 3:1   b. 9:1   c. 1:2:1   d. 9:3:3:1

7. Crosses between F_1 pea plants resulting from the cross $AABB \times aabb$ lead to F_2 phenotypic ratios close to _____ .
   a. 1:2:1   b. 3:1   c. 1:1:1:1   d. 9:3:3:1

8. Match each example with the most suitable description.
   ____ dihybrid cross   a. bb
   ____ monohybrid cross   b. $AABB \times aabb$
   ____ homozygous condition   c. Aa
   ____ heterozygous condition d. $Aa \times Aa$

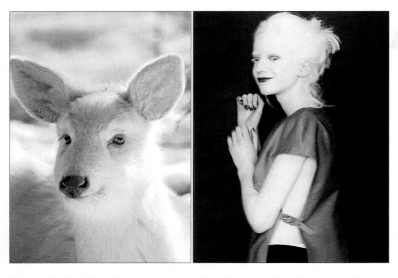

Figure 10.18 Two albino organisms. By not posing his subjects as objects of ridicule, the photographer of human albinos is attempting to counter the notion that there is something inherently unbeautiful about them.

Genetics Problems

Answers in Appendix IV

1. A certain recessive allele c is responsible for *albinism*, an inability to produce or deposit melanin, a brownish-black pigment, in body tissues. Humans and a number of other organisms can have this phenotype. Figure 10.18 shows two stunning examples. In cases of albinism, what are the possible genotypes of the father, the mother, and their children?
   a. Both parents have normal phenotypes; some of their children are albino and others are unaffected.
   b. Both parents are albino and have albino children.
   c. The woman is unaffected, the man is albino, and they have one albino child and three unaffected children.

2. One gene has alleles A and a. Another has alleles B and b. For each genotype, what type(s) of gametes will form? Assume that independent assortment occurs.
   a. $AABB$ c. $Aabb$
   b. $AaBB$ d. $AaBb$

3. Refer to Problem 2. What will be the genotypes of offspring from the following matings? Indicate the frequencies of each genotype among them.
   a. $AABB \times aaBB$ c. $AaBb \times aabb$
   b. $AaBB \times AABb$ d. $AaBb \times AaBb$

4. Certain dominant alleles are so essential for normal development that an individual who is homozygous recessive for a mutant recessive form can't survive. Such recessive, *lethal alleles* can be perpetuated in the population by heterozygotes.
   Consider the Manx allele (M^L) in cats. Homozygous cats ($M^L M^L$) die when they are still embryos inside the mother cat. In heterozygotes ($M^L M$), the spine develops abnormally. The cats end up with no tail (Figure 10.19).
   Two $M^L M$ cats mate. What is the probability that any one of their *surviving* kittens will be heterozygous?

5. In one experiment, Mendel crossed a pea plant that bred true for green pods with one that bred true for yellow pods. All the F_1 plants had green pods. Which form of the trait (green or yellow pods) is recessive? Explain how you arrived at your conclusion.

6. Return to Problem 2. Assume you now study a third gene having alleles C and c. For each genotype listed, what type(s) of gametes will be produced?
   a. $AABBCC$ c. $AaBBCc$
   b. $AaBBcc$ d. $AaBbCc$

7. Mendel crossed a true-breeding tall, purple-flowered pea plant with a true-breeding dwarf, white-flowered plant. All F_1 plants were tall and had purple flowers. If an F_1 plant self-fertilizes, then what is the probability that a randomly selected F_2 offspring will be heterozygous for the genes specifying height and flower color?

8. *DNA fingerprinting* is a method of identifying individuals by locating unique base sequences in their DNA molecules (Section 15.4). Before researchers refined the method, attorneys often relied on the ABO blood-typing system to settle disputes over paternity. Suppose that you, as a geneticist, are asked to testify during a paternity case in which the mother has type A blood, the child has type O blood, and the alleged father has type B blood. How would you respond to the following statements?

a. Attorney of the alleged father: "The mother's blood is type A, so the child's type O blood must have come from the father. My client has type B blood; he could not be the father."

b. Mother's attorney: "Because further tests prove this man is heterozygous, he must be the father."

9. Suppose you identify a new gene in mice. One of its alleles specifies white fur. A second allele specifies brown fur. You want to determine whether the relationship between the two alleles is one of simple dominance or incomplete dominance. What sorts of genetic crosses would give you the answer? On what types of observations would you base your conclusions?

10. Your sister gives you a purebred Labrador retriever, a female named Dandelion. Suppose you decide to breed Dandelion and sell puppies to help pay for your college tuition. Then you discover that two of her four brothers and sisters show *hip dysplasia*, a heritable disorder arising from a number of gene interactions. If Dandelion mates with a male Labrador known to be free of the harmful alleles, can you guarantee to a buyer that her puppies will not develop the disorder? Explain your answer.

11. A dominant allele *W* confers black fur on guinea pigs. A guinea pig that is homozygous recessive (*ww*) has white fur. Fred would like to know whether his pet black-furred guinea pig is homozygous dominant (*WW*) or heterozygous (*Ww*). How might he determine his pet's genotype?

12. Red-flowering snapdragons are homozygous for allele R^1. White-flowering snapdragons are homozygous for a different allele (R^2). Heterozygous plants (R^1R^2) bear pink flowers. What phenotypes should appear among first-generation offspring of the crosses listed? What are the expected proportions for each phenotype?

a. $R^1R^1 \times R^1R^2$ c. $R^1R^2 \times R^1R^2$

b. $R^1R^1 \times R^2R^2$ d. $R^1R^2 \times R^2R^2$

(In cases of incomplete dominance, alleles are usually designated by superscript numerals, as shown here, not by the uppercase letters for dominance and lowercase letters for recessiveness.)

13. Two pairs of genes affect comb type in chickens (Figure 10.12). When both genes are recessive, a chicken has a single comb. A dominant allele of one gene, *P*, gives rise to a pea comb. Yet a dominant allele of the other (*R*) gives rise to a rose comb. An *epistatic* interaction occurs when a chicken has at least one of both dominants, *P__ R__*, which gives rise to a walnut comb.

Predict the ratios resulting from a cross between two walnut-combed chickens that are heterozygous for both genes (*PpRr*).

14. As Section 3.6 explains, a single mutant allele gives rise to an abnormal form of hemoglobin (Hb^S instead of Hb^A). Homozygotes (Hb^SHb^S) develop sickle-cell anemia. Heterozygotes (Hb^AHb^S) show few obvious symptoms.

Suppose a woman's mother is homozygous for the Hb^A allele. She marries a male who is heterozygous for the allele, and they plan to have children. For *each* of her pregnancies, state the probability that this couple will have a child who is:

a. homozygous for the Hb^S allele
b. homozygous for the Hb^A allele
c. heterozygous Hb^AHb^S

Figure 10.19 The Manx, a breed of cat that has no tail.

IMPACTS, ISSUES   *Strange Genes, Tortured Minds*

"This man is brilliant," was the entire text of a letter of recommendation from Richard Duffin, a mathematics professor at Carnegie Mellon University. Duffin wrote it in 1948 on behalf of John Forbes Nash, Jr. (Figure 11.1), who was twenty years old at the time and applying to Princeton University's graduate school.

In the next decade, Nash made brilliant contributions to the field of mathematics and was considered to be

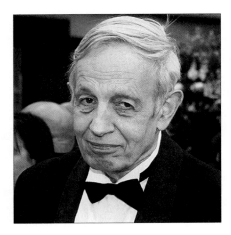

Figure 11.1   John Forbes Nash, Jr., a prodigy who solved problems that had long baffled some of the greatest minds in mathematics. His early work in economic game theory won him a Nobel Prize. He is shown here at a premier of "A Beautiful Mind," an award-winning film based on his long, tormented battle with schizophrenia.

one of the nation's top scientists. Apart from his social awkwardness, which is common among highly gifted people, there was no warning that paranoid schizophrenia would debilitate him in his thirtieth year. Nash had to abandon his position at the Massachusetts Institute of Technology. Two decades would pass before he would return to his work in mathematics.

Of every hundred people worldwide, one is affected by *schizophrenia*, which is characterized by delusions, hallucinations, disorganized speech and behavior, and social dysfunction. Many researchers have speculated that extraordinary creativity is linked to schizophrenia and other neurobiological disorders (NBDs) including depression, bipolar disorder (manic depression), and autism.

Certainly not every individual with high IQ shows such a link, but a higher percentage of geniuses have NBDs compared to the general population. Creative writers alone are eighteen times more suicidal, ten times more likely to be depressed, and twenty times more likely to have bipolar disorder.

We now have evidence that highly creative, healthy people have more personality traits in common with the mentally ill than with normal, less creative people, particularly in their sensitivity to environmental stimuli. People with NBDs belong to an illustrious crowd that includes Socrates, Newton, Beethoven, Darwin, Lincoln, Poe, Dickens, Tolstoy, Van Gogh, Freud, Churchill, Einstein, Picasso, Woolf, Hemingway, and Nash.

the big picture

Focus on Chromosomes   Males and females differ in their sex chromosomes, but all other homologous chromosomes are the same in both. All of their chromosomes are subject to crossing over and other changes, which diagnostic tools can detect.

Human Inheritance Patterns   Family pedigrees often reveal patterns of autosomal dominant, autosomal recessive, and sex-linked inheritance. Such patterns underlie many genetic abnormalities and disorders.

Abnormal brain biochemistry underlies NBDs. For instance, people with *bipolar disorder* show extreme swings in mood, thoughts, energy, and behavior. In their brain cells, expression of some mitochondrial genes that control aerobic respiration and protein breakdown is markedly low.

Change in any step of a crucial biochemical pathway could impair the brain's wiring. Therefore, we can expect that alterations in genes contribute to the abnormal neurochemistry in NBDs. Indeed, NBDs tend to run in families. Geniuses and individuals with one or more types of NBD often appear in the same family.

We already know about several mutant genes that predispose individuals to neural disorders. We also know that their bearers do not always show severe symptoms. Individuals who push the envelope of human creativity walk a razor's edge of mental stability, and it may take interplays of gene products and environmental factors to knock them off.

This brief account of neurobiological disorders is a glimpse into the world of modern genetics research. It invites you to think about how far you have come in this unit of the book. You first looked at cell division, the starting point of inheritance. You looked at how chromosomes and the genes they carry are shuffled during meiosis, then at fertilization. You also mulled over Mendel's insights into patterns of inheritance and some exceptions to his conclusions. Turn now to the chromosomal basis of inheritance.

✓ How Would You Vote?

Diagnostic tests for predisposition to neurobiological disorders will soon be available. Individuals might use knowledge of their susceptibility to modify life-style choices. Insurance companies and employers might also use that information to exclude predisposed but otherwise healthy individuals. Would you support legislation governing these tests? See the Media Menu for details, then vote online.

Chromosome Abnormalities   Certain genetic disorders arise from structural alterations of chromosomes or from abnormal changes in the chromosome number. Some changes occur spontaneously, and others result from exposure to harmful agents in the environment.

Prospects in Human Genetics   Some genetic disorders are treatable. Prospective parents who are at risk of transmitting a gene for a severe disorder often request genetic counseling or screening options, including prenatal and preimplantation diagnosis.

11.1 The Chromosomal Basis of Inheritance

You already know about chromosomes and what happens to them during meiosis. Now we'll start correlating chromosome structure to human inheritance patterns.

A REST STOP ON OUR CONCEPTUAL ROAD

Before driving on to the land of human inheritance, take a few minutes to check the following road map. It offers perspective on six important concepts:

1. A *gene*, again, is a unit of information about a heritable trait. The genes of eukaryotic cells are distributed among a number of chromosomes. Each gene has its own location, or locus, in one type of chromosome.

2. A cell with a diploid chromosome number, or 2*n*, has *pairs of homologous chromosomes*. All but one pair are normally the same in length, shape, and order of genes. The exception is a pairing of nonidentical sex chromosomes, such as X with Y in humans. Each chromosome becomes aligned with its homologous partner at metaphase I of meiosis.

3. Genes mutate, so a pair of genes on homologous chromosomes may or may not be the same. All of the slightly different molecular forms of a gene that occur among individuals of a population are called *alleles*.

4. A *wild-type* allele is a gene's most common form, in either a natural population or in a standardized, laboratory-bred strain of the species. A less common form of the gene is a *mutant* allele.

5. All genes on the same chromosome are physically connected. The farther apart any two genes are along the length of a chromosome, the more vulnerable they are to *crossing over*. By this event, a chromatid of one chromosome and a chromatid of its homologue swap corresponding segments (Figure 11.2). Crossing over between nonsister chromatids is a form of *genetic recombination* that introduces novel combinations of alleles in chromosomes.

6. On rare occasions, the *structure* of a chromosome or the parental *chromosome number* changes in mitosis or meiosis. Such chromosomal abnormalities can have severe phenotypic consequences.

AUTOSOMES AND SEX CHROMOSOMES

Some species show separation into sexes, and it all starts with genes on chromosomes. Say the species has a diploid chromosome number, so that body cells have pairs of homologous chromosomes. All but one pair are alike in length, shape, and gene sequence. A unique chromosome occurs in either females *or* males of many species, but not both.

For instance, a diploid cell in a human female has two X chromosomes (XX). A diploid cell in a human male has one X and one Y chromosome (XY). This is a common inheritance pattern among mammals, fruit flies, and many other animals. It is not the only one. In butterflies, moths, birds, and certain fishes, the two sex chromosomes are identical in males, and they are not identical in females.

Human X and Y chromosomes differ physically. The Y is a lot shorter, almost a remnant of the other in appearance. The two also differ in which genes they carry. They still synapse (zipper together briefly) in a small region. That bit of zippering allows them to interact as homologues during meiosis.

Human X and Y chromosomes fall into the more general category of **sex chromosomes**. When inherited in certain combinations, sex chromosomes determine a new individual's gender—whether a male or female will develop. All other chromosomes in a cell are the same in both sexes. We categorize them as **autosomes**.

Read Me First!

and watch the narrated animation on crossing over and genetic recombination

a A pair of duplicated homologous chromosomes (two sister chromatids each). In this example, nonidentical alleles occur at three gene loci (*A* with *a*, *B* with *b*, and *C* with *c*).

b In prophase I of meiosis, a crossover event occurs: Two nonsister chromatids exchange corresponding segments.

c What is the outcome of the crossover? Genetic recombination between nonsister chromatids (which are shown here, after meiosis, as two unduplicated, separate chromosomes).

Figure 11.2 Review of crossing over. As shown in Figure 9.6, this event occurs in prophase I of meiosis.

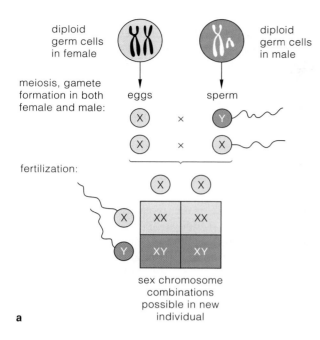

diploid germ cells in female

diploid germ cells in male

meiosis, gamete formation in both female and male:

eggs   sperm

X × Y

X × X

fertilization:

X   X

sex chromosome combinations possible in new individual

a

Figure 11.3 (**a**) Punnett-square diagram showing the sex determination pattern in humans.

(**b**) Early on, a human embryo is neither male nor female. Then tiny ducts and other structures that can develop into male *or* female reproductive organs start forming. In an XX embryo, ovaries form *in the absence of the SRY gene on the Y chromosome*. In an XY embryo, the gene product triggers the formation of testes. A hormone secreted from testes calls for development of male traits. (**c**) External reproductive organs in human embryos.

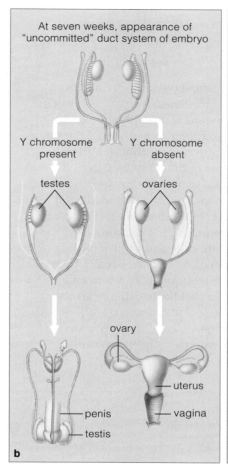

At seven weeks, appearance of "uncommitted" duct system of embryo

Y chromosome present   Y chromosome absent

testes   ovaries

ovary

uterus

penis   vagina

testis

b

At seven weeks, appearance of structures that will give rise to external genitalia

Y chromosome present   Y chromosome absent

10 weeks   10 weeks

penis   vaginal opening

birth approaching

c

XX

XY

SEX DETERMINATION IN HUMANS

Each normal egg produced by a human female has one X chromosome. Half the sperm cells formed in a male carry an X chromosome, and half carry a Y. Say an X-bearing sperm fertilizes an X-bearing egg. The individual will develop into a female. If the sperm carries a Y chromosome, the individual will develop into a male (Figure 11.3*a*).

SRY, one of 330 genes in a human Y chromosome, is the master gene for male sex determination. Its expression in XY embryos triggers the formation of testes, the primary male reproductive organs (Figure 11.3*b*). Testes make testosterone, and this sex hormone governs the emergence of male sexual traits.

An XX embryo has no *SRY* gene, so primary female reproductive organs—ovaries—form instead. Ovaries make estrogens and other sex hormones that govern the development of female sexual traits.

The human X chromosome carries 2,062 genes. Like other chromosomes, it carries some genes associated with sexual traits, such as the distribution of body fat

and hair. But most of its genes deal with *nonsexual* traits, such as blood-clotting functions. Such genes can be expressed in males as well as in females. Males, remember, also inherit one X chromosome.

Diploid cells have pairs of genes, on pairs of homologous chromosomes. The alleles (alternative forms of a gene) at a given locus may be identical or nonidentical.

As a result of crossing over and other events, offspring inherit combinations of alleles not found on parental chromosomes.

Abnormal events at meiosis or mitosis can change the structure and number of chromosomes.

Autosomes are pairs of chromosomes that are the same in males and females of a species. One other pairing, the sex chromosomes, differs between males and females.

The SRY gene on the human Y chromosome dictates that a new individual will develop into a male. In the absence of the Y chromosome (and the gene), a female develops.

11.2 Karyotyping Made Easy

Karyotyping is a diagnostic tool that allows us to check images of the structure and number of chromosomes in an individual's somatic cells.

How do we know so much about an individual's autosomes and sex chromosomes? *Karyotyping* is one diagnostic tool. A **karyotype** is a preparation of an individual's metaphase chromosomes, sorted out by their defining visual features. Any abnormalities in chromosome structure or number can be detected by comparing a standard karyotype for the species.

Chromosomes are in their most condensed form and easiest to identify when a cell enters metaphase. Technicians don't count on finding dividing cells in the body—they culture cells and induce mitosis artificially. They put a sample of cells, usually from blood, into a solution that stimulates growth and mitotic cell division. They add colchicine to arrest

the cell cycle at metaphase. Colchicine, recall, is a microtubule poison that blocks spindle formation.

The cell culture is centrifuged to isolate all the metaphase cells (Figure 11.4). A hypotonic solution makes the cells swell, by osmosis. The cells, along with their chromosomes, move apart. Then they are mounted on slides, fixed, and stained.

The chromosomes are viewed and photographed through a microscope. The photograph is cut, either with scissors or on a computer, and the individual chromosomes are lined up by their size and shape.

Spectral karyotyping uses a range of colored fluorescent dyes that bind to specific regions of chromosomes. Analysis of the resulting rainbow-hued karyotype often reveals crossovers and abnormalities that would not be otherwise visible. You will see an example of a multicolor spectral karyotype in Section 11.7.

Read Me First!

and watch the narrated animation on karyotype preparation

Figure 11.4 Karyotyping. With this type of diagnostic tool, an image of metaphase chromosomes is cut apart. Individual chromosomes are aligned by their centromeres and arranged according to size, shape, and length.

(**a**) A sample of cells from an individual is added to a medium that stimulates cell growth and mitotic cell division. The cell cycle is arrested at metaphase, with colchicine. (**b**) The culture is subjected to *centrifugation*, which works because cells have greater mass and density than the solution bathing them. A centrifuge's spinning force moves the cells farthest from the center of rotation, so they collect at the base of the centrifuge tubes.

(**c**) The culture medium is removed; a hypotonic solution is added. As the cells swell, the chromosomes move apart. (**d**) The cells are mounted on a microscope slide, fixed by air-drying, and stained. Chromosomes show up.

(**e**) A photograph of one cell's chromosomes is cut up and organized, as in the human karyotype in (**f**), which shows 22 pairs of autosomes and 1 pair of sex chromosomes—XX *or* XY. Scissors or computers do the cuts.

11.3 Impact of Crossing Over on Inheritance

Crossing over between homologous chromosomes is one of the main pattern-busting events in inheritance.

We now know there are many genes on each type of autosome and sex chromosome. All the genes on one chromosome are called a **linkage group**. For instance, the fruit fly (*Drosophila melanogaster*) has four linkage groups, corresponding to its four pairs of homologous chromosomes. Indian corn (*Zea mays*) has ten linkage groups, corresponding to its ten pairs, and so on.

If linked genes stayed connected through meiosis, then there would be no surprising mixes of parental traits. You could expect parental phenotypes among, say, F_2 offspring of dihybrid crosses to show up in a predictable ratio. As early experiments with fruit flies made clear, however, plenty of genes on the same chromosomes do *not* stay together through meiosis.

In one experiment, mutant female flies that bred true for white eyes and a yellow body were crossed with wild-type males (red eyes and gray body). As expected, 50 percent of the F_1 offspring had one or the other parental phenotype. However, 129 of the 2,205 F_2 offspring were recombinants! They had white eyes and a gray body, or red eyes and a yellow body.

Why? Some alleles tend to stay together more often than others through meiosis. They are closer together along the length of a chromosome and therefore less vulnerable to a crossover. *The probability that crossing over will disrupt the linkage between any two gene loci is proportional to the distance between them.*

If, say, genes *A* and *B* are twice as far apart as genes *C* and *D*, we can expect crossing over to disrupt the linkage between *A* and *B* far more frequently:

Two genes are very closely linked when the distance between them is small. Their combinations of alleles nearly always end up in the same gamete. Linkage is more vulnerable to crossing over when the distance between two gene loci is greater (Figure 11.5). When two loci are far apart, crossing over is so frequent that the genes assort independently into gametes.

Human gene linkages were identified by tracking phenotypes in families over generations. One thing is clear from such studies: Crossovers are not rare. For most eukaryotes, meiosis cannot even be completed properly until at least one crossover occurs between each pair of homologous chromosomes.

> All of the genes at different locations along the length of a chromosome belong to the same linkage group. They do not all assort independently at meiosis.
>
> Crossing over between homologous chromosomes disrupts gene linkages and results in nonparental combinations of alleles in chromosomes.
>
> The farther apart two genes are on a chromosome, the greater will be the frequency of crossing over and genetic recombination between them.

a Full linkage between two genes; no crossing over. Half of the gametes have one parental genotype, and half have the other. Genes that are very close together along the length of a chromosome typically stay together in gametes.

Most gametes have parental genotypes

A smaller number have recombinant genotypes

b Incomplete linkage; crossing over affected the outcome. Genes that are far apart along the length of a chromosome are more vulnerable to crossing over.

Figure 11.5 Examples of outcomes of crossing over between two gene loci.

11.4 Human Genetic Analysis

Some organisms, including pea plants and fruit flies, are ideal for genetic analysis. They do not have a lot of chromosomes. They grow and reproduce fast in small spaces, under controlled conditions. It doesn't take long to track a trait through many generations. Humans, however, are another story.

Unlike fruit flies in the laboratory, we humans live under variable conditions in diverse environments, and we live as long as the geneticists who study our traits. Most of us select our own mates and reproduce if and when we want to. Most families are not large, which means there are not enough offspring available for researchers to make easy inferences.

Geneticists often gather information from several generations to increase the numbers for analysis. If a trait follows a simple Mendelian inheritance pattern, they can be confident about predicting the probability of its showing up again. The pattern also can be a clue to the past (Figure 11.6).

Such information is often displayed in **pedigrees**, or charts of genetic connections among individuals. Standardized methods, definitions, and symbols that represent different kinds of individuals are used to

Figure 11.6 An intriguing pattern of inheritance. Eight percent of the men in Central Asia carry nearly identical Y chromosomes, which implies descent from a shared ancestor. If so, then 16 million males living between northeastern China and Afghanistan—close to 1 of every 200 men alive today—belong to a lineage that started with the warrior and notorious womanizer Genghis Khan. In time, his offspring ruled an empire that stretched from China all the way to Vienna.

construct these charts. Figure 11.7 gives an example. Those who analyze pedigrees rely on their knowledge of probability and Mendelian inheritance patterns that may yield clues to a trait. As you will see, they have traced many genetic abnormalities and disorders to a dominant or recessive allele, even to its location on an autosome or a sex chromosome.

Bear in mind, a genetic *abnormality* is simply a rare or uncommon version of a trait, as when a person is born with six digits on each hand or foot instead of the usual five. Whether we view such a condition as disfiguring or merely interesting is subjective; there is nothing inherently life-threatening about it. A **genetic disorder**, however, is a heritable condition that sooner or later gives rise to mild to severe medical problems. A set of symptoms, or **syndrome**, characterizes each abnormality or disorder. Table 11.1 gives examples.

You might be thinking that a disease, too, has a set of symptoms that arises from an abnormal change in how the body functions. However, each **disease** is an illness that results from an infection, dietary problems, or environmental factors—*not* from a heritable mutation. It might be appropriate to call it a *genetic* disease only when such factors alter previously workable genes in a way that disrupts body functions.

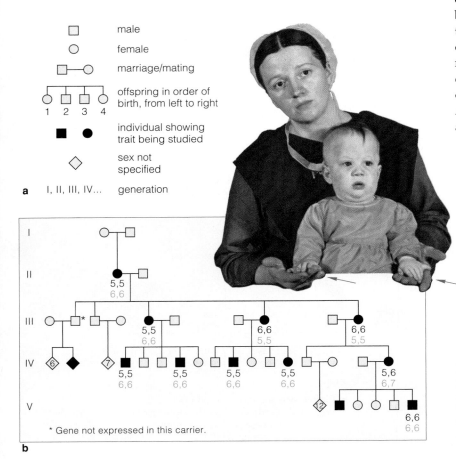

a

□	male
○	female
□─○	marriage/mating
	offspring in order of birth, from left to right
■ ●	individual showing trait being studied
◇	sex not specified
I, II, III, IV...	generation

b

* Gene not expressed in this carrier.

Figure 11.7 (**a**) Some standardized symbols used in pedigrees. (**b**) A pedigree for *polydactyly*, characterized by extra fingers, toes, or both. *Black* numerals signify the number of fingers on each hand; *blue* numerals signify the number of toes on each foot. This condition recurs as one symptom of Ellis–van Creveld syndrome.

Table 11.1 Examples of Human Genetic Disorders and Genetic Abnormalities

Disorder or Abnormality	Main Symptoms	Disorder or Abnormality	Main Symptoms
Autosomal recessive inheritance		**X-linked recessive inheritance**	
Albinism	Absence of pigmentation	Androgen insensitivity syndrome	XY individual but having some female traits; sterility
Blue offspring	Bright blue skin coloration	Color blindness	Inability to distinguish among some or all colors
Cystic fibrosis	Excessive glandular secretions leading to tissue, organ damage		
Ellis–van Creveld syndrome	Extra fingers, toes, short limbs	Fragile X syndrome	Mental impairment
Fanconi anemia	Physical abnormalities, bone marrow failure	Hemophilia	Impaired blood-clotting ability
		Muscular dystrophies	Progressive loss of muscle function
Galactosemia	Brain, liver, eye damage	X-linked anhidrotic dysplasia	Mosaic skin (patches with or without sweat glands); other effects
Phenylketonuria (PKU)	Mental impairment		
Sickle-cell anemia	Adverse pleiotropic effects on organs throughout body	**Changes in chromosome number**	
		Down syndrome	Mental impairment; heart defects
Autosomal dominant inheritance		Turner syndrome	Sterility; abnormal ovaries, abnormal sexual traits
Achondroplasia	One form of dwarfism		
Camptodactyly	Rigid, bent fingers	Klinefelter syndrome	Sterility; mild mental impairment
Familial hypercholesterolemia	High cholesterol levels in blood; eventually clogged arteries	XXX syndrome	Minimal abnormalities
		XYY condition	Mild mental impairment or no effect
Huntington disease	Nervous system degenerates progressively, irreversibly	**Changes in chromosome structure**	
Marfan syndrome	Abnormal or no connective tissue	Chronic myelogenous leukemia (CML)	Overproduction of white blood cells in bone marrow; organ malfunctions
Polydactyly	Extra fingers, toes, or both		
Progeria	Drastic premature aging	Cri-du-chat syndrome	Mental impairment; abnormally shaped larynx
Neurofibromatosis	Tumors of nervous system, skin		

Alleles that give rise to severe genetic disorders are rare in populations, because they put their bearers at risk. Why don't they disappear? Rare mutations introduce new copies of the alleles into populations. Also, in heterozygotes, a normal allele is paired with a harmful one and may cover its functions, in which case the harmful allele can be transmitted to offspring.

With these qualifications in mind, we turn next to examples of chromosomal inheritance patterns in the human population. Figure 11.8 is an early introduction to one of these examples—an autosomal dominant disorder called Huntington disease.

Pedigree analysis often reveals simple Mendelian inheritance patterns. From such patterns, specialists infer the probability that offspring will inherit certain alleles.

A genetic abnormality is a rare or less common version of a heritable trait. A genetic disorder is a heritable condition that results in mild to severe medical problems.

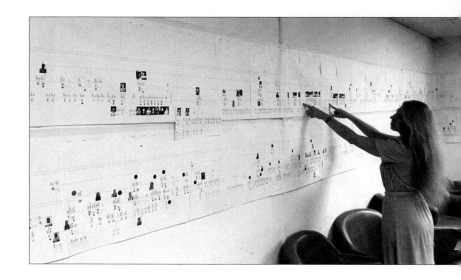

Figure 11.8 Pedigree for *Huntington disease*, a progressive degeneration of the nervous system. Researcher Nancy Wexler and her team constructed this extended family tree for nearly 10,000 Venezuelans. Their analysis of unaffected and affected individuals revealed that a dominant allele on human chromosome 4 is the culprit. Wexler has a special interest in the disease; it runs in her family.

11.5 Examples of Human Inheritance Patterns

Some human phenotypes arise from a dominant or recessive allele on an autosome or X chromosome that is inherited in simple Mendelian patterns.

AUTOSOMAL DOMINANT INHERITANCE

Figure 11.9*a* shows a typical inheritance pattern for an autosomal dominant allele. If one of the parents is heterozygous and the other homozygous, any child of theirs has a 50 percent chance of being heterozygous. The trait usually appears in every generation because the allele is expressed even in heterozygotes.

Achondroplasia is a classic example. This autosomal dominant disorder affects approximately 1 in 10,000 people. The homozygous dominants often die before birth, but heterozygotes can still reproduce. Skeletal cartilage does not form properly in achondroplasiacs. Adults have abnormally short arms and legs relative to other body parts. They are about 4 feet, 4 inches tall, as in Figure 11.9*a*. The dominant allele often has no other phenotypic effects.

In *Huntington disease*, the nervous system slowly deteriorates, and involuntary muscle action becomes more frequent. Symptoms may not start until past age thirty; those affected die in their forties or fifties. Many unknowingly transmit the gene to children, before the onset of symptoms. The mutation causing the disorder changes a protein necessary for normal development of brain cells. It is an *expansion* mutation, which ends up as multiple repeats in the same DNA segment. The repeats disrupt gene function.

A few dominant alleles that cause severe problems persist in populations because expression of the allele may not interfere with reproduction, or affected people reproduce before the symptoms become severe. Rarely, spontaneous mutations reintroduce some of them.

AUTOSOMAL RECESSIVE INHERITANCE

For some traits, inheritance patterns reveal clues that point to a recessive allele on an autosome. First, if both parents are heterozygous, any child of theirs will have a 50 percent chance of being heterozygous and a 25 percent chance of being homozygous recessive, as in Figure 11.9*b*. Second, if they are both homozygous recessive, any child of theirs will be, also.

About 1 in 100,000 newborns is homozygous for a recessive allele that causes *galactosemia*. They do not have working copies of one of the enzymes that digest lactose, so a reaction intermediate builds up to toxic levels. Normally, lactose is converted to glucose and galactose, then glucose–1–phosphate (which is broken down by glycolysis or converted to glycogen). The conversion is blocked in galactosemics (Figure 11.10).

High galactose levels can be detected in urine. The excess causes malnutrition, diarrhea, vomiting, and damage to the eyes, liver, and brain. When untreated, galactosemics typically die early. If they are quickly placed on a restricted diet excluding dairy products, they grow up symptom-free.

Figure 11.9 (**a**) A case of autosomal dominant inheritance. A dominant allele (coded *red*) is fully expressed in carriers. The three males shown above have achondroplasia, an autosomal dominant disorder. At center is Verne Troyer, known as Mini Me in the Mike Myers spy movies. Verne stands two feet, eight inches tall.

(**b**) An autosomal recessive pattern. In this case, both parents are heterozygous carriers of the recessive allele (coded *red*).

Figure 11.10 Blocked metabolic pathway in galactosemics.

Figure 11.11 One pattern for X-linked recessive inheritance. In this case, the mother carries a recessive allele on one of her X chromosomes (*red*).

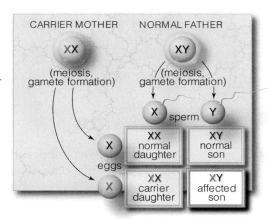

Figure 11.12 One of many standardized tests that can reveal color blindness. If you cannot see the red "29" inside this circle, then you may have some form of red–green color blindness.

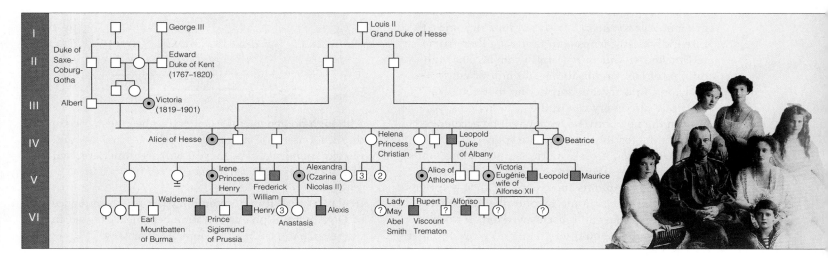

X-LINKED INHERITANCE

An X-linked gene is found only on the X chromosome. In X-linked genetic disorders, females are not affected as often as males, because a dominant allele on their other X chromosome can mask a recessive one (Figure 11.11). A son cannot inherit a X-linked allele from his father, but a daughter can. When she does, each of her sons has a 50 percent chance of inheriting it.

Color blindness is an inability to distinguish among some or all colors. It results from several common recessive disorders associated with X-linked genes. Mutant forms of the genes change the light-absorbing capacity of sensory receptors inside the eyes.

Normally, humans can detect differences among 150 colors. A person who is red–green color blind sees fewer than 25 colors; some or all of the receptors that respond to visible light of red and green wavelengths are weakened or absent. Others confuse red and green colors or see shades of gray instead of green. Tests can identify affected people (Figure 11.12).

The trait is more common in men, but heterozygous women also show symptoms. Can you explain why?

Figure 11.13 Partial pedigree for Queen Victoria's descendants, including carriers and affected males who inherited the X-linked allele for hemophilia A. At one time, the recessive allele was present in eighteen of Victoria's sixty-nine descendants, who sometimes intermarried. Of the Russian royal family members shown, the mother was a carrier; Crown Prince Alexis was hemophilic.

Hemophilia A, a blood-clotting disorder, is one case of X-linked recessive inheritance. Normally, a clotting mechanism quickly stops bleeding from minor injuries. Some clotting proteins are products of genes on the X chromosome. Bleeding is prolonged in males with one of these mutant X-linked genes. About 1 in 7,000 males is affected. In heterozygous females, clotting time is close to normal.

The frequency of hemophilia A was high in royal families of nineteenth-century Europe, in which close relatives often married (Figure 11.13).

Genetic analyses of family pedigrees have revealed simple Mendelian inheritance patterns for certain traits, as well as for many genetic disorders that arise from expression of alleles on an autosome or X chromosome.

11.6 Too Young, Too Old

FOCUS ON HEALTH

Sometimes textbook examples of the human condition seem a bit abstract, so take a moment to think about two boys who were too young to be old.

Imagine being ten years old with a mind trapped in a body that is getting a bit more shriveled, more frail—*old*—every day. You are barely tall enough to peer over the top of the kitchen counter; you weigh less than thirty-five pounds. Already you are bald and have a crinkled nose. Maybe you have a few more years to live. Would you, like Mickey Hayes and Fransie Geringer, still be able to laugh?

Of every 8 million newborn humans, one will grow old far too soon. On one of its autosomes, that rare individual carries a mutant gene that gives rise to *Hutchinson–Gilford progeria syndrome*. Through billions of DNA replications and mitotic cell divisions, information encoded in that gene was distributed to every cell in the growing embryo, then in the newborn. Its legacy will be accelerated aging and a terribly reduced life span.

The mutation grossly disrupts interactions among genes that bring about growth and development. Observable symptoms start to materialize before age two. Skin that should be plump and resilient starts to thin. Skeletal muscles weaken. Tissues in limb bones that should lengthen and grow stronger soften. Hair loss is pronounced; premature baldness is inevitable (Figure 11.14). There are no documented cases of progeria running in families, so we suspect it arises from spontaneous mutations. Probably the mutated gene is dominant over a normal allele on the homologous chromosome.

Most progeriacs expect to die in their early teens, from strokes or heart attacks. These final insults are brought on by a hardening of the wall of arteries, a condition typical of advanced age. When Mickey turned eighteen, he was the oldest living progeriac. Fransie was seventeen when he died.

Figure 11.14 Two boys who met at a gathering of progeriacs at Disneyland, California, when they were not yet ten years old.

11.7 Altered Chromosomes

Rarely, chromosome structure changes spontaneously or by exposure to chemicals or irradiation. Some changes can be detected. Many have severe or lethal outcomes.

THE MAIN CATEGORIES OF STRUCTURAL CHANGE

DUPLICATION Even normal chromosomes have gene sequences that have been repeated several to many thousands of times. These are **duplications**:

normal chromosome

one segment repeated

three repeats

Although no genetic information has been lost, certain duplications cause a variety of neural problems and physical abnormalities. As you will see, others proved useful over evolutionary time.

INVERSION With an **inversion**, part of the sequence of DNA within the chromosome becomes oriented in the reverse direction, with no molecular loss:

segments G, H, I become inverted

An inversion is not a problem if it does not disrupt a crucial gene region. But it mispairs during meiosis, so it can lead to chromosome deletions in gametes. Some people don't even know they have an inverted chromosome region until they have kids.

DELETION Whether it happens as a consequence of inversion or of an attack by an environmental agent, a **deletion** is the loss of a portion of a chromosome:

segment C deleted

In mammals, most deletions cause serious disorders, or are lethal. Why? Missing or incomplete genes disrupt the body's program of growth, development, and maintenance activities. For example, one deletion from human chromosome 5 results in an abnormally shaped larynx and mental impairment. When affected infants cry, they produce sounds like a cat's meow. Hence the name of the disorder, *cri-du-chat*, which means cat-cry in French. Sounds become normal later on. Figure 11.15 shows an affected boy.

TRANSLOCATION   In **translocation**, a broken part of a chromosome is attached to a different chromosome. Most translocations are reciprocal; both chromosomes exchange broken parts:

If the chromosome's genetic information does not get garbled, translocations may not pose a threat to the individual or its offspring. However, translocations can cause severe problems, including some sarcomas, lymphomas, myelomas, and leukemias. One notorious reciprocal translocation results in the Philadelphia chromosome, which is named after the city in which it was discovered. It is a killer (Figure 11.16).

DOES CHROMOSOME STRUCTURE EVOLVE?

Alterations in the structure of chromosomes generally are not good and tend to be selected against. Even so, over evolutionary time, many alterations with neutral effects became built into the DNA of all species.

We can expect that some of the duplications turned out to be adaptive. Perhaps some copies continued to specify an unaltered gene product even as others underwent modification. Think back on hemoglobin's polypeptide chains (Section 3.6). In humans and other primates, several globin genes are strikingly similar. They may have evolved as an outcome of duplications, mutations, and transpositions of the same gene. With small structural differences, the different globins have slightly different capacities to bind and then transport oxygen under a range of cellular conditions.

In addition, alterations in chromosome structure may have contributed to differences among closely related organisms, such as apes and humans. Consider this: Eighteen of the twenty-three pairs of human chromosomes are almost identical with chimpanzee

Figure 11.15   (**a**) Male infant who developed cri-du-chat syndrome. His ears are low on the side of the head relative to the eyes. (**b**) Same boy, four years later. By this age, affected humans stop making mewing sounds typical of the syndrome.

Figure 11.16   A reciprocal translocation, as revealed by spectral karyotyping. The Philadelphia chromosome is longer than its normal counterpart, human chromosome 9.

By chance, chromosomes 9 and 22 broke in a stem cell in bone marrow. Each broken part was reattached on the wrong one. At the broken end of chromosome 9, a gene with a role in cell division fused with the control region of a gene at chromosome 22's broken end. Overexpression of the mutant gene leads to uncontrolled divisions of white blood cells. A type of cancer, chronic myelogenous leukemia (CML), is the outcome.

and gorilla chromosomes. The other five chromosomes differ only at inverted and translocated regions.

On rare occasions, a segment of a chromosome may become duplicated, inverted, moved to a new location, or deleted.

Most chromosome changes are harmful or lethal. Others have been conserved over evolutionary time; they confer adaptive advantages or have had neutral effects.

11.8 Changes in the Chromosome Number

Occasionally, abnormal events occur before or during cell division, and gametes and new individuals end up with the wrong chromosome number. Consequences range from minor to lethal physical changes.

In **aneuploidy**, cells usually have one extra or one less chromosome. Autosomal aneuploidy is usually fatal for humans and is linked to most miscarriages. In **polyploidy**, cells have three or more of each type of chromosome. Half of all species of flowering plants, some insects, fishes, and other animals are polyploid.

Nearly all changes in chromosome number arise through **nondisjunction**, whereby one or more pairs of chromosomes don't separate as they should during mitosis or meiosis. Figure 11.17 shows an example.

The chromosome number also may change during fertilization. Suppose a normal gamete fuses by chance with an $n+1$ gamete, with one extra chromosome. The new individual will be trisomic ($2n+1$), with three of one type of chromosome and two of every other type. If an $n-1$ gamete fuses with a normal n gamete, the new individual will be monosomic ($2n-1$). Mitotic divisions perpetuate the mistake when the embryo is growing in size and developing.

AUTOSOMAL CHANGE AND DOWN SYNDROME

A few trisomics are born alive, but only trisomy 21 individuals reach adulthood. A newborn with three chromosomes 21 will develop *Down syndrome*. This autosomal disorder is the most frequent type of altered chromosome number in humans; it occurs once in every 800 to 1,000 births. It affects more than 350,000 people in the United States. Figure 11.17*b* shows a karyotype for a trisomic 21 female. About 95 percent of all cases arise through nondisjunction at meiosis.

Affected individuals have upward-slanting eyes, a fold of skin that starts at the inner corner of each eye, a deep crease across each palm and foot sole, one (not two) horizontal furrows on their fifth fingers, and somewhat flattened facial features. Not all of these symptoms develop in every individual.

That said, trisomic 21 individuals have moderate to severe mental impairment and heart defects. Their skeleton develops abnormally, so older children have shorter body parts, loose joints, and misaligned hip, finger, and toe bones. Muscles and reflexes are weak. Speech and other motor skills develop slowly. With medical care, they live fifty-five years, on average.

The incidence of nondisjunction rises as mothers become older (Figure 11.18). It may originate with the father, but less often. Trisomy 21 is one of hundreds

b

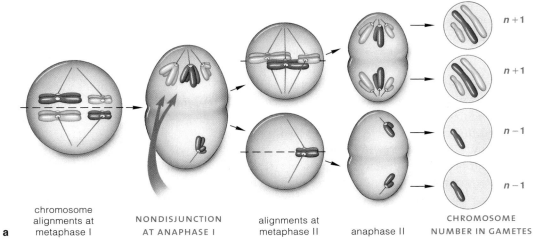

Figure 11.17 (**a**) One example of how nondisjunction arises. Of the two pairs of homologous chromosomes shown here, one fails to separate during anaphase I of meiosis. The chromosome number is altered in the gametes that form after meiosis.

(**b**) An actual case of nondisjunction. This karyotype reveals the trisomic 21 condition of a human female.

a chromosome alignments at metaphase I

NONDISJUNCTION AT ANAPHASE I

alignments at metaphase II

anaphase II

CHROMOSOME NUMBER IN GAMETES

$n+1$

$n+1$

$n-1$

$n-1$

of conditions that can be detected through prenatal diagnosis (Section 11.9). With early special training and medical intervention, individuals still can take part in normal activities. As a group, they tend to be cheerful and sociable.

CHANGES IN THE SEX CHROMOSOME NUMBER

Nondisjunction also causes most of the alterations in the number of X and Y chromosomes. The frequency of such changes is 1 in 400 live births. Most often they lead to difficulties in learning and motor skills, such as speech, although problems can be so subtle that the underlying cause is not even diagnosed.

FEMALE SEX CHROMOSOME ABNORMALITIES *Turner syndrome* individuals have an X chromosome and no corresponding X or Y chromosome (XO). About 1 in 2,500 to 10,000 newborn girls are XO. Nondisjunction originating with the father accounts for 75 percent of the cases. Yet cases are few, compared to other sex chromosome abnormalities. At least 98 percent of XO embryos may spontaneously abort early in pregnancy.

Despite the near lethality, XO survivors are not as disadvantaged as other aneuploids. On average, they are only four feet, eight inches high, but they are well proportioned (Figure 11.19). Most can't make enough sex hormones; they don't have functional ovaries. This affects development of secondary sexual traits, such as breast enlargement. A few eggs form in ovaries but are destroyed by the time these girls are two years old.

Another example: A few females inherit three, four, or five X chromosomes. The *XXX syndrome* occurs at a frequency of about 1 in 1,000 live births. Adults are fertile. Except for slight learning difficulties, most fall within the normal range of social behavior.

MALE SEX CHROMOSOME ABNORMALITIES About one of every 500 to 2,000 males inherits one Y and two or more X chromosomes, mainly through nondisjunction. Most have an XXY mosaic genotype. About 67 percent of those affected inherited the extra chromosome from their mother.

The resulting *Klinefelter syndrome* develops during puberty. XXY males tend to be overweight and tall. The testes and the prostate gland usually are smaller than average. Many XXY males are within the normal range of intelligence, although some have short-term memory loss and learning disabilities. They make less testosterone and more estrogen than normal males, with feminizing effects. Sperm counts are low. Hair is sparse, the voice is pitched high, and the breasts are a

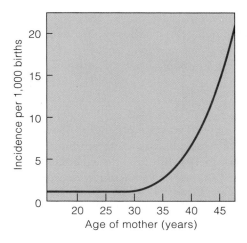

Figure 11.18 Relationship between the frequency of Down syndrome and mother's age at childbirth. The data are from a study of 1,119 affected children. The risk of having a trisomic 21 baby rises with the mother's age. This may seem odd, because about 80 percent of trisomic 21 individuals are born to women not yet 35 years old. However, these women are in the age categories with the highest fertility rates, and they simply have more babies.

bit enlarged. Testosterone injections starting at puberty can reverse the feminized traits.

About 1 in 500 to 1,000 males has one X and two Y chromosomes, an *XYY condition*. They tend to be taller than average, with mild mental impairment, but most are phenotypically normal. XYY males were once thought to be genetically predisposed to a life of crime. This misguided view was based on a sampling error—too few cases among narrowly selected groups, such as prison inmates. The same researchers gathered the karyotypes *and* personal histories. Fanning the stereotype was a report that a mass murderer of young nurses was XYY. He wasn't.

In 1976 a Danish geneticist reported on a study of 4,139 tall males, all twenty-six years old, who had reported to their draft board. Besides giving results of physical examinations and intelligence tests, those draft records held clues to their social and economic status, education, and criminal convictions, if any. Twelve of the males were XYY, which meant there were more than 4,000 males in the control group. The only finding was that mentally impaired, tall males who engage in criminal activity are just more likely to get caught—irrespective of karyotype.

The majority of XXY, XXX, and XYY children may not even be properly diagnosed. Some are dismissed unfairly as being underachievers.

Figure 11.19 One young girl with Turner syndrome.

Nondisjunction in germ cells, gametes, or early embryonic cells changes the number of autosomes or the number of sex chromosomes. The change affects development and the resulting phenotypes.

Nondisjunction at meiosis causes most sex chromosome abnormalities, which typically lead to subtle difficulties with learning, and speech and other motor skills.

11.9 Prospects in Human Genetics

With the arrival of their newborn, parents typically ask, "Is our baby normal?" Quite naturally, they want their baby to be free of genetic disorders, and most babies are. What are the options when they are not?

BIOETHICAL QUESTIONS

Humans do not view diseases and genetic disorders the same way. We attack diseases with antibiotics, surgery, and other tactics. But how do we attack a heritable "enemy" that can be transmitted to our own offspring?

Should we institute regional, national, or global programs to identify people who may carry harmful alleles? Should they be told that they are "defective" and might bestow some disorder on their children? Who decides which alleles are bad? Should society bear the cost of treating all genetic disorders of all individuals, before and after birth? If so, should society also have a say in whether an embryo that bears harmful alleles will be born at all, or aborted? An **abortion** is the expulsion of a pre-term embryo or fetus from the uterus.

As you most likely have learned by now, such questions are the tip of an ethical iceberg.

SOME OF THE OPTIONS

GENETIC SCREENING Through large-scale screening programs, affected individuals or carriers of some harmful allele can be detected early enough to start preventive measures before symptoms develop. For instance, most hospitals in the United States routinely screen newborns for PKU, described next, so we now see fewer individuals with symptoms of the disorder.

PHENOTYPIC TREATMENTS The symptoms of a number of genetic disorders can be minimized or alleviated by surgery, drugs, hormone replacement therapy, or in some cases by controlling diet.

For instance, dietary control works for individuals affected by *phenylketonuria*, or PKU. In this case, a homozygous recessive mutation impairs an enzyme that converts the amino acid phenylalanine to tyrosine. Phenylalanine builds up and is diverted into other pathways. Compounds that impair brain function form as a result. Affected people can lead relatively normal lives by restricting phenylalanine intake. For example, they can avoid soft drinks and other food products sweetened with aspartame, a compound that contains phenylalanine.

Figure 11.20 Amniocentesis, a prenatal diagnostic tool.

A pregnant woman's doctor holds an ultrasound emitter against her abdomen while drawing a sample of amniotic fluid into a syringe. He monitors the path of the needle with an ultrasound screen, in the background. Then he directs the needle into the amniotic sac that holds the developing fetus and withdraws twenty milliliters or so of amniotic fluid. The fluid contains fetal cells and wastes that can be analyzed for genetic disorders.

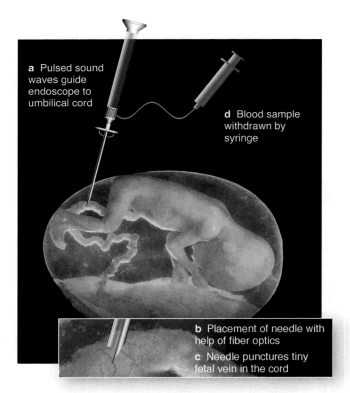

a Pulsed sound waves guide endoscope to umbilical cord

d Blood sample withdrawn by syringe

b Placement of needle with help of fiber optics

c Needle punctures tiny fetal vein in the cord

Figure 11.21 Fetoscopy for prenatal diagnosis.

PRENATAL DIAGNOSIS   Methods of *prenatal diagnosis* are used to determine the sex of embryos or fetuses and to screen for more than 100 genetic abnormalities. *Prenatal* means before birth. *Embryo* is a term that applies until eight weeks after fertilization, after which the term *fetus* is appropriate.

Suppose a forty-five-year-old woman is pregnant and worries about Down syndrome. Between 8 and 12 weeks after conception, such women often request *amniocentesis* (Figure 11.20). A tiny sample of fluid inside the amnion, a membranous sac enclosing the fetus, is withdrawn. Some cells shed by the fetus are suspended in the sample. The cells are analyzed for many genetic disorders, such as Down syndrome, sickle-cell anemia, and cystic fibrosis.

Chorionic villi sampling (CVS) is another procedure. A clinician withdraws a few cells from the chorion, a membranous sac that encloses the amnion and helps form the placenta. Unlike amniocentesis, CVS can yield results as early as eight weeks into pregnancy.

A developing fetus can be seen with an endoscope, a fiber-optic device. During *fetoscopy*, pulsed sound waves are used to scan the uterus and locate parts of the fetus, umbilical cord, or placenta (Figure 11.21). A sample of fetal blood can be drawn, so fetoscopy also is useful to diagnose blood cell disorders such as sickle-cell anemia and hemophilia.

There are risks to a fetus associated with all three procedures, including punctures or infection. Also, if the amnion does not reseal itself fast, too much fluid can leak out and endanger the fetus. Amniocentesis raises the risk of miscarriage by 1 to 2 percent. With CVS, placental development may be compromised, and 0.3 percent of newborns will have missing or underdeveloped fingers and toes. Fetoscopy raises the risk of a miscarriage by 2 to 10 percent.

GENETIC COUNSELING   Parents-to-be can seek *genetic counseling* to compare risks of diagnostic procedures against the risk that their child will be affected by a severe genetic disorder. But they also should be told about the small overall risk of 3 percent that *any* child might have some kind of birth disorder. And they should also consider whether the risk becomes greater with increased age of the potential mother or father.

Suppose a first child or close relative has a severe disorder. Genetic counseling may involve diagnosis of parental genotypes, pedigrees, and genetic testing for known disorders. Using this information, geneticists can predict the risk for disorders in future children. Counselors should remind prospective parents that the same risk usually applies to each pregnancy.

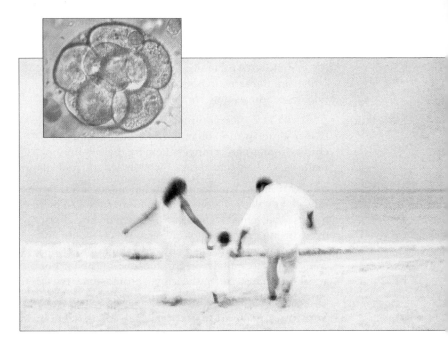

Figure 11.22   Eight-cell and multicelled stages of human development.

REGARDING ABORTION   What happens after prenatal diagnosis reveals a serious problem? Do prospective parents opt for an induced abortion? We can only say here that they must weigh awareness of the severity of the disorder against their ethical and religious beliefs. Worse, they must play out their personal tragedy on a larger stage dominated by a nationwide battle between highly vocal "pro-life" and "pro-choice" factions. We return to this topic in Section 38.12.

PREIMPLANTATION DIAGNOSIS   This procedure relies on **in vitro fertilization**. Sperm and eggs taken from prospective parents are mixed in a sterile culture medium. One or more eggs may be fertilized. If so, within forty-eight hours, mitotic cell divisions may convert it into a ball of eight cells (Figure 11.22 and Section 38.12). According to one view, the tiny, free-floating ball is a *pre*-pregnancy stage. Like unfertilized eggs discarded monthly from a woman, it has not attached to the uterus. All of its cells have the same genes; all are not yet committed to being specialized for any organ. Doctors take one of the undifferentiated cells and analyze its genes. If it has no detectable genetic defects, the ball is inserted into the uterus.

Some couples at risk of passing on cystic fibrosis, muscular dystrophy, or other genetic disorders have opted for the procedure. Many of the resulting "*test-tube*" babies have been born in good health.

Summary

Section 11.1 Human somatic cells are diploid ($2n$), with twenty-three pairs of homologous chromosomes. One is a pairing of sex chromosomes—XX in females, XY in males. All other chromosomes are autosomes, which are the same in both sexes. Crossing over in mitosis produces new combinations of alleles.

Section 11.2 Karyotyping is a diagnostic tool that makes a preparation of an individual's metaphase chromosomes and arranges them by their defining features, such as centromere location, length, and shape.

Section 11.3 The theory of independent assortment does not explain all gene combinations, because pairs of homologous chromosomes swap segments by crossing over during prophase I of meiosis. This frequent and expected event invites genetic recombination, or new combinations of alleles in chromosomes. The farther apart two genes are on the same chromosome, the more likely they will undergo crossing over.

Section 11.4 Geneticists often use pedigrees, or charts of genetic connections in a lineage over time, to estimate probabilities that offspring will inherit a given trait.

Sections 11.5, 11.6 Dominant or recessive alleles on either an autosome or X chromosome can be tracked when they are inherited in simple Mendelian patterns.

Section 11.7 On rare occasions, a chromosome's structure changes. A segment is deleted, inverted, moved to a new location (translocated), or duplicated. Most alterations are harmful or lethal. However, many have accumulated in the chromosomes of all species over evolutionary time. Either they had neutral effects or they later proved to be useful.

Section 11.8 The parental chromosome number can change, as by nondisjunction during meiosis. Aneuploids have one extra or one less chromosome; most autosomal aneuploids die before birth. About half of all flowering plants and some insects, fishes, and other animals are polyploid (three or more of each type of chromosome). More often, changes in number cause genetic disorders.

Section 11.9 Phenotypic treatments, genetic screening, genetic counseling, prenatal diagnosis, and preimplantation diagnosis are some options available for potential parents at risk of having children who will develop a genetic disorder.

Self-Quiz
Answers in Appendix III

1. The probability of a crossover occurring between two genes on the same chromosome is _____ .
   a. unrelated to the distance between them
   b. increased if they are close together
   c. increased if they are far apart

2. Genes on the same chromosome are _____ .
   a. linked   c. homologous   e. all of the
   b. identical alleles   d. autosomes   above

3. Chromosome structure can be altered by a _____ .
   a. deletion   c. inversion   e. all of the
   b. duplication   d. translocation   above

4. A recognized set of symptoms that characterize a specific disorder is a _____ .
   a. syndrome   b. disease   c. pedigree

5. Most genes for human traits are located on _____ .
   a. the X chromosome   c. autosomes
   b. the Y chromosome   d. dominant chromosomes

6. Nondisjunction at meiosis can result in _____ .
   a. karyotyping   c. duplications
   b. crossing over   d. aneuploidy

7. Turner syndrome (XO) is an example of _____ .
   a. dominance   c. aneuploidy
   b. polyploidy   d. gene linkage

8. Match the chromosome terms appropriately.
   ___ crossing over   a. number and defining
   ___ deletion   features of an individual's
   ___ nondisjunction   metaphase chromosomes
   ___ translocation   b. segment of a chromosome
   ___ karyotype   moves to a nonhomologous
   ___ linkage group   chromosome
                         c. disrupts gene linkages
                         d. one outcome: gametes with
                            wrong chromosome number
                         e. a chromosome segment lost
                         f. all genes on a chromosome

Genetics Problems
Answers in Appendix IV

1. Human females are XX and males are XY.
   a. Does a male inherit the X from his mother or father?
   b. With respect to X-linked alleles, how many different types of gametes can a male produce?
   c. If a female is homozygous for an X-linked allele, how many types of gametes can she produce with respect to that allele?
   d. If a female is heterozygous for an X-linked allele, how many types of gametes might she produce with respect to that allele?

2. Marfan syndrome follows a pattern of autosomal dominant inheritance. What is the chance that any child will inherit the dominant allele if one parent does not carry the allele and the other is heterozygous for it?

3. Somatic cells of individuals with Down syndrome usually have an extra chromosome 21; they contain forty-seven chromosomes.
   a. At which stages of meiosis I and II could a mistake alter the chromosome number?
   b. A few individuals have forty-six chromosomes, including two normal-appearing chromosomes 21 and a longer-than-normal chromosome 14. Speculate on how this chromosome abnormality may have arisen.

4. Much of what we know about human genetics comes from studies of experimental organisms. For example, the embryologist Thomas Morgan discovered a genetic basis for a relationship between sex determination and some nonsexual traits in *Drosophila melanogaster*. This fruit fly

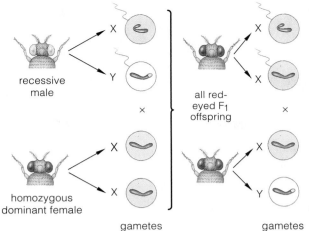

Figure 11.23 The common fruit fly *Drosophila melanogaster*, with (**a**) wild-type red eyes and (**b**) white eyes. (**c**) One experimental search for a *Drosophila* sex-linked gene.

recessive male

×

homozygous dominant female

gametes

all red-eyed F_1 offspring

×

gametes

1/2

1/2

1/2

1/2

1/4 ♂

1/4 ♀

1/4 ♂

1/4 ♀

White-eyed males show up in F_2 generation.

can live in small bottles on agar, cornmeal, molasses, and yeast. A female lays hundreds of eggs in a few days and offspring reproduce in less than two weeks. In a single year, Morgan was able to follow traits through nearly thirty generations of thousands of flies.

At first all the flies were wild-type for eye color; they had red eyes. Then mutation gave rise to a recessive allele for eye color, and a white-eyed male turned up. Morgan established true-breeding strains of white-eyed males and females for *reciprocal crosses*. In the first of such paired crosses, one parent displays the trait of interest. In the second cross, the other parent displays it.

Morgan let white-eyed males mate with homozygous red-eyed females. All F_1 offspring had red eyes, and some F_2 males had white eyes (Figure 11.23). Then Morgan mated true-breeding red-eyed males with white-eyed females. Half the F_1 offspring were red-eyed females, and half were white-eyed males. Also, of the F_2 offspring, 1/4 were red-eyed females, 1/4 white-eyed females, 1/4 red-eyed males, and 1/4 white-eyed males.

Test results pointed to a relationship between an eye-color gene and sex determination. Was the locus on a sex chromosome? Which one? Before answering, think about male and female sex chromosome pairings, and how a dominant allele can mask a recessive one.

5. Does the phenotype indicated by red circles and squares in this pedigree show a Mendelian inheritance pattern that's autosomal dominant, autosomal recessive, or X-linked?

6. One of the *muscular dystrophies*, a category of genetic disorders, is due to a recessive X-linked allele. Usually, symptoms start in childhood. Gradual, progressive loss of muscle function leads to death, usually by age twenty or so. Unlike color blindness, the disorder is nearly always restricted to males. Suggest why.

7. In the human population, mutation of two genes on the X chromosome causes two types of X-linked hemophilia (A and B). In a few cases, a woman is heterozygous for both mutant alleles (one on each of the X chromosomes). All of her sons should have either hemophilia A or B.

However, on very rare occasions, one of these women gives birth to a son who does not have hemophilia, and his one X chromosome does not have either mutant allele. Explain how such an X chromosome could arise.

Media Menu

Student CD-ROM

Impacts, Issues Video
Strange Genes, Tortured Minds
Big Picture Animation
Human inheritance and genetic variations
Read-Me-First Animation
Crossing over and genetic recombination
Karyotype preparation
Other Animations and Interactions
Autosomal dominant inheritance
Autosomal recessive inheritance
X-linked recessive inheritance

InfoTrac
- Not Seeing Red (or Blue or Green). *Bioscience*, August 2000.
- Discovering the Genetics of Autism. *USA Today*, January 2003.
- Origins: A Gene for Speech. *Time*, August 2002.

Web Sites
- Online Mendelian Inheritance in Man: www.ncbi.nlm.nih.gov/Omim
- Genetic Pathology Gallery: www.pathology.washington.edu:80/Cytogallery/
- Eugenics Archive: www.eugenicsarchive.org

How Would You Vote?

Soon there will be diagnostic tests that detect mutations in genes associated with neurobiological disorders. Because environmental factors can influence the development and progression of many NBDs, people with a positive result could make informed life-style choices that may prevent them from becoming affected. Insurance companies and employers might also use that information to deny benefits or employment to individuals at high risk for developing NBDs. Would you support regulation of such diagnostic tests in order to protect individual privacy?

IMPACTS, ISSUES *Goodbye, Dolly*

In 1997, geneticist Ian Wilmut made headlines when he coaxed part of a specialized cell from an adult sheep into becoming part of the first cell of an embryo. His team had been removing the nucleus from unfertilized eggs and slipping a nucleus from a specialized cell of an adult animal into them. Of hundreds of modified eggs, one developed into a whole animal.

The cloned lamb, named Dolly, grew up (Figure 12.1). In time she gave birth to six lambs of her own. Since then, researchers all over the world have cloned other kinds of mammals, including mice, cows, pigs, rabbits, a mule, and cats.

Wilmut and Dolly were back in the limelight in early 2002 with bad news. By age five, Dolly had become fat and arthritic. Sheep usually don't get old-age symptoms until they are about ten years old. In 2003, Dolly developed a progressive lung infection and was put to sleep.

Did Dolly develop health problems simply because she was a clone? Earlier studies of her telomeres had raised suspicions. Telomeres are short segments that cap the ends of chromosomes and stabilize them. They get shorter and shorter as an animal ages. When Dolly was only two years old, her telomeres were as short as those of a six-year-old sheep—the exact age of the animal from which she had been cloned.

Cloning mammals is difficult. Not many nuclear transfers are successful. Most individuals that develop from modified eggs die before birth or shortly afterward.

Figure 12.1 (**a**) Where the molecular revolution started— James Watson and Francis Crick posing in 1953 by their newly unveiled structural model of DNA, the molecule of inheritance in all living cells. (**b**) The now-deceased Dolly. She helped awaken society to the implications of where the molecular revolution is taking us.

the big picture

DNA's Function Early experiments with viruses and bacteria revealed that DNA is the hereditary material in living things. Biochemical analysis helped show that DNA has two strands, helically coiled in a precise pattern.

DNA's Structure In all living things, the structure of DNA arises from base pairing between adenosine and thymine, guanine and cytosine. The sequence of bases along its two strands is unique for each species. It is the foundation for life's diversity.

It took almost seven hundred attempts to get one live clone of a guar, a wild ox on the endangered species list. Less than two days after his foster mother gave birth, he died from complications following an infection.

Surviving clones typically have health problems. Like Dolly, many become unusually overweight as they age. Other clones are exceptionally large from birth or have some enlarged organs. Cloned mice develop lung and liver problems, and almost all die prematurely. Cloned pigs have heart problems, they limp, and one never did develop a tail or, worse still, an anus.

Physically moving a nucleus from one cell to another is just part of the challenge. Most genes in an adult cell are inactive. They have to be reprogrammed or switched on in controlled ways in an unfertilized egg. So far, not all genes in clones are being properly activated.

Some people want to put a stop to cloning complex animals, saying the risk of bringing defective ones into the world troubles them deeply. Others want research to continue because the potential benefits are enormous. However, nearly all agree that cloning humans would be an outrage.

Think about these issues as you read through the rest of the chapters in this unit of the book. They deal with how cells replicate and repair their DNA, how genes are expressed, and what happens when things go wrong. They also invite you to reflect on how researchers are programming these molecular events in previously unimaginable ways.

 How Would You Vote?

Animal cloning experiments often produce abnormal animals, but cloning research may also result in new drugs and organ replacements for human patients. Should animal cloning be banned? See the Media Menu for details, then vote online.

How DNA Is Replicated   During the DNA replication process, the two strands of a double-stranded DNA molecule unwind from each other. Each serves as the template for assembly of a new strand with a complementary base sequence.

Reprogramming DNA   With recent advances in nuclear transfer techniques, researchers have been able to reprogram the DNA of specialized cells to make exact genetic replicas, or clones, of adult mammals.

12.1 The Hunt for Fame, Fortune, and DNA

With this chapter, we turn to investigations that led to our understanding of DNA. The chapter is more than a march through details of DNA's structure and function. It also reveals how ideas are generated in science.

EARLY AND PUZZLING CLUES

In the 1800s, Johann Miescher was collecting cells from the pus of open wounds and sperm cells from a fish. Such cells have little cytoplasm, which makes it easier to isolate their nuclear material. Miescher, a physician, wanted to identify the composition of the nucleus. In time, he isolated an acidic compound that had a bit of phosphorus. He had discovered what became known as **deoxyribonucleic acid**, or **DNA**.

Now fast-forward to 1928. An army medical officer, Frederick Griffith, wanted to develop a vaccine against the bacterium *Streptococcus pneumoniae*, a major cause of pneumonia. He did not succeed, but he isolated and cultured two strains that unexpectedly shed light on mechanisms of heredity. Colonies of one strain had a rough surface appearance; colonies of the other strain appeared smooth. Griffith designated the strains *R* and *S*. He then used them in a series of four experiments, as shown in Figure 12.2.

First, he injected mice with live *R* cells. The mice did not develop pneumonia. *The R strain was harmless.*

Second, he injected other mice with live *S* cells. The mice died. Blood samples from them teemed with live *S* cells. *The S strain was pathogenic; it caused the disease.*

Third, he killed *S* cells by exposing them to high temperature. *Mice injected with dead S cells did not die.*

Fourth, he mixed live *R* cells with heat-killed *S* cells. He injected them into mice. The mice died—*and blood samples drawn from them teemed with live S cells!*

What went on in the fourth experiment? Maybe heat-killed *S* cells in the mix weren't really dead. But if that were so, then the mice injected with just the heat-killed *S* cells in experiment 3 would have died. Or maybe the harmless *R* cells had mutated into a killer strain. But if that were so, then mice injected with just *R* cells in experiment 1 would have died.

The simplest explanation was this: *Heat had killed the S cells but did not destroy their hereditary material— including the part that specified "how to cause infection."* Somehow, that material had been transferred from the dead *S* cells into living *R* cells, which put it to use.

After later tests, it was clear that the transformation was permanently heritable. Even after a few hundred generations, *S* cell descendants were still infectious.

What was the hereditary material that caused the transformation? Scientists started looking in earnest, but most were thinking "proteins." Because heritable traits are diverse, they assumed the molecules of inheritance had to be structurally diverse, too. Proteins, they said, could be built from unlimited mixes of twenty kinds of amino acids. Other molecules just seemed too simple.

But Griffith's results intrigued Oswald Avery, who went on to transform the harmless bacterial cells with extracts of killed pathogens. When he added protein-digesting enzymes to the extracts, bacterial cells were still transformed. However, when he added an enzyme that digests DNA but not protein to the extracts, the cells were *not* transformed. DNA was looking good.

CONFIRMATION OF DNA FUNCTION

By the 1950s, molecular detectives were using viruses for experiments. **Bacteriophages**, which infect certain bacteria, were the viruses of choice. These infectious particles contain information on how to build new virus particles. At some point after they infect a host cell, viral enzymes take over its metabolic machinery. And then the machinery starts synthesizing substances necessary to make more virus particles.

As researchers knew, some bacteriophages consist only of DNA and a protein coat. Also, micrographs revealed that the coat remains on the *outer surface* of infected cells. Were the viruses injecting only hereditary material into cells? If so, was the material protein, DNA, or both? Figure 12.3 outlines just two of many experiments that pointed to DNA.

❶ Mice injected with live cells of harmless strain *R*

❷ Mice injected with live cells of killer strain *S*

❸ Mice injected with heat-killed *S* cells

❹ Mice injected with live *R* cells plus heat-killed *S* cells

Mice don't die. No live *R* cells in their blood

Mice die. Live *S* cells in their blood

Mice don't die. No live *S* cells in their blood

Mice die. Live *S* cells in their blood

Figure 12.2 Summary of results of Fred Griffith's experiments with *Streptococcus pneumoniae* and laboratory mice.

Read Me First!
and watch the narrated animation on
the Hershey–Chase experiments

virus particle
labeled with ^{35}S

DNA (*blue*)
being injected
into bacterium

^{35}S remains
outside cells

a

virus particle
labeled with ^{32}P

DNA (*blue*)
being injected
into bacterium

^{32}P remains
inside cells

b

DNA
inside
protein
coat

tail
fiber

hollow
sheath

micrograph of virus particles
injecting DNA into an *E. coli* cell

Figure 12.3 Example of the landmark experiments that tested whether genetic material resides in bacteriophage DNA, proteins, or both. As Alfred Hershey and Martha Chase knew, sulfur (S) but not phosphorus (P) is present in proteins, and phosphorus but not sulfur is present in DNA.

(**a**) In one experiment, bacterial cells were grown on a culture medium with a tracer, the radioisotope ^{35}S. The cells used the ^{35}S when they built proteins. Bacteriophages infected the labeled cells, which started making viral proteins. So the proteins, and new virus particles, became labeled with the ^{35}S. The labeled virus particles infected a new batch of unlabeled cells. The mixture was whirred in a kitchen blender. Whirring dislodged the viral protein coats from infected cells. Chemical analysis revealed the presence of labeled protein in the solution but only traces of it inside the cells.

(**b**) In another experiment, bacteriophages infected cells that had taken up the radioisotope ^{32}P. The infected cells used the ^{32}P when they built viral DNA. The DNA became labeled, as did new virus particles. Later, the labeled viruses infected bacteria in solution, then were dislodged from them. Most of the labeled viral DNA stayed in the cells—evidence that DNA is the genetic material of this virus.

Then Linus Pauling did something no one had done before. With his training in biochemistry, a talent for model building, and a dose of intuition, he deduced the structure of a protein—collagen. His discovery was electrifying. If someone could pry open the secrets of proteins, then why not DNA? And if DNA's structural details were deduced, wouldn't they hold clues to how it functions? *Someone could go down in history as having discovered the secret of life!*

ENTER WATSON AND CRICK

Having a shot at fame and fortune quickens the pulse of men and women in any profession, and scientists are no exception. However, science is a community effort. Individuals share not only what they find but also what they do not understand. Even if an experiment does not yield an expected result, it may turn up something that others can use or raise questions others can answer.

And so scientists all over the world started sifting through all the clues. Among them were James Watson, a postdoctoral student from Indiana University, and Francis Crick, a researcher at Cambridge University. They spent hours arguing over everything they read about DNA's size, shape, and bonding requirements. They fiddled with cardboard cutouts and badgered chemists to help them identify possible bonds they may have overlooked. They built models from thin bits of metal connected with suitably angled "bonds" of wire.

In 1953, they built a model that fit all the pertinent biochemical rules and insights they had gleaned from other sources. They had discovered DNA's structure. As you will see, the structure's breathtaking simplicity helped Crick answer another enormous riddle—*how life can show unity at the molecular level and still give rise to spectacular diversity at the level of whole organisms.*

DNA functions as the cell's treasurehouse of inheritance. Its molecular structure encodes the information required to reproduce parental traits in offspring.

12.2 The Discovery of DNA's Structure

Long before the bacteriophage studies were under way, biochemists knew that DNA contains only four kinds of nucleotides that are the building blocks of nucleic acids. But how were the nucleotides arranged in DNA?

DNA'S BUILDING BLOCKS

Each **nucleotide** consists of a five-carbon sugar (which, in DNA, is deoxyribose), a phosphate group, and one of the following nitrogen-containing bases:

adenine	guanine	thymine	cytosine
A	G	T	C

In all four types of nucleotides, the component parts are organized the same way (Figure 12.4). But T and C are pyrimidines, with a carbon backbone arranged as a single ring. A and G are purines, which are larger, bulkier molecules; they have two carbon rings.

By 1949, the biochemist Erwin Chargaff had shared with the scientific community two crucial insights into the composition of DNA. First, the amount of adenine relative to guanine differs from one species to the next. Second, the amounts of thymine and adenine in a DNA molecule are exactly the same, and so are the amounts of cytosine and guanine. We may show this as A = T and G = C. The symmetrical proportions of the four kinds of nucleotides had to mean something. Was this a tantalizing clue to how the nucleotides are arranged in the DNA molecule?

The first convincing evidence of that arrangement emerged from Maurice Wilkins's research laboratory in England. Rosalind Franklin, one of his colleagues, made good **x-ray diffraction images** of DNA. (Maybe for reasons given in Section 12.3, her contribution has only recently been acknowledged.) X-ray diffraction images are made after directing a beam of x-rays at a molecule, which scatters the x-rays in a pattern that can be captured on film. The pattern consists only of dots and streaks; in itself, it is not the structure of the molecule. Researchers use it to calculate the positions of the molecule's atoms.

Also, DNA must be processed first. A suspension of DNA molecules has to be spun rapidly, spooled onto a rod, and gently pulled into gossamer fibers, like cotton candy. When dry, the fibers twist and turn into two forms, which makes x-ray diffraction images too complicated to decipher. Wet fibers have only one.

Franklin was the first to make a spectacularly clear x-ray diffraction image of wet DNA fibers, as Figure 12.5 shows. She used it to calculate that a molecule of DNA is long and thin, with a 2-nanometer diameter. And she calculated that some molecular configuration is repeated every 0.34 nanometer along its length, and another every 3.4 nanometers.

Figure 12.5 Rosalind Franklin's superb x-ray diffraction image of DNA fibers.

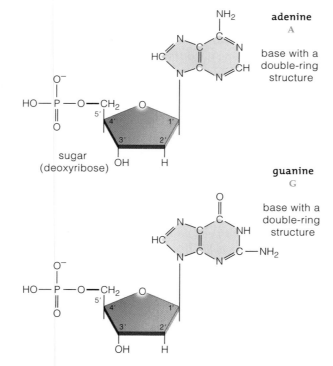

Figure 12.4 All of the chromosomes in a cell contain DNA. What does DNA contain? Four kinds of nucleotides: A, G, T, and C.

Could the sequence of DNA's nucleotide bases be twisting up in a repeating pattern, a bit like a circular stairway? Certainly Pauling thought so. After all, he discovered a helical shape in collagen. Like everyone else—including Wilkins, Watson, and Crick—he was thinking "helix." Watson later wrote, "We thought, why not try it on DNA? We were worried that *Pauling* would say, why not try it on DNA? Certainly he was a very clever man. He was a hero of mine. But we beat him at his own game. I still can't figure out why."

Pauling, it turned out, made a big chemical mistake. His model had all the negatively charged phosphate groups inside the DNA helix instead of outside. If they were that close together, they would repel each other too much to be stable.

PATTERNS OF BASE PAIRING

Franklin filed away the image of wet fibers, and then Watson and Crick took the lead. They perceived that DNA must consist of two strands of nucleotides, held together at their bases by hydrogen bonds (Figure 12.6). Such bonds form when the two strands run in opposing directions and twist to form a double helix. Two kinds of base pairings form along the molecule's length: A—T and G—C.

The bonding pattern accommodates variation in the order of bases. For instance, a stretch of DNA from a rose, a human, or any other organism might be:

All DNA molecules show the same bonding pattern, but each species has a number of unique DNA base sequences. *This molecular constancy and variation among species is the foundation for the unity and diversity of life.*

Intriguingly, computer simulations show that if you want to pack a string into the least space, coil it into a helix. Was a space-saving advantage a factor in the molecular evolution of the DNA double helix? Maybe.

The pattern of base pairing between the two strands in DNA is constant for all species—A with T, and G with C. However, each species has a number of unique sequences of base pairs along the length of their DNA molecules.

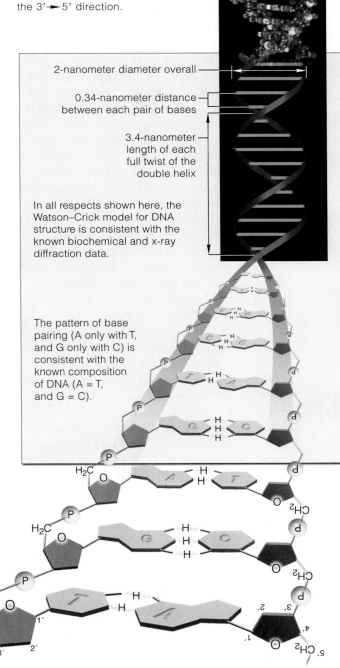

Figure 12.6 Composite of different models for a DNA double helix. The two sugar–phosphate backbones run in *opposing* directions. Think of the sugar units (deoxyribose) of one strand as being upside down.

By comparing the numerals used to identify each carbon atom of the deoxyribose molecule (1', 2', 3', and so on), you see that one strand runs in the 5'→3' direction and the other runs in the 3'→5' direction.

2-nanometer diameter overall

0.34-nanometer distance between each pair of bases

3.4-nanometer length of each full twist of the double helix

In all respects shown here, the Watson–Crick model for DNA structure is consistent with the known biochemical and x-ray diffraction data.

The pattern of base pairing (A only with T, and G only with C) is consistent with the known composition of DNA (A = T, and G = C).

12.3 Rosalind's Story

Watson and Crick got the attention of the world, in part by using Rosalind Franklin's data. She got cancer, most likely because of her intensive work with x-rays.

When Rosalind Franklin started at King's Laboratory of Cambridge University, she already had impressive credentials. She developed a refined x-ray diffraction method while studying the structure of coal. She took a new mathematical approach to interpreting x-ray diffraction images and, like Pauling, had built three-dimensional molecular models. At Cambridge, she was asked to create and run a state-of-the-art x-ray crystallography laboratory. Her assignment was to investigate the structure of DNA.

No one bothered to tell Franklin that, down the hall, Maurice Wilkins was working on the puzzle. Even the graduate student assigned to assist her didn't mention it. No one bothered to tell Wilkins about Franklin's assignment; he assumed she was a technician hired to do his x-ray crystallography work because he didn't know how to do it himself. And so the clash began. To Franklin, Wilkins seemed inexplicably prickly. To Wilkins, Franklin displayed an appalling lack of the deference that technicians usually show researchers.

Wilkins had a prized cache of crystalline DNA fibers —each with parallel arrays of hundreds of millions of DNA molecules—which he gave to his "technician." Five months later, Franklin gave a talk on what she had learned. DNA, she said, may have two, three, or four parallel chains twisted in a helix, with phosphate groups projecting outward.

With his crystallography background, Crick would have recognized the significance of her report—*if* he had been there. (A *pair* of chains oriented in opposing directions would be the same even if flipped 180 degrees. Two pairs of chains? No. DNA's density ruled that out. But one pair of chains? Yes!) Watson was in the audience but didn't know what Franklin was talking about.

Later, Franklin produced her outstanding x-ray diffraction image of wet DNA fibers. It fairly screamed *Helix!* She also worked out DNA's length and diameter. But she had been working with dry fibers for a long time and didn't dwell on her new data. Wilkins did.

In 1953, he let Watson see that image and reminded him of what Franklin had reported more than a year before. When Watson and Crick did focus on her data, they had the final bit of information they needed to build a DNA model—one with two helically twisted chains running in opposing directions.

Figure 12.7 Portrait of Rosalind Franklin arriving at Cambridge in style, from Paris.

12.4 DNA Replication and Repair

The discovery of DNA structure was a turning point in studies of inheritance. Crick understood at once how cells duplicate their DNA before they divide.

Until Watson and Crick presented their model, no one could explain **DNA replication**, or how the molecule of inheritance is duplicated before a cell divides.

Enzymes easily break the hydrogen bonds between the two nucleotide strands of a DNA molecule. When enzymes and other proteins act on the molecule, one strand unwinds from the other and exposes stretches of its nucleotide bases. Cells contain stockpiles of free nucleotides that can pair with the exposed bases.

Each parent strand stays intact, and a companion strand gets assembled on each one according to this base-pairing rule: A to T, and G to C. As soon as a stretch of a new, partner strand forms on a stretch of the parent strand, the two twist together into a double helix. Because the parent DNA strand is conserved during the replication process, half of every double-stranded DNA molecule is "old" and half is "new." Figures 12.8 and 12.9 describe this process, which we call *semiconservative* replication.

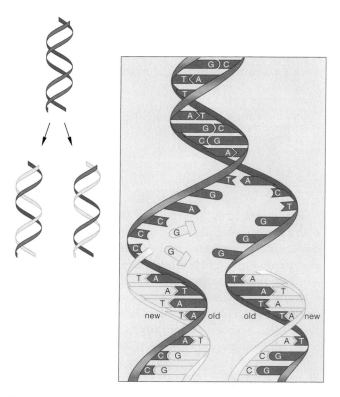

Figure 12.8 A simple picture of the semiconservative nature of DNA replication. The original two-stranded DNA molecule is coded *blue*. Each parent strand remains intact. One new strand (*gold*) is assembled on each of the parent strands.

Read Me First!
and watch the narrated animation on
DNA replication

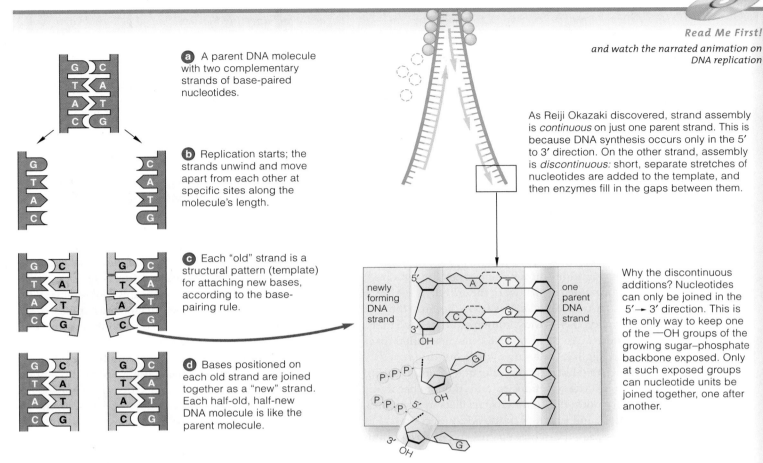

a A parent DNA molecule with two complementary strands of base-paired nucleotides.

b Replication starts; the strands unwind and move apart from each other at specific sites along the molecule's length.

c Each "old" strand is a structural pattern (template) for attaching new bases, according to the base-pairing rule.

d Bases positioned on each old strand are joined together as a "new" strand. Each half-old, half-new DNA molecule is like the parent molecule.

As Reiji Okazaki discovered, strand assembly is *continuous* on just one parent strand. This is because DNA synthesis occurs only in the 5' to 3' direction. On the other strand, assembly is *discontinuous*: short, separate stretches of nucleotides are added to the template, and then enzymes fill in the gaps between them.

newly forming DNA strand

one parent DNA strand

Why the discontinuous additions? Nucleotides can only be joined in the 5'→ 3' direction. This is the only way to keep one of the —OH groups of the growing sugar–phosphate backbone exposed. Only at such exposed groups can nucleotide units be joined together, one after another.

Figure 12.9 A closer look at DNA replication.

DNA replication uses a team of molecular workers. In response to cellular signals, replication enzymes become active along the length of the DNA molecule. Along with other proteins, some enzymes unwind the strands in both directions and prevent them from rewinding. Enzyme action jump-starts the unwinding but is not required to unzip hydrogen bonds between the strands; hydrogen bonds are individually weak.

Now **DNA polymerases**, a class of enzymes, attach short stretches of free nucleotides to unwound parts of the parent template. Free nucleotides themselves drive the strand assembly. Each has three phosphate groups. A DNA polymerase splits off two, releasing energy that drives the attachments.

DNA ligases fill in the tiny gaps between the new short stretches and form one continuous strand. Then enzymes wind up the template and complementary strands together, forming a DNA double helix.

Sometimes a molecule of DNA breaks. Sometimes it is replicated incorrectly so that it does not exactly match the parent molecule; it acquired mismatched bases or has missing or extra segments. DNA repair processes minimize such damage. Ligases can fix the breaks, and specialized DNA polymerases can correct the mismatched base pairs or replace mutated ones.

The repair processes confer survival advantage on cells. Why? Mistakes in DNA can result in the altered or diminished function of encoded proteins and thus disrupt how cells operate. This is what happens with genetic disorders. Also, a broken strand of DNA may block replication and cause a cell to commit suicide by issuing signals for its own death.

DNA is replicated prior to cell division. Enzymes unwind its two strands. Each strand remains intact throughout the process—it is conserved—and enzymes assemble a new, complementary strand on each one.

Mistakes happen. Repair systems fix mismatched base pairs and help maintain the integrity of genetic information. They also bypass breaks, which helps keep replication from shutting down.

12.5 Reprogramming DNA To Clone Mammals

Knowledge of DNA structure and function opened up exciting, and troubling, research avenues—including cloning adult mammals from little more than DNA and a cell stripped of its nucleus.

Geneticists started out cloning some embryos from *in vitro* fertilization. Briefly, a sperm fertilizes an egg in a petri dish. Mitotic divisions produce two cells, then four, then eight. The eight-cell stage is gently split into two- or four-cell clusters that are implanted in surrogate mothers. There they grow and develop into genetically identical animals; they are clones.

Splitting such early stages, or *artificial twinning*, gives us clones that are identical to one another but not to sexually reproducing parents; they have genes from both parents. Such clones must grow up before researchers can find out whether genes for a desired maternal or paternal trait were inherited.

Using a differentiated cell is faster, because the desired genotype is already known. You may wonder what "differentiated" means. All cells descended from a fertilized egg have the same DNA, but different lineages start making unique selections from it during development. The selections commit them to becoming liver cells, blood cells, or other specialists in structure, composition, and function in the adult (Section 14.3).

A differentiated cell must be tricked into rewinding the clock. *Its DNA must be reprogrammed into starting over again and directing the development of a whole individual.* Nuclear transfer is one way to trick it. The nucleus of a differentiated cell from an animal to be cloned replaces an unfertilized egg's nucleus (Figure 12.10). Chemicals or electric shocks may induce the cell to divide. If all goes well, a cluster of embryonic cells forms and can be implanted in a surrogate mother.

In Dolly's case, the nucleus came from a cell in a sheep's udder. Nuclei from cumulus cells were used to clone mice and CC, the first cloned cat. Cumulus cells surround immature eggs in mammalian ovaries. Genetic tests confirmed that CC (short for *Carbon Copy*) is a clone, even though her coat patterning differs from that of the genetic donor (Figure 12.10). Here is visible evidence that environmental factors (in this case, in the uterus) can alter gene expression.

Variations in gene expression are less obvious but more of a problem in other clones. About 4 percent of all the genes tested in cloned mice were expressed at abnormal levels. Gene expression also was disturbed in cloned mice that had received genetic material from cells of entirely different tissues. Does this mean that nuclear transfer procedures invite defects? Probably.

Abnormal gene expression in clones isn't surprising. Genes switch on and off during normal development. Researchers are not yet rewinding the clock to cover all of the ticks—minutes or hours—between nuclear transfer and the first cell divisions.

① A microneedle is about to remove the nucleus from an unfertilized sheep egg (center).

② The microneedle has now emptied the sheep egg of its own nucleus, which held the DNA.

③ DNA from a donor cell is about to be deposited in the enucleated egg.

④ An electric spark will stimulate the egg to enter mitotic cell division. After a few rounds of divisions, the ball of cells will be implanted inside the womb of a surrogate female sheep (ewe).

the first cloned sheep

Figure 12.10 Steps in the nuclear transfer process that led to Dolly, and a gallery of famous firsts in the brave new world of mammalian cloning.

the first cloned mice

the first cloned pigs

CC (*left*), and her genetic donor, Rainbow.

Summary

Section 12.1 Experimental tests that used bacteria and bacteriophages offered the first solid evidence that DNA is the hereditary material in living organisms.

Sections 12.2, 12.3 DNA consists only of nucleotides, each with a five-carbon sugar (deoxyribose), a phosphate group, and one of four kinds of nitrogen-containing bases: adenine, thymine, guanine, or cytosine.

A DNA molecule consists of two nucleotide strands twisted together into a double helix. Bases of one strand hydrogen-bond with bases of the other.

Bases of the two DNA strands pair in a constant way. Adenine pairs with thymine (A–T), and guanine with cytosine (G–C). Which base follows another along a strand varies among species. The DNA of each species incorporates some number of unique stretches of base pairs that set it apart from the DNA of all other species.

Section 12.4 In DNA replication, enzymes unwind the two strands of a double helix and assemble a new strand of complementary sequence on each parent strand. Two double-stranded DNA molecules result. One strand of each molecule is old (is conserved); the other is new.

During replication, repair systems fix base-pairing mistakes and mutated bases. They also repair breaks in DNA strands that could shut down replication. DNA ligase and special DNA polymerases are involved.

Section 12.5 Embryo splitting and nuclear transfers are two methods that produce clones, or individuals that have identical DNA. Clones can show phenotypic differences if they are exposed to different factors that affect gene expression during development.

Self-Quiz

Answers in Appendix III

1. Which is *not* a nucleotide base in DNA?
   a. adenine   c. uracil   e. cytosine
   b. guanine   d. thymine   f. All are in DNA.

2. What are the base-pairing rules for DNA?
   a. A–G, T–C   c. A–U, C–G
   b. A–C, T–G   d. A–T, G–C

3. One species' DNA differs from others in its _____ .
   a. sugars   c. base sequence
   b. phosphates   d. all of the above

4. When DNA replication begins, _____ .
   a. the two DNA strands unwind from each other
   b. the two DNA strands condense for base transfers
   c. two DNA molecules bond
   d. old strands move to find new strands

5. DNA replication requires _____ .
   a. free nucleotides   c. many enzymes
   b. new hydrogen bonds   d. all of the above

6. Cell differentiation involves _____ .
   a. cloning   c. selective gene expression
   b. nuclear transfers   d. both b and c

Critical Thinking

1. Matthew Meselson and Frank Stahl's experiments supported a semiconservative model of DNA replication. These researchers obtained "heavy" DNA by growing *Escherichia coli* in a medium enriched with ^{15}N, a heavy isotope of nitrogen. They also prepared "light" DNA by growing *E. coli* in the presence of ^{14}N, the more common isotope. An available technique helped them identify which replicated molecules were heavy, light, or hybrid (one heavy strand and one light). Use pencils of two colors, one for heavy strands and one for light. Starting with a DNA molecule having two heavy strands, arrange them to show how daughter molecules would form after replication in a ^{14}N-containing medium. Show the four DNA molecules that would form if daughter molecules are replicated a second time in the ^{14}N medium.

2. Mutations, permanent changes in DNA base sequences, are the original source of genetic variation and the raw material of evolution. Yet how can mutations accumulate, given that cells have repair systems that fix structurally altered or discontinuous DNA strands during replication?

IMPACTS, ISSUES   *Ricin and Your Ribosomes*

In 2003, police acting on an intelligence tip stormed a London apartment, where they collected castor oil beans (Figure 13.1) and laboratory glassware. They arrested several young men and reminded the world of ricin's potential as a biochemical weapon.

Ricin is a protein product of the castor oil plant (*Ricinus communis*). A dose the size of a grain of salt can kill you; only plutonium and botulism toxin are more deadly. Researchers knew about it as long ago as 1888. Later, when Germany unleashed mustard gas against allied troops during World War I, the United States and England feverishly investigated whether ricin, too, could be used as a weapon. Both countries shelved the research when the war ended.

Fast-forward to 1969. Georgi Markov, a Bulgarian writer, defected to the West at the height of the Cold War.

Figure 13.1 Castor oil plant seeds, source of the ribosome-busting ricin.

As he strolled down a London street, an assassin used a modified umbrella to poke a tiny ball laced with ricin into one of his legs. Markov died in agony three days later.

Ricin is on stage once again. Traces of ricin showed up in hastily abandoned Afghanistan caves in 2001 and on a Chechen fighter killed in Moscow in 2003. In early 2004, traces turned up in a United States Senate mailroom, in a State Department building, and in an envelope addressed to the White House.

Each year, a lot of ricin-rich wastes form during castor oil production. The entire castor oil plant is poisonous, but ricin is most concentrated in seeds. How does ricin exert its deadly effects? It inactivates ribosomes, the cell's protein-building machinery.

Ricin is a protein. One of its two polypeptide chains, shown in the filmstrip, helps ricin insert itself into cells. The other chain, an enzyme, wrecks part of the ribosome where amino acids are joined together. It yanks adenine subunits from an RNA molecule that is a crucial part of the ribosome. The ribosome's three-dimensional shape unravels, protein synthesis stops, and cells spiral toward death. So does the individual; there is no antidote.

It's possible to get on with your life without knowing what a ribosome is or what it does. It also is possible to recognize that protein synthesis is not a topic invented to torture biology students. It is something worth knowing about and appreciating for how it keeps us alive—and for appreciating anti-terrorism researchers who are working to keep us that way.

the big picture

Making the Transcripts   It takes two steps to get from DNA's protein-building information to a new protein molecule. In the first step, a strand of mRNA is transcribed from a gene region, a sequence of nucleotide bases in an unwound part of a DNA molecule.

Readers of the Genetic Code   Every three ribonucleotide bases along the length of an mRNA transcript is a "word" corresponding to a particular amino acid. Two other classes of RNAs recognize a range of these base triplets, which represent a genetic code.

So start with what you know about DNA, the book of protein-building information in cells. The alphabet used to write the book is simple enough—just A, T, G, and C, for the nucleotide bases adenine, thymine, guanine, and cytosine. But how do you get from an alphabet to a protein? The answer starts with the order, or sequence, of those four nucleotide bases in a DNA molecule.

As you already know, the two strands unwind from each other entirely when a cell is replicating its DNA. At other times, however, cells selectively unwind the two strands in certain regions and thereby expose the base sequences we call genes. Most of the genes encode information on building particular proteins.

You'll see from this chapter that it takes two steps, transcription and translation, to do something with the information in a gene. In all eukaryotic cells, the first step proceeds in the nucleus. A newly exposed DNA base sequence functions as a structural pattern, or a template, for making a strand of ribonucleic acid (RNA) from the cell's pool of free nucleotides.

The RNA then moves into the cytoplasm, where it is translated. In this second step of protein synthesis, the RNA guides the assembly of amino acids into a new polypeptide chain. The new chains become folded into the three-dimensional shapes of specific proteins.

In short, DNA guides the synthesis of RNA, then RNA guides the synthesis of proteins:

$$\text{DNA} \xrightarrow{\text{transcription}} \text{RNA} \xrightarrow{\text{translation}} \text{PROTEIN}$$

How Would You Vote?

Ricin is difficult to disperse through the air and is unlikely to be used in a large-scale terrorist attack. However, ricin powder did turn up in a Senate office building. Scientists are working to develop a vaccine against ricin. If mass immunizations were to be offered, would you sign up to be vaccinated? See the Media Menu for details, then vote online.

Translating the Transcripts   In the second step of protein synthesis, tRNAs and rRNAs interact to translate the mRNA transcript of a gene region into a polypeptide chain. The chain grows as one of the rRNA components of ribosomes catalyzes the bonding between amino acids.

Mutations and Proteins   Mutations change the genetic code words in the messages that specify particular proteins. When the protein is an essential part of cell architecture or metabolism, we can expect the outcome to be an abnormal cell.

13.1 How Is RNA Transcribed From DNA?

In transcription, the first step in protein synthesis, the base sequence in an unwound DNA region becomes a template for assembling a strand of ribonucleic acid, or RNA.

It takes three classes of RNA molecules to synthesize proteins. Most genes are transcribed into **messenger RNA**, or **mRNA**—the only kind that carries *protein-building* instructions. Other genes are transcribed into **ribosomal RNA**, or **rRNA**, a component of ribosomes.

A ribosome is a large molecular structure upon which polypeptide chains are assembled. Transcription of still other genes yields **transfer RNA**, or **tRNA**, which can deliver amino acids one at a time to a ribosome.

THE NATURE OF TRANSCRIPTION

An RNA molecule is almost but not quite like a single strand of DNA. It has four kinds of ribonucleotides, each with the five-carbon sugar ribose, a phosphate group, and a base. Three bases—adenine, cytosine, and guanine—are the same as those in DNA. In RNA, however, the fourth base is **uracil**, not thymine. Like thymine, uracil can pair with adenine. This means a new RNA strand can be built according to the same base-pairing rules as DNA (Figure 13.2).

Transcription *differs* from DNA replication in three respects. Part of a DNA strand, not the whole molecule, is used as the template. The enzyme **RNA polymerase**, not DNA polymerase, adds ribonucleotides one at a time to the end of a growing strand of RNA. And transcription results in one free strand of RNA, not a hydrogen-bonded double helix.

The many coding regions in DNA are transcribed separately, and each has its own START and STOP signal. A **promoter** is one sequence of bases in DNA that signals the start of a gene. RNA synthesis gets going as soon as RNA polymerases and other proteins attach to it. Each polymerase moves along the DNA strand, joining one ribonucleotide after another on

a   sugar (ribose)   b   sugar (deoxyribose)

Figure 13.2 (**a**) Uracil, one of four ribonucleotides in RNA. The other three—adenine, guanine, and cytosine—differ only in their bases. Compare uracil to (**b**) thymine, a DNA nucleotide. (**c**) Base pairing of DNA with RNA at transcription, compared to base pairing during DNA replication.

DNA — RNA   base pairing during transcription

DNA — DNA   base pairing during DNA replication

promoter region

RNA polymerase, the enzyme that catalyzes transcription

newly forming RNA transcript

DNA template at selected transcription site

DNA template winding up

DNA template unwinding

a RNA polymerase initiates transcription at a promoter region in DNA. It recognizes a base sequence located next to the promoter as a template. It will link the nucleotides adenine, cytosine, guanine, and uracil into a strand of RNA, in the order specified by DNA.

b All through transcription, the DNA double helix becomes unwound in front of the RNA polymerase. Short lengths of the newly forming RNA strand briefly wind up with its DNA template strand. New stretches of RNA unwind from the template (and the two DNA strands wind up again).

Figure 13.3 Gene transcription. By this process, an RNA molecule is assembled on a DNA template. (**a**) Gene region of DNA. The base sequence along one of DNA's two strands (not both) is used as the template. (**b–d**) Transcribing that region results in a molecule of RNA.

the DNA as a template (Figure 13.3). It soon reaches another base sequence in the DNA that signals "the end," and the new RNA is released as a free transcript.

FINISHING TOUCHES ON mRNA TRANSCRIPTS

In eukaryotic cells, each new mRNA molecule is not in final form. Just as a dressmaker may snip off some threads or put bows on a dress before it leaves the shop, so do these cells tailor their "pre-mRNA." For instance, some enzymes attach a modified guanine "cap" to the start of each pre-mRNA transcript. Others attach about 100 to 300 adenine ribonucleotides as a tail to the other end. Hence its name, poly-A tail.

Later, in the cytoplasm, the cap will help bind the mRNA to a ribosome. Enzymes will begin to nibble away at the tail. Each tail's length determines how long a particular mRNA molecule will last. It helps keep protein-building messages intact for as long as the cell requires them.

The protein-coding parts of eukaryotic genes are **exons**. In between them are one or more **introns**, or sequences that are removed before a mature mRNA transcript is translated. The introns are sites where the protein-building information can be snipped apart and spliced back together in more than one way. This *alternative splicing* lets cells use the same gene to make proteins that differ slightly in form and function. It may be a way to increase DNA's information-storing capacity, hence the capacity to make diverse proteins.

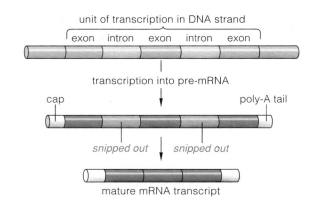

Figure 13.4 Pre-mRNA transcript processings. Some or all introns are removed before the transcript leaves the nucleus.

As Figure 13.4 shows, introns are transcribed along with exons. They are snipped out before the mature mRNA transcript leaves the nucleus. Think of it this way: The *ex*ons are *ex*ported from the nucleus, and *in*trons stay *in* the nucleus, where they are recycled.

> *In gene transcription, a sequence of exposed bases on one of the two strands of a DNA molecule serves as a template.*
>
> *Using that template, RNA polymerase assembles a single strand of RNA from the four ribonucleotides, A, U, C, and G.*
>
> *Before leaving the nucleus, each new mRNA transcript, or pre-mRNA, undergoes modification into final form.*

C What happened at the assembly site? RNA polymerase catalyzed the assembly of ribonucleotides, one after another, into an RNA strand, using exposed bases on the DNA as a template. Many other proteins assist this process.

d At the end of the gene region, the last stretch of the new transcript is unwound and released from the DNA template. Shown below is a model for a transcribed strand of RNA.

13.2 Deciphering mRNA Transcripts

Like a strand of DNA, an mRNA transcript is a linear sequence of nucleotides. What are the protein-building "words" in its sequence? Each is a three-letter word, composed of three nucleotide bases.

THE GENETIC CODE

Marshall Nirenberg, Philip Leder, Severo Ochoa, and Gobind Korana deduced the correspondence between genes and proteins. Picture an mRNA transcript that was built on a DNA template, as in Figure 13.5a. To translate it into the amino acid sequence of a protein, you have to know just how many letters—nucleotide bases—correspond to a word (an amino acid). That is what the researchers figured out.

As you will see, after mRNA docks at a ribosome, its ribonucleotide bases are "read" *three at a time*, as triplets. Base triplets in mRNA transcripts were given the name **codons**. Figure 13.5b shows how the order of different codons in mRNA determines the order in which different kinds of amino acids will follow one another in a growing polypeptide chain.

Count the different codons in Figure 13.6, and you see there are sixty-four possible choices. That's a lot more than the twenty kinds of amino acids. But some amino acids are encoded by more than one codon. For instance, both GAA and GAG call for glutamate.

In most species, the first AUG in a transcript is a START signal for translating "three-bases-at-a-time." This means methionine is the first amino acid in all new polypeptide chains. UAA, UAG, and UGA do not call for any amino acid. They are STOP signals that block further additions of amino acids to a new chain.

The set of sixty-four different codons is a **genetic code**. Protein synthesis adheres to this code in nearly all cases. Mitochondria are one of the exceptions. They have their own "mitochondrial code," which includes a few unique codons. This is one clue that supports a theory of endosymbiotic origins for eukaryotic cells. By this theory, ancient aerobic bacteria were ingested by other cells. They resisted digestion, then evolved into mitochondria inside host cells (Section 18.4).

DNA

mRNA

a

mRNA codons

amino acids

b threonine proline glutamate glutamate lysine

Figure 13.5 Example of the correspondence between genes and proteins. (**a**) An mRNA transcript (*brown*) of a gene region of DNA (*blue*). Three nucleotide bases, equaling one codon, specify one amino acid. This series of codons (base triplets) specifies the sequence of amino acids shown in (**b**).

Figure 13.6 The near-universal genetic code. Each codon in mRNA is a set of three ribonucleotide bases. Sixty-one of these base triplets encode specific amino acids. Three are signals that stop translation.

The *left* vertical column (*dark brown*) lists choices for the first base of a codon. The *top* horizontal row (*light tan*) lists the second choices. The *right* vertical column (*dark tan*) lists the third.

To give three examples, reading from left to right, the triplet U G G corresponds to tryptophan. Both U U U and U U C correspond to phenylalanine.

first base	second base				third base
	U	C	A	G	
U	phenylalanine	serine	tyrosine	cysteine	U
	phenylalanine	serine	tyrosine	cysteine	C
	leucine	serine	STOP	STOP	A
	leucine	serine	STOP	tryptophan	G
C	leucine	proline	histidine	arginine	U
	leucine	proline	histidine	arginine	C
	leucine	proline	glutamine	arginine	A
	leucine	proline	glutamine	arginine	G
A	isoleucine	threonine	asparagine	serine	U
	isoleucine	threonine	asparagine	serine	C
	isoleucine	threonine	lysine	arginine	A
	methionine (or START)	threonine	lysine	arginine	G
G	valine	alanine	aspartate	glycine	U
	valine	alanine	aspartate	glycine	C
	valine	alanine	glutamate	glycine	A
	valine	alanine	glutamate	glycine	G

Figure 13.8 (**a**) Ribbon model for the large subunit of a bacterial ribosome. It has two rRNA molecules (*gray*) and thirty-one structural proteins (*gold*), which stabilize the structure. At one end of a tunnel through the large subunit, rRNA catalyzes polypeptide chain assembly. This is an ancient, highly conserved structure. Its role is so vital that the corresponding subunit of the eukaryotic ribosome, which is larger, may be similar in structure and function. (**b**) Model for the small and large subunits of a eukaryotic ribosome.

codon in mRNA transcript

anticodon in tRNA

amino acid

a

Figure 13.7 Model for a tRNA. The icon shown to the right is used in following illustrations. The "hook" at the lower end of this icon represents the binding site for a specific amino acid.

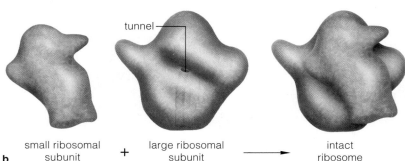

tunnel

small ribosomal subunit + large ribosomal subunit → intact ribosome

b

THE OTHER RNAs

In the cytoplasm of all cells are pools of free amino acids and tRNA molecules. Each tRNA has a molecular "hook," an attachment site for an amino acid. It has an **anticodon**, a ribonucleotide base triplet that can pair with an mRNA codon (Figure 13.7). When tRNAs bind to mRNA on a ribosome, their amino acid cargo will become automatically positioned in the order that the codons specify.

There are sixty-four codons but not as many kinds of tRNAs. How do tRNAs match up with more than one type of codon? According to base-pairing rules, adenine pairs with uracil, and cytosine with guanine. However, in codon–anticodon interactions, these rules can loosen for the third base in a codon. This freedom in codon–anticodon pairing at a base is known as the "wobble effect."

To give one example, AUU, AUC, and AUA specify isoleucine. All three codons can base-pair with one type of tRNA that hooks on to isoleucine.

Again, interactions between the tRNAs and mRNA take place at ribosomes. A ribosome has two subunits (Figure 13.8). They are built from rRNA and structural proteins in the nucleus, then shipped separately to the cytoplasm. There, a large and small subunit converge into an intact, functional ribosome only when mRNA is to be translated.

The genetic code is a set of sixty-four codons, which are ribonucleotide bases in mRNA that are read in sets of three. Different amino acids are specified by different codons.

Only mRNA carries DNA's protein-building instructions from the nucleus into the cytoplasm.

tRNAs deliver amino acids to ribosomes. Their anticodons base-pair with codons in the order specified by mRNA.

Polypeptide chains are built on ribosomes, each consisting of a large and small subunit made of rRNA and proteins.

13.3 Translating mRNA Into Protein

DNA's hereditary information must be stored intact, safely, in one place. Think of mRNA transcripts as intermediaries that deliver messages from DNA to ribosomes, which translate them into the polypeptide chains of proteins.

Translation has three stages: initiation, elongation, and termination. *Initiation* requires an initiator tRNA, the only one that can start translation. It binds to a small ribosomal subunit. Then mRNA's START codon, AUG, joins with that tRNA's anticodon. A large ribosomal subunit now joins with the small one (Figure 13.9*a–c*). Together, the ribosome, mRNA, and initiator tRNA are an initiation complex. The next stage can begin.

In *elongation*, a polypeptide chain is synthesized as the mRNA passes between the two ribosomal subunits, like a thread passing through the eye of a needle. tRNA molecules move amino acids to the ribosome, where they bind to the mRNA in the order specified by its codons. Part of an rRNA molecule at the center of the large ribosomal subunit functions as an enzyme. It catalyzes peptide bond formation between the amino acids (Figure 13.9*d–f*).

Figure 13.9*g* shows how one peptide bond forms between the most recently attached amino acid and the next one brought to the ribosome. Here, you might look once more at Section 3.5, which includes a step-by-step description of peptide bond formation during protein synthesis.

During the last stage of translation, *termination*, the ribosome reaches the mRNA's STOP codon. No tRNA

elongation

binding site for mRNA

P (first binding site for tRNA)　　*A* (second binding site for tRNA)

c Initiation ends when a large and small ribosomal subunit converge and bind together. In elongation, the second stage of translation, mRNA occupies a binding site at one end of a tunnel through the large subunit (Figure 13.8). tRNAs that deliver amino acids to the intact ribosome will occupy two other binding sites.

b *Initiation*, the first stage of translating mRNA, will start when an initiator tRNA binds to a small ribosomal subunit. The small subunit/tRNA complex will attach to the start of the mRNA, move along the transcript, and scan it for the START codon AUG.

initiation

amino acid 1

amino acid 2

d The initiator tRNA binds to the ribosome. Its anticodon matches up with the mRNA START codon AUG, and it has the amino acid methionine attached to it. A second tRNA binds with the next codon (here it is GUG).

amino acid 1　amino acid 2

e One of the rRNA molecules that make up the large ribosome catalyzes formation of a peptide bond between the amino acids (here, methionine and valine).

a A mature mRNA transcript leaves the nucleus through a pore in the nuclear envelope. It enters the cytoplasm, which has many free amino acids, tRNAs, and ribosomal subunits.

Figure 13.9 Translation, the second step of protein synthesis. Here we track a mature mRNA transcript that formed inside the nucleus of a eukaryotic cell. It passes through pores across the nuclear envelope and enters the cytoplasm, which contains pools of many free amino acids, tRNAs, and ribosomal subunits.

has a corresponding anticodon. Proteins called release factors bind to the ribosome. Binding triggers enzyme activity that detaches the mRNA *and* the polypeptide chain from the ribosome (Figure 13.9*i–k*).

Unfertilized eggs and other cells that rapidly make many copies of different proteins usually stockpile mRNA transcripts in their cytoplasm. In cells that are quickly using or secreting proteins, you often observe many clusters of ribosomes (polysomes) on an mRNA transcript, all translating it at the same time.

Many newly formed polypeptide chains carry out their functions in the cytoplasm. Many others have a shipping label, a special sequence of amino acids. The label lets them enter the ribosome-studded, flattened sacs of rough ER (Section 4.7). In the organelles of the endomembrane system, the chains will take on final form before shipment to their ultimate destinations as structural or functional proteins.

Translation is initiated when a small ribosomal subunit and an initiator tRNA arrive at an mRNA transcript's START codon, and then a large ribosomal subunit binds to them.

tRNAs deliver amino acids to a ribosome in the order dictated by the linear sequence of mRNA codons. A polypeptide chain lengthens as peptide bonds form between the amino acids.

Translation ends when a STOP codon triggers events that cause the polypeptide chain and the mRNA to detach from the ribosome.

Read Me First!
and watch the narrated animation on translation

f The first tRNA is released, and the ribosome moves to the next codon position.

g A third tRNA binds with the next codon (here it is UUA). The ribosome catalyzes peptide bond formation between amino acids 2 and 3.

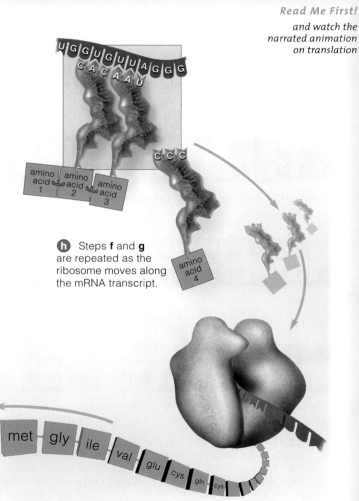

h Steps **f** and **g** are repeated as the ribosome moves along the mRNA transcript.

termination

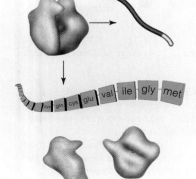

i A STOP codon moves into the area where the chain is being built. It is the signal to release the mRNA transcript from the ribosome.

j The new polypeptide chain is released from the ribosome. It is free to join the pool of proteins in the cytoplasm or to enter rough ER of the endomembrane system.

k The two ribosomal subunits now separate, also.

13.4 Mutated Genes and Their Protein Products

When a cell taps its genetic code, it is making proteins with precise structural and functional roles that keep it alive. If something changes a gene, the protein that it encodes may change. If the protein has an essential role, we can expect the outcome to be an abnormal cell.

Gene sequences can change. Sometimes one base gets substituted for another in the nucleotide sequence. At other times, an extra base is inserted or one is lost. Such small-scale changes in the nucleotide sequence of a DNA molecule are **gene mutations**.

There is some leeway here, because more than one codon specifies the same amino acid. If UCU were changed to UCC, for example, it probably would not have dire effects, because both codons specify serine. However, many mutations give rise to proteins that function in altered ways or not at all. Repercussions are sometimes harmful or lethal.

COMMON MUTATIONS

In gene mutations called **base-pair substitutions**, one base is copied incorrectly during DNA replication. Its outcome? A protein may incorporate the wrong amino acid, or its synthesis may have been cut off too soon. In the example in Figure 13.10*b*, adenine *replaced* one thymine in a gene for beta hemoglobin. The mutant gene's product has a single amino acid substitution that causes sickle-cell disease (Section 3.6).

Figure 13.10*c* shows a different gene mutation, one in which a single base—thymine—was *deleted*. Again, DNA polymerases read base sequences in blocks of three. A deletion is one of the *frameshift* mutations; it shifts the "three-bases-at-a-time" reading frame. An altered mRNA is transcribed from the mutant gene, so an altered protein is the result.

Frameshift mutations fall in the broader categories of **insertions** and **deletions**. One or more base pairs become inserted into DNA or are deleted from it.

Other mutations arise from transposable elements, or **transposons**, that can jump around in the genome. Geneticist Barbara McClintock found that these DNA segments or copies of them move spontaneously to a new location in a chromosome or even to a different chromosome. When transposons land in a gene, they alter the timing or duration of its activity, or block it entirely. Their unpredictability can give rise to odd variations in traits. Figure 13.11 gives an example.

part of DNA template

mRNA transcribed from DNA

a THREONINE | PROLINE | GLUTAMATE | GLUTAMATE | LYSINE — resulting amino acid sequence

base substitution in DNA

altered mRNA

b THREONINE | PROLINE | VALINE | GLUTAMATE | LYSINE — altered amino acid sequence

deletion in DNA

altered mRNA

c THREONINE | PROLINE | GLYCINE | ARGININE — altered amino acid sequence

Figure 13.10 Gene mutation. (**a**) Part of the gene, the mRNA, and the resulting amino acid sequence of the hemoglobin beta chain. (**b**) A base substitution in DNA replaces a thymine with an adenine. When the altered mRNA transcript is translated, valine replaces glutamate as the sixth amino acid of the new polypeptide chain. Sickle-cell anemia is the eventual outcome.

(**c**) Deletion of the same thymine would be a frameshift mutation. The reading frame for the rest of the mRNA shifts, a different protein product forms, and it causes thalassemia, a different type of red blood cell disorder.

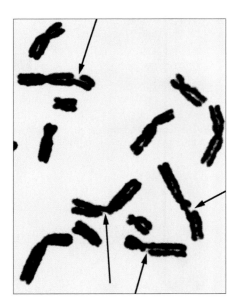

Figure 13.12 Chromosomes from a human cell exposed to gamma rays, a form of ionizing radiation. We can expect such broken pieces (*arrows*) to be lost in interphase, when DNA is being replicated. The extent of chromosome damage in an exposed cell typically depends on how much radiation it absorbed.

Figure 13.11 Barbara McClintock, who won a Nobel Prize for her research. She proved that transposons slip into and out of different locations in DNA. The curiously nonuniform coloration of kernels in strains of Indian corn (*Zea mays*) sent her on the road to discovery.

Several genes govern pigment formation and deposition in corn kernels, which are a type of seed. Mutations in one or more of these genes produce yellow, white, red, orange, blue, and purple kernels. However, as McClintock realized, *unstable* mutations can cause streaks or spots in *individual* kernels.

All of a corn plant's cells have the same pigment-encoding genes. But a transposon invaded a pigment-encoding gene before the plant started growing from a fertilized egg. While a kernel's tissues were forming, its cells couldn't make pigment. But the same transposon jumped back out of the pigment-encoding gene in some of its cells. Descendants of *those* cells could make pigment. The streaks and spots in individual kernels are evidence of those cell lineages.

HOW DO MUTATIONS ARISE?

Many mutations happen spontaneously while DNA is being replicated. This is not surprising, given the swift pace of replication (about twenty bases per second in humans and a thousand bases per second in certain bacteria). DNA polymerases can repair most of the mistakes (Section 12.4). Sometimes, however, they go on assembling a new strand right over an error. The bypass can result in a mutated DNA molecule.

Not all mutations are spontaneous. A number arise after DNA is exposed to mutation-causing agents. For instance, x-rays and similar high-energy wavelengths of **ionizing radiation** break chromosomes into pieces (Figure 13.12). Such radiation also indirectly damages DNA because it penetrates living tissue, leaving in its wake a potentially destructive trail of free radicals. Because of this, doctors and dentists use the lowest possible doses of x-rays in order to minimize damage to their patients' DNA.

Nonionizing radiation boosts electrons to a higher energy level. DNA absorbs one form, ultraviolet (UV) radiation. Two types of nucleotides in DNA, cytosine and thymine, are most vulnerable to excitation that can change base-pairing properties. For example, UV light can induce two thymine bases to pair (T–T, not A–T). At least seven gene products interact as a DNA repair mechanism to remove this error, which is a thymine dimer. Thymine dimers form in skin cells after exposure of unprotected skin to sunlight.

When they are not repaired, thymine dimers cause DNA polymerase to make additional errors during the next cycle of replication. They are the source of mutations that lead to certain cancers.

Natural and synthetic chemicals accelerate the rates of spontaneous mutations. **Alkylating agents** are one example. They transfer charged methyl or ethyl groups to reactive sites in DNA. At these sites, DNA is more vulnerable to base-pair changes that invite mutation. Cancer-causing agents in cigarette smoke and many substances exert their effects by alkylating DNA.

THE PROOF IS IN THE PROTEIN

When a mutation arises in a somatic cell, its good or bad effects do not endure, because it cannot be passed on to offspring. When a mutation arises in a germ cell or a gamete, however, it may enter the evolutionary arena. It also may do so if it arises during asexual reproduction. Either way, *a protein product of a heritable mutation will have harmful, neutral, or beneficial effects on the individual's ability to function in the prevailing environment.* The effects of uncountable mutations in millions of species have had spectacular evolutionary consequences. And that is a topic of later chapters.

A gene mutation is a change involving one or more bases in the nucleotide sequence of DNA. The most common types are base-pair substitutions, insertions, and deletions.

Exposure to harmful radiation and chemicals in the environment can cause mutations in DNA.

A protein specified by a mutated gene may have harmful, neutral, or beneficial effects on the ability of an individual to function in the environment.

Summary

Section 13.1   Transcription and translation are two steps of a process leading from genes to proteins.

$$DNA \xrightarrow{\text{transcription}} RNA \xrightarrow{\text{translation}} PROTEIN$$

In eukaryotic cells, transcription occurs in the nucleus (Figure 13.13). DNA's two strands unwind from each other at a selected region. RNA polymerases use the exposed DNA bases as a template on which an RNA molecule is built from free ribonucleotides (adenine, guanine, cytosine, and uracil). The mRNA transcript becomes modified before it leaves the nucleus.

Section 13.2   In translation, three types of RNAs interact to build polypeptide chains. Messenger RNA (mRNA) carries DNA's transcribed protein-building information from the nucleus to the cytoplasm. Its genetic message is written in codons, or sets of three nucleotides along an mRNA strand that specify an amino acid. There are sixty-four codons, a few of which act as START or STOP signals for translation. Ribosomes consist of ribosomal RNA (rRNA) and proteins that stabilize it. Transfer RNA (tRNA) molecules bind free amino acids in the cytoplasm and deliver them to ribosomes during protein synthesis. Different tRNAs bind different amino acids.

Section 13.3   During translation, amino acids are joined in the order specified by codons in mRNA.

Translation proceeds through three stages. In initiation, ribosomal subunits, an initiator tRNA, and an mRNA join to form an initiation complex. In elongation, tRNAs deliver amino acids to ribosomes, which synthesize a polypeptide chain from them. Part of the rRNA in the ribosomes catalyzes peptide bond formation between the amino acids. At termination, a STOP codon and other factors trigger the release of the mRNA and the new polypeptide chain, and they make the ribosome's subunits separate from each other.

Section 13.4   Gene mutations are heritable, small-scale changes in the base sequence of DNA. Major types are base-pair substitutions, insertions, and deletions. Many arise spontaneously as DNA is being replicated. Some occur when transposons jump around in a gene or after DNA is exposed to ionizing radiation or to chemicals in the environment. Mutations may cause changes in protein structure, protein function, or both.

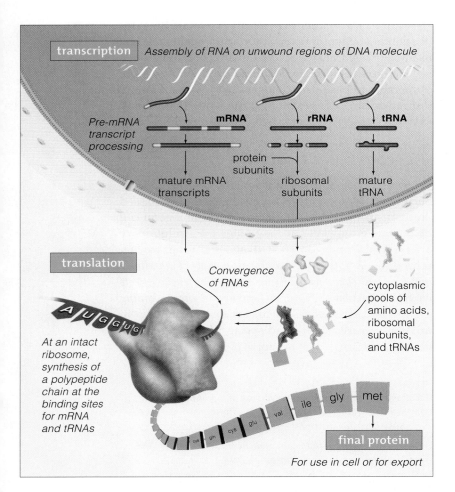

Figure 13.13   Summary of protein synthesis in eukaryotes. DNA is transcribed into RNA in the nucleus. RNA is translated in the cytoplasm. Prokaryotic cells don't have a nucleus; transcription and translation proceed in their cytoplasm.

Self-Quiz
Answers in Appendix III

1. DNA contains many different genes that are transcribed into different _____ .
    a. proteins   c. mRNAs, tRNAs, rRNAs
    b. mRNAs only   d. all of the above

2. An RNA molecule is typically _____ .
    a. a double helix   c. double-stranded
    b. single-stranded   d. triple-stranded

3. An mRNA molecule is synthesized by _____ .
    a. replication   c. transcription
    b. duplication   d. translation

4. Each codon specifies a (an) _____ .
    a. protein   c. amino acid
    b. polypeptide   d. mRNA

5. Anticodons pair with _____ .
    a. mRNA codons   c. RNA anticodons
    b. DNA codons   d. amino acids

6. Match the terms with the most suitable description.
    ____ alkylating   a. parts of mature mRNA
         agent   transcript
    ____ chain   b. base triplet for amino acid
         elongation   c. second stage of translation
    ____ exons   d. base triplet; pairs with codon
    ____ genetic code   e. one environmental agent that
    ____ anticodon   induces mutation in DNA
    ____ introns   f. set of 64 codons in mRNA
    ____ codon   g. the parts removed from a
                                pre-mRNA transcript before
                                translation

Critical Thinking

1. *Antisense drugs* may help us fight cancer and viral diseases, including SARS. They are short mRNA strands that are complementary to the mRNAs associated with these illnesses. Speculate on how these drugs work.

2. A DNA polymerase made an error while a crucial gene region of DNA was being replicated. DNA repair enzymes didn't detect or repair the damage. Here is the part of the DNA strand that contains the error:

..AATTCCGACTCCTATGG
..TTAAGGTTGAGGATACC

After the DNA molecule is replicated, two daughter cells form. One daughter cell is carrying the mutation and the other cell is normal. Develop a hypothesis to explain this observation.

3. *Neurofibromatosis* is a human autosomal dominant disorder caused by mutations in the *NF1* gene. It is characterized by the formation of soft, fibrous tumors in the peripheral nervous system and skin as well as abnormalities in muscles, bones, and internal organs (Figure 13.14).

   Because the gene is dominant, an affected child usually has an affected parent. Yet in 1991, scientists reported on a boy who had neurofibromatosis whose parents did not. When they examined both copies of his *NF1* gene, they found the copy he had inherited from his father contained a transposon. Neither the father nor the mother had a transposon in any of the copies of their own *NF1* genes. Explain the cause of neurofibromatosis in the boy and how it arose.

4. Cigarette smoke is mostly carbon dioxide, nitrogen, and oxygen. The rest contains at least fifty-five different chemicals that have been identified as carcinogenic, or cancer-causing, by the International Agency for Research on Cancer (IARC). When these carcinogens enter the bloodstream, enzymes convert them to a series of chemical intermediates in an attempt to make them easier to excrete. Some of the intermediates bind irreversibly to DNA. Speculate on one mechanism by which smoking causes cancer.

5. Using the data in Figure 13.6, translate the following mRNA segment into an amino acid sequence:

   5'-GGTTTCTTCAAGAGA-3'

6. The termination of DNA transcription by prokaryotic RNA polymerases depends in some cases on the structure of the newly forming RNA transcript. The terminal end of an mRNA chain often folds back on itself and makes a hairpin-looped structure like the one shown to the right.

   Why do you suppose a "stem-loop" structure such as this one causes RNA polymerases to stop transcription when they reach it?

```
              C
          U — C
          G — C
          A — U
          C — G
          C — G
          G — C
          C — G
          C — G
 ...CCCACAG — CAUUUUU...
```

Figure 13.14 Soft skin tumors on a person with neurofibromatosis, an autosomal dominant disorder.

IMPACTS, ISSUES   *Between You and Eternity*

You are in college, your whole life ahead of you. Your risk of developing cancer is as remote as old age, an abstract statistic that is easy to forget.

"There is a moment when everything changes—when the width of two fingers can suddenly be the total distance between you and eternity." Robin Shoulla wrote those words after being diagnosed with breast cancer. She was only seventeen. At an age when most young women are thinking about school, parties, and potential careers, Robin was dealing with a radical mastectomy, pleading with her oncologist not to use her jugular vein for chemotherapy, wondering if she would survive through the next year (Figure 14.1).

Robin became an annual statistic—one of 10,000 or so females and males under age forty who develop breast cancer. About 180,000 new cases are diagnosed each year in the United States population at large.

Cancers are as diverse as their underlying causes, but several gene mutations predispose individuals to developing certain kinds. Either the mutant genes are inherited or they mutate spontaneously in individuals after being assaulted by environmental agents, such as toxic chemicals and ultraviolet radiation.

One gene on chromosome 17 encodes *Her2*, a type of membrane receptor. *Her2* is part of a control pathway that governs the cell cycle—that is, when and how often cells divide. It also is one of the **proto-oncogenes**. When mutated or overexpressed, such genes help bring about cancerous transformations. Cells of about

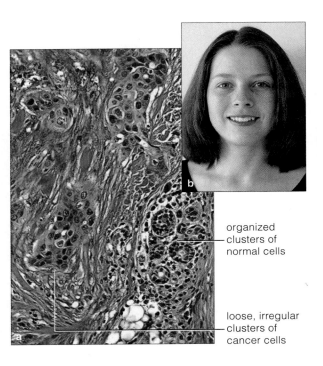

organized clusters of normal cells

loose, irregular clusters of cancer cells

Figure 14.1   Breast cancer. (**a**) Infiltrating ductal carcinoma cells form irregular clusters in breast tissue. (**b**) Robin Shoulla. Diagnostic tests revealed the presence of such cells in her body.

the big picture

Types of Control Mechanisms   Whether and how a gene is expressed depends on regulatory proteins that interact with DNA, RNA, proteins, and one another. It also depends on the attachment and detachment of certain functional groups to the DNA.

Gene Control in Prokaryotes   Being tiny single cells and fast reproducers, prokaryotic cells make rapid responses to short-term changes in nutrient availability and other aspects of the environment. They compensate for the changes by quickly adjusting gene transcription rates.

25 percent of breast cancer patients have too many *Her2* receptors or extra copies of the gene itself. They divide too fast, and abnormal masses of cells result.

Proteins encoded by two different genes, *BRCA1* and *BRCA2*, are among the tumor suppressors that help keep benign or cancerous cell masses from forming. The filmstrip shows part of one of the proteins, which are crucial for DNA repair processes. When *BRCA1* or *BRCA2* is mutated, a cell's capacity to fix breaks in DNA or correct replication errors is compromised. Diverse mutations are free to accumulate throughout the DNA, and such an accumulation leads to cancer.

BRCA1 and *BRCA2* are known as *breast cancer genes*, because cancerous breast cells often hold mutated versions of them. A female in which a *BRCA* gene has mutated in one of three especially dangerous ways has about an 80 percent chance of developing breast cancer before reaching seventy.

Robin Shoulla survived. She may never know which mutation caused her cancer. Thirteen years later, she has what she calls a normal life—a career, husband, children. Her goal is to grow very old with grey hair and spreading hips, smiling.

Robin's story invites you to enter the world of **gene controls**, the molecular mechanisms that govern when and how fast specific genes will be transcribed and translated, and whether gene products will be switched on or silenced. You will be returning to the impact of such controls in many chapters throughout the book.

 How Would You Vote?

Some females at high risk of developing breast cancer opt for prophylactic mastectomy, the surgical removal of one or more breasts even before cancer develops. Many of them would never have developed cancer. Should the surgery be restricted to cancer treatment? See the Media Menu for details, then vote online.

Gene Control in Eukaryotes Like prokaryotes, eukaryotic cells control short-term shifts in diet and activity. Unlike prokaryotes, they also control a long-term program of development, which is based largely on selective gene expression and cell differentiation.

Researching Gene Controls In complex, multicelled eukaryotes, cascades of gene controls guide development of a single fertilized egg into a complete individual with a predictable body plan. A century of research with the common fruit fly has yielded clues about how these controls work.

14.1 Some Control Mechanisms

When, how, and to what extent any gene is expressed depends on the type of cell, its functions, its chemical environment, and signals from the outside.

Diverse mechanisms control gene expression through interactions with DNA, RNA, and new polypeptide chains or the final proteins. Some respond to rising or falling concentrations of specific substances in a cell. Others respond to external signaling molecules.

Control agents include **regulatory proteins** that intervene before, during, or after gene transcription or translation. They include signaling molecules such as hormones, which initiate changes in cell activities when they dock at suitable receptors.

With **negative control**, regulatory proteins slow or stop gene action; with **positive control**, they promote or enhance it. Some DNA base sequences that don't encode proteins are sites of transcriptional control. A **promoter** is a common type of noncoding sequence that marks where to start transcription. **Enhancers** are binding sites for some activator proteins.

Chemical modification offers more control. With methylation, for example, methyl groups ($-CH_3$) get attached to specific regions of newly replicated DNA and prevent access to them. Many heavily methylated genes are activated when the groups are stripped off. Attachment of acetyl groups to histones that organize DNA also exerts control (Section 8.1 and Figure 14.2).

Figure 14.2 How loosening of the DNA–histone packaging in chromosomes may expose genes for transcription. Attachment of an acetyl group to a histone makes it loosen its grip on the DNA that is wound around it. Enzymes that are associated with transcription attach or detach acetyl groups.

You'll read about major signaling mechanisms later. Here, we will sample the events they set in motion.

Gene expression is controlled by regulatory proteins that interact with one another, with control elements built into the DNA, with RNA, and with newly synthesized proteins.

Control also is exerted through chemical modifications that inactivate or activate specific gene regions or the histones that organize the DNA.

14.2 Prokaryotic Gene Control

Think about the dot of the letter "i." About a thousand bacterial cells would stretch side by side across the dot— and each depends as much on gene controls as you do!

Prokaryotic cells grow and divide fast when nutrients are plentiful and other conditions also favor growth. At such times, controls promote the rapid synthesis of enzymes for nutrient absorption and other growth-related metabolic events. Genes that specify enzymes for a metabolic pathway often occur as a linear set in the DNA. And they all may be transcribed together, in a single RNA strand.

NEGATIVE CONTROL OF THE LACTOSE OPERON

With this bit of background, consider an example of how one kind of prokaryote responds to the presence or absence of lactose. *Escherichia coli* lives in the gut of mammals, where it dines on nutrients traveling past. Milk typically nourishes mammalian infants. It does not contain glucose, the sugar of choice for *E. coli*. It does contain lactose, a different sugar.

After being weaned, infants of most species drink little (if any) milk. Even so, *E. coli* cells can still use lactose if and when it shows up in the gut. They can activate a set of three genes for lactose-metabolizing enzymes. A promoter precedes all three genes in *E. coli* DNA, and operators flank it. An **operator** is a binding site for a repressor, a regulatory protein that stops transcription. Such an arrangement, in which a promoter and a set of operators control more than one bacterial gene, is called an **operon** (Figure 14.3).

In the absence of lactose, a repressor molecule binds to a set of operators. Binding causes the DNA region that contains the promoter to twist into a loop, as in Figure 14.3*b*. RNA polymerase, the workhorse that transcribes genes, can't bind to a looped-up promoter. So operon genes aren't used when they aren't required.

When lactose *is* in the gut, *E. coli* converts some of it to allolactose. This sugar binds to the repressor and changes its molecular shape. The altered repressor can't bind to operators. The looped DNA unwinds and RNA polymerase transcribes the genes, so lactose-degrading enzymes are produced when required.

POSITIVE CONTROL OF THE LACTOSE OPERON

E. coli cells pay far more attention to glucose than to lactose. They transcribe genes for its breakdown faster, and continuously. Even when lactose is in the gut, the lactose operon is not used much—unless there is no glucose. Such conditions call for an **activator** protein

Read Me First!
*and watch the narrated animation
on the lactose operon*

a Lactose operon: a set of operators, a promoter, three genes that specify three enzymes, and a binding site for a second messenger (*white*). A different gene upstream from the operon specifies a repressor protein that can block access to the lactose operon.

b In the absence of lactose, the repressor binds to two operators in DNA. It makes the DNA loop out in a way that blocks operon gene transcription; it stops RNA polymerase from binding to its promoter.

Figure 14.3 Negative control of the lactose operon. The operon's first gene codes for an enzyme that splits lactose, a disaccharide, into glucose and galactose. The second gene codes for an enzyme that helps transport lactose into cells. The third gene's product helps metabolize certain sugars.

c When lactose is present, some is converted to a form that binds to the repressor and alters its shape. The altered repressor can't bind to operators, so RNA polymerase is free to transcribe the operon genes.

known as CAP (short for catabolite activator protein). This activator exerts positive control over the lactose operon by making a promoter far more inviting to RNA polymerase. But CAP can't issue the invitation until it is bound to a chemical messenger called cAMP (cyclic adenosine monophosphate). When cAMP and this activator join together and bind to the promoter, they make it far easier for RNA polymerase to start transcribing genes.

When glucose is plentiful, ATP forms by glycolysis, but synthesis of an enzyme necessary to make cAMP is blocked. Blocking is lifted when glucose is scarce and lactose becomes available. cAMP accumulates, CAP–cAMP complexes form, and the lactose operon genes are transcribed. The gene products allow lactose to be used as an alternative energy source.

Unlike cells of *E. coli*, many of us develop *lactose intolerance*. Cells making up the lining of our small intestine make and then secrete lactase into the gut. As many people age, however, concentrations of this lactose-digesting enzyme decline. Lactose accumulates and ends up in the large intestine (colon), where it promotes population explosions of resident bacteria. As the bacteria digest the lactose, a gaseous metabolic product accumulates, distends the colon, and causes pain. Short fatty acid chains released by the bacteria also lead to diarrhea, which can be severe.

Transcription rates of bacterial genes for nutrient-digesting enzymes are quickly adjusted downward and upward by control systems that respond to nutrient availability.

14.3 Eukaryotic Gene Control

Like bacteria, eukaryotic cells control short-term shifts in diet and in levels of activity. If those cells happen to be among hundreds or trillions of cells in a multicelled organism, long-term controls also enter the picture. They orchestrate gene interactions during development.

SAME GENES, DIFFERENT CELL LINEAGES

Later in the book, you will be reading about how you and other complex organisms developed from a single cell. For now, tentatively accept this premise: All cells of your body started out life with the same genes, because every one arose by mitotic cell divisions from the same fertilized egg. And they all transcribe many of the same genes, because they are alike in most aspects of structure and basic housekeeping activities.

In other ways, however, *nearly all of your body cells became specialized in composition, structure, and function.* This process of **cell differentiation** occurs during the development of all multicelled organisms. Differences arise among cells that use different subsets of genes. Specialized tissues and organs are the result.

For example, nearly all of your cells continually transcribe genes for the enzymes of glycolysis. Only immature red blood cells can transcribe the genes for hemoglobin. Your liver cells transcribe genes required to make enzymes that neutralize certain toxins, but

they are the only ones that do. When your eyes first formed, certain cells accessed the genes necessary for synthesizing crystallin. No other cells can activate the genes for this protein, which helps make transparent fibers of the lens in each eye.

WHEN CONTROLS COME INTO PLAY

Ultimately, gene expression is all about controlling the amounts and kinds of proteins present in a cell in any specified interval. Just imagine the coordination that goes into making, stockpiling, using, exporting, and degrading thousands of types of proteins in the same moment of cellular time.

Most genes in complex, multicelled organisms are switched off, either permanently or part of the time. Expression of the rest is adjusted up and down. Why? Cells continually deliver and secrete substances into tissue fluids—the body's internal environment—and withdraw substances from it. Inputs and outputs cause slight shifts in the concentrations of nutrients, signaling molecules, metabolic products, and other solutes. Homeostasis is maintained as cells respond to these changes by adjusting gene expression.

Controls over gene expression work at certain stages before, during, and after transcription and translation. Figure 14.4 introduces the main control points.

a Chemical modification of DNA restricts access to genes. Genes can be duplicated or rearranged.

b Pre-mRNA spliced in alternative ways can lead to different forms of a protein. Other modifications affect whether a transcript reaches the cytoplasm.

c Transport protein binding determines whether an mRNA becomes delivered to the correct region of cytoplasm for local translation.

d How long an mRNA lasts depends on the proteins that are attached to it and the length of its poly-A tail.

e Translation can be blocked. mRNA cannot attach to a ribosome when proteins bind to it. Initiation factors can be inactivated.

f Processing of new polypeptides may activate or disable them. Control here indirectly affects other activities.

Figure 14.4 Controls that influence whether, when, and how a eukaryotic gene will be expressed.

Figure 14.5 Polytene chromosomes. To sustain their rapid growth rate, *Drosophila* (fruit fly) larvae eat continuously and use a lot of saliva. Giant chromosomes in their salivary gland cells are produced by repeated mitotic DNA replication without cell division. Each strand contains many copies of the same chromosome, aligned side-by-side.

An insect hormone, ecdysone, serves as a regulatory protein; it promotes gene transcription. In response to the hormonal signal, these chromosomes loosen, and they puff out in the regions where genes are being transcribed. Puffs are largest and most diffuse where transcription is most intense.

CONTROLS BEFORE TRANSCRIPTION Remember how many histones and other proteins organize the DNA in a eukaryotic chromosome (Section 8.1)? They affect whether RNA polymerase can access genes and start transcription. Methyl, acetyl, and other functional groups attached to the DNA also can block access to genes. In diploid cells, either the maternal or paternal allele at a gene locus may get methylated, which can block the maternal or paternal influence on a trait.

Also before transcription, some controls trigger the duplication or rearrangement of gene sequences. In immature amphibian eggs and gland cells of certain insect larvae, chromosomes are copied repeatedly in an undividing cell. These multiple DNA replications produce *polytene* chromosomes that have hundreds or thousands of side-by-side gene copies (Figure 14.5).

CONTROL OF TRANSCRIPT PROCESSING After genes are transcribed, several mechanisms control what the cell does with the RNA. Transcript processing steps dictate whether, when, and how pre-mRNA becomes translated (Section 13.1). For instance, in different kinds of muscle cells, enzymes remove different parts of the pre-RNA transcript for troponin, a contractile protein. After the remaining exons are spliced together, their protein-building message is unique in one tiny region. In each cell type, the resulting protein works in a distinctive way, which helps account for subtle variations in how different kinds of muscles function.

The nuclear membrane is a barrier between a new mRNA and the cellular machinery that can translate it. Only after the mRNA binds to certain proteins will nuclear pore complexes let it cross to the cytoplasm.

Once in the cytoplasm, an mRNA is guided about according to base sequences in its untranslated ends, which are like zip codes. A transport protein bound to a zip code region delivers the mRNA to a particular area of the cell, where it will be translated or stored. In immature eggs, uneven distribution of "maternal messages" and their protein products determines the head-to-tail polarity of the future developing embryo. Control over mRNA localization occurs in the form of binding proteins that attach to the zip code region. These delay or block delivery of an mRNA.

Some transcripts are shelved when the cytoplasm has too many Y-box proteins. When phosphorylated, these proteins bind and help stabilize an mRNA, but if too many of them become attached they block its translation. Thus, phosphorylation of Y-box proteins is a control point for mRNA inactivation. The mRNA stored in unfertilized eggs is bound to Y-box proteins.

CONTROLS AT TRANSLATION The greatest range of controls over eukaryotic gene expression operates at translation. This process depends on the coordinated participation of many kinds of molecules, including ribosomal subunits and a host of initiation factors. Each kind of molecule is regulated independently of the other kinds.

The stability of mRNA transcripts is also a control point. The more stable a given transcript is, the more proteins can be produced from it. Enzymes typically start digesting mRNA within minutes, nibbling away at its poly-A tail (Section 13.1). How fast they do the deed depends on the tail's length, on base sequences in untranslated regions and other sequences in the coding region, and on attached proteins.

CONTROLS AFTER TRANSLATION Lastly, control over gene expression is exerted when the protein products are modified, as when phosphate groups are attached to Y-box proteins. Diverse controls activate, inhibit, and stabilize enzymes and other molecules used in protein synthesis. A case in point is allosteric control of tryptophan synthesis, as Section 5.4 describes.

Cell differentiation arises when diverse populations of cells activate or suppress genes in selective, unique ways.

In the cells of complex, multicelled species, gene expression is controlled by mechanisms that govern events before, during, and after transcription and translation.

Most controls over gene expression occur at translation.

14.4 Examples of Gene Controls

Cells rarely use more than 5 to 10 percent of their genes at a given time; controls silence most of them. Which genes are active depends on the type of organism, the stage of growth and development it's passing through, and the controls operating at that stage.

The preceding section introduced you to an important idea. All differentiated cells in a complex, multicelled body use most of their genes in much the same way, but each type also uses a fraction of those genes in a unique, selective way. **Selective gene expression** has made each kind distinctive in one or more aspects of their structure, composition, and function. Here we consider two examples of the controls that guide their selections during embryonic development.

HOMEOTIC GENES AND BODY PLANS

Whether a particular gene gets transcribed depends in part on the action of regulatory proteins, which can bind with promoters, enhancers, or one another. For example, **homeotic genes** are a class of master genes in most eukaryotic organisms. They are transcribed in specific locations in the developing embryo, so their products form in local tissue regions. By interacting with one another and with other control elements, homeotic genes guide formation of organs and limbs by turning on other genes in precise areas, according to a basic body plan.

Homeotic genes were discovered through mutations that cause cells in a *Drosophila* embryo to develop into a body part that belongs somewhere else. For instance, the *antennapedia* gene is supposed to be transcribed where a thorax, complete with legs, should form. In all other regions, cells normally don't transcribe this gene. But Figure 14.6*a* shows what happens when a mutation allows the gene to be wrongly transcribed in the body region destined to become a head.

Do animals alone have homeotic genes? No. In corn plants, for instance, a different homeotic gene guides the formation of all leaf veins in straight, parallel lines. If the gene mutates, the veins will twist.

Homeotic genes code for regulatory proteins that include a "homeodomain," a sequence of about sixty amino acids. This sequence binds to control elements in promoters and enhancers (Figure 14.6*b*). More than 100 homeotic genes have been identified in diverse eukaryotes—and the same mechanisms control their transcription. Many of the genes are interchangeable among species as evolutionarily distant as yeasts and humans, so we can expect that they evolved in the most ancient eukaryotic cells. Their protein products often differ only in *conservative* substitutions. In other words, one amino acid has replaced another, but it has similar chemical properties.

X CHROMOSOME INACTIVATION

Diploid cells of female humans and female calico cats have two X chromosomes. One is in threadlike form. The other stays scrunched up, even during interphase. This scrunching is a programmed shutdown of all but about three dozen genes on *one* of two homologous X chromosomes. The shutdown is called **X chromosome inactivation**, and it happens in female embryos of all placental mammals and their marsupial relatives.

Figure 14.7*a* shows one condensed X chromosome in the nucleus of a cell at interphase. We also call this condensed structural form a Barr body (after Murray Barr, who first identified it).

One X chromosome gets inactivated when embryos are still a tiny ball of cells. In placental mammals, the shutdown is random, in that *either* chromosome could become condensed. The maternal X chromosome may be inactivated in one cell; the paternal or the maternal X chromosome may be inactivated in a cell next to it.

Figure 14.6 (**a**) Experimental evidence of controls over where body parts develop. In *Drosophila* larvae, activation of genes in one group of cells normally results in antennae on the head. A mutation that affects *antennapedia* gene transcription puts legs on the head. This is one of the genes controlled by regulatory proteins that have homeodomains. (**b**) Stick model for the binding of a homeodomain sequence to a transcriptional control site in DNA.

Unaffected skin with normal sweat glands

Affected skin with no normal sweat glands

Figure 14.7 (**a**) In the somatic cell nucleus of a human female, a condensed X chromosome, also called a Barr body (*arrow*). The X chromosome in cells of human males is not condensed this way. (**b**) A mosaic tissue effect that shows up in anhidrotic ectodermal dysplasia.

Figure 14.8 Why is this female cat "calico"? In her body cells, one of her two X chromosomes has a dominant allele for the brownish-black pigment melanin. Expression of the allele on her other X chromosome results in orange fur. When this cat was an embryo, one X chromosome was inactivated at random in each cell that had formed by then.

Patches of different colors reflect which allele was inactivated in cells that formed a given tissue region. White patches are an outcome of an interaction that involves a different gene, the product of which blocks melanin synthesis.

Once that random molecular decision is made in a cell, all of that cell's descendants make the exact same decision as they go on dividing to form tissues. What is the outcome? *A fully developed female has patches of tissue where genes of the maternal X chromosome are being expressed, and patches of tissue where genes of the paternal X chromosome are being expressed.* She has become a "mosaic" for X chromosome expression!

When alleles on two homologous X chromosomes are not identical, differences may occur among patches of tissues throughout the body. Mosaic tissues can be observed in human females who are heterozygous for a rare recessive allele that causes an absence of sweat glands. Sweat glands formed in only parts of their skin. Where the glands are absent, the recessive allele is on the X chromosome that was not shut down. This mosaic effect is one symptom of *anhidrotic ectodermal dysplasia* (Figure 14.7b), a heritable disorder that is characterized by abnormalities in the skin and in the structures derived from it, including teeth, hair, nails, and sweat glands.

A different mosaic tissue effect shows up in female calico cats, of the sort shown in Figure 14.8. These cats are heterozygous for a certain coat color allele on their X chromosomes.

The shutdown isn't an accident of evolution; it has an important function. In mammals, recall, males have one X and one Y chromosome. This means the females have twice as many X chromosome genes. Inactivating one of their X chromosomes balances gene expression between the sexes. The normal development of female embryos depends on this type of control mechanism, which is called **dosage compensation**.

How, in a single nucleus, does one X chromosome get shut down while the other does not? Several molecules participate, including histone methylases and an X chromosome gene called *XIST*. The *XIST* product, a large RNA molecule, binds chromosomal DNA like a gene-masking paint. Although the *XIST* gene is found on both X chromosomes, only one of them expresses it. As it does, it gets fully painted with RNA, and so its genes become inactivated. The other X chromosome does not express *XIST*, and does not become painted. Only the genes on this chromosome remain active and may be transcribed. Why only one of the two X chromosomes expresses *XIST* is not yet fully understood.

Controls over when, how, and to what extent a gene is expressed depend on the type of cell and its functions, on the cell's chemical environment, and on signals for change. Homeotic gene expression and dosage compensation are examples of control mechanisms in eukaryotic cells.

14.5 There's a Fly in My Research

Structural patterns emerge as the embryos of animals and plants develop, and they are both beautiful and fascinating. Researchers have correlated those patterns with the expression of specific genes at particular times, in particular tissues, for diverse organisms.

DROSOPHILA!

For about a hundred years, *Drosophila melanogaster* has been the fly of choice for laboratory experiments. It costs almost nothing to feed and house this tiny fruit fly. *D. melanogaster* reproduces fast in bottles, it has a short life cycle, and disposing of spent bodies after an experiment is a snap. Thanks to automated gene sequencing, we now know how its 13,601 genes are distributed among its four pairs of chromosomes.

Studies of *Drosophila* at the anatomical, cytological, biochemical, and genetic levels continue to reveal much about gene controls over how animal embryos develop. They also yield insights into the evolutionary connections among groups of animals.

CLUES TO GENE CONTROLS

Over the past ten years, *Drosophila* researchers have made remarkable discoveries about how embryos develop, especially through **knockout experiments**. In such experiments, individual genes are deleted from wild-type experimental organisms. Differences between the engineered and wild-type organisms, either morphological or behavioral, are clues to the function of the missing gene.

Knockout experiments have identified many hundreds of *Drosophila* genes, which tend to be named after what happens when they are missing. Many turned out to be homeotic genes. For instance, *eyeless* is a control gene expressed in fruit fly embryos. In its absence, no eyes form. Other named genes include *dunce* (a regulatory protein required for learning and memory), *wingless*, *wrinkled*, *tinman* (necessary for heart development), *minibrain*, and *groucho* (which, among other things, prevents overproduction of whisker bristles). Figures 14.9 and 14.10 show a small sampling of mutant flies.

More ambitious *Drosophila* experiments with deleted genes yield intriguing information about the controls over development. By adding special promoters to a gene, researchers can control its expression with external cues, such as temperature. They also can delete genes from one part of the *Drosophila* genome and insert them into another. This molecular sleight-of-hand with the *eyeless* gene demonstrated that its expression can induce an eye to form not only on the fly's head, but also on the legs, wings, and antennae (Figure 14.10).

Astonishingly, the *eyeless* gene has counterparts in humans (a gene named *Aniridia*), mice (*Pax-6*), and squids (also *Pax-6*). Humans who have no

Figure 14.9 A few *Drosophila* mutants. (**a**) Wild-type (normal) fruit fly. The photograph above it can give you an idea of a fruit fly's size relative to the surface of a peach. (**b**) Yellow miniature. (**c**) Curly wings. (**d**) Vestigial wings.

Figure 14.10 Two more *Drosophila* mutations. *Left*, an eye that formed on a fruit fly leg. *Right*, a fruit fly with a double thorax, the outcome of a homeobox gene mutation.

Figure 14.11 Genes and *Drosophila*'s segmented body plan. (**a**) Fate map for the surface of a *Drosophila* zygote. Such maps indicate where each differentiated cell type in the adult originated. The pattern starts with the polar distribution of maternal mRNA and proteins in the unfertilized egg. This polarity dictates the future body axis. A series of segments will develop along this axis. Genes specify whether legs, wings, eyes, or some other body parts will develop on a particular segment.

Briefly, here's how it happens: Maternal gene products prompt expression of gap genes. Different gap genes become activated in regions of the embryo with higher or lower concentrations of different maternal gene products. Gap gene products influence each other's expression as well. They form a primitive spatial map.

Depending where they occur relative to the concentrations of gap gene products, embryonic cells express different pair-rule genes. Products of pair-rule genes accumulate in seven transverse stripes that mark the onset of segmentation (**b**). They activate other genes, the products of which divide the body into units (**c**). These interactions influence the expression of homeotic genes, which collectively govern the identity of each segment.

functional *Aniridia* genes have eyes without irises. *Aniridia* or *Pax-6* inserted into an *eyeless* mutant fly has the same effect as the *eyeless* gene—it induces eye formation wherever it is expressed. Here is evidence that animals as evolutionarily distant as insects, cephalopods, and mammals are connected by a shared ancestor.

GENES AND PATTERNS IN DEVELOPMENT

Different cells become organized in different ways in a new embryo. They divide, differentiate, and live or die; they migrate or stick to cells of the same type in tissues. Descendant cells fill in the details in orderly patterns, in keeping with a master body plan.

Such master plans consist of genes expressed in certain places at certain times during development. The regional and temporal gene expression generates a three-dimensional map of many overlapping proteins, most of which are transcription factors.

As an embryo develops, certain proteins induce undifferentiated cells to develop into different body tissues, depending on where the cells start out on the map. One example is the development of segments in *Drosophila* embryos (Figure 14.11).

Figure 14.12 *Left*, seven spots in the embryonic wing of a moth larva identify the presence of a gene product that will induce the formation of seven "eyespots" in the wing of the adult (*right*).

Pattern formation is the name for the emergence of embryonic tissues and organs in predictable patterns, in places where we expect them to be. Figures 14.11 and 14.12 are graphic examples. In Section 38.5, you will be taking a closer look at the controlled gene interactions that fill in details of the body plan.

Summary

Section 14.1 Whether, when, and how a gene gets expressed depends on controls over transcription and translation, and on modifications to protein products. Regulatory proteins (e.g., activators and inhibitors of transcription) and hormones are examples of control agents. These controls interact with one another, with control elements built into DNA molecules, with RNA, and with gene products.

With negative control mechanisms, regulatory proteins slow or curtail gene activity. With positive control mechanisms, they promote gene activity.

Section 14.2 Like all cells, prokaryotes respond quickly to short-term shifts in nutrients and other environmental conditions. Most of their gene control mechanisms adjust transcription rates in response to nutrient availability. Bacterial operon systems are examples of prokaryotic gene regulation.

Section 14.3 All cells of complex multicelled eukaryotes inherit the same genes, but each cell type selectively activates or suppresses a fraction of the genes in ways that lead to one or more unique aspects of structure, composition, and function.

At any time, most genes in a eukaryotic cell are shut off, unused. Those that the cell uses for housekeeping purposes, ongoing metabolic functions, are switched on all the time, at low levels. Expression of the other genes is adjustable. When control mechanisms come into play depends on cell type, prevailing chemical conditions, and signals from other cell types that can change a target cell's activities.

Gene expression within a cell changes in response to external conditions and is subject to long-term controls over growth and development. Eukaryotic cells control gene expression at key points, including transcription, RNA processing, RNA transport, mRNA degradation, translation, and protein activity. Translation is the major control point for most eukaryotic genes because so many participating molecules are regulated.

Section 14.4 Selective gene expression is the basis of cell differentiation during growth and development. It gives rise to cells that differ from one another in structure and function.

Complex eukaryotic body plans are influenced by homeotic genes, the master genes that control the emergence of the basic body plan during development.

X-chromosome inactivation is an example of dosage compensation, a control mechanism that maintains a crucial balance of gene expression between the sexes.

Section 14.5 Experiments with *Drosophila* identified a host of control genes. In embryo development, spatial maps of regulatory proteins guide the formation of tissues and organs in expected patterns.

Self-Quiz   *Answers in Appendix III*

1. The expression of a given gene depends on the _____ .
    a. type of cell and its functions c. environmental signals
    b. chemical conditions   d. all of the above

2. Hormones may _____ gene transcription in target cells.
    a. promote   c. participate in
    b. inhibit   d. both a and b

3. Eukaryotic genes guide _____ .
    a. fast short-term activities c. development
    b. overall growth   d. all of the above

4. Gene expression adjusts in response to changing _____ .
    a. nutrient availability   c. signals from other cells
    b. solute concentrations   d. all of the above

5. Cell differentiation _____ .
    a. occurs in all complex multicelled organisms
    b. requires unique genes in different cells
    c. involves selective gene expression
    d. both a and c
    e. all of the above

6. Regulatory proteins interact with _____ .
    a. DNA   c. gene products
    b. RNA   d. all of the above

7. An operon typically governs _____ .
    a. bacterial genes   c. genes of all types
    b. eukaryotic genes   d. DNA replication

8. In prokaryotic cells but not eukaryotic cells, a(n) _____ is a type of base sequence that precedes genes of an operon.
    a. lactose molecule   c. operator
    b. promoter   d. both b and c

9. A nucleotide sequence that signals the start of a gene is a(n) _____ .
    a. promoter   b. operator   c. enhancer   d. activator

10. Eukaryotic cells in complex organisms regulate gene expression by controlling different processes in _____ .
    a. transcription   e. mRNA degradation
    b. RNA processing   f. protein activity
    c. translation   g. a and d
    d. RNA transport   h. all of the above

11. X chromosome inactivation _____ .
    a. is dosage compensation   c. makes calico cats
    b. balances gene expression   d. both a and b

12. Homeotic genes _____ .
    a. are part of a bacterial operon
    b. control eukaryotic body plans
    c. control X chromosome inactivation
    d. both a and c

13. A cell with a Barr body is _____ .
    a. prokaryotic   c. from a female mammal
    b. from a male mammal   d. infected by the Barr virus

14. Match the terms with the most suitable description.
    _____ homeotic gene   a. binding site for repressor
    _____ operator   b. specialization during
    _____ proto-oncogene   development
    _____ differentiation   c. inactivated X chromosome
    _____ Barr body   d. can cause cancer
                             e. body plan development

Critical Thinking

1. Distinguish between:
   a. repressor protein and activator protein
   b. promoter and operator

2. Define three types of gene controls. Do they work for both prokaryotic and eukaryotic cells?

3. Unlike most rodents, guinea pigs are well developed at the time of birth. Within a few days, they can eat grass, vegetables, and other plant material. Suppose a breeder decides to separate baby guinea pigs from their mothers after three weeks. He wants to keep the males and females in different cages. However, he has trouble identifying the sex of young guinea pigs. Suggest how a microscope can help him identify their sex.

4. A plant, a fungus, and an animal consist of diverse cell types. How might this diversity arise, given that body cells in each of these organisms inherit the same set of genetic instructions? As part of your answer, define cell differentiation and the general way that selective gene expression brings it about.

5. In what fundamental way do negative and positive controls of transcription differ? Is the effect of one or the other form of control (or both) reversible?

6. If all cells in your body start out life with the same inherited information on how to build proteins, then what caused the differences between a red blood cell and a white one? Between a white blood cell and a nerve cell?

7. *Duchenne muscular dystrophy*, a genetic disorder, affects boys almost exclusively. Early in childhood, muscles begin to atrophy (waste away) in affected individuals, who typically die in their teens or early twenties.

   Muscle biopsies of a few women who carry an allele that is associated with the disorder identified some body regions of atrophied muscle tissue. They also showed that muscles adjacent to a region of atrophy were normal or even larger and more chemically active, as if to compensate for the weakness of the adjoining region.

   Form a hypothesis about the genetic basis of Duchenne muscular dystrophy that includes an explanation of why it might appear in some body regions but not others.

8. The closer a mammalian species is to humans in its genetic makeup, the more useful information it yields in laboratory studies of the mechanisms of cancer. Do you support the use of any mammal for cancer research? Why or why not?

9. Geraldo isolated an *E. coli* strain in which a mutation has hampered the capacity of CAP to bind to a region of the lactose operon, as it would do normally. How will this mutation affect transcription of the lactose operon when the *E. coli* cells are exposed to the following conditions? Briefly state your answers:
   a. Lactose and glucose are both available.
   b. Lactose is available but glucose is not.
   c. Both lactose and glucose are absent.

10. Calico cats are almost always female. A male calico cat is usually sterile. Briefly explain why.

11. The *Drosophila* embryo in Figure 14.13 displays a repeating pattern of gene expression. Reflect on Figure 14.11, then think about the gene product that made the red rings. What type of gene specified this product?

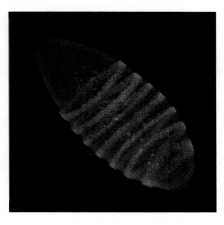

Figure 14.13 *Drosophila* zygote. The fluorescent rings are evidence that a gene is being expressed in certain regions.

IMPACTS, ISSUES *Golden Rice or Frankenfood?*

Not too long ago, the World Health Organization made a conservative estimate that 124 million children around the world show vitamin A deficiencies. These children may become permanently blind and develop other disorders. Researchers began working on a solution. They transferred three genes into rice plants. The genes directed the plants to make beta-carotene, a yellow pigment that is a precursor for vitamin A. Eating just 300 grams per day of the new "golden rice" might be enough to prevent vitamin A deficiency.

No one wants children to suffer or die. But many people oppose the idea of genetically modified foods, including golden rice. Possibly they are unaware of the history of agrarian societies. It isn't as if our ancestors were twiddling their green thumbs. For thousands of years, their artificial selection practices coaxed new plants and new breeds of cattle, cats, dogs, and birds from wild ancestral stocks. Meatier turkeys, seedless watermelons, big juicy corn kernels from puny hard ones—the list goes on (Figure 15.1).

And we're newcomers at this! During the 3.8 billion years before we even made our entrance, nature busily conducted uncountable numbers of genetic experiments by way of mutation, crossing over, and gene transfers

Figure 15.1 Snapshots from our food spectrum: Too little food in Ethiopia and lots of Indonesian rice plants. *Above*, an artificial selection success story—a big kernel from a modern strain of corn next to tiny kernels of an ancestral corn species discovered in a prehistoric cave in Mexico.

the big picture

The Genome Project   The discovery of the structure of DNA in 1953 sparked intense interest in creating technologies to manipulate that structure. Fifty years later, the entire sequence of bases in the human genome was completed.

Tools of the Trade   Scientists use DNA technologies to cut, identify, isolate, clone, copy, sequence, compare, and manipulate the DNA of any organism they wish to study. They put these technologies to practical use in genetic engineering.

between species. These processes introduced changes in the molecular messages of inheritance, and today we see their outcomes in the sweep of life's diversity.

Maybe the unsettling thing about the more recent human-directed changes is that the pace has picked up, hugely. We're getting pretty good at tinkering with the genetics of many organisms. We do this for pure research and for useful, practical applications.

Some say we must never alter the DNA of anything. The concern is that we as a species simply do not have the wisdom to bring about genetic changes without causing irreparable harm. One is reminded of our very human tendency to leap before we look.

And yet, we dream of the impossible. Something about the human experience gave us a capacity to imagine wings of our own making, and that capacity carried us to the frontiers of space. Someone else dreamed of turning plain rice into more nutritious food that might keep some children from going blind.

Many economic questions also remain unanswered. For example, will the patents on golden rice translate into higher production costs for the rural farmers of developing countries that urgently need the rice? Will transfer of beta-carotene genes disrupt a rice plant's messages of inheritance in some unexpected way?

The questions confronting you are these: Should we be more cautious, believing the risk takers may go too far? What do we stand to lose if risks are not taken?

☑ How Would You Vote?

Nutritional labeling is required on all packaged food in the United States, but genetically modified food products may be sold without labeling. Should food distributors be required to label all products made from genetically modified plants or livestock? See the Media Menu for details, then vote online.

Genetic Engineering   Normal or modified genes are being inserted into individual organisms both for research and practical applications. Genetically modified organisms help farmers produce food more efficiently. They are also a source of biomaterials and pharmacologic products.

Bioethics   The human genome has been sequenced, and the findings are being used for gene therapy and other applications. Many ethical and social issues remain unresolved as objections to the use of genetic engineering continue.

15.1 Tinkering With the Molecules of Life

In this unit, you started with cell division mechanisms that allow parents to pass on DNA to new generations. You moved to the chromosomal and molecular basis of inheritance, then on to gene controls that guide life's continuity. The sequence parallels the history of genetics. And now, you have arrived at the point where geneticists hold molecular keys to the kingdom of inheritance.

EMERGENCE OF MOLECULAR BIOLOGY

In 1953, James Watson and Francis Crick unveiled their model of the DNA double helix and ignited a global blaze of optimism about genetic research. The very book of life seemed to open up for scrutiny. In reality, it dangled just beyond reach. Major scientific breakthroughs are seldom accompanied by the simultaneous discovery of the tools necessary to study them. New methods of DNA research had to be invented before that book could become readable.

In 1972, Paul Berg and his associates were the first to make **recombinant DNA**. They fused fragments of DNA from one species into the genetic material from a different species, which they had grown in the laboratory. Their new recombinant DNA technique allowed them to isolate and replicate manageable subsets of DNA from any organism they wanted to study. The science of molecular biology was born, and suddenly everybody was worried about it.

Although researchers knew that DNA was not toxic, they could not predict with absolute certainty what would happen every time they fused genetic material from different organisms. Would they create new super-pathogens by accident? Could DNA from normally harmless organisms be fused to create a new form of life? What if their creation escaped into the environment and transformed other organisms?

In a remarkably quick and responsible display of self-regulation, scientists reached a consensus on safety guidelines for DNA research. Adopted at once by the National Institutes of Health (NIH), the guidelines listed laboratory procedural precautions. They covered the design and use of host organisms that could survive only under the narrow range of conditions that occur in the laboratory. Researchers stopped using the DNA from pathogenic or toxic organisms for recombination experiments until proper containment facilities were developed.

A golden age of recombinant DNA research soon followed. The emphasis had shifted from DNA's chemical and physical properties to its specific molecular structure. In 1977, Allan Maxam, Walter Gilbert, and Fred Sanger developed a method for determining the nucleotide sequence of cloned DNA fragments. The tools for reading the book of life, opened more than twenty years before, were now available for everyone to use.

DNA sequencing was cool, a visually rewarding, data-rich technique that entranced more than a few scientists. Unbelievable amounts of sequence data accumulated, from unbelievably diverse organisms. Computer technology at the time was advancing simultaneously, but it was barely keeping pace with

Figure 15.2 A few bases of the human genome—and a few of the supercomputers used to sequence it—at Celera Genomics in Maryland.

the tremendous demand for sequence data analysis and storage. In 1982, the NIH provided three million dollars to fund the first large-scale DNA database in the United States, one accessible to the public.

THE HUMAN GENOME PROJECT

Around 1986, everyone seemed to be arguing about sequencing the human genome. A **genome** is all the DNA in a haploid number of chromosomes. Many scientists insisted that the benefits for medicine and pure research would be incalculable. Others said the mapping would divert funding from other studies that had greater urgency as well as more likelihood of success.

At the time, sequencing three billion bases was a daunting and seemingly impossible task. With the techniques available at the time, it would have taken a worldwide consortium at least fifty years just to identify the sequence, even before deciphering what it meant. But techniques were getting better every year, and more bases were being sequenced and analyzed in less time. Automated (robotic) sequencing had just been invented, as had PCR, the polymerase chain reaction. Although both techniques were still cumbersome, expensive, and far from standardized, many sensed their potential for molecular biology.

Sequencing was still laborious. Waiting for faster methods seemed to be the most efficient means of sequencing the human genome, but exactly when was the technology going to be fast *enough*? Who would decide?

It was during this heated debate in 1987 that several independent organizations launched their own versions of the Human Genome Project. Among them was a company started by Walter Gilbert. He declared that his company would not only sequence the human genome, it would also patent the genome.

In early 1988, the NIH effectively annexed the entire Human Genome Project by hiring Watson as its head and providing researchers with 200 million dollars per year. A consortium formed between the NIH and other institutions working on different versions of the project.

Watson set aside 3 percent of the funding to study ethical and social issues arising from the research. He resigned in 1992 because of a disagreement with the NIH about patenting partial gene sequences. Francis Collins replaced him in 1993.

Amid ongoing squabbles over patent issues, the bulk of genome project sequencing in the United States continued at the NIH until 1998, when the

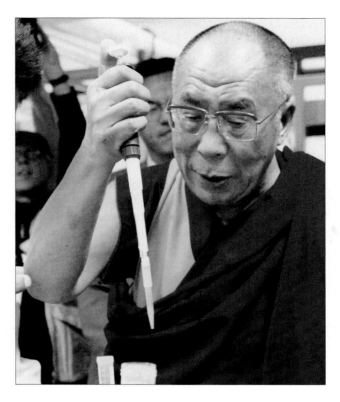

Figure 15.3 Everyone pitches in. The Dalai Lama prepares mouse DNA for sequencing during his 2003 visit to Whitehead Institute/MIT Center for Genome Research in Cambridge, Massachusetts.

scientist Craig Venter started Celera Genomics (Figure 15.2). Venter declared that *his* new company would be the first to complete the genome sequence. His challenge prompted the United States government to move its sequencing efforts into high gear.

Sequencing of the human genome was officially completed in 2003—fifty years after the discovery of DNA structure. About 99 percent of the coding regions in human DNA have been deciphered with a high degree of accuracy. A number of other genomes also have been fully sequenced (Figure 15.3).

What do we do with this vast amount of data? The next step is to investigate questions about precisely what that sequence means—where the genes are, where they are not, what the control mechanisms are and how they operate.

Recently, more than 21,000 human genes were identified. This doesn't mean we know what all those genes encode; it only means we know they are definitely genes. One of the many interesting discoveries is that the first intron and the last exon of most gene sequences are longer than the others. They may actually be part of an as yet undiscovered transcriptional control mechanism.

15.2 A Molecular Toolkit

Analysis of genes starts with manipulation of DNA. With molecular tools, researchers can cut DNA from different sources, then splice the fragments together.

THE SCISSORS: RESTRICTION ENZYMES

In 1970, Hamilton Smith and his colleagues were studying viral infection of *Haemophilus influenzae*, a species of bacteria. The bacteria protected themselves from infection by cutting up the invading viral DNA before it inserted itself into the bacterial chromosome.

Smith isolated one of the bacterial enzymes that was cutting up the viral DNA, the first known **restriction enzyme**. In time, several hundred strains of bacteria and a few eukaryotic cells yielded thousands more. Each restriction enzyme cuts double-stranded DNA at a specific base sequence between four and eight base pairs in length. Most of these recognition sites contain the same nucleotide sequence on both strands of the DNA. For instance, GAATTC is recognized on both strands and cut by the enzyme EcoRI.

Many restriction enzymes make staggered cuts that put a single-stranded "tail" on DNA fragments. Such cuts have a "sticky" end—a single-stranded tail. That tail can base-pair with a tail of another DNA molecule cut by the same enzyme, because the sticky ends of both fragments will match up (Figure 15.4a).

Tiny nicks remain when the fragments base-pair. A different enzyme, **DNA ligase**, seals the nicks, which results in a recombinant DNA molecule (Figure 15.4b). Recombinant DNA can consist of base sequences from different organisms of the same or different species.

CLONING VECTORS

Bacterial cells, recall, have only one chromosome—a circular DNA molecule. But many also have plasmids. A **plasmid** is a small circle of extra DNA with just a few genes (inset, *left*). It gets replicated right along with the bacterial chromosome. Bacteria normally can live without plasmids. Even so, some plasmid genes are useful, as when they confer resistance to antibiotics.

Under favorable conditions, bacteria divide often, so huge populations of genetically identical cells form swiftly. Before each division, replication enzymes copy both the chromosomal DNA *and* the plasmid DNA, in some cases repeatedly. This gave researchers an idea. Why not try to insert a fragment of foreign DNA into a plasmid and see if a bacterial cell replicates it?

A modified plasmid that accepts foreign DNA and slips into a host bacteria, yeast, or some other cell is a **cloning vector**. Cloning vectors usually have multiple cloning sites, which are several unique restriction enzyme sequences clustered in one part of the vector. As you'll see later, the vector also has genes that help researchers identify which cells it slips into, such as genes for antibiotic resistance (Figure 15.5).

A cell that takes up a cloning vector may found a huge population of descendant cells, each containing an identical copy of the vector and the foreign DNA inserted into it. Collectively, all of the identical cells hold many "cloned" copies of the foreign DNA.

Such DNA cloning is a tool that helps researchers amplify and harvest unlimited amounts of particular DNA fragments for their studies (Figure 15.6).

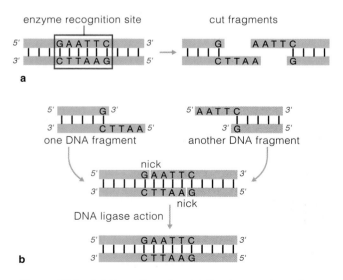

Figure 15.4 (**a**) Formation of restriction fragments and (**b**) splicing fragments into a recombinant DNA molecule.

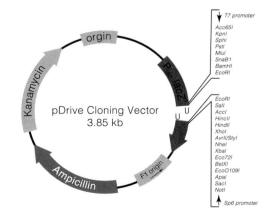

Figure 15.5 A commercially available cloning vector, with its useful restriction enzyme sites listed at right. This vector includes antibiotic resistance genes (*blue*) and the bacterial *lacZ* gene (*red*). These genes help researchers identify cells that take up recombinant molecules.

Read Me First!
and watch the narrated animation
on DNA recombination

a A restriction enzyme cuts a specific base sequence everywhere it occurs in DNA.

b The DNA fragments have sticky ends.

c The same enzyme cuts the same sequence in plasmid DNA.

d The plasmid DNA also has sticky ends.

e The DNA fragments and the plasmid DNA are mixed with DNA ligase.

f The result? A collection of recombinant plasmids that incorporate foreign DNA fragments.

g Host cells that can divide rapidly take up the recombinant plasmids.

Figure 15.6 (**a–f**) Formation of recombinant DNA—in this case, a collection of DNA fragments sealed into bacterial plasmids. (**g**) Recombinant plasmids are inserted into host cells that can rapidly amplify the foreign DNA of interest.

Figure 15.7 How to make cDNA. Reverse transcriptase catalyzes the assembly of a single DNA strand on an mRNA template, forming an mRNA–cDNA hybrid molecule. Next, DNA polymerase replaces the mRNA with another DNA strand. The result is double-stranded DNA.

cDNA CLONING

Chromosomal DNA usually contains introns (Section 13.1). Sometimes it's impossible to tell whether a gene sequence is part of an intron or exon or to pinpoint where it starts and ends. A researcher investigating gene products or gene expression focuses on mRNA, because the introns have already been snipped out of it. All that's left is coding sequence and some small signal sequences. Any time a gene is being expressed, mRNA is being transcribed, so cells that are actually using a gene will also contain the mRNA it encodes.

Restriction enzymes do not cut RNA, so mRNA cannot be cloned until it has been translated first into DNA. Replication enzymes isolated from viruses or bacteria can be used to translate the mRNA *in vitro*, or inside a test tube. **Reverse transcriptase** is a viral enzyme that uses the mRNA as a template. Using free nucleotides, it assembles a single strand of **cDNA**, or *complementary* DNA, on the template (Figure 15.7). A hybrid molecule is the outcome; one strand of mRNA and one strand of cDNA are base-paired together.

DNA polymerase added to the mix strips the RNA from the hybrid molecule as it copies the first strand of cDNA into a second strand. The result is a double-stranded DNA copy of the original mRNA. And that copy may be used for cloning.

Molecular biologists manipulate DNA and RNA. Restriction enzymes cut DNA from organisms of the same or different species, and ligases glue the fragments into plasmids.

A recombinant plasmid is a cloning vector; it can slip into bacteria, yeast, or other cells that divide rapidly. Host cells make multiple, identical copies of the foreign DNA.

Reverse transcriptase uses mRNA as a template to make cDNA.

15.3 Haystacks to Needles

Any genome consists of thousands of genes. E. coli has 4,279; humans have about 30,000. To study or modify any one of those genes, researchers must first find it among all others in the genome, and it's like searching for a needle in a haystack. Once found, it must be copied many times to make enough material for experiments.

ISOLATING GENES

Each **gene library** is a collection of bacterial cells that house different cloned fragments of DNA. We call the cloned fragments of an entire genome a *genomic* library. By contrast, a *cDNA library* is derived from mRNA.

A particular gene of interest must be isolated from millions of other genes. Clones containing that gene are mixed up in a library with thousands or millions of others that do not. A **probe**, a short stretch of DNA labeled with a radioisotope (or sometimes a pigment), may be used to find a one-in-a-million clone. Probes distinguish one DNA sequence from all of the others in a library of clones or any other collection of mixed DNA. A radiolabeled probe base-pairs with DNA in the gene region of interest, then researchers can find it with devices that detect radiation. Such base-pairing between DNA (or RNA) from more than one source is known as **nucleic acid hybridization**.

How do researchers make a suitable probe? If they already know the desired gene sequence, they can use it to design and build an oligomer (a short stretch of nucleotides). Or they can use DNA from a closely related gene as a probe even if it isn't an exact match.

Figure 15.8 shows steps of one probe hybridization technique. Bacterial cells containing a gene library are spread apart on the surface of a solid growth medium, usually enriched agar, in a petri dish. Individual cells undergo repeated divisions, which form large clusters, or colonies, of genetically identical bacterial cells.

Press a piece of nylon or nitrocellulose paper on top of the petri dish and some of the cells from each colony will stick to it, mirroring the distribution of all colonies on the dish. Soaking the paper in an alkaline solution ruptures the cells, which releases their DNA. The solution also denatures the DNA, separating it into single strands that stick to the paper in the spots

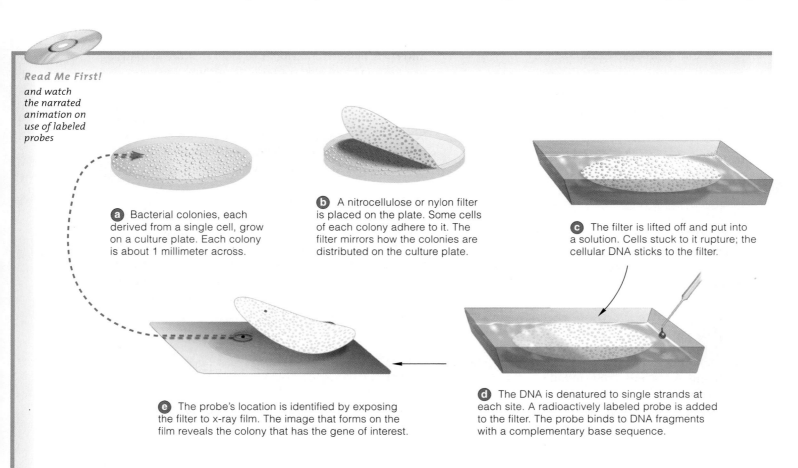

Read Me First!
and watch the narrated animation on use of labeled probes

a Bacterial colonies, each derived from a single cell, grow on a culture plate. Each colony is about 1 millimeter across.

b A nitrocellulose or nylon filter is placed on the plate. Some cells of each colony adhere to it. The filter mirrors how the colonies are distributed on the culture plate.

c The filter is lifted off and put into a solution. Cells stuck to it rupture; the cellular DNA sticks to the filter.

d The DNA is denatured to single strands at each site. A radioactively labeled probe is added to the filter. The probe binds to DNA fragments with a complementary base sequence.

e The probe's location is identified by exposing the filter to x-ray film. The image that forms on the film reveals the colony that has the gene of interest.

Figure 15.8 Use of a radioactive probe to identify a bacterial colony that contains a targeted gene.

Read Me First!
and watch the narrated animation on PCR

where the colonies were. When the probe is washed over the paper, it hybridizes with, or sticks to, only the DNA that has the target sequence. The hybridized probe makes a radioactive spot that can be detected with x-ray film. The position of the spot on the film reflects the position of the original colony on the petri dish. Cells from that colony alone are cultured to isolate the cloned gene of interest.

PCR

Researchers may replicate a gene, or part of it, with **PCR** (*Polymerase Chain Reaction*). PCR uses primers and a heat-tolerant polymerase for a hot–cold cycled reaction that replicates targeted DNA fragments. And it can replicate them by a billionfold. This technique can transform one needle in a haystack, that one-in-a-million DNA fragment, into a huge stack of needles with a little hay in it.

Figure 15.9 shows the reaction steps. **Primers** are synthetic nucleotide oligomers, usually between ten and thirty bases long. They are designed to base-pair with specific nucleotide sequences on either end of the fragment of interest. In a PCR reaction, researchers mix primers, DNA polymerase, nucleotides, and the DNA, which will act as a template for replication. Then the researchers expose the mixture to repeated cycles of high and low temperatures. The two strands of a DNA double helix separate into single strands at high temperature. When the mixture is cooled, some of the primers will hybridize with the DNA template.

The elevated temperatures required to separate DNA strands destroy typical DNA polymerases. But the heat-tolerant DNA polymerase employed for PCR reactions is from *Thermus aquaticus*, a bacterium that lives in superheated water springs (Chapter 19). Like all DNA polymerases, it recognizes primers bound to DNA as places to initiate synthesis.

Synthesis proceeds along the DNA template until the temperature cycles up and the DNA strands are separated again. When the temperature cycles down, primers rehybridize, and the reactions start all over. With each round of temperature cycling, the number of copies of targeted DNA can double. PCR quickly and exponentially amplifies even a tiny bit of DNA.

Probes may be used to help identify one particular gene among many in gene libraries.

The polymerase chain reaction (PCR) is a method of rapidly and exponentially amplifying DNA samples of interest.

a Primers, free nucleotides, and DNA template are mixed with heat-tolerant DNA polymerase.

b When the mixture is heated, the DNA denatures. When it is cooled, some primers hydrogen-bond to the DNA template.

c *Taq* polymerase uses the primers to initiate synthesis, copying the DNA template. The first round of PCR is completed.

d The mixture is heated again. This denatures all the DNA into single strands. When the mixture is cooled, some of the primers hydrogen-bond to the DNA.

e *Taq* polymerase uses the primers to initiate synthesis, copying the DNA. The second round of PCR is complete. Each successive round of synthesis can double the number of DNA molecules.

Figure 15.9 Two rounds of the polymerase chain reaction, or PCR. A bacterium, *Thermus aquaticus*, is the source of the *Taq* polymerase. Thirty or more cycles of PCR may yield a billionfold increase in the number of starting DNA template molecules.

15.4 First Just Fingerprints, Now DNA Fingerprints

Except for identical twins, no two people have exactly the same base sequence in their DNA. Scientists can distinguish one person from another on the basis of differences in those sequences.

Each human has a unique set of fingerprints. Like all other sexually reproducing species, each also has a **DNA fingerprint**—a unique array of DNA sequences inherited in a Mendelian pattern from parents. More than 99 percent of the DNA is the same in all humans, but the other 1 percent is unique to each individual. These unique stretches of DNA are sprinkled through the human genome as **tandem repeats**—many copies of the same short DNA sequences, positioned one after the other along the length of a chromosome.

For example, one person's DNA might contain four repeats of the bases TTTTC in a certain location. Another person's DNA might have them repeated fifteen times in the same location. One person might have five repeats of CGG, and another might have fifty. Such repetitive sequences slip spontaneously into the DNA during replication, and their numbers grow or shrink over time. The mutation rate is high in these regions.

DNA fingerprinting reveals differences in the tandem repeats among individuals. A restriction enzyme cuts their genomic DNA into an assortment of fragments. The sizes of those fragments are unique to the individual. They reveal genetic differences between individuals, and they can be detected as RFLPs (restriction fragment length polymorphisms).

The differences show up with **gel electrophoresis**. In this technique, an electric field pulls a sample of DNA fragments through a slab of a semisolid matrix, such as an agar gel. Fragments of different sizes migrate through the matrix at different rates. Larger molecules are hindered by the matrix more than smaller ones, much as elephants are slower than tigers at slipping between trees in a dense forest. In short, gel electrophoresis separates the fragments of DNA according to their length. After a time, the different fragments separate into distinct bands.

A banding pattern of genomic DNA fragments is the DNA fingerprint unique to the individual. It is identical only between identical twins. Otherwise, the odds of two people sharing an identical DNA fingerprint are one in three trillion.

PCR can also be used to amplify tandem-repeat regions. Differences in the size of the resulting amplified DNA fragments are again detected with gel electrophoresis. A few drops of blood, semen, or cells from a hair follicle at a crime scene or on a suspect's clothing yield enough DNA to amplify with PCR, and then generate a fingerprint.

DNA fingerprints help forensic scientists identify criminals, victims, and innocent suspects. Figure 15.10 shows some tandem repeat RFLPs that were separated by gel electrophoresis. Those samples of DNA had been taken from seven people and from a bloodstain left at a crime scene. One of the DNA fingerprints matched.

Defense attorneys initially challenged the use of DNA fingerprinting as evidence in court. Today, however, the procedure has been firmly established as accurate and unambiguous. DNA fingerprinting is routinely submitted as evidence in disputes over paternity, and it is being widely used to convict the guilty and exonerate the innocent. To date, such evidence has already helped release 143 innocent people from prison.

DNA fingerprint analysis also has confirmed that human bones exhumed from a shallow pit in Siberia belonged to five individuals of the Russian imperial family, all shot to death in secrecy in 1918. It also was used to identify the remains of those who died in the World Trade Center on September 11, 2001.

① ② ③   FROM BLOOD AT CRIME SCENE   ④ ⑤ ⑥ ⑦

Figure 15.10 Damning comparison of the DNA fingerprints from a bloodstain left behind at a crime scene and from blood samples of seven suspects (the circled numbers). Can you point out which of the seven is a match?

15.5 Automated DNA Sequencing

Sequencing reveals the order of nucleotides in DNA. This technique uses DNA polymerase to partially replicate a DNA template. In current research labs, manual methods have been replaced largely by automated techniques.

Automated DNA sequencing can reveal the sequence of a stretch of cloned or PCR-amplified DNA in just a few hours. Researchers use four standard nucleotides (T, C, A, and G). They also use four modified versions, which we represent as T*, C*, A*, and G*. Each of the four types of modified nucleotide has become labeled with a pigment that will fluoresce in a particular color when it passes through a laser beam.

Researchers mix all eight kinds of nucleotides with a single-stranded DNA template, a primer, and DNA polymerase. The polymerase uses the primer to copy the template DNA into new strands of DNA. One by one, it adds nucleotides in the order dictated by the sequence of the DNA template.

Each time, the polymerase randomly attaches one of the standard *or* one of the modified nucleotides to the DNA template. When one of the modified nucleotides becomes covalently bonded to the newly forming DNA strand, it stops further synthesis of that strand. After enough time passes, there will be some new strands that stop at each base in the DNA template sequence.

Eventually the mixture holds millions of copies of DNA fragments, all fluorescent-tagged. The fragments are now separated by gel electrophoresis, which is part of an automated sequencer. Shortest fragments migrate fastest and reach the end of the block of gel first; the longest fragment is last. Fragments of the same length migrate through the gel at the same speed, and they form observable bands (Figure 15.11*a*).

Each fragment passes through a laser beam, and the modified nucleotide attached to its tail end makes it fluoresce a certain color. The sequencer detects and records the fluorescent colors as the fragments pass through the end of the gel. Because each color codes for a particular nucleotide, the order of colored bands is the DNA sequence. The machine itself assembles the sequence data.

Figure 15.11*b* shows partial results from a run through an automated DNA sequencer. Each peak in the tracing represents the detection of one fluorescent color as the fragments reached the end of the gel. The sequence is shown beneath the graph line.

> *With automated DNA sequencing, the order of nucleotides in a DNA fragment that has been cloned or amplified can be determined rapidly.*

Read Me First!
and watch the narrated animation on automated DNA sequencing

Figure 15.11 Automated DNA sequencing. (**a**) DNA fragments are synthesized using a template and fluorescent nucleotides. The bands are separated by gel electrophoresis. (**b**) The order of the fluorescent bands that appear in the gel is detected by the automated sequencer, and indicates the template DNA sequence. Today, researchers use sequence databases that are accessible globally via the Internet.

15.6 Practical Genetics

Even the tiniest living organisms are able to make complex organic compounds. Researchers harness this ability for practical purposes in genetic engineering.

DESIGNER PLANTS

As crop production expands to keep pace with human population growth, it puts unavoidable pressure on ecosystems everywhere. Irrigation leaves mineral and salt residues in soils. Tilled soil erodes, taking topsoil with it. Runoff clogs rivers, and fertilizer in it causes algae to grow so much that fish suffocate. Pesticides harm humans, other animals, and beneficial insects.

Pressured to produce more food at lower cost and with less damage to the environment, some farmers are turning to genetically engineered crop plants.

Genetic engineering is the process of changing the genetic makeup of an organism, often with intent to alter one or more aspects of phenotype. Researchers may accomplish this by transferring a gene from one species into another species, or by modifying a gene and inserting it into an organism of the same species.

Cotton plants with a built-in insecticide gene kill only the insects that eat it, so farmers that grow them don't have to use as many pesticides. Genetically modified wheat has double yields per acre. Certain transgenic tomato plants survive in salty soils that wither other plants; they also absorb and store excess salt in their leaves, thus purifying saline soil for future crops. *Transgenic* simply refers to an organism into which DNA from another species has been inserted, as in Figures 15.12 and 15.13.

a   b

Figure 15.12 Transgenic plants. (**a**) Cotton plant (*left*), and cotton plant with a gene for herbicide resistance (*right*). Both were sprayed with weed killer. (**b**) Genetically engineered aspen seedlings in which lignin biosynthesis has been partially blocked. Unmodified seedling is on the left.

The cotton plants in Figure 15.12*a* were genetically engineered for resistance to a relatively short-lived herbicide. Spraying fields with this herbicide will kill all weeds but not the engineered cotton plants. The practice means farmers can use fewer and less toxic chemicals. They also don't have to till the soil as much to control weeds, so there is less river-clogging runoff.

Aspen tree seedlings in which a lignin biosynthesis pathway has been modified still make lignin, but not as much—and root, stem, and leaf growth are greatly enhanced (Figure 15.12*b*). Wood from lignin-deficient trees makes it easier to manufacture paper and clean-burning fuels such as ethanol.

Read Me First!
and watch the narrated animation on gene transfer

a A bacterial cell contains a Ti plasmid (*purple*) that has a foreign gene (*blue*).

b The bacterium infects a plant and transfers the Ti plasmid into it. The plasmid DNA becomes integrated into one of the plant's chromosomes.

c The plant cell divides. Its descendant cells form an embryo, which may develop into a mature plant that can express the foreign gene.

d

e Example of a young plant with a fluorescent gene product.

Figure 15.13 (**a–d**) Gene transfer from *Agrobacterium tumefaciens* to a plant cell using a Ti plasmid. (**e**) A transgenic plant expressing a firefly gene for the enzyme luciferase.

Figure 15.14 Two genetically modified animals: (**a**) Mira, a goat transgenic for human antithrombin III, an anticlotting factor. (**b**) This transgenic mouse has been engineered to produce green fluorescent protein (GFP).

(**c**) A featherless chicken breed developed by traditional cross-breeding methods in Israel. They thrive in desert environments where cooling systems are not an option. Chicken farmers in the United States have lost millions of feathered chickens at a time when temperatures skyrocketed.

Engineering plant cells starts with vectors that can carry genes into plant cells. *Agrobacterium tumefaciens* is a bacterial species that infects eudicots, including beans, peas, potatoes, and other vital crops. Its plasmid genes cause tumor formation on these plants; hence the name Ti plasmid (*Tumor-inducing*). The Ti plasmid is used as a vector for transferring new or modified genes into plants.

Researchers excise the tumor-inducing genes, then insert a desired gene into the plasmid (Figure 15.13). Some plant cells cultured with the modified plasmid take it up. Whole plants may be regenerated.

Modified *A. tumefaciens* bacteria deliver genes into monocots that also are food sources, including wheat, corn, and rice. Researchers can also transfer genes into plants by way of electric shocks, chemicals, or blasts of microscopic particles coated with DNA.

BARNYARD BIOTECH

The first mammals enlisted for experiments in genetic engineering were laboratory mice. Transgenic mice appeared on the research scene in 1982 when scientists built a plasmid containing a gene for rat somatotropin (also known as growth hormone). They injected the recombinant DNA into fertilized mouse eggs, which were subsequently implanted into female mice. One-third of the resulting offspring grew much larger than their littermates—the rat gene had become integrated into their DNA and was being expressed.

Transgenic animals are now used routinely for medical research. The function and regulation of many gene products have been discovered using "knockout mice," in which targeted genes are inactivated. Defects in the resulting mice give clues about the gene. Strains of mice engineered to be susceptible to human diseases allow researchers to study both the diseases and their cures without experimenting on humans.

Genetically engineered animals are also sources of pharmacological and other valuable proteins. As a few examples, goats produce CFTR protein (for treating cystic fibrosis) and TPA protein (to counter the effects of a heart attack). Rabbits produce human interleukin-2, a protein that stimulates divisions of immune cells (T-lymphocytes). Cattle, too, may soon be producing human collagen that can be used to repair cartilage, bone, and skin. Goats make spider silk protein that might be used to make bullet-proof vests, medical supplies, and space equipment. Other goats make human antithrombin, used to treat people with blood clotting disorders (Figure 15.14*a*).

Genetic engineering has also given us dairy goats with healthier milk, pigs whose manure is easier on the environment, freeze-resistant salmon, extra-hefty sheep, low-fat pigs, mad cow disease-resistant cows, and even allergen-free cats.

Tinkering with the genetics of animals for the sake of human convenience does raise some serious ethical issues, particularly because failed experiments can have gruesome results. However, is transgenic animal research simply an extension of thousands of years of acceptable barnyard breeding practices (Figure 15.14*c*)? The techniques have changed, but not the intent. Like our ancestors, we continue to have a vested interest in improving our livestock.

Transgenic plants help farmers grow crops more efficiently and with less impact on the environment.

Transgenic animals are widely used in medical research. Some are sources of medically valued proteins and other biomaterials. Food animals are being altered to be more nutritious, disease resistant, or easier to raise.

15.7 Weighing the Benefits and Risks

We as a society continue to work our way through the ethical implications of DNA research even while we are applying the new techniques to medicine, industry, agriculture, and environmental remediation.

WHO GETS WELL?

More than 15,500 genetic disorders affect between 3 and 5 percent of all newborns, and they cause 20 to 30 percent of all infant deaths per year. They account for about 50 percent of the mentally impaired and nearly 25 percent of all hospital admissions.

Rhys Evans (Figure 15.15*a*) was born with a severe immune deficiency known as SCID-X1, which stems from mutations in gene *IL2RG*. Children affected by this disorder can live only in germ-free isolation tents, because they cannot fight infections.

In 1998, a virus was used to insert nonmutated copies of *IL2RG* into stem cells taken from the bone marrow of eleven boys with SCID-X1. *Stem* cells are still "uncommitted" and have the potential to differentiate into other types, including white blood cells of the immune system. Each child's modified stem cells were infused back into his bone marrow. Months afterward, ten of the children left their isolation tents for good. Their immune systems had been repaired by the gene therapy. Since then, many other SCID-X1 patients, including Rhys Evans, have been cured in other gene therapy trials.

In 2002, two children from the initial experiment in 1998 developed leukemia. Their illness surprised researchers, who had anticipated that any cancer related to the therapy would be extremely rare. An overproduction of white blood cells (T-lymphocytes) caused the leukemia in both children. The very gene targeted for repair work— *IL2RG*—may be the problem, particularly when combined with the viral vector used in the gene therapy. No other children in any gene therapy experiments for SCID-X1 have developed leukemia. Even so, our understanding of the human genome clearly lags behind our ability to modify it.

WHO GETS ENHANCED?

Modifying the human genome has profound ethical implications even beyond the unexpected risks. To many of us, human gene therapy to correct genetic disorders seems like a socially acceptable goal. Now take this idea one step further. Is it acceptable to change some genes of a normal human in order to alter or enhance traits?

Through gene transfers, researchers have already engineered strains of mice with enhanced memory and learning abilities. Maybe their work heralds help for Alzheimer's disease patients, perhaps even for those who just want more brain power.

The idea of being able to select desirable human traits is referred to as *eugenic engineering*. Yet who decides which forms of a trait are the most desirable? Realistically, cures for many severe but rare genetic disorders will not be pursued because the payback for research is not financially attractive. Eugenics, however, might turn a profit. Just how much would potential parents pay to engineer tall or blue-eyed or fair-skinned children? Would it be okay to engineer

Figure 15.15 Experimental gene therapy patients. (**a**) Rhys Evans was born with a gene that causes SCID-X1. His immune system never developed in a way that could fight infections. A gene transfer freed him from life in a germ-free isolation tent. (**b**) Max Randell smiles at his mother after receiving gene therapy in 2001 for Canavan's disease. This is a degenerative and fatal disease of the central nervous system. At the time of this writing, Max is alive and doing well.

"superhumans" who have breathtaking strength or intelligence? How about an injection that would help you lose extra weight, and keep it off permanently? The borderline between interesting and abhorrent is not the same for everyone.

In a survey conducted not long ago in the United States, more than 40 percent of those interviewed said it would be fine to use gene therapy to make smarter and better looking babies. In one poll of British parents, 18 percent were willing to use genetic enhancement to prevent their children from being aggressive, and 10 percent were willing to use it to keep them from growing up to be homosexual.

KNOCKOUT CELLS AND ORGAN FACTORIES

Each year, about 75,000 people are on waiting lists for an organ transplant, but human donors are in short supply. There is talk of harvesting organs from pigs (Figure 15.16), because pig organs function very much like ours do. Transferring an organ from one species into another is called **xenotransplantation**.

The human immune system battles anything that it recognizes as "nonself." It rejects a pig organ at once. A certain sugar molecule occurs on the plasma membrane of cells that make up a pig organ's blood vessels. Antibodies circulating in human blood latch on to that sugar quickly and doom the transplant. Within a few hours, cascading reactions lead to massive coagulation inside the organ's blood vessels, and failure is swift. Drugs can suppress this immune response, but there's a serious side effect: the drugs make organ recipients vulnerable to infections.

Pig DNA contains two copies of *Ggta1*, the gene for alpha-1,3-galactosyltransferase. This enzyme catalyzes a key step in biosynthesis of alpha-1,3-galactose, the pig sugar that human antibodies recognize. Researchers succeeded in knocking out both copies of the *Ggta1* gene in transgenic piglets. Without the gene, the pigs lack alpha-1,3-galactose. If one of their organs or tissues is transplanted, the human immune system might not recognize it. The tissues and organs from such animals could benefit millions of people, including those who suffer from diabetes or Parkinson's disease.

Critics of xenotransplantation are concerned that, among other things, pig–human transplants would invite pig viruses to cross species and infect humans, perhaps catastrophically. Their concerns are not unfounded. In 1918, an influenza pandemic killed twenty million people worldwide. It originated with a swine flu virus—in pigs.

Figure 15.16 Inquisitive transgenic pig at the Virginia Tech Swine Research facility.

REGARDING "FRANKENFOOD"

Genetically engineered food crops are widespread in the United States. At least 45 percent of cotton crops, 38 percent of soybean crops, and 25 percent of corn crops are now engineered to withstand weedkillers or to make their own pesticides. For years, modified corn and soybeans have been used in tofu, breakfast cereals, soy sauce, vegetable oils, beer, and soft drinks. They are fed to farm animals. Engineered crop plants hold down food production costs, reduce dependence on pesticides and herbicides, and enhance crop yields. Food plants are being designed for flavor, nutritional value, and extended shelf life.

In Europe especially, public resistance to modified food runs high. Besides arguing that modified foods might be toxic and have lower nutritional value, many people worry that designer plants might cross-pollinate wild plants and produce "superweeds."

The chorus of critics in Europe may provoke a trade war with the United States. The outcome is not small potatoes, so to speak. In 1998, the value of American agricultural exports was about 50 billion dollars. Restrictions will profoundly impact agriculture in the United States, and inevitably the impact will trickle down to what you eat and how much you pay for it.

All of which invites you to read scientific research and form your own opinions. The alternative is to be swayed by media hype (the term "Frankenfood," for instance), or by potentially biased reports from other groups that might have a different agenda (such as chemical pesticide manufacturers).

15.8 Brave New World

The structural and comparative analysis of genomes is yielding information about evolutionary trends as well as potential therapies for genetic diseases.

GENOMICS

Research into genomes of humans and other species has converged into a new research field—**genomics**. The *structural* genomics branch deals with the actual mapping and sequencing of genomes of individuals. The *comparative* genomics branch is concerned with finding evolutionary relationships among groups of organisms. Comparative genomics researchers analyze similarities and differences among genomes.

Comparative genomics has practical applications as well as potential for research. The basic premise is that the genomes of all existing organisms are derived from common ancestors. For instance, pathogens share some conserved genes with human hosts even though the lineages diverged long ago. Shared gene sequences, how they are organized, and where they differ may hold clues to where our immune defenses against pathogens are strongest or the most vulnerable.

Genomics has potential for **human gene therapy**—the transfer of one or more normal or modified genes into a person's body cells to correct a genetic defect or boost resistance to disease. However, even though the human genome is fully sequenced, it is not easy to manipulate within the context of a living individual.

Experimenters employ stripped-down viruses as vectors that inject genes into human cells. Some gene therapies deliver modified cells into a patient's tissue. In many cases, therapies make a patient's symptoms subside even when the modified cells are producing just a small amount of a required protein.

However, no one can yet predict where virus-injected genes will end up in a person's chromosomes. The danger is that the insertion will disrupt other genes, particularly those controlling cell division and growth. One-for-one gene swaps with recombination methods are possible but still experimental.

DNA CHIPS

Analysis of genomes is now advancing at a stunning pace. Researchers pinpoint which genes are silent and which are being expressed with the use of **DNA chips**. These are microarrays of thousands of gene sequences representing an entire genome—all stamped onto a glass plate about the size of a smallish business card.

A fluorescent labeled cDNA probe is made using mRNA, say, from cells of a cancer patient. Only the

Figure 15.17 Complete yeast genome array on a DNA chip that is about 19 millimeters (3/4 inch) across. Green spots pinpoint the genes that are active during fermentation. Red spots pinpoint the ones used during aerobic respiration. Yellow spots indicate genes active during both pathways.

genes expressed at the time the cells are harvested will be making mRNA, so they alone will make up the resulting probe population. The labeled probe is then incubated with a chip made from genomic DNA. Wherever the probe binds with complementary base sequences on the chip, there will be a spot that glows under fluorescent light. Analysis of which spots on the chip are glowing reveals which of the thousands of genes inside the cells are active and which are not.

DNA chips are being used to compare different gene expression patterns between cells. Examples are yeasts grown in the presence and absence of oxygen, and different types of cells from the same multicelled individual. RNA from one set of cells is transformed into green fluorescent cDNA, and RNA from the other set into red fluorescent cDNA. The cDNAs are mixed and incubated with a genomic DNA chip. Green or red fluorescence indicates expression of genes in the different cell types. Yellow is a mixture of both red and green, and it indicates that both genes were being expressed at the same time in a cell (Figure 15.17).

In genomics, new techniques such as DNA chips allow researchers to rapidly evaluate and compare genome-spanning expression patterns.

Summary

Section 15.1 Discovery of DNA's double helical structure sparked interest in deciphering its genetic messages. A global race to complete the sequence of the human genome spurred rapid development of new techniques to study and manipulate DNA. The entire human genome has been sequenced and is now being analyzed. Genomes of other organisms have been sequenced as well.

Section 15.2 Recombinant DNA technology uses restriction enzymes that cut DNA into fragments. The fragments may be spliced into cloning vectors by using DNA ligase. Recombinant plasmids are taken up by rapidly dividing cells, such as bacteria, to make multiple, identical copies of the foreign DNA. Reverse transcriptase copies mRNA into cDNA for cloning.

Section 15.3 A gene library is a mixed collection of cells that have taken up cloned DNA. A particular gene can be isolated from a library by using a probe, a short stretch of DNA that can base-pair with the gene and that is traceable with a radioactive or pigment label. Probes help researchers identify one particular clone among millions of others. Base-pairing between nucleotide sequences from different sources is called nucleic acid hybridization.

The polymerase chain reaction (PCR) is a way to rapidly copy particular pieces of DNA. A sample of a DNA template is mixed with nucleotides, primers, and a heat-resistant DNA polymerase. Each round of PCR proceeds through a series of temperature changes that amplifies the number of DNA molecules exponentially.

Section 15.4 Tandem repeats are multiple copies of a short DNA sequence that follow one another along a chromosome. The number and distribution of tandem repeats, unique in each person, can be revealed by gel electrophoresis; they form a DNA fingerprint.

Section 15.5 Automated DNA sequencing rapidly reveals the order of nucleotides in DNA fragments. As DNA polymerase is copying a template DNA, progressively longer fragments stop growing when one of four different fluorescent nucleotides becomes attached. Electrophoresis separates the resulting labeled fragments of DNA into bands according to length. The order of the colored bands as they migrate through the gel reflects which fluorescent base was added to the end of each fragment, and so indicates the template DNA base sequence.

Section 15.6 Genetic engineering is the directed modification of the genetic makeup of an organism, often with intent to modify its phenotype. Researchers insert normal or modified genes from one organism into another of the same or different species. Gene therapies also reinsert altered genes into individuals.

Genetically engineered bacteria produce medically valued proteins. Transgenic crop plants help farmers produce food more efficiently. Genetic engineering of animals allows commercial production of human proteins, as well as research into genetic disorders.

Section 15.7 Human gene therapy and modification of animals for xenotransplantation are examples of developing technologies. As with any new technology, potential benefits must be weighed against potential risks, including ecological and social repercussions.

Section 15.8 Genomics, the study of human and other genomes, is shedding light on evolutionary relationships and has practical uses. Human gene therapy transfers normal or modified genes into body cells to correct genetic defects. Gene chips are used to compare patterns of gene expression.

Self-Quiz

Answers in Appendix III

1. _____ is the transfer of normal genes into body cells to correct a genetic defect.
   - a. Reverse transcription
   - b. Nucleic acid hybridization
   - c. PCR
   - d. Gene therapy

2. DNA is cut at specific sites by _____ .
   - a. DNA polymerase
   - b. DNA probes
   - c. restriction enzymes
   - d. reverse transcriptase

3. Fill in the blank: A _____ is a small circle of bacterial DNA that is not part of the bacterial chromosome.

4. By reverse transcription, _____ is assembled on a(n) _____ template.
   - a. mRNA; DNA
   - b. cDNA; mRNA
   - c. DNA; ribosomes
   - d. protein; mRNA

5. PCR stands for _____ .
   - a. polymerase chain reaction
   - b. polyploid chromosome restrictions
   - c. polygraphed criminal rating
   - d. politically correct research

6. By gel electrophoresis, fragments of DNA can be separated according to _____ .
   - a. sequence
   - b. length
   - c. species

7. Automated DNA sequencing relies on _____ .
   - a. supplies of standard and labeled nucleotides
   - b. primers and DNA polymerases
   - c. gel electrophoresis and a laser beam
   - d. all of the above

8. _____ can be used to insert genes into human cells.
   - a. PCR
   - b. Modified viruses
   - c. Xenotransplantation
   - d. DNA microarrays

9. Match the terms with the most suitable description.
   - ____ DNA fingerprint
   - ____ Ti plasmid
   - ____ nature's genetic experiments
   - ____ nucleic acid hybridization
   - ____ eugenic engineering

   - a. selecting "desirable" traits
   - b. mutations, crossovers
   - c. used in some gene transfers
   - d. a person's unique collection of tandem repeats
   - e. base pairing of nucleotide sequences from different DNA or RNA sources

Figure 15.18 (**a**) ANDi, the first transgenic primate; his cells incorporate a jellyfish gene for bioluminescence. (**b**) The same gene was transferred into these zebrafish.

Media Menu

Student CD-ROM

Impacts, Issues Video
   Golden Rice or Frankenfood?
Big Picture Animation
   DNA technology, genetic engineering, and human applications
Read-Me-First Animation
   DNA recombination
   Use of labeled probes
   Polymerase chain reaction (PCR)
   Automated DNA sequencing
   Gene transfer
Other Animations and Interactions
   Action of restriction enzymes
   Making cDNA

InfoTrac

- Speed Reader. *Forbes*, November 2002.
- New Tools, Moon Tigers, and the Extinction Crisis. *BioScience*, September 2001.
- The Terminator's Back: Controversial Scheme Might Prevent Transgenic Spread. *Scientific American*, September 2002.
- Human Gene Therapy: Harsh Lessons, High Hopes. *FDA Consumer*, September 2000.

Web Sites

- National Center for Biotechnology Information: www.ncbi.nlm.nih.gov
- National Human Genome Research Institute: www.genome.gov
- Ag BioTech InfoNet: www.biotech-info.net
- Issues in Biotechnology: www.actionbioscience.org/biotech

How Would You Vote?

The United States is the world's leading producer and consumer of genetically modified organisms. Some people are uneasy about genetic engineering and would like to avoid products based on this technology. Should food that contains genetically modified plants or livestock be clearly identified on product labels?

Critical Thinking

1. What if it were possible to create life in test tubes? This is the question behind recent attempts to model and eventually create *minimal organisms*, which we define as living cells having the smallest set of genes required to survive and reproduce.

Craig Venter and Claire Fraser recently found that *Mycoplasma genitalium*, a bacterium that has 517 genes (and 2,209 transposons), is a good candidate for genetic research. By disabling its genes one at a time in the laboratory, they discovered that it may have only 265–350 essential protein-coding genes.

What if those genes were to be synthesized one at a time and inserted into an engineered cell consisting only of a plasma membrane and cytoplasm? Would the cell come to life? The possibility that it might prompted Venter and Fraser to seek advice from a panel of bioethicists and theologians. No one on the panel objected to synthetic life research. They felt that much good might come of it, provided scientists didn't claim to have found "the secret of life." The 10 December 1999 issue of *Science* includes an essay from the panel and an article on *M. genitalium* research. Read both, then write down your thoughts about creating life in a test tube.

2. Lunardi's Market put out a bin of tomatoes having vine-ripened redness, flavor, and texture. A sign identified them as genetically engineered produce. Most shoppers selected unmodified tomatoes in the adjacent bin even though those tomatoes were pale pink, mealy-textured, and tasteless. Which tomatoes would you pick? Why?

3. The sequence of the human genome has been completed, and knowledge about a number of the newly discovered genes is already being used to detect genetic disorders. Many women have refused to take advantage of genetic screening for a gene that is associated with the development of breast cancer.

Should medical records about people participating in genetic research and in genetic clinical services be made readily available to insurance companies, potential employers, and others? If not, how can such information be protected?

4. Scientists at Oregon Health Sciences University produced Tetra, the first primate clone. They also made the first transgenic primate by inserting a jellyfish gene into a fertilized egg of a rhesus monkey. (The gene encodes a bioluminescent protein that fluoresces green.) The egg was implanted in a surrogate monkey's uterus, where it developed into a male that was named ANDi (Figure 15.18).

The long-term goal of this gene transfer project is not to make glowing-green monkeys. It is the transfer of human genes into the primates whose genomes are most like ours. Transgenic primates could be studied to gain insight into genetic disorders, which might lead to the development of cures for those who are affected and vaccines for those at risk.

However, something more controversial is at stake. Will the time come when foreign genes can be inserted into human embryos? Would it be ethical to transfer a chimpanzee or monkey gene into a human embryo to cure a genetic defect? Or to bestow immunity against a potentially fatal disease such as AIDS?

III Principles of Evolution

Two male frigate birds (Fregata minor) in the Galápagos Islands, hundreds of miles off the coast of Ecuador. Each male inflates a gular sac, a balloon of red skin at his throat, in a display that may catch the eye of a female. The males lurk in the bushes together, gular sacs inflated, until a female flies overhead. Then they wag their head back and forth and call seductively to her. We can expect that, like other male structures used exclusively in courtship displays, the gular sac is an outcome of sexual selection—with the females being the selective agents.

IMPACTS, ISSUES   *Rise of the Super Rats*

Slipping in and out of the pages of human history are rats—*Rattus*—the most notorious of mammalian pests. One kind or another has distributed pathogens and parasites that cause bubonic plague, typhus, and other deadly infectious diseases (Figure 16.1). The death toll from fleas that bit infected rats, then people, has exceeded the dying in all wars combined.

In contrast, rats thrive around us. By one estimate, there is one rat for every person in urban and suburban centers of the United States. Besides spreading diseases, they chew their way through the walls and wires of our homes and cities. In any given year, the economic losses they inflict typically approach nineteen billion dollars.

People have been fighting back for years with traps, ratproof storage facilities, and toxic poisons, including arsenic and cyanide. During the 1950s, scientists devised baits laced with warfarin. This chemical interferes with blood clotting mechanisms. Rats that ate the baits died within days from internal bleeding or minor scratches that would not heal.

Warfarin was extremely effective. Compared to other rat poisons, it had a lot less impact on harmless species. It quickly became the rodenticide of choice.

In 1958, a Scottish researcher reported that some rats were indifferent to the poison. Similar reports followed from other European countries. About twenty years

Figure 16.1   Medieval attempts to deal with a bubonic plague pandemic—the Black Death that may have killed half the people in Europe alone. Not knowing that rats spread the disease agent, Europeans tried to protect themselves by praying and dancing 'til they dropped; physicians wore bird masks. For the next 300 years, anyone accused of causing the sporadic outbreaks of the plague—no matter how absurd the evidence—was burned alive.

In this century, by using ever more potent poisons on rats, we unwittingly promoted the rise of super rats. Three centuries from now, how will people be viewing us?

the big picture

Evolutionary Views Emerge   The world distribution of species, similarities and differences in body form, and the fossil record gave early evidence of evolution—of changes in lines of descent. Darwin and Wallace had an idea of how those changes occur.

Variation and Adaptation   An adaptation is any heritable aspect of form, function, behavior, or development that promotes survival and reproduction. An outcome of microevolution, it enhances the fit between the individual and prevailing conditions.

later, 10 percent of the urban rats caught in the United States were warfarin resistant. What happened?

To find out, researchers compared warfarin-resistant rat populations with still-vulnerable rats. They traced the difference to a gene on one of the rat chromosomes. A dominant allele at that locus was common among warfarin-resistant rat populations but rare among the vulnerable ones. And the product of the dominant allele neutralizes warfarin's effect on blood clotting.

"What happened" was evolution. As warfarin started to exert pressure on rat populations, the previously rare dominant allele abruptly proved adaptive. The lucky rats that inherited the allele survived and produced more rats. The unlucky ones that inherited the recessive allele had no defense; they died. Over time, the dominant allele's frequency increased in all populations of rats exposed to the poison.

Of course, selection pressures can and often do change. In response to increasing warfarin resistance, people stopped using warfarin. And guess what: The dominant allele's frequency declined. Now the latest worry is the evolution of "super rats," which the newer and even more potent rodenticides can't seem to kill.

The point is, if you hear someone question whether life evolves, remember the word simply means heritable change in lines of descent. The actual mechanisms that bring about changes in organisms are the focus of this chapter. Later chapters highlight how these mechanisms contribute to the evolution of new species.

 How Would You Vote?

Antibiotic-resistant strains of bacteria are becoming dangerously pervasive. Standard animal husbandry practice includes continually dosing healthy animals with antibiotics—the same antibiotics prescribed for people. Should this practice stop? See the Media Menu for details, then vote online.

Think Gene Pools Individuals don't evolve; populations do. All individuals of a population represent a pool of genes, the frequencies of which can shift over the generations. Such shifts can change the characteristics that define the population, and the species.

Microevolutionary Processes Mutation, genetic drift, natural selection, and gene flow are microevolutionary processes. By changing allele frequencies in a population, they change the observable characteristics that define the population and, more broadly, the species.

16.1 Early Beliefs, Confounding Discoveries

Prevailing beliefs can influence how we interpret clues to natural processes and their observable outcomes.

QUESTIONS FROM BIOGEOGRAPHY

At one time Europeans viewed nature as a great Chain of Being extending from the "lowest" forms of life to humans, and on to spiritual beings. Each kind of being, or **species** as it was called, was one separate link in the chain. All the links had been designed and forged at the same time at one center of creation. They had not changed since. Once all the links were discovered and described, the meaning of life would be revealed.

Then Europeans embarked on their globe-spanning explorations and discovered the world is a lot bigger than Europe. Tens of thousands of unique plants and animals turned up in Asia, Africa, the New World, and the Pacific islands.

In 1590, the naturalist Thomas Moufet attempted to sort through the bewildering arrays. He simply gave up and wrote such gems as this description of locusts and grasshoppers: "Some are green, some black, some blue. Some fly with one pair of wings; others with more; those that have no wings they leap; those that cannot fly they walk. Some there are that sing, others are silent." It was not a work of breathtaking insight.

Later on, Alfred Wallace and a few other naturalists saw *patterns* in the distribution of species. Not content with cataloging species, they looked for forces that shaped similarities and differences among them. They were pioneers in **biogeography**—the study of patterns in the geographic distribution of species. They found clues to the ecological and evolutionary forces that influence individual species and entire communities.

Some patterns were intriguing. For instance, many plants and animals are unique to islands in the middle of the ocean and other remote places. Similar-looking species often live in the same kinds of habitats, even when vast expanses of open ocean or high, impassable mountain ranges separate them.

Consider: Flightless, long-necked, long-legged birds are native to three continents (Figure 16.2a–c). Why are they so much alike? The plants in Figure 16.2d,e live on separate continents. Both have spines, tiny leaves, and short fleshy stems. Why are *they* so much alike?

Compare the global locations of the flightless birds and you'll find they all live in flat, open grasslands in dry climates. American and African desert plants are about the same distance from the equator, in regions where water is scarce. Their fleshy stems store water and have a thick cuticle, which keeps water in. Their rows of sharp spines deter thirsty, hungry animals.

Figure 16.2 Species that resemble one another, strikingly so, even though they are native to distant geographic realms.

(**a**) South American rhea, (**b**) Australian emu, and (**c**) African ostrich. All three types of birds live in similar habitats. They are unlike most birds in several traits, most notably in their inability to get airborne.

(**d**) A spiny cactus native to the hot deserts of the American Southwest. (**e**) A spiny spurge native to southwestern Africa.

coccyx

ankle
bone

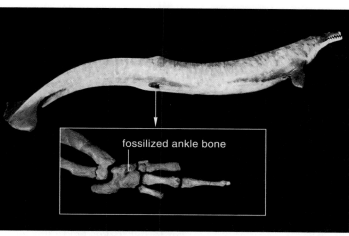

fossilized ankle bone

Figure 16.3 Body parts with no apparent function. *Above*, reconstruction of an ancient whale (*Basilosaurus*). This giant predator's head was as long as a sofa. The whale was fully aquatic, so it did not use its hindlimbs to support body weight as you do—yet it had ankle bones. We use our ankles, but not our coccyx bones.

Figure 16.4 (**a**) Fossilized ammonites and (**b**) a cutaway view of its shell, compared to that of a chambered nautilus (**c**). Ammonites, now extinct, lived hundreds of millions of years ago. Like the chambered nautilus, which exists today, ammonites were marine predators.

If all birds and plants were created in one place, then how did such similar kinds end up in the same kind of habitat in such remote and distant places?

QUESTIONS FROM COMPARATIVE MORPHOLOGY

The similarities and differences in body plans among groups of organisms raised questions that started yet another line of inquiry—**comparative morphology**. For example, bones of a human arm, whale flipper, and bat wing differ in size, shape, and function. As you'll see in Chapter 17, these bones are similarly located in the body and are constructed of the same kinds of tissues arranged in much the same patterns. And they develop in much the same way in embryos. Naturalists who deduced all of this wondered: How can there be animals that differ hugely in certain features yet are so much alike in other features?

By one hypothesis, body plans are so perfect there was no need to make a new design for each organism at the time of its creation. Yet if that were so, then why were useless body parts created? For instance, an ancient, ocean-dwelling whale had ankle bones but didn't walk (Figure 16.3). Why the bones? Our coccyx is like some tailbones in many other mammals. We do not have a tail. Why do we have parts of one?

QUESTIONS ABOUT FOSSILS

Geologists had been mapping layers of rock exposed by erosion or quarrying, and their discoveries added to the confusion. They found the same kinds of layers in different parts of the world. They thought the layers formed when sediments slowly collected, year after year, on the floor of ancient rivers and seas. If that were so, then the deepest layers were the oldest. And if that were so, then fossils embedded in successive layers could be clues to the past. **Fossils** came to be recognized as stone-hard evidence of earlier forms of life. Figures 16.3 and 16.4 show examples.

A puzzle: Many deep layers held fossils of simple marine life. Some layers above them contained fossils that were structurally similar but more intricate. In higher layers, fossils were like living species. What did the sequences in complexity among fossils of a given type mean? Were they evidence of lines of descent?

Taken as a whole, the findings from biogeography, comparative morphology, and geology did not fit with prevailing beliefs. Scholars floated novel hypotheses. If a simultaneous dispersal of all species from a center of creation was unlikely, *then perhaps species originated in more than one place.* If species had not been created in a perfect state—and fossil sequences and "useless" body parts among species implied they had not—*then perhaps species had become modified over time.* Awareness of evolution was in the wind.

Awareness of biological evolution emerged over centuries through the cumulative observations of many naturalists, biogeographers, comparative anatomists, and geologists.

16.2 A Flurry of New Theories

Nineteenth-century naturalists found themselves trying to reconcile the evidence of change with a traditional conceptual framework that simply did not allow for it.

SQUEEZING NEW EVIDENCE INTO OLD BELIEFS

A respected anatomist, Georges Cuvier, was among those trying to make sense of the growing evidence for change. For years Cuvier had compared body plans of fossils and living organisms. He knew about fossils in sedimentary layers. He was the first to recognize that abrupt shifts in the fossil record mark mass extinctions.

By Cuvier's hypothesis, the changes resulted from periodic natural disasters, which he called revolutions. Except for those episodes, Earth was an unchanging stage for the human drama. A single time of creation had populated the world with all species. Many died in periodic global natural disasters, such as floods and earthquakes, and then the survivors repopulated the world. There never were any *new* species; naturalists simply hadn't yet found all of the fossils that would date back to the time of creation.

The hypothesis enjoyed support for some time. It even became elevated to the rank of theory, one that later became known as **catastrophism**.

Even so, many scholars kept at the puzzle. Jean Lamarck proposed a different hypothesis, known as inheritance of *acquired* characteristics. Environmental pressures and internal "needs" induced permanent changes in an individual's body form and functioning,

then the individual passed on the changes, acquired during its lifetime, to offspring. Life, created long ago in a simple state, slowly improved as a result. The force for change was a drive toward perfection, up the Chain of Being. It was centered in nerves that directed an unknown "fluida" to body parts in need of change.

Apply his hypothesis to modern giraffes. Say they had a short-necked, hungry ancestor that had to keep stretching its neck to browse upon leaves beyond the reach of other animals. Big stretches directed fluida to its neck and made it lengthen permanently. Offspring inherited the longer neck, then stretched their necks, too. Generation after generation of straining to reach ever loftier leaves led to the modern giraffe.

As Lamarck correctly inferred, the environment *is* a factor in evolution. However, his hypothesis as well as others made at the time have not been supported by experimental tests. Environmental factors can alter an individual's traits, as when serious strength training builds huge muscles. But offspring of a muscle-bound parent won't be born muscle-bound. They can inherit genes, but not increased muscle mass.

VOYAGE OF THE *BEAGLE*

In 1831, in the midst of the confusion, Charles Darwin was twenty-two years old and wondering what to do with his life. Ever since he was eight, he had wanted to hunt, fish, collect shells, or just watch insects and birds—anything but sit in school. Later, at his father's

Figure 16.5 (**a**) Charles Darwin. (**b**) Replica of the *Beagle* sailing off a rugged coastline of South America. During one of his trips, Darwin ventured into the Andes. He discovered fossils of marine organisms in rock layers 3.6 kilometers above sea level.

(**c–e**) The Galápagos Islands are isolated in the ocean, far to the west of Ecuador. They arose by volcanic action on the seafloor about 5 million years ago. Winds and currents carried organisms to the once lifeless island. All the island's native species are descended from those travelers. At far right, a blue-footed booby, one of many species Darwin observed during his voyage.

insistence, he did attempt to study medicine in college. The crude, painful procedures being used on patients at that time sickened him. His exasperated father urged him to become a clergyman, so he packed for Cambridge. His grades were good enough to earn a degree in theology. Yet he spent most of his time with faculty members who embraced natural history.

John Henslow, a botanist, perceived Darwin's real interests. He hastily arranged for Darwin to become ship's naturalist aboard the *Beagle*, which was about to embark on a five-year voyage around the world. The young man who hated schoolwork, and who had no formal training as a naturalist, suddenly started to work with enthusiasm.

The *Beagle* sailed first to South America to finish work on mapping the coastline (Figure 16.5). During the Atlantic crossing, Darwin collected and studied marine life. He read Henslow's parting gift, the first volume of Charles Lyell's *Principles of Geology*. He saw diverse species in environments ranging from sandy shores of remote islands to high mountains. And he started circling the question of evolving life, which was now on the minds of many respected individuals.

Darwin started by mulling over a radical theory. As Lyell and other geologists were arguing, erosion and other gradual, natural processes of change had more impact on Earth history than rare catastrophes. Geologists for years had chipped away at sandstones, limestones, and other rocks that form after sediments accumulate in the beds of rivers and seas. They saw how those beds often consist of a number of stacked layers. If, they hypothesized, deposition took place as slowly in the past as it did in their own era, then it must have taken many millions of years for thick stacks to form. They even incorporated earthquakes and other infrequent events into their theory. After all, immense floods, over a hundred big earthquakes, and twenty or so volcanic eruptions occur in a typical year, so catastrophes aren't that unusual.

The idea that gradual, repetitive change shaped the Earth became known as the **theory of uniformity**. It challenged the prevailing views of the Earth's age.

The theory bothered scholars who firmly believed that Earth could not be more than 6,000 years old. They thought people had recorded everything that happened during those 6,000 years—and in all that time, no one had ever mentioned seeing a species evolve. Yet, by Lyell's calculations, it must have taken millions of years to sculpt out the present landscape. *Wasn't that enough time for species to evolve in diverse ways?* Later, Darwin thought so. But exactly *how* did they evolve? He would end up devoting the rest of his life to that burning question.

Prevailing beliefs can influence how we interpret clues to natural processes and their observable outcomes.

Darwin's observations during a global voyage helped him think about species in a novel way.

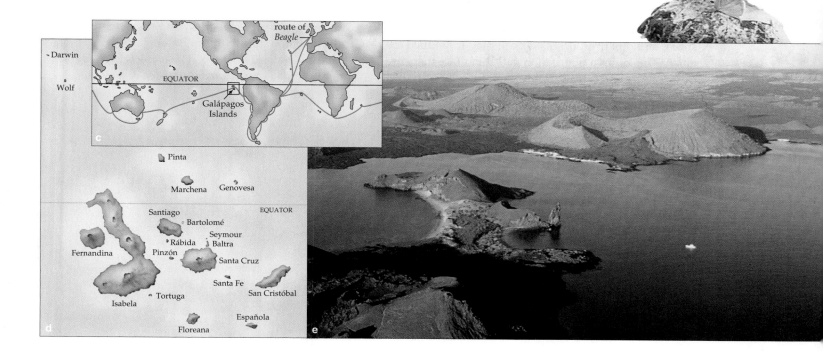

16.3 Darwin's Theory Takes Form

Darwin's observations of thousands of species in different parts of the world helped him see how species may evolve.

OLD BONES AND ARMADILLOS

When Darwin returned to England, he had volumes of notes and thousands of specimens, although he had not done a great job of correlating them with habitats. Other naturalists helped him fill in some important blanks—and in time he arrived at an explanation of how all of that diversity evolved.

In Argentina, for example, Darwin found fossils of glyptodonts, now extinct. Of all animals on Earth, only living armadillos are like them (Figure 16.6). Of all places on Earth, armadillos live only in the same places where glyptodonts had lived. If the two kinds of animals had been created at the same time, lived in the same place, and were so much alike, why is only one still alive? Would it be reasonable to assume that glyptodonts were the early ancestors of armadillos? Many of their shared traits might have been retained over many thousands of generations. Other traits may have been modified in the armadillo branch of a family tree. *Descent with modification*—it did seem possible. What, then, could be the driving force for evolution?

A KEY INSIGHT—VARIATION IN TRAITS

While Darwin assessed his notes, an influential essay by Thomas Malthus, a clergyman and economist, made him consider a topic of social interest. Malthus had correlated population size with famine, disease, and war. Humans, said Malthus, run out of food, living space, and other resources because they reproduce too much. The larger a population gets, the more there are to reproduce. Resources dwindle, and struggles to live intensify. Many starve, get sick, or engage in war and other forms of competition for dwindling resources.

Darwin thought that *any* population has a capacity to produce more individuals than the environment can support. Case in point: A sea star can release 2,500,000 eggs a year, but the seas don't fill up with sea stars. Predators eat many of the eggs and larvae. Most of the survivors die of disease and other assaults.

Darwin also reflected on populations he had seen during his voyage. Their individuals were not alike down to the last detail. They varied in size, color, and other traits. *It dawned on him that variations in traits could affect an individual's ability to secure resources—and to survive and reproduce in particular environments.*

Did the Galápagos Islands reveal evidence of this? Between the islands and the South American coast are 900 kilometers of open ocean. The islands have diverse shoreline, desert, and mountain habitats. Nearly all of the species live nowhere else, although they resemble mainland species. Did wind and water put colonizing species on the Galápagos? Were species on different islands island-hopping descendants of the colonizers?

Try correlating variations in the Galápagos finches with different environmental challenges. Imagine yourself watching a large-billed, seed-cracking type (Figure 16.7). A few birds with a stronger bill crack seeds too tough for the others. If most of the seeds that form during a given season have hard coats, then the strong-billed birds will have a competitive edge and a better chance of surviving and producing offspring. If the trait is heritable, then the advantage will help their descendants, also. If environmental factors continue to "select" the most adaptive version of this trait, then the population may end up being mostly strong-billed birds. *And a population is evolving when forms of heritable traits change over the generations.*

Figure 16.6 (**a**) A modern armadillo. (**b**) A Pleistocene glyptodont, about as big as a Volkswagen Beetle, and now extinct. Glyptodonts shared unusual traits and a restricted distribution with the existing armadillos. Yet the two kinds of animals are widely separated in time. Their similarities were a clue that helped Darwin develop a theory of evolution by natural selection.

a

b

Figure 16.7 Four of thirteen finch species on the Galápagos Islands. (**a**) *Geospiza conirostris* and (**b**) *G. scandens*. Both eat cactus flowers and fruits; other finches eat its seeds. (**c**) *Certhidea olivacea*, a tree-dweller, uses its slender bill to probe for food. (**d**) *Camarhynchus pallidus* eats wood-boring insects. It learns to break and use cactus spines and twigs as probes. We now know that bill shape depends on when and where the protein BMP4, a signaling molecule, is switched on in bird embryos. Mutations in the gene for this protein, are an evolutionary source for the differences.

NATURAL SELECTION DEFINED

Here are Darwin's key observations and conclusions about evolution, expressed in modern terms:

1. *Observation:* Natural populations have an inherent reproductive capacity to increase in numbers through successive generations.

2. *Observation:* No population can indefinitely grow in size, because its individuals will run out of food, living space, and other resources.

3. *Inference:* Sooner or later, individuals will end up competing for dwindling resources.

4. *Observation:* Those individuals generally have the same genes encoding the same shared traits. Genes are the population's pool of heritable information.

5. *Observation:* Mutations have given rise to alleles, or slightly different molecular forms of genes, which are a source of differences in phenotypic details.

6. *Inferences:* Some phenotypes are better than others at helping an individual compete for resources, and to survive and reproduce. Alleles for those phenotypes increase in the population, and other alleles decrease. In time the genetic changes lead to increased **fitness**—an increase in adaptation to the environment.

7. *Conclusions:* **Natural selection** among individuals of a population is an outcome of variation in traits that affect which individuals survive and reproduce in each generation. This microevolutionary process results in adaptation, or increased fitness to the environment.

Darwin kept on looking for patterns in his data and filling in gaps in his reasoning. He waited too long. More than ten years after he wrote but did not publish his theory, Alfred Wallace sent him a letter (Figure 16.8). Wallace had been doing impressive work in the Amazon Basin and Malay Archipelago, Madagascar, and elsewhere. He had written earlier to Lyell and Darwin about the causes of species distributions. And he had arrived at the same theory!

Though daunted, Darwin encouraged Wallace to publish. Wallace and other colleagues thought Darwin should get credit. At a scientific meeting in 1858,

Figure 16.8 Alfred Wallace, one of the first to study island biogeography. For a stimulating view of the Darwin–Wallace story, read David Quamman's *The Song of the Dodo.*

with neither Darwin nor Wallace in attendance, the theory was presented and attributed to both. The next year, Darwin published *On the Origin of Species*, with detailed evidence in support of the theory.

You may have heard that Darwin's book fanned an intellectual firestorm, but most scholars were quick to accept the idea that diversity is the result of evolution. The theory of natural selection *was* fiercely debated. Decades passed before experimental evidence from a new field, genetics, led to its widespread acceptance.

Natural selection is an outcome of differences in survival and reproduction among members of a population that vary in some traits. It increases adaptation to the environment, as measured by fitness—the relative genetic contribution of an individual to future generations.

16.4 The Nature of Adaptation

A word of caution: Observable traits are not always easy to correlate with conditions in an organism's environment.

"Adaptation" is one of those words that have different meanings in different contexts. An individual plant or animal often can quickly adjust its form, function, and behavior. Junipers in inhospitably windy places grow less tall than junipers of the same species in more sheltered places. A clap of thunder may make you lurch the first time you hear it, but then you may get used to the sound over time and ignore it. These are examples of *short-term* adaptations, because they last only as long as the individual does.

Over the long term, an **adaptation** is some heritable aspect of form, function, behavior, or development that improves the odds for surviving and reproducing in a given environment. It is an *outcome* of microevolution —natural selection especially—an enhancement of the fit between the individual and prevailing conditions.

Figure 16.9 (**a**) Severe, rapid wilting of one commercial tomato plant (*Solanum lycopersicum*) that absorbed salty water. (**b**) Galápagos tomato plant, *S. cheesmanii*, which stores most absorbed salts in its leaves, not in its fruits.

SALT-TOLERANT TOMATOES

As an example of long-term adaptation, compare how tomato species handle salty water. Tomatoes evolved in Ecuador, Peru, and the Galápagos Islands. The type sold most often in markets, *Solanum lycopersicum*, has eight close relatives in the wild. If you mix ten grams of table salt with sixty milliliters of water, then pour it into the soil around *S. lycopersicum*'s roots, the plant will wilt severely in less than thirty minutes (Figure 16.9*a*). Even if the soil has only 2,500 parts per million of salt, this species will grow poorly.

Yet the Galápagos tomato (*S. cheesmanii*) survives and reproduces in seawater-washed soils. We know that its salt tolerance is a heritable adaptation. How? F_1 crosses of a wild species with the commercial one result in a small, edible hybrid. The hybrid tolerates irrigation water that is two parts fresh and one part salty. It's gaining interest in places where fresh water is scarce and where salts have built up in croplands.

It may take modification of only a few traits to get new salt-tolerant plants. Revving up just one gene for a sodium–hydrogen ion transporter helps the tomato plants use salty water and still bear edible fruits.

NO POLAR BEARS IN THE DESERT

You can safely bet that a polar bear (*Ursus maritimis*) is finely adapted to the icy Arctic, and that its form and function would be a flop in a desert (Figure 16.10). You

Figure 16.10 Which adaptations of a polar bear (*Ursus maritimus*) won't help in a desert? Which ones help an oryx (*Oryx beisa*)? For each animal, make a tentative list of possible structural and functional adaptations to the environment. Later, after you finish reading Unit VI, see how you can expand the list.

Figure 16.11 Adaptation to what? A heritable trait is an adaptation to specific environmental conditions. Hemoglobin of llamas, which live at high altitudes, has a high oxygen-binding affinity. However, so does hemoglobin of camels, which live at lower elevations.

might be able to make some educated guesses about why that is so. However, detailed knowledge of its anatomy and physiology might make you view it—or any other animal or plant—with respect. How does a polar bear maintain its internal temperature when it sleeps on ice? How can its muscles function in frigid water? How often must it eat? How does it find food? Conversely, how can an oryx walk about all day in the blistering heat of an African desert? How does it get enough water when there is no water to drink? You will find some answers, or at least ideas about how to look for them, in the next three units of this book.

ADAPTATION TO WHAT?

Bear in mind, it isn't always easy to identify a direct relationship between adaptation and the environment. For instance, the prevailing environment may be very different from the one in which a trait evolved.

Consider the llama. It is native to the cloud-piercing peaks of the Andes in western South America (Figure 16.11). The llama lives 4,800 meters (16,000 feet) above sea level. Compared to humans at lower elevations, its lungs have more air sacs and blood vessels. The llama heart has larger chambers, so it pumps larger volumes of blood. Llamas don't have to make extra blood cells, as people do when they move permanently from the lowlands to high elevations. (Extra cells make blood "stickier," so the heart has to pump harder.) But the most publicized adaptation is this: Llama hemoglobin is better than ours at latching on to oxygen. It picks up oxygen in the lungs far more efficiently.

Superficially, at least, the oxygen-binding affinity of llama hemoglobin appears to be an adaptation to thin air at high altitudes. Is it? Apparently not.

Llamas belong to the same family as dromedary camels. Both share camelid ancestors that evolved in the Eocene grasslands and deserts of North America. Later, the ancestors of camels and llamas went their separate ways. Camel forerunners moved into Asia's low-elevation grasslands and deserts by a land bridge, which later submerged when the sea level rose. Llama forerunners moved in a different direction—down the Isthmus of Panama, and on into South America.

Intriguingly, a dromedary camel's hemoglobin also shows a high oxygen-binding capacity. So if the trait arose in a shared ancestor, then how was it adaptive at *low* elevations? We know camels and llamas didn't just *happen* to evolve in the same way. They are close kin, and their most recent ancestors lived in very different environments with different oxygen concentrations.

Who knows why the trait was originally favored? Eocene climates were alternately warm and cool, and hemoglobin's oxygen-binding capacity does go down as temperatures go up. Did it prove adaptive during a long-term shift in climate? Or were its effects neutral at first? What if the mutant gene for the trait became fixed in some ancestral population simply by chance?

Use all of these "what-ifs" as a reminder to think carefully about connections between form and function. Identifying such connections takes a lot of intuition, research, and experimental tests.

A long-term adaptation is any heritable aspect of form, function, behavior, or development that contributes to the fit between an individual and its environment.

An adaptive trait improves the odds of surviving and reproducing, or at least it did so under conditions that prevailed when genes for the trait first evolved.

16.5 Individuals Don't Evolve, Populations Do

Evolution starts with changes in the gene pool of a population, as brought about by mutation, natural selection, gene flow, and genetic drift.

VARIATION IN POPULATIONS

As Darwin and Wallace perceived, *individuals do not evolve; populations do*. By definition, a **population** is a group of individuals of the same species occupying a given area. To understand how it evolves, start with variation in the features that characterize it.

Individuals of a population share basic features. Jays have two feathered wings, three toes forward, one toe back, and so on. These are *morphological* traits (*morpho–*, form). The body parts of all individuals work much the same way in metabolism, growth, and reproduction. Such *physiological* traits help the body function in its environment. Individuals respond the same way to basic stimuli, as when babies imitate adult facial expressions. These are *behavioral* traits.

Especially for sexually reproducing species, details of most traits vary among individuals of populations. Pigeon feathers and butterfly wings differ in colors or patterns. Figure 16.12 hints at human variations in skin color and distribution, color, texture, and amount of hair. Almost every trait of every species is variable.

Many traits show *qualitative* differences; they have two or more distinct forms, or morphs. (Remember the purple or white pea plant flowers?) Two forms is a dimorphism; three or more is polymorphism. Other traits, such as height and eye color, show *quantitative* differences, which are small and incremental.

THE "GENE POOL"

Genes hold information on heritable traits. In general, a population's individuals all have the same number and kinds of genes. We say "in general" because in sexually reproducing species, the sex chromosomes of males and females are not alike.

Think of all the genes in a population as the **gene pool**—a pool of genetic resources that the individuals of the population, and their offspring, share. Often, each kind of gene in the pool is present in two or more slightly different molecular forms, or **alleles**.

Individuals have different combinations of alleles. This leads to variations in *phenotype*, or differences in details of traits. Whether you have black, brown, red, or blond hair depends upon the certain alleles that you inherited from your two parents.

Also, don't forget that offspring inherit genes, not phenotypes. Environmental conditions often alter gene expression (Section 10.6). Variation resulting from their effects lasts no longer than the individual.

Which alleles end up in a gamete and then the new individual? Five events, described in earlier chapters, shape the outcome. We summarize them here:

1. Gene mutation (produces new alleles)

2. Crossing over during meiosis I (introduces novel combinations of alleles in chromosomes)

3. Independent assortment at meiosis I (puts mixes of maternal and paternal chromosomes in gametes)

4. Fertilization (combines alleles from two parents)

5. Change in chromosome number or structure (leads to the loss, duplication, or repositioning of genes)

Only mutation *creates* new alleles. Other events shuffle *existing* alleles into different combinations—but what a shuffle! For example, a human gamete gets one of 10^{600} possible combinations. Not even 10^{10} humans are alive today. Unless you are an identical twin, it is extremely unlikely that any other person with your precise genetic makeup has ever lived, or ever will.

STABILITY AND CHANGE IN ALLELE FREQUENCIES

Researchers typically track **allele frequencies**, or the abundance of certain alleles in a population. They start at **genetic equilibrium**, a theoretical reference

Figure 16.12 A sampling of variation in human and snail populations.

STARTING POPULATION	THE NEXT GENERATION	THE NEXT GENERATION

490 *AA* butterflies
dark-blue wings

490 *AA* butterflies
dark-blue wings

490 *AA* butterflies
dark-blue wings

420 *Aa* butterflies
medium-blue wings

420 *Aa* butterflies
medium-blue wings

420 *Aa* butterflies
medium-blue wings

90 *aa* butterflies
white wings

90 *aa* butterflies
white wings

90 *aa* butterflies
white wings

Figure 16.13 How to find out if a population is evolving. Start with five assumptions of the *Hardy–Weinberg rule*: A population's allele frequencies don't change when there is no mutation, the population is infinitely large and isolated from others of the species, mating is random with respect to the alleles of interest, and all individuals survive and reproduce sexually (no selection).

Track a hypothetical pair of alleles that influence butterfly wing color. Butterflies are sexually reproducing organisms, with pairs of genes on pairs of homologous chromosomes. Allele *A* is associated with dark-blue, *a* with white, and *Aa* with medium-blue wings. At genetic equilibrium, the proportions of the wing-color genotypes are

$$p^2\,AA + 2pq\,Aa + q^2\,aa = 1.0$$

where *p* and *q* are frequencies of alleles *A* and *a*. The frequencies of *A* and *a* must add up to 1.0. Example: If *A* occupies half of all loci for this gene in the population, then *a* must occupy the other half (0.5 + 0.5 = 1.0). If *A* occupies 90 percent of all the loci, then *a* must occupy 10 percent (0.9 + 0.1 = 1.0). No matter what the proportions,

$$p + q = 1.0$$

At meiosis, the alleles of each pair segregate and end up in different gametes. So the proportion of gametes having the *A* allele is *p*, and the proportion having the *a* allele is *q*. This Punnett square shows the genotypes possible in the next generation (*AA*, *Aa*, and *aa*):

	p Ⓐ	*q* ⓐ
p Ⓐ	*AA* (p^2)	*Aa* (pq)
q ⓐ	*Aa* (pq)	*aa* (q^2)

The frequencies add up to 1.0: $p^2 + 2pq + q^2 = 1.0$.

Let's say the population has 1,000 individuals, and that each one produces two gametes:

490 *AA* individuals produce 980 *A* gametes
420 *Aa* individuals produce 420 *A* and 420 *a* gametes
90 *aa* individuals produce 180 *a* gametes

The frequency of alleles *A* and *a* among the 2,000 gametes is

$$A = \frac{2 \times 490 + 420}{2,000 \text{ alleles}} = \frac{1,400}{2,000} = 0.7 = p$$

$$a = \frac{2 \times 90 + 420}{2,000 \text{ alleles}} = \frac{600}{2,000} = 0.3 = q$$

At fertilization, gametes combine at random, forming a new generation. If the population stays at 1,000, you have 490 *AA*, 420 *Aa*, and 90 *aa* individuals. Because the allele frequencies for dark-blue, medium-blue, and white wings are the same as in the original gametes, they will give rise to the same phenotypic frequencies as the second generation.

As long as the five conditions of the Hardy–Weinberg rule hold, this pattern will persist. If traits show up in different proportions from one generation to the next, then one or more of the five assumptions is not being met. *The hunt can begin for one or more specific evolutionary forces driving the change.*

point. At this point, a population is *not* evolving with respect to the allele frequencies being studied. Why? Five conditions are prevailing: No mutations have occurred; the population is infinitely large; it is fully isolated from all other populations of the species; the product of the allele being studied has no effect on survival or reproduction; and all mating is random.

Figure 16.13 is an example. If you choose to pass on reading it, just know that the five conditions hardly ever prevail at the same time in the natural world. Why? Mutations are rare, but inevitable. Also, three processes—*natural selection, gene flow,* and *genetic drift* —drive such populations out of genetic equilibrium.

Microevolution refers to small-scale changes in population allele frequencies that arise from mutation, natural selection, gene flow, and genetic drift.

A population or species is characterized by morphological, physiological, and behavioral traits, most of which are heritable. Details of these traits vary among its individuals.

Different combinations of alleles among individuals of a population give rise to variations in phenotype—that is, to differences in the details of their shared structural, functional, and behavioral traits.

In sexually reproducing species, a population's individuals share a pool of genetic resources—that is, a gene pool.

Few populations ever achieve genetic equilibrium. Natural selection, gene flow, and genetic drift change a gene pool's allele frequencies.

16.6 Mutations Revisited

Reflect, for a moment, on the statement that mutations are the original source of alleles. These heritable changes in DNA usually give rise to altered gene products, and evolution starts with them.

We can't predict when or in which individual a given mutation will occur. We do know that each gene has a **mutation rate**, the probability of its mutating during or between DNA replications. For each gamete the rate is between 10^{-5} and 10^{-6} mutations per gene.

Mutations may give rise to structural, functional, or behavioral alterations that reduce an individual's chances of surviving and reproducing. Even a single biochemical change can be devastating. For instance, skin, bones, tendons, lungs, blood vessels, and many other vertebrate organs incorporate collagen. When the collagen gene mutates, drastic changes all through the body follow. Any mutation that severely changes phenotype usually causes death; it is a **lethal mutation**.

By comparison, a **neutral mutation** doesn't help or hurt. Natural selection neither increases nor decreases its frequency in a population, because it won't have a discernible effect on whether an individual survives and reproduces. If you carry a mutant gene that keeps earlobes attached to your head instead of swinging freely, this in itself shouldn't stop you from surviving and perhaps reproducing just as well as anybody else.

Every so often, a mutation proves useful. A mutant gene product that influences growth may make a corn plant grow larger or faster and thus give it the best access to sunlight and to nutrients. A neutral mutation may prove helpful after conditions in the environment change. Even when it bestows only a small advantage, chance events or natural selection might preserve the mutant gene in the DNA and favor its representation in the next generation.

Mutations are so rare they usually have little or no immediate effect on a population's allele frequencies. But both beneficial and neutral mutations have been accumulating in diverse lineages for billions of years. All that time, they have functioned as raw material for evolutionary change, for the staggering range of biodiversity, past and present. From an evolutionary view, the reason you don't look like a bacterium or an avocado or earthworm or even your neighbors down the street began with mutations that arose at different times, in different lines of descent.

Go ahead. Hit me with a pesticide.

> *Mutations occur at a predictable rate, but are themselves unpredictable. They may alter an individual's phenotype.*

16.7 Directional Selection

We turn now to mechanisms of natural selection. In some cases, the range of values for a trait shifts in one direction. At other times, an existing range of values may be stabilized or disrupted, depending on conditions.

In cases of **directional selection**, allele frequencies that give rise to a range of variation in phenotype tend to shift in a consistent direction. The shift is a response to directional change in the environment or to one or more new conditions. A new mutation that benefits individuals also may cause a directional shift. Either way, forms at one end of a phenotypic range become more common than midrange forms (Figure 16.14).

PESTICIDE RESISTANCE

As the chapter introduction made clear, pesticides can bring about directional selection. A few individuals usually survive the initial applications. Some heritable aspect of body structure, physiology, or behavior helps them resist the chemicals. The most resistant ones are favored, so resistance becomes more common. Today, 450 species resist one or more pesticides.

Pesticides also kill natural predators of pests. Freed from natural constraints, resistant populations cause more damage than before. This outcome of directional selection is called *pest resurgence*. Crops of genetically engineered plants can resist pests. But these plants, too, will start exerting selection pressure on populations of pests. Individuals that can overcome the engineered defenses tend to be favored.

ANTIBIOTIC RESISTANCE

Natural and synthetic antibiotics can fight bacterial diseases. Natural **antibiotics** are toxins that some microorganisms release to kill bacterial competitors for nutrients. Streptomycins, for instance, inhibit bacterial protein synthesis. The penicillins disrupt covalent bonds that hold a bacterial cell wall together.

Antibiotics should be prescribed with restraint and care. Besides performing their intended function, they often disrupt resident populations in the intestines and vagina. Imbalances lead to secondary infections.

Antibiotics also have been overprescribed, often for simple infections that most people can fight on their own. Genetic variation in bacterial gene pools allows individuals with certain genotypes to survive while others do not. Overuse of antibiotics has thus favored resistant populations, which does not bode well for the millions of people who each year contract cholera, tuberculosis, and other bacterial diseases.

COAT COLOR IN DESERT MICE

Researchers Michael Nachman, Hopi Hoekstra, and Susan D'Agostino reported directional selection among rock pocket mice (*Chaetodipus intermedius*) living in the same part of Arizona's Sonoran Desert (Figure 16.15). Of more than eighty genes known to affect coat color in mice, they pinpointed one that governs a difference between two populations of this mouse species.

Rock pocket mice are small mammals that spend the day in underground burrows. At night they forage for seeds, scampering over the tawny-colored granite outcroppings. Here, individuals with tawny fur are camouflaged from predators (Figure 16.15*b*).

A smaller population of pocket mice lives in the same region, but these mice scamper over dark basalt of ancient lava flows. They have dark coats, so they, too, are camouflaged from predators (Figure 16.15*c*).

We can expect that night-flying predatory birds are selective agents that affect fur color. Earlier studies, for example, demonstrated that owls have an easier time seeing mice with fur that does not match the rocks. Nachman drew on genetic data about laboratory mice to formulate a hypothesis on coat color differences in the two populations. He predicted that a mutation of either the *Mclr* or *agouti* gene causes the difference.

He collected DNA from dark-colored pocket mice at a lava flow and from light-colored mice at adjacent granite outcroppings. DNA analysis showed that the *Mclr* gene sequence differs between the groups. The gene sequence for all dark-fur mice differed by four nucleotides from that of their light-furred neighbors.

With directional selection, allele frequencies underlying a range of variation tend to shift in a consistent direction in response to some change in the environment.

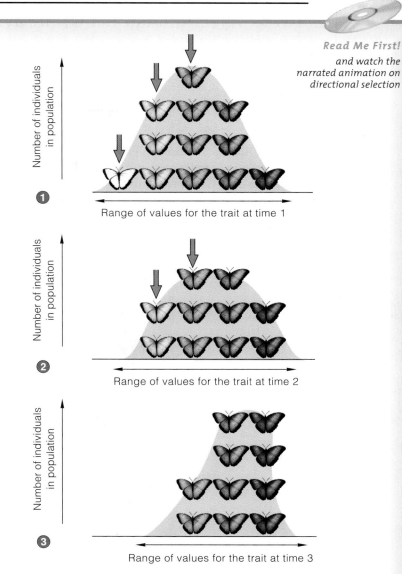

Read Me First!
and watch the narrated animation on directional selection

Figure 16.14 Directional selection in a butterfly population. This bell-shaped curve signifies a range of continuous variation in a wing color trait. *Medium-blue*, the most common form, is between two extremes—*white* and *dark purple*. The *orange* arrows signify which forms are being selected against over time.

Figure 16.15 Visible evidence of directional selection between two neighbor populations of rock pocket mice. (**a**) Lava basalt flow at the study site. The two color morphs of rock pocket mice, each posed on two different backgrounds: (**b**) tawny fur and (**c**) dark fur.

16.8 Selection Against Or in Favor of Extreme Phenotypes

Consider now two additional modes of natural selection. One works against phenotypes at the fringes of a range of variation; the other favors them (Figure 16.16).

STABILIZING SELECTION

With **stabilizing selection**, intermediate forms of a trait in a population are favored and alleles for the extreme forms are not. This mode of selection can counter mutation, gene flow, and genetic drift. It preserves the most common phenotypes.

As an example, prospects are not good for human babies who weigh far more or far less than average at birth. Also, pre-term instead of full-term pregnancies increase the danger, as reflected in Figure 16.17. Newborns weighing less than 5.51 pounds or born before thirty-eight weeks of pregnancy are completed tend to develop high blood pressure, diabetes, and heart disease when they are adults. The mother's blood concentration of cortisol, a stress hormone, may be linked to low birth weight and the illnesses that develop later in life.

Read Me First!

and watch the narrated animation on stabilizing and disruptive selection

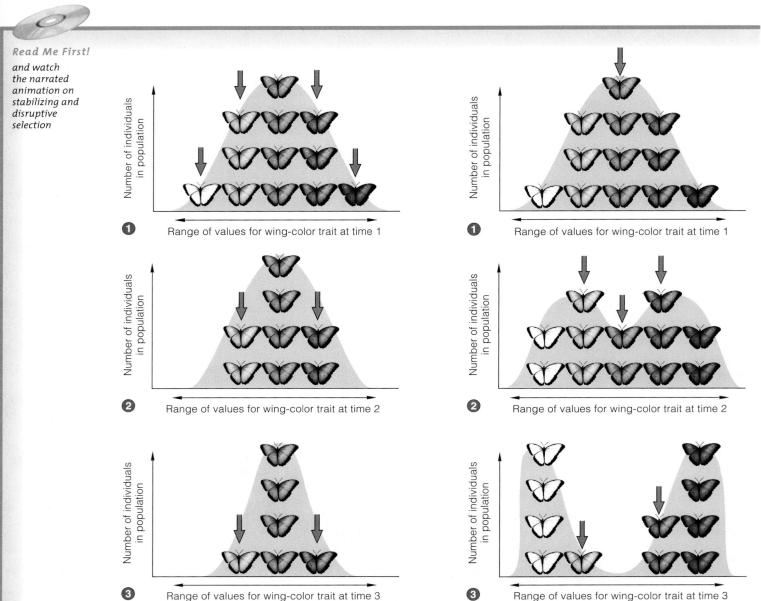

Figure 16.16 Selection against or in favor of extreme phenotypes, with a population of butterflies as the example. *Left*, stabilizing selection and *right*, disruptive selection. The *orange* arrows show forms of the trait being selected against.

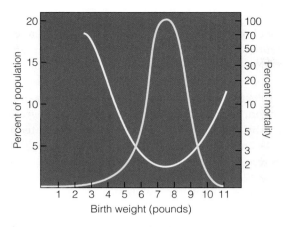

Figure 16.17 Weight distribution for 13,730 human newborns (*yellow* curve) correlated with death rate (*white* curve).

Figure 16.18 Adult sociable weaver (*Philetairus socius*), a native of the African savanna. These birds cooperate in constructing and using large communal nests in a region where trees and other good nesting sites are scarce.

Rita Covas and her colleagues gathered evidence of stabilizing selection on the body mass of juvenile and adult sociable weavers (*Philetairus socius*), as in Figure 16.18. Between 1993 and 2000, they captured, measured, tagged, released, and recaptured 70 to 100 percent of birds present in communal nests during the breeding season. Their field studies supported a prediction that body mass is a trade-off between risks of starvation and predation, giving intermediate-mass birds the selective advantage. Foraging is not easy in this habitat, and lean birds don't store enough fat to avoid starvation. We can expect that fat ones are more attractive to predators and not as good at escaping.

DISRUPTIVE SELECTION

With **disruptive selection**, forms at both ends of the range of variation are favored and intermediate forms are selected against.

Consider the black-bellied seedcracker (*Pyrenestes ostrinus*) of Cameroon. Females and males of these African finches have large or small bills—but no sizes in between (Figure 16.19). It's like everyone in Texas being four feet *or* six feet tall, with no one in between.

The pattern holds all through the geographic range. If unrelated to gender or geography, what causes it? If only two bill sizes persist, then disruptive selection may be eliminating birds with intermediate-size bills. Which factors affect feeding performance? Cameroon's swamp forests flood during the wet season, and fires sparked by lightning burn during the dry season. Two kinds of sedges dominate these forests. Sedges are fire-resistant, grasslike plants. One species produces hard seeds and the other, soft.

Obviously, a seedcracker's ability to crack seeds directly affects survival. It turns out that birds with

lower bill 12 mm wide   lower bill 15 mm wide

Figure 16.19 Disruptive selection in African finch populations. Selection pressures favor birds with bills that are about 12 *or* 15 millimeters wide. The difference is correlated with competition for scarce food resources during the dry season.

small bills prefer to eat soft seeds in the habitat, and birds with large bills are better at cracking the hard ones. In the dry season, all seeds are scarce and birds compete fiercely for them. Limited availability of two types of seeds during recurring periods of famine has had a disruptive effect on bill size in the seedcracker population; birds with intermediate sizes are selected against and all bills are 12 or 15 millimeters wide.

In the seedcracker, bills of a particular size have a genetic basis. In experimental crosses between two birds with the two optimal bill sizes, all offspring had a bill of one size or the other, nothing in between.

With stabilizing selection, intermediate phenotypes are favored and extreme phenotypes at both ends of the range of variation are eliminated.

With disruptive selection, intermediate forms of traits are selected against; extreme forms in the range of variation are favored.

16.9 Maintaining Variation in a Population

Natural selection theory helps explain diverse aspects of nature, including male–female differences, and the relationship between sickle-cell anemia and malaria.

SEXUAL SELECTION

The individuals of many sexually reproducing species show a distinct male or female phenotype, or **sexual dimorphism** (*dimorphos,* having two forms). Often the males are larger and flashier than females. Courtship rituals and male aggression are common.

These adaptations and behaviors seem puzzling. All take energy and time away from an individual's survival activities. Why do they persist if they do not contribute directly to survival? The answer is **sexual selection**: a form of natural selection in which the genetic winners are the ones that outreproduce others of the population. The most adaptive traits help individuals defeat same-sex rivals for mates or are the ones most attractive to the opposite sex.

By choosing mates, one gender acts as an agent of selection on its own species. For example, the females of some species shop among a clustering of males, which differ in appearance and courtship behavior. The selected males, as well as the females making the selection, pass on their alleles to the next generation.

Flashy structures and behaviors are correlated with species in which males have little or nothing to do with raising offspring. The female chooses a male by observable signs of his health and vigor, which might improve the odds of producing healthy, vigorous offspring (Figure 16.20).

You might be wondering whether we can correlate genes with specific forms of sexual behavior. One of the most amazing demonstrations of this comes from sexual deception as practiced by an Australian orchid. The flowers of *Chiloglottis trapeziformis* attract male wasps by making a compound—a sex pheromone—that is identical to one released by the female wasps, the point being to get pollinated as the male is busy doing what otherwise would perpetuate its genes.

This orchid is stingy. It gives a male wasp nothing in return, not a single drop of nectar, even though it is the orchid's sole pollinator. The wingless female wasps hatch in soil. When males don't lift and carry them to a food source, they starve to death.

When *C. trapeziformis* puts out blooms, the male wasps waste precious time and metabolic energy trying to find females. Evolutionary biologist Florian Schiestl suggests that selection pressure is afoot for wasps that can make a new sex pheromone, one that the orchid can't duplicate.

And while the interaction exploits males, Wittko Francke thinks it might put pressure on their brains to evolve. In an orchid patch, the average tiny-brained male wasp copulates blindly with whatever smells right. It will try to copulate even with the head of a pin that has a few micrograms of pheromone sprayed on it. However, a few wasps with a slightly less robotic brain might be able to identify the females by other cues, such as visual ones. Alternatively, both species could face extinction, another pattern in nature.

Figure 16.20 One male bird of paradise is engaged in a flashy courtship display. He caught the eye (and, perhaps, the sexual interest) of the smaller, less colorful female. The males of this species compete fiercely for females, which are the selective agents. (Why do you suppose drab-colored females have been favored?) (*Paradisaea raggiana*) is engaged in a flashy courtship display. He caught the eye (and, perhaps, the sexual interest) of the smaller, less colorful female.

SICKLE-CELL ANEMIA—LESSER OF TWO EVILS?

With *balancing* selection, two or more alleles for a trait are being maintained at frequencies above 1 percent in the population. Their persistence is called **balanced polymorphism** (*polymorphos,* having many forms). The allele frequencies might shift slightly, but they often return to the same values over the long term. We often see this balance when conditions favor heterozygotes.

In some way, their nonidentical alleles for a given trait give them higher fitness than homozygotes, which, recall, have identical alleles for the trait.

Consider the environmental pressures that favor an Hb^A/Hb^S pairing in humans. The Hb^S allele codes for a mutant form of hemoglobin, an oxygen-transporting protein in blood. Homozygotes (Hb^S/Hb^S) develop *sickle-cell anemia*, a genetic disorder (Section 3.6).

The Hb^S frequency is highest in subtropical and tropical regions of Asia and Africa. Often, Hb^S/Hb^S homozygotes die in their early teens or early twenties. Yet, in these same regions, heterozygotes (Hb^A/Hb^S) make up nearly a third of the population! Why is this combination maintained at such high frequency?

The balancing act is most pronounced in areas that have the highest incidence of *malaria* (Figure 16.21). Mosquitoes transmit the parasitic agent of malaria, *Plasmodium*, to human hosts. The parasite multiplies in the liver and, later, in red blood cells. The target cells rupture and release new parasites during severe, recurring bouts of infection (Section 20.3).

It turns out that the mutant hemoglobin interferes with the life cycle of the parasitic agent, so Hb^A/Hb^S heterozygotes are more likely to survive malaria than people who produce normal (Hb^A/Hb^A) hemoglobin. There are several potential survival mechanisms. In heterozygotes, infected cells have a sickle shape under normal conditions. The abnormal shape marks them as targets for the immune system, which proceeds to destroy them. Also, heterozygotes have one normal hemoglobin allele. Although they are not completely healthy, they produce enough normal hemoglobin to support body functions. As a result, they are more likely than the Hb^S/Hb^S homozygotes to survive and reach reproductive age.

So the persistence of the "harmful" Hb^S allele is a matter of relative evils. Natural selection has favored the Hb^A/Hb^S combination in malaria-ridden areas because heterozygotes show more resistance to the disease. In such environments, the combination has more survival value than either Hb^S/Hb^S or Hb^A/Hb^A. And malaria has been a selective force for thousands of years in tropical and subtropical areas of Asia, the Middle East, and Africa.

With sexual selection, some version of a gender-related trait gives the individual an advantage in reproductive success. Sexual dimorphism is one outcome of sexual selection.

In a population showing balanced polymorphism, natural selection is maintaining two or more alleles at frequencies greater than 1 percent over the generations.

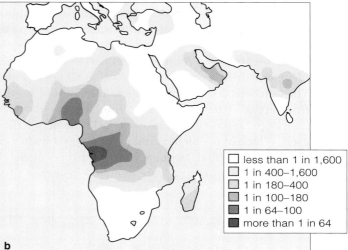

☐	less than 1 in 1,600
☐	1 in 400–1,600
☐	1 in 180–400
▨	1 in 100–180
▨	1 in 64–100
■	more than 1 in 64

Figure 16.21 (**a**) Distribution of malaria cases in Africa, Asia, and the Middle East in the 1920s, before the start of programs to control mosquitoes, the vector for *Plasmodium*. (**b**) Distribution and frequency of people with the sickle-cell trait. Notice the close correlation between the maps. (**c**) Physician searching for *Plasmodium* larvae in Southeast Asia.

16.10 Genetic Drift—The Chance Changes

Random changes in allele frequencies can lead to a loss of genetic diversity in a population. The drift in those frequencies is greatest for small populations.

Genetic drift is a random change in allele frequencies over time, brought about by chance alone. It tends to have minor impact in very large populations. Even so, it increases the likelihood that an allele will become more or less prevalent when the population is small.

Sampling error, a rule of probability, helps explain the difference. By this rule, you are less likely to come closer to an expected outcome of some event if that event doesn't happen very often. For instance, flip a coin. With each flip, there is a 50 percent chance the coin will turn up heads. With ten flips, the odds are low that it will turn up heads half the time. With a thousand flips, you are more likely to come close to 500 heads and 500 tails. Sampling error also applies each time random mating and fertilization take place in a population.

Figure 16.22 is a computer simulation of the effect of genetic drift in one large and one small population. The outcomes are a simple way to think about results from actual experiments. The simulation starts with nine populations of 25 flies each and nine of 500 flies each. Which ones entered either group was a matter of chance. The initial frequency of wild-type allele *A* was 0.5. Some offspring were removed in each of fifty

generations to maintain population size. In the end, *A* wasn't the only allele left in large groups. But it was fixed in five of the small groups. **Fixation** means only one kind of allele remains at a locus in a population. All individuals have become homozygous for it.

Thus, *in the absence of other forces, random change in allele frequencies leads to the homozygous condition and a loss of genetic diversity over the generations.* It happens in all populations. It just happens faster in small ones. Once alleles from a parent population are fixed, their frequencies will not change again unless mutation or gene flow introduces new alleles.

BOTTLENECKS AND THE FOUNDER EFFECT

Genetic drift is pronounced when a few individuals rebuild a population or start a new one. This happens after a **bottleneck**, a drastic reduction in population size brought about by severe pressure or a calamity. Suppose contagious disease, habitat loss, or hunting nearly wipes out a population. Even if a moderate number of individuals survive the bottleneck, allele frequencies will have been altered at random.

In the 1890s, hunters killed all but twenty of a large population of northern elephant seals. Government restrictions since then have allowed them to recover to a current population of about 130,000. Each of them is homozygous at every gene locus examined so far.

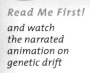

Read Me First!
and watch the narrated animation on genetic drift

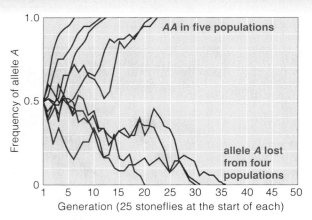

a The size of nine populations of flies was held constant at 25 breeding individuals in each generation, through fifty generations. The five graph lines reaching the top of this diagram tell you that allele *A* became fixed in five of these small populations. The four lines plummeting off the bottom of the diagram tell you that it was lost from four of them. As you can see, *alleles can be fixed or lost even in the absence of selection.*

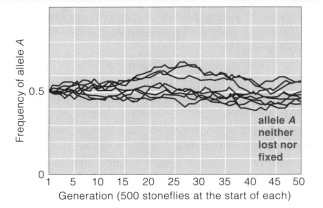

b The size of nine different populations was kept at 500 individuals in each generation, through fifty generations. In these larger populations, allele *A* did not become fixed. The magnitude of genetic drift was much less in each generation than in the small populations tracked in (**a**).

Figure 16.22 Computer simulation of genetic drift's effect on allele frequencies in small and large populations of flies. Equal fitness is assumed for three simulations (*AA* = 1, *Aa* = 1, and *aa* = 1).

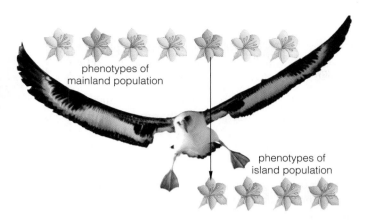

phenotypes of
mainland population

phenotypes of
island population

Figure 16.23 Founder effect. This wandering albatross carries seeds, stuck to its feathers, from the mainland to a remote island. By chance, most of the seeds carry an allele for orange flowers that are rare in the original population. Without further gene flow or selection for color, genetic drift will fix the allele on the island.

Genetic outcomes also can be unpredictable after a few individuals establish a new population. This form of bottlenecking is a **founder effect**. By chance, allele frequencies of the founders may differ in the original population. If there is no further gene flow, genetic drift will work on the small population. The effect can be pronounced on isolated islands (Figure 16.23).

GENETIC DRIFT AND INBRED POPULATIONS

Genetic drift is pronounced in an inbred population. **Inbreeding** is nonrandom mating among very close relatives, which share many identical alleles. It leads to the homozygous condition. It also lowers fitness if harmful recessive alleles are increasing in frequency.

Most human societies forbid or discourage incest (inbreeding between parents and children or siblings). But inbreeding among other close relatives is common in geographically or culturally isolated small groups. The Old Order Amish of Pennsylvania, for instance, are a highly inbred group having distinct genotypes. One outcome of inbreeding is a high frequency of the recessive allele that causes *Ellis–van Creveld syndrome*. Affected individuals have extra fingers, toes, or both and short limbs (Section 11.4). The allele might have been rare when a few founders entered Pennsylvania. Currently, about 1 in 8 individuals of the community are heterozygous and 1 in 200 are homozygous for it.

Genetic drift is the random change in allele frequencies over the generations, brought about by chance alone. The magnitude of its effect is greatest in small populations, such as one that endures a bottleneck.

16.11 Gene Flow

Individuals, and their alleles, move into and out of populations, and this physical flow counters changes introduced by other microevolutionary processes.

Individuals of the same species don't always stay put. A population will lose alleles whenever an individual permanently leaves it, an event called *emigration*. The population gains alleles whenever new individuals permanently move in, an event called *immigration*. In both cases, there is a **gene flow**—a physical flow of alleles between two or more populations. Gene flow tends to counter genetic differences that we expect to see developing by way of mutation, natural selection, and genetic drift. It helps keep separated populations genetically similar.

Think of the acorns that blue jays disperse when they gather nuts for the winter. Each fall the jays visit acorn-bearing oak trees repeatedly, then bury acorns in the soil of home territories that may be as much as a mile away (Figure 16.24). Alleles flowing in with the "immigrant acorns" help decrease genetic differences between stands of oak trees.

Figure 16.24 Blue jay, a mover of acorns that helps keep genes flowing between separate oak populations.

Or think of the millions of people from politically explosive, economically bankrupt countries who seek a more stable home. The scale of their emigrations is unprecedented, but the flow of genes is not. Human history is rich with cases of gene flow that minimized many of the genetic differences among geographically separate groups. Remember Genghis Khan? His genes flowed from China to Vienna (Figure 11.6). Similarly, the armies of Alexander the Great brought the genes for green eyes from Greece all the way to India.

Gene flow is the physical movement of alleles into and out of a population, through immigration and emigration. It tends to counter the effects of mutation, natural selection, and genetic drift.

Summary

Section 16.1 Awareness of evolution, or changes in lines of descent over time, emerged long ago from biogeography, comparative morphology, and geology.

Section 16.2 Prevailing cultural belief systems influence interpretation of natural events. In the nineteenth century, naturalists worked to reconcile traditional belief systems with a growing body of physical evidence in support of evolution.

Section 16.3 Charles Darwin and Alfred Wallace proposed a novel theory that natural selection in populations results in evolution. The theory of natural selection is this: Populations increase in size until resources dwindle and individuals compete for them. When individuals have forms of traits that make them more competitive, they tend to produce more offspring. Over generations, those forms increase in frequency. Nature "selects" variations in traits that are more effective at helping individuals survive and reproduce in particular environments.

Section 16.4 Long-term adaptations are heritable aspects of form, function, behavior, or development that improve the chance of surviving and reproducing.

Section 16.5 Individuals of a population generally have the same number and kinds of genes for the same traits. Mutations are the source of *new* alleles (different molecular forms of genes). Individuals who inherit different allele combinations vary in details of one or more traits. An allele at any locus may become more or less common relative to other kinds or may be lost.

Microevolution refers to changes in allele frequencies of a population brought about by mutation, natural selection, gene flow, and genetic drift (Table 16.1).

At genetic equilibrium, a population is not evolving. By the Hardy–Weinberg rule, this occurs only if there is no mutation, the population is infinitely large and isolated from other populations of the species, there is no selection, mating is random, and members survive and reproduce equally. Deviations from this theoretical baseline indicate microevolution is at work.

Section 16.6 Mutations are rare for any given individual, but they happen at a predictable rate. Where a mutation does occur is unpredictable.

Section 16.7 Natural selection acts on phenotypic variation within a population. Directional selection favors the forms at one end of the phenotypic range.

Section 16.8 Intermediate forms of a trait are favored by stabilizing selection. In disruptive selection, forms at both ends of a range of variation are favored over the intermediate forms.

Section 16.9 Sexual selection, by females or males, leads to forms of traits that favor reproductive success. Persistence in phenotypic differences between males and females (sexual dimorphism) is one outcome.

Selection may result in balanced polymorphism, with nonidentical alleles for a trait being maintained over time at frequencies greater than 1 percent.

Section 16.10 Genetic drift is a random change in allele frequencies over time due to chance alone. The random changes tend to lead to the homozygous condition and loss of genetic diversity through the generations. The effect of genetic drift is greatest in small populations, such as ones that pass through a bottleneck or arise from a small group of founders.

Section 16.11 Gene flow shifts allele frequencies by physically moving alleles into a population (by way of immigration) and out of it (by emigration). It tends to keep different populations of the same species alike by countering mutation, natural selection, and genetic drift.

Table 16.1	Microevolutionary Processes
Mutation	A heritable change in DNA
Natural selection	Change or stabilization of allele frequencies; an outcome of differences in survival and reproduction among variant individuals of a population
Genetic drift	Random fluctuation in allele frequencies over time due to chance occurrences alone
Gene flow	Individuals, and their alleles, move into and out of populations; the physical flow counters the effects of the other microevolutionary processes

Self-Quiz   *Answers in Appendix III*

1. Individuals don't evolve, _____ do.

2. Biologists define evolution as _____ .
   a. equilibrium in allele frequencies
   b. heritable change in a line of descent
   c. acquiring traits during the individual's lifetime
   d. both a and b

3. _____ is the original source of new alleles.
   a. Mutation   c. Genetic drift   e. All give rise
   b. Natural selection   d. Gene flow   to new alleles

4. Natural selection may occur when there are _____ .
   a. differences in forms of traits
   b. differences in survival and reproduction among individuals that differ in one or more traits
   c. both a and b

5. Directional selection _____ .
   a. eliminates uncommon forms of alleles
   b. shifts allele frequencies in a consistent direction
   c. favors intermediate forms of a trait
   d. works against adaptive traits

6. Disruptive selection _____ .
   a. eliminates uncommon forms of alleles
   b. shifts allele frequencies in a consistent direction
   c. doesn't favor intermediate forms of a trait
   d. both b and c

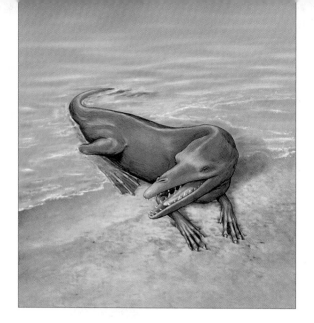

Figure 16.25 Reconstruction, based on fossils discovered in Pakistan, of *Rodhocetus*. This cetacean lived 47 million years ago, along the shores of the Tethys Sea. Its ankle bones indicate a close evolutionary link between early whales and hoofed land mammals.

7. _____ tends to reduce allelic differences among populations of a species.
   a. Genetic drift   c. Mutation
   b. Gene flow   d. Natural selection

8. Match the evolution concepts.
   _____ gene flow   a. source of new alleles
   _____ natural   b. changes in a population's allele
   selection   frequencies due to chance alone
   _____ mutation   c. allele frequencies change owing to
   _____ genetic   immigration, emigration, or both
   drift   d. outcome of differences in survival,
                            reproduction among individuals
                            that vary in forms of shared traits

Critical Thinking

1. Martha is studying a population of tropical birds. Male birds have brightly colored tail feathers and the females don't. She suspects this difference is maintained by sexual selection. Design an experiment to test her hypothesis.

2. A few families in a remote region of Kentucky show a high frequency of *blue offspring*, an autosomal recessive disorder. Skin of affected individuals appears bright blue. Homozygous recessives lack an enzyme that maintains hemoglobin in its normal molecular form. Without it, a blue form of hemoglobin accumulates in blood and shows through the skin. Formulate a hypothesis to explain why the blue offspring trait recurs among a cluster of families but is rare in the human population at large.

3. For some time, evolutionists accepted that the ancestors of whales were four-legged animals that walked on land, then took up life in water about 55 million years ago. Fossils show gradual changes in skeletal features that made an aquatic life possible. But which four-legged mammals were its ancestors? The answer came from Philip Gingerich and Iyad Zalmout. While digging in Pakistan, they found fossils of early aquatic whales. Intact, sheep-like ankle bones *and* archaic whale skull bones were in the same fossilized skeletons (Figures 16.3 and 16.25).

Ankle bones of fossilized, early whales from Pakistan have the same form as the unique ankle bones of extinct and modern artiodactyls. Modern cetaceans no longer have even a remnant of an ankle bone. Here is evidence of an evolutionary link between certain aquatic mammals and a major group of mammals on land.

No one was around to witness the transition. Yet the fossils are real, just as the morphology and molecular makeup of living organisms are real. As you'll see in the next chapter, radiometric dating assigns fossils to places in time.

Because there were no witnesses, do you think there can be absolute proof of evolution? Is the circumstantial evidence of fossil morphology enough to convince you that the theory is valid?

IMPACTS, ISSUES *Measuring Time*

How do you measure time? Is your comfort level with the past limited to your own generation? Probably you can relate to a few centuries of human events. But geologic time? Comprehending the distant past requires a huge intellectual leap from the familiar to the unknown.

Consider this: Asteroids are rocky, metallic bodies hurtling through space. They are a few meters to 1,000 kilometers across. When our solar system's planets were forming, their gravitational force swept most asteroids from the sky. At least 6,000 asteroids, including the one shown in Figure 17.1a, still orbit the sun in a belt between Mars and Jupiter. Millions more frequently zip past Earth. They are hard to spot because they don't emit light. We don't discover most of them unless they pass close by. Some have passed too close for comfort.

Big asteroid impacts altered the course of evolution. For instance, researchers found a thin layer of iridium around the world, and it dates to a mass extinction that wiped out the last of the dinosaurs (Figure 17.1b). Iridium is rare on Earth but not in asteroids. Above the layer, there are no more fossils of dinosaurs, anywhere.

It has only been about 100,000 years since the first modern humans (*Homo sapiens*) evolved. We know that dozens of humanlike species evolved in Africa during the 5 million years before our species even showed up. So why are we the only ones left?

Unlike today's large, globally dispersed populations of humans, the early species lived in small bands. What

Figure 17.1 (**a**) An asteroid 19 kilometers (about 12 miles) long, still hurtling through space. (**b**) Two views of the last few minutes of the Cretaceous. The filmstrip at far right shows a sample of the worldwide, iridium-rich layer of sediment (*black*) that dates precisely to the K–T boundary. It's evidence of an asteroid impact.

the big picture

Evidence of Evolution Fossils are direct evidence of ancient life. Biogeography and comparisons of body form, developmental patterns, and biochemistry are helping us piece together and interpret the fossil record.

How Species Originate Microevolution and plain luck have both contributed to the origin of species. Reproductively isolated subpopulations of a species diverge genetically. A new species is recognized when divergences are great enough to prevent successful interbreeding.

if most were casualties of the twenty asteroids that struck when they were alive? What if *our* ancestors were just plain lucky? About 2.3 million years ago, one huge object from space hit the ocean, west of what is now Chile. If it had collided with the rotating Earth just a few hours earlier, our ancestors in southern Africa might have been incinerated.

Now that we know what to look for, we are seeing more and more craters in satellite images of the Earth. Less than 4,000 years ago, in what is now Iraq, an impact released energy that was equivalent to the detonation of hundreds of nuclear weapons.

If we can figure out what an asteroid impact will do to us, we can figure out how impacts affected life in the past. We *can* comprehend life long before our own. This chapter introduces some tools and evidence used to interpret patterns, trends, and rates of change among life's major lineages. Along the way you will read about their causes, including good and bad cosmic luck.

This leap through time starts with the premise that any aspect of the natural world, past as well as present, has one or more underlying causes. We look for clues by studying physical and chemical aspects of the Earth, analyzing fossils, and comparing the morphology and biochemistry of species. We test our hypotheses with experiments, models, and advancing technologies. This shift from experience to inference—from the known to what can only be surmised—has given us astonishing glimpses into the past.

How Would You Vote?

A major asteroid impact could obliterate civilization and much of Earth's biodiversity. Should nations around the world contribute resources to searching for and tracking asteroids? See the Media Menu for details, then vote online.

Big Evolutionary Events All species share genetic connections through ancient lineages that have changed over evolutionary time. New species emerged in response to opportunities that opened up following often-catastrophic challenges.

Organizing the Evidence Naming and classifying species helps us manage data on biodiversity. Evolutionary classification systems group species with respect to derived traits, which evolved only once in the most recent shared ancestor of two or more groups.

17.1 Fossils—Evidence of Ancient Life

The fossil record helps explain the connections between Earth's evolution and life's evolution.

About 500 years ago, Leonardo da Vinci was puzzled by seashells entombed in rocks of northern Italy's high mountains, hundreds of kilometers from the sea. How did they get there? By the prevailing belief, water from a stupendous, divinely invoked flood had surged up against the mountains, where it deposited the shells. But many of the shells were thin, fragile, and intact. If they had been swept across such great distances, then wouldn't they be battered to bits?

Leonardo also brooded about the rocks. They were stacked like cake layers. Some layers had shells, others had none. Then he remembered how large rivers swell with spring floodwaters and deposit silt in the sea. Did such depositions happen in ancient seasons? If so, then shells in the mountains could be evidence of layered communities of organisms that once lived in the seas!

By the 1700s, **fossils** were accepted as the remains and impressions of organisms that lived in the past. (*Fossil* comes from a Latin word for "something that was dug up.") People were still interpreting fossils through the prism of cultural beliefs, as when a Swiss naturalist unveiled the remains of a giant salamander and excitedly announced that they were the skeleton of a man who drowned in the great flood.

By midcentury, though, scholars were questioning these interpretations. Why? Mining, quarrying, and excavations for canals were under way. Diggers were finding similar rock layers and similar sequences of fossils in distant places, such as the nearshore cliffs on both sides of the English Channel. If those layers had been deposited with the passing of time, then the vertical sequence of fossils in them might be a record of past life—*a fossil record*.

HOW DO FOSSILS FORM?

Most fossils discovered so far are bones, teeth, shells, seeds, spores, and other hard parts (Figure 17.2). Fossilized feces (coprolites) hold residues of species that were eaten in ancient times. Imprints of leaves, stems, tracks, burrows, and other *trace* fossils provide further indirect evidence of past life.

Fossilization is a slow process that starts when an organism or traces of it become covered by volcanic ash or sediments. Water slowly infiltrates the remains, and metal ions and other inorganic compounds that are dissolved in it replace the minerals in bones and other hardened tissues. As sediments accumulate, they exert increasing pressure on the burial site. In time, the pressure and mineralization processes transform those remains into stony hardness.

Remains that become buried quickly are less likely to be obliterated by scavengers. Preservation is also favored when a burial site stays undisturbed. Usually, however, erosion and other geologic assaults deform, crush, break, or scatter the fossils. This is one reason fossils are relatively rare.

Figure 17.2 Representatives of more than 250,000 ancient species known from the fossil record. *Left,* fossilized parts of the oldest known land plant (*Cooksonia*). Its stems were less than seven centimeters tall. *Right,* fossilized skeleton of an ichthyosaur. This marine reptile lived 200 million years ago.

Figure 17.3 A slice through time—Butterloch Canyon, Italy, once at the bottom of a sea. Its sedimentary rock layers slowly formed over hundreds of millions of years. Later, geologic forces lifted the stacked layers above sea level. Later still, the erosive force of river water carved the canyon walls and exposed the layers. Scientists Cindy Looy and Mark Sephton are climbing to reach the Permian–Triassic boundary layer, where they will look for fossilized fungal spores.

Other factors affect preservation. Organic materials cannot decompose in the absence of free oxygen, for instance. They might endure if sap, tar, ice, mud, or another air-excluding substance protects them. Insects in amber and frozen woolly mammoths are examples.

FOSSILS IN SEDIMENTARY ROCK LAYERS

Stratified (stacked) layers of sedimentary rock formed long ago from deposits of volcanic ash, silt, sand, and other materials. Sand and silt piled up after rivers transported them from land to the sea, as Leonardo suspected. Sandstones formed from sand, and shales from silt. Depositions were sometimes interrupted, in part because the sea level changed as ice ages began. Tremendous volumes of water froze in glaciers, rivers dried up, and the depositions ended in some regions. Later in time, when the climate warmed and glaciers melted, the depositions resumed.

The formation of sedimentary rock layers is called **stratification**. The first to form are now the deepest layers, and those closest to the surface were the last. Most formed horizontally, as in Figure 17.3, because particles tend to settle in response to gravity. You may see tilted or ruptured layers, as along a road cut into a mountainside. Major crustal movements or upheavals caused them, much later in time.

We find most fossils in sedimentary rock. When you understand how rock layers form, it is obvious that the fossils in a particular layer formed at a given time in Earth history. Specifically, *the older the layer, the older the fossils.* Given that rock layers formed in sequence, their fossils are unique to sequential ages.

INTERPRETING THE FOSSIL RECORD

We have fossils for more than 250,000 known species. Judging from the current range of biodiversity, there must have been many, many millions more. Yet the fossil record will never be complete. Why is this so?

The odds are against finding signs of an extinct, ancient species. At least one individual had to be gently buried before it decomposed or something ate it. The burial site had to escape erosion, lava flows, and other geologic forces. The fossil had to end up in a place where someone could actually find it. Fossils often are found on the side of a canyon carved out by a river that exposed the layers of sedimentary rock.

Fossils did not form in many habitats, and most species didn't lend themselves to preservation. Unlike bony fishes and hard-shelled mollusks, jellyfishes and soft worms don't show up as much in the fossil record. Yet they probably were just as common, or more so.

Also think about population density and body size. One plant population might release millions of spores in a single season. The earliest humans lived in small bands and raised few offspring. What are the odds of finding even one fossilized human bone compared to spores of plants that lived at the same time?

Finally, imagine one line of descent, a **lineage**, that vanished when its habitat on a remote volcanic island sank into the sea. Or imagine two lineages, one lasting only briefly and the other for billions of years. Which is more likely to be represented in the fossil record?

Fossils are physical evidence of life in the remote past. Those embedded in sedimentary rock layers are a historical record of life. The deepest layers generally contain the oldest fossils.

The fossil record is incomplete. Geologic events obliterated much of it. The record is slanted toward species that had large bodies, hard parts, dense populations, and wide distribution, and that persisted for a long time.

Even so, the fossil record is now substantial enough for us to reconstruct patterns and trends in the history of life.

17.2 Dating Pieces of the Puzzle

How do we assign fossils to a place in time? In other words, how do we know how old they really are?

RADIOMETRIC DATING

For as long as they've been digging up rocks, people have been coming across fossils. At one time, they could assign only *relative* ages to their treasures, not absolute ones. For instance, a fossilized mollusk in a rock layer was said to be younger than a fossil below it and older than a fossil above it, and so on.

Things changed with **radiometric dating**. This is a way to measure the proportions of a daughter isotope and the parent radioisotope of some element trapped inside a rock since the time the rock formed. Again, a radioisotope is a form of an element with an unstable nucleus (Section 2.2). Radioactive atoms decay, or lose energy and subatomic particles until they reach a more stable form.

It is not possible to predict the exact instant of one atom's decay, but a predictable number of an isotope's atoms will decay over a period of time. Like the ticking of a perfect clock, the characteristic rate of decay for each isotope is constant. In other words, changes in pressure, temperature, or chemical state

don't alter it. The time it takes for half of a quantity of a radioisotope's atoms to decay is its **half-life** (Figure 17.4*a*). For instance, uranium 238 has a half-life of 4.5 billion years. It decays into thorium 234, which in turn decays into something else, and so on through a series of intermediate daughter isotopes. The final, stable daughter element is lead. By measuring the ratio of uranium 238 to lead in the oldest rocks, geologists estimated that Earth formed more than 4.6 billion years ago.

Radiometric dating doesn't work for sedimentary rock. It works for volcanic rock or ashes, which hold the most fossils. The ratio of carbon 14 to carbon 12 is used to date recent fossils that still contain some carbon (Figure 17.4*b–d*). The only way to date older fossils is to determine their position relative to any volcanic rocks in the same area. This dating method has an error factor of less than 10 percent.

PLACING FOSSILS IN GEOLOGIC TIME

Early geologists carefully counted backward through layers of sedimentary rock, then used their counts to construct a chronology of Earth history as a **geologic time scale** (Figure 17.5). By comparing evidence from

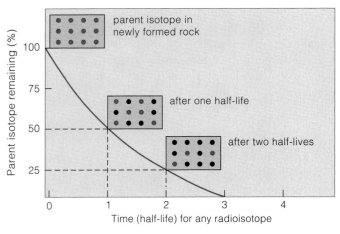

a A simple way to think about the decay of a radioisotope to a more stable form, as plotted against time.

Figure 17.4 (**a**) The decay of radioisotopes at a fixed rate to more stable forms. The half-life of each kind of radioisotope is the time it takes for 50 percent of a sample to decay. After two half-lives, 75 percent of the sample has decayed, and so on.

(**b–d**) Radiometric dating of a fossil. Carbon 14 (^{14}C) forms in the atmosphere. There, it combines with free oxygen, the result being carbon dioxide. Along with far greater quantities of its more stable isotopes, trace amounts of carbon 14 enter food webs by way of photosynthesis. All organisms incorporate carbon into body tissues.

b Long ago, trace amounts of ^{14}C and a lot more ^{12}C were incorporated into tissues of a living mollusk. The carbon was part of the organic compounds making up the tissues of its prey. As long as it lived, the proportion of ^{14}C to ^{12}C in its tissues remained the same.

c When the mollusk died, it stopped gaining carbon. Over time, proportion of ^{14}C to ^{12}C in its remains declined because of the radioactive decay of ^{14}C. Half of the ^{14}C had decayed in 5,370 years, half of what remained was gone in another 5,370 years, and so on.

d Fossil hunters find the fossil. They measure its $^{14}C/^{12}C$ ratio to determine the half-life reductions since death. The ratio turns out to be one-eighth of the $^{14}C/^{12}C$ ratio in living organisms. Thus the mollusk lived about 16,000 years ago.

Eon	Era	Period	Epoch	Millions of Years Ago	Major Geologic and Biological Events That Occurred Millions of Years Ago (mya)
PHANEROZOIC	CENOZOIC	QUATERNARY	Recent	0.01	1.8 mya to present. Major glaciations. Modern humans evolve. The most recent *extinction crisis* is under way.
			Pleistocene	1.8	
		TERTIARY	Pliocene	5.3	65–1.8 mya. Major crustal movements, collisions, mountain building. Tropics, subtropics extend poleward. When climate cools, dry woodlands, grasslands emerge. *Adaptive radiations* of flowering plants, insects, birds, mammals.
			Miocene	22.8	
			Oligocene	33.7	
			Eocene	55.5	
			Paleocene	65	65 mya. Asteroid impact; *mass extinction* of all dinosaurs and many marine organisms.
	MESOZOIC	CRETACEOUS	Late		99–65 mya. Pangea breakup continues, inland seas form. Adaptive radiations of marine invertebrates, fishes, insects, and dinosaurs. Origin of angiosperms (flowering plants).
				99	
			Early		145–99 mya. Pangea starts to break up. Marine communities flourish. *Adaptive radiations* of dinosaurs.
				145	145 mya. Asteroid impact? Mass extinction of many species in seas, some on land. Mammals, some dinosaurs survive.
		JURASSIC		213	248–213 mya. *Adaptive radiations* of marine invertebrates, fishes, dinosaurs. Gymnosperms dominate land plants. Origin of mammals.
		TRIASSIC		248	248 mya. *Mass extinction*. Ninety percent of all known families lost.
	PALEOZOIC	PERMIAN			286–248 mya. Supercontinent Pangea and world ocean form. On land, *adaptive radiations* of reptiles and gymnosperms.
				286	360–286 mya. Recurring ice ages. On land, *adaptive radiations* of insects, amphibians. Spore-bearing plants dominate; cone-bearing gymnosperms present. Origin of reptiles.
		CARBONIFEROUS			
				360	360 mya. *Mass extinction* of many marine invertebrates, most fishes.
		DEVONIAN			410–360 mya. Major crustal movements. Ice ages. *Mass extinction* of many marine species. Vast swamps form. Origin of vascular plants. *Adaptive radiation* of fishes continues. Origin of amphibians.
				410	
		SILURIAN			440–410 mya. Major crustal movements. *Adaptive radiations* of marine invertebrates, early fishes.
				440	
		ORDOVICIAN			505–440 mya. All land masses near equator. Simple marine communities flourish until origin of animals with hard parts.
				505	
		CAMBRIAN			544–505 mya. Supercontinent breaks up. Ice age. *Mass extinction*.
				544	
PROTEROZOIC					2,500–544 mya. Oxygen accumulates in atmosphere. Origin of aerobic metabolism. Origin of eukaryotic cells. Divergences lead to eukaryotic cells, then protists, fungi, plants, animals.
				2,500	
ARCHEAN AND EARLIER					3,800–2,500 mya. Origin of photosynthetic prokaryotic cells. 4,600–3,800 mya. Origin of Earth's crust, first atmosphere, first seas. Chemical, molecular evolution leads to origin of life (from proto-cells to anaerobic prokaryotic cells).

Figure 17.5 Geologic time scale. Major boundaries mark the times of the greatest mass extinctions. If the time spans listed were to the same scale, the Archean and Proterozoic portions would run off this page. Compare Figure 17.6.

around the world, they found four abrupt transitions in fossil sequences and used them as boundaries for four great intervals. They named the first interval the Proterozoic, after finds that predate fossils of animals. They named other intervals the Paleozoic, Mesozoic, and the "modern" era, the Cenozoic.

The current geologic time scale now correlates with **macroevolution**, or major patterns, trends, and rates of change among lineages. Also, being more immense than early researchers suspected, the Proterozoic is now subdivided into more intervals. Life originated in one of those intervals, the Archean eon.

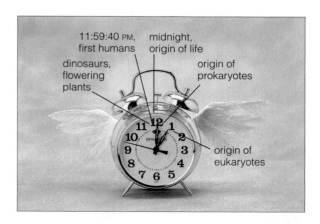

Figure 17.6 A geologic time clock. Think of the spans as minutes on a clock that runs from midnight to noon. If we say that life started at midnight, then the Paleozoic began at 10:04 A.M., the Mesozoic at 11:09 A.M., and the Cenozoic at 11:47 A.M. The recent epoch of the Cenozoic era started after the last 0.1 second before noon. And where does that put you?

17.3 Evidence From Biogeography

By clinking their hammers against the rocks, geologists discovered that the "solid" Earth hasn't stayed put.

As the early geologists were discovering fossils and mapping the record of Earth history, Charles Darwin was so taken with one of their theories that it helped shape his view of life. The **theory of uniformity** held that mountain building and erosion had repeatedly worked over the Earth's surface in precisely the same ways through time (Section 16.2). However, as more fossils were found, geologists realized that repetitive change was only part of the picture. Like life, Earth itself had changed irreversibly.

AN OUTRAGEOUS HYPOTHESIS

For instance, the Atlantic coasts of South America and Africa seemed to "fit" like jigsaw puzzle pieces. Were all continents once part of a bigger one that had split into fragments and drifted apart? One model for the proposed supercontinent—**Pangea**—took into account the world distribution of fossils and existing species. It also took into account glacial deposits, which held clues to ancient climate zones.

Most scientists did not accept the continental drift theory. Continents drifting about on their own across the Earth's mantle seemed to be an outrageous idea, and they preferred the theory of uniformity.

But more evidence kept piling up. Iron-rich rocks are molten when they form. Iron particles in them orient north–south in response to Earth's magnetic poles, and stay that way after the rocks harden. Yet in North and South America, the tiny iron compasses in rocks that had formed 200 million years ago did *not* point pole to pole. So scientists came up with a map. They made a north–south alignment work by joining North America and western Europe—and orienting them in a way that sure didn't look like modern maps.

More puzzles! Deep-sea probes showed that the seafloor is spreading away from mid-oceanic ridges (Figure 17.7). Molten rock spewing from a ridge flows sideways in both directions, then it hardens into new crust. The spreading new crust forces older crust into deep trenches elsewhere in the seafloor. All the ridges and trenches are actually edges of thin but enormous plates, like pieces of a cracked eggshell. They move with almost imperceptible slowness. However, over time, land masses take up new positions.

These findings put continental drift into a broader explanation of crustal movements, now known as the **plate tectonic theory**. Researchers soon found ways to use the new theory's predictive power.

a

b

island arc | oceanic crust | oceanic ridge | trench | continental crust

lithosphere (solid layer of mantle) | hot spot | athenosphere (plastic layer of mantle) | subducting plate

Figure 17.7 Some forces of geologic change.

(**a**) Present configuration of Earth's crustal plates. These immense, rigid parts of the crust split, drift apart, and collide at almost imperceptible rates. On the seafloor, *red* signifies the newest crust (less than 10 million years old), and *blue* is the oldest; it formed 180 million years ago.

(**b**) Huge plumes of molten material drive the movement. They well up from the interior, spread laterally under the crust, and rupture it at mid-oceanic ridges. At these deep ridges, molten material seeps out, cools, and slowly forces the seafloor away from the rupture, which displaces the plates from ridges.

The leading edge of one plate commonly plows under an adjoining plate and uplifts it. *Blue* lines in (**a**) show where this is now happening. The Cascades, Andes, and other great mountain ranges paralleling the coasts of continents formed this way.

Long ago, superplumes violently ruptured the crust at what are now called "hot spots" in the mantle. The Hawaiian Archipelago has been forming this way. Continents also rupture. Deep rifting and splitting are happening now in Missouri, at Lake Baikal in Russia, and in eastern Africa.

a 420 mya   b 260 mya   c 65 mya   d 10 mya

Figure 17.8 Reconstructions of drifting continents. (**a**) Gondwana (*yellow*) 420 million years ago. (**b**) Later in time, all major land masses had collided to form a supercontinent, Pangea. (**c**) Positions of the fragments after Pangea split apart 65 million years ago, and (**d**) their positions 10 million years ago.

About 260 million years ago, seed ferns and other plants lived nowhere except on the part of Pangea that had once been Gondwana. So did the therapsids, or mammal-like reptiles. (**e**) Fossilized leaf of one of the seed ferns, *Glossopteris*. (**f**) *Lystrosaurus*, a therapsid about 1 meter (3 feet) long. This tusked herbivore fed on the fibrous plants of dry floodplains.

For instance, the same series of glacial deposits, coal seams, and basalt are found in Africa, India, Australia, and South America. All four southern land masses hold fossils of *Glossopteris*, a seed fern. They also hold fossils of *Lystrosaurus*, a mammal-like reptile (Figure 17.8). Neither the plant's heavy seeds nor the reptile could have floated across a vast, open ocean. Researchers suspected they had evolved together on **Gondwana**, a supercontinent that preceded Pangea.

Antarctica, too, formed after Gondwana broke up. A geologist predicted that fossils of *Glossopteris* and *Lystrosaurus* would be discovered there, in a series of glacial deposits, coal seams, and basalt just like that on the other southern continents. Sure enough, Antarctic explorers did find the same series and fossils, which supported the prediction and plate tectonics theory.

DRIFTING CONTINENTS, CHANGING SEAS

Let's take stock. In the remote past, crustal movement put huge land masses on collision courses. In time, the masses converged to form supercontinents, which later split at deep rifts and formed new ocean basins. Gondwana drifted south from the tropics, across the south pole, then north until it slowly piled into other land masses. The outcome? A supercontinent, Pangea, that extended from pole to pole, with a single world ocean lapping against its coasts. All the while, erosive forces of water and wind resculpted the land surface. Asteroids and meteorites smacked into the crust. The impacts and their aftermath had long-term effects on global temperature and climate.

Such changes on land and in the ocean and atmosphere influenced life's evolution. Imagine early life in shallow, warm waters along continents. Shorelines vanished as continents collided and wiped out many lineages. Yet, even as habitats vanished, new ones opened up for survivors—and evolution took off in new directions.

Over the past 3.8 billion years, gradual as well as catastrophic events changed the Earth's crust, the atmosphere, and the ocean—and influenced the evolution of life.

17.4 Evidence From Comparative Morphology

Read Me First!
and watch the narrated animation
on morphological divergence

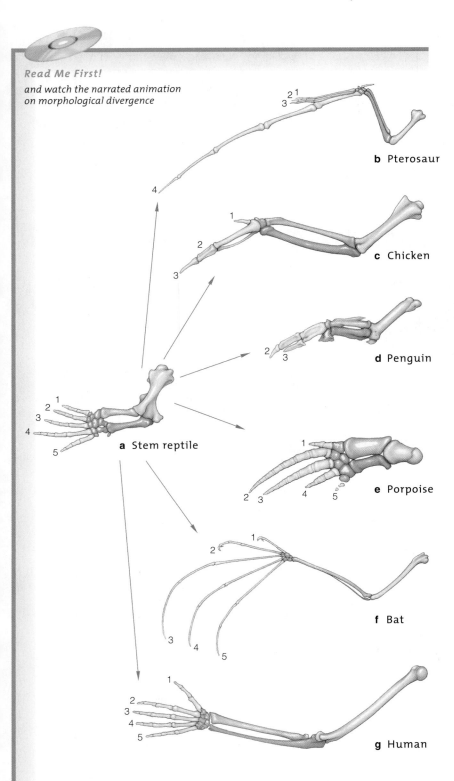

b Pterosaur

c Chicken

d Penguin

a Stem reptile

e Porpoise

f Bat

g Human

Figure 17.9 Morphological divergence among vertebrate forelimbs, starting with bones of a stem reptile. Similarities in the number and position of skeletal elements were preserved when diverse forms evolved. Some bones were lost over time (compare the numbers 1 through 5). The drawings are not to the same scale.

Evolution, remember, simply means heritable changes in lines of descent. Comparisons of the body form and structures of major groups of organisms yield clues to evolutionary trends.

Comparative morphology is the study of body forms and structures of major groups of organisms, such as vertebrates and flowering plants. Often it reveals similarities in one or more body parts that suggest inheritance from a common ancestor. Such body parts are **homologous structures** (*homo*– means the same). In such cases, genetically based similarities are there, even when different kinds of organisms are using the structures for different functions.

MORPHOLOGICAL DIVERGENCE

Populations of a species genetically diverge when gene flow ends between them (Chapter 16). In time, some morphological traits that help to define their species commonly diverge, also. Change from the body form of a common ancestor is a major macroevolutionary pattern called **morphological divergence**. The Greek *morpho*– means body form.

Even if the same body part of two related species became dramatically different, some other aspects of the species may remain alike. A careful look beyond unique modifications may reveal the shared heritage. For example, all vertebrates on land are descendants of the first amphibians. Divergences led to what we generally call reptiles, then to birds and mammals. We know about cotylosaurs, the "stem reptiles" that probably were ancestral to all of those groups. Their fossilized, five-toed limb bones tell us the cotylosaurs crouched low to the ground (Figure 17.9a). Their descendants diversified into many new land habitats. We now know a few kinds adapted to land returned to the seas when environmental conditions changed.

A five-toed limb was evolutionary clay. It became molded into different kinds of limbs with different functions. In lineages that eventually led to penguins and porpoises, it became modified into flippers used in swimming. In the lineage leading to modern horses, it became modified into long, one-toed limbs suitable for running fast. Among moles, it became stubby and useful for burrowing into dirt. Among elephants, it became strong and pillarlike, suitable for supporting a great deal of weight.

The five-toed limb also became modified into the human arm and hand. Later, a thumb evolved in opposition to the four fingers of the human hand; it was the basis of stronger and more precise motions.

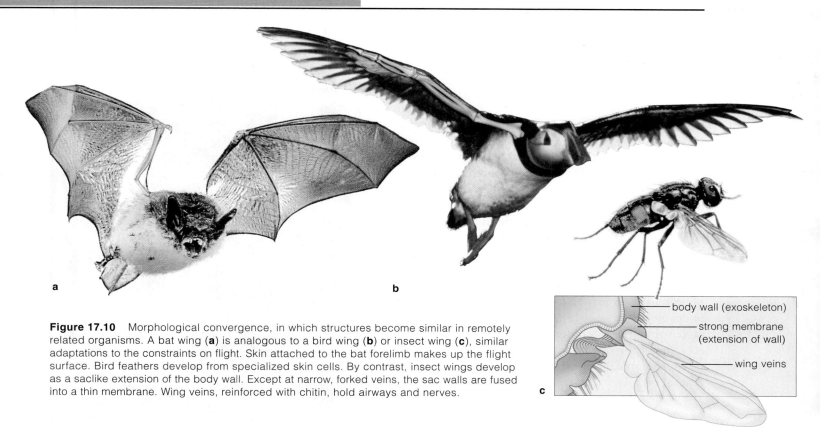

Figure 17.10 Morphological convergence, in which structures become similar in remotely related organisms. A bat wing (**a**) is analogous to a bird wing (**b**) or insect wing (**c**), similar adaptations to the constraints on flight. Skin attached to the bat forelimb makes up the flight surface. Bird feathers develop from specialized skin cells. By contrast, insect wings develop as a saclike extension of the body wall. Except at narrow, forked veins, the sac walls are fused into a thin membrane. Wing veins, reinforced with chitin, hold airways and nerves.

Even though vertebrate forelimbs are not the same in size, shape, or function from one group to the next, they clearly are alike in the structure and positioning of their bony elements. Also, the forelimbs are alike in the internal arrangement of nerves, blood vessels, and muscles that develop inside them. In addition, other comparisons of early vertebrate embryos reveal strong resemblances in patterns of bone development. Such similarities point to common ancestry.

MORPHOLOGICAL CONVERGENCE

Body parts with similar form or function in different lineages aren't *always* homologous. Sometimes they evolved independently in remote lineages. Parts that differed at first might have become similar because organisms were subjected to similar environmental pressures. **Morphological convergence** refers to cases where dissimilar body parts evolved in similar ways in evolutionarily distant lineages.

For instance, you just read about the homologous forelimbs of birds and bats. Bones aside, are bird and bat wings homologous, too? No. The flight surface of birds evolved as a sweep of feathers, all derived from skin. The forelimb structurally supports it. The flight surface of bats is a thin membrane, an extension of the skin itself. The bat wing is attached to reinforcing bony elements inside the forelimb (Figure 17.10*a,b*).

The insect wing, too, resembles bird and bat wings in its function—flight. Is it homologous with them? No. This wing develops as an extension of an outer body wall reinforced with chitin. It has no underlying bony elements to support it (Figure 17.10*c*).

The differences between bat, bird, and insect wings are evidence that each of these animal groups adapted independently to the same physical constraints that govern how a wing can function in the environment. The wings of all three are **analogous structures**. They are not modifications of comparable body parts in different lineages. They are three different responses of dissimilar parts to the same challenge. The Greek *analogos* means similar to one another.

With morphological divergence, comparable body parts became modified in different ways in different lines of descent from a common ancestor.

Such divergences resulted in homologous structures. Even if these body parts differ in size, shape, or function, they have an underlying similarity because of shared ancestry.

In morphological convergence, dissimilar body parts became similar in independent lineages that are not closely related.

Such body parts are analogous structures. They became similar only as a result of similar pressures; ancestry had nothing to do with it.

17.5 Evidence From Patterns of Development

Comparing the patterns of embryonic development often yields evidence of evolutionary relationships.

Over evolutionary time, mutations in lines of descent have resulted in built-in constraints on how a plant or an animal embryo can grow and develop. That is why most mutations and changes in chromosomes tend to be selected against. During embryonic development, certain steps can't occur unless earlier steps precede them in expected ways. Some mutations disrupt key steps. Every so often, though, a neutral or beneficial change moves a lineage past a constraint.

Let's look at how a developmental step might shift. Homeotic genes, remember, orchestrate how the body plan of multicelled organisms develops (Chapter 14). Mutations in homeotic genes can cause an organism to develop differently, sometimes drastically so. Most of the time the alteration causes real problems, but occasionally an altered form offers a survival benefit.

For example, homeotic genes guide the timing and pattern of flower formation. A single mutation in one of these genes, *Apetala1*, causes field mustard flowers to develop male floral structures, or anthers, where there are supposed to be petals (Figure 17.11*a*). In the laboratory at least, the abundantly anthered mutant is notably fertile. This master gene has been found in many other flowering plants (Figure 17.11*d*).

As another example, embryos of some vertebrate lineages are alike in early stages. Tissues form in similar ways as cells divide, differentiate, and interact. The gut and heart, bones, skeletal muscles, and other parts

Figure 17.11 How a single mutation in a plant homeotic gene affects flower form and function.

(**a**) Mutation in the *Apetala1* gene in field mustard (*Brassica oleracea*) results in the formation of a truly distorted flower. (**b**) Normal *Brassica* flower. (**c**) The *Apetala1* mutation in mouse-ear cress (*Arabidopsis thaliana*) results in flowers that have no petals. (**d**) Normal *Arabidopsis* flower.

Read Me First!
and watch the narrated animation on mutation and proportional changes

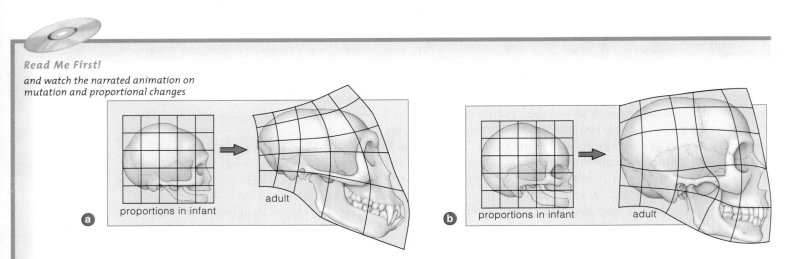

Figure 17.12 Differences between two primates, a possible outcome of mutations that changed the timing of steps in the body's development. The skulls are depicted as paintings on a rubber sheet divided into a grid. Stretching both sheets deforms the grid in a way that corresponds to differences in growth patterns between these primates. (**a**) Proportional changes in chimpanzee skull, and (**b**) human skull.

Figure 17.13 How many legs? Mutations in ancestral genes may help explain why animals differ in the number of legs and other appendages. *Dll* is a homeotic gene that initiates limb development, and other genes control its expression. Fluorescent *green* reveals *Dll* expression in (**a**) a velvet walking worm, and (**b**) a sea star; and (**c**) *blue* in a mouse embryo's foot. (**d**) Cambrian legs.

develop in orderly spatial patterns that are strikingly similar among these groups.

How, then, did adults of different groups get to be so different? We can expect that heritable changes in the onset, rate, or completion of developmental steps led to many of the differences. Some changes could have increased or decreased relative sizes of tissues and organs. Some changes could have put a stop to growth during a juvenile stage. In some cases, adults with some juvenile traits are still functional.

Altered growth rates might have caused the major proportional differences between the chimpanzee and human skull bones (Figure 17.12). For humans, facial bones and skull bones around the brain grow at fairly consistent rates, from infant to adult. The growth rate is faster for chimp facial bones, so the proportions of an infant skull and an adult skull differ significantly.

Did transposons cause some of the variation among lineages? As you read in Section 13.4, these short DNA segments can spontaneously and repeatedly slip into new places in genomes. Depending on where they end up, they can have powerful effects on gene expression.

Only primates carry the 300 base-pair transposons called *Alu* elements, and they have done so for at least 30 million years. *Alu* elements are noncoding, yet they have sequences that resemble intron–exon splice signals. When inserted into coding regions of DNA, they cause alternative splicing of genes. *Alu* elements have had profound effects on the expression of genes for estrogen, thyroid hormones, and other essential proteins that control growth and development, so we can expect that they were pivotal in primate evolution.

About one million *Alu* elements make up more than 10 percent of the human genome. Ancestors of humans and chimpanzees diverged between 6 and 4 million years ago, but more than 98 percent of human and chimpanzee DNA remains identical. Something about the remaining 2 percent accounts for the differences. Uniquely positioned *Alu* elements may be factors.

As a final example, appendages as diverse as crab legs, beetle legs, butterfly wings, sea star arms, fish fins, and mouse feet start out as buds from the body surface. The buds form wherever the *Dll* gene product is expressed. This product is a signal for clusters of dividing embryonic cells to "stick out from the body" in an expected pattern, as in Figure 17.13. Normally, *Hox* genes help sculpt the body by suppressing *Dll* expression where appendages aren't supposed to form.

The *Dll* gene is expressed in similar ways across many phyla, which is a strong case for its ancient origin. Indeed, some Cambrian fossils suggest that its early expression was unrestricted (Figure 17.13*d*). Over time, layers of gene controls evolved, resulting in the variable numbers and locations of appendages we see today. Control extends to all complex animals, including humans and other vertebrates.

Similarities in patterns of development are often clues to an evolutionary relationship among plant and animal lineages.

Heritable changes that alter key steps in a developmental program may be enough to bring about major differences between adult forms of related lineages. Transposons and single gene mutations can bring about such changes.

17.6 Evidence From Biochemistry

All species are a mix of ancestral and novel traits, which include biochemical traits. The kinds and numbers of traits they do or don't share are clues to relationships.

Each species, recall, has its own DNA base sequence, which encodes instructions for making RNAs and then proteins. We can expect that a number of genes have mutated over time in each line of descent. In addition, we can expect that fewer mutations have accumulated in lineages that originated recently compared to those that evolved much earlier in time. Because of this, the RNA and proteins of closely related species will be more similar than those of distantly related ones.

Identifying biochemical similarities and differences among species is now rapid, thanks to methods of automated gene sequencing (Section 15.5). Extensive sequence data of many genomes and proteins are compiled in internationally accessible databases. With such data, we know (for example) that 31 percent of the 6,000 yeast cell genes have counterparts in our genome. So do 40 percent of the 19,023 roundworm genes and 50 percent of the fruit fly genes.

PROTEIN COMPARISONS

When two species have many proteins with similar or identical amino acid sequences, we can expect them to be close relatives. When most of the sequences differ a lot, many mutations have accumulated, so a long time must have passed since the two shared a common ancestor.

A few essential genes have evolved very little; they are highly *conserved* across diverse species. One of them encodes cytochrome *c*. This protein component of electron transfer chains occurs in species that range from aerobic bacterial species to humans. In humans, its primary structure consists of only 104 amino acids. Figure 17.14 shows the striking similarity between the entire amino acid sequences for cytochrome *c* from a yeast, a plant, and an animal. And think about this: The *entire* amino acid sequence of human cytochrome

c is identical to that of chimpanzee cytochrome *c*. It differs by only 1 amino acid in rhesus monkeys, 18 in chickens, 19 in turtles, and 56 in yeasts. With this biochemical information in hand, would you predict that humans are more closely related to chimpanzees or to rhesus monkeys? Chickens or yeast?

NUCLEIC ACID COMPARISONS

Mutations that cause structural differences between species are often dispersed through the nucleotide sequences of their DNA. Some unique alterations have accumulated in each lineage.

Nucleic acid hybridization refers to base-pairing between DNA strands from different sources (Section 15.3). In a hybrid molecule, more hydrogen bonds form between matched bases than mismatched bases; the more matched bases, the stronger the association between the strands. The amount of heat required to separate two strands of a hybrid can be used as a comparative measure of their similarity. It takes more heat to disrupt hybrid DNA of closely related species.

Evolutionary distances are still being measured by DNA–DNA hybridizations, although automated DNA sequencing now gives faster, more quantifiable results. In a version of DNA fingerprinting (Section 15.4), restriction fragments of DNA from different species can be compared after they are separated by gel electrophoresis. All of these techniques are used to compare DNA isolated from nuclei, mitochondria, and chloroplasts (Figure 17.15).

Mitochondrial DNA (mtDNA) can be used to estimate diversity in eukaryotic populations because it mutates quickly. It is inherited in entirety from one parent in sexually reproducing species (typically, the mother), so any changes between maternally related individuals are due to mutations, not recombination. Genes encoding ribosomal RNA (rDNA) are used to compare species.

Computer programs quickly compare collections of DNA sequencing data. Often, evolutionary trees

$^+$NH$_3$-gly asp val glu lys gly lys lys ile phe ile met lys cys ser gln cys his thr val glu lys gly gly lys his lys thr gly pro asn leu his gly leu phe gly arg lys thr gly gln ala pro gly tyr ser

$^+$NH$_3$-ala ser phe ser glu ala pro pro gly asn pro asp ala gly ala lys ile phe lys thr lys cys ala gln cys his thr val asp ala gly ala gly his lys gln gly pro asn leu his gly leu phe gly arg gln ser gly thr thr ala gly tyr ser

$^+$NH$_3$-thr glu phe lys ala gly ser ala lys lys gly ala thr leu phe lys thr arg cys leu gln cys his thr val glu lys gly gly pro his lys val gly pro asn leu his gly ile phe gly arg his ser gly gln ala glu gly tyr ser

Figure 17.14 Comparison of the primary structure of cytochrome *c* from a yeast (*top row*), wheat plant (*middle*), and primate (*bottom*). This protein is a vital component of electron transfer chains in cells. Its amino acid sequence has been highly conserved, even in these three evolutionarily distant lineages. The parts highlighted in *gold* are identical in all three. The probability that such a pronounced molecular resemblance resulted by chance alone is extremely low.

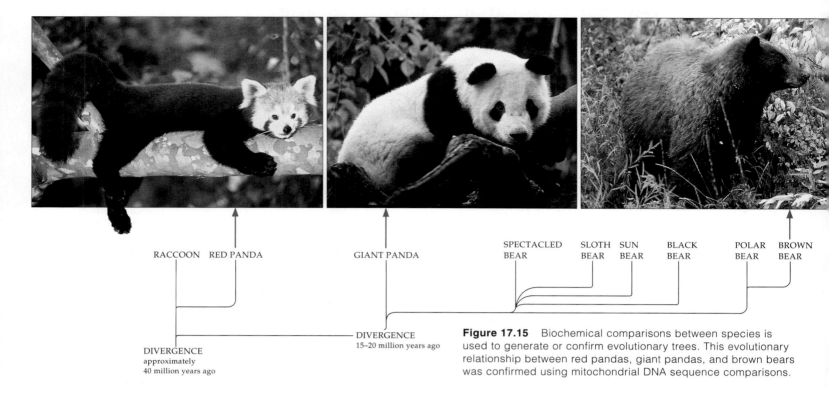

RACCOON   RED PANDA

GIANT PANDA

SPECTACLED BEAR | SLOTH BEAR | SUN BEAR | BLACK BEAR | POLAR BEAR | BROWN BEAR

DIVERGENCE
approximately
40 million years ago

DIVERGENCE
15–20 million years ago

Figure 17.15 Biochemical comparisons between species is used to generate or confirm evolutionary trees. This evolutionary relationship between red pandas, giant pandas, and brown bears was confirmed using mitochondrial DNA sequence comparisons.

based on this comparative analysis have reinforced morphological findings and the fossil record.

However, gene transfers between species can slant the results. For example, after hybridization between two different species of plants, hybrid offspring may cross back to either parental species, thus transferring genes from one species into the other. Gene swapping is rampant among prokaryotic species.

MOLECULAR CLOCKS

Some researchers estimate the timing of divergence by comparing the numbers of neutral mutations in highly conserved genes (Section 13.4). Because such mutations have little or no effect on the individual's survival or reproduction, we can expect that neutral mutations have accumulated in conserved genes at a fairly constant rate.

The addition of neutral mutations to the DNA of a given lineage has been likened to the predictable ticks of a **molecular clock**. Turn the hands of such a clock back, so that the total number of ticks unwind down through past geologic intervals. Where the last tick stops, that is the approximate time when molecular, ecological, and geographic events put the lineage on its unique evolutionary road.

How are molecular clocks calibrated? The number of differences in DNA base sequences or amino acid sequences between species can be plotted against a series of branch points inferred from the fossil record. Graphs like this may reflect relative divergence times among species, phyla, and other groups.

Biochemical similarity is greatest among the most closely related species and smallest among the most remote.

ala ala asn lys asn lys gly ile ile trp gly glu asp thr leu met glu tyr leu glu asn pro lys lys tyr ile pro gly thr lys met ile phe val gly ile lys lys lys glu glu arg ala asp leu ile ala tyr leu lys lys ala thr asn glu-COO⁻

ala ala asn lys asn lys ala val glu trp glu glu asn thr leu tyr asp tyr leu leu asn pro lys lys tyr ile pro gly thr lys met val phe pro gly leu lys lys pro gln asp arg ala asp leu ile ala tyr leu lys lys ala thr ser ser-COO⁻

asp ala asn ile lys lys asn val leu trp asp glu asn asn met ser glu tyr leu thr asn pro lys lys tyr ile pro gly thr lys met ala phe gly gly leu lys lys glu lys asp arg asn asp leu ile thr tyr leu lys lys ala cys glu-COO⁻

17.7 Reproductive Isolation, Maybe New Species

Speciation is a macroevolutionary event based on reproductive isolation and microevolution of a population or subpopulation. Several mechanisms prevail.

Species is a Latin word that simply means "kind," as in "one particular kind of duck." Long ago, naturalists defined species mainly in terms of morphological traits, but common sense told them to consider other factors as well. For instance, individuals of the same species often look different because they grew under different conditions. The plant in Figure 17.16 is an example.

Actually, individuals of *most* species vary greatly in morphological details. But they are all more closely related to one another than to any other species. We now identify unique traits by using morphological or DNA sequence comparisons, and by identifying the basic functions that isolate a species from others.

Reproduction is such a basic, defining function. It is the core of the **biological species concept**. Ernst Mayr, an evolutionary biologist, phrased it this way: Species are groups of interbreeding natural populations that are reproductively isolated from other such groups. In his view, it doesn't matter how much the phenotypes vary. Populations belong to the same species as long as individuals share traits that let them interbreed and produce fertile offspring.

Mayr's species concept applies only to sexually reproducing organisms. If we subscribe to his view, speciation is the attainment of reproductive isolation. Isolation does not happen on purpose. *Any structural, functional, or behavioral trait that favors reproductive isolation is simply a by-product of genetic change.*

Gene flow alone counters genetic changes between populations of a species. **Gene flow**, remember, is the movement of alleles into and out of a population by immigration and emigration. This microevolutionary process helps maintain a shared pool of alleles.

If gene flow between populations or subpopulations of a species ends, **genetic divergence** follows. By this process, gene pools of isolated populations slowly or quickly diverge, for mutation, natural selection, and genetic drift happen independently in each one.

Regardless of differences in the duration or details, **reproductive isolating mechanisms** kick in at some point. These heritable aspects of body form, function, or behavior prevent interbreeding between divergent populations. Some prevent successful pollination or mating between individuals, so hybrid zygotes do not form. Others prevent gametes from forming or block fertilization. Still others kill hybrids or make them weak or infertile (Figures 17.17 and 17.18). Let's start with the *prezygotic* mechanisms listed in Figure 17.18a,b.

Figure 17.16 Morphological differences between plants of the same species (*Sagittaria sagittifolia*) growing (**a**) in water and (**b**) on land. The leaf shapes are responses to different environmental conditions, not to different genetic programs.

Figure 17.17 (**a**) Mechanical isolation. Notice the fit between the reproductive parts of this zebra orchid and the wasp body. Few pollinating insects fit as precisely on this flower's landing platform.

(**b**) Behavioral isolation. Courtship displays precede sex among blue-footed boobies. Individuals recognize tactile, visual, and acoustical signals, including a prancing dance followed by back arching, skyward pointing bill and exposed throat, and wing spreading.

(**c**) Temporal isolation. *Magicicada septendecim*, a periodical cicada that matures underground and emerges to reproduce every 17 years. Its populations often overlap the habitats of a sibling species (*M. tredecim*), which reproduces every 13 years. Adults live only a few weeks.

Mechanical isolation. Incompatibility between body parts keeps potential mates or pollinators mechanically isolated. Two sage species keep pollen to themselves by attracting different insect pollinators. The pollen-bearing stamens of one extend from a nectar cup, above petals that form a big platform for the big pollinators. Small bees landing on it don't often touch the stamens and pick up pollen. The platform of the other sage species is too small to hold the big pollinators. Figure 17.17*a* shows another example.

Behavioral isolation. Behavioral differences bar gene flow between related species in the same vicinity. For instance, before male and female birds copulate, they often engage in courtship displays (Figure 17.17*b*). A female is genetically prewired to recognize distinctive singing, head bobbing, wing spreading, or prancing by a male of her species as an overture to sex. Females of different species usually ignore his behavior.

Temporal isolation. Interbreeding might be possible in diverging populations, but not if they differ in timing of reproduction. Three 17-year cicada species often occupy the same habitat in the eastern United States. They differ in form and behavior, but they all mature underground and feed on juicy roots. Every 17 years, they emerge to reproduce (Figure 17.17*c*). Each species has a *sibling* species—one that resembles it in form and behavior. The siblings emerge on a 13-year cycle. Thus, each species and its sibling can't get together except once in every 221 years!

Ecological isolation. Populations occupying different microenvironments in a habitat might be ecologically isolated. In the open forests of seasonally dry foothills of the Sierra Nevada are two manzanita species. One lives at elevations between 600 and 1,850 meters, the other between 750 and 3,350 meters. These species rarely hybridize, and only where the ranges overlap. Water-conserving mechanisms help them through dry seasons. But one species is adapted to sheltered sites where water stress isn't intense. The other species lives in drier, more exposed sites on rocky hillsides, so cross-pollination is unlikely.

Gamete mortality. Gametes of different species may have molecular incompatibilities. Example: If pollen lands on a plant of another species, it usually does not even recognize the molecular signals that trigger the germination of same-species pollen.

Postzygotic isolating mechanisms may act while an embryo is developing. Unsuitable interactions among genes or gene products cause early death, sterility, or weak hybrids with low survival rates. Some hybrids are sturdy but sterile. Mules, the offspring of a female horse and male donkey, are examples of such hybrids.

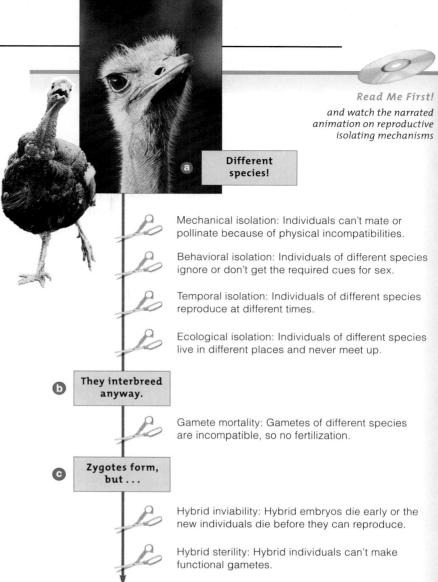

Read Me First!
and watch the narrated animation on reproductive isolating mechanisms

a Different species!

Mechanical isolation: Individuals can't mate or pollinate because of physical incompatibilities.

Behavioral isolation: Individuals of different species ignore or don't get the required cues for sex.

Temporal isolation: Individuals of different species reproduce at different times.

Ecological isolation: Individuals of different species live in different places and never meet up.

b They interbreed anyway.

Gamete mortality: Gametes of different species are incompatible, so no fertilization.

c Zygotes form, but . . .

Hybrid inviability: Hybrid embryos die early or the new individuals die before they can reproduce.

Hybrid sterility: Hybrid individuals can't make functional gametes.

No offspring, sterile offspring, or weak offspring that die before reproducing

Figure 17.18 When reproductive isolating mechanisms halt interbreeding. There are barriers to (**a**) getting together, mating, or pollination, (**b**) successful fertilization, and (**c**) survival, fitness, or fertility of hybrid embryos or offspring.

A species is one or more populations of individuals having a unique common ancestor. Its individuals share a gene pool, produce fertile offspring, and remain reproductively isolated from individuals of other species.

Speciation is the process by which daughter species form from a population or subpopulation of a parent species. The process varies in its details and duration.

In two or more isolated populations, mutation, natural selection, and genetic drift operate independently. The buildup of genetic differences may lead to speciation.

Reproductive isolating mechanisms may evolve simply as by-products of the genetic changes. They are heritable traits that, one way or another, prevent interbreeding.

17.8 The Main Model for Speciation

Three models for speciation differ in their basic premise of how populations become reproductively isolated.

THE NATURE OF ALLOPATRIC SPECIATION

If we assume that genetic changes leading to a new species require *physical separation* between populations, then allopatry might be the main speciation route. By the model for **allopatric speciation**, a physical barrier arises and stops gene flow between two populations or subpopulations of a species. (*Allo*– means different; *patria* can be taken to mean homeland.) Reproductive isolating mechanisms evolve in the two populations. In time, speciation is over. Interbreeding is no longer possible even when the new species moves back to the area still occupied by the parent species.

Whether a geographic barrier can effectively block gene flow depends on an organism's means of travel (deliberate or accidental), how fast it can travel, and whether it is inclined to disperse. Some measurable distance often separates populations of a species, so gene flow among them is an intermittent trickle, not a steady stream. Barriers can arise abruptly and shut off the trickles. In the 1800s, a major earthquake buckled part of the Midwest. The Mississippi River changed course; it cut through habitats of insects that couldn't swim or fly. Gene flow across the previously adjacent habitats stopped in an instant of geologic time.

As the fossil record suggests, geographic isolation also occurs over great spans of time. This happened after glaciers advanced down into North America and Europe during the ice ages and cut off populations of plants and animals from one another. When glaciers retreated, the descendants of the separate populations met up. Although related by descent, some were no longer reproductively compatible. They had evolved into new species. Genetic divergence was not as great between other separated populations, so descendants are still interbreeding. Reproductive isolation in their case was incomplete; speciation did not follow.

Also, remember how Earth's crust is fractured into gigantic plates? Slow but colossal movements of the plates have altered the configurations of land masses. As Central America was forming, part of an ancient ocean basin was uplifted and became a land bridge—the Isthmus of Panama. Some camelids moved across it into South America. Geographic separation led to new species, the llamas and vicunas (Figure 17.19).

ALLOPATRIC SPECIATION ON ARCHIPELAGOS

An **archipelago** is an island chain some distance from a continent. Some are so close to the mainland that gene flow is more or less unimpeded; there is little if any speciation. The Florida Keys are like this. But the isolated ones are another matter (Figure 17.20). The Hawaiian Archipelago is nearly 4,000 kilometers west from the California coast, and the Galápagos Islands, 900 kilometers or so from the Ecuadoran coast. The islands of both chains are only the tops of immense

Figure 17.19 Allopatric speciations. The earliest camelids, no bigger than a jackrabbit, evolved in the Eocene grasslands and deserts of North America. By the end of the Miocene, they included the now-extinct *Procamelus*. The fossil record and comparative studies indicate that this may have been the common ancestral stock for llamas (**a**), vicunas (**b**), and camels (**c**). One of the descendant lineages dispersed into Africa and Asia and evolved into modern camels. A different lineage, ancestral to the llamas and vicunas, dispersed into South America after gradual crustal movements formed a land bridge between the two continents.

Late Eocene paleomap, before a land bridge formed between North and South America

Read Me First!
and watch the narrated animation on allopatric speciation

a A few individuals of a species on the mainland reach isolated island 1. Speciation follows genetic divergence in a new habitat.

b Later in time, a few individuals of the new species colonize nearby island 2. In this new habitat, speciation follows genetic divergence.

c Speciation may also follow colonization of islands 3 and 4. And it may follow invasion of island 1 by genetically different descendants of the ancestral species.

The shared ancestor of all of Hawaii's honeycreepers probably looked like this housefinch (*Carpodacus*).

Akepa (*Loxops coccineus*) · Akekee (*L. caeruleirostris*) · Nihoa finch (*Telespyza ultima*) · Palila (*Loxioides bailleui*)

Maui parrotbill (*Pseudonestor xanthrophrys*) · Alauahio (*Paroreomyza montana*) · Kauai Amakihi (*Hemignathus kauaiensis*) · Akiapolaau (*H. munroi*)

d Akohekohe (*Palmeria doli*) · Apapane (*Himatione sanguinea*) · Iiwi (*Vestiaria coccinea*)

Figure 17.20 (a–c) One example of allopatric speciation on an isolated archipelago. Can you envision other possibilities? (d) Some of the more than twenty species of Hawaiian honeycreepers. Their bills are adapted to diverse foods, such as insects, seeds, fruits, and floral nectar cups.

volcanoes that formed long ago on the seafloor. When the volcanoes first broke the surface of the sea, their fiery surfaces were devoid of life.

In one view, flotsam or winds brought mainland finches to one of the Galápagos Islands. Descendants colonized other islands in the chain, where habitats and selection pressures differed. Divergences within and between islands fostered episodes of allopatric speciation. Later, new species returned to the islands of their ancestors. The distances between islands are enough to foster divergences, but not enough to stop the occasional colonizers.

The youngest island in the Hawaiian Archipelago, Hawaii, formed less than a million years ago. Here alone we see diverse habitats, ranging from cooled lava beds, rain forests, and alpine grasslands to snow-capped volcanoes. When the first birds arrived, they found a buffet of fruits, seeds, nectars, tasty insects, and not many competitors for them. The near absence of competition spurred rapid speciations into vacant adaptive zones. Figure 17.20*d* hints at the variation among Hawaiian honeycreepers. Today, these and thousands of other species of animals and plants that originated in the archipelago are found nowhere else. As another example of their speciation potential, the Hawaiian Islands combined make up less than 2 percent of the world's land masses. Yet they are home to 40 percent of all fruit fly (*Drosophila*) species.

By an allopatric speciation model, some type of physical barrier intervenes between populations or subpopulations of a species and prevents gene flow among them. It favors genetic divergence that ends in speciation.

17.9 Other Speciation Models

There is evidence that some species have arisen and are maintained by less common mechanisms in which environmental barriers do not play a role.

SYMPATRIC SPECIATION

By the model for **sympatric speciation**, a species may form *within* the home range of an existing species, in the absence of a physical barrier. (*Sym–* means together with, as in "together with others in the homeland.")

EVIDENCE FROM CICHLIDS IN AFRICA In Cameroon, West Africa, cichlid populations may have undergone sympatric speciation in lakes, the basins of which are the collapsed cones of small volcanoes. Many cichlid species coexist in each lake. They probably colonized the lakes before volcanic action severed connections with a nearby river system.

Figure 17.21 A small, isolated crater lake in Cameroon, West Africa, where different species of cichlids may have originated by way of sympatric speciation.

Researchers analyzed mitochondrial DNA from all eleven cichlid species living in one lake. They did the same for all nine species coexisting in the other lake as the basis for comparison. The nucleotide sequences are all similar. The sequences from related cichlid species in nearby lakes and rivers are not as similar.

Allopatric populations could not have evolved. The physical and chemical conditions, even shorelines, are uniform in all lakes (Figure 17.21). Gene flow is absent; tiny fish-free creeks trickling from higher elevations are the only way in. Also, cichlids are good swimmers; individuals of different species often meet up.

Because clusters of species are more closely related to one another than to species anywhere else, we can expect them to share a common ancestor. Because of their restricted distribution, we can expect that they formed in the same lake. They do show small degrees of *ecological* separation. Feeding preferences put species in different places. Some feed in open waters, others at the lake bottom. Yet they all *breed* close to the lake bottom, in sympatry. Was the small-scale ecological separation enough to promote sexual selection among potential mates? Maybe. Over time, it may have led to reproductive isolation, then speciation.

POLYPLOIDY'S IMPACT Sympatric speciation has been common among flowering plants; about half of all known species are polyploid. In **polyploidy**, somatic cells have three or more of each type of chromosome characteristic of the species (Figures 17.22 and 17.23). Such changes can happen when chromosomes do not separate properly at meiosis or mitosis (Section 11.8). Changes also happen when a germ cell replicates its DNA but fails to divide, then functions as a gamete.

Polyploidy induces rapid speciation in plants. Often a polyploid plant has fertile offspring when crossed with identical polyploids; a new species has formed.

Speciation may have been fast for many flowering plants that self-fertilize or hybridize. If their offspring were polyploid and extra chromosomes paired with

Figure 17.22 Love those polyploids! Among them are five cotton species, marigolds, watermelons, sugarcane, and coffee plants, which have 22, 33, 66, or 88 chromosomes.

Triticum monococcum (einkorn) x Unknown species of wild wheat CROSS-FERTILIZATION, FOLLOWED BY SPONTANEOUS CHROMOSOME DOUBLING *T. turgidum* (wild emmer) x *T. tauschii* (a wild relative) *T. aestivum* (one of the common bread wheats)

14AA x 14BB → 14AB ──────→ 28AABB x 14DD → 42AABBDD

a By 11,000 years ago, humans were cultivating wild wheats. Einkorn has a diploid chromosome number of 14 (two sets of 7). It probably hybridized with another wild wheat species having the same number of chromosomes.

b About 8,000 years ago, polyploidy arose in plants that may have evolved from sterile, self-fertilizing AB hybrid wheat. (Wild emmer is tetraploid, or AABB; it has two sets of 14 chromosomes). The polyploids are fertile; the A and B chromosomes have partners at meiosis.

c An AABB plant probably hybridized with *T. tauschii*, a wild relative of wild emmer. Its diploid chromosome number must have been 14 (two sets of 7 DD). Hybrids now include common bread wheats. One has a chromosome number of 42 (six sets of 7 AABBDD).

Figure 17.23 Presumed sympatric speciation in wheat by polyploidy and hybridizations. Wheat grains 11,000 years old and diploid wild wheats have been found in the Near East.

each other during meiosis, the extra set of genes may have done no harm. Bread wheat may have arisen by way of hybridization and polyploidy (Figure 17.23).

Polyploid mammals are rare, maybe because of failures in dosage compensation. By this normal event, genes on sex chromosomes are expressed at the same levels in females and males. Remember, one of two X chromosomes in all female mammals gets inactivated (Section 14.4), so only one X chromosome is active in males *and* females. Polyploidy changes that. Alleles on extra sex chromosomes are active, which upsets the balance of gene expression necessary for normal functioning. Typically this has lethal consequences. Plants generally do not have sex chromosomes, so polyploidy does not always cause bad imbalances.

PARAPATRIC SPECIATION

Parapatric speciation might happen when different selection pressures operating across a broad region affect populations that are in contact along a common border. Hybrids that form in the contact are less fit than individuals on either side of the contact zones. Because hybrids are selected against, they appear in the hybrid zone only (Figure 17.24).

By a sympatric speciation model, daughter species arise from a group of individuals within an existing population. Polyploid flowering plants probably formed this way.

By a parapatric speciation model, populations maintaining contact along a common border evolve into distinct species.

T. barretti
hybrid zone
T. anophthalmus

Figure 17.24 Parapatric speciation in rare Tasmanian velvet worms that live in adjoining regions of northeastern Tasmania, as shown in the map. (**a**) The giant velvet worm, *Tasmanipatus barretti* and (**b**) the blind velvet worm, *T. anophthalmus*.

(**c**) The habitats of the two kinds of walking worms overlap very little in a hybrid zone. The hybrids are sterile, which may be the main reason these two species are maintaining their separate identities in the absence of an obvious physical barrier.

17.10 Patterns of Speciation and Extinction

All species, past and present, are related by descent. They share genetic connections through lineages that extend back in time to the molecular origin of the first proto-cells. Later chapters focus on evidence in support of this view. Here, simply start thinking about ways to interpret the large-scale histories of species.

BRANCHING AND UNBRANCHED EVOLUTION

The fossil record reveals two patterns of evolutionary change, one branching, the other unbranched. The first is known as **cladogenesis** (from *klados*, branch; and *genesis*, origin). In this pattern, a lineage splits, the populations become genetically isolated, and so they diverge. Biologists generally accept cladogenesis as the most probable pattern of speciation, and it is the one we described earlier in the chapter.

In the second pattern, **anagenesis**, changes in allele frequencies and morphology accumulate in one line of descent. (In this context, *ana–* means renewed.) Because gene flow never stops among populations, directional changes are confined to that lineage alone.

EVOLUTIONARY TREES AND RATES OF CHANGE

Evolutionary trees summarize information about the continuity of relationships among groups. The sketch below is one way to start thinking about how these diagrams are constructed. Each branch represents a single line of descent from a common ancestor. Each *branch point* represents a time of genetic divergence and speciation, as brought about by natural selection or other microevolutionary processes:

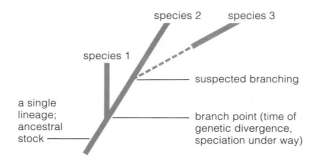

When plotted against time, a branch that ends before the present (the "treetop") signifies that the lineage is extinct. A dashed line signifies we know something about a lineage but not exactly where it fits in the tree.

The **gradual model of speciation** holds that species originate by small morphological changes over long

Figure 17.25 A series of fossilized foraminiferan shells from a vertical sequence of sedimentary rock layers. The first shell (*bottom*) is 64.5 million years old. The most recent (*top*) is 58 million years old. Detailed analysis of shell patterns confirmed the evolutionary order matches the geological sequence.

spans of time. The model fits with many fossil sequences. As one example, sedimentary rock layers often hold vertical sequences of fossilized shells of foraminiferans, and these reflect gradual morphological change (Figure 17.25).

The **punctuation model of speciation** is a different way to view patterns of speciation. Most morphological changes are said to evolve in a rather brief period when populations are just starting to diverge— within hundreds or thousands of years at most. Bottlenecks, the founder effect, directional selection, genetic drift, or some combination of all these processes promote rapid speciation. Daughter species recover fast from the adaptive wrenching, then don't change much over the next 2 to 6 million years.

Stability has indeed prevailed for all but 1 percent of the history of most lineages. And many parts of the fossil record reveal episodes of abrupt change. But both models help explain speciation patterns. Changes have been gradual, abrupt, or both. Species originated at different times and have differed in how long they last. Some have not changed much over millions of years; others were the start of adaptive radiations.

ADAPTIVE RADIATIONS

An **adaptive radiation** is a burst of divergences from a single lineage that led to many new species, each adapted to an unoccupied or new habitat, or to using novel resources. The honeycreepers in Figure 17.20 are examples. In the past, species of one lineage typically radiated into vacant **adaptive zones**. Think of each adaptive zone as a way of life, such as "burrowing in deep seafloor sediments" or "catching winged insects in the air at night." Species must have

platypus, other monotremes
kangaroos, other marsupials
elephants, other proboscideans
manatees
anteaters
armadillos
shrews, other insectivores; bats
carnivores
whales, dolphins
deer, other artiodactyls
horses, other perissodactyls
primates
rodents
rabbits

CENOZOIC

MESOZOIC

ancestral mammal

Figure 17.26 Adaptive radiation of mammals. Branch widths indicate the range of biodiversity at different times. Mammals arose 220 million years ago but did not start a great radiation until after the K–T impact removed the last of the dinosaurs. Not all lineages are shown. The 4,000 existing species include tiny shrews, flying bats, and giant whales.

The photograph shows a fossil of *Eomaia scansoria* (Greek for ancient mother climber). About 125 million years ago, this insectivore lived in low shrubs and branches. At this writing, it is the earliest placental mammal we know about.

physical, evolutionary, or ecological access to new adaptive zones before they can occupy them (Figure 17.26). What does this mean?

Physical access means a lineage happens to be there as adaptive zones open up. For example, mammals were once distributed in uniformly tropical regions of a great continent. The continent broke up into huge land masses that slowly drifted apart. Habitats and resources changed in many different ways in different places, setting the stage for independent radiations.

Evolutionary access means a lineage has changed in structure or function in such a way that it can exploit the environment in a new or more efficient manner. Such modifications are known as **key innovations**. As an example, when some vertebrate forelimbs evolved into wings, novel adaptive zones opened for ancestors of birds and bats.

Ecological access means a lineage has the means to enter one or more vacated adaptive zones, or that it is competitive enough to displace resident species.

EXTINCTIONS—THE END OF THE LINE

An **extinction** is an irrevocable loss of a species. The fossil record gives evidence of twenty or more **mass** **extinctions**—catastrophic losses of entire families or other major groups. Past extinctions differed in size. For example, 250 million years ago, 95 percent of all known species were abruptly lost; in other times, a few groups were lost. Afterward, biodiversity slowly recovered as new species filled vacant adaptive zones.

Luck had a lot to do with it. Many species were devastated by global changes in climate; given their adaptations, they had nowhere else to go. And when one asteroid struck and wiped out the last dinosaurs, mammals were among the survivors that radiated into vacated adaptive zones. Asteroids, imperceptibly drifting continents, climatic change—all contributed to past patterns of extinctions and recoveries.

Lineages have changed gradually, abruptly, or both. Their member species originated at different times and have differed in how long they have persisted.

Adaptive radiations are bursts of divergences from a single lineage that gave rise to many new species, each adapted to a vacant or new habitat or to using a novel resource.

Repeated and often large extinctions happened in the past. After times of reduced biodiversity, new species formed and occupied new or vacated adaptive zones.

17.11 How Can We Organize the Evidence?

Biodiversity patterns are being revealed by taxonomy, phylogenetic reconstruction, and classification. This type of organization makes it easier for us to assimilate information about species and their heritage.

NAMING, IDENTIFYING, AND CLASSIFYING SPECIES

Taxonomists are biologists who systematically identify and classify species, then assign each kind a two-part name. Bacteria, junipers, vanilla orchids, houseflies, humans—all have a scientific name that gives people all over the world a way to know they are talking about the same organism. The name's first part, the **genus** (plural, genera), identifies a group of similar species (Chapter 1). The second part of the name is the specific epithet. Together with the generic name, it designates one species alone. For instance, *Ursus maritimus* is the polar bear, *U. arctos* the brown bear, and *U. americanus* is the black bear. Figure 17.27 has other examples.

All **classification systems** are organized systems of retrieving information about how species fit in the big picture. Ever more inclusive groupings of species are called **higher taxa** (singular, taxon). Family, order, class, phylum, and kingdom are examples. Taxonomists use phylogeny to assign species to higher taxa. **Phylogeny** is the systematic study of evolutionary relationships among species, from the most ancestral through the divergences that led to all descendant species.

At one time, many biologists tentatively accepted a **six-kingdom system** as a way to retrieve information about species. This system recognizes the first great divergence after life originated. That divergence led to the archaea and bacteria, the only prokaryotic cells. Soon afterward, another divergence led to ancestors of single-celled eukaryotes, collectively known as protists. Later on, some protists gave rise to multicelled plants, fungi, and animals. Figure 17.28 shows the groupings.

Consensus is growing to subsume the six-kingdom system in a **three-domain system**, which better reflects evolutionary relationships (Figure 17.29).

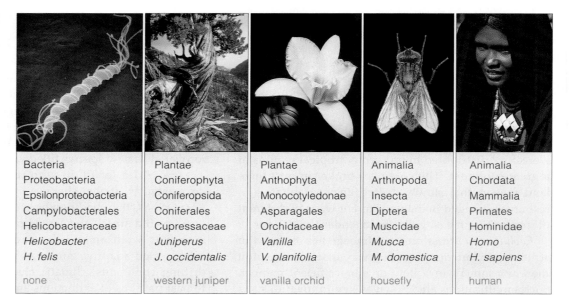

Figure 17.27 Taxonomic classification of five organisms. Each is assigned to ever more inclusive categories (higher taxa)— in this case, from species to kingdom.

KINGDOM	Bacteria	Plantae	Plantae	Animalia	Animalia
PHYLUM	Proteobacteria	Coniferophyta	Anthophyta	Arthropoda	Chordata
CLASS	Epsilonproteobacteria	Coniferopsida	Monocotyledonae	Insecta	Mammalia
ORDER	Campylobacterales	Coniferales	Asparagales	Diptera	Primates
FAMILY	Helicobacteraceae	Cupressaceae	Orchidaceae	Muscidae	Hominidae
GENUS	*Helicobacter*	*Juniperus*	*Vanilla*	*Musca*	*Homo*
SPECIES	*H. felis*	*J. occidentalis*	*V. planifolia*	*M. domestica*	*H. sapiens*
COMMON NAME	none	western juniper	vanilla orchid	housefly	human

Figure 17.28 Six-kingdom system of classification. In time, protists may be divided into more kingdoms.

Figure 17.29 A recent three-domain system of classification. Protists, plants, fungi, and animals share features that unite them in domain Eukarya.

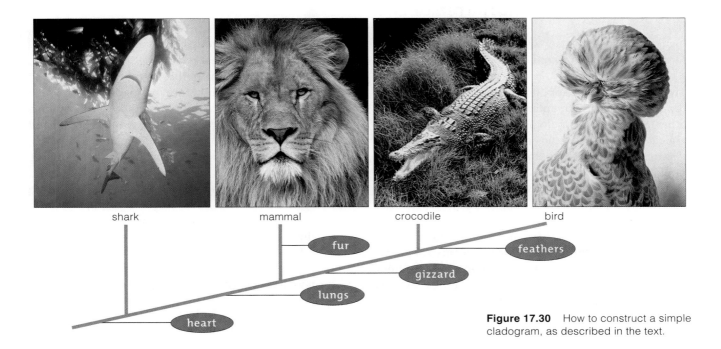

shark   mammal   crocodile   bird

fur

feathers

gizzard

lungs

heart

Figure 17.30 How to construct a simple cladogram, as described in the text.

WHAT'S IN A NAME? A CLADISTICS VIEW

Clues to phylogenetic relationships abound in fossils, morphology, biochemistry, and Earth's history. Degrees of morphological divergence help us construct tree diagrams in which branch points are the measure of relatedness. Traditional classification systems group species based on overall similarities, relying on the judgment and experience of taxonomists. In *cladistics*, species are grouped by derived traits.

A **derived trait** is a novel feature that evolved in just one species and is shared only by its descendants. In the evolutionary tree diagrams called **cladograms**, taxa are grouped by their shared derived traits. All descendants from an ancestral species in which a trait first evolved are a **monophyletic group**, which means "single tribe." A cladogram is like a time bar without absolute dates. It does not directly convey "who came from whom." Rather, it uses the position of branch points from the last shared ancestor to convey inferred evolutionary relationships (phylogenies) between taxa.

As a very simple example, sharks, crocodiles, birds, and mammals have a heart. Sharks do not have lungs, but crocodiles, birds, and mammals do. All crocodiles and birds have a specialized part of the gut called a gizzard, but no mammal has a gizzard. Birds alone have feathers; mammals alone have fur. You can use these traits to make a cladogram, as in Figure 17.30.

In this case, the bird is more closely related to the crocodile than it is to the mammal. The crocodile isn't its *ancestor*; it's a modern organism, too. But birds share

an ancestor with crocodiles that is more recent than the one they share with mammals. The more ancient ancestor makes birds, crocodiles, and mammals closer to one another than to the shark.

Cladograms reflect information from comparative biochemistry. In the next section, you will find an evolutionary tree of life based on such evidence. Bear in mind that the simple branchings don't convey how many thousands of morphological and biochemical traits were analyzed to work out the connections. For example, it incorporates information from detailed comparisons of ribosomal RNA and protein-encoding genes of major groups of organisms. More importantly, it correlates well with parts of the fossil record. Even as you read this, researchers are rethinking how parts of the record are interpreted.

Taxonomists systematically identify species and classify them into ever more inclusive groupings—the higher taxa.

Classification systems organize and simplify the retrieval of information about species. Phylogenetic systems attempt to reflect evolutionary relationships among species.

Reconstructing the evolutionary history of a given lineage is based on detailed understanding of the fossil record, morphology, life-styles, and habitats of its representatives, and on biochemical comparisons with other groups.

Recent evidence, especially from comparative biochemistry, favors grouping organisms into a three-domain classification system—the archaea, bacteria, and eukarya (protists, plants, fungi, and animals).

17.12 An Evolutionary Tree of Life

In preparation for the next unit on life's diversity, take a moment to review this evolutionary tree.

The evolutionary tree of life in Figure 17.31 is a work in progress. Armed with automated gene sequencers and other advanced technologies, evolutionary biologists are clarifying phylogenies at a remarkable pace. Once researchers have identified a monophyletic group, they rank it in the context of other known relationships.

Taxonomic ranks have only relative meaning. Within the Eukarya, plants are a monophyletic group, a *clade*. Within the plant clade are many subclades, including mosses, liverworts, and vascular plants. These in turn have subclades. Within the "vascular plants" alone are several monophyletic groups, such as ferns, conifers, cycads, and the flowering plants. Within the flowering plants are subclades such as monocots, magnoliids,

and eudicots. Sister groups, such as echinoderms and chordates, are the closest relatives.

The important point is that ranking should reflect *sets within sets*. If Eukarya form a kingdom, then green plants could be ranked a phylum, and vascular plants a class. If Eukarya form a domain, green plants could be ranked a kingdom, and vascular plants a phylum. The "kingdom" of protists is about to be split into multiple groups. This is not a problem if we accept that absolute rankings are arbitrary. Focus instead on phylogenetic patterns—sets within sets—then even a shift of that magnitude won't snap against the other branches of a reconstructed tree of life.

> *Relationships between clades or monophyletic groups continue to be clarified as a pattern of sets within sets.*

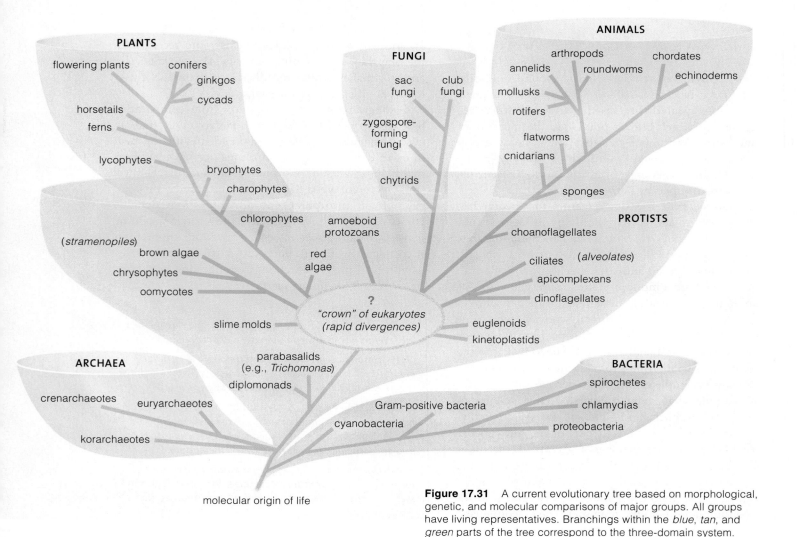

Figure 17.31 A current evolutionary tree based on morphological, genetic, and molecular comparisons of major groups. All groups have living representatives. Branchings within the *blue*, *tan*, and *green* parts of the tree correspond to the three-domain system.

Summary

Section 17.1 Fossils are physical evidence of organisms that lived in the distant past. Many are found in stacked layers of sedimentary rock, with the oldest generally near the bottom and the most recent on top. Although the fossil record is incomplete, it reveals many patterns in the history of life.

Section 17.2 Examining fossil layers provides a relative time scale that helps explain macroevolution: patterns, rates, and trends in evolution (Table 17.1). The geologic time scale uses abrupt transitions in the fossil record as boundaries. Radiometric dating has allowed us to assign absolute dates to the scale with a very small margin of error.

Section 17.3 The global distribution of land masses and fossils, magnetic patterns in volcanic rocks, and evidence of seafloor spreading from mid-oceanic ridges support the plate tectonic theory. Movements of the Earth's crustal plates shifted continents in ways that profoundly influenced life's evolution.

Section 17.4 Comparative morphology reveals evidence of evolution. It has identified homologous structures: comparable body parts that were modified in different ways in different lines of descent from a shared ancestor. Researchers must distinguish these parts from analogous structures: dissimilar body parts that became similar in independent lineages as a response to similar environmental pressures.

Section 17.5 Similarities in patterns and structures of embryonic development suggest common ancestry. Even minor genetic changes can alter the onset, rate, and completion time of developmental stages. They can have major impact on the adult form.

Section 17.6 Automated sequencing methods now help clarify evolutionary relationships. DNA, RNA, or proteins from different species are compared. Neutral mutations may accumulate in DNA at a constant rate; like ticks of a molecular clock, they help researchers calculate the divergence of one lineage from another.

Section 17.7 All populations of a species share a unique common ancestor and a gene pool, and they can interbreed and produce fertile offspring under natural conditions. If and when gene flow between them stops, reproductive isolating mechanisms typically evolve, because mutation, natural selection, and genetic drift operate independently in each population. Such divergence may give rise to a new species.

Section 17.8 By the allopatric speciation model, a geographic barrier cuts off gene flow between two or more populations and promotes genetic divergence, reproductive isolation, and finally speciation.

Section 17.9 By a sympatric speciation model, the divergence starts while populations are in physical contact. By a parapatric speciation model, it occurs between populations that share a common border.

Section 17.10 The time and duration of speciation has differed among lineages. Most species have become extinct. Mass extinctions, slow recoveries, and adaptive radiations are major macroevolutionary patterns.

Section 17.11 Each species has a unique, two-part scientific name. Taxonomy deals with identifying and naming species. Phylogenetic reconstruction uses analytical methods to infer evolutionary relationships. Classification is a way to organize information about related species.

The currently favored three-domain classification system is based largely on phylogenetic evidence. It recognizes three domains: bacteria, archaea, and eukarya. The eukarya include protists, plants, fungi, and animals.

Section 17.12 The evolutionary tree is a work in progress. More phylogenies are now being rapidly clarified. New groups are ranked in the scheme according to their relationship to other groups.

Table 17.1 Summary of Processes and Patterns of Evolution		
Microevolutionary Processes		
Mutation	Original source of alleles	Stability or change in a species is the outcome of balances or imbalances among all of these processes, the effects of which are influenced by population size and by the prevailing environmental conditions.
Gene flow	Preserves species cohesion	
Genetic drift	Erodes species cohesion	
Natural selection	Preserves or erodes species cohesion, depending on environmental pressures	
Macroevolutionary Processes		
Genetic persistence	Basis of the unity of life. The biochemical and molecular basis of inheritance extends from the origin of first cells through all subsequent lines of descent.	
Genetic divergence	Basis of life's diversity, as brought about by adaptive shifts, branching, and radiations. Rates and times of change varied within and between lineages.	
Genetic disconnect	End of the line for a species. Mass extinctions are catastrophic events in which major groups abruptly and simultaneously are lost.	

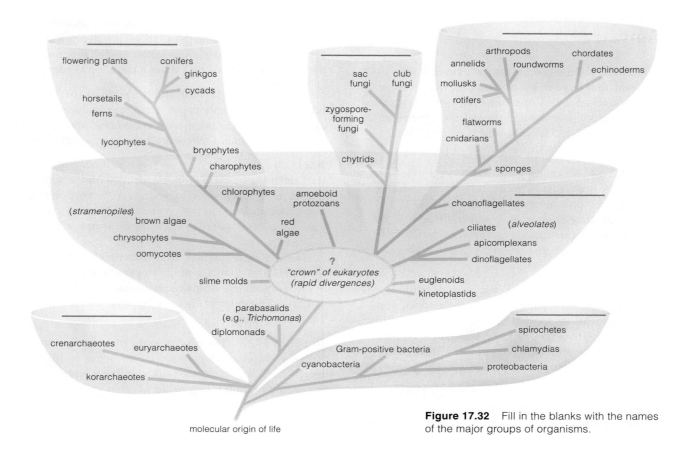

Figure 17.32 Fill in the blanks with the names of the major groups of organisms.

Labels in figure:

flowering plants, conifers, ginkgos, cycads, horsetails, ferns, lycophytes, bryophytes, charophytes, chlorophytes, amoeboid protozoans, *(stramenopiles)*, brown algae, chrysophytes, oomycotes, red algae, slime molds, *"crown" of eukaryotes (rapid divergences)*, parabasalids (e.g., *Trichomonas*), diplomonads, crenarchaeotes, euryarchaeotes, korarchaeotes, cyanobacteria, Gram-positive bacteria, molecular origin of life

sac fungi, club fungi, zygospore-forming fungi, chytrids

arthropods, chordates, annelids, roundworms, echinoderms, mollusks, rotifers, flatworms, cnidarians, sponges, choanoflagellates, ciliates, *(alveolates)*, apicomplexans, dinoflagellates, euglenoids, kinetoplastids

spirochetes, chlamydias, proteobacteria

Self-Quiz

Answers in Appendix III

1. Morphological convergences may lead to _____ .
   a. analogous structures
   b. homologous structures
   c. divergent structures
   d. both a and c

2. Sexually reproducing individuals of a species _____ .
   a. can interbreed
   b. have fertile offspring
   c. share genetic history
   d. all of the above

3. In evolutionary trees, a branch point represents a _____ , and a branch that ends represents _____ .
   a. single species; incomplete data on lineage
   b. single species; extinction
   c. time of divergence; extinction
   d. time of divergence; speciation complete

4. *Pinus banksiana, Pinus strobus,* and *Pinus radiata* are _____ .
   a. three families of pine trees
   b. three different names for the same organism
   c. three species belonging to the same genus
   d. both a and c

5. Heritable changes in DNA underlying morphological differences between lineages may have _____ .
   a. been caused by transposons
   b. affected the onset, rate, and time of development
   c. both a and b

6. Reproductive isolating mechanisms _____ .
   a. stop interbreeding
   b. stop gene flow
   c. reinforce genetic divergence
   d. all of the above

7. Individuals of a monophyletic group _____ .
   a. are all descended from an ancestral species
   b. demonstrate morphological convergence
   c. share a unique derived trait
   d. both a and c

8. A(n) _____ classification system reflects presumed evolutionary relationships.
   a. epigenetic
   b. tectonic
   c. phylogenetic
   d. both b and c

9. Increasingly inclusive taxa range from _____ to _____ .
   a. kingdom; genera and species
   b. kingdom; genera and domain
   c. genera; domain and kingdom
   d. species; kingdom and domain

10. Fill in the blanks in Figure 17.32 for the six major groups of organisms. You will soon trace the branching within each group.

11. Match these terms suitably.
   ____ phylogeny
   ____ fossils
   ____ stratification
   ____ homologous structures
   ____ cladogram
   ____ analogous structures
   ____ adaptive radiation

   a. evidence of life in distant past
   b. branching lineages
   c. similar body parts in different lineages with shared ancestor
   d. insect wing and bird wing
   e. evolutionary relationships among species, from ancestors through descendants
   f. layers of sedimentary rock
   g. burst of divergences into new habitats

Critical Thinking

1. At the end of your backbone is a coccyx, a few small bones that are fused together. Could the human coccyx be a *vestigial* structure—all that's left of the tail of some distant vertebrate ancestors? Or is it the start of a newly evolving structure? Formulate a hypothesis, then design a way to test predictions based on the hypothesis.

2. You notice several duck species in the same lake habitat, with no physical barriers hampering the ducks' movements. All the females of the various species look quite similar to one another. But the males differ in the patterning and coloration of their feathers. Speculate on which forms of reproductive isolation may be keeping each species distinct. How does the appearance of the male ducks provide a clue to the answer?

3. Richard Lenski uses bacterial populations in culture tubes to develop model systems for studying evolution. Bacteria produce several generations in a day. Researchers can store them in the deep freeze, then bring them back to active form, unaltered, to directly compare ancestors and their descendants. Are bacterial models relevant to any evolutionary studies of sexually reproducing organisms? Before you answer, read a short article by P. Raine and M. Travisano entitled "Adaptive Radiation in a Heterogeneous Environment" (*Nature*, 2 July 1998, 69–72).

4. Shannon thinks there are too many kingdoms and sees no reason to make another one for something as small as bacteria. "Keep them with the other prokaryotes!" she says. Taxonomists would call her a "lumper." But Andrew is a "splitter." He sees no reason to withhold kingdom status from bacteria simply because they are part of a microscopic world that not many people know about. Which may be the most useful: more or fewer boundaries between groups? Explain your answer.

5. *Rama the cama*, a llama-camel hybrid, was born in 1997 (Figure 17.33). Camels and llamas have a shared ancestor but have been separated for 30 million years. Veterinarians collected semen from a male camel that weighed close to 1,000 pounds, then used it to artificially inseminate a female llama one-sixth his weight. The idea was to breed an animal having a camel's strength and endurance and a llama's gentle disposition.

   Instead of being large, strong, and sweet, Rama is smaller than expected and has a camel's short temper. Rama resembles both parents, with a camel's long tail and short ears but no hump, and llama-like hooves rather than camel footpads. Now old enough to mate, he is too short to get together with a female camel and too heavy to mount a female llama. He has his eye on Kamilah, a female cama born in early 2002, but will have to wait several years for her to mature. The question is, will any offspring from such a match be fertile?

   What does Rama's story tell you about the genetic changes required for irreversible reproductive isolation in nature? Explain why a biologist might not view Rama as evidence that llamas and camels are the same species.

6. A key innovation, again, is some modification in structure or function that permitted a species to exploit the environment in a more efficient or novel way, compared to ancestral species. Speculate on what might have been a key innovation in human evolution. Describe how that innovation might be the basis of an adaptive radiation in environments of the distant future.

Figure 17.33 Rama the cama displaying his unexpected short temper.

Media Menu

Student CD-ROM

Impacts, Issues Video
   Measuring Time
Big Picture Animation
   Evidence of evolution and macroevolution
Read-Me-First Animation
   Morphological divergence
   Mutation and proportional changes
   Reproductive isolating mechanisms
   Allopatric speciation
Other Animations and Interactions
   The geologic time scale
   Continental drift
   Sympatric speciation in wheat

InfoTrac

- Speciation Genes Stick Together. *Applied Genetics News*, September 2001.
- Species: Life's Mystery Packages. *U.S. News & World Report*, July 2002.
- Unlocking the Mystery of Extinction. *Knight Ridder/Tribune News Service*, December 2002.
- Describing the "Tree of Life": Attainable Goal or Stuff of Dreams? *BioScience*, October 2002.

Web Sites

- Talk.Origins–Speciation: www.talkorigins.org/faqs/faq-speciation.html
- USGS Geologic Time: pubs.usgs.gov/gip/geotime
- NASA Asteroid and Comet Impact Hazards: impact.arc.nasa.gov/
- BBC Extinction Files: http://www.bbc.co.uk/education/darwin/exfiles/index.htm

How Would You Vote?

NASA's Spaceguard Survey is spending over 3 million dollars a year to map all near-Earth asteroids and comets larger than one kilometer in diameter. The survey will warn of a potentially catastrophic asteroid impact decades in advance and, it is hoped, allow us to prevent it. Weighing the certainty of catastrophe in the event of an asteroid impact against its low likelihood in the near future, do you support continued funding for this project? Should other nations contribute?

IMPACTS, ISSUES *Looking for Life in All the Odd Places*

In the 1960s, microbiologist Thomas Brock looked into hot springs and pools in Yellowstone National Park (Figure 18.1). He found an ecosystem of microscopic cells, including *Thermus aquaticus*. This prokaryote lives on simple carbon compounds dissolved in *really* hot water, on the order of 80°C (176°F)! *T. aquaticus* is one of the thermophiles, or "heat lovers."

Brock's work had two unexpected results. First, it sent researchers down paths that led them to a great domain of life, the Archaea. Second, it led to a faster

way to copy DNA and end up with useful amounts of it. *T. aquaticus* happens to make a heat-resistant enzyme that can catalyze the polymerase chain reaction—PCR. Enlisting synthetic versions of this enzyme triggered a revolution in biotechnology.

So *bioprospecting* became the new game in town. Companies started exploring extreme environments for species that might yield valuable products. And they found a dazzling array of life forms that withstand extraordinary levels of temperature, acidity, alkalinity, salinity, and pressure.

To extreme thermophiles, Yellowstone's hot springs would be relatively chilly. Many of these prokaryotic species evolved on the seafloor, near hydrothermal vents that spew superheated, mineral-rich water. One species even grows and reproduces at 121°C (249°F).

Other prokaryotes live in acidic springs, where the pH approaches zero, or in highly alkaline soda lakes. In polar regions, some types cling to life in salt ponds that never freeze and glacial ice that never melts.

It's not just prokaryotes. Populations of snow algae tint mountain glaciers red. Some of the transparent

Figure 18.1 Cells of *Thermus aquaticus*, a heat-loving prokaryotic species discovered in thermal springs in Yellowstone National Park. The filmstrip (*far right*) shows nanobes, found miles below Earth's surface.

the big picture

Molecular Beginnings We can expect that the first chemical step toward life was the spontaneous formation of complex organic compounds from simpler substances present on the early Earth. A low-oxygen atmosphere favored this process.

On to Proto-Cells Laboratory studies and computer simulations have yielded indirect evidence that self-assembly of membranes, combined with chemical and molecular evolution, gave rise to the structural and functional forerunners of cells.

invertebrates called salps form long chains in frigid polar seas. One red alga, *Cyanidium caldarium*, grows in hot acidic springs. Some diatoms inhabit extremely salty lakes, where the hypertonicity would make cells of most organisms shrivel up and die.

The nanobes shown in the filmstrip were discovered 3.8 kilometers (3 miles) below Earth's surface in truly hot rocks—170°C (338°F). Being one-tenth the size of most bacteria, nanobes cannot be observed without electron microscopes. They look like the simplest fungi. Some researchers think nanobes are too small to be alive, to hold the metabolic machinery that runs life processes. Yet nanobes do have DNA. And they appear to grow. Are they like proto-cells, which preceded the origin of the first living cells? Maybe.

What's the point of all these examples? Simply this: *Life can take hold in almost any environment that has sources of carbon and energy.*

This chapter invites you to think about life in new ways, starting with theories about its origin and early evolution. Together with the next unit, it is a sweeping slice through time, one that can give you perspective on how life's spectacular range of diversity came about.

The road map is incomplete. Yet evidence from many avenues of research points to a principle that helps us organize information about an immense journey: *Life is a magnificent continuation of the physical and chemical evolution of the universe, and of the planet Earth.*

 ☑ *How Would You Vote?*

Private companies make millions of dollars selling an enzyme first isolated from cells in Yellowstone National Park. Should the federal government let private companies bioprospect within the boundaries of national parks, as long as it shares in the profits from any discoveries? See the Media Menu for details, then vote online.

The First Cells   The first cells were anaerobic prokaryotes. Some gave rise to bacteria, others to archaea and ancestral eukaryotic cells. Evolution of the noncyclic pathway of photosynthesis in some bacteria added oxygen to the atmosphere, which became a new selection pressure.

Origin of Organelles   The nucleus and ER of eukaryotic cells may have evolved through infoldings of the plasma membrane. Mitochondria and chloroplasts are probably descended from bacteria that were engulfed but not digested by other predatory cells.

18.1 In the Beginning . . .

Life originated when Earth was a thin-crusted inferno, so we probably will never find evidence of the first cells. Yet we can arrive at a plausible explanation of how they emerged by considering three questions. What were the prevailing conditions? Do known principles of physics, chemistry, and evolution disprove the hypothesis that the first cells could have originated through chemical and molecular evolution? Can we design experiments to test that hypothesis? Let's take a look.

Think about how you rewind a videotape on a VCR, then imagine "rewinding" the universe. As you do, the galaxies start moving closer together. After 12 to 15 billion years of rewinding, all galaxies, all matter, and space are compressed into a hot, dense volume at one single point. You have arrived at time zero.

That incredibly hot, dense state lasted only for an instant. What happened next is called the **big bang**, the nearly instantaneous distribution of all matter and energy throughout the universe. Within minutes, the temperature dropped a billion degrees. Nuclear fusion reactions created most of the light elements, including helium, which still are the most abundant elements. With radio telescopes, we have now detected radiation left over from the beginning of time.

Over the next billion years, uncountable numbers of gaseous particles collided, and gravitational forces condensed them into the first stars. When stars were massive enough, nuclear reactions ignited inside them and gave off tremendous light and heat, and heavier elements formed. All stars have a life history, from birth to an often explosive death. Heavier elements released from dying ones were swept up when new stars formed and helped form even heavier elements.

When explosions of dying stars ripped through our galaxy, they left behind a dense cloud of dust and gas that extended trillions of kilometers in space. As the cloud cooled, countless bits of matter gravitated toward one another. By 5 billion years ago, the shining star of our solar system—the sun—was born.

CONDITIONS ON THE EARLY EARTH

Figure 18.2 shows part of one of the vast clouds in the universe. It is mostly hydrogen gas, along with water, iron, silicates, hydrogen cyanide, ammonia, methane, formaldehyde, and other small inorganic and organic substances. Between 4.6 billion and 4.5 billion years ago, the cloud that became our solar system probably was like it in composition. Clumps of minerals and ice at the cloud's perimeter grew more massive. They eventually became planets; one of particular interest was the early Earth.

Four billion years ago, hot gases blanketed patches of Earth's thin, fiery crust. We suspect that this first atmosphere was a mix of gaseous hydrogen, nitrogen, carbon monoxide, and carbon dioxide. Did it include free oxygen? Probably not much. Geologic evidence reveals that the free oxygen level in the atmosphere was relatively low until about 2.2 billion years ago.

If that atmosphere had not been relatively free of oxygen, organic compounds necessary to make cells would not have been able to form wherever they were exposed to open air. Free oxygen would have attacked and destroyed the compounds as they assembled.

Figure 18.2 Part of the Eagle nebula, a hotbed of star formation. Each pillar is wider than our solar system. New stars shine brightly on the tips of gaseous streamers.

Figure 18.3 To the right, a diagram of the type of reaction chamber that Stanley Miller used to test whether small organic compounds form spontaneously under conditions thought to have prevailed on the early Earth.

At first, any water falling on the molten surface must have evaporated. However, after the crust cooled and became solid, rainfall and runoff eroded mineral salts from the rocks. Over millions of years, salty waters collected in crustal depressions and formed the earliest seas. If this liquid water had not accumulated, membranes could not have formed, because they take on their bilayer structure in water. No membrane, no cell. Life at its most basic level *is* the cell.

ABIOTIC SYNTHESIS OF ORGANIC COMPOUNDS

Cells appeared less than 200 million years after the crust solidified, so complex carbohydrates and lipids, proteins, and nucleic acids must have formed by then. We know that meteorites, Mars, and Earth all formed at the same time, from the same cosmic cloud. Their rocks contain simple sugars, fatty acids, amino acids, and nucleotides, so we can expect that the precursors of biological molecules were on the early Earth, too.

As you know, synthesis of biological molecules is an uphill run. So which energy sources spontaneously drove the assembly of the simple organic compounds into the first molecules of life? The sun, lightning, or heat from hydrothermal vents could have given off enough energy to drive the reactions.

Stanley Miller was the first to test this hypothesis. He placed water, methane, hydrogen, and ammonia inside a reaction chamber, as in Figure 18.3. He kept the mixture circulating while zapping it with sparks to simulate lightning. In less than a week, amino acids and other small organic compounds had formed.

Recent geologic evidence suggests that Earth's early atmosphere was not quite like Miller's mixture. But simulations that used other gases have also yielded different organic compounds, including some types that can act as nucleotide precursors of nucleic acids.

By another hypothesis, simple organic compounds formed in outer space. Researchers detect amino acids in interstellar clouds and in some of the carbon-rich meteorites that have landed on Earth. One meteorite discovered in Australia contains eight amino acids identical to those in all living organisms.

What about proteins, DNA, and the other *complex* organic compounds? Where did they form? In open water, hydrolysis reactions would have broken them apart as fast as they assembled. By one hypothesis, the clay of tidal flats bound and protected the newly forming polymers. Certain clays contain mineral ions that attract amino acids or nucleotides. Experiments show that once some of these molecules stick to clay, other molecules bond to them, forming chains that resemble the proteins or nucleic acids in living cells.

Another hypothesis that is currently getting a lot of attention is this: The first biological molecules were synthesized near hydrothermal vents. The depths of ancient seas were certainly oxygen-poor. Experiments show that amino acids, at least, will condense into proteinlike structures when heated in water.

Experiments provide indirect evidence that the complex organic molecules characteristic of life could have formed under conditions that probably prevailed on the early Earth.

18.2 How Did Cells Originate?

Metabolism refers to all reactions by which cells harness and use energy. In the first 600 million years or so of Earth history, enzymes, ATP, and other crucial organic compounds probably assembled spontaneously. If they did so in the same places, their close association might have promoted the start of metabolic pathways.

ORIGIN OF AGENTS OF METABOLISM

However the first proteins formed, their molecular structure dictated their behavior. If some behaved like weak enzymes and promoted reactions, they would snap up more amino acids and enzyme helpers, such as metal ions. Chemical selection could have been at work before cells appeared, favoring the evolution of enzymes and other complex organic compounds.

Imagine an ancient estuary, where seawater mixes with mineral-rich water being drained from the land. Beneath the sun's rays, organic molecules stick to clay (Figure 18.4*a*). At first there are quantities of an amino acid; call it D. Molecules of D get incorporated into proteins—until D starts to run out. Nearby, however, is a weakly catalytic protein. This protein promotes formation of D from a plentiful, simpler substance C.

By chance, clumps of organic molecules include the enzyme-like protein. Such clumps have an edge in any chemical competition for starting materials. Say the C molecules become scarce. The advantage tilts to molecular clumps that promote formation of C from simpler substances B and A. Suppose that B and A are carbon dioxide and water. The atmosphere and seas contain unlimited amounts of both. And so chemical selection has favored a synthetic pathway:

$$A + B \longrightarrow C \longrightarrow D$$

Were some clumps better at absorbing and using energy? Think back on chlorophyll *a* (Section 6.1). A group of rings in this pigment absorbs light and gives up electrons. The same kind of ring structures also occur in electron transfer chains in all photosynthetic and aerobically respiring cells. Such structures form spontaneously from formaldehyde, one of the legacies of cosmic clouds. Were similar structures transferring electrons in early metabolic pathways? Probably.

ORIGIN OF THE FIRST PLASMA MEMBRANES

All living cells have an outer membrane that controls which substances move into and out of a cell (Section 4.1). By one hypothesis, proto-cells were transitional forms between chemical evolution and the first living cells. The **proto-cells** were simple membrane-bound sacs that enclosed basic metabolic machinery and that were self-replicating.

Experiments have shown that membrane-like sacs can form spontaneously. Under conditions that could

Figure 18.4 Where did the first cells originate? Two candidates: (**a**) Clay templates in tidal flats, as explained in the text, and (**b**) iron sulfide-rich rocks at hydrothermal vents. Iron-sulfide rocks at hydrothermal vents are honeycombed with microscopic, cell-sized chambers. (**c**) Similar iron-sulfide structures have formed in the laboratory.

Did such chambers serve as protected microenvironments that favored the assembly of the first membranes? In experiments simulating conditions near vents, iron sulfides catalyzed the synthesis of short peptide chains and other reactions that resemble metabolism. Many reactions in living cells use iron-sulfide cofactors. Are these cofactors a metallic legacy from a deep-sea ancestor? Perhaps.

20 μm

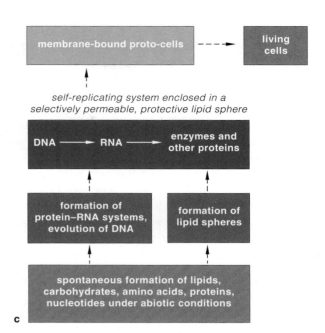

Figure 18.5 Laboratory-formed proto-cells. (**a**) Selectively permeable vesicles with a protein outer membrane were formed by heating amino acids, then wetting the resultant protein chains. (**b**) RNA-coated clay (*red*) surrounded by a membrane of fatty acids and alcohols (*green*). The mineral-rich clay catalyzes RNA polymerization and promotes the formation of membranous vesicles.

(**c**) Model for the steps that may have led to living cells.

have occurred on ancient sunbaked tidal flats, amino acids do form chains, then assemble like a membrane around a fluid interior (Figure 18.5a). Fatty acids and alcohols can spontaneously form vesicles, especially when clays rich in minerals are present (Figure 18.5b).

Or proto-cells may have formed near hydrothermal vents. Chimney-like rocks that pile up around vents are honeycombed with microscopic chambers (Figure 18.4b,c). By one hypothesis, such chambers were three-dimensional molds for the first cell membranes. Iron sulfides in the rocks could have served as catalysts for early metabolic reactions.

ORIGIN OF SELF-REPLICATING SYSTEMS

Another defining characteristic of life is a capacity for reproduction, which now starts with protein-building instructions in DNA. As you know, the translation of DNA into proteins requires the participation of RNA, enzymes, and other molecules.

Coenzymes and metal ions assist most enzymes—and certain coenzymes are structurally identical with RNA subunits. When you mix and heat RNA subunits with very short chains of phosphate groups, they self-assemble into strands of RNA. Simple self-replicating systems of RNA, enzymes, and coenzymes have been made in laboratories. So we know RNA can serve as an information-storing template for making proteins.

Also, an rRNA in ribosomes does catalyze protein synthesis (Figure 13.8). Ribosomes, recall, are highly conserved structures. The ribosomes in a eukaryotic cell are not that different from those in prokaryotes. It's likely that rRNA's catalytic behavior evolved early in Earth history.

Did an **RNA world** in which RNA was a template for protein synthesis *precede* the origin of DNA? It's possible. Remember, RNA and DNA are a lot alike. Three of four bases are identical, and RNA's uracil differs from DNA's thymine by one functional group.

Why would DNA be any better than RNA? *Because DNA is more stable than RNA, it can form longer chains that can carry far more information.* Computer analyses also show that helical coiling of a strandlike molecule is the best way to pack subunits in the smallest space.

Until we identify the chemical ancestors of RNA and DNA, the story of life's origin will be incomplete. But details are filling in. For instance, researchers fed data about inorganic compounds and energy sources into a supercomputer. They programmed the computer to simulate random chemical competition and natural selection among selected compounds, as might have happened untold billions of times in the distant past. They ran the program again and again. The outcome was always the same. *Simple precursors evolved. Then they spontaneously organized themselves as large, complex molecules. And they began to interact as complex systems.*

There are many gaps in the story of life's origin. But diverse laboratory experiments and computer simulations show that chemical evolution can produce the organic molecules and structures that we think of as characteristic of life.

18.3 The First Cells

The first cells probably evolved during the **Archaean**, an eon that started 3.8 billion years ago. Until that time, constant bombardment by comets and asteroids almost certainly made the planet inhospitable to life.

Biologists generally accept that the first cells were like existing prokaryotic cells; they had no nucleus. Because free oxygen was absent, we can expect that they obtained energy by anaerobic pathways such as fermentation. Plenty of "food" was available; geologic processes had already enriched the seas with organic compounds and minerals. There were no predators, and cells were not exposed to oxygen attacks.

By analyzing the phylogenies of living prokaryotes, we have evidence that some populations diverged not long after life originated. One lineage gave rise to the bacteria. Another gave rise to the shared ancestors of archaea and eukaryotic cells.

Microscopic fossils in 3.5-billion-year-old rocks hint at what some of those earliest prokaryotes may have looked like (Figure 18.6a). Other fossils clearly show that chemoautotrophic forms had become established near deep-sea hydrothermal vents by 3.2 billion years ago. In some groups, pigments probably detected the type of weak infrared radiation (heat) that has been measured at hydrothermal vents. Pigments may have helped cells avoid boiling water, as they do for some existing hydrothermal vent species.

At some point, mutations occurred independently in some groups and altered their radiation-sensitive pigments, electron transfer chains, and other bits of metabolic machinery. They favored a novel mode of nutrition—the cyclic pathway of photosynthesis. The new groups were photoautotrophic; they tapped into sunlight, an unlimited energy source (Section 6.2).

Populations of photosynthesizers formed flattened mats, which trapped sediments dissolved in water. In time, continuous deposition of the sediments and cell remains formed layered, dome-shaped rocks, with the most recent mats of still-living cells on top. Some of these structures, called **stromatolites**, are more than one meter high (Figure 18.7). Fossilized ones provide the most abundant evidence of early prokaryotic life.

Stromatolites were abundant by the dawn of the Proterozoic, 2.7 billion years ago. A noncyclic pathway of photosynthesis had evolved among cyanobacteria. Its by-product, free oxygen, slowly accumulated in the sea surface waters, then in air. And here we return to events sketched out earlier, in Chapters 4 and 6.

20 µm

Figure 18.6 A sampling of early life. (**a**) A strand of what may be walled prokaryotic cells dates back 3.5 billion years. (**b**) One of the oldest eukaryotic species, *Grypania spiralis*, a colonial eukaryote that formed macroscopic colonies 2.1 billion years ago. (**c**) *Tawuia*, another early Proterozoic eukaryotic megafossil. (**d**) *Bangiomorpha pubescens*, a sexually reproducing multicellular red alga, from 1.2 billion years ago.

Figure 18.7 Stromatolites. (**a**) A painting of how one shallow sea might have looked during the early Proterozoic. (**b**) A cut through a stromatolite reveals many layers of fossilized cyanobacterial mats, hardened by calcium carbonate deposits. (**c**) In Shark Bay, Australia, are mounds 2,000 years old. Structurally, they are similar to some of the stromatolites dated at more than 3 billion years old.

An atmosphere enriched with free oxygen had two irreversible effects. First, *it stopped the further chemical origin of living cells*. Except in a few anaerobic habitats, complex organic compounds could no longer form spontaneously and stay intact; they could not escape attacks by free oxygen. Second, *aerobic respiration now evolved and in time became the dominant energy-releasing pathway*. In many prokaryotic lineages, selection had favored this pathway, which neutralized oxygen by *using* it as an electron acceptor. Aerobic respiration was a key innovation that contributed to the rise of multicelled eukaryotes.

Eukaryotic cells also evolved in the Proterozoic. Traces of the kinds of lipids that are made by existing eukaryotic cells have been isolated from rocks that are 2.8 billion years old. But the first complete eukaryotic fossils are about 2.1 billion years old (Figure 18.6*b,c*). These species had organelles.

As you know, organelles are the defining features of eukaryotic cells. Where did they come from? The next section presents a few plausible hypotheses.

We still don't know how the earliest eukaryotes fit in evolutionary road maps. The earliest known form we can assign to a modern group is a filamentous red alga, *Bangiomorpha pubescens*. This alga was alive 1.2 billion years ago. It is the first multicelled species we know about, and its cells were differentiated. Some

cells in the strandlike body functioned as anchoring structures (holdfasts). Others produced two types of sexual spores—which makes *B. pubescens* the earliest known sexually reproducing organism.

By 1.1 billion years ago the first supercontinent—Rodinia—had formed. Stromatolites flourished along its vast shorelines. About 300 million years after that, however, they were in decline. Were newly evolved predators—animals, perhaps—feeding on them? The earliest fossils that are unambiguously "animals" are 600 million years old. We can interpret this to mean that simpler animals originated before then. It was not until 570 million years ago, when oxygen in the atmosphere was approaching modern levels, that the first adaptive radiations of animals began.

The first living cells evolved by about 3.8 billion years ago, during the Archaean eon. All were prokaryotic. We can expect that they made ATP by fermentation pathways. Early on, ancestors of archaea and eukaryotic cells diverged from the lineage that led to modern bacteria.

After the noncyclic pathway of photosynthesis evolved, free oxygen accumulated in the atmosphere and put an end to the further spontaneous chemical origin of life. That atmosphere set the stage for the evolution of eukaryotic cells.

18.4 Where Did Organelles Come From?

Thanks to globe-hopping microfossil hunters, we have considerable evidence of early life, including the fossil treasures shown in Sections 4.4 and 18.3. Today, most descendant species contain a profusion of organelles. Where did the organelles come from?

ORIGIN OF THE NUCLEUS AND ER

Prokaryotic cells, recall, do not have an abundance of organelles. Some do have infoldings of their plasma membrane, which contains many enzymes and other metabolic workers (Figure 18.8a). By applying the theory of natural selection, we can hypothesize that such infoldings first evolved among the ancestors of eukaryotic cells, and they became narrow channels in which nutrients, organic compounds, and other substances became concentrated. What selective advantage did infoldings offer? More membrane meant more metabolic machinery; reactions, transport tasks, and waste disposal were completed faster.

Did ER channels evolve this way? And did they also protect the cell's metabolic machinery from uninvited guests? Possibly. From time to time, metabolically "hungry" foreign cells do enter the cytoplasm of existing prokaryotic cells.

By this scenario, infoldings also surrounded the DNA and became a nuclear envelope (Figure 18.8b). A nuclear envelope was favored because it helped protect the cell from invasions of foreign DNA. Bacteria and simple eukaryotic cells called yeasts do transfer bits of DNA among themselves. Early

eukaryotic cells equipped with a nuclear envelope could copy and use their messages of inheritance, free from metabolic competition from a potentially disruptive hodgepodge of foreign DNA.

ORIGIN OF MITOCHONDRIA AND CHLOROPLASTS

Early in the history of life, cells became food for one another. Heterotrophs started feeding on autotrophs and other heterotrophs. In some cases, the meals turned into uninvited guests. They lingered on and slowly evolved into mitochondria, chloroplasts, and other organelles.

Such novel partnerships are one premise of a theory of **endosymbiosis**, as championed by Lynn Margulis and others. (*Endo–* means within; *symbiosis* means living together.) One species (the symbiont) lives out its life inside another species (the host), and the interaction benefits one or both.

According to this theory, eukaryotic cells evolved after the noncyclic pathway of photosynthesis had emerged and changed the atmosphere. By 2.1 billion years ago, remember, certain prokaryotic cells had adapted to the accumulation of free oxygen and were carrying out aerobic respiration.

Aerobic bacteria were among the meals ingested by the ancestor of eukaryotic cells. At some point, a meal resisted digestion. It stayed alive, grew, and reproduced in the sheltered, nutrient-rich cytoplasm. The host reproduced, too. After many generations,

a

b

Figure 18.8 (**a**) Sketch of a bacterial cell (*Nitrobacter*) that lives in soil. Cytoplasmic fluid bathes permanent infoldings of the plasma membrane. (**b**) Model for the origin of the nuclear envelope and endoplasmic reticulum. In prokaryotic ancestors of eukaryotic cells, infoldings of the plasma membrane may have evolved into these organelles.

Figure 18.9 Diverse organelles are hallmarks of eukaryotic cells. *Cyanophora paradoxa* is one of the flagellated protists called glaucophytes. Its mitochondria resemble aerobic bacteria in size and structure. Its photosynthetic structures resemble cyanobacteria—they even have a wall like that of cyanobacteria.

its descendants started using ATP formed by the aerobic symbiont; in time, some depended on it.

The aerobic symbiont no longer had to gather raw materials or build proteins when the host was doing the work for it. Host DNA regions that specified the proteins were free to mutate and lose their function. In time, both types of cells became incapable of independent life.

EVIDENCE OF ENDOSYMBIOSIS

Is such a theory far-fetched? A chance discovery in Jeon Kwang's laboratory suggests otherwise. In 1966, a rod-shaped bacterium had infected his culture of *Amoeba discoides*. Some infected cells died right away. Others grew more slowly, and they were smaller and vulnerable to starving to death. Kwang maintained the infected culture. Five years later, the infection's bad effects were gone. Amoebas stuffed with bacteria were thriving. Exposing these infected amoebas to antibiotics actually killed them.

Infection-free cells stripped of their nucleus died after receiving the nucleus from an infected cell. Yet, more than 90 percent survived when a few bacteria tagged along with the transplant. In other studies, infected amoebas lost their ability to synthesize an essential enzyme. They depended on the bacterium to make it for them! The invading bacterial cells had become symbiotic with the amoebas.

When you think about it, mitochondria do resemble bacteria in size and structure. Each has its own DNA and divides independently of cell division. The inner membrane of a mitochondrion resembles a bacterial cell's plasma membrane. Its DNA has just a few genes (thirty-seven in human mitochondrial DNA). And its genetic code differs a bit from the genetic code shared by all other organisms.

We can expect that chloroplasts, too, originated by endosymbiosis. By one hypothesis, photosynthetic cells were engulfed by predatory aerobic bacteria and escaped digestion. They started absorbing nutrients from the host's cytoplasm and continued to function. They also released oxygen during photosynthesis. By releasing oxygen in their aerobically respiring hosts, they acted as agents favoring endosymbiosis.

In metabolism and overall nucleic acid sequence, chloroplasts resemble the type of prokaryotes called cyanobacteria. Chloroplast DNA and cellular DNA replicate independently of each other. Chloroplasts also divide independently of the cell's division.

Finally, consider the glaucophytes. This intriguing protist group has unique photosynthetic organelles. The organelles closely resemble cyanobacteria; they even are surrounded by a cell wall (Figure 18.9).

However they arose, the first eukaryotic cells had a nucleus, an endomembrane system, mitochondria, and, in some lineages, chloroplasts. They were the world's first protists.

18.5 Timeline for Life's Origin and Evolution

hydrogen-rich anaerobic atmosphere

atmospheric oxygen, 10%

archaean lineage

d ancestors of eukaryotes

h endomembrane system and nucleus

noncyclic pathway of photosynthesis

cyclic pathway of photosynthesis **f**

e

c bacterial lineage

a **b**

origin of prokaryotes

g aerobic respiration

| 3.8 billion years ago | 3.2 billion years ago | 2.5 billion years ago |

Chemical and Molecular Evolution

a Between 5 billion and 3.8 billion years ago, as an outcome of chemical and molecular competition, complex carbohydrates, lipids, proteins, and nucleic acids evolved from simple organic compounds present on the early Earth.

Origin of Prokaryotic Cells

b The first living cells probably had evolved by 3.8 billion years ago. All were prokaryotic; they lacked a nucleus or other organelles. Atmospheric oxygen was low and the early cells made ATP by anaerobic pathways.

Three Domains of Life

c The first major divergence gave rise to bacteria and to the common ancestor of the archaea and all eukaryotic cells.

d Not long after, ancestors of archaea and eukaryotic cells diverged.

Photosynthesis, Aerobic Respiration Evolve

e A cyclic pathway of photosynthesis evolved in some bacterial groups.

f An oxygen-releasing noncyclic pathway evolved later in the cyanobacteria and, over time, changed the atmosphere.

g Aerobic respiration evolved in many bacterial groups.

Origin of Endomembrane System, Nucleus

h Cell sizes and the amount of genetic information continued increasing in ancestors of what would become the eukaryotic cells. The endomembrane system, including the nuclear envelope, arose through the modification of cell membranes.

Figure 18.10 Milestones in the history of life. As you read the next unit on life's past and present diversity, refer to this visual overview. It can serve as a simple reminder of the evolutionary connectedness of all groups of organisms, from the structurally simple to the most complex.

Read Me First!
and watch
the narrated animation on
the history of life

atmospheric oxygen, 20%; the ozone layer slowly develops

ARCHAEA

Extreme thermophiles

Extreme thermophiles and mesophiles

Halophiles and methanogens

EUKARYOTES

Animals

Fungi

Heterotrophic protists

Photosynthetic protists with chloroplasts that evolved from red and green algae

Red and green algae; their chloroplasts evolved from cyanobacterial symbionts

Plants

BACTERIA

Oxygen-releasing photosynthetic bacteria (cyanobacteria)

Other photosynthetic bacteria

Heterotrophic bacteria, including chemoheterotrophs

k origin of animals

j origin of eukaryotes, the first protists

k origin of fungi

origin of mitosis, meiosis

i endosymbiotic origin of mitochondria

j endosymbiotic origin of chloroplasts

k origin of lineage leading to plants

l

Aerobic species becomes endosymbiont of anaerobic forerunner of eukaryotes.

1.2 billion years ago

900 million years ago

435 million years ago

Endosymbiotic Origin of Mitochondria

i Before about 1.2 billion years ago, aerobic bacterial species and an anaerobic ancestor of eukaryotic cells entered into close symbiotic interaction. The endosymbiont evolved into the mitochondrion.

Endosymbiotic Origin of Chloroplasts

j Cyanobacteria entered into a close symbiotic interaction with early protists and evolved into chloroplasts. Later, photosynthetic protists would evolve into chloroplasts inside other protist hosts.

Plants, Fungi, and Animals Evolve

k By 900 million years ago, all major lineages—including fungi, animals, and the algae that would give rise to plants—had evolved along shorelines of the first supercontinent.

Lineages That Made It to the Present

l Today, all regions of the Earth's waters, crust, and atmosphere hold organisms. They are related by descent and share certain common traits. But each lineage encountered different selective pressures, and each has evolved its own characteristic traits.

Summary

Section 18.1 Earth formed more than 4 billion years ago. Indirect evidence suggests that complex organic compounds characteristic of life could have formed spontaneously on early Earth.

Section 18.2 Proto-cells may have arisen through chemical evolution and the self-assembly of lipid or protein spheres. RNA probably preceded DNA as the template for protein synthesis.

Section 18.3 The first prokaryotic cells arose about 3.8 billion years ago. They were anaerobic. An early divergence separated bacteria from the ancestors of archaea and eukaryotes. Evolution of the noncyclic pathway of photosynthesis in cyanobacteria resulted in an accumulation of free oxygen in the atmosphere. The increase paved the way for aerobic respiration and the evolution of eukaryotic cells.

Media Menu

Student CD-ROM	*Impacts, Issues Video*
	   Looking for Life in All the Odd Places
	Big Picture Animation
	   Life's origin and evolution
	Read-Me-First Animation
	   The history of life
	Other Animations and Interactions
	   Miller's experimental apparatus

InfoTrac	• How Life Began: Microbes at the Extremes May Tell Us. *Time*, July 2002. • Nanobes? How Small Is the Smallest Life? *Odyssey*, February 2002. • Life on Earth: Scientists Probe the Early Days. *Astronomy*, November 2001. • From Proteins to Protolife. *Science News*, July 1994.

Web Sites	• Life Over Time:
	  www.fmnh.org/exhibits/exhibit_sites/lot/LOT1.htm
	• Center for Studies of the Origins of Life:
	  www.rpi.edu/~straca/NSCORT/origin.html
	• ActionBioScience New Frontiers:
	  actionbioscience.org/newfrontiers/index.html

How Would You Vote?	PCR uses a DNA polymerase first isolated from a prokaryote that was discovered in Yellowstone National Park. Sales of the enzyme, which has been patented, exceed 500 million dollars annually. Yellowstone receives none of this revenue. Now park managers want to charge companies to "bioprospect" in thermal pools and to share in profits from any new products. Do you support this use of our national parks?

Section 18.4 The internal membranes of eukaryotic cells may have evolved through infoldings of the cell membrane. Mitochondria and chloroplasts most likely evolved by endosymbiosis, at the times indicated in the Section 18.5 visual summary.

Self-Quiz

Answers in Appendix III

1. An abundance of _____ in the atmosphere would have prevented the spontaneous (abiotic) assembly of organic compounds on the early Earth.
   a. hydrogen   b. methane   c. oxygen   d. nitrogen

2. The prevalence of iron-sulfide cofactors in living organisms may be evidence that life arose _____ .
   a. in outer space   c. near deep-sea vents
   b. on tidal flats   d. in the upper atmosphere

3. The evolution of _____ resulted in an increase in the levels of atmospheric oxygen.
   a. sexual reproduction
   b. aerobic respiration
   c. the noncyclic pathway of photosynthesis
   d. the cyclic pathway of photosynthesis

4. Mitochondria may be descended from _____ .
   a. chloroplasts   c. early protists
   b. bacteria   d. archaea

5. Chronologically arrange the evolutionary events, with 1 being the earliest and 6 the most recent.

   _____ 1
   _____ 2
   _____ 3
   _____ 4
   _____ 5
   _____ 6

   a. emergence of the noncyclic pathway of photosynthesis
   b. origin of mitochondria
   c. origin of proto-cells
   d. emergence of the cyclic pathway of photosynthesis
   e. origin of chloroplasts
   f. the big bang

Critical Thinking

1. What if it were possible to create life in test tubes? This is the question behind attempts to model and perhaps create minimal organisms: living cells having the smallest set of genes required to survive and reproduce.

Craig Venter and Claire Fraser recently found that *Mycoplasma genitalium*, a bacterium that has 517 genes (and 2,209 transposons), is a good candidate for such an experiment. By disabling its genes one at a time in the laboratory, they discovered that *M. genitalium* may have only 265–350 essential protein-coding genes.

What if those genes were synthesized one at a time and inserted into an engineered cell consisting only of a plasma membrane and cytoplasm? Would the cell come to life? The possibility that it might prompted Venter and Fraser to seek advice from a panel of bioethicists and theologians. No one on the panel objected to synthetic life research. They said that much good might come of it, provided scientists didn't claim to have found "the secret of life." The December 10, 1999, issue of *Science* includes an essay from the panel and an article on *M. genitalium* research. Read both, then write down your thoughts about "creating" life in a test tube.

IV Evolution and Biodiversity

From the Green River formation near Lincoln, Wyoming, the stunning fossilized remains of a bird trapped in time. During the Eocene, some 50 million years ago, sediments that had been slowly deposited in layers at the bottom of a large inland lake became its tomb. In this same formation, fossils of palms, cattails, sycamore, and other plants tell us that the climate was warm and moist. Fossils from places all around the world yield major clues to life's early history.

IMPACTS, ISSUES   *West Nile Virus Takes Off*

In 336 BC, when he was twenty years old, Alexander the Great ascended to the throne of Macedonia (Figure 19.1). During his reign, he carved out an empire that stretched across the Middle East and into northern India. He died twelve years later while visiting Babylon, the site of modern-day Baghdad. According to one scribe, a flock of ravens announced Alexander's arrival in the city. The birds behaved strangely, and some fell dead at his feet. Soon thereafter, Alexander suffered a sudden onset of back pain. Fever and chills, weakness, delirium, paralysis, and finally death followed.

An infectious disease expert, Charles Calisher, and epidemiologist John Marr connected the dots between the birds' behavior and Alexander's symptoms. As they hypothesize, Alexander died as an outcome of West Nile encephalitis. A flavivirus, first isolated in 1937, causes this disease, which results in severe inflammation of the brain. Researchers had discovered that the virus was once prevalent in Africa, West Asia, and the Middle East. Was there an outbreak of West Nile encephalitis during Alexander's time? Possibly.

Until the summer of 1999, no one knew that the virus had entered the Western Hemisphere. Then people in and around New York City started wondering about the dead and dying crows—and the mysteriously sickened horses and humans. West Nile virus particles turned up in tissue samples from infected individuals. Sixty-two people were infected that summer. Seven died.

The virus hitchhiked, inside birds, across North and Central America. By 2003, nearly 9,000 human cases of West Nile encephalitis or fevers had been reported; and more than 200 people had died. Cases popped up in all states except Maine, Washington, and Oregon, also in Canada and the Cayman Islands. Birds and horses in

Figure 19.1   A clue to viral history? Alexander the Great may have died of West Nile encephalitis. Ravens and crows are highly susceptible to this viral disease. Today these birds are being monitored to assess the spread of the disease through the Americas.

the big picture

to ancestors of eukaryotes

ARCHAEA

BACTERIA

chemical and molecular evolution of life

Prokaryotic Cell Features   Prokaryotic cells generally are the smallest and, structurally, the simplest organisms. They show great reproductive potential and, collectively, the greatest metabolic diversity.

Major Prokaryotic Groups   The first cells were prokaryotic. Some gave rise to archaea, others to bacteria and to ancestral eukaryotic cells. One kind of prokaryote or another is found in almost all environments.

Mexico showed signs of infection. Most likely, the virus will spread throughout all of the Americas.

West Nile virus clearly is pathogenic. A **pathogen** is an infectious, disease-causing agent that can invade a host organism and multiply in or on it. Disease follows when metabolic activities of its descendants damage tissues and interfere with how the host body works.

As far as we know, you can't "catch" West Nile virus from a dog, cat, or classmate. Like other flaviviruses, this one usually travels in mosquitoes that can suck blood from an infected host, then deposit it into another host. In North America, at least forty-three different kinds of mosquitoes have become vectors for West Nile virus.

Infections may have widespread ecological impact. In North America, the virus has been detected in more than 150 bird and mammal species, as well as alligators. Many of these species harbor the virus without getting sick. They act as reservoirs, from which the virus may be transmitted to others that are harmed by infection. If an endangered species can be infected and sickened by West Nile virus, it may be pushed toward extinction.

This chapter can start you thinking about the unseen multitudes at the boundary between nonliving and living things. Viruses hover near the boundary, and the prokaryotes—structurally the simplest forms of life— are right inside it. From the human perspective, some kinds of prokaryotes and viruses are dangerous, others are beneficial. As you'll soon see, even the bad ones have relatives that impact positively on human affairs.

 How Would You Vote?

Eliminating mosquitoes is the best defense against West Nile virus. Many local agencies are spraying pesticides wherever mosquitoes are likely to breed. Some people fear ecological disruptions and bad effects on health, and say spraying will never get rid of all mosquitoes anyway. Would you support a spraying program in your community? See the Media Menu for details, then vote online.

Viruses   Viruses are infectious particles that require a living host to replicate themselves. Each consists of little more than DNA or RNA and a few viral enzymes inside a protein coat.

The Bad Bunch   Pathogenic bacteria and viruses have coevolved with specific hosts. In evolutionary terms, like any other organism, the pathogen that leaves the most descendants wins.

19.1 Characteristics of Prokaryotic Cells

Of all organisms, prokaryotic cells are the smallest and most far-flung, abundant, and metabolically diverse. These cells have the longest evolutionary history. They arose about 3.8 billion years ago, and all living things are their descendants.

BODY PLANS, SHAPES, AND SIZES

Prokaryotic cells arose before nucleated cells did. (*Pro–* means before; *karyon* is taken to mean nucleus.) A rare few have simple organelles but no nucleus (Table 19.1). Reactions typically proceed at the plasma membrane or in the cytoplasm; for example, at ribosomes.

Figure 19.2 shows a typical prokaryotic body plan and common shapes. The structural simplicity doesn't mean the cells are inferior to eukaryotic cells. These are tiny, fast reproducers, so they don't require a lot of complexity. A semirigid, permeable cell wall usually surrounds the plasma membrane and helps a cell hold its shape and resist rupturing as internal fluid pressure increases. A capsule or slimy layer of polysaccharides, polypeptides, or both often encloses the wall. It helps cells stick to surfaces and resist phagocytes.

Many species have one or more bacterial flagella, as in Figure 19.3a. These whip-shaped motile structures differ from eukaryotic flagella. They do not have a 9+2 array of microtubules inside, and they don't bend; they rotate like a propeller around the attachment site.

Pili (singular, pilus) often project above a cell wall. These protein filaments help cells adhere to surfaces, such as the vaginal lining. A sex pilus (Figure 19.3b) serves in **bacterial conjugation**. By this mechanism, a donor cell's sex pilus latches onto a recipient cell and then retracts, pulling it close. The shortened sex pilus

serves as a conjugation bridge, the means by which the donor transfers a plasmid to the recipient. A **plasmid**, recall, is a small, self-replicating circle of DNA.

The tiniest cells barely show up in light microscopes. *E. coli* is more typical; it is 1–2 micrometers long and 0.5–1.0 micrometers wide. *Thiomargarita namibiensis*, the largest prokaryote known, is visible to the naked eye. At 750 micrometers, it's larger than some eukaryotic cells.

METABOLIC DIVERSITY

Collectively, prokaryotes show the most diversity in how they get food. *Photoautotrophs* make their own by a photosynthetic pathway that uses sunlight and CO_2 (as a carbon source). *Chemoautotrophs* are self-feeders, too. They use CO_2 and get energy by oxidizing organic compounds or inorganic ones, such as iron and sulfur. *Photoheterotrophs* are not self-feeders. They do tap the sun's energy, but they break down organic compounds for carbon. The parasitic *chemoheterotrophs* pirate energy and carbon from a living host; saprobic types get them from organic remains or wastes.

Table 19.1 Characteristics of Prokaryotic Cells
1. No membrane-bound nucleus.
2. Generally a single chromosome (a circular DNA molecule) with no associated proteins; many species also contain plasmids.
3. Cell wall present in most species.
4. Reproduction mainly by prokaryotic fission.
5. Collectively, great metabolic diversity among species.

coccus

bacillus

spirillum

a

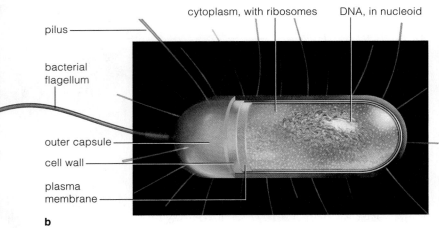

pilus

bacterial flagellum

outer capsule

cell wall

plasma membrane

cytoplasm, with ribosomes   DNA, in nucleoid

b

Figure 19.2 (**a**) The three most common shapes among prokaryotic cells: spheres, rods, and spirals. (**b**) Generalized body plan of a prokaryotic cell.

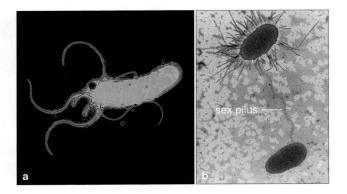

sex pilus

a   **b**

Figure 19.3 (**a**) *Helicobacter pylori*, a flagellated pathogen. It can cause peptic ulcers. (**b**) A sex pilus that has connected two *Escherichia coli* cells.

Read Me First!
and watch the narrated
animation on prokaryotic fission

a The bacterial chromosome is attached to the plasma membrane prior to DNA replication.

b Replication starts and proceeds in two directions from a certain site in the bacterial chromosome.

c The DNA copy becomes attached at a membrane site near the attachment site of the parent DNA molecule.

d Then the two DNA molecules are moved apart by membrane growth between the two attachment sites.

e Lipids, proteins, and carbohydrates are built for new membrane and new wall material. Both get inserted across the cell's midsection.

f The ongoing, orderly deposition of membrane and wall material at the midsection cuts the cell in two.

Figure 19.4 Reproduction by way of prokaryotic fission. Only bacteria and archaea reproduce by this cell division mechanism.

GROWTH AND REPRODUCTION

We measure prokaryotic growth as an increase in the number of cells in a population. Each cell divides in two, the two become four, four become eight, and so forth. Before the cell divides, it nearly doubles in size. After division, each daughter cell has a single bacterial chromosome: a circular double-stranded molecule of DNA that has very few proteins attached to it.

In some species, a daughter cell merely buds from a parent cell. Most often, a cell reproduces by a division mechanism called **prokaryotic fission** (Figure 19.4). Briefly, a parent cell replicates its DNA molecule, and the replica docks at the plasma membrane next to the parent molecule. New membrane gets added between them and so moves the two apart. The membrane and cell wall extend into the cell's midsection and divide it into two genetically equivalent daughter cells.

The reproductive potential of prokaryotic cells is staggering. Some types divide every twenty minutes. If that rate held, a single cell would give rise to nearly a billion descendants in ten hours! So why don't the unseen multitudes take over the world? Their rapidly growing populations use up all available nutrients and pollute their habitats with metabolic wastes. They change the very conditions required for reproduction. Also, viral pathogens keep their populations in check.

CLASSIFICATION

Once, reconstructing the evolutionary history of the prokaryotes seemed an impossible task. Except for the stromatolite builders (Section 18.3), the most ancient groups are not well represented in the fossil record. It's likely that most early forms are not represented at all. Early classification systems had to be constructed by **numerical taxonomy**. This process compares traits of an unidentified cell with those of a known group, such as its shape, metabolic patterns, nutrition, and staining attributes of its wall (if any). The more traits the two have in common, the closer the inferred relatedness.

Today, automated gene sequencing and other tools of comparative biochemistry are helping to reveal the phylogenies. A recent discovery that most prokaryotic genomes contain hereditary material from more than one source does complicate the search. These mixed genomes arise by **lateral gene transfer**: movement of genetic information between cells, often of different species, by conjugation or some other mechanism.

The mostly microscopic prokaryotes are the most abundant and far-flung organisms on Earth. Enormous reproductive capacity and diverse modes of metabolism among these groups contributed to their evolutionary success.

19.2 The Bacteria

There may be millions of prokaryotic species. By far, most of the 400 named genera are bacteria. The few that cause human disease often get the spotlight, but most have vital ecological roles. A single gram of rich soil may contain thousands of species and billions of individuals. Figure 19.5 shows the main groups, which we can only sample here.

REPRESENTATIVE DIVERSITY

The **cyanobacteria** are photoautotrophs that help cycle carbon, oxygen, nitrogen, and other key nutrients. Like plants, they contain chlorophyll *a* and release oxygen during photosynthesis. Chloroplasts arose from early species, through endosymbiosis. Nearly all atmospheric oxygen can probably be traced to free-living cyanobacteria or to the chloroplasts descended from them.

We find most cyanobacteria in aquatic habitats and soil. After division, daughter cells often stay attached as mucus-sheathed chains, which can form dense, slimy mats (Figure 19.6*a*). *Spirulina* is a commercially grown "health food." A few species of cyanobacteria even are found in lichens, as symbionts with fungi.

Anabaena, an aquatic species, is one of the nitrogen fixers. When nitrogen is scarce, some cells in *Anabaena* chains develop into heterocysts, which convert nitrogen gas to ammonia (Figure 19.6*b*). The heterocysts share nitrogen compounds with other cells in the chain and get carbohydrates in return.

Gram-positive bacteria are so named because Gram staining tints their thick, multilayered walls purple. Most are chemoheterotrophs. The *Lactobacillus* species that form lactate during fermentation are used to make yogurt and other popular foods. *L. acidophilus* lowers the pH of skin and of intestinal and vaginal linings, which helps stop pathogens from becoming established. Antibiotics can cripple or kill *L. acidophilus* and other symbionts. Diarrhea or vaginitis may follow.

Clostridium and *Bacillus* are Gram-positive bacteria that survive hostile conditions when an **endospore**, a

Figure 19.6 Cyanobacteria. (**a**) A huge population on the surface of a nutrient-rich pond. (**b**) Some cells become resting spores or heterocysts when conditions restrict growth.

type of resting structure, forms inside the cell body. An endospore encloses the bacterial chromosome and a bit of cytoplasm (Figure 19.7). It resists heat, irradiation, drying out, acids, disinfectants, and boiling water. The dormant state lasts until favorable conditions return. Then the endospore germinates; a bacterium emerges from it, and normal function resumes.

Endospores that enter the human body can be lethal. In 2001, someone put *Bacillus anthracis* endospores in envelopes and mailed them. A few people opened the mail, breathed in endospores, and died of anthrax. The endospores of *Clostridium tetani* can pass across broken skin and cause tetanus, another disease (Section 32.7). Endospores of *C. botulinum* often taint canned food. Toxins produced by bacteria that emerge from these germinating endospores cause botulism, a dangerous form of food poisoning.

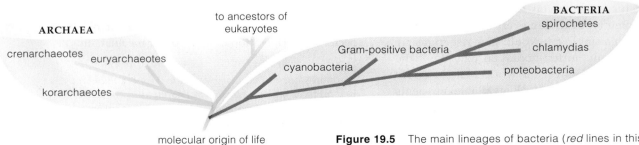

Figure 19.5 The main lineages of bacteria (*red* lines in this cladogram).

Figure 19.7 Endospore forming in *Clostridium tetani*.

spore coat — DNA

Figure 19.8 *Thiomargarita namibiensis*, the largest bacterium. The cells can be seen with the naked eye and are mostly a giant nitrate-filled vacuole.

a b c

Figure 19.9 (**a**) *Borrelia burgdorferi*, a spirochete that causes Lyme disease, now the most prevalent tick-borne disease in the United States. (**b**) Deer ticks (*Ixodes*) are the common vector for this bacterium. Bottom to top on this finger, a female, male, and nymph. (**c**) Typical bull's-eye rash, the reaction to a tick bite.

Proteobacteria make up the largest, most diverse bacterial group. Gram staining tints their wall pink, so they also are known as Gram-negative bacteria. Many attack humans and other animals. The most common strain of *E. coli* lives inside the mammalian gut. This chemoheterotroph makes vitamin K, helps digest fat, and normally keeps foodborne pathogens in check by outcompeting them for nutrients.

There are photosynthetic proteobacteria, but they don't release oxygen and they don't use chlorophylls. Their bacteriochlorophylls are a similar pigment, with slightly different structure and absorption spectra.

Many proteobacteria are players in nutrient cycles. *T. namibiensis* helps connect sulfur and nitrogen cycles in the seas (Figure 19.8). It strips electrons from sulfur compounds as an energy source and also stores nitrate (in a big vacuole) for use as a final electron acceptor. *Rhizobium* causes nitrogen-fixing nodules to form on roots of peas and many other legumes. *Agrobacterium tumefaciens* causes crown gall tumors in plants, but it also is used in genetic engineering (Chapter 15).

Chlamydias are cocci that can't make ATP; they get it by slipping into animal cells, as parasites. *Chlamydia trachomatis* causes chlamydia, a sexually transmitted disease. Infection by this intracellular parasite also can result in trachoma, an eyelid disease that has blinded about 6 million people in developing countries.

Spirochetes look like stretched-out springs (Figure 19.9). All are motile. The group includes free-living species, symbionts, and parasites. *Pillotina* and other spirochetes live in the gut of termites and help them digest wood. The pathogenic *Borrelia burgdorferi* uses ticks as vectors and causes Lyme disease.

REGARDING THE "SIMPLE" BACTERIA

Bacteria are small. Their insides are not elaborate. *But bacteria are not simple.* They sense and move toward areas where nutrients are more plentiful and where other conditions also favor growth. Aerobes move to oxygen; anaerobes move away from it. Photosynthetic types move toward light but away from light that's too intense. Many species avoid toxins or predators.

Magnetotactic bacteria migrate in the ocean, and they grow best in oxygen-poor seawater. They contain a compass—chains of magnetic particles that respond to Earth's magnetic field. For instance, in the Northern Hemisphere, geomagnetic north points downward at a slight angle, so the migrating cells move into deeper water, which has less oxygen than surface waters.

Some bacteria show a collective behavior, as when millions of myxobacteria cells come together and form a "predatory" colony. These cells lack flagella but can glide about. They secrete enzymes that digest "prey," which include other bacterial cells. What's more, they migrate, change direction, and move as a single unit toward areas where there may be food. Finally, many myxobacteria colonies interact to form spore-bearing structures (fruiting bodies). Some cells differentiate to form a slime stalk. Others form branches or clusters of spores. Each spore holds a single living cell that can germinate and give rise to a new colony.

Bacteria are the most common and diverse prokaryotic cells. They include free-living cells with vital ecological roles, as well as pathogens of plants, animals, and humans.

19.3 The Archaea

Archaea are prokaryotic, but in other respects they are as similar to eukaryotes as they are to bacteria. Many survive in extreme habitats. Others are turning up almost everywhere. None are vertebrate pathogens.

THE THIRD DOMAIN

It's easy to see why archaea were once lumped with bacteria. Both have the same basic sizes and shapes. Both lack a nucleus. They differ a bit in metabolism and structure, but the differences just didn't seem to be significant enough to divide them into two groups.

Things changed in the 1970s. Molecular biologist Carl Woese began comparing the ribosomal RNAs of prokaryotes. Even though the rRNAs are necessary for protein synthesis, they can change a bit in certain base sequences without losing their function. Many small changes that have accumulated in rRNAs of different lineages can be measured directly.

To Woese's surprise, lineages that are adapted to high temperature or that produce methane gas formed a distinct group of organisms. They differed as much from bacteria as from eukaryotes. Woese proposed a new classification scheme in which Archaebacteria are equivalent to domains Bacteria and Eukaryota. Today we call the "third" domain Archaea. The Greek prefix *archae–* means "ancient." Biologists suspect that some archaea may resemble the first cells on Earth.

Through ongoing rRNA comparisons, we have now divided the archaea into three groups: Euryarchaeota, Crenarchaeota, and the newly discovered Korarchaeota (Figure 19.10). More data may reveal other groups.

HERE, THERE, EVERYWHERE

Physiologically, most of the archaea are **methanogens**, (methane makers), **extreme halophiles** (salt lovers), or **extreme thermophiles** (heat lovers). Methanogens and halophiles are euryarchaeotes, and most of the thermophiles are crenarchaeotes.

a

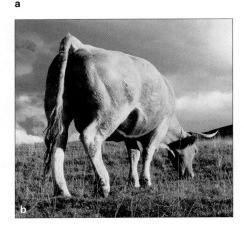

Figure 19.11 Methanogenic archaea. (**a**) *Methanocaldococcus jannaschii*, isolated near a deep-sea hydrothermal vent, grows at 85°C. (**b**) Three genera of archaea have been isolated from the cattle gut.

Methanogens have been isolated from marsh mud, Antarctic lakes, hydrothermal vents, and rocks deep below Earth's surface. Others are symbionts that live in the gut of termites or a few other animals (Figure 19.11). Some bacteria are also methanogens.

All archaean methanogens are strict anaerobes; free oxygen kills them. They strip electrons from hydrogen gas (H_2) or acetate in ATP-forming reactions that use carbon (from CO_2) as the final electron acceptor. In these reactions, methane (CH_4) forms as a by-product. Collectively, methanogens release about 2 billion tons of methane annually, so they have profound effects on the global cycling of carbon (Section 41.8).

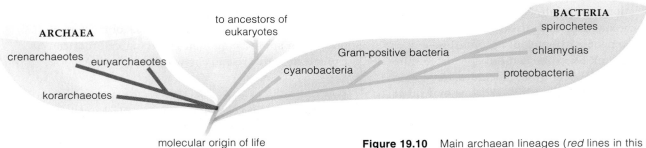

Figure 19.10 Main archaean lineages (*red* lines in this cladogram).

Figure 19.12 Life in extreme environments. (**a**) In saltwater evaporation ponds in Utah's Great Salt Lake, extreme halophiles (certain archaea and red algae) tint the water pink. (**b**) Extreme thermophiles live in Yellowstone National Park's hot springs. (**c**) The parasitic *Nanoarchaeum equitans* (smaller *blue* spheres) grows only when attached to *Ignicoccus* (larger spheres). Both archaea were isolated from 100°C water near a hydrothermal vent.

5 μm

Extreme halophiles live in the Dead Sea, Great Salt Lake, saltwater evaporation ponds, and other highly salty habitats (Figure 19.12a). Most get ATP by aerobic reactions but they also harness light energy when free oxygen is scarce. Their plasma membrane incorporates bacteriorhodopsin, a unique light-activated pigment. When this pigment absorbs sunlight energy, it changes shape in a way that pumps protons (H^+) out from the cell. Then H^+ flows back in, through an ATP synthase, and drives ATP formation.

Many extreme thermophiles live in sulfur-rich hot springs where water temperatures exceed 80°C (Figure 19.12b). Like methanogens, they are strict anaerobes. Unlike them, they use sulfur as an electron acceptor or donor in ATP-forming reactions. *Sulfolobus* cells grow in acidic hot springs. They can endure the same harsh conditions as the bacterium *Thermus aquaticus*, source of heat-stable DNA polymerases that are the basis of PCR, a DNA amplification method in biotechnology.

Other extreme thermophiles are the start of food webs in sediments around hydrothermal vents, where water temperatures exceed 110°C. They use hydrogen sulfide escaping from the vents as an electron source. Their presence at these vents adds to the evidence that life may have originated on the seafloor.

Nanoarchaeum equitans turned up during exploration of deep-sea hydrothermal vents off the coast of Iceland. This crenarchaeote is the smallest known cell and the only parasitic archaean (Figure 19.12c). Its genome is sequenced; to date, it is the smallest genome known.

Seawater collected near the coasts of Antarctica and California holds an abundance of archaea (primarily crenarchaeotes). So do mounds of methane hydrate on the ocean floor and sediments from the Great Lakes. And so do soils of agricultural fields, grasslands, boreal forests, and Siberian tundra. Biologists are now finding new species almost everywhere they look.

Today, Woese compares the discovery of archaea to the discovery of a new continent. He and many others are exploring and mapping this domain. By screening samples for archaean rRNA, they came to realize that members of this domain are diverse and widespread. They also realize that archaea and bacteria living in the same places may swap genes—and may do so often.

Archaea is the most recently defined domain of life. Many species thrive in extreme environments. Others are being discovered alongside bacteria in more hospitable habitats.

19.4 Viruses and Viroids

In ancient Rome, "virus" meant poison or venom. In the late 1800s, this rather nasty word was bestowed on newly discovered pathogens, smaller than the bacteria being studied by Louis Pasteur and others. Many viruses deserve the name. They attack animals, fungi, plants, protists, bacteria—you name it, there are viruses that infect it.

Today, we define a **virus** as a noncellular infectious particle that consists of DNA or RNA enclosed within a protein coat. The coat protects the genetic material as the virus journeys to a new host cell. Different viral coats consist of one or more types of protein subunits organized into a rodlike or many-sided shape (Figure 19.13). Complex viruses have sheaths, tail fibers, and other accessory structures attached to the coat. Some of the coat proteins on each type of virus chemically recognize and bind to specific receptors on host cells.

An outer envelope wraps around the coat of some viruses. It is composed mostly of membrane remnants from the previously infected cell, and it bristles with spikes of glycoproteins.

Cells of the vertebrate immune system can recognize viral proteins. However, genes for many of the proteins mutate often, so a modified virus might slip past the immune fighters. "Flu shots" prepare the human body to detect and battle some influenza viruses. Flu shots

are tailored each year to the modified coat proteins of particular strains of influenza viruses.

Each kind of virus multiplies only in specific hosts and cannot be studied easily except in living cells. This is why researchers study bacteria and bacteriophages, the class of viruses that infect them. Unlike the cells of complex multicelled organisms, both can be cultured easily and quickly. Bacteriophages were used in early experiments to determine the function of DNA. They are still used as research tools in genetic engineering.

Animal viruses contain double- or single-stranded DNA or RNA. In size, they range from parvoviruses (18 nanometers) to poxviruses (350 nanometers). Many cause diseases, including the common cold, influenza, measles, and some cancers. HIV causes AIDS. Section 34.10 describes how it weakens the immune system.

What happens during viral infections? Once host cells are infected, different viruses multiply in different ways. Even so, nearly all multiplication cycles proceed through five basic steps, as outlined here:

1. *Attachment.* A virus recognizes and binds to specific molecular groups at the host cell surface. Each type of virus can only infect cells having receptors for it.

2. *Penetration.* The entire virus particle or just its genetic material enters the cytoplasm of the host cell.

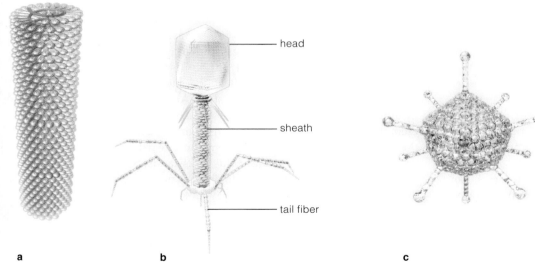

a   b   c   d

Figure 19.13 Body plans of viruses. (**a**) Protein subunits bound to nucleic acid form a spiral in rod-shaped viruses, including this tobacco mosaic virus. (**b**) Complex viruses, such as this bacteriophage, have structures attached to the coat. (**c**) Polyhedral (many-sided) viruses include this adenovirus. (**d**) A lipid envelope derived from membrane fragments of a lysed host cell surrounds HIV and other enveloped viruses. A capsid of regularly arrayed protein subunits encloses the viral RNA and a few enzymes. (**e**) Influenza virus. Spikes of glycoprotein project above its envelope. (**f**) One of the herpes viruses. Its envelope is pulled aside for this image.

3. *Replication and protein synthesis.* Viral DNA or RNA hijacks the host cell's machinery, which then churns out copies of viral nucleic acids and proteins.

4. *Assembly.* The new viral nucleic acids and proteins become organized as new infectious particles.

5. *Release.* Virus particles are released from the host cell by one mechanism or another.

Figure 19.14 shows two pathways that are common among bacteriophages. Steps from attachment through assembly are fast in a *lytic* pathway. Viral particles are released when damage to a host's plasma membrane, wall, or both lets cytoplasm dribble out. The host cell assists in its own **lysis**, or disintegration. It actually builds the viral enzyme that causes the damage.

In a *lysogenic* pathway, a latent period extends the cycle. The virus doesn't kill a host cell outright. A viral enzyme cuts the host chromosome, then integrates the viral genes into it. Before the cell divides, it replicates the recombinant molecule. Miniature time bombs thus await all of the host's descendants. Later, a molecular signal or some other stimulus may reactivate the cycle.

Latency is typical of many viruses, such as Type I *Herpes simplex*, which causes cold sores (fever blisters). This virus lies hidden in facial tissues. Emotional stress, even sunburn, can reactivate it; painful skin eruptions are the outcome. Like other enveloped viruses, this one enters a new host cell by a version of endocytosis, and it escapes by budding from the cell surface.

In multiplication cycles of RNA viruses, the viral RNA serves as a template for making mRNA or DNA. One enzyme of HIV and other retroviruses catalyzes DNA formation by reverse transcription.

Stripped-down as they are, viroids are simpler still. **Viroids** are circles of RNA with no protein coat and no protein-coding genes. Many viroids are pathogens that kill millions of dollars' worth of crop plants annually. Only one type has been implicated in a human disease. This viroid exerts its effect only when an individual is infected with the hepatitis D virus at the same time.

Viroids actually resemble introns, those noncoding portions of eukaryotic DNA you read about in Section 13.1. Are self-splicing introns that once escaped from DNA molecules? Or are they molecular remnants of an ancient RNA world? We don't know yet.

> *Viruses consist of nucleic acids enclosed in a protein coat, and some types have an outer envelope. Viroids are snippets of RNA. Both kinds of infectious agents can multiply only by hijacking the metabolic machinery of their hosts.*

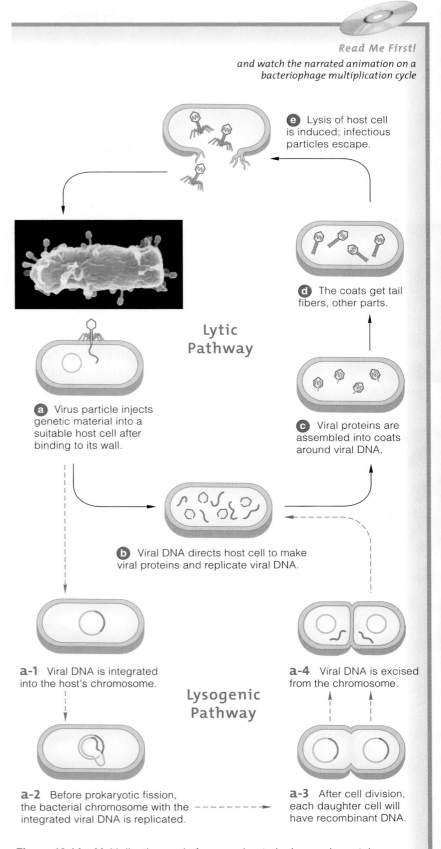

Read Me First!
and watch the narrated animation on a bacteriophage multiplication cycle

e Lysis of host cell is induced; infectious particles escape.

d The coats get tail fibers, other parts.

Lytic Pathway

a Virus particle injects genetic material into a suitable host cell after binding to its wall.

c Viral proteins are assembled into coats around viral DNA.

b Viral DNA directs host cell to make viral proteins and replicate viral DNA.

a-1 Viral DNA is integrated into the host's chromosome.

a-4 Viral DNA is excised from the chromosome.

Lysogenic Pathway

a-2 Before prokaryotic fission, the bacterial chromosome with the integrated viral DNA is replicated.

a-3 After cell division, each daughter cell will have recombinant DNA.

Figure 19.14 Multiplication cycle for some bacteriophages. In certain viruses, the lytic pathway may expand to include a lysogenic phase.

19.5 Evolution and Infectious Diseases

Just by being human, you are a potential host for diverse pathogenic viruses, protists, fungi, and parasitic worms. Before leaving the chapter, reflect on what this means from an evolutionary perspective.

THE NATURE OF DISEASE

Infection is the invasion of a cell or multicelled body by a pathogen (or the pathogen's genetic material). **Disease** follows when the pathogen multiplies and the metabolic activities of its descendants interfere with body activities. Table 19.2 gives some examples.

It takes direct contact with an infected person or their secretions for a contagious disease to spread. *Sporadic* diseases, such as whooping cough, occur irregularly among very few people. *Endemic* diseases, such as tuberculosis, are more or less always present in a population but are confined to a small part of it.

During an *epidemic*, a disease quickly spreads through part of a population for just a limited time, then it subsides. *Pandemic* refers to epidemics of the same disease breaking out in several countries in the same interval of time.

AIDS is a pandemic with no end in sight. A 2003 outbreak of SARS (short for *Severe Acute Respiratory Syndrome*) was a brief pandemic. It started in China, and travelers quickly brought it to countries around the world. Before government-ordered quarantines halted its spread, thousands were sickened and 774 died. A previously unknown coronavirus causes SARS. It has reappeared in China and may not subside entirely.

Now consider disease in terms of the pathogen's prospects for survival. A pathogen stays around only for as long as it has access to outside sources of energy and raw materials. To a tiny bacterial or viral pathogen, a human body is the jackpot. With bountiful resources, a pathogen may multiply to astounding population sizes. The point? Evolutionarily speaking, the ones that leave the most descendants win.

Two barriers prevent pathogens from taking over the world. First, any species with a history of being attacked by a specific pathogen has coevolved with it and has built-in defenses against it. Vertebrates, with their immune systems, are examples. Second, if a pathogen kills its individual host too quickly, it might vanish along with the individual. This is one reason most pathogens have less-than-fatal effects. After all, infected individuals who can live longer spread more germs and contribute to the pathogen's reproductive or replicative success.

Usually, the individual will die only if it becomes host to overwhelming numbers of a pathogen, if it is a novel host with no coevolved defenses, or if a mutant pathogenic strain has emerged and has breached the current defenses.

DRUG-RESISTANT STRAINS

Antibiotics are compounds synthesized by one organism that can kill another, and we have learned to use them as weapons against bacterial pathogens. Penicillin, our first widely used antibiotic, has been mass produced since 1943. Penicillin and other antibiotics have saved millions of lives. However, as explained in Chapter 15, these drugs act as agents of directional selection. As a result, resistant forms of many bacteria are on the rise.

Lateral gene transfers among bacterial cells of the same or different species contribute to the problem. Many bacterial pathogens swap genes like teens swap music files. The genes that confer antibiotic resistance move about in plasmids or as a free piece of DNA.

Antiviral drugs are also losing their punch. Here's an example. In one study, about 27 percent of newly diagnosed patients with AIDS had been infected by an HIV strain resistant to at least one antiretroviral drug. Multiple classes of drugs didn't work in more than 10 percent of these patients. Resistant HIV strains had evolved in other individuals, before the patients in the study had taken any of the drugs.

FOODBORNE DISEASES AND MAD COWS

According to Centers for Disease Control estimates, contaminated food sickens as many as 80 million Americans annually. Most vulnerable are the very

SARS VIRUS

MYCOBACTERIUM TUBERCULOSIS

Table 19.2	The Eight Deadliest Infectious Diseases		
Disease	Main Agents	Estimated New Cases per Year	Estimated Deaths per Year
Acute respiratory infection*	Bacteria, viruses	1 billion	4.7 million
Diarrhea**	Bacteria, viruses, protozoans	1.8 billion	3.1 million
Tuberculosis	Bacteria	9 million	3.1 million
Malaria	Sporozoans	110 million	2.7 million
AIDS	Virus (HIV)	5.6 million	2.6 million
Measles	Viruses	200 million	1 million
Hepatitis B	Virus	200 million	1 million
Tetanus	Bacteria	1 million	0.5 million

* Includes pneumonia, influenza, and whooping cough.
** Includes amoebic dysentery, cryptosporidiosis, and gastroenteritis.

Figure 19.15 (**a**) Charlene Singh, the first reported case of an American resident affected by vCJD, being cared for by her mother. The disease has since claimed Charlene's life. (**b**) Section through a brain damaged by bovine spongiform encephalopathy (BSE). The light "holes" are areas where tissue has been destroyed. (**c**) Model of a normal prion protein. The vCJD-causing version misfolds into a different three-dimensional structure.

young or old, or those with a compromised immune system. *Campylobacter*, *Salmonella*, *Staphylococcus aureus*, and *Listeria* are frequent contaminants of dairy products, meat, and poultry. All can cause abdominal pain, nausea, and diarrhea. *Listeria* infection during pregnancy can cause miscarriage or premature birth.

0157:H7 is one of the pathogenic strains of *E. coli*, a normally harmless bacterial species that lives in your gut. Tons of ground beef contaminated with 0157:H7 have been recalled from vendors. Even vegetarians are at risk, because 0157:H7 has been found in fruit juices, lettuce, green onions, and alfalfa sprouts.

Some simple practices can help you avoid bacterial food poisoning. Thoroughly cook meats and poultry. Avoid unpasteurized dairy products, and don't eat raw eggs. Kitchen sponges can be sanitized by a trip through the dishwasher. So can utensils that have been in contact with uncooked meat or poultry.

Bacteria aren't the only foodborne threats. The first case of bovine spongiform encephalopathy (BSE) in the United States was reported in 2003. A single cow in Washington State developed this "mad cow disease." Meat from the herd was recalled and consumers were assured that none had entered the human food supply. What if it had? Before 2000, thousands of cattle with BSE entered food chains, mainly in Great Britain. So far, that meat has caused about 153 cases of a fatal brain disease, variant Creutzfeldt–Jakob disease (vCJD).

Charlene Singh was one of those vCJD cases (Figure 19.15). She probably contracted the disease when she was growing up in Britain, before moving to Florida. In 2001, the mood swings, balance problems, and

memory loss started. She soon became bedridden and could not control her body functions or communicate. Charlene died in the summer of 2004.

According to the most widely accepted hypothesis, prions cause BSE, vCJD, and some other degenerative diseases of humans and other animals, including deer and elk. **Prions** are infectious proteins. Normal prions, which are present in tissues of the nervous system, are folded into a characteristic configuration. Infectious prions are misfolded (Figure 19.15c). In some way, they induce normal prions to become misfolded, also.

Misfolded prions accumulate and form massive deposits in the brain. Brain tissue samples taken from individuals who died from one of these diseases are riddled with holes (Figure 19.15b). The holes give the tissue a spongiform (sponge-like) appearance.

The prions that cause BSE and vCJD are nearly identical. They may be mutated strains of prions that cause scrapie in sheep. Apparently some prions were transferred from sheep to cattle by way of livestock feed that contained animal parts. Governments have now banned the use of such feed.

Prions are not alive, so they can't be killed. It is not easy to destroy them through denaturation; they resist boiling, baking, irradiation, and most disinfectants.

Should you worry about eating American beef? The USDA says no. If you are a beef lover and a skeptic, avoid eating brains or meat cuts that contain spinal cord, such as oxtails. Ground beef and sausage may accidentally contain nervous tissue, so watch how it is being made or make your own. You may also want to look for meat from grass-fed cattle.

Summary

Section 19.1 Bacteria and archaea are prokaryotic cells; they do not have a nucleus. Some species have membrane infoldings and other structures in the cytoplasm. None has a profusion of organelles. As a group, prokaryotes show great diversity in modes of acquiring energy and carbon. Prokaryotic fission is a cell division mechanism used only by prokaryotic cells. Replication of the bacterial chromosome is followed by the division of a parent cell into two genetically equivalent daughter cells. Prokaryotes transfer DNA by exchanging plasmids (conjugation).

Section 19.2 Bacteria are the most common and diverse prokaryotes. The cyanobacteria are oxygen-releasing photoautotrophs. Gram-positive bacteria have thick cell walls. Most are chemoheterotrophs. Some are pathogens. Gram-negative proteobacteria are the most diverse bacterial group, with autotrophs and heterotrophs, free-living species, symbionts, and pathogens. The chlamydias are all intracellular parasites. The spring-shaped spirochetes include symbionts, free-living species, and pathogens.

Section 19.3 The archaea are prokaryotic, but in some respects they resemble eukaryotes as much as they do bacteria. They are now placed in a separate domain from the bacteria. Distribution and diversity of the group are highly active research areas.

Comparative rRNA studies give evidence of three archaean groups: euryarchaeotes, crenarchaeotes, and korarchaeotes. Most of the methanogens (methane makers) and halophiles (salt lovers) are in the first group, and most extreme thermophiles belong to the second group. Archaea and bacteria coexist in many habitats and apparently engage in gene transfers.

Section 19.4 Viruses and viroids are noncellular infectious agents. Each virus particle consists of a core of DNA or RNA and a protein coat that in some species is enclosed in a lipid envelope. All viral multiplication cycles go through five steps: attachment to a host; then penetration into it; replication and protein synthesis; assembly of new particles; then their release. Cycles differ in their details. An example is the lytic pathway of bacteriophages, which may include a latent phase.

Viroids are circles of infectious RNA with no protein or protein-coding genes. They resemble introns. Many are plant pathogens. Only one kind infects humans.

Section 19.5 Infection is the invasion of a cell or multicelled body by a pathogen, which causes disease when its activities interfere with body functions. Hosts and pathogens coevolve. Antibiotics and antiviral drugs select for drug-resistant strains. Many bacteria cause food poisoning. Prions are infectious proteins that kill by destroying the brain and nervous tissue.

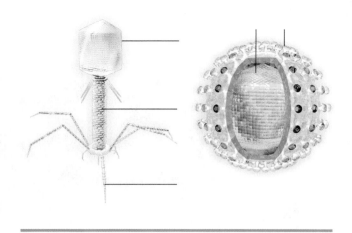

Self-Quiz

Answers in Appendix III

1. Label the components of the viruses shown above.

2. Only _____ are prokaryotic.
   - a. archaea
   - b. bacteria
   - c. viroids
   - d. a and b are correct

3. None of the _____ are human pathogens.
   - a. bacteria
   - b. viruses
   - c. archaea
   - d. prions

4. Bacteria transfer plasmids by _____ .
   - a. prokaryotic fission
   - b. endospore formation
   - c. conjugation
   - d. the lytic pathway

5. The _____ are all oxygen-releasing photoautotrophs.
   - a. spirochetes
   - b. chlamydias
   - c. cyanobacteria
   - d. proteobacteria

6. The symbiotic *E. coli* in your gut are _____ .
   - a. spirochetes
   - b. chlamydias
   - c. cyanobacteria
   - d. proteobacteria

7. All _____ are intracellular parasites of vertebrates.
   - a. spirochetes
   - b. chlamydias
   - c. cyanobacteria
   - d. proteobacteria

8. Some Gram-positive bacteria (e.g., *Bacillus anthracis*) survive harsh conditions by forming a(n) _____ .
   - a. pilus
   - b. heterocyst
   - c. endospore
   - d. plasmid

9. DNA or RNA may be the genetic material of _____ .
   - a. bacteria
   - b. a virus
   - c. archaea
   - d. a prion

10. Is this statement true or false: All viruses consist of genetic material, a protein coat, and an outer envelope.

11. Bacteriophages can multiply by _____ .
   - a. prokaryotic fission
   - b. the lytic pathway
   - c. the lysogenic pathway
   - d. both b and c

12. Only _____ reproduce by prokaryotic fission.
   - a. viruses
   - b. archaea
   - c. bacteria
   - d. b and c are correct

13. Match the terms with their most suitable description.
   - ___ archaea
   - ___ bacteria
   - ___ virus
   - ___ plasmid
   - ___ extreme halophile
   - ___ viroid
   - ___ prion
   - a. infectious protein
   - b. nonliving infectious particle; nucleic acid core, protein coat
   - c. salt lover
   - d. prokaryotes that most closely resemble eukaryotes
   - e. most common prokaryotic cells
   - f. small circle of bacterial DNA
   - g. small circle of RNA

Figure 19.16 A radura, international symbol for irradiated food. It means that a product has been exposed to high-energy radiation to kill foodborne pathogens. Irradiation cannot make the food itself radioactive.

Figure 19.17 Chinese civet cats. In 2004, a waitress who worked in a restaurant that serves their meat was one of the first reported cases of SARS. Thousands of these animals were slaughtered to halt the spread of the virus.

Critical Thinking

1. Perhaps you think the world would be better off if there were no viruses. If so, consider this. Curtis Suttle at the University of British Columbia studies interactions among certain viruses, bacterial communities, and algal communities in ocean water. He selectively removed viruses from seawater and found that algae stopped growing. Further investigation revealed that they were dependent on nutrients released by dying bacteria that lysed after viral infection. Make a list of other ways in which viruses might play vital roles in ecosystems.

2. One way to prevent bacterial food poisoning is by *food irradiation*, exposing food to high-energy rays or beams that kill pathogenic bacteria. The process also slows spoilage and prolongs shelf life. Some think this is a safe way to protect consumers. Others think the treatment could produce harmful chemicals. They say irradiation does not kill endospores that can cause botulism. In their view, the best way to prevent food poisoning is to tighten and enforce food safety standards.

Irradiated meat, poultry, and fruits are now available in many supermarkets. By law, they must be marked with a special symbol (Figure 19.16). Would seeing this symbol on a package make you more or less likely to purchase the product? Explain your answer.

3. How did the 2003 SARS pandemic start? Probably with Chinese civet cats (Figure 19.17) and other wildlife coveted as food delicacies in China. These animals harbor a coronavirus. The viral DNA sequence is nearly identical with a sequence that causes SARS in humans.

Gene sequencing studies show that the animal virus has a unique stretch of twenty-nine added nucleotides. Because of the additions, researchers suspect that the SARS virus probably jumped from wildlife to humans, rather than in the opposite direction. Review Section 11.7, then explain their reasoning.

4. *Thermotoga maritima* was discovered in geothermically heated marine sediments and grows best at about 80°C. Analysis of rRNA puts it squarely in the bacterial group, but sequencing of the entire genome presents a far more complicated picture. About 50 percent of the *T. maritima* genes resemble bacterial genes, about 25 percent of its genes are unique, and about 25 percent resemble those of archaea, especially the thermophile *Pyrococcus horikoshii*. Suggest two possible explanations for the similarity between these members of different domains.

Media Menu

Student CD-ROM

Impacts, Issues Video
    West Nile Virus Takes Off
Big Picture Animation
    Prokaryotes, viruses, and disease
Read-Me-First Animation
    Prokaryotic fission
    A bacteriophage multiplication cycle
Other Animations and Interactions
    Prokaryotic body plan
    Body plans of viruses

InfoTrac

- Germs Make the Man: Your Body Is Teeming with Trillions of Infectious Microbes. That's a Very Good Thing. *Fortune*, January 2003.
- What the Bugs Can Do For You. *Time*, July 2002.
- Caught Off Guard: SARS Reveals Gaps in Global Disease Defense. *Scientific American*, June 2003.
- The Virus Hunter. *British Medical Journal*, October 2003.
- Flying Fever. *Audubon*, July 2000.

Web Sites

- MicrobeWorld: www.microbeworld.org
- Intimate Strangers: www.pbs.org/opb/intimatestrangers
- Bad Bug Book: vm.cfsan.fda.gov/~mow/intro.html
- Institute for Molecular Virology: virology.wisc.edu/IMV
- National Center for Infectious Diseases: www.cdc.gov/ncidod

How Would You Vote?

West Nile virus is carried by mosquitoes. Many localities are spraying pesticides to kill these insect vectors. The pesticides have environmental and possibly health side effects. It is unlikely that spraying will successfully eliminate all of the virus-carrying mosquitoes. Would you support a spraying program in your community?

IMPACTS, ISSUES *Tiny Critters, Big Impacts*

Go for a swim just about anywhere and you will be sharing the water with multitudes of protists. Like their most ancient ancestors, almost all of these eukaryotic species are aquatic. Many kinds are photosynthesizers or predators in freshwater or the seas. Others live in cells or moist tissues of host species. Name a fungus, plant, or animal, and most likely a symbiotic, parasitic, or pathogenic protist associates with it.

You'll also find symbionts, parasites, and pathogens among the fungi, but far more kinds are free-living decomposers. Unlike protists, most fungi live on land.

Structurally, protists and fungi are the simplest of all eukaryotes. Most are microscopically small. And yet, collectively, some have big impacts on our world.

For instance, coccolithophores and foraminiferans drift about in the world ocean. These tiny cells have a shell or plates hardened with calcium carbonate. Long ago, calcium-rich remains of ancient species started to accumulate on the seafloor. Gradually, the deposits were compressed and then tectonic forces lifted them above sea level. We know the deposits as more familiar materials—chalk and limestone.

Dover, England, is renowned for its white chalk cliffs (Figure 20.1). The cliffs are a monument to individual coccolithophores that drifted down to the seafloor and formed 0.5-millimeter-thick deposits every year for millions of years. Austin, Texas, sits on 400-meter-thick chalk deposits, formed in another ancient sea.

Figure 20.1 Two examples of protistan architecture. *Left*, towering 300 feet above the English Channel, the white chalk cliffs of Dover, England. *Right*, Hagar Qim temple, Malta. The filmstrip at *far right* shows a fossilized shell of one kind of foraminiferan, *Textularia*, that lived in Miocene times.

the big picture

branch leading to fungi

?
"crown" of eukaryotes (rapid divergences)

What Is a Protist? Structurally, protists are the simplest eukaryotes and the ones most like the first eukaryotic cells on Earth. They were already diverging rapidly when fungi arose. Researchers are discovering how divergent lineages are related and are reclassifying major groups.

Single-Celled Types A dazzling variety of flagellated, ciliated, and amoeboid cells arose from heterotrophic ancestors. Modern groups live in almost every aquatic or moist environment, including tissues of humans and other animals.

In 3200 BC on Malta, foraminiferan-rich limestone deposits were quarried to build the Hagar Qim temple, now a World Heritage site (Figure 20.1). A few thousand years later, other limestones were used to build the Egyptian pyramids and Sphinx. Look closely at the walls of the Empire State Building, the Pentagon, or Chicago's Tribune Tower, and you'll find foraminiferan shells.

Tiny cells also can make their presence known in a big way. One of the water molds, *Phytophthora infestans*, infects potatoes and causes the disease late blight. In the late 1840s, this pathogen destroyed potato crops that were the main food staple in Europe. Millions of people starved to death or emigrated to other countries.

Today, related species threaten forests in the United States, Europe, and Australia. *P. ramorum* started an epidemic of sudden oak death in California. Tens of thousands of oaks have died, and the pathogen has now jumped to madrone, redwoods, and other novel hosts. Cascading ecological changes will diminish sources of food and shelter for forest species.

Contrasting these two examples can help you keep the global perspective in mind. Some simple eukaryotes do cause disease or damage, and we tend to assign "value" to them in terms of their direct impact on our lives. There is nothing wrong with battling the harmful kinds and admiring the beneficial ones. Even so, don't lose sight of how protists or fungi or any other kind of organism fit in nature's larger picture.

 How Would You Vote?

The pathogen that causes sudden oak death has been found infecting twenty-six kinds of plants in California and Oregon. Some infected species are commonly sold as nursery stock. Should states that are free of the pathogen be allowed to prohibit the shipping of plants from states that are affected? See the Media Menu for details, then vote online.

Mostly Multicelled Types   Stramenopiles and groups informally called algae also are structurally simple eukaryotes. They range from single-celled species to giant kelps and include the closest relatives of plants. Chloroplasts hint at endosymbiotic origins for some photosynthetic species.

The Fabulous Fungi   Nearly all fungi are multicelled heterotrophs. Most species are decomposers that help cycle nutrients in communities. Others are symbionts, pathogens, and parasites. We recognize three major lineages, mainly by the type of sexual spores they produce.

20.1 Characteristics of Protists

Traditionally, kingdom Protista includes photoautotrophs, decomposers, parasites, and predators. Most are single-celled, but nearly every lineage also has multicelled forms.

Of all existing species, **protists** are the most like the first eukaryotic cells. Unlike prokaryotes, protist cells have a nucleus. Most also have mitochondria, ER, and Golgi bodies. Their ribosomes are larger than those in bacteria and they have more than one chromosome, each consisting of DNA with many proteins attached. They have a cytoskeleton that includes microtubules. Many cells have chloroplasts. And unlike prokaryotes, protist cells divide by way of mitosis, meiosis, or both.

Of course, these also are defining features of plant, fungal, and animal cells. Until recently, protists were defined largely by what they are *not*—not bacteria, not plants, not fungi, not animals. They were lumped into a kingdom meant to signify an evolutionary crossroad between prokaryotes and "higher" forms of life.

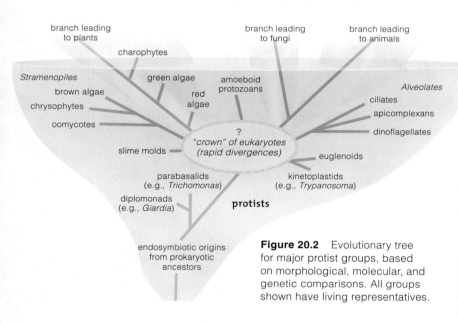

Figure 20.2 Evolutionary tree for major protist groups, based on morphological, molecular, and genetic comparisons. All groups shown have living representatives.

Phylogenetic threads are still being untangled with the help of molecular comparisons. So consider Figure 20.2, an evolutionary tree for these organisms, as a work in progress. "Protists" are being sorted out into a number of monophyletic groups. Here we can only sample their astounding diversity.

> *Protists are organisms thought to be most like the earliest eukaryotes. Most are single cells, but nearly all lineages include multicelled forms. Different lineages appear to have evolutionary connections with plants, fungi, and animals.*

20.2 The Most Ancient Groups

Eukaryotic species arose hundreds of millions of years ago. Studying the single-celled forms that are their descendants can help give us an idea of how cellular complexity evolved.

FLAGELLATED PROTOZOANS

Flagellated protozoans are heterotrophs having one or more flagella and a **pellicle**. This thin, translucent, protein-rich outer covering imparts shape to the cell. Soil, seawater, and freshwater often hold free-living, bacteria-eating species. The termite gut has mutualist species that make cellulose-degrading enzymes. The human body can be host to parasitic types.

Giardia lamblia intrigues evolutionary biologists. It may be a representative of one of the first eukaryotic lineages because it has no mitochondria or lysosomes. *G. lamblia* parasitizes humans, cattle, and wild animals (Figure 20.3a). It can survive outside the host body, as cysts, in feces-contaminated water. Generally, a **cyst** is a protective covering formed by cell secretions. Ingesting *G. lamblia* cysts invites *giardiasis*. The symptoms of this disease can range from mild cramps to severe diarrhea that lasts for weeks.

Another group, the trichomonads, also lack mitochondria. Worldwide, *Trichomonas vaginalis* infects about 170 million people. The pathogen is transmitted mainly during sexual intercourse. It can damage the urinary and reproductive tracts of hosts unless they receive prompt treatment.

Trypanosomes form another group of flagellated protozoans. They have one large mitochondrion and a kinetoplast, a unique structure that houses the DNA. Biting insects spread species that can infect humans. *Trypanosoma brucei* causes African sleeping sickness (Figure 20.3b). *T. cruzi* causes Chagas disease in South and Central America. Both diseases are fatal. During Iraq conflicts, sand fleas that are vectors for *Leishmania donovani* bit some American soldiers and caused cases of skin leishmaniasis. The soldiers are being treated with surgery and experimental drugs.

EUGLENOIDS

Euglenoids live in freshwater ponds and lakes. Like many other freshwater protists, they have a contractile vacuole that expels excess water in the cell body. The euglenoids are related to trypanosomes, and all have more than one mitochondrion. They also have a cluster of pigments, called an eyespot, that partially shields a light-sensitive receptor. When activated, the receptor helps direct the cell toward locations that are suitable for photosynthesis. Figure 20.4 shows the body plan.

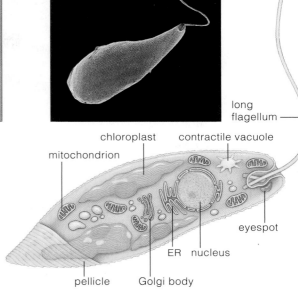

long flagellum

chloroplast   contractile vacuole

mitochondrion

eyespot

ER   nucleus

pellicle   Golgi body

Figure 20.3 Two flagellated protozoans. (**a**) Scanning electron micrograph of *Giardia lamblia*. The cell adheres to intestinal epithelium by means of its sucking disk. (**b**) Light micrograph of *Trypanosoma brucei*, the cause of African sleeping sickness. The tsetse fly is its insect vector.

Figure 20.4 Body plan and light micrograph of *Euglena*.

The earliest euglenoids were colorless heterotrophs that ingested dissolved organic matter. Some still are, but about half are photoautotrophs. Their chloroplasts probably arose by endosymbiosis after some ancient euglenoid engulfed green algal cells. The chloroplasts, like those of green algae, hold chlorophylls *a* and *b*.

AMOEBOID PROTOZOANS

The first **amoeboid protozoans** had permanent motile structures, but these structures were lost as the group evolved. Pseudopods, or "false feet," became their means of motility. Naked amoebas, radiolarians, and foraminiferans are in this group.

Each pseudopod, again, is a lobe of cytoplasm that forms and retracts as microtubules assemble and then disassemble (Sections 4.10 and 5.8). These cytoskeletal elements are especially evident in naked amoebas and foraminiferans. Naked amoebas have soft bodies and no symmetry (Figure 20.5). Most types are free-living phagocytes in freshwater or in the soil. One species, *Entamoeba histolytica*, can live in the human intestines, where it may cause amoebic dysentery.

Perforated shells protect living foraminiferans and radiolarians. Prey-capturing pseudopods project from the fine perforations, as in Figure 20.6*b*. Nearly all of the radiolarians and some foraminiferans drift with ocean currents, as components of plankton. **Plankton** refers to aquatic communities of mostly microscopic organisms that drift or swim weakly in water. More species of foraminiferans live on the ocean floor, and a few have been discovered in lake sediments.

Among the living representatives of the most ancient protists are a variety of heterotrophic cells and one lineage that apparently acquired its chloroplasts by endosymbiosis.

Figure 20.5 *Amoeba proteus*, a naked amoeboid protozoan that is a favorite for experiments in biology classes. Compared to other species, it has stubbier pseudopods.

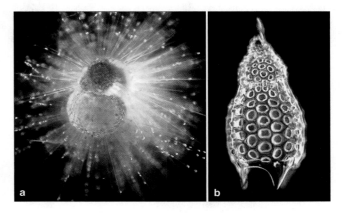

Figure 20.6 Amoeboid protozoans. (**a**) A live foraminiferan. The many tiny yellow spheres on its thin pseudopods are golden algal symbionts. This protist is about as small as the eye of a needle. (**b**) Radiolarian shell.

20.3 The Alveolates

Ciliates, apicomplexans, and dinoflagellates are a monophyletic group—the alveolates. All of these single cells have tiny, membrane-bound sacs just under their outermost membranes. The sacs, called alveoli (singular, alveolus), may help stabilize the cell surface.

CILIATED PROTOZOANS

Ciliated protozoans, or **ciliates**, live in freshwater, saltwater, and soil. Most are free-living predators, but some attach to substrates or form colonies. About 30 percent are endosymbionts of diverse animals. The majority have many cilia that beat in synchrony and function in motility, in directing food toward an oral cavity, or both. Ciliates eat bacteria, algae, and one another. The chloroplasts of algal prey are not always digested, and the cell that ingested them often uses part of the compounds that the chloroplasts continue to make. Some ciliates even have bacterial symbionts.

Figure 20.7 shows *Paramecium*, a freshwater species. A mature cell is about 200 micrometers long. Outer membranes form a pellicle. An oral cavity that opens on the surface leads to a gullet. Ingested food gets enclosed in enzyme-filled digestive vesicles. Like a number of other protists, the ciliates have contractile vacuoles that get rid of excess water in the cell body.

Most ciliates can reproduce sexually and asexually. Each typically has two types of nuclei and distributes them to daughter cells in a complex way. Some engage in binary fission, an asexual reproductive mode that divides the cell body into two parts.

APICOMPLEXANS

Apicomplexans are parasitic alveolates equipped with a unique microtubular device that can attach to and penetrate a host cell. The tip of the device protrudes from the cell's anterior end. Plastids much like the ones in red algae occur in many species. They may be the legacy of past endosymbiotic shenanigans.

More than 4,000 apicomplexan species are known to parasitize animals ranging from insects to humans. Here we will focus on four species of *Plasmodium* that cause the disease *malaria*. These species have infected more than 300 million people in tropical parts of the world, especially sub-Saharan Africa.

The bite of an infected female *Anopheles* mosquito puts the parasite in a host. Sporozoites, a motile stage, travel the bloodstream to the liver. They reproduce asexually in liver cells until the cells burst and release another stage—the merozoites. Some of the merozoites reinfect the liver. Others enter red blood cells, where they develop into types of immature gametes known as male and female gametocytes (Figure 20.8).

Amazingly, *Plasmodium* is sensitive to temperature and oxygen levels of its hosts. The gametocytes cannot mature in humans, because humans are warm-bodied and there is little free oxygen in human blood; most is bound to hemoglobin. Gametocytes must mature in a female mosquito, which has a lower body temperature and takes in oxygen from air along with blood. In this insect's gut, zygotes form, divide, and give rise to sporozoites. This infective stage then migrates to the mosquito's salivary glands, there to await a new bite.

Malaria symptoms start after the infected liver cells rupture and release merozoites, metabolic wastes, and cellular debris into the bloodstream. Shaking, chills, a burning fever, and drenching sweats follow. After one episode, symptoms usually subside for a few weeks or months. Infected individuals might feel healthy, but they should expect relapses. Jaundice, kidney failure, convulsions, and coma are outcomes of the disease.

Many *Plasmodium* strains are now resistant to older antimalarial drugs. Artemisinin, a compound isolated from sweet wormwood (*Artemisia*), shows promise. It lowers fever and reduces the number of parasites in the blood, so it can slow the course of infection.

Efforts to design a vaccine are ongoing. At this writing, a new vaccine based on a merozoite surface protein is being tested in African adults. If the clinical trials go smoothly and the vaccine proves safe and effective, widespread clinical trials will probably start in 2006 or 2007. The sooner the better—every thirty seconds, one African child dies of malaria.

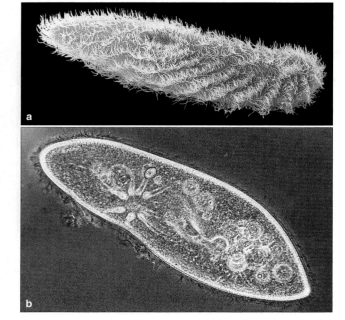

Figure 20.7
(**a**) Surface of *Paramecium*, a single-celled predator. You can see the start of its oral cavity, the depression in the cell surface.
(**b**) A look at *Paramecium* organelles.

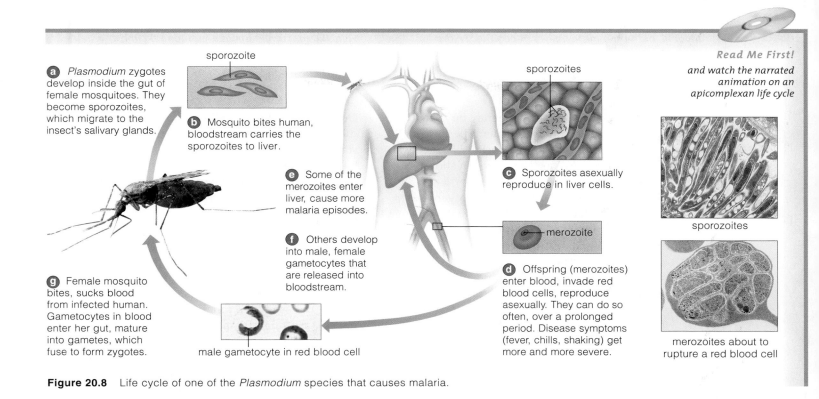

Read Me First!
and watch the narrated
animation on an
apicomplexan life cycle

a *Plasmodium* zygotes develop inside the gut of female mosquitoes. They become sporozoites, which migrate to the insect's salivary glands.

sporozoite

b Mosquito bites human, bloodstream carries the sporozoites to liver.

sporozoites

c Sporozoites asexually reproduce in liver cells.

sporozoites

e Some of the merozoites enter liver, cause more malaria episodes.

merozoite

f Others develop into male, female gametocytes that are released into bloodstream.

d Offspring (merozoites) enter blood, invade red blood cells, reproduce asexually. They can do so often, over a prolonged period. Disease symptoms (fever, chills, shaking) get more and more severe.

merozoites about to rupture a red blood cell

g Female mosquito bites, sucks blood from infected human. Gametocytes in blood enter her gut, mature into gametes, which fuse to form zygotes.

male gametocyte in red blood cell

Figure 20.8 Life cycle of one of the *Plasmodium* species that causes malaria.

DINOFLAGELLATES

Dinoflagellates are single-celled alveolates that most often have cellulose plates just beneath the plasma membrane plus one flagellum at the posterior end and another in a groove that runs around the cell body. About half of the species in this diverse group are key producers in freshwater and marine habitats. Some prey on bacteria and algae, others parasitize marine or freshwater fish and invertebrates.

Remember the theory of endosymbiosis, explained in Section 18.4? Dinoflagellate chloroplasts appear to be descended from red or green algae that evolved as symbionts with cells that ingested them. In addition, many dinoflagellates are endosymbionts of corals.

At high densities, bioluminescent dinoflagellates in tropical seas can have dramatic effects. Disturb them and the water shimmers (Figure 20.9). The light might be like a burglar alarm that attracts help, such as larger predators that can eat the disturber. Less pretty are the fish kills that result from high densities of the toxic species. We turn to this topic in the next section.

Alveolates are diverse single-celled eukaryotes that swim or drift in lakes, seas, and the bloodstream of animals. Ciliates, apicomplexans, and dinoflagellates are in this group.

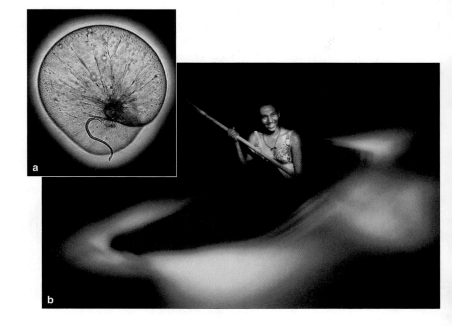

Figure 20.9 (**a**) *Noctiluca scintillans*, a bioluminescent dinoflagellate tinted red by symbionts living inside it. (**b**) Near Vieques Island, Puerto Rico, a liter of water holds about 5,000 of these cells, which flash with blue light when agitated. Like fireflies and other bioluminescent species, *Noctiluca scintillans* makes luciferase, an enzyme that destabilizes a luciferin molecule by catalyzing the transfer of a phosphate group to it. When the luciferin restabilizes, it releases energy as fluorescent light.

20.4 Algal Blooms

FOCUS ON THE ENVIRONMENT

Algal blooms are huge increases in population sizes of aquatic protists, most often dinoflagellates. They tend to develop in warm, shallow water. Nutrient-rich pollutants, as in agricultural run-off, fan the population growth.

During a typical algal bloom, there may be as many as 8 million cells in each liter of water. The cells of some species tint the water red, especially near coasts; hence the name *red tide*. Vast blooms even show up in images from space satellites.

The blooms can sicken or kill aquatic organisms by polluting the water with metabolic wastes and sometimes potent toxins. The enormous numbers of algal cells inevitably die, and aerobic bacteria go to work on the remains. Their metabolic action depletes oxygen from the water, so aquatic animals suffocate.

As an example, populations of *Karenia brevis* often bloom near the coasts of the Gulf of Mexico and the Atlantic seaboard (Figure 20.10). This dinoflagellate makes brevetoxin, which binds to sodium pumps in cell membranes and disrupts nerve cell function. *K. brevis* blooms have killed billions of fishes. They have also sickened and killed seabirds, bottlenose dolphins, sea turtles, and manatees.

What about humans? Eat shellfish contaminated by brevetoxins and you may end up with a case of *neurotoxic shellfish poisoning* (NSP). The toxin rarely is dangerous. It does cause gut irritation, tingling, dizziness, and muscle cramps that can last for days. Even swimming in toxin-tainted water may irritate your skin and eyes, and walking on a beach near a *K. brevis* bloom can irritate airways to your lungs.

Figure 20.10 (**a**) *Karenia brevis*, a toxin-producing dinoflagellate. Its two flagella beat in such a way that they spin the cell through water. (**b**) In the autumn of 2003, blooms of *K. brevis* and related species near Naples, Florida, killed hundreds of fishes. Dead fishes washed up on beaches, including this one.

20.5 The Stramenopiles

Flagellated stramenopiles are among the most ancient eukaryotic lineages. Even multicelled types have flagellated spores. Unique filaments project like tinsel from one of two flagella.

Oomycotes and the mostly photosynthetic species called chrysophytes and brown algae belong to the stramenopile lineage. All oomycotes are colorless, but the other groups make chlorophylls *a* and *c* and fucoxanthin, a distinctive brown pigment.

OOMYCOTES

The water molds, downy mildews, and white rusts are common **oomycotes**. Their name means egg fungus; it refers to an egg cell that forms by way of meiosis in a parent cell. Unlike others in the group, oomycotes have no chloroplasts. Like fungi, these heterotrophs absorb nutrients through filaments that form in the life cycle. Many are decomposers of aquatic habitats. *Saprolegnia*, an opportunistic parasite, is quick to infect freshwater fishes with damaged tissues. Fuzzy white patches on fish skin are a sign of infection. The oomycotes called downy mildews are pathogens. A notorious species, *Plasmopara viticola*, infects grapevines and fruits.

The *Phytophthora* species introduced at the start of the chapter live up to their name—plant destroyer. All are plant pathogens. *P. infestans* can destroy a field of potatoes in a single day. *P. ramorum* is killing many of California's native trees and woody shrubs. Dark oozing from cankers on trunks is a telltale symptom of infection (Figure 20.11).

Figure 20.11 Dead trees on a hill near Big Sur, California, are part of an ongoing epidemic of sudden oak death. Despite the disease's name, the oomycote causing it is also killing many other species, including Douglas fir and redwoods. Dieback from the crown and oozing from bark are symptoms.

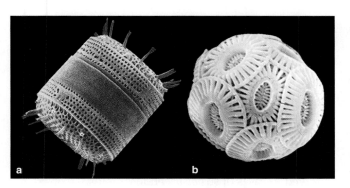

Figure 20.12 Shell of a diatom (**a**) and coccolithophore (**b**).

Figure 20.13 *Right*: Representative brown algae. (**a**) *Postelsia*, a brown alga along coasts from central California to Vancouver Island. Its blades top a resilient stipe fastened to rocks. (**b**) Photograph and (**c**) body plan of a giant kelp, *Macrocystis*. Giant kelps form dense underwater forests.

CHRYSOPHYTES

Golden algae, yellow-green algae, diatoms, and the coccolithophores are grouped as **chrysophytes**. Most are free-living photosynthetic cells of aquatic habitats.

A silica shell encasing diatoms has two parts that overlap like a hatbox (Figure 20.12*a*). For 100 million years, finely crushed diatom shells accumulated at the bottom of lakes and seas. We now use the sediments for filters, insulation, and abrasives. Living diatoms are key producers of many oceanic food chains. Land plants and diatoms release about the same amount of free oxygen into the atmosphere every year.

Coccolithophores secrete calcium carbonate plates beneath their plasma membrane (Figure 20.12*b*). They produce about 1.5 million tons of calcium carbonate per year. As described in the chapter opening, such plates accumulated over time, and formed chalk and limestone deposits. Most species of coccolithophores drift through calm, nutrient-poor seas. In nutrient-rich water, diatoms usually are the dominant competitors.

BROWN ALGAE

Most species of **brown algae** live in temperate or cool seas, from intertidal zones into the open ocean. Their accessory pigments tint them olive-green, golden, or dark brown (Figure 20.13). Brown algae range from microscopic filaments to plantlike forms thirty meters tall. All are multicelled photoautotrophs that reproduce sexually and asexually. Large sporophytes that make walled spores alternate with gametophytes (gamete-producing bodies) during the life cycle.

Macrocystis, *Laminaria*, and other giant kelps are the largest protists. The biggest seaweeds are sporophytes that have stipes (stemlike parts), blades (leaflike parts), and holdfasts (structures that anchor the alga). Hollow, gas-filled bladders along the stipes impart buoyancy and keep blades upright, near the sunlit ocean surface. Spores form in specialized blades near the base.

Maybe you've heard of the Sargasso Sea. This sea without shores, in the middle of the western North Atlantic, is named after vast, free-floating mats, as much as nine meters thick of *Sargassum*. Giant rafts of this kelp also drift off the southeastern United States. They are the crucial breeding grounds for American eels, nurseries for loggerhead turtles, and habitats for diverse fishes and invertebrates.

Sargassum, *Macrocystis*, and a few other species have commercial uses. Algal extracts become ingredients in ice creams, puddings, jelly beans, salad dressings, toothpaste, cosmetics, and other products. Alginic acid from the cell walls of some species is used to make algins, which act as thickeners, emulsifiers, and suspension agents in many commercial products.

Oomycotes (heterotrophs) and chrysophytes and brown algae (photoautotrophs) are stramenopiles. A tinsel-type flagellum is the shared trait that unites these diverse groups. The brown algae include the kelps, the largest of the protists.

20.6 Red Algae

Red algae offer us a classic story of endosymbiosis. Their chloroplasts contain reddish pigments and are probably derived from cyanobacteria. Red alga-like plastids in other protists point to an episode of endosymbiosis during their evolution.

Nearly all of the 6,000 known species of **red algae** live in warm marine currents and clear tropical seas. Of all algae, they survive at the greatest depths. A few encrusting types help build coral reefs. Along with chlorophyll *a*, red alga chloroplasts hold phycobilins. These red accessory pigments absorb green and blue-green wavelengths, which penetrate deep water. The more phycobilins an alga has, the redder it appears. Shallow-water species tend to have little phycobilin and look green. Deep dwellers are almost black.

By the theory of endosymbiosis, chloroplasts of red algae may have evolved from ancient cyanobacteria, which were engulfed by predatory cells but resisted digestion. The cyanobacteria became an endosymbiont, then evolved into chloroplasts (Section 18.4).

Ancient, single-celled red algae may also have been involved in cases of *secondary* endosymbiosis. Consider that plastids in apicomplexans and chloroplasts in some dinoflagellates have more than two external membranes. Did these protists get their chloroplasts "secondhand," after they engulfed photosynthetic red algal cells that became endosymbionts? Probably.

Some single-celled red algae persist, but most are multicelled. They have a branching or sheetlike growth pattern but no true tissues (Figures 20.14 and 20.15). Life cycles are diverse, with asexual phases as well as sexual phases that do not involve flagellated gametes.

Agar is an inert gel extracted from cell walls of a few species. It keeps baked goods and cosmetics moist, firms up jellies, and is used as a culture medium and as soft capsules for drugs. Derivatives stabilize paints. Carrageenan thickens sauces and dairy foods. *Porphyra* species are cultivated worldwide, more than 130,000 tons are harvested annually (Figure 20.15).

Most red algae are multicellular and marine. Of all algae, they survive at the greatest depths. The phycobilins that allow this are the legacy of a cyanobacterial endosymbiont.

Read Me First!
and watch the narrated animation on a red alga life cycle

zygote

sporophyte (2n)

a

fertilization

Diploid Stage
Haploid Stage

meiosis

b

germinating spore (n)

d

c

male gametes

female gametes

gametophyte (n)

Figure 20.14 Life cycle of a red alga (*Porphyra*). For centuries, Japanese fishermen cultivated and harvested a red alga in early fall. The rest of the year, it seemed to vanish. Kathleen Drew-Baker studied its sheetlike form in the laboratory. She saw gametes forming in packets near the sheet margins. She also studied gametes in a petri dish. After zygotes formed, individuals developed into tiny, branching filaments on bits of shell in the dish. That was how the alga spent most of the year!

People had known about the pinkish growths on shells, but no one figured out the growths were algal sporophytes. *Porphyra* species could be cultivated on shells or other calcium-rich surfaces! Within a few years, researchers worked out the life cycle of *P. tenera*, a species used for nori. (It is pressed into thin sheets and used for seasoning or as a sushi wrapper.) By 1960, cultivation was a billion-dollar industry.

Figure 20.15 A red alga, *Antithamnion plumula*. The filamentous, branching growth pattern is the most common among red algae.

20.7 Green Algae

Green algae resemble plants so closely that many botanists now put them inside the plant kingdom. Evolutionarily, some green algae are certainly closer to plants than they are to other green algae. Single-celled and multicelled forms abound, mainly in freshwater habitats.

All species commonly called "green algae" are similar to plants in several traits. Their chloroplasts contain chlorophylls *a* and *b*, they store sugars as starch, and their cell walls incorporate cellulose. Most species are now classified as **chlorophytes**. A smaller group of green algae, **charophytes**, includes the closest living relatives of plants. Genetic analysis reveals that some charophytes have more in common with plants than they do with other green algae.

Figure 20.16*a–c* is a sampling of chlorophytes, which have sheetlike, filamentous, and colonial forms. Some can't be seen without a microscope. Chlorophytes live mostly in freshwater, although certain species live in the seas, in soil, and on rocks, snow, bark, and other organisms. A few kinds are even symbionts with fungi (in lichens) and with some invertebrates.

Melvin Calvin employed a single-celled green alga (*Chlorella*) to track the steps in the light-independent reactions of photosynthesis. *Volvox* is a colonial form, a whirling sphere of 500 to 60,000 flagellated cells that's common in freshwater ponds. White, powdery beaches in the tropics formed in part from remnants of *Udotea* cells. The tiny cells built calcified walls, died, and then disintegrated. *Ulva* is distributed worldwide, and it is one of the most conspicuous seaweeds. Blades of some species are as much as 65 centimeters long, but they are seldom more than 40 microns thick. Many people in Scotland and Japan consider these wispy sheets of "sea lettuce" as highly nutritious food.

Charophytes include microscopic cells such as the desmids, multicelled stoneworts, and filamentous forms (Figure 20.16*d,e*). Desmids are diverse and widespread. They live mainly in freshwater, but some live in the ocean or on its surface; others live on snow or ice.

Some green algae might help astronauts on long missions in outer space. They could provide the crew with oxygen (released in photosynthesis) and dispose of carbon dioxide wastes. Many single-celled forms are easily cultured. They reproduce asexually and rapidly reach high densities. The next section offers a close look at a representative life cycle.

Green algae are photoautotrophs now classified as either chlorophytes or charophytes, the closest relatives of plants. Their forms and life cycles are diverse.

Figure 20.16 Chlorophytes: (**a**) Sea lettuce (*Ulva*) grows in estuaries and attaches to kelps in the seas. Reproductive cells form within the margins of a sheetlike body and are released from it. (**b**) *Udotea cyathiformis*, one of more than 100 known *Udotea* species in tropical and subtropical waters. Many are calcified. (**c**) *Volvox*, a colony of interdependent cells that resemble the free-living, flagellated *Chlamydomonas* cells described in the next section. This colony is rupturing. Each released daughter cell may found a new colony.

Charophytes: (**d**) *Chara*, a stonewort known as muskweed, after its skunky odor. (**e**) *Micrasterias*, a type of desmid. Two new cells are forming from two parts of the now-smaller parent cell (the bow-tie shape in the *middle*).

20.8 Environmental Escape Artists

Imagine you are a single-celled protist. Life is good. There's plenty of food, and your habitat is just right. Then, without warning, things change. Nutrients start to run out. With your limited mobility, you can't get away. Is it all over? Perhaps not, if you belong to one of the groups that can wait out hard times by forming resting spores.

CONSIDER A GREEN ALGA

Green algae engage in diverse reproductive modes. A classic example is *Chlamydomonas*, a single-celled alga about twenty-five micrometers wide. It is able to reproduce sexually. Most of the time, however, it engages in asexual reproduction. As many as sixteen daughter cells form by mitotic cell division within the confines of the parent cell wall. Daughter cells may live in this cell for a while, but sooner or later they leave by secreting enzymes that digest what's

left of their parent. When nitrogen is scarce, haploid cells of opposite mating types fuse and form a diploid zygote having a tough outer wall. Inside the wall, this dormant cell can withstand drying out and freezing. After conditions improve, meiosis occurs, and then germination releases the haploid cells (Figure 20.17).

CONSIDER A SLIME MOLD

As a final example, consider the **slime molds**. These protists are common heterotrophs living in temperate regions. None has chloroplasts. All types spend part of their life cycle as a single cell and part as a larger, cohesive aggregation. Unfavorable conditions prompt the formation of spore-bearing structures.

In *plasmodial slime molds*, the most conspicuous stage of the life cycle is a multinucleated mass called the plasmodium. It arises from a single diploid cell

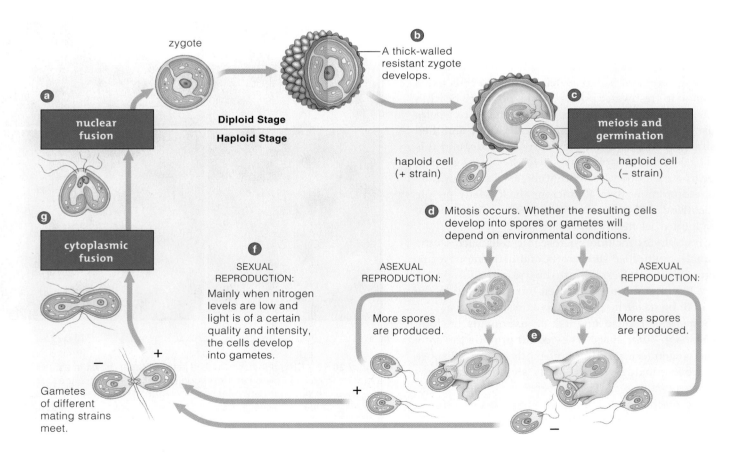

Figure 20.17 Life cycle of a species of *Chlamydomonas*, a common green algae of freshwater habitats. This single-celled species reproduces asexually most of the time, but two mating strains (designated + and –) can engage in sexual reproduction.

that undergoes many rounds of mitosis without any cytoplasmic division. A typical plasmodium fans out in what looks like a network of veins (Figure 20.18*a*). As it streams along rotting wood or the forest floor, it engulfs microbes and decaying organic matter. When stressed, as by dwindling food, the mass gives rise to many spore-bearing fruiting bodies (Figure 20.18*b*).

Cellular slime molds spend most of their life cycle as amoeboid (amoebalike) cells. *Dictyostelium discoideum* is typical of the group. The cells feed individually on bacteria and reproduce asexually, by way of mitotic cell divisions. When food runs out, as many as 100,000 of these cells aggregate into a mound.

Environmental cues, including moisture gradients, induce the mass to crawl about as a "slug" (Figure 20.19*a*–*c*). While the slug is migrating, cells start to differentiate into two types. Some will form a stalk; others will form spores. The slug migrates along until

Figure 20.18 *Physarum*, a plasmodial slime mold. (**a**) Feeding plasmodium on a log. (**b**) Spore-bearing structure.

it reaches a favorable spot. There, an elongating stalk elevates nonmotile spores, as in Figure 20.19*d*–*f*. In this way, a mature fruiting body forms. Spores float away from it, on currents of air. When they germinate, each releases a single diploid amoeboid cell, and the cycle begins anew.

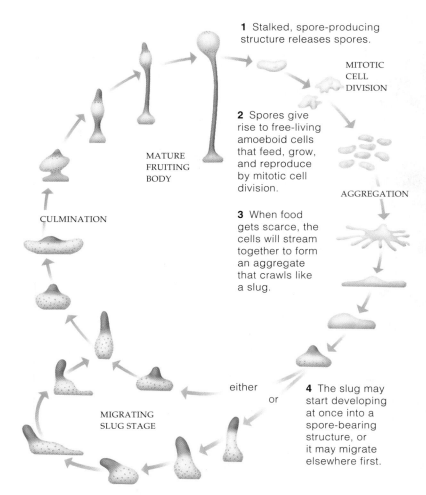

1 Stalked, spore-producing structure releases spores.

MITOTIC CELL DIVISION

2 Spores give rise to free-living amoeboid cells that feed, grow, and reproduce by mitotic cell division.

AGGREGATION

3 When food gets scarce, the cells will stream together to form an aggregate that crawls like a slug.

MATURE FRUITING BODY

CULMINATION

either

or

MIGRATING SLUG STAGE

4 The slug may start developing at once into a spore-bearing structure, or it may migrate elsewhere first.

a　Life cycle of *Dictyostelium discoideum*.

Several amoeboid cells streaming together:

Figure 20.19 (**a**) Life cycle of *Dictyostelium discoideum*, one of the cellular slime molds. Spores give rise to free-living amoeboid cells, which grow and divide asexually until food dwindles.

(**b**) When starving, cells secrete cyclic AMP, a signaling molecule that induces them to stream together. Cells become sticky and adhere to one another. They secrete a cellulose sheath around themselves, thus forming a "slug" two to four millimeters long (**c**).

In a migrating slug, cells differentiate into prestalk (*red*), prespore (*white*), and anterior-like cells (*brown dots*).

The prestalk and prespore cells differentiate and form a stalked, spore-bearing structure (**d**–**f**). Anterior-like cells may help elevate nonmotile spores for dispersal from the spore-bearing structure.

20.9 Characteristics of Fungi

Fungi, too, arose from the "crown" of single-celled eukaryotes that evolved more than 900 million years ago. These are ancient lineages, closer on the family tree to animals than to plants.

Do you think "mushrooms" when you think of fungi? Mushrooms are just spore-bearing structures (fruiting bodies) of only a few of the 56,000 fungal species we know about, and there may be at least a million more fungi that we haven't even found yet!

When plants invaded land about 435 million years ago, fungal species accompanied them. Less than 100 million years later, three lineages—zygomycetes, sac fungi, and club fungi—were well established. Many other species are an informal group, called "imperfect fungi," until gene sequencing methods can point to where they belong. The vast majority of species in all of these groups are multicelled (Figure 20.20).

It takes a list of features to define **fungi**, because other groups are like them in many ways. All fungi are heterotrophs. Like some bacteria, they decompose organic compounds in a distinctive way. Fungal cells growing in or on organic matter secrete digestive enzymes, then individual cells absorb digested bits. The extracellular digestion helps plants, which absorb some of the released nutrients and metabolic products. Thus they help cycle nutrients through food webs.

Most fungi are saprobes, or organisms that absorb nutrients from nonliving organic material and cause its decay. Many are parasites; they feed on tissues of living hosts. As you will see, others are symbionts.

Most fungi reproduce sexually at opportune times and, more often, asexually. Either way, spores form. A **spore** is a single-celled or multicelled resting structure. Most fungal spores have a rigid protective wall. In multicelled fungi, spores germinate and start a **hypha** (plural, hyphae), a cellular filament encased in chitin. Mitotic cell divisions in hyphae produce a **mycelium** (plural, mycelia). This mesh of branching filaments has a high surface-to-volume ratio. As the mycelium grows over or into organic matter, hyphal cells absorb nutrients. Nutrients flow easily through the mesh; no walls form between the cytoplasm of adjoining cells.

Life cycles of fungi are diverse. We'll sample just a few species, starting with a mushroom-forming **club fungus**. For most of its life cycle, this fungus consists of a haploid mycelium buried in soil or growing into decaying wood. Figure 20.21 illustrates how sexual reproduction starts with fusion of two mating strains of hyphae. Cytoplasmic fusion produces a mycelium in which each cell contains two haploid nuclei ($n+n$). Under favorable conditions, specialized hyphae start to intertwine and form fruiting bodies—in this case, mushrooms. Nonmotile, haploid spores form in club-shaped cells on the fruiting body. After air currents disperse them, germinating spores start new mycelia.

a PURPLE CORAL FUNGUS *Clavaria* **b** RUBBER CUP FUNGUS *Sarcosoma*

c BIG LAUGHING MUSHROOM *Gymnophilus*

d TRUMPET CHANTERELLE *Craterellus* **e** SCARLET HOOD *Hygrophorus*

Figure 20.20 From southeastern Virginia, a small sampling of the spectacular range of fungal diversity. Most of these fruiting bodies belong to species of club fungi. Only (**b**) shows fruiting bodies of a sac fungus.

Figure 20.21 Generalized life cycle of many club fungi. When the hyphal cells of two compatible mating strains make contact, their cytoplasms fuse but the nuclei do not. Mitotic cell divisions yield a mycelium in which each cell has two nuclei. When the conditions are right, this mycelium forms a mushroom.

Club-shaped, spore-bearing structures form on gills, the mushroom cap's leaflike inner surface. Inside each structure, the two nuclei fuse. The result is a diploid zygote. With zygote formation, the cycle starts again.

The scanning electron micrograph (*below*) shows part of a mycelium, the underground portion of a club fungus that absorbs water and dissolved nutrients.

Read Me First!
and watch the narrated animation on a club fungus life cycle

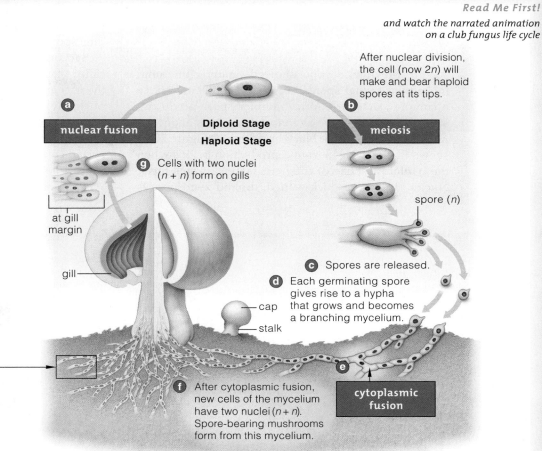

After nuclear division, the cell (now 2*n*) will make and bear haploid spores at its tips.

a nuclear fusion

Diploid Stage
Haploid Stage

b meiosis

g Cells with two nuclei (*n* + *n*) form on gills

spore (*n*)

at gill margin

gill

c Spores are released.

d Each germinating spore gives rise to a hypha that grows and becomes a branching mycelium.

cap

stalk

e

f After cytoplasmic fusion, new cells of the mycelium have two nuclei (*n* + *n*). Spore-bearing mushrooms form from this mycelium.

cytoplasmic fusion

one hyphal cell among the many hyphae that make up a mycelium

Figure 20.22 Fly agaric mushroom (*Amanita muscaria*), a hallucinogenic species. In Central America, India, and Russia, it was used in ancient rituals to induce trances.

Club fungi are the most diverse fungal group. Many are known as mushrooms, puffballs, stinkhorns, and shelf fungi. Most are decomposers that grow through soil, but many closely associate plant roots. Rusts and smuts are parasites that attack plants. The cultivation of *Agaricus bisporus*—the mushroom of grocery store fame—is a multimillion-dollar business.

So which organism is the oldest and largest? The honey mushroom (*Armillaria ostoyae*) may well be the winner. One individual has been spreading through the soil of an Oregon forest for 2,400 years. It extends across 2,200 or so acres to an average depth of 1 meter. Imagine 1,665 football fields side by side and you get an idea of its amazing size. *A. ostoyae* definitely is not one of the beneficial fungi. Its hyphae penetrate and clog tree roots, which eventually kills the tree.

Do *not* gather and eat wild mushrooms. Identifying mushrooms is tricky. Most of the edible species look extremely similar to poisonous ones. Many fungi make psychoactive alkaloids, and some of these can be deadly (Figure 20.22). It is said that there are old mushroom hunters and bold mushroom hunters, but no old, bold mushroom hunters.

Fungi are single-celled or multicelled heterotrophs that digest organic material outside the cell body. They help cycle nutrients when plants take up some of the released nutrients.

20.10 Fungal Diversity

A fungus has a thing about spores. It produces sexual spores, asexual spores, or both. The major fungal groups are defined by the ways that they produce sexual spores.

As the previous section showed, many club fungi make their sexual spores in fruiting bodies that have a complex structure. Much simpler spore-producing structures occur in the **zygomycetes**. Think about the *Rhizopus stolonifer*, the black bread mold. Figure 20.23 shows its life cycle. Sexual reproduction begins when the haploid hyphae of different mating strains meet. Nuclei fuse and a thick-walled, diploid zygospore is

formed. It undergoes meiosis, then germinates, giving rise to a hypha that will form a spore sac at its tip. This spore sac releases many haploid spores.

Sac fungi form sexual spores in sac-shaped cells called asci (singular, ascus). In multicelled sac fungi, the asci are enclosed in reproductive structures made of tightly woven hyphae. Different kinds are shaped like flasks, globes, and shallow cups (Figure 20.24*b,c*).

Among the sac fungi are the truffles. These highly-prized culinary delicacies grow as symbionts with tree roots. Their fruiting bodies are formed undergound. When spores mature and are ready to be released, the

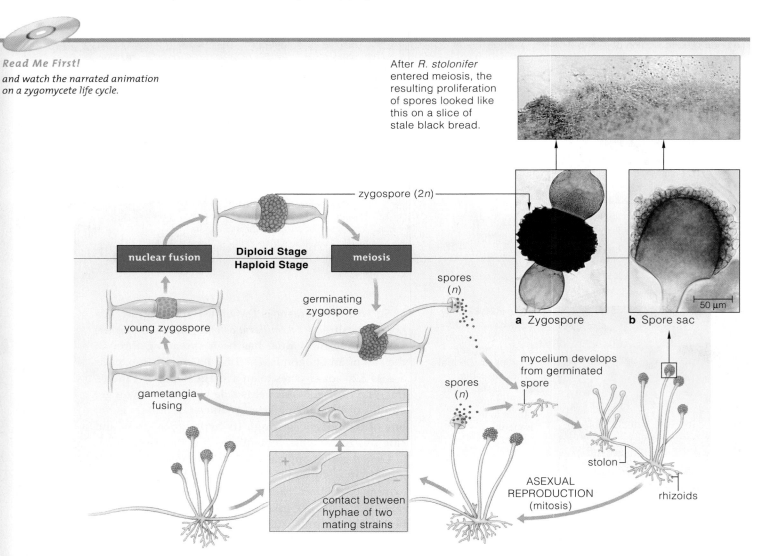

Read Me First!

and watch the narrated animation on a zygomycete life cycle.

After *R. stolonifer* entered meiosis, the resulting proliferation of spores looked like this on a slice of stale black bread.

zygospore (2n)

Diploid Stage
nuclear fusion | **Haploid Stage** | meiosis

young zygospore

gametangia fusing

germinating zygospore

spores (n)

mycelium develops from germinated spore

spores (n)

contact between hyphae of two mating strains

stolon

ASEXUAL REPRODUCTION (mitosis)

rhizoids

a Zygospore **b** Spore sac

50 µm

Figure 20.23 Life cycle of the black bread mold *Rhizopus stolonifer*. Asexual phases are common. Also, different mating strains (+ and −) reproduce sexually. Either way, haploid spores form and give rise to mycelia. Chemical attraction between a + hypha and a − hypha makes them fuse. Two gamete-producing structures (gametangia) form, each with several haploid nuclei. The nuclei pair up. Each pair fuses and forms a zygote. Some zygotes disintegrate. Others become thick-walled zygospores and may be dormant for several months. Meiosis occurs as the zygospore germinates, and asexual spores form.

Figure 20.24 Sexual spores. (**a**) Spores on the gill margin of a club fungus. (**b,c**) *Sarcoscypha coccinea*, a scarlet cup fungus. On the cup's inner surface, many spores form in saclike cells called asci.

fruiting body produces a complex mixture of aromatic compounds. This gives the ripe truffle its distinctive taste and smell. Trained dogs or pigs are often used to sniff them out underground. Several kinds of truffles sell for hundreds of dollars an ounce.

Most food-spoiling molds are multicelled sac fungi. Some of the single-celled fungi we call yeasts are also sac fungi (others are club fungi). One packet of baking yeast holds thousands of tiny spores of *Saccharomyces cerevisiae*. In a warm, moist spot, such as bread dough set out to rise, these spores germinate. The cells that emerge can reproduce asexually by budding. Aerobic respiration in these cells releases carbon dioxide and causes the bread dough to rise. Fermentation by other yeast species is used to make alcoholic beverages. Yeasts also can be genetically engineered and grown in large vats, as drug-producing factories.

Unique sexual spores characterize the major fungal lineages. In zygomycetes, spores form at the tip of a specialized hypha that germinates from a zygospore. Many club fungi and sac fungi produce distinctive sexual spores in specialized hypha that differentiate as part of a complex fruiting body.

20.11 The Unloved Few

You know you are a serious student of biology when you view organisms objectively in terms of their role in nature, not in terms of their impact on humans in general and you in particular. As a student you hail saprobic fungi as decomposers and praise parasitic fungi that can keep populations of bad insects and weeds in check. And then you cross paths with a bad fungus.

Have you ever opened the fridge to get a bowl of fruit and find a fungus beat you to it? Has a fungus made the tissues of your warm, damp toes scaly, red, and cracked (Figure 20.25a)? Think about it. Which home gardeners wax poetic about powdery mildew on roses? Which farmers happily hand over millions of dollars a year to sac fungi that infect corn, wheat, peaches, and apples? Who rejoices that a sac fungus, *Cryphonectria parasitica*, blitzed nearly all chestnut trees in eastern North America?

Who willingly inhales airborne spores of *Ajellomyces capsulatus*? In soil, the spores give rise to mycelia. In moist lung tissues, they form yeastlike cells that can cause *histoplasmosis*, a respiratory disease. The body's macrophages normally can eliminate the direct threat, but debris from the battle damages lung tissue. Heavy exposure to spores invites pneumonia.

And household molds! Thank them for sinus, ear, and lung infections, for rashes, and for increasing asthma attacks 300 percent over the past two decades.

Some fungi have even tweaked human history. One species, *Claviceps purpurea*, is a parasite of rye and other cereal grains (Figure 20.25b). Give it credit; we use some of its alkaloid by-products to treat migraines and to stop hemorrhages after childbirth by shrinking the uterus. However, its alkaloids can be toxic. Eat bread made with spoiled rye and you may end up with *ergotism*. Disease symptoms include vomiting, diarrhea, hallucinations, hysteria, and convulsions. Untreated, ergotism turns limbs gangrenous and causes death.

Ergotism epidemics were common in the Middle Ages in Europe, when rye was a key crop. They thwarted Peter the Great, a Russian czar who was obsessed with conquering ports along the Black Sea for his nearly landlocked empire. Soldiers laying siege to the ports ate mostly rye bread and fed rye to their horses. The soldiers convulsed, and horses went into "blind staggers." Ergotism outbreaks might have provided an excuse to launch witchhunts in the early American colonies.

Figure 20.25 Love those fungi! (**a**) Athlete's foot, courtesy *Epidermophyton floccosum*. (**b**) Spores of a historically notable fungus, *Claviceps purpurea*, on infected rye plant parts.

20.12 Fungal Symbionts

*"Symbiosis" refers to species that interact closely during their life cycles. In **mutualism**, their interaction benefits both or does no harm to either one. Fungal endophytes, lichens, and mycorrhizae generally are like this.*

FUNGAL ENDOPHYTES

Endophytic fungi are symbionts that live inside the leaves and stems of almost all plants, with helpful or neutral effects. Tall fescue (*Festuca arundinacea*) offers an example. Lush clumps of this grass are vulnerable to herbivores, but fungi living in the tissues of most *F. arundinacea* strains produce alkaloids that are toxic to the grazers, which quickly learn to leave the plants alone. Diverse fungal endophytes also help protect host plants from pathogens, such as other fungi and the oomycote *Phytophthora*.

LICHENS

A **lichen** consists of a fungal species intertwined with one or more photoautotrophs, most often green algae or cyanobacteria. It forms after a hypha's tip binds to a host cell and both lose their wall. Their cytoplasm fuses or the hypha induces the host cell to cup around it. Both start dividing; they give rise to a multicelled and often layered vegetative body that is flattened, erect, leaflike, or pendulous (Figure 20.26). The fungus usually makes up most of the lichen; its partner lives in or on it. The photoautotroph supplies the fungus with nutrients and gets a protected habitat in return.

The cyanobacteria in lichens help cycle nitrogen by converting nitrogen gas in the air to forms that plants can absorb. Lichens get minerals from airborne particles and are highly vulnerable to air pollution.

Lichens can colonize places that are too hostile for most organisms. For example, when a glacier retreats, they colonize newly exposed bedrock. Their metabolic products help form new soil or enrich whatever soil is already present. As conditions improve, other species move in and typically replace the pioneers. This may be what happened when plants first invaded the land.

MYCORRHIZAE

Many soil fungi live as mutualists on or in tree roots. Those truffles in Figure 20.24 are an example. They interact as a **mycorrhiza** (plural, mycorrhizae), which means fungus-root. Underground parts of the fungus grow through soil and provide a large surface area for absorbing nutrients. When ions of phosphorus and other minerals are abundant, the fungus takes them up, and it releases some to the plant when ions are scarce. In return, the fungus withdraws some sugars from the plant. The plant's loss is a trade-off, because it doesn't grow as efficiently without a fungal partner. Figure 20.27 shows an example of the dramatic effects of a mycorrhizal fungus on plant growth.

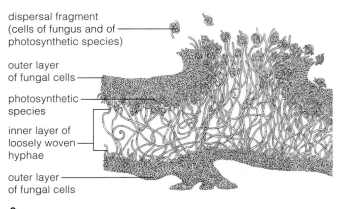

dispersal fragment (cells of fungus and of photosynthetic species)

outer layer of fungal cells

photosynthetic species

inner layer of loosely woven hyphae

outer layer of fungal cells

e

Figure 20.26 (**a**) One of the encrusting lichens, growing on a sunlit rock. (**b**) *Cladonia rangiferina*, an erect, branching lichen. (**c**) *Lobaria oregana*, a leaflike lichen that includes nitrogen-fixing cyanobacteria. (**d**) *Usnea*, a pendant lichen commonly called old man's beard. (**e**) Body plan of a stratified (layered) lichen, cross-section.

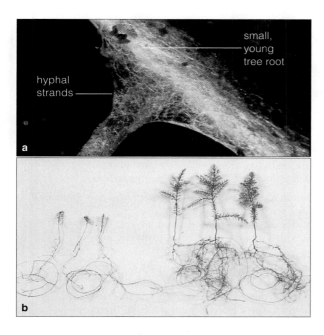

Figure 20.27 (**a**) Mycorrhiza of a young hemlock tree. (**b**) Effects of the presence or absence of mycorrhizae on plant growth in sterilized, phosphorus-poor soil. The juniper seedlings at left were controls; they grew without a fungus. Seedlings at right, the experimental group, are six months old. They were grown with a partner fungus.

AS FUNGI GO, SO GO THE FORESTS

Since the early 1900s, collectors have been counting the numbers and kinds of wild mushroom populations in European forests. The populations clearly are declining. Mushroom gatherers are not the cause; inedible as well as edible species are vanishing. However, the decline does correlate with rising levels of air pollution.

Now consider this: Myccorhizal fungi benefit nearly all trees. Trees do have hairlike absorptive structures projecting from young roots, but fungal hyphae are smaller and better at penetrating small crevices in soil. As a tree ages, one mycorrhizal fungus gives way to another, in predictable patterns. When fungi die, trees lose a vital support system and are more vulnerable to damage from severe frosts, droughts, and acid rain. As a result, many forests are deteriorating.

In mutualism—a symbiotic interaction—one or both partners derive benefits. Endophytes are protective fungi living inside leaves or stems. In lichens, one or more photoautotrophs are sheltered by a fungus. In mycorrhizae, a fungus living in or on a plant's roots shares mineral ions it has absorbed. All photosynthetic partners give up some sugars to the fungus.

Summary

Section 20.1 Structurally, protists are the simplest eukaryotes and the ones most like the first eukaryotic cells. The groups are undergoing major reclassification.

Section 20.2 Flagellated protozoans are among the most ancient eukaryotic cells. Some of these single cells do not have mitochondria. Amoeboid protozoans lost their flagella as they evolved. Foraminiferans and radiolarians are amoebas with secreted shells or plates. Euglenoids are freshwater flagellates; most acquired chloroplasts by endosymbiosis.

Section 20.3 The ciliates, apicomplexans, and dinoflagellates are all alveolates. These single cells have membrane-bound sacs beneath the plasma membrane. Beating cilia help ciliates move or direct food into their gullet. Apicomplexans are intracellular parasites with host-penetrating microtubular devices; malaria-causing *Plasmodium* species are examples. Dinoflagellates are flagellated photoautotrophs or heterotrophs.

Section 20.4 Nutrient-enriched water fuels algal blooms: huge increases in population sizes of aquatic protists that can harm other organisms.

Section 20.5 Stramenopiles have tinselly flagella and are heterotrophs (oomycotes) or photoautotrophs (single-celled diatoms and coccolithophores, and multicelled brown algae). Different oomycotes are decomposers, parasites, or pathogens. Diatoms have silica shells and coccolithophores have calcium plates. Brown algae include kelps, the largest and most structurally complex protists.

Section 20.6 Red algae are colored by phycobilins that mask chlorophylls and that can capture light even at great depths. Most are multicelled and live in marine habitats. Single-celled species may have given rise to plastids in some other protists, by endosymbiosis.

Section 20.7 The "green algae" are single cells or multicelled. Like plants, they have chlorophylls a and b, they store sugars as starch, and their cell walls contain cellulose. Most are chlorophytes. The charophytes are more closely related to plants than to other green algae.

Section 20.8 Many protists are escape artists; they survive adverse conditions by forming resting spores. *Chlamydomonas* and slime molds are examples.

Section 20.9 Fungi are heterotrophs that engage in extracellular digestion and absorption: Individual cells secrete digestive enzymes on organic matter and then absorb breakdown products. Most are decomposers that help ecosystems; some are pathogenic or parasitic.

In multicelled fungi, germinating spores give rise to a mycelium (a mesh of filamentous hyphae). Hyphal cells have chitin-stiffened walls but not at abutting ends, so cells making up each filament share their cytoplasm.

Section 20.10 Besides asexual spores, the three major lineages of fungi (zygomycetes, sac fungi, and club fungi) form unique sexual spores that help us classify them. "Imperfect fungi" is an informal group of fungal species awaiting classification.

Section 20.11 Most fungi are beneficial, but others destroy crops, spoil food, and cause diseases.

Section 20.12 Endophytes, lichens, and mycorrhizae are cases of mutualism. Endophytic fungi live in most stems and leaves. Lichens consist of a fungal symbiont and a cyanobacterium or green alga. A mycorrhiza consists of a fungus associating with young roots.

Self-Quiz

Answers in Appendix III

1. Some of the flagellated protozoans lack _____ .
   a. a nucleus   c. DNA
   b. mitochondria   d. a plasma membrane

2. The euglenoids are _____ .
   a. flagellated   c. marine   e. all are correct
   b. single cells   d. a and b

3. The _____ are parasitic alveolates, such as the organism that causes malaria.
   a. trypanosomes   c. oomycotes
   b. apicomplexans   d. zygomycetes

4. Silica reinforces the cell walls of _____ .
   a. coccolithophores   c. foraminiferans
   b. diatoms   d. fungi

5. The giant kelp *Macrocystis* and the plant pathogen *Phytophthora* are both _____ .
   a. brown algae   c. oomycotes
   b. chrysophytes   d. stramenopiles

6. Algal blooms are most commonly population explosions of _____ .
   a. diatoms   c. euglenoids
   b. radiolarians   d. dinoflagellates

7. Phycobilins are the signature pigments of _____ .
   a. red algae   c. green algae
   b. brown algae   d. all algae

8. The closest relatives of *Chara* and other charophytes are the _____ .
   a. oomycotes   c. plants
   b. chrysophytes   d. fungi

9. When a spore of a cellular slime mold germinates, it gives rise to a _____ .
   a. free-living amoeboid   c. mycelium
   b. fruiting body   d. mycorrhiza

10. When a spore of a multicelled fungus germinates, it gives rise to a _____ .
    a. free-living amoeboid   c. mycelium
    b. fruiting body   d. mycorrhiza

11. A "mushroom" is _____ .
    a. the food-absorbing part of a fungal body
    b. composed entirely of haploid cells
    c. a reproductive structure
    d. the longest lived part of a fungal life cycle

12. Match these terms suitably.
    ____ trypanosome   a. single-celled fungus
    ____ foraminiferan   b. flagellated protozoan
    ____ diatom   c. "fungus-root"
    ____ *Paramecium*   d. brown alga
    ____ oomycote   e. calcium-carbonate shell
    ____ charophyte   f. two-part silica shell
    ____ kelp   g. fungus plus photoautotroph
    ____ yeast   h. "egg fungus"
    ____ mycorrhiza   i. ciliated protozoan
    ____ lichen   j. closest relative of plants

Table 20.1 Comparison of Prokaryotes With Eukaryotes

	Prokaryotes	Eukaryotes
Organisms represented:	Archaea, Bacteria	Protists, fungi, plants, and animals
Ancestry:	Two major lineages that evolved more than 3.5 billion years ago	Equally ancient prokaryotic ancestors gave rise to forerunners of eukaryotes, which evolved more than 1.2 billion years ago
Level of organization:	Single-celled	Protists, single-celled or multicelled. Nearly all others multicelled; division of labor among differentiated cells, tissues, and often organs
Typical cell size:	Small (1–10 micrometers)	Large (10–100 micrometers)
Cell wall:	Most with distinctive wall	Cellulose or chitin; none in animal cells
Membrane-bound organelles:	Rarely; no nucleus, no mitochondria	Typically profuse; nucleus present; most with mitochondria, many with chloroplasts
Modes of metabolism:	Both anaerobic and aerobic	Aerobic modes predominate
Genetic material:	One chromosome; plasmids in some	Chromosomes of DNA plus many associated proteins in a nucleus
Mode of cell division:	Prokaryotic fission, mostly; some reproduce by budding	Nuclear division (mitosis, meiosis, or both) associated with one of various modes of cytoplasmic division

Figure 20.28 A drift of snow on a glacier. It has been tinted red by an abundance of snow algae, including *Chlamydomonas nivalis* (*right*). Notice the red footprint.

Critical Thinking

1. Take a moment to consider Table 20.1, which shows the general features of prokaryotic and eukaryotic cells. Consider that, as you learned in this and the previous chapter, some of the prokaryotes exceed eukaryotes in size, some primitive eukaryotes lack mitochondria, and many eukaryotes carry within them the descendants of endosymbiotic prokaryotes. Recall also that archaea are as closely related to eukaryotes as they are to bacteria. Because of these and other discoveries, some scientists have suggested that the division of organisms into prokaryotes and eukaryotes is not particularly useful. Do you agree or disagree with this argument? Suggest how these taxonomic categories might help us interpret biodiversity and how they might be misleading.

2. Runoff from heavily fertilized cropland, animal waste, and raw sewage promotes algal blooms that can result in massive kills of aquatic species, birds, and other forms of wildlife. Are such kills an unfortunate but inevitable side effect of human life?

   If you find the environmental cost unacceptable, how would you stop the pollution? Bear in mind that we now absolutely depend on high-yield (and heavily fertilized) crops. How do you suggest we dispose of the waste generated by farms and cities?

3. The most common "snow alga," *Chlamydomonas nivalis,* lives in the world's glaciers (Figure 20.28). Although it's a green alga, an abundance of carotenoid pigments colors it red. Think about the wavelengths of radiation striking its icy habitats. Besides their role in photosynthesis, what other function might the carotenoids be serving? (*Hint:* Review Section 6.1.)

4. The cellular slime mold *Dictyostelium discoideum* is now being studied by more than 660 biologists. Each year, hundreds of papers about its biology appear in scientific journals. It is one of the "model organisms" that help us learn about cell differentiation and other developmental processes. Speculate on why this simple soil organism is such a desirable experimental organism.

5. The fungus *Fusarium oxysporum* is a plant pathogen. Some view it as a potential weapon in the war on drugs. Why? Strains of the fungus attack and kill only specific plants—including coca plants used to produce cocaine. Proposals to spray *F. oxysporum* to kill marijuana plants in Florida were abandoned after public outcries.

   What concerns might you have about allowing the use of such natural mycoherbicides to kill off plants that are sources of the drugs favored by substance abusers?

Media Menu

Student CD-ROM

Impacts, Issues Video
    Tiny Critters, Big Impacts
Big Picture Animation
    Evolution and diversity of simple eukaryotes
Read-Me-First Animation
    An apicomplexan life cycle
    A red alga life cycle
    A club fungus life cycle
    A zygomycete life cycle
Other Animations and Interactions
    A green alga life cycle
    A cellular slime mold life cycle

InfoTrac

- Fatal Attraction. *Natural History,* April 2001.
- Parasites from Another Kingdom. *BioScience,* December 2002.
- Hunting Slime Molds. *Smithsonian,* March 2001.
- Penicillin: From Discovery to Product. *Bulletin of the World Health Organization,* August 2001.
- Algal Research. *The Scientist,* October 29, 2001.

Web Sites

- Protist Image Database: megasun.bch.umontreal.ca/protists/protists.html
- Malaria Foundation International: malaria.org
- The Harmful Algae Page: www.whoi.edu/redtide
- Tom Volk's Fungi: botit.botany.wisc.edu/toms_fungi
- Lichens of North America: www.lichen.com
- Dr. Fungus: www.doctorfungus.org

How Would You Vote?

Phytophthora ramorum, the pathogen that causes sudden oak death, has been detected in twenty-six kinds of plants in California and Oregon. A few of the infected plants belong to species that are raised in nurseries and sold nationwide. Some states that are free of the pathogen want to ban the shipment of some or all plants from affected states. Should those states be allowed to institute such a ban?

IMPACTS, ISSUES *Beginnings and Endings*

Change is the way of life. About 300 million years ago, in the Carboniferous, swamp forests carpeted the wet, warm lowlands of continents. Tree-high ancestors of today's tiny club mosses and horsetails were dominant. Then things changed. As the long-term global climate became cooler and drier, moisture-loving plants declined. Hardier plants—cycads, ginkgos, and conifers—rose to dominance. They were the gymnosperms, and they would flourish for millions of years.

Things changed again. Other plants that had been evolving had more potential to radiate into diverse environments. The angiosperms—flowering plants—started to take over. They still dominate most regions. But in the far north, at high elevations, and in parts of the Southern Hemisphere, some conifers retained the competitive edge and held on as great forests.

Things changed yet again. Humans learned to grow some flowering plants as crops, which became the foundation for human population growth. Human populations grew spectacularly, and they required more resources than crops. Conifers had the bad luck to become the premier sources of lumber, paper, resins, and other forest products. They became vulnerable to deforestation—the removal of all trees from large tracts of land.

In California, only 4 percent of the original coastal redwood forest is still undisturbed. In Maine, an area the size of Delaware has been logged over in just the past fifteen years. Many logs end up

Figure 21.1 A clear-cut mountainside in British Columbia. This province supplies about half the Canadian softwood exported to the United States. The forest on the facing page is in Washington State. Similar scenes of deforestation can be found worldwide, in just about every coniferous forest.

the big picture

flowering plants
conifers
ginkgos
cycads
horsetails
ferns
lycophytes
bryophytes
charophytes
chlorophytes

Evolutionary Trends Nearly all plants are multicelled photoautotrophs. Their lineage apparently started as a branching from a green algal group. Through modifications in structure and life cycles, different kinds became adapted to higher and drier habitats on land.

sporophyte (diploid)

gametophyte (haploid)

Nonvascular Plants Mosses and other bryophytes are small plants that have no internal transport tissues. A gamete-forming body dominates the life cycle. A spore-forming body forms on it and remains attached even at maturity.

as paper or pulp. In the United States, the amount of wood processed in pulp mills between 1920 and 1997 increased about twentyfold. The United States exports many logs to foreign lumbermills. At the same time, it imports timber from Canada and tropical rain forests as far away as New Zealand (Figure 21.1).

Clearing the world's forests has drastic ecological effects. Nutrients wash away from exposed soils, and sediments clog streams. Herbicides applied to stop non-timber species from taking over also prevent ecosystem recovery and push species toward extinction. Almost 750 species of plants in the United States alone are now on the endangered list.

And with this bit of perspective on change, we turn to the origins, and end of the line, of the major groups of plants. With rare exceptions, plants are multicelled photoautotrophs. These metabolic wizards make organic compounds by absorbing energy from the sun, carbon dioxide from the air, and minerals from water. As they do, they split water molecules and release free oxygen. Their oxygen-releasing pathway of photosynthesis influences the atmosphere even while sustaining the growth of multicelled forms as tall as redwoods and as vast as an aspen forest that is one continuous clone.

We know of at least 295,000 kinds of existing plants. Be glad their ancient ancestors left the water. Without those pioneers in a new world, we humans and other land-dwelling animals never would have made it onto the evolutionary stage.

How Would You Vote?

Recycling paper saves trees, slows the filling of landfills, and reduces water pollution, although it is often more expensive than making new paper. Plans are afoot to give tax breaks to companies that recycle paper. Would you support such a plan? See the Media Menu for details, then vote online.

Seedless Vascular Plants Dominating the life cycles of seedless vascular plants is a spore-forming body with vascular tissues, roots, and other adaptations to dry habitats. Without free water, flagellated male gametes of these plants cannot reach eggs for fertilization.

The Rise of Seed Plants Gymnosperms and angiosperms are vascular plants that make pollen grains and seeds. These unique traits contributed to their reproductive success. Only angiosperms produce flowers. They are the most widespread and diverse group of existing plants.

21.1 Evolutionary Trends Among Plants

We share the world with at least 295,000 kinds of plants. Within this tremendously diverse group, we find many recurring structures that evolved in response to present and past environmental challenges.

Five hundred million years ago, the invasion of land was under way. Why then? Astronomical numbers of photosynthetic cells had come and gone, and oxygen-producing types had changed the atmosphere. High above Earth, the sun's energy had converted much of the oxygen into a dense ozone layer, a shield against lethal doses of ultraviolet radiation. Until then, life had not ventured above the surface of water and mud.

Algae were evolving at the water's edge, and one group—probably the charophytes—gave rise to plants. *Cooksonia*, a simple branching plant a few centimeters tall, evolved by 430 million years ago. It took another 160 million years for the taller *Psilophyton* to evolve. Then the evolutionary pace picked up. It took only 60 million years for plants to radiate from the swampy lowlands to high mountains and nearly all places in between. They did so through modifications in their structure, function, and reproductive modes.

ROOTS, STEMS, AND LEAVES

Underground absorptive structures evolved as plants colonized the land, and in some lineages they became systems of roots. Most roots had a large surface area that helped plants absorb more water and dissolved mineral ions. As they do today, young roots probably

COOKSONIA

PSILOPHYTON

associated with fungal mycorrhizal symbionts, which help plants obtain water and dissolve nutrients. In many lineages, older roots started anchoring the plant.

Above ground, shoot systems evolved. Their stems and leaves captured energy from the sun and carbon dioxide from the air. Stems became erect, taller, and branched when plants developed a capacity to make and incorporate lignin, a glue-like polymer, in their cell walls. Lignin-strengthened walls supported the stems as they grew upward and outward, in patterns that increased the light-intercepting surface of leaves.

Vascular tissues, xylem and phloem, first appeared in plants called rhyniophytes. As happens now, xylem distributed water and mineral ions through the plant; phloem distributed photosynthetic products.

Also, having enough water for metabolism was not a problem in most aquatic habitats. It was a challenge for life on land. A waxy coat—a cuticle—evolved and helped conserve water inside shoots on hot, dry days. Water and carbon dioxide could cross the cuticle only at tiny gaps called stomata. When stomata evolved, so did control over water loss (Section 6.6). These tissue specializations endure in most existing land plants.

FROM HAPLOID TO DIPLOID DOMINANCE

In plant life cycles, remember, a gametophyte phase alternates with a sporophyte phase (Section 9.5). Each **gametophyte** is a haploid body that produces haploid gametes. Male and female gametes fuse at fertilization, and the resulting zygotes are the start of **sporophytes**. In these multicelled vegetative bodies, haploid spores form by way of meiosis. **Spores** are resting structures, typically walled, that help a new generation wait out harsh environmental conditions. After they germinate, plant spores grow and develop into gametophytes.

Many algae live in places where conditions do not change much, so making resting spores would be a big waste of energy. As you might expect for these algae, gamete production dominates the life cycles. Haploid spores often develop right from the zygote itself, not from a multicelled vegetative body (Figure 21.2).

When plants moved to dry land, spore production became crucial—and sporophytes came to dominate most life cycles. Shrubs, trees, and other sporophyte forms having waxy cuticles, complex vascular tissues, and spore-producing capsules (sporangia) evolved. The timing of fertilization and spore dispersal became adapted to the seasons. A land plant could now retain, nourish, and protect its gametophytes and its embryo sporophytes as they formed and developed, right up to the least risky time to leave home.

green algae   bryophytes   ferns   gymnosperms   angiosperms

Figure 21.2 One evolutionary trend in plant life cycles. Algae and bryophytes put the most energy into making gametophytes. The other groups evolved in seasonally dry habitats on land. They put the most energy into structures that produce spores and retain, nourish, and protect gametes through harsh times.

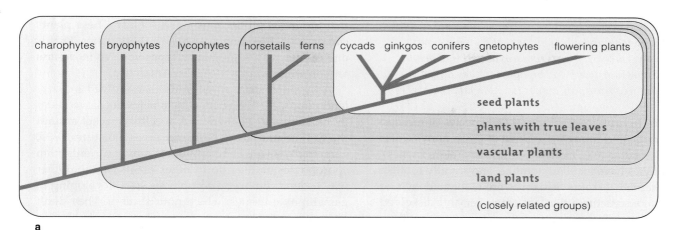

Figure 21.3 (**a**) Evolutionary tree for plants, with nested monophyletic groups. (**b**) Milestones in the evolution of plants.

EVOLUTION OF POLLEN AND SEEDS

Figure 21.3*a* is an evolutionary tree for major plant groups. All but 24,000 of the 295,000 existing species are *vascular*, with internal tissue systems that conduct water and solutes through roots, stems, and leaves. Compared to the *nonvascular* groups—bryophytes—they became success stories on land. The lycophytes, horsetails, and ferns are among the *seedless* vascular plants. The cycads, ginkgos, gnetophytes, and conifers are all gymnosperms, a group of *seed-bearing* vascular plants. Angiosperms are vascular, seed-bearing plants as well, but they alone make flowers. The flowering plants are the largest and most diverse group.

Most seedless vascular plants make only one type of spore; they are *homo*sporous. Some seedless and all seed-bearing vascular plants make two types; they are *hetero*sporous. In gymnosperms and flowering plants, the two are designated megaspores and microspores.

Megaspores divide and form female gametophytes, which make female gametes (call them eggs). Far smaller microspores give rise to **pollen grains**, which are like well-packed suitcases. The protective suitcase wall encloses a few cells that will eventually develop into a mature, sperm-bearing, male gametophyte.

Typically, air currents or animals such as insects deliver pollen grains to eggs. Environmental water is not required, as it is for the male gametes of algae.

Pollen grains were a key innovation that helped seed-bearing plants radiate into high, dry habitats.

Another innovation: Embryo sporophytes became packaged in nutritive tissues and a tough, waterproof coat. The term **seed** refers to the whole package. Over time, the development and dispersal of seeds became attuned to environmental change—for instance, a dry season alternating with a warm, wet season favoring germination and growth. It was no coincidence that seed plants rose to dominance in the Permian, a time of extreme shifts in the global climate.

The plant kingdom includes multicelled, photosynthetic species called bryophytes, seedless vascular plants, and seed-bearing vascular plants.

Most plant lineages became structurally adapted to life on land. They have root and shoot systems, a waxy cuticle, stomata, vascular tissues, and lignin-reinforced tissues.

Sporophytes with well-developed roots, stems, and leaves came to dominate the life cycles of most land plants. Parts of these complex sporophytes nourish and protect the new generation until conditions favor dispersal and growth.

Some plants started to produce two types of spores, not one. This led to the evolution of male gametes that could be dispersed without liquid water and to the evolution of seeds.

21.2 The Bryophytes—No Vascular Tissues

The first plants were bryophytes. The fossil record of this group is sketchy, and whether the first ones were more like hornworts or liverworts remains a matter of debate. We do know that, like living bryophytes, they lacked the specialized transport tissues found in later plants.

Modern **bryophytes** include 24,000 species of mosses, liverworts, and hornworts. None of these nonvascular plants is taller than twenty centimeters (eight inches). Bryophytes have leaflike, stemlike, and usually rootlike parts named rhizoids. **Rhizoids** are elongated cells or threadlike structures that absorb water and dissolved mineral ions. They also attach gametophytes to soil.

Bryophytes don't have specialized vascular tissues because they don't need them. Most grow in habitats that are moist all the time or seasonally. These plants can dry out, then revive after absorbing water. That is one reason why some hardy types survive in deserts and on windswept plateaus in Antarctica.

All groups share three traits that evolved in early land plants. *First*, a waxy cuticle helps conserve water in aboveground parts. *Second*, a cellular jacket around the gamete-producing parts conserves moisture. *Third*, gametophytes are large and don't draw nutrients from sporophytes as they do in other plants. It's the other way around. Embryo sporophytes start developing in gametophyte tissues, which nourish *them*. They don't let go when they reach maturity. They stay attached to the gamete-producing body and draw nutrients from it through at least part of the life cycle (Figure 21.4).

Read Me First!
and watch the narrated animation on a moss life cycle

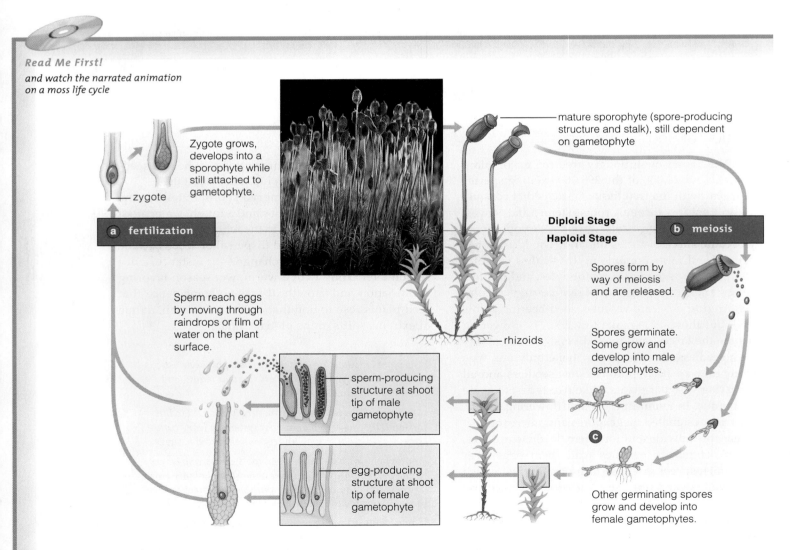

Figure 21.4 Life cycle of a moss (*Polytrichum*), one of the bryophytes. The sporophyte stays attached to the gametophyte, which provides it with nutrients and absorbed water.

sporophyte

gametophyte

Figure 21.5 (**a**) Moss plants on damp rocks. (**b**) Peat bog in Ireland. This family is cutting blocks of peat and stacking them to dry as a home fuel source. Most bryophytes grow slowly. Peat mosses grow fast enough to yield thirteen tons of organic matter per hectare annually, about twice the yield from corn plants. Nova Scotia alone exports 200,000 bales of peat per year to Japan and the United States. Nearly all harvested peat is burned to generate electricity in power plants, which releases fewer pollutants compared to coal burning. (**c**) Peat moss (*Sphagnum*). A few sporophytes—brown, jacketed structures on white stalks—are attached to a pale gametophyte.

Mosses are the bryophytes that most people know about (Figure 21.5*a*). Like others of this group, they are sensitive to air pollution. Where air quality is poor, mosses are few or absent. Most gametophytes have a low mounded or feathery form. Gametes develop at their shoot tips, in vessels enclosed in a tissue jacket. Sperm swim to eggs through water droplets clinging to the plants. Moss sporophytes consist of a stalk and jacketed structure in which haploid spores develop.

Figure 21.5*c* shows *Sphagnum*, one of 350 kinds of peat mosses. Their compressed, moist, organic remains form **peat bogs**, which can be as acidic as vinegar. In cold and temperate regions, bogs cover an area equal to half of the United States. Only highly acid-tolerant plants, such as cranberries, live in them.

Acidic metabolic products of bryophytes hamper growth of bacterial and fungal decomposers. Well-preserved bodies about 2,000 to 3,000 years old have turned up in peat bogs in Europe. The bogs may have been sites of ceremonial human sacrifices, and acidity kept the bodies from decomposing. Also, compared to cotton, *Sphagnum* soaks up five times more water, which enters dead cells in the leaflike parts. Because of the antiseptic properties and absorbency of peat moss, World War I doctors used it for bandages.

Although hornworts and liverworts are less well known, one or the other group may resemble the first plants. Even now, they are among the hardy pioneer

Figure 21.6 (**a**) One hornwort. The hornlike sporophyte is attached to a gametophyte. Liverworts, including *Marchantia*, reproduce sexually as well as asexually. Unlike other types, this genus has (**b**) female and (**c**) male reproductive parts on different plants.

species that are the first to colonize barren habitats. Their gametophytes are mostly ribbonlike or "leafy" forms that are often flat and thin, as in Figure 21.6.

Bryophytes are small, structurally simple nonvascular plants most of which grow in fully or seasonally moist habitats. Their flagellated sperm require environmental water to swim to the eggs.

The sporophyte grows from gametophyte tissues and draws nutritional support from it. Only among bryophytes does the sporophyte remain attached to a larger gametophyte.

21.3 Seedless Vascular Plants

The earliest seedless vascular plants were branching photosynthetic forms that lacked roots and leaves. A spectacular radiation produced leafy and even treelike forms. Diversity peaked during the Carboniferous.

Figure 21.7 depicts early seedless vascular plants in a Devonian swamp. Some of their descendants, which include the **lycophytes**, **horsetails**, and **ferns**, are with us today. Like their ancestors, they differ in big ways from bryophytes: The sporophytes have xylem and phloem, they are not attached to gametophytes, and they are the larger, longer lived phase of the life cycle.

Figure 21.7 What a Devonian swamp may have looked like, based on fossil evidence. Section 21.1 showed fossils of *Cooksonia* and *Psilophyton*, two of the first seedless vascular plants that evolved in such habitats.

The seedless vascular plants are sometimes called "amphibians" of the plant kingdom. Like frogs, they require a moist habitat for the sexual phase of the life cycle. Like bryophytes and lichens, a few species do live in dry habitats, but they reproduce sexually only during brief pulses of seasonal rain. The resurrection plant, *Selaginella lepidophylla*, shrivels up completely and goes dormant, then quickly revives. It grows in Mexico, New Mexico, and Texas.

LYCOPHYTES

Tree-sized lycophytes lived in Carboniferous swamp forests. The 1,100 or so modern forms are a great deal smaller. The best known types are club mosses, which range from tundra to the tropics (Figure 21.8a). Most club mosses have vascularized stems and roots, which grow from a branching, underground rhizome.

Microphylls—tiny leaves, each with an unbranched vein—are a defining trait of lycophytes. Chambers for spore production develop at the base of some of these leaves. In certain species, such leaves are organized around a central stem, as a **strobilus** (plural, strobili). This general term refers to any conelike reproductive structure derived from modified leaves. Some of the *Lycopodium* species are informally called ground pines, partly because they look a bit like miniature Christmas trees with "cones." Strobili also form on sporophytes of other groups, such as horsetails and gymnosperms.

Figure 21.8 A few seedless vascular plants. (**a**) Club moss (*Lycopodium*), a lycophyte. Notice its strobili, conelike clusters of modified leaves with spore-forming chambers. (**b**) Horsetail (*Equisetum*) vegetative stem. (**c**) Strobilus of a nonphotosynthetic, fertile *Equisetum* stem. (**d**) Tree fern (*Cyathea*). Tree fern forests once cloaked much of New Zealand. Most have been cleared for farming.

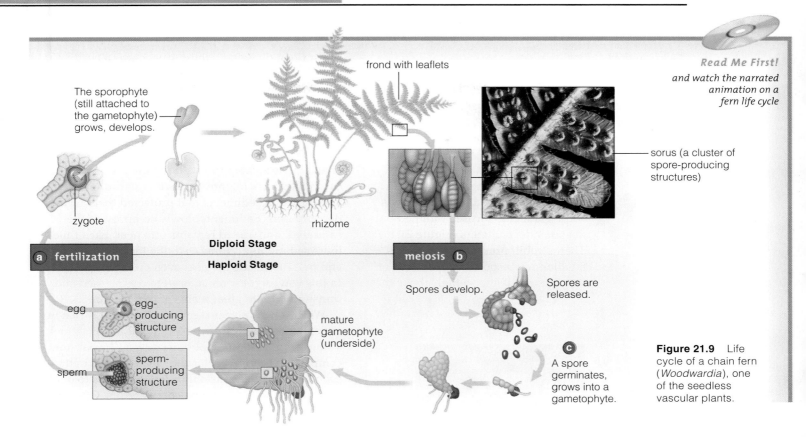

frond with leaflets

The sporophyte
(still attached to
the gametophyte)
grows, develops.

Read Me First!
and watch the narrated
animation on a
fern life cycle

sorus (a cluster of
spore-producing
structures)

zygote

rhizome

Diploid Stage

a fertilization

meiosis b

Haploid Stage

Spores develop.

Spores are
released.

egg

egg-
producing
structure

sperm

sperm-
producing
structure

mature
gametophyte
(underside)

c
A spore
germinates,
grows into a
gametophyte.

Figure 21.9 Life
cycle of a chain fern
(*Woodwardia*), one
of the seedless
vascular plants.

HORSETAILS

Seedless vascular plants called sphenophytes grew as tall as trees in Carboniferous swamp forests. About thirty smaller species are with us today. They are the "horsetails" (*Equisetum*) you may notice growing near many streams, railroad tracks, roads, and vacant lots.

The sporophytes of most horsetails have rhizomes, hollow stems, and scalelike leaves at the stem nodes. A cylinder of xylem and phloem runs parallel with the stems. The stems have horizontal ribs reinforced with silica granules, so they feel like sandpaper. American colonists and pioneers traveling west gathered these stems for use as disposable pot scrubbers.

Figure 21.8*b,c* shows the photosynthetic vegetative stems and pale, fertile stems of the sporophyte of one *Equisetum* species. The strobilus at each fertile stem's tip releases spores. When spores germinate, they give rise to free-living gametophytes no bigger than one millimeter to one centimeter across.

FERNS

With 12,000 or so species, the ferns are the largest and most diverse group of seedless vascular plants. All but about 380 species are native to the tropics, but we find them in homes and gardens all over the world (Figure 21.8*d*). And talk about sizes! Leaves of some floating ferns are less than 1 centimeter wide. Some tropical tree ferns grow 25 meters (82 feet) tall.

In most fern species, roots and leaves grow from vascularized rhizomes. Exceptions include tropical tree ferns and species growing as epiphytes. *Epiphyte* refers to any aerial plant that grows attached to tree trunks or branches. Young fern leaves, or fronds, develop in a coiled pattern that looks a bit like a fiddlehead, and are uncoiled by maturity. Fronds of many species, such as chain ferns, are divided into leaflets (Figure 21.9).

Rust-colored patches are distributed on the lower surface of most fern fronds. Each patch, a tiny cluster of spore-forming chambers, is called a sorus (plural, sori). Most chamber walls are only one cell thick, and when they pop open, spores are catapulted through the air. After germination, each spore gives rise to a heart-shaped gametophyte, a few centimeters across (Figure 21.9).

Seedless vascular plants include lycophytes, horsetails, and ferns. All require environmental water to complete the sexual phase of the life cycle. Large, independent sporophytes with xylem, phloem, and true leaves dominate the life cycle.

In lycophytes and horsetails, clusters of modified leaves form conelike protective structures in which spores are produced.

21.4 Ancient Carbon Treasures

Three hundred million years or so ago, in the middle of the Carboniferous, mild climates prevailed and swamp forests carpeted the wet lowlands of continents. The absence of pronounced seasonal swings in temperature favored plant growth through much of the year. Plants with lignin-reinforced tissues and well-developed root and shoot systems now had the competitive edge.

In vast forests of the Carboniferous, massively stemmed lycophyte trees —the giant club mosses—topped out at almost forty meters (Figure 21.10). Their strobili produced as many as 8 billion microspores or hundreds of megaspores. Being so high above the forest floor, dispersing these spores was a cinch. The giant horsetails, including species of *Calamites*, were close to twenty meters tall. Their aboveground stems often grew quickly from ever spreading underground rhizomes, and they formed dense thickets.

As it happened, the sea level rose and fell fifty times during the Carboniferous. Each time the sea receded, the swamp forests flourished. When the sea moved back in, forest trees became submerged and buried in sediments that protected them from decomposers. Sediments slowly compressed the saturated, undecayed remains into peat. Each time more sediments accumulated, the increased heat and pressure made the peat even more compact. In this way, organic remains were compressed and transformed into great seams of **coal** (Figure 21.10).

With its high percentage of carbon, coal is rich in energy and one of our premier "fossil fuels." It took a staggering amount of photosynthesis, burial, and compaction to form each major seam of coal in the ground. It has taken us only a few centuries to deplete much of the world's known coal deposits. Often you will hear about annual production rates for coal or some other fossil fuel. But how much do we really produce each year? None. We simply *extract* it from the ground. Coal is a nonrenewable source of energy.

Lepidodendron

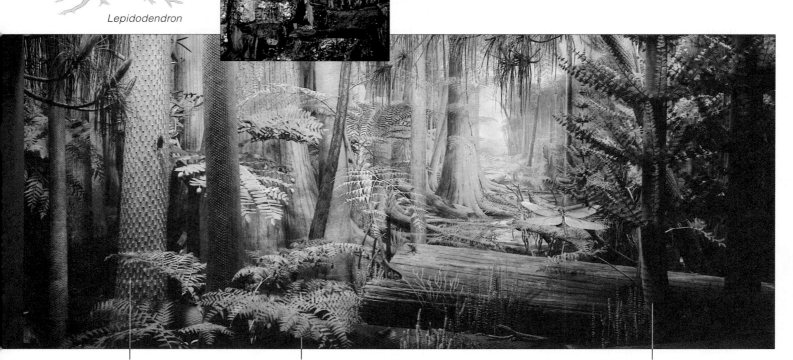

stem of a giant lycophyte (*Lepidodendron*)

seed fern (*Medullosa*), one of the early seed-bearing plants

stem of a giant horsetail (*Calamites*)

Figure 21.10 Reconstruction of a Carboniferous forest. The boxed inset shows part of a seam of coal.

21.5 The Rise of Seed-Bearing Plants

In diversity, numbers, and distribution, seed producers became the most successful groups of the plant kingdom.

Seed-bearing plants made their entrance in the late Devonian. The first kinds were close relatives of ferns, but seeds, in some cases as big as walnuts, formed on their fernlike fronds (Figure 21.10). Those "seed ferns" flourished through the Carboniferous, and a few made it through a mass extinction that ended the Permian. Long before the last ones vanished, cycads, conifers, and other gymnosperms started adaptive radiations. They became the dominant plants of Mesozoic times. (What some folks call the Age of Dinosaurs, botanists call the Age of Cycads.) However, flowering plants originated during the late Jurassic or early Cretaceous. By 120 million years ago, they were diversifying even as gymnosperms were on the decline. Their adaptive radiations continued, and flowering plants have now assumed supremacy in nearly all land habitats.

Think back on the factors that promoted the rise of land plants. Structural modifications, such as a water-conserving cuticle, were one reason they endured in seasonally dry, often cold habitats. Just as important were the extraordinarily adaptive ways in which two kinds of spores evolved in the seed-bearers.

Today, we can see how their **microspores** divide and grow into pollen grains—walled structures that protect immature male gametophytes from the time they are released to the time they start journeying on their own to eggs. We see how they afford protection when conditions are dry and cold. Pollen grains are tiny, and easily dispersed by air currents or animals. Regardless of the dispersing agent, **pollination** refers to the arrival of pollen on female reproductive parts of a seed plant. The evolution of pollen grains meant that it didn't matter whether water was available or not; sperm could travel to eggs without it.

In contrast to microspores, the **megaspores** of seed plants form in ovules. Each **ovule** starts out as a tiny mass of sphorophyte tissue. The megaspore part of it gives rise to a female gametophyte, with an egg cell. Other parts form nutritious tissue and a multilayered coat. Sperm fertilize the egg and an embryo develops inside an ovule. *What we call a "seed" is a mature ovule.* Its tough coat protects the embryo when conditions force a parent sporophyte to stop growing and enter dormancy. Nutrient-rich tissue packaged inside, the equivalent of a power bar, jump-starts the embryo's renewed growth once conditions favor germination.

A footnote to this story: Seed plants appeal to more than pollinators. They appealed even to early human species. Half a million years ago, for example, *Homo erectus* was stashing nuts and rose hips in caves, and roasting seeds. By 11,000 years ago, modern humans were domesticating seed plants as reliable sources of food. We now recognize 3,000 or so species as edible, and we plant staggering numbers of about 200 species as vital crops (Figure 21.11). In this sense, agricultural practices are contributing to the ongoing evolutionary success of at least some seed-bearing plants.

pine
pollen grains

> *Seed plants originated late in the Devonian. Two descendant groups, gymnosperms and flowering plants, radiated through dry land habitats with the help of structural modifications, pollen grains, and ovules that mature into seeds.*

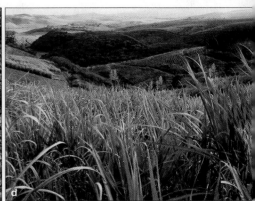

Figure 21.11 Edible treasures from flowering plants. (**a**) Fruits, which function in seed dispersal. (**b**) Mechanized harvesting of wheat, *Triticum*. (**c**) Indonesians picking shoots of tea plants (*Camellia sinensis*). Leaves of plants on hillsides in moist, cool regions have the best flavor. Only the terminal bud and two or three youngest leaves make the finest teas. (**d**) From Hawaii, a field of sugarcane, *Saccharum officinarum*. Sap extracted from its stems is boiled to make table sugar and syrups.

21.6 Gymnosperms—Plants With Naked Seeds

With this bit of history behind us, we turn to a survey of some modern gymnosperms. Their seeds are perched, in exposed fashion, on a spore-producing structure. Gymnos means naked; sperma is taken to mean seed.

The 600 species of **conifers** are woody trees and shrubs. Many have needlelike or scalelike leaves. Only conifers have female **cones**: reproductive structures with ovules wedged between clusters of papery or woody scales. Most conifers shed some leaves all year long but stay leafy, or *evergreen*. A few shed all leaves in autumn; they are *deciduous*. The most abundant trees of the Northern Hemisphere (pines), the tallest trees (redwoods), and the oldest (bristlecone pine) are conifers. Figures 21.12a and 21.13 show two kinds. Also in this group are firs, spruces, junipers, cypresses, larches, and podocarps.

About 130 species of **cycads** made it to the present. Pollen-bearing strobili and seed-bearing strobili form on separate plants (Figure 21.12b). Pollen travels from male to female plants on air currents or as "dust" on beetle pollinators. Most cycads evolved in the tropics or subtropics. Many have become ornamental plants, but some species in the wild face extinction.

Ginkgos (Figure 21.12d–f) were diverse in dinosaur times. The maidenhair tree, *Ginkgo biloba*, is the only surviving species. Like a few other gymnosperms, it is deciduous. Ginkgos were widely planted in China a few thousand years ago. Their natural populations nearly vanished, maybe because firewood was scarce. Ginkgos have attractive fan-shaped leaves and resist insects, disease, and air pollutants. Once again these hardy trees are widely planted—male trees, at least. The thick, fleshy seeds of female trees are stinky.

Gnetophytes include tropical trees, leathery leafed vines, and desert shrubs. An herbal stimulant, now banned, was extracted from photosynthetic stems of *Ephedra* (Figure 21.12g). A single *Welwitschia* species lives in African deserts. Its sporophyte has a deep taproot and a woody stem with strobili. Its two strap-shaped leaves may grow five meters long. A mature plant looks ragged because its leaves split lengthwise repeatedly during growth (Figure 21.12h).

What about life cycles? Consider a mature pine tree. This sporophyte has female cones with megaspore-containing ovules on tier after tier of woody scales (Figure 21.13). The ovules are exposed, not embedded

Figure 21.12 Gymnosperm diversity. (**a**) Bristlecone pine (*Pinus longaeva*) in the Sierra Nevada. Cycads: (**b**) Male strobilus of *Dioon* and (**c**) ovule-bearing, loose-leafed strobilus of *Cycas*. (**d**) The seeds of *Ginkgo biloba*. Existing ginkgo leaves (**e**) look like fossilized ones (**f**). (**g**) *Ephedra viridins* and (**h**) *Welwitschia mirabilis*.

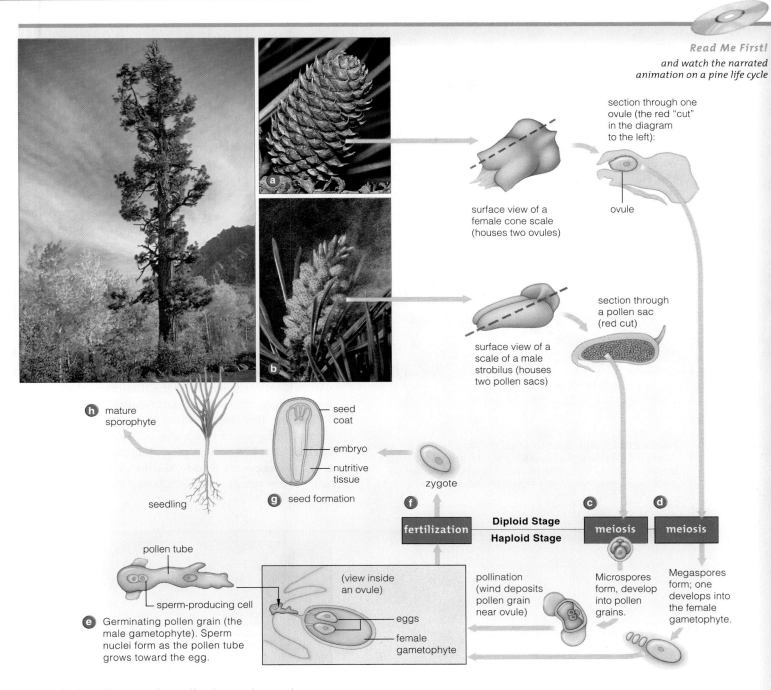

Read Me First!
and watch the narrated animation on a pine life cycle

section through one ovule (the red "cut" in the diagram to the left):

ovule

surface view of a female cone scale (houses two ovules)

section through a pollen sac (red cut)

surface view of a scale of a male strobilus (houses two pollen sacs)

h mature sporophyte

seed coat

embryo

nutritive tissue

zygote

seedling

g seed formation

f fertilization

Diploid Stage

Haploid Stage

c meiosis

d meiosis

pollen tube

(view inside an ovule)

pollination (wind deposits pollen grain near ovule)

Microspores form, develop into pollen grains.

Megaspores form; one develops into the female gametophyte.

sperm-producing cell

eggs

female gametophyte

e Germinating pollen grain (the male gametophyte). Sperm nuclei form as the pollen tube grows toward the egg.

Figure 21.13 Life cycle of a conifer, the ponderosa pine.

in woody tissue. The pine tree also has male strobili (not cones) in which microspores form and become pollen grains. At a suitable time, millions of pollen grains drift away on air currents, and some land on ovules of the same sporophyte or a different one. After this pollination event, a pollen grain germinates, and part of it starts growing as a tubular structure. This is the sperm-bearing male gametophyte. It grows slowly through tissues of the ovule for about a

year. After it penetrates the female gametophyte, its sperm reaches the egg, fertilization occurs, and a new pine tree zygote starts down a developmental road.

Existing gymnosperms include conifers, ginkgos, cycads, and gnetophytes. Like their Permian ancestors, all are adapted to seasonally dry climates. Depending on the group, seeds form on exposed surfaces of strobili or female cones.

21.7 Angiosperms—The Flowering Plants

What does "angiosperm" mean? Sperma, recall, refers to seeds. The angio– part of the name, derived from a Greek word for vessel, refers to a flower's ovary. Seeds of flowering plants form inside this closed vessel.

We find flowering plants almost everywhere, from icy tundra to deserts and remote oceanic islands. What accounts for their distribution and diversity? Consider the **flower**, a specialized reproductive shoot (Figure 21.14). When plants invaded the land some 435 million years ago, insects that ate plant parts and spores were not far behind. After pollen-producing plants evolved, many kinds of insects apparently made the connection between "plant parts with pollen" and "food."

Even as flowering plants gave up some pollen to hungry insects, they gained a reproductive edge. How? Insects crawling on plants became dusted with pollen, which they deposited directly on female plant parts. Plants coevolved in ways that became more enticing to insects, as with flowers, fragrances, and nectar. Instead of wasting energy on random searches, many insects coevolved with specific plants; they became specialists in finding and collecting pollen or nectar from them—and coincidentally dispersing pollen for them.

Coevolution refers to two or more species jointly evolving because of their close ecological interactions. Heritable changes in one exert selection pressure on the other, which evolves also. That is how **pollinators**—agents that deliver pollen of one species to female parts of the same species—coevolved with seed plants. They include insects, bats, and birds. By recruiting pollinators to assist in their reproduction, flowering plants have managed to be the dominant plant group over the past 100 million years (Figure 21.15).

Did the first angiosperms look like *Archaefructus*? Complete fossils of this plant, including flowers, date back at least 125 million years (Figure 21.14). Or were early angiosperms woody shrubs or trees? We really don't know yet, and research is ongoing.

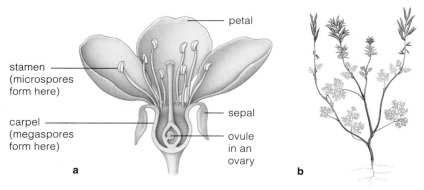

Figure 21.14 (**a**) Floral structures. (**b**) *Archaefructus sinensis*, one of the earliest known flowering plants, apparently lived in shallow lakes. As in modern flowers, it had male and female parts (stamens and carpels), but no petals or sepals.

Figure 21.15 Vascular plant diversity in Mesozoic times. Conifers and other gymnosperms began to decline even before flowering plants started an adaptive radiation. The painting above showcases *Archaeanthus linnenbergeri*, which lived in the Mesozoic. In many traits, living magnolias resemble this now-extinct flowering plant.

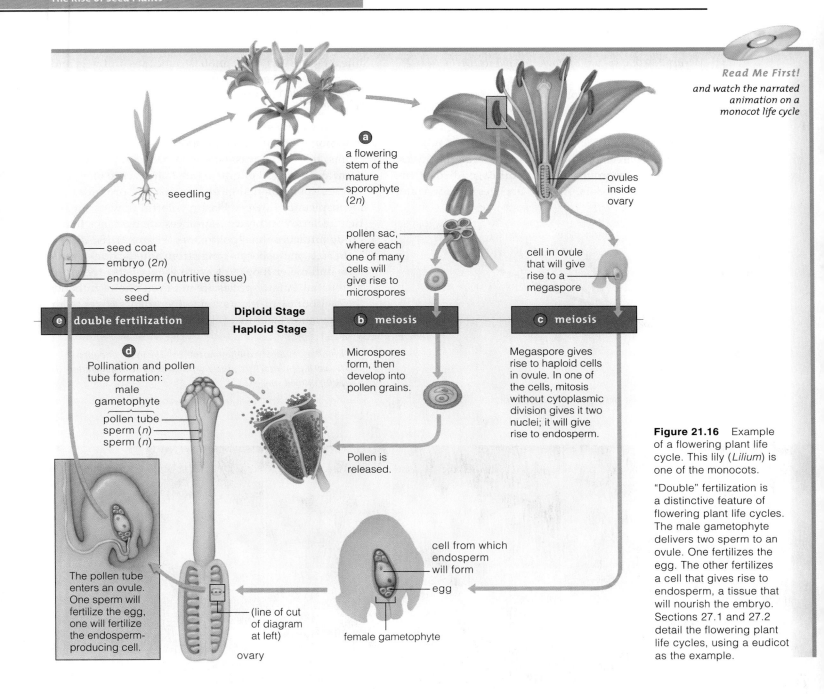

Read Me First!
and watch the narrated animation on a monocot life cycle

a a flowering stem of the mature sporophyte (2n)

seedling

ovules inside ovary

seed coat
embryo (2n)
endosperm (nutritive tissue)
seed

pollen sac, where each one of many cells will give rise to microspores

cell in ovule that will give rise to a megaspore

Diploid Stage		
e double fertilization	**b** meiosis	**c** meiosis
Haploid Stage		

d Pollination and pollen tube formation: male gametophyte

pollen tube
sperm (n)
sperm (n)

Microspores form, then develop into pollen grains.

Pollen is released.

Megaspore gives rise to haploid cells in ovule. In one of the cells, mitosis without cytoplasmic division gives it two nuclei; it will give rise to endosperm.

Figure 21.16 Example of a flowering plant life cycle. This lily (*Lilium*) is one of the monocots.

"Double" fertilization is a distinctive feature of flowering plant life cycles. The male gametophyte delivers two sperm to an ovule. One fertilizes the egg. The other fertilizes a cell that gives rise to endosperm, a tissue that will nourish the embryo. Sections 27.1 and 27.2 detail the flowering plant life cycles, using a eudicot as the example.

The pollen tube enters an ovule. One sperm will fertilize the egg, one will fertilize the endosperm-producing cell.

(line of cut of diagram at left)

ovary

cell from which endosperm will form
egg

female gametophyte

The adaptive radiations of flowering plants were triggered by asteroid impacts, rupturing and collisions of land masses, and changes in sea level and the global climate. Given these severe environmental challenges, and given their protected seeds and all that help from pollinators, flowering plants displaced gymnosperms and older lineages in most habitats (Figure 21.15).

Chapter 27 in the next unit offers a closer look at the structure and function of flowering plants. For now, start thinking about the monocot life cycle in Figure 21.16 and how a flower-producing sporophyte dominates it. You'll see a eudicot life cycle in Section 27.2. Seeds of both groups become packaged with a nutrient-rich tissue (endosperm), and parts of ovaries mature into structures called fruits. Fragrant, colorful, tiny, large, hard-shelled, winged, sticky—no matter what the features, all fruits function in dispersing the new generation of sporophytes.

Angiosperms are the most successful plants. They alone produce flowers. These specialized reproductive structures have pollen-producing male parts and female parts with ovaries, protected chambers in which seeds develop.

21.8 A Glimpse Into Flowering Plant Diversity

At least 260,000 species of flowering plants live in diverse habitats. They bloom in meadows and forests, in parched deserts, and on windswept mountaintops. Some thrive in lakes and streams. A few kinds have adapted to saltmarshes or to shallow marine habitats.

Almost 90 percent of living plants are angiosperms. The three major groups are **magnoliids**, **eudicots** (true dicots), and **monocots**. Ancient lineages include water lilies (Figure 21.17). Magnolias, avocados, nutmeg, and pepper plants are among the 9,200 kinds of magnoliids.

The 170,000 or so eudicot species include cabbages, daisies, and most other herbaceous (nonwoody) plants; most flowering shrubs and trees, and cacti.

Among 80,000 or so monocot species are orchids, palms, lilies, and grasses. Rye, sugarcane, rice, wheat, corn, and some other grasses are valued crop plants.

Wind-pollinated angiosperms usually produce tiny, inconspicuous flowers. Flashy, colorful flowers, sugar-rich nectars, and heady fragrances are evidence that a group attracts animal pollinators (Figure 21.18a–d).

In size, angiosperms range from aquatic duckweeds (one millimeter long) to towering *Eucalyptus* trees 100 meters tall. A few species are not even photosynthetic; they withdraw nutrients from other plants or get them from mychorrizal fungi (Figure 21.18e,f).

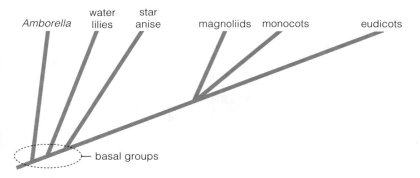

basal groups

Figure 21.17 Evolutionary tree diagram for flowering plants.

> *Magnoliids, eudicots, and monocots are the main lineages of flowering plants. The largest group, the eudicots, includes 170,000 named species.*

Figure 21.18 (**a**) Examples of flowers, reproductive structures that function in pollination and seed formation. Their colors, patterns, and shapes attract pollinators. (**b**) This hummingbird pollinator is withdrawing nectar from the flower of a columbine (*Aquilegia*) plant.

(**c**) Sacred lotus (*Nelumbo nucifera*), an aquatic species. The flower's radial pattern is typical of ancient lineages.

(**d**) More recently evolved lineages, including these pansies (*Viola*), display a bilateral pattern, with roughly equivalent left and right parts.

(**e**) Indian pipe (*Monotropa uniflora*), a nonphotosynthetic species, pilfers nutrients from mycorrhizae of some photosynthetic plants. (**f**) Dwarf mistletoe (*Arceuthobium*). It parasitizes trees directly and stunts their growth.

21.9 Deforestation Revisited

Paralleling the huge increases in human population size are increasing demands for wood as fuel, lumber, and other forest products, and for pastures and croplands. More and more people compete for dwindling resources and for economic profit, but also because alternative ways of life simply are not available to most families.

Seed-bearing plants dominate the world's existing great forests, and they profoundly influence life. Like enormous sponges, the watersheds of these forested regions absorb, hold, then gradually release water. By mediating downstream flows, they slow erosion, flooding, and sedimentation that can disrupt rivers, lakes, and reservoirs. Strip the forests, and exposed soil is more vulnerable to leaching of nutrients and to erosion, especially on steep slopes.

Deforestation means mass removal of all trees from large tracts for logging, agriculture, and grazing. This chapter started with a look at what is going on in the once-sweeping temperate forests of the United States, Canada, New Zealand, and elsewhere. But tropical forests are the most threatened. For 10,000 years or more, these forests have been home to 50 to 90 percent of all land-dwelling species. In less than four decades, we have cleared more than half of them.

Deforestation is now greatest in Brazil, Indonesia, Colombia, and Mexico. If it continues at present rates, only Brazil and Zaire will still have large tropical forests in the year 2010. By 2035, most of their forests will be gone as well.

In tropical regions, clearing forests for agriculture sets the stage for long-term losses in productivity. The irony is that tropical forests are one of the worst places to grow crops or raise animals. Deep topsoils simply cannot form. Nutrient-rich litter and organic remains cannot build up because high temperatures and heavy, frequent rainfall promote their rapid decomposition. As fast as decomposers release nutrients, trees and other plants take them up.

Long before mechanized logging became pervasive, many people were carrying out *shifting cultivation*, once referred to as slash-and-burn agriculture. They cut and burned trees, then tilled ashes into the soil. Nutrient-rich ashes can sustain crops only for a few seasons, at most. Later, people abandon cleared plots, because heavy leaching results in infertile soil.

If a small human population practices shifting cultivation on small, widely spaced plots, they may not damage a forest much. But soil fertility plummets when human populations become more dense. Why? Larger areas are cleared, then plots are cleared again after shorter and shorter intervals.

Figure 21.19 Tropical rain forest in Southeast Asia.

Deforestation even can alter regional climates. It affects the rates of evaporation, runoff, and maybe even the pattern of rainfall. Trees release 50 to 80 percent of the water vapor in tropical forests. In a logged area, annual precipitation declines, and rain swiftly drains away from the exposed, nutrient-poor soil. A logged region gets hotter and drier—and soil fertility and moisture levels fall. In time, sparse, dry grasslands or deserts may come to prevail where the formerly dense forest once stood.

Also, tropical forests absorb much of the sunlight striking equatorial regions. Deforested land is shinier, so to speak, and reflects more incoming energy back into space. The combined photosynthetic activity of so many trees affects the global cycling of carbon and oxygen. Extensive tree harvesting and the burning of tree biomass release stored carbon, as carbon dioxide. Some researchers suspect that deforestation might be amplifying the greenhouse effect for that reason.

Conservation biologists are working to protect the forests. As three examples, a coalition of 500 groups is dedicated to preserving Brazil's remaining tropical forests. In India, women have already constructed and installed 300,000 inexpensive, smokeless wood stoves. In the past decade, the stoves saved more than 182,000 metric tons of trees by cutting demands for fuelwood. In Kenya, women planted millions of trees as a source of wood and to counter erosion.

Table 21.1   Comparison of Major Plant Groups

Bryophytes	24,000 species. Moist, humid habitats.

Nonvascular land plants. Fertilization requires standing water. Haploid dominance. Cuticle and stomata present in some.

Club mosses	1,100 species; simple leaves. Mostly wet or shady habitats.
Horsetails	25 species of single genus. Swamps and disturbed habitats.
Ferns	12,000 species. Wet, humid habitats in mostly tropical and temperate regions.

Seedless vascular plants. Fertilization requires standing water. Diploid dominance. Cuticle and stomata present in all.

Conifers	600 species, mostly evergreen, woody trees, shrubs with pollen- and seed-bearing cones. Widespread distribution.
Cycads	130 slow-growing tropical, subtropical species.
Ginkgo	1 species, a tree with fleshy-coated seeds.
Gnetophytes	70 species. Limited to some deserts, tropics.

Gymnosperms—vascular plants with "naked seeds." Water not required for fertilization. Diploid dominance. Cuticle and stomata present in all.

Flowering plants

Monocots	80,000 species. Floral parts often arranged in threes or in multiples of three; one seed leaf; parallel leaf veins common; pollen with one furrow.
Eudicots (true dicots)	At least 170,000 species. Floral parts often arrayed in fours, fives, or multiples of these; two seed leaves; often net-veined leaves; pollen with three or more pores and/or furrows.
Magnoliids and basal groups	9,200 species. Many spirally arranged floral parts or fewer, in threes; two seed leaves; net-veined leaves common; pollen with one furrow.

Angiosperms—vascular plants with flowers, protected seeds. Water not required for fertilization. Diploid dominance. Cuticle and stomata in all.

Table 21.2   Evolutionary Trends Among Plants

Bryophytes	Ferns	Gymnosperms	Angiosperms
Nonvascular	Vascular		
Haploid dominance	Diploid dominance		
Spores of one type	Spores of two types		
Motile gametes		Nonmotile gametes*	
Seedless		Seeds	

* Require pollination by wind, insects, animals, etc.

Summary

Section 21.1   Plants are multicelled eukaryotes. Nearly all are photoautotrophs on land. Table 21.1 summarizes the main characteristics of major groups.

Comparisons among lineages reveal trends in plant evolution. Plants on land became structurally modified; a waterproof cuticle, stomata, and vascular tissues evolved in most groups. The haploid phase, dominant in algal life cycles, gave way to diploid dominance. Complex sporophytes that hold, nourish, and protect spores and gametophytes arose. Two types of spores, not one, evolved. For gymnosperms and angiosperms, heterospory paved the way for the evolution of pollen grains and seeds. Table 21.2 summarizes these trends.

Section 21.2   Mosses, liverworts, and hornworts are bryophytes. These nonvascular plants do not have xylem and phloem. Their flagellated sperm reach eggs only by swimming through films or droplets of water. The sporophyte begins to develop inside gametophyte tissues. It remains attached to and dependent upon the gametophyte even when mature.

Section 21.3   Lycophytes, horsetails, and ferns are seedless vascular plants. Sporophytes with leaves and vascular tissues are the larger, longer lived phase of the life cycle. Like bryophytes, their flagellated sperm must swim through water to reach the eggs.

Section 21.4   Giant bryophytes dominated swamp forests during the Carboniferous. Coal is the energy-rich, compressed organic remains of these forests.

Section 21.5   Gymnosperms and flowering plants (angiosperms) are seed-bearing vascular plants. They produce microspores that give rise to pollen grains, from which sperm-bearing male gametophytes develop. They also produce megaspores that give rise to female gametophytes (with eggs). This happens in ovules, other parts of which form nutritive tissue and a seed coat. A seed is a mature ovule.

Section 21.6   Gymnosperms, including conifers, cycads, ginkgos, and gnetophytes, are adapted to dry climates. Their ovules form on exposed surfaces of strobili or, in the case of conifers, of female cones.

Section 21.7   Angiosperms alone produce flowers, and their seeds form inside floral ovaries. Some ovary tissues later develop into fruits.

Section 21.8   Angiosperms are the most diverse and widespread group of plants. Most species coevolved with animal pollinators. The two largest groups are monocots and eudicots. Magnoliids and basal groups are a more ancient lineage.

Section 21.9   Deforestation threatens ecosystems worldwide. Forests play a role in the global cycling of water, carbon dioxide, and oxygen.

Self-Quiz

1. Which of the following statements is *not* correct?
   a. Gymnosperms are the simplest vascular plants.
   b. Bryophytes are nonvascular plants.
   c. Lycophytes and angiosperms are vascular plants.
   d. Only angiosperms produce flowers.

2. Which does *not* apply to gymnosperms or angiosperms?
   a. vascular tissues   c. single spore type
   b. diploid dominance   d. cuticle with stomata

3. Of all land plants, bryophytes alone have independent
   _____ and attached, dependent _____ .
   a. sporophytes;   c. rhizoids; zygotes
      gametophytes   d. rhizoids; stalked
   b. gametophytes;   sporangia
      sporophytes

4. Lycophytes, horsetails, and ferns are classified as
   _____ plants.
   a. multicelled aquatic   c. seedless vascular
   b. nonvascular seed   d. seed-bearing vascular

5. The _____ produce flagellated sperm.
   a. ferns   c. monocots
   b. conifers   d. a and c

6. A seed is _____ .
   a. a female gametophyte   c. a mature pollen tube
   b. a mature ovule   d. an immature embryo

7. Match the terms appropriately.
   ____ gymnosperm   a. gamete-producing body
   ____ sporophyte   b. help control water loss
   ____ lycophyte   c. "naked" seeds
   ____ ovary   d. only plant that produces
   ____ bryophyte   flowers
   ____ gametophyte   e. spore-producing body
   ____ stomata   f. nonvascular land plant
   ____ angiosperm   g. seedless vascular plant
                      h. ovules form in it

Critical Thinking

1. Figure 21.20 shows a Nahmint Valley forest in British Columbia after logging. It also shows the wood frames of homes being built. Would you chain yourself to a tree in such a forest to stop loggers? If your answer is yes, would you also give up the chance to own a wood-frame home (as most homes are, in developed countries)? What about forest products, such as paper tablets, newspapers, toilet tissue, wood furniture, and fireplace wood?

2. With respect to question 1, multiply each of your answers by 6.3 billion people and describe what might happen when we run out of trees. Research the pros and cons of tree farms—say, of a single species of pine.

3. The 2004 Nobel Peace Prize was awarded to Wangari Maathai, who founded the Green Belt Movement in 1977 (Figure 21.21). Since then, group members—mostly poor rural women—have planted more than 25 million trees. Reforestation efforts help halt soil erosion, provide sources of fruits, food, and firewood, and empower women to work together for the long term good of their communities. Are any nonprofit groups working to protect forests in your area? Do you support their goals and methods?

Figure 21.20 Part of a logged-over forest in British Columbia.

Figure 21.21 Kenyan Wangari Maathai, who in 2004 won a Nobel Peace Prize for her work in support of women's rights and sustainable development.

Media Menu

Student CD-ROM

Impacts, Issues Video
   Beginnings and Endings
Big Picture Animation
   Origin and evolution of the plants
Read-Me-First Animation
   Moss life cycle
   Fern life cycle
   Pine life cycle
   Monocot life cycle
Other Animations and Interactions
   Evolutionary trends in major plant groups
   Evolutionary tree for major plant groups

InfoTrac

- Doing the Peat Bog Two-Step. *Mother Earth News*, June 2001.
- A Fern with a Taste for Toxic Waste. *Environment*, May 2001.
- Rocks Yield Clues to Flower Origins. *Science News*, April 21, 2001.
- The Future of Our Forests. *Audubon*, January 2001.
- Flower Power. *Odyssey*, March 2003.

Web Sites

- Botanical Society of America: www.botany.org
- Smithsonian Bryophytes: bryophytes.plant.siu.edu
- American Fern Society: amerfernsoc.org
- Wayne's Word: waynesword.palomar.edu/wayne.htm
- USDA Plants Database: plants.usda.gov

How Would You Vote?

In the United States, the average person uses about 750 pounds of paper each year, twice the amount in other developed countries. Recycling paper saves trees, lessens the impact on landfills, and reduces water pollution. However, recycled paper is often more expensive to produce than new paper. The government could encourage use of recycled paper by giving companies tax breaks for using it. Would you support such a program?

IMPACTS, ISSUES *Old Genes, New Drugs*

Other than Samoans, few of us prize cone snails (*Conus*) as tasty treats. However, the intricate patterns of cone snail shells obviously have been prized down through the ages. Archaeologists exploring past cultures in Peru, Arizona, Micronesia, and Iran uncovered bracelets, rings, and ritual objects made from *Conus* shells. Collectors still covet the shells, the most beautiful of which still end up in jewelry and other forms of decorative art.

Cone snails also fascinate biologists, for different reasons. More than 500 species of these predatory mollusks inhabit the seas, and all of them subdue small prey with *conotoxins*. These paralytic secretions pack a wallop that can kill even large animals. Seventy percent of the people who were harpooned by a cone snail and were not treated promptly ended up dead.

Conotoxins are peptides that can bind to channel proteins of cell membranes. Any cone snail species can make 100 to 300 conotoxins, each with a highly specific molecular target. What's so fascinating about toxin diversity and specificity of targets? It means cone snails are potential sources of many new drugs.

One *C. magnus* conotoxin is 1,000 times more potent than morphine—yet nonaddictive. It works by blocking channels where nerve cells release signaling molecules that can alter activities of neighboring cells. A synthetic version is now being tested in clinical trials. The test groups consist of cancer and AIDS patients who are in unbearable but otherwise untreatable pain.

C. geographicus, shown in Figure 22.1, secretes a toxin that one day might help epileptics. While studying this

Figure 22.1 *Conus geographicus* engulfing a small fish. This mollusk's siphon (extended straight up) detects disturbances in the water, as happens when small fishes and other prey swim within range. *C. geographicus* impaled this fish with a harpoon-like device, then pumped paralyzing conotoxins into it. The filmstrip (*far right*) shows just a tiny sampling of the diverse patterns of *Conus* shells.

the big picture

So What Is an Animal? Animals are alike in many respects, but differ in their type of body symmetry, whether they have a head and segments, and in their type of gut and of body cavity, if any. The invertebrate animals evolved first and far outnumber vertebrates.

Getting Along Without Organs Sponges, an ancient lineage, have endured with no body symmetry, tissues, or organs. Cnidarians—the jellyfishes and their relatives—have a radial type of body symmetry, and also tissues, but no organs or definite head.

species, University of Utah researchers discovered a gene that has been conserved in the DNA of different species for a long, long time. In cone snails, the gene codes for an enzyme, gamma-glutamyl carboxylase (GGC), that catalyzes a step in the conotoxin synthesis pathway. Oddly, humans make GGC, but it functions in blood clotting processes. Fruit flies make it even though they don't secrete toxins and their blood doesn't clot. To date, no one knows what GGC does in fruit flies.

We can expect that the GGC gene arose more than 500 million years ago, before genetic divergences from a common ancestor gave rise to mollusks, insects, and vertebrates. As an outcome of different mutations in the independently evolving lineages, the enzyme product of the ancestral gene took on different functions.

The point is this: Carefully look back through time and you will see how organisms interconnect. *At every branch point in the animal family tree, microevolutionary processes gave rise to changes in heritable traits—in biochemistry, body plans, and behavior.* Your uniquely human traits emerged through modification of traits that evolved in countless generations of vertebrates and, before them, in ancient invertebrate forms.

This chapter and the next compare the major groups of animals, from the structurally simplest to the most complex. Don't assume that the more ancient lineages are somehow evolutionarily stunted or "primitive." As you will see, even the simplest among them are exquisitely adapted to their environment.

 How Would You Vote?

Cone snails are diverse, but most kinds have a limited geographic range, which makes them highly vulnerable to extinction. We don't know how many are harvested, because no one monitors the trade. Should the United States push to extend regulations on trade in endangered species to cover any species captured from the wild? See the Media Menu for details, then vote online.

Protostomes   Most animals have tissues and organ systems, bilateral symmetry, and a head. Many of them also have a type of body cavity called a coelom. These features emerge in a particular way in the embryos of protostomes, one of two major lineages of bilateral animals.

Invertebrate Deuterostomes   Deuterostomes are bilateral animals that share some features with protostomes, but these features develop in a different way in their embryos. The deuterostome lineage includes chordates as well as the radial invertebrates called echinoderms.

22.1 Overview of the Animal Kingdom

We already know of more than 2 million animal species, and no simple definition can cover that much diversity.

GENERAL CHARACTERISTICS

Animals share many traits with other organisms, and no one trait sets them apart. For that reason, we must define **animals** by a list of their characteristics:

1. All animals are multicelled, and most have tissues, organs, and organ systems. They have no walled cells.

2. Animals are aerobic heterotrophs that ingest other organisms or withdraw nutrients from them.

3. Animals typically reproduce sexually; many also reproduce asexually. Their embryos usually develop through a series of stages.

4. Most animals are motile; they actively move about during all or part of the life cycle.

These shared traits arose in a common ancestor. Later on, different traits accumulated in dozens of lineages.

Mammals, birds, reptiles, amphibians, and fishes are all **vertebrates**, with a backbone. These are the most familiar lineages. Yet all but 50,000 of 2 million named species are **invertebrates**! Invertebrates have many features, but a backbone isn't one of them (Table 22.1).

CLUES IN BODY PLANS

How might we get a conceptual handle on animals as different as flatworms and dinosaurs, hummingbirds and humans? Comparing similarities and differences with respect to six basic features is one way to start. These features are body symmetry, embryonic tissue layers, cephalization, type of gut, type of body cavity, and segmentation.

BODY SYMMETRY, CEPHALIZATION Nearly all animal bodies are radial or bilateral. With **radial symmetry**, body parts are arranged in a regular pattern around a central axis, like spokes of a bike wheel (Figure 22.2). Radial animals live in water, and their body plan is adapted to intercept food that is equally likely to drift or swim toward them from any direction.

Most animals have **bilateral symmetry**. In this body type, there is an *anterior* end (front) and *posterior* end (back). The body has right and left sides, a *ventral* surface (the underside) and *dorsal* surface (upper side). Early on, bilateral animals moving along in a forward direction apparently favored **cephalization**: Being the first to encounter food, danger, and other stimuli, their anterior end evolved into a distinct *head*, in which many nerve and sensory cells became concentrated. Some cells were the start of sensory organs and a complex brain in many bilateral lineages.

TISSUE LAYERS IN EMBRYOS Section 8.4 introduced the nature of cleavage, the first stage of animal development. Mitotic cell divisions transform an animal zygote into a tiny ball of cells in this stage. After this, primary tissue layers form in all animals except sponges, and they will give rise to adult tissues and organs. In structurally simple radial animals, two layers form.

Group	Some Representatives	Existing Species
Table 22.1	**Major Animal Groups**	
Poriferans	Sponges	8,000
Cnidarians	Hydras, jellyfishes	11,000
Flatworms	Turbellarians, flukes, tapeworms	15,000
Annelids	Leeches, earthworms, polychaetes	15,000
Mollusks	Snails, slugs, squids, octopuses	110,000
Arthropods	Crustaceans, spiders, insects	1,000,000+
Roundworms	Pinworms	20,000
Echinoderms	Sea stars, sea urchins, sea cucumbers	6,000
Chordates	Invertebrate chordates: Tunicates, lancelets	2,100
	Vertebrate chordates:	
	Jawless fishes	84
	Jawed fishes	21,000
	Amphibians	4,900
	Reptiles	7,000
	Birds	8,600
	Mammals	4,500

radial symmetry

bilateral symmetry

Figure 22.2 Simple way to think about radial and bilateral symmetry in animal body plans.

The outer **ectoderm** is the start of a tissue that lines body surfaces. The inner **endoderm** develops into a gut lining. Ectoderm and endoderm are the source of additional body parts in bilateral animals. In the embryos of these animals, a third layer—**mesoderm**—forms between the other two. This third layer is the source of the muscles and many other organs in all animals more complex than jellyfish and their relatives.

TYPE OF GUT   A **gut** is either a sac with one opening or a tube with two openings to the outside. Food is digested in the gut, then absorbed into the internal environment. A saclike gut is an *incomplete* digestive system, and it was the first kind to evolve. Cnidarians have such a gut.

A tubular gut, called a *complete* digestive system, starts at a mouth and ends at an anus. It forms in the embryos of two major lineages of bilateral animals, but not the same way. In both cases, it starts at or near the first opening (a blastopore) that develops on the embryo's surface. In **protostomes**, such as flatworms, mollusks, annelids, roundworms, and arthropods, the opening persists as the mouth. It becomes the anus in the echinoderms, chordates, and other **deuterostomes**. We return to these two lineages at the chapter's end, after tracing pivotal steps in their evolution.

BODY CAVITIES   Flatworms, like a few other groups of invertebrates, have no body cavity; they just have tissues and organs packed between the gut and body wall (Figure 22.3*a*). But most bilateral animals have a body cavity of the type known as a **coelom**. It lies between the gut and body wall and has a lining—the *peritoneum*—derived from mesoderm (Figure 22.3*c*).

The coelom develops differently in the embryos of protostomes and deuterostomes. Either way, it was a key innovation. Organs that formed inside this cavity were cushioned in fluid, which helped protect them from shocks. The organs also were free to move, grow, and develop independently of the body wall.

Some protostomes have a reduced coelom or none at all. The coelom apparently was reduced or lost as a result of mutations in certain genes that affected early development. Roundworms, as one example, have a **pseudocoel** (false coelom). Their main body cavity is only partly lined with tissue derived from mesoderm (Figure 22.3*b*).

SEGMENTATION   One of the other characteristics that help define bilateral groups is the presence or absence of segments. **Segmentation** refers to the division of a body into interconnecting units that are repeated one

Read Me First! and watch the narrated animation on types of body cavities

a No coelom (*acoelomate* animals)
epidermis
gut cavity
organs packed between gut and body wall

b Pseudocoel (*pseudocoelomate* animals)
epidermis
gut cavity
unlined body cavity around gut

c Coelom (*coelomate* animals)
epidermis
gut cavity
peritoneum
body cavity with lining that holds internal organs in place

Figure 22.3   Type of body cavity (if any) in different groups of animals.

after the next along the main body axis. This body plan evolved early in some groups. Most annelids have pronounced segmentation; other groups have fused segments or segments with specialized appendages. We see clues to our segmented origins in human embryos.

Animals are multicelled heterotrophs. Nearly all have tissues, organs, and organ systems. They reproduce sexually, and many can reproduce asexually. Embryos develop in stages. Most species are motile for at least part of the life cycle.

Animals differ in body symmetry, cephalization, type of gut, type of body cavity, and segmentation. The vast majority have no backbone; they are invertebrates.

22.2 Animal Origins

Judging from genetic evidence and radiometric dating of fossilized tracks, burrows, and pinhead-sized embryos, divergences from protist lineages gave rise to animals somewhere between 1.2 billion and 670 million years ago.

How did animals arise? Perhaps they evolved from colonies of flagellated cells. Over time, some of the cells could have mutated and taken on reproduction and other specialized tasks. Such a division of labor is typical of multicellularity. The choanoflagellates offer support for this theory. They are single-celled protists that form colonies. Each cell has a ring of absorptive structures (microvilli) around the base of a flagellum. As you will see, body cells of sponges have a similar structure. Comparisons of DNA sequences also point to choanoflagellates as the group most likely to have been involved in the early stages of animal evolution.

Suppose a colony of cells flattened out over time and started creeping along on the seafloor. Their lifestyle could have favored the evolution of cell layers, such as those of *Trichoplax adhaerens*. This tiny, soft-bodied marine animal is the only living placozoan, and it can be viewed as the simplest animal (Figure 22.4*a*). Its several thousand cells are arranged in two layers. When *Trichoplax* glides over food, gland cells in the lower layer secrete digestive enzymes, and other cells absorb breakdown products.

By 610 million years ago, tiny, soft-bodied animals shaped like fronds, disks, and blobs were living on or in seafloor sediments (Figure 22.4*b,c*). We call them the **Ediacarans**. Some were similar to living groups and may be related to them. Most Ediacarans had a flattened body form, suitable for absorbing dissolved organic compounds. So far, fossils show no evidence of predatory species.

Things got rougher in the Cambrian, when animals started a great adaptive radiation. Large, aggressive predators emerged, as did spiny, shelled, and armored prey. Before the Cambrian ended, all major groups of animals had originated in the seas. What caused that Cambrian explosion of diversity? Changes in the land masses, sea level, and climates may have triggered it. Also, new predators and new prey exerted selection pressure on one another.

We have much to learn about animal origins, but we know all existing lineages are one monophyletic group. All are descended from a common ancestor, as outlined in the evolutionary tree in Figure 22.5.

a
b
c
d

Figure 22.4 (**a**) The only living placozoan; the soft-bodied *Trichoplax adhaerens*. Of all animals, it has the smallest genome. Fossilized Ediacarans, (**b**) *Spriggina* and (**c**) *Dickensonia*. The oldest known Ediacarans lived between 610 million and 510 million years ago. (**d**) Fossil of one of the earliest trilobites.

Animal life dates back more than 600 million years, but an adaptive radiation in the Cambrian gave rise to the ancestors of all major groups with living representatives.

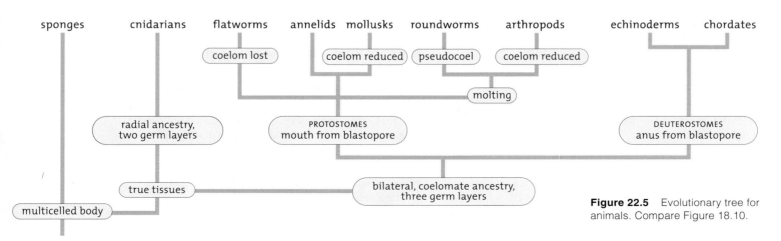

Figure 22.5 Evolutionary tree for animals. Compare Figure 18.10.

22.3 Sponges—Success in Simplicity

Yes, you can endure through time with simplicity, as demonstrated by the sponges.

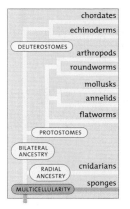

The **sponges**—animals with no symmetry, tissues, or organs—are one of nature's true success stories. They have thrived since precambrian times. Most live in shallow tropical seas, but some live in deep-sea trenches, cold Antarctic seas, and freshwater. Some sponges are large enough to sit in, others are as small as a fingernail. Different kinds have flattened and sprawling, lobed, compact, or tubelike shapes, as in Figure 22.6. Cells line all body surfaces but don't form distinct tissues, as seen in more complex animals.

Sponges are typically "filter feeders." Water flows into a sponge through tiny pores and chambers in the body, then out through one or more large openings. It does so because of the beating of the flagella of collar cells. These cells make up the lining of the chambers. A ring of microvilli—long, thin extensions from the cell body—acts as a food-trapping "collar" (Figure 22.6b). The cells engulf bacteria or organic debris flowing past. They release some of it to amoeba-like cells that prowl between the linings. These cells process food and carry out other tasks that benefit the body.

Sea slugs are among the relatively few predators that eat sponges. Defenses include protein fibers and sharp spicules of calcium carbonate or silica that stiffen the sponge body. Predators learn that sampling a sponge is like eating a mouthful of glass splinters. Some species coat themselves in slime. Others stink.

As a group, sponges have a large chemical arsenal against predators, parasites, and pathogens. They also harbor symbionts that secrete chemicals and help get rid of wastes. The chemicals intrigue researchers, who view them as potential drugs to fight inflammation, viruses, microbes, and tumors.

Mature sponges don't move about, so how do they deal with sex? Some release sperm into water, which might carry the sperm to another sponge. Sperm of some sponges can fertilize eggs produced by the same colony, which they then retain in the body for a time. Zygotes develop into ciliated larva. A **larva** (plural, larvae) is a free-living, sexually immature stage of animal development, one that precedes the adult form. Sponge larvae have flagella which enable them to briefly swim about until they settle on a suitable spot and grow up to be adults.

Many sponges reproduce asexually by budding or fragmentation. Small buds or pieces break away and grow into new sponges. Under stressful conditions, some freshwater species make gemmules; they encase clumps of cells in a hardened coat. Gemmules survive oxygen-poor water, drying out, and freezing. Later on, when conditions improve, each may be the start of a new sponge.

Sponges have no symmetry, tissues, or organs; they are at the cellular level of construction. Yet their lineage has endured since precambrian times. Fibers and spicules in the body wall and chemical defenses help keep predators away.

Figure 22.6 (**a**) An encrusting sponge growing on a ledge in a temperate sea. (**b**) A vase-shaped sponge. Flagellated, phagocytic cells line canals and chambers in the body wall. Each has a collar of food-trapping structures (microvilli). Fine filaments connect microvilli to one another, forming a collarlike "sieve." The sieve strains food particles from water flowing through it. Cells at the collar's base engulf trapped food.

section through body wall:

- glasslike structural elements
- amoeboid cell
- pore
- semifluid matrix
- flattened surface cells

collar cell with a central flagellum ringed by microvilli

22.4 Cnidarians—Simple Tissues, No Organs

All cnidarians are at the tissue level of organization. Their forms are diverse, but all are variations on two radial body plans—the medusa and polyp.

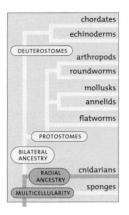

The radial, tentacled animals called **cnidarians** include sea anemones, hydroids, corals, and jellyfishes. Most species live in the seas. They have two tissue layers but no internal organs. They alone make capsules with dischargeable threads stuffed inside. These **nematocysts** stud the body surface (Figure 22.7). Different kinds release toxins, entangle prey, or pierce them with barbs. Most nematocyst toxins are harmless to humans. But some do irritate tissues, and a few can be deadly.

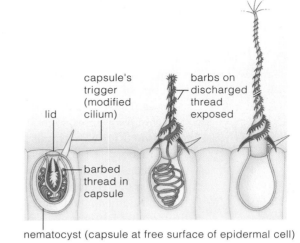

nematocyst (capsule at free surface of epidermal cell)

Figure 22.7 Nematocyst before and after prey (not shown) touched its trigger and made the capsule "leakier" to water. As water diffused in, turgor pressure built up and forced the thread to turn inside out. The barbed tip pierces prey.

Figure 22.8 Cnidarian body plans: (**a**) medusa and (**b**) polyp. (**c**) Sea wasp (*Chironex*), a jellyfish with tentacles about three meters long. Its toxin can kill you within minutes. (**d**) Sea anemone and (**e**) part of a colony of reef-building corals. The external walls of reef-building corals interconnect. (**f**) Portuguese man-of-war (*Physalia utriculus*), a colonial cnidarian equipped with a float.

Read Me First!
and watch the narrated
animation on a
cnidarian life cycle

feeding polyp

medusa

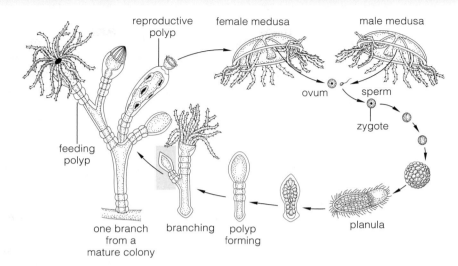

Figure 22.9 Life cycle of a hydroid (*Obelia*). The medusa is a sexual stage; it produces eggs or sperm. After zygotes form, they develop into planulas, a type of ciliated larva. After crawling about briefly, the planula settles down and develops into a polyp. More polyps form by asexual reproduction. A mature colony often contains thousands of feeding polyps, as well as reproductive polyps that give rise to medusae.

The most common body forms, the medusa and polyp, have a saclike gut. Depending on the group, medusae, polyps, or both form during the life cycle. Medusae look a bit like bells or saucers with tentacles around the rim, and they move freely through water (Figure 22.8*a,c*). Polyps are tube shaped, and usually one end attaches to a substrate. The gut opening is at the polyp's other end, and a ring of tentacles around it captures prey from the water (Figure 22.8*b,d,e*).

Like all other animals except sponges, cnidarians have **epithelia** (singular, epithelium): sheetlike tissues that line the free surfaces of body parts. A specialized type lines the cnidarian gut cavity; its gland cells secrete digestive enzymes. Another type, epidermis, covers the body's outer surfaces.

Nerve cells threading through cnidarian epithelia form a *nerve net*. This simple nervous system interacts with contractile cells embedded in the epithelia. By controlling contraction, the nerve net causes simple changes in shape that make the body move. How? Between the linings is mesoglea, a secreted gelatinous mass with a few scattered cells. It acts as a buoyant, deformable skeleton; the jellyfish bell moves when contractions squeeze water out from under it.

Any fluid-filled cavity or cell mass that contractile cells act against is a *hydrostatic* skeleton. It is one of several kinds of skeletons, all of which interact with contractile cells to bring about movement.

Although alike in body plans, cnidarians differ in their life cycles. Sea anemones produce only polyps, which for them is a sexual stage with gonads (gamete-producing reproductive organs). The hydroid *Obelia* produces polyps and medusae (the sexual stage), and planulas develop from zygotes (Figure 22.9). A planula is a bilateral, usually ciliated larva, and it moves about briefly before developing into an adult.

Besides this, many cnidarians grow as colonies. For instance, polyps of colonial corals are cemented side by side in their own "houses" of calcium carbonate secretions (Figure 22.8*e*). Their compressed remains help form tropical reefs. Coral polyps are among the cnidarians with dinoflagellate symbionts. They offer their photoautotrophic guests safe habitats and access to dissolved carbon dioxide and minerals. The guests supply their hosts with oxygen and recycle mineral wastes. They also contribute to house-building; their metabolic activities alter the pH of seawater in a way that promotes the deposition of calcium carbonate.

Finally, consider the Portuguese man-of-war. A blue, gas-filled float keeps this colonial cnidarian near the sea surface. Groups of polyps and medusae interact in feeding, reproduction, and other tasks (Figure 22.8*f*).

Cnidarians are the simplest living animals with cells that are structurally and functionally organized in tissue layers.

22.5 Flatworms—The Simplest Organ Systems

Beyond the cnidarians are animals ranging from simple worms to humans. Most are bilateral; all have organs.

Organs are structural units of two or more tissues arrayed in certain patterns and interacting in a common task. Two or more organs interacting chemically, physically, or both in common tasks make up an **organ system**. The flatworms are the simplest animals with organs that form from three tissue layers in the embryo—ectoderm, endoderm, and *mesoderm*. They include the mostly free-living turbellarians, and parasitic flukes and tapeworms. Although these bilateral, cephalized animals have no coelom, they probably had coelomate ancestors. In body form, the flatworms slightly resemble cnidarian planulas.

Most of the 15,000 flatworms are hermaphrodites. Each has male and female gonads and a penis. The turbellarians classified as planarians can reproduce asexually by *transverse* fission. The body splits in two, and each piece regrows the missing parts. Planarians respond similarly to injury. Cut one in pieces and you can end up with several planarians.

Most planarians live in freshwater. Like the marine turbellarians, they prey on or scavenge tiny animals. Figure 22.10 shows organ systems of one planarian. The pharynx is a muscular tube between the mouth and saclike gut. It sucks up food and expels wastes.

The head has light-detecting eyespots and clusters of nerves that form a rudimentary brain. Dorsal nerve cords relay messages through the body. A system of tubular protonephridia (singular, protonephridium) helps control the composition and volume of internal fluids. It extends from pores at the body surface to bulb-shaped flame cells. When excess water diffuses into flame cells in the body, a tuft of cilia "flickering" in the bulb drives water through tubes to the outside.

Flukes and tapeworms are parasites that live in or on one to four different hosts during their life cycle. Generally, we find a parasite's mature stage in the *definitive* host. Immature stages live in one or more *intermediate* hosts. As most parasites do, flukes and tapeworms withdraw nutrients from cells or tissues of a host. An infected host loses weight, weakens, and becomes vulnerable to other parasites and pathogens.

Tapeworms have a scolex with barbs or hooks at one end. They use it to attach to the wall of a vertebrate's intestine. None has a mouth or a gut; nutrients diffuse across the body wall. The chemical secretions of some species slow the muscle contractions that propel food

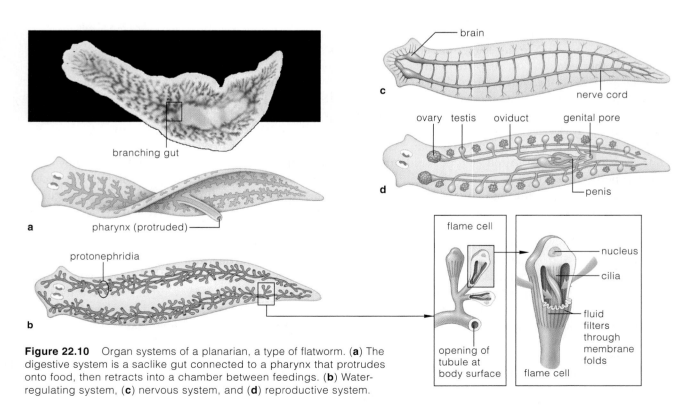

Figure 22.10 Organ systems of a planarian, a type of flatworm. (**a**) The digestive system is a saclike gut connected to a pharynx that protrudes onto food, then retracts into a chamber between feedings. (**b**) Water-regulating system, (**c**) nervous system, and (**d**) reproductive system.

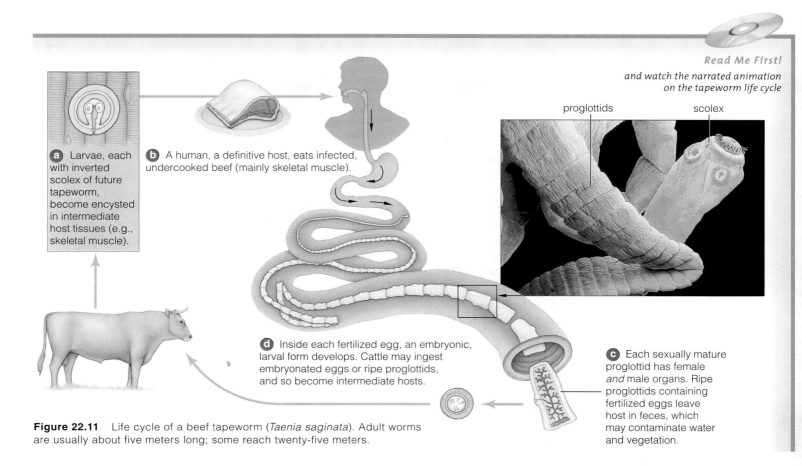

Read Me First!
and watch the narrated animation
on the tapeworm life cycle

proglottids   scolex

a Larvae, each with inverted scolex of future tapeworm, become encysted in intermediate host tissues (e.g., skeletal muscle).

b A human, a definitive host, eats infected, undercooked beef (mainly skeletal muscle).

d Inside each fertilized egg, an embryonic, larval form develops. Cattle may ingest embryonated eggs or ripe proglottids, and so become intermediate hosts.

c Each sexually mature proglottid has female *and* male organs. Ripe proglottids containing fertilized eggs leave host in feces, which may contaminate water and vegetation.

Figure 22.11 Life cycle of a beef tapeworm (*Taenia saginata*). Adult worms are usually about five meters long; some reach twenty-five meters.

through the host's gut. Their action prolongs the time the tapeworm spends in a nutrient-rich "habitat." The chemicals may be potential sources of drugs to slow contractions in humans who have an overactive gut.

Some tapeworm larvae enter humans in tissues of raw, undercooked, or improperly pickled pork, beef, or fish (Figure 22.11). New units of the tapeworm body, called proglottids, bud behind the scolex. Each unit is hermaphroditic; it transfers sperm to other units of the same tapeworm or a different one. Proglottids farthest from the scolex store the fertilized eggs. They detach from the tapeworm and are expelled from the body in feces. Fertilized eggs can survive for months outside an intermediate host.

Figure 22.12 shows the life cycle of a blood fluke that causes *schistosomiasis*. Larvae bore into the skin but don't cause obvious symptoms. Later, when white blood cells attack the masses of fluke eggs, the liver, spleen, bladder, and kidneys can be damaged.

Flatworms are among the simplest bilateral, cephalized animals with organ systems, which arise from three primary tissue layers that form in their embryos.

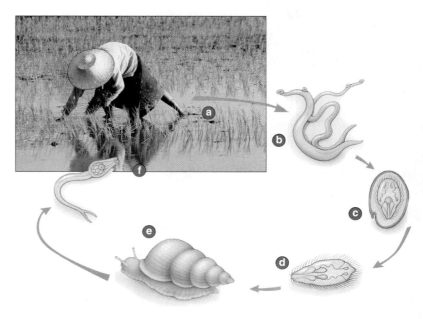

Figure 22.12 Life cycle of *Schistosoma japonicum*. (**a**) This blood fluke grows, matures, and mates in human hosts. (**b,c**) Fertilized eggs exit in feces, then hatch as ciliated, swimming larvae. (**d**) Larvae burrow into an aquatic snail and multiply asexually. (**e,f**) Fork-tailed, swimming larvae develop, leave the snail, then bore into the skin of a human host. They migrate to thin-walled intestinal veins, and a new cycle starts.

22.6 Annelids—Segments Galore

Annelids have so many segments, it makes you wonder about controls over how the animal body plan develops.

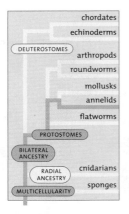

Did you ever see earthworms wriggling out of their flooded burrows after a rain? They are among 15,000 or so species of bilateral and highly segmented animals: the **annelids**. Leeches and the less well known but far more diverse polychaetes are annelids as well. Figures 22.13 through 22.15 show examples of their structure and function.

Segmented bodies are most pronounced in annelids. Except in leeches, nearly all segments have pairs or clusters of chitin-reinforced bristles on each side. Setae or chaetae are the formal names for the bristles; hence the names oligochaete for earthworms, which have a few bristles per segment; and polychaete for marine worms, which usually have a lot. (*Oligo–* means few; *poly–*, many.) Bristles shoved into soil or sediments afford traction for crawling and burrowing.

ADVANTAGES OF SEGMENTATION

Segmented bodies had great evolutionary potential, given all the diversity in annelid (and arthropod) body plans. Most earthworm segments are alike, but leeches have suckers at both ends (Figure 22.15), and polychaetes have an elaborate head and fleshy-lobed parapods ("closely resembling feet"). Existing species hint at the modifications in body plans that favored increases in size, more complex internal organs, and body segments adapted for specialized tasks.

ANNELID ADAPTATIONS—A CASE STUDY

Let's use an earthworm as our representative annelid. Its body has a flexible, permeable cuticle that is good for gas exchange but not for conserving body water. It is restricted to moist habitats, just as other annelids are restricted to aquatic habitats. The earthworm has a muscular pharynx that draws soil rich in detritus (decaying bits of organic matter) into the gut. It can scavenge the equivalent of its own weight each day, and it aerates the soil as it burrows through it.

Partitions divide the body into a series of coelomic chambers, each with muscles, blood vessels, nerves, and other organs (Figure 22.14). A complete digestive system with specialized regions extends through all chambers. In each segment's wall is a layer of circular muscles. As longitudinal muscles bridging several of the segments contract, circular muscles relax, all the way down the long body. They all exert force against fluid-filled coelomic chambers—a type of hydrostatic skeleton. Forward movement is complicated business; contractions are coordinated with each segment's bristles being plunged into soil and then released.

head end

on both sides of the similar segments, bristles used in locomotion

Figure 22.13 (**a**) Earthworm, a familiar annelid. (**b**) All body segments look alike on the outside, but regional specializations occur inside, along the main body axis.

(**c**) A tube-dwelling polychaete with featherlike, mucus-coated structures on its head. Polychaetes are one of the most common animals along coasts. (**d**) As a group, they have a dizzying variety of modified segments. Many are predators or scavengers; others dine on algae.

Read Me First!

and watch the narrated animation on the earthworm body plan

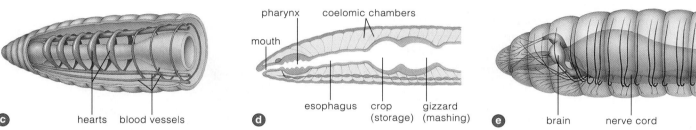

circular muscles · longitudinal muscles · dorsal blood vessel · coelom · cuticle

coelom · cuticle

seta (retracted) · ventral blood vessel · nerve cord · nephridium

a

bladderlike storage region of nephridium · nephridium's thin loop reabsorbs some solutes, relinquishes them to blood

blood vessels

body wall

funnel (coelomic fluid with waste enters here) · external pore (fluid containing wastes discharged here)

b

hearts · blood vessels

c

pharynx · coelomic chambers

mouth

esophagus · crop (storage) · gizzard (mashing)

d

brain · nerve cord

e

Figure 22.14 Earthworm body plan. (**a**) Midbody, transverse section. (**b**) A nephridium, one of many functional units that help maintain the volume and the composition of body fluids. (**c**) Portion of the closed circulatory system, which is functionally linked to nephridia. (**d**) Part of the digestive system, near the worm's head end. (**e**) Part of the nervous system.

The earthworm has a rudimentary **brain**, actually a fused pair of ganglia (singular, ganglion) in its head. Ganglia are clusters of nerve cell bodies. Connected to it are a pair of **nerve cords**, lines of communication that help the brain coordinate activities of different segments. Within each segment, the cords connect to smaller ganglia that help control local activity.

Nephridia (singular, nephridium) are units of an organ system that help regulate the composition and volume of body fluids. Usually, each nephridium has a funnel that collects fluid from a coelomic chamber and then drains it into tubes in the chamber behind it (Figure 22.14*b*). The fluid enters a tiny bladder, which delivers it to a pore at the body surface of the next segment in line.

Like most annelids, earthworms also have a closed circulatory system with multiple hearts. Contractions of the heart and of muscularized blood vessels keep blood flowing in one direction. Smaller vessels service the gut, nerve cord, and body wall (Figure 22.14*c*).

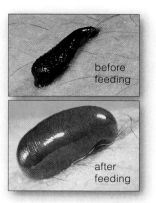

before feeding

after feeding

Figure 22.15 Leeches. Most leeches are aquatic scavengers or predators with sharp jaws and a blood-sucking device. *Hirudo medicinalis*, the freshwater species shown here, has been used for at least 2,000 years as a blood-letting tool to "cure" nosebleeds, obesity, and some other conditions.

Leeches are still being used. They draw off pooled blood after a doctor reattaches a severed ear, lip, or fingertip. A patient's body can't do this on its own until blood circulation routes are reestablished.

Annelids are bilateral, coelomate, segmented worms that have digestive, nervous, excretory, and circulatory systems. Some species show the degree of specialization that became possible after segmented body plans evolved.

22.7 The Evolutionarily Pliable Mollusks

Is there a "typical" mollusk? No. The group has more than 110,000 named species, including tiny snails in treetops, burrowing clams, and giant predators of the open seas.

Mollusks are bilateral, soft-bodied animals with a small coelom. Those with a distinct head have eyes and tentacles, but not all have a head. Many have a shell, or a reduced one. A radula, a tongue-like organ that rasps at food, is common. Only mollusks have a cloaklike extension of the body mass—a *mantle*—that drapes back over itself. The respiratory organ is a gill with thin-walled leaflets.

Gastropods, chitons, bivalves, and cephalopods are among the better known mollusks. The gastropods, the largest and most diverse group, include snails, slugs, and nudibranchs (Figure 22.16). Their name, meaning "belly foots," refers to the underside of the gastropod body mass. The muscular "foot" allows the animals to glide about over surfaces or to burrow.

Maybe it was their fleshy, soft bodies, so forgiving of chance evolutionary changes, that gave the early mollusks the potential to diversify so many ways in so many different habitats. Let's explore this idea by focusing on a few representative examples.

HIDING OUT, OR NOT

If you were small, soft of body, and *tasty*, a shell could discourage predators. A shell protects chitons. When something disturbs it, a chiton hunkers down under its eight-piece shell (Figure 22.16*d*). Foot muscles pull down as the mantle presses like a suction cup against a rock. A shell protects scallops, oysters, mussels, and other bivalves; its two hinged parts (valves) snap shut. When the parts clap together, they force water out, which propels the scallop backward—usually enough to get away from a hungry sea star.

If you were small, soft of body, and *toxic*, a shell would be superfluous if you advertised your toxicity to predators. Some nudibranchs do just that. Gland cells of certain species secrete sulfuric acid and other bad-tasting substances. Other nudibranchs graze on

a

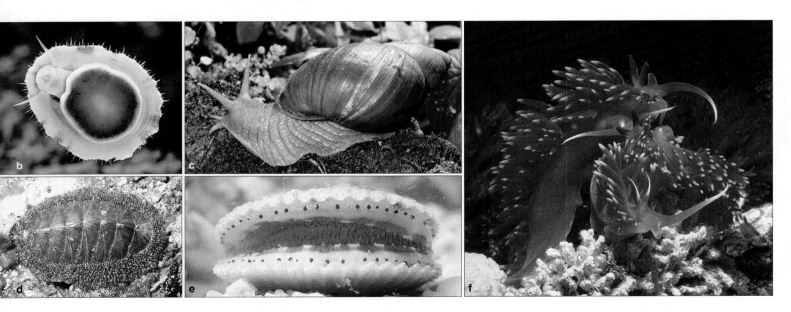

Figure 22.16 Mollusks. (**a**) Body plan and (**b**) ventral view of an aquatic snail crawling on aquarium glass. (**c**) Land snail. (**d**) Chiton. (**e**) Scallop, with light-sensitive "eyes" (blue dots) fringing its shell. (**f**) Two Spanish shawl nudibranchs (*Flabellina iodinea*). These hermaphrodites exchange sperm and then deposit chains of fertilized eggs into their partner.

Figure 22.17 (**a**) Artist's reconstruction of some animals that lived in the vast, tropical seaways of the Ordovician. Shown here, trilobites and the cephalopods called nautiloids, along with crinoids that looked like stalked plants.

(**b**) Generalized body plan of a cuttlefish, one of the cephalopods.

(**c**) Chambered nautilus, a living descendant of Ordovician nautiloids.
(**d**) Diver and squid (*Dosidicus*) inspecting each other.

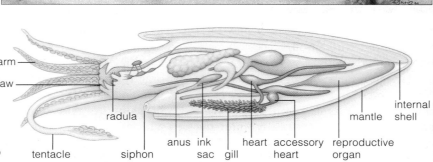

arm
jaw
radula
tentacle
siphon
anus ink sac gill heart accessory heart reproductive organ
mantle
internal shell

cnidarians, then incorporate ingested nematocysts in their own tissues. Spanish shawl nudibranchs flash red, nematocyst-studded respiratory organs as they mate (Figure 22.16*f*). Any predators attracted to the colorful couple get stung with the first bite and learn to avoid the species. Unlike most mollusks, which have separation into sexes, all of the nudibranchs are hermaphrodites.

A CEPHALOPOD NEED FOR SPEED

Some 500 million years ago, the cephalopods were supreme predators of the seas (Figure 22.17*a*). There were thousands of species with a chambered shell. And yet, in all but one of their modern descendants, the shell is reduced or gone (Figure 22.17*b,c*).

What happened? We can correlate its loss with an adaptive radiation of fishes that hunted cephalopods or were competitors for the same prey. In what might have been a long-term race for speed and wits, most cephalopods lost the external shell and became active, streamlined and, for invertebrates, smart.

For all cephalopods, *jet propulsion* became the name of the game. They learned to move by forcing a water jet through a funnel-shaped siphon, out from beneath the mantle cavity. We see how this works in modern species. Mantle muscles relax, which draws water into the cavity. They contract while the mantle's free edge closes over the head. The rapid squeeze shoots water through the siphon. The brain controls the siphon's action, hence the direction of escape or pursuit.

We also can correlate increases in speed with the evolution of more efficient respiratory and circulatory systems. Cephalopods are the only mollusks with a closed circulatory system. Their heart pumps blood to two gills, then two accessory (booster) hearts pump blood for oxygen uptake and carbon dioxide removal in metabolically active tissues, muscles especially.

Cephalopods now include the fastest invertebrates (squids), the largest (giant squid), and the smartest (octopuses). Of all mollusks, members of this group have the largest brain relative to body size, and they show the most complex behavior.

Lively stories emerge when we use evolutionary theory to interpret the fossil record and the nature of existing species diversity, as we have done here for the mollusks.

22.8 Roundworms

Roundworms are among the most abundant living animals. Sediments in shallow water often hold a million per square meter; a cupful of topsoil teems with them.

Roundworms make up a group known formally as nematodes. At least 12,000 kinds are named. Many are microscopic; a typical one isn't even five millimeters long. Their cylindrical body has bilateral features and tapered ends. Its false coelom is packed mostly with reproductive organs (Figure 22.18). A roundworm has a flexible cuticle, which it sheds repeatedly as it grows.

Most roundworms are free-living species that are decomposers. Many play vital roles in the cycling of nutrients through ecosystems.

The free-living *Caenorhabditis elegans* is a favorite experimental organism among biologists. It has the same general body plan and tissue types as more complex animals, but it is tiny and transparent, with only 959 somatic cells. Researchers can monitor the developmental fate of each cell. Also, the generation time is short, and the genome is one-thirtieth the size of the human genome. Self-fertilizing hermaphroditic forms can be used to get populations of offspring that are homozygous for desired traits.

Many roundworms infect roots of crop plants; they are significant agricultural pests. Others are internal parasites of animals that include humans, dogs, and insects. *Ascaris lumbricoides*, a very large roundworm, infects more than 1 billion people worldwide, mostly in Latin America and Asia (Figure 22.19a).

Pigs or game animals can carry *Trichinella spiralis*. Adults of this parasite live attached to the intestinal lining. Eggs develop into juveniles, which migrate through blood vessels to muscles, where they become encysted (Figure 22.19b). *Trichinosis*, the disease that results from infection, can be fatal.

Repeated infections by the roundworm *Wuchereria bancrofti* can result in *elephantiasis*, or a severe case of edema. Adult parasites get lodged in lymph nodes, organs that filter excess tissue fluid being sent back to the bloodstream. Worms obstruct the flow, so fluid backs up in tissue spaces and enlarges the legs, feet, and other body parts (Figure 22.19c). Mosquitoes serve as intermediate hosts.

The pinworms (*Enterobius vermicularis*) are tiny nematodes that live in the human rectum. They are most commonly parasites of young children.

Roundworms are abundant bilateral animals with a false coelom and a complete digestive system. They include free-living decomposers and assorted parasites.

pharynx   intestine   eggs in uterus   gonad   anus

false coelom   muscularized body wall

Figure 22.18 Body plan of *Caenorhabditis elegans*.

Figure 22.19 (**a**) Living roundworms (*Ascaris lumbricoides*). Infection by these intestinal parasites causes stomach pain, vomiting, and appendicitis. (**b**) *Trichinella spiralis* juveniles in the muscle tissue of a host animal. (**c**) A case of elephantiasis brought on by the roundworm *Wuchereria bancrofti*.

22.9 Arthropods—The Most Successful Animals

Evolutionarily speaking, "success" means having the most offspring, the most species in the most habitats, fending off threats and competition efficiently, and exploiting the greatest amounts and kinds of food. These are the features that characterize arthropods.

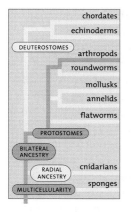

Arthropods are bilateral, with a complete gut, a much reduced coelom, and an *open* circulatory system, in which blood flows out of simple vessels or hearts, bathes tissues, and flows back. More than a million species are known. One lineage (trilobites) is extinct. Surviving groups are the crustaceans and chelicerates, insects, and myriapods (both centipedes and millipedes).

We consider representatives in sections that follow. For now, start thinking about six major adaptations that contributed to the success of arthropods in general and insects in particular.

A hardened exoskeleton. Arthropods have a cuticle of chitin, proteins, and waxes, often stiffened by calcium carbonate deposits. The cuticle is a protective external skeleton, or an **exoskeleton**. Exoskeletons might have evolved as defenses against predation. They took on added functions when some arthropods invaded the land. They support a body no longer made buoyant by water, and their waxes block water loss.

This exoskeleton does not restrict increases in size, because arthropods grow in spurts—and they **molt**. Molting is a process that is under hormonal control. At some stages of the life cycle, hormones induce a soft new cuticle to form under the old one, which is then shed (Figure 22.20). The body mass increases by the uptake of air or water and by rapid, continuous cell divisions before the new cuticle hardens. Molting is a trait that arthropods share with roundworms.

Jointed appendages. If uniformly hardened, a cuticle would restrict movement. Arthropod cuticle, however, thins where it spans *joints*, or regions where two body parts abut. Contraction of muscles makes a cuticle region bend, which shifts the position of body parts with respect to one another. A jointed exoskeleton was a key innovation; it allowed the evolution of diverse specialized appendages, including wings, antennae, and legs. ("Arthropod" means jointed leg.)

Specialized segments and fused-together segments. The first arthropods were segmented, and the segments were more or less similar. In most lineages that made it to the present, some of the segments became fused together or modified in specialized ways. As just one example, compare the nearly identical segments of a centipede with the winged body of a butterfly.

Respiratory structures. Many freshwater and marine arthropods use gills as respiratory organs. Among insects and other arthropods on land, air-conducting tubes called tracheas evolved. Insect tracheas start at pores on the body surface. They branch into tubes that let oxygen diffuse directly into tissues, which supports the aerobic respiration so essential for flight and other energy-demanding activities.

Specialized sensory structures. For many arthropods, diverse sensory structures also contributed to success. For instance, some evolved into eyes that can sample a big part of the visual field and provide the brain with information from many directions.

Specialized developmental stages. Many arthropods divide the task of surviving and reproducing among different stages of development. The division of labor is most common among insects. For some species, the new individual is a *juvenile* —a miniaturized version of the adult that simply grows in size until sexually mature. Other species undergo **metamorphosis**. The body changes in often drastic ways between the embryo and adult forms. Hormones help direct this tissue reorganization and remodeling.

Eating and growing fast are specialties of many immature stages; the adult form specializes in reproduction and in dispersal of the new generation. (Think of caterpillars chewing up leaves, then think of butterflies mating and laying eggs.) Different developmental stages are adapted to specific environmental conditions, especially seasonal shifts in food sources, water supplies, and availability of potential mates.

Figure 22.20 Molting by a red-orange centipede wriggling out of its old exoskeleton.

As a group, the arthropods are notably abundant and widespread, and they have enormously diverse life-styles.

The success of the arthropod lineage as a whole arises largely from having a body with a hardened, jointed exoskeleton; fused, modified body segments; specialized appendages; and specialized respiratory and sensory structures.

In many species—insects especially—success also arises from a division of labor among different stages of the life cycle, such as larvae, juveniles, and adults.

22.10 A Look at the Crustaceans

Shrimps, lobsters, crabs, barnacles, pillbugs, and other crustaceans got their name because they have a hard yet flexible "crust"—an external skeleton—but so do nearly all arthropods. The vast majority live in marine habitats, where they are so abundant they are dubbed "insects of the seas." Humans harvest many edible types.

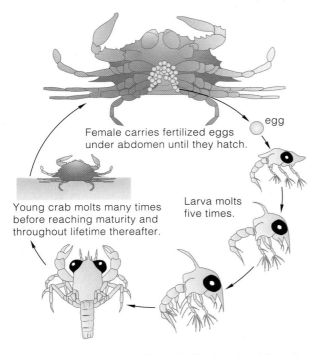

Figure 22.21 Life cycle of a crab. The larval and juvenile stages molt repeatedly and grow in size.

egg

Female carries fertilized eggs under abdomen until they hatch.

Young crab molts many times before reaching maturity and throughout lifetime thereafter.

Larva molts five times.

Like other arthropods, crustaceans repeatedly molt and shed their exoskeleton during a life cycle. Figure 22.21 depicts a crab's larval and juvenile stages.

Most crustaceans have sixteen to twenty segments; some have more than sixty. Their head has pairs of antennae, and jawlike and food-handling appendages. In the simplest groups, all appendages are similar. In other groups, specialized claws and pincers evolved. The lobsters, shrimps, and crabs have five pairs of walking legs. A dorsal cuticle extends back from the head and covers the thoracic segments (Figure 22.22a).

Of all arthropods, only barnacles have a "shell," made of calcium secretions, that protects them from predators, drying winds, and surf (Figure 22.22b). Adult barnacles spend their lives cemented to piers, rocks, and even other animals. Once attached they can't move, so you would think that mating might be tricky. But barnacles tend to cluster in groups, and most are hermaphrodites. They extend their penis, which can be several times their body length, out to nearby neighbors.

Copepods are among the more numerous animals in aquatic habitats. Most of them are less than two millimeters long (Figure 22.22c). Some are parasites of invertebrates and fishes. Most ingest phytoplankton. Copepods in turn serve as food for invertebrates, fishes, and baleen whales.

Crustaceans differ greatly in the number and kind of appendages. They grow in stages and repeatedly replace their external skeleton by molting.

Figure 22.22 (**a**) Lobster in the Caribbean Sea, off the coast of Belize. (**b**) Goose barnacles. Adults cement themselves to one spot. You might mistake them for mollusks, but when they open their hinged shell to filter-feed, you see jointed appendages—the hallmark feature of arthropods. (**c**) One of the copepods. Different kinds are free-living filter feeders, predators, and parasites.

22.11 Spiders and Their Relatives

Chelicerates originated in shallow seas early in the Paleozoic. They are named for unique appendages called chelicerae. Horseshoe crabs are among the few living marine species. Of species on land, we might say this: Never have so many been loved by so few.

Horseshoe crabs (Figure 22.23a) haven't changed much since the Devonian. Their horseshoe-shaped carapace (a hardened dorsal shield) protects their ten-legged body. Their closest relatives on land are arachnids—spiders, scorpions, ticks, and chigger mites.

The spiders and scorpions are predators, and most sting, bite, or stun prey with venom, which they inject from fanged chelicerae. Spiders also have appendages that spin silk threads for webs and egg cases.

Our world would be overrun with insects without the predatory efforts of spiders. Even so, a few kinds that can harm humans and other mammals have given the whole group a bad name. For example, all spiders of genus *Loxosceles* are venomous. The brown recluse (*L. reclusa*) has a violin-shaped marking on its thorax (Figure 22.24b). Its bite ulcerates skin, which may heal poorly. In a few cases, amputations or skin grafts have been necessary. Bites of the black widow spiders (*Latrodectus*) kill about five people a year in the United States. Neurotoxin in their venom triggers potentially dangerous muscle spasms. Especially for children, prompt treatment is essential.

Mites are among the smallest, most diverse, and most widely distributed arachnids. Of an estimated 500,000 species, a few are parasites. Ticks are larger than most mites. Some transmit pathogens. One is the

Figure 22.23 (**a**) Horseshoe crab (*Limulus*). Its long spine helps steer the body. (**b**) Fat-bodied scorpion of Australia.

vector for *Borrelia burgdorferi*, the bacterial agent of Lyme disease. Rocky Mountain spotted fever, scrub typhus, tularemia, and encephalitis are a few of the many other tickborne diseases. Ticks don't fly or jump. They crawl onto grasses or shrubs, then crawl onto animals brushing past, then into scalp, groin, or other hairy hidden body parts. What about house dust mites? They can't kill you, but they can provoke allergies in susceptible people (Figure 22.24d).

Spiders, scorpions, and their relatives have appendages specialized for predatory or parasitic life-styles.

Figure 22.24 Arachnids. (**a**) Tarantula with a fearless female. Like all spiders, it helps keep insects in check. Its bite does not harm humans. (**b**) Brown recluse, with six eyes and a violin-shaped mark on its forebody. It is native to some central and southern parts of the United States. (**c**) Female black widow, with a red hourglass-shaped marking on the underside of a shiny black abdomen. Black widows live in North America. Males are smaller and don't bite. (**d**) Dust mite. Mite feces and corpses commonly cause allergies in susceptible people.

chelicerae

22.12 A Look at Insect Diversity

Insects are the most diverse animal group. There are more species of dragonflies alone than there are species of mammals. Many insects also produce staggering numbers of offspring. Ants and termites alone may make up a third of the biomass of all animals on land.

In most adult insects, the segmented body is divided into three distinct parts: a head, thorax, and abdomen. The head has paired sensory antennae and paired mouthparts that are specialized for biting, chewing, or other tasks (Figure 22.25). The thorax has three pairs of legs and, usually, two pairs of wings. Insects are the only winged invertebrates.

Insects have a complete digestive system divided into a foregut, a midgut where food is digested, and a hindgut where water is reabsorbed. A system of *malpighian tubules* disposes of wastes and maintains ion concentrations in body fluids. After proteins are digested, residues diffuse from blood into the tubes. There, enzymes convert them to crystals of uric acid, which are eliminated in feces. The tubules help land-dwelling insects get rid of potentially toxic wastes without also losing precious water.

Larvae, nymphs, and pupae are immature stages of many insect life cycles (Figure 22.26). Like human children, some insect nymphs have the adult's form, but in miniature. Unlike children, however, these immature stages molt. Most insect larvae undergo metamorphosis, in which the drastic reorganization and remodeling of tissues necessary for the transition to adulthood take place.

Figure 22.27 shows a few of the more than 800,000 species. If numbers and distribution are the measure, the most successful kinds are small and reproduce often. Huge numbers grow and reproduce on a plant that would be only an appetizer for another animal. By one estimate, if all the offspring of a single female fly survived and reproduced for six more generations, she would have more than 5 trillion descendants!

The most successful insects are winged. They move among food sources that are too widely scattered to be exploited by other kinds of animals. The capacity for flight contributed hugely to their success on land.

The very features that contribute to insect success also make them our most aggressive competitors for crops and natural products, such as paper or lumber.

Figure 22.25 Examples of insect appendages. Head parts typical of (**a**) grasshoppers, which chew food; (**b**) flies, which sponge up nutrients; (**c**) butterflies, which siphon nectar; and (**d**) mosquitoes, which pierce hosts and suck up blood.

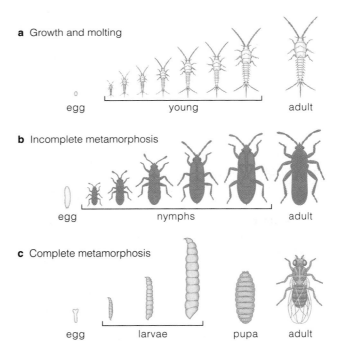

a Growth and molting

egg   young   adult

b Incomplete metamorphosis

egg   nymphs   adult

c Complete metamorphosis

egg   larvae   pupa   adult

Figure 22.26 Insect development. (**a**) Young silverfish change little except in size and proportion as they mature into adults. (**b**) True bugs show *incomplete* metamorphosis, or gradual change from the immature form until the last molt. (**c**) Fruit flies show *complete* metamorphosis. Tissues of larvae are destroyed and replaced before an adult emerges.

Some insects transmit pathogens that sicken us. We fight "bad" insects by spraying many millions of pounds of pesticides each year. Yet most play vital roles as decomposers or as pollinators of flowering plants—including almost all crop plants. Also, many "good" insects are predators or parasites of the ones we would rather do without. Think about it.

Like other arthropods, insects have a hardened cuticle that forms an exoskeleton; jointed appendages; fused, modified body segments; respiratory structures; and specialized sensory organs. Many have a division of labor among stages that form in the life cycle. The most successful are winged.

In terms of distribution, number of species, population sizes, competitive adaptations, and exploitation of diverse foods, insects are the most successful animals.

Figure 22.27 Examples of insects. (**a**) Mediterranean fruit fly. Its larvae destroy citrus fruit and other crops. (**b**) Duck louse. It eats bits of feathers and skin. (**c**) European earwig, one of the common household pests. The long curved forceps at the tail end show this is a male.

(**d**) Flea, with strong legs good for jumping onto and off of its animal hosts. (**e**) Stinkbugs, newly hatched. (**f**) In the center of this group of honeybee workers, one bee is signaling the position of a food source to her hive mates by dancing, as described in Section 43.4.

With more than 300,000 species of beetles, Coleoptera is the largest order in the animal kingdom. (**g**) Ladybird beetles swarming. (**h**) *Cyclommatus*, a staghorn beetle.

(**i**) Swallowtail butterfly and (**j**) luna moth. (**k**) An adult damselfly, which preys on flying insects. Its larvae are voracious predators of aquatic invertebrates and small fishes.

22.13 The Puzzling Echinoderms

Reflect, for a moment, on Figure 22.5. Ancestors of the roundworms, arthropods, annelids, and mollusks you just read about were at a major branch point in time. At that point, ancestors of echinoderms and chordates set out on a separate evolutionary journey.

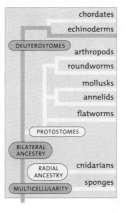

The name **echinoderm** means spiny-skinned. It refers to the interlocking calcium carbonate spines or plates in the body wall. All echinoderms live in the seas. They include sea stars, brittle stars, sea urchins, sea cucumbers, feather stars, sand dollars, sea biscuits, and sea lilies, or crinoids. Figure 22.28 shows some representatives.

Most adults are radial with a few bilateral features, and most of them have bilateral larvae. Did the group evolve from a bilateral ancestor? DNA studies now indicate that this is most likely the case.

Almost all adult echinoderms are bottom dwellers. Some sea lilies live attached to the seafloor by a stalk, although most echinoderms creep about. They lack a brain, but their decentralized nervous system allows them to respond to food, predators, and so forth that appear from different directions. For instance, any arm of a sea star that touches a tasty mussel's shell can become the leader, directing the rest of the body to move toward the prey.

Echinoderms have tube feet. These are fluid-filled, muscular structures, sometimes with suckerlike disks at the tips (Figure 22.28*a*). Sea stars use their tube feet for walking, burrowing, clinging to rocks, or gripping a clam or snail about to become a meal. Tube feet are components of a **water–vascular system**, unique to echinoderms. In sea stars, that system includes a main canal in each arm. Short side canals extend from them and deliver water to the tube feet (Figure 22.29). Inside a tube foot is a fluid-filled, muscular structure shaped like the rubber bulb on a medicine dropper. As the bulb contracts, it forces fluid into the foot and causes it to lengthen.

Figure 22.28 A few echinoderms. (**a**) Sea star, with a close-up of its little tube feet. (**b**) Sea urchins, which can move about on spines and a few tube feet. (**c**) Brittle star. Its slender arms (rays) make rapid, snakelike movements. (**d**) Sea cucumber, with rows of tube feet along its elongated body.

part of the water–vascular system

sieve
plate
ring
canal
ampulla

Figure 22.29 Water–vascular system of a sea star. Combined with many tube feet, it is the basis of locomotion.

Tube feet change shape continually when muscle action redistributes fluid through the water–vascular system. Hundreds of tube feet may move at a time. After being released, each tube foot swings forward, reattaches to the substrate, then swings backward and is released before swinging forward once again. Their motions are splendidly coordinated, so sea stars glide smoothly rather than lurch along.

Most sea stars are active predators with a feeding apparatus on their ventral surface. Some swallow prey whole. Others push part of their stomach outside the mouth and around their prey, then start digesting their meal even before swallowing it. Sea stars get rid of coarse, undigested residues through the mouth, even though they do have a small anus.

In sea urchins, the calcium carbonate plates form a stiff, rounded "test" from which the spines protrude (Figure 22.28b). Sea urchin roe (eggs) are an ingredient in some sushi. Each year, Japan imports about 20 million pounds of sea urchins that were captured off the California coast. Harvests are largely unregulated and are probably contributing to the decline of the most prized species.

Sea cucumbers also are a favorite food in Asia, and some species are declining. These echinoderms have the softest body; their plates have been reduced to microscopic spicules (Figure 22.28d). They creep about on the ocean floor, picking up prey and bits of organic debris with sticky tentacles. When a predator attacks, a sea cucumber shoots its internal organs out through the region of the anus to distract it. If this succeeds, the sea cucumber regenerates its missing parts.

With their odd combination of radial and bilateral traits, echinoderms make a case for this point: There are exceptions to the major trends that we observe in animal evolution.

Summary

Section 22.1 Animals cannot be defined in a simple way; they share many traits with other organisms. Instead, they can be characterized by a list of features:

a. Animals are multicelled heterotrophs. Their cells engage in aerobic respiration and have no secreted wall.

b. Except for sponges, animals have tissues. Most also have organs and organ systems.

c. Animals typically reproduce sexually, but many also can reproduce asexually. Their life cycles often involve a series of stages.

d. Immature stages, adults, or both are usually motile (able to move from place to place).

Vertebrates have a backbone. The invertebrates, which have none, vastly outnumber them.

Animals also can be characterized with respect to six features of their basic body plan: body symmetry, embryonic tissue layers, cephalization, type of gut, type of body cavity, and segmentation. Not all of these features occur in every group.

a. Except for sponges, body parts of animals are arranged radially or bilaterally with respect to the main body axis.

b. During the evolution of bilateral animals, many sensory structures and nerve cells became concentrated in a head (a process called cephalization).

c. Two or three primary tissue layers—ectoderm, endoderm, and mesoderm—form in animal embryos. They give rise to all tissues and organs of adults.

d. Except for placozoans and sponges, animals have a saclike or a tubular gut. A tubular gut has a mouth and anus, and its formation in embryos is one of many defining traits for the two major lineages of bilateral animals. In protostomes, the mouth forms first. In deuterostomes, the anus forms first.

e. Protostomes and deuterostomes differ in whether they have a coelom, a main body cavity between their gut and body wall, and in how that cavity develops. In some lineages the coelom was lost or reduced; such animals are called acoelomates and pseudocoelomates.

f. Bilateral animals differ in whether they are segmented, with repeats of body units, and in how pronounced segmentation has become.

Section 22.2 Ediacarans are the earliest animals known from the fossil record. Most modern groups arose during an adaptive radiation in the Cambrian. *Trichoplax*, a type of tiny soft-bodied animal with two cell layers, is the simplest living animal.

Section 22.3 Sponges are simple animals with no body symmetry, true tissues, or organs. Certain flagellated cells lining canals and chambers in their bodies move water and take up microscopic food particles. Chemical and structural features make sponges unappealing to most predators.

Section 22.4 The radially symmetrical cnidarians, such as jellyfishes, sea anemones, corals, and hydroids, are the simplest animals with tissues. They have an outer epidermis, inner gastrodermis with gland cells, and a jellylike substance in between that has only a few scattered cells. They are the only animals that make nematocysts, or capsules with dischargeable threads that can sting, entangle, or ensnare prey.

Section 22.5 Flatworms are among the simplest protostomes. They are bilateral. Three primary tissue layers form in their embryos and give rise to simple organ systems. Flatworms have a saclike gut, and no coelom, but it appears they had coelomate ancestors. Turbellarians, tapeworms, and flukes are in this group, which includes major parasites of humans.

Section 22.6 Annelids are highly segmented worms. They include the oligochaetes (such as earthworms), polychaetes (marine worms), and leeches. The first two groups have bristles along their body length that work in conjuction with muscle contractions to bring about body movements.

Section 22.7 With 110,000 species, mollusks are one of the largest groups. Only mollusks have a mantle, a sheetlike part of the body mass draped back on itself. All are soft-bodied and often shelled. Examples are the gastropods and cephalopods. The largest, swiftest, and smartest invertebrates are cephalopods.

Section 22.8 Roundworms (nematodes) have a cylindrical body with bilateral features, a flexible cuticle, a complete gut, and a false coelom. At least 12,000 kinds are free-living decomposers or parasites.

Section 22.9 We know of more than a million kinds of bilateral, coelomate animals called arthropods. The major groups are chelicerates, crustaceans, centipedes, millipedes, and insects. Features that contributed to their success are a hardened exoskeleton, jointed appendages, specialized and fused segments, efficient respiratory and sensory structures, and developmental stages adapted to specific environmental conditions.

Section 22.10 The mostly marine crustaceans include crabs, lobsters, barnacles, and copepods. Specialized appendages underlie their diverse life-styles.

Section 22.11 Chelicerates include horseshoe crabs and arachnids (spiders, scorpions, ticks, and mites). They are predators, parasites, or scavengers.

Section 22.12 Insects are the most successful group of animals, and the only winged invertebrates. They are our major competitors for food.

Section 22.13 Echinoderms, such as sea stars, are invertebrates of the deuterostome lineage. They have an exoskeleton of spines, spicules, or plates of calcium carbonate. Adults are radial, but bilateral ancestry is evident in their larval stages and other features.

Figure 22.30 A soft, branching coral (*Telesto*).

Self-Quiz *Answers in Appendix III*

1. All animals _____ .
   a. are motile for at least some stage in the life cycle
   b. consist of tissues arranged as organs
   c. can reproduce asexually as well as sexually
   d. both a and b

2. A coelom is a _____ .
   a. lined body cavity   c. sensory organ
   b. resting stage   d. type of bristle

3. The cnidarians alone have _____ .
   a. nematocysts   c. a hydrostatic skeleton
   b. tissues   d. radial symmetry

4. Flukes are most closely related to _____ .
   a. planarians   c. arthropods
   b. roundworms   d. echinoderms

5. Which group has the greatest number of species?
   a. crustaceans   c. mollusks
   b. insects   d. roundworms

6. The nephridia of earthworms perform a function most similar to the _____ .
   a. gemmules of sponges   c. flame cells of planarians
   b. chelicerae of spiders d. tube feet of echinoderms

7. The _____ are coelomate and radial as adults.
   a. cnidarians   c. roundworms
   b. echinoderms   d. both a and c

8. Chordates are most closely related to _____ .
   a. cnidarians   c. mollusks
   b. echinoderms   d. arthropods

9. Match the organisms with their appropriate descriptions.
   _____ Ediacarans   a. complete gut, false coelom
   _____ sponges   b. simplest organ systems
   _____ cnidarians   c. no tissues, no organs
   _____ flatworms   d. jointed exoskeleton
   _____ roundworms   e. mantle over body mass
   _____ annelids   f. segmented worms
   _____ arthropods   g. tube feet, spiny skin
   _____ mollusks   h. nematocyst producers
   _____ echinoderms   i. known from fossils only

Figure 22.31 To which group might this burrowing marine animal belong?

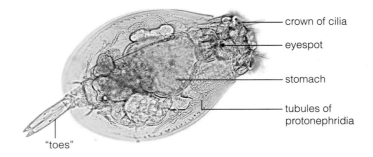

"toes"

crown of cilia

eyespot

stomach

tubules of protonephridia

Figure 22.32 A rotifer (*Euchlanis*).

Critical Thinking

1. You are diving in the calm, warm waters behind a large tropical reef. You see something that looks like a plant with tentacles (Figure 22.30). This is a branching, soft coral with many polyps. You will not see it growing on reef surfaces exposed to the open sea. Formulate a hypothesis that might explain the animal's distribution. Describe how you would go about testing this hypothesis.

2. The marine worm in Figure 22.31 has a segmented body. Most segments are similar; they have bristles on their sides that the worm uses to burrow in sandy marine sediments. What group do you think this worm belongs to? What other features could you examine to confirm your suspicions?

3. The flatworms, roundworms, and annelids all include species that are parasites of mammals, whereas sponges, cnidarians, mollusks, and echinoderms do not. What do you think explains the difference?

4. In the summer of 2000, only 10 percent of the lobster population of Long Island Sound survived after a massive die-off. Many lobstermen believe the deaths followed heavier spraying of pesticides, used to control mosquitoes that carry West Nile virus. Speculate about why a chemical designed to target mosquitoes might also harm lobsters.

5. There are many invertebrate groups besides the major ones described in this chapter. For example, rotifers are protostomes that are usually less than a millimeter long. All but about five percent live in freshwater. Most have a crown of cilia at the head end that assists in swimming and in wafting food toward the mouth (Figure 22.32). A jawlike structure known only in this group is used to grind up food. The digestive system lies inside a pseudocoelom and is complete. Paired gastric glands secrete enzymes that aid in digestion. Fluid balance and waste removal are controlled by protonephridia. Nerve cell bodies clustered in the head integrate body activities. Certain rotifers have clusters of pigments that serve as simple "eyes."

Even with all this complexity, rotifers do not have any special circulatory or respiratory organs. Do you consider this surprising? Explain your reasoning.

Media Menu

Student CD-ROM

Impacts, Issues Video
  Old Genes, New Drugs
Big Picture Animation
  Origin and evolution of the invertebrates
Read-Me-First Animation
  Types of body cavities
  Cnidarian life cycle
  Tapeworm life cycle
  Earthworm body plan
Other Animations and Interactions
  Life cycle of a crab
  Examples of insect appendages

InfoTrac

- Fossil Finder: Complex Life Emerged with a Bang 540 Million Years Ago. Andrew Knoll Wants to Know Why. *Time*, August 2001.
- Jellies on a Roll. *Audubon*, May–June 2002.
- The Worm and the Parasite: Some Tropical Scourges Call for a Defense Against an Entire Micro-Ecosystem. *Natural History*, February 2003.
- The Perfect Limpet. *Natural History*, February 2001.

Web Sites

- Animal Diversity Web: animaldiversity.ummz.umich.edu
- Introduction to the Metazoa: www.ucmp.berkeley.edu/phyla/phyla.html
- Wadsworth's Wonderful Worms: rigel.umdnj.edu
- Bugbios: www.insects.org

How Would You Vote?

Harvesting wild animals is usually not regulated until a species is threatened by extinction. In the United States, some conservationists want to expand rules that govern trade in endangered species. They want the rules to cover *any* species in the wild. Critics argue that such an approach is too costly and would hurt developing countries. Should the United States push for tighter regulations governing international trade of any species that is caught rather than farmed?

IMPACTS, ISSUES *Interpreting and Misinterpreting the Past*

In Charles Darwin's time, major groups of organisms were already identified. A big obstacle to acceptance of his theory of evolution by natural selection was the seeming lack of transitional forms. If new species evolve from older ones, then where were the "missing links" in the fossil record—forms with intermediate traits that bridge the groups? Ironically, workmen at a limestone quarry in Germany had already unearthed one.

The pigeon-sized fossil looked like a small meat-eating dinosaur (Figure 23.1). It had three long clawed fingers on each forelimb, a long bony tail, and a heavy jaw with short, spiky teeth. Later, diggers found another one. Later still, someone noticed the feathers. If those were birds, why did they have teeth and a bony tail? If dinosaurs, what were they doing with feathers? The specimen was named *Archaeopteryx*, meaning ancient winged one.

Between 1860 and 1988, six *Archaeopteryx* specimens and a fossilized feather were found. Anti-evolutionists tried to dismiss them as forgeries. Someone, they said, had pressed modern bird bones and feathers against wet plaster. The dried, imprinted plaster casts merely looked like fossils. Microscopic examination confirmed the fossils are real. Further confirmation of their age came from the remains of obviously ancient jellyfishes, worms, and other species in the same limestone layers.

Radiometric dating shows that *Archaeopteryx* lived 150 million years ago. Why are its remains preserved so well after so much time? *Archaeopteryx* lived in tropical forests bordering a lagoon. The lagoon was large, warm, and stagnant; coral reefs barred inputs of oxygenated water from the sea. So it was not an inviting place for scavenging animals that might have eaten and wiped

| *Dromaeosaurus* | *Archaeopteryx* | *Confuciusornis* |

Figure 23.1 Morphological comparison of *Archaeopteryx* and *Dromaeosaurus*, a duck-sized, two-legged dinosaur. Birds and dinosaurs evolved from early reptiles. *Right*, reconstruction of another "missing link" between reptiles and birds—*Confuciusornis sanctus*, based on fossils from China.

the big picture

Origins, Evolutionary Trends   The first chordates were invertebrates. A cranium, then a vertebral column, jaws, paired fins, and paired lungs were key innovations in the evolution of fishes, amphibians, reptiles, birds, and mammals.

Diversification of Fishes   Fishes were the first vertebrates. They were and still are the most diverse group. Most have jaws and an internal skeleton of cartilage or bone. The paired fins of fish are homologous to the limbs of land vertebrates.

out the remains of *Archaeopteryx* and other organisms that fell from the sky or drifted offshore.

Fine sediments driven over the reefs by each storm surge gently buried the carcasses on the bottom of the lagoon. Over time, soft, muddy sediments were slowly compacted and hardened. They became a limestone tomb for over 600 species, including *Archaeopteryx*.

Is *Archaeopteryx* the only transitional form linking dinosaurs and birds? No. *Confuciusornis sanctus* lived about the same time as *Archaeopteryx*. It is the earliest known bird with a bill (Figure 23.1). Fossils show how the clawed limbs of its four-legged ancestor became modified into wings. Each wing had three digits, one a reduced claw with flight feathers attached.

No one was around to witness such transitions in the history of life. But fossils are real, just as the morphology, biochemistry, and molecular makeup of living organisms are real. Radiometric dating assigns fossils to their place in time. *So evolution is not "just a theory."* Remember, a scientific theory differs from speculation because its predictive power has been tested in nature many times, in many different ways. If a theory stands, there is a high probability that it is not wrong.

Evolutionists argue all the time among themselves. They argue over how to interpret the evidence and which mechanisms and events can explain life's history. At the same time, they do not ignore evidence—which is there for us to gather and interpret. Here is one account of vertebrate evolution, including our own origins.

 How Would You Vote?

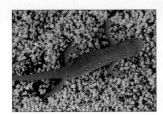

The Move Onto Land

Amphibians were the first animals to spend much of their life on land. They still need to keep their skin moist for respiration and require standing water to lay their eggs.

Enter the Amniotes

Reptiles, birds, and mammals are amniotes. They are fully adapted to life on land. Eggs of amniotes are fertilized internally and embryos develop inside a series of membranes.

Human Evolution

Human traits arose through modifications in earlier mammalian, primate, and then early hominid lineages. Diverse humanlike and early human forms arose in Africa.

23.1 The Chordate Heritage

We start our look at vertebrates and their chordate origins with a reminder: Each kind of organism, humans included, is a mosaic of traits. Many traits have been conserved from remote ancestors, and others are unique to its branch on the family tree.

CHORDATE CHARACTERISTICS

Most of the **chordates** are coelomate, bilateral animals characterized in part by four features, as illustrated in Figure 23.2. *First,* a rod of stiffened tissue—not bone or cartilage—helps support the body. It is a notochord. *Second,* a nerve cord runs parallel with the notochord and gut; its anterior end develops into a brain. Unlike the nerve cords of invertebrates, this one is dorsal. *Third,* **gill slits** punctuate the wall of a **pharynx**—a muscular tube that functions in feeding, respiration, or both. *Fourth,* a tail extends past an anus. The four features develop in all chordate embryos, although they are not always present in the adult form.

INVERTEBRATE CHORDATES

Many of the animals that preceded vertebrates were like the simplest living chordates—the urochordates. This group includes sea squirts and other tunicates. Adult tunicates secrete a jellylike or leathery "tunic" around the pharynx. Figure 23.3 shows translucent blue ones. Most species are a few centimeters long and live alone or in colonies from intertidal zones to the deep ocean. Adults usually attach to substrates. Sea squirts are the most common; they spurt water through a siphon when something irritates them.

For our story, a key point to remember about the tunicates is that their larvae are bilateral swimmers. A firm, flexible notochord acts like a torsion bar. It interacts with bands of muscles that lie just under the epidermis. When muscles on one side of the tail or the other contract, the notochord bends, then springs back as muscles relax. Strong, side-to-side motion propels a larva forward. Most fishes use their muscles and their backbone for the same kind of motion.

Another point: Tunicates are **filter feeders**. They filter food from seawater that flows in through one siphon, past gill slits in the pharynx, then out by way of another siphon. The pharynx is also a respiratory organ. It takes up dissolved oxygen from seawater. Oxygen diffuses into blood vessels near the pharynx. Carbon dioxide from aerobic respiration diffuses into the water flowing out.

Our simplest relatives also include the fish-shaped lancelets (cephalochordates). Lancelets are only three to seven centimeters long. Their body tapers at both ends and displays the chordate features (Figure 23.2). Like vertebrates, lancelets have a head and a simple brain that develops from one end of their nerve cord (*cephalo–*, head). Their nerve cord forms as it does in vertebrates. Nerve cells (neurons) in the brain control reflex responses to light and other stimuli.

Also like vertebrates, the lancelets have segmented muscles. Contractile units in muscle cells run parallel with the body's long axis. Contractile force directed against the notochord allows side-to-side swimming motions. Even so, lancelets spend most of the time burrowing in shallow sediments and filter-feeding. Coordinated beating of cilia that line numerous gill slits move water into their pharynx, where food gets caught in mucus before moving to the rest of the gut.

A gill-slitted pharynx, a simple brain, a nerve cord, and bands of muscles that parallel the body's long axis—as you will see, these features turned out to have amazing evolutionary possibilities for vertebrates.

Read Me First!
and watch the narrated animation on the chordate body plan

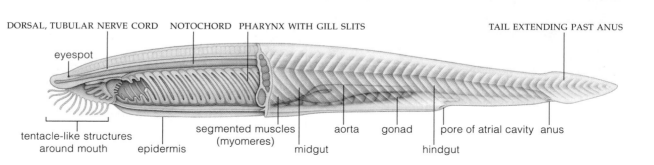

DORSAL, TUBULAR NERVE CORD NOTOCHORD PHARYNX WITH GILL SLITS TAIL EXTENDING PAST ANUS

eyespot

tentacle-like structures around mouth epidermis segmented muscles (myomeres) midgut aorta gonad hindgut pore of atrial cavity anus

Figure 23.2 Lancelet body plan. This small filter-feeder shows the four defining features of chordates.

Figure 23.3 (**a**) Generalized larval form of a sea squirt, a tunicate. A new larva swims about briefly, then it undergoes drastic tissue remodeling and reorganization. This metamorphosis starts when its head end attaches to a substrate. Its tail, notochord, and most of the nervous system are resorbed (that is, recycled and used to form new tissues). Many gill slits form in the pharynx wall. Organs rotate until openings through which water flows into and out from the pharynx are pointing away from the substrate, as in (**b**). (**c**) Adult sea squirts (*Rhopalaea crassa*).

Figure 23.4 Hagfish body plan. The two photographs to the right show one hagfish before and after it secretes slimy mucus from its pores.

A WORD OF CAUTION

We are describing chordate features that will help give you insight into the evolution of vertebrates, but don't assume tunicates and lancelets are missing links between vertebrates and the earliest chordates. These groups do share some characteristics that preceded the origin of vertebrates. However, each group also has unique traits that put it on separate branches of the chordate family tree.

Similarly, the hagfishes are like jawless vertebrates called lampreys, but some traits that evolved early on are theirs alone. Unlike vertebrates, for instance, their brain is not enclosed in a hardened, bony chamber. Even so, some cartilage helps protect it.

And don't consider hagfishes evolutionary dead-ends. A hagfish is less than a meter long and nearly blind, but it easily finds food with sensory tentacles (Figure 23.4). Its rasping, tonguelike structure draws small invertebrate prey and tissues of a dead or dying fish into its mouth. Hagfishes are not active animals;

they generally burrow in marine sediments. However, low metabolic rates work for them. They can wait out an absence of food for more than half a year.

If a hagfish is threatened, it secretes up to a gallon of sticky mucus. Fishermen who snag hagfishes view this behavior as disgusting. Even so, sliming potential predators has served this soft-bodied and otherwise vulnerable lineage quite well.

All chordate embryos have a notochord, a tubular dorsal nerve cord, a pharynx with gill slits in its wall, and a tail that extends past the anus. These traits are legacies from a shared ancestor, an early invertebrate chordate.

Existing filter-feeding lancelets and tunicates are among the groups closest to the earliest chordates.

Hagfishes represent the next level in chordate complexity. A portion of their brain is encased in protective cartilage.

23.2 Evolutionary Trends Among the Vertebrates

The first vertebrates were jawless fishes that appeared during the Cambrian. Later in time, their body plans were modified in diverse ways during an adaptive radiation that led to all modern vertebrate groups.

Figure 23.5 (**a**) Reconstruction of an early craniate (*Myllokunmingia*). (**b**) Larva of a lamprey, an existing jawless fish. Most of the lampreys are parasitic. They attach to prey with a suckerlike oral disk and rasp away at it with their horny mouthparts (**c**). We often observe several lampreys attached to the same host. In the 1800s, sea lampreys probably entered the Hudson River, then canals built for commerce. By 1946, they were established in all of the Great Lakes of North America. Trout, salmon, catfishes, and other natives have no coevolved defenses against them.

EARLY CRANIATES

Fishes, amphibians, reptiles, birds, and mammals are **craniates**. A cranium is a chamber of cartilage or bone that encloses all or part of a brain. Craniates evolved by 530 million years ago. Like lancelets, they had a notochord and segmented muscles; they were active swimmers. Adults resembled the larvae of lampreys, a group of existing jawless fishes (Figure 23.5).

Thirty million years later, ostracoderms and other jawless fishes had evolved. They had armorlike plates of bony tissue and dentin, a hard tissue still found in teeth. The armor may have worked against pincers of giant sea scorpions. But it didn't stop **jaws**—hinged, bony feeding structures. Jawed craniates had evolved. Most jawless fishes disappeared during the adaptive radiation of jawed craniates. Hagfishes and lampreys are the only groups that made it to the present.

Placoderms were among the earliest jawed fishes. Armor plates protected their head, but not much else (Figure 23.6). They didn't have teeth, but razor-sharp bony plates served the same function. The placoderms must have been major agents of natural selection. With their arrival, an arms race of offensive and defensive body structures began that continues to this day.

Read Me First!

and watch the narrated animation on the evolution of jaws

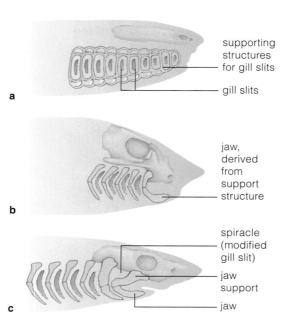

supporting structures for gill slits

gill slits

jaw, derived from support structure

spiracle (modified gill slit)

jaw support

jaw

Figure 23.6 Comparison of gill-supporting structures. (**a**) In early jawless fishes, supporting elements reinforced a series of gill slits on both sides of the body. (**b**) In placoderms and other early jawed fishes, the first elements in the series were modified, and they functioned as jaws. Cartilage reinforced the mouth's rim. (**c**) Sharks and other modern jawed fishes have strong jaw supports. *Above*, the size of a human compared to an extinct jawed fish, the placoderm *Dunkleosteus*.

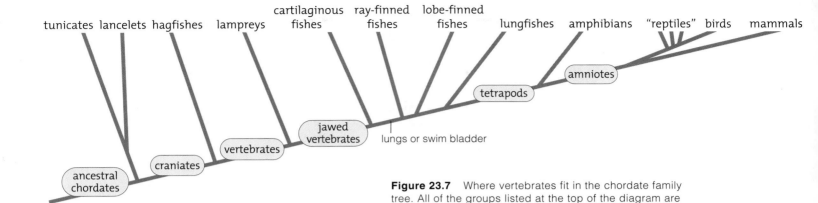

Figure 23.7 Where vertebrates fit in the chordate family tree. All of the groups listed at the top of the diagram are multicelled, bilateral, coelomate animals.

KEY INNOVATIONS

Vertebrates profoundly changed the course of animal evolution. Unlike their nearly brainless, filter-feeding relatives, they have bones—organs made of mineral-hardened secretions. **Vertebrae** (singular, vertebra), a series of bony or cartilaginous segments, replaced the notochord as the partner of muscles. The vertebral column, part of an inner skeleton, was more flexible and less clunky than external armor plates. The union of vertebrae and muscles promoted maneuverability, and hard bones invited more forceful contractions. The outcome? *Agile, fast-moving jawed fishes came to be the dominant animals in the seas.*

Jaws sparked another trend. They had evolved from the first of a series of hard parts that structurally supported gill slits in the pharynx (Figure 23.6). They were a key innovation, one that invited novel feeding possibilities. As noted earlier, they also triggered the evolution of defenses and of ways to overcome them. For example, the fishes with better eyes for detecting predators or bigger brains to plan escapes now had a competitive edge. *A trend toward more complex sensory organs and nervous systems started among ancient fishes and continued among vertebrates on land.*

Another trend started when pairs of fins evolved. Fins are appendages that help propel, stabilize, and guide the body through water. Some fish groups had fleshy ventral fins with internal skeletal supports that became forerunners of limbs. *Paired, fleshy fins were a starting point for all legs, arms, and wings that evolved among amphibians, reptiles, birds, and mammals.*

A shift in respiration set another trend in motion. In lancelets, oxygen and carbon dioxide diffuse across skin. In most early vertebrates, paired gills evolved. **Gills** are respiratory organs having moist, thin folds serviced by blood vessels. They have a large surface area that exchanges gases between the environment and the body. Oxygen diffuses *from* water inside the mouth into blood vessels within each gill as carbon dioxide diffuses *into* the surrounding water.

Gills became more efficient in larger, more active fishes. But gills can't function out of water; their thin surfaces stick together unless flowing water moistens them. In fishes ancestral to land vertebrates, two tiny outpouchings developed on the gut wall and evolved into **lungs**: internally moistened sacs for gas exchange. Lungs supplemented and then replaced gills during the invasion of land. In a related trend, modifications to the heart enhanced the uptake of oxygen and removal of carbon dioxide. *The first vertebrates on land relied more on a pair of lungs than on gills. Increased efficiency of circulatory systems accompanied the evolution of lungs.*

In the Carboniferous, placoderms and other jawed fishes became extinct after faster, brainier predators replaced them in the seas. Through adaptation, key innovations, and plain luck, the descendants of some jawless and jawed groups made it to the present.

MAJOR VERTEBRATE GROUPS

Figure 23.7 shows how vertebrates fit in the chordate family tree. Besides lampreys, the major groups are known informally as cartilaginous fishes, bony fishes, amphibians, reptiles, birds, and mammals. The rest of this chapter introduces their defining characteristics.

Vertebrates arose in the Cambrian. All of the major modern groups evolved during an adaptive radiation of jawed fishes.

Jaws, a vertebral column, paired fins, and lungs were pivotal developments in the evolution of lineages that gave rise to fishes, amphibians, reptiles, birds, and mammals.

23.3 Jawed Fishes and the Rise of Tetrapods

Unless you study underwater life, you may not know that fishes are the world's dominant vertebrates. Their numbers exceed those of all other vertebrate groups combined, and they show far more diversity in their body plans.

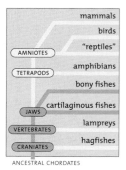

Of four major groups of living fishes, one has a skeleton of cartilage, the others a skeleton of bone. Their body form and behavior hint at the challenges of living in water. Water is 800 times more dense than air and resists movements through it. One response to this constraint, a streamlined shape, reduced friction during fast pursuits and escapes. For instance, sharks and other predators of the open ocean have a long, trim body with strong tail muscles for forward propulsion (Figure 23.8*a*). If a species doesn't make high-speed runs, the body tends to be flattened and adapted for concealment. Bottom dwellers or crevice lurkers are like this (Figure 23.8*b*). Fish scales were another response. These small, bony plates protect the body without weighing it down.

A trout suspended in shallow water shows another adaptation to water's density. It can maintain neutral buoyancy with its swim bladder. This is an adjustable flotation device that exchanges gases with the blood. When a trout is gulping air at the water's surface, it is adjusting the volume of its swim bladder.

CARTILAGINOUS FISHES

Sharks and rays are well-known **cartilaginous fishes**. These predators have prominent fins, a skeleton of cartilage, and five to seven gill slits on both sides. At fifteen meters in length, some sharks rank among the largest living vertebrates. They continually shed and replace their teeth, which are modified scales with a dentin core and an outer enamel layer. Jaws are used to grab and rip chunks of prey. The sharks have been predators for hundreds of millions of years, but rare attacks on humans give the group a bad reputation.

Figure 23.8 Representative cartilaginous fishes. (**a**) Galápagos shark and (**b**) stingray. Notice the pairs of gill slits. Representative bony fishes: (**c**) Coral grouper, one of many species that have been overfished as food for humans. (**d**) Flying fish. Like bird wings, modified pectoral fins help these fishes fly out of the water when chased by predators. Evolutionarily speaking, they are not yet equipped to take off again from the deck of boats that happen to cross their flight path. (**e**) Perch fins.

"BONY FISHES"

More than 400 million years ago, even before sharks appeared, ray-finned fishes, lobe-finned fishes, and lungfishes evolved. With more than 21,000 known species in freshwater and marine habitats, the ray-finned fishes are the most diverse in shape, size, and coloration. Their fins are bony elements in a web of skin. Figure 23.8c–e shows just a few representatives.

Coelacanths (*Latimeria*) are the only surviving group of lobe-finned fishes. We know about two populations that may be separate species. The ventral fins of these fishes are fleshy extensions of the body, with skeletal support elements inside (Figure 23.9a). Like the ray-finned fishes, coelacanths have gills.

Lungfishes (Figure 23.9b) have gills *and* one or two small, modified outpouchings of the gut wall. These sacs help take in oxygen and remove carbon dioxide. Lungfishes must surface and gulp air; they drown if held under water. When streams dry out seasonally, lungfishes encase themselves in slimy secretions and mud. They stay inactive until the rainy season starts.

Do such fishes share a common ancestor with four-legged walkers—**tetrapods**? Probably. Like tetrapods, a lungfish has a separate blood circuit to the lungs. Its skullbones are arranged the same way. It has the same tooth enamel. Also, the fins of both lobe-finned fishes and lungfishes are about the same as tetrapod limbs in structure, position, and sizes. A lungfish even uses its limbs to move forward on underwater substrates.

Limb joints, digits, and other traits of fossils imply that walking originated in water, not on land. By the late Devonian, some swimming and crawling species were using limbs and digits (Figure 23.9c–e).

Ray-finned fishes are now the most diverse and abundant vertebrates. The lobe-finned fishes have fleshy ventral fins reinforced with skeletal parts. The lungfishes have simple lunglike sacs that supplement respiration by gills.

Walking probably started under water during the Devonian, among the aquatic forerunners of tetrapods on land.

inside lobed fins, bony or cartilaginous structures undergoing modification

Figure 23.9 (**a**) Living coelacanth (*Latimeria*), of the only lineage of surviving lobe-finned fishes. (**b**) Australian lungfish. (**c**) Painting of Devonian tetrapods. *Acanthostega*, submerged, and *Ichthyostega* crawling on land. Both had a fishlike skull, caudal tail, and fins. Unlike fishes, both had a short neck and four limbs with digits. (**d,e**) Proposed evolution of skeletal elements inside fins into the limb bones of early amphibians.

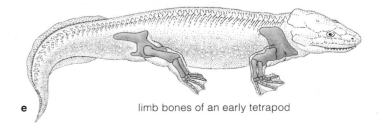

limb bones of an early tetrapod

23.4 Amphibians—The First Tetrapods on Land

Even a small genetic change could have transformed lobed fins into limbs. Remember those enhancers that control gene transcription? One of them influences how digits form on limb bones. Even a single mutation in one master gene can lead to a big change in morphology.

ANCESTRAL CHORDATES

Amphibians are the group of vertebrates that were the first tetrapods on land. They either have four legs or a four-legged aquatic ancestor, and the body plan and reproductive modes are somewhere between fishes and "reptiles."

What drove certain aquatic tetrapods onto land? Consider this: Asteroids hit the Earth at least five times during the Devonian. One of the last impacts coincided with a mass extinction of marine life. If it triggered the release of methane hydrates, much of the oxygen dissolved in the open seas and swampy coastal habitats would have been displaced by carbon dioxide (Section 3.2). However it happened,

tetrapods with lungs had the advantage; they could get enough oxygen by gulping air. Fossils show that some tetrapods were semiaquatic and were spending time on land when the Devonian came to a close.

Life in the new, drier habitats was both dangerous and promising. Temperatures shifted more on land than in the water, air didn't support the body as well as water did, and water was not always plentiful. But air is richer in oxygen. Amphibian lungs continued to be modified in ways that enhanced oxygen uptake. The heart, newly divided into three chambers, pumped more oxygen-rich blood to cells. Increased reliance on aerobic respiration, a pathway that makes enough ATP to sustain more active life-styles, was now possible.

New sensory information also challenged the early amphibians. Swamp forests supported vast numbers of insects and other small invertebrate prey. On land, vision, hearing, and balance turned out to be highly advantageous senses. They promoted the expansion of the brain regions that received, interpreted, and responded to the richly novel sensory input.

All living amphibians are descendants of the first tetrapods. None has escaped the water entirely. Even

fish locomotion

salamander locomotion

Figure 23.10 Familiar amphibians. (**a**) Red-spotted salamander (*Notophthalamus*). Coordinated controls over its legs move the body the same way that controls over fins move fishes. (**b**) A frog jumping and displaying its tetrapod heritage. (**c**) American toad.

Figure 23.11 One of the caecilians, a legless tropical amphibian. How do you suppose it burrows into soil?

species that have gills or lungs also exchange oxygen and carbon dioxide across the surface of their thin skin. Respiratory surfaces must be moist at all times, however, and skin dries easily in air.

Some amphibians spend their entire life in water. Most release eggs in water, where their larvae can develop. Most species adapted to dry habitats lay eggs in moist places. A few protect their embryos as they develop inside the moist adult body.

Frogs and toads are among the most well-known amphibians, and with more than 4,800 named species, they are also the most diverse (Figure 23.10b,c). Most are active predators; they capture insect prey with an extensible, sticky-tipped tongue. They generally lay eggs in water, and these hatch into tadpoles, a type of legless larvae with gills and a tail. Later on, the gills and tail disappear, and four legs form.

The 500 or so salamander species have a long body, and a tail that persists in adults. A few species, such as the mudpuppy and axolotl, retain other juvenile amphibian features, including external gills. As most salamanders walk, they bend from side to side like fishes (Figure 23.10a). Probably the first tetrapods on land walked this way.

As some amphibians evolved, they lost their limbs and vision, but not their jaws. This group gave rise to caecilians (Figure 23.11). Most of the 165 or so species burrow through moist soil, using the senses of touch and smell as they pursue insects. A few are aquatic predators that use electrical cues to locate prey.

In body form and behavior, amphibians show resemblances to aquatic tetrapods and reptiles. Most still rely on access to aquatic or moist habitats to complete their life cycle. Frogs, toads, and salamanders are the most familiar groups.

23.5 Vanishing Acts

FOCUS ON THE ENVIRONMENT

Amphibians are survivors. They originated before the dinosaurs and outlasted them. They have been around a thousand times longer than humans, but now human activities are putting many species at risk.

There is no question that amphibians are in trouble. Of about 5,500 known species, population sizes of at least 200 are plummeting. The declines have been well-documented in North America and Europe, but things look bad worldwide.

Six frog, four toad, and eleven salamander species are currently listed as threatened or endangered in the United States and Puerto Rico. One of them, the California red-legged frog, is the largest frog native to the western United States. It is the species that inspired Mark Twain's classic short story, "The Celebrated Jumping Frog of Calaveras County."

Many declines correlate with shrinking habitats and deteriorating habitat conditions. Developers and farmers often fill low-lying ground that became seasonal pools of standing water. Nearly all species of amphibians have trouble breeding unless they can deposit their eggs in water.

Also contributing to the declines are introductions of new species in amphibian habitats, long-term changes in climate, increases in ultraviolet radiation, and the spread of fungal and parasitic diseases into new habitats (Figure 23.12). Contamination of aquatic habitats by pesticides and other chemicals is another contributing factor. Chapter 31 offers a closer look at the effects of habitat contamination.

Figure 23.12 (a) Example of frog deformities. (b) A parasitic fluke (*Ribeiroia*) that burrows into frog tadpole limb buds and physically or chemically alters their cells. Infected tadpoles grow extra legs or none at all. Where *Ribeiroia* population densities are great, the number of tadpoles that successfully complete metamorphosis is low. Trematode cysts have been found in deformed frogs and salamanders in California, Oregon, Arizona, and New York.

23.6 The Rise of Amniotes

Amniotes were the first vertebrates to adapt to dry land habitats. They did so through modifications in their organ systems, behavior, and eggs, which have four membranes that conserve water and support embryonic development.

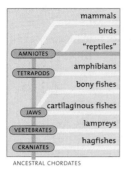

ANCESTRAL CHORDATES

THE "REPTILES"

Late in the Carboniferous, some amphibian species gave rise to **amniotes**, the only vertebrates that produce eggs having four membranes. These membranes conserve water and also protect the developing embryo before hatching or being born (Figure 23.13). All amniotes have dry, tough, or scaly skin that restricts water loss. They also have a pair of efficient water-conserving kidneys.

Synapsids and sauropsids are two major groups of amniotes. *Synapsids* are mammals and their extinct relatives. *Sauropsids* are "**reptiles**," and birds. Reptiles are no longer recognized as a formal taxonomic group because they are not monophyletic. The name persists only as a way to refer to several lineages that show the basic amniote features but not the derived traits that define birds or mammals.

Compared to amphibians, the early reptiles chased prey with more cunning and speed. With well-muscled jaws and sharp teeth, they could seize and crush prey with sustained force. The four limbs of many species were better at supporting the body's trunk on land. Gas exchange across the skin, vital for amphibians, was abandoned; lungs became better developed. The brainier reptiles engaged in behaviors not seen among the amphibians (Figure 23.13*a*).

THE AGE OF DINOSAURS

The first **dinosaurs** descended from Triassic reptiles. They were not much bigger than a turkey. They might have had high metabolic rates, and it is possible they were warm-blooded. Many moved about on two legs.

Adaptive zones opened up for dinosaurs about 213 million years ago. Huge fragments from an asteroid or comet fell upon the spinning Earth, creating craters in what is now France, Quebec, Manitoba, and North Dakota. Nearly all animals that survived the impacts were small, had high rates of metabolism, and could tolerate big temperature changes.

Descendants of the surviving dinosaurs dominated the land for 125 million years, even as ichthyosaurs and other groups flourished in the seas (Figure 23.14). Many dinosaurs were lost in a mass extinction at the end of the Jurassic, and others in a pulse of Cretaceous extinctions. Some lineages did recover, and new ones evolved in forests, swamps, and open habitats.

By the **K–T asteroid impact theory**, the Cretaceous ended when an asteroid as huge as Mount Everest struck Earth (Chapter 17). Dinosaurs vanished, along with most life in the seas. Afterward, gradual crustal movements repositioned the impact site. Researchers discovered it in the Yucatán peninsula of Mexico.

The asteroid struck at 160,000 kilometers per hour and made a crater wider than Connecticut. Monstrous waves 120 meters high raced across the ocean, wiped out life on islands, then slammed into continents. By a **global broiling hypothesis**, it was as if 100 million nuclear bombs had been detonated. Trillions of tons

Figure 23.13 Some amniote eggs. (**a**) Painting of the nest of a duck-billed, plant-eating dinosaur (*Maiasaura*), found in Montana. This species, which lived 80 million years ago, showed social behavior; it traveled in herds, protected eggs, and cared for the young. (**b**) Eastern hognose snakes emerging from leathery-shelled eggs.

Figure 23.14 Jurassic scene. (**a**) *Archaeopteryx* gliding in the foreground. Left to right, the plant eaters *Stegosaurus* and *Apatosaurus*, the carnivore *Saurophaganax* ("king of reptile eaters"), and *Camptosaurus*, a beaked plant eater. A climbing mammal looks out from the shrubbery. *Apatosaurus* was one of the largest land animals that ever lived. This tiny-brained plant eater grew as long as 27 meters (90 feet) and weighed 33 to 38 tons.

(**b**) *Temnodontosaurus*, an ichthyosaur that hunted large squid, ammonites, and other prey in the warm, shallow seaways. Fossils measuring 9 meters (30 feet) long have been found in England and Germany.

of vaporized debris rose in a colossal fireball. As the debris fell back to Earth, atmospheric temperatures rose by thousands of degrees. In one horrific hour, nearly all plants erupted in flames. All animals out in the open, not in burrows, were broiled alive.

Some reptiles survived; their descendants include the crocodilians, turtles, tuataras, snakes, and lizards. So did a group of feathered dinosaurs—birds.

Tough, scaly, or dry skin and other ways of conserving water, eggs with membranes that support the embryos, and active life-styles helped amniotes radiate into dry land habitats.

One amniote lineage, the sauropsids, includes reptiles and birds. Another, the synapsids, includes mammals and the now-extinct mammal-like reptiles.

Much of amniote history was a matter of luck, of species with particular traits being in the right or wrong places on a changing geologic stage.

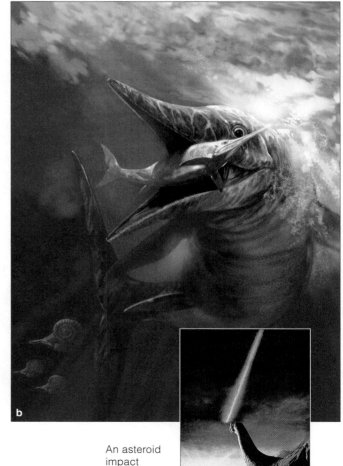

An asteroid impact ended the golden age of dinosaurs.

23.7 Existing Reptilian Groups

The name reptile is derived from the Latin repto, *which means to creep. Some existing forms do creep. Others race, lumber, or swim about. These diverse lineages include all living amniotes that are not birds or mammals.*

More than 8,160 species of reptiles flourish in diverse habitats on land, in fresh water, and in the oceans. All are cold-blooded. Fertilization is internal; eggs are usually laid on land, even by aquatic species. Both males and females have a cloaca, an opening used in excretion as well as reproduction. Figure 23.15 shows a body plan that is typical of one group.

About 305 existing species of turtles live inside a shell attached to their skeleton. If threatened, turtles can pull their head and limbs inside (Figure 23.16a,b). Their body plan works well; it has been around since the Triassic. Only among sea turtles and other highly mobile types has the shell been reduced in size.

Instead of teeth, turtles have horny plates that are used to grip and chew food. Strong jaws and a fierce disposition help deter predators. But turtle eggs are vulnerable to many predators on land. Add hunting and destruction of nesting sites to the mix, and most sea turtles are now poised at the brink of extinction.

With about 4,710 species, the lizards are the most diverse reptiles. The largest, *Varanus komodoensis*, is a monitor lizard that grows ten feet long. This ambush predator lurches out of the bushes at deer, wild boar, an occasional human, and other prey. The smallest lizard, *Sphaerodactylus ariasae*, can fit on a dime (*left*). Like most predatory lizards, it snags prey with sharp, peglike teeth. Chameleons (page 475) use accurate flicks of a sticky tongue, which can be longer than the length of their body. Iguanas, including the marine iguanas of the Galápagos, are herbivores.

Being small themselves, most lizards are prey for other animals. Some attempt to startle predators by flaring a throat fan; others try to outrun them (Figure 23.16c,d). Many will give up their tail when a predator grabs them. The detached tail wriggles for a bit and may be distracting enough to permit a getaway.

The two surviving species of tuataras live on small islands near New Zealand (Figure 23.16e). They are the descendants of a group that had its heyday in the Triassic. They are reptiles, yet they resemble modern amphibians in their brain and locomotion. Like some lizards, the tuataras have a third "eye" under the skin, complete with retina, a rudimentary lens, and links to the brain. It may detect changes in daylength and light intensity and affect hormonal control of reproduction.

During the Cretaceous, short-legged, long-bodied lizards gave rise to elongated, limbless snakes. Some snakes retain bony remnants of ancestral hindlimbs. Most move their body in S-shaped waves, much like salamanders do.

All 2,955 living snake species are carnivores with flexible skull bones and jaws. Many can swallow prey wider than they are. Rattlesnakes and other fanged types bite and subdue their prey with venom (Figure 23.16f). Snakes in general are not aggressive toward humans. Even so, venomous snakes bite about 7,000 people annually in the United States; 15 of them die. About 40 percent of those bitten reported that they were handling or otherwise disturbing the snake at the time. Worldwide, the annual death toll from snake bites is estimated at 30,000–40,000.

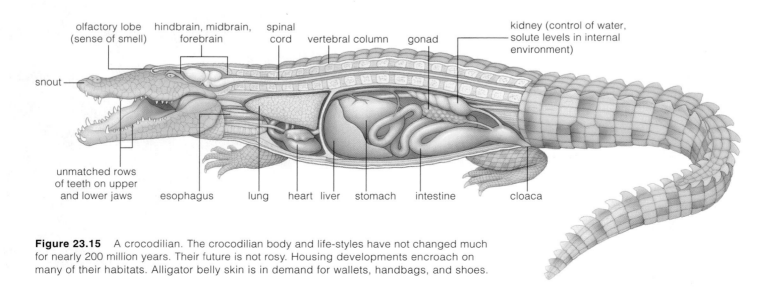

Figure 23.15 A crocodilian. The crocodilian body and life-styles have not changed much for nearly 200 million years. Their future is not rosy. Housing developments encroach on many of their habitats. Alligator belly skin is in demand for wallets, handbags, and shoes.

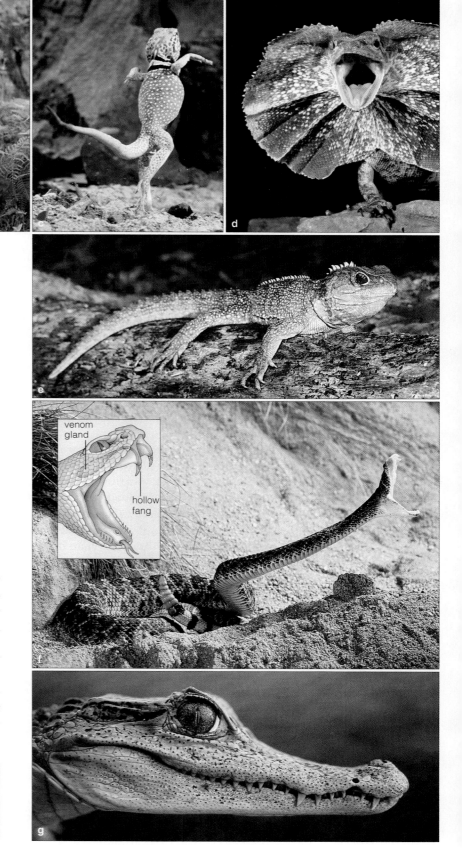

vertebral column

hard shell

b

Figure 23.16 (**a**) Galápagos tortoise. (**b**) Turtle shell and skeleton. (**c,d**) Lizard fleeing, lizard confronting. (**e**) Tuatara (*Sphenodon*). (**f**) Rattlesnake. (**g**) Spectacled caiman, a crocodilian. It has two unmatched rows of peglike teeth.

The twenty-three modern crocodilian species are the most advanced reptiles and the closest living relatives of birds. Crocodiles, alligators, and caimans all live in or near water and are predators that have powerful jaws, a long snout, and sharp teeth (Figure 23.16*g*). They drag prey into the water, tear it apart as they spin around and around, then gulp down the chunks. The crocodilians are the only reptiles with a four-chambered heart, like mammals and birds. Like most birds, they show parental behavior, such as guarding nests and assisting the hatchlings.

Turtles and tuataras have body plans that have changed very little since the Triassic. Lizards and snakes are the most diverse reptilians. Crocodilians have the most complex brain and circulatory system and are the closest relatives of birds.

23.8 Birds—The Feathered Ones

Birds are warm-blooded amniotes with feathers. They are descended from dinosaurs and still retain a number of reptilian traits. Internal and external modifications in body plans led to the capacity for flight.

mammals
birds
"reptiles"
AMNIOTES
amphibians
TETRAPODS
bony fishes
cartilaginous fishes
JAWS
lampreys
VERTEBRATES
hagfishes
CRANIATES

ANCESTRAL CHORDATES

The capacity for flight evolved in four groups—the insects, pterosaurs (extinct), birds, and bats. Of these groups, feathers are a trait unique to **birds** and some extinct dinosaurs. These lightweight structures, derived from scales, are used in flight, as body insulation, and often in courtship displays.

There are about 28 orders of birds and close to 9,000 named species. They vary in size, proportions, coloration, and capacity for flight. The smallest, the bee hummingbird, weighs 1.6 grams (0.6 ounces). The ostrich, the largest living bird, weighs as much as 150 kilograms (330 pounds). It cannot fly, but its powerful leg muscles make it an impressive sprinter. Feather colors and patterns can be especially spectacular among warblers, parrots, and some other perching species that show complex social behaviors. Bird song and other forms of behavior are among the topics of Chapter 43.

Birds originated when Mesozoic reptiles began an adaptive radiation. They diverged from a lineage of small theropod dinosaurs, a type of carnivore that ran about on two legs. Bird feathers evolved from highly modified reptilian scales. *Archaeopteryx* was in or near that lineage. As you read earlier, it had reptilian *and* avian traits, including feathers. Birds still share many traits with their closest relatives, the dinosaurs and crocodilians. For instance, like them, birds have scales on their legs, and they have some of the same internal parts. Like them, birds fertilize eggs internally and have a cloaca used in excretion and reproduction.

Unlike reptiles, birds closely regulate their body temperature. Most heat gains come from metabolism, not the outside environment. Feathers help conserve body heat when external temperatures drop. When temperatures rise, elastic sacs connected to bird lungs help dissipate excess metabolic heat. The sacs enhance air flow and can move warmed air out of the body.

The evolution of bird flight involved more than feathers. It involved modification of the entire body, from internal bone structure to highly efficient modes of respiration and circulation. The high metabolic rates that sustain flight depend on a strong flow of oxygen through the body. The elastic sacs that connect to the lungs greatly enhance oxygen uptake (Section 35.4). Like mammals, birds have a large, four-chambered heart. It pumps oxygen-rich blood to the lungs and to the rest of the body in separate circuits, as it probably did in the reptilian ancestors of birds.

Flight also demands an airstream, low weight, and a powerful downstroke that can provide lift—a force at right angles to an airstream. Each wing, a modified forelimb, consists of feathers and lightweight bones attached to powerful muscles. Its bones do not weigh much, owing to profuse air cavities in the bone tissue (Figures 23.17*b* and 23.18*d*). The flight muscles attach

yolk sac   embryo   amnion   chorion   allantois

hardened shell   albumin ("egg white")

a

Figure 23.17 (**a**) Sketch of one type of amniote egg. (**b**) Body plan of a typical bird, a pigeon. Flight muscles attach to the large, keeled breastbone (sternum).

skull

radius

ulna

pectoral girdle

humerus

internal structure of bird limb bones

pelvic girdle

sternum

two main flight muscles attached to keel of sternum

b

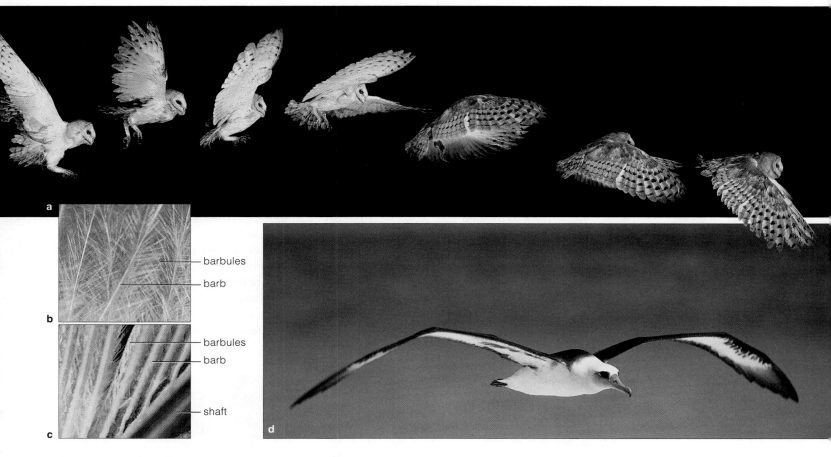

Figure 23.18   Bird flight. (**a**) Of all living animals, only birds and bats fly by *flapping* their wings. A bird wing is an intricate system of lightweight bones and feathers. (**b**) Innermost down feathers function in insulation. (**c**) A hollow central shaft and an interlocked lattice of barbs and barbules strengthen flight feathers and increase the wing flight surface without imposing much weight. (**d**) Laysan albatross. Its wingspan is more than two meters, yet this seabird weighs less than 10 kilograms (22 pounds). It is so at home in the air, it even sleeps while riding the winds. One bird monitored by researchers flew 24,843 miles in ninety days as it searched for food and brought it back to its nestlings. That is the equivalent of a trip around the globe.

to an enlarged breastbone (sternum) and to the upper limb bones attached to it. When the muscles contract, they produce a powerful downstroke.

With their long flight feathers, wings are like the airfoils on planes. A bird normally spreads out long feathers on a downstroke, thereby increasing the area of the surface pushing against the air. Then, on the upstroke, the bird folds its feathers a bit, so each wing surface decreases and offers less resistance to air.

We see one of the most dramatic forms of behavior among birds that migrate with the changing seasons. **Migration** is a recurring pattern of movement from one region to another in response to environmental rhythms. Seasonal change in daylength is one factor that influences the internal timing mechanisms that we call "biological clocks." It causes physiological

and behavioral changes that induce migratory birds to make round trips between breeding grounds and wintering grounds. Arctic terns migrate the farthest. They spend summers in arctic regions and winters in antarctic regions. One monitored bird took off from an island near Great Britain and landed three months later in Melbourne, Australia. It racked up more than 22,000 kilometers (14,000 miles) on that one journey.

Of all animals, birds alone produce feathers, which they use in flight, heat conservation, and socially significant communication displays.

A bird's feathers, lightweight bones, and highly efficient respiratory and circulatory systems sustain flight.

23.9 Mammals

Remember the premise of cause and effect? Mammals evolved before dinosaurs and outlasted them. They were artful dodgers, sidelined only until the dinosaurs' luck ran out. Then they radiated, spectacularly, into the vacated adaptive zones.

They alone make hair, but they differ in its amount, type, and distribution. Female mammals alone secrete milk for their young. This nutrient-rich fluid forms in mammary glands that have ducts to the body surface.

Jaws and teeth also help set mammals apart. They alone have a lower jawbone on each side. Most have four types of upper and lower teeth that match up (Figure 23.19). Reptiles have a hinged two-part jaw, and all teeth are one type only and don't match up.

Most mammals care for offspring for an extended time. Young mammals typically have the capacity to learn and repeat behaviors with survival value. Some species show stunning *behavioral flexibility*—a capacity to expand on basic activities with novel behaviors.

Where did mammals come from? Remember, the extinct mammal-like reptiles and all extinct and living species of mammals are synapsids. The defining trait of this group is an opening in a particular location on each side of the skull that functions in the attachment and alignment of jaw muscles. The trait arose in the Carboniferous. Later on, in the Permian, the synapsids underwent adaptive radiations and became the major land vertebrates. At the close of the Permian, many groups died out in a mass extinction.

Fortunately for humans, the group called *therapsids* endured. Some became more like mammals during the Triassic. For instance, limbs became positioned upright under their trunk. Their skeletal arrangement made it easier to walk erect, but a trunk higher from

Figure 23.19 Distinctly mammalian traits. (**a**) A human baby, already with a mop of hair, being nourished by milk from mammary glands. Humans are placental mammals. (**b**) Four types of teeth; upper and lower ones match up. (**c**) Platypus, an egg-laying mammal, with live, unfinished offspring. (**d**) Koala, a pouched mammal. The smallest and largest living placental mammals: (**e**) Kitti's hognosed bat, 1.5 grams; and (**f,g**) blue whale, 200 tons.

Read Me First!
and watch the narrated animation on
mammalian radiations

a About 150 million years ago, during the Jurassic, the first monotremes and marsupials evolved and migrated throughout the supercontinent Pangea.

b Between 130 and 85 million years ago, in the Cretaceous, placental mammals emerged and started to spread. Monotremes and marsupials of the southern supercontinent evolved in isolation from placental mammals.

c About 20 million years ago, in the Miocene, placental mammals expanded in range and diversity. On Antarctica, mammals vanished. Marsupials and early placental mammals displaced monotremes in South America.

d About 5 million years ago, in the Pliocene, advanced placental mammals invaded South America. And they drove most marsupials and early placental species to extinction.

Figure 23.20 (a–d) Adaptive radiations of mammals. *Right*, one consequence—a case of convergent evolution: **(e)** Australia's spiny anteater, one of two living monotremes. **(f)** Africa's aardvark and **(g)** South America's giant anteater. Compare the specialized ant-snuffling snouts.

the ground was not as stable. Not coincidentally, their cerebellum, a brain region that controls balance and spatial positioning, increased in size.

The earliest known groups of modern mammals evolved early on in the Jurassic (Figure 23.20*a*). They were mouselike forms, and they coexisted with the dinosaurs. By 130 million years ago, three separate lineages had evolved. We call them **monotremes** (egg-laying mammals), **marsupials** (pouched mammals), and **eutherians** (placental mammals).

At the end of the Jurassic, mammals evolved into the adaptive zones vacated, under extreme duress, by dinosaurs. In the Cenozoic, the supercontinent Pangea split apart. Some monotremes and marsupials drifted off on what would become Australia and Antarctica. They became isolated from placental mammals that were rapidly expanding their range and diversity on other continents (Figure 23.20*b,c*).

On a fragment that would become South America, monotremes were replaced by marsupials and early placental mammals. When a land bridge joined South and North America in the Pliocene, highly evolved placental mammals radiated southward and rapidly replaced many of their previously isolated relatives

(Figure 23.20*d*). Only the opossums and a few other species invaded land masses in the other direction.

What gave placental mammals a competitive edge? They had higher metabolic rates, more precise ways to control body temperature, and a more efficient way to nourish embryos. You will read about them in later chapters. For now, note that the adaptive radiation of the advanced placental mammals came about at the expense of their less competitive relatives.

One more point: Although the lineages originally evolved on separate continents, many lived in similar habitats. In time, they came to resemble one another in form and function. For example, Australia's spiny anteater, South America's giant anteater, and Africa's aardvark all have similar snouts adapted to preying on ants (Figure 23.20*e–g*). Such species are examples of *convergent* evolution, as described in Section 17.4.

Only female mammals produce milk. Mammals also are the only animals with hair.

The three major groups are the egg-laying monotremes, the pouched marsupials, and the placental mammals. Placental mammals are now the dominant group in most regions.

23.10 From Early Primates to Hominids

So far, you have traveled 570 million years through time, from tiny flattened animals to craniates, then on to the jawed fishes with a backbone. You encountered some early tetrapods, then amniotes—reptiles, birds, and mammals. Now you are about to access what we know about the evolutionary events that led to modern humans.

TRENDS IN PRIMATE EVOLUTION

Living **primates** are prosimians, tarsioids, and the anthropoids (Figure 23.21). The first prosimians were tree-dwellers of northern forests for millions of years until monkeys and apes nearly displaced them. The gibbon, siamang, orangutan, gorilla, chimpanzee, and bonobo are classified as apes. All monkeys, apes, and humans are anthropoids. In biochemistry and body form, apes and humans are close relatives. They, and their closest extinct ancestors, are the only hominoids. All humanlike and human species, past and present, are known as the **hominids**.

Most of the primates live in the trees of forests, woodlands, or savannas (grasslands with a scattering of trees). No one feature sets them apart from other mammals, and each lineage has its defining traits. The five trends that define the lineage leading to humans were set in motion among the first tree-dwellers.

First, a reliance on the sense of smell became less important than daytime vision. *Second*, skeletal changes promoted **bipedalism**—upright walking—which freed hands for novel tasks. *Third*, bone and muscle changes led to diverse hand motions. *Fourth*, teeth became less specialized. *Fifth*, the evolution of the brain, behavior, and culture became interlocked. **Culture** is the sum of a social group's behavioral patterns, passed on through the generations by learning and symbolic behavior. In brief, *"uniquely" human traits emerged by modification of traits that had evolved earlier, in ancestral forms.*

Enhanced daytime vision. Early primates had an eye on each side of their head. Later on, many species had forward-directed eyes, which were better at sampling shapes and movements in three-dimensional space. Visual systems became more and more responsive to variations in light intensity (dim to bright) and colors.

Upright walking. How a primate walks depends on arm length and the shape and position of the shoulder blades, pelvic girdle, and the backbone (Figure 23.21*d*). With arm and leg bones of about the same length, a monkey climbs and runs on four legs. A gorilla walks with weight on two legs and the knuckles of its long arms. Humans have the shortest and the most flexible backbone, one curved in an S shape. These structural features are signs of bipedalism.

Figure 23.21 A few primates. (**a**) A tarsier, small enough to fit on a human hand, is a climber and leaper. (**b**) Spider monkey, an agile four-legged climber, leaper, and runner. (**c**) Male gorilla, a knuckle-walker. (**d**) Comparison of skeletal organization and stances. These images are not at the same scale. A monkey has long, thin limbs relative to its torso, and its long, thin digits are good at gripping, not at supporting the body's weight. Gorillas use their forelimbs to climb and support body weight when walking on knuckles and two legs. Humans are two-legged walkers.

Power grip and precision grip. So how did we get our versatile hands? Early mammals spread their toes apart to support the body as they walked or ran on four legs. In ancient tree-dwelling primates, hand bone modifications allowed them to curl fingers around objects (*prehensile* movements) and touch a thumb to fingertips (*opposable* movements). In time, hands were freed from load-bearing functions. Later, as hominids evolved, so did power and precision grips:

power grip   precision grip

Versatile hand positions gave human ancestors the capacity to make and use tools. They were a basis for unique technologies and cultural development.

Teeth for all occasions. Modifications to the jaws and teeth accompanied a shift from eating insects, to fruits and leaves, then to a mixed diet. Rectangular jaws and lengthy canines evolved in monkeys and apes. A bow-shaped jaw evolved in early hominid lineages. So did teeth that were smaller and all about the same length.

Brains, behavior, and culture. Shifts in reproductive and social behavior accompanied a shift to life in the trees. In many lineages, parents put more effort in fewer offspring and maternal care became intense. The young required more extended periods of dependency and learning before setting out on their own.

New forms of behavior promoted new connections in brain regions that deal with sensory inputs. A brain having more intricate wiring favored new behavior. The brain, behavior, and culture evolved together through enriched learning and symbolic behaviors, especially language. A capacity for language arose among hominids. In addition to a bigger, more complex brain, it required a bigger chamber to hold it.

ORIGINS AND EARLY DIVERGENCES

The first primates arose between 85 million and 65 million years ago, in tropical Paleocene forests. Like the tree shrews they resembled (Figure 23.22), they had huge appetites and foraged at night for eggs, insects, seeds, and buds under trees. They had a long snout and a good sense of smell for snuffling out food or predators. They clambered about among stems and branches, although not with much speed or grace.

By the Eocene, the prosimians were living higher in the trees. They had a shorter snout, a larger brain,

Figure 23.22 Tropical forest of Southeast Asia and one of its tiny inhabitants, a tree shrew (*Tupaia*).

better daytime vision, and better ways to grasp and manipulate objects. *How did these traits evolve?*

Trees offered food and safety from ground-dwelling predators. They also were zones of uncompromising selection. Picture an Eocene morning, leaves swaying in the breeze, colorful fruit, and predatory birds. An odor-sensitive, long snout would not have been useful; breezes disperse odors. A brain that assessed motion, depth, shape, and color, along with skeletal changes, must have been favored. For example, eye sockets became positioned in front instead of on the sides of the skull—a better arrangement for depth perception.

By 36 million years ago, tree-dwelling anthropoids had evolved. They were on or close to the lineage that led to monkeys and apes. One had forward-directed eyes, a flattened face, and an upper jaw that sported shovel-shaped front teeth. It probably used its hands to grasp food. Some of the early anthropoids lived in swamps infested with predatory reptiles. Was that one reason why it became so imperative to think fast, grip strongly, and avoid the ground? Possibly.

Between 23 and 5 million years ago, in the Miocene, apelike forms—*the first hominoids*—evolved and spread through Africa, Asia, and Europe. Shifts in land masses and ocean circulation were causing long-term change in climate. Africa became cooler and drier, with more seasonal change. Tropical forests, with their edible soft fruits, leaves, and abundant insects, gave way to open woodlands and later to grasslands. Food had become drier, harder, and more difficult to find.

Hominoids that had evolved in lush forests had two options—move into new adaptive zones or die out. Most died out, but not the common ancestor of apes and, later in time, modern humans. By 7 million years ago, the *hominid* lineage had become established.

Complex, forward-directed vision; bipedalism; refined hand movements; generalized teeth; and interlocked elaboration of brain regions, behavior, and culture were key adaptations on the road from arboreal primates to modern humans.

23.11 Emergence of Early Humans

In central, eastern, and southern Africa, the Miocene through the Pliocene was a "bushy" time of hominid evolution, meaning many forms rapidly evolved during that span. We still don't know how they are related.

Sahelanthropus tchadensis was one species that evolved in Central Africa about 6 or 7 million years ago, when human ancestors were becoming distinct from apes (Figure 23.23*a*). One fossilized braincase is no bigger than that of a chimpanzee. Yet, like brainier hominids that evolved later, its owner had a shorter, flattened face with a prominent brow ridge, and smaller canines. Was it an ape or a hominid? We don't know.

Australopithecus and *Paranthropus* were hominids informally called **australopiths**. Figure 23.23*a* shows examples. Like apes, australopiths had a large face relative to their braincase, which indicates that they were small-brained. Also like apes, australopiths had protruding jaws. However, they differed in critical respects from earlier hominoids. For one thing, their thick-enameled molars could grind harder food. And these hominids walked upright. We know this from

fossilized hip bones and limb bones. We also have footprints. About 3.7 million years ago, an *A. afarensis* individual walked across newly fallen volcanic ash. A light rain transformed the ash to quick-drying cement and preserved its tracks (Figure 23.23*c*).

In the late Miocene, bipedal hominids were still adapted to forest life. Their hands could make strong, precision grips. Their descendants eventually left the trees for life on the ground but were not specialized for running fast on all fours. Rather, they became fully upright and used manipulative skills to advantage. We can expect that they used their hands to hang on to offspring and to carry precious food during their foraging expeditions out in the open.

What do fossilized fragments tell us about human origins? The record is still too sketchy to reveal how all the diverse forms were related, let alone which were our ancestors. Besides, which traits should we use to define **humans**—members of the genus *Homo*?

Well, what about brains? Our brain is the basis of analytical and verbal skills, complex social behavior, and technological innovation. This sets us apart from

Sahelanthropus tchadensis
7–6 million years

Australopithecus afarensis
3.6–2.9 million years

A. africanus
3.2–2.3 million years

A. garhi
2.5 million years
(first tool user?)

Paranthropus boisei
2.3–1.4 million years
(huge molars)

P. robustus
1.9–1.5 million years

H. habilis
1.9–1.6 million years

a

  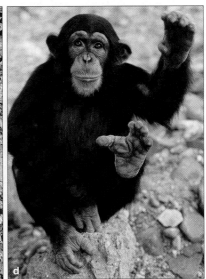

Figure 23.23 (**a**) African hominid fossils. (**b**) Fossilized bones of "Lucy" (*Australopithecus afarensis*), a two-legged walker who lived 3.2 million years ago.

(**c**) At Laetoli in Tanzania, Mary Leakey found these footprints, made in soft, damp, volcanic ash 3.7 million years ago. The arch, big toe, and heel marks of these footprints are signs of bipedal hominids. Unlike apes, the early hominids did not have a splayed-out big toe, of the sort being held up by the chimpanzee in (**d**).

apes that have a far smaller skull volume and brain size. Yet this feature alone cannot tell us when some hominids made the evolutionary leap to becoming human. Their brain size fell within the range of apes. They did use simple tools, but so do chimps and some birds. We have no clues to their social behavior.

We are left to speculate on evidence of physical traits, such as bipedalism, manual dexterity, a small face, a large braincase, and small, thickly enameled teeth. These traits emerged late in the Miocene and are features of what might be the first humans—*Homo habilis*. The name means "handy man."

Between 2.4 and 1.6 million years ago, early forms of *Homo* lived in woodlands bordering the savannas of eastern and southern Africa (Figure 23.24). Fossil teeth show that they ate hard-shelled nuts, dry seeds, soft fruits and leaves, and insects—all seasonal foods. *H. habilis* had to think ahead, to plan when to gather and store foods for cool, dry seasons ahead.

H. habilis shared its habitat with carnivores such as saber-tooth cats, which could impale prey and tear flesh but could not crush open the marrow bones. Carcasses with meat shreds clinging to bones offered nutrients in nutrient-stingy places. *H. habilis* was not one of the full-time carnivores. It may have enriched its diet opportunistically, by scavenging carcasses.

Who was the first tool maker? The earliest stone tools, from a site in Ethiopia, date to about 2.5 million years ago. Bits of volcanic rock, chipped to create a sharp edge, were found alongside bones of animals that appear to have been butchered with similar tools. Fossils of *A. garhi* were found nearby and date to the same period. Did *A. garhi* make the tools? Perhaps.

The layers of Tanzania's Olduvai Gorge document later refinements in toolmaking skills (Figure 23.25). The deepest of these sedimentary rock layers date to about 1.8 million years ago and hold crudely chipped pebbles similar to the earliest ones in Ethiopia. They also hold the fossil remains of australopiths and of *H. habilis*. We don't know which group created the tools. More recently deposited layers hold somewhat more complex tools. But innovations in tool form were limited. The same types appear in layers that formed slowly, over a period of about 1 million years.

Australopiths were a group of early hominids that walked upright. Some coexisted with the first human species, Homo habilis, which had a larger brain than earlier hominids.

The formation and use of stone tools may have started with one of the australopiths. H. habilis was another toolmaker.

Figure 23.24 Reconstruction of *Homo habilis* in an East African woodland. Two australopiths are in the distance.

Figure 23.25 From Olduvai Gorge in Africa, a sampling of stone tools. *Left to right,* crude chopper, more refined chopper, hand ax, and cleaver.

23.12 Emergence of Modern Humans

Ancestors of modern humans stayed put in Africa until Homo erectus evolved. Then cultural evolution took off.

EARLY BIG-TIME WALKERS

Homo erectus is a species related to modern humans (Figure 23.26). Its name means "upright man." Its forerunners were upright walkers, but some *H. erectus* populations did the name justice. They walked out of Africa, turned left into Europe, and also right, all the way to China. Middle East and Southeast Asia fossils date from 1.8 million to 1.6 million years ago. So *H. erectus* had to adapt to brutally cold climates of that period, and to vast ice sheets that advanced several times into northern Europe and North America.

Whatever pressures triggered the far-flung travels, this was a time of physical changes, as in skull size and leg length. It was a time of cultural lift-off for the human lineage. *H. erectus* had a larger brain and was a more creative toolmaker. Its social organization and communication skills must have been well developed. How else can we explain its grand dispersal? From southern Africa into England, different populations used the same kinds of hand axes and other tools. They met environmental challenges by building fires and using furs for clothing. Clear evidence of fire use dates from an early Pleistocene ice age.

A new fossil discovered in Ethiopia shows *Homo sapiens* had evolved by 160,000 years ago. Compared to *H. erectus*, the new species had smaller teeth, facial bones, and jawbones. Many individuals also had a novel feature—a chin. They had a higher, rounder skull, a bigger brain, and perhaps complex language.

A different group, the massively built and large-brained Neandertals, lived in Europe and the Near East from 200,000 to 30,000 years ago (Figure 23.26). Their extinction coincided with the arrival of modern humans in the same regions (Figure 23.27). Did they interbreed or war with those new arrivals? We do not know. But Neandertal DNA has unique sequences that are not found in present-day gene pools. They may not have modern descendants.

WHERE DID MODERN HUMANS ORIGINATE?

If we are interpreting the fossil record correctly, then Africa was the cradle for us all. At this writing, no one has found *Homo* fossils older than 2 million years *except* in Africa. Some *H. erectus* groups dispersed into the grasslands, forests, and mountains of Europe and Asia. Between 2 million and 500,000 years ago, they left Africa in waves. Some still lived in Southeast Asia between 53,000 and 37,000 years ago.

Figure 23.26
Fossils of two species of early humans. Evidence indicates that both species coexisted for a time with fully modern humans, *Homo sapiens.*

H. erectus
2 million–53,000? years

H. neanderthalensis
200,000–30,000 years

H. sapiens
fossil from Ethiopia,
160,000 years old

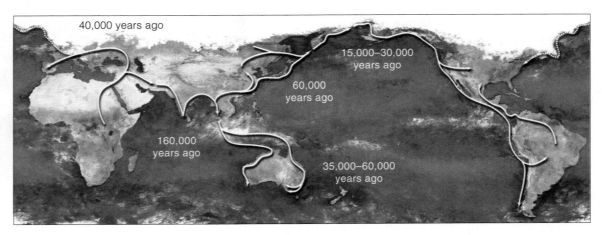

Figure 23.27 Estimated times when early *H. sapiens* colonized different parts of the world, based on radiometric dating of fossils. White lines show the presumed dispersal routes. Molecular data show that very few populations left Africa. Genetically, all modern human populations are extremely similar.

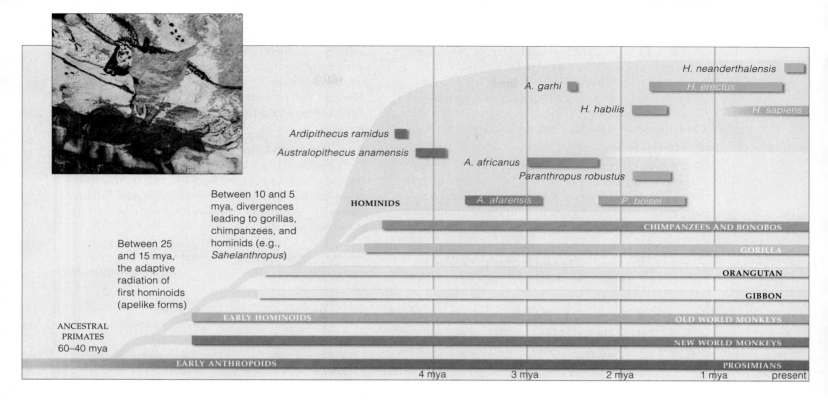

Figure 23.28 Summary of presumed evolutionary branchings in the primate family tree, including lineages on or near the road that led to modern humans. Also shown is one of the cave paintings by modern early humans at Lascaux, France.

But when did *H. sapiens* originate? *Here we find one case of how the same body of evidence can be interpreted in different ways.* The *multiregional* model and the *African emergence* model explain the distribution of early and modern humans differently, yet they both rely on measurements of genetic distance between existing populations. Biochemical and immunological studies show the greatest genetic distance is between human populations native to Africa and all others.

According to the multiregional model, data show that modern humans are the descendants of *H. erectus* populations that left Africa and spread through many regions by about one million years ago. In each area, the selective pressure on a population was different. Over time, subpopulations (races) of *H. sapiens* arose as a result of different selection pressures.

The African emergence model does not dispute fossil evidence that *H. erectus* migrated out of Africa and evolved further in different regions. However, it has *H. sapiens* arising in sub-Saharan Africa, then migrating out to other regions (Figure 23.27). By this model, as modern humans spread out of Africa, they replaced the archaic *H. erectus* populations that had preceded them. Only later did the regional phenotypic differences called races evolve.

In support of this model, the oldest of the *H. sapiens* fossils are from Africa. In Zaire, intricately wrought barbed-bone tools indicate that African populations were as skilled at making tools as *Homo* populations in Europe. Gene patterns from forty-three ethnic Asian groups suggest modern humans moved from Central Asia, along India's coast, then on into Southeast Asia and southern China. They later dispersed north and west into China and Siberia, then into the Americas.

Whatever the case, for the past 40,000 years, our cultural evolution has outpaced the macroevolution of the human species. One point to remember: Humans spread rapidly through the world by devising *cultural* means to deal with diverse environments. Hunters and gatherers persist in areas even as other groups moved from "stone-age" technology to high tech. The evolutionary road to this adaptable species was long, as summarized in Figure 23.28.

Cultural evolution has outpaced the biological evolution of the only remaining human species, H. sapiens. Today, humans everywhere rely on cultural innovation to adapt rapidly to a broad range of environmental challenges.

Summary

Section 23.1 Four features help define chordates. Their embryos develop a notochord, a dorsal hollow nerve cord (which becomes a brain and spinal cord), a pharynx with gill slits, and a tail extending past the anus. Some or all of these features persist in adults. The tunicates and the lancelets are invertebrate filter-feeding chordates.

Section 23.2 Craniates are chordates with a brain chamber of cartilage or bone. Vertebrates are craniates with a vertebral column of mainly bony segments. Jaws, paired fins, and later, lungs, were key innovations that led to the adaptive radiation of vertebrates.

Section 23.3 Cartilaginous fishes and bony fishes are aquatic vertebrates. The most diverse vertebrates are the bony fishes. Late in the Devonian, bony fishes having lobed fins and lungs gave rise to the tetrapods.

Sections 23.4, 23.5 Amphibians, the first tetrapods on land, share traits with aquatic tetrapods and reptiles. Most life cycles still go through aquatic stages. Existing groups include salamanders, frogs, and toads. As a result of habitat loss and other factors, many species are now declining.

Section 23.6 Amniotes, the first vertebrates to escape reproductive dependence on outside water, have water-conserving skin and kidneys, and amniote eggs. Synapsids (mammals and extinct mammal-like species) are one major group. Sauropsids ("reptiles" and birds) are another major group.

Section 23.7 Modern reptilian groups live in diverse habitats on land, in fresh water, and in the seas. Eggs are fertilized inside the body and usually laid on land. Lizards and snakes are the most diverse. Tuataras have some amphibian-like traits. Crocodilians are the closest relatives of birds.

Section 23.8 Birds are warm-blooded animals and the only existing ones with feathers. In most species, the body plan has been highly modified for flight.

Section 23.9 Mammals have hair and distinctive jaws and teeth. They are the only animals in which females nourish the young with milk from mammary glands. Three lineages are egg-laying mammals (monotremes), pouched mammals (marsupials), and placental mammals (eutherians). Placental mammals are the most diverse group.

Section 23.10 Living primates are prosimians, tarsioids, and anthropoids (monkeys, apes, and humans). The first primates were tiny shrewlike animals. In the lineage leading to humans, there was a trend toward better daytime vision, upright walking, more refined hand movements, smaller teeth, bigger brains, and increased parental care.

Section 23.11 The hominoid branch of the primate family tree evolved in Africa as the climate became cooler and drier. It includes apes and the hominids, humans and their most recent ancestors. Australopiths were early hominids, and they walked upright. The relationships among the diverse forms are sketchy. The first known members of the genus *Homo* lived in Africa between 2.4 million and 1.6 million years ago.

Section 23.12 Some *H. erectus* populations moved from Africa into Europe and Asia. Neandertal were early humans. DNA evidence suggests that they did not contribute to the gene pool of modern humans. *H. sapiens* had evolved by 160,000 years ago. Exactly where different populations of modern humans arose is unresolved.

Self-Quiz
Answers in Appendix III

1. All chordates have _____ .
   a. a backbone   c. jaws
   b. a notochord   d. both b and c

2. Vertebrate gills function in _____ .
   a. respiration   d. water regulation
   b. blood circulation   e. both a and c
   c. food trapping

3. Amphibians are descended from _____ .
   a. bony fishes   c. cartilaginous fishes
   b. lizards   d. lobe-finned fishes

4. Which of the following is an amniote?
   a. a shark   c. a turtle
   b. a toad   d. a sea squirt

5. Reptiles are adapted to life on land by _____ .
   a. tough skin   d. amniote eggs
   b. internal fertilization   e. none of the above
   c. good kidneys   f. all of the above

6. _____ are the closest living relatives of birds.
   a. Crocodilians   c. Prosimians
   b. Tuataras   d. Lizards

7. Only birds have _____ .
   a. a cloaca   c. feathers
   b. a four-chambered heart d. all of the above

8. A chimpanzee is a _____ .
   a. craniate   d. amniote
   b. vertebrate   e. placental mammal
   c. hominoid   f. all of the above

9. *Homo erectus* _____ .
   a. was the earliest member of the genus *Homo*
   b. was one of the australopiths
   c. evolved in Africa and spread to diverse regions
   d. was the first to make stone tools

10. Match the organisms with the appropriate description.
    _____ fishes   a. first land tetrapods
    _____ amphibians   b. feathered amniotes
    _____ reptiles   c. egg-laying mammals
    _____ birds   d. humans and close relatives
    _____ monotremes   e. cold-blooded amniotes
    _____ marsupials   f. pouched mammals
    _____ hominids   g. most diverse vertebrates

1 cm

Figure 23.29 This sea squirt, *Ciona savignyi*, is native to the seas near Japan, but has invaded waters along the west coast of the United States, including the San Francisco Bay.

Figure 23.30 The duck-billed platypus (*Ornithorhynchus anatinus*) lives in Australia and Tasmania. In addition to other unusual traits, it is the only mammal that produces venom.

Critical Thinking

1. The National Human Genome Research Institute is currently funding a project to sequence the genome of *Ciona savignyi*, a sea squirt (Figure 23.29). What might these studies tell us about the human genome?

2. In 1798, a stuffed platypus specimen was delivered to the British Museum. Many were sure it was a fake. Soft brown fur and beaverlike tail put the animal firmly in the mammalian camp. But a ducklike bill and webbed feet suggested an affinity with birds (Figure 23.30). Reports that the animal laid eggs only added to the confusion.

We now know that platypuses burrow in riverbanks and forage for prey under water. The webbing on their feet can be retracted to reveal claws. The highly sensitive bill allows the animal to detect prey even with its eyes and ears tightly shut.

To modern biologists, a platypus is clearly a mammal. Like other mammals, it has fur and the females produce milk. Young animals have more typical mammalian teeth that are replaced by hardened pads as the animal matures. Why do you think modern biologists can more easily accept the idea that a mammal can have some reptilelike traits, such as laying eggs? What do they know that gives them an advantage over scientists living in 1798?

3. Reflect on the flight muscles of birds and their demands for oxygen and ATP energy. Which organelle would you expect to be profuse in flight muscles? Why?

4. During the Triassic, hundreds of species of tuataralike reptiles roamed Pangea. Now only two species survive on a few small cold islands off the coast of New Zealand. Speculate on why this relic population has survived there, despite disappearing from the rest of its former range.

5. In early *H. sapiens*, the average braincase volume was about 1,200 cubic centimeters. Today, it averages about 1,400 cubic centimeters. What selective pressure fueled this increase? According to one theory, female preference for clever mates led to selection for increasingly brainier males. Females are thought to have benefited secondarily, because they share the genes that affect intelligence. Do you find this theory plausible? What other selective agents might have favored increased intelligence? What kind of data could researchers gather to support or disprove the sexual selection hypothesis?

Media Menu

Student CD-ROM

Impacts, Issues Video
Interpreting and Misinterpreting the Past
Big Picture Animation
Origin and evolution of the chordates
Read-Me-First Animation
Chordate body plan
The evolution of jaws
Mammalian radiations
Other Animations and Interactions
Vertebrate evolution
Crocodilian body plan
A typical amniote egg

InfoTrac

- High Regard for Sea Squirt's Genome. *BioWorld Week*, December 2002.
- How the West Was Swum. *Natural History*, June 2001.
- Skull of Earliest Human Relative Discovered in African Nation of Chad. *Knight Ridder/Tribune News Service*, July 2002.

Web Sites

- Animal Diversity Web: animaldiversity.ummz.umich.edu
- Introduction to the Metazoa: www.ucmp.berkeley.edu/phyla/phyla.html
- Amphibian Declines and Deformities: www.usgs.gov/amphibians.html
- Cornell Lab of Ornithology: birds.cornell.edu/
- Becoming Human: www.becominghuman.org

How Would You Vote?

Fossils are among the best evidence we have of the evolution of life. Fossils of vertebrates are particularly sought after, rare, and valuable. In response to poaching of such fossils from public lands, some people have proposed that ownership and sale of vertebrate fossils should be banned. Others say that private collectors help ensure preservation of the fossil record. Would you support a ban on private collecting and ownership of vertebrate fossils?

IMPACTS, ISSUES   *Too Hot To Handle*

Heat can kill. In the summer of 2001 Korey Stringer, a football player for the Minnesota Vikings, collapsed after morning practice. His team had been working out in full uniform on a field where temperatures were in the high 90s. High humidity put the heat index above 100°F.

With an internal body temperature of 108.8°F and blood pressure too low to record, Korey was rushed to the hospital. Doctors immersed him in an icewater bath, then wrapped him in cold, wet towels. It was too late.

Korey's blood clotting mechanism shut down and he started to bleed internally. Then his kidneys faltered and he was placed on dialysis. He stopped breathing on his own and was put on a respirator. His heart gave up.

Less than twenty-four hours after football practice had started, Korey Stringer was pronounced dead. He was twenty-seven years old.

All organisms function best within a limited range of internal operating conditions. For humans, "best" is when the body's internal temperature remains between about 97°F and 100°F. Past 104°F, blood is transporting a great deal of metabolically generated heat. Transport controls divert heat from the brain and other internal organs to the skin (Figure 24.1). Skin transfers heat to the outside environment—as long as it is not too hot outside. Profuse sweating can dissipate more heat, but not on hot, humid days.

Figure 24.1   Temperature control mechanisms that have worked for millions of years. *Left*, scanning electron micrograph of a sweat gland pore at the skin surface. When water in sweat moves out through such pores, it helps dissipate excess body heat. Blood circulating through the body reaches fine capillaries in skin (*right*). When skin tissues are not as warm as blood, heat is transferred into them, then into the air— if the air is cooler still. In Korey Stringer's case, cooling could not happen fast enough.

the big picture

Multicellular Organization   Each cell of a multicelled organism engages in activities that help assure its own survival. The structural and functional organization of all the cells collectively sustains the whole body.

Recurring Challenges   Plants and animals both require mechanisms of gas exchange, nutrition, internal transport, salt–water balance between the body and the external environment, and defense.

These normal cooling mechanisms fail above 105°F. The body cannot sweat as much, and its temperature starts to climb rapidly. The heart beats faster and the individual becomes confused or faints. These are signs of *heat stroke*. The high internal temperature that causes it can denature the enzymes and other proteins that keep us alive. When heat stroke is not countered fast enough, brain damage or death is the expected outcome.

We use this sobering example as our passport to the world of anatomy and physiology. *Anatomy* is the study of body form—that is, its morphology. *Physiology* is the study of patterns and processes by which an individual survives and reproduces in the environment. It deals with how structures are put to use and how metabolism and behaviors are adjusted when conditions change. It also deals with how, and to what extent, physiological processes can be controlled.

You can use a lot of this information to interpret what goes on in your own body. More broadly, you can use it to perceive commonalities among all animals and plants. Regardless of the species, *the structure of a given body part almost always has something to do with a present or past function*. Most aspects of form and function are long-standing adaptations that evolved as responses to environmental challenges.

That said, nothing in the evolutionary history of the human species suggests that the organ systems making up our body are finely tuned to handle intense football practice on hot, humid days.

How Would You Vote?

Heat sickness can affect a person's ability to think clearly. Should a coach who doesn't stop team practice or a game when the heat index soars be held responsible if a player dies from heat stroke? See the Media Menu for details, then vote online.

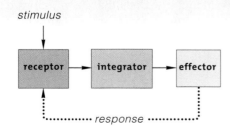

Homeostasis   In plants as well as in animals, cells, tissues, and organs interact in maintaining the body's internal operating conditions. Feedback control mechanisms guide much of this activity.

Cell-to-Cell Communication   Signaling mechanisms control and integrate body activities. Cells send messages that other cells receive and transduce, then change activities in response.

403

24.1 Levels of Structural Organization

We introduce important concepts in this chapter. They have broad application across the next two units of the book. Become familiar with them; they can deepen your sense of how plants and animals function under environmental conditions, both favorable and stressful.

FROM CELLS TO MULTICELLED ORGANISMS

Most plants and animals have cells, tissues, organs, and organ systems that split up the task of survival. A separate cell lineage gives rise to body parts that will function in reproduction. Said another way, the plant or animal body shows a division of labor.

A **tissue** is a community of cells and intercellular substances that are interacting in one or more tasks. For example, wood and bone are tissues that function in structural support. An **organ** has at least two tissues that are organized in certain proportions and patterns and that perform one or more common tasks. A leaf adapted for photosynthesis and an eye that responds to light in the surroundings are examples. An **organ system** has two or more organs interacting physically, chemically, or both in the performance of one or more common tasks. A plant's shoot system, with organs of photosynthesis and reproduction, is like this (Figure 24.2). So is an animal's digestive system, which takes in food, breaks it up into bits of nutrients, absorbs the bits, and expels the unabsorbed leftovers.

GROWTH VERSUS DEVELOPMENT

A plant or animal body becomes structurally organized during growth and development. Generally, **growth** of a multicelled organism means that its cells increase in number, size, and volume. **Development** is a succession of stages in which specialized tissues, organs, and organ systems form. In other words, we measure growth in *quantitative* terms and development in *qualitative* terms.

STRUCTURAL ORGANIZATION HAS A HISTORY

The structural organization of each tissue, organ, and organ system has an evolutionary history. Think back on how plants invaded land and you have some idea of how their structure relates to function. Plants that left aquatic habitats found an abundance of sunlight and carbon dioxide for photosynthesis, oxygen for aerobic respiration, and mineral ions. As they dispersed farther away from their aquatic cradle, however, they faced a new challenge—how to keep from drying out in air.

Remember that challenge when micrographs show the internal structure of plant roots, stems, and leaves (Figure 24.2). Pipelines conduct streams of water from soil to leaves. Stomata, the small gaps across a leaf's epidermis, open and close in ways that help conserve water (Section 26.4). Cells that have lignin-reinforced walls collectively support upright growth of stems. Remember that same challenge when you come across

shoot system (aboveground parts)

root system (belowground parts, mostly)

reproductive organ (flower)

stem tissues, cross-section, for support, storage, water distribution, food distribution

root water-conducting cells, longitudinal section

Figure 24.2 Morphology of a tomato plant (*Solanum lycopersicon*). Different cell types make up vascular tissues that conduct water, dissolved mineral ions, and organic compounds. These tissues thread through others that make up most of the plant body. Another tissue covers all surfaces exposed to the surroundings.

Figure 24.3 Some of the structures that function in human respiration. Their cells carry out specialized tasks. Airways to a pair of organs called lungs are lined with ciliated cells that whisk away bacteria and other airborne particles that might cause infections. Inside the lungs are tubes (blood capillaries), a fluid connective tissue called blood, and thin air sacs (epithelial tissue), all with roles in gas exchange.

ciliated cells and mucus-secreting cells that line respiratory airways

lung tissue (tiny air sacs) laced with blood capillaries—one-cell-thick tubular structures that hold blood, which is a fluid connective tissue

organs (lungs), part of an organ system (the respiratory tract) of a whole organism

examples of root systems. Any soil region with plenty of water and dissolved mineral ions stimulates root growth in its direction; hence the branching roots.

Similarly, the respiratory systems of land animals are adaptations to life in air. Gases can only move into and out of the animal body by diffusing across a moist surface. This is not a problem for organisms in water, but moist surfaces dry out in air. Land animals have moist sacs for gas exchange *inside* their body (Figure 24.3), which brings us to the body's own environment.

THE BODY'S INTERNAL ENVIRONMENT

To stay alive, plant and animal cells must be bathed in a fluid that delivers nutrients and carries away metabolic wastes. In this they are no different from free-living single cells. But each plant or animal has thousands to many trillions of living cells that must draw nutrients from the fluid bathing them and dump wastes into it.

Body fluid *not* inside cells—extracellular fluid—is an **internal environment**. Changes in its composition and volume affect cell activities. The type and number of ions are vital; they must be kept at concentrations compatible with metabolism. It makes no difference whether the plant or animal is simple or complex. *It requires a stable fluid environment for all of its living cells.* This concept is central to understanding how plants and animals work.

START THINKING "HOMEOSTASIS"

The next two units describe how a plant or an animal carries out the following functions: The body works to maintain favorable conditions for all of its living cells. It acquires water, nutrients, and other raw materials, distributes them throughout the body, and disposes of wastes. It actively as well as passively defends itself against attack. It has the capacity to reproduce. Parts of the body nourish and protect gametes, and often the embryonic stages of the next generation.

Each cell engages in metabolic activities that ensure its own survival. Collectively, however, the activities of cells in tissues, organs, and organ systems sustain the body as a whole. Their interactions keep operating conditions of the internal environment within tolerable limits, a state we call **homeostasis**.

The structural organization of plants and animals emerges during growth and development.

Cells, tissues, and organs require a favorable internal environment, and they collectively maintain it. All body fluids not contained in cells make up that environment.

Acquiring diverse substances and distributing them to cells, disposing of wastes, protecting cells and tissues, reproducing, and often nurturing offspring are basic body functions.

24.2 Recurring Challenges to Survival

Plants and animals have such diverse body plans that we sometimes forget how much they have in common.

GAS EXCHANGE IN LARGE BODIES

How often do you think of similarities between, say, Tina Turner and a tulip (Figure 24.4)? Connections are there. Consider: Cells in Tina or the tulip would die if oxygen and carbon dioxide stopped diffusing across their surface. Like most other heterotrophs, Tina supplies her aerobically respiring cells with a lot of oxygen and disposes of carbon dioxide products. Like most other autotrophs, the plant secures carbon dioxide for its photosynthetic cells. At the same time, it retains enough oxygen for *its* aerobically respiring cells, and disposes of any excess.

All multicelled species respond, structurally and functionally, to the same challenge. *They must quickly move gaseous molecules to and from individual cells.*

Remember **diffusion**? It is a net movement of ions or molecules of a substance from a region where they are concentrated to an adjacent region where they are less concentrated. Animals and plants have ways to keep gases diffusing in directions most suitable for metabolism and cell survival. How? That question will lead you to the stomata at leaf surfaces (Chapter 26) and to the circulatory and respiratory systems of animals (Chapters 33 and 35).

INTERNAL TRANSPORT IN LARGE BODIES

Metabolic reactions happen fast. If it takes too long for reactants to diffuse through the body or to and from its surface, systems will shut down. This is one reason cells and multicelled species have the sizes and shapes that they do. As they grow, their volume increases in three dimensions; in length, width, and depth. But their surface area increases in two dimensions only. That is the essence of the **surface-to-volume ratio**. If a body were to develop into a solid, dense mass, it would not have enough surface area for fast, efficient exchanges with the environment (Section 4.1).

When a body or some body part is thin, as it is for the flatworm and lily pads in Figure 24.5, substances can easily diffuse between individual cells and the environment. In massive bodies, individual cells that are far from an exchange point with the environment depend on systems of rapid internal transport.

Most plants and animals have vascular tissues, or systems of tubes that move substances to and from cells. In most plants, xylem distributes soil water and mineral ions to all plant parts, and phloem distributes sugars from leaves. Each leaf vein has long strands of xylem and phloem (Figure 24.5c).

In large animals, vascular tissues extend from a surface facing the environment to living cells. Each time Tina belts out a song, she has to move oxygen into her lungs and carbon dioxide out of them. Blood vessels thread through her lung tissue where gases are exchanged. Other vessels transport oxygen to other body regions and branch into tiny capillaries, where interstitial fluid and cells exchange gases (Figure 24.5d).

In both plants and animals, the vascular system also transports nutrients and water. It also functions in the transport of chemical messages called hormones and in internal defense. In animals, vascular tissues transport infection-fighting white blood cells and assorted chemical weapons. In many plants, a wound triggers synthesis of defensive chemicals that are distributed throughout the plant by way of phloem.

MAINTAINING THE WATER–SOLUTE BALANCE

Plants and animals continually gain and lose water and solutes. They routinely produce metabolic waste. Given all the inputs and outputs, how are the volume and composition of their internal environment kept within a tolerable range? Plants and animals differ hugely in this respect. Yet we still can find common responses among them, when we zoom down to the level of molecular movement.

Figure 24.4 What do these organisms have in common besides their good looks?

Figure 24.5 (**a**) A flatworm gliding along in water, (**b**) waterlily leaves floating on water, (**c**) veins visible in a decaying dicot leaf, and (**d**) human veins and capillaries. These body plans and structures are responses to the same constraint. What is it?

Substances tend to follow concentration gradients when moving into and out of the body or from one body compartment into another. At such interfaces, the cells of sheetlike tissues passively and actively transport substances (Section 5.6). **Active transport**, recall, is a pumping of substances *against* gradients.

Active transport mechanisms in roots help control which solutes move into the plant. In leaves, they help control water loss and gas exchange by closing and opening stomata at different times. In animals, such mechanisms occur in kidneys and other organs. As you will see again and again in the next two units, *active and passive transport help maintain the internal environment and metabolism itself by adjusting the kinds, amounts, and directional movements of substances.*

CELL-TO-CELL COMMUNICATION

Plants and animals show another major similarity in their structure and function. A number of their cells release signaling molecules that coordinate and also control events inside the body as a whole. Different signaling mechanisms guide how the plant or animal body grows, develops, and maintains and protects itself. Section 24.5 has a good example of this.

ON VARIATIONS IN RESOURCES AND THREATS

A **habitat** is the place where individuals of a species normally live. Each has different resources and poses a unique set of challenges. What are its physical and chemical characteristics? Is water plentiful, with the right kinds and amounts of solutes? Is the habitat rich or poor in nutrients? Is it sunlit, shady, or dark? Is it warm or cool, hot or icy, windy or calm? How much does the outside temperature vary from day to night? How much do conditions vary with the seasons?

And what about biotic (living) components of the habitat? Which producers, predators, prey, pathogens, and parasites live there? Is competition for resources and reproductive partners fierce? Such ever changing variables promote diversity in form and function.

Even with all that diversity, however, we still may observe similar responses to similar challenges. Sharp cactus spines or sharp porcupine quills help deter most of the other animals that might eat the cactus or porcupine. Both develop from specialized epidermal cells. Vascular tissues and other body parts helped nurture those cells. And by contributing to building spines and other defensive structures at the body's surface, the epidermal cells help protect the body as a whole against environmental threats.

Plant and animal cells function in ways that help ensure survival of the body as a whole. At the same time, tissues and organs that make up the body function in ways that allow the continued survival of individual living cells.

The connection between each cell and the body as a whole is evident in the requirements for—and contributions to—gas exchange, nutrition, internal transport, stability in the internal environment, and defense.

24.3 Homeostasis in Animals

In preparation for your trek through the next two units, focus here on what homeostasis means to survival.

Like other adult humans, your body has more than 65 trillion living cells. Each must draw nutrients from and dump wastes into the same fifteen liters of fluid (less than four gallons). The fluid not inside cells is extracellular fluid. Much is **interstitial fluid**, meaning it fills the spaces between cells and tissues. The rest is **plasma**, the fluid portion of blood. Interstitial fluid exchanges substances with cells and with blood.

Homeostasis, again, is a state in which the body's internal environment is being maintained within a range that cells can tolerate. In nearly all animals, sensory receptors, integrators, and effectors interact in maintaining this state. They detect, process, and respond to information on *how things are* compared to preset points of *where they are supposed to be.*

Sensory receptors are cells or cell parts that detect stimuli, which are specific forms of energy. A kiss, for example, is a form of mechanical energy that changes pressure on the lips. Receptors in lip tissues translate the stimulus into signals that flow to the brain. The brain is a major **integrator**, a central command post that receives and processes information about stimuli. It also issues signals to the body's **effectors**—muscles, glands, or both—that carry out suitable responses.

NEGATIVE FEEDBACK

Feedback mechanisms are key homeostatic controls. By **negative feedback mechanisms**, an activity changes some condition, and when the condition shifts past a

Figure 24.7 A climber near the summit of Mount Everest, where the normal act of breathing isn't enough.

certain point, it triggers a response that reverses the change (Figure 24.6).

Think of a furnace with a thermostat. A thermostat senses the surrounding air temperature relative to a preset point on a thermometer built into the furnace's control system. When the temperature falls below the preset point, it signals a switching mechanism, which turns on the furnace. When the air heats enough and matches the preset level, the thermostat senses the match and signals the switch to shut off the furnace.

Now think about how we breathe to take in oxygen and get rid of carbon dioxide. Most of us live at low elevations, where 1 in 5 molecules is oxygen. There, the brain pays attention to receptors that are sensitive to how much carbon dioxide is dissolved in blood. It compares receptor signals against a set point for what the carbon dioxide level is supposed to be, and calls for tweaks in breathing and other activities that can put the carbon dioxide level in line with the set point.

Above 3,300 meters (10,000 feet), oxygen is scarce (Figure 24.7). Its level in blood plummets. Receptors sensitive to this steep decline signal the brain, which responds by making us hyperventilate (breathe faster and far more deeply than usual). If the level stays low, shortness of breath, heart palpitations, headaches, nausea, and vomiting will follow. All are last-ditch warnings that cells are screaming for oxygen.

Read Me First!
and watch the narrated animation on negative feedback

STIMULUS input into the system

receptor	integrator	effector
(such as a free nerve ending in the skin)	(such as the brain or spinal cord)	(a muscle or gland)

RESPONSE to stimulus causes change. The change is "fed back" to receptor. In *negative* feedback, the system's response cancels or counters the effect of the original stimulus.

Figure 24.6 Three components of negative feedback systems in multicelled organisms. Figure 24.7 gives a specific example.

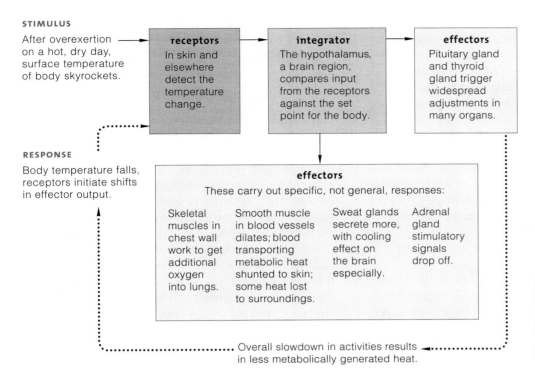

STIMULUS

After overexertion on a hot, dry day, surface temperature of body skyrockets.

receptors
In skin and elsewhere detect the temperature change.

integrator
The hypothalamus, a brain region, compares input from the receptors against the set point for the body.

effectors
Pituitary gland and thyroid gland trigger widespread adjustments in many organs.

RESPONSE

Body temperature falls, receptors initiate shifts in effector output.

effectors

These carry out specific, not general, responses:

Skeletal muscles in chest wall work to get additional oxygen into lungs.

Smooth muscle in blood vessels dilates; blood transporting metabolic heat shunted to skin; some heat lost to surroundings.

Sweat glands secrete more, with cooling effect on the brain especially.

Adrenal gland stimulatory signals drop off.

Overall slowdown in activities results in less metabolically generated heat.

Figure 24.8 Major homeostatic controls over a human body's internal temperature. The *solid* arrows signify the main control pathways. The *dotted* arrows signify the feedback loop. In Korey Stringer's case, environmental conditions overwhelmed the control pathways.

Similarly, feedback mechanisms work to keep the internal body temperature of humans and many other mammals near 98.6°F (37°C) even during hot or cold weather. When a football player runs around on a hot summer day, the body becomes hot. Receptors trigger changes that should slow down the entire body *and* its cells (Figure 24.8). Under typical conditions, these and other control mechanisms counter overheating. They curb activities that naturally generate metabolic heat, and give up excess heat to the surrounding air.

POSITIVE FEEDBACK

In some situations, **positive feedback mechanisms** operate. These controls initiate a chain of events that *intensify* change from an original condition, and after a limited time, the intensification reverses the change. Positive feedback mechanisms are usually associated with instability in a system.

For instance, rapid flows of sodium ions across the plasma membrane of neurons help the body send and receive signals quickly. A suitable signal opens a few sodium channels across a neuron's membrane. Ions flow into the neuron, and the increased concentration inside makes more channels open, then more, until sodium ions flood inside. Such ion flows start the "messages" that travel along a neuron's surface.

As another example, during labor, the fetus exerts pressure on the wall of the chamber enclosing it, the uterus. Pressure induces the production and secretion of a hormone, oxytocin, that makes muscle cells in the wall contract. Contractions exert pressure on the fetus, which puts more pressure on the wall, and so on until the fetus is expelled from the mother's body.

We have been introducing a pattern of detecting, evaluating, and responding to a flow of information about the internal and external environment. During this activity, organ systems work together. Soon, you will be asking these questions about their operation:

1. Which physical or chemical aspects of the internal environment are organ systems working to maintain?

2. How are organ systems kept informed of change?

3. By what means do they process the information?

4. What mechanisms are set in motion in response?

As you will read in Unit Six, organ systems of nearly all animals are under neural and endocrine control.

Homeostatic controls, such as negative and positive feedback mechanisms, help maintain physical and chemical aspects of the body's internal environment within ranges that its individual cells can tolerate.

24.4 Does Homeostasis Occur in Plants?

Plants differ from animals in important respects. Even so, they, too, have controls over their internal environment.

Direct comparisons between plants and animals are not always possible. In young plants, new tissues arise only at the tips of actively growing roots and shoots. In animal embryos, tissues form all through the body. Plants do not have a centralized integrator that serves as the equivalent of an animal brain. They do have some decentralized mechanisms that protect the internal environment and work to keep the body as a whole functioning. Later chapters explain how, but two simple examples here will make the point.

WALLING OFF THREATS

Unlike people, trees consist mostly of dead and dying cells. Also unlike people, trees cannot run away from attacks. When a pathogen infiltrates their tissues, trees cannot unleash infection-fighting phagocytic cells in response, because they have none. Some trees live in habitats that are too harsh or remote to favor most pathogens, so they have been able to grow, slowly, for thousands of years. Remember those bristlecone pines in Section 21.6? They spend most of the year in snow and the rest under intense radiation from the sun.

Most trees protect their internal environment by building thickened cell walls around wounds. They unleash phenols and other toxic compounds. Many secrete resins. The heavy flow of gooey compounds saturates and protects bark and wood at an attack site. It also can seep into soil around roots.

Some toxins are so potent that they also kill cells of the tree itself. As a result, compartments form around injured, infected, or poisoned sites, and new tissues grow right over them. This plant response to attack is called **compartmentalization**.

Drill holes in a tree species that makes a strong compartmentalization response and it quickly walls off that wound (Figure 24.9). In a species that makes a moderate response, decomposers that cause wood to decay expand farther, into regions more distant from the holes. Drill holes in a weak compartmentalizer, and decomposers will cause massive decay.

Even strong compartmentalizers live only so long. When they form too many walls, they shut off the flow of water and solutes to living cells.

SAND, WIND, AND THE YELLOW BUSH LUPINE

Anyone who has tiptoed barefoot across sand near the coast on a hot, dry day has a tangible clue to why few plants grow in it. One of the exceptions is the yellow bush lupine, *Lupinus arboreus* (Figure 24.10).

L. arboreus is native to warm, dry areas of Central and Southern California. It is a hardy colonizer of soil exposed by fires or abandoned after being cleared for agriculture. Like all other legumes, this species has nitrogen-fixing symbionts in its young roots. The mutualistic interaction gives it a competitive edge in nitrogen-deficient soil. In the early 1900s, this species was planted in northern California to stabilize coastal dunes. Unfortunately, it is too successful in its new habitat. It outgrows and displaces native plants.

A big environmental challenge near the beach is the lack of fresh water. Leaves of a yellow bush lupine

strong

moderate

weak

Figure 24.9 You think it's easy being a tree? Drilling patterns for an experiment to test compartmentalization responses. From top to bottom, decay patterns (*green*) in stems of three tree species that make strong, moderate, or weak compartmentalizing responses.

Figure 24.10 Yellow bush lupine, *Lupinus arboreus*, in a sandy shore habitat. On hot, windy days, its leaflets fold up longitudinally along the crease that runs down their center. This helps minimize evaporative water loss.

1 A.M.   6 A.M.   NOON   3 P.M.   10 P.M.   MIDNIGHT

are structurally adapted for water conservation. Each leaf has a surprisingly thin cuticle, but a dense array of fine epidermal hairs projects above it, especially on lower leaf surfaces. Collectively, the hairs will trap moisture escaping from stomata. Trapped, moist air slows evaporation and helps maintain water levels inside the leaf at optimal levels for metabolism.

These leaves make homeostatic responses to the environment. They fold along their length, like two parts of a clam shell, and resist the moisture-sucking force of the wind. Each folded, hairy leaf is better at stopping moisture loss from stomata (Figure 24.10).

Leaf folding by *L. arboreus* is a controlled response to changing conditions. When winds are strong and potential for water loss is greatest, these leaves fold tightly. The least-folded leaves are near the plant's center or on the side most sheltered from the wind. Folding is a response to heat as well as to wind. When air temperature is highest during the day, leaves fold at an angle that helps reflect the sun's rays from their surface. The response minimizes heat absorption.

ABOUT RHYTHMIC LEAF FOLDING

In case you think leaf folding couldn't possibly be a coordinated response, take a look at the bean plant in Figure 24.11. Like some other plants, it holds its leaves

Figure 24.11 Observational test of the rhythmic movements of leaves of a young bean plant (*Phaseolus*). The investigator, Frank Salisbury, kept this plant in complete darkness for twenty-four hours. Its leaves continued to fold and unfold independently of sunrise (6 A.M.) and sunset (6 P.M.).

horizontally during the day but folds them closer to a stem at night. Keep the plant in full sun or darkness for a few days and it will continue to move its leaves in and out of the "sleep" position, independently of sunrise and sunset. The response might help reduce heat loss at night, when air cools, and so maintain the plant's internal temperature within tolerable limits.

Rhythmic leaf movements are just one example of a **circadian rhythm**, a biological activity repeated in cycles that each last for close to twenty-four hours. Circadian means "about a day." As you will see in a later chapter, a pigment molecule called phytochrome may be part of a control mechanism over leaf folding.

Homeostatic mechanisms are at work in plants, although they are not governed from central command posts as they are in most animals.

Compartmentalization and rhythmic leaf movements are two examples of responses to specific environmental challenges.

24.5 How Cells Receive and Respond to Signals

Signal reception, transduction, and response—this is a fancy way of saying cells chatter among themselves in ways that bring about changes in their activities.

Reflect on Section 4.11, the overview of how adjoining cells communicate as by plasmodesmata in plants and gap junctions in animals. Also think back on how free-living *Dictyostelium* cells issue signals that induce them to converge and make a spore-bearing structure. They do so in response to dwindling supplies of food, an environmental cue for change (Figure 20.19).

In large multicelled organisms, one cell type signals others in response to cues from both the internal and external environment. Local and long-distance signals lead to local and regional changes in metabolism, gene expression, growth, and development.

The molecular mechanisms by which cells "talk" to one another evolved early in the history of life. They often involve three events. *First*, a specific receptor is activated, perhaps by the reversible binding of some signaling molecule. *Second*, the signal is transduced— it is changed into a form that can operate inside the cell. *Third*, the cell makes the functional response.

Most receptors are membrane proteins of the sort shown in Section 4.3. When activated, the receptor starts signal transduction. It might activate an enzyme that in turn activates many molecules of a different enzyme, which activates many molecules of another kind, and so on. These chains of cascading reactions inside the cell greatly amplify the original signal.

In the next two units, you will come across diverse cases of signal reception, transduction, and response. For now, consider this example to get a sense of the kinds of events that signals set in motion.

The very first cell of a new multicelled individual holds marching orders that will guide its descendants through growth, development, and reproduction, and often death. As part of that program, many cells heed calls to self-destruct when they finish their prescribed function. **Apoptosis** is the process of programmed cell death. It begins with signals that unleash enzymes of self-destruction, which each body cell synthesizes and stockpiles (Figure 24.12).

Proteases are part of the stockpile. They chop apart structural proteins, including the building blocks of cytoskeletal elements and of proteins that keep DNA structurally organized. Nucleases, enzymes that snip up nucleic acids, are also stockpiled.

During apoptosis, a cell shrinks away from other cells in a tissue. Its surface bubbles outward and inward. Its chromosomes bunch up near the nuclear envelope. The nucleus, then the cell, break apart. Phagocytic cells that patrol and protect body tissues engulf dying cells and their remnants. Lysosomes inside the phagocytes digest the engulfed bits, which are recycled.

Cells were busily committing suicide when your own hands were developing. The human hand starts out as a paddlelike structure (Figure 24.13a). When development proceeds normally, apoptosis causes the

Signal to die docks at receptor.

Signal leads to activation of protein-destroying enzymes.

Figure 24.12 Artist's depiction of a suicidal cell. Normally, body cells self-destruct when they finish their functions or become infected or cancerous, which could threaten the body as a whole.

Figure 24.13 Formation of human fingers with the help of apoptosis. (**a**) Forty-eight days after fertilization, webs of tissue connect embryonic digits. (**b**) Three days later, after cells between the digits commit suicide, there are no more webs. (**c**) Digits stay attached when the cells do not commit suicide on cue.

paddle to split apart into individual fingers in a few days (Figure 24.13*b*).

While embryos develop, the timing of cell death is usually predictable. As one example, keratinocytes are skin cells that live about three weeks, then die and form protective layers that are continually sloughed off from the surface of skin. These cells can be induced to die ahead of schedule. So can other cells that are supposed to last a lifetime. All it takes is sensitivity to molecular signals that can call up the enzymes of death.

Control genes suppress or trigger programmed cell death. One gene, *bcl-2*, helps keep normal body cells from dying before their time. In some cancers, this gene has mutated and cells do not respond to signals to die. Apoptosis is no longer a control option.

What about plants? The controlled death of xylem cells creates pipelines that carry water throughout the plant. Also, when a tissue is under attack, signals may bring about the controlled death of nearby cells in a pattern that walls off the threat.

Plant and animal cells communicate with one another by secreting signaling molecules into extracellular fluid and selectively responding to signals from other cells.

Communication involves receiving signals, transducing them, and bringing about change in a target cell's activity.

Signal transduction requires membrane receptors and other membrane proteins. It often involves enzymes that induce a cascade of reactions that amplify the initial signal.

Summary

Section 24.1 Anatomy is the study of body form. Physiology is the study of how the body functions in the environment. The structure of most body parts correlates with current or past functions.

A plant or animal's structural organization emerges during growth and development. Each living cell performs metabolic functions that keep it alive. Cells are organized in tissues, organs, and often organ systems. These structural units function in coordinated ways in the performance of specific tasks.

Section 24.2 Plants and animals have responded in similar ways to certain environmental challenges. All exchange gases, transport substances to and from cells, and maintain water and solute concentrations within the body at tolerable limits. All have mechanisms for integrating and controlling body parts in ways that ensure survival of the whole organism. They also have mechanisms for responding to signals from other cells and to signals or cues from the outside environment.

Section 24.3 Homeostasis is a stable state in which the body's internal environment is being maintained within a range that its cells can tolerate. In animal cells, sensory receptors, integrators, and effectors interact to maintain tolerable conditions, often by feedback loops.

With negative feedback mechanisms, a change in some condition, such as body temperature, triggers a response that results in reversal of the change. With positive feedback mechanisms, a change leads to a response that intensifies the condition that caused it.

Section 24.4 Plants have decentralized mechanisms that work to maintain the internal environment and ensure their survival. Some trees wall off threats by secreting resins and toxins, a defensive response called compartmentalization. Some plants respond to changes in environmental conditions by folding their leaves. Rhythmic leaf folding is a circadian rhythm that may help a plant maintain its internal temperature.

Section 24.5 Cell-to-cell communication involves signal reception, signal transduction, and a response by a target cell. Many signals are transduced by membrane proteins that trigger reactions in the cell. Reactions may alter the activity of genes or enzymes that take part in cellular events. An example is a signal that unleashes the protein-cleaving enzymes of apoptosis; a target cell self-destructs.

Self-Quiz   *Answers in Appendix III*

1. An increase in the number, size, and volume of plant cells or animal cells is called _____ .
   a. growth   c. differentiation
   b. development   d. all of the above

Figure 24.14 (**a**) Consuelo De Moraes studies plant stress responses. (**b,c**) A caterpillar chewing on tobacco causes the plant to secrete chemicals that attract a wasp. (**d**) The wasp grabs the caterpillar and lays an egg inside it.

2. The internal environment consists of _____ .
   a. all body fluids   c. all body fluids outside cells
   b. all fluids in cells   d. interstitial fluid

3. _____ influences the concentrations of water and solutes in the internal environment.
   a. Diffusion   c. Passive transport
   b. Active transport   d. all are correct

4. Cell communication typically involves signal _____ .
   a. reception   c. response
   b. transduction   d. all are correct

5. Match the terms with their most suitable description.
   ___ circadian rhythm   a. programmed cell death
   ___ homeostasis   b. 24-hour or so cyclic activity
   ___ apoptosis   c. stable internal environment
   ___ negative   d. an activity changes some
         feedback   condition, then the change
                         triggers its own reversal

Critical Thinking

1. Many plants protect themselves with thorns or toxins or nasty-tasting chemicals that deter plant-eating animals. Some get help from wasps.

Consuelo De Moraes, currently at Pennsylvania State University, studies interactions among plants, caterpillars, and parasitoid wasps. *Parasitoids* are a special class of parasites; their larvae eat a host from the inside out.

When a caterpillar chews on a tobacco plant leaf, it secretes a lot of saliva. Some chemicals in its saliva are an external signal that triggers a chemical response from leaf cells. The cells release certain molecules that diffuse through the air. Parasitoid wasps follow the concentration gradient to the stressed leaves. They attack a caterpillar, and each deposits one egg inside it. The eggs grow and develop into caterpillar-munching larvae (Figure 24.14).

As De Moraes discovered, plant responses are highly specific. Leaf cells release different chemicals in response to different caterpillar species. Each chemical attracts only the wasps that parasitize the particular kind of caterpillar that triggers the chemical's release.

Are the plants "calling for help"? Not likely. Give a possible explanation for this plant–wasp interaction in terms of cause, effect, and natural selection theory.

2. The Arabian oryx (*Oryx leucoryx*), an endangered antelope, evolved in the harsh deserts of the Middle East. Most of the year there is no free water, and temperatures routinely reach 117°F (47°C). The most common tree in the region is the umbrella thorn tree (*Acacia tortilis*). List common challenges that the oryx and the acacia face. Also research and report on the morphological, physiological, and behavioral responses of both organisms.

3. In the summer of 2003, a record heat wave lasted for weeks and caused the death of more than 5,000 people in France. The elderly and very young were at greatest risk. High humidity increases the chance of heat-related illness because it reduces the rate of evaporative cooling. So does tight clothing that doesn't "breathe," like the uniform Korey Stringer wore during his last practice. Other risk factors are obesity, poor circulation, dehydration, and alcohol intake. Using Figure 24.8 as a reference, briefly suggest how each factor may interfere with homeostatic controls over the body's internal temperature.

V How Plants Work

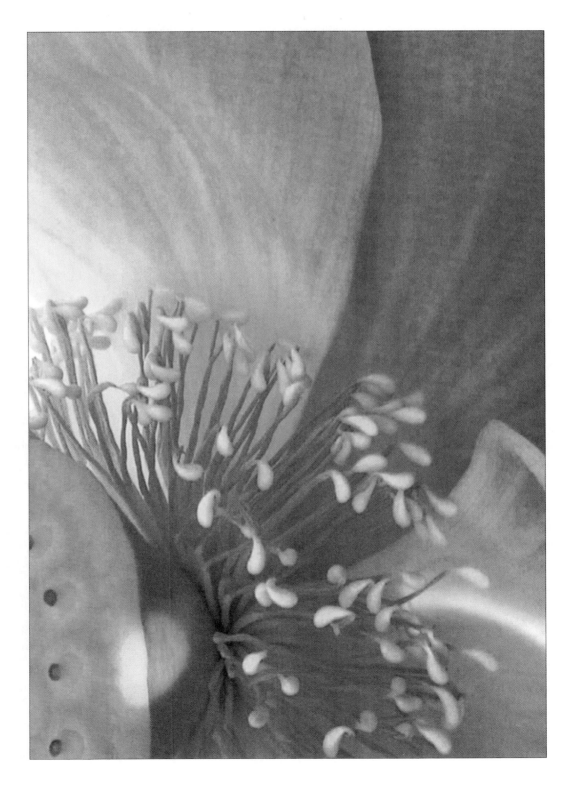

The sacred lotus, Nelumbo nucifera, busily doing what its ancestors did for well over 100 million years—flowering spectacularly during the reproductive phase of its life cycle.

IMPACTS, ISSUES *Droughts Versus Civilization*

The more we dig up records of past climates, the more we wonder about what's happening now. In any given year there are bad *droughts*, or significantly less rainfall than we expect in a given region. Drought is a normal climatic cycle, but it's been worse than usual in North America, Australia, Africa, and elsewhere—enough to cause mass starvation, cripple economies, even to make hungry North Koreans rattle their nuclear weapons. And what is the long-term forecast? More of the same.

Humans built the whole of modern civilization on a vast agricultural base. Today we reel from droughts that last two, five, seven years or so. Imagine a drought lasting 200 *years*. It happened. About 3,400 years ago, rainfall dried up and brought an end to the Akkadian civilization in northern Mesopotamia. Samples taken from the seafloor, lake bottoms, and ice cores point to a worldwide change in climate that also brought an end to societies elsewhere.

Look beyond the architectural marvels of the Maya civilization (Figure 25.1). A drought that lasted more than 150 years contributed to its collapse in the ninth century. Or look to Afghanistan, recently scorched by

Figure 25.1 Why study plants? We depend absolutely on the adaptations by which they acquire and use resources from the environment. In the absence of any of those resources, including water, communities fall apart. *Left*, reminder of the failed Maya civilization. *Above*, from modern-day Guatemala, an atrophied corn cob, one outcome of a drought that caused widespread crop failures.

the big picture

Basic Body Plans Two major classes of flowering plants, eudicots and monocots, consist of the same tissues arranged in somewhat different ways. A dermal tissue system covers exposed parts of the body, the bulk of which is a ground tissue system through which vascular tissues thread.

Types of Plant Tissues Parenchyma, collenchyma, and sclerenchyma are simple plant tissues consisting of only one of a variety of cell types. Xylem, phloem, epidermis, and other complex plant tissues consist of more than one cell type.

the worst drought in a hundred years. Subsistence farmers make up 70 percent of the population. Six years of droughts wiped out harvests, dried up wells, and killed livestock. In desperation, the rural families sold their land, their possessions, and their daughters. Despite crop imports, starvation has affected about 37 percent of the population. And the world's leading exporters of wheat and other crops cannot do much for others when they themselves experience droughts.

Even brief episodes of drought reduce photosynthesis and crop production. Plants conserve water by closing stomata, which of course stops carbon dioxide from moving in. No carbon dioxide, no sugars. A stressed flowering plant also makes fewer flowers. The flowers it does make might not open fully, so they might not get pollinated. Even if they do, its seeds and fruits may fall off the plant before they ripen.

This unit focuses on seed-bearing vascular plants, with emphasis on flowering types that are intimately interconnected with human life. You will be taking a look at how these plants function and consider their patterns of growth, development, and reproduction. Their structure and physiology help them get through short-term droughts and other hostile times—but not through long-term water deprivation.

This is a huge question mark for your future. Which nations will stumble during long-term climate change? Which ones will make it through a severe drought that doesn't end any time soon?

 How Would You Vote?

Large-scale farms and large cities compete for clean, fresh water, which is becoming scarcer as population growth skyrockets. Should cities restrict urban growth to reduce conflicts over water supplies? See the Media Menu for details, then vote online.

Primary Growth Patterns   Each growing season, tissues of stems, leaves, and other shoots lengthen, and so do roots. The lengthening, or primary growth, originates only at apical meristems in shoot and root tips. Meristems are regions of undifferentiated cells that retain the capacity to divide.

Secondary Growth Patterns   Also during the growing season, tissues of the stems and roots of many plants thicken. This secondary growth originates at lateral meristems inside the stems and roots. When wood forms this way, plants grow taller and often broader, and better able to compete for access to sunlight.

25.1 Overview of the Plant Body

With 260,000 species, flowering plants dominate the plant kingdom, as surveyed in Chapter 21. Its major groups are the magnoliids, eudicots (true dicots), and monocots. We focus here on the eudicots and monocots.

Many flowering plants have a body plan similar to the one in Figure 25.2. Above ground are **shoots**, such as stems, leaves, and reproductive shoots called flowers. Stems are structural frameworks for upright growth, which helps photosynthetic cells harvest energy from the sun. **Roots** are plant structures that typically grow downward and outward through the soil. Roots absorb water and dissolved minerals, and often anchor parts above ground. Their cells store and release food for their own use and often for distribution to the rest of the plant body.

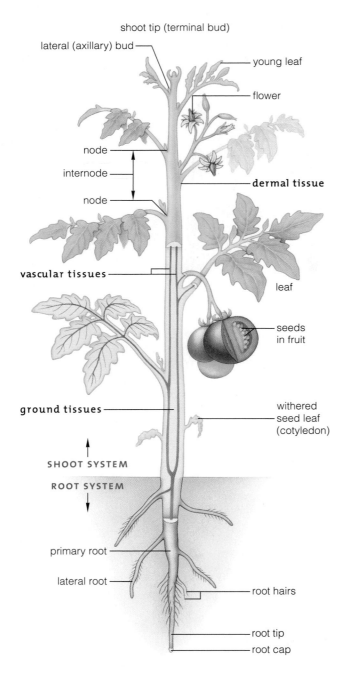

Figure 25.2 legend labels:
shoot tip (terminal bud)
lateral (axillary) bud
young leaf
flower
node
internode
node
dermal tissue
vascular tissues
leaf
seeds in fruit
ground tissues
withered seed leaf (cotyledon)
SHOOT SYSTEM
ROOT SYSTEM
primary root
lateral root
root hairs
root tip
root cap

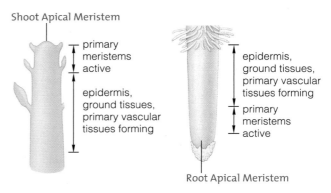

Shoot Apical Meristem

primary meristems active

epidermis, ground tissues, primary vascular tissues forming

epidermis, ground tissues, primary vascular tissues forming

primary meristems active

Root Apical Meristem

a The cellular descendants of *apical meristems* divide, grow, and differentiate. They form three *primary meristems*, the activity of which lengthens shoots and roots:

Protoderm ──────► epidermis
Ground meristem ──────► ground tissues
Procambium ──────► primary vascular tissues

vascular cambium

cork cambium

thickening

Lateral Meristems

b In woody plants, activity of two *lateral meristems* thickens older stems and roots; they result in secondary growth:

Vascular cambium ──────► secondary vascular tissues
Cork cambium ──────► periderm

Figure 25.2 Body plan for the commercially grown tomato plant (*Solanum lycopersicum*). Vascular tissues (*purple*) conduct water, dissolved minerals, and organic substances. They thread through ground tissues, which make up most of the plant body. Epidermis, a type of dermal tissue, covers the root and shoot surfaces.

Figure 25.3 (**a**) Apical meristems, sources of primary growth—that is, a lengthening of shoot and root tissues. (**b**) In many plants, other meristems form in older stems and roots. They are sources of secondary growth—a thickening of tissues that leads to increases in diameter.

THREE PLANT TISSUE SYSTEMS

Plant stems, branches, leaves, and roots have the same three tissue systems (Figure 25.2). The *ground tissue* system has basic functions, such as food storage. The *vascular tissue* system threads through ground tissue and delivers water and solutes through the plant. The *dermal tissue* system protects all exposed surfaces.

Parenchyma, collenchyma, and sclerenchyma are the simplest tissues in these systems; they are made of one type of cell. Xylem, phloem, and dermal tissues are complex, with organized arrays of two or more types.

Most plant tissues originate at **meristems**, each a localized region of dividing cells. In the new tissues in back of a growing meristem, cellular descendants are maturing or are already mature. *Apical* meristems are in shoot and root tips, where all young plant parts start lengthening. Populations of cells that form here are immature forerunners of epidermis, ground tissue, and primary vascular tissues. Taken as a whole, apical meristem activity and the lengthening of shoots and roots are the plant's primary growth (Figure 25.3a).

Also in the growing season, older stems and roots of many plants thicken. Increases in girth start by cell divisions from *lateral* meristems, cylindrical arrays of cells that form in stems and roots (Figure 25.3b).

EUDICOTS AND MONOCOTS—SAME TISSUES, DIFFERENT FEATURES

Maples, roses, buttercups, and beans are among the true dicots, or eudicots. Lilies, orchids, ryegrass, corn, and wheat are typical monocots. Both eudicots and monocots are made of the same tissues, organized in different ways. For instance, eudicot seeds have two cotyledons, and monocot seeds have one. **Cotyledons** are leaflike structures commonly called seed leaves. They form in seeds as part of the plant embryo, and store or absorb food for it. They wither after the seed germinates and new leaves start to make food. Figure 25.4 shows some other structural and organizational differences between eudicots and monocots.

Most vascular plants have stems for upright growth, leaves that function in photosynthesis, specialized reproductive shoots called flowers, and roots that absorb water and solutes. Vascular tissue threads all through the plant body.

Shoots and roots lengthen as descendants of apical meristems divide and develop into new, primary tissues. In many species, older stems and roots put on secondary growth, or thicken, when lateral meristems are active.

Read Me First!
and watch the narrated animation on eudicots versus monocots

ⓐ Eudicots

Inside seeds, two cotyledons (seed leaves of embryo)

Usually four or five floral parts (or multiples of four or five)

Leaf veins usually in a netlike array

Three pores or furrows in the pollen grain surface

Vascular bundles organized as a ring in ground tissue

ⓑ Monocots

Inside seeds, one cotyledon (seed leaf of embryo)

Usually three floral parts (or multiples of threes)

Leaf veins usually running parallel with one another

One pore or furrow in the pollen grain surface

Vascular bundles distributed throughout ground tissue

Figure 25.4 How eudicots and monocots differ. (Magnoliids are dicots with different features than those listed. Compare Table 21.1.)

25.2 Two Categories of Plant Tissues

A variety of tissues classified as either simple or complex make up the flowering plant body.

Growth, remember, is an increase in the number, size, and volume of cells that make up a multicelled body. Plants don't grow all over. They grow longer only in back of shoot and root tips, as descendants of apical meristems divide, lengthen, and differentiate. In stems and roots that thicken, new tissues form only on top of older tissues that are no longer growing. Before considering the tissues that form, review Figure 25.5. It will help you interpret our micrographs of them.

SIMPLE TISSUES

Figures 25.6 and 25.7 are examples of simple tissues. Parenchyma tissue makes up most of the soft, moist, primary growth of roots, stems, leaves, and flowers. Parenchyma cells tend to be thin-walled, pliable, and many-sided. Unlike some other plant cells, they are alive at maturity and have a capacity to divide. Their mitotic divisions can repair plant wounds. Mesophyll is a photosynthetic parenchyma in leaves. Air spaces between its cells favor gas exchange. Parenchyma also functions in storage, secretion, and other tasks. Some cells also are present in vascular tissue systems.

Collenchyma, with unevenly thickened cell walls, forms as patches or rings in many lengthening stems and in leaf stalk ribs. Pectin imparts flexibility to the tissue. This pliable polysaccharide cements cellulose strands together inside the walls (Figure 25.7*a*).

Sclerenchyma gains strength to resist compression from lignin in its cell walls. Lignin also thwarts some fungal attacks, and waterproofs pitted walls of xylem's

radial   tangential   transverse

Figure 25.5 Terms that identify how tissue specimens are cut from a plant. Along the radius of a stem or a root, longitudinal cuts give *radial* sections. Cuts at right angles to a root or stem radius give *tangential* sections. Cuts perpendicular to the long axis of a stem or root give *transverse* sections (cross-sections).

stem epidermis   simple and complex tissues inside the stem

Figure 25.6 Locations of simple and complex tissues inside the stem of a yellow buttercup (*Ranunculus*), transverse sections.

Figure 25.7 More simple tissues. (**a**) Collenchyma from a supporting strand in a celery stem, transverse section. Sclerenchyma: (**b**) a strong flax stem, tangential view. (**c**) Stone cells, a type of sclereid in pears, transverse section.

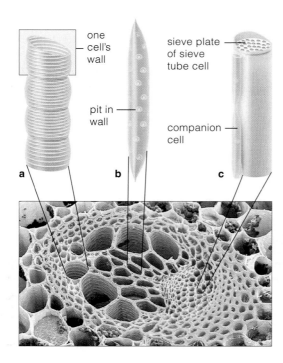

one cell's wall

pit in wall

sieve plate of sieve tube cell

companion cell

a   b   c

Figure 25.8 From xylem, (**a**) part of a column of vessel members and (**b**) one tracheid. These cells are dead at maturity; only the pitted walls remain as interconnected water-conducting tubes. (**c**) One of the living cells that interconnect as sieve tubes in phloem, which transport sugars and other solutes.

leaf surface   cuticle   epidermal cell   photosynthetic cell

Figure 25.9 A typical plant cuticle, with epidermal cells and photosynthetic cells just beneath it.

water-conducting cells. Upright plants could not have evolved on land without the mechanical support that lignin provides them (Section 21.1).

Fibers and sclereids are typical sclerenchyma cells. *Fibers* are long, tapered cells in the vascular tissues of some stems and leaves (Figure 25.7*b*). They flex, twist, and resist stretching. We use some fibers as materials for cloth, rope, paper, and other commercial products. The far stubbier *sclereids* structurally support mature tissues, such as an acorn's hard coat, coconut shells, peach pits, and "gritty" pear flesh (Figure 25.7*c*).

COMPLEX TISSUES

VASCULAR TISSUES Xylem and phloem are vascular tissues. They have long conducting cells that are often sheathed in fibers and parenchyma.

Xylem conducts water and dissolved mineral ions, and structurally supports plants. Its conducting tubes consist of interconnected *vessel members* and *tracheids* (Figure 25.8*a*,*b*). Both cell types have lignified, pitted secondary walls and are dead at maturity. The lignin, laid down in various patterns, stiffens the walls. The

walls of these dead xylem cells interconnect as sturdy water-conducting tubes. Many of the pits in adjoining walls match up, so that water moves laterally beween tubes as well as upward through the xylem.

Phloem conducts sugars and other organic solutes. Its main cells, *sieve-tube members*, are alive at maturity. They connect end to end, at their sieve plates (Figure 25.8*c*). Special, living parenchyma cells, the *companion cells*, load sugars and other photosynthetic products into these transport tubes, which deliver them to cells in all growing or storage regions.

DERMAL TISSUES The first dermal tissue to form on a plant is **epidermis**, which is usually a single layer of cells. The epidermal cells secrete waxes and cutin onto their outward-facing wall. The secretions form a body covering, a **cuticle**, that helps the plant conserve water and deflect attacks by some pathogens (Figure 25.9).

Leaf epidermis and young stems have specialized cells. For instance, a tiny gap across the epidermis, or **stoma** (plural, stomata), forms when paired guard cells plump and move apart. Most plants have many stomata. Each is a point where the diffusion of water vapor, oxygen, and carbon dioxide across epidermis is controlled. Periderm, a different tissue, replaces the epidermis in woody stems and roots.

Most of the plant body—the ground tissue system—consists of parenchyma, collenchyma, and sclerenchyma. Each of these simple tissues has only one of a variety of cell types.

Xylem and phloem are vascular tissues. In xylem, tubes of interconnected tracheids and vessel members conduct water and dissolved ions. In phloem, sieve-tube members interact with companion cells to distribute organic compounds.

Epidermis covers the surfaces of the primary plant body. Periderm forms on older, woody stems and roots.

25.3 Primary Structure of Shoots

Ground, vascular, and dermal tissues of monocot and dicot stems become organized in characteristic patterns during growth.

BEHIND THE APICAL MERISTEM

Each new flowering plant has its structure already mapped out by the time it is an embryo sporophyte inside a seed coat. It already has a primary root and shoot. In the primary shoot's tip, apical meristem and descendant tissues are laying out orderly frameworks for the stem primary structure (Figure 25.10). Below the shoot tip, new tissues become specialized as cells divide at different rates in prescribed directions, and then differentiate in size, shape, and function. These tissues make up distinctive stem regions, leaves, and lateral (axillary) buds from which lateral shoots will develop. The lateral shoots give rise to a plant's side branches and reproductive structures.

Notice, in Figure 25.10*a*, how tiny bulges of tissue develop along the sides of the apical meristem. Each bulge is one immature leaf. As growth continues, the stem lengthens between tier after tier of new leaves. Each part of the stem where one or more leaves have formed is a node. As shown in Figure 25.2, the stem region between two successive nodes is an internode.

Buds develop in the upper angle where the leaves attach to the stem. Each **bud** is an undeveloped shoot of mostly meristematic tissue, often protected by bud scales (modified leaves). Buds are the start of new stems, leaves, and flowers.

INSIDE THE STEM

In most flowering plants, primary xylem and phloem develop within the same cylindrical sheath of cells. They are the conducting lines of **vascular bundles**, or multistranded cords that thread lengthwise through a

b Sketch of the shoot tip, corresponding to (**a**)

c Same stem region later on, after shoot lengthened above it

d Same stem region, later still

immature leaf

shoot apical meristem

lateral bud forming

cortex
vascular tissues

pith

Figure 25.10 Stem of *Coleus*, a eudicot. (**a**) The light micrograph is a tangential cut through the stem center. The top tiers of the plant in the photograph above it formed according to this pattern of development. (**b–d**) Successive stages of the stem's primary growth, starting at the shoot apical meristem.

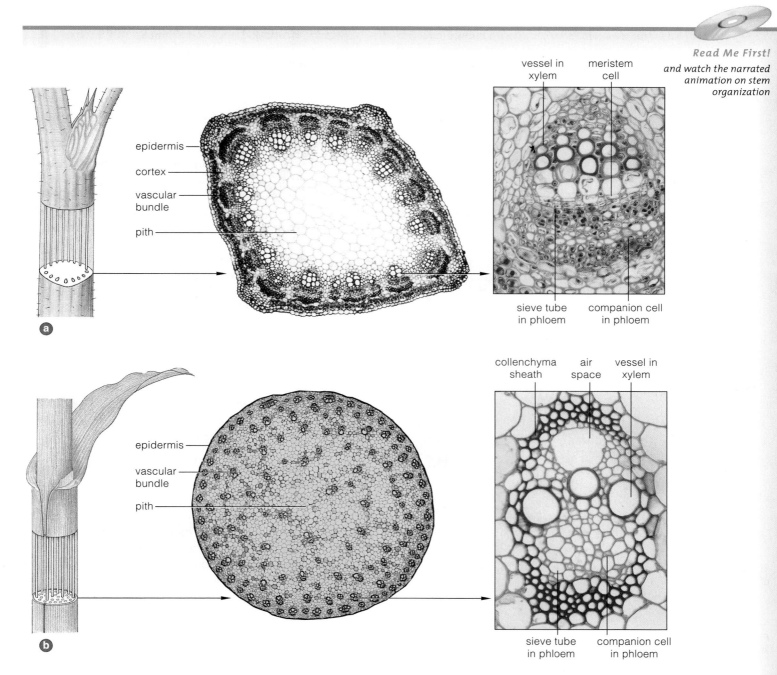

Read Me First!
*and watch the narrated
animation on stem
organization*

Figure 25.11 Internal stem organization. (**a**) Alfalfa (*Medicago*), a eudicot.
(**b**) Corn (*Zea mays*), a monocot.

ground tissue system of primary and lateral shoots. They commonly develop in two genetically dictated patterns. In most eudicot stems, long vascular bundles form a ring, which divides the ground tissue into a cortex and pith (Figure 25.11*a*). The cortex is a region between the bundles and the epidermis. The pith is the central region, inside the ring of vascular bundles. Ground tissue of the roots becomes similarly divided, into root cortex and pith.

Most monocot and some magnoliid stems have a different arrangement. Their long vascular bundles are distributed through the ground tissue (Figure 25.11*b*).

Shoot apical meristems give rise to the primary plant body, which develops a distinctive internal structure, as in the pattern in which its vascular bundles are arranged.

25.4 A Closer Look at Leaves

Each leaf is a metabolic factory with many photosynthetic cells. Even so, leaves vary greatly in size, shape, surface specializations, and internal structure.

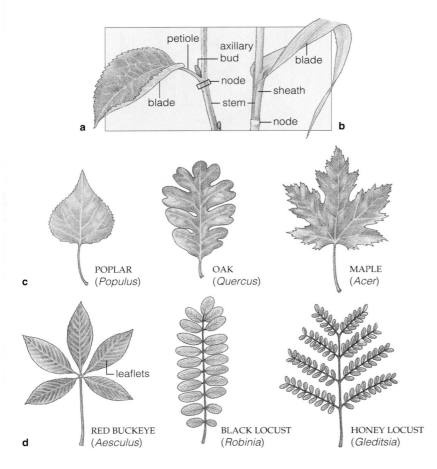

Figure 25.12 Common leaf forms of (**a**) dicots and (**b**) monocots. Examples of (**c**) simple leaves and (**d**) compound leaves.

Figure 25.13 Example of leaf specializations. Leaves of the cobra lily (*Darlingtonia californica*) form a "pitcher." The pitcher becomes partially filled with plant secretions that hold digestive enzymes. And it also releases chemical odors that some insects find irresistible. Insects lured in often cannot find the way back out; light shining through the patterned dome of the pitcher's leaves confuses them. They just wander around and down, adhering to downward-pointing leaf hairs—which are slickened with wax above the potent vat.

LEAF SIMILARITIES AND DIFFERENCES

Talk about diversity! A duckweed leaf is no more than 1 millimeter (0.04 inch) across; the leaves of one palm (*Attalea*) are 12 meters across. Leaves are shaped like needles, blades, spikes, cups, tubes, and feathers. They differ hugely in color, odor, and edibility; many form toxins. Leaves of birches and other *deciduous* species wither and drop away from stems with the approach of winter. Leaves of camellias and other *evergreen* plants also drop, but not all at the same time.

A typical leaf has a flat blade and a petiole, or stalk, attached to the stem (Figure 25.12). *Simple* leaves are undivided, although many are lobed. *Compound* leaves have blades divided as leaflets. Most monocots such as ryegrass and corn, have flat blades, the base of which encircles and sheathes the stem.

Many leaves have shapes and orientations that are good at intercepting sunlight and exchanging gases. Most are thin, with a high surface-to-volume ratio, and most reorient themselves during the day and remain perpendicular to the sun's rays. Leaves often project out from a stem in patterns that keep one another in the sun.

LEAF FINE STRUCTURE

LEAF EPIDERMIS Cuticle-protected epidermis covers all leaf surfaces exposed to the air. It may be smooth, sticky, or slimy, with "hairs," scales, spikes, hooks, glands, and other surface specialties (Figure 25.13). Most leaves have many more stomata on their lower epidermal surface. In arid or cold habitats, the water-conserving stomata and thick epidermal hairs often are more protected, in depressions in the leaf surface.

MESOPHYLL—PHOTOSYNTHETIC GROUND TISSUE A leaf's internal structure, too, is adapted to intercept sunlight and promote gas exchange. Inside the leaf, cells of **mesophyll**, a photosynthetic parenchyma, are exposed to air spaces (Figure 25.14). Carbon dioxide reaches them by diffusing in through stomata; oxygen diffuses the opposite way. Cells exchange substances at plasmodesmata. These junctions, which match up across their adjoining walls, permit the exchange of substances between their cytoplasm (Section 4.11).

Leaves oriented perpendicular to the sun have two mesophyll regions. Attached to the upper epidermis is *palisade* mesophyll—columnar parenchymal cells with more chloroplasts and more photosynthetic potential, compared to cells of the *spongy* mesophyll layer below them (Figure 25.14). Monocot leaves grow vertically

Read Me First!
and watch the narrated
animation on leaf structure

a leaf blade, leaf vein, stem

Leaf Vein
(one vascular bundle) — cuticle

Upper Epidermis

Palisade Mesophyll

Spongy Mesophyll

Lower Epidermis

xylem

Water, dissolved mineral ions from roots and stems move into leaf vein (blue arrow).

phloem

Photosynthetic products (pink arrow) enter vein, will be transported throughout plant body.

b

Oxygen and water vapor diffuse out of leaf at stomata.

Carbon dioxide in outside air enters leaf at stomata.

cuticle-coated cell of lower epidermis

one stoma (opening across epidermis)

c   50 μm

d

Figure 25.14   Leaf organization for *Phaseolus*, a bean plant. (**a**) Foliage leaves. (**b**,**c**) Leaf fine structure. (**d**) Plumped stomata.

and intercept light from all directions. As you might expect, their mesophyll is not organized as two layers. The blades of lawn grasses are like this.

VEINS—THE LEAF'S VASCULAR BUNDLES   Leaf **veins** are vascular bundles, usually strengthened with fibers. Their continuous strands of xylem rapidly move water and dissolved nutrients to all mesophyll cells, and the continuous strands of phloem carry photosynthetic products—especially sugars—away from them. In most eudicots, veins branch lacily into a number of minor veins embedded in the mesophyll. In most monocots, the veins are more or less similar in length and run parallel with the leaf's long axis (Figure 25.15).

A leaf's shape, orientation, and structure typically function in sunlight interception, gas exchange, and distribution of water and solutes to and from its living cells. Its epidermis encloses photosynthetic parenchyma (mesophyll) and veins.

Figure 25.15   Typical venation patterns in flowering plants. (**a**) A netlike array of leaf veins is common among eudicots. In many species, a stiffened midrib runs from the petiole to the leaf tip, and ever smaller veins branch from it. (**b**) A pronounced parallel orientation of veins is typical of monocot leaves. Like our own blood-transporting veins, leaf veins are conduits for dissolved substances. Like an umbrella's ribs, the stiffened veins also help maintain the leaf's shape.

25.5 Primary Structure of Roots

Roots function mainly in providing plants with a large surface area for absorbing water and essential mineral ions dissolved in it.

Unless tree roots start to buckle a sidewalk or choke off a sewer line, most of us do not pay much attention to flowering plant root systems. While we are walking above them, roots are busily mining the soil for water and minerals, and most reach a depth of 2 to 5 meters. In hot deserts, where free water is scarce, one hardy mesquite shrub sent its roots down 53.4 meters (175 feet) near a stream bed. Some cacti have shallow roots radiating outward for 15 meters.

Someone once measured the roots of a young rye plant that had been growing for four months in 6 liters of soil water. If the surface area of that root system were laid out as one sheet, it would occupy more than 600 square meters!

So how do these remarkable structures form, and what structural patterns does their growth follow? It all starts when a seed germinates. A primary root is the first structure to poke through the seed coat. In nearly all eudicot seedlings, the primary root thickens as it grows downward. Later, lateral roots, which branch sideways, form in the primary root's tissues, perpendicular to its axis, then erupt through epidermis.

Read Me First!
and watch the
narrated animation
on root organization

VASCULAR CYLINDER
endodermis
pericycle
xylem
phloem

cortex

epidermis

root hair

Vessel members are mature; root hairs are about to form.

New root cells lengthen, sieve tubes mature, vessel members start forming.

Most cells have stopped dividing.

Meristem cells are dividing fast.

No cell division is occurring here.

root tip

root cap

(a)

epidermis

root cortex

root cortex

endodermis

pericycle

primary xylem

primary phloem

(b) Vascular cylinder, cross-section

Figure 25.16 **(a)** Organization of a typical primary root and the zones where cells divide, lengthen, and then differentiate into primary tissues. Oldest root cells are farthest from apical meristem at the tip, which a root cap protects. **(b)** *Ranunculus* vascular cylinder.

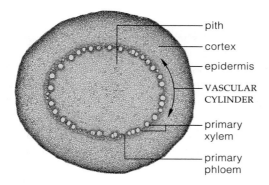

pith
cortex
epidermis
VASCULAR CYLINDER
primary xylem
primary phloem

Figure 25.17 Root of corn (*Z. mays*), transverse section. Its vascular cylinder divides the ground tissue into two zones—cortex and pith.

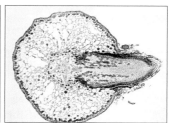

Figure 25.18 Lateral root formation from the pericycle, a cylindrical sheet of cells, one cell thick, inside the endodermis. Transverse sections.

Figure 25.16*a* shows the meristems in one root tip. The cellular descendants of the root apical meristem give rise to the three primary meristems. These cells divide, enlarge, elongate, and differentiate to become the dermal, ground, and vascular tissue systems. Some cells of the apical meristem also give rise to a root cap. This dome-shaped mass of cells helps protect the young root as it lengthens through soil.

Root epidermis is the plant's absorptive interface with soil. Some epidermal cells send out extensions called **root hairs**. Collectively, root hairs enormously increase the surface area available for taking up water and dissolved oxygen and mineral ions. We'll look at the nutritional role of root hairs in the next chapter. For now, simply note that an abundance of spaces between the cells of the ground tissue system allows oxygen to diffuse easily through roots. Like other living cells in the plant, root cells make ATP by aerobic respiration and require oxygen for the reactions.

Root apical meristem also gives rise to the ground tissue system and to the **vascular cylinder**. A vascular cylinder consists of primary xylem and phloem and one or more layers of cells called the *pericycle*. Figures 25.16*b* and 25.17 show transverse sections of a vascular cylinder of a eudicot and a monocot root.

When water enters a root, it moves from cell to cell until it reaches the endodermis, a layer of cells around the vascular cylinder. Where the endodermal cells abut, their walls are waterproofed, which forces incoming water to pass through their cytoplasm. This way, cellular mechanisms can control the flow of water and dissolved substances into the vascular cylinder.

The pericycle is inside the endodermis. Some of its cells divide repeatedly and form lateral roots, which erupt through the cortex and epidermis (Figure 25.18).

All of this growth activity results in one of two kinds of root systems. In eudicots, a **taproot system**

a   b

Figure 25.19 (**a**) Taproot system of the eudicot California poppy. (**b**) Fibrous root system of a grass plant, a monocot.

consists of a primary root and its lateral branchings. Dandelions, carrots, oak trees, and poppies are among the plants having this system (Figure 25.19*a*). By comparison, the primary root of most monocots, such as grasses, is short-lived. Adventitious roots arise from the stem in its place, and then lateral roots branch from them. (*Adventitious* refers to any plant structure that forms at an uncommon place.) Lateral roots are more or less alike in diameter and length. Such roots form a **fibrous root system** (Figure 25.19*b*).

Roots provide a plant with a tremendous surface area for absorbing water and solutes. Inside each is a vascular cylinder, with long strands of primary xylem and phloem.

Taproot systems consist of a primary root and lateral branchings. Fibrous root systems consist of adventitious roots that replace the primary root.

25.6 Secondary Growth—The Woody Plants

Thicker roots and stems—secondary growth—starts with two lateral meristems. One gives rise to secondary xylem and phloem, the other to a sturdier surface covering.

Flowering plant life cycles extend from germination to seed formation, then death. Annuals finish a cycle in one growing season and usually are "nonwoody," or herbaceous. For carrots and other biennials, roots, stems, and leaves form in one season; flowers form, seeds form, and the plant dies in the next. Perennials continue growth and seed formation year after year.

Plants that add secondary growth in two or more growing seasons are woody plants. They include all gymnosperms, some monocots, and many eudicots. They form lateral meristems—**vascular cambium** and

cork cambium. Each spring, primary growth resumes at buds on twigs and roots. New tissues form inside a bud from cell divisions of an apical meristem.

In older parts of the plant, secondary xylem forms on the inner face of the vascular cambium, and secondary phloem forms on the outer face (Figure 25.20). Vascular bundles thread through gaps in these secondary tissues at leaf and branch nodes. Most woody roots have xylem, not pith, at their center.

Some meristematic cells of vascular cambium are the start of vascular tissue that extends longitudinally through the stem. Other meristematic cells give rise to horizontal rays of parenchyma. These conduct water "sideways" in the stem, in a pattern like bike wheel spokes. Without secondary vascular tissues, water and

Read Me First!

and watch the narrated animation on secondary growth

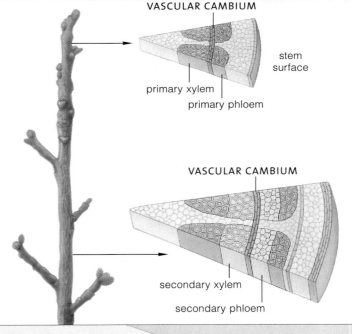

(a) Two lateral meristems in older stems and roots of woody plants produce secondary growth, or increases in diameter. *Vascular cambium* gives rise to secondary vascular tissues. *Cork cambium* gives rise to periderm.

(b) Pattern of activity at vascular cambium. Reading left to right, ongoing cell divisions enlarge the inner core of secondary xylem and displace vascular cambium toward the stem or root surface.

Figure 25.20
(a) Two meristems that become active during secondary growth. **(b)** Twig from a walnut tree in winter, after leaves dropped. In spring, *primary* growth resumes at terminal and lateral buds. *Secondary* growth resumes at vascular cambium.

One of the cells of vascular cambium at the start of secondary growth.

One of the two daughter cells differentiates into a xylem cell (coded *blue*), and the other remains meristematic.

One of the two daughter cells differentiates into a phloem cell (coded *pink*), and the other remains meristematic.

The same pattern of cell division and differentiation into xylem and phloem cells continues through the growing season.

Figure 25.21 (**a**) One type of woody stem. (**b**) Early and late wood cut from red oak. (**c**) A stem with three annual rings. The first year, the stem put on primary and some secondary growth; the next two years, it put on secondary growth.

dissolved substances would not be able to travel up, down, and sideways through woody stems.

As seasons pass and the plant ages, its inner core of xylem expands outward, and the meristematic cells are displaced toward root and stem surfaces. Some cortex and epidermis split away. A new cover, *periderm*, forms from cork cambium. Periderm, along with secondary phloem, make up **bark** (Figure 25.21).

Cell divisions at cork cambium produce **cork**. This tissue has densely packed rows of cell walls thickened with suberin, a fatty substance. Only the innermost cells are kept alive by xylem and phloem. Cork's suberized layers protect, waterproof, and insulate the stem or root surface. Cork also forms over wounds.

Wood's appearance and function change as a stem or root becomes older. The core becomes **heartwood**, a dry tissue that no longer transports water and solutes. It helps a plant defy gravity and also is a dumping ground for metabolic wastes. In time, resins, tannins, and oils darken and strengthen the oldest, crushed primary xylem, and make it more aromatic.

Sapwood is all secondary growth—the functioning xylem—between heartwood and the vascular cambium (Figure 25.21*a*). In spring, dissolved sugars that had been stored in roots travel through secondary xylem to buds. The sugar-rich fluid is **sap**.

Vascular cambium becomes inactive for part of the year in regions having cool winters or prolonged dry spells. *Early* wood, formed at the start of the growing season, has large-diameter, thin-walled xylem cells. By contrast, *late* wood forms in the drier summer months of the growing season, and cells have smaller diameters and thicker walls (Figure 25.21*b*).

Trunk cross-sections reveal bands of early and late wood, which alternate and reflect light differently. The differences are **growth rings** or, more informally, "tree rings" (Figure 25.21*c*). The trees in temperate zones add a growth ring every year. Such rings don't form often in the tropics, which don't have pronounced seasons.

Oak, hickory, and other eudicot trees that evolved in temperate and tropical zones are hardwoods; their xylem has vessels, tracheids, and fibers. Conifers such as redwoods and pines are softwood trees; their xylem has tracheids and rays of parenchyma, but no vessels. They are weaker and less dense than the hardwoods.

What is wood's selective advantage? Like all other organisms, plants compete for resources. Plants having taller stems or broader canopies that can defy gravity intercept more energy streaming in from the sun. With more energy for photosynthesis, they have a metabolic means to form large root and shoot systems. With larger systems, they are more competitive in acquiring resources. And so they ultimately can be successful, in reproductive terms, in particular habitats.

Secondary vascular tissues often form in stems and roots of woody plants. Wood is mostly an accumulation of secondary xylem.

With their sturdier tissues, woody plants defy gravity and grow taller and broader. Where competition for sunlight is intense, the ones that intercept the most sunlight win.

Wood may be classified by its location and functions (as in heartwood versus sapwood) and by the type of plant (many eudicots form hardwood, and conifers form softwood).

Summary

Section 25.1 Angiosperms (flowering plants) now dominate the plant kingdom. Their aboveground shoot systems include stems, leaves, and reproductive parts called flowers. Most species have root systems that grow downward and outward through soil.

A vascular tissue system distributes nutrients and water through the plant. A dermal system protects plant surfaces. The ground tissue makes up the bulk of the plant body. All tissues originate at meristems: regions of cells that are undifferentiated and retain a capacity for division. Primary growth at apical meristems lengthens shoots and roots. Secondary growth thickens older stems and roots, often accumulating as wood.

Monocots and eudicots have the same tissues, but these are organized in somewhat different ways.

Section 25.2 Parenchyma, sclerenchyma, and collenchyma are the simple tissues; each has only one cell type that may vary in form and function (Table 25.1).

Parenchyma cells, alive and metabolically active at maturity, make up the bulk of ground tissue systems.

Sclerenchyma cells die at maturity, leaving behind their lignin-reinforced walls to support the plant body. Collenchyma supports growing plant parts.

Complex plant tissues have two or more types of cells (Table 25.1). Complex tissues include vascular tissues (xylem and phloem) and dermal tissues (epidermis and periderm). Vessel members and tracheids are water-conducting cells of xylem. They are not alive at maturity, but their pitted, interconnected walls remain to serve as pipelines for water and dissolved minerals.

Sieve-tube members, phloem's main conducting cells, are alive at maturity. Adjoining cells interconnect and transport sugars. Companion cells assist them.

Epidermis covers and protects the outer surfaces of primary plant parts. Periderm replaces epidermis on plants showing extensive secondary growth.

Section 25.3 Stems support upright growth, and vascular bundles (xylem and phloem) thread through them. In most eudicot stems, a ring of bundles divides the ground tissue into cortex and pith. Monocot stems often have vascular bundles distributed throughout the ground tissue.

Figure 25.22
Flower of (**a**) St. John's wort (*Hypericum*) and (**b**) an iris (*Iris*).

Table 25.1 Summary of Flowering Plant Tissues

Simple tissues

Parenchyma	Parenchyma cells
Collenchyma	Collenchyma cells
Sclerenchyma	Fibers or sclereids

Complex tissues

Xylem	Conducting cells (tracheids, vessel members); parenchyma cells; sclerenchyma cells
Phloem	Conducting cells (sieve-tube members); parenchyma cells; sclerenchyma cells
Epidermis	Undifferentiated cells; also guard cells (stomata) and other specialized cells
Periderm	Cork; cork cambium; new parenchyma

Section 25.4 Leaves are photosynthesis factories. They have veins (vascular bundles) and mesophyll (photosynthetic parenchyma) between upper and lower epidermis. Air spaces around mesophyll cells enhance gas exchange. Water vapor, oxygen, and carbon dioxide cross the cuticle-covered epidermis at stomata.

Section 25.5 Roots absorb water and dissolved mineral ions, which become distributed to aboveground parts. Most anchor plants and store food. Some help support aboveground parts. Eudicots typically have a taproot system, and many monocots have a fibrous root system, with many lateral branchings.

Section 25.6 Activity at lateral meristems thickens many older stems and roots. Wood is classified by location and function (as in heartwood or sapwood). Bark consists of secondary phloem and periderm.

Self-Quiz   *Answers in Appendix III*

1. Which of the two stem sections below is common among eudicots? Which is common among monocots? Label the main tissue regions of both sections.

2. Roots and shoots lengthen through activity at _____ .
   a. apical meristems   c. vascular cambium
   b. lateral meristems   d. cork cambium

3. In many plant species, older roots and stems thicken by activity at _____ .
   a. apical meristems   c. vascular cambium
   b. cork cambium   d. both b and c

4. Is the plant with the yellow flower in Figure 25.22*a* a eudicot or a monocot? What about the plant with the purple flower (Figure 25.22*b*)?

Figure 25.23 **Figure 25.23** Examples of tree rings.

5. _____ conducts water and ions; _____ conducts food.
   a. Phloem; xylem   c. Xylem; phloem
   b. Cambium; phloem   d. Xylem; cambium

6. Mesophyll consists of _____ .
   a. waxes and cutin   c. photosynthetic cells
   b. lignified cell walls   d. cork but not bark

7. In phloem, organic compounds flow through _____ .
   a. collenchyma cells   c. vessels
   b. sieve tubes   d. tracheids

8. Xylem and phloem are _____ tissues.
   a. ground   b. vascular   c. dermal   d. both b and c

9. Label the components of the section from the three-year-old tree pictured above.

10. In early wood, cells have _____ diameters, _____ walls.
    a. small; thick   c. large; thick
    b. small; thin   d. large; thin

11. Match the plant parts with the suitable description.
    _____ apical meristem   a. masses of xylem
    _____ lateral meristem   b. source of primary growth
    _____ xylem   c. distribution of sugars
    _____ phloem   d. source of secondary growth
    _____ vascular cylinder   e. distribution of water
    _____ wood   f. central column in roots

Critical Thinking

1. Think about the conditions that might prevail in a hot desert in New Mexico or Arizona; on the floor of a shady, moist forest in Georgia or Oregon; and in the arctic tundra in Alaska. Then "design" a flowering plant that could do well in one of those places, and explain why.

2. Fitzgerald, widely known for his underdeveloped sense of nature, sneaks into a forest preserve at night and then maliciously girdles an old-growth redwood. Girdling a tree means cutting away the bark and cambium in a ring around its girth, and can kill it by interrupting the circulation of water and nutrients. He settles down and watches the tree, waiting for it to die. But the tree will not die that night or any time soon. Explain why.

3. Sylvia lives in Santa Barbara, where droughts are common. She replaced most of her lawn with drought-tolerant plants and waters the remaining lawn twice a week after the sun goes down, soaking it to a depth of several inches. Why is her strategy good for lawn grasses?

4. Oscar and Lucinda meet in a tropical rain forest and fall in love, and he carves their initials into the bark of a tiny tree. They never do get together, though. Ten years later, still heartbroken, Oscar searches for the tree. Given what you know about primary and secondary growth, will he find the carved initials higher relative to ground level? If he goes berserk and chops down the tree, what kinds of growth rings will he see?

5. Figure 25.23 shows growth layers, or tree rings, of (**a**) oak and (**b**) elm, two hardwood eudicots that are durable and strong; (**c**) shows growth layers for pine, a softwood. Pine is lightweight, resists warping, and grows faster than the hardwoods. It is commercially farmed as a source of relatively inexpensive lumber. Why, then, are hardwoods more expensive and more favored for furniture, wall panels, and kitchen cabinets?

IMPACTS, ISSUES   *Leafy Clean-Up Crews*

From World War I until the 1970s, the United States Army used a weapons testing and disposal site at Aberdeen Proving Grounds in Maryland (Figure 26.1). When explosives and chemical weapons became obsolete, they were burned in open pits at the site, along with plastics and other wastes.

Toxic levels of lead, arsenic, mercury, and other metals contaminate soil and water at the site. Dozens of organic compounds, including ethylene (TCE), seeped into groundwater. Today, the toxic brew is approaching nearby marshes and the Chesapeake Bay. Parts of the bay are already so polluted that they are a dead zone, where no marine life survives.

Together with the Environmental Protection Agency, the Army is repairing the site, which is known as J-Field. Workers cannot dig up and cart off the contaminated soil; there is too much of it. Instead, they have planted hybrid poplars (*Populus trichocarpa x deltoides*), which are cleansing the groundwater of TCE and other organic solvents. By processes of *phytoremediation*, workers are using plants to take up and concentrate or degrade contaminants in a blighted region.

The poplar roots take up contaminants as well as mineral ions dissolved in the groundwater. Living cells inside the tree tissue degrade some of them. They release gaseous forms of other contaminants into the

Figure 26.1 J-Field, once a weapons testing and disposal site. Today, hybrid poplars are helping to remove substances that contaminate the field's soil and groundwater.

the big picture

Plant Nutrients and Soil Vascular plants have structures and mechanisms that distribute water, dissolved mineral ions, and organic compounds to individual cells. How well a given soil supports plant growth depends on its composition and properties.

Root Function The uptake of water and dissolved mineral ions depends partly on the tissue organization of roots and on the selective transport functions of certain root cells. Symbionts living in or on young roots help many plants grow more efficiently.

surrounding air. Sometimes airborne contaminants can be the lesser of two evils. For example, TCE persists for a long time in groundwater, but it breaks down faster in polluted air.

Different plants are taking up different classes of contaminants in other blighted areas. At some sites, plants break down the targeted compounds or release them into the air, as J-Field poplars are doing. At other sites, microbial symbionts with plants break them down, and plant roots take up the waste products.

Also, certain plants can take up contaminants and store them in their tissues. Workers then remove the plants from the site for convenient disposal.

The best plants for phytoremediation take up many contaminants, grow fast, and grow large. Not many plants in nature can do so, but genetically engineered ones might. For instance, the filmstrip at right shows alpine pennycress (*Thlaspi caerulescens*). This species absorbs ions of dissolved minerals from soil, including zinc and cadmium. Unlike most plants, its living cells move the potentially toxic zinc and cadmium into their central vacuole, where they are stored. Researchers are attempting to transfer a gene that confers this storage capacity to other plants.

In this chapter, we focus on how vascular plants function in their environment. Many of the same processes that allow them to take up and transport water and solutes through the plant body are the ones that come in handy in phytoremediation.

 How Would You Vote?

Phytoremediation using genetically engineered plants can increase the efficiency with which a contaminated site is cleaned up. Do you support planting genetically engineered plants for such projects? See the Media Menu for details, then vote online.

Water Transport and Conservation After water and solutes enter roots, interconnected water-conducting tubes of xylem distribute them through the plant body, all the way to leaves. Plants conserve water by reversibly closing many stomata, the openings across the cuticle-covered epidermis.

Translocation In phloem, living cells that abut end to end form transport tubes. They distribute dissolved products of photosynthesis to all plant parts where cells are growing or storing food. Solute concentration and pressure gradients drive the bulk flow.

26.1 Plant Nutrients and Availability in Soil

*We've mentioned nutrients in passing. But exactly what are they? A **nutrient** is any element that is essential in an organism's life because no other element can indirectly or directly fulfill its metabolic role.*

Plant growth requires the sixteen essential elements listed in Table 26.1. Soil water usually contains ionic forms of them, such as calcium (Ca^{++}) and potassium (K^+). Nine of the sixteen are *macro*nutrients. They are required in amounts above 0.5 percent of the plant's dry weight (weighed after all water has been removed

from the body). Seven are *micro*nutrients; they make up traces (usually a few parts per million) of the dry weight. Deficiencies in any nutrient cause problems (Tables 26.1 and 26.2).

PROPERTIES OF SOIL

Soil consists of mineral particles mixed with variable amounts of decomposing organic material, or **humus**. The minerals form by the weathering of hard rocks. Humus forms from dead organisms and organic litter (fallen leaves, feces, and so on). Water and air occupy spaces between the particles and organic bits.

Soils differ in the proportions and compaction of mineral particles, mostly sand, silt, and clay. The biggest sand grains are 0.05 to 2 millimeters across. You can see individual grains by dribbling beach sand through your fingers. When you rub silt between your fingers, you can't see individual particles; they are only 0.002 to 0.05 millimeter across. Clay particles are the finest of all.

How suitable is a given soil for plant growth? Is it gummy when wet because it does not have enough air spaces? Does it form hard clods when dry? The answer depends partly on its proportions of sand, silt, and clay. The more clay, the finer the soil's texture.

In a clay particle, sheets of water molecules are interspersed among thin, stacked layers of negatively charged aluminosilicate crystals. Negative charges in the particle attract and reversibly bind the positively charged ions dissolved in soil water. This is how clay latches on to nutrients even while water is percolating past and draining away.

Soil with too much clay isn't good for most plants. Without enough sand and silt, clay particles pack so tightly they exclude air, so root cells can't get oxygen for aerobic respiration. In addition, water does not soak into heavy clay soils; it tends to flow along the surface, as runoff. Runoff ends up in streams, taking with it soil and dissolved minerals. The soils having the best oxygen and water penetration are **loams**, with roughly equal proportions of sand, silt, and clay.

Humus supports plant growth. This decomposing organic matter releases nutrients that plants can take up. Like clay, humus has an abundance of negatively charged organic acids that attract positively charged mineral ions. Humus also quickly swells with water, then shrinks by quickly releasing it. The alternation causes spaces in the soil that air can penetrate; in other words, humus helps aerate the soil.

In general, soils with 10 to 20 percent humus are best for plants. The worst, as in swamps and bogs,

Table 26.1	Plant Nutrients and Symptoms of Deficiencies
Carbon Hydrogen Oxygen	No symptoms; all three macronutrients are available in abundance from water and carbon dioxide
Nitrogen	Stunted growth; young leaves turn yellow and die (these are symptoms of chlorosis)
Potassium	Reduced growth; curled, mottled, or spotted older leaves; burned leaf edges; weakened plant
Calcium	Terminal buds wither; deformed leaves; stunted roots
Magnesium	Chlorosis; drooped leaves
Phosphorus	Purplish veins; stunted growth; fewer seeds, fruits
Sulfur	Light-green or yellowed leaves; reduced growth
Chlorine	Wilting; chlorosis; some leaves die
Iron	Chlorosis; yellow, green striping in leaves of grasses
Boron	Terminal buds, lateral branches die; leaves thicken, curl, become brittle
Manganese	Dark veins, but leaves whiten and fall off
Zinc	Chlorosis; mottled or bronzed leaves; abnormal roots
Copper	Chlorosis; dead spots in leaves; stunted growth
Molybdenum	Pale green, rolled or cupped leaves

Table 26.2	What to Ask When Home Gardening Hits a Wall

1. What are the symptoms? (e.g., brown, yellow, curled, wilted, chewed leaves)
2. What is the species? Is part of one plant, a whole plant, or many plants affected?
3. Is the planting soil loose or compact? Were amendments added? Are fertilizers used, and how often?
4. Is watering by hand, hose, sprinklers, drip system? When and how often?
5. Is the plant indoors? Outdoors, in full sun or partial or full shade? In wind?
6. Dig gently to expose a few small feeder roots. Are they black and mushy (overwatering), brown and dry (not enough water), or white with a crisp "snap"?
7. Do you see insects, or insect droppings, webs, cast skins, or slime?
8. Some unique symptoms of infections rather than nutrient deficiencies:
   Viral: Leaves or petals stunted, with mottling, colored rings, distorted shapes.
   Bacterial: Tissues have a soaked, slimy texture, often a rotting smell.
   Fungal: Leaves with dry texture, discolored spots with distinct margins, usually with concentric rings (usually tan at the center, then brown, then light yellow at edge of infection).

O HORIZON
Fallen leaves and other
organic material littering
the surface of mineral soil

A HORIZON
Topsoil, with decomposed organic
material; variably deep (only a few
centimeters in deserts, elsewhere
extending as far as thirty centimeters
below the soil surface)

B HORIZON
Compared with A horizon, larger soil
particles, not much organic material,
more minerals; extends thirty to sixty
centimeters below soil surface

C HORIZON
No organic material, but partially
weathered fragments and grains of
rock from which soil forms;
extends to underlying bedrock

BEDROCK

Figure 26.2 (**a**) Example of soil horizons. These horizons developed in a habitat in Africa.

(**b**) Erosion. In Chiapas State, Mexico, forests that once sponged up water from soil were cut down. Out-of-control runoff from rains cut deeper and wider gullies in the land and carried away topsoil. (**c**) Why the Great Plains of North America became known as the Dust Bowl. Drought and strong winds prevail in this region. In the 1850s, native prairies were plowed under for farms. In time, a third of the once-deep topsoil and half the nutrients were blown away. By the 1930s, monstrous clouds of dust were forming. Not surprisingly, they ushered in erosion control practices.

either have less than 10 percent humus or have more than 90 percent humus.

Soils develop slowly, over thousands of years, and they are in different stages of development in different places. They generally form in layers that are distinct in color and other properties. A soil profile identifies three layers, called horizon types. The A horizon, for instance, is **topsoil**. This top layer is the most essential for plant growth. It is far deeper in some places than others. You can see examples in Figure 18.15 and in Section 42.8, which shows soil profiles for some of the world's major land regions.

LEACHING AND EROSION

Leaching is the downward percolation of water, and small quantities of dissolved nutrients, through soil. Leaching is fastest in sandy soils, which are not as good as clay at binding nutrients. During spring rains and the resulting runoff, leaching happens more in forests than in grasslands. Why? Grass plants grow faster and take up more of the water.

Soil erosion is a loss of soil under the force of wind and water. Strong winds, fast-moving water, and poor vegetation cover cause the greatest losses. For example, erosion from croplands puts about 25 billion metric tons of topsoil into the Mississippi River, then the Gulf of Mexico each year. Figure 26.2*b,c* are other examples.

Either way, the nutrient losses affect plant growth and all organisms that depend on plants for survival.

Nutrients are essential elements. No other element can substitute for their direct or indirect roles in the metabolic activities that sustain growth and survival.

Soils consist mainly of particles ranging from large-grained sand to silt and fine-grained clay. The particles, especially clay, reversibly bind water molecules and dissolved mineral ions, which thereby become more accessible for uptake by plant roots.

Soil contains humus, a reservoir of organic material that is rich in organic acids and in different stages of decay. Most plants grow best in soils having equal proportions of sand, silt, and clay, as well as 10 to 20 percent humus.

26.2 The Root of It All

Energetically, building extensive root systems is costly, yet new roots keep replacing old ones as they grow outward and downward through soil. They aren't "exploring." Patches of soil having greater concentrations of water and mineral ions simply stimulate outward growth.

SPECIALIZED ABSORPTIVE STRUCTURES

A mature corn plant absorbs as much as three liters of water per day. Plants in general require a lot of water and dissolved nutrients. Root hairs, mycorrhizae, and root nodules enhance the uptake.

Root hairs are thin extensions of specialized root epidermal cells that greatly increase the surface area available for absorption (Figure 26.3). A young plant may develop billions of root hairs during its growth.

A **mycorrhiza** (plural, mycorrhizae) is a symbiotic interaction between a young root and fungus. As you read in Chapter 20, fungal hyphae grow as a velvety cover around a root or penetrate its cells. Collectively, the hyphae have a far larger surface area and absorb minerals from a larger volume of soil than roots can do on their own. The root cells give up some sugars and nitrogen-rich compounds to the fungus. They get scarce minerals in return (Figure 26.4). Any two-way flow of benefits between species is called **mutualism**.

Some bacteria in soil also are mutualists with clover, peas, string beans, and other legumes which, like all other plants, can't grow without nitrogen. Gaseous nitrogen ($N \equiv N$, or N_2) is plentiful in air, but plants just can't break its three covalent bonds. The bacteria can. They make enzymes that split all three bonds, then they use the atoms to form ammonia. This metabolic conversion of gaseous nitrogen to ammonia is known as **nitrogen fixation**. In other reactions, the ammonia is converted to forms that plants take up (Chapter 41).

Nitrogen-fixing bacteria infect plant roots, then they become symbionts in localized swellings called **root nodules** (Figure 26.5). The bacterial cells pilfer some sugars and other products of photosynthesis. But the plants absorb some of the nitrogen that the bacterial cells assimilated from the air.

Figure 26.3
Profusion of root hairs on a young root of a marjoram plant.

Figure 26.4 One view of a lodgepole pine (*Pinus contorta*) seedling, with the soil carefully pushed away from its root system. Notice the mycorrhizae's growth compared to that of the shoot system, which is only four centimeters tall.

a Root nodule

b

Figure 26.5 (**a**) Soybean root nodules. Infected plant cells and bacterial cells inside them divide repeatedly, forming a swollen mass that becomes a root nodule. The bacteria fix nitrogen after plant cell membranes surround them. The plant gets some of the nitrogen; the bacteria get some photosynthetic products. (**b**) Soybean plants growing in nitrogen-poor soil show the effect of root nodules on growth. Only the plants in the rows at *right* were inoculated with *Rhizobium* bacteria and formed nodules.

Figure 26.6 Control of the uptake of water and dissolved nutrients in roots. (**a,b**) Inside the roots of most flowering plants is an endodermis (a sheet of single cells wrapped around the vascular cylinder) and an exodermis (a cell layer just beneath the epidermis).

(**c**) Abutting walls of cells of both layers contain a waxy Casparian strip. The waxes stop water from passing around cells. Water may enter the vascular cylinder only by moving through the cytoplasm of endodermal cells. (**d**) Transport proteins in the plasma membrane of endodermal cells selectively control water and nutrient uptake.

(**d**) Waxy, water-impervious Casparian strip (*gold*) in abutting walls of endodermal cells that control water and nutrient uptake

HOW ROOTS CONTROL WATER UPTAKE

Section 25.5 introduced you to the tissue organization of typical roots. Turn now to how roots carry out their absorptive function. Water molecules in soil are only weakly bound to clay particles, so they readily move across root epidermis and continue on to a column of vascular tissue. This is the root's **vascular cylinder**. A cylindrical sheet of single cells is all that intervenes between the root cortex and the cylinder's xylem and phloem—the pipelines to the rest of the plant.

Wherever endodermal cells abut, we find a band of waxy deposits. This **Casparian strip** is a barrier to the unrestricted flow of water and solutes into the vascular cylinder. It forces water and solutes to cross only at unwaxed wall regions and through a cell. They move across the plasma membrane facing the cortex, then through cytoplasm, then across the plasma membrane facing the vascular cylinder (Figure 26.6).

Like all living cells, endodermal cells have a lot of transport proteins embedded in the plasma membrane. The proteins let some solutes but not others cross the membrane. *The transport proteins of endodermal cells are control points where a plant can adjust the quantity and types of solutes absorbed from soil water.*

Roots of many plants also have an **exodermis**, a cell layer just beneath their surface (Figure 26.6a). Walls of exodermal cells commonly have a Casparian strip that functions like the one next to the root vascular cylinder.

Root hairs, root nodules, and mycorrhizae greatly enhance the uptake of water and dissolved nutrients.

The roots of flowering plants control their uptake of water and dissolved nutrients at the vascular cylinder's endodermis and exodermis. A waxy Casparian strip around cells prevents water from penetrating anywhere except through cytoplasm.

26.3 Water Transport Through Plants

By now, you have a sense of how the distribution of water and dissolved mineral ions to all living cells is central to plant growth and functioning. Turn now to a model that explains how water actually moves through a plant.

TRANSPIRATION DEFINED

Plants secure only a fraction of the water they absorb for growth and metabolism. Most water is lost, mainly through the many stomata in leaves. Evaporation of water molecules from leaves, stems, and other plant parts is a process called transpiration. Far from being a waste, the process is vital for cell functioning.

COHESION—TENSION THEORY

Start with a basic question: How does water actually move from the soil, into the roots, and all the way up into leaves? What propels individual water molecules to the uppermost tips of plants, including redwoods and other trees that may be more than 100 meters tall? Inside vascular tissues, water moves through a complex tissue called xylem. Section 25.1 introduced you to **tracheids** and **vessel members**, the water-conducting cells of xylem. Figure 26.7 provides a

closer look at them. These cells are dead at maturity; only their lignin-impregnated walls are left. This means that the conducting cells in xylem cannot be expending energy to pull water "uphill."

Some time ago, the botanist Henry Dixon came up with an explanation of how water is transported in plants. By his **cohesion–tension theory**, water inside xylem is pulled upward by air's drying power, which creates a continuous negative pressure called tension. The tension extends all the way from leaves to roots. Figure 26.8 illustrates Dixon's theory. As you review this figure, consider the following points.

First, air's drying power causes **transpiration**: the evaporation of water from all parts of the plant that are exposed to the air, but most notably at stomata. The force of transpiration puts water molecules that are inside the waterproof conducting tubes of xylem into a state of tension. The tension extends from veins inside leaves, down through the stems, and on into young roots where water is being absorbed.

Second, the fluid columns of water show *cohesion*: they resist breaking into droplets as they are pulled up under tension. (Here you may wish to reflect on Section 2.5.) The collective strength of all the hydrogen bonds between water molecules imparts cohesion.

pits in tracheid

vessel member

perforation plate

Tracheids have tapered, unperforated end walls. Pits in adjoining tracheid walls match up.

Three adjoining members of a vessel. Thick, finely perforated walls of these dead cells connect as long vessels, another type of water-conducting tube in xylem.

Perforation plate at the end wall of one type of vessel member. Perforated ends allow water to flow unimpeded.

Figure 26.7 A few types of tracheids and vessel members from xylem. Interconnected, pitted walls of cells that died at maturity form these water-conducting tubes. Lined with pectin, the pits may control water distribution to specific regions. According to researchers, when hydrated, the pectins swell and close off water flow. During drought conditions, they shrink, and water moves freely through wide-open pits toward the leaves above.

Read Me First!
and watch the
narrated animation on
cohesion–tension theory

mesophyll (photosynthetic cells)   vein   upper epidermis

stoma

The driving force of evaporation in air

(a) Transpiration is the evaporation of water molecules from aboveground plant parts, especially at stomata. The process puts the water in xylem in a state of tension that extends from roots to leaves.

xylem   vascular cambium   phloem

**Cohesion in root, stem, leaf xylem
plus water uptake in growth regions**

(b) The collective strength of hydrogen bonds among water molecules, which are confined within the narrow water-conducting tubes in xylem, imparts cohesion to water. Hence the narrow columns of water in xylem can resist rupturing under the continuous tension.

vascular cylinder endodermis cortex   water molecule root hair cell

Figure 26.8 Key points of the cohesion–tension theory of water transport in vascular plants.

Ongoing water uptake at roots

(c) For as long as water molecules continue to escape by transpiration, that tension will drive the uptake of replacements from soil water.

Third, for as long as individual molecules of water escape from a plant, the continuous tension inside the xylem permits more molecules to be pulled upward from the roots, and therefore to replace them.

Hydrogen bonds are strong enough to hold water molecules together inside the water-conducting tubes of xylem. However, the bonds are not strong enough to stop the water molecules from breaking away from one another during transpiration and then escaping from leaves, through stomata.

Transpiration is the name for the process of evaporation from plant parts.

By the cohesion–tension theory of water transport, this process is the key source of water tension in xylem.

The tension extends downward from the highest leaves to the roots. It allows columns of water molecules that are hydrogen-bonded to one another to be pulled upward through the plant body.

26.4 How Plants Conserve Water

At least 90 percent of the water transported from roots to a leaf evaporates right out. Only about 2 percent gets used in photosynthesis, membrane functions, and other activities, but that amount must be maintained.

Think of a young plant cell, with a soft primary wall. As it grows, water diffuses in, fluid pressure increases on the wall and makes it expand, and the cell grows in volume. Expansion ends when there is enough *turgor* pressure, or internal fluid pressure against the wall, to counterbalance the water uptake. When soil dries out or gets too salty, however, the balance can end, with bad effects on the cell (Figure 26.9). Yet plants are not entirely at the mercy of their surroundings. They have a cuticle, and they have stomata.

THE WATER-CONSERVING CUTICLE

Even mildly water-stressed plants would wilt and die without a cuticle (Figure 26.10). Epidermal cells secrete this translucent, water-impermeable layer, which coats cell walls exposed to air. Cuticle is made of waxes, pectin, and cellulose fibers embedded in cutin, an insoluble lipid polymer. A cuticle does not stop light rays from reaching photosynthetic tissues. It does restrict water loss. It also restricts the *inward* diffusion of carbon dioxide necessary for photosynthesis, and *outward* diffusion of oxygen formed as a by-product.

CONTROLLED WATER LOSS AT STOMATA

Photosynthesis in plants, recall, uses carbon dioxide and releases free oxygen. Both gases diffuse across the cuticle-covered epidermis at tiny, collapsible openings called stomata (singular, stoma). Water moves out, but when soil holds enough water, roots can replace the losses. Stomata of most plants close at night, CAM plants being notable exceptions (Section 6.6). Water is conserved, and carbon dioxide collects in leaves as cells make ATP by way of aerobic respiration.

A pair of specialized parenchyma cells define each stoma; we call them guard cells (Figure 26.11). When the pair swell with water, turgor pressure bends them and a gap—the stoma—forms in between them. When the pair lose water and turgor pressure declines, they collapse against each other, so the gap closes.

A water-stressed plant closes stomata in response to the hormone abscisic acid (ABA). Remember the Section 24.5 preview of signal reception, transduction, and response? ABA is a signal. It binds to receptors on a guard cell's plasma membrane, which causes gated channels across the membrane to open. Calcium ions flow into the cells. They cause other channels to open, so potassium ions and other substances flow out from the cytoplasm. When many ions leave, water follows

Figure 26.9 Plant wilting. When a plant with soft green leaves is growing well, the soil water is dilute (hypotonic) compared to fluids in its living cells. Water moves osmotically into its cells, and internal fluid pressure builds up against the cell walls. Water also is squeezed out when this turgor pressure becomes great enough to counter the attractive force of cytoplasmic fluid. (Cytoplasm usually has more solutes compared to soil water.)

In soft, erect plant parts, as much water is moving into cells as is moving out. The constant pressure keeps cells plump. When the soil dries or gets too salty, water's concentration gradient reverses and the cells lose water. Osmotically induced shrinkage of cytoplasm in the young cells results in wilting.

(**a**) Cells from an iris petal, plump with water. Their cytoplasm and central vacuole extend to the cell wall. (**b**) Cells from a wilted iris petal. The loss of turgor pressure resulted in plasmolysis; their cytoplasm and central vacuole shrank, and the plasma membrane moved away from the wall.

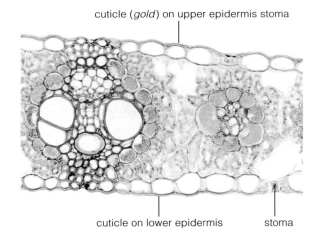

cuticle (*gold*) on upper epidermis stoma

cuticle on lower epidermis   stoma

Figure 26.10 Upper and lower cuticle on a basswood leaf.

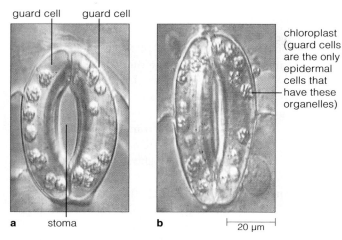

guard cell   guard cell

chloroplast (guard cells are the only epidermal cells that have these organelles)

a   stoma

b   20 µm

Figure 26.11 Stomata in action. Whether a stoma is open or closed at any given time depends on the shape of two guard cells that define this small gap across a cuticle-covered leaf epidermis. (**a**) This stoma is open. Water entered collapsed guard cells, which swelled under turgor pressure and moved apart. The stoma formed this way. (**b**) This stoma is closed. Water left the two swollen guard cells, which collapsed against each other and closed the stoma.

K^+

ABA signal

Ca^{++}

malate

K^+

Ca^{++}

malate

a Stoma is open; water has moved in.

b Stoma is closed; water has moved out.

Figure 26.12 Hormonal control of stomatal closure. (**a**) When a stoma is open, high solute concentrations in the cytoplasm of both guard cells have raised the turgor pressure, keeping the cells plumped up. (**b**) In a water-stressed plant, the hormone abscisic acid binds to receptors on the guard cell plasma membrane. It activates a signal transduction pathway that lowers solute concentrations inside the cells, which lowers the turgor pressure and closes the stoma.

its gradient and moves out of the guard cells. As the water exits, the stoma closes. It opens when the ions move in reverse, into the guard cells (Figure 26.12).

Whether stomata are open or closed is influenced by other cues, such as carbon dioxide's concentration inside a leaf, incoming light, and temperature. For example, photosynthesis starts after the sun comes up. As the morning progresses, carbon dioxide levels fall in all photosynthetic cells, including guard cells. The decrease stimulates active transport of potassium ions into the cells. Blue light is another signal. It is intercepted by a carotenoid, which transduces it into ion gradients. Water follows the ions into the guard cells, and the stoma opens.

As this section makes clear, plant survival depends on stomatal function. Think about it when you are out and about on smog-shrouded days (Figure 26.13).

Water-dependent events in plants are severely disrupted when water loss exceeds uptake at roots for extended periods. Wilting is one observable outcome.

Transpiration and gas exchange occur mainly at stomata. These numerous small openings span the waxy cuticle, which covers all plant epidermal surfaces exposed to air.

Plants open and close stomata at different times to control water loss, carbon dioxide uptake, and oxygen disposal, all of which affect rates of photosynthesis and plant growth.

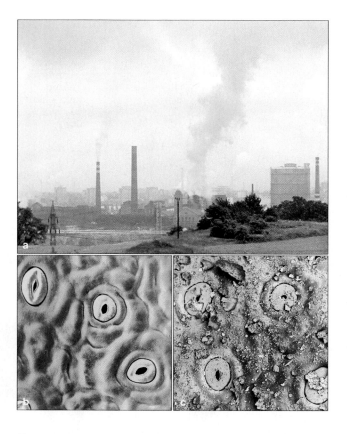

Figure 26.13 (**a**) Smog in Central Europe. (**b**) Stomata at a holly leaf surface. (**c**) A holly plant growing in industrialized regions becomes covered with gritty airborne pollutants that clog stomata and prevent sunlight from reaching photosynthetic cells in the leaf.

26.5 Distribution of Organic Compounds in Plants

Xylem distributes water and minerals through plants. The vascular tissue called phloem distributes organic products of photosynthesis.

Like xylem, phloem has many conducting tubes, fibers, and strands of parenchyma cells. Unlike xylem, it has **sieve tubes** through which organic compounds rapidly flow. *Living* cells form these long tubes. The cells are positioned side by side and end to end. Their abutting end walls, known as sieve plates, are porous (Figure 26.14a,b). **Companion cells** are pressed against

the tubes. These cells help load organic compounds into neighboring sieve tubes.

Some organic products of photosynthesis are used inside the leaf cells that make them. The rest move to roots, stems, buds, flowers, and fruits. In most cells, carbohydrates are stored as starch. Starch molecules are too large for transport across the plasma membrane of cells and too insoluble for transport. Cells convert them to sucrose, which is more easily transportable.

Experiments with insects show that sucrose is the main carbohydrate transported in phloem. Aphids were anesthetized by exposing them to high levels of carbon dioxide as they were feeding on the juices in phloem's conducting tubes (Figure 26.14c). Then researchers detached the body from the mouthparts, which they left attached to the plant. They collected and analyzed the exuded fluid. For most of the plants studied, sucrose was the most abundant carbohydrate in the fluid that was being forced out of the tubes.

Translocation is the formal name for the process that transports sucrose and other organic compounds through phloem. High pressure drives the process. Often, the pressure inside phloem's conducting tubes is five times higher than the air pressure inside an automobile tire.

Phloem translocates photosynthetic products along declining pressure and solute concentration gradients. The *source* of the flow is any region of the plant where organic compounds are being loaded into sieve tubes. Common sources are mesophylls—the photosynthetic tissues in leaves. The flow ends at a *sink*, which is any plant region where the products are being unloaded, used, or stored. Example: While flowers and fruits are growing during development, they are sink regions.

Why do organic compounds flow from a source to a sink? According to the **pressure flow theory**, internal pressure builds up at the source end of the sieve tube system and *pushes* the solute-rich solution on toward any sink, where solutes are being removed.

one of a series of living cells that abut, end to end, and form a sieve tube

companion cell (in the background, pressed right against the sieve tube)

perforated end plate of sieve tube cell, of the sort shown in (**b**)

a

b

c

Figure 26.14 (**a**) Part of a sieve tube in phloem. Arrows point to perforated ends of individual tube members.

(**b**) Scanning electron micrograph of the sieve plate on the end of two side-by-side sieve tube members.

(**c**) Honeydew exuding from an aphid after this insect's mouthparts penetrated a sieve tube. High pressure in phloem forced this droplet of sugary fluid out through the terminal opening of the aphid gut.

Read Me First!

and watch the narrated animation on translocation in phloem

SOURCE (e.g., mature leaf cells)

interconnected sieve tubes

WATER

a Active transport mechanisms move solutes into the sieve tube, against concentration gradients.

b As a result of the increased solute concentration, the water potential is decreased in the sieve tube, and water moves in, increasing turgor pressure.

c The pressure then pushes solutes by bulk flow between a source and a sink, with water moving into and out of the system all along the way.

bulk flow

d Both pressure and solute concentrations gradually decrease between the source and the sink.

e Solutes are unloaded into sink cells, and the water potential in those cells is lowered. Water moves out of the sieve tube and into sink cells.

SINK (e.g., developing root cells)

Steps in the translocation process

upper leaf epidermis

photosynthetic cell

sieve tube in leaf vein

companion cell next to sieve tube

lower leaf epidermis

Typical source region
Photosynthetic tissue in a leaf

sieve tube

Typical sink region
Actively growing cells in a young root

Figure 26.15 Translocation of organic compounds. Review Section 6.5 to get an idea of how translocation relates to photosynthesis in vascular plants.

You can use Figure 26.15 to track what happens after sucrose moves from the photosynthetic cells into small veins in leaves. In energy-requiring reactions, companion cells in the veins load sucrose into sieve tube members. As the sucrose concentration increases in the tubes, water also moves in, by osmosis. More internal fluid pressure is exerted against the sieve tube walls. When pressure increases, it pushes the sucrose-laden fluid out of the leaf, into the stem, and on toward the sink.

Plants store carbohydrates as starch but distribute them in the form of sucrose and other small, water-soluble units.

Translocation is the distribution of organic compounds to different plant regions. It depends on concentration and pressure gradients in the sieve tube system of phloem.

Gradients last as long as companion cells load compounds into sieve tubes at sources, such as mature leaves, and as long as compounds are unloaded at sinks, such as roots.

Summary

Section 26.1 Nutrients are elements essential for survival and growth; no other element can replace their metabolic role. Plants require sixteen nutrients, including carbon, oxygen, hydrogen, nitrogen, and phosphorus, which they get from air, water, and soil. Soil characteristics greatly affect the accessibility of water, oxygen, and nutrients to plants.

Section 26.2 Root systems take up water and nutrients, which often are scarce in soil. Root hairs increase the absorptive surface area. Mycorrhizae and bacteria in root nodules are mutualists that assist plants in nutrient uptake.

Some control over water and nutrient uptake occurs in the roots of flowering plants. A waxy Casparian strip prevents water from entering roots except through the cytoplasm of endodermal and exodermal cells. There, membrane transport proteins control the uptake.

Section 26.3 Plants distribute water and dissolved mineral ions through water-conducting tubes of xylem, a vascular tissue. Cells (tracheids and vessel members) form the tubes but are dead at maturity. The remaining waterproofed cell walls connect as narrow pipelines.

Transpiration is the evaporation of water from plant parts, mainly at stomata, into air. By the cohesion–tension theory, it is a force that pulls water upward in plants. Transpiration inside xylem's water-conducting cells causes continuous negative pressure (tension) that extends from leaves to roots.

Even as water molecules are escaping from leaves, replacements are pulled into the leaf under tension. The collective strength of hydrogen bonds among the water molecules imparts cohesion, which allows water to be pulled up as continuous fluid columns. The "cohesion" part of the theory refers to the capacity of the hydrogen bonds between water molecules to resist rupturing.

Section 26.4 A cuticle is a waxy, waterproof cover on all plant surfaces exposed to the surroundings. It helps the plant conserve water on hot, dry days.

Gas exchange and transpiration occur at stomata, openings across the cuticle-covered epidermis of leaves and many stems. In response to environmental cues, ions and water move into and out of paired guard cells, which alternately plump up or collapse against each other, opening and closing the gap between them. When stomata are open, they permit gas exchange. Plants conserve water when stomata are closed.

Section 26.5 By an energy-requiring process called translocation, sucrose and other organic compounds are distributed throughout plants. Sugars produced by photosynthesis are loaded into vessels of phloem with the help of companion cells. They are unloaded at the plant's actively growing regions or at storage regions.

Self-Quiz

1. Carbon, hydrogen, and oxygen are examples of _____ for plants.
   - a. macronutrients
   - b. micronutrients
   - c. trace elements
   - d. essential elements
   - e. both a and d

2. A _____ strip in endodermal cell walls forces water and solutes to move through root cells, not around them.
   - a. cutin
   - b. lignin
   - c. Casparian
   - d. cellulose

3. The nutrition of some plants depends on a root–fungus association known as a _____ .
   - a. root nodule
   - b. mycorrhiza
   - c. root hair
   - d. root hypha

4. In phloem, organic compounds flow through _____ .
   - a. collenchyma cells
   - b. sieve tubes
   - c. vessels
   - d. tracheids

5. Water evaporation from plant parts is called _____ .
   - a. translocation
   - b. expiration
   - c. transpiration
   - d. tension

6. Water transport from roots to leaves occurs by _____ .
   - a. pressure flow
   - b. differences in source and sink solute concentrations
   - c. the pumping force of xylem vessels
   - d. cohesion–tension among water molecules

7. In daytime, most plants lose _____ and take up _____ .
   - a. water; carbon dioxide
   - b water; oxygen
   - c. oxygen; water
   - d. carbon dioxide; water

8. At night, most plants conserve _____ , and _____ accumulates.
   - a. carbon dioxide; oxygen
   - b. water; oxygen
   - c. oxygen; water
   - d. water; carbon dioxide

9. Match the concepts of plant nutrition and transport.
   - _____ stomata
   - _____ nutrient
   - _____ sink
   - _____ root system
   - _____ hydrogen bonds
   - _____ transpiration
   - _____ translocation

   - a. evaporation from plant parts
   - b. harvesting soil nutrients
   - c. balance water loss with carbon dioxide requirements
   - d. cohesion in water transport
   - e. sugars unloaded from sieve tubes
   - f. organic compounds distributed through the plant body
   - g. element with roles in metabolism that no other element can fulfill

Critical Thinking

1. Home gardeners, like farmers, must be sure that their plants have access to nitrogen from either nitrogen-fixing bacteria or fertilizer. Insufficient nitrogen stunts plant growth; leaves yellow and die. Which biological molecules incorporate nitrogen? How would nitrogen deficiency affect biosynthesis and cause these symptoms?

2. You just returned home from a three-day vacation. Your severely wilted plants tell you they weren't watered before you left. Being aware of the cohesion–tension theory of water transport, explain what happened to them.

3. When moving a plant from one location to another, it helps to include some native soil around the roots. Explain why, in terms of mycorrhizae and root hairs.

c 1587–1589   1606–1612

Figure 26.16 (**a**) Location of two of the early American colonies. (**b**) Bald cypress swamp. (**c**) Bald cypress growth layers, cross-section. This tree was living when English colonists first settled in North America. Narrow annual rings mark severe drought years, which stunted tree growth. Compare Section 25.6.

 4. In the sketch at *left*, label the stoma. Now think about Henry, who discovered a way to keep all of a plant's stomata open at all times. He also figured out how to keep those of another plant closed all the time. Both plants died. Explain why.

5. Allen is studying the rate of transpiration from tomato plant leaves. He notices that several environmental factors, including wind and relative humidity, affect the rate. Explain how they might do so.

6. At the middle fork of the Salmon River in Idaho, clear water from a wilderness area converges on brown water. The brown water is filled with silt from a wilderness area disturbed by cattle ranching and other human activities. Knowing what you do about the nature of topsoil, what is probably happening to plant growth in the disturbed habitat? Knowing how fishes acquire oxygen for aerobic respiration, how might the silt be affecting their survival?

7. In 1587, about 150 English settlers arrived at Roanoke Island off the coast of North Carolina. When ships arrived in 1589 to resupply the colony, they found the island had been abandoned. Searches up and down the coast failed to turn up the missing colonists. About twenty years later, the English established a colony at Jamestown, Virginia. Although this colony survived, the initial years were hard. In the summer of 1610 alone, more than 40 percent of the colonists died, many of starvation.

Scientists examined wood cores from bald cypress (*Taxodium distichum*) that had been growing at the time the Roanoke and Jamestown colonies were founded. Growth rings revealed the colonists were in the wrong place at the wrong time (Figure 26.16). They arrived at Roanoke just in time for the worst droughts in eight hundred years. Nearly a decade of severe drought struck Jamestown.

We know the corn crop of the Jamestown colony failed totally. Drought-related crop failures probably occurred at Roanoke. The settlers also had trouble finding fresh water. Jamestown was established at the head of an estuary; when river levels dropped, the water became brackish.

If secrets locked in trees intrigue you, look into a field of study called *dendroclimatology*. For example, find out what growth layers reveal about fluctuations in climate where you live. See if you can correlate them with human events at the time the changes were being recorded.

Media Menu

 Student CD-ROM
Impacts, Issues Video
    Leafy Clean-Up Crews
Big Picture Animation
    Plant tissues, growth, and function
Read-Me-First Animation
    Root functioning
    Cohesion–tension theory
    Translocation in phloem

 InfoTrac
• High and Dry. *Natural History*, October 2000.
• Friendly Fungi Help Trees Grow. *Applied Genetics News*, May 2000.
• How Plants Conserve Water. *Applied Genetics News*, July 2001.
• Ancient Tree Rings Reveal Past Climate. *Science News*, March 2001.

 Web Sites
• Wayne's Word: waynesword.palomar.edu
• Botany Basics: extension.orst.edu/mg/botany
• The Virtual Foliage Home Page: botit.botany.wisc.edu
• Ultimate Tree-Ring Web Pages: web.utk.edu/~grissino
• USDA Soils Site: soils.usda.gov

 How Would You Vote?
Some plants have an amazing ability to absorb metals and other chemical contaminants. Scientists are now transferring genes of those plants into fast-growing species, for use in phytoremediation. Some people worry that growing genetically modified plants in areas filled with toxic chemicals could have unforeseen outcomes. Should we allow the planting of genetically engineered plants for phytoremediation?

IMPACTS, ISSUES *Imperiled Sexual Partners*

Imagine a world without chocolate. No chocolate bars, no cocoa, no brownies. Too far-fetched? Chocolate comes from seeds of *Theobroma cacao* (Figure 27.1). This tree doesn't have to grow in a tropical rain forest. It grows well in places where rain forests were cleared to make way for plantations. There's just one problem. Trees in sunny plantations don't form many seeds.

As it turns out, *T. cacao*'s only pollinators are midges, a type of fly about as tiny as a pin head. This species relies exclusively on midges to carry pollen from flower to flower. Midges happen to live and breed only in the deep, damp shade of tropical rain forests. No forests, no midges. That is a problem for chocolate production.

Now, some growers are planting cacao trees in the shade of rain forests. Besides keeping trees in midges,

they are preserving otherwise threatened habitat for rain-forest birds and other species.

Dwindling numbers of pollinators are a worldwide problem. By some estimates, about three-fourths of the world's crop plants complete their life cycles with the help of pollinators—mostly insects, and especially bees, butterflies, flies, moths, and beetles.

Honeybees (*Apis mellifera*) pollinate most crops in the United States. Beekeepers easily pack up and move artificial hives containing thousands of bees from field to field. However, diseases, parasitic mites, pesticides, and other threats are wiping out the honeybees.

Different bees are essential pollinators of one or a few crop plants. Alfalfa plants use *Megachile rotundata*, the alfalfa leaf-cutting bee, as the pollinator of choice.

Figure 27.1 Flowers, fruits, and seeds of *Theobroma cacao*, a plant pollinated by tiny insects. As many as forty seeds ("beans") are packed inside each fruit. The seeds are processed into cocoa butter and essences of chocolate, which are mildly addictive. An average American buys 8 to 10 pounds of chocolate annually. An average Swiss citizen craves a whopping 22 pounds. With that kind of demand, *T. cacao* trees have the good luck to have humans interested in their survival.

the big picture

Reproductive Modes
Flowering plants coevolved with pollinators. Their life cycles alternate between sporophytes and gametophytes and often have asexual phases. Pollen, eggs, then seeds and fruits develop on floral shoots.

Developmental Cues
An embryo sporophyte that has developed inside a seed resumes growth after germinating. A combination of species-specific genetic programs and environmental cues guide its growth and development in expected patterns.

The southeastern blueberry bee (*Habropoda laboriosa*) is the main pollinator of blueberry plants. The blueberry bee is being clobbered by habitat loss, pesticides, and (whoops) accidentally imported ants that eat its larvae.

More than 60 percent of all flowering plant species aren't making as many seeds, and pollinator shortages are a factor. In United States, about 570 plant species are listed as endangered and another 140 as threatened. Some of North America's most beautiful wildflowers, such as the prairie white-fringed orchid (*Platanthera*) shown in the filmstrip, are skittering toward extinction. The impact ripples through communities; seeds and fruits are staples in the diet of an astounding number of animal species, including songbirds.

It won't be easy to protect the pollinators. Declines in pollinator species often follow an accumulation of many small, local environmental assaults. Especially at risk are the migratory ones. Pollinators as diverse as monarch butterflies, rufous hummingbirds, and the lesser long-nosed bat move seasonally between the United States and their wintering grounds in Mexico. Food sources all along the route sustain their vast migrations. Lose any of those sources, and their migration might come to a screeching halt.

Flowering plants, again, rose to dominance through coevolutionary reliance on animal species. Those same interactions now put many at risk. Recognizing their vulnerability is our first step toward finding workable ways to protect them.

 How Would You Vote?

Microencapsulated pesticides are easy to apply and are effective for long periods. But they are about the size of pollen grains and are a tempting but toxic threat to certain pollinators. Should we restrict their use? See the Media Menu for details, then vote online.

Chemical Signals At least five classes of plant hormones interact with one another in the control of growth and development. Chemical signals also play a role in plant defenses and in growth responses to environmental conditions, such as sunlight and gravity.

Seasonal Changes Plants sense and respond to seasonal changes in daylight, temperature, and other environmental factors. Such changes can stimulate different plants to respond in specific ways, as by putting out flowers, losing leaves, and becoming dormant.

27.1 Sexual Reproduction in Flowering Plants

Sexual reproduction is the primary reproductive mode of flowering plant life cycles. It involves the formation of spores and gametes, both of which develop inside specialized reproductive shoots called flowers.

REGARDING THE FLOWERS

Each **flower** is a specialized reproductive shoot that develops on a sporophyte. A **sporophyte**, remember, is a spore-producing body that grows by mitotic cell divisions from a fertilized egg. Two kinds of haploid spores form inside flowers, and these spores are the start of **gametophytes**—structures that produce male or female gametes (Figure 27.2).

Sperm develop in male gametophytes, and eggs develop in female gametophytes. Fusion of a haploid sperm with a haploid egg at fertilization results in a diploid zygote that grows into a new sporophyte.

A flower develops as whorls of modified leaves attached to the receptacle, or top of the floral shoot. As leaves differentiate, most form an outermost whorl of sepals around a ring of nonfertile petals. The petals surround stamens and carpels, a flower's fertile parts.

Stamens are male reproductive parts of a flower. They usually consist of an anther atop a thin stalk, or filament. Inside a typical anther are two to six pairs of pouches called pollen sacs. Tiny walled, haploid spores form in the pouches and give rise to **pollen grains**, a type of male gametophyte.

Pollen grains often have sticky or spiked, sculpted walls that stick to hairs, fur, or feathers of a pollinator (Figure 27.3). The wall structure also helps the pollen to adhere to and germinate on a receptive stigma.

One or more **carpels** are female reproductive parts, at the center of a flower. The upper portion of a carpel or of a fused group of carpels is known as a stigma. The stigma's sticky or hairy surface tissue gently traps pollen grains long enough for them to germinate. The carpel's lower, swollen portion is an **ovary**. Inside the ovary are one or more **ovules**, structures in which an egg-producing female gametophyte forms. If an egg inside gets fertilized, the ovule becomes a seed.

Read Me First!

and watch the narrated animation on floral structure and function

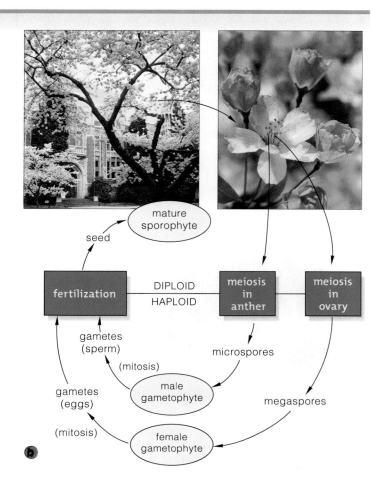

Figure 27.2 (**a**) Structure of one type of flower, a blossom from a cherry (*Prunus*) tree. (**b**) Overview of flowering plant life cycles, using a cherry tree as the example.

Perfect flowers have both stamens and carpels. But some species have *imperfect* flowers, with stamens *or* carpels; male and female flowers form on the same plant or on different ones. For example, cucumbers and corn have separate male and female flowers on the same plant. Asparagus and kiwi have male and female flowers on different plants.

COEVOLUTION WITH POLLINATORS

Only 10 percent or so of all flowering plant species (and most gymnosperms) simply release pollen to the wind. Their flowers usually are inconspicuous and do not make nectar. They make a lot of pollen.

The overwhelming majority of flowering species preferentially attract **pollinators**: insects, birds, and other agents that directly deliver pollen from male to female floral reproductive parts of the same species (Figure 27.4). Flowers lure them in with suitable cues and reinforcements, the outcomes of millions of years of coevolution. Petals of particular colors, sizes, and shapes are such cues. Nectar and nutritious pollen are among the positive reinforcements.

Many flowers have bright colors and compelling fragrances. Floral ground tissue cells make pigments, including carotenoids (yellow to red-orange) and anthocyanins (red to blue). Cells of petals and other parts make fragrant oils. The petals often incorporate light-refracting crystals that make them shimmer in the light of the sun and moon.

Birds have acute daytime vision but a notably poor sense of smell. Bird-pollinated plants typically make flowers with vivid red components but no perfume. Beetles fly clumsily and land better on heavy-petaled

Figure 27.3 Pollen grains of (**a**) grass, (**b**) chickweed, and (**c**) ragweed plants. Pollen grains of most families of plants differ in size, wall details, and the number of wall pores.

flowers, preferably the ones that smell like fermenting fruit or dung. Daisies and other composite flowers consist of many tiny flowers, all fused into a landing platform sturdy enough for butterflies. Night-flying, nectar-sipping bats home in on flowers with intensely fruity or musky odors and white or pale petals, which show up best in moonlight (Figure 27.4). Light-colored flowers with sweet odors lure the moths.

Bees visit sweet-smelling flowers with yellow, blue, or purple petals. Often the petals have pigments that absorb and reflect ultraviolet light in patterns that bees use as visual guides to nectar. Humans can't see the patterns without special camera filters; our eyes don't contain receptors that respond to ultraviolet light.

Flowering plant life cycles alternate between formation of sporophytes (spore-producing bodies) and gametophytes (gamete-producing bodies). The sporophytes produce flowers, specialized reproductive shoots. The overwhelming majority of flowering plants coevolved with pollinators.

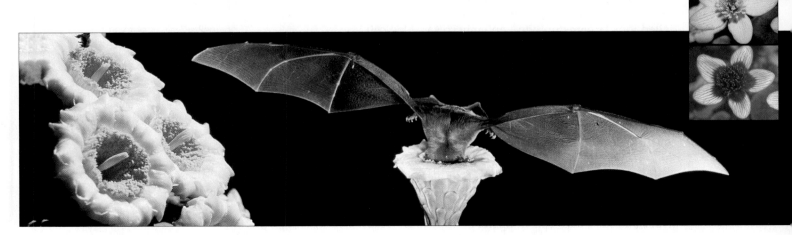

Figure 27.4 Pollinators. As birds and insects sip nectar from the large, white flowers of the giant saguaro by day, and bats by night, pollen grains stick to their body. At far right, shine ultraviolet light on a gold-petaled marsh marigold (*above*) and the nectar guide for bees shows up (*below*).

27.2 A New Generation Begins

In flowering plants alone, fertilization is a two-step process. One step brings together the male and female gametes to form a zygote. The other produces a tissue (endosperm) that will nourish the zygote as it develops.

MICROSPORE AND MEGASPORE FORMATION

Recall, from Chapter 21, that flowering plants make two types of spores. In anthers of a cherry blossom, masses of spore-producing cells form by mitotic cell divisions. Walls develop around each mass, forming a pouchlike pollen sac (Figure 27.5*a*). Inside the sacs, cells undergo meiosis and cytoplasmic divisions, thus forming haploid **microspores**. A microspore has its own wall, which includes species-specific molecules.

Pollen grains form from microspores that undergo one or two mitotic cell divisions. Once formed, their growth is arrested until their release from the anther.

In flowering plants, only the *nucleus* of a sperm is a male gamete. Depending on the species, it forms as a gametophyte develops inside a pollen grain or after the pollen grain germinates. So different pollen grains are mature *or* immature male gametophytes.

The second type of spore forms in ovules (Figure 27.5*e*). An ovule starts out as a cell mass on the inner wall of an ovary, in the base of a carpel. As the ovule matures, protective cell layers, or integuments, form around it. Inside each ovule, a cell undergoes meiosis and cytoplasmic division. Four haploid spores, called **megaspores**, are the outcome. Megaspores are usually larger than the microspores that form in anthers.

Three of the megaspores typically disintegrate. The remaining cell undergoes three rounds of mitosis but not cytoplasmic division. The result is a cell with eight nuclei. Each nucleus migrates to a prescribed location, then the cytoplasm divides to form a seven-celled female gametophyte—an embryo sac. One cell, the endosperm mother cell, has two nuclei ($n + n$). It will give rise to a nutritive tissue. The other cells are all haploid, and one of them is the egg.

POLLINATION AND FERTILIZATION

Pollination occurs when pollen grains released from an anther land on a receptive stigma. In response to cues from the stigma, the pollen grain germinates; it resumes metabolic activity. It becomes a pollen tube, which grows down through the stigma and into the ovary's tissues. Two sperm nuclei are inside it. When the pollen tube reaches an ovule, it grows right into the female gametophyte. The tube's tip ruptures and two sperm nuclei are released (Figure 27.5*i*).

"Fertilization" usually means the fusion of a sperm and egg. But flowering plants have a unique trait; they undergo **double fertilization**. During this process, one sperm nucleus from the pollen tube fuses with the egg nucleus, which forms a zygote. The other sperm nucleus fuses with both of the nuclei in the

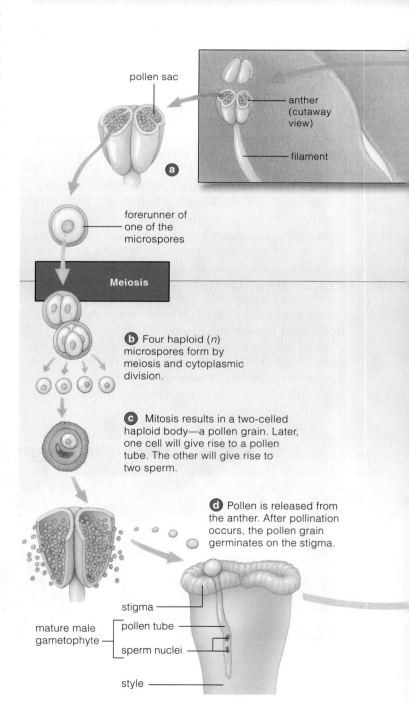

pollen sac

anther (cutaway view)

filament

a

forerunner of one of the microspores

Meiosis

b Four haploid (*n*) microspores form by meiosis and cytoplasmic division.

c Mitosis results in a two-celled haploid body—a pollen grain. Later, one cell will give rise to a pollen tube. The other will give rise to two sperm.

d Pollen is released from the anther. After pollination occurs, the pollen grain germinates on the stigma.

mature male gametophyte

stigma

pollen tube

sperm nuclei

style

endosperm mother cell. The outcome is a cell with a triploid (3*n*) nucleus. This cell will give rise to the **endosperm**, a nutritive tissue found only in seeds of the flowering plants. Right after a seed germinates, endosperm will sustain the sporophyte seedling until photosynthesis kicks in.

In flowering plants, sperm form in pollen grains, the male gametophytes. Female gametophytes (with egg cells) develop in the ovules of ovaries. After pollination, double fertilization starts the formation of an embryo and endosperm.

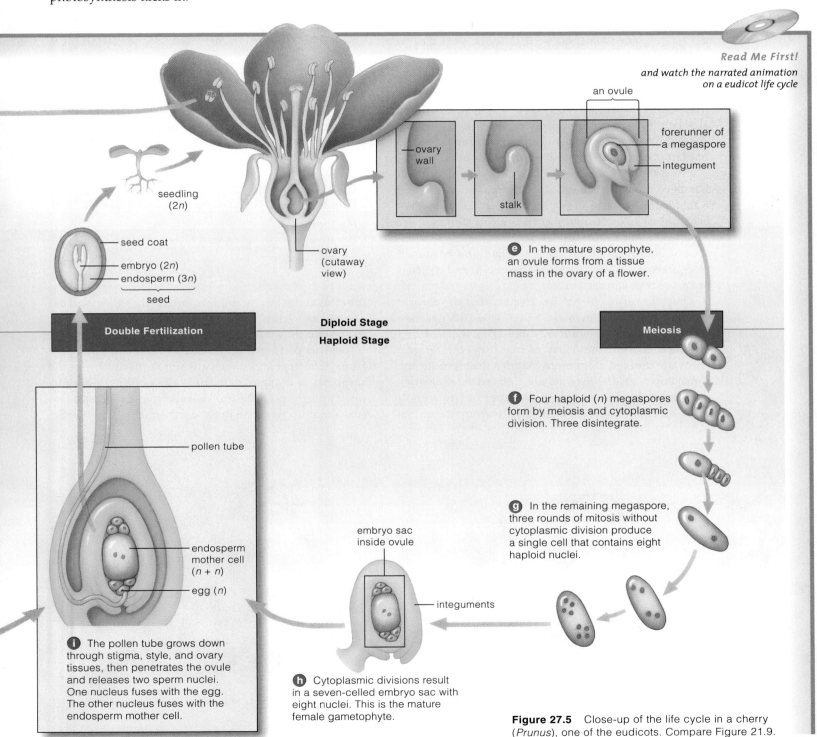

Read Me First!
and watch the narrated animation on a eudicot life cycle

an ovule

ovary wall

forerunner of a megaspore

integument

stalk

seedling (2*n*)

seed coat

embryo (2*n*)

endosperm (3*n*)

seed

ovary (cutaway view)

e In the mature sporophyte, an ovule forms from a tissue mass in the ovary of a flower.

Double Fertilization

Diploid Stage

Haploid Stage

Meiosis

f Four haploid (*n*) megaspores form by meiosis and cytoplasmic division. Three disintegrate.

pollen tube

endosperm mother cell (*n* + *n*)

egg (*n*)

embryo sac inside ovule

integuments

g In the remaining megaspore, three rounds of mitosis without cytoplasmic division produce a single cell that contains eight haploid nuclei.

i The pollen tube grows down through stigma, style, and ovary tissues, then penetrates the ovule and releases two sperm nuclei. One nucleus fuses with the egg. The other nucleus fuses with the endosperm mother cell.

h Cytoplasmic divisions result in a seven-celled embryo sac with eight nuclei. This is the mature female gametophyte.

Figure 27.5 Close-up of the life cycle in a cherry (*Prunus*), one of the eudicots. Compare Figure 21.9.

27.3 From Zygotes to Seeds in Fruits

*The new zygote embarks on mitotic cell divisions and becomes an embryo sporophyte, which develops in an ovule. Outside the ovule, the ovary wall and sometimes neighboring tissues mature into what we call **fruit**.*

THE EMBRYO SPOROPHYTE

Flowering plant embryos develop in different ways, but let's focus on a shepherd's purse (*Capsella*) embryo. Its **cotyledons**, or seed leaves, develop from two tiny lobes of meristematic tissue (Figure 27.6). All eudicot embryos have two cotyledons. All monocot embryos have one. Embryo sporophytes of *Capsella*, a eudicot, absorb nutrients from endosperm and store them in cotyledons. By contrast, most monocot embryos do not tap nutritive tissue until after germination.

From the time a zygote forms in the ovule until it has become a mature embryo, the parent plant will transfer nutrients to the ovule tissues. Food reserves accumulate in endosperm or in the cotyledons. Over time, the ovule wall pulls away from the ovary wall. A seed coat forms from the thickened and hardened integuments. The embryo, food reserves, and the coat are a **seed**—a mature ovule.

Only flowering plants have seeds that develop in ovaries, and only they make fruits. Botanists describe fruits by the way they form (Table 27.1). *Simple* fruits are derived from the ovary of one flower. *Aggregate* fruits are derived from many ovaries that form in the same flower. *Multiple* fruits are derived from ovaries of many separate flowers. Many types of fruits in all three categories are derived largely from structures

other than the ovary, and they are further classified as "accessory" fruits. All apples and strawberries, such as the one shown in Figure 27.7, are like this.

Capsella pods, pea pods, and acorns are simple, dry fruits. Cherries and olives are simple fleshy fruits. A raspberry is an aggregate fruit. Most of the flesh of a pineapple, a multiple fruit, forms from the receptacle (the shoot tip to which floral parts attach). An apple is

Table 27.1 Categories of Fruits
Simple Fruits From one ovary of one flower.
Dry types:
Dehiscent. Fruit wall splits along definite seams to release seeds. Legume (e.g., pea, bean), poppy
Indehiscent. Fruit wall does not split on seams. Acorn, grains (e.g., corn), sunflower, carrot, maple
Fleshy types:
Berry. Compound or simple ovary, many seeds. Tomato, grape, banana
Pepo. Ovary wall has hard rind. Watermelon, cucumber
Hesperidium. Wall has leathery rind. Orange, lemon
Drupe. One or two seeds. Thin skin, part of flesh around seed enclosed in hardened ovary tissue. Cherry, almond
Aggregate Fruits
Many ovaries of one flower attached to same receptacle. Many seeds. Raspberry, blackberry (not really "berries")
Multiple Fruits
Combined from ovaries of many flowers. Pineapple, fig

fruit

One fruit — ovule — wall of ovary

Embryo forming

embryo

Heart-shaped embryo

shoot tip

cotyledons

root tip

endosperm

Torpedo-shaped embryo

Mature embryo

Figure 27.6 Embryonic development of shepherd's purse (*Capsella*), a eudicot.

Figure 27.7 Examples of fruits. (**a**) Simple, dry wind-dispersed fruits of maple (*Acer*). (**b**) Cedar waxwing eating mountain ash berries, a simple fleshy accessory fruit. (**c**) Cocklebur, a prickly accessory fruit that sticks to the fur of animals. (**d**) Strawberry (*Fragaria*). This aggregate fruit is mostly accessory tissues. The fleshy red portion is derived from the receptacle. When mature, the receptacle has many individual fruits on its surface.

wall of one fruit

enlarged receptacle tissues

is a simple, fleshy, accessory fruit of a type called a pome. The core is derived from the ovary of a single flower; the rest of the fruit is an enlarged floral tube.

SEED DISPERSAL

All fruits function in seed dispersal. They are adapted to specific dispersing agents, such as currents of air or water, or animals passing by. Winds disperse fruits of maples (*Acer*). Part of the fruit wall extends out, like wings (Figure 27.7*a*). When a fruit breaks in half and drops, air currents spin it sideways. It may whirl far from the parent plant, which outcompetes seedlings for water, minerals, and sunlight.

Brisk winds transport some fruits many kilometers from a parent tree. Air currents lift a dandelion fruit by its "parachute." The same currents readily disperse orchid seeds, which are as fine as dust particles.

Many mammals, birds, insects, and other animals are seed dispersers. As a strawberry passes through an animal gut, the embryo sporophytes escape digestive enzymes, which actually digest only a part of the seed coat. The thinned-down regions make it easier for the embryos to break through their hard seed coat during germination. This happens some time after the seeds are expelled from the animal body, in feces.

Many fruits have hooks, barbs, sticky substances, or other surface specializations that can adhere to fur, feathers, and other animal parts. Fruits of bur clover, cocklebur, and bedstraw are examples (Figure 27.7*c*).

Some water-dispersed fruits have heavy wax coats, and others have sacs of air that help them float. With its thick, water-repelling tissue, a coconut palm fruit is adapted to bob along in the open ocean. It can drift hundreds of kilometers before saltwater, which could kill the embryo, penetrates the fruit.

Humans are terrific agents of dispersal. We have transported seeds all over the world. Cacao, oranges, corn, and many other plants were domesticated and coaxed to reproduce in new places, in huge numbers.

A mature ovule, which encases an embryo sporophyte and food reserves inside a protective coat, is a seed. A mature ovary, with or without additional floral parts that have become incorporated into it, is a fruit.

Fruits are dry or fleshy. Different kinds of fruits develop from one or more ovaries, and many incorporate other tissues in addition to those of the ovary wall.

Seeds and fruits are structurally adapted for dispersal by air currents, water currents, and many kinds of animals.

27.4 Asexual Reproduction of Flowering Plants

Sexual reproduction is the predominant reproductive mode among flowering plants. But many species also reproduce asexually, which permits rapid production of genetically-identical offspring. We put the alternative to use when we propagate plants that interest us.

ASEXUAL REPRODUCTION IN NATURE

Many flowering plant species reproduce asexually by modes of **vegetative growth**, as listed in Table 27.2. In essence, the new roots and shoots grow right out from extensions or fragments of parent plants. Such asexual reproduction proceeds by way of mitosis. This means a new generation of offspring is genetically identical to the parent; it is a clone.

A "forest" of quaking aspen (*Populus tremuloides*) is a spectacular example of vegetative reproduction. Figure 27.8 shows shoot systems of one individual plant. That parent plant's root system gives rise to new shoots. One aspen forest in Colorado stretches across hundreds of acres and includes 47,000 shoots.

Barring rare mutations, trees at the north end of this clone are genetically identical to those at the south end. Water travels from roots near a lake all the way to shoot systems living in much drier soil. Dissolved ions travel in the opposite direction.

As long as environmental conditions favor growth and regeneration, such clones are about as close as an organism can get to being immortal. No one knows how old the aspen clones are. The oldest known clone might be a ring of creosote bushes (*Larrea divaricata*) growing in the Mojave Desert. It has been around for the past 11,700 years.

The reproductive possibilities are amazing. Watch strawberry plants send out horizontal, aboveground stems (runners), then watch the new roots and shoots develop at every other node. Most navel oranges we eat are produced by the descendants of one particular California tree that reproduced by **parthenogenesis**. This is a process by which an embryo develops from an unfertilized egg.

Parthenogenesis is sometimes stimulated when a pollen grain contacts a stigma but a pollen tube does not grow through the style. In some cases, hormones released by the stigma or pollen grains diffuse to the unfertilized egg and stimulate it to develop into an embryo. That embryo is $2n$ as a result of fusion of the products of mitotic cell division in the egg. A $2n$ cell outside the gametophyte also may be stimulated to develop into an embryo. In all cases, the plant that is produced by parthenogenesis has the same genes as the parent plant.

Table 27.2	Asexual Reproductive Modes of Flowering Plants	
Mechanism	**Examples**	**Characteristics**
Vegetative Reproduction on Modified Stems		
1. Runner	Strawberry	New plants arise at nodes along aboveground horizontal stems.
2. Rhizome	Bermuda grass	New plants arise at nodes of underground horizontal stems.
3. Corm	Gladiolus	New plants arise from axillary buds on short, carbohydrate-storing, underground stems.
4. Tuber	Potato	New shoots arise from axillary buds (tubers are enlarged tips of slender underground rhizomes).
5. Bulb	Onion, lily	New bulbs arise from axillary buds on short underground stems.
Parthenogenesis		
	Orange, rose	Embryo develops without nuclear or cellular fusion (for example, from an unfertilized haploid egg or by developing adventitiously, from tissue surrounding the embryo sac).
Vegetative Propagation		
	Jade plant, African violet	New plant develops from tissue or structure (a leaf, for instance) that drops from the parent plant or gets separated from it.
Tissue Culture Propagation		
	Orchid, lily, wheat, rice, corn, tulip	New plant induced to arise from a parent plant cell that is not irreversibly differentiated.

INDUCED PROPAGATION

Most houseplants, woody ornamentals, and orchard trees are clones. People typically propagate them from cuttings or fragments of shoot systems. For example, with suitable encouragement, a severed African violet leaf may form a callus from which adventitious roots develop. A callus is one type of meristem. Remember, a meristem is a localized region of young plant cells that retain the potential for mitotic cell division.

A twig or bud from one plant may be grafted onto a different variety of some closely related species. As one example, many vintners in France now graft their prized but disease-prone grapevines onto disease-resistant root stock imported from America.

Frederick Steward was one of the pioneers in the field of **tissue culture propagation**. He placed small bits of phloem from the roots of carrot plants (*Daucus carota*) in flasks along with a liquid growth medium. The liquid included sucrose, minerals, and vitamins.

Figure 27.8 Quaking aspen (*Populus tremuloides*), a plant that reproduces asexually by runners, which give rise to a stand of trees that is an impressive clone.

a Protoplast. This is a "naked" living plant cell without a wall, which digestive enzymes have stripped away.

b Protoplast regenerates the missing cell wall, and then mitotic cell divisions begin.

c Repeated divisions result in a clump of living cells. All of these cells are not yet differentiated.

d Cell divisions continue and result in a callus, a mass of undifferentiated cells that can be induced to give rise to a plant embryo.

e One somatic embryo. Many asexually formed embryos can be produced at the same time in tanks (bioreactors).

f Young plant that grew from a somatic embryo. Clones of thousands of plants are produced by such micropropagation methods.

Figure 27.9 One method of tissue culture propagation.

It also contained coconut milk, which Steward knew was rich in a then-unidentified chemical that induces cell divisions. As the flasks rotated, individual cells torn away from the tissue bits divided and formed multicelled clumps. Some clumps gave rise to new plants (Figure 27.9). These experiments were among the first to show that cells of specialized plant tissues still retain the genetic instructions that are required to produce a whole individual.

Researchers now use shoot tips and other parts of individual plants for tissue culture propagations. The techniques often prove useful when an advantageous mutant arises. For instance, a mutant may show better resistance to a disease that can cripple other plants of the same species. Tissue culture propagation is used to produce hundreds or thousands of identical plants from a single promising mutant. These techniques are being employed to improve food crops, including corn, wheat, soybeans, and rice. They also are being used commercially to increase production of hybrid orchids, lilies, and other prized ornamentals.

Flowering plants reproduce asexually by a variety of processes. Clonal propagation has produced some of the world's largest individual plants, and the oldest.

Plants can also be propagated asexually by culturing bits of plant tissue in the laboratory.

Asexual propagation yields genetically uniform offspring.

27.5 Patterns of Early Growth and Development

The growth of the embryo sporophyte idles before or after a seed has been dispersed from the parent plant. Later, if all goes well, the seed germinates. This is the first phase of the new plant's growth and development, one that requires genetically prescribed programs and environmental cues.

HOW DO SEEDS GERMINATE?

A flowering plant grows and develops according to a predetermined pattern. The pattern starts while the embryo sporophyte is still inside the seed coat. Figure 27.10 shows such an embryo in a grain of corn. A "grain" is simply one type of dry fruit. At a suitable time, it may germinate. **Germination** is the process by which an immature stage in the life cycle resumes its growth after a time of arrested development.

Germination depends on seasonal factors, such as the number of daylight hours, temperature, and levels of oxygen and moisture in the soil. Mature seeds, for instance, don't hold enough water to support cell expansion or metabolism. Where water is ample only on a seasonal basis, germination coincides with the spring rains. At that time, water molecules infiltrate seeds; the water is attracted to hydrophilic groups of proteins stored in endosperm or cotyledons. After a seed swells enough with water, its coat ruptures.

Once the seed coat splits, more oxygen reaches the embryo, and aerobic respiration moves into high gear. The embryo's meristematic cells now divide rapidly. Germination is finished when the primary root breaks out of the seed coat.

GENETIC PROGRAMS, ENVIRONMENTAL CUES

Have you ever noticed the buds along a young stem? Each holds a community of cells, a meristem, much like tissues of stem cells in your body. These cells can divide again and again, with descendants giving rise to more primary tissues and organs. The growth and development of new tissues are under gene controls, and environmental cues influence gene expression. Sunlight, gravity, temperature changes, and seasonal shifts in hours of darkness versus daylight are some cues that commonly affect development.

You might be surprised that growth, development, and survival of a plant require extensive coordination among cells, as they do in animals. Plant cells, too, communicate with other cells that are some distance away in the body. As you will see, hormones function as the main signals for plant cell communication.

Figure 27.11 shows two patterns of germination, growth, and development. The patterns have a genetic basis. All cells in a plant are descended from the same cell—a zygote—so they generally have inherited the same genes. But unequal divisions put daughter cells in different positions in the young plant body, and the cells start using different genes selectively.

For instance, activated apical meristematic cells start synthesizing a growth-stimulating hormone, which diffuses down into new tissues below. In those tissue regions, it stimulates cells to divide and elongate. The cells and their descendants do so in a direction that lengthens the shoot. Figure 4.24 shows the two basic patterns of enlargement available to plant cells.

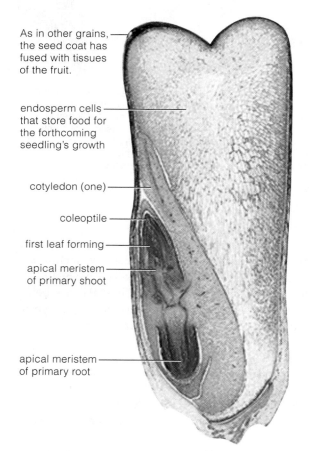

As in other grains, the seed coat has fused with tissues of the fruit.

endosperm cells that store food for the forthcoming seedling's growth

cotyledon (one)

coleoptile

first leaf forming

apical meristem of primary shoot

apical meristem of primary root

Figure 27.10 Embryo sporophyte and its food reserves inside a grain of corn (*Zea mays*).

Plant growth involves cell divisions and cell enlargements.

Plant development requires cell differentiation, as brought about by selective gene expression.

Interactions among genes, hormones, and the environment govern how each individual plant grows and develops.

Read Me First!
and watch the narrated animation
on plant development

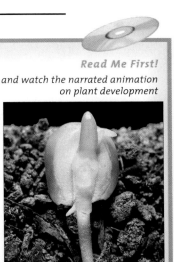

coleoptile

branch root

primary root

primary root

germinating seed   new seedling

first foliage leaf

adventitious root

branch root

primary root

prop roots that form on corn seedlings and that afford additional support for the rapidly growing stem

first foliage leaf

coleoptile

a Corn (*Zea mays*), a monocot. After a corn grain germinates, the seedling grows through the soil. A thin sheath (coleoptile) protects new leaves. Adventitious roots develop at the coleoptile's base.

one foliage leaf (this type is divided into three leaflets)

primary leaf

seed coat   primary root

cotyledons (two)

hypocotyl

primary leaf

withered cotyledon

primary leaf

branch roots

primary root

root nodule

germinating seed   new seedling

b Common bean plant (*Phaseolus vulgaris*), a eudicot. After a seed germinates, a hypocotyl (a hook-shaped shoot region below the cotyledons) forces a channel through soil for the cotyledons. Sunlight makes the hook straighten. Photosynthetic cells make food for several days in the cotyledons, which then wither and fall off. Foliage leaves take over photosynthesis.

Figure 27.11 Patterns of growth and development for a eudicot and a monocot.

27.6 Cell–Cell Communication in Plant Development

As a plant grows and develops, its cell types increase in number, size, and volume, and specialized tissues form. These events require chemical communication among different cell types, which secrete and respond to hormones and growth regulators in response to different genetic and environmental cues.

Plants put on new growth by repeated cell divisions at meristems (Chapter 25). Young plant cells enlarge as they take up water. The fluid pressure (*turgor* pressure) stretches their primary wall. Just as a soft balloon inflates easily, soft-walled cells expand under pressure.

About half of the meristematic cells don't increase in size, but they retain the capacity to divide. The rest divide and enlarge, then differentiate according to an inherited program that hormones and other molecules carry out. Cells interact in ways that give rise to plant parts with specific shapes and functions.

MAJOR TYPES OF PLANT HORMONES

Hormones, remember, are key signaling molecules that alter activities in target cells. After one binds to a receptor on a target, the signal is transduced. Signal transduction often alters the receptor's shape. This may affect gene expression, enzyme activity, or ion concentrations in the target cell's cytoplasm.

Table 27.3 lists the five major classes of hormones, called gibberellins, auxins, cytokinins, ethylene, and abscisic acid.

Figure 27.12 Gibberellin experiment. Controls (*left*) were two untreated cabbage plants. Three gibberellin-treated plants are at *right*.

Figure 27.12 shows one effect of **gibberellins**. These acidic compounds form in flowering plants, mosses, gymnosperms, ferns, and some fungi. For example, the fungus *Gibberella fujikuroi* is one pathogen that infects rice seedlings. Infection makes stems lengthen so much that they become spindly. The plants eventually topple and die. In the 1930s, researchers in Japan discovered they could cause the same effect by applying extracts from the fungus to healthy plants. They later purified the compound that triggered stem elongation and gave it the name gibberellin, after the fungus.

Remember the dwarf pea plants that Mendel used in some studies of heredity (Section 10.1)? Their short stems resulted from a mutation in a gene that affects gibberellin action. Gibberellins help seeds germinate by stimulating cell division and elongation. They also help induce flowering in some species.

Auxins form in the apical meristems of shoots and coleoptiles. A coleoptile is a sheath around the first primary shoot and keeps it from shredding as it grows through soil. Downward transport of auxins causes both structures to elongate by cell divisions (Figure 27.13*a–e*). Auxins also induce the cell divisions and differentiation that make vascular cambium. As seeds develop, an auxin stimulates fruit formation.

However, auxins also have inhibitory effects. For example, secretion of auxin by a shoot tip prevents the growth of lateral buds along the lengthening stem. This effect is known as *apical dominance*. Gardeners routinely pinch off shoot tips to end the supply of auxin in a main stem, so lateral buds are free to give rise to lateral branches. Auxins also stop abscission, a dropping of leaves, flowers, and fruits. They inhibit the growth of main roots, yet enhance the growth of roots from stems of plant cuttings (Figure 27.13*f*).

Table 27.3	Major Classes of Plant Hormones and Their Effects	
Hormone	Source and Mode of Transport	Stimulatory or Inhibitory Effects
Gibberellins	Young tissues of shoots, seeds, possibly roots. May travel in xylem and phloem	Make stems lengthen greatly through cell division and elongation. Help seeds germinate. Help induce flowering in some plants
Auxins	Apical meristems of shoots, coleoptiles, seeds. Transport in one direction: downward	Lengthen shoots, coleoptiles. Induce vascular cambium formation, fruit formation. Block lateral bud formation. Role in abscission
Cytokinins	Mainly the root tip. Travels from roots to shoots inside xylem	Stimulate cell division in roots and shoots, leaf expansion. Inhibit leaf aging. Applications release buds from apical dominance
Ethylene (a gas)	Most tissues undergoing ripening, aging, or stress. Diffusion in all directions	Stimulates fruit ripening, senescence of leaves, flowers; leaf and fruit abscission
Abscisic acid (ABA)	Mature leaves in response to water stress; may be produced in seeds	Stimulates stomatal closure. Induces transport of photosynthetic products from leaves to seeds. Stimulates embryo formation in seeds. May induce and maintain dormancy in some species

Read Me First!

and watch the narrated animation on auxin's effects

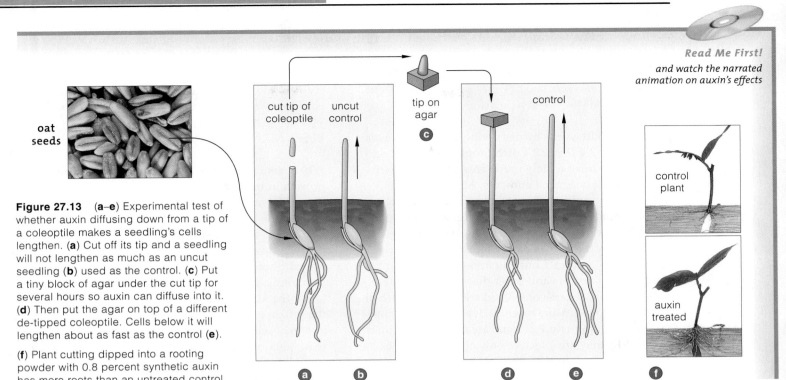

oat seeds

cut tip of coleoptile | uncut control

tip on agar

c

control

control plant

auxin treated

Figure 27.13 (a–e) Experimental test of whether auxin diffusing down from a tip of a coleoptile makes a seedling's cells lengthen. (**a**) Cut off its tip and a seedling will not lengthen as much as an uncut seedling (**b**) used as the control. (**c**) Put a tiny block of agar under the cut tip for several hours so auxin can diffuse into it. (**d**) Then put the agar on top of a different de-tipped coleoptile. Cells below it will lengthen about as fast as the control (**e**).

(**f**) Plant cutting dipped into a rooting powder with 0.8 percent synthetic auxin has more roots than an untreated control.

a **b** **d** **e** **f**

IAA (indoleacetic acid) is the most pervasive auxin in nature. Applying it in an orchard in springtime to thin flowers results in fewer but larger fruits. We also use it to keep mature fruit from dropping from trees. This way all the fruit can be picked at the same time. Synthetic auxins are used as herbicides, a subject we'll consider in the section to follow.

Cytokinins stimulate rapid cell division in root and shoot meristems and in maturing fruits. They oppose auxin's effects; they make lateral buds grow. They also keep leaves from aging and are used to prolong the shelf life of cut flowers and other horticultural prizes.

Ethylene, the only gaseous plant hormone, helps control the growth of most tissues, fruit ripening, and leaf dropping and other aging responses. Today, food distributors use ethylene to ripen tomatoes and other green fruits after shipping them to grocery stores. Fruit picked green does not bruise or deteriorate as fast. Ethylene is used to brighten rinds of citrus fruits.

Abscisic acid (ABA) was misnamed; it is a growth inhibitor that has little to do with abscission. ABA causes stomata to close in water-stressed plants. It induces and maintains dormancy in buds and seeds. ABA also stimulates transport of some photosynthetic products to seeds, and in the seeds it triggers protein synthesis and embryo formation. Growers apply ABA to nursery stock before it is shipped in order to induce dormancy. Plants that are in a dormant state bruise less easily when they are handled.

OTHER SIGNALING MOLECULES

Besides these main hormones, plants synthesize other growth regulators. The *brassinolides* resemble animal steroids in structure. They promote cell elongation and division; plants that lack them are dwarfed. *Jasmonates* are derived from fatty acids. They interact with other hormones in defense, germination of seeds, and root growth. *Salicylic acid*, the active ingredient in aspirin, was first discovered in willows. It helps plants defend themselves against pathogens. *Systemin*, a peptide, is produced when plant tissues are attacked by insects. It travels to other parts of the plant, where it turns on genes for substances that can interfere with an insect's ability to digest proteins. *Florigen* is the name given to a still unidentified chemical that stimulates flowering. Botanists suspect there are many more than these.

Interactions among hormones and other kinds of signaling molecules govern the normal growth, development, daily functioning, and reproduction of plants.

Plant hormones are signaling molecules secreted by specific cells that alter activities in target cells. At specific times, they stimulate or inhibit cell division, elongation, differentiation, and other events.

The main classes of plant hormones are gibberellins, auxins, cytokinins, abscisic acid, and ethylene.

27.7 Herbicides!

FOCUS ON THE ENVIRONMENT

Plants are plants, but when they grow where we don't want them, we call them weeds. Weeds compete with crops for sunlight, nutrients, and other resources. They slip into gardens and lawns. And so we unleash weedkillers.

Herbicides kill weeds by disrupting their metabolism and growth. *2,4-D* (for 2,4 dichlorophenoxyacetic acid) is one of the most widely used. This synthetic auxin is effective only against eudicots. Application of 2,4-D causes rapid, uncontrolled, and disorganized growth that leads to death of the targeted plant.

2,4-D was in the chemical mix in *Agent Orange*, a herbicide used to clear heavily forested areas during the Vietnam conflict. Contaminants in Agent Orange are linked to cancers and other disorders in Vietnam veterans, but 2,4-D is considered relatively safe when properly applied. More recently developed herbicides, such as Triclopyr and Picloram, are also auxin mimics. Like 2,4-D, they are effective only against eudicots.

The most widely used herbicide in agriculture is *atrazine*. After plant roots take it up in soil water, it moves through xylem all the way into leaves. There, it interferes with photosynthesis by disrupting electron transfers (Section 6.3). Atrazine has contaminated groundwater in many parts of the United States. Its use and possible effects on animal hormones are the focus of the opening essay in Chapter 31.

Glyphosate kills just about any plant it contacts by inhibiting an enzyme required for amino acid synthesis. Roundup, a highly popular herbicide produced by Monsanto, is largely a glyphosate salt.

Monsanto also produces genetically engineered plants that are resistant to application of Roundup. Roundup Ready soybeans, canola, cotton, and corn are examples. Once planted, these crops can be kept almost weed-free by blanketing them with periodic applications of Roundup. In contrast, using Roundup with conventional crops requires painstaking care to avoid accidentally treating the crop plants themselves.

All herbicides have the potential to kill nontargeted plants when applied indiscriminately. They may have other detrimental effects, as when they harm aquatic organisms. Prudent users avoid the most persistent and toxic types and apply others in minimal required amounts, and with great care. Even so, herbicides are a frequent contaminant of many ecosystems.

2,4-D   Atrazine   DDT

27.8 About Those Tropisms

Plant hormones orchestrate "tropic" responses of roots and shoots to cues from the environment. They guide the direction of growth toward or away from light, the Earth's gravitational force, or some other stimulus.

RESPONDING TO LIGHT

When stems or leaves adjust the rate and direction of growth in response to light, they show **phototropism**. The adjustments position plant parts so they can better capture the light required for photosynthesis.

Charles Darwin saw the response by a coleoptile growing toward light. Later, in the 1920s, Fritz Went linked phototropism with redistribution of a growth-promoting substance. He named the substance auxin (after the Greek *auxein*, "to increase"). Went showed that auxin moves from the tip of a coleoptile into cells less exposed to light and makes them elongate faster than cells on the illuminated side. The difference in growth rates made the coleoptile bend (Figure 27.14).

Blue wavelengths of light are known to induce the strongest phototropic responses. These wavelengths are absorbed by assorted *flavoproteins*. These yellow pigments might be the receptors that transduce light energy into the phototropic bending mechanism.

a Rays of sunlight strike one side of a coleoptile.   **b** The coleoptile bends after the auxin diffuses from tip to cells on shaded side.

c

Figure 27.14 Examples of phototropism in plants. In (**a**,**b**), hormone-mediated differences in the rates of cell elongation induce a coleoptile to bend toward light. In (**c**), seedlings of flowering shamrock (*Oxalis*) are busily demonstrating the response to light.

RESPONDING TO GRAVITY

Consider an embryo sporophyte in a corn seed. Under normal conditions, the first root to break through the seed coat always curves down, and the coleoptiles and primary shoot curve up. Flip the new seedling upside down. It adjusts; its primary root curves down and the shoot curves up (Figure 27.15*a*). A growth response to Earth's gravitational force is a form of **gravitropism**.

Turn the seedling on its side in a dark room and it still makes a gravitropic response. Its shoot curves up even in the absence of light. How does a plant sense which way is up? In that horizontally oriented stem, cell elongations slow down greatly on the side facing up, but elongations on the lower side increase fast. Different rates of elongation on opposing sides of the shoot make it bend up. Something makes cells on the "bottom" of the shoot that's been turned on its side *more* sensitive to a hormone, and those on top *less* so.

Auxin, together with a growth-inhibiting hormone in roots, may trigger such responses. Proteins that transport auxin across the cell membrane also play a role. Figure 27.15*b* shows seedlings that were placed perpendicular to the Earth's surface and then exposed to an inhibitor of auxin transporters. Unlike untreated control seedlings, they did not bend.

Gravity-sensing mechanisms of many organisms are based on **statoliths**. Plant statoliths are modified plastids made denser by clusters of starch grains. In response to gravity, statoliths in specialized root cap cells settle downward until they rest in the lowest cytoplasmic region (Figure 27.16). If a plant's roots become reoriented with respect to gravity, shifting of statoliths may cause *auxin* redistribution inside these cells. The change initiates the gravitropic response.

RESPONDING TO CONTACT

In **thigmotropism**, plants shift a direction of growth as they contact objects. Auxin and ethylene may guide the response. Vines, or stems too slender or soft to grow upright without support, make such a response. So do tendrils (modified leaves). As they grow against a stem, cells on the contact side stop elongating and the vine or tendril curls around it, often more than once (Figure 27.17). Afterward, the cells on both sides resume growth at the same rate.

Plants adjust the direction and rate of growth in response to environmental stimuli.

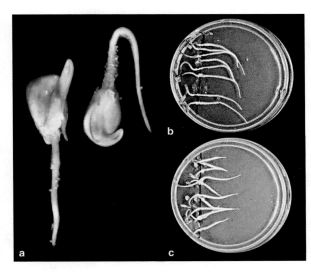

b Control group (petri dish of seedlings oriented perpendicular to gravitational force of Earth)

c Experimental group (seedlings oriented the same way as in (**b**) but exposed to an inhibitor of auxin transport protein)

Figure 27.15 (**a**) Gravitropic responses of a corn primary root and shoot growing in a normal orientation (*left*) and turned upside down (*right*).

(**b**) Experimental test of whether these responses require auxin transport proteins in root cell membranes. In primary roots turned on their side, auxin is laterally transported to the down-facing side, where it stops cells from lengthening. Cells of the upward-facing side continue to lengthen, so the roots bend downward. (**c**) In an experimental group exposed to auxin transport inhibitors, downward bending stopped. Mutations in genes that encode auxin transport proteins also stop the bending.

a   statoliths   **b**   statoliths

Figure 27.16 Effect of gravity on statolith distribution. (**a**) Normal orientation of gravity-sensing cells in a corn root cap. Statoliths in the cells settle downward. (**b**) Turn the root sideways. Ten minutes later, statoliths have settled to the new "bottom" of the cells. A gravity-sensing mechanism using statoliths may induce auxin redistribution in root tips. A difference in auxin concentrations makes cells on the top side of a horizontal root lengthen faster than cells on the bottom. Differences in elongation rates make the root tip curve downward.

Figure 27.17 Passion flower (*Passiflora*) tendril busily twisting thigmotropically.

Figure 27.18 Experiments showing that plants flower in response to night length. Each horizontal bar represents 24 hours. *Yellow* signifies daylight; *blue-gray* signifies night. (**a**) Long-day plants flower when night is *shorter* than a length critical for flowering. (**b**) Short-day plants flower when night is *longer* than a critical value.

(**c**) When an intense red flash interrupts a long night, both plants respond as if it were a short night. (**d**) A short pulse of far-red light after the red flash cancels the disruptive effect of the red flash by inactivating phytochrome.

(**e**) The flowering responses of a short-day plant, chrysanthemum, to (*left*) short-day exposure and (*right*) long-day exposure.

Figure 27.19 Conversion of the phytochrome molecule from inactive form (Pr) to active form (Pfr). The active form reverts to the inactive form in darkness. When active, phytochrome affects genes involved in growth and development, by mechanisms that are still not understood.

27.9 Control of Floral Development

Environmental cues, especially light levels, trigger the shift from making leaves to producing flowers. Development of flowers is controlled by a set of master switches that turn other genes on and off in ways that shape floral parts.

HOW DO PLANTS KNOW WHEN TO FLOWER?

You have probably noticed that different species of plants flower at different times of the year. Summer-blooming irises and spinach are *long-day* plants. The cockleburs and chrysanthemums are *short-day* plants; they flower in spring or autumn. *Day-neutral* plants, such as sunflowers, form flowers when they become mature, without regard to the season.

Actually, those names are misleading, because the cue that triggers flower formation is *night length*, not daylength. "Short-day" plants will flower only when nights are *longer* than a critical value, and "long-day" plants flower when nights are *shorter* than a critical value (Figure 27.18*a,b*).

Photoperiodism is the name for such a biological response to change in the length of daylight relative to darkness over twenty-four hours. **Phytochrome**, a type of receptor, helps a plant detect changes in light. It absorbs red and far-red wavelengths, with different results. Red ones activate phytochrome; far-red ones inactivate it (Figure 27.19). The relative amounts of red and far-red light vary with the time of day and through the seasons.

Simple experiments demonstrate its effect. When a dark interval is interrupted by a pulse of red light, phytochrome becomes activated, and a short-day plant won't flower (Figure 27.18*c*). But if the red light is followed by a pulse of far-red, phytochrome is inactivated and plants behave as if the dark period was uninterrupted (Figure 27.18*d*).

Phytochrome may affect the secretion of hormones that inhibit or induce a switch from leaves to flowers. But no one has yet identified the elusive hormone(s). Experiments indicate that leaves make signals that reach floral buds. In one experiment, all leaves but one were removed from a cocklebur (a short-day plant). The remaining leaf was enclosed in black paper for 8-1/2 hours a day to simulate long nights. The plant soon flowered. Moreover, when this leaf was grafted onto a cocklebur that had *not* been exposed to long periods of darkness, that plant flowered as well.

GENES THAT CONTROL FLOWER FORMATION

When it is time for a plant to flower, populations of cells that would otherwise give rise to leaves instead

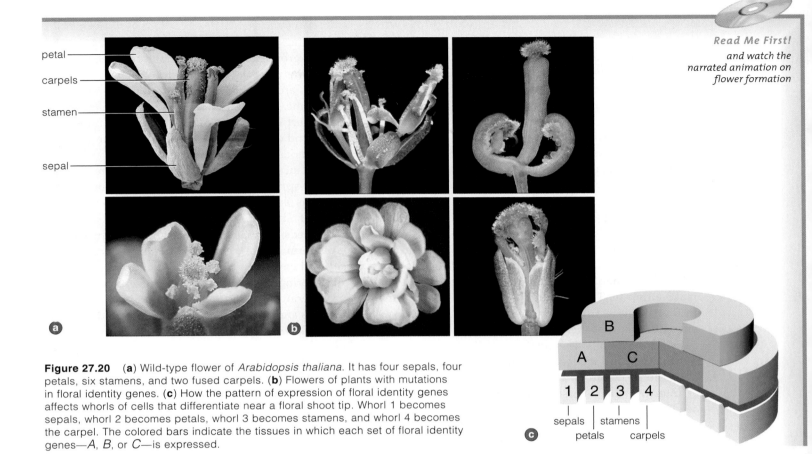

Read Me First!
and watch the narrated animation on flower formation

Figure 27.20 (**a**) Wild-type flower of *Arabidopsis thaliana*. It has four sepals, four petals, six stamens, and two fused carpels. (**b**) Flowers of plants with mutations in floral identity genes. (**c**) How the pattern of expression of floral identity genes affects whorls of cells that differentiate near a floral shoot tip. Whorl 1 becomes sepals, whorl 2 becomes petals, whorl 3 becomes stamens, and whorl 4 becomes the carpel. The colored bars indicate the tissues in which each set of floral identity genes—*A*, *B*, or *C*—is expressed.

differentiate into floral parts—sepals, petals, stamens, and carpels. How does the switch happen? The study of mutants of common wall cress, *Arabidopsis thaliana*, supports an **ABC model** for flowering. By this model, three groups of master switch genes designated *A, B,* and *C* control the formation of floral parts.

Divisions of meristematic cells at the tip of a floral shoot yield whorls of cells arranged one inside the next. The fate of cells in each whorl is determined by which components of the *ABC* group are turned on. In the outermost whorl, only the *A* genes are activated. These cells give rise to sepals. Moving inward, cells in the next whorl express *A* and *B* genes; they give rise to petals. Farther in are cells that turn on *B* and *C* genes; their descendants form stamens. Innermost is a whorl of cells that switch on *C* genes only. They give rise to two fused carpels. Figure 27.20*c* shows which genes govern which parts.

Evidence supporting the model comes from plants with mutations in genes of the *ABC* group. Mutation in one of the *A* group genes alters the two outermost whorls. The resulting flower has stamens and carpels but no petals. A mutation in the *B* group influences the second and third whorls, so petals are replaced by

sepals, and stamens by carpels. A mutation in the *C* group changes the innermost whorls. The result is a lovely but sterile flower—additional petals form in place of the stamens and carpels.

These *ABC* genes are analogous to the homeotic genes that control the mapping of body parts during the development of animals. *ABC* genes themselves are controlled by the product of another gene known as *Leafy*. Some mutations in *Leafy* result in shoots with no flowers. And what switches on *Leafy*? Evidence now points to a steroid hormone.

Phytochrome, a blue-green pigment, is part of a switching mechanism for photoperiodic responses to light of red and far-red wavelengths. Photoperiodic responses to red and far-red light involve phytochrome. The activated form of this photoreceptor, Pfr, may trigger secretion of one or more hormones that induce and inhibit flowering at different times of year.

The main environmental cue for flowering is the length of night—the hours of darkness, which vary seasonally. Different kinds of plants form flowers at different times of year, depending on their phytochrome mechanism.

Products of a set of master genes govern flower formation by selectively acting on meristematic cells of floral shoots.

Chapter 27 Plant Reproduction and Development   463

27.10 Life Cycles End, and Turn Again

In spring and summer, auxin from leaf cells moves into stems, where it interacts with cytokinins and gibberellins to maintain growth. With dwindling hours of daylight in autumn, leaf growth shuts down as nutrients are channeled to storage sites in twigs, stems, and roots.

In autumn, plants shed flowers, leaves, and fruits by a process called **abscission**. At the base of a leaf stalk or some other part about to drop, an abscission zone of parenchyma forms. Figure 27.21 shows one example.

Senescence is the sum of all processes that lead to death of a plant or some of its parts. The recurring cue is a decrease in daylength. But other factors, such as drought, wounds, and nutrient scarcity, also bring it about. Auxin production in leaves and fruits slows. A molecular signal stimulates abscission zone cells to make ethylene. The cells enlarge and deposit suberin in their walls. At the same time, the cells synthesize enzymes that digest cellulose and pectin in the layers between their walls. The outcome? Cells separate from one another, and the leaf or some other part above the abscission zone drops away.

Interrupt the diversion of nutrients into flowers, seeds, or fruits, and you slow the aging of a plant's other parts. For example, remove a plant's flowers or seed pods, and its leaves and stems remain vigorous and green much longer. Gardeners routinely remove faded flowers to maintain vegetative growth.

For many plants, growth shuts down in autumn. It does so even when the temperature is mild, the sky is bright, and water remains plentiful. A plant that has stopped growing under conditions that seem (to us) suitable for growth has entered **dormancy**; metabolic activities are idling. Ordinarily, the plant's buds will not resume growth until environmental cues emerge at the start of spring. Cold night temperatures, long nights, and dry, nitrogen-poor soil are some cues that can promote the onset of dormancy.

Dormancy-breaking mechanisms operate between fall and spring. Temperatures often become milder, and rains and nutrients become available. With the return of favorable conditions, life cycles turn once more. Seeds germinate, buds resume growth, and new leaves form. These events require gibberellins, abscisic acid, and environmental cues.

Flowering is often a response to seasonal changes in temperature. For instance, unless the buds of some biennials and perennials are exposed to cold air in the winter, flowers will not form on their stems in spring. Low-temperature stimulation of flowering is called **vernalization** (from *vernalis*, which means "to make springlike"). Figure 27.22 shows an experiment that provided evidence of this effect.

Multiple cues from the environment influence hormonal secretions that stimulate or inhibit processes of plant growth and development. Changes in daylength, temperature, nutrient availability, and moisture are among the cues.

Figure 27.21 In autumn, the changing colors of leaves of a horse chestnut tree (*Aesculus hippocastanum*). The leaf scar at *right* defines an abscission zone. Before the leaf detached from the tree, a tissue formed in a horseshoe-shaped zone (hence the tree's name). The tissue helps prevent water loss and infection after plant parts drop away.

Branch exposed to cold outside air flowered.

Potted plant grown inside a greenhouse did not flower.

Figure 27.22 Localized effect of cold temperature on dormant buds of a lilac (*Syringa*). For this experiment, which started during the cold winter months, a lilac plant was kept inside a greenhouse, where the temperature was warmer than the outside air. But one of its branches protruded outside the greenhouse. Only the buds on that branch resumed growth in spring.

Summary

Section 27.1 A complex diploid sporophyte, usually with roots, stems, and leaves, is the dominant phase of flowering plant life cycles. They form flowers, which have nonfertile parts (sepals and petals), fertile male reproductive parts (stamens), and female reproductive parts (carpels). Pollinators are insects and other agents that transfer pollen from male to female reproductive parts of flowers of the same species. Most flowers coevolved with one or more specific pollinators.

Section 27.2 Two kinds of haploid spores form in flowers and give rise to male and female gametophytes.

In pollen sacs of anthers, haploid microspores form by way of meiosis. Each will become a pollen grain, with a male gametophyte inside that produces sperm nuclei.

A flower has one or more carpels often fused together. Each carpel has an ovary. One or more cell masses form on the ovary wall and become ovules. One cell in each mass, a megaspore, gives rise to a female gametophyte. One gametophyte cell gives rise to an egg. Another has two nuclei; it helps form endosperm, a nutritive tissue.

Pollination is the arrival of pollen grains on a receptive stigma (the upper part of a carpel). After a pollen grain germinates, one of its cells develops as a pollen tube, which has two sperm nuclei inside. A pollen tube grows down through tissues of the ovary, toward the egg. With double fertilization, one of its nuclei fertilizes the egg, forming a zygote. The other nucleus fuses with the two nuclei of the cell that will become endosperm.

Section 27.3 A seed is a mature ovule. It consists of the embryo sporophyte and its food reserves inside a protective seed coat. Eudicot embryos have two seed leaves, or cotyledons. Monocot embryos have one. The ovary wall and sometimes other tissues form fruits around seeds. Fruits function in seed dispersal.

Section 27.4 Many flowering plants can reproduce asexually by sending out runners or by other types of vegetative growth. Tissue culture propagation is a laboratory technique used to propagate thousands to millions of plants from vegetative cells.

Section 27.5 Embryo sporophytes are dormant in seeds until conditions favor germination (resumption of growth). Enzymes are activated and the embryo's meristematic cells begin to divide rapidly. Genes, signaling molecules, and environmental cues interact and guide growth and development.

Section 27.6 Plant hormones are signaling molecules secreted by specific cells that alter activities in target cells. Along with growth regulators, hormones interact in guiding the rates and direction of growth and development. There are five main classes:

Gibberellins make stems lengthen greatly, help seeds germinate, and stimulate flowering in long-day plants and in biennials.

Auxins make stems and coleoptiles lengthen, promote vascular cambium and fruit formation, and help plants make phototropic and gravitropic responses.

Cytokinins interact with auxins to stimulate growth of roots, shoots, lateral buds, and fruits.

Ethylene induces tissue growth, the ripening of fruits, and abscission.

Abscisic acid promotes stomatal closure, induces and maintains bud and seed dormancy, and has major roles in food storage and embryo formation in seeds.

Section 27.7 Herbicides are applied to protect crop plants. They interfere with plant growth by a variety of mechanisms, including mimicking the hormone auxin.

Section 27.8 Plant parts make tropic responses to light, gravity, and other environmental conditions by altering rates and direction of growth. In gravitropism, roots grow downward and stems grow upward in a response to gravity. In phototropism, stems and leaves bend and grow toward light. In thigmotropism, they grow in response to contact with some solid object.

Section 27.9 The main cue for flowering is the number of hours of darkness, which varies seasonally. The pigment phytochrome acts in these photoperiodic responses. Flowers form when a set of master switches induces descendants of meristematic cells at shoot tips to differentiate into floral structures instead of leaves.

Section 27.10 With senescence, processes lead to the death of a plant or plant part. With dormancy, a perennial or biennial stops growing even when conditions appear suitable for continued growth.

Self-Quiz

Answers in Appendix III

1. The _____ , which bears flowers, roots, stems, and leaves, dominates the life cycle of flowering plants.
   a. sporophyte
   b. gametophyte
   c. megaspore
   d. microspore

2. Seeds are mature _____ ; fruits are mature _____ .
   a. ovaries; ovules
   b. ovules; stamens
   c. ovules; ovaries, mostly
   d. stamens; ovaries, mostly

3. Haploid _____ form by meiosis in a pollen sac.
   a. megaspores
   b. microspores
   c. seeds
   d. endosperm cells

4. Cotyledons develop as part of _____ .
   a. endosperm
   b. stamens
   c. embryo sporophytes
   d. ovaries

5. A whole plant grows from a tissue or structure that became separated from a parent plant. This is _____ .
   a. double fertilization
   b. vernalization
   c. vegetative propagation
   d. photoperiodism
   e. senescence

6. Plant hormones _____ .
   a. interact with one another
   b. are influenced by environmental cues
   c. are active in plant embryos within seeds
   d. are active in adult plants
   e. all of the above

Figure 27.23 Field of sunflowers (*Helianthus*) that are busily demonstrating solar tracking.

7. _____ stops lateral buds from branching out.
   a. Gibberellin   c. Cytokinin
   b. Auxin   d. Ethylene

8. In Michigan, a long-day plant will flower _____ .
   a. in July   c. all year round
   b. in April   d. whenever the air warms up

9. Match the terms with the most suitable description.
   _____ phototropism
   _____ photoperiodism
   _____ double
         fertilization
   _____ ovule
   _____ carpel
   _____ abscission

   a. flowers, leaves, fruits drop from a plant in autumn
   b. formation of zygote and first cell of endosperm
   c. starts out as cell mass in ovary; may become a seed
   d. female floral reproductive structure
   e. light triggers adjusted rate and direction of growth
   f. change in length of daylight relative to darkness causes it

Media Menu

Student CD-ROM

Impacts, Issues Video
   Imperiled Sexual Partners
Big Picture Animation
   Plant life cycles and development
Read-Me-First Animation
   Floral structure and function
   Eudicot life cycle
   Plant development
   Auxin's effects
   Flower formation
Other Animations and Interactions
   Seed development in *Capsella*
   Phototropism
   Gravitropism

InfoTrac

- On the Nectar Trail. *Audubon*, March 2001.
- Birds, Bees, and Even Nectar-Feeding Bats Do It. *Smithsonian*, April 2000.
- Cloning for Conservation. *Audubon*, November–December 2001.
- Ghost Stories from the Ice Age. *Natural History*, September 2001.
- Springing to Life. *Audubon*, March–April 2002.

Web Sites

- Electronic Field Trip to an Orchard:
  http://www.ca.uky.edu/agc/dl/ftrips/orchard/orchact.htm
- How a Corn Plant Grows:
  maize.agron.iastate.edu/corngrows.html
- Plants-in-Motion:
  sunflower.bio.indiana.edu/~rhangart/plantmotion

How Would You Vote?

Microencapsulated pesticides are a special threat to bees. The tiny pellets are designed for long-term release of some poison. They closely resemble pollen in size. When dusted on flowers of crops, the pellets are gathered up and carried to the hive, where they can kill an entire colony. Should we prohibit the use of these pesticides on flowering crops that are attractive to honeybees?

Critical Thinking

1. Photosynthesis sustains plant growth, and sunlight energy sustains photosynthesis. Why, then, do seedlings germinated in a fully darkened room grow taller than seedlings of the same species that germinated in the sun?

2. Most plants evolved under selection pressure of attacks by insects and other organisms and often repel attackers with natural toxins. A *toxin* is an organic compound, a normal metabolic product of one species, but its chemical effects harm or kill a different organism that contacts it. Humans, too, encounter traces of natural toxins, even in such familiar foods as hot peppers, potatoes, figs, celery, rhubarb, and alfalfa sprouts. Still, we do not die in droves. Propose a few reasons why the toxins of "edible" plants don't kill us.

3. How we protect crop plants is just one aspect of what must be done to support a human population that now exceeds 6.2 billion. Should vital crop plants be genetically engineered? Can we, and should we, make them better at fixing nitrogen or tolerating salts? What about bruise resistance? Built-in pesticide resistance?

4. *Solar tracking* refers to the observation that many plants maintain the flat blades of their leaves at right angles to the sun throughout the day (Figure 27.23). This tropic response enhances photosynthesis. Name one type of molecule that might be involved in the response.

5. Belgian scientists isolated a mutated gene in wall cress (*Arabidopsis thaliana*) that makes too much auxin. Predict what some of the resulting phenotypic traits might be.

6. Elaine, a plant physiologist, succeeded in introducing genes for pest resistance into petunia cells. How might she use tissue culture propagation to produce many petunia plants with those genes?

7. All but one of New Zealand's native large-billed birds are extinct. The kereru, the survivor, is the only dispersal agent for several native trees that have big seeds and fruits. One, the Puriri (*Vitex lucens*) is a highly valued hardwood. Explain, in terms of natural selection theory, why we might expect to see no new Puriri trees in the near future.

VI How Animals Work

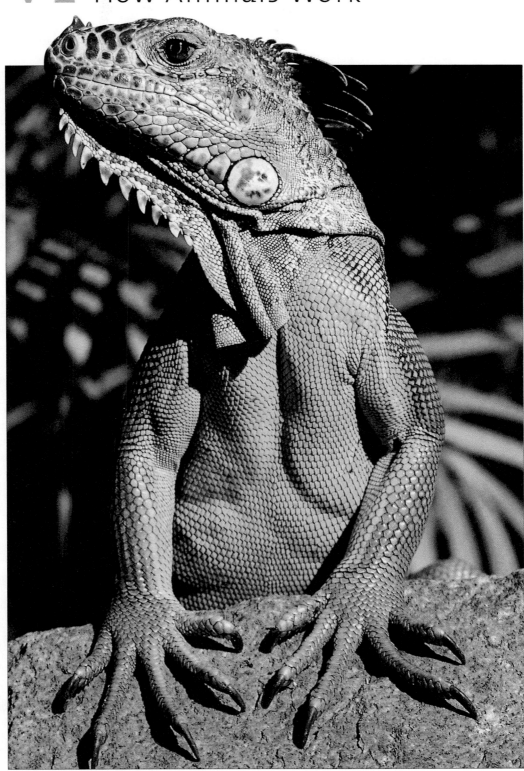

How many and what kinds of body parts does it take to function as a lizard in a tropical forest? Make a list of what comes to mind as you start reading Unit VI, then see how resplendent the list can become at the unit's end.

IMPACTS, ISSUES *Open or Close the Stem Cell Factories?*

Each year in the United States, about 10,000 people injure their spinal cords. In 1995 actor Christopher Reeve, known for his role as Superman, was one of them. A fall from a horse left him paralyzed and unable to breathe on his own. He became an advocate for the disabled and a champion of human embryonic stem cell research before his death in 2004.

Stem cells are any undifferentiated cells that give rise to specialized ones. The first cells to form when a fertilized animal egg divides are like this. Embryonic stem cells cultured in the laboratory can be coaxed to divide again and again. Their descendants often can be coaxed into developing into blood cells, muscle cells—

Figure 28.1 Christopher Reeve before and after a severe spinal cord injury left him paralyzed.

and, of interest to Reeve, nerve cells. Stem cells may one day help people affected by paralysis, leukemia, damage from heart attacks, Parkinson's disease, and other ailments.

Researchers have already used embryonic stem cells to fix spinal cord injuries in laboratory rats. Dividing and differentiating cells replaced damaged tissues, then the rats regained limited use of their legs.

Some people argue it is unethical to use human embryonic cells for any reason. Embryonic stem cell lineages already established were taken from aborted embryos, from clusters of cells that formed by *in vitro* fertilization but were never used, or from blood inside discarded umbilical cords, as in the filmstrip.

Adults retain patches of stem cells in bone marrow, fat, and other tissues. Researchers at Duke University have been experimenting with stem cells from fat that was removed by liposuction. So far, they have prodded descendant cells to become bone, cartilage, or nerve cells. Adult stem cells, however, are not as versatile as embryonic ones. They form fewer kinds of tissues and often lose the ability to differentiate when maintained in culture for long periods.

Stem cell research invites you to start thinking more closely about this unit's topics. You will be looking at how animals are put together (their anatomy) and how they function in the environment (their physiology).

Amazingly, it takes only four basic types of tissues to construct nearly all animals—epithelial, connective,

the big picture

Sheetlike Linings Sheetlike epithelial tissues line the body's outer surface and its inner cavities and tubes. They serve protective and secretory functions. Junctions structurally and functionally connect adjoining cells.

Internal Tissues Most types of internal tissues are connective such as cartilage, bone, and blood. They help support, organize, protect, and often insulate the other tissues—epithelia, muscle tissues, and nervous tissue.

muscle, and nervous tissues. A *tissue*, remember, is a community of cells and intercellular substances that are interacting in the performance of one or more tasks, such as contraction by muscle tissue. Each *organ* consists of different tissues organized in proportions and patterns necessary to carry out particular tasks. The vertebrate heart, for instance, has all four types of tissues, in predictable proportions and arrangements. In *organ systems*, two or more organs and structures interact physically, chemically, or both in a common task, as when interconnected vessels transport blood, driven by the force of a beating heart.

Even normal activities of cells, tissues, and organs change the composition of the internal environment, which in most animals is blood and interstitial fluid. Homeostasis, remember, is the state in which internal operating conditions stay within ranges that individual cells can tolerate. It is maintained by interactions that range from the level of cells to organ systems.

Whether you look at a flatworm or salmon, a bird or human, these are the challenges the body faces:

1. Function in ways that ensure homeostasis.

2. Acquire water, nutrients, and other materials; distribute them; and dispose of wastes.

3. Protect tissues against injury or attack.

4. Reproduce and, in many species, help nourish and protect the offspring through their early growth and development.

 How Would You Vote?

Many researchers believe that embryonic stem cell studies would greatly benefit medical science. Others object to the use of any cells derived from human embryos. Should researchers be allowed to start embryonic stem cell lines from human embryos that were frozen but not used for in vitro fertilization? See the Media Menu for details, then vote online.

Tissue Research Sometimes the body cannot replace damaged tissues or synthesize vital substances. Researchers are already making laboratory-grown replacement tissues. They are also experimenting with designer organs.

Organ Systems Vertebrates have the same assortment of basic organ systems. Each system has specialized functions, but all interact and contribute to the survival of the body as a whole. Skin is an example of the integumentary system.

469

28.1 Epithelial Tissue

*Most of what you see when you look in the mirror—
your skin, hair, and nails—is epithelial tissue or structures
derived from it. Sheetlike epithelium also lines internal
tubes and cavities, including airways, blood vessels, and
the digestive, urinary, and reproductive tracts.*

WHAT IS EPITHELIUM?

Like other tissues, **epithelium** (plural, epithelia) is an
organized array of cells and intercellular substances.
This type is sheetlike, with one free surface exposed
to the outside environment or a body fluid (Figure
28.2). The opposite surface is attached to the tissue by
a noncellular, protein-rich basement membrane. The
cells are close together in an epithelial sheet, with little
extracellular material between them. Some cell types
are specialized for absorbing or secreting substances.

Squamous (flattened), cuboidal, or columnar cells
are the most common in epithelia. *Simple* epithelium
is a single cell layer thick. *Stratified* epithelium has two
or more cell layers. The outer layer of your own skin
is mainly stratified squamous epithelium. Figure 28.2
shows examples.

GLANDULAR EPITHELIUM

In all animals, cells in some epithelia secrete products
that are unrelated to their metabolism and are used
elsewhere. Most animals also have **glands**: secretory

free surface
of epithelium

a

simple
epithelium

basement
membrane

underlying connective tissue

b

Figure 28.2 Characteristics of
epithelium, an animal tissue with
a free surface exposed to a body
fluid or the outside environment.

(**a**) Section near the surface of
stratified squamous epithelium.
It consists of multiple layers and
cells that become flattened near
the free surface. (**b**) In all epithelia,
the opposite surface rests on a
basement membrane that binds it
to an underlying connective tissue.

(**c**) Light micrographs and sketches
of three simple epithelia, showing
three of the most common shapes
of cells in this tissue.

c

TYPE: Simple squamous
DESCRIPTION: Friction-reducing
slick, single layer of flattened cells
COMMON LOCATIONS: Lining of
blood and lymph vessels, heart;
air sacs of lungs; peritoneum
FUNCTION: Diffusion, filtration,
secretion of lubricants

TYPE: Simple cuboidal
DESCRIPTION: Single layer of
squarish cells
COMMON LOCATIONS: Ducts,
secretory part of small glands;
retina; kidney tubules; ovaries,
testes; bronchioles
FUNCTION: Secretion, absorption

TYPE: Simple columnar
DESCRIPTION: Single layer of tall cells;
free surface may have cilia, mucus-
secreting glandular cells, microvilli
COMMON LOCATIONS: Glands, ducts;
gut; parts of uterus; small bronchi
FUNCTION: Secretion, absorption;
ciliated types move substances

Figure 28.3 Section through the glandular epithelium of a frog. This frog, of the genus *Dendrobates*, makes one of the most lethal glandular secretions known. Natives of one tribe in Colombia use its exocrine gland secretion to poison tips of darts, which they shoot out of blowguns.

Pigment-rich cells that extend all through this epithelium impart color to the skin. The skin of all poisonous frogs has distinctive colors and patterns that evolved as a clear warning signal. In essence, it says to predators, "Don't even think about it."

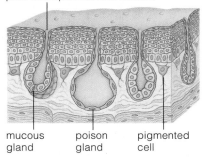

pore that opens at skin surface

mucous gland poison gland pigmented cell

Read Me First!

and watch the narrated animation on animal cell junctions

organs derived from epithelia. Bear in mind, *secretion* is not the same as *excretion*, or the concentrating and removal of substances of no use to the body.

Exocrine glands secrete mucus, saliva, earwax, oil, milk, tears, digestive enzymes, and other products. They are released onto the free surface of epithelium by way of ducts or tubes (Figures 28.2*c* and 28.3). The **endocrine glands** have no ducts. They secrete their products, hormones, directly into a body fluid. The hormones typically enter the bloodstream and travel to target cells some distance away. Chapter 31 takes a look at endocrine glands and their secretions.

CELL JUNCTIONS

In epithelia, as in most tissues, cell junctions connect adjoining cells. **Tight junctions** stop most substances from leaking across a tissue. Rows of proteins form tight seals by fusing each cell to its neighbors (Figure 28.4*a*). Substances must pass *through* epithelial cells to get to the other side. The plasma membrane of these cells includes transport proteins that selectively allow only certain molecules or ions across (Section 5.6).

Think about the acidic fluid in your stomach. If it were to leak across the stomach's epithelial lining, it would digest proteins of your body instead of the proteins you eat. That is what happens in people who get peptic ulcers. The epithelial lining of the stomach is breached, typically by a bacterial infection, and so strong acid leaks into the internal environment.

Figure 28.4 Cell junctions in animal tissues. (**a**) In tight junctions, rows of proteins running parallel with a tissue's free surface prevent leaks between adjoining cells. (**b**) Adhering junctions are spot welds or collars between adjoining cells. Many intermediate filaments of the cytoskeleton anchor them to the plasma membrane. (**c**) Gap junctions are cylindrical arrays of proteins paired up between the plasma membranes of two cells. They are open channels for rapid signaling between the cytoplasm of adjoining cells.

Adhering junctions are like spot welds that lock cells together (Figure 28.4*b*). They are profuse in skin and other body parts subjected to ongoing abrasion. **Gap junctions** permit ions and small molecules to flow between the cytoplasm of abutting cells (Figure 28.4*c*). They are communication channels in smooth muscle, the heart, and other tissues in which activities of adjacent cells must be coordinated.

Epithelia are sheetlike tissues covering the body surface and lining its cavities, ducts, and tubes. Epithelia have one free surface exposed to a body fluid or the outside environment. Glands are secretory organs derived from epithelium. Cell junctions structurally and functionally link adjoining cells.

28.2 Connective Tissues

Connective tissues are the most abundant and widely distributed animal tissues. They range from soft types to cartilage, bone tissue, adipose tissue, and blood. Different kinds organize, strengthen, functionally support, and sometimes insulate other tissues.

All connective tissues except blood have cells called fibroblasts, which make and secrete structural fibers of collagen and elastin. Cells also secrete modified polysaccharides, which accumulate between cells and fibers as the tissue's "ground substance."

SOFT CONNECTIVE TISSUES

Soft connective tissues all have the same components but different proportions of them. **Loose connective tissue** is the most common connective tissue in the body. Its fibroblasts and fibers are very loosely arranged in a semifluid ground substance. Figure 28.5*a* shows an example. Loose connective tissue often is present beneath epithelia. White blood cells patrol it and make early counterattacks against pathogens that breach the epithelial linings.

Dense connective tissues have an abundance of fibers in the extracellular matrix. In **dense, irregular connective tissue**, fibroblasts and collagen fibers are oriented every which way (Figure 28.5*b*). This tissue is part of skin. It also supports intestinal muscles, and it forms protective capsules around kidneys and other organs that do not stretch much.

Dense, regular connective tissue has orderly rows of fibroblasts between parallel bundles of fibers, as in Figure 28.5*c*. This organization helps keep the tissue from being torn apart under mechanical stress. All tendons, which connect skeletal muscle to bones, are primarily dense, regular connective tissue. So are the ligaments that attach one bone to another.

SPECIALIZED CONNECTIVE TISSUES

In vertebrate embryos, **cartilage** forms as specialized cells secrete a solid but pliable intercellular material, which eventually imprisons them (Figure 28.5*d*). This rubbery tissue resists compression. It maintains the shape of the outer ear, the nose, and some other body parts. It protects and cushions the joints between limb bones, between bones of the vertebral column, and elsewhere. Blood vessels do not run through cartilage. Individual cells imprisoned inside it are serviced by diffusion only, which moves substances to and from them. In adults, cartilage cells do not divide very often. Damage to cartilage is notoriously slow to heal.

collagenous fiber
fibroblast
elastic fiber
a

collagenous fibers
b

collagenous fibers
fibroblast
c

ground substance with very fine collegen fibers
cartilage cell (chondrocyte)
d

TYPE: Loose connective tissue
DESCRIPTION: Fibers, fibroblasts, other cells loosely arranged in extensive ground substance
COMMON LOCATIONS: Beneath skin and most epithelia
FUNCTION: Elasticity, diffusion

TYPE: Dense, irregular connective tissue
DESCRIPTION: Collagen fibers, fibroblasts occupy most of the ground substance
COMMON LOCATIONS: In skin and in capsules around some organs
FUNCTION: Structural support

TYPE: Dense, regular connective tissue
DESCRIPTION: Collagen fibers bundled in parallel, long rows of fibroblasts, little ground substance
COMMON LOCATIONS: Tendons, ligaments
FUNCTION: Strength, elasticity

TYPE: Cartilage
DESCRIPTION: Chondrocytes inside pliable, solid ground substance
COMMON LOCATIONS: Nose, ends of long bones, airways, skeleton of cartilaginous fish, vertebrate embryo
FUNCTION: Support, flexion, low-friction surface for joint movements

Figure 28.5 Characteristics of connective tissue.

cartilage on knobby end of a long bone

compact bone tissue

spaces in spongy bone tissue

Figure 28.6 Cartilage and bone tissue. Spongy bone tissue has needlelike hard parts with spaces between. Compact bone tissue is more dense. Bone, a load-bearing tissue, resists being compressed. It favored increases in size as many land vertebrates, including the ancestors of giraffes, evolved. It gives large animals selective advantages. They can ignore most predators and roam farther for food and water. They gain or lose heat more slowly than smaller animals do; they have a lower surface-to-volume ratio.

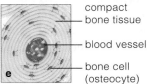

compact bone tissue

blood vessel

bone cell (osteocyte)

e

nuclues

cell bulging with fat droplet

f

TYPE: Bone tissue
DESCRIPTION: Collagen fibers, osteocytes occupying extensive calcium-hardened ground substance
LOCATION: Bones of all vertebrate skeletons
FUNCTION: Movement, support, protection

TYPE: Adipose tissue
DESCRIPTION: Large, tightly packed fat cells occupying most of ground substance
COMMON LOCATIONS: Under skin, around the heart and kidneys
FUNCTION: Energy storage, insulation, padding

white blood cell   platelet

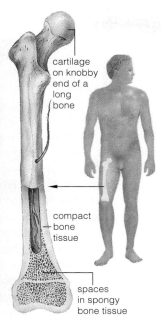

red blood cell

Figure 28.7 Cellular components of human blood. Many diverse proteins, nutrients, oxygen and carbon dioxide, and other substances also are dissolved in plasma, blood's straw-colored fluid portion.

Bone tissue is a hardened connective tissue with living cells imprisoned in their mineralized secretions (Figure 28.5e). It is the main tissue of bones: organs that support and protect the soft internal tissues and organs and interact with muscles to move the body. Figure 28.6 shows one of the limb bones, which have load-bearing functions. As explained in Sections 32.2 and 33.3, some bones are sites of blood cell formation.

Adipose tissue is an energy reservoir. Fat droplets form inside many cells as they are converting excess carbohydrates and lipids to fats. But cells of adipose tissue get so swollen with stored fat that the nucleus gets pushed off to one side (Figure 28.5f). This tissue has little extracellular matrix but lots of blood vessels that rapidly move fats to and from individual cells. Fat deposits under the skin form an insulating layer and cushion certain body parts. They also accumulate around some organs, such as the kidneys and heart.

Blood is generally classified as connective tissue because its cellular components form from stem cells in bone, a connective tissue. Blood is a mostly fluid transport medium. The fluid, plasma, is mostly water with diverse proteins, gases, ions, sugars, and other substances dissolved in it. It also transports red blood cells, white blood cells, and platelets (Figure 28.7). Red blood cells deliver oxygen to metabolically active tissues, and remove carbon dioxide. A variety of white blood cells defend and repair tissues. Platelets have a role in blood clotting.

Connective tissues organize, support, protect, and insulate other tissues. Except for blood, they have characteristic proportions and arrangements of fibroblasts and fibers embedded in a ground substance of their own making.

Soft connective tissues have their components arranged densely in irregular or regularly parallel patterns.

Cartilage, bone, blood, and adipose tissue are specialized connective tissues. Cartilage and bone are both structural materials. Adipose tissue is a reservoir of stored energy. Blood is a fluid connective tissue with transport functions.

28.3 Muscle Tissues

In muscle tissues cells contract, or forcefully shorten in response to stimulation, then relax and passively lengthen. These tissues consist of many cells in parallel, in tight or loose arrays. Coordinated contractions of layers or rings of muscles move the whole body or its components.

The three broad categories of muscle tissue are called skeletal muscle, cardiac muscle, and smooth muscle tissues. Each type has unique properties that reflect its specific functions.

In chordates ranging from lampreys to humans, **skeletal muscle tissue** attaches to skeletal elements and helps the body move and maintain positions. The tissue has parallel arrays of long cylindrical muscle cells, or *muscle fibers* (Figure 28.8*a*). Inside each cell are long strands, or myofibrils, divided transversely into many contractile units. Skeletal muscle has a striated (striped) appearance because of these repeating units. Section 32.4 describes the fine structural organization of these units and explains how the units work.

Skeletal muscle tissue makes up about 40 percent of the weight of an average human. Reflexes activate it, but we also make it contract by thinking about it, as is happening in Figure 28.8*a*. That's why skeletal muscles are sometimes called "voluntary" muscle.

Only the heart wall contains **cardiac muscle tissue** (Figure 28.8*b*). Its cells contract as a unit in response to signals that flow through gap junctions between their abutting plasma membranes. Each branching cardiac cell has a single nucleus. It has far more mitochondria than other types of muscle cells, because the heart's incessant contractions demand a lot of ATP on an uninterrupted basis.

Cardiac muscle tissue also is striated, although less conspicuously so. Cardiac muscle and smooth muscle are both said to be "involuntary" muscle; we usually can't make them contract just by thinking about it.

We find layers of **smooth muscle tissue** in the wall of many soft internal organs, including the stomach, bladder, and uterus. It has unbranching cells, tapered at both ends, with one centrally positioned nucleus. Contractile units are not arranged in orderly repeating fashion, as they are in skeletal and cardiac muscle tissue. In other words, smooth muscle tissue is *not* striated (Figure 28.8*c*). Its contractions are slower than those of skeletal muscle, but they can be sustained much longer. Contractions drive many internal events, as when they propel material through the gut, shrink the diameter of arteries, and close sphincters.

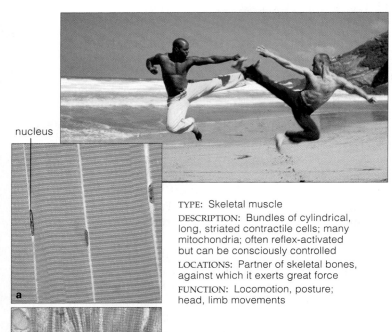

nucleus

TYPE: Skeletal muscle
DESCRIPTION: Bundles of cylindrical, long, striated contractile cells; many mitochondria; often reflex-activated but can be consciously controlled
LOCATIONS: Partner of skeletal bones, against which it exerts great force
FUNCTION: Locomotion, posture; head, limb movements

a

junction between abutting cells

TYPE: Cardiac muscle
DESCRIPTION: Unevenly striated, fused-together cylindrical cells that contract as a unit owing to signals at gap junctions between them
LOCATIONS: Heart wall
FUNCTION: Pump blood forcefully through circulatory system

b

TYPE: Smooth muscle
DESCRIPTION: Contractile cells tapered at both ends; not striated
LOCATIONS: Wall of arteries, sphincters, stomach, intestines, urinary bladder, many other soft internal organs
FUNCTION: Controlled constriction; motility (as in gut); arterial blood flow

c

Figure 28.8 (**a**) Skeletal muscles, the functional partners of bones, move the vertebrate body. Row after row of contractile units give skeletal muscle cells a striated appearance. (**b**) Striated cells of cardiac muscle tissue. The dark horizontal bands where cells abut hold communication junctions. (**c**) Smooth muscle tissue, showing tapered cells and no striations.

Muscle tissue contracts in response to stimulation. It helps move the body and its component parts. Contraction requires ATP, and all muscle tissues are rich in mitochondria.

Skeletal muscle attaches to and moves skeletal elements such as bones. Cardiac muscle is present only in the heart wall. Smooth muscle tissue is present in many soft internal organs.

28.4 Nervous Tissue

Of all animal tissues, the one with the communication lines made of neurons exerts the most control over how the body senses and responds to changing conditions.

Nervous tissue is composed of neurons and a variety of cells, collectively called neuroglia, that structurally and functionally support them. The **neurons** make up the communication lines in most nervous systems.

All cells respond to stimulation, but a neuron (and muscle cell) is highly excitable in a specific way. It can propagate a message, without decay, along its plasma membrane, all the way to an output zone. There, the message causes the release of signaling molecules that can enhance or suppress the activity of a neighboring neuron or muscle cell.

More than 100 billion neurons form communication lines in your own body. The sensory neurons detect specific changes in conditions. Interneurons in your brain and spinal cord integrate the information and coordinate immediate and long-term responses. Motor neurons, of the sort shown in Figure 28.9, deliver the commands from the brain and spinal cord to muscles and glands. And that is a topic of later chapters.

> Neurons are the basic units of communication in nervous tissue. Different kinds detect specific stimuli, integrate information, and issue or relay commands to other tissues.
>
> Also in nervous tissue are neuroglial cells that structurally and functionally support the neurons.

Figure 28.9 A motor neuron relaying signals to the muscle cells involved in rapidly unrolling a chameleon's tongue precisely in the direction and to the distance of the stimulus, a tasty fly.

28.5 Tissue Engineering   *FOCUS ON SCIENCE*

Growing new tissues and organs in the laboratory was once a pipe dream. Today, tissue engineers have had some success doing just that.

As each new animal grows and develops, its cells interact in prescribed ways to give rise to the body's tissues. Cells of new tissue synthesize specific gene products that are crucial for normal body functions.

For decades, medical researchers worked to make tissues in a laboratory. Lab-grown epidermis (Figure 28.10a) for treating skin ulcers is one of the successes. One company makes "skin" by growing cells from foreskins that were discarded from circumcised male infants. They grow cells on a layer of collagen from cattle tendons. Lab-grown skin is simplified—it lacks some cell types and glands —but it helps protect the underlying tissues and helps wounds heal.

A more challenging project is a laboratory-grown bladder. At present, when a person has their bladder removed, surgeons fashion a replacement by using a segment of the intestine. This is a less than optimum solution; intestinal cells in the replacement organ absorb materials that should be excreted.

Using another approach, Anthony Atala and his colleagues made a replacement organ with a lining of bladder cells. As an experiment, they grew epithelial cells from the lining of a dog's bladder on the inner surface of a biodegradable plastic shell. They grew smooth muscle cells on the shell's outer surface.

When the experimental bladders were inserted into dogs, blood vessels and nerves grew into them as the shells degraded (Figure 28.10b). Nerves made the proper connection and restored bladder functions. If tests in humans are successful, bioartificial bladders might become available in less than ten years.

Also on the horizon are *designer organs*, each type containing cells that synthesize hormones, enzymes, growth factors, or other substances patients cannot make themselves. If a bit of the patient's epithelium encloses the cells, then the immune system might not attack the designer organ. Biotechnologists are now close to making molecular cues that will help designer organs remain in suitable locations inside the body and also will help them function properly.

Figure 28.10 (a) A large sheet of epidermis grown in the laboratory. (b) X-ray revealing the position of a bioartificial bladder that was surgically inserted into a dog.

28.6 Overview of Major Organ Systems

Organ systems perform specialized functions, such as gas exchange, blood circulation, and locomotion. At least some of their activities contribute to day-to-day survival. Their formation and specialized functions are outcomes of a long-term program of growth and development.

Figure 28.11 has an overview of eleven organ systems of a typical vertebrate, an adult human. It lists some of the terms we use when describing positions of organs. It also shows major body cavities in which a number of important organs are located.

The amazing thing is that the internal environment stays within tolerable limits even though all of these organ systems are putting various amounts and kinds of substances into it and withdrawing substances from it as they grow, develop, and maintain themselves. Talk about clockwork!

As the skeletal system and muscular system move us, say, toward sources of nutrients and water, their cells are withdrawing oxygen from interstitial fluid the circulatory system delivered to it. Their movement helps keep the blood circulating, as when contractions of leg muscles help drive blood in veins back to the heart. At the same time, blood is picking up carbon dioxide wastes from the interstitial fluid bathing their cells. Also at the same time, the respiratory system is replenishing oxygen for them by pulling in more from the air even as it forces out the carbon dioxide wastes —and so it goes, throughout the entire body.

The clock started ticking when germ cells, a type of immature reproductive cell, gave rise to a sperm and an egg. (The body's other cells, remember, are somatic, after a Greek word for body.) At fertilization, a zygote formed, then mitotic cell divisions turned it into an embryo. Then embryonic cells arranged themselves into three primary tissues—ectoderm, mesoderm, and endoderm. In all vertebrates, these are the embryonic forerunners of all adult tissues and organs. **Ectoderm** gives rise to the skin's outer layer and to the nervous system's tissues. **Mesoderm** is the source of muscles, bones, and most of the circulatory, reproductive, and urinary systems. **Endoderm** gives rise to linings of the digestive tract and organs derived from it. The parts have been interacting ever since (Chapter 38).

Many organ systems serve specialized functions, such as gas exchange, blood circulation, or locomotion.

Vertebrate tissues, organs, and organ systems arise from three primary tissues that form in the early embryo. These tissues are ectoderm, mesoderm, and endoderm.

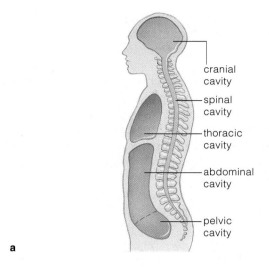

cranial cavity

spinal cavity

thoracic cavity

abdominal cavity

pelvic cavity

a

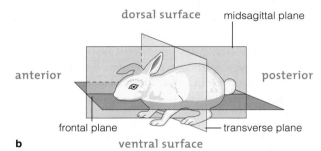

dorsal surface

midsagittal plane

anterior

posterior

frontal plane

transverse plane

b

ventral surface

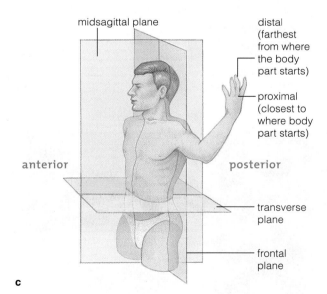

midsagittal plane

distal (farthest from where the body part starts)

proximal (closest to where body part starts)

anterior

posterior

transverse plane

frontal plane

c

Figure 28.11 (**a**) Major cavities in the human body. (**b,c**) Directional terms and planes of symmetry for the body. Most vertebrates move with the main body axis parallel to Earth's surface. For them, *dorsal* refers to their back (upper surface) and *ventral* to the opposite, or lower surface. In humans, who are upright, *anterior* refers to the front of a standing person; it corresponds to ventral. *Posterior*, the back, is equivalent to dorsal in the rabbit. (**d**) Human organ systems and their functions.

Integumentary System

Protects body from injury, dehydration, and some pathogens; controls its temperature; excretes certain wastes; receives some external stimuli.

Nervous System

Detects external and internal stimuli; controls and coordinates the responses to stimuli; integrates all organ system activities.

Muscular System

Moves body and its internal parts; maintains posture; generates heat by increases in metabolic activity.

Skeletal System

Supports and protects body parts; provides muscle attachment sites; produces red blood cells; stores calcium, phosphorus.

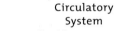

Circulatory System

Rapidly transports many materials to and from cells; helps stabilize internal pH and temperature.

Endocrine System

Hormonally controls body functioning; works with nervous system to integrate short-term and long-term activities.

d

Lymphatic System

Collects and returns some tissue fluid to the bloodstream; defends the body against infection and tissue damage.

Respiratory System

Rapidly delivers oxygen to the tissue fluid that bathes all living cells; removes carbon dioxide wastes of cells; helps regulate pH.

Digestive System

Ingests food and water; mechanically, chemically breaks down food and absorbs small molecules into internal environment; eliminates food residues.

Urinary System

Maintains the volume and composition of internal environment; excretes excess fluid and blood-borne wastes.

Reproductive System

Female: Produces eggs; after fertilization, affords a protected, nutritive environment for the development of new individuals. *Male:* Produces and transfers sperm to the female. Hormones of both systems also influence other organ systems.

28.7 Human Skin—Example of an Organ System

Go back in time to the Cambrian seas, when jawed predatory fishes were evolving. Some prey species had heavy armor-plated integuments. Although protective, armor plates do not lend themselves to moving fast or making hairpin turns. Among many bony fishes, a thinner, layered integument—skin—evolved.

Skin, your body's largest organ, can stretch, conserve body water, and repair small cuts or burns. It helps make vitamin D and helps cool you by dissipating metabolic heat. Its sensory receptors help the brain assess the outside world. Skin even contains white blood cells that help defend the body.

Two key features of skin evolved when vertebrates moved onto dry land. First, keratinocytes evolved in the outer layer. These cells secreted keratin, a tough, water-resistant protein, and they made skin more durable and more waterproof. Second, glands with ducts became recessed below the surface.

Vertebrate skin has two layers, with the epidermis above attached to a lower dermis (Figure 28.12). The **epidermis** is a stratified squamous epithelium made up mainly of keratinocytes. These cells are linked by adhering junctions. Ongoing mitotic division of cells in the deepest epidermal layers pushes previously formed cells toward the skin's surface. Abrasion and pressure exerted by the growing cell mass flatten and kill cells before they reach the surface. Dead ones are continually rubbed off or flaking away.

Langerhans cells are phagocytes that migrate through epidermis. When they encounter bacteria or virus particles, they swiftly engulf them. Then they sound molecular alarms that mobilize the immune system (Chapter 34). When ultraviolet (UV) radiation damages Langerhans cells, the skin becomes much more vulnerable to viral outbreaks, such as the cold sores caused by *Herpes* virus.

The **dermis**, mainly a dense connective tissue, has fibers of stretch-resistant elastin and supportive collagen. Blood vessels, lymph vessels, and sensory receptors thread through it. The hypodermis (not a part of skin) anchors the dermis to tissues below. It consists mainly of loose connective tissue and adipose tissue, which insulates and cushions some body parts (Figure 28.12).

Variations in skin color reflect differences in the distribution and activity of melanocytes. These cells make **melanin**, a brownish-black pigment, which they donate to the keratinocytes. Melanin shields the body from harmful UV radiation. In pale skin, little melanin forms. Such skin often appears pink because hemoglobin's red color shows through thin-walled blood vessels and epidermis. An orange pigment, carotene, also contributes to skin color.

Human skin has many exocrine glands, including mammary glands, oil glands, and about 2.5 million sweat glands. Secretions from sweat glands help the body dissipate heat. Secretions from the oil glands lubricate and soften hair and skin, and kill bacteria.

An average human scalp has about 100,000 hairs, made mostly of keratin (Figure 28.13). Hairs form by mitotic divisions of specialized epidermal cells. Genes, nutrition, and hormones affect hair growth. Protein deficiency thins hair (amino acids are required for keratin synthesis). High fever, emotional stress, and excess vitamin A intake also cause hair to thin.

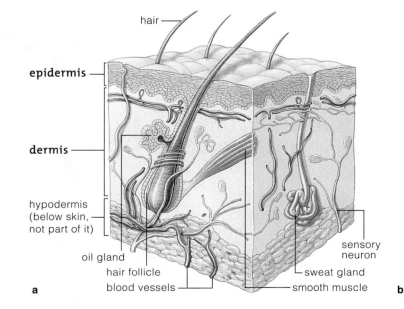

Figure 28.12 (**a**) Structure of skin that is typical of body parts with hair, oil glands, and sweat glands. (**b**) A section through part of human skin.

hair

epidermis

dermis

hypodermis (below skin, not part of it)

oil gland
hair follicle
blood vessels
smooth muscle
sweat gland
sensory neuron

a

outermost epidermal layer (all dead cells)

keratinized cells being flattened

rapidly dividing cells of epidermis

dermis

b

Figure 28.13 Hair fine structure. Dead, flattened hair cells accumulate and form a tubelike cuticle around a hair's shaft. They are derived from modified skin cells that make the polypeptide chains of the protein keratin. Disulfide bridges link three chains into thin fibers, which become bundled into larger fibers. The fibers almost fill the cells, which in time die off. Figure 28.14 shows how people play with the disulfide bridges.

Figure 28.14 (a) Curly or straight hair? It depends on the disulfide bridges between keratin chains. Bridges break when exposed to chemicals. Their keratin chains are held in new positions by ironing (flattening out) hair or rolling it around curlers. Exposure to a different chemical makes new disulfide bridges form between different sulfur-bearing amino acids. Displaced bonding locks the chains in new positions. That is how many women, including actress Nicole Kidman (b,c), straighten their naturally curly hair.

THE VITAMIN CONNECTION

When you sit out in the sun, exposure to UV light stimulates epidermal cells to make cholecalciferol, a precursor of vitamin D. ("Vitamin D" is a generic name for steroid-like compounds that help the body absorb calcium from food.) At the same time, UV exposure causes the breakdown of a B vitamin, folate. Among other things, an embryo's nervous system cannot develop normally without folate.

By one hypothesis, variations in skin color among human populations are adaptations to differences in exposure to sunlight. Humans first evolved beneath the intense sun of the African savanna. Even with melanin-rich skin that helped them keep enough folate, people could still synthesize plenty of vitamin D. What happened after some populations migrated to colder regions, where they spent a lot of time bundled up or indoors? We can hypothesize that embryo-protecting dark skin was less favored and increased UV exposure was more favored. In such regions, paler skin offered a selective advantage.

This hypothesis was tested by comparing skin color of people native to more than fifty countries with annual UV levels. The data supported the hypothesis. Increasingly lower levels of annual UV light correlate with increasingly lighter skin.

OF SUNTANS AND SHOE-LEATHER SKIN

Melanocytes that UV light stimulates into making melanin produce the "tan" that many light-skinned people covet. Melanin production accelerates slowly, then peaks about ten days after the first exposure. Before then, a light-skinned person with unprotected skin can become mildly to severely burned.

Dark-skinned people are better protected, but only at first. Prolonged or repeated UV exposure makes elastin fibers in the dermis clump up and damages collagen. Chronically tanned skin loses its resiliency and starts to look like shoe leather. UV also damages DNA in skin cells, increasing the risk of skin cancer.

As we age, epidermal cells divide less often. Skin thins and is more vulnerable to injury. Glandular secretions that kept it soft and moistened dwindle. Elasticity fades as collagen and elastin fibers become sparse. Wrinkles deepen. Many people needlessly accelerate the aging process by indulging in tanning or smoking, which shrinks the skin's blood supply.

how to get shoe-leather skin in a tanning salon

Summary

Section 28.1 A tissue is an aggregation of cells and intercellular substances that perform one or more common tasks. Epithelial tissues cover external body surfaces and line internal cavities and tubes. Epithelium has one free surface exposed to some body fluid or the environment. Gland cells and glands are derived from epithelium. Endocrine glands are ductless and secrete hormones. Exocrine glands secrete other products, such as sweat or milk, through ducts to body surfaces.

Animal tissues have a variety of cell-to-cell junctions. Tight junctions prevent substances from leaking across the tissue. Adhering junctions cement neighboring cells together. Gap junctions are open channels that connect the cytoplasm of abutting cells. They permit the rapid transfer of ions and small molecules between cells.

Section 28.2 Connective tissues bind, organize, support, strengthen, protect, and insulate other tissues. Soft connective tissues have different proportions and arrangements of protein fibers, fibroblasts, and other cells in ground substance. Specialized connective tissues include cartilage, bone tissue, adipose tissue, and blood.

Section 28.3 Muscle tissues contract (shorten), then passively lengthen. They help move the body and its component parts. The three types are skeletal muscle, smooth muscle, and cardiac muscle tissue. Skeletal muscle and cardiac muscle are striated.

Section 28.4 Neurons make up the communication lines of nervous tissue. Different kinds detect, integrate, and assess internal and external conditions, and deliver commands to the glands and muscles that carry out responses. Neuroglial cells in nervous tissue protect and support the neurons.

Section 28.5 Researchers have successfully grown some human tissues and replacement organs in the laboratory. Such bioengineered products are not as complex as their natural counterparts.

Section 28.6 An organ is composed of different tissues combined in definite proportions and patterns that allow them to perform one or more common tasks. An organ system is two or more organs that interact chemically, physically, or both in tasks that help keep individual cells as well as the whole body functioning.

Tissues, organs, and organ systems interact to provide the stable internal environment required for survival of individual cells and the body as a whole.

All adult tissues and organs arise from three tissues that form during development: endoderm, mesoderm, and ectoderm.

Section 28.7 The organ system called skin functions in protection, temperature control, detection of shifts in external conditions, vitamin D production, and defense.

Self-Quiz   *Answers in Appendix III*

1. _____ tissues are sheetlike with one free surface.
   a. Epithelial   c. Nervous
   b. Connective   d. Muscle

2. _____ function in cell-to-cell communication.
   a. Tight junctions   c. Gap junctions
   b. Adhering junctions d. all of the above

3. In most animals, glands are located in _____ tissue.
   a. epithelial   c. muscle
   b. connective   d. nervous

4. Most _____ have many collagen and elastin fibers.
   a. epithelial tissues c. muscle tissues
   b. connective tissues d. nervous tissues

5. _____ is mostly plasma.
   a. Adipose tissue   c. Cartilage
   b. Blood   d. Bone

6. Your body converts excess carbohydrates and proteins to storage fats, which accumulate in _____ .
   a. epithelial tissue   c. adipose tissue
   b. dense connective tissue d. both b and c are correct

7. In your body, cells of _____ can shorten (contract).
   a. epithelial tissue   c. muscle tissue
   b. connective tissue   d. nervous tissue

8. Only _____ muscle tissue has a striated appearance.
   a. skeletal   c. cardiac
   b. smooth   d. a and c are correct

9. _____ detects and integrates information about changes and controls responses to those changes.
   a. Epithelial tissue   c. Muscle tissue
   b. Connective tissue   d. Nervous tissue

10. Match the terms with the most suitable description.
    ____ exocrine gland   a. strong, pliable; like rubber
    ____ endocrine gland   b. secretion through duct
    ____ cartilage   c. outermost primary tissue
    ____ ectoderm   d. contracts, not striated
    ____ smooth muscle   e. cements cells together
    ____ blood   f. fluid connective tissue
    ____ adhering   g. ductless secretion
         junction

Critical Thinking

1. *Porphyria*, a genetic disorder, occurs in about one in 25,000 individuals. Affected people lack enzymes of a metabolic pathway that forms heme, the iron-containing group of hemoglobin. Accumulation of porphyrins, which are intermediates of the pathway, results in frightening symptoms, especially after sunlight exposure. Lesions and scars form on skin (*right*). Hair grows thick on the face and hands. Gums retreat, so the canines take on a fanglike appearance. Ingesting alcohol, garlic, and some other substances makes symptoms worse. Affected individuals have to avoid sunlight and aggravating substances. They also can get injections of heme from normal red blood cells.

Figure 28.15 Assaults on the integument. Metal bits embedded in the skin of an individual with an unusual sense of personal style.

Figure 28.16 In the Kalahari Desert, gray meerkats (*Suricata suricatta*) face the sun's warming rays, as they do each morning. This simple behavior helps them maintain internal body temperature. How animals function in their environment is the focus of this unit.

Porphyria may have started vampire stories that were being told even before the Middle Ages. Superstitious folk didn't have a clue to the basis of the condition, but molecular biologists do. Research its possible causes.

2. The nose, lips, tongue, navel, and private parts are often targets for *body piercing:* cutting holes into the body so jewelry can be threaded through them (Figure 28.15). *Tattooing,* or using permanent dyes to make patterns in skin, is another fad. Besides being painful, both skin invasions can invite bacterial infections, chronic viral hepatitis, AIDS, and other diseases if the piercers and tattooers reuse unsterilized needles, dye, razors, gloves, swabs, and trays. Months may pass before any problems develop, so the cause-and-effect connection isn't always obvious. If despite the risks you think tissue invasions are okay, how can you be sure the equipment used is sterile?

3. Adipose tissue and blood are often said to be atypical connective tissues. Compared to other types of connective tissues, which features are *not* typical?

4. Opponents of using animals to test the safety of cosmetic products argue that alternative methods are available. Laboratory grown tissues are already used for some studies of this sort. Given what you have learned in this chapter about how tissues and organs interact, discuss the advantages and disadvantages of tissue-specific methods compared to tests using living animals.

5. After a cold night in Africa's Kalahari Desert, animals small enough to fit inside a coat pocket emerge stiffly from burrows. These "meerkats" are a type of mongoose. They stand on tiny hind legs and face east, exposing their chilled bodies to the warm rays of the morning sun (Figure 28.16). Once meerkats warm up, they fan out and search for food. Into the meerkat gut go insects and the occasional lizard. These are pummeled, dissolved, and digested into glucose and other nutritious bits small enough to move across the gut wall, into blood, and on to the body's cells.

Off the top of your head, name as many tissues as you can that might have roles in (1) keeping the meerkat body warm, (2) moving the body, as during foraging and heart-thumping flights from predators, and (3) digestion and absorption of nutrients, and elimination of the residues.

Media Menu

Student CD-ROM

Impacts, Issues Video
 Open or Close the Stem Cell Factories?
Big Picture Animation
 Tissues and organ systems
Read-Me-First Animation
 Animal cell junctions
 Human organ systems
Other Animations and Interactions
 Structure of epithelium
 Examples of connective tissue
 Directional terms and planes of symmetry

InfoTrac

- The Human Body Shop. *Technology Review*, April 2001.
- The Child Within. *Scientific American*, June 2002.
- Spring in Your Step? The Forces in Cartilage. *Science News*, July 2002.
- Skin Deep. *Scientific American*, June 2003.
- Hair Today. *BioScience*, April 2000.

Web Sites

- HistoWeb: www.kumc.edu/instruction/medicine/anatomy/histoweb
- Stem Cells Primer: stemcells.nih.gov/
- AMA Body Atlas: www.ama-assn.org/ama/pub/category/7140.html
- BBC Human Body and Mind: www.bbc.co.uk/science/humanbody/

How Would You Vote?

Many biologists believe that embryonic stem cells offer the most potential for developing ways to treat hundreds of human diseases. Stem cell lineages have already been established (aborted fetuses and cells cultured but not used for in vitro fertilization were sources). Should new stem cell lineages be started from these same sources?

IMPACTS, ISSUES *In Pursuit of Ecstasy*

Ecstasy is a drug that makes you feel socially accepted, relieves anxiety, and sharpens the senses while giving you a mild high. It also can leave you dying in a hospital bed, foaming at the mouth and bleeding from every orifice as your temperature skyrockets. It can send your family and friends spiraling into horror and disbelief as they watch you stop breathing. Lorna Spinks ended life that way. She was nineteen years old (Figure 29.1).

Her anguished parents released the photograph at far right, taken minutes after her death. They wanted others to know what Lorna did not: Ecstasy can kill.

Ecstasy's active chemical ingredient, MDMA, is related to amphetamine, or "speed." Among other effects, it interferes with the function of serotonin, a signaling molecule that works in the brain. MDMA causes neurons to release too much serotonin. Instead of being cleared away, as normally happens, serotonin molecules saturate receptors on the target cells, which cannot be released from overstimulation. With luck, the serotonin surge results in feelings of increased energy, empathy, and euphoria.

Reality does not always coincide with expectations. MDMA can interfere with the body's ability to control its internal temperature. According to Lorna's friends, she became dizzy, incoherent, and flushed after taking a tablet of Ecstasy. She died after her body temperature rose high enough to shut down vital organ systems.

Few MDMA overdoses do end in death. Ecstasy's most commonly reported short-term side effects are panic attacks and temporary psychosis. The long-term effects are still not known. Today's Ecstasy users are unwitting guinea pigs for unscripted experiments.

Figure 29.1 Increased use of Ecstasy at all-night raves and other parties correlates with a rise in Ecstasy-related visits to emergency rooms. At right, Lorna Spinks, who was pursuing a college degree in sociology. Irma Perez did not get even that far. In 2004, the fourteen-year-old from Belmont, California, took one Ecstasy tablet at a slumber party, fell ill, and died in a hospital the next day.

the big picture

How Neurons Work Messages that travel swifty through the nervous system are based on electrical disturbances across the plasma membrane of individual neurons, and on signals sent between neurons and other cells.

Information Flow Chemical signals transmit information between neurons or from neurons to glands and muscles. Nerves are communication lines that carry messages back and forth between the brain and spinal cord and the rest of the body.

We do know that neurons do not rebound quickly when serotonin stores are depleted. We know that repeated experimental doses of MDMA change the number and structure of serotonin-secreting neurons in laboratory animals. And we know that depressed serotonin levels in humans contribute to a loss of concentration, memory problems, and depression. Ecstasy use has been linked to depression, but it is not clear which comes first in the individual. It may be that depressed individuals are more likely to use the drug than those who are emotionally balanced.

The effects on memory are better understood. As many studies show, the extent of memory loss increases with a rise in the frequency of drug use. At least over the short term, the capacity for memory seems to be restored to normal when Ecstasy use stops, although undoing the neural imbalances can take many months.

How do we function as individuals? It depends a lot on whether we decide to nurture, ignore, or abuse our nervous system—which is still forming its delicate connections well into our twenties.

In earlier chapters, you traced the evolution of nervous systems over hundreds of millions of years, first among invertebrates, then through vertebrate lineages ancestral to humans. Through that legacy, we alone measure our lives against a sense of history and a sense of destiny. As individuals, we alone can ask how we came to be, and where we are headed from here.

Think about it.

 ## How Would You Vote?

In one view, people caught using illegal drugs should not be jailed. In an opposing view, the threat of jail time in itself makes some individuals think twice before experimenting with possibly dangerous drugs. Should users of illegal drugs be compelled to enter mandatory drug rehabilitation programs as an alternative to jail? See the Media Menu for details, then vote online.

Nervous Systems Nearly all nervous systems have signal-conducting and information-processing communication lines. Information about conditions inside and outside the body are detected and assessed, then muscles and glands are commanded to make suitable responses.

The Human Brain As in all other vertebrates, the human brain incorporates ancient tissues that control reflex responses to stimuli. Conscious behavior and uniquely human abilities originate in the brain's more recently evolved layerings of neurons.

29.1 Neurons—The Great Communicators

Nervous systems evolved as a way to sense and respond with increasing precision to changing conditions inside and outside the body. They do so by way of communication lines that are made of excitable cells called neurons.

NEURONS AND THEIR FUNCTIONAL ZONES

Responding to sounds and sights, to odors, to hunger and passion, fear and rage—just about everything that an animal does in life starts with information flowing along communication lines of the nervous system. Are these lines silent unless they receive signals from the outside, much as telephone lines wait to carry calls from all over the country? No. For instance, before you were born, newly forming cells called neurons were starting to organize themselves in embryonic tissues. They began chattering among themselves. All through your life, in times of danger and reflection, excitement and sleep, their chattering has continued—and it will continue for as long as you do.

Compare the single neuron in Figure 29.2 with the communication line in Figure 29.3. Each neuron is

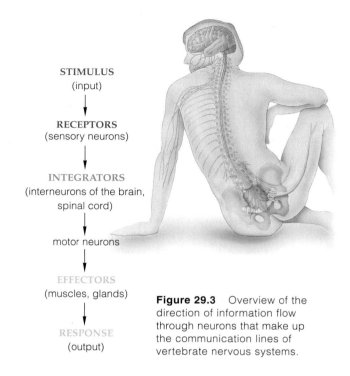

STIMULUS
(input)

↓

RECEPTORS
(sensory neurons)

↓

INTEGRATORS
(interneurons of the brain,
spinal cord)

↓

motor neurons

↓

EFFECTORS
(muscles, glands)

↓

RESPONSE
(output)

Figure 29.3 Overview of the direction of information flow through neurons that make up the communication lines of vertebrate nervous systems.

Read Me First!
and watch the narrated animation on neuron structure and function

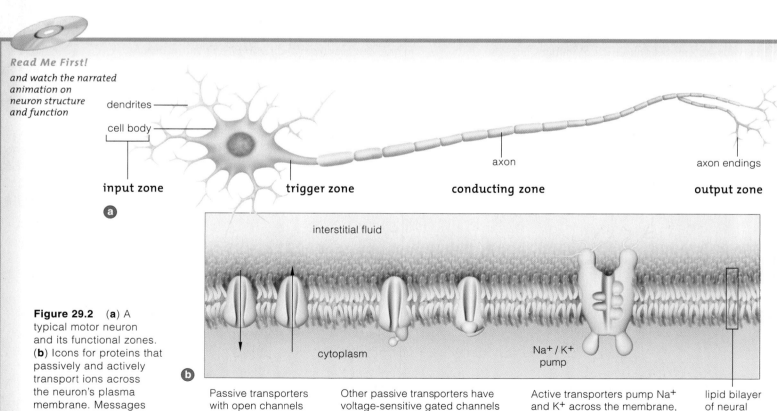

Figure 29.2 (**a**) A typical motor neuron and its functional zones. (**b**) Icons for proteins that passively and actively transport ions across the neuron's plasma membrane. Messages travel from the neuron's input zone to its output zone with the help of these proteins.

dendrites

cell body

input zone

trigger zone

axon

conducting zone

axon endings

output zone

interstitial fluid

cytoplasm

Na^+ / K^+
pump

Passive transporters with open channels let ions continually leak across the membrane.

Other passive transporters have voltage-sensitive gated channels that open and shut. They assist diffusion of Na^+ and K^+ across the membrane, down their concentration gradients.

Active transporters pump Na^+ and K^+ across the membrane, against their concentration gradients. They counter ion leaks and restore resting membrane conditions.

lipid bilayer of neural membrane

like one person in a bucket brigade. In such brigades, people in a line pass along buckets of water, hand to hand, to put out a fire. Neurons hand off information, not buckets. And in all animals with a brain, it takes different kinds of neurons to pass information along, from stimulus to response.

Sensory neurons detect and relay the information to the brain and, if the animal has one, a spinal cord. A *stimulus* is a specific form of energy, such as light, that a specific receptor has detected. Only the brain and spinal cord house **interneurons**, which do more than receive sensory information and issue calls for response. Interneurons also interact to integrate, store, and retrieve information. The **motor neurons** deliver commands from other neurons to muscles or glands, the body's effectors, which carry out responses.

All neurons have cytoplasmic extensions from the cell body, but these differ in number and in length. The cell body and slender extensions called **dendrites** are often *input* zones for information. Here, incoming signals cause a small or big ruckus. When big, the disturbance spreads to the *trigger* zone, an adjoining patch of plasma membrane where self-propagating signals arise. These signals travel down an **axon**, a slender, typically long extension that is the neuron's *conducting* zone. Axon endings, which are commonly branched, are *output* zones where information may be passed to other cells (Figure 29.2a).

Neurons are so essential for survival, they get a lot of help. They are metabolically assisted, structurally supported, and protected by an enormous number of diverse cells collectively called *neuroglia*.

A NEURON'S PLASMA MEMBRANE

When a neuron is "at rest," or not being stimulated, a small electric gradient is being maintained across its plasma membrane. The cytoplasmic fluid next to the membrane is a bit negatively charged compared to interstitial fluid outside. We measure such charges in millivolts. The amount of energy it takes to maintain this steady voltage difference (often −70 millivolts) is known as the **resting membrane potential**.

All cells carry a slight negative charge inside, but only in *excitable* cells can the electric gradient across the plasma membrane fleetingly reverse in response to adequate stimulation. The reversal in charge is an **action potential**. It sets in motion a whole series of fleeting reversals, just like itself, that travels from the trigger zone all the way down to an output zone in almost no time at all.

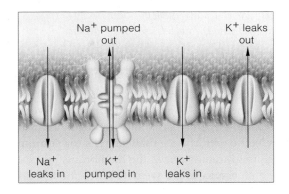

Figure 29.4 Pumping and leaking of sodium and potassium ions across a resting membrane.

Membrane crossings of particular ions are the key. Like other ions, ions of potassium (K^+) and sodium (Na^+) cannot cross the lipid bilayer of the neural membrane. They must be assisted by transport proteins that span the bilayer—where controls can be exerted over how many cross at a given time. Some transporters are like channels that never shut; they let a bit of Na^+ leak in and K^+ leak out all the time (Figures 29.2 and 29.4). Others are channels with molecular gates. Gates are shut tight in a motor neuron at rest. You will soon see what happens when a lot of them open at once.

If nothing were to counter the tiny back-and-forth leaks, the Na^+ and K^+ gradients across the resting membrane would in time vanish. However, **sodium–potassium pumps** maintain the gradients. They also restore them after every action potential. Remember, this type of pump is an active transporter (Section 5.6). When activated by a phosphate-group transfer from ATP, the ones that span the neural membrane move Na^+ out and K^+ in at the same time, against their concentration gradients.

With this bit of background, let's turn next to how an action potential arises and propagates itself.

Sensory neurons, interneurons, and motor neurons are excitable cells, the communication units of nervous systems.

In response to adequate stimulation, action potentials arise and propagate themselves from a trigger zone to an output zone of a neuron's plasma membrane.

An action potential is a fleeting reversal in the voltage difference across the plasma membrane. It occurs as certain ions are induced to flow through transport proteins that span the membrane's lipid bilayer.

29.2 How Is an Action Potential Propagated?

Action potential propagation is not hard to follow when you already know that ions move back and forth across the neural membrane through transport proteins.

Signals arriving at a neuron's input zone can be weak, intense, or something in between. Said another way, their magnitude varies, depending on the strength of the stimulus and how long it lasts. Suppose a weak signal arrives at the input zone of a motor neuron. It causes an electric disturbance that changes the sodium and potassium gradients across the membrane, but not much. The disturbance is too weak to reach the trigger zone, so an action potential doesn't happen. Why not? *Many gated channels for sodium must be opened to trigger an action potential, and input zones do not have them.*

Signals about an intense or long-lasting stimulus *can* spread along the membrane to the trigger zone. Here, sodium channels having voltage-sensitive gates are abundant. If the voltage difference across a resting membrane changes by a certain amount—a threshold level—then an action potential will inevitably follow.

Remember, the cytoplasm of a neuron at rest is more negatively charged than interstitial fluid outside (Figure 29.5a). At threshold, gated sodium channels in the trigger zone start to open. Sodium ions are free to follow their electric and concentration gradients across the membrane, and they start flowing into the

neuron. The cytoplasm becomes less negative, which compels more gates to open; more sodium flows in, which causes even more gates to open, and so it goes:

The ever increasing flow of sodium ions is a case of **positive feedback**; it intensifies as an outcome of its own occurrence. Stimulus strength no longer matters. The inward rush of positively charged ions past all the open gates reverses the voltage difference across the membrane (Figure 29.5b).

Figure 29.6 has a recording of the voltage across a neuron before, during, and after an action potential. At threshold, the voltage change always spikes with the same intensity, as an *all-or-nothing* response.

Each spike lasts just a millisecond or so. Why? The gates on sodium channels close and stop the inflow of sodium at the membrane patch where the charge was reversed. Also, halfway through the reversal, gated channels for *potassium* opened, and potassium flowed out. These events restored the voltage difference at the membrane patch, but not the original gradients (Figure 29.5c). The sodium–potassium pumps actively transport sodium back outside and potassium inside.

a Membrane at rest; inside of neuron negative with respect to the outside. An electrical disturbance (*yellow* arrow) spreads from an input zone to an adjacent trigger region of the membrane, which has a great number of gated sodium channels.

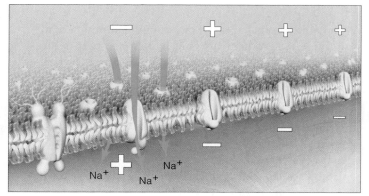

b A strong disturbance initiates an action potential. Sodium gates open. The sodium inflow decreases the negativity inside the neuron. The change causes more gates to open, and so on until threshold is reached and the voltage difference across the membrane reverses.

Figure 29.5 Propagation of an action potential along the axon of a motor neuron.

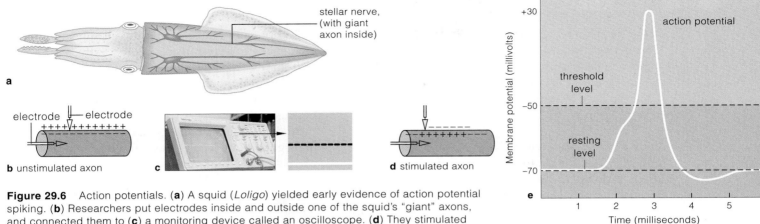

Figure 29.6 Action potentials. (**a**) A squid (*Loligo*) yielded early evidence of action potential spiking. (**b**) Researchers put electrodes inside and outside one of the squid's "giant" axons, and connected them to (**c**) a monitoring device called an oscilloscope. (**d**) They stimulated the axon. The voltage change showed up as a deflection in a beam of light across the screen of the oscilloscope. (**e**) This *white* line shows a waveform typical of action potentials.

Then most potassium gates close, and sodium gates are ready to open with the arrival of a suitable signal.

How does an action potential move along an axon? The sodium inflow alters the charge distribution in an adjacent patch of membrane, gated channels open, then channels open in the *next* patch, and so on in a self-propagating, positive feedback event that does not weaken with time. Also, the action potential spreads *away* from the original point of stimulation. Why? After sodium rushes across the patch of membrane, the voltage-gated gates on sodium channels stay shut.

> A neuron's trigger zone has many sodium channels with voltage-sensitive gates. Action potentials are initiated when the gates open in an ever accelerating way. The inward rush of sodium briefly reverses the voltage difference across the membrane. Sodium–potassium pumps restore the original ion gradients at the membrane patch.
>
> An action potential causes a voltage reversal at the adjacent membrane patch, and so on down to the output zone. It self-propagates away from the trigger zone because sodium gates stay shut for a bit after each inward rush of sodium.

Read Me First!

and watch the narrated animation on action potential propagation

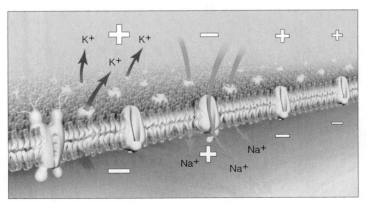

(**c**) With the reversal, sodium gates shut and potassium gates open (*red* arrows). Potassium follows its gradient out of the neuron. Voltage is restored. The disturbance triggers an action potential at the adjacent site, and so on, away from the point of stimulation.

(**d**) Following each action potential, the inside of the plasma membrane becomes negative once again. However, the sodium and potassium concentration gradients are not yet fully restored. Active transport at sodium–potassium pumps restores them.

29.3 How Neurons Send Messages to Other Cells

Action potential propagation typically stops after traveling from a trigger zone to an output zone of a neuron. There, a different signaling mechanism takes over, one that bridges the gap between the neuron endings and another cell.

CHEMICAL SYNAPSES

Only a thin cleft, a **chemical synapse**, separates the output zone of a neuron from a neighboring neuron or a muscle cell or gland cell. The *pre*synaptic neuron's plasma membrane has many gated channels for ions of calcium. When its membrane is resting, the ions are more concentrated in the extracellular fluid than they are inside. Near the membrane are many vesicles that store one or more kinds of **neurotransmitters** (Figure 29.7a,b). These are a type of signaling molecule that diffuses across chemical synapses.

When an action potential reaches the output zone, the gates on the calcium channels open. Calcium ions diffuse inside, and they induce synaptic vesicles to fuse with the plasma membrane. When this happens, neurotransmitter is released into the synaptic cleft.

Receptors on the *post*synaptic cell's membrane can bind specific neurotransmitters that diffuse across the cleft. Binding changes the shape of the receptors, and a passageway opens through their interior. Ions now enter the postsynaptic cell by diffusing through the passageway (Figure 29.7c).

And how does the postsynaptic cell respond? Will the signaling molecules have an *excitatory* effect that can help drive the membrane toward the threshold of an action potential? Or will they have an *inhibitory* effect that pulls it away from the threshold? It depends on the type and number of receptors, neurotransmitter molecules, and gated channels across the membrane.

Because the factors vary, a given neurotransmitter can have different effects on different cells. Consider what happens at a neuromuscular junction, a type of synapse between a motor neuron and a muscle cell (Figure 29.8). Acetylcholine (ACh) released from the motor neuron diffuses across the synaptic cleft and binds to membrane receptors on the muscle cell. It has an excitatory effect, and so it may help initiate muscle contraction (Section 32.4). But some interneurons in

Read Me First!
and watch the narrated animation on chemical synapses

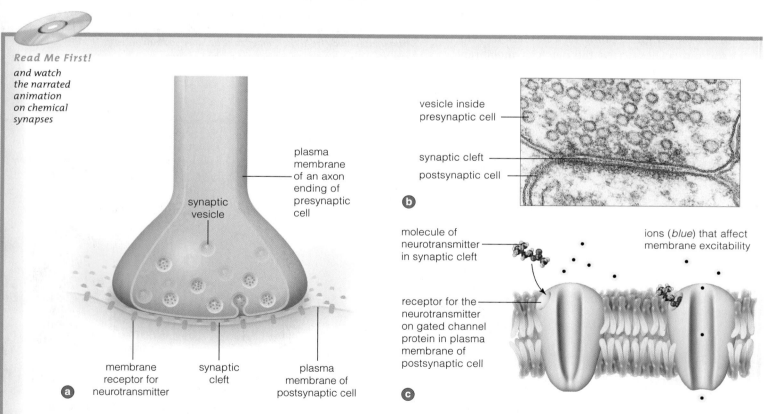

Figure 29.7 Example of a chemical synapse. (**a,b**) Only a thin cleft separates a presynaptic cell from a postsynaptic cell. Arrival of an action potential at an axon ending triggers exocytosis of molecules of neurotransmitter from the presynaptic cell. (**c**) The molecules diffuse across the cleft. They bind to receptors that are part of gated proteins of the postsynaptic cell membrane. The gates open; ions flow in, causing a local electrical disturbance at the membrane site.

muscle fiber

axon ending

neuromuscular junction

motor neuron axons from spinal cord to skeletal muscle cells

transverse slice of spinal cord

part of a skeletal muscle

Figure 29.8 A common type of chemical synapse: a neuromuscular junction. It forms between the axon endings of motor neurons and muscle cells (muscle fibers).

Figure 29.9 Synaptic integration. Many excitatory and inhibitory signals typically arrive at a postsynaptic neuron's input zone at the same time, and they compete for control of it.

If an excitatory signal of a certain magnitude arrived by itself, you would see the *yellow* waveform. This one was not strong enough to drive the membrane to threshold, so an action potential didn't happen. If an inhibitory signal arrived all by itself, you would see the *purple* waveform. If the two arrived at the same time, you would see the *red* waveform. It shows that the inhibitory signal pulled the excitatory signal's effect even farther below threshold. Summation of the two did not result in an action potential (*white* dashed line).

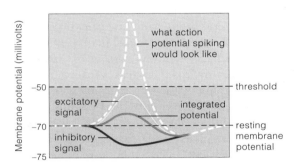

what action potential spiking would look like

threshold

excitatory signal

integrated potential

resting membrane potential

inhibitory signal

Membrane potential (millivolts)

−50

−70

−75

the brain also release ACh. In that setting, one of the effects is inhibition of cells concerned with memory.

Many other neurotransmitters have excitatory or inhibitory effects in the brain. Serotonin, for instance, influences sleeping, sensory perception, emotions, and body temperature. Norepinephrine affects emotions, dreaming, and alertness. Without dopamine, there is no sense of pleasure or positively reinforced learning. GABA (gamma aminobutyric acid) has widespread inhibitory effects. Anti-anxiety drugs, such as Valium, enhance binding of GABA to receptors in the brain.

SYNAPTIC INTEGRATION

Each neuron in your brain is on the receiving end of thousands of inhibitory and excitatory signals, which can vary in intensity. By a process known as **synaptic integration**, a postsynaptic neuron sums all signals from different communication lines that reach its input zone at the same time. Many signals dampen, suppress, or reinforce one another in ways that drive the postsynaptic membrane closer to or farther away from threshold. Figure 29.9 gives two examples.

For signals to keep flowing along communication lines, neurotransmitter molecules must be removed promptly from synaptic clefts after they have done

their work. Some simply diffuse away from the cleft. Enzymes get rid of others, as when acetylcholinesterase breaks apart ACh molecules in the cleft. Transport proteins actively pump neurotransmitter molecules into neuroglial cells or back into the presynaptic cell. Drugs often interfere with neurotransmitter uptake.

What happens if neurotransmitter is not removed? Start thinking about one example: Cocaine blocks the uptake of dopamine, which keeps on stimulating its target cells. At first, the abnormal stimulation invites a sense of euphoria, or intense pleasure. Later on, bad things happen, as you will read in Section 29.9.

Neurotransmitters are signaling molecules that cross a synaptic cleft between two neurons or between a neuron and a muscle cell or gland cell.

The type and amount of neurotransmitter influence whether a signal will have excitatory or inhibitory effects. So do the type and amount of receptors and gated ion channels on the postsynaptic cell membrane.

Synaptic integration is the summation of all excitatory and inhibitory signals that happen to arrive at the same time at a postsynaptic cell's input zone. Summation allows messages to be reinforced or downplayed, sent onward or suppressed. It is essential for normal body functioning.

29.4 Paths of Information Flow

By synaptic integration, messages arriving at a neuron may be reinforced and sent on to its neighbors. What determines the direction in which a message travels? It depends on how neurons are organized in the body.

BLOCKS AND CABLES OF NEURONS

As in other complex animals, information generally flows through your body from sensory receptors, to interneurons, then to motor neurons. However, many signals loop about in amazing ways.

Consider the billions of interneurons in your brain. Thousands engage in something like block parties that are really hopping. In *diverging* circuits, dendrites and axons of neurons fan out from one block and communicate with other blocks. In *converging* circuits, signals from many neurons zero in on just a few. In some circuits, neurons synapse back on themselves, repeating signals like gossip that just won't go away. As one example, such *reverberating* circuits make your eye muscles twitch rhythmically while you sleep.

Information also flows rapidly through **nerves**, the long-distance cables between body regions. In nerves, long axons of many sensory neurons, motor neurons, or both are bundled inside connective tissues. Figure 29.10a is an example of this structural organization.

Schwann cells, a type of neuroglial cell, wrap like jelly rolls around most nerve axons. Collectively, they form a myelin sheath. This electric insulator enhances action potential propagation (Figure 29.10b). How? Ions cannot move back and forth across the sheath, so the disturbance caused by a signal spreads down an axon's cytoplasm until it reaches a tiny exposed node between two jelly rolls. There, a new action potential is initiated, and this happens at each node in line. By jumping node to node, a message can travel as fast as 120 meters per second down a nerve's long axon.

In *multiple sclerosis* (MS), certain white blood cells wrongly identify a type of protein in myelin sheaths as foreign and destroy it. This autoimmune response leads to inflammation of axons in the brain and spinal cord. Some individuals are genetically predisposed to develop this disorder, but a viral infection may set it in motion. Either way, information flow is disrupted, and weakened muscles, fatigue, pain, dizziness, visual problems, and other symptoms follow. About 500,000 people in the United States now have MS.

REFLEX ARCS

Reflexes are the most ancient paths of information flow. A **reflex** is a movement or some other response

a

myelin
sheath
axon

blood vessels

nerve fascicle
(a number of axons
bundled inside
connective tissue)

the nerve's outer
wrapping

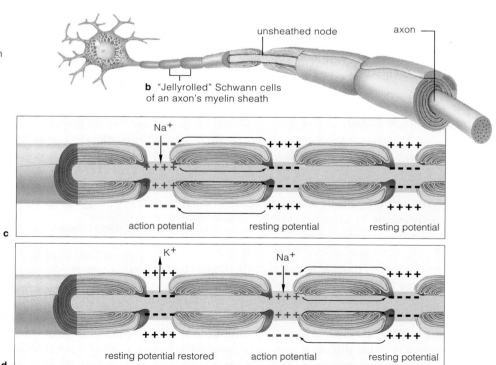

unsheathed node

axon

b "Jellyrolled" Schwann cells
of an axon's myelin sheath

c

Na⁺

action potential   resting potential   resting potential

d

K⁺   Na⁺

resting potential restored   action potential   resting potential

Figure 29.10 (**a**) Structure of a nerve. (**b–d**) In axons with myelin sheaths, ions flow across the neural membrane only at nodes, or tiny unsheathed gaps where many gated channels for sodium ions are exposed. When a disturbance caused by an action potential reaches each node, sodium gates open and start a new action potential. The disturbance spreads swiftly to the next node, where it triggers a new action potential, and so on in line.

Read Me First!

and watch the narrated animation on the stretch reflex

a Fruit being loaded into a bowl puts weight on an arm muscle and stretches it. Will the bowl drop? NO! Muscle spindles in the muscle sheath also are stretched.

STIMULUS
Biceps stretches.

b Stretching stimulates sensory receptor endings in this muscle spindle. Action potentials are propagated toward spinal cord.

c In the spinal cord, axon endings of the sensory neuron release a neurotransmitter that diffuses across a synaptic cleft and stimulates a motor neuron.

d The stimulation is strong enough to generate action potentials that self-propagate along the motor neuron's axon.

e Axon endings of the motor neuron synapse with muscle fibers in the stretched muscle.

f ACh released from the motor neuron's axon endings stimulates cells making up muscle fibers.

RESPONSE
Biceps contracts.

g Stimulation makes the stretched muscle contract. Ongoing stimulations and contractions hold the bowl steady.

muscle spindle   neuromuscular junction

Figure 29.11 Stretch reflex. In skeletal muscle, stretch-sensitive receptors of sensory neurons are in muscle spindles. Stretching generates action potentials.

to a stimulus that happens automatically, no thought required. In the simplest reflex arcs, sensory neurons synapse directly on motor neurons. In more complex reflexes, sensory neurons interact with one or more interneurons that can activate or suppress all motor neurons required for a coordinated response.

The *stretch reflex* is a simple reflex arc that causes a muscle to contract after gravity or some other load stretches it. Suppose you hold a bowl while someone puts peaches into it. The load makes your hand drop a bit, which stretches an arm muscle called a biceps. The biceps stretches in response, which stretches the receptor endings of sensory organs inside it. These organs are **muscle spindles**. Their endings, enclosed in a sheath running parallel with the muscle, actually are the input zones of sensory neurons (Figure 29.11).

In the spinal cord, axons of these sensory neurons synapse with motor neurons, the axons of which lead back to the stretched muscle. In the axon endings of the motor neurons, action potentials trigger the release of ACh, which makes the biceps contract. Contraction helps steady your arm against the peach load.

In complex animals, neurons are organized in blocks and in cables, some of which arc back to the point of stimulation.

Nerves are long-distance cables between body regions. Most long axons of their sensory neurons, motor neurons, or both have a myelin sheath, which speeds signal propagation.

Reflex arcs, in which sensory neurons synapse directly on motor neurons, are the simplest paths of information flow.

29.5 Types of Nervous Systems

All animals above the sponge level of organization have a nervous system, with signal-conducting and information-processing lines through the body.

REGARDING THE NERVE NET

To appreciate the diversity of nervous systems, start by recalling that animals first evolved in the seas. We still find animals with the simplest nervous systems in aquatic habitats. They are cnidarians, such as sea anemones and jellyfishes, and all of them show *radial* symmetry (Sections 22.1 and 22.4).

Cnidarians have a **nerve net**. This loosely arranged mesh of nerve cells threads around epithelial cells (Figure 29.12*a*). When nerve, sensory, and contractile cells interact in reflex pathways, tissues are moved in simple stereotyped ways. As an example, consider the reflex pathway that controls a sea anemone's feeding actions. It extends from sensory receptors in tentacles to nerve cells that control contractions of cells that surround the opening to the gut.

A nerve net extends through the body but does not focus information. It does little more than bring about weak contractions of the body wall or cause tentacles to bend slowly. Its owners will never dazzle you with speed or acrobatics. Even so, their nerve net lets them capture food that randomly drifts into their tentacles and get it into the gut. The radial system is equally responsive to tidbits arriving from any direction.

ON THE IMPORTANCE OF HAVING A HEAD

Flatworms are the simplest animals having a bilateral nervous system. Their two cordlike nerves branch repeatedly in a ladderlike array (Figure 29.12*b*). The head end of some species has ganglia. A **ganglion** (plural, ganglia) is a cluster of nerve cell bodies acting as a local integrating center. In flatworms, the ganglia integrate signals from paired sensory organs, such as eyespots. Some exert local control over the nerves.

Bilateral nervous systems may have evolved from nerve nets. Remember, bilateral larvae called planulas form in some cnidarian life cycles (Section 22.5). In one hydrozoan planula (*Podocoryne*), a nerve net develops in a serially repeated pattern, from front to back, that helps the planula move in the forward direction. The nerve net becomes meshlike when tissues of the next stage (the polyp) expand during development. But in its initial repeating pattern, and in its control over forward motion, this planula reminds us of flatworms.

Visualize a planula on the Cambrian seafloor. By one scenario, a mutation blocked the transition to the adult form, but reproductive organs still matured. The planula, then its offspring, could still reproduce. Other mutations led to a concentration of sensory cells in the "head" end, which met up with food or danger first. Up-front control meant faster response time.

Did **cephalization**—the formation of a head—and bilateral symmetry evolve this way? Possibly. We see

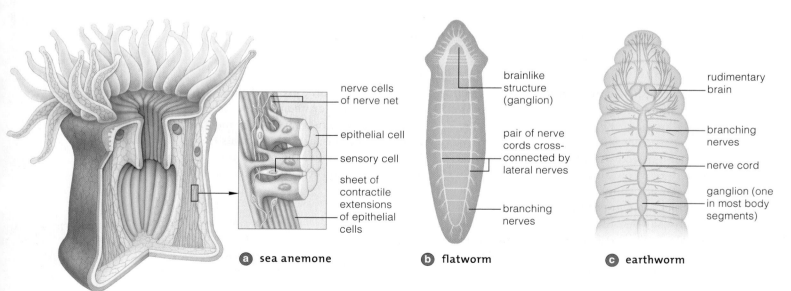

(a) sea anemone   **(b) flatworm**   **(c) earthworm**

nerve cells of nerve net
epithelial cell
sensory cell
sheet of contractile extensions of epithelial cells

brainlike structure (ganglion)
pair of nerve cords cross-connected by lateral nerves
branching nerves

rudimentary brain
branching nerves
nerve cord
ganglion (one in most body segments)

Figure 29.12 (**a**) Nerve net of a sea anemone, a cnidarian. Interacting nerve cells, sensory cells, and contractile cells thread through epithelial cells of the epidermis. They face the jellylike midlayer, the mesoglea. (**b,c**) Like most invertebrates, the flatworms and earthworms have a bilateral nervous system.

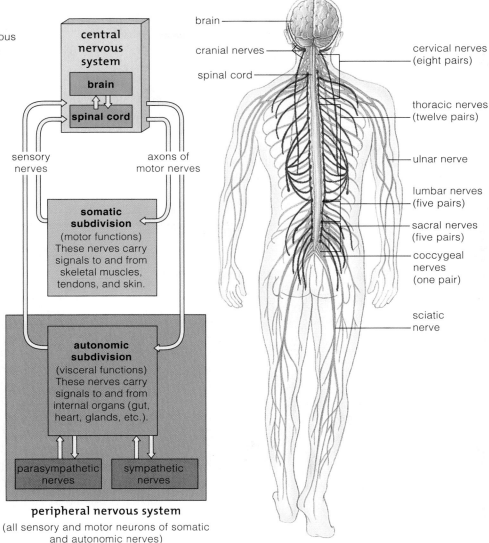

Figure 29.13 Functional divisions of the human nervous system. These divisions are typical of all vertebrates.

cephalization and bilateral symmetry in most invertebrates. We see them in the rudimentary brain and paired ganglia of earthworms (Figure 29.12c). They even are apparent in the paired sensory structures and brain centers, the paired nerves, and paired skeletal muscles of all vertebrates, including humans.

EVOLUTION OF THE SPINAL CORD AND BRAIN

Hundreds of millions of years ago, as the first fishlike vertebrates were evolving, a hollow, tubular nerve cord was evolving into a brain and spinal cord (Chapter 23). Simple reflex pathways still dominated; sensory neurons synapsed right on motor neurons that directly signaled muscles to contract. Later, fast-moving predators and prey that could better assess and respond to food or danger gained a competitive edge. Starting with amphibians, the senses of hearing, vision, and balance became increasingly important on land. Maneuverability and dexterity improved through modifications in bones and muscles. New tissues thickened the brain.

The oldest regions of the vertebrate brain still deal with reflex control of respiration and other essential functions. But far more interneurons now synapse on sensory and motor neurons of the ancient pathways and on one another.

The vertebrate nervous system has two functional divisions, one central and one peripheral. The spinal cord and brain are the *central* nervous system (Figure 29.13). The *peripheral* nervous system is mostly nerves that extend through the rest of the body; they carry signals into and out of the central nervous system.

Inside the brain and spinal cord, communication lines are called tracts, not nerves. Tracts that make up *white* matter have axons with glistening white, fatty myelin sheaths; they afford rapid signal transmission. The *gray* matter consists of cell bodies, dendrites, and unmyelinated axons of neurons, along with neuroglia.

Cnidarians have a nerve net. This diffuse mesh of nerve cells extends through their radial body and interacts with sensory and contractile cells in simple reflex pathways.

Cephalized, bilateral animals have a nervous system with a brain or ganglia at the head end, paired nerves, and paired sensory structures.

The central nervous system of vertebrates consists of a brain and spinal cord. The peripheral nervous system consists mostly of nerves that thread through the rest of the body and carry signals into and out of the central region.

29.6 What Are the Major Expressways?

Let's now take a look at the peripheral nervous system and the spinal cord. The two interconnect as the main expressways for information flow through the body.

PERIPHERAL NERVOUS SYSTEM

THE SOMATIC AND AUTONOMIC SUBDIVISIONS In humans, the peripheral nervous system has thirty-one pairs of *spinal* nerves, which connect with the spinal cord. It also has twelve pairs of *cranial* nerves, which connect directly with the brain. We can further classify cranial and spinal nerves by their function. The ones carrying information about moving the head, trunk, and limbs are **somatic nerves**. Their sensory axons carry messages from receptors in the skin, skeletal muscles, and tendons to the central nervous system. Their motor axons deliver commands from the brain and the spinal cord to skeletal muscles. **Autonomic nerves** are spinal and cranial nerves that service the viscera—smooth muscle, cardiac muscle, glands, and other internal organs and structures.

SYMPATHETIC AND PARASYMPATHETIC NERVES Two kinds of autonomic nerves are called sympathetic or

Read Me First!

and watch the narrated animation on autonomic nerves

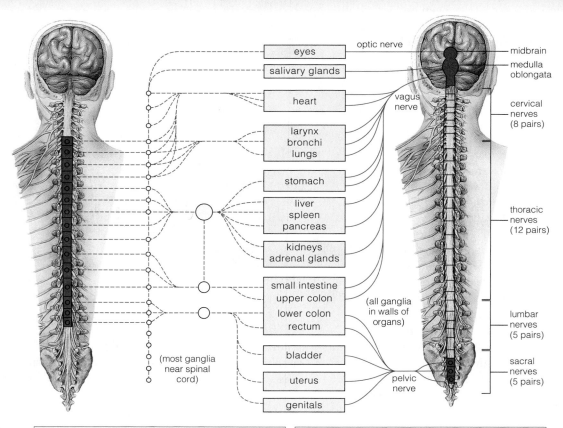

Some Responses to Sympathetic Outflow	Some Responses to Parasympathetic Outflow
Heart rate increases	Heart rate decreases
Pupils of eyes dilate (widen, let in more light)	Pupils of eyes constrict (keep more light out)
Glandular secretions decrease in airways to lungs	Glandular secretions increase in airways to lungs
Salivary gland secretions thicken	Salivary gland secretions become dilute
Stomach and intestinal movements slow down	Stomach and intestinal movements increase
Sphincters (rings of muscle) contract	Sphincters (rings of muscle) relax

(a) sympathetic outflow from the spinal cord

(b) parasympathetic outflow from the spinal cord and brain

Figure 29.14 (**a**) Major sympathetic nerves and (**b**) parasympathetic nerves of the central nervous system. These nerves are paired; the body's right and left halves contain one of each. The ganglia (clusters of nerve cell bodies) are local control centers, not part of the central nervous system; their axons are bundled inside nerves.

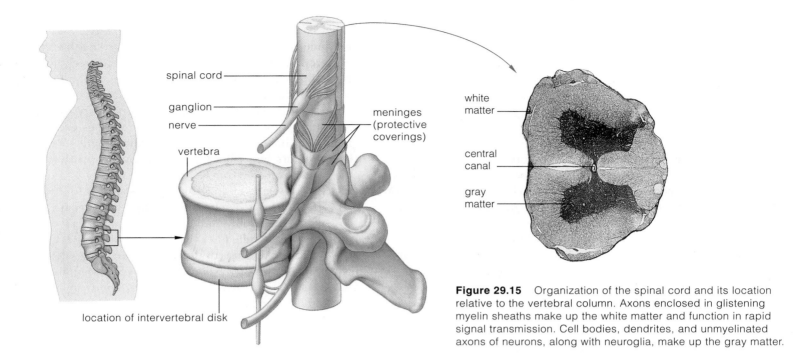

spinal cord

ganglion

nerve

meninges (protective coverings)

vertebra

white matter

central canal

gray matter

location of intervertebral disk

Figure 29.15 Organization of the spinal cord and its location relative to the vertebral column. Axons enclosed in glistening myelin sheaths make up the white matter and function in rapid signal transmission. Cell bodies, dendrites, and unmyelinated axons of neurons, along with neuroglia, make up the gray matter.

parasympathetic (Figure 29.14). Both carry excitatory and inhibitory signals to internal organs. However, they normally work antagonistically, meaning signals from one oppose signals from the other. Synaptic integration of their opposing signals leads to precise adjustments in an organ's level of activity.

Signals from the **sympathetic nerves** dominate in times of stress, excitement, or danger. Housekeeping tasks, such as digestion, are set aside and awareness is heightened as the body prepares to fight or escape from threats or to frolic, as in play or sexual behavior.

Parasympathetic nerves dominate when the body is not receiving much outside stimulation. They tend to slow down overall activity and divert the body's energies to housekeeping tasks, such as digestion.

What happens when something startles or scares you? Parasympathetic input slows down, giving way to sympathetic signals that raise your heart rate and blood pressure, make you sweat more, and increase your rate of breathing. The signals also act on adrenal glands, which secrete epinephrine. Formerly known as adrenaline, this neurotransmitter puts you into a state of intense arousal—primed to fight or get away fast. Hence the term *fight–flight response*.

SPINAL CORD

By definition, the **spinal cord** is a vital expressway for signals between the peripheral nervous system and the brain. But direct reflex connections also are made

in the cord. Such connections exert some control in limb movements (remember the stretch reflex?), and in bladder emptying and some other basic tasks.

In amphibians, spinal reflexes play a greater role in motor activity. Between the frog brain and spinal cord are circuits that make bent legs straighten. Surgically cut these circuits near the brain, and the legs will be paralyzed—but only for about sixty seconds. Reflex pathways in the spinal cord will recover fast and have the frog hopping again soon. Humans and all other primates rely far more on brain centers, so they show little or no recovery from a similar injury.

Notice, in Figure 29.15, that the spinal cord threads through a canal made of bones of the vertebral column. Bones and ligaments protect it. So do three meninges, or coverings, that also cloak the brain. In *meningitis*, a viral or bacterial infection inflames these coverings. Fever, severe headaches, a stiff neck, and nausea are among the symptoms that follow.

Nerves of the peripheral nervous system connect the brain and spinal cord with the rest of the body.

The somatic nerves of the peripheral nervous system deal with skeletal muscle movements. Its autonomic nerves deal with smooth muscle, cardiac muscle, and glands.

The spinal cord is a vital expressway for signals between the brain and peripheral nerves. Some of its interneurons also exert direct control over certain reflex pathways.

29.7 The Vertebrate Brain

The spinal cord is continuous with the brain, the body's master control center. The brain receives, integrates, stores, retrieves, and issues information. It coordinates responses to sensory input. As is the case for the spinal cord, bones and membranes protect the brain.

THE BRAIN'S SUBDIVISIONS

The hollow, tubular nerve cord that forms in every chordate embryo becomes a brain and spinal cord in vertebrates. The same genes that control segmentation in most invertebrates subdivide the tube into forebrain, midbrain, and hindbrain regions (Figure 29.16). The most ancient nervous tissue, the **brain stem**, persists in all regions and is continuous with the spinal cord.

The hindbrain's **medulla oblongata** houses reflex centers for respiration, circulation, and other vital tasks. It integrates motor responses and many complex reflexes, such as coughing, and it affects sleep. The **cerebellum** uses inputs from the eyes, ears, muscle spindles, and forebrain regions to help control motor skills and posture. Axons from its two halves reach the **pons** (meaning bridge). The pons is like a traffic officer; it controls the flow of signals between the cerebellum and integrating centers in the forebrain.

Fishes and amphibians have the most pronounced midbrain. It coordinates most of their sensory input and starts motor responses. In all vertebrates except mammals, the midbrain includes a pair of optic lobes, brain centers for visual input. As mammals evolved, the forebrain took over responsibility for integrating visual stimuli; the midbrain just relays signals to it.

Vertebrates first evolved in water, where chemical odors diffusing from predators, prey, and mates were vital cues. They relied heavily on their olfactory lobes and on paired outgrowths from the brain stem, where olfactory input and responses to it were integrated. Later on, the outgrowths expanded, especially in land vertebrates. They became two halves of the **cerebrum**; the two cerebral hemispheres.

The **thalamus** became a forebrain center for sorting out sensory input and relaying it to the cerebrum. The **hypothalamus** ("under the thalamus") evolved into the main center for homeostatic control of the internal environment. It assesses and governs all behaviors related to internal organ activities, such as thirst, sex, and hunger. It also governs related emotions, such as sweating with passion and vomiting from fear.

PROTECTION AT THE BLOOD–BRAIN BARRIER

The neural tube's lumen (the space inside it) persists in adult vertebrates as a system of cavities and canals filled with cerebrospinal fluid (Figure 29.17). This clear extracellular fluid cushions the brain and spinal cord against potentially jarring movements.

A **blood–brain barrier** protects the brain and spinal cord of vertebrates from harmful substances. It exerts some control over which solutes enter cerebrospinal

b Brain at 7 weeks

c Brain at 9 weeks

d Brain at birth

Figure 29.16 How the anterior end of the neural tube develops into three brain subdivisions in a human embryo. (**a**) This sketch flattens out the tube to list the dominant brain regions that develop along its length. The tube actually curves forward in the embryo, as in (**b–d**).

Figure 29.17 Cerebrospinal fluid (*blue*). This extracellular fluid fills four interconnected ventricles (cavities) in the brain and in the spinal cord's central canal.

Figure 29.18 (**a**) Major brain regions for five vertebrates, dorsal views. The sketches are not to the same scale. (**b**) This photograph shows more detail for the right half of the human brain, sagittal view. Meninges protect the brain tissue but these coverings were removed for instructional purposes.

olfactory lobe

forebrain

midbrain

hindbrain

a

FISH
shark

AMPHIBIAN
frog

REPTILE
alligator

BIRD
goose

MAMMAL
human

fluid. No other part of extracellular fluid has solute concentrations kept within such narrow limits. Even changes brought on by eating or exertion are limited. Why? Some hormones and other substances in blood can alter neural function. Also, shifts in some ions can alter the threshold for action potentials.

The barrier works at the wall of blood capillaries that service the brain. In most brain regions, tight junctions fuse all abutting cells that make up the wall of capillaries, so water-soluble substances must pass *through* the cells to reach the brain. Transport proteins in the plasma membrane of these cells allow glucose, other vital nutrients, and some ions to cross. They bar many toxins and wastes, such as urea. The barrier can't stop a fat-soluble substance, such as oxygen, carbon dioxide, alcohol, caffeine, and nicotine. Inflammation, toxins, and traumatic blows can destroy this barrier and compromise neural function.

corpus callosum   hypothalamus   thalamus   pineal gland location

part of optic nerve

midbrain
cerebellum
pons
medulla oblongata

b

THE HUMAN BRAIN

An average-sized human brain weighs 1,300 grams (3 pounds) and has about 100 billion interneurons. More than half of its volume is neuroglia. Compared with all other vertebrate brains, including the ones shown in Figure 29.18, the human midbrain is smaller. The hindbrain's cerebellum is larger. It is about the size of a fist and has more interneurons than all other brain regions combined. As in other vertebrates, it deals with the sense of balance and coordination. But it took on added functions during human evolution. For example, it assists in the learning of motor skills and mental skills, such as language.

What about the forebrain? A fissure divides its cerebrum into two halves, the cerebral hemispheres (Figure 29.18). The next section takes a look at their thin outer layers, the cerebral cortex. Each half deals mainly with input from the opposite side of the body. For instance, signals about pressure on the right arm

go to the left hemisphere. The activities of both halves are coordinated by two-way signals through a thick band of nerve tracts, the corpus callosum.

The pineal gland is located near the hypothalamus. The hypothalamus receives signals about light levels from the retina and relays them to this light-sensitive endocrine gland. Chapter 31 explores the functions of both of these forebrain components.

The vertebrate brain develops from a hollow neural tube, the lumen of which persists in adults as a system of cavities and canals filled with cerebrospinal fluid. The fluid cushions nervous tissue from sudden, jarring movements.

The nervous tissue is subdivided into a hindbrain, forebrain, and midbrain. The brain stem is the most ancient tissue in all three divisions. The forebrain has the most complex integrating centers.

29.8 Consider the Cerebral Cortex

Our "humanness" starts in the outer layer of gray matter of the cerebral cortex, which governs conscious behavior. The cerebral cortex processes and coordinates responses to sensory input. It interacts with the limbic system, which governs emotions and contributes to memory.

FUNCTIONAL AREAS OF THE CORTEX

The two halves of our cerebrum look alike, but each has functional specializations. The left half typically deals with most analytical skills, mathematics, and speech. The right half assesses visual images, judges spatial relations, and interprets music. Both have four subdivisions (frontal, occipital, temporal, and parietal lobes). In the layer of gray matter at their surface—the cerebral cortex—distinctive areas receive and process specific kinds of information. The *motor* areas affect voluntary motor activity. *Sensory* areas assist in the perception of what specific sensations mean. Diverse *association* areas integrate the information that brings about conscious actions. All of these areas interact.

The body is spatially mapped out in the primary motor cortex of each hemisphere's frontal lobe. This area controls the coordinated movements of skeletal muscles. Much of it deals with thumb, finger, and tongue muscles. Figure 29.19 can give you an idea of the control required for voluntary hand movements and verbal expression. The motor cortex of each lobe controls muscles on the opposite side of the body.

The frontal lobe includes the premotor cortex and Broca's area. The premotor cortex governs learned motor skill patterns. Dribble a basketball, play a piano, type on a keyboard—the repetitive movements are evidence that simultaneous and sequential actions of different muscle groups are being coordinated. Broca's area helps translate thoughts into speech by controlling tongue, throat, and lip muscles. It is the basis of a capacity to understand and create complex sentences. In most people, Broca's area is located in the left cerebral hemisphere (Figure 29.20*a*).

The primary somatosensory cortex, in the front of each parietal lobe, is mapped out to correspond to the human body. As you will see in the next chapter, it is a receiving center for sensory input from the skin and joints. Another sensory area in this lobe deals with taste perception. At the back of the occipital lobe, the primary visual cortex gets signals from the eyes. The perception of sound and odor arises in sensory areas in each temporal lobe.

a

b

Figure 29.19 (**a**) A slice of the primary motor cortex, through the region indicated in (**b**). The sizes of body parts draped over the diagrammatic slice are distorted to show which ones get the most precise control.

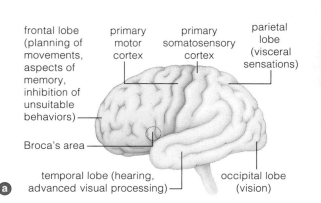

frontal lobe (planning of movements, aspects of memory, inhibition of unsuitable behaviors)

primary motor cortex

primary somatosensory cortex

parietal lobe (visceral sensations)

Broca's area

temporal lobe (hearing, advanced visual processing)

occipital lobe (vision)

a

Motor cortex activity when speaking

Prefrontal cortex activity when generating words

Visual cortex activity when seeing written words

b

Figure 29.20 (**a**) Primary receiving and integrating centers of the human cerebral cortex. Association areas coordinate and process sensory input from diverse receptors. (**b**) Three PET scans identifying which areas were active when an individual performed three different tasks.

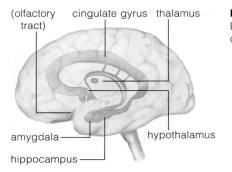

(olfactory tract)   cingulate gyrus   thalamus

amygdala

hippocampus

hypothalamus

Figure 29.21 Limbic system components.

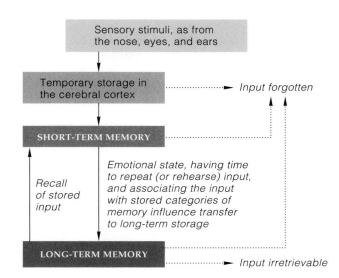

Sensory stimuli, as from the nose, eyes, and ears

Temporary storage in the cerebral cortex ···········▶ *Input forgotten*

SHORT-TERM MEMORY ···········

Recall of stored input

Emotional state, having time to repeat (or rehearse) input, and associating the input with stored categories of memory influence transfer to long-term storage

LONG-TERM MEMORY ···········

···········▶ *Input irretrievable*

Figure 29.22 Stages of memory processing, starting with the temporary storage of sensory inputs in the cerebral cortex.

The cortex has many association areas, but not in the primary motor and sensory areas. Each integrates many inputs (Figure 29.20a). For instance, a visual association area surrounds the primary visual cortex and compares what we see with visual memories. The most recently evolved area, the prefrontal cortex, is the start of our personality and intellect, of abstract thought, judgment, planning, and concern for others.

CONNECTIONS WITH THE LIMBIC SYSTEM

Encircling the upper brain stem is the **limbic system**, which controls emotions and has roles in memory. It includes the hypothalamus, parts of the thalamus, the cingulate gyrus, the hippocampus, and amygdala (Figure 29.21). The cingulate (belt-shaped) gyrus is a fold in the brain tissue that lies just above the corpus callosum. It affects motivation and is more active in extroverts and risk-takers than it is in the introverted or cautious. The hypothalamus is a link between our emotions and visceral activities. The almond-shaped amygdala is essential for emotional stability and for interpreting social cues. It also becomes active when the individual is frightened or anxious.

The limbic system is evolutionarily related to the olfactory lobes and still deals with the sense of smell. That is one reason you might feel warm and fuzzy when you recall the scent of a special person.

The limbic system helps correlate organ activities with self-gratifying behavior, such as eating and sex. That is why it is called our emotional–visceral brain. It can put a heart on fire with passion or a stomach on fire with indigestion. Generally, the prefrontal cortex can override or dampen such "gut reactions."

MAKING MEMORIES

Memory is a capacity to process, store, and retrieve information about past sensory experience. Without it, learning and behavioral modification would not be possible. Information is stored in stages. In *short-term* storage, excitation lasts seconds to a few hours. This stage is limited to a few bits of sensory information— numbers, words of a sentence, and so on. *Long-term* storage files vast amounts of information in almost permanent ways (Figure 29.22).

The brain processes facts and skills separately, by memory circuits that handle different kinds of input. A circuit for facts starts as inputs at the sensory cortex are shuttled to the amygdala and hippocampus. The amygdala is the gatekeeper; it connects the sensory cortex with areas in the thalamus and hypothalamus.

The hippocampus is the first brain region affected in *Alzheimer's disease*, which usually starts late in life. Affected people often recall long-known facts but have trouble moving new facts into long-term memory. As the cerebral cortex becomes affected, patients become confused, depressed, and incoherent.

Skill memory also starts at the sensory cortex, but a circuit routes sensory input to a brain region dealing with motor responses. Motor skills involve muscle conditioning. This circuit connects to the cerebellum, the brain region that coordinates motor activity.

The cerebral cortex, each hemisphere's outermost layer of gray matter, contains motor, sensory, and association areas. Interactions among these areas govern conscious behavior.

The cerebral cortex also interacts with the limbic system, which affects emotions and contributes to memory.

29.9 Drugging the Brain

Psychoactive drugs are substances that affect the function or action of neurotransmitters. Some induce the release of a neurotransmitter. Others prevent its breakdown or uptake. Still others block receptors on the membrane of a postsynaptic cell where it would bind.

ROOTS OF ADDICTION

Psychoactive drugs, both legal and illegal, are often taken to mediate pain, mental illness, or stress. Some are also used and abused by people who simply find their effects pleasurable. Either way, drug users often develop *tolerance*; they need larger or more frequent doses to obtain the same effect. Tolerance is a sign that drug addiction may be next. Addicts abruptly deprived of drugs undergo biochemical upheaval, which causes physical pain and mental anguish. Table 29.1 lists the warning signs of drug addiction.

All major addictive drugs stimulate the release of dopamine, the neurotransmitter directly involved in the sense of pleasure. For example, most people who smoke think this is just a habit. Actually, they smoke because of the dopamine spike when nicotine hits the brain's neurons. *All* addicts specifically crave the spike and the vast majority can't stop the craving.

EFFECTS OF PSYCHOACTIVE DRUGS

STIMULANTS   Stimulants make you more alert, then depress you. The *caffeine* in coffee, tea, chocolate, and many soft drinks is one. Low doses act at the cerebral cortex and can increase alertness, but they harm fine motor coordination and increase nervousness. *Nicotine* in tobacco is a stimulant that mimics ACh. It affects

Figure 29.23   Want to be one of the brain-dead? (**a**) Normal brain activity revealed by a PET scan. (**b**) PET scan of a comparable section that reveals cocaine's long-term effect. *Red* marks regions of greatest activity, followed by *yellow*, *green*, and *blue* for the least activity.

diverse sensory receptors. Chapter 35 gets into the impacts of nicotine addiction on health.

As many as 3.7 million Americans may be *cocaine* abusers. This stimulant fans pleasure by blocking the reabsorption of norepinephrine, dopamine, and other neurotransmitters. Postsynaptic cells are not released from stimulation. Blood pressure and sexual appetite rise. In time, the molecules diffuse away, but the body can't synthesize replacements fast enough. The sense of pleasure vanishes permanently as hypersensitized postsynaptic cells demand and cannot get stimulation. Figure 29.23 highlights one long-term effect.

Abusers inhale a granular form of cocaine or burn crack cocaine and inhale the smoke. This extremely addictive drug has huge social and economic costs. In 2001, it put more than 100,000 people in emergency rooms. There is no antidote for overdoses, which can cause seizures, respiratory failure, or heart failure.

Amphetamines induce massive release of dopamine and norepinephrine. Addicts often smoke, snort, inject, or gulp a form called *crank*. It triggers euphoria, sexual arousal, heart pounding, agitation, dry mouth, tremors, and paranoia. Crank kills the appetite; some get addicted after using it to lose weight. Dopamine and norepinephrine synthesis decline over time, and the brain depends more on artificial stimulation. The outcomes? Malnutrition, psychosis, depression, loss of memory, and brain, heart, lung, and liver damage.

You already read about MDMA (*Ecstasy* or XTC). It is only one of many *club drugs* used at bars, clubs, and raves. Sexual predators use these settings to sneak drugs into the drinks of people they wish to assault.

DEPRESSANTS, HYPNOTICS   Depressants sedate, and hypnotics induce sleep (*not* a hypnotic trance). Their effects depend on the dose and on physiological and emotional states. They can invite calm, drowsiness,

Table 29.1	Warning Signs of Drug Addiction*

1. Tolerance—it takes increasing amounts of the drug to get the same effect.

2. Habituation—it takes continued drug use over time to maintain the self-perception of functioning normally.

3. Inability to stop or curtail drug use, even if the desire to do so persists.

4. Concealment—not wanting others to know of the drug use.

5. Extreme or dangerous actions to get and use a drug, as by stealing, by asking more than one doctor for prescriptions, or by jeopardizing employment by using drugs at work.

6. Deterioration of professional and personal relationships.

7. Anger and defensiveness if someone suggests there may be a problem.

8. Drug use preferred over previous customary activities.

*Three or more of these signs may be cause for concern.

FOCUS ON HEALTH

sleep, coma, or death. Low doses impact inhibitory synapses the most, so users get excited or euphoric at first. Larger doses suppress excitatory synapses and lead to depression. Both drugs amplify each other, as when alcohol and barbiturates heighten depression.

Alcohol, or ethyl alcohol, alters cell functions. Like nicotine and cocaine, it easily crosses the blood–brain barrier. Its potency has put some who went on a binge and overindulged into comas and coffins. Even low intake can cause diminished judgment, disorientation, and uncoordinated movements. Long-term addiction can lead to liver failure, as described in Chapter 5.

ANALGESICS When severe stress leads to emotional or physical pain, your brain produces endorphins and enkephalins. These are natural painkillers—*analgesics*. The narcotic analgesics *codeine* and *heroin* are two of the most highly addictive drugs (Figure 29.24). Withdrawal causes chills, fever, anxiety, hyperactivity, severe vomiting, and diarrhea.

PSYCHEDELICS, HALLUCINOGENS These drugs skew sensory perception by interfering with ACh, norepinephrine, or serotonin. *LSD* alters serotonin's roles in inducing sleep, maintaining temperature, and mediating sensory perception. Even small doses can warp perceptions. Some users thought they could fly and "flew" off buildings.

The hallucinogen *marijuana* is made from parts of the plant *Cannabis*. In low doses it is like a depressant. It slows down but does not impair motor activity. It relaxes the body and usually elicits a mild euphoria, but can cause an acute panic attack in some people. It often disrupts short-term memory. As alcohol does, it skews the performance of complex tasks, such as driving a car. Smoking marijuana, like smoking tobacco, can increase the risk of respiratory ailments and of cancers of the mouth, throat, and lungs.

Figure 29.24 Are we having fun yet? Shooting up heroin on the street. Sharing needles puts users at risk of AIDS, viral-induced hepatitis B, and bacterial infections.

Summary

Section 29.1 Neurons are a type of excitable cell, which means an adequate stimulus can cause an abrupt reversal in the voltage difference across their plasma membrane. Sensory neurons relay signals into the spinal cord and brain. Interneurons within the spinal cord and brain integrate information. Motor neurons carry signals away from the spinal cord and brain.

Pumping and leaking of ions across a neuron's plasma membrane sets up a resting membrane potential. Signals at a neuron's input zones can spread to a trigger zone, where action potentials arise. These signals travel along the membrane to axon endings (an output zone).

Section 29.2 An action potential is an abrupt, brief reversal in voltage across a patch of plasma membrane. It occurs only if stimulation exceeds a certain threshold level. Gated, voltage-sensitive channels for sodium open in an ever accelerating way. Sodium rushes into the neuron; potassium flows out. The disturbance triggers an action potential, which triggers another at an adjacent membrane patch, and on to the neuron's output zone. Sodium–potassium pumps restore the ion gradients necessary for the next round of action potentials.

Section 29.3 Neurons interact at chemical synapses. An action potential arriving at the axon endings of a presynaptic cell triggers the release of neurotransmitter molecules. These diffuse across a tiny cleft and bind to receptors on a postsynaptic cell, which may respond by changing its activities. Competing signals arriving at the same time are integrated, which helps determine whether a message will be sent on or suppressed.

Section 29.4 Nerves are long-distance cables, with bundles of axons that carry messages through the body. Myelin sheaths enclose most of the axons and speed the flow of information. Reflexes are simple, automatic responses to stimulation. In the simplest reflex arcs, a sensory neuron synapses directly on a motor neuron.

Section 29.5 Radial animals have simple nerve nets. Most animals have a bilateral nervous system with a rudimentary or well-developed brain at their head end. Vertebrate nervous systems are functionally divided into central and peripheral regions. The brain and spinal cord make up the central nervous system. The peripheral nervous system has paired nerves connecting the brain and spinal cord to the rest of the body.

Section 29.6 The peripheral system's somatic nerves service skeletal muscles. Its autonomic nerves service the viscera (soft internal organs). Its sympathetic and parasympathetic subdivisions often work in opposition to control organ activity. Sympathetic signals dominate in times of heightened awareness or danger. When the body is not receiving much outside stimulation, parasympathetic signals dominate.

Section 29.7 In vertebrate embryos, a neural tube develops into a spinal cord and to three brain regions: a hindbrain, midbrain, and forebrain. The brain stem is the most evolutionarily ancient neural tissue. It persists in all three brain regions and governs many reflex centers dealing with vital tasks.

A blood–brain barrier, at cells making up blood capillaries servicing the brain, keeps many harmful substances from reaching brain cells.

The thin surface layer of gray matter of the cerebral hemispheres is the most recently evolved tissue, and the most complex in its functions.

Table 29.2 summarizes the components and functions of the central nervous system.

Section 29.8 The cerebral cortex is functionally divided into motor, sensory, and association areas. This thin layer of gray matter on the cerebrum controls conscious behavior. Its centers also interact with the limbic system to control emotion and memory.

Section 29.9 Psychoactive drugs act at synapses, where they mimic neurotransmitters or affect their release or uptake. Habitual use of psychoactive drugs can rewire the brain and cause addiction.

Table 29.2 Summary of the Central Nervous System*

FOREBRAIN	Cerebrum	Localizes and processes sensory inputs; initiates and controls skeletal muscle activity. Governs memory, emotions, and abstract thought in the most complex vertebrates
	Olfactory lobe	Relays sensory input from the nose to olfactory centers of cerebrum
	Thalamus	Has relay stations for conducting sensory signals to and from cerebral cortex; has role in memory
	Hypothalamus	With the pituitary gland, a homeostatic control center that adjusts the volume, composition, and temperature of internal environment. Governs behaviors affecting organ functions (e.g., thirst, hunger, and sex) and expression of emotion
	Limbic system	A complex of brain structures that governs emotions; has roles in memory
	Pituitary gland (Chapter 31)	With hypothalamus, provides endocrine control of metabolism, growth, and development
	Pineal gland (Chapter 31)	Helps control some circadian rhythms; also has role in mammalian reproductive physiology
MIDBRAIN	Roof of midbrain (tectum)	In fishes and amphibians, its centers coordinate sensory input (as from optic lobes) and motor responses. In mammals, its centers (mainly reflex centers) swiftly relay sensory input to forebrain
HINDBRAIN	Pons	Is a "bridge" of tracts between the cerebrum and cerebellum; its other tracts connect spinal cord with forebrain. Works with the medulla oblongata to control the rate and depth of respiration
	Cerebellum	Coordinates motor activity for moving limbs and maintaining posture, and for spatial orientation
	Medulla oblongata	Its tracts relay signals between the spinal cord and pons; its reflex centers help control heart rate, adjustments in blood vessel diameter, respiratory rate, vomiting, coughing, other vital functions
SPINAL CORD		Makes reflex connections for limb movements. Many of its tracts carry signals between the brain and the peripheral nervous system

*The reticular formation extends from the spinal cord to the cerebral cortex.

Self-Quiz
Answers in Appendix III

1. _____ relay messages from the brain and spinal cord to muscles and glands.
   a. Motor neurons   c. Interneurons
   b. Sensory neurons   d. Neuroglia

2. When a neuron is at rest _____ .
   a. it is at threshold potential
   b. gated sodium channels are open
   c. the sodium–potassium pump is operating
   d. both a and c

3. Action potentials occur when _____ .
   a. a neuron receives adequate stimulation
   b. sodium gates open in an ever accelerating way
   c. sodium–potassium pumps kick into action
   d. both a and b

4. Neurotransmitters are released by _____ .
   a. axon endings   c. dendrites
   b. the cell body   d. myelin sheaths

5. You would find the axons of interneurons _____ .
   a. in the brain   c. in autonomic nerves
   b. in the spinal cord   d. both a and b

6. Skeletal muscles are controlled by _____ .
   a. sympathetic nerves   c. somatic nerves
   b. parasympathetic nerves   d. both a and b

7. When you sit quietly on the couch reading, output from the _____ system prevails.
   a. sympathetic   c. both
   b. parasympathetic   d. neither

8. Compared to lower vertebrates, mammals have a relatively small _____ .
   a. forebrain   c. hindbrain
   b. midbrain   d. both a and c

9. The cerebrum is part of the _____ .
   a. forebrain   c. hindbrain
   b. midbrain   d. brain stem

10. Learning new facts involves the _____ .
    a. cerebellum   c. primary motor cortex
    b. amygdala   d. brain stem

11. Match each item with its description.
    _____ muscle spindle   a. embryonic structure
    _____ neurotransmitter   b. controls basic reflexes
    _____ limbic system   c. protects brain and spinal
    _____ brain stem   cord from some toxins
    _____ cerebral cortex   d. signaling chemical
    _____ neural tube   e. stretch-sensitive receptor
    _____ blood–brain   f. roles in emotion, memory
          barrier   g. governs higher thought

Critical Thinking

1. The neuron at right has a bushy crown of dendrites, and a single axon below the cell body. Would you expect to find such a neuron in the brain, the spinal cord, or one of the peripheral nerves?

2. The sea hare *Aplysia californica* is a marine mollusk about the length of your hand (Figure 29.25). It is a favorite research organism for many neurobiologists. Its nervous system only has 20,000 or so neurons, some with large cell bodies that make it easy for scientists to place electrodes or inject dyes.

The sea hare's limited behavior includes responses to food and to touch. Some researchers found dopamine affects its feeding behavior. This neurotransmitter affects human reward-seeking behavior, including addiction to drugs. How might studies of dopamine's effects on sea hares help shed light on treating drug addiction?

3. The blood–brain barrier in human newborns is not fully developed, especially in premature ones. Why is this one reason to pay careful attention to their early diet?

4. Brightly colored frogs of the genus *Dendrobates* secrete a potent toxin through their skin (Section 28.1). The structure of the toxin molecule is similar to that of the neurotransmitter ACh. Speculate on how this toxin might interfere with the nervous system.

5. When Jennifer was six years old, a man lost control of his car and hit a tree in front of her house. She ran over to him and screamed when she saw blood from a wound dripping on a bush of red roses. Thirty-five years later, someone gave her a bottle of *Tea Rose* perfume. When she sniffed it, she became frightened and extremely anxious. A few minutes later she also had a vivid recollection of the accident. Explain this incident in terms of what you learned about memory, olfaction, and the limbic system.

6. Eric typically drinks one cup of coffee nearly every hour, all day long. And by midafternoon, he has trouble concentrating on his studies, and he feels tired and more than a little clumsy. Drinking another cup of coffee doesn't make him more alert. Explain how the caffeine in coffee might produce such symptoms.

7. Steve Wade (Figure 29.26) abused methamphetamine for five years. One night he downed a fifth of 151-proof rum, injected meth five times, and snorted it ten times. His temperature skyrocketed to 110°F, hearing in one ear shut down, and his cognitive skills and reasoning dissolved.

Five years later, he shakes too much to shave with a razor. For him, objects and rooms won't stop spinning. He forgets words in midsentence. Therapists help him practice simple expressions—"How?" "What?" "Are you okay?" Steve, now in his thirties, is irrevocably disabled.

Even before many adolescents and teenagers start experimenting with drugs, they have heard about or know about people like him, yet many still don't stop. Why do you suppose they don't connect the dots to their own behavior? As a class project, do a web search for accounts of drug addicts in their own words. Compare reports in class. Can you identify common social and physiological factors that came into play?

Figure 29.25 Sea hare (*Aplysia*). This brainless mollusk is a boon to researchers who want to study the neural basis of behavior in a simple organism.

Figure 29.26 Portrait of Steve Wade, now irrevocably disabled.

Media Menu

Student CD-ROM

Impacts, Issues Video
   In Pursuit of Ecstasy
Big Picture Animation
   Neurons and nervous systems
Read-Me-First Animation
   Neuron structure and function
   Action potential propagation
   Chemical synapses
   The stretch reflex
   Autonomic nerves

InfoTrac

- Solving the MS Puzzle. *U.S. News & World Report*, October 2002.
- How We Think. *Maclean's*, May 1, 2000.
- Making the Connection. *Biological Sciences Review*, January 2001.
- Alzheimer's: Searching for a Cure. *FDA Consumer*, July–August 2003.

Web Sites

- Society for Neuroscience: web.sfn.org/
- Neuroscience for Kids: faculty.washington.edu/chudler/neurok.html
- Comparative Mammalian Brain Collections: www.brainmuseum.org
- Dana Foundation: www.dana.org
- National Institute on Drug Abuse: www.nida.nih.gov

How Would You Vote?

Of all prison inmates in the United States, close to 25 percent are serving time for nonviolent drug offenses. The cost of keeping them locked up exceeds 9 billion dollars a year. Some people argue that mandatory drug rehabilitation programs would be a better use of the money. Would you support state legislation mandating that convicted individuals enter treatment programs instead of going to jail?

IMPACTS, ISSUES *A Whale of a Dilemma*

Put yourself in the sensory world of a whale moving 200 meters below the ocean surface. Little sunlight penetrates beyond that depth, so eyes do not help much. Fishes use dissolved chemicals as navigational cues, but the whale has few chemical receptors, and its olfactory lobes are greatly reduced.

So how does a whale sense where it is going? It uses sounds—acoustical cues. Water is an ideal medium for transmitting sound waves, which move five times faster in water than in air. Also, compared to mammals on land, whales have far more neurons that collect and integrate auditory information. Unlike humans, whales do not have external flaps that intercept sound waves.

Some species have no opening to their ears. In others, wax fills the ear canal. How, then, do whales hear? Their *jaws* pick up vibrations, which travel through a layer of fat to the middle ear.

Whales use sound to communicate, locate food, and map out the contours of their surroundings. A killer whale or another toothed species uses *echolocation*. It emits high-pitched sounds and listens as echoes from them bounce off objects, including prey. Its ears are especially sensitive to high-frequency sounds. A baleen whale, such as a humpback, makes very low-pitched sounds that can travel across an entire ocean basin. Its ears are attuned to sounds in this range.

The ocean is becoming a lot noisier, and the superb acoustical adaptations of whales may now be working against them. In 2001, for instance, whales beached themselves near an area where the United States Navy was testing a low-frequency sonar system (Figure 30.1). The system is designed to detect new missile-launching submarines that are capable of running in near-silence. We cannot hear its intense, low-frequency emissions, but they may be bad news for whales.

Figure 30.1 A few children drawn to one of the whales that beached itself during military testing of a new sonar system. Of sixteen stranded whales, six died on the beach. Volunteers pushed the others out to sea and their fate is unknown.

the big picture

What Is a Sensory Pathway? Sensory systems are front doors of the nervous system. Nearly all have specific sensory receptors, nerve pathways from receptors to a brain, and brain regions that receive and process sensory information.

Somatic Sensations The somatic sensations start at sensory receptors located in tissues throughout the body, not in sensory organs. They are the basis for responses to touch, pressure, pain, temperature, and positional changes in body parts.

As autopsies showed, the beached whales had blood in their ears and acoustic fat. Were their ears damaged by sounds as intense as 235 decibels? Did the sounds frighten them into surfacing too fast, so that rapid pressure changes damaged the internal tissues? We don't know yet.

Public outcry halted deployment of the new sonar system. Testing continues, however, because the threat of stealth attacks against the United States has not gone away.

Besides, commercial shipping may be a far more pervasive problem for whales. Tankers generate low-frequency sounds that frighten whales or drown out acoustical cues on a daily basis. Realistically, the global shipping of oil and other resources that nations require is not going to stop. If research shows that whales are at risk, will those same nations be willing to design and deploy new tankers that minimize the damage?

With this chapter we turn to sensory systems, the means by which animals receive signals from inside and outside the body, then decode them in ways that give rise to awareness of sounds, sights, odors, and other sensations. Most systems contain some sensory neurons, nerve pathways, and specialized brain regions. Depending on the type and the numbers of sensory receptors they have, animals sample the environment in different ways and differ in their perception of it. How to meet pressing human needs while respecting the sensory diversity can be something of a dilemma.

 How Would You Vote?

Shipping and other human activities generate an underwater ruckus. To what extent should we limit these activities to protect whales against potential harm? Would you support banning activities that exceeded a certain noise level from United States territorial waters? If so, how would you get other nations to do the same? See the Media Menu for details, then vote online.

Taste and Smell The senses of taste and smell start with sensory receptors that bind molecules that are dissolved in the fluid bathing them. Some species have these chemoreceptors on antennae, tentacles, or legs. Other species have them in a mouth, nose, or both.

Balance, Hearing, and Vision In land vertebrates, balance, hearing, and vision became highly evolved senses. Balance and hearing start with acoustical receptors that respond to vibrations. Vision starts with arrays of light-sensitive receptors, such as the rods and cones in your own eyes.

30.1 Overview of Sensory Pathways

*In most animals, a **sensory system** consists of pathways from sensory neurons and nerves to specific brain regions.*

Each **stimulus** (plural, stimuli) is a form of energy that activates sensory receptors, the peripheral endings of a certain type of sensory neuron. **Mechanoreceptors** detect changes in pressure, position, or acceleration, all forms of mechanical energy. **Pain receptors** detect tissue damage. **Thermoreceptors** are sensitive to heat and cold. **Chemoreceptors** detect chemical energy of substances dissolved in the fluid that bathes them. **Osmoreceptors** detect changes in the solute levels of some body fluid. **Photoreceptors** detect differences in the energy of visible and ultraviolet light.

The *somatic* sensations arise from sensory receptors located in tissues throughout the body. They differ from the *special senses*, such as taste, smell, balance, hearing, and vision, which rely on receptors that are restricted to certain body parts, such as ears and eyes.

Regardless of the differences, all sensory receptors transduce stimulus energy into action potentials. But action potentials aren't like an ambulance siren; they don't vary in amplitude. So how does the brain assess a stimulus? It considers (1) *which* nerve pathways are carrying action potentials, (2) the *frequency* of action potentials traveling on each axon in the pathway, and (3) the *number* of axons recruited by the stimulus.

First, an animal's brain is prewired, or genetically programmed, to interpret action potentials in certain ways. That is why you can "see stars" after an eye gets poked, even in a dark room. Many photoreceptors in the eye were mechanically disturbed and sent signals down one of two optic nerves to the brain, which will interpret any signals from an optic nerve as "light."

Second, receptors fire action potentials more often and longer when a stimulus is stronger (Figure 30.2). The same receptor might detect the sound of a throaty whisper or a wild screech. The brain interprets the difference from variations in the frequency of signals.

Third, a strong stimulus recruits far more sensory receptors, compared to a weak stimulus. Gently tap a spot of skin on one of your arms and you will activate a few receptors. Press harder on the same spot and you will activate more receptors in a larger area. A big disturbance translates into action potentials in many sensory axons. And the brain interprets the combined activity as an increase in stimulus intensity.

The frequency of action potentials may decline or stop even when a stimulus is maintained at constant strength. A reduced response to an ongoing stimulus is called **sensory adaptation**. For instance, right after you put on clothing, awareness of its pressure on your skin stops. Some skin mechanoreceptors adapt fast to sustained stimulation; they can signal the onset or the removal of a stimulus. Other receptors adapt more slowly or not at all. They help the brain continually monitor the presence or absence of specific stimuli.

The gymnast in Figure 30.2 is sensing any motions of his limbs and his body's position in space with the help of mechanoreceptors in skeletal muscle, joints, tendons, ligaments, and skin. Those stretch receptors of muscle spindles you read about in Section 29.4 are among them. Their activation depends on how much and how fast a muscle stretches. In this way, the brain is continually informed of changes in muscle length, which helps maintain balance and posture.

A sensory system has sensory receptors for specific stimuli, nerve pathways that conduct information from receptors to the brain, and brain regions that receive the information.

The brain assesses a stimulus according to which nerve pathway is signaling it, the frequency of action potentials from the pathway, and the number of axons recruited.

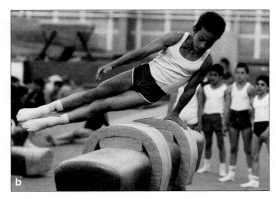

Figure 30.2 (**a**) Recordings of action potentials from some pressure receptor endings in a human hand. A thin rod was pressed against skin with a certain amount of pressure. The vertical bars above the thick horizontal line show individual action potentials. Increases in their frequency correspond to the increases in the stimulus strength.

(**b**) Example of how we make use of sensory pathways—in this case, from receptors in muscles throughout the body to the spinal cord and brain.

Pressure (grams) / Time (seconds)

30.2 Sensing Touch, Pressure, Temperature, and Pain

Somatic sensations start with receptors near the body surface, in skeletal muscles, and in the wall of internal organs. These receptors are most highly developed in birds and mammals. Fishes and amphibians have only a few.

Somatic sensations arise in the cerebral cortex—the cerebrum's outer layer of gray matter. In its sensory areas, interneurons are arrayed like maps for parts of the body surface, just as they are for the motor cortex (Figure 30.3). Maps with the largest areas represent body parts with the greatest sensory acuity and fine motor control, such as fingers, thumbs, and lips.

You and other mammals have receptors for touch, pressure, cold, warmth, and pain (Figure 30.4). The parts having the most of these receptors, including fingertips, are the most sensitive. Parts having fewer, such as the back of the hand, are less sensitive.

Skin and internal tissues have free nerve endings, which act as mechanoreceptors, thermoreceptors, and pain receptors. The branched endings are "naked" or thinly myelinated. Some free nerve endings help us feel prickling pain, as when a finger gets jabbed with a pin. One type makes skin or epithelial linings feel itchy or warm after exposure to irritating chemicals.

Encapsulated receptors are common near the body surface. Meissner's corpuscle adapts very slowly to vibrations of low frequency. It is abundant in the lips, fingertips, eyelids, nipples, and genitals. The bulb of Krause detects cold. Below 10°C, it starts the painful sensations of freezing. Ruffini endings detect steady touch, pressure, and heat. The Pacinian corpuscle is widely distributed deep in the skin, where it responds to fine textures. It is located near freely movable joints and in some internal organs. Its onionlike membrane layers alternate with fluid-filled spaces around the sensory endings. The array is responsive to the rapid pressure changes caused by touch and vibrations.

As described in Section 29.4, stretch receptors in muscle spindle fibers increase their firing rate when the muscle stretches. Along with receptors in the skin and near joints, they send signals to the brain about limb positions.

Pain is a perception of injury to some body region. Sensations of *somatic* pain start with signals from pain receptors in the skin, skeletal muscles, tendons, and joints. Sensations of *visceral* pain are associated with internal organs. Excess chemical stimulation, muscle spasms, muscle fatigue, gross distension of the gut, and poor blood flow to organs are common triggers.

Sensory receptors near the body surface and in internal organs detect touch, pressure, temperature, pain, motion, and positional changes of body parts.

Figure 30.3 Differences in representation of different body parts in the primary somatosensory cortex. This region is a strip of cerebral cortex, about an inch wide, from the top of the human head to above the ear.

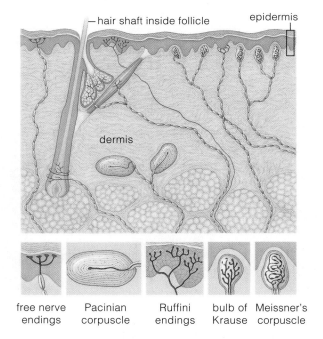

free nerve endings   Pacinian corpuscle   Ruffini endings   bulb of Krause   Meissner's corpuscle

Figure 30.4 A sampling of the mechanoreceptors in human skin. Different kinds are more common in some body regions than others, as the text explains.

30.3 Sampling the Chemical World

In the remainder of this chapter, we will sample the special senses, which have receptors in special organs. They include smell, taste, balance, hearing, and vision.

The sensory pathways of olfaction (smell) and taste begin with chemoreceptors, which become activated by binding molecules of a substance that is dissolved in the fluid bathing them. In both sensory pathways, this input can trigger signals that travel along nerve expressways through the thalamus. The signals end up in the cerebral cortex. There, perceptions of certain stimuli take shape and undergo fine-tuning. Sensory information also reaches the limbic system, which acts to integrate it with emotional states and with stored memories (Section 29.8).

Olfactory receptors fire off signals when they are exposed to water-soluble or volatile (easily vaporized) chemicals. The lining of your own nose has about 5 million of these receptors. That of a bloodhound nose has more than 200 million. The receptor axons lead into one of two olfactory bulbs. In these small brain structures, they synapse with groups of cells that sort out components of a scent. From there, information flows along the olfactory tract to the cerebrum, where it is further processed (Figure 30.5).

Many animals use olfactory cues to navigate, locate food, and communicate socially, as with pheromones. Pheromones, again, are signaling molecules secreted by one individual that influence the social behavior of other individuals of its species. For example, olfactory receptors on antennae of a male silk moth allow him to locate a pheromone-secreting female more than a kilometer upwind.

In reptiles and most mammals (not primates), a cluster of sensory cells forms a *vomeronasal organ* that detects pheromones. In humans, a reduced version of this organ is located about 12.7 millimeters (half an inch) from the entrance to the nose. Whether it is still functional, and what role it might play in human behavior, is a matter of continued debate (Chapter 43).

Different animals have **taste receptors** on antennae, legs, or tentacles, or inside the mouth. On the surface of your mouth, throat, and especially the upper part of the tongue, chemoreceptors are positioned in about 10,000 sensory organs called taste buds (Figure 30.6). Fluids in the mouth reach the receptors by entering a pore in each taste bud.

You perceive many different tastes, but they all are some combination of five primary sensations: sweet (as elicited by glucose and other simple sugars), sour (acids), salty (NaCl or other salts), bitter (plant toxins, including alkaloids), and *umami* (elicited by amino acids such as glutamate, which has a savory taste typical of aged cheese and meat). With its strong savory taste and no odor or texture of its own, MSG (monosodium glutamate) has become a common flavor enhancer.

olfactory tract from receptors to the brain

olfactory bulb

bony plate

ciliated endings of olfactory receptor that project into mucus inside nose

Figure 30.5 Pathway from sensory endings of olfactory receptors in the human nose to the cerebral cortex and the limbic system. Receptor axons pass through holes in a bony plate between the lining of the nasal cavities and the brain.

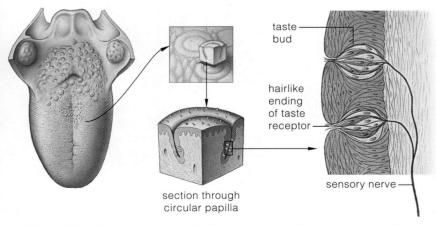

taste bud

hairlike ending of taste receptor

sensory nerve

section through circular papilla

Figure 30.6 Taste receptors in the human tongue. The structures called circular papillae ring epithelial tissue that contains taste buds. A human tongue has approximately 10,000 of these sensory organs, each of which has as many as 150 chemoreceptors.

The senses of smell and taste start at chemoreceptors. Both involve sensory pathways that lead to processing regions in the cerebral cortex and in the limbic system.

30.4 Balancing Acts

All animals assess and respond to displacements from equilibrium, when the body is balanced in relation to gravity, acceleration, and other forces that may affect its positions and its movement. Even brainless jellyfishes right themselves when turned upside down.

The first organs of equilibrium evolved in fishes. You have a pair of them, one in each of your inner ears. A **vestibular apparatus** inside each ear consists of two sacs—the utricle and the saccule—along with three semicircular canals (Figure 30.7). The sacs and canals interconnect as a continuous, fluid-filled system. In this system, many mechanoreceptors are stimulated when you move about or rotate your head.

Inside the bulging base of a semicircular canal is an organ of *dynamic* equilibrium. It has a gelatinous mass with hair cells—a type of mechanoreceptor—projecting into it. Any rotation of the head displaces the fluid in these canals. The pressure exerted by the fluid movement shifts the gelatinous mass, which in turn bends the hair cells. The bending causes signals to flow from sensory neurons to the vestibular nerve. This nerve then carries the signals to the brain.

Inside each utricle and saccule is an organ of *static* equilibrium. These organs send messages to the brain about how the head is oriented relative to the ground. In each, a thick membrane rests on top of hair cells that project upward from the floor of the sac. That membrane contains a mass of crystals that weigh it down. When your head is held upright, the weighted membrane presses down on hair cells, which bends them slightly. This sends a steady stream of action potentials from the sensory neurons along the vestibular nerve to the brain. If your posture changes or if you speed up or slow down movement in one direction, the position of the membrane above the hair cells will shift. Depending on the position, a hair cell will now have more or less weight on it, so signaling from sensory cells will step up or slow down. To interpret the body's posture and movement, the brain constantly compares and integrates the signals from sensory cells in all organs of static and dynamic equilibrium in the inner ears on both sides of the head.

Figure 30.7 The vestibular apparatus inside the human ear. Organs of dynamic and static equilibrium inside this fluid-filled organ contribute to the sense of balance.

At the same time, the brain evaluates input from receptors in the skin, joints, and tendons. Integration of all the information allows your brain to control eye muscles and keep the visual field in focus, even as you shift position or your head rotates. Integration also helps maintain awareness of the body's position and motion in space, as demonstrated by champion figure skater Sarah Hughes.

A stroke, an inner ear infection, or loose particles in the semicircular canals cause *vertigo*, a sensation that the world is moving or spinning around. Vertigo also arises from conflicting sensory inputs, as when you stand on the top viewing deck of a skyscraper and look down. The vestibular apparatus reports that you are motionless, while your eyes report that your body is floating in space far above the ground.

Mismatched signals also give rise to the dizziness and nausea of *motion sickness*. Passengers in a vehicle moving fast on a hilly, curvy road experience changes in acceleration and direction that scream "motion" to each vestibular apparatus. At the same time, signals from the eyes about objects inside of the vehicle tell the brain that the body is standing still. Being the driver minimizes motion sickness because a driver is forced to focus on the scenery rushing past, so visual signals are consistent with vestibular signals.

Organs of equilibrium help keep the body balanced in relation to gravity, velocity, acceleration, and other forces that influence its position and movement.

30.5 Making Sense of Sounds

Many arthropods and most vertebrates perceive sounds; they have a sense of hearing. Land vertebrates have ears that capture and sort out sound waves traveling in air.

PROPERTIES OF SOUND

Sounds are waves of compressed air, or one form of mechanical energy. Clapping your hands compresses air and creates pressure variations that can be shown as wave forms. The *amplitude* of a sound corresponds to its loudness, or intensity. We measure amplitude in decibels. The human ear can detect a one-decibel difference between sounds. The *frequency* of a sound is the number of wave cycles per second. Each cycle extends from the peak (or trough) of one wave to the corresponding peak (or trough) on the next wave in line (Figure 30.8).

Unlike a tuning fork's pure tone, most sounds are combinations of waves of different frequencies. Their timbre, or sound quality, varies. When you perceive one person's voice as nasal and another as deep, it is because of differences in timbre. Variations in timbre help you recognize many individuals by their voices.

EVOLUTION OF VERTEBRATE HEARING

Water readily transfers vibrations to body tissues, so fishes do not require elaborate outer ears to capture them. The transfer of sounds from air to the body is more dicey, because sounds spread out in air. Land vertebrates have special structures that trap, amplify, and process sound waves.

Figure 30.8 Wavelike properties of sound.

Consider the three regions of your ears (Figure 30.9a). The *outer* ear is adapted for gathering sounds from air, as it is for most mammals. Its pinna is a much-folded flap of cartilage, sheathed in skin, that projects from the side of your head. An auditory canal leads from the pinna into the middle ear.

The *middle* ear amplifies and transmits air waves to the inner ear. It has an eardrum, which first evolved in early reptiles as a tiny depression on each side of the head. The "drum" is a thin membrane that can vibrate fast in response to pressure waves. Between the eardrum and the inner ear is an air-filled cavity and one or more small bones. These are the hammer, anvil, and stirrup inside the mammalian ear (Figure 30.9b). They interact to transmit the force of sound waves from the eardrum inward, to a still smaller surface known as the oval window.

Figure 30.9 Components of the human ear.

The oval window is an elastic membrane in front of a spiral, pea-sized, fluid-filled cochlea. The cochlea and the vestibular apparatus described earlier are the major parts of the *inner* ear. Figure 30.9c shows how a cochlea would look if it were uncoiled. Inside are the mechanoreceptors that respond to sound waves.

As the stirrup pushes against the oval window, it starts pressure waves in the fluid inside two cochlear ducts (scala vestibuli and scala tympani). The waves reach another membrane, the round window, causing it to bow inward and outward in response.

The cochlear duct, the inner ear's third duct, sorts out pressure waves. Its *basilar* membrane is stiff and narrow near the oval window, then it broadens out and becomes more flexible deeper into the coil. These structural differences cause areas of the membrane to respond to different frequencies. High-pitched sounds cause the first part of the coil to vibrate; low-pitched

Read Me First!

and watch the narrated animation on ear structure and function

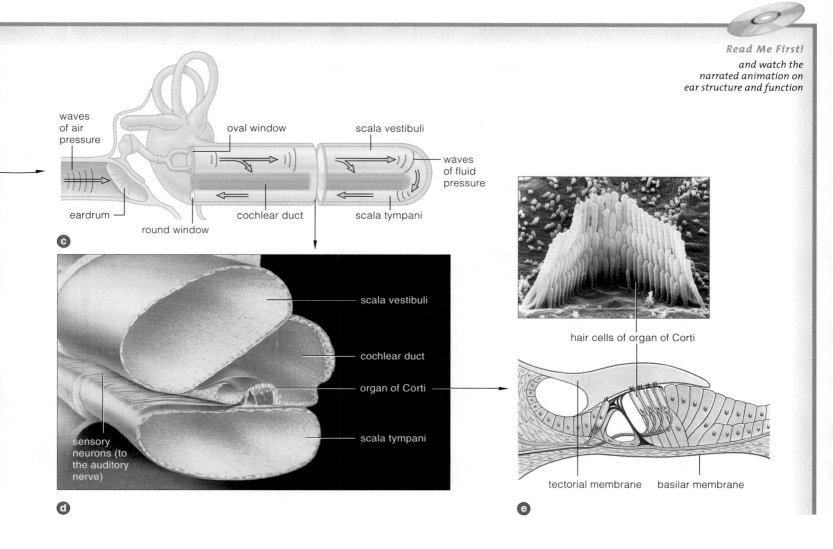

waves of air pressure

oval window

scala vestibuli

waves of fluid pressure

eardrum

cochlear duct

scala tympani

round window

c

scala vestibuli

cochlear duct

organ of Corti

scala tympani

sensory neurons (to the auditory nerve)

d

hair cells of organ of Corti

tectorial membrane   basilar membrane

e

sounds affect parts deeper within. How? Attached to the inner basilar membrane is an organ of Corti, with arrays of vibration-sensitive acoustical receptors.

The mechanoreceptors are a type of hair cell, with modified cilia. They project into a *tectorial* membrane draped over them (Figure 30.9*d,e*). Their cilia bend as pressure waves displace the basilar membrane. This bending transduces the mechanical energy of pressure into action potentials, which travel along an auditory nerve into the brain. And think about this: Hearing losses follow when hair cells are damaged, as from long-term exposure to intense sounds (Figure 30.10).

Ears of land vertebrates collect, amplify, and sort out sound waves. In the inner ear, sound waves produce fluid pressure variations and trigger action potentials in hair cells.

Figure 30.10   Results of an experiment on the effect of intense sound on the inner ear. *Left:* From a guinea pig ear, two rows of hair cells that normally project into the tectorial membrane in the organ of Corti. *Right:* Hair cells inside the same organ after twenty-four hours of exposure to noise levels comparable to extremely loud music.

To give you a sense of how sound intensity is measured, a ticking watch measures 10 decibels. Normal conversation is about 60 decibels, a food blender operating at high speed is about 90 decibels, and an amplified rock concert is about 120 decibels. The perceived loudness of a sound depends not only on its intensity, but also on its frequency.

30.6 Do You See What I See?

Shine a light on a single-celled amoeba and it abruptly stops moving. Sunflowers track the sun as it arcs overhead. Most organisms are sensitive to light, but few "see" as you do.

REQUIREMENTS FOR VISION

At the minimum, the sense of **vision** requires eyes and image perception in brain centers, which receive and assess patterns of visual information on brightness, shapes, positions, and movements. **Eyes** are sensory organs with a dense array of photoreceptors. And all photoreceptors have this in common: *They incorporate pigment molecules that can absorb photon energy, which is converted to excitation energy in sensory neurons.*

Earthworms and some other eyeless animals have photoreceptors scattered in the integument or grouped in a depression in it. They use light to orient the body, detect shadows, or adjust biological clocks.

Complex eyes evolved among swift predators and prey, which had to discriminate among objects in a fast-changing visual field. A **visual field** is the part of the outside world that an animal sees. Forming images of it is easier with a lens, a transparent body that bends all light rays from any point in the visual field so they converge on photoreceptors. Rays of light bend at the boundaries between substances of different densities. Bending sends them off in new directions.

Spiders, abalones, and some other animals have *simple* eyes with one lens (Figure 30.11). Crustaceans and insects have *compound* eyes with many lensed units. Squids, octopuses, and vertebrates have *camera* eyes. Like cameras, their eyeball contains a darkened chamber. Light enters it through a pupil, an opening in an iris. The iris is a ring of contractile tissue equivalent to a camera's diaphragm. In back of the pupil, a lens focuses light onto a **retina**—a tissue densely packed with photoreceptors (equivalent to light-sensitive film in the camera). Sensory nerve axons start at the retina and converge as an optic tract to the brain.

THE HUMAN EYE

As in most vertebrates, a human eyeball has a three-layered wall (Figure 30.12). Its outer layer is the sclera and cornea. The sclera is the densely fibrous "white" part; it protects most of the eyeball. A cornea made of transparent collagen fibers covers the rest of it.

The middle layer has a choroid, ciliary body, iris, and pupil. The choroid is a darkly pigmented tissue, which absorbs light wavelengths that photoreceptors missed and stops them from scattering in the eye.

Suspended just behind the cornea is the doughnut-shaped, pigmented iris. Light enters the dark "hole" in the center. In bright sunlight, the circular muscles embedded in the iris contract, so the pupil's diameter shrinks. In dim light, however, radial muscles in the iris contract, making the pupil enlarge and allowing more light to enter the eye.

Once past the pupil, light rays pass through a lens kept moist by a clear fluid called the aqueous humor. A jellylike vitreous body fills the area behind the lens. At the very back of this chamber is the retina.

Because the cornea and lens have a curved surface, light rays coming from a particular point in the visual

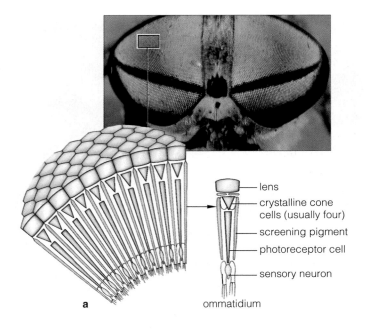

a

lens
crystalline cone cells (usually four)
screening pigment
photoreceptor cell
sensory neuron

ommatidium

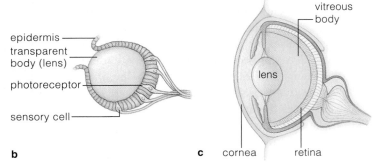

epidermis
transparent body (lens)
photoreceptor
sensory cell

b

vitreous body
lens
c cornea retina

Figure 30.11 A few invertebrate eyes. (**a**) Compound eye of a deerfly. In each of many units, a lens directs light onto a crystalline cone, which focuses light onto one photoreceptor cell. (**b**) Abalone eye, complete with a lens. (**c**) Camera eye typical of some cephalopods (squids and octopuses, as shown here) and vertebrates. Cephalopods and vertebrates are not closely related, so similarities in eye structure and function may be a case of convergent evolution. By one theory, a group of genes that once affected development of the nervous system were appropriated for the task of eye formation.

Read Me First!
and watch the narrated animation on
eye structure and function

Wall of eyeball	(three layers)
Fibrous Tunic	*Sclera*. Protects eyeball
	Cornea. Focuses light
Vascular Tunic	*Choroid*. Blood vessels nutritionally support wall cells; pigments prevent light scattering
	Ciliary body. Its muscles control lens; shape; its fine fibers hold lens upright
	Iris. Adjusting iris controls incoming light
	Pupil. Serves as entrance for light
Sensory Tunic	*Retina*. Absorbs, transduces light energy
	Fovea. Increases visual acuity
	Start of optic nerve. Carries signals to brain

Interior of eyeball	
Lens	Focuses light on photoreceptors
Aqueous humor	Transmits light, maintains pressure
Vitreous body	Transmits light, supports lens and eyeball

Figure 30.12 Structure and component parts of the human eye.

field strike them at different angles, so the trajectories change. The newly angled trajectories make light rays converge at the back of the retina, in a pattern upside-down and reversed left to right relative to the original source of the light rays (Figure 30.13*a*).

The closer any object is to the eye, the more rays of light coming from it will be diverging from one another when they enter the eye. But the shape of the lens can be adjusted in ways that focus light rays on the retina, regardless of how far they have traveled.

Lens adjustments are called **visual accommodation**. A ciliary muscle encircles the lens and attaches to it by fibers (Figures 30.12 and 30.13). Focus on something close up, and the ciliary muscle contracts and the lens bulges. Rays of light that are diverging from the object are focused onto the retina (Figure 30.13*b*). When an object is farther away, light rays do not have to be bent as much to be focused. The ciliary muscles can relax a bit, which allows the lens to flatten out (Figure 30.13*c*). When you spend long periods staring at a computer screen, reduce eyestrain by taking a break once in a while and focusing on more distant objects.

Vision requires eyes with a dense array of photoreceptors and image formation in brain centers that can receive and process patterns of visual information on brightness, shapes, positions, and movements.

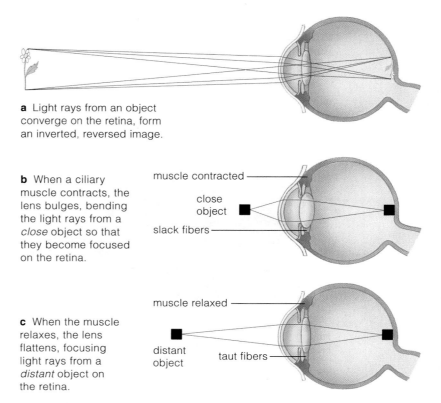

a Light rays from an object converge on the retina, form an inverted, reversed image.

b When a ciliary muscle contracts, the lens bulges, bending the light rays from a *close* object so that they become focused on the retina.

muscle contracted

close object

slack fibers

c When the muscle relaxes, the lens flattens, focusing light rays from a *distant* object on the retina.

muscle relaxed

distant object

taut fibers

Figure 30.13 (**a**) Pattern of retinal stimulation in a human eye. A curved, transparent cornea in front of the pupil changes the trajectories of light rays as they enter the eye. (**b,c**) Two focusing mechanisms use a ciliary muscle that encircles the lens and attaches to it.

30.7 A Closer Look at the Retina

Each eye is an outpost of the brain, busily collecting and analyzing information about the distance, shape, position, brightness, and movement of visual stimuli. Its sensory pathway starts at the retina and ends in the brain.

In between the retina and the choroid is a basement layer, a pigmented epithelium. Resting on this layer are densely packed arrays of **rod cells** and **cone cells**, two classes of photoreceptors that are derived from cilia (Figure 30.14). Rod cells detect very dim light. At night or in dark places, they respond to changes in light intensity across the visual field. Their signals are the start of coarse perception of motion. Cone cells detect bright light. Their signals are the start of sharp vision and perception of color in the daytime.

A *rod* cell's outer segment is folded into several hundred membrane disks, each containing a hundred million molecules of the visual pigment rhodopsin. Membrane stacking and the high pigment density increase the likelihood of intercepting packets of light energy—photons. Action potentials triggered by the absorption of even a few photons make us conscious of dimly lit objects, as in darkened rooms or late at night.

The sense of color and of daytime vision starts with photon absorption by *cone* cells. There are three types, and each has a different pigment. One is best at absorbing red wavelengths, another blue, and a third green. Normal vision requires all three types.

Several layers of cells lie above the rods and cones, where they accept information and process signals from them (Figure 30.15). Input from about 125 million

rods and cones synapses on bipolar cells, which send signals to just one million ganglion cells, the axons of which start an optic nerve. Signals that reach ganglion cells have been dampened or enhanced by amacrine cells and horizontal cells. Thus, *signal integration and processing start in the retina.*

The retinal cells are organized into receptive fields, or restricted areas that affect the activity of individual sensory neurons. For example, the field for a ganglion cell is a small circle. Different cells respond to rapid changes in light intensity, a spot of one color, motion, and so on in their tiny receptive field.

Humans have two optic nerves, one from the retina in each eye. The optic nerve leading away from each eye delivers signals about a stimulus from the *left* visual field to the right cerebral hemisphere. It delivers signals from the *right* visual field to the left hemisphere (Section 29.8).

Optic nerve axons end in a layered brain region that has a map corresponding to the receptive fields that deal with each kind of visual stimulus, such as form, movement, depth, color, and texture. After early processing, signals rapidly and simultaneously reach different parts of the visual cortex. Final integration organizes the signals into the sensation of sight.

Organization of visual signals begins at receptive fields in the retina. Signals are further processed in a layered brain region, and finally integrated in the visual cortex.

Figure 30.14 Scanning electron micrograph and sketches of rod cell and cone cell structure.

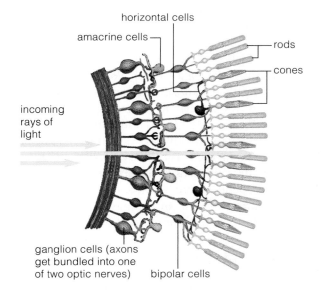

Figure 30.15 Organization of the photoreceptors and related sensory cells in the retina.

30.8 Visual Disorders

Vision is impaired when light is not focused properly, when photoreceptors do not work as they should, or when visual processing breaks down.

SO NEAR, SO FAR

Sometimes the eyeball is not shaped quite right and the lens cannot be adjusted enough to make the focal point match up with the retina. In *nearsightedness*, the eyeball is too long or the ciliary muscle that adjusts the lens contracts too strongly. Light from distant objects gets focused too soon—in front of the retina instead of right on it (Figure 30.16*a*).

In *farsightedness*, the eyeball is not long enough or the lens is "lazy." The focal point for light from close objects is behind the retina (Figure 30.16*b*). This often happens after age forty; the lens usually loses some of its natural flexibility as we grow older.

Most cases of farsightedness and nearsightedness can be corrected with glasses or by surgically altering the shape of the cornea. Today, lasers are used to reshape the cornea. The most common procedure, LASIK, was performed on 1.8 million Americans in 2002. When all goes well, laser surgery eliminates the need to wear glasses during most activities, although those over age forty may still need reading glasses. Results vary, and chronic eye irritation is common.

COLOR BLINDNESS

Individuals who lack one or more types of cones are *color-blind*. Remember *red–green color blindness*? This X-linked, recessive abnormality appears most often in males (Section 11.5). Some or all of the cone cells that detect light of red or green wavelengths are missing. Most affected people have trouble distinguishing red from green in dim light, and some cannot distinguish between the two even in bright light.

AGE-RELATED DISORDERS

Near the retina's center is the macula, an area that is rich in photoreceptors (Figure 30.17). It surrounds the fovea. This funnel-shaped depression has the greatest density of photoreceptors and contributes the most to visual acuity. Cone cells in the fovea discriminate the most precisely between adjacent points in space.

Macular degeneration, one of the most common causes of blindness, usually starts as a person ages. In the United States, about 13 million people have age-related macular degeneration (AMD). Destruction of photoreceptor cells in the macula and fovea clouds

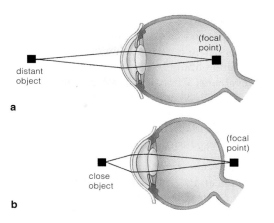

Figure 30.16 Focusing problems. In nearsightedness (**a**) images from distant objects converge in front of the retina. In farsightedness (**b**) images from close objects have not yet converged when they reach the retina.

the center of the visual field more than the periphery. The causes of AMD are not fully known, although there is a genetic component. Smoking, obesity, and high blood pressure increase the risk.

AMD cannot be cured, but its progression can be slowed with eye drops, vitamins, and laser therapy. An experimental treatment filters a patient's blood and removes cholesterol and certain proteins. So far, it has helped about 30 percent of the patients.

Glaucoma results when too much aqueous humor accumulates inside the eyeball. This damages blood vessels and ganglion cells, and it can interfere with peripheral vision and visual processing. Glaucoma can be treated with medication and surgery.

A *cataract* is a gradual clouding of the lens. It alters the amount of light that reaches the retina and where light is focused. The clouded lens can be replaced with an artificial implant. Each year, millions of people undergo cataract surgery.

Figure 30.17 Part of the retina at the back of the eye. The fovea contains only cones. A photoreceptor-rich area, the macula, surrounds it. The optic nerve is made up of the bundled axons of ganglion cells.

Summary

Section 30.1 Sensory receptors convert stimulus energy, such as light and pressure, into action potentials. Mechanoreceptors, pain receptors, chemoreceptors, osmoreceptors, and photoreceptors are the main types.

The brain evaluates action potentials from sensory receptors based on which nerve delivers them, their frequency, and the number of axons firing in a given interval. Continued stimulation of a receptor may lead to a lessened response (sensory adaptation).

Section 30.2 Receptors for somatic sensations, such as touch, pressure, and warmth, are not localized in a special organ or tissue. They signal sensory areas of the cerebral cortex, where interneurons are organized like maps for individual parts of the body surface. Receptors for the special senses—taste, smell, hearing, balance, and vision—are located in special sensory organs.

Section 30.3 The senses of taste and smell involve pathways from chemoreceptors to processing regions in the cerebral cortex and limbic system. Taste receptors are concentrated in taste buds in the tongue and mouth. Olfactory receptors line the nasal passages. Pheromones are chemical signals secreted by many animals that also have specialized organs or structures to detect them.

Section 30.4 The vestibular apparatus is a fluid-filled organ of equilibrium in the vertebrate inner ear. It detects gravity, acceleration, and other forces that affect the body's position and movement.

Section 30.5 Sound is the detection of pressure waves that vary in amplitude and frequency. In vertebrates, the outer ear collects sound waves. The middle ear amplifies sound waves and transmits them to the fluid-filled inner ear. Inside the cochlea of the inner ear, pressure waves are sorted out according to their frequency and converted to action potentials that are relayed along auditory nerves to the brain.

Section 30.6 All organisms are sensitive to light, but vision requires eyes and brain centers that can process visual information. An eye is a sensory organ that has a dense array of photoreceptors.

Like squids and octopuses, humans have camera eyes, each with an iris that adjusts incoming light, and a lens that focuses light on the retina in the back of the eyeball chamber. The retina has densely packed photoreceptors.

Section 30.7 Rods detect dim light, and cones detect bright light and colors. These photoreceptors interact with other retinal cells to process visual information even before sending it on to the brain. Two optic nerves carry signals that eventually reach the cerebral cortex.

Section 30.8 Abnormalities in eye shape, in the lens, and with retinal cells often impair vision, especially with advancing age.

Self-Quiz
Answers in Appendix III

1. _____ is defined as a decrease in the response to an ongoing stimulus.
   a. Perception   c. Sensory adaptation
   b. Visual accommodation d. Somatic sensation

2. Which is a somatic sensation?
   a. taste   d. hearing
   b. smell   e. both a and c
   c. touch   f. all of the above

3. Chemoreceptors play a role in the sense of _____ .
   a. taste   d. hearing
   b. smell   e. both a and b
   c. touch   f. all of the above

4. In the _____ neurons are arranged like maps that correspond to different parts of the body surface.
   a. cerebral cortex   c. basilar membrane
   b. retina   d. all of the above

5. Mechanoreceptors in the _____ send signals to the brain about the body's position relative to gravity.
   a. eye   c. tongue
   b. ear   d. nose

6. The middle ear functions in _____ .
   a. detecting shifts in body position
   b. amplifying and transmitting sound waves
   c. sorting sound waves out by frequency
   d. both b and c

7. Match each structure with its description.
   ____ rod cell   a. type of mechanoreceptor
   ____ cochlea   b. function in balance
   ____ lens   c. detects color
   ____ hair cell   d. detects dim light
   ____ cone cell   e. contains chemoreceptors
   ____ taste bud   f. focuses light
   ____ vestibular   g. sorts out sound waves
        apparatus

Critical Thinking

1. Figure 30.18 shows pits above and below a python's mouth. Inside are thermoreceptors that detect body heat, or infrared energy. What kinds of organisms are the likely prey of this snake? Also, what kinds of otherwise edible organisms might slip past it?

Figure 30.18 Here's heat-sampling you!

Figure 30.19 Here's looking at you!

Figure 30.20 Inner ear structures from (**a**) a bushbaby, a highly agile land mammal, (**b**) a forerunner of whales (*Ichthyolestes*) that lived on land 50 million years ago, (**c**) a marine whale (*Indocetus*) that lived 45 million years ago, and (**d**) a dolphin. These computer reconstructions are adjusted to account for differences in body size.

2. In humans, photoreceptors are most concentrated at the very back of the eyeball. In birds of prey, such as owls and hawks, the greatest density of photoreceptors is in a region closer to the eyeball's roof. When these birds are on the ground, they can't see objects even slightly above them unless they turn their head almost upside down (Figure 30.19). How is this retinal organization adaptive?

3. The vestibular apparatus of all modern whales and dolphins is much smaller, relative to their body size, than it is in other mammals (Figure 30.20). Whales, remember, are descended from land-dwelling mammals (page 257). Fossils of early whales show that the vestibular apparatus became modified as their ancestors made the transition back to a life in the seas. Think about how the sense of balance in an aquatic animal might differ from that of a land mammal that walks about on four legs. Speculate on why selection might have favored a rapid reduction of the vestibular apparatus during the transition from a terrestrial to an aquatic life-style.

4. The strength of Earth's magnetic field and its angle relative to the surface vary with latitude. Diverse species sense these differences and use them as cues for asssessing their location and direction of movement. Behavioral experiments have shown that sea turtles, salamanders, and spiny lobsters use information from Earth's magnetic field during their migrations. Whales and certain burrowing rodents also seem to have a magnetic sense. Evidence about humans is contradictory. Is it likely that humans have such a sense? Suggest an experiment that might support or disprove the possibility.

5. Laura loves broccoli and brussels sprouts. Lionel can't stand them. Everyone has the same five kinds of taste receptors, so what is going on? Is Lionel just being difficult? Perhaps not. Studies have shown that the number and distribution of receptors that respond to bitter substances vary and that some of this variation is genetic. People having the greatest number of these receptors find many fruits and vegetables unpalatable. These *supertasters* make up about 25 percent of the general population. They tend to be slimmer than average but are more likely to have colon polyps and colon cancer. How might Lionel's highly sensitive taste buds put him at increased risk for colon cancer?

Media Menu

Student CD-ROM

Impacts, Issues Video
    A Whale of a Dilemma
Big Picture Animation
    Sensory neurons, systems, and senses
Read-Me-First Animation
    Ear structure and function
    Eye structure and function
Other Animations and Interactions
    Recordings of action potentials
    Wavelike properties of sound
    Organization of the retina

InfoTrac

- Taking a New Look at Pain: Why Do We Hurt? *Newsweek*, May 2003.
- What's So Hot About Peppers? *Odyssey*, March 2003.
- Seeing How the Brain Hears. *The Scientist*, May 2002.
- How the Retina Works. *American Scientist*, January–Feburary 2003.
- Saving Your Sight: Early Detection Is Critical. *FDA Consumer*, March–April 2002.

Web Sites

- Converging Pathways of Pain Research: history.nih.gov/exhibits/pain/
- Olfaction: www.leffingwell.com/olfaction.htm
- Seeing, Hearing, and Smelling the World: www.hhmi.org/senses/
- Virtual Tour of the Ear: ctl.augie.edu/perry/ear/hearmech.htm
- National Eye Institute: www.nei.nih.gov/

How Would You Vote?

Whales and dolphins are highly dependent on acoustical cues. It is possible—although not certain—that sounds from supertankers, undersea mining, and other human activities may be pushing some species toward the brink of extinction. There are ways to lessen undersea noise, but quieter engines and other noise abatement methods are expensive. Should the United States prohibit activities that exceed a certain noise level in its waters?

IMPACTS, ISSUES *Hormones in the Balance*

Atrazine has been widely used as an herbicide for more than forty years. Each year, people in the United States apply about 60 million tons of it to croplands, home gardens, and lawns in parks and golf course fairways. Where does it all go after that? Into soil and water. Atrazine molecules break down fast, but they are still turning up in well water and groundwater, especially in the American Midwest.

Enter Tyrone Hayes. As this University of California biologist suspects, atrazine may be one of the known synthetic organic compounds that are inadvertently acting as *endocrine disrupters*. They mimic, block, or boost the effects of natural hormones in ways that interfere with reproduction and development.

Hayes, shown in the filmstrip, looked at atrazine's effects on African clawed frogs (*Xenopus laevis*). In his laboratory, many male tadpoles were exposed to the herbicide. They became hermaphrodites, with male *and* female reproductive organs. Was this happening in nature? Hayes collected native leopard frogs (*Rana pipiens*) from ponds and ditches across the Midwest. Male frogs from every atrazine-contaminated site he sampled had abnormal sex organs. In the pond with the highest atrazine level, 92 percent of males were

Figure 31.1
Not a good time to be a frog.

the big picture

The Endocrine System In all animals, hormones and other signaling molecules support metabolism, growth, development, and reproduction. Glands and cells of the vertebrate endocrine system secrete hormones.

Signaling Mechanisms Signaling molecules secreted from one type of cell bind to and activate receptors on target cells. The signals are converted to forms that may alter cell activities. A few cell types are targets of some signals; cells throughout the body are targets of others.

abnormal. Only one site had no detectable atrazine—and it had no abnormal male frogs.

Scientists funded by herbicide industries dispute his conclusions, but the Environmental Protection Agency found his work persuasive. This agency regulates the agricultural application of atrazine and has called for more research on atrazine's effects on amphibians.

Meanwhile, chemical pollutants may be having negative effects on other species in natural aquatic habitats. In some pesticide-contaminated lakes in Florida, alligators have low testosterone levels and an abnormally small penis. In New York, species patterns of colored spots and blotches on salamander skin became discombobulated after a golf course was developed next to their breeding pond.

What about humans? Hayes thinks atrazine levels as low as 0.1 part per billion harm frogs. As much as 3 parts per billion are allowed in our drinking water. Some researchers are conducting experiments to test the hypothesis that chemicals in the environment are contributing to an earlier onset of puberty and to low sperm counts. Others dismiss the hypothesis as junk science. Ongoing detective work on both sides of the controversy should clarify the picture.

This chapter focuses on hormones—their sources, targets, and interactions. If the details start to seem remote, remember the endocrine disrupters. What you learn here can help you evaluate research reports that may affect your health and life-style.

✓ How Would You Vote?

Crop yields that sustain the human population currently depend on agricultural pesticides, some of which may disrupt hormone function in frogs and other untargeted species. Should the pesticides that are under suspicion remain in use while researchers investigate them? See the Media Menu for details, then vote online.

From Signaling to Response   Close interactions between the hypothalamus and the pituitary gland adjust the secretions of many endocrine glands. Some endocrine glands and cells secrete signaling molecules in response to local chemical changes.

Environmental Effects   Environmental cues affect endocrine functions. So does the introduction of natural and synthetic compounds into habitats, which may affect sperm counts and other aspects of reproduction or disrupt growth and development.

31.1 Introducing the Vertebrate Endocrine System

Hormones and other signaling molecules secreted from one cell type cause the activities of target cells to change in ways that affect the body's form, metabolism, health, and behavior. In complex animals, they have long-term roles in reproduction, growth, and development.

COMPARISON OF SIGNALING MOLECULES

In every complex organism, *signaling molecules* help integrate cell activities in ways that keep the body working smoothly. Hormones, neurotransmitters, local signaling molecules, and pheromones are signaling molecules in animals. Each type acts on target cells, the "target" being any cell that bears receptors for it. Different kinds of receptors are located at the plasma membrane, in the cytoplasm, or in the nucleus.

In animals, **hormones** are the secretory products of endocrine glands, endocrine cells in tissues, and a few kinds of neurons. Insulin is an example of an animal hormone. Certain pancreatic cells secrete molecules of this protein hormone into interstitial fluid. Like most of the other animal hormones, insulin diffuses into blood capillaries, then the bloodstream transports it throughout the body. Insulin molecules diffuse out of the blood into all tissues, but the hormone acts only on cells that have functioning receptors for it.

You already know about the **neurotransmitters**. In Chapter 29, you saw how these signaling molecules are released from the axon endings of neurons, then swiftly activate or inhibit target cells after diffusing across the tiny synaptic cleft between them.

Also, many cells release **local signaling molecules**. These change chemical conditions in local tissues, as when the action of some prostaglandins restricts or enhances blood flow to a tissue.

Pheromones diffuse through water or air to target cells in other individuals of the same species. They help integrate social behavior. For example, soldier termites secrete one pheromone that acts as an alarm signal when ants attack their colony. In Chapter 43, we will consider these nearly odorless, hormonelike signals and their roles in animal communication. Here, our focus will be on the functions of hormones and local signaling molecules.

OVERVIEW OF THE ENDOCRINE SYSTEM

The first hormones were discovered in the early 1900s by physiologists W. Bayliss and E. Starling. They were trying to find out what triggers secretion of pancreatic juices when food is traveling through the canine gut. As they knew, acids mix with food inside the stomach,

model for the hormone insulin

and the pancreas secretes an alkaline solution after the acidic mixture enters the small intestine. But what was it that stimulated the pancreatic response? Was it a type of neural stimulation or something else?

To find an answer, Bayliss and Starling severed the nerves that carry signals to and from the upper small intestine of a laboratory dog. They did not sever the small intestine's blood vessels. Later, when acidic food entered the small intestine, the pancreas still secreted the alkaline solution.

The researchers also injected juices extracted from the intestinal lining—a *glandular* epithelium—into the dog's blood. The extract induced the same response. Glandular cells of the lining were probably producing a pancreas-stimulating substance.

That substance is now called secretin. Identifying its mode of action supported a hypothesis that dated back centuries: *The bloodstream carries internal secretions that influence the activities of organs inside the body.* Starling coined the term "hormone" for glandular secretions (after *hormon*, meaning to set in motion).

Later, researchers found that all vertebrates make hormones. Even jawless lampreys make a few. Figure 31.2 uses the human body to provide an overview of hormones produced by all jawed vertebrates. It also shows where the major sources of these hormones are located. Gene duplications and modifications may account for the diversity in these signaling molecules and in the receptors for them.

All of the sources of hormones inside the vertebrate body came to be viewed as the **endocrine system**. The name implies that there is a separate control system for the body, apart from the nervous system. (*Endon* means "within." *Krinein* is taken to mean "secrete.") Later, biochemical research and electron microscopy studies revealed that most endocrine sources and the nervous system interact in intricate ways.

In all animals, signaling molecules integrate cell activities. Hormones, neurotransmitters, local signaling molecules, and pheromones are the main categories. Different kinds function in day-to-day activities, and others in long-term programs of growth, development, and reproduction.

A signaling molecule serves in cell-to-cell communication and may cause target cells to alter their activities.

Endocrine glands or cells produce animal hormones. Blood transports most hormones to nonadjacent target cells.

Most vertebrates have the same types of hormones, produced by similar glandular sources. The endocrine system interacts closely with the nervous system.

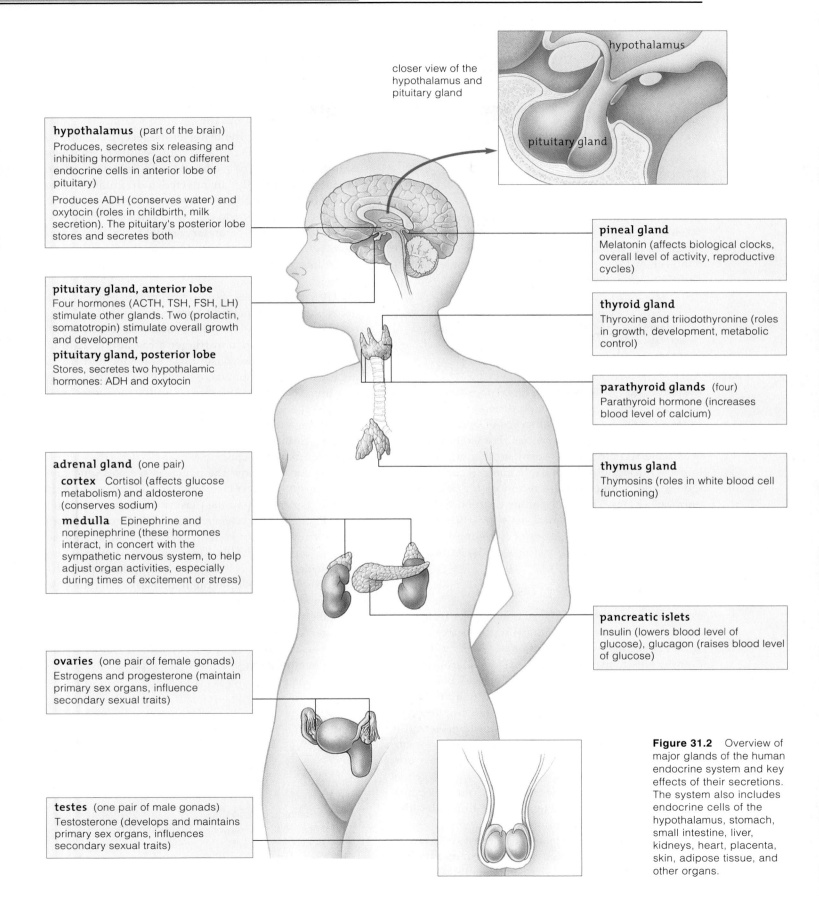

closer view of the hypothalamus and pituitary gland

hypothalamus

pituitary gland

hypothalamus (part of the brain)
Produces, secretes six releasing and inhibiting hormones (act on different endocrine cells in anterior lobe of pituitary)

Produces ADH (conserves water) and oxytocin (roles in childbirth, milk secretion). The pituitary's posterior lobe stores and secretes both

pituitary gland, anterior lobe
Four hormones (ACTH, TSH, FSH, LH) stimulate other glands. Two (prolactin, somatotropin) stimulate overall growth and development

pituitary gland, posterior lobe
Stores, secretes two hypothalamic hormones: ADH and oxytocin

adrenal gland (one pair)

  cortex Cortisol (affects glucose metabolism) and aldosterone (conserves sodium)

  medulla Epinephrine and norepinephrine (these hormones interact, in concert with the sympathetic nervous system, to help adjust organ activities, especially during times of excitement or stress)

ovaries (one pair of female gonads)
Estrogens and progesterone (maintain primary sex organs, influence secondary sexual traits)

testes (one pair of male gonads)
Testosterone (develops and maintains primary sex organs, influences secondary sexual traits)

pineal gland
Melatonin (affects biological clocks, overall level of activity, reproductive cycles)

thyroid gland
Thyroxine and triiodothyronine (roles in growth, development, metabolic control)

parathyroid glands (four)
Parathyroid hormone (increases blood level of calcium)

thymus gland
Thymosins (roles in white blood cell functioning)

pancreatic islets
Insulin (lowers blood level of glucose), glucagon (raises blood level of glucose)

Figure 31.2 Overview of major glands of the human endocrine system and key effects of their secretions. The system also includes endocrine cells of the hypothalamus, stomach, small intestine, liver, kidneys, heart, placenta, skin, adipose tissue, and other organs.

31.2 The Nature of Hormone Action

Like most other signaling molecules, hormones induce target cells to transcribe genes, make more proteins or fewer, to hurry along with glucose uptake, to step up or slow down certain secretions, to grow or get ready to divide, to commit suicide—all in the service of the body's programs of growth, maintenance, and reproduction.

SIGNAL RECEPTION, TRANSDUCTION, AND RESPONSE

Recall, from Section 24.5, that response to a hormone takes three steps: (1) activation of a receptor for that hormone, (2) transduction of the receptor's signal into a molecular form that can exert effects inside the cell, and (3) the functional response:

Responses to hormones vary, largely for two reasons. *First,* different hormones act on different mechanisms in target cells. *Second,* not all types of cells are able to respond to a particular hormone. For example, many cells have receptors for cortisol, so this hormone has impacts throughout the body. In contrast, only certain kidney and bone cells have receptors for parathyroid hormone (PTH). These cells alone respond to PTH.

With these points in mind, let's look briefly at the effects of two main categories of signaling molecules, the steroid and peptide hormones (Table 31.1).

HORMONE SIGNALING MECHANISMS

STEROID HORMONES   In vertebrates, cells inside the adrenal glands and the primary reproductive organs (gonads) secrete **steroid hormones**. Cells build these lipid-soluble molecules from stores of cholesterol.

How do the steroid hormones operate? Being lipid soluble, many diffuse directly across the lipid bilayer of a cell's plasma membrane, then bind to a receptor in the cytoplasm or nucleus. This hormone–receptor complex interacts with DNA in a specific gene region. Most often, it stimulates transcription of genes into mRNA. The transcripts are translated into enzymes and other protein molecules that carry out a response to the hormonal signal. Figure 31.3*a* illustrates this mechanism of steroid hormone action.

Some steroid hormones also can exert their effects without entering the nucleus. These hormones bind to receptors at a target cell's plasma membrane. Their binding causes a fast response—either by altering ion channels in the membrane itself or by affecting some regulatory proteins in the target cell's cytoplasm.

Testosterone is a steroid hormone that controls the development of sperm and of male secondary sexual traits. The gene for the testosterone receptor is on the X chromosome. One mutated form results in complete *androgen insensitivity syndrome,* an X-linked recessive disorder. Genetically, an affected individual is male (XY). His testes produce and secrete testosterone, but testosterone's targets do not have workable receptors. He develops female secondary sexual traits. Symptoms vary, because different mutations in the gene lead to variations in receptor structure and sensitivity.

PEPTIDE HORMONES   Some new polypeptide chains get modified into peptide hormones inside the Golgi complex. Nearly all **peptide hormones** bind to cell membrane receptors and trigger a cascading chain of reactions. By analogy, suppose someone e-mails a great joke to a friend, who e-mails it to 10 others, each of whom sends it to 10 others. The 100 recipients mail it to 10 others, the 1,000 send it to 10 others, the 10,000 send it on, and so it goes down the line. By the end of the day, millions of people all over the country have been stimulated to chuckle.

Now consider how the pancreas secretes glucagon when the glucose level in blood falls. Receptors for this peptide hormone span the plasma membrane of target cells (Figure 31.3*b*). When activated, glucagon receptors signal an enzyme to convert ATP to cyclic adenosine monophosphate (cAMP). This is a **second messenger**, a molecule within a cell that mediates a signal from the outside. cAMP calls up protein kinases. Each molecule of this enzyme signals molecules of another enzyme, the molecules of which activate others, and so on until glycogen stored in the cell is broken down to glucose.

Or consider what happens in a muscle cell when the blood level of glucose rises and the pancreas secretes

Table 31.1	Two Main Categories of Hormones

	Examples
Steroid hormones	Estrogens; progestins; androgens, such as testosterone; aldosterone; cortisol
Peptide-based hormones	
Peptides	Glucagon, ADH, oxytocin, TRH
Proteins	Insulin, somatotropin, prolactin
Glycoproteins	FSH, LH, TSH

Read Me First!

and watch the narrated animation on hormone signaling mechanisms

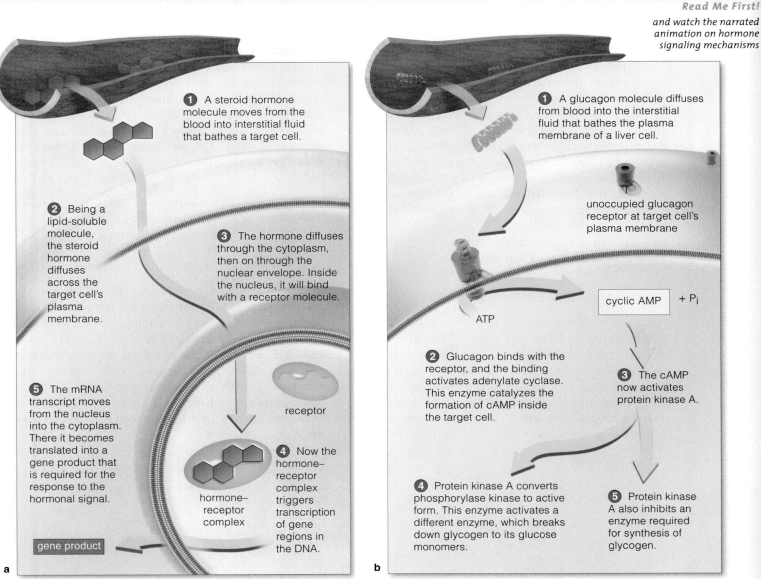

1 A steroid hormone molecule moves from the blood into interstitial fluid that bathes a target cell.

2 Being a lipid-soluble molecule, the steroid hormone diffuses across the target cell's plasma membrane.

3 The hormone diffuses through the cytoplasm, then on through the nuclear envelope. Inside the nucleus, it will bind with a receptor molecule.

5 The mRNA transcript moves from the nucleus into the cytoplasm. There it becomes translated into a gene product that is required for the response to the hormonal signal.

receptor

4 Now the hormone–receptor complex triggers transcription of gene regions in the DNA.

hormone–receptor complex

gene product

a

1 A glucagon molecule diffuses from blood into the interstitial fluid that bathes the plasma membrane of a liver cell.

unoccupied glucagon receptor at target cell's plasma membrane

ATP

cyclic AMP $+ P_i$

2 Glucagon binds with the receptor, and the binding activates adenylate cyclase. This enzyme catalyzes the formation of cAMP inside the target cell.

3 The cAMP now activates protein kinase A.

4 Protein kinase A converts phosphorylase kinase to active form. This enzyme activates a different enzyme, which breaks down glycogen to its glucose monomers.

5 Protein kinase A also inhibits an enzyme required for synthesis of glycogen.

b

Figure 31.3 (**a**) Example of a mechanism by which a *steroid* hormone initiates change in a target cell's activities. (**b**) Example of a mechanism by which a *peptide* hormone initiates change in a target cell's activities. Here, glucagon binds to a receptor and initiates reactions inside the cell. Cyclic AMP, a type of second messenger, relays the signal into the cell interior.

insulin. When this protein hormone binds to muscle cell receptors, free molecules of glucose transporters in the cytoplasm are induced to move to the cell's plasma membrane and insert themselves into it. With more transporters, the cell takes up glucose faster.

Bear in mind, there are other hormone categories, including catecholamines. One of the catecholamines —epinephrine—also binds with membrane receptors, and cAMP helps transduce this signal.

Hormones reversibly bind with receptors on target cells. The receptors for different hormones are located at the surface of the plasma membrane, in the cytoplasm, or in the nucleus.

Formation of a hormone–receptor complex leads to the transduction of the hormone's signal into a molecular form that works inside the cell. Responses include adjusting specific metabolic reactions and the uptake of certain substances.

31.3 The Hypothalamus and Pituitary Gland

The hypothalamus and the pituitary gland interact as a major brain center that controls activities of other organs, many of which also have endocrine functions.

The **hypothalamus**, recall, is part of the forebrain that helps govern the internal environment, the viscera, and emotional states. Neurotransmitters are secreted by many of its neurons. However, peptide hormones are secreted by others, the axons of which extend into a slender stalk at the base of the hypothalamus, then into a lobed, pea-sized **pituitary gland**.

The pituitary gland's *posterior* lobe is a distribution point for two hypothalamic hormones. The *anterior* lobe itself responds to others. In many vertebrates (not humans), the pituitary also has an *intermediate* lobe. Its hormonal secretion controls reversible changes in the skin or coat color of many species.

POSTERIOR LOBE SECRETIONS

Some hypothalamic neurons end at a capillary bed in the pituitary's posterior lobe (Figure 31.4). They make ADH (antidiuretic hormone) and oxytocin, then store them in axon endings. Capillaries take up the secreted hormones, then blood transports them to target cells. ADH affects urine formation in the kidneys. Oxytocin induces muscle contraction in the lining of the uterus and in the wall of milk ducts.

ANTERIOR LOBE SECRETIONS

The axons of other hypothalamic neurons end near a capillary bed in the stalk *above* the anterior pituitary (Figure 31.5). Most of these neurons make and secrete **releasers**, hypothalamic hormones that stimulate their targets. A few secrete **inhibitors**, which slow secretion from target cells. These hormones enter the nearby capillaries, then flow into a *second* capillary bed, then diffuse out. They stimulate or inhibit the secretion of the following anterior lobe hormones:

ACTH	Corticotropin
TSH	Thyrotropin
FSH	Follicle-stimulating hormone
LH	Luteinizing hormone
PRL	Prolactin
STH (or GH)	Somatotropin (growth hormone)

In turn, ACTH induces the release of cortisol from the adrenal glands. TSH stimulates the release of thyroid hormones. FSH and LH function in gamete formation and other aspects of sexual reproduction. STH—also known as growth hormone—affects the metabolism in most tissues. It triggers secretions from liver cells that affect the growth of bone and soft tissues. Prolactin helps initiate and maintain the production of milk in mammary glands after other hormones have primed the tissues.

Read Me First!
and watch the narrated animation on the posterior lobe of the pituitary gland

ⓐ Cell bodies of secretory neurons in hypothalamus synthesize ADH or oxytocin.

ⓑ The ADH or oxytocin moves downward inside the axons of the secretory neurons and accumulates in the axon endings.

ⓒ Action potentials trigger the release of these hormones, which enter blood capillaries in the posterior lobe of the pituitary.

ⓓ Small blood vessels deliver the hormone molecules to the general circulation.

oxytocin

ADH

muscles in uterus wall

mammary glands

nephrons in kidneys

Figure 31.4 Functional links between the hypothalamus and the posterior lobe of the pituitary gland. Targets of posterior lobe secretions are shown at right.

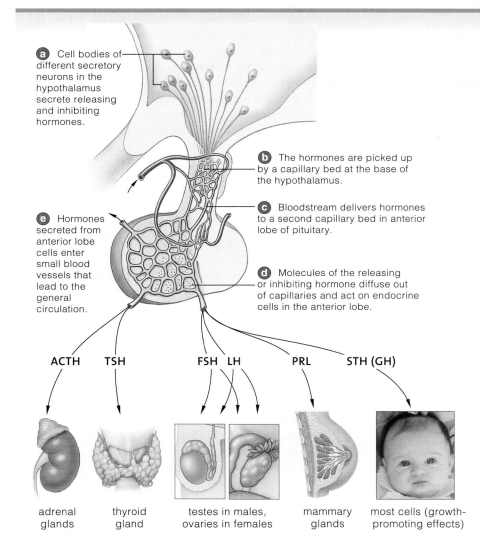

a Cell bodies of different secretory neurons in the hypothalamus secrete releasing and inhibiting hormones.

b The hormones are picked up by a capillary bed at the base of the hypothalamus.

c Bloodstream delivers hormones to a second capillary bed in anterior lobe of pituitary.

e Hormones secreted from anterior lobe cells enter small blood vessels that lead to the general circulation.

d Molecules of the releasing or inhibiting hormone diffuse out of capillaries and act on endocrine cells in the anterior lobe.

ACTH   TSH   FSH LH   PRL   STH (GH)

adrenal glands

thyroid gland

testes in males, ovaries in females

mammary glands

most cells (growth-promoting effects)

Read Me First!
and watch the narrated animation on the anterior lobe of the pituitary gland

Figure 31.5 Functional links between the hypothalamus and the pituitary gland's anterior lobe. Also shown are the main targets of anterior lobe secretions. The photograph above shows one case of abnormal output of a posterior pituitary hormone. This male is twelve years old. He is affected by pituitary gigantism, caused by overproduction of STH. At six feet five inches tall, he towers over his mother.

ABNORMAL PITUITARY OUTPUTS

Hormones are not secreted in huge quantities. Roger Guillemin and Andrew Schally sifted through 7 metric tons of hypothalamic tissue and purified only a single milligram of TSH, the first known releaser. Even so, tiny amounts of hormones have big impacts. So does their undersecretion or oversecretion. When controls over secretion fail, the body's form, function, or both will be altered.

Overproduction of STH leads to *pituitary gigantism.* Affected adults have the same body form as average people but proportionally are far larger (Figure 31.5). Underproduction of somatotropism leads to *pituitary dwarfism.* Affected adults have the form of an average person but are proportionally much smaller. What if STH secretion continues at high levels after limb bones stop lengthening? *Acromegaly* is the result. Cartilage, bone, and other connective tissues in the jaws, feet,

and hands thicken. So do epithelial tissues of the skin, nose, eyelids, lips, and tongue. A final example: ADH secretion may slow or end if the posterior pituitary is damaged, as by a blow to the head. This is one cause of *diabetes insipidus.* The body loses too much water in urine. The result can be life-threatening dehydration.

ADH and oxytocin are two hormones produced by the hypothalamus but dispersed from the posterior lobe of the pituitary gland. Both have a narrow range of targets.

Other hypothalamic hormones stimulate or inhibit the secretion of anterior pituitary hormones.

The anterior pituitary hormones are ACTH, TSH, FSH, LH, PRL, and STH. They trigger secretion of hormones from different glands, which have diverse effects on the body.

31.4 Feedback Loops in Hormone Secretions

Reflect on Figure 31.2, which lists hormones from endocrine sources other than the hypothalamus and pituitary gland. Here are a few examples of controls over their output.

The preceding section might lead you to believe that the road from signal to response is always a straight line. This is not always the case. *First*, hormones often do not act alone. One or more kinds may dampen or enhance the effect of another. One might be required to prime cells to respond to another. *Second*, feedback mechanisms adjust the secretion of many hormones. *Third*, a target cell might not react the same way at all times. How concentrated are the hormone molecules? Are receptor binding sites vacant and exposed? Have they become altered? Such factors affect the outcome. *Fourth*, environmental events often intervene in both expected and unexpected ways.

Consider the types of feedback loops that control secretion. When a hormone's concentration increases or decreases in a body region, the change can trigger events that respectively suppress or stimulate further secretion. This is what happens in feedback loops from a target to the hypothalamus and pituitary, which interact as an endocrine control center. As hormone secretion slows or steps up, the concentration in blood changes. The information is "fed back" to the control center, which compares it to a set point. In *negative* feedback, the rise inhibits further secretion of the hormone. In *positive* feedback, a rise stimulates it.

NEGATIVE FEEDBACK AND THE ADRENAL CORTEX

Humans have a pair of adrenal glands, one above each kidney (Figure 31.6). Some cells of the **adrenal cortex**, the outer layer of this type of gland, secrete cortisol. Normally, a negative feedback loop to the anterior pituitary and hypothalamus governs cortisol secretion. Negative feedback keeps the cortisol level in blood from rising or falling too far.

Figure 31.6 shows what happens when the level of cortisol in blood falls below a set point. The decline triggers the secretion of CRH (corticotropin releasing hormone) from the hypothalamus. CRH prods cells in the anterior pituitary to secrete ACTH (corticotropin). In turn, ACTH stimulates cells of the adrenal cortex to secrete cortisol.

Secretion continues until the blood level of cortisol rises above the set point. Then, the hypothalamus and anterior pituitary issue signals that inhibit secretion of CRH and ACTH. And so cortisol secretion slows.

Cortisol helps maintain the blood level of glucose when food is not being absorbed from the gut. Most notably, it induces liver cells to break down stored glycogen, and it suppresses glucose uptake by other cells. Its action indirectly helps keep a steady supply of glucose reaching the brain. It also causes adipose cells to break down fats to fatty acids, and skeletal muscle cells to break down proteins to amino acids. These substances enter the blood. Later, other body cells can take them up.

With injury, illness, or anxiety, the nervous system overrides the feedback loop and lets the blood level of cortisol soar. Over the short term, this is an adaptive response. It helps get enough glucose to the brain at a time when food supplies are likely to be low. Cortisol also helps keep inflammatory responses in check.

Chronic stress, which continues unabated for a long time, invites problems. An abnormally high level of cortisol interferes with the production and release of other hormones, including somatotropin and the sex hormones required for growth. A chronically elevated

Read Me First!
and watch the narrated animation on feedback control of cortisol secretion

(a) STIMULUS
Decline in blood level of cortisol past a set point

(b) hypothalamus — CRH → anterior pituitary — ACTH → adrenal cortex — cortisol

RESPONSE

(f) Hypothalamus and pituitary detect increase in cortisol level in blood, inhibit further secretion

adrenal cortex
adrenal medulla
adrenal gland
kidney

(c) Blood glucose uptake inhibited in many tissues, especially muscles (not the brain).

(d) Proteins degraded in many tissues, especially muscles. Free amino acids converted to glucose, also used to synthesize or repair cell structures.

(e) Fats in adipose tissue degraded to fatty acids, which are released to the blood as alternative energy sources. Helps conserve blood glucose for brain.

Figure 31.6 Structure of the human adrenal gland. One gland rests on top of each kidney. The diagram shows a negative feedback loop that governs cortisol secretion from the adrenal cortex.

Figure 31.7 (**a**) Human thyroid gland and four parathyroid glands. (**b**) A mild case of goiter, displayed by Maria de Medici. A neck this rounded was viewed as a sign of beauty during the late Renaissance. Goiter is common in parts of the world where supplies of iodine are insufficient for normal thyroid function.

level of cortisol also invites hunger and can lead to weight gain. Also, over the long term, suppression of inflammatory responses can open the door to disease.

LOCAL FEEDBACK AND THE ADRENAL MEDULLA

The **adrenal medulla**, the inner region of the adrenal gland, also houses neurons. They secrete epinephrine and norepinephrine, which are neurotransmitters in some contexts and hormones in others. Signals reach them by sympathetic nerves from the hypothalamus. Suppose the signals trigger norepinephrine secretion. Molecules of this hormone accumulate in the synaptic cleft between target cells in the adrenal medulla and axon endings from the nerve. When there is too much norepinephrine, a local negative feedback mechanism kicks in. Norepinephrine binds to receptors for it on the axon endings and inhibits its further release.

When you are excited or stressed, epinephrine and norepinephrine adjust blood flow rates and fat and carbohydrate metabolism. They boost the heart rate, make arteriole diameters widen or narrow in different regions, and widen airways. More of the total blood volume—and oxygen—flows to heart and muscle cells, which are demanding a great deal of energy. These are features of the *fight–flight response* (Section 29.4).

SKEWED FEEDBACK FROM THE THYROID

Humans have a **thyroid gland** at the base of the neck in front of the trachea, or windpipe (Figure 31.7*a*). When stimulated by TSH from the anterior pituitary, the thyroid secretes thyroxine and triiodothyronine. These hormones have essential roles in development.

In addition, they regulate metabolic rates of warm-blooded animals. A few examples will highlight the importance of feedback control of their secretion.

Thyroid hormone production requires iodide, a form of iodine. Iodine-deficient diets can lead to an enlarged thyroid or *simple goiter* (Figure 31.7*b*). This occurs after low levels of thyroid hormones cause the anterior pituitary to secrete TSH. The thyroid grows in response, but it still cannot make any hormones. A feedback loop to the pituitary keeps calling for TSH, which invites more enlargement of the thyroid.

Hypothyroidism is a low level of thyroid hormones. Without treatment, affected adults are often sluggish, overweight, intolerant of cold, dry-skinned, confused, and depressed. This condition is no longer common in regions where people use iodized salt.

Graves' disorder and other *toxic goiters* result from an excess of thyroid hormones or *hyperthyroidism*. The symptoms typically include anxiety, tremors, trouble sleeping, heat intolerance, protruding eyes, and an erratic heartbeat. Some cases arise when antibodies wrongly stimulate the thyroid. Other cases arise after inflammation or tumors develop in this gland.

Hormone secretion and its effects depend on hormone interactions, feedback mechanisms, and variations in the state of the hormone's target cells.

Feedback mechanisms control secretions from endocrine glands. In many cases, feedback loops to the hypothalamus, pituitary, or both govern the secretory activity.

Negative feedback slows down the further secretion of a hormone. Positive feedback enhances its further secretion.

31.5 Direct Responses to Chemical Change

Some endocrine glands and cells are not under the control of the hypothalamus and pituitary. They respond directly to chemical changes in the internal environment.

SECRETIONS FROM PARATHYROID GLANDS

Young children who have a vitamin D deficiency have trouble absorbing calcium and phosphorus from the gut, so their rapidly growing bones don't develop as they should. The resulting bone abnormality, *rickets*, causes bowed legs, a malformed pelvis, and often a malformed skull and rib cage (Figure 31.8).

On the back surface of the thyroid gland are four **parathyroid glands** (Figure 31.7). They help control calcium levels in blood with secretions of parathyroid hormone (PTH). Calcium ions have roles in muscle contraction, enzyme action, blood clotting, and other activities. PTH is secreted when the calcium level in blood falls. Its secretion slows when the level rises.

PTH acts on cells in the skeleton and the kidneys. It induces living bone cells to secrete enzymes that digest bone tissue, which releases ions of calcium and other minerals into interstitial fluid. Calcium enters the blood, then small tubes in the kidneys. There, PTH enhances calcium reabsorption. It also induces kidney cells to secrete enzymes, which act on precursors of an active form of vitamin D_3. The activated form stimulates intestinal cells to absorb calcium from the food in the gut. That is why children who don't have enough vitamin D run the risk of bone disorders.

EFFECTS OF LOCAL SIGNALING MOLECULES

Changes in the chemical environment make cells alter their activity. Many types respond by secreting local signaling molecules. The signals have very localized effects, because target cells take up most of the molecules so fast that few enter the general circulation.

For example, prostaglandins form in many tissues. The production and secretion of these local signaling molecules can be stepped up when tissue conditions change. Some redirect blood flow by constricting or dilating the smooth muscle inside the wall of arterioles. Others make pain receptors more responsive to noxious chemicals. Still others influence inflammation, intestinal wall motility, and stomach acid secretion. Prostaglandins also function in overall endocrine controls. Some are used in therapeutic treatments of asthma, hypertension, and heart attacks.

Figure 31.8 A child affected by rickets.

SECRETIONS FROM PANCREATIC ISLETS

The pancreas has exocrine and endocrine functions. Its *exocrine* cells secrete digestive enzymes into the small intestine. Its *endocrine* cells are grouped in 2 million or so clusters, known as the **pancreatic islets**. Each islet has three types of hormone-secreting cells:

1. *Alpha* cells inside the pancreas secrete the hormone glucagon. Between meals, cells throughout the body take up and use free glucose, which lowers the blood level of glucose. Glucagon secretion causes liver cells to convert amino acids and stored glycogen to glucose, which enters the blood. *Thus, glucagon raises the level of glucose in blood.*

2. *Beta* cells, the most abundant cells in the pancreatic islets, secrete insulin. After meals, when glucose levels in the blood become high, insulin stimulates uptake of glucose, especially by muscle cells and adipose cells. Insulin promotes the synthesis of proteins and fats, and it inhibits the conversion of proteins to glucose. *Insulin lowers the level of glucose in blood.*

3. *Delta* cells secrete somatostatin. This hormone helps control the digestion and absorption of nutrients. Also, it can block secretion of insulin and glucagon.

Figure 31.9 shows how these pancreatic hormones interact to maintain the level of glucose in blood even when the amount and timing of food intake vary. Bear in mind, insulin is the *only* hormone that causes cells to take up and store glucose, which is a form of instant energy. Its central importance in the metabolism of carbohydrates, fats, and proteins becomes clear when we study people who cannot produce enough insulin or whose target cells cannot respond to it.

For example, insulin deficiency might lead to one form or another of *diabetes mellitus*. In this disorder, cells have trouble taking up glucose, which builds up in blood, then in urine. Affected people must urinate frequently. This alters the body's water–solute balance and causes abnormal thirst. Without a steady supply of glucose, body cells start depleting structural fats and proteins as alternative energy sources. Weight loss is one outcome. Another is ketone accumulation in blood and urine. Ketones are normal acidic products of fat breakdown, but when they accumulate, the result is *ketoacidosis*. Such changes in the acid–base balance and in solute levels may interfere with normal brain function. Extreme disruptions can result in death.

In *type 1 diabetes*, the body mounts an autoimmune response against its insulin-secreting beta cells. White blood cells mistakenly identify the beta cells as foreign

Read Me First!
and watch the narrated animation on hormones and glucose metabolism

Figure 31.9 Some of the homeostatic controls over glucose metabolism.

(**a**) *Following* each meal, glucose enters the blood faster than cells can take it up. The blood glucose level increases. (**b**) The chemical change stimulates pancreatic beta cells to secrete insulin. (**c,d**) The main targets, liver cells and muscle cells, use glucose at once and store excess amounts as glycogen. (**e**) The outcome? Insulin has *lowered* the blood level of glucose.

(**f**) *Between* meals, the blood glucose level decreases. (**g**) The chemical change stimulates pancreatic alpha cells to secrete glucagon. (**h**) Target cells with receptors for this hormone convert glycogen back to glucose, which enters the blood. (**i**) The outcome? Glucagon has *raised* the blood level of glucose.

The nervous system helps control glucose metabolism. For example, as you read earlier, the hypothalamus induces the adrenal medulla to secrete glucocorticoids when the body is stressed and depleting glucose. Two outcomes: Glycogen is degraded to glucose units in the liver, and glycogen synthesis slows, in liver and muscle tissue especially.

insulin

d Muscle cells especially take up glucose and then use it as an instant energy source or convert the excess to glycogen.

c In the liver, excess glucose is converted to glycogen, fats, and proteins.

b Beta cells of pancreas release insulin into blood.

e Blood glucose level declines to a set point, so stimulus calling for insulin diminishes.

a **Stimulus**
Glucose is absorbed following a meal.

ELEVATED blood glucose

Homeostasis: normal blood glucose levels (90 mg/100 ml)

LOW blood glucose

i Blood glucose level increases to a set point, so stimulus calling for glucagon diminishes.

f **Stimulus**
Cells use or store glucose between meals.

h Liver converts glycogen to glucose and stops synthesizing glycogen.

glucagon

g Alpha cells of pancreas release glucagon into blood.

and kill them. Environmental factors add to a genetic predisposition to the disorder. This form of diabetes mellitus is not as common as others, but it is more dangerous in the short term. Usually, symptoms start appearing in childhood and adolescence (the disorder is also called juvenile-onset diabetes). Type 1 diabetic patients require insulin injections.

In *type 2 diabetes*, insulin levels are normal or even high. But target cells do not respond to the hormone as they should, and blood sugar levels remain high. Symptoms typically appear in middle age, as insulin production declines. Genetics is a factor, but obesity increases the risk. Diet, exercise, and oral medications can control many cases of type 2 diabetes. However, if glucose levels are not lowered, pancreatic beta cells are constantly stimulated. Eventually, they falter, and so does insulin production. When that happens, a type 2 diabetic also requires insulin injections.

Secretions from some endocrine glands and endocrine cells are direct homeostatic responses to chemical changes in the internal environment.

31.6 Hormones and the Environment

This section of the chapter invites you to reflect on a key point. An individual's growth, development, and reproduction start with genes and hormones, and so does behavior. But certain environmental factors often influence activation and rates of hormonal secretion.

DAYLENGTH AND THE PINEAL GLAND

Deep within the vertebrate brain is the **pineal gland** (Figure 31.2). This tiny gland secretes melatonin, but only when there is little or no light. The amount of light varies from day to night and with the seasons, and so does melatonin secretion. In many species, the variation in the blood level of melatonin helps control the growth and development of gonads, the primary reproductive organs. How? Melatonin is part of an internal timing mechanism, or **biological clock**. The variations in its secretion affect reproductive cycles as well as sleep and other behaviors.

Think of a male songbird in winter, when there are more hours of darkness than in summer. Its elevated blood level of melatonin indirectly suppresses sexual activity. In spring, there are more hours of daylight, and the brakes on melatonin secretion lift. Now the gonads begin secreting the hormones responsible for territorial singing behavior and courtship behavior.

Does melatonin affect human behavior? Possibly. As clinical observations and studies indicate, low melatonin levels may trigger **puberty**, the time when reproductive organs and structures start to mature. We have reports of early puberty among patients who do not have a functional pineal gland.

Melatonin also influences neurons that can lower body temperature and make us drowsy in low light. Just after sunrise, less melatonin is secreted, so body temperature rises and we wake up. A biological clock responsive to daylight exposure governs the sleep/ arousal cycle. That is why travelers are advised to spend time in the sun to minimize jet lag.

Seasonal affective disorder (SAD) hits some people in winter. They become exceedingly depressed, binge on carbohydrates, and crave sleep. These "winter blues" might develop when a biological clock is out of sync with the seasonal decline in daylight hours. Clinically administered melatonin can make symptoms worse. Exposure to artificial light, which shuts down pineal activity, can cause dramatic improvement.

THYROID FUNCTION AND FROG DEFORMITIES

At the start of this chapter, you read about possible effects of hormone disrupters on sex organs in frogs. Researchers also have jarringly visible evidence that deformities of amphibians in general are on the rise (Figure 31.10). Parasites and increasing levels of UV radiation have been implicated. Studies suggest that hormone disrupters also play a role.

In one experiment, researchers exposed African clawed frog (*X. laevis*) embryos to water from lakes in Minnesota and Vermont. Half of the water samples came from lakes where deformity rates were low. The other half came from "hot spots" that had as many as twenty kinds of dissolved pesticides—and high frog deformity rates. Tadpoles that developed in hot-spot water had a bent spine and other abnormalities. Some did not fully metamorphose into adults. Tadpoles that formed in water from other lakes were normal.

A surge in thyroid hormone output stimulates a tadpole to metamorphose. Are pesticides interfering with thyroid functions? Perhaps. When embryos in hot-spot water were given extra thyroid hormones, they displayed few symptoms, or none at all.

CHEMICAL SOUPS AND SPERM COUNTS

Are frogs too far removed from your interests? Chemicals that disrupt endocrine function may also be affecting men. Between 1938 and 1990, the sperm counts of males in Western countries declined about 40 percent (Figure 31.11).

What causes this? Estrogen-like compounds in the environment? Possibly. Sperm counts in human males who have taken a natural estrogen are lower. Many pesticides bind to estrogen receptors all through the

a   b

Figure 31.10 (a) Evidence of a connection between environmental effects on the thyroid gland and developmental abnormalities in *Xenopus laevis*. The uppermost *X. laevis* tadpole was raised in water taken from a lake that held few deformed frogs. The three tadpoles below it were raised in water from three "hot-spot" lakes in which increasingly higher concentrations of natural and synthetic organic compounds were dissolved.

(b) X-ray of a deformed adult frog. Such deformities are especially common sights in pesticide-contaminated water.

Figure 31.11 Normal and low sperm counts. Grid lines help clinicians make counts. (**a**) Normally, there are about 113 million sperm per milliliter of semen per ejaculate. (**b**) A low sperm count of 60 million to 70 million sperm per milliliter. Sperm counts have been declining among human males.

body. They include DDT and Kepone, both estrogen-mimicking insecticides that have since been banned in the United States. Kepone has been linked to cancer as well as to low sperm counts in males exposed to it.

Environmental estrogens might also interfere with sperm function. For one study, mouse sperm were exposed to environmental estrogens. They matured too fast and were unable to fertilize eggs.

COMPARATIVE LOOK AT INVERTEBRATES

This chapter focused on vertebrates, but all organisms produce signaling molecules of one sort or another. Let's look at hormonal control of **molting**, a periodic discarding and replacement of a hardened cuticle that otherwise would limit increases in body mass. Molting occurs in the life cycle of all insects, crustaceans, and other invertebrates with thick cuticles.

Although details vary among groups, molting is largely under the control of **ecdysone**. This steroid hormone is derived from cholesterol and is chemically related to many vertebrate hormones. In insects and crustaceans, the molting glands synthesize and store ecdysone and then release it for distribution through the body at molting time. Hormone-secreting neurons in the brain seem to control its release. They respond to a combination of environmental cues, such as light and temperature, as well as internal signals.

Figure 31.12 is an example of the control steps in crabs and other crustaceans. Right before and during the molting periods, coordinated interactions among ecdysone and other hormones bring about structural and physiological changes. The interactions make the old cuticle separate from the epidermis and muscles. They induce changes that dissolve inner layers of the cuticle and recycle the remnants. The interactions also

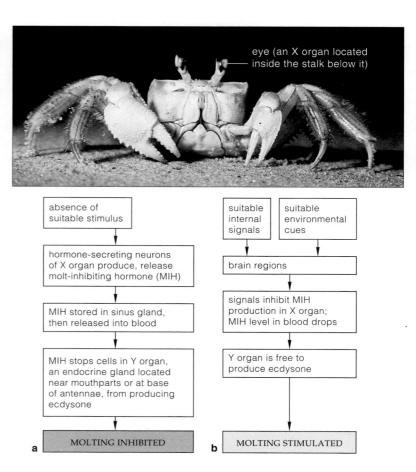

Figure 31.12 Steps in hormonal control of molting in crabs and other crustaceans. Ecdysone is a hormone that stimulates molting. (**a**) In the absence of environmental cues for molting, an X organ in each crab eye stalk produces a hormone that inhibits ecdysone synthesis. (**b**) Right before and during molts, signals from the brain turn off cell activities in the X organ, and so ecdysone is produced and secreted.

trigger changes in metabolism and in the composition of the internal environment. They promote the rapid cell divisions, secretions, and pigment formation that help make a new cuticle. At the same time, hormonal interactions control the heart rate, muscle action, color changes, and other processes.

The steps differ a bit in insects, which do not have a molt-inhibiting hormone. Rather, stimulation of an insect brain causes a cascade of signals that trigger the production of molt-inducing ecdysone.

Environmental cues, such as seasonal changes in the number of daylight hours, influence the secretion of some hormones. Melatonin secretion is a classic example.

Environmental chemicals can disrupt endocrine functions.

Summary

Section 31.1 Hormones, neurotransmitters, local signaling molecules, and pheromones are the main kinds of signaling molecules. They are chemical secretions from one cell type that adjust the behavior of other, target cells. Any cell is a target if its plasma membrane, cytoplasm, or nucleus has the necessary receptors. In target cells, hormones influence gene activation, protein synthesis, and the structure and function of enzymes, membranes, and other cellular components.

All vertebrates have an organ system of endocrine glands and cells. In most cases, the hormonal secretions travel the bloodstream to nonadjacent targets. Tables 31.2 and 31.3 summarize the hormonal sources, targets, and actions in the vertebrates called mammals.

Section 31.2 There are two main categories of hormones. Steroid hormones are lipid soluble and derived from cholesterol. Some steroid hormones enter a target cell and interact directly with DNA. Others bind to the cell's plasma membrane, altering membrane properties. Peptide hormones are water soluble. They bind to membrane receptors at the cell surface and then activate enzyme systems. Binding may lead to formation of a second messenger, such as cAMP. The second messenger relays the signal into the cytoplasm, where it causes a cascade of enzyme activations.

Section 31.3 The hypothalamus, a forebrain region, is a major center for homeostatic control of the internal environment, the viscera, and emotions. It interacts closely with the adjacent, pea-sized pituitary gland.

Some hypothalamic neurons make ADH or oxytocin, two hormones that the posterior pituitary disperses for them. ADH acts on kidneys. Oxytocin acts on milk ducts.

Other hypothalamic neurons make six releasers and inhibitors that act on target cells in the anterior pituitary. These releaser and inhibitor hormones control secretion of ACTH, TSH, FSH, LH, PRL, and STH.

These pituitary hormones in turn control secretion of hormones from other endocrine glands. ACTH acts on the adrenal cortex, TSH on the thyroid, FSH and LH on male and female gonads, and PRL on mammary glands and the uterus. STH (somatotropin, or growth hormone) has growth-promoting effects on cells throughout the body.

Section 31.4 Secretion from many glands depends on feedback loops to the hypothalamus, pituitary gland, or both. For example, when the cortisol level is low, the hypothalamus stimulates the pituitary to secrete ACTH, which induces the adrenal cortex to secrete cortisol. When the cortisol level rises and returns to a set point, the hypothalamus responds by issuing signals that inhibit further cortisol secretion.

This is a case of negative feedback. In times of stress, the central nervous system can override the feedback controls, allowing cortisol levels to rise.

Table 31.2   Hormones Released From the Mammalian Pituitary Gland

Pituitary Lobe	Secretions	Designation	Main Targets	Primary Actions
POSTERIOR Nervous tissue (extension of hypothalamus)	Antidiuretic hormone	ADH	Kidneys	Induces water conservation as required during control of extracellular fluid volume and solute concentrations
	Oxytocin	OCT	Mammary glands	Induces milk movement into secretory ducts
			Uterus	Induces uterine contractions during childbirth
ANTERIOR Glandular tissue, mostly	Corticotropin	ACTH	Adrenal cortex	Stimulates release of cortisol, an adrenal steroid hormone
	Thyrotropin	TSH	Thyroid gland	Stimulates release of thyroid hormones
	Follicle-stimulating hormone	FSH	Ovaries, testes	In females, stimulates estrogen secretion, egg maturation; in males, helps stimulate sperm formation
	Luteinizing hormone	LH	Ovaries, testes	In females, stimulates progesterone secretion, ovulation, corpus luteum formation; in males, stimulates testosterone secretion, sperm release
	Prolactin	PRL	Mammary glands	Stimulates and sustains milk production
	Somatotropin (or growth hormone)	STH (GH)	Most cells	Promotes growth in young; induces protein synthesis, cell division; roles in glucose, protein metabolism in adults
INTERMEDIATE* Glandular tissue, mostly	Melanocyte-stimulating hormone	MSH	Pigmented cells in skin and other integuments	Induces color changes in response to external stimuli; affects some behaviors

* An intermediate pituitary lobe is present in most vertebrates, but not in humans. MSH is associated with the human anterior lobe.

Negative feedback governs secretion of epinephrine and norepinephrine by the adrenal medulla. Secretion of thyroid hormones requires iodide. In iodide's absence, feedback control of the thyroid is disrupted.

Section 31.5 Some glands are not under the control of the pituitary and hypothalamus. They respond to changes in the conditions in their local environment. The parathyroid glands respond to changes in the blood level of calcium. Prostaglandins are local signaling molecules and are secreted in response to local tissue conditions.

Insulin and glucagon are pancreatic hormones. They are secreted in response to shifts in the concentration of glucose in blood. Insulin is secreted by beta cells when the glucose level is high. It stimulates glucose uptake by muscle and liver cells and thus lowers the blood level of glucose. Glucagon secreted by alpha cells stimulates the release of glucose when the blood level is too low.

Section 31.6 Secretion of some hormones depends on signals from the external environment, such as seasonal variations in daylight hours. Melatonin secretion by the vertebrate pineal gland is an example. It is part of the biological clock, an internal timing mechanism. In arthropods, molting is controlled by hormones that are responsive to changes in environmental conditions.

Natural and synthetic compounds in the environment alter endocrine function in animals, including humans.

Table 31.3 Hormone Sources Other Than the Mammalian Hypothalamus and Pituitary Gland

Source	Secretion(s)	Main Targets	Primary Actions
ADRENAL CORTEX	Glucocorticoids (including cortisol)	Most cells	Promote breakdown of glycogen, fats, and proteins as energy sources; thus help raise blood level of glucose
	Mineralocorticoids (including aldosterone)	Kidney	Promote sodium reabsorption (sodium conservation); help control the body's salt–water balance
ADRENAL MEDULLA	Epinephrine (adrenaline)	Liver, muscle, adipose tissue	Raises blood level of sugar, fatty acids; increases heart rate and force of contraction
	Norepinephrine	Smooth muscle of blood vessels	Promotes constriction or dilation of certain blood vessels; thus helps control the flow of blood volume to different body regions
THYROID	Triiodothyronine, thyroxine	Most cells	Regulate metabolism; have roles in growth, development
	Calcitonin	Bone	Lowers calcium level in blood
PARATHYROIDS	Parathyroid hormone	Bone, kidney	Elevates calcium level in blood
GONADS			
Testes (in males)	Androgens (including testosterone)	General	Required in sperm formation, development of genitals, maintenance of sexual traits, growth, and development
Ovaries (in females)	Estrogens	General	Required for egg maturation and release; preparation of uterine lining for pregnancy and its maintenance in pregnancy; genital development; maintenance of sexual traits; growth, development
	Progesterone	Uterus, breasts	Prepares, maintains uterine lining for pregnancy; stimulates development of breast tissues
PANCREATIC ISLETS	Insulin	Liver, muscle, adipose tissue	Promotes cell uptake of glucose; thus lowers glucose level in blood
	Glucagon	Liver	Promotes glycogen breakdown; raises glucose level in blood
	Somatostatin	Insulin-secreting cells	Inhibits digestion of nutrients, hence their absorption from gut
THYMUS	Thymopoietin	T lymphocytes	Poorly understood regulatory effect on T lymphocytes
PINEAL	Melatonin	Gonads (indirectly)	Influences daily biorhythms, seasonal sexual activity
STOMACH, SMALL INTESTINE	Gastrin, secretin, etc.	Stomach, pancreas, gallbladder	Stimulate activities of stomach, pancreas, liver, gallbladder; required for food digestion, absorption
LIVER	Somatomedins	Most cells	Stimulate cell growth and development
KIDNEYS	Erythropoietin	Bone marrow	Stimulates red blood cell production
	Angiotensin*	Adrenal cortex, arterioles	Helps control secretion of aldosterone (hence sodium reabsorption, and blood pressure)
	1,25-hydroxyvitamin D_6* (calcitriol)	Bone, gut	Enhances calcium reabsorption from bone and calcium absorption from gut
HEART	Atrial natriuretic hormone	Kidney, blood vessels	Increases sodium excretion; lowers blood pressure

* Kidneys produce *enzymes* that modify precursors of this substance, which enters the general circulation as an activated hormone.

Self-Quiz

Answers in Appendix III

1. _____ are signaling molecules released from one type of cell that can alter target cell activities.
   - a. Hormones
   - b. Neurotransmitters
   - c. Pheromones
   - d. Local signaling molecules
   - e. both a and b
   - f. a through d

2. ADH and oxytocin are hormones produced in the hypothalamus but distributed from the _____ .
   - a. anterior lobe of pituitary
   - b. posterior lobe of pituitary
   - c. pancreas
   - d. pineal gland

3. In the diagrams below, fill in the blanks with the names of the appropriate secretions from the pituitary.

4. Overproduction of _____ causes gigantism.
   - a. somatotropin
   - b. ADH
   - c. insulin
   - d. melatonin

5. Which do not stimulate hormone secretions?
   - a. neural signals
   - b. local chemical changes
   - c. hormonal signals
   - d. environmental cues
   - e. All of the above can stimulate secretion.

6. _____ lowers blood sugar levels; _____ raises it.
   - a. Glucagon; insulin
   - b. Insulin; glucagon
   - c. Gastrin; insulin
   - d. Gastrin; glucagon

7. The pituitary detects a rising hormone concentration in blood and inhibits the gland secreting the hormone. This is a _____ feedback loop.
   - a. positive
   - b. negative
   - c. long-term
   - d. b and c

8. The _____ has endocrine and exocrine functions.
   - a. liver
   - b. pancreas
   - c. pineal gland
   - d. parathyroid gland

9. Match the hormone source listed at left with the most suitable description at right.
   - _____ adrenal medulla
   - _____ thyroid gland
   - _____ parathyroid glands
   - _____ pancreatic islets
   - _____ pineal gland
   - _____ prostaglandin
   - a. affected by daylength
   - b. potent local effects
   - c. raises blood calcium level
   - d. epinephrine source
   - e. insulin, glucagon
   - f. hormones require iodide

Critical Thinking

1. In the late 1990s, Marcia Herman-Giddens asked this question of pediatricians throughout the United States: At what age are female patients developing secondary sexual traits? They reported that girls are developing breasts a little more than one year earlier than had been reported before.

Herman-Giddens's data were criticized, because the girls were not a random sample; they were brought in to see a pediatrician. Nevertheless, many researchers suspect that her findings reflect a real trend. What might be the cause of this *precocious breast development*?

Evidence that chemicals may play a role comes from the island of Puerto Rico, which has the world's highest incidence of premature breast development (Figure 31.13). A recent study found that most affected girls started to develop breasts when they were between six and twenty-four months old! Sixty-eight percent had high blood levels of phthalates—chemicals used in the manufacture of some plastics and pesticides. By comparison, none of the normal girls that researchers examined had high phthalate levels.

What type of information might be used to test the hypothesis that exposure to phthalates may be contributing to early breast development among young girls in the mainland United States?

2. Many abnormally small but normally proportioned people live in a remote village in Pakistan. On average, the men are about 130 centimeters (a little over 4 feet) tall, and the women about 115 centimeters (3 feet, 6 inches) tall. Northwestern University researchers interviewed the

Figure 31.13 A case of the physical impact of hormone disrupters? This Puerto Rican girl, *twenty-three months old*, shows premature breast development.

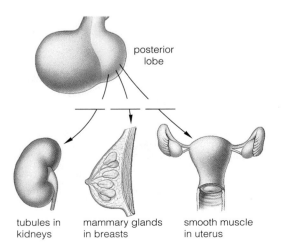

tubules in kidneys   mammary glands in breasts   smooth muscle in uterus

posterior lobe

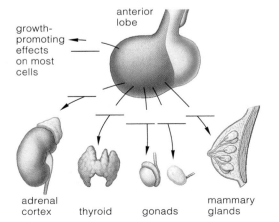

anterior lobe

growth-promoting effects on most cells

adrenal cortex   thyroid   gonads   mammary glands

Figure 31.14 Hiralal Maheshwari, with two males who show a heritable form of dwarfism.

Figure 31.15 A blue crab that has just molted its old shell. For twelve hours or so, it will be a soft-shelled crab, considered a delicacy by many seafood lovers.

people. They found out that all were part of an extended family. The abnormality, a type of *dwarfism*, is one case of autosomal recessive inheritance (Figure 31.14).

Biochemical and genetic analysis show that affected individuals have a low level of somatotropin but a normal gene that codes for this growth hormone. The abnormality starts with a mutant gene that specifies a receptor for one of the hypothalamic releasing hormones. That receptor is on cells in the anterior lobe of the pituitary gland. Explain how a mutant receptor could cause the abnormality.

3. In winter, with its fewer daylight hours compared to summer, Maxine became exceedingly depressed, craved carbohydrate-rich foods, and stopped exercising regularly. She put on a great deal of weight. Her doctor diagnosed her condition as *seasonal affective disorder*, the winter blues. He advised Maxine to purchase a cluster of intense, broad-spectrum lights and sit near them for at least an hour each day. The treatment quickly lifted the cloud of depression. Use your knowledge of pineal gland secretion to explain why Maxine's symptoms appeared and why the prescribed treatment worked.

4. Maya is affected by *type 1 insulin-dependent diabetes*. One day, she miscalculates and injects herself with too much insulin. She begins to shake and to feel confused. Maya knows that if she does not get treatment, she can fall into a coma and die. She quickly injects herself with glucagon that her doctor prescribed for emergencies, then calls for medical assistance. An ambulance arrives and she is given dextrose (a sugar) intravenously. What caused her symptoms? How did the glucagon injection help? Why would a sugar injection help her recover fast?

5. The blue crab (*Callinectes sapidus*) is found in estuaries along the Gulf and Atlantic coasts of the United States. Like other arthropods, these crabs molt as they grow (Figure 31.15). After molting, it takes about twelve hours for the new shell to harden. During this interval, the crab is vulnerable to natural predators and to human crabbers who can sell it as an expensive "soft-shelled" crab. Blue crab populations have declined in some areas. Chemical pollutants that bind to and block ecdysone receptors may be a factor. How might a chemical that affects ecdysone action interfere with the crab's life cycle?

Media Menu

Student CD-ROM

Impacts, Issues Video
  Hormones in the Balance
Big Picture Animation
  Hormone function and effects
Read-Me-First Animation
  Hormone signaling mechanisms
  Posterior lobe of the pituitary gland
  Anterior lobe of the pituitary gland
  Feedback control of cortisol secretion
  Hormones and glucose metabolism
Other Animations and Interactions
  Overview of human endocrine system
  Hypothalamus–pituitary interactions

InfoTrac

- Hormone Disrupters: A Clue to Understanding the Environmental Causes of Disease. *Environment*, June 2001.
- The Emerging Diabetes Epidemic: Science has now proven that diet and exercise can prevent diabetes, but motivating the public is the biggest challenge. *Saturday Evening Post*, May–June 2002.

Web Sites

- The Endocrine Society: www.endo-society.org
- Pituitary Network Association: www.pituitary.com
- American Diabetes Association: www.diabetes.org
- Fooling with Nature: www.pbs.org/wgbh/pages/frontline/shows/nature/

How Would You Vote?

Evidence suggests that agricultural chemicals can act as endocrine disrupters. Some environmentalists call atrazine a threat to both humans and wildlife. Several European countries have banned its use. The manufacturer says atrazine is safe when used properly and that it controls weeds at lower cost than the alternatives. Should the United States continue to allow use of an agricultural chemical, such as atrazine, while investigations into its effects and safety are ongoing?

IMPACTS, ISSUES *Pumping Up Muscles*

Want to be more muscular and stronger, with a lot more endurance? Just swallow our pills or powders and be like the guy in Figure 32.1. That is the message sent to body builders and other athletes in advertisements of many dietary supplements. Such products are available in health food stores and on the Internet. Because the Food and Drug Administration (FDA) does not classify supplements as drugs, testing for effectiveness and long-term side effects has been negligible. And quality control of commercial production is largely unregulated.

Consider androstenedione, or "andro." It got a big boost in 1998, when Mark McGwire admitted he used it during his successful attempt to break Major League

Baseball's home-run record. Androstenedione forms naturally in the body as an intermediate in the synthesis of testosterone. Testosterone *does* have tissue-building (anabolic) effects that are well documented. Andro is said to raise the blood level of testosterone, which in turn boosts the rate of protein synthesis in muscles.

Does andro work? Probably not. In controlled studies, males taking the supplement did not increase muscle mass or strength any more than males who took a placebo. At most, andro raised testosterone levels for a few hours.

Andro also is an intermediate in the synthesis of estrogen, a sex hormone with feminizing effects. Its known side effects on males include shrunken testicles, the development of female-like breasts, and hair loss. Users in both sexes risk liver damage, acne, and a lower blood level of "good" cholesterol (HDL). Females can develop masculinized patterns of hair growth and speech, and their menstrual cycle can be disrupted.

In early 2004, the FDA issued an advisory that taking andro supplements has serious side effects. They told companies to stop distributing products containing andro immediately.

Figure 32.1 An overabundance of contractile tissue—the goal of some supplement users.

the big picture

Types of Skeletons Animals have a hydrostatic skeleton, exoskeleton, or endoskeleton. Different species move the body and its individual parts by applying the force of contraction against an enclosed fluid or against structural elements.

Bones and Joints Bones are hardened organs that help move the body, protect and support soft organs, and store minerals. Some are sites of blood cell formation. Tendons attach bones to each other; ligaments bridge the joints between bones.

Creatine, a short chain of amino acids, is another supplement being touted as a performance enhancer. The body normally makes some creatine and gets more from food. When called upon to contract hard and fast, muscles draw on phosphorylated creatine as an instant energy source.

Unlike andro, creatine supplements might work. In several controlled studies, they improved performance during brief, high-intensity exercise. Clinical trials are under way to determine whether creatine may benefit individuals affected by muscular dystrophy and other muscle disorders. However, excessive creatine intake puts a strain on the kidneys, and we do not yet know what long-term side effects might be. No regulatory agency checks to see how much creatine is actually present in any given commercial product.

With this chapter, we turn to interactions that give most animal species their shape and capacity to move. Regardless of the species, movement of the body or parts of it requires contractile cells and some medium or structure against which contractile force is applied. Among different animal groups, hydrostatic skeletons, exoskeletons, and endoskeletons receive the applied force. We humans have an endoskeleton of many bones attached to many deep and superficial muscles. The structure of our skeletal and muscular systems, and how they work, has an evolutionary history—one that can help you evaluate just how far both systems can be pushed in the pursuit of enhanced performance.

 How Would You Vote?

Dietary supplements are largely unregulated. Should they be placed under the jurisdiction of the Food and Drug Administration so they would be subject to more stringent testing for effectiveness and for safety? See the Media Menu for details, then vote online.

Bones and Muscles Interact When skeletal muscles contract, they shorten and move the bones to which they are attached. They are like a lever system, with rigid bones moving about a fixed attachment point. Many muscle groups work in opposable pairs.

actin   myosin   actin

How Muscles Work Muscle cells, or fibers, contain a repeating series of contractile units called sarcomeres. Each unit contracts by ATP-driven sliding interactions between organized arrays of actin and myosin filaments. As sarcomeres shorten, cells shorten, and so does the whole muscle.

32.1 What Is a Skeleton?

Animal skeletons are structural elements or a contained fluid against which contractile force is applied.

There are three types of skeletons. With a **hydrostatic skeleton**, muscle contractions redistribute a volume of fluid contained within a body cavity. Soft-bodied invertebrates, such as earthworms and sea anemones, have a hydrostatic skeleton (Figure 32.2).

With an **exoskeleton**, a cuticle, shell, or some other rigid, *external* body part receives the applied force of muscle contraction. Arthropods, recall, have a hinged exoskeleton. Muscle sets are attached to its hard parts and move them, like levers. Remember the arthropod cuticle? It extends over body segments and over gaps between them. Being pliable, it acts like a hinge at the gaps when contracting muscles act on body parts, such as a thorax with its attached wings (Figure 32.3).

Like other vertebrates, you have an **endoskeleton** (Figure 32.4). Its *internal* body parts—bones—accept the contractile force. How did that bony framework evolve? Think back on Chapter 23. From notochord to vertebral column, from gill supports to jaws, from the structural elements in lobed fins to limbs—these key evolutionary trends emerged when some vertebrates left the water and adapted to life on land.

Think of the changes that were required to support body weight deprived of water's buoyancy. In fishes, two sets of bones—the pelvic and pectoral girdles—are a stable base for moving fins, which propel, guide, and stabilize the body in water. For early vertebrates, the air was not dense enough to support body weight, which was now transferred to four legs. Bones of the girdles were modified. Repositioning the limbs helped support the body's center of gravity. Bones enclosed the heart, lungs, and other soft organs that would have collapsed under the weight of a body out of water.

Now compare a human's endoskeleton with that of an ancient land vertebrate, as in Figure 32.4. Notice the pectoral girdle (at the shoulders), the pelvic girdle (at the hips), and the paired arms, hands, legs, and feet. These are legacies of early tetrapods. So are the flat shoulder blades and slender collarbone of their pectoral girdles. These work well for fishes. But it is easy for a "fish out of water" to dislocate shoulder blades or fracture a collarbone—the one broken most often—after falling hard on an outstretched arm.

The rest of a human skeleton—jaws and other skull bones, twelve pairs of ribs, a breastbone, and twenty-six vertebrae—are the legacies of ancient craniates and jawed vertebrates. **Vertebrae** (singular, vertebra), the bony segments of a backbone, extend from the skull to the pelvic girdle. The backbone's **intervertebral disks** are shock absorbers and flex points made of cartilage.

What does your backbone do? Its bony parts have attachment sites for paired muscles and form a canal that protects the delicate spinal cord. It also transmits the weight of your upright torso to your lower limbs.

Our hominid ancestors started walking on two legs more than 4 million years ago. The switch to upright posture put a notably S-shaped curve in the backbone. Ever since, bones and disks of the vertebral column have been stacked against gravity. A severe or rapid shock can force a disk to slip out of place or rupture—*herniate*—and cause chronic pain. No longer aquatic, no longer four-legged, we now live with costs as well as benefits of getting around in the world on two legs.

Nearly all animals have a hydrostatic skeleton, exoskeleton, or endoskeleton. The skeleton of land vertebrates reflects ancient adaptations to life in water and later adaptations to a life deprived of water's buoyancy.

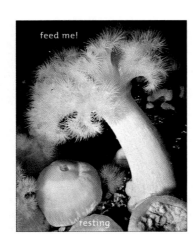

Figure 32.2 Hydrostatic skeleton. Some contractile cells in a sea anemone's body wall are longitudinal to the main body axis; others run radially around the gut. When this cnidarian rests, the radial cells are relaxed and the longitudinal ones are contracted. When it is feeding, radial cells are contracted and longitudinal ones are relaxed.

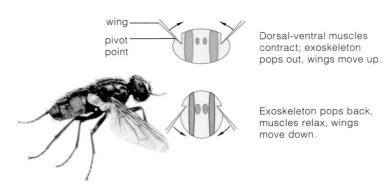

Figure 32.3 Fly wing motion. An insect cuticle forms a pliable hinge across gaps between body segments. Contracting muscles change the shape of the thorax, causing wings attached to it to move up and down.

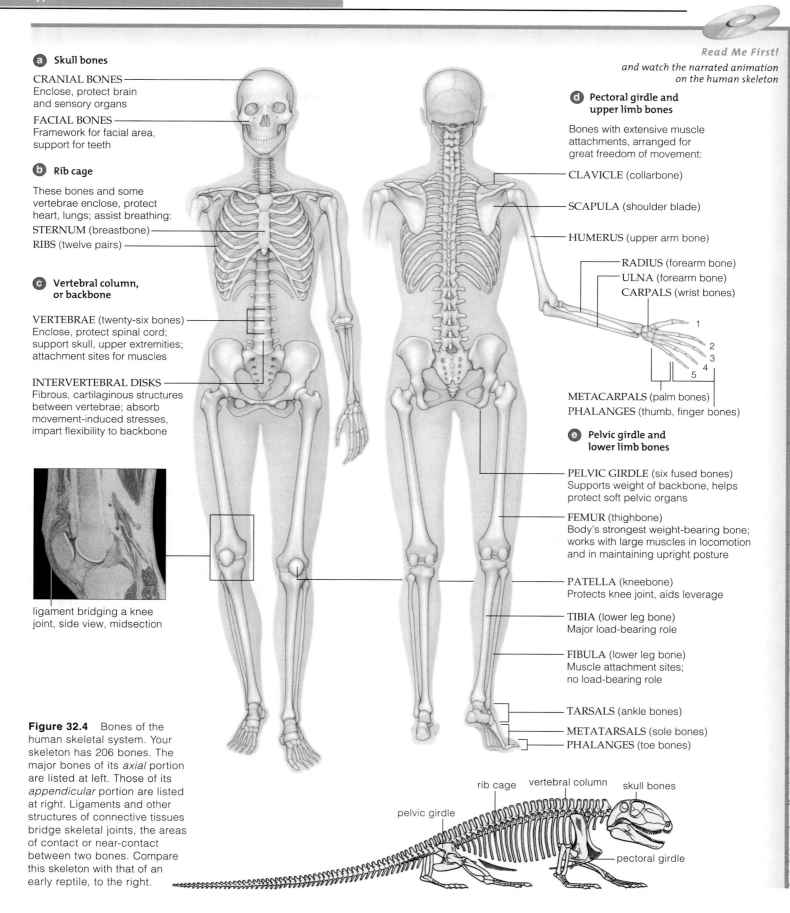

a Skull bones

CRANIAL BONES
Enclose, protect brain
and sensory organs

FACIAL BONES
Framework for facial area,
support for teeth

b Rib cage

These bones and some
vertebrae enclose, protect
heart, lungs; assist breathing:

STERNUM (breastbone)

RIBS (twelve pairs)

c Vertebral column,
or backbone

VERTEBRAE (twenty-six bones)
Enclose, protect spinal cord;
support skull, upper extremities;
attachment sites for muscles

INTERVERTEBRAL DISKS
Fibrous, cartilaginous structures
between vertebrae; absorb
movement-induced stresses,
impart flexibility to backbone

ligament bridging a knee
joint, side view, midsection

Read Me First!
*and watch the narrated animation
on the human skeleton*

d Pectoral girdle and
upper limb bones

Bones with extensive muscle
attachments, arranged for
great freedom of movement:

CLAVICLE (collarbone)

SCAPULA (shoulder blade)

HUMERUS (upper arm bone)

RADIUS (forearm bone)
ULNA (forearm bone)
CARPALS (wrist bones)

1
2
3
4
5

METACARPALS (palm bones)
PHALANGES (thumb, finger bones)

e Pelvic girdle and
lower limb bones

PELVIC GIRDLE (six fused bones)
Supports weight of backbone, helps
protect soft pelvic organs

FEMUR (thighbone)
Body's strongest weight-bearing bone;
works with large muscles in locomotion
and in maintaining upright posture

PATELLA (kneebone)
Protects knee joint, aids leverage

TIBIA (lower leg bone)
Major load-bearing role

FIBULA (lower leg bone)
Muscle attachment sites;
no load-bearing role

TARSALS (ankle bones)

METATARSALS (sole bones)

PHALANGES (toe bones)

Figure 32.4 Bones of the
human skeletal system. Your
skeleton has 206 bones. The
major bones of its *axial* portion
are listed at left. Those of its
appendicular portion are listed
at right. Ligaments and other
structures of connective tissues
bridge skeletal joints, the areas
of contact or near-contact
between two bones. Compare
this skeleton with that of an
early reptile, to the right.

pelvic girdle

rib cage   vertebral column   skull bones

pectoral girdle

32.2 Zooming In on Bones and Joints

Bones of a vertebrate skeleton are organs with diverse roles (Table 32.1). They function in movement, protection of soft internal parts, and as reservoirs for mineral ions. Blood cells originate from stem cells in some of the bones.

BONE STRUCTURE AND FUNCTION

Human bones range in size from middle ear bones smaller than lentils to clublike thighbones or femurs (Figure 32.4). Bone tissue consists of bone cells in a calcium-hardened organic matrix rich with collagen fibers. There are three types of bone cells. *Osteoblasts* are bone-forming cells that secrete components of the matrix. We find huge populations of osteoblasts on the outer surfaces and internal cavities of adult bones. *Osteocytes* are osteoblasts that became imprisoned in tiny gaps in their own secretions. They are the most common bone cells in adult bones. *Osteoclasts* are different cells that break down bone. They secrete acids and enzymes into the hardened matrix.

The femur in Figure 32.5 has two types of bone tissue. Its *compact* bone tissue resists mechanical shock. This tissue's matrix is laid down as dense concentric rings around canals for nerves and blood vessels. Tiny gaps between any two rings house osteocytes. A femur's shaft and ends also contain *spongy* bone tissue. This tissue is strong without weighing much because its hardened matrix is pocketed with open spaces.

Red marrow, the major site of blood cell formation, fills the spaces in spongy bone. The central cavity of the femur and most mature bones of adults is filled with **yellow marrow**. This marrow is mostly fat, but it can be converted to blood cell-producing red marrow in times of severe blood loss.

BONE FORMATION AND REMODELING

Early in development, all vertebrates have a skeleton made of cartilage, which persists in adult cartilaginous fishes. In all other vertebrates, however, osteoblasts invade the cartilage model and transform it into bone. Cartilage in the bone shaft breaks down and a marrow cavity opens up (Figure 32.6).

The total bone mass in healthy young adults does not change much, yet mineral ions and osteocytes are being simultaneously removed and replaced all the time. Mineral deposits and removals help maintain the blood levels of calcium and phosphorus, and they adjust bone strength. In an ongoing process known as **bone remodeling**, osteoblasts deposit bone even as osteoclasts break it down. Osteoclast action liberates mineral ions that enter interstitial fluid and then the bloodstream, which transports them through the body.

Calcium ions are the most prevalent mineral ions stored in and released from bones. Neural function, muscle contraction, and many other activities require them. The calcium level in blood is a tightly controlled

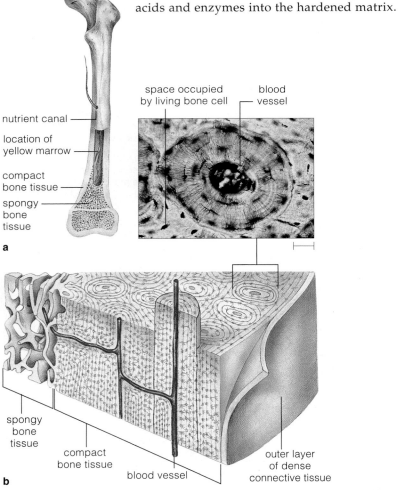

space occupied by living bone cell

blood vessel

nutrient canal

location of yellow marrow

compact bone tissue

spongy bone tissue

a

spongy bone tissue

compact bone tissue

blood vessel

outer layer of dense connective tissue

b

Figure 32.5 (**a**) Structure of a human thighbone (femur), and (**b**) a section through its spongy and compact bone tissues.

Table 32.1	Functions of Bone

1. *Movement.* Bones interact with skeletal muscle and change or maintain the position of the body and its parts.

2. *Support.* Bones support and anchor muscles.

3. *Protection.* Many bones are organized as hard compartments that enclose and protect soft internal organs.

4. *Mineral storage.* Bones are a reservoir for calcium and phosphorus ions. Deposits and withdrawals of these minerals help maintain essential ion concentrations in body fluids.

5. *Blood cell formation.* Only certain bones contain regions where blood cells form.

Embryo:
cartilage model of
future bone in embryo

Fetus:
blood vessel invades
model; osteoblasts
start producing bone
tissue; marrow
cavity forms

Newborn:
remodeling and
growth continue;
secondary bone-
forming centers
appear at knobby
ends of bone

Adult:
mature bone

Figure 32.6 Long bone formation, starting with osteoblast activity in a cartilage model formed earlier in the embryo. The bone-forming cells are active first in the shaft region, then at the knobby ends. In time, cartilage is left only at the ends.

Figure 32.7
(**a**) Normal bone tissue. (**b**) Bone affected by osteoporosis.

aspect of metabolism. Bones and teeth store all but 1 percent or so of the body's calcium.

Negative feedback loops from the blood to glands govern calcium release and uptake. When the calcium level rises, the thyroid gland releases calcitonin. This hormone suppresses osteoclast action, so calcium's release into blood slows. When the calcium level falls, parathyroid glands release parathyroid hormone, or PTH. This hormone stimulates calcium release from bones and calcium reabsorption from the kidneys. It enhances osteoclast action and also activates vitamin D, which stimulates calcium absorption by the gut.

Until humans are about twenty-four years old, the osteoblasts are secreting more matrix than osteoclasts can break down, and bone mass increases. So bone becomes denser and stronger. After this, osteoblast activity slows, and bones gradually weaken.

A decline in bone density past a certain point is called *osteoporosis* (Figure 32.7). Calcium or vitamin D deficiency, lack of physical activity, and parathyroid problems increase the risk. So can the decline in female sex hormones that occurs with menopause. Smoking, excessive alcohol intake, and prolonged use of steroids can also slow bone deposition.

WHERE BONES MEET—SKELETAL JOINTS

Connective tissue bridges **joints**, the areas of contact or near-contact between bones. **Ligaments** are straps of dense connective tissue at many joints, such as knees. They attach one bone to another, yet let both move freely. The breastbone, vertebrae, and ribs move only a bit. Their joints of cartilage are pliable enough to cushion abutting bones and help absorb shocks.

Knee joints are especially vulnerable to injury. They let you swing, bend, and turn the long bones below them, and can absorb the force of your weight. Abruptly stretch or twist a knee joint too far and you *strain* it. Tear its ligaments or tendons and you *sprain* it. Move the wrong way and you may even dislocate attached bones. In collision sports such as football, a blow to the knee can sever ligaments. The ligaments must be surgically put back together within ten days. Why? Phagocytic white blood cells patrol the fluid in the joint, where they clean up everyday wear and tear. Their action can turn torn ligaments to mush.

Joint inflammation and degenerative disorders are collectively called arthritis. In *osteoarthritis*, cartilage at freely movable joints wears away. Joints in fingers, knees, hips, and the backbone are affected most. With *rheumatoid arthritis*, joint membranes become inflamed and thicken, cartilage degenerates, and bone deposits accumulate because of an autoimmune response. A bacterial or viral infection may trigger it, but genetics may be a factor. Smoking raises the risk. Rheumatoid arthritis can develop at any age, but symptoms of the disorder usually appear before age fifty.

Bones are collagen-rich, mineralized organs that function in movement, support, protection, storage of calcium and other minerals, and in blood cell formation.

A balance between ongoing calcium deposits and withdrawals maintains bone mass and calcium concentrations in blood.

32.3 Skeletal–Muscular Systems

Skeletal muscles are the functional partners of bones. They contract (shorten) in response to stimulation, then passively return to a resting position (lengthen).

Skeletal muscle cells are not your typical cells. Before they can differentiate and mature in embryos, groups of them fuse together into one multinucleated **muscle fiber.** Then dense connective tissue forms a sheath that encloses and extends past bundles of such fibers. The extension, a cordlike or straplike **tendon**, attaches the resulting organ, a skeletal muscle, to bone.

Most of these attachment sites are like a gearshift. *They act as a lever system, in which a rigid rod is attached to a fixed point and moves about it.* Muscles connect to bones (rigid rods) near a joint (fixed point). As they contract, they transmit force that makes bones move.

The skeletal muscles also interact with one another. Some work in pairs or groups in ways that promote a movement. Others work in opposition; the action of one opposes or reverses the action of another. As an example, extend your right arm forward, then place your left hand over the biceps in your upper right arm and slowly bend the elbow, as in Figure 32.8. Feel the biceps contract? Even when a biceps contracts just

a bit, it causes a large motion of the bone connected to it. This is the case for most leverlike arrangements as the frog in Figure 32.9 obligingly illustrates.

Bear in mind, only *skeletal* muscle is the functional partner of bone. As mentioned earlier, smooth muscle is mainly a component of soft internal organs, such as the stomach. Cardiac muscle forms only in the heart wall. Later chapters will consider the structure and function of smooth muscle and cardiac muscle.

The human body has close to 700 skeletal muscles, some near the surface, others deep in the body wall (Figure 32.10). The trunk has muscles of the thorax, backbone, abdominal wall, and pelvic cavity. Other muscles attach to the upper and lower limb bones.

c The first muscle group in the frog's upper hindlimb contracts again and draws the leg back toward the body.

b An opposing muscle group attached to the limb forcefully contracts and pulls the limb back. The contractile force, applied against the ground, propels the frog forward.

a A frog muscle attached to each upper hindlimb contracts and pulls the leg slightly forward relative to the body's main axis.

Figure 32.9 A leaping frog demonstrating how a small change in the length of a contracting muscle can cause big movements.

Read Me First!

and watch the narrated animation on opposing muscle action

Triceps relaxes.

Biceps contracts at the same time, and pulls forelimb up.

Triceps contracts, pulls the forelimb down.

At the same time, biceps relaxes.

Figure 32.8 Two opposing muscle groups in human arms. (**a**) When the triceps relaxes and its opposing partner (biceps) contracts, the elbow joint flexes and the forearm is pulled upward. (**b**) When the triceps contracts and the biceps relaxes, the forearm is extended down.

TRICEPS BRACHII
Straightens
the forearm
at elbow

PECTORALIS MAJOR
Draws the arm forward
and in toward the body

SERRATUS ANTERIOR
Draws shoulder blade
forward, helps raise arm,
assists in pushes

EXTERNAL OBLIQUE
Compresses the abdomen,
assists in lateral rotation
of the torso

RECTUS ABDOMINIS
Depresses the thoracic
(chest) cavity, compresses
the abdomen, bends the
backbone

ADDUCTOR LONGUS
Flexes, laterally rotates,
and draws the thighs
toward the body

SARTORIUS
Bends the thigh at the hip,
bends lower leg at the
knee, rotates the thigh in
an outward direction

QUADRICEPS FEMORIS
Flexes the thigh at hips,
extends the leg at the knee

TIBIALIS ANTERIOR
Flexes the foot toward
the shin

BICEPS BRACHII
Bends the forearm at
the elbow

DELTOID
Raises the arm

TRAPEZIUS
Lifts the shoulder blade,
braces the shoulder,
draws the head back

LATISSIMUS DORSI
Rotates and draws the
arm backward and
toward the body

GLUTEUS MAXIMUS
Extends and rotates the
thigh outward when
walking, running, and
climbing

BICEPS FEMORIS
(Hamstring muscle)
Draws thigh backward,
bends the knee

GASTROCNEMIUS
Bends the lower leg at
the knee when walking,
extends the foot when
jumping

a

muscle
tendon
bursae
synovial
cavity

b

Later chapters explain how skeletal muscles assist
respiration and circulation. For now, let's turn to the
mechanisms that bring about their contraction.

*Cordlike or straplike tendons of dense connective tissue
attach skeletal muscles to bones. Only skeletal muscles
transmit contractile force to bones.*

Many skeletal muscles work as opposable groups.

Figure 32.10 (**a**) Major skeletal muscles of the human skeletal–muscular
system. Not all skeletal muscles are shown, but these are the ones that
body builders are most familiar with. (**b**) A typical tendon. Bursae form
between tendons and bones (or some other structure). Each bursa is a
flattened sac filled with synovial fluid. It helps reduce friction between
body parts during movements.

32.4 How Does Skeletal Muscle Contract?

Bones of a dancer or any other human in motion move in some direction when skeletal muscles attached to them shorten. A muscle shortens when its fibers, and individual contractile units inside them, shorten.

The basic units of contraction are called **sarcomeres**. In both skeletal and cardiac muscles, these are highly oriented arrays of actin, myosin, and other proteins. The orientation starts with long, slender muscle fibers that run parallel with a muscle's long axis. Inside each fiber are parallel arrays of **myofibrils**, or threadlike, cross-banded structures. Staining reveals a transverse, light–dark pattern, which gives the fibers a striped, or striated, look (Figure 32.11*b*).

Two dense arrays of cytoskeletal elements, called Z bands, define the ends of the sarcomere and anchor its parts (Figure 32.11*c*). Anchored to each Z band are parallel arrays of thin filaments that project toward the sarcomere's center. The filaments consist mainly of repeating units of **actin**, a globular protein. Not attached to the Z bands is a parallel array of thicker filaments that partially overlaps the thin filaments (Figure 32.12*a*). The thick filaments are composed of **myosin**, a motor protein with a clublike head. The thick and thin filaments run parallel with myofibrils. The myosin heads are a few nanometers away from the thin actin filaments.

The muscle bundles, muscle fibers, myofibrils, and filaments have the same parallel orientation. What is the point? *The repetitive orientation focuses the force of contraction, so that all sarcomeres in all fibers of a muscle work together to pull a bone in the same direction.*

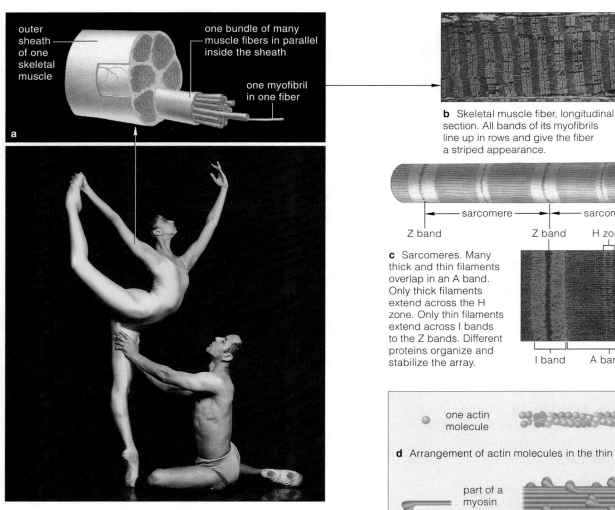

a

outer sheath of one skeletal muscle

one bundle of many muscle fibers in parallel inside the sheath

one myofibril in one fiber

one myofibril inside fiber:

b Skeletal muscle fiber, longitudinal section. All bands of its myofibrils line up in rows and give the fiber a striped appearance.

sarcomere — sarcomere

Z band   Z band   H zone   Z band

c Sarcomeres. Many thick and thin filaments overlap in an A band. Only thick filaments extend across the H zone. Only thin filaments extend across I bands to the Z bands. Different proteins organize and stabilize the array.

I band   A band   I band

one actin molecule — part of a thin filament

d Arrangement of actin molecules in the thin filaments

part of a myosin molecule — part of a thick filament

e Arrangement of myosin molecules in the thick filaments

Figure 32.11 From the Dance Theatre of Harlem, an example of exquisite control of skeletal muscle movements. **(a–e)** Zooming down through skeletal muscle from a biceps down to molecules with contractile properties.

Read Me First!
and watch the narrated animation
on the sliding-filament model

actin   **myosin**   **actin**

a Sarcomere between contractions

b Same sarcomere, contracted

Figure 32.12 (**a**,**b**) The arrangement of actin and myosin filaments that interact to reduce a sarcomere's width. (**c**–**g**) Sliding-filament model for sarcomere contraction in a myofibril of a muscle cell. For clarity, we show the action of just two of the many myosin heads. Many myosin heads bind repeatedly and then slide all of the actin filaments. Their collective action makes the sarcomere shorten (contract).

— myosin head

one of many myosin binding sites on actin

c One myosin filament in a muscle at rest. The myosin heads were energized earlier by the binding of ATP, which they hydrolyzed to ADP and inorganic phosphate.

cross-bridge   cross-bridge

d Release of calcium from a cellular storage system allows formation of cross-bridges; myosin binds to sites on actin filaments.

e Binding makes each myosin head tilt toward the center of the sarcomere and slide the bound actin filaments along with it.

f ADP and phosphate are released as the myosin heads drag the actin filaments inward, pulling the Z lines closer together.

ATP   ATP

g New ATP binds to the myosin heads and they detach from actin. Hydrolysis of the ATP will return the myosin heads to their original orientation, ready to act again.

SLIDING-FILAMENT MODEL FOR CONTRACTION

How do sarcomeres, and a skeletal or cardiac muscle, shorten? By the **sliding-filament model**, short, ATP-driven movements of the myosin heads pull neighbor actin filaments toward the center of a sarcomere. The myosin stays in place as actin slides over it. It is like two pocket doors closing—but pulling the walls to which they are attached with them. As myosin pulls on the actin, the actin pulls the two Z bands inward, and so the sarcomere shortens (Figure 32.12*b*).

Each myosin head connects repeatedly to binding sites on a nearby actin filament. Part of the head acts as an enzyme. It catalyzes a phosphate-group transfer from ATP that provides energy to drive contraction.

The myosin head forms a cross-bridge to the actin when a change in the local concentration of calcium ions exposes a binding site for it. Binding makes the head tilt and pull the actin filament toward the center of the sarcomere. Another ATP boost breaks the grip on actin, and the myosin head reverts to the original

position (Figure 32.12*c*–*e*). Depending on calcium and ATP levels, the head may attach to the next binding site on the actin filament, tilt in another stroke, and so on. Just a single contraction of one sarcomere requires hundreds of myosin heads, all performing a series of short strokes down the length of the actin filaments.

Contracting skeletal muscles move the body or its parts in response to signals from the nervous system. Remember the neuromuscular junctions (Section 29.3)? At these synapses, signals from motor neurons can excite a muscle fiber in ways that cause a release of calcium that brings about contraction.

Sarcomeres are the basic units of contraction in skeletal and cardiac muscle. The parallel orientation of all of the muscle components focuses contractile force on a bone to be moved.

By energy-driven interactions between myosin and actin filaments, the many sarcomeres of a muscle cell shorten and collectively account for the muscle's contraction.

32.5 Energy for Contraction

When a muscle fiber at rest is called upon to contract, phosphate donations from ATP increase 20 to 100 times. Where does all this energy come from?

A muscle fiber has only a small supply of ATP when it starts contracting. It can produce more very quickly by transferring one phosphate group from **creatine phosphate** to ADP (Figure 32.13). A muscle fiber has about six times as much creatine phosphate as ATP. Even so, the supply can only fuel about 15 seconds of contraction. After the call for contraction is over, the fiber's supply of creatine phosphate is restored; ATP donates phosphate to creatine.

Figure 32.13 Three metabolic pathways by which ATP forms inside muscles in response to the demands of physical exercise.

During prolonged but moderate exercise, aerobic respiration typically supplies most of the energy for contraction. A muscle fiber taps its store of glycogen for glucose (the starting substrate) in the first five to ten minutes. After that, glucose and fatty acid that are delivered by the blood sustain the next half hour or so of activity. Fatty acids become the main fuel source for any further contraction.

Not all fuel is burned in aerobic respiration. Even in resting muscle, some of the pyruvate produced by glycolysis is converted to lactate by a fermentation pathway. Lactate production increases with exercise. Remember that the net ATP yield from this anaerobic process is small. But lactate serves as a transportable fuel. It can be transported out of the skeletal muscle fibers and picked up by the blood, which delivers it to other muscle fibers or tissues. There, it can be used as fuel or stored as glycogen (Section 31.5).

> *Muscle contraction requires high concentrations of ATP.*
>
> *Muscle fibers make ATP by dephosphorylation of creatine phosphate, by aerobic respiration, and by glycolysis.*

32.6 Properties of Whole Muscles

Muscle contractions vary in duration and intensity. Exercise can strengthen or fatigue muscles. Age and certain genetic disorders can weaken them.

TYPES OF CONTRACTIONS

A motor neuron has many axon endings that synapse on different fibers inside each muscle. A motor neuron and all muscle fibers it controls constitute one **motor unit**. When a motor neuron is briefly stimulated, all fibers in the motor unit contract. A fleeting contractile force is generated that lasts just a few milliseconds. That contraction is a **muscle twitch** (Figure 32.14a).

When a new stimulus is applied before a response ends, the muscle twitches once again. When a motor unit is stimulated repeatedly during a short interval, the twitches all run together. The result is a sustained contraction called **tetanus**. The contraction produces three or four times the force of a single twitch. Figure 32.14c shows a recording of tetanic contraction.

Muscle tension is the mechanical force exerted by a muscle on an object. It is affected by the number of fibers recruited into action. Opposing this force is a load, either an object's weight or gravity's pull on the muscle. Only when muscle tension exceeds opposing forces does a stimulated muscle shorten. *Isotonically* contracting muscles shorten and move a load (Figure 32.15a). *Isometrically* contracting muscles do develop tension but cannot shorten, as when you try to lift something that is too heavy (Figure 32.15b).

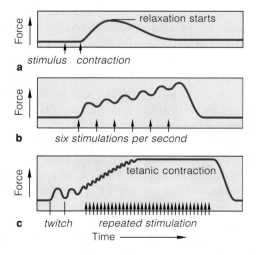

Figure 32.14 Three recordings of twitches in muscles exposed to artificial stimulation of different frequencies. (**a**) A single twitch. (**b**) Summation of twitches after six stimulations per second. (**c**) Tetanic contraction after twenty stimulations per second.

WHAT IS MUSCLE FATIGUE?

When ongoing, strong stimulation keeps a muscle in a state of tetanic contraction, muscle fatigue follows. **Muscle fatigue** is a decrease in a muscle's capacity to generate force, a decline in tension.

After a few minutes of rest, a fatigued muscle will contract again in response to stimulation. The extent of recovery depends largely upon how long and how often the muscle was stimulated previously. Muscles trained by a pattern of brief, intense exercise fatigue and recover fast. This happens during weight lifting. Muscles used in prolonged, moderate exercise fatigue slowly. They take longer to recover, often up to a day. Exactly what causes muscle fatigue is unknown, but glycogen depletion is one factor.

A *muscle cramp* is a sudden involuntary, painful contraction that resists release. Any skeletal muscle can cramp, but problems occur most often in muscles of the calves and thighs. Motion usually aggravates cramping; gentle stretching, massage, and heat often may relieve it. To avoid cramps, stretch regularly and avoid overexertion and dehydration.

WHAT ARE MUSCULAR DYSTROPHIES?

Muscular dystrophies are a class of genetic disorders in which muscles progressively weaken and degenerate. Duchenne muscular dystrophy is the most common form among children. Myotonic muscular dystrophy is the most common form among adults.

A single mutant gene on the X chromosome causes Duchenne muscular dystrophy. The gene encodes one of a group of proteins found in extensions of a muscle fiber's plasma membrane. The protein group attaches the plasma membrane to actin filaments at Z bands of sarcomeres. Unless actin binds firmly to proteins at the bands, sarcomeres cannot contract. Like all other X-linked disorders (Section 11.8), muscular dystrophy occurs mostly in males, at a frequency of about 1 in 3,500. Generally, affected individuals are confined to a wheelchair by their teens. They can expect to survive only into their twenties. There is no cure.

MUSCLES, EXERCISE, AND AGING

No matter how much you exercise, you will not make any more muscle fibers. Existing ones just get bigger, more active, and more resistant to fatigue. Think of *aerobic exercise*—not intense, but long in duration. It increases the number of mitochondria in muscles and the number of blood capillaries servicing them. These

Figure 32.15 (**a**) Isotonic contraction. The load is less than a muscle's peak capacity to contract. The muscle can contract, shorten, and lift the load. (**b**) Isometric contraction. The load exceeds a muscle's peak capacity, so the muscle contracts but cannot shorten.

physiological changes improve endurance. By contrast, *strength training* (brief but very intense exercise, such as weight lifting) makes the fibers thicken. Also, it promotes the synthesis of enzymes used for glycolysis. Bulging muscles form, but they do not have much endurance. They fatigue fast.

As people age, the number and size of their muscle fibers decline. Tendons that attach muscles to bone stiffen and are more likely to tear. Older people may exercise long and intensely, but their muscles don't change as much as when they were young. Even so, aerobic exercise can improve blood circulation, and modest strength training can slow the loss of muscle tissue.

As a beneficial side effect, aerobic exercise among the middle-aged and elderly can lift major depression as well as drugs do. It may improve memory and the capacity to plan and carry out complex tasks. Also, exercise increases blood flow and helps nerves make new connections. In short, exercise is good for more than muscles. It is good for the brain.

Cross-bridges in sarcomeres collectively exert a mechanical force called muscle tension. Muscles shorten only if this tension exceeds other opposing forces.

Muscle tension is related to the number of fibers that have been stimulated. A signal from a motor neuron stimulates all the fibers of the same motor unit.

Exercise, aging, and some genetic disorders affect the ability of muscles to generate tension.

32.7 Oh, *Clostridium!*

FOCUS ON HEALTH

We conclude this chapter with a look at two bacterial infections that interfere with how muscles work.

Clostridium botulinum is an anaerobic soil bacterium that can cause disease in humans. Improperly canned food may contain its endospores. When these spores germinate, they start making an odorless, tasteless botulinum toxin. In a person who has consumed toxin-tainted food, the poison enters motor neurons and stops the release of acetylcholine (ACh). Without this neurotransmitter, muscles cannot contract. They become more and more flaccid and paralyzed. These are symptoms of *botulism*, a type of food poisoning. Affected people must be treated with antitoxin as quickly as possible. Muscles in the heart and those involved with breathing must be protected.

A related bacterium, *C. tetani*, lives in the gut of horses, cattle, and other grazing animals, and the gut of many people. If sunlight and oxygen do not reach them, the endospores can endure for years in soil, manure-rich soils especially. They resist disinfectants, heat, and boiling water. After entering the body at a deep puncture or cut, they can germinate in necrotic tissues. The bacterial cells don't spread from these anaerobic, dead tissues. But they make a toxin that blood or nerves deliver to the spinal cord and brain.

In the spinal cord, the toxin acts on nerve cells that help control motor neurons. It blocks release of two neurotransmitters, GABA and glycine, so the motor neurons are freed from normal inhibitory control. That is when symptoms of the disease *tetanus* begin.

After four to ten days, the overstimulated muscles stiffen and go into spasm. They cannot be released from contraction. Prolonged, spastic paralysis follows. Fists and jaws may remain clenched. (Lockjaw is the other name for the disease.) The backbone may become frozen into an arch. If respiratory and cardiac muscles become paralyzed, death nearly always follows.

Vaccines were not available for soldiers of early wars, when *C. tetani* lurked in dead cavalry horses and manure on battlefields (Figure 32.16). Vaccines have all but eradicated tetanus in the United States. Worldwide, the annual death toll is about 50,000.

Figure 32.16
Painting of a young victim of a contaminated battle wound, as he lay dying of tetanus in a military hospital.

Summary

Section 32.1 Three categories of skeletal systems are common in animals—hydrostatic skeletons, exoskeletons, and endoskeletons. Humans, like other land vertebrates, have an endoskeleton with skull bones, a pelvic girdle, a pectoral girdle, and paired limbs. Evolutionarily, the move of our vertebrate ancestors from water to land, then upright walking, involved many skeletal modifications.

Section 32.2 Bones are collagen-rich, mineralized organs. They function in mineral storage, movement, and protection and support of soft organs. Blood cells form in bones that contain red marrow. Ongoing mineral deposits and removals help maintain the blood levels of mineral ions, and also adjust bone strength. Hormones secreted by parathyroid glands promote bone breakdown and calcium release. Ligaments and cartilage hold bones together at joints.

Section 32.3 Tendons attach skeletal muscles to bones. When skeletal muscles contract, they transmit force that makes the bones move. Many muscles work as opposing pairs. The internal organization of skeletal and cardiac muscle promotes strong, directional contraction.

Section 32.4 In each skeletal muscle fiber (and cardiac muscle cell), many threadlike structures called myofibrils are divided into sarcomeres, the basic unit of contraction. A sarcomere has parallel arrays of actin and myosin filaments. ATP-driven sliding interactions between actin and myosin shorten sarcomeres in response to signals from a motor neuron.

Section 32.5 ATP energy for muscle contraction comes from dephosphorylation of creatine phosphate, aerobic respiration, and glycolysis.

Section 32.6 Muscle tension is a mechanical force caused by cross-bridge formation. A muscle shortens only when muscle tension exceeds an opposing load. A motor neuron and all muscle fibers that form junctions with its endings are a motor unit. Repeated stimulation of a motor unit results in tetanus, or sustained contraction.

Section 32.7 Bacterial toxins can interfere with nervous signals that stimulate muscle contraction.

Self-Quiz

Answers in Appendix III

1. Bone functions in _____ .
   a. mineral storage   c. blood cell formation
   b. movement   d. all of the above

2. The _____ is the basic unit of contraction.
   a. osteoblast   c. muscle fiber
   b. sarcomere   d. myosin filament

3. ATP for muscle contraction can be formed by _____ .
   a. aerobic respiration   c. creatine phosphate
   b. glycolysis   breakdown
                           d. all of the above

4. Bones move when _____ muscles contract.
   a. cardiac   c. smooth
   b. skeletal   d. all of the above

5. A part of the _____ molecule can bind and break down ATP to release energy.
   a. actin   c. both
   b. myosin   d. neither

6. A sarcomere shortens when _____ .
   a. thick filaments shorten
   b. thin filaments shorten
   c. both thick and thin filaments shorten
   d. none of the above

7. Match the words with their defining feature.
   ____ osteoblast   a. actin's partner
   ____ muscle twitch   b. all in the hands
   ____ muscle tension   c. blood cell production
   ____ joint   d. decline in tension
   ____ myosin   e. bone-forming cell
   ____ marrow   f. motor unit response
   ____ metacarpals   g. force exerted by cross-bridges
   ____ myofibrils   h. area of contact between bones
   ____ muscle fatigue   i. muscle fiber's threadlike parts

Critical Thinking

1. The genetic disorder *osteogenesis imperfecta* (OI) arises from a mutant form of collagen. As bones develop, this protein forms a scaffold for the deposition of mineralized bone tissue. The scaffold forms improperly in children with OI, who end up with fragile, brittle, easily broken bones (Figure 32.17).

   OI is rare; most doctors never come across it. Parents are often accused of child abuse when they take an OI child with multiple bone breaks to an emergency room. Prepare a brief handout about OI that could be used to alert the staff in emergency rooms to the condition. You may wish to start with resources on the Osteogenesis Imperfecta Foundation web site at www.oif.org.

2. Lydia is training for a marathon. While training, she plans to take creatine supplements because she heard that they give athletes extra energy. She asks you whether you think this is a good idea. What is your response?

3. Compared with most people, long-distance runners have more mitochondria in skeletal muscles. In sprinters, skeletal muscle fibers have more of the enzymes necessary for glycolysis, but not as many mitochondria. Think about the differences between the two forms of exercise. Suggest why marathon runners and sprinters have pronounced differences in their muscles.

4. In adults, the nuclei of muscle fibers are stuck in G1 of the cell cycle and cannot undergo division. Lots of adults increase the size of muscles by lifting weights. What accounts for the increased bulk of the muscles?

5. Botulinum toxin can kill you if you eat it, but some people pay hundreds of dollars to have it injected. These *botox injections* prevent certain facial muscles from contracting and pulling on the skin in ways that cause wrinkling. Most commonly, botox is injected into muscles between the eyebrows. It makes eyelids droop for weeks in 3 percent of the patients. Explain why.

Figure 32.17 (**a**) X-ray of an arm bone deformed by osteogenesis imperfecta (OI). (**b**) Tiffany is affected by OI. She was born with multiple fractures in her arms and legs. Now six years old, she has undergone surgery to correct more than 200 bone fractures and to place steel rods in her legs. Every three months she receives intravenous infusions of an experimental drug that may help strengthen her bones.

Media Menu

Student CD-ROM

Impacts, Issues Video
  Pumping Up Muscles
Big Picture Animation
  Skeletons and muscles
Read-Me-First Animation
  The human skeleton
  Opposing muscle action
  The sliding-filament model
Other Animations and Interactions
  How a fly wing operates
  Structure of the human thigh bone
  Structure of a skeletal muscle
  Energy sources for muscle contraction

InfoTrac
- Use Them or Lose Them: Exercise to Keep Bones Strong. *Environmental Nutrition*, March 2002.
- The Age of Arthritis. *Time*, December 2002.
- Facts About Metabolic Diseases of Muscle. *Muscular Dystrophy Association*, June 2000.

Web Sites
- Bone Biology: depts.washington.edu/bonebio
- Human Anatomy Online: www.innerbody.com/htm/body.html
- Hypermuscle: www.med.umich.edu/lrc/Hypermuscle/Hyper.html
- American Academy of Orthopaedic Surgeons: orthoinfo.aaos.org

How Would You Vote?
Many dietary supplements have effects on metabolism, but because they are not sold as drugs, they have not been regulated or tested for side effects. Many believe that bringing dietary supplements under the jurisdiction of the Food and Drug Administration would help consumers sort fact from fiction and protect them against possibly dangerous products. Supplement makers oppose the idea and say it will decrease access to products and increase their cost. Do you support increased regulation of dietary supplements?

IMPACTS, ISSUES   *And Then My Heart Stood Still*

In the nineteenth century, physiologist Augustus Waller put the feet of his pet bulldog Jimmie in bowls of salty water, connected by wires to a crude recording device. Salt water is a good conductor of electricity. It picked up faint electric signals from Jimmie's beating heart, and the recording device scratched out the world's first graph of a heartbeat—an electrocardiogram.

Now fast-forward to the present. A graph of your heart's normal activity would look much the same as Jimmie's. The pattern emerged a few weeks after you started growing from a fertilized egg and developing into an embryo. Early on, some of the embryonic cells differentiated into cardiac muscle cells and started to contract on their own. The first patch of cells to start contracting set the pace for the others and have been the pacemaker ever since. If all goes well, that patch of cardiac muscle cells will continue to contract and make your heart beat steadily, about seventy times a minute, until the day you die.

However, the heart abruptly stops beating in a lot of people who are not expecting it, in a lot of places besides hospitals. We call this *sudden cardiac arrest*. It happens more than 250,000 times a year in the United States alone. It happened to Tammy Higgins before her thirtieth birthday. She had just left church and abruptly collapsed and stopped breathing.

Her husband could not find her pulse. As a former lifeguard, he knew oxygen-rich blood had to resume flowing to her brain, and fast. He immediately started cardiopulmonary resuscitation, or *CPR*. This method of mouth-to-mouth respiration and chest compressions kept Tammy alive until an ambulance arrived.

The emergency crew used a portable defibrillator, a device that delivers an electric shock to the chest, and they jump-started her heart (Figure 33.1). In the hospital, Tammy found out that her heartbeat is abnormal. She also found out that she was pregnant.

Surgeons implanted a tiny defibrillator in her chest wall. It constantly monitors her heart and shocks it when it stops beating. The implant saved her during pregnancy and once more during the three years that followed.

Figure 33.1 The two defibrillator paddles that may shock a heart into beating again after cardiac arrest. At the *far right*, Jimmie the bulldog taking part in a painless experiment that yielded the first electrical recording of a beating heart.

the big picture

What Circulatory Systems Do   The circulatory system of large animals has a muscular pump that drives blood through vessels of different diameters. The large vessels function in rapid transport; tiny ones are zones of leisurely diffusion between blood and interstitial fluid.

What Is Blood?   Blood is a fluid connective tissue, mostly water with cellular components and dissolved nutrients, proteins, and other substances. Red and white blood cells and platelets with specialized functions arise from stem cells in bone marrow.

Tammy was lucky; her brain was not damaged, and she did not die. Her husband and a few bystanders knew how to perform CPR. The chance of surviving sudden cardiac arrest increases by 50 percent when CPR starts within four to six minutes. Also, for every minute that passes without defibrillation, the chance of staying alive decreases by about 7 to 10 percent.

AEDs (automated external defibrillators) have been distributed to many senior centers and shopping malls, hotels, and other public places. About the size of a laptop computer, an AED checks for a heartbeat and shocks the heart if required. Its electronic voice commands trained bystanders to carry out appropriate steps. The American Heart Association, the American Red Cross, community groups, and adult education programs offer CPR and AED training to the public. In short, we have come a long way from Jimmie in checking out the heart (*below*) and in helping the hearts that falter.

And with this bit of history, we turn to the structure and function of a major organ system, one that steadily circulates blood with the help of a durable pump.

How Would You Vote?

CPR can make the difference between life and death after a cardiac arrest or a heart attack. Should public high schools in your state require all students to take a course in CPR? Is such a course worth diverting time and resources from the basic curriculum? See the Media Menu for details, then vote online.

The Cardiovascular System   The cardiovascular system of humans has a four-chambered heart that pumps blood through two separate circuits. A natural pacemaker in the heart wall stimulates cardiac contraction. Contraction of ventricles drives blood flow through the system.

The Lymphatic System   Fluid leaks out from gaps between cells in the smallest blood vessels. It enters the lymph vascular system, which returns it to blood. The lymphoid tissues and organs, including lymph nodes, can cleanse blood of pathogens and other threats to the body.

33.1 The Nature of Blood Circulation

Imagine an earthquake closing off your neighborhood's highway. Grocery trucks can't enter; waste-disposal trucks can't leave. Your food supplies dwindle and garbage piles up. Cells would be similarly stressed if something were to disrupt the flows along the body's highways.

The **circulatory system** moves substances to and from cellular neighborhoods. **Blood**, its transport medium, typically flows inside tubular vessels under pressure generated by a muscular pump, a **heart**. Blood makes exchanges with a fluid that fills tissue spaces beween cells. This **interstitial fluid** then exchanges substances with cells. Therefore, blood and interstitial fluid are functionally inseparable as an *internal* environment for the body's cells. Many organ systems interact to keep the composition, volume, and temperature of the environment within tolerable ranges for cell activities.

Circulatory systems are open or closed. In the *open* systems of most mollusks and all arthropods, blood flows out from muscular vessels, or hearts, mingles with fluid in body tissues, then moves back in (Figure 33.2*a,b*). You and many other animals have a *closed* circulatory system. It confines blood inside walls of a heart and blood vessels that vary in thickness and in diameter. The earthworm circulatory system, partly shown in Figure 33.2*c,d,* is one example. Figure 33.3 includes sketches of a few others.

Reflect on the overall "design" of a closed system. The total *volume* of blood is pumped continuously out of the heart, through vessels, then back to the heart. It travels fastest in large-diameter transport vessels. It slows in specific body regions, thereby giving blood enough time to exchange substances with interstitial fluid. The slowdowns take place in beds consisting of many small-diameter blood vessels called capillaries.

EVOLUTION OF VERTEBRATE CIRCULATION

All vertebrates have a closed circulatory system, but fishes, amphibians, birds, and mammals differ in the pumps and plumbing. These differences evolved over hundreds of millions of years after some vertebrates left the water for land. The earliest vertebrates, recall, had gills. Like all respiratory structures, gills have a thin, moist surface that oxygen and carbon dioxide can diffuse across. In time, *internally moistened* sacs called lungs evolved and supported the move to dry land. So did modifications that helped the circulatory systems move blood faster to and from the lungs.

In fishes, blood still flows in a single circuit (Figure 33.3*a*). The contractile force of a two-chambered heart drives blood through two capillary beds. Blood flows through gill capillary beds into a large vessel, through capillary beds in body tissues and organs, and back to

Figure 33.2 Comparison of open and closed circulatory systems. (**a,b**) In a grasshopper's open system, a heart (not like yours) pumps blood through a vessel (an aorta). From there, blood moves into tissue spaces. It mingles with fluid bathing cells, then reenters the heart at openings in the heart wall. (**c,d**) The closed system of an earthworm confines blood in pairs of muscular hearts near the head end and inside many blood vessels.

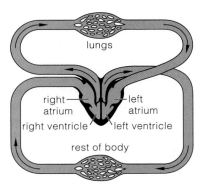

a In fishes, a two-chambered heart (atrium, ventricle) pumps blood in one circuit. Blood picks up oxygen in gills, delivers it to rest of body. Oxygen-poor blood flows back to heart.

b In amphibians, a heart pumps blood through two partially separate circuits. Blood flows to lungs, picks up oxygen, returns to heart. But it mixes with oxygen-poor blood still in the heart, flows to rest of body, returns to heart.

c In birds and mammals, the heart is fully partitioned into two halves. Blood circulates in two circuits: from the heart's right half to lungs and back, then from the heart's left half to oxygen-requiring tissues and back.

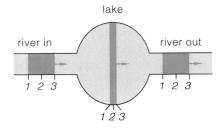

d Why have capillary beds? Picture two fast rivers flowing into and out from a lake. The flow *rate* is the same in all three places, with an identical volume of water moving from point *1* to point *3* during the same interval. But flow *velocity* decreases within the lake. Why? The volume spreads out through a larger cross-sectional area and moves forward a shorter distance during the specified interval.

Figure 33.3 Comparison of the blood flow circuits in the closed circulatory systems of fishes, amphibians, birds, and mammals.

the heart. Blood is not under much pressure when it moves from the gill capillaries so it travels slowly to the rest of the body tissues.

The heart of early amphibians became partitioned into three chambers. Two atria emptied into a single ventricle. Oxygenated blood and oxygen-poor blood flowed in two circuits, but they still mixed somewhat in the ventricle (Figure 33.3*b*).

In birds and mammals, the heart is partitioned into two halves—two chambers each—that pump blood in two separate circuits (Figure 33.3*c*). In the **pulmonary circuit**, oxygen-poor, carbon dioxide rich blood flows from the *right* half of the heart to the lungs. The blood picks up oxygen, gives up carbon dioxide, and flows into the left half of the heart.

In the **systemic circuit**, the *left* half of the heart pumps oxygenated blood to all tissues where oxygen is used and carbon dioxide forms. After giving up the oxygen and taking on carbon dioxide, blood flows into the heart's right half, then back to the lungs.

The double circuit is a fast, efficient mode of blood delivery. It supports the high levels of activity typical of vertebrates whose ancestors evolved on land.

LINKS WITH THE LYMPHATIC SYSTEM

The heart's pumping pressure forces some water, nutrients, and a few proteins dissolved in blood into interstitial fluid. Excess fluid and reclaimable solutes move back into capillaries or enter lymphatic vessels, which return them to the blood. As you will see in Section 33.10, the **lymphatic system** consists of these vessels and other structures that cleanse the fluid.

Circulatory systems move substances to and from interstitial fluid that bathes all living cells in a multicelled body. A typical closed system confines blood in a heart and blood vessels.

Blood flows fastest in large-diameter vessels between the vertebrate heart and capillary beds, where it slows enough for efficient exchanges with interstitial fluid.

In fishes and amphibians, oxygen-poor and oxygenated blood mix a bit. In birds and mammals, blood flows in two separate circuits, through a heart divided into two side-by-side pumps. The double circuit keeps oxygenated blood separate, which helps support higher levels of activity.

33.2 Characteristics of Blood

Tumbling along in the fluid portion of blood are cells and substances that do more than move gases and nutrients to and from cellular neighborhoods. Many defend the body and help maintain internal operating conditions.

FUNCTIONS OF BLOOD

Blood is a fluid connective tissue that moves oxygen, nutrients, and other solutes to cells. It carries away metabolic wastes and secretions, including hormones. Blood helps stabilize internal pH and is a highway for cells and proteins that defend and repair tissues. In birds and mammals, it helps keep body temperature within tolerable limits by moving excess heat to skin, where it can be dissipated into the surroundings.

red blood cell

white blood cell

platelets

BLOOD COMPOSITION AND VOLUME

The volume of blood depends on body size and on the concentrations of water and solutes. In average-sized humans, it is about four or five quarts (6 to 8 percent of the total body weight). In all vertebrates, blood is a viscous fluid, thicker than water and slower flowing. Its components are plasma, red and white blood cells, and platelets. The cells and platelets arise from stem cells in bone marrow. Any **stem cell** is unspecialized, and it retains a capacity for mitotic cell division. Some portion of its daughter cells divide and differentiate into specialized types.

PLASMA   About 50 to 60 percent of the total blood volume is plasma (Figure 33.4). **Plasma** is 90 percent water. Besides being the transport medium for blood cells and platelets, it acts as a solvent for hundreds of different plasma proteins, other molecules, and ions. Some of the proteins transport lipids and fat-soluble vitamins through the body. Others function in blood clotting. Other solutes include glucose, lipids, amino acids, vitamins, hormones, as well as oxygen, carbon dioxide, and gaseous nitrogen.

RED BLOOD CELLS   Erythrocytes, or **red blood cells**, are biconcave disks about 8 micrometers in diameter and 2 micrometers thick (Figure 33.5). They transport oxygen from lungs to aerobically respiring cells and carry carbon dioxide wastes from them. When oxygen first diffuses into blood, it binds to hemoglobin in red

Figure 33.4 Typical components of human blood. The scanning electron micrograph above shows a few of the cellular components.

The sketch of a test tube shows what happens when you keep a blood sample from clotting. The sample separates into the straw-colored plasma, which floats on a reddish-colored cellular portion.

Components	Relative Amounts	Functions
Plasma Portion (*50%–60% of total volume*):		
1. Water	91%–92% of plasma volume	Solvent
2. Plasma proteins (albumin, globulins, fibrinogen, etc.)	7%–8%	Defense, clotting, lipid transport, roles in extracellular fluid volume, etc.
3. Ions, sugars, lipids, amino acids, hormones, vitamins, dissolved gases	1%–2%	Roles in extracellular fluid volume, pH, etc.
Cellular Portion (*40%–50% of total volume*):		
1. Red blood cells	4,800,000–5,400,000 per microliter	Oxygen, carbon dioxide transport
2. White blood cells:		
Neutrophils	3,000–6,750	Phagocytosis during inflammation
Lymphocytes	1,000–2,700	Immune responses
Monocytes (macrophages)	150–720	Phagocytosis in all defense responses
Eosinophils	100–360	Defense against parasitic worms
Basophils	25–90	Secrete substances for inflammatory response and for fat removal from blood
3. Platelets	250,000–300,000	Roles in clotting

Figure 33.5 The cellular components of blood and where they originate.

Labels in figure:
natural killer cells — eosinophils — neutrophils — basophils — mast cells — B lymphocytes (mature in bone marrow) — T lymphocytes (mature in thymus) — forerunners of the white blood cells (leukocytes) — stem cells that multiply and differentiate in bone marrow — committed cell (proerythroblast) — red blood cells (erythrocytes) — monocytes (immature phagocytes) — dendritic cells — mature macrophages — megakaryocytes — platelets

blood cells. You learned about this protein in Section 3.6. Stored hemoglobin fills about 98 percent of red blood cells. It makes both the cells and oxygenated blood appear bright red. Oxygen-poor blood is darker red and appears blue through blood vessel walls near the body surface.

Mature red blood cells no longer have a nucleus, nor do they require it. They have enough hemoglobin, enzymes, and other proteins to last about 120 days. The body's phagocytes engulf the oldest red blood cells and the ones that are already dead. Normally, ongoing replacements keep the cell count fairly stable.

A **cell count** is a measure of the quantity of cells of a specified type in a microliter of blood. The average number of red blood cells per cubic centimeter is 5.4 million in human males and 4.8 million in females.

WHITE BLOOD CELLS Diverse leukocytes, or **white blood cells**, function in housekeeping and defense. Some patrol tissues and engulf damaged or dead cells and anything else chemically recognized as "nonself." Some sound the alarm that tissues are under siege by specific viruses, bacteria, or other threats. Many form highly organized masses inside lymph nodes and the spleen, which are organs of the lymphatic system.

White blood cells differ in size, nuclear shape, and staining traits (Figure 33.5). Their numbers depend on whether an individual is active, healthy, or under attack. We consider these cells in the next chapter, but a few examples can give you a sense of what they do. Neutrophils and basophils are phagocytes with roles in inflammation. Macrophages and dendritic cells are phagocytes that call up immune responses to specific threats. Lymphocytes are cells that carry out immune responses. There are two types: B cells and T cells. Natural killer cells directly kill body cells that have turned cancerous or have been infected by viruses.

PLATELETS Megakaryocytes originate from some stem cells. They later shed membrane-wrapped cytoplasmic fragments of themselves, as **platelets**. Each platelet lasts five to nine days, but hundreds of thousands are always circulating in blood. When activated, they release substances that initiate blood clotting.

Blood, a fluid connective tissue, functions as a transport medium and solvent. Diverse proteins, gases, sugars, and other substances are dissolved in it. Its cellular components include red blood cells, white blood cells, and platelets.

33.3 Blood Disorders

The body continually replaces blood cells, for good reason. Besides aging and dying off regularly, blood cells typically encounter a variety of pathogens that use them as places to complete their life cycle. Besides this, sometimes blood cells malfunction as a result of gene mutations.

RED BLOOD CELL DISORDERS   Too few red blood cells or deformed ones result in the disorders collectively called **anemias**. The oxygen level in blood becomes too low to support normal metabolism. Shortness of breath, fatigue, and chills follow. *Hemorrhagic* anemias follow sudden blood loss, as from a severe wound; *chronic* anemias result from suppressed red blood cell production or a slight but persistent blood loss, as happens from a bleeding ulcer.

Certain bacteria and protozoans replicate in blood cells. They cause some *hemolytic* anemias; they kill off infected cells when they escape by lysis. Insufficient iron in the diet results in *iron deficiency* anemia; red blood cells cannot make enough normal hemoglobin without iron. *Sickle-cell anemia* arises from a mutation that alters hemoglobin (Section 3.6). *Thalassemias* are outcomes of mutations that disrupt or stop synthesis of globin chains of hemoglobin. Too few red blood cells form. Those that do form are thin and fragile.

Polycythemias (having far too many red blood cells) make blood flow sluggish. So does *blood doping*. Some athletes withdraw and store their own red blood cells, then reinject them a few days before they are to compete in strenuous events. The withdrawal triggers red blood cell formation because the body attempts to replace "lost" cells. When withdrawn cells are put back in the body, the cell count goes up. The idea is to increase oxygen-carrying capacity and endurance. Blood doping elevates blood pressure and increases the viscosity of blood. Some athletes also artificially elevate red cell counts by injecting hormones that stimulate red cell production.

WHITE BLOOD CELL DISORDERS   An Epstein–Barr virus causes *infectious mononucleosis*, a highly contagious disease. Too many monocytes and lymphocytes form. After a few weeks of fatigue, aches, low-grade fever, and a chronic sore throat, patients usually recover.

Recovery is dicey for the *leukemias*. These cancers originate in the bone marrow and can suppress or impair white blood cell formation. Radiation therapy and chemotherapy kill cancer cells, but side effects are often severe. Remissions (symptom-free periods in chronic illness) may last for months or years. New gene therapies are putting some leukemias into long-term remission with only mild side effects.

33.4 Blood Typing

Blood from donors can be transfused into patients who are affected by severe blood loss or a blood disorder. Such blood transfusions cannot be hit-or-miss. Red blood cells of a potential donor and recipient must bear the same kinds of recognition proteins at their surface.

Among the many recognition proteins on the body's red blood cells are "self" markers. As the next chapter explains, they identify cells as belonging to one's own body. If a blood donor's cells have a different marker, a recipient's immune system will recognize them as foreign and attack them.

Figure 33.6 shows what happens when blood from incompatible donors and recipients intermingles. In a defense response called **agglutination**, proteins called antibodies act against foreign cells and make them clump together for faster disposal. Free antibodies circulating in the recipient's plasma bind to nonself

Figure 33.6   Light micrographs showing (**a**) an absence of agglutination in a mixture of two different yet compatible blood types and (**b**) agglutination in a mixture of incompatible types. (**c**) Responses in blood types A, B, AB, and O when mixed with samples of the same and different types.

Read Me First!

and watch the narrated animation on Rh and pregnancy

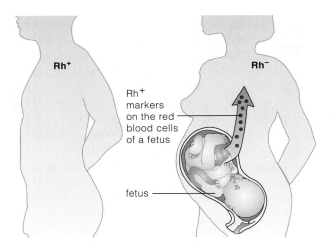

a A forthcoming child of an Rh⁻ woman and Rh⁺ man inherits the gene for the Rh⁺ marker. During childbirth, some of its cells bearing the marker may leak into the maternal bloodstream.

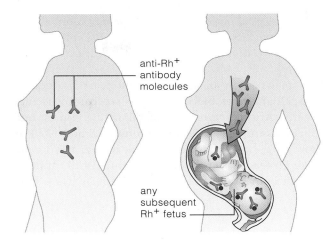

b The foreign marker stimulates antibody formation. If the same woman becomes pregnant again and if her second fetus (or any other) inherits the gene for the marker, her circulating anti-Rh⁺ antibodies will act against it.

Figure 33.7 Antibody production in response to Rh⁺ markers on red blood cells of a fetus.

markers on the introduced cells and target them for destruction by the immune system. The clumped cells can clog up small blood vessels and damage tissues. Without treatment, death may follow. The same thing can happen during a pregnancy if a mother and child differ in some red blood cell markers. The mother's immune system will attack the child's cells.

Blood typing—the analysis of the surface markers on red blood cells—can help prevent such problems.

ABO BLOOD TYPING

Molecular variations in one kind of self marker on red blood cells are analyzed with **ABO blood typing**. The genetic basis of such variation is described in Section 10.4. People with one form of the marker have type A blood; those with another form have type B blood. Those with both forms of the marker on their red blood cells have type AB blood. Others do not have either form of the marker; they have type O blood.

If you are type A, your immune system will treat blood with type B markers as foreign. If you are type B, your body will react against type A markers. If your blood is type AB, your body is familiar with both markers and won't react against either type; you are able to receive blood from anyone. If you are type O, both A and B markers are perceived as foreign. You can only receive blood from others who are also type O. But you can donate blood to anyone.

Rh BLOOD TYPING

Rh blood typing is based on the presence or absence of the Rh marker (first identified in blood of *Rh*esus monkeys). If you are type Rh⁺, your blood cells bear this marker. If you are type Rh⁻, they do not. Usually people do not have antibodies against Rh markers, because their body has never been exposed to them. But an Rh⁻ recipient of transfused Rh⁺ blood will make antibodies, and they will remain in the blood.

If an Rh⁻ woman is impregnated by an Rh⁺ man, there is a chance the fetus will be Rh⁺. Some fetal red blood cells usually leak into a woman's blood during childbirth and her body makes antibodies against Rh (Figure 33.7). If she gets pregnant again, some of her Rh antibodies may enter the bloodstream of her fetus. If the fetus is Rh⁺, her antibodies will cause its red blood cells to swell, rupture, and release hemoglobin.

Erythroblastosis fetalis can be the outcome. When too many red cells die, so does the fetus. Diagnosing the mismatch before birth can help. The newborn's blood is slowly replaced by transfusions that are free of Rh antibodies. A known Rh⁻ woman can be treated after a first pregnancy to protect her next fetus.

Markers on red cell surfaces are the basis for ABO and Rh blood typing. Incompatible blood types cause problems in transfusions and in some pregnancies.

33.5 Human Cardiovascular System

"Cardiovascular" comes from the Greek kardia *(for heart) and Latin* vasculum *(vessel). In the human cardiovascular system, a muscular heart pumps blood to the lungs and the tissues by way of two completely separate circuits.*

In humans, as in all mammals, the heart is a double pump that drives blood through two cardiovascular circuits. Arteries, arterioles, capillaries, venules, and veins make up each circuit (Figures 33.8 and 33.9). A short loop, the pulmonary circuit, oxygenates blood. Again, it leads from the heart's right half to capillary beds in both lungs, then back to the heart's left half.

The systemic circuit is a longer loop. The heart's left half pumps oxygenated blood into the main artery, the **aorta**. That blood gives up oxygen in all tissues, then oxygen-poor blood flows back to the heart's right half.

Most blood flows through only one capillary bed. Some flows through two. Blood picks up glucose and other substances at capillaries in the intestines, then flows to and through a capillary bed inside the liver. This organ has vital roles in metabolizing and storing nutrients. It also neutralizes many toxins.

As Figure 33.10 shows, the cardiovascular system distributes oxygen, nutrients, and other substances

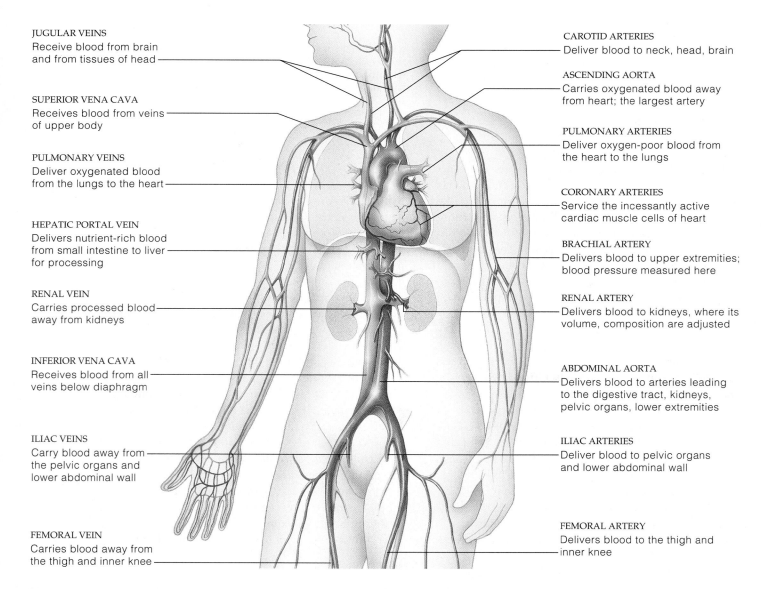

JUGULAR VEINS
Receive blood from brain and from tissues of head

SUPERIOR VENA CAVA
Receives blood from veins of upper body

PULMONARY VEINS
Deliver oxygenated blood from the lungs to the heart

HEPATIC PORTAL VEIN
Delivers nutrient-rich blood from small intestine to liver for processing

RENAL VEIN
Carries processed blood away from kidneys

INFERIOR VENA CAVA
Receives blood from all veins below diaphragm

ILIAC VEINS
Carry blood away from the pelvic organs and lower abdominal wall

FEMORAL VEIN
Carries blood away from the thigh and inner knee

CAROTID ARTERIES
Deliver blood to neck, head, brain

ASCENDING AORTA
Carries oxygenated blood away from heart; the largest artery

PULMONARY ARTERIES
Deliver oxygen-poor blood from the heart to the lungs

CORONARY ARTERIES
Service the incessantly active cardiac muscle cells of heart

BRACHIAL ARTERY
Delivers blood to upper extremities; blood pressure measured here

RENAL ARTERY
Delivers blood to kidneys, where its volume, composition are adjusted

ABDOMINAL AORTA
Delivers blood to arteries leading to the digestive tract, kidneys, pelvic organs, lower extremities

ILIAC ARTERIES
Deliver blood to pelvic organs and lower abdominal wall

FEMORAL ARTERY
Delivers blood to the thigh and inner knee

Figure 33.8 Major blood vessels of the human cardiovascular system. This drawing is simplified for clarity. For example, each arm has a complete set of the arteries and veins listed, and so does each leg. Humans, remember, have bilateral ancestry.

Read Me First!
and watch the narrated animation
on human blood circulation

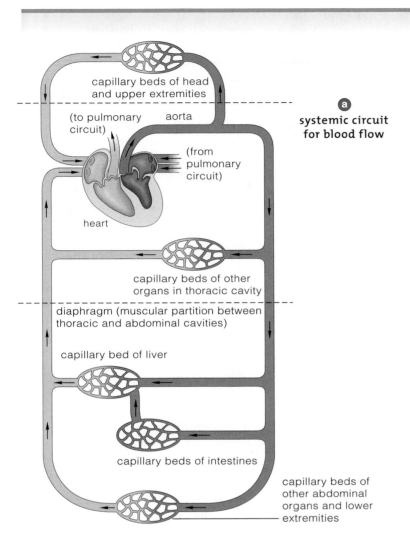

a systemic circuit for blood flow

capillary beds of head and upper extremities

(to pulmonary circuit)   aorta

(from pulmonary circuit)

heart

diaphragm (muscular partition between thoracic and abdominal cavities)

capillary beds of other organs in thoracic cavity

capillary bed of liver

capillary beds of intestines

capillary beds of other abdominal organs and lower extremities

b pulmonary circuit for blood flow

right pulmonary artery   left pulmonary artery

capillary bed of right lung   pulmonary trunk   capillary bed of left lung

(to systemic circuit)

(from systemic circuit)   pulmonary veins

heart

Figure 33.9 (**a,b**) Systemic and pulmonary circuits for blood flow through the human cardiovascular system. Blood vessels carrying oxygenated blood are color-coded *red*. Those carrying oxygen-poor blood are color-coded *blue*.

that enter the body by way of the respiratory and digestive systems. The system also transports carbon dioxide and other waste products to the respiratory and urinary systems for disposal. Interactions among organ systems help keep operating conditions of the internal environment within tolerable ranges. That is the state we call homeostasis (Sections 24.1 and 24.3).

In the pulmonary circuit of a human cardiovascular system, oxygen-poor blood flows from the heart's right half, through both lungs, then flows back to the heart. In the lungs, it takes up oxygen and gives up carbon dioxide.

In the systemic circuit, oxygenated blood flows from the heart's left half and aorta to capillary beds of all metabolically active regions. There it gives up oxygen and takes up carbon dioxide, then moves to the heart's right half.

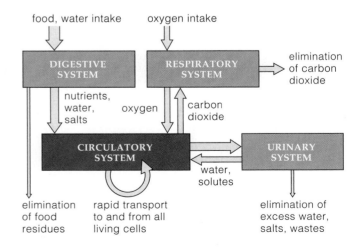

food, water intake   oxygen intake

DIGESTIVE SYSTEM   RESPIRATORY SYSTEM   elimination of carbon dioxide

nutrients, water, salts   oxygen   carbon dioxide

CIRCULATORY SYSTEM   URINARY SYSTEM

water, solutes

elimination of food residues   rapid transport to and from all living cells   elimination of excess water, salts, wastes

Figure 33.10 Functional links between the circulatory system and other organ systems that maintain the internal environment.

33.6 The Heart Is a Lonely Pumper

The human heart is a durable pump that spontaneously beats 2.5 billion times in a seventy-year life span.

HEART STRUCTURE

The human heart's durability arises from its structure. Outermost is a pericardium, a tough, double-layered, connective tissue sac that anchors the heart to nearby structures (Figure 33.11). In between the two layers is a fluid that lubricates the heart during its perpetual wringing motions. The sac's inner layer is part of the heart wall. Most of this wall is myocardium—cardiac muscle cells tethered to elastin and collagen fibers. The densely crisscrossed fibers serve as a "skeleton" against which contractile force is applied. A type of epithelial sheet called endothelium lines the heart's inner wall. Coronary arteries branch off the aorta and deliver nutrients and oxygen to a capillary bed that services only cardiac muscle cells.

Each half of the heart has two chambers: an atrium (plural, atria) that receives blood, and a ventricle that pumps it out. Between each atrium and ventricle is an atrioventricular (AV) valve. Between each ventricle and the artery leading away from it is a semilunar valve. Both types are *one-way* valves. Fluid pressure alternately forces them open and shut—which helps keep the blood moving in the forward direction only.

The heart's four chambers go through a sequence of contraction and relaxation called the **cardiac cycle**. First, relaxed atria fill. AV valves open under the fluid pressure. Blood flows into the ventricles, which fill as the atria contract (Figure 33.12). When the ventricles contract, the fluid pressure rises so sharply above the pressure in the great arteries that both semilunar valves open and blood flows out. As ventricles relax, the already filling atria start a new cycle. Thus, atrial contraction only helps fill the ventricles. *Contraction of the ventricles is the driving force for blood circulation.*

a

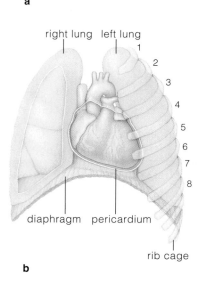

right lung left lung

1
2
3
4
5
6
7
8

diaphragm pericardium

rib cage

b

superior vena cava (from head, upper limbs)

arch of aorta

trunk of pulmonary arteries

right semilunar valve (shown closed); to the pulmonary trunk

left semilunar valve (shown closed); to aorta

right pulmonary veins (from lungs)

left pulmonary veins (from lungs)

right atrium

left atrium

right AV valve (shown open)

left AV valve (shown open)

right ventricle

left ventricle

(muscles that prevent valve from everting)

endothelium and underlying connective tissue

inferior vena cava (from trunk, legs)

septum (partition between heart's two halves)

heart's apex

inner layer of pericardium

myocardium

c

Figure 33.11 (**a**) Photograph of the human heart and (**b**) its location in the thoracic cavity. (**c**) Cutaway view of the heart, showing its wall and internal organization.

HOW DOES CARDIAC MUSCLE CONTRACT?

Cardiac muscle (heart muscle) has orderly arrays of sarcomeres, or contractile units. Like skeletal muscle, it contracts by a sliding-filament mechanism driven by ATP, which many mitochondria provide (Sections 28.3 and 32.4). Unlike skeletal or smooth muscle, its cells are short and abut at their branching ends. Near these ends, gap junctions allow signals that call for contraction to spread swiftly from one cell to the next in cardiac muscle tissue (Figure 33.13*a*).

A small number of specialized cardiac muscle cells *do not* contract. Seventy or so times a minute, these cells initiate and propagate waves of excitation that travel from atria to ventricles. The cells' synchronized excitation directs the heart's pumping.

An electric disturbance starts at the SA (sinoatrial) node, a clump of specialized cells in the right atrium's wall (Figure 33.13*b*). It reaches the AV node, another cluster of cell bodies. This is the only electric bridge between the heart's atria and its ventricles, which are insulated everywhere else by connective tissue.

From the AV node, excitation travels along special conducting cells that run through the heart's septum, the partition between its two halves (Figure 33.13*b*). Excitation is conducted from the heart apex right up the ventricle walls. Ventricles respond by contracting in an upward twisting from the apex. The contraction ejects blood into the aorta and pulmonary arteries.

The SA node is the first to fire signals to contract, so it functions as the **cardiac pacemaker**. Its rhythmic, spontaneous signals are the basis for the normal rate of heartbeat. The nervous system can only *adjust* the rate and strength of contractions set by this natural pacemaker. Even when all nerves leading to the heart are severed, as by a bad injury, the pacemaker keeps the heart beating. When the SA node is not working properly, cardiac arrest may follow. A defibrillator works by resetting the SA node.

A double-layer sac that holds lubricating fluid wraps around the outer heart wall. Endothelium lines the inner wall.

Under spontaneous, rhythmic signals from the cardiac pacemaker, the branching, abutting cardiac muscle cells contract in synchrony, almost as if they were a single unit.

A partition (septum) divides the four-chambered heart into two halves, each with an atrium and a ventricle. Ventricular contraction drives blood flow away from the heart.

The SA node, the cardiac pacemaker, sets the rate at which the heart beats. The nervous system can only adjust the rate of its rhythmic, spontaneous signals to contract.

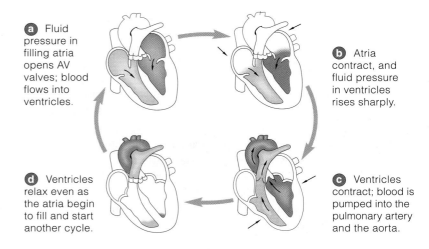

a Fluid pressure in filling atria opens AV valves; blood flows into ventricles.

b Atria contract, and fluid pressure in ventricles rises sharply.

c Ventricles contract; blood is pumped into the pulmonary artery and the aorta.

d Ventricles relax even as the atria begin to fill and start another cycle.

Figure 33.12 Cardiac cycle. Blood and heart movements make a "lub-dup" sound at the chest wall. At each "lub," AV valves are closing while the ventricles are contracting. At each "dup," the semilunar valves are closing as the ventricles are relaxing.

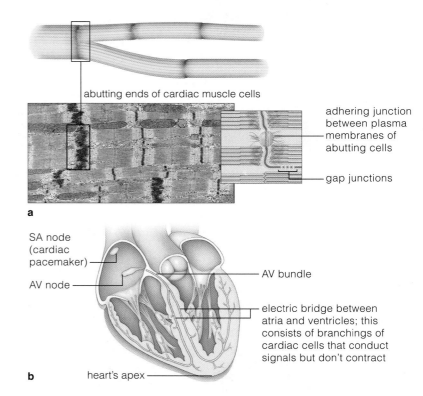

abutting ends of cardiac muscle cells

adhering junction between plasma membranes of abutting cells

gap junctions

a

SA node (cardiac pacemaker)

AV node

AV bundle

electric bridge between atria and ventricles; this consists of branchings of cardiac cells that conduct signals but don't contract

heart's apex

b

Figure 33.13 (**a**) Cardiac muscle cells, structurally held together by tight junctions and adhering junctions at their abutting ends. Gap junctions near the abutting ends offer low electrical resistance and enhance the rapid spread of excitation from cell to cell.

(**b**) Location of noncontracting cardiac muscle cells, which initiate and propagate signals that make the heart beat.

33.7 Pressure, Transport, and Flow Distribution

During any specified interval, the pressure driving blood through your body depends on how fast and how strongly the heart is beating. It depends also on the total resistance to flow through the vascular system.

Figure 33.14 compares the structure of blood vessels. **Arteries** are rapid-transport vessels for blood pumped out of the heart's ventricles. They grade into **arterioles**, smaller vessels where controls over the distribution of blood flow operate. These grade into **capillaries**, the small vessels that form diffusion zones. **Venules** are also small vessels between capillaries and veins. **Veins** are large vessels that transport blood to the heart; they also are blood volume reservoirs.

a Artery

b Arteriole

c Capillary

d Vein

Figure 33.14 Comparison of the structure of human blood vessels. Capillaries grade into venules (not shown), which are functionally like them. These drawings are not drawn to the same scale.

Blood pressure is fluid pressure imparted to blood by ventricular contractions. Pressure is highest in the contracting ventricles, still high at the start of arteries, and lowest in the relaxed atria (Figure 33.15). In both the pulmonary and systemic circuits, the difference in pressure between two points sets the flow rate. The beating heart establishes high pressure at the start of a circuit. Blood rubs against the vessel walls, and that friction impedes its flow. Resistance to flow depends mainly on how *wide* the blood vessels are. Simply put, resistance rises as tubes narrow. A twofold decrease in the radius of a blood vessel will increase the resistance to flow sixteenfold.

RAPID TRANSPORT IN ARTERIES

With their large diameter and low resistance to flow, arteries are fast, efficient transporters of oxygenated blood. They also are pressure reservoirs that smooth out pulsations in pressure that are generated by each cardiac cycle. An artery's thick, muscular, elastic wall bulges from the large volume of blood forced into it by ventricular contraction. As the elastic wall recoils, it forces oxygen-rich blood farther through the circuit between contractions.

DISTRIBUTING BLOOD FLOW

No matter what you are doing, your lungs normally receive all blood from the heart's right half. But the flow distribution to one organ or another along the systemic circuit can vary. When you are just lounging around, the flow distribution from the heart's left half is probably close to the values shown in Figure 33.16.

What happens when you go for a run? A greater volume of blood is diverted to your skeletal muscles

Figure 33.15 Plot of the decline in fluid pressure for a volume of blood moving through the systemic circuit.

relative to the volume being distributed to your skin, kidneys, and gut. Like traffic cops, arterioles guide the flow according to instructions from a command post. The nervous and endocrine systems send signals that act on rings of smooth muscle cells in arteriole walls (Figure 33.14b). Specific signals make these cells relax, causing **vasodilation**; the diameter of the blood vessel enlarges. Other signals make them contract, causing **vasoconstriction**; the diameter shrinks. Dilate all arterioles in one tissue, and more blood flows to it. Constrict them, and less blood flows to the tissue.

Arterioles also responds to metabolic events that shift concentrations of substances in a tissue. Such local chemical changes are "selfish" in that they invite or divert blood flow to meet a tissue's own metabolic needs. For instance, as you run, your skeletal muscle cells use up oxygen, and the levels of carbon dioxide, hydrogen and potassium ions, and other solutes rise. The ion levels cause arterioles in the skeletal muscle tissues to dilate. More blood flows past, delivering more raw materials and carting away cell products and metabolic wastes. When skeletal muscles relax, demand for oxygen declines. Now the oxygen level rises and the arterioles constrict.

CONTROLLING BLOOD PRESSURE

We generally measure blood pressure at an upper arm's brachial artery (Figure 33.17). In each cardiac cycle, *systolic* (peak) pressure is exerted by contracting ventricles against the artery wall. *Diastolic* pressure, the lowest arterial blood pressure of the cardiac cycle, is reached when ventricles are relaxed. An average resting blood pressure is recorded as 120/80 (that is the systolic pressure/diastolic pressure).

Blood pressure depends on the total blood volume, how much blood is being pumped out of the heart (cardiac output), and arteriole resistance. *Baroreceptors* in the wall of some arteries, such as the carotids, can signal a control center in the brain when pressure rises or falls. In response, the brain calls for changes in cardiac output and arteriole diameter. This reflex response is the main short-term control over blood pressure. Long-term controls, exerted at the kidneys, adjust the total volume and composition of the blood.

Blood pressure depends on the rate and strength of heartbeat and on resistance to flow through blood vessels. It is greatest in contracting ventricles and at the start of arteries.

The distribution of blood flow is controlled mainly through adjustments in the diameter of arterioles in different tissues.

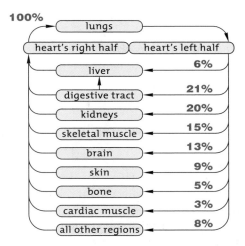

Figure 33.16 Distribution of the heart's output in people napping. How much flows through a given region is adjusted through selective vasodilation and vasoconstriction at many arterioles all along the systemic circuit.

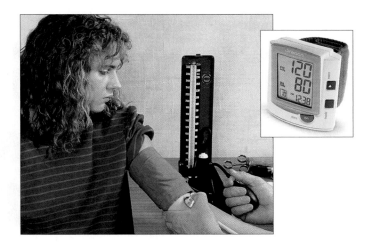

Figure 33.17 Measuring blood pressure. A hollow inflatable cuff attached to a pressure gauge is wrapped around the upper arm. A stethoscope is placed over the brachial artery, just below the cuff.

The cuff is inflated with air to a pressure above the highest pressure of the cardiac cycle, when ventricles contract. Above this pressure, you will not hear sounds through the stethoscope, because no blood is flowing through the vessel.

Air in the cuff is slowly released until the stethoscope picks up soft tapping sounds. The sounds are caused by blood spurting into the artery under the force of the strongest ventricular contraction. When sounds start, the gauge's value is typically about 120 mm Hg. That measured amount of pressure would force mercury (Hg) to move up 120 millimeters in a narrow glass column of standardized size.

More air is released from the cuff. Eventually the sounds stop. Blood is now flowing continuously, even when the ventricles are the most relaxed. The pressure when the sounds stop is the lowest during a cardiac cycle, usually about 80 mm Hg.

Monitors, such as the one shown in the inset, are now available that automatically record the systolic/diastolic blood pressure.

33.8 Diffusion at Capillaries, Then Back to the Heart

Every capillary bed is a diffusion zone for exchanges between blood and interstitial fluid. Living cells in any body tissue die quickly when deprived of the exchanges. Brain cells start dying off within four minutes.

CAPILLARY FUNCTION

Between 10 billion and 40 billion capillaries service the human body. Collectively, they offer a tremendous surface area for gas exchanges. At least one is next to living cells in nearly all tissues. The proximity is vital, because diffusion distributes molecules and ions very slowly, and not very far.

Red blood cells are eight micrometers across. They have to squeeze single file through the capillaries. The squeeze puts oxygen-transporting red blood cells and solutes in plasma in direct contact with the exchange area—the capillary wall—or only a very short distance away from it.

Each capillary is a cylindrical sheet of endothelial cells, organized as a single layer inside a basement membrane. So how do oxygen and carbon dioxide get from blood, across these endothelial cells, and into interstitial fluid? Like many other small, lipid-soluble substances, they diffuse through the lipid bilayer of each cell's plasma membrane, through its cytoplasm, then through the bilayer on the other side. Certain proteins cross by way of endocytosis and exocytosis.

In most capillaries, the adjoining cells have narrow clefts between them. Ions and other small, water-soluble substances enter and leave through the clefts between endothelial cells. So do white blood cells, as the next chapter explains. In capillaries inside the brain, tight junctions join endothelial cells, and clefts are nonexistent. Substances cannot "leak" in between the cells; they have to pass through them. The junctions help maintain the blood–brain barrier, described in Section 29.7.

Figure 33.18 Bulk flow in a capillary bed. Fluid crosses a capillary wall by way of ultrafiltration and reabsorption. (**a**) At the capillary's arteriole end, a difference between blood pressure and interstitial fluid pressure forces out some plasma but few plasma proteins. Bulk flow moves plasma through clefts between endothelial cells of the capillary wall. Ultrafiltration is the bulk flow of fluid *out* of a capillary.

(**b**) Reabsorption is the osmotic movement of some interstitial fluid *into* the capillary. It happens when the water concentration between interstitial fluid and the plasma differs. Plasma, with its dissolved proteins, has a greater solute concentration and therefore a lower water concentration. Reabsorption near the end of a capillary bed tends to balance ultrafiltration at the start of it. Normally there is only a small *net* filtration of fluid, which the lymphatic system returns to the blood.

ARTERIOLE END OF CAPILLARY BED	
Outward-Directed Pressure:	
Hydrostatic pressure of blood in capillary:	35 mm Hg
Osmosis due to interstitial proteins:	28 mm Hg
Inward-Directed Pressure:	
Hydrostatic pressure of interstitial fluid:	0
Osmosis due to plasma proteins:	3 mm Hg
Net Ultrafiltration Pressure:	
(35 − 0) − (28 − 3) = 10 mm Hg	
ULTRAFILTRATION FAVORED	

VENULE END OF CAPILLARY BED	
Outward-Directed Pressure:	
Hydrostatic pressure of blood in capillary:	15 mm Hg
Osmosis due to interstitial proteins:	28 mm Hg
Inward-Directed Pressure:	
Hydrostatic pressure of interstitial fluid:	0
Osmosis due to plasma proteins:	3 mm Hg
Net Reabsorption Pressure:	
(15 − 0) − (28 − 3) = −10 mm Hg	
REABSORPTION FAVORED	

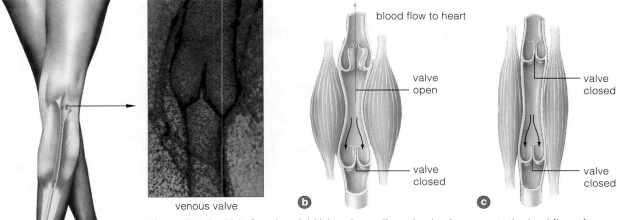

Figure 33.19 Vein function. (**a**) Valves in medium-sized veins prevent the backflow of blood. Neighboring skeletal muscles helps raise fluid pressure inside a vein. (**b**) These muscles bulge into the vein when they contract, which increases the pressure that keeps blood flowing forward. (**c**) When muscles relax, venous valves shut and stop backflow.

Diffusion is not the only process acting at capillary beds. **Bulk flow** is the movement of water and solutes in response to a difference in fluid pressure between regions. It does not affect diffusion much. But it helps maintain the fluid distribution between blood and interstitial fluid. Figure 33.18 illustrates how water and solutes in blood and interstitial fluid influence the direction of bulk flow. At the start of a capillary bed, a small amount of protein-free plasma is pushed out in bulk through clefts in the capillary wall. This process is called **ultrafiltration**. Farther on, tissue fluid moves back into capillaries through the clefts in the capillary wall. This process is called **capillary reabsorption**.

Normally, there is a tiny *net* outward flow from a capillary bed, which the lymphatic system returns to blood. High blood pressure increases the amount of fluid leaving capillaries. Too much fluid in interstitial spaces is called *edema*. Commonly, gravity causes the fluid to pool in foot and ankle tissues. Edema often follows physical exercise, because arterioles in many tissues have dilated. It also can result from obstructed veins or heart failure. It is dramatic in *elephantiasis,* when roundworm infection slows the return of fluid to lymphatic vessels (Section 22.8).

VENOUS PRESSURE

Blood moving out of capillary beds enters venules, or "little veins," which merge into large-diameter veins. Functionally, venules are a bit like capillaries. Certain solutes diffuse across their thin walls.

Veins are large-diameter, low-resistance transport tubes that carry blood toward the heart. Within many veins, especially in the leg, valves prevent backflow (Figure 33.19). When gravity causes blood in the vein to reverse direction, the valves are pushed shut.

The vein wall can bulge greatly under pressure, far more so than an arterial wall. Thus, veins can act as reservoirs for variable volumes of blood. About 60 percent of the total blood volume is in the veins.

The vein wall contains some smooth muscle. When blood must circulate faster, as during exercise, the smooth muscle contracts. The vein wall stiffens, so the vessel cannot bulge as much. As pressure in the lumen increases, more blood is driven toward the heart. Also, when limbs move, contracting skeletal muscles bulge against nearby veins. This helps raise the venous pressure that moves blood toward the heart (Figure 33.19). Rapid breathing also contributes to an increased venous pressure. Inhaling pushes air down on internal organs, which alters the pressure gradient between the heart and the veins.

Sometimes venous valves lose their elasticity. The veins become enlarged and bulge at the skin surface. Such *varicose veins* are common in legs. In the area around the anus, they are called *hemorrhoids.* Exercise and maintaining an ideal weight reduce the risk.

Capillary beds are diffusion zones for exchanges between blood and interstitial fluid. Here, bulk flow contributes to the fluid balance between blood and interstitial fluid.

Venules overlap somewhat with capillaries in function. Veins are highly distensible blood volume reservoirs, where the flow volume back to the heart can be adjusted.

33.9 Cardiovascular Disorders

A stroke results from a blood clot or ruptured blood vessel in the brain. A heart attack results when blood vessels that service the heart are narrowed or clogged. High blood pressure and atherosclerosis increase the risks, many of which can be minimized by changes in life-styles.

GOOD CLOT, BAD CLOT

Small blood vessels are vulnerable to ruptures, cuts, and other injuries. **Hemostasis** is a multistep process that stops blood loss and lays down a framework for repairs. In the vascular phase of the process, smooth muscle in a damaged vessel wall goes into spasm, an automatic contractile response. Vasoconstriction may briefly cut off the blood flow. In the platelet phase, platelets temporarily close the breach by clumping together. They release substances that prolong the spasm and attract more platelets. In the coagulation phase, plasma proteins help convert blood to a gel. The fibrinogens (rodlike proteins) react to collagen fibers in the damaged vessel. They stick together as long, insoluble threads, which stick to the exposed collagen. The fibers all form a net that traps blood cells and platelets, as in Figure 33.20. The entire mass is a *blood clot.* The clot retracts and forms a compact mass that seals the breach in the blood vessel wall.

Clot formation is vital for repairing blood vessels. But clots cause problems when they completely block blood flow through a vessel, as you will see shortly.

Figure 33.20 Fibrous protein net that helps blood clots form.

A SILENT KILLER—HYPERTENSION

Hypertension is a chronically high blood pressure even during rest. Blood pressure stays above 140/90, often for unknown reasons. This chronic condition is called a "silent killer," because outward symptoms do not always show up. Heredity may be a factor; the disorder tends to run in families. Diet and lack of regular exercise are factors; high salt intake raises blood pressure and adds to the heart's workload in some people. The heart may enlarge and fail to pump efficiently. Also, high blood pressure may add to the "hardening" of arterial walls, which slows delivery of oxygen to the brain, heart, and other vital organs.

Of 23 million hypertensive Americans, most do not seek treatment. About 180,000 die each year.

ATHEROSCLEROSIS

In *arteriosclerosis*, arteries thicken and lose elasticity. With *atherosclerosis*, the condition worsens as lipids accumulate in the arterial wall and narrow the inside diameter. You have probably heard that cholesterol plays a role in this "hardening of the arteries." Your body uses some cholesterol to make cell membranes, myelin sheaths, bile salts, and steroid hormones. The liver makes enough cholesterol for cell structure and function, but more cholesterol is absorbed from food in the gut. People differ in how the excess is handled.

In the blood, most cholesterol is bound to proteins as *low-density lipoproteins*, or LDLs. Most cells take up this form of cholesterol. A smaller amount is bound as *high-density lipoproteins*, or HDLs, which the liver takes up. HDL is metabolized in the liver, and the residues are eliminated by way of bile, secreted back to the gut. As levels of LDLs rise, so does the risk of atherosclerosis. The first sign of trouble is a build-up of lipids in an artery's endothelium (Figure 33.21*b*). As lipids accumulate, smooth muscle proliferates, the

wall of artery, cross-section

unobstructed lumen of normal artery

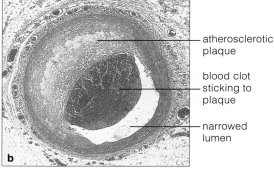

atherosclerotic plaque

blood clot sticking to plaque

narrowed lumen

Figure 33.21 Sections from (**a**) a normal artery and (**b**) an artery with a lumen narrowed by an atherosclerotic plaque. A clot clogged this one.

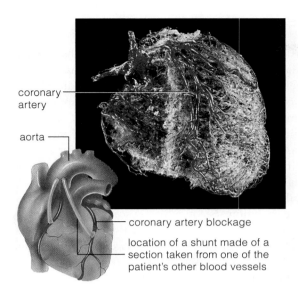

coronary artery

aorta

coronary artery blockage

location of a shunt made of a section taken from one of the patient's other blood vessels

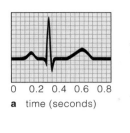

0   0.2 0.4 0.6 0.8

a   time (seconds)

Figure 33.23 **(a)** An ECG of a single, normal beat of the human heart. **(b–d)** Three arrhythmia recordings.

bradycardia (here, 46 beats per minute)

b

tachycardia (here, 136 beats per minute)

c

ventricular fibrillation

d

Figure 33.22   Coronary arteries and other blood vessels of the heart. Resins were injected into them, then the rest of the cardiac tissues were dissolved, leaving this accurate, three-dimensional corrosion cast. The sketch shows two coronary bypasses (artificially colored *green*), which extend from the aorta and past two clogged parts of the coronary arteries.

site becomes inflamed, and fibrous connective tissue is laid down. The resulting *atherosclerotic plaque* makes the arterial wall bulge inward, which narrows this blood vessel's diameter.

Hardened plaque can rupture the artery wall. If the resulting blood clot stays put, it is a *thrombus*. If it is dislodged and travels in blood, it is an *embolus*, which can clog small vessels. Coronary arteries are at high risk. If they narrow by 25 percent, they invite *angina pectoris* (chest pain) or a heart attack. During an attack, some of the cardiac muscle cells die from lack of blood. A first attack is fatal about 30 percent of the time. Survivors often have a badly weakened heart.

With *coronary bypass surgery*, a section of blood vessel from elsewhere in the body is stitched to the aorta and to the coronary artery below a clogged region (Figure 33.22). With *laser angioplasty*, laser beams directed at the plaques vaporize them. With *balloon angioplasty*, a small balloon inflated within a blocked artery flattens the plaques.

RHYTHMS AND ARRYTHMIAS

As described in Section 33.6, electric signals from the SA node stimulate cardiac muscle contraction. An *electrocardiogram*, or ECG, records the electrical activity of a beating heart (Figure 33.23*a*).

ECGs can reveal *arrhythmias*, or abnormal rhythms in heartbeat (Figure 33.23*b–d*). Arrhythmias are not always dangerous. Endurance athletes may experience *bradycardia*, a below-average resting cardiac rate. In response to ongoing exercise, their nervous system has adjusted the cardiac pacemaker's rate of firing downward. Exercise or stress commonly causes 100 or more heartbeats a minute, or *tachycardia*.

In *atrial fibrillation*, the atria quiver instead of contracting normally, which increases the chance of blood clots and stroke. *Ventricular fibrillation* is the most dangerous arrhythmia. It caused the collapse of Tammy Higgins, as described at the start of this chapter. Ventricles flutter and their pumping action falters or halts. The individual loses consciousness and faces death. A shock from a defibrillator might be able to restore the heart's normal rhythm.

RISK FACTORS

Cardiovascular disorders are the leading cause of death in the United States. Each year, about 40 million people experience cardiovascular disorders, and each year about a million die. Nine factors top the list of risks, and tobacco smoking tops them all. Other factors include genetic predisposition to a heart attack, hypertension, a high blood level of cholesterol, obesity, and diabetes mellitus, a condition described in Section 31.6. Advancing age also is a risk factor; the older you get, the greater the risk. Physical inactivity increases the risk. Regular exercise can lower it, even if the activity is not particularly strenuous. Gender is a factor; until about age fifty, males are at greater risk.

33.10 Connections With the Lymphatic System

We conclude this chapter with a brief look at how the lymphatic system interacts with blood circulation. Think of this section as a bridge to the next chapter, on immunity, because the lymphatic system also helps defend the body against injury and attack.

LYMPH VASCULAR SYSTEM

A portion of the lymphatic system, called the **lymph vascular system,** consists of many tubes that collect and deliver water and solutes from interstitial fluid to ducts of the circulatory system. Its main components are lymph capillaries and vessels (Figure 33.24). Tissue fluid that moves into these vessels is called **lymph**.

The lymph vascular system serves three functions. First, its vessels are drainage channels for water and plasma proteins that have leaked out from the blood at capillary beds and must be delivered back to the blood circulation. Second, the system takes up fats that the body has absorbed from the small intestine and delivers them to the general circulation (Section 36.4). Third, it delivers pathogens, foreign cells, and cellular debris from tissues to the lymph vascular system's disposal centers, the lymph nodes.

lymph capillary   interstitial fluid

flaplike "valve" made of overlapping cells at tip of lymph capillary

b   capillary bed

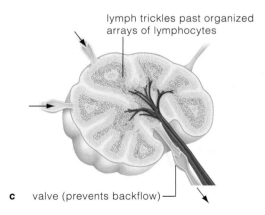

lymph trickles past organized arrays of lymphocytes

c   valve (prevents backflow)

TONSILS
Defense against bacteria and other foreign agents

RIGHT LYMPHATIC DUCT
Drains right upper portion of the body

THYMUS GLAND
Site where certain white blood cells acquire means to chemically recognize specific foreign invaders

THORACIC DUCT
Drains most of the body

SPLEEN
Major site of antibody production; disposal site for old red blood cells and foreign debris; site of red blood cell formation in the embryo

SOME OF THE LYMPH VESSELS
Return excess interstitial fluid and reclaimable solutes to the blood

SOME OF THE LYMPH NODES
Filter bacteria and many other agents of disease from lymph

BONE MARROW
Marrow in some bones is production site for infection-fighting blood cells (as well as red blood cells and platelets)

a

Figure 33.24 (**a**) Components of the human lymphatic system and their functions. Not shown are the patches of lymphoid tissue in the small intestine and in the appendix. (**b**) Diagram of lymph capillaries at the start of a drainage network, the lymph vascular system. (**c**) Cutaway view of a lymph node. Its inner compartments are packed with organized arrays of infection-fighting white blood cells.

The lymph vascular system starts at all capillary beds. There, excess fluid enters the lymph capillaries. These capillaries have no obvious entrance; water and solutes move into clefts between cells that form their flaplike valves. As you can see from Figure 33.24b, the endothelial cells overlap here. Lymph capillaries merge and become larger diameter lymph vessels. The lymph vessels have some smooth muscle in their wall and valves that prevent backflow. These vessels converge into collecting ducts that drain into veins in the lower neck.

LYMPHOID ORGANS AND TISSUES

The other portion of the lymphatic system has roles in the body's defense responses to injury and attack. We call them *lymphoid* organs and tissues. The major types are the lymph nodes, spleen, and thymus, as well as the tonsils, and patches of tissue in the wall of the small intestine and appendix.

Lymph nodes are strategically located at intervals along lymph vessels (Figure 33.24c). Before entering blood, the lymph trickles through at least one node for filtration. Masses of lymphocytes take up stations in the nodes after they form in bone marrow. When they contact cellular debris or an invader, they divide rapidly and form large armies that destroy it.

The **spleen** is the largest lymphoid organ. In an average adult, it is about the size of a fist. The spleen functions as a site of red blood cell formation only in embryos. After birth, it filters pathogens and used-up red cells and platelets from branching blood vessels that thread through it. Phagocytic white cells in the spleen engulf and digest defunct cells and alert the body to invaders. Lymphocytes in the spleen produce antibodies. Even so, we can survive without it. If the spleen is damaged by trauma and must be removed, a heightened risk of infection is the only ill effect.

In the **thymus gland**, T lymphocytes differentiate in ways that allow them to recognize and respond to particular pathogens. The thymus also produces the hormones that influence these actions. It is central to immunity, the focus of the next chapter.

The lymph vascular portion of the lymphatic system consists of many vessels that start in capillary beds. They return water and solutes from tissue fluid to blood, deliver absorbed fats to the bloodstream, and deliver pathogens to lymph nodes.

The lymphatic system also includes lymph nodes and other lymphoid organs that have specific roles in defending the body against tissue damage and infectious diseases.

Summary

Section 33.1 A circulatory system moves substances to and from the interstitial fluid that bathes cells. All vertebrates have a closed circulatory system, with blood confined in a heart and blood vessels. As vertebrates moved onto land and evolved lungs, the circulatory system changed in ways that made oxygen delivery to cells more efficient. Fluid leaking from blood vessels enters the lymphatic system, which filters it and then returns it to the circulatory system.

Section 33.2 Blood consists of plasma, blood cells, and platelets. Plasma is mostly water with dissolved ions and molecules. Red blood cells, packed with hemoglobin, function in the transport of oxygen and, to a lesser extent, carbon dioxide. Many kinds of white blood cells function in housekeeping and defense. Platelets release substances that initiate blood clotting. All blood cells and platelets arise from stem cells in bone marrow.

Section 33.3 In a blood disorder, an individual has too many, too few, or abnormal red or white blood cells.

Section 33.4 Markers on the surface of red blood cells determine an individual's ABO and Rh blood type. The ABO blood types of a donor and recipient must be compatible to avoid blood transfusion problems. Rh type incompatibilities can cause problems for offspring.

Section 33.5 The human heart pumps blood through two separate circuits. In the pulmonary circuit, oxygen-poor blood from the heart's right half flows to the lungs, picks up oxygen, then flows to the heart's left half. In the systemic circuit, the oxygen-rich blood flows from the heart's left half to all body tissues, then oxygen-poor blood flows to the heart's right half. Most blood flows through one capillary system. In one route, it flows through intestinal capillaries, then liver capillaries. This gives the liver time to metabolize and store nutrients. The liver also neutralizes some bloodborne toxins.

Section 33.6 The human heart is a double pump that beats constantly. Each half has two chambers (atrium and ventricle). Blood flows from veins into atria, ventricles, then the aorta and pulmonary arteries. Ventricular contraction is the driving force for blood circulation through the body.

The cardiac conduction system is the basis of the heart's beating. Its SA node spontaneously produces rhythmic electric signals for contraction; it is the cardiac pacemaker. Waves of excitation wash over the atria, then down the heart's septum, then up the ventricles. The nervous system can adjust the rate and the strength of contractions, but it does not initiate them.

Section 33.7 Blood pressure is highest in contracting ventricles, slightly lower in the arteries, and falls more as blood flows through the systemic or pulmonary circuit. It is lowest in relaxed atria. It depends on the strength and rate of heartbeat and on resistance to flow in different blood vessels. Flow distribution to different tissue regions depends mainly on adjustments in arteriole diameter.

Section 33.8   Capillary beds are diffusion zones where blood and interstitial fluid exchange substances. Ultrafiltration pushes a small amount of fluid out of capillaries. Fluid moves back in by capillary reabsorption. Normally, both processes are almost balanced, with just a small net outward flow of fluid from a capillary bed.

Venules overlap capillaries in function. Veins are rapid-transport vessels and a blood volume reservoir for adjusting the flow volume back to the heart.

Section 33.9   Hemostasis is a multistep process that stops blood flow from small vessels after injury. It results in clot formation, which normally is a good thing but can be bad in some cardiovascular disorders. The most common disorders are hypertension (chronic high blood pressure), atherosclerosis, heart attacks, strokes, and arrythmias. Regular exercise, maintaining a normal body weight, and not smoking lower the risks for most people.

Section 33.10   The vascular portion of the lymphatic system takes up excess water and plasma proteins from interstitial fluid, as well as absorbed fats, and transports them to blood. It transports pathogens and foreign material to lymph nodes and the spleen.

The lymphoid organs and tissues are sites of maturation for some white blood cells. Some are battlegrounds where organized arrays of these cells screen lymph and battle disease-causing agents.

Self-Quiz   *Answers in Appendix III*

1. Cells directly exchange substances with _____ .
   a. blood vessels   c. interstitial fluid
   b. lymph vessels   d. both a and b

2. All vertebrates have _____ .
   a. an open circulatory system
   b. a closed circulatory system
   c. a four-chambered heart
   d. both b and c

3. Which are not found in the blood?
   a. plasma
   b. blood cells and platelets
   c. gases and dissolved substances
   d. All of the above are found in blood.

4. A person who has type O blood _____ .
   a. can receive a transfusion of blood of any type
   b. can donate blood to a person of any blood type
   c. can donate blood only to a person of type O
   d. cannot be a blood donor
   e. both a and b

5. In the blood, most oxygen is transported _____ .
   a. in red blood cells   c. bound to hemoglobin
   b. in white blood cells   d. both a and c

6. Blood flows directly from the left atrium to _____ .
   a. the aorta   c. the right atrium
   b. the left ventricle   d. the pulmonary arteries

7. Contraction of _____ drives the flow of blood through the aorta and pulmonary arteries.
   a. the atria   c. the ventricles
   b. arterioles   d. skeletal muscle

8. Blood pressure is highest in the _____ and lowest in the _____ .
   a. arteries, veins   c. veins, arteries
   b. arterioles, venules   d. capillaries, arterioles

9. At rest, the largest volume of blood is in the _____ .
   a. arteries   c. veins
   b. capillaries   d. arterioles

10. Which is not a function of the lymphatic system?
    a. filters out pathogens
    b. returns fluid to the circulatory system
    c. helps certain white blood cells mature
    d. distributes oxygen to the tissues

11. Match the components with their functions.
    ____ capillary bed   a. filters out pathogens
    ____ lymph node   b. cardiac pacemaker
    ____ blood   c. main blood volume
    ____ ventricle   reservoir
    ____ SA node   d. largest artery
    ____ veins   e. fluid connective tissue
    ____ aorta   f. zone of diffusion
                           g. contractions drive
                              blood circulation

12. Label the components of the heart diagram below.

13. Identify the blood vessels below and describe the function of each.

Critical Thinking

1. The highly publicized deaths of a few airline travelers have led to warnings about *economy-class syndrome*. The theory is that sitting motionless for long periods on flights allows blood to pool and clots to form in the legs. Low oxygen levels in airline cabins may increase clotting. If a clot becomes large enough to block blood flow or breaks free and is carried to the lungs or the brain, the outcome can be deadly.

There might be a time lag between clot formation and health problems, so a connection to air travel might easily have been overlooked. Studies are now under way to determine whether economy-class travel represents a significant risk. Given what you know about blood flow in the veins, explain why periodically getting up and moving around in the plane's cabin during a long flight may lower the risk of clot formation.

2. Some molecules of the bacterium called *Streptococcus pyrogenes* are similar to molecules on plasma membranes of cells in connective tissues throughout the body. When the bacterium causes throat infections, the weapons called antibodies go to work against the invader. But they also go to work on connective tissues of the heart, joints, and elsewhere. Prolonged inflammation over the course of a few years or even decades leads to *rheumatic heart disease*. The heart valves become damaged or deformed. Explain how this disease affects the heart's function and what problems might arise as a consequence.

3. Consider the micrograph in Figure 33.25. It shows red blood cells moving through a blood vessel. What type of vessel is this? Explain how you came to this conclusion.

4. Like other insects, the fruit fly (*Drosophila melanogaster*) has an open circulatory system. The transport medium is hemolymph, not blood. Contraction of one of the main vessels of the system drives the flow of hemolymph through the tiny fly body.

In 1993, researchers found that normal development of the fly "heart" requires the presence of a gene they called *tinman*. The gene's name is a reference to the character in *The Wizard of Oz* who had no heart. If the *tinman* gene is mutated, no heart forms in the embryonic fly, which dies. Genes having a similar base sequence have been found in zebrafish, the African clawed toad, chickens, mice, and humans. In all of these evolutionarily distant species, mutant versions result in abnormal heart development. In humans, mutations cause some genetic defects in the partition dividing the heart chambers and in the cardiac conduction system. Does it surprise you that a single gene would have such a major effect on the development of hearts or heartlike organs in so many different organisms? Explain your answer.

5. Miranda is nineteen, a college student, and a dedicated runner. About a month ago, a friend convinced her to switch to a *strict vegan diet*. She avoids all foods and other products that contain materials from animals, including eggs and milk. As she began training for a race, she found herself feeling unusually fatigued. Her friends commented that she looked pale. When she visited her doctor, she was told that the iron stores in her body were being depleted. Explain how iron deficiency would affect her blood and could cause her symptoms.

10 µm

Figure 33.25 Light micrograph of a branching blood vessel.

Media Menu

Student CD-ROM

Impacts, Issues Video
  And Then My Heart Stood Still
Big Picture Animation
  Blood, circulation, and cardiovascular function
Read-Me-First Animation
  Circulatory systems
  Rh and pregnancy
  Human blood circulation
Other Animations and Interactions
  Major blood vessels
  Cardiac cycle
  Transfusion interaction
  Measuring blood pressure
  The lymphatic system

InfoTrac

- A Whole New Heart. *Mechanical Engineering-CIME*, August 2002.
- How to Keep Your Heart Healthy. *FDA Consumer*, November–December 2003.
- Murmurs of the Heart. *Discover*, March 2000.
- The Hidden Unity of Hearts. *Natural History*, April 2000.

Web Sites

- National Heart, Lung, and Blood Institute: www.nhlbi.nih.gov
- American Heart Association: www.americanheart.org
- The Heart: sln.fi.edu/biosci/heart.html
- Red Gold: www.pbs.org/wnet/redgold/index.html

How Would You Vote?

If a person's heart stops beating because of cardiac arrest or a heart attack, CPR can keep oxygenated blood flowing to the brain until the condition can be treated. The sooner CPR is started, the better the chances for survival. Some health professionals would like to see all students in public high schools trained in CPR. Would you support a requirement that all high school students in your state be required to complete a training course in CPR before they graduate?

IMPACTS, ISSUES *The Face of AIDS*

Thirteen-year-old Chedo Gowero spends all of her days gathering firewood and working in her neighbors' homes and fields. With both parents dead from AIDS, she was forced to leave school to support herself, her ten-year-old brother, and her grandmother. Despite hardships, she is proud that her efforts have so far allowed her younger brother to continue his education.

Chedo and her brother are among the estimated 800,000 children in Zimbabwe alone who became

orphans after their parents died from AIDS (Figure 34.1). HIV, the virus that causes this disease, has infected more than one-third of the adult population. In this African nation, heterosexual intercourse is the most common mode of transmission.

The situation is getting worse. Infection and death rates have been rising for decades. Young women are especially vulnerable, because the virus is more easily transmitted from male to female rather than the other way around. Having few options, many of the orphaned girls become prostitutes to support themselves and their siblings. At present, twenty-five percent of female Zimbabweans between ages fifteen and twenty-four test positive for the virus.

Also, a long-term drought that is reducing crop yields is contributing to the AIDS crisis in Zimbabwe. In 2003, 70 percent of those living on farms tested HIV positive. Many people are trapped in a vicious cycle of infection *and* malnutrition. Malnourished individuals are more easily infected. Once infected, they often are too weak to tend crops and livestock.

Figure 34.1 Putting a face on one of the social casualties of AIDS—Chedo Gowero at her village in Zimbabwe. The filmstrip at far right shows a micrograph of HIV, the viral agent that causes AIDS. This virus killed Chedo's parents.

the big picture

Evolutionary Background The vertebrate body has physical, chemical, and cellular defenses against coevolved threats. Targets include diverse viruses, bacteria, and other pathogens, as well as cancerous cells and other altered body cells.

First Line of Defense Skin and mucous membranes are major barriers to the internal environment. Enzymes in saliva, tears, and other secretions kill some pathogens. Harmless microorganisms living in or on the body stop many pathogens from becoming established.

The same vicious cycle persists in other countries, but it is pronounced in sub-Saharan Africa. By the United Nations estimate, as many as 46 million people were infected worldwide when 2003 ended. Of that staggering total, 70 percent live in sub-Saharan Africa.

When confronted by such statistics, you might be tempted to walk away, believing that the problem is insurmountable. However, keep in mind that most Africans—including Chedo and her young brother—have *not* been infected. And there are reasons for hope.

In some African countries, the infection rates are declining as a result of educational and prevention programs. Volunteers from developed countries are working with African agencies to get information even into rural areas. The United States government has promised 15 billion dollars in funding for AIDS research and treatments over a five-year period. Other developed countries are also funding these efforts. Drugs that can help patients survive longer are available, although few undeveloped countries can afford them.

AIDS is a new threat on the evolutionary stage. So are many newly emerging pathogens, including the mutated coronavirus that causes SARS, the antibiotic-resistant strains of pathogens, and anthrax strains used as a biological weapon. How does a human body counter such threats? What happens when it cannot respond against them successfully? Such questions invite you into the world of immune fighters, and into the kinds of research that might provide answers.

 How Would You Vote?

AIDS drugs can extend a life span. Their high cost puts them out of reach of most developing countries, which urgently need them. Should the United States government offer incentives to American drug companies to sell AIDS drugs to developing countries at a discount? What about AIDS patients at home? Who actually ends up paying? See the Media Menu for details, then vote online.

Second Line of Defense Inflammation is a local response to tissue damage. It involves white blood cells, chemical signals, and plasma proteins. Cascading activation of complement proteins occurs both in inflammation and during immune responses.

Third Line of Defense If infection persists, white blood cells make immune responses against an agent they chemically recognize as nonself. They rapidly divide and differentiate into huge cell populations sensitized to destroy that agent and to intercept any subsequent attacks.

34.1 Overview of the Body's Defenses

Vertebrates continually cross paths with astoundingly diverse viruses, bacteria, fungi, protozoans, parasitic worms, and other disease agents. They coevolved with most of them, and three lines of defense are now the evolutionary outcome.

INTACT SKIN AND MUCOUS MEMBRANES

Pathogens cannot do damage unless they first breach surface barriers and enter the internal environment. These barriers—intact skin and specialized epithelial linings of the body's tubes and cavities—are the first line of defense against tissue damage.

NONSPECIFIC INTERNAL DEFENSES

When surface barriers are breached, a second line of defense forms in the tissues under siege. Certain kinds of white blood cells (lymphocytes) and antimicrobial chemicals make both rapid and prolonged responses to the agent doing damage. Figure 34.2 shows these cells, which all arise from stem cells in bone marrow. Like SWAT teams, some react fast to any threat. The most abundant are **neutrophils**. Each can ingest and digest about twenty bacterial cells. **Eosinophils** secrete enzymes that punch holes in parasitic worms. The **basophils** and **mast cells** make and secrete chemical substances that help keep inflammation going. The next section describes this inflammatory response.

Although slower to act, the white blood cells called **macrophages** are "big eaters" (Figure 34.2). Each can engulf and digest a hundred bacterial cells or almost anything else chemically recognized as foreign to the body. They also clean up tissues. Immature forms of these cells (monocytes) circulate in blood. Like the **dendritic cells**, macrophages do more than patrol and engulf; they also alert immune cells to threats.

Natural killer cells (NK cells) don't need a wake-up call. They patrol the body and bind to tumor cells, virus-infected cells, and bacterial cells. Some of their secretions poke holes in the targeted cells, which then die. Other NK secretions enhance inflammation.

EVOLUTION OF ADAPTIVE IMMUNITY

Your immune system, the third and most recent line of defense, has a story to tell. It is one of increasingly precise defenses based on cell-to-cell communication and self-recognition. It started hundreds of millions of years ago, when multicelled animals were evolving from single-celled eukaryotes. At the least, their cells had membrane proteins that helped them chemically recognize one another and stay together as a unit. To do this, they had to chemically distinguish *self* from *nonself*. And this capacity had selective advantages. How do we know? Early on, some body cells started recognizing and destroying pathogens—which bear molecules that do *not* belong to the animal body.

Phagocytic cells were the earliest infection fighters. We find scavenging types in the most ancient animal lineages, such as sponges, cnidarians, and flatworms. It was a tiny evolutionary step from ingesting bacteria as food to digesting harmful ones that infiltrated the body. The enzymes called **lysozymes** were especially good at digesting cell walls of many bacteria.

Later, nonself cues triggered signal transduction pathways. Cells recognizing the cues made chemical weapons that walled off the threat or punched holes in it. Later still, **cytokines** evolved in invertebrates. This major class of signaling molecules coordinates the interactions among cells that defend the body. In all vertebrates, three types—interleukins, interferons, and tumor necrosis factors—became part of the arsenal.

But each individual had an unchanging capacity to recognize certain potential threats. Adapting to *new* threats was not possible in the individual's lifetime.

By 450 million years ago, something new had come along. White blood cells called **B lymphocytes** and **T lymphocytes** had evolved in jawed fishes. Only these cells can make proteins that chemically recognize a truly immense number of nonself configurations. More than this, *during each individual's life, B cell and T cell populations specifically tailor a defense response to one of potentially limitless kinds of pathogens.* We call this capacity **adaptive immunity**.

neutrophil

eosinophil

basophil

mast cell

B lymphocyte

T lymphocyte

NK cell

dendritic cell

macrophage

Figure 34.2 Gallery of cellular defenders of the vertebrate body, all stained to reveal structural details. The scanning electron micrograph shows a macrophage caught in the act of engulfing a pathogenic yeast cell. Some of these white blood cells are like rapid-deployment SWAT teams. Others divide and are the start of huge subpopulations that hang in there over the long term.

Table 34.1 Three Vertebrate Lines of Defense

I. Barriers at Body Surfaces (*nonspecific* targets)

Intact skin; mucous membranes at other body surfaces

Infection-fighting chemicals (e.g., lysozyme in tears, saliva)

Normally harmless microorganisms of skin and other body surfaces that can outcompete pathogenic visitors

Flushing effect of tears, saliva, urination, and gut motility

II. Nonspecific Responses (*nonspecific* targets)

Phagocytes and cytotoxic cells

  a. Fast-acting white blood cells (neutrophils, eosinophils, mast cells, basophils)

  b. Macrophages (also take part in immune responses)

  c. Natural killer cells

Inflammation

  a. Histamine, prostaglandins, blood-clotting proteins

  b. Antimicrobial proteins (e.g., complement, interferons)

Fever

III. Adaptive Immunity (*specific* targets only)

T cells and B lymphocytes, alerted by dendritic cells, macrophages

Communication signals (e.g., interleukins and interferons)

Chemical weapons (e.g., antibodies, perforins)

Now reflect on the evolutionary trends among the vertebrates that invaded land. As circulation of body fluids became ever more efficient, so did the capacity to defend the body. Phagocytic cells now had a faster way of reaching embattled tissues and intercepting pathogens, as they trickled along vascular highways. So did chemical weapons dissolved in plasma.

In short, *internal defenses against threats are in place before damage is inflicted.* Standing guard are agents of fast, nonspecific responses. Ready and waiting are B and T cells that counter specific threats (Table 34.1).

Surface and chemical barriers, nonspecific responses to tissue damage, and immune responses protect vertebrate bodies.

Recognition of nonself evolved among the earliest animals, which responded generally to threats with phagocytosis and chemical weapons. Signaling pathways that recruited cellular defenders evolved among invertebrates.

Vertebrates still require these nonspecific defenses. They also have B and T lymphocytes that tailor immune responses to potentially limitless kinds of threats.

34.2 First Line of Defense

A pathogen can only cause infection when it enters the internal environment. It must penetrate skin or slip across mucous membranes. Both are formidable barriers.

Vertebrate skin has tough surface layers made of dead, keratin-packed epithelial cells bound tightly together (Section 28.7). The surface of this body covering is oily, waterproof, salty, and acidic. Even so, it teems with normally harmless bacteria and yeasts. Many are permanently established in such large numbers that pathogenic types typically cannot compete with them and become entrenched.

Pathogens can enter through wounds in the skin or through cavities and tubes that open onto the body surface. Think of your eyes, nose, mouth, and urinary and genital openings. All are portals for pathogens, and all are heavily protected. For one thing, white blood cells patrol the layers of living epithelial cells that continually form beneath the dead ones.

Besides this, eyes are continually washed by tears that hold bacteria-killing lysozyme. Lysozyme-laden mucus in airways to the lungs protects against inhaled bacteria that slipped past fine nasal hairs. Ciliated cells in the airways sweep out the debris. Coughs and sneezes put bacteria back outside.

To colonize the mouth, a pathogen must get past lysozyme in saliva and harmless microbes already living in the warm, moist, nutrient-rich habitat. Microbes that do reach the stomach usually die from the low pH. Others die from exposure to bile salts in the small intestine and the constant churning of material in the gut. The few hardy types that make it to the large intestine are confronted by more than 100 resident species that can restrict their growth.

Urine's low pH and its excretion from the body keep pathogens from taking hold in the urinary tract. Lactate wastes of *Lactobacillus* populations keep the vaginal pH low. Related species inhabit the mouth and the large intestine. Taking antibiotics can kill off some of these "good" bacteria and invite pathogenic opportunists to replace them. Also, when a female's estrogen levels decline during menopause, bacterial lactate production in the vagina declines, and her risk of vaginal infections increases.

The vertebrate body's first line of defense against pathogens includes intact skin, mucous membranes, ciliated cells, lysozyme, and resident bacteria at body surfaces and in the gut lumen.

34.3 Second Line of Defense—Nonspecific Internal Responses

What happens when pathogens do slip into the internal environment? They typically trigger an inflammatory response, the release of antimicrobial proteins, and fever.

INFLAMMATORY RESPONSE

Suppose pathogenic bacterial cells enter a tissue and start releasing metabolic products into the interstitial fluid. By a mechanism known as **acute inflammation**, fast-acting phagocytes leave the bloodstream at the tissue's capillary beds. So do complement proteins, clotting factors, and other plasma proteins. Redness, warmth, swelling, and pain are localized signs that inflammation is under way. What is going on?

In response to the tissue disturbance, mast cells in connective tissue release **histamine** and other local signaling molecules that make local arterioles dilate. Vasodilation, recall, widens a blood vessel's diameter by making smooth muscle in its wall relax. As blood engorges the arterioles, the tissue reddens and warms with blood-borne metabolic heat (Figure 34.3).

Histamine also makes thin-walled capillaries more permeable. It induces endothelial cells of their wall to pull apart more at narrow clefts between them. The capillaries are now "leaky" to plasma proteins that normally do not leave blood. As some proteins leak out, osmotic pressure rises in interstitial fluid. Also, blood pressure rises because selective controls over arterioles are distributing more of the blood flow to the damaged tissue. More fluid leaves the capillaries and less gets reabsorbed. The fluid balance across the capillary wall shifts, causing edema. The tissue swells with fluid and presses on its pain receptors.

Within hours of the body's first response to tissue damage, neutrophils are squeezing across capillary walls. They swiftly go to work. Monocytes arrive and differentiate into macrophages. As these cells engulf invaders and digest cellular debris, they secrete local signaling molecules that attract more phagocytes.

The plasma proteins leaking into the tissue include clotting factors, which phagocyte secretions activate. Clotting factors, remember, stimulate the formation of

Read Me First!
and watch the narrated animation on the inflammatory response

a Bacteria invade a tissue and directly kill cells or release metabolic products that damage tissue.

b Mast cells in tissue release histamine, which then triggers arteriole vasodilation (hence redness and warmth) as well as increased capillary permeability.

c Fluid and plasma proteins leak out of capillaries; localized edema (tissue swelling) and pain result.

d Complement proteins attack bacteria. Clotting factors wall off inflamed area.

e Neutrophils, macrophages, and other phagocytes engulf invaders and debris. Macrophage secretions also kill targets, attract more phagocytes, and call for fever as well as T and B cell proliferation.

Figure 34.3 Acute inflammation in response to a bacterial invasion. Fast-acting white blood cells and diverse plasma proteins enter an irritated and damaged tissue. Collectively, they inactivate, destroy, and wall off the invaders, remove chemicals and cellular debris, and prepare the tissue for repair. The micrograph shows a white blood cell squeezing out through a cleft between endothelial cells of a blood capillary.

a In some responses, complement proteins are activated when antibodies (the Y-shaped molecules) bind to them. These antibodies have already latched on to the bacterial cell.

b Binding directly to a bacterium activates some types of complement proteins.

c By cascading reactions, huge numbers of different complement proteins form. They assemble into many molecules, which assemble into many attack complexes.

d The attack complexes become inserted into the target cell's lipid envelope or plasma membrane. Each complex makes a large pore form across it.

e The pores bring about lysis of the cell, which dies because of the severe structural disruption.

Figure 34.4 What complement proteins do. One reaction pathway starts as these proteins bind to microbes during nonspecific responses. Another operates during immune responses. Both lead to the formation of membrane pore complexes, which induce lysis. The micrograph shows the surface of one kind of pathogenic cell. Membrane attack complexes made the holes.

fibrin threads. These stick to exposed collagen fibers in the tissue. All of the fibers form a net that traps blood cells and platelets. The entire mass is a clot.

You read about this clotting response earlier, in Section 33.9. By walling off inflamed areas, clots may prevent or delay the spread of invaders and toxins to other tissues. When the threat ends, anticlotting factors that escaped from the capillaries dissolve the clots.

ANTIMICROBIAL PROTEINS

Complement proteins circulate in blood in an inactive form. During nonspecific responses, the complement system is primed to attack microbes or interfere with their reproduction. The proteins continue working in an immune response. Some become part of membrane attack complexes that bring about lysis; they grossly disrupt a target cell's plasma membrane (Figure 34.4). Other complement proteins bind to pathogens and tag them for destruction by white blood cells. Virus-infected cells that are dying secrete interferons. These chemical signals bind to nearby cells and make them more resistant to viral infection.

FEVER

A *fever* is a core temperature that exceeds the body's normal set point. Even some cold-blooded reptiles get a fever when infected. Fevers of about 39°C (100°F) in

humans probably enhance immune functions. Some interferons, the *endogenous pyrogens*, promote fever by stimulating the secretion of a prostaglandin from the brain. This local signaling molecule causes a rise in the set point of a hypothalamic thermostat that controls the body's core temperature. Aspirin lowers a fever by inhibiting prostaglandin synthesis.

Some white blood cells and chemical weapons are responses to threats in general, not to specific targets.

Inflammation develops as part of nonspecific and specific defenses. The tissue under attack warms and reddens as local arterioles vasodilate. Edema, caused by a fluid imbalance across the wall of capillaries, swells the tissue and causes pain.

Macrophages and other phagocytes engulf invaders and debris, and secrete chemical mediators. Plasma proteins, such as complement proteins, target invaders for destruction. Clotting factors wall off the inflamed tissue.

A fever is a temporary elevation in the set point for body temperature. It helps the body fight infection.

34.4 Third Line of Defense—Adaptive Immunity

Sometimes surface barriers, fever, and inflammation are not enough to end a threat, and an infection becomes established. Dedicated white blood cells of the immune system are now called into action.

Four features characterize the **immune system** of all jawed vertebrates:

1. Certain white blood cells of the system, the B and T lymphocytes, distinguish *self* from *nonself*. They ignore normal body cells but *not* pathogens, parasites, and infected or altered body cells, all of which have foreign or altered molecular configurations at their surface.

2. B and T cells mount attacks after they have become sensitized to a *specific* threat to the body. They are not activated by tissue damage in general.

3. Collectively, B and T cells show staggering diversity in receptors for signals that can unleash them against just one of virtually unlimited kinds of threats.

4. The immune system has *memory*. A portion of the B cell and T cell populations that form during a first-time encounter with a specific threat are set aside for future battles against the same threat.

Recognition of nonself molecules triggers mitotic cell divisions that give rise to huge populations, all sensitized to the same threat. Any molecule that incites divisions of B and T cells and becomes their target is classified as an **antigen**. Most antigens are proteins.

antigen fragments

MHC molecule

antigen–MHC complex

Figure 34.5 Dendritic cell. Such cells are one of the major classes of antigen-presenting cells. Dendritic cells patrol blood, internal organs, and skin. They ingest, process, and display antigen next to MHC molecules. Langerhans cells of skin, described in Section 28.7, are a type of dendritic cell.

Some subpopulations of new B and T cells become *effectors*. These differentiated cells act at once, in a primary response to antigen. However, others become *memory* cells. These enter a resting phase, but they all have receptors for that antigen. If it shows up again, memory cells are already available and primed to take part in a faster, *secondary* response to the same threat.

THE DEFENDERS AND THEIR MAIN TARGETS

In every immune response, antigen-presenting cells, cytotoxic T cells, and activated helper T cells interact. Some of these cells process and display antigen, and others recognize and attack the antigen bearers.

Antigen that becomes paired with MHC markers on a body cell is the call to arms. MHC markers are specific recognition proteins at the plasma membrane of the body's own cells. Their name refers to a gene complex encoding them (*Major Histocompatibility Complex*).

Diverse **antigen-presenting cells** process antigen, then display fragments of it next to MHC markers. Macrophages, B cells, and dendritic cells can do this. They engulf free antigen or the whole cell or virus bearing it. Once antigen is housed in a cytoplasmic vesicle, lysosomal enzymes swiftly cleave it into bits. The fragments bind to MHC markers already built in the cytoplasm. Antigen–MHC complexes are moved to the plasma membrane. There they are displayed at the surface, like emergency flags (Figure 34.5).

Odds are that some **helper T cell** will be receptive to this particular flag. It will issue signals that can induce any B and T cell sensitized to the same antigen to start the divisions and differentiations that form armies of effector and memory cells (Figure 34.6).

The effector **B cells** immediately execute *antibody-mediated* responses. These are the only cells that can make and secrete **antibodies**, a type of molecule with regions that can bind specific antigens. Neutrophils, macrophages, and NK cells attack any foreign cell or virus that has antibodies attached to it. Antibodies do more. They inactivate antigen-bearing cells and toxins circulating in blood or interstitial fluid. They promote inflammation and switch on complement proteins.

Antigen-presenting cells and helper T cell signals also activate **cytotoxic T cells** that bear receptors for the same antigen. Effector cytotoxic T cells carry out *cell-mediated* responses. They do not respond to freely circulating antigen. But they bind to any body cell infected by viruses or by intracellular pathogens or parasites. They recognize and kill cancer cells. These are the cells that attack cells of a transplanted organ and can bring about its rejection.

Read Me First!
and watch the narrated animation
on immune responses

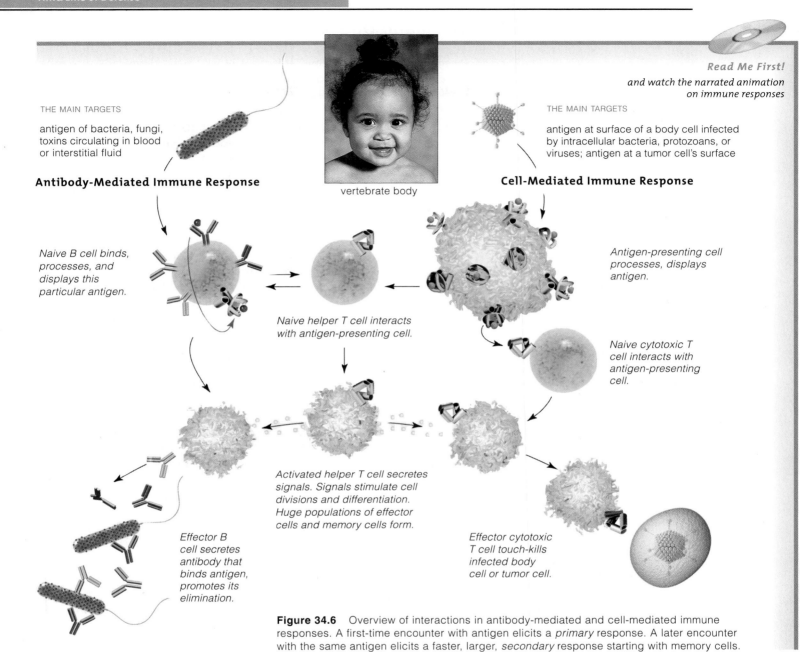

THE MAIN TARGETS

antigen of bacteria, fungi, toxins circulating in blood or interstitial fluid

Antibody-Mediated Immune Response

vertebrate body

THE MAIN TARGETS

antigen at surface of a body cell infected by intracellular bacteria, protozoans, or viruses; antigen at a tumor cell's surface

Cell-Mediated Immune Response

Naive B cell binds, processes, and displays this particular antigen.

Antigen-presenting cell processes, displays antigen.

Naive helper T cell interacts with antigen-presenting cell.

Naive cytotoxic T cell interacts with antigen-presenting cell.

Activated helper T cell secretes signals. Signals stimulate cell divisions and differentiation. Huge populations of effector cells and memory cells form.

Effector B cell secretes antibody that binds antigen, promotes its elimination.

Effector cytotoxic T cell touch-kills infected body cell or tumor cell.

Figure 34.6 Overview of interactions in antibody-mediated and cell-mediated immune responses. A first-time encounter with antigen elicits a *primary* response. A later encounter with the same antigen elicits a faster, larger, *secondary* response starting with memory cells.

CONTROL OF IMMUNE RESPONSES

Helper T cells recognize antigen *only* if a macrophage, dendritic cell, or B cell presents it. They themselves divide only after contact with it. *The resulting clone of helper T cells drives sustained immune responses.*

Immune responses subside once antigen is cleared away. The tide of battle turns as effector cells and their secretions kill most antigen-bearing agents. With less antigen present, fewer immune fighters are recruited. Complement proteins help in the cleanup. They clear immune complexes from blood and deposit them into the liver and spleen for disposal.

Recognition of antigen bound to MHC markers on body cells triggers immune responses. Huge subpopulations of effector and memory T and B cells form, all sensitized to the one kind of antigen that stimulated their formation.

In antibody-mediated immune response, activated B cells make and secrete antibodies that target specific antigen.

In cell-mediated immune response, cytotoxic T cells act to directly kill body cells that are already infected by intracellular pathogens. They also can kill tumor cells.

34.5 How Lymphocytes Form and Do Battle

Your body is continually exposed to a mind-boggling array of antigens. However, it continually produces B and T cells that collectively bear an equally mind-boggling array of novel receptors, capable of recognizing about a billion kinds of antigens! How does such receptor diversity arise, and where do the immune fighters respond to it?

B CELLS, T CELLS, AND ANTIGEN RECEPTORS

A gene for an antigen-binding receptor is not a single, continuous sequence. Different parts of the sequence are located in different places in a B or T cell's DNA. Which parts get transcribed in a new cell is random. Then the transcripts are snipped and spliced in one of several ways before they are translated (Figure 34.7). Such random recombinations are the source of the diversity among B and T antigen receptors.

Consider antibody synthesis. The antigen-binding region varies among antibody molecules. It folds into a unique array of bumps and grooves with their own distribution of charge. For each antibody, that region will bind only the antigen that bears complementary grooves, bumps, and distribution of charge.

A B cell synthesizes its unique, typically Y-shaped antibody molecules before leaving bone marrow. That still-maturing cell is already making many copies of these antigen receptors, which move to the plasma membrane. The tail end of each becomes positioned in the lipid bilayer, and the two arms project above it.

In time, the B cell bristles with antigen receptors—bound antibodies. It is now a "naive" B cell, meaning it has not yet met up with antigen complementary to its receptors. When membrane-bound, the antibodies can bind antigen. Helper T cells chemically recognize that bound antigen.

What about T cells? After forming in bone marrow, they migrate to the thymus gland. There they mature and acquire **TCRs** (*T*-Cell *R*eceptors), antigen-binding receptors unique to them. A naive T cell also bristles with antigen receptors when it leaves the thymus and enters blood. Its TCRs will not recognize free antigen or MHC markers unadorned with antigen. They will recognize and bind antigen–MHC complexes.

In short, each B cell or T cell bears many copies of only one type of antigen receptor. By a theory of *clonal selection*, antigen "chooses" only the B or T cell that bears the receptor able to bind with it (Figure 34.8*a*). Descendants of the activated cell form by mitotic cell divisions. They are a lineage of genetically identical cells—a clone—all with the same antigen receptor.

The clonal selection theory also explains how each individual acquires *immunological memory* of a first encounter with antigen. The term refers to the body's capacity to mount a secondary immune response to any later encounter with the same type of antigen that provoked the primary response (Figure 34.8*b*).

Memory cells formed in a primary response do not attack. They circulate and patrol tissues for years or decades. A secondary response to the same antigen is faster than the first because these cells quickly give rise to effector populations. Infection often ends before a person gets sick. Even greater numbers of memory cells form the second time around (Figure 34.9).

In evolutionary terms, these advance preparations against encounters with the same pathogen bestow a survival advantage on the individual.

WHERE ARE THE BATTLEGROUNDS?

Where is antigen intercepted and dealt with? In the skin, mucous membranes, and linings of organs, macrophages and dendritic cells

a As a B cell matures, different segments of antibody-coding genes recombine at random into a final gene sequence.

b The final sequence is transcribed into mRNA.

c Processing yields a mature mRNA transcript (e.g., introns excised, exons spliced).

d mRNA is translated into one of the polypeptide chains of an antibody molecule.

Figure 34.7 How antigen receptor diversity arises, using the antibody molecule as an example.

Antibodies are proteins. Genes encode instructions for synthesizing them. Instructions for an antibody molecule's variable regions are not one continuous stretch of chromosome; they are divided up in different segments along its length. Here we show different kinds of V, J, and C segments on a chromosome (see Figure 34.11*a*).

In this region, a recombination event occurs while each B cell is maturing. Any one of the V segments may be joined to any one of the J segments, which are all fairly long. The randomly arranged sequence is attached to a C (*C*onstant) segment. This particular gene will be present in all of the B cell's descendants.

Antigen binds only to antibody specific to it on a naive B cell.

clonal population of effector B cells

Effector B cells secrete antibodies.

a

First exposure to antigen provokes primary immune response.

naive T or B cell

effector cells

memory cells

Another exposure to the same antigen provokes secondary response.

effector cells

memory cells

b

Figure 34.8 (**a**) Clonal selection of a B cell with the only antibody able to combine with a specific antigen. (**b**) Not all B and T cells are activated in a primary immune response to antigen. Many go on circulating as memory cells, which become activated during a secondary immune response.

first exposure to antigen

subsequent exposure to same antigen

Relative concentrations of antibody

Response time (weeks)

Figure 34.9 Example of differences in the magnitude and duration between a primary and secondary immune response to the same antigen. This primary response peaked 24 days after it started. The secondary response peaked after 7 days (the span shown between weeks 5 and 6). In the secondary response, antibody concentration was 100 times greater.

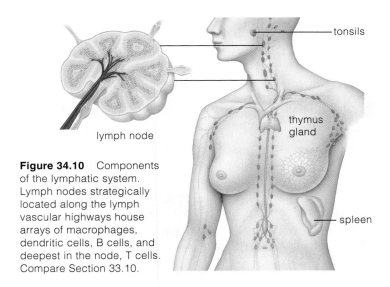

tonsils

lymph node

thymus gland

spleen

Figure 34.10 Components of the lymphatic system. Lymph nodes strategically located along the lymph vascular highways house arrays of macrophages, dendritic cells, B cells, and deepest in the node, T cells. Compare Section 33.10.

are an early line of defense. They take up antigen by way of phagocytosis or endocytosis, then migrate to lymph nodes. There, they present antigen to naive B and T cells positioned inside (Figure 34.10).

Free antigen in interstitial fluid enters the lymph vascular system. Because lymph returns to the blood, there is always the chance that antigen will circulate to all body tissues. However, lymph nodes trap most of it. Antigen that does manage to enter the blood is filtered out as blood flows through the spleen.

In a lymph node, free antigen percolates through an array of B cells, macrophages, and dendritic cells. It is bound, processed, and presented to helper T cells. Other dendritic cells bear receptors for antibodies and complement proteins. These cells can activate B cells, which multiply and mature. Deeper inside the node, T cells also multiply and mature.

During infection, white blood cells and antibody molecules accumulate inside lymph nodes. Hence the swollen "lumps" you may have noticed from time to time under the jaw and elsewhere when you are sick.

Genes for antigen receptors are not one continuous sequence in the DNA. When a B or T cell is maturing, different parts of the sequence are transcribed and spliced together randomly. Each cell ends up with a shuffled gene sequence for one of a billion possible antigen receptors.

If "chosen" by antigen, all of that cell's descendants will bind that antigen only.

Guided by activated helper T cells, immune responses by macrophages, dendritic cells, B cells, and cytotoxic T cells play out in lymph nodes and the spleen.

34.6 Antibody-Mediated Response

The plasma membrane of a naive B cell bristles with antibodies. All of its antibody molecules are alike. They all have receptors that can bind to the same antigen.

antigen binding site

variable region of heavy chain

antigen binding site

variable region of light chain

flexible hinge region

constant region of light chain

constant region of heavy chain (*bright green*) that includes a hinged region

a

antigen on bacterial cell

binding site on one kind of antibody molecule for a specific antigen

b

antigen on virus particle

Figure 34.11 Antibody structure. (**a**) An antibody molecule has four polypeptide chains, often bonded in a Y-shaped configuration.

(**b,c**) At antigen binding sites of the molecule, antigen fits into the grooves and onto protrusions.

c

binding site on another kind of antibody molecule for a different antigen

ANTIBODY FUNCTION

Figure 34.11 shows antigen-binding sites of a typical antibody molecule. If antigen binds to one of these sites, it will move into the cell by receptor-mediated endocytosis (Section 5.8). After antigen is digested into fragments, antigen–MHC complexes form and move to the B cell surface, where they are displayed.

Suppose a responsive helper T cell meets up with the B cell's antigen–MHC complex (Figure 34.12). The cells exchange a costimulatory signal, then disengage. When that B cell once again encounters unprocessed antigen, its antibodies will bind it. That binding, in combination with interleukins secreted by activated helper T cells nearby, drives the B cell into mitosis.

Effector B cells form. (These also are called plasma cells.) They synthesize and secrete large numbers of antibodies. When freely circulating antibody binds to it, antigen becomes a target that is recognized by the phagocytes and complement proteins.

Remember, the main targets of antibody-mediated responses are extracellular pathogens and toxins that are freely circulating in extracellular fluid. *Antibodies cannot bind to pathogens already inside a host cell.*

CLASSES OF IMMUNOGLOBULINS

There are five classes of antibodies, formally called the **immunoglobulins**, or **Igs**. These protein products of gene shufflings form when B cells mature and engage in immune responses. We abbreviate them as IgM, IgD, IgG, IgA, and IgE. Each type has antigen-binding sites. The secreted molecules of IgA and IgM form larger structures, which clump antigen-bearing threats together for elimination from the body:

IgG, IgD, IgE   IgA   IgM

IgM is the first to be secreted in a primary response and the first made by newborns. It may help activate helper T cells. IgM molecules form a structure with ten antigen binding sites. They can bind to clumped targets, including masses of virus particles. IgM and *IgD* are the most common antibodies on naive B cells.

IgG makes up 80 percent of the immunoglobulins in blood. It is the best at activating complement and

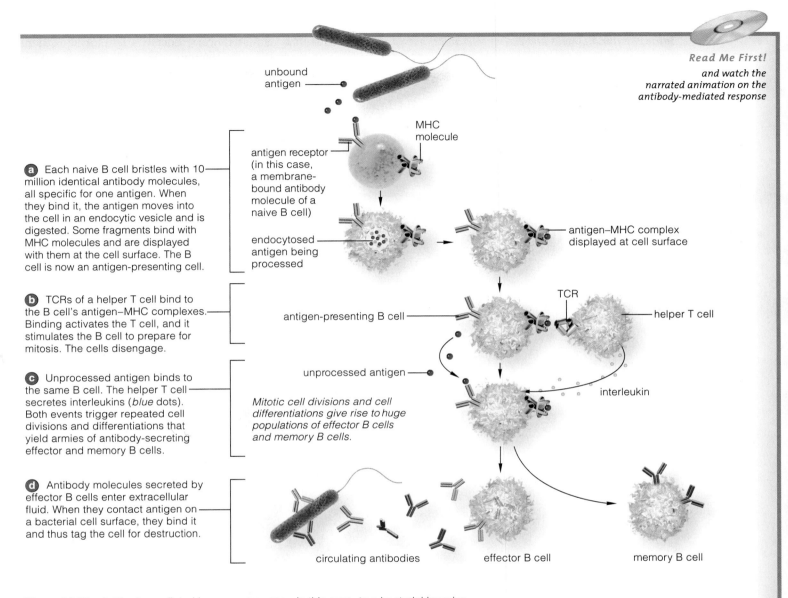

Read Me First!
*and watch the
narrated animation on the
antibody-mediated response*

a Each naive B cell bristles with 10 million identical antibody molecules, all specific for one antigen. When they bind it, the antigen moves into the cell in an endocytic vesicle and is digested. Some fragments bind with MHC molecules and are displayed with them at the cell surface. The B cell is now an antigen-presenting cell.

b TCRs of a helper T cell bind to the B cell's antigen–MHC complexes. Binding activates the T cell, and it stimulates the B cell to prepare for mitosis. The cells disengage.

c Unprocessed antigen binds to the same B cell. The helper T cell secretes interleukins (*blue* dots). Both events trigger repeated cell divisions and differentiations that yield armies of antibody-secreting effector and memory B cells.

d Antibody molecules secreted by effector B cells enter extracellular fluid. When they contact antigen on a bacterial cell surface, they bind it and thus tag the cell for destruction.

unbound antigen

MHC molecule

antigen receptor (in this case, a membrane-bound antibody molecule of a naive B cell)

endocytosed antigen being processed

antigen–MHC complex displayed at cell surface

TCR

antigen-presenting B cell

helper T cell

unprocessed antigen

Mitotic cell divisions and cell differentiations give rise to huge populations of effector B cells and memory B cells.

interleukin

circulating antibodies

effector B cell

memory B cell

Figure 34.12 Antibody-mediated immune response—in this case, to a bacterial invasion.

neutralizing many toxins. IgG can cross the placenta and helps protect a fetus with the mother's acquired immunities. It is secreted into early milk.

IgE triggers inflammation after attacks by parasitic worms and other pathogens. It attaches to basophils and mast cells with its tail. Binding of antigen to its receptors induces the cells to release histamine. IgE is involved in allergic reactions, as described later.

IgA is the main immunoglobulin in exocrine gland secretions, including tears, saliva, and milk. It also is present in the respiratory, digestive, and reproductive tract linings. Like IgM, it forms larger structures that bind big antigens. Bacteria and viruses cannot bind to cells when secreted IgA binds to them first.

Monoclonal antibodies have useful roles in research, medicine, and industry (Section 34.8). First, hybrid cell lines are produced by fusing animal tumor cells with B cells from animals that were exposed to a specific antigen. Cultured hybrid cells are induced to multiply and synthesize large amounts of antibody against one specific antigen.

Antibodies secreted by B effector cells bind to antigens of extracellular pathogens or toxins. The antibodies tag the invaders for attack by phagocytes and complement.

Immunoglobulins are classes of antibody molecules that defend against diverse threats in different body regions.

34.7 Cell-Mediated Response

Intracellular pathogens and cancer cells are targets of the cell-mediated immune response.

Certain kinds of viruses, bacteria, fungi, and protists slip inside body cells. Once inside, they are invisible to antibodies. Helper T cells, cytotoxic T cells, NK cells, macrophages, neutrophils, and eosinophils can battle them (Figure 34.13). These immune fighters also act against cancers. Locally high concentrations of interleukins, interferons, and other signals get the attention of these executioners. They induce them to divide, differentiate, and attack. So do antibodies that picked up antigen released from dead or dying cells.

Cytotoxic T cells respond directly to body cells that bear antigen–MHC complexes and altered proteins. Once certain molecules at their surface make contact

Read Me First!

and watch the narrated animation on cell-mediated response

a A virus particle infects a macrophage. The host cell's pirated metabolic machinery makes viral proteins, which are antigens. Macrophage enzymes cleave antigen into fragments.

b Some antigen fragments bind to class I MHC molecules, which occur on all nucleated cells. On the infected cells, antigen–MHC complexes form and move to the cell surface, where they are displayed.

e Meanwhile, antigen–MHC complexes on the first macrophage bind to TCRs of a cytotoxic T cell. Binding, in combination with interleukin signals from the helper T cell, stimulate the cytotoxic T cell to divide and differentiate, forming huge populations of effector and memory T cells.

g An effector cytotoxic T cell touch-kills the infected cell by releasing perforins and toxic chemicals (*green* dots) onto it.

h The effector disengages and reconnoiters for more targets. Its perforins punch holes in the infected cell's plasma membrane; its toxins disrupt the target's organelles and DNA, so the infected cell dies.

virus particles

antigen–MHC complex

TCR

c Particles of the same virus also are engulfed by *another* macrophage, which processes and displays viral antigen.

d A responsive helper T cell binds to antigen–MHC complexes on this macrophage, which secretes interleukins (*yellow* dots) in response. These signaling molecules stimulate the helper T cell to secrete different interleukins (*blue* dots).

f Meanwhile, inside the respiratory tract, the same virus infects an epithelial cell, which processes and displays antigen at its surface, too.

Figure 34.13 Diagram of a T cell–mediated immune response.

cytotoxic
T cell

tumor
cell

Figure 34.14 Cytotoxic T cell caught in the act of touch-killing a tumor cell.

with infected or altered body cells, their touch-kill mechanism takes over (Figure 34.14). The cytotoxic cell secretes perforins and other molecules that punch holes in the target, something like those in Figure 34.4.

Remember **apoptosis**, or cell suicide (Section 24.5)? Signaling molecules secreted by bound cytotoxic T cells can induce apoptosis in the target. After making its lethal hit, the cytotoxic T cell lets go of the target and resumes reconnoitering.

Cytotoxic T cells also are the cause of rejections of tissue and organ transplants. When MHC markers on donor cells differ from those on a recipient's cells, cytotoxic T cells recognize them as antigens. MHC typing procedures and matching donors to recipients help minimize the risk.

Antigen–MHC complexes do not activate the NK cells. These "natural killers" attack infected cells that are tagged for destruction by antibodies. They also detect stress markers on infected cells or on cells that are becoming cancerous. A stressed cell that still has normal MHC markers is not killed. When the markers are altered or fewer than normal, the NK cell puts holes in it and induces apoptosis.

If cancer cells have altered surface proteins that make them targets of cytotoxic T cells and NK cells, how do they survive? Some release many copies of altered protein, which saturate the antigen receptors of both cytotoxic cells. Also, after tumor cells reach a critical mass, they can overwhelm defenses.

Intracellular pathogens and altered body cells are targets of cell-mediated immune responses.

Helper T cells and cytotoxic T cells are antigen-specific defenders. NK cells, macrophages, neutrophils, eosinophils, and other cells take part in these responses. All cytotoxic types touch-kill infected cells and tumor cells.

34.8 Immunotherapy

FOCUS ON
SCIENCE

Despite the body's natural defenses, cancers can arise and infections can persist. Enhancing the body's responses is the goal of new immunotherapies.

All blood cells, including those involved in defense, arise from cells in bone marrow. By turning up the production of these cells, we can enhance the body's defensive capacity. *Hematopoietic growth factors* are signaling molecules that stimulate the production of blood cells in the bone marrow. When given to patients undergoing radiation and chemotherapy, they can encourage the replacement of "good" cells that are also killed by the treatment.

Cytokines, such as interferons and interleukins, are signaling molecules that stimulate defenses. Natural and recombinant alpha interferons are used in treating cancers and chronic viral infections, such as hepatitis and genital herpes. Interleukin-2, which stimulates B and T cell production and action, is used to treat some metastatic cancers.

Overexpressed proteins that are displayed on the surface of cancer cells can be targeted by monoclonal antibodies (Section 34.6). For example, some kinds of metastatic breast cancer cells have an overabundance of HER2 proteins at their surface. The drug Herceptin (Trastuzumab) is a monoclonal antibody that binds to these proteins and invites the attention of NK cells. There are side effects. HER2 proteins also occur in fewer numbers on some normal body cells, so normal cells also may be attacked.

Monoclonal antibodies can be bound to poisons to make *immunotoxins*. Remember ricin (Chapter 13)? This toxin stops protein synthesis in cells by disabling ribosomes. Researchers have attached a genetically engineered ricin A chain to monoclonal antibodies. The antibodies bind to antigen that is abundant on certain cancer cells. After the immunotoxin binds to antigen at a cancer cell's surface, the ricin chain enters the cell and shuts down translation. Other researchers have bound ricin to monoclonal antibodies that can specifically target cells infected by HIV. So far, these treatments are only experimental.

Also on the horizon are *cancer vaccines*, which may prime the body to recognize and fight cancer cells. Such vaccines are being tested against malignant melanoma. This deadliest of skin cancers can spread to lungs and other body regions. In some ongoing clinical trials, patients' cancer cells are removed and treated with a substance that makes them appear foreign. The treated cells are reinjected into a patient, the idea being that they will sound the alarm for immune responses against the cancer.

34.9 Defenses Enhanced or Compromised

Sometimes immune responses are not strong enough or are misdirected or compromised. Here are examples of what we can and cannot do about it.

IMMUNIZATION

Immunization refers to processes that may promote immunity against diseases. In *active* immunization, an antigen-containing preparation known as a **vaccine** is taken orally or is injected into the body (Figure 34.15). A first injection elicits a primary immune response. A subsequent injection, a booster, elicits the secondary response. More effector cells and memory cells form rapidly and contribute to long-lasting protection.

Many vaccines are made from weakened or killed pathogens or from inactivated natural toxins. Others consist of harmless viruses that have had genes from three or more pathogens inserted into their DNA or RNA. After vaccination, body cells make the antigen encoded by those genes, and immunity is established.

Passive immunization helps those who are already battling hepatitis B, tetanus, rabies, and a few other diseases. It requires an injection of antibody purified from the blood of someone who already fought the disease, or from another source. The antibodies fight the infection but do not activate the immune system; memory cells do not form. When the body disposes of injected antibody molecules, the protection ends.

Vaccines occasionally fail or have bad side effects. There are a few reports of vaccines that have caused immunological problems. There is also some concern that mercury used in some vaccines could harm the nervous system. Mercury-containing vaccines are no longer administered to children. However, if you are considering getting vaccinated, it may be a good idea to assess the risks and benefits of the procedure.

ALLERGIES

In millions of people, exposure to harmless proteins triggers inflammation, excess secretion of mucus, and immune responses. Any substance that ordinarily is harmless and yet provokes such responses is called an **allergen**. Hypersensitivity to them is called an **allergy**. Drugs, foods, pollen, dust mites, fungal spores, and venom from bees, wasps, and other insects are among the most common allergens.

Some people are genetically predisposed to having allergies. Infections, emotional stress, or changes in air temperature can trigger reactions. Exposure to an allergen stimulates secretion of IgE, which binds to mast cells. An inflammatory response causes copious amounts of mucus to be secreted. Airways constrict. Stuffed-up sinuses, labored breathing, sneezing, and a drippy nose are typical symptoms of *asthma* and *hay fever* among many millions of people (Figure 34.16).

Recommended Vaccines	Recommended Ages
Hepatitis B	Birth–2 months
Hepatitis B booster 1	1–4 months
Hepatitis B booster 2	6–18 months
Hepatitis B assessment	11–12 years
DTP (Diphtheria; Tetanus; and	
Pertussis, or whooping cough)	2, 4, and 6 months
DTP booster 1	15–18 months
DTP booster 2	4–6 years
DT	11–16 years
HiB (*Haemophilus influenzae*)	2, 4, and 6 months
HiB booster	12–15 months
Polio	2 and 4 months
Polio booster 1	6–18 months
Polio booster 2	4–6 years
MMR (Measles, Mumps, Rubella)	12–15 months
MMR booster	4–6 years
MMR assessment	11–12 years
Pneumococcal	2, 4, and 6 months
Pneumococcal booster 1	12–15 months
Pneumococcal booster 2	1–18 years
Varicella	12–18 months
Varicella assessment	11–12 years
Hepatitis A (in selected areas)	1–12 years

Figure 34.15 Centers for Disease Control and Prevention guidelines for immunizing children as of December 2003. Doctors routinely immunize infants and children. Low-cost or no-cost vaccinations are available at many community clinics and health departments.

Figure 34.16 One of the effects of ragweed pollen, an allergen, on sensitive people.

Figure 34.17 A case of severe combined immunodeficiency (SCID). Cindy Cutshwall was born with a deficient immune system. She carries a mutated gene for ADA (adenosine deaminase). Without this enzyme, her cells could not break down adenosine, so a reaction product that is toxic to white blood cells accumulated in her body. High fevers, severe ear and lung infections, diarrhea, and an inability to gain weight were outcomes.

In 1991, when Cindy was nine years old, she and her parents consented to one of the first human gene therapies. Genetic engineers spliced the normal ADA gene into the genetic material of a harmless virus. The modified virus was the vector that delivered copies of the normal gene into her bone marrow cells. Some cells incorporated the gene in their DNA and started making the missing enzyme.

Now in her early twenties, Cindy is doing well. She still requires weekly injections to supplement her ADA production. Other than that, she is able to live a normal life. She is a strong advocate of gene therapy.

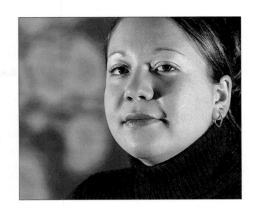

Antihistamines, or anti-inflammatory drugs, often relieve symptoms of allergies. Also, in desensitization programs, skin tests are used to identify the allergens. Increasing doses of allergens are slowly administered. Each time, the body produces more IgG, which freely circulates as an allergen neutralizer.

Anaphylactic shock is a life-threatening response to an allergen. For instance, someone allergic to wasp or bee venom can die within minutes of a sting. Airways constrict, making it hard to breathe. As fluids leak out of permeable capillaries, blood pressure plummets. The circulatory shock may lead quickly to death.

AUTOIMMUNE DISORDERS

When self-recognition fails, the immune fighters start an attack against the body. The misdirected attack on one's own body is called an **autoimmune response**. It can severely damage tissues.

In some cases, B and T cells become hyperactive. For example, *rheumatoid arthritis* occurs as the result of chronic inflammation of joints and often the heart, blood vessels, and lungs. Auto-antibodies typically collect in joints. They invite cascades of reactions that make complement, which results in T cell responses and chronic inflammation. In time, the joints become immobilized as a result.

In other cases, certain antibodies bind to hormone receptors and stimulate cell division. *Graves' disease* (thyrotoxicosis) is like this. Antibodies bind to some receptors in the thyroid gland, causing it to speed up thyroid hormone synthesis. Overproduction of these hormones alters metabolic rates in tissues throughout the body. Feedback loops that govern their synthesis do not affect antibody production. Antibodies keep on acting, and thyroid production and metabolism spin out of control. Among the symptoms are weight loss, arrhythmias, sleeplessness, pronounced mood swings, and excessive sweating.

A common neurological disorder, *multiple sclerosis*, is another example. It develops after autoreactive T cells attack myelin sheaths on axons in nerves and then enter the cerebrospinal fluid. Numbness, paralysis, and blindness are symptoms. At least eight genes heighten susceptibility to multiple sclerosis, but the activation of white blood cells against viral infections may be the trigger for the disorder.

Immune responses tend to be stronger in women than in men, and autoimmunity is far more frequent in women. We know that the receptor for estrogen is involved in controls over gene expression throughout the body. By one hypothesis, the estrogen receptor is wrongly occupied in some autoimmune responses.

DEFICIENT IMMUNE RESPONSES

Loss of immune function can have a lethal outcome. *Primary* immune deficiencies are present at birth and result from altered genes or abnormal developmental steps. Severe combined immunodeficiency (SCID) is like this. A genetic defect called adenosine deaminase deficiency (ADA) is one type (Figure 34.17). *Secondary* immune deficiencies are losses of immune function after exposure to some outside agent, such as a virus. Severe immune deficiencies make individuals more vulnerable to infections by opportunistic agents that are otherwise harmless to those in good health.

AIDS, or acquired immunodeficiency syndrome, is the most common secondary immune deficiency. The next section describes its cause and its effects.

Immunization programs are designed to boost immunity to specific diseases.

Some heritable disorders, developmental abnormalities, and attacks by viruses and other outside agents result in misdirected, compromised, or nonexistent immunity.

34.10 AIDS Revisited—Immunity Lost

Worldwide, HIV infection rates continue to skyrocket. An effective vaccine still eludes us. For now, the best protection is avoiding unsafe behavior.

AIDS is a constellation of disorders that develop after infection by HIV (*H*uman *I*mmunodeficiency *V*irus). The virus cripples the immune system and makes the body highly susceptible to infections and rare forms of cancer. Worldwide, HIV has infected an estimated 34 million to 46 million individuals (Table 34.2).

There is no way to rid the body of known forms of the virus, HIV-I and HIV-II. *There is no cure for those already infected.* At first an infected person appears to be in good health, maybe fighting "a bout of the flu." But symptoms that foreshadow AIDS emerge: fever, many enlarged lymph nodes, fatigue, chronic weight loss, and drenching night sweats. Then opportunistic infections strike. They include yeast infections of the mouth, esophagus, vagina, and elsewhere, and a form of pneumonia caused by *Pneumocystis carinii*. Painless colored lesions often erupt, especially on the legs and feet (Figure 34.18). The lesions are signs of *Kaposi's sarcoma*, a form of cancer common in AIDS patients.

HIV INFECTION—A TITANIC STRUGGLE BEGINS

HIV, a retrovirus, has a lipid envelope, a bit of plasma membrane acquired as it budded from an infected cell. Proteins spike out from the envelope, span it, and line its inner surface. Inside the envelope, more viral proteins enclose two RNA strands and several copies of reverse transcriptase, a retroviral enzyme.

HIV infects cells having the receptor CD4 at their surface. Macrophages and helper T cells, the only ones that do, are also known as **CD4 lymphocytes.**

Once the viral enzyme enters a CD4 lymphocyte, it uses viral RNA as a template to make DNA, which is inserted into a host chromosome. In some cells, the inserted genes are silent until activated in a later round of infection. Transcription yields copies of viral RNA. Some are translated into proteins. Others are enclosed in a protein coat, as the viral RNA of new particles. The particles bud from the plasma membrane, ready to infect more cells (Figure 34.19).

Infection marks the onset of a titanic battle between HIV and the immune system. B cells start making antibodies to antigenic HIV proteins. Armies of helper and cytotoxic T cells form. In some infection phases, the virus infects about 2 billion helper T cells, and 100 million to 1 billion virus particles

Figure 34.18 The lesions that are a sign of Kaposi's sarcoma.

Table 34.2	Global Cases of HIV and AIDS*	
Region	AIDS Cases	New HIV Cases
Sub-Saharan Africa	25,000,000	3,000,000
South/Southeast Asia	4,600,000	610,000
Latin America	1,300,000	⋅120,000
Central Asia/East Europe	1,200,000	180,000
North America	790,000	36,000
East Asia and Pacific	700,000	150,000
Middle East/North Africa	550,000	43,000
Western Europe	520,000	30,000
Caribbean Islands	350,000	45,000
Australia/New Zealand	12,000	700

*Conservative global estimates as of December 2003, *www.unaids.org*

are built every day. Every two days, half the virus particles are destroyed and half the helper T cells lost in battle are replaced. HIV reservoirs and masses of infected T cells collect in the lymph nodes.

Gradually, the number of virus particles in blood rises and the battle tilts. The body produces fewer replacement helper T cells. It may take a decade or more, but erosion of helper T cell counts destroys the capacity to initiate effective immune responses.

Some viruses, including the measles virus, make far more particles in a given day, but the immune system usually wins. Other viruses, including herpes viruses, remain in the body for a lifetime, but the immune system keeps them under control. With HIV, the immune system is destroyed. Secondary infections and tumors kill the individual.

HOW IS HIV TRANSMITTED?

In North America, about 36,000 new HIV infections occur annually. Most often, the mode of transmission is having unprotected sex with an infected partner. HIV in semen and vaginal secretions enters a partner through the epithelial lining of the penis, vagina, rectum, or (rarely) the mouth. Epithelial linings already damaged, as by other sexually transmitted diseases, rough sex, or anal intercourse, increase the likelihood of transmitting the virus from one person to another.

Infected mothers can transmit HIV to their children during vaginal birth and by breast-feeding. HIV also travels in tiny amounts of infected blood in syringes that are shared and reused by intravenous drug abusers and by patients in hospitals of poor countries. This virus is not effectively transmitted by way of food, air, water, casual contact, or insect bites.

Read Me First!

*and watch the narrated animation
on HIV replication*

viral coat   viral enzyme
proteins   (reverse transcriptase)

25–30 µm

a Viral RNA enters a CD4 lymphocyte.

b Viral RNA forms by reverse transcription of viral RNA.

c The viral DNA becomes integrated into host cell's DNA.

d DNA, including the viral genes, is transcribed.

nucleus

viral DNA

viral RNA

viral proteins

viral RNA

viral RNA   lipid envelope with proteins

f Virus particles that bud from the infected cell may attack a new one.

e Some transcripts are new viral RNA, others are translated into proteins. Both self-assemble as new virus particles.

Figure 34.19 Replication cycle of HIV, the retrovirus that causes AIDS. The micrograph at the left shows one viral particle.

WHAT ABOUT DRUGS AND VACCINES?

Current drugs cannot cure people; we cannot go after HIV genes already inserted in their DNA. Also, given the huge number of replications in infected people, HIV genes mutate at high rates. Natural selection is also in play; drug resistance is favored in patients undergoing drug therapy.

AZT (azidothymidine) is one of the drugs that can block reverse transcription. It replaces normal nucleotide phosphates in DNA. Over time, however, such drugs also become incorporated in the DNA of the body's cells and kill them.

Other drugs act as protease inhibitors. Translation of HIV genes results in a single long polypeptide strand that enzymes cut up into proteins. Protease inhibitors prevent the cutting and thus disrupt the assembly. But they also affect host cell proteases and, like all AIDS drugs, have debilitating side effects.

The cost of current treatments—as much as 15,000 dollars a year—puts them beyond the reach of most people in developing countries, where the AIDS pandemic is raging. At present, the best option for halting the spread of the pandemic appears to be the development of a safe, effective, low-cost vaccine.

Researchers are designing AIDS vaccines. They want to make a synthetic virus, one with enough viral components to activate the immune system against HIV but not enough to threaten health. At this writing, they have developed more experimental vaccines than ever. About a dozen are in the first phases of clinical testing. Uninfected volunteers are receiving vaccinations to gauge safety and potential side effects. The second phase tests how a vaccine triggers production of antibodies against the virus.

One vaccine has reached the final phase, which evaluates its ability to prevent infection. AIDSVAX is a genetically engineered mix of surface proteins from two HIV strains. Starting in 1998, about 5,400 at-risk but uninfected volunteers were vaccinated. Results were disappointing. The vaccine did not appear to stop infection. It did produce a slight degree of protection among African and Asian males. This finding may point to variation in the infection process or immune response among ethnic groups.

Another phase III trial is under way in Thailand, with results due in 2008. In this case, the AIDSVAX is being tested in combination with another vaccine.

Summary

Section 34.1 All animals defend their body (self) from nonself (foreign cells and viruses). Phagocytic white blood cells engulf foreign cells and debris (Table 34.3). Cytokines coordinate responses. Lysozymes break down bacterial cell walls. These are nonspecific responses to threats. Most jawed vertebrates show adaptive immunity, also. Their B and T cells can recognize potentially billions of kinds of threats and specifically zero in on each one.

Section 34.2 Physical and chemical barriers at body surfaces, as well as established populations of harmless microorganisms, are the first line of defense.

Section 34.3 Inflammation is a fast, nonspecific response to tissue damage that also continues in immune responses. Histamine increases blood flow and makes capillaries leaky. Phagocytes, complement and other plasma proteins, and clotting factors escape from blood and work in the damaged tissue. A fever is a deviation in the set point for body temperature and helps stem infection.

Section 34.4 An immune system shows self/nonself recognition, sensitivity to specific threats, and memory. Antigen is any molecular configuration that can elicit the formation of effector and memory B and T cells. Pathogens, parasites, and altered or cancerous body cells bear unique antigen. Macrophages and other white blood cells engulf and process antigen, then display it in combination with MHC markers at their surface, as a call to arms.

Section 34.5 B cells form and express unique antigen receptors in bone marrow. T cells also form there but get unique antigen-binding receptors in the thymus. Each cell's receptors can bind one kind of antigen. After the primary response to antigen, a portion of memory B and T cells persists, giving the body immunological memory.

Antigen-presenting cells throughout the body move to lymph nodes, where B and T cells are activated and where most immune responses are carried out.

Section 34.6 B cells carry out antibody-mediated responses. Naive B cells exposed to both antigen and interleukins secreted by a sensitized helper T cell are stimulated to divide and differentiate. They form armies of antibody-secreting effector B cells and memory B cells. Different effector B cells secrete five classes of antibodies called immunoglobulins. Antibodies bind and tag specific extracellular pathogens for elimination.

Section 34.7 T cells and NK cells make cell-mediated responses against intracellular pathogens and altered body cells. Sensitized helper T cells and cytotoxic T cells exposed to antigen–MHC complexes are stimulated to divide. Their descendants differentiate into effector and memory cells. Effector cytotoxic cells touch-kill their targets with protein weapons, including perforins that punch holes in the target and cause lysis.

Section 34.8 Immunotherapies are new methods of enhancing the immune response to fight viral infection or cancers. Manufactured cytokines, cancer vaccines, and monoclonal antibodies are examples.

Section 34.9 Immunization provokes immunity. A weakened or stripped-down version of a pathogen is used as a vaccine to trigger a primary response. Allergic reactions are immune responses to normally harmless substances. In an autoimmune disorder, the body's own cells are wrongly viewed as foreign. Immune deficiency is a weakened capacity to mount an immune response.

Section 34.10 HIV, a retrovirus, causes AIDS. It infects and destroys helper T cells and macrophages (also called CD4 lymphocytes). Sex is the most common mode of transmission. The disease can be treated, at great cost, with drugs that have negative side effects. AIDS cannot be cured. Attempts to create a vaccine are ongoing.

Table 34.3	Summary of White Blood Cells and Their Functions
Cell Type	Main Characteristics
Macrophage	Phagocyte; acts in nonspecific and specific responses; presents antigen to T cells, cleans up and helps repair tissue damage
Neutrophil	Fast-acting phagocyte; takes part in inflammation, not in sustained responses, and is most effective against bacteria
Eosinophil	Secretes enzymes that attack certain parasitic worms
Basophil and mast cell	Secrete histamine, other substances that act on small blood vessels to produce inflammation; also contribute to allergies
Dendritic cell	Processes and directly presents antigen to helper T cells
Lymphocytes:	*All take part in most immune responses; after antigen recognition, clonal populations of effector and memory cells form.*
B cell	Effectors secrete five types of antibodies
Helper T cell	Effectors secrete interleukins that stimulate rapid division and differentiation of both B cells and T cells
Cytotoxic T cell	Effectors touch-kill infected cells, tumor cells, and foreign cells
Natural killer (NK) cell	Cytotoxic cell of undetermined affiliation; kills virus-infected cells, bacteria, and altered body cells by a touch-kill mechanism

Self-Quiz

Answers in Appendix III

1. _____ are nonspecific defenses against threats.
   - a. Intact skin and mucous membranes
   - b. Tears, saliva, gastric fluid
   - c. Resident bacteria
   - d. Urine flow
   - e. all of the above

2. Complement proteins _____ .
   - a. neutralize toxins
   - b. enhance resident bacteria
   - c. promote inflammation
   - d. form pore complexes
   - e. both a and b
   - f. both c and d

3. _____ trigger immune responses.
   - a. Interleukins
   - b. Lysozymes
   - c. Immunoglobulins
   - d. Antigens
   - e. Histamines
   - f. all of the above

4. Recognition of specific threats is based on _____ .
   a. antigen receptor diversity c. mast cell proliferation
   b. recombinations of   d. both a and b
      receptor genes   e. all of the above

5. Localized _____ is a sign of inflammation.
   a. nausea   d. warmth
   b. swelling   e. b through d
   c. redness   f. all of the above

6. Antibody-mediated responses work against _____ .
   a. intracellular pathogens d. both a and c
   b. extracellular pathogens e. both b and c
   c. extracellular toxins   f. all of the above

7. Immunoglobulin(s) _____ promote(s) antimicrobial
   activity in mucus-coated surfaces of some organ systems.
   a. IgA   b. IgE   c. IgG   d. IgM   e. IgD

8. _____ are targets of effector cytotoxic T cells.
   a. Extracellular virus particles in blood
   b. Virus-infected body cells or tumor cells
   c. Parasitic flukes in the liver
   d. Bacterial cells in pus
   e. Pollen grains in nasal mucus

9. Match the immunity concepts.
   ____ inflammation   a. neutrophil
   ____ antibody secretion   b. effector B cell
   ____ fast-acting phagocyte   c. nonspecific response
   ____ immunological   d. immune response
         memory   against own body
   ____ autoimmunity   e. secondary responses

Critical Thinking

1. Pigs are considered the most likely animal candidates as future organ donors for humans. As a first step, some researchers are developing transgenic pigs—animals that received and are expressing foreign genes (Figure 34.20). Which type of human genes might be inserted to stop the human immune system from attacking a pig-to-human transplant? Explain your reasoning.

2. As Section 34.5 explains, generation of tremendously diverse B cell and T cell antigen receptors is the basis of pathogen-specific immune responses. The genetic roots of this adaptive immunity lie deep in the chordate family tree. Remember the lancelets (Section 23.1)? Researchers have found that these invertebrate chordates have genes that resemble the V-region genes of vertebrates. However, these genes cannot undergo rearrangements in lancelets, as they do in vertebrates. Lancelets can make nonspecific defense responses, but not immune responses. Why?

3. Elena developed *chicken pox* when she was in first grade. Later in life, when her children developed chicken pox, she remained healthy even though she was exposed to countless virus particles daily. Explain why.

4. Vaccines hold promise for preventing or treating some cancer. *Therapeutic* vaccines are designed to help the body fight an existing cancer. They activate the immune system against antigens specific to tumor cells. In typical clinical trials, some patients receive a known cancer treatment and others the new vaccine. If you had a life-threatening cancer, would you take part in a clinical trial of a vaccine without knowing what its side effects might be?

Figure 34.20 Two piglets developed from the same cell lineage at the University of Missouri. The yellow hooves and snout on the piglet at the left are evidence that it is transgenic.

Media Menu

IMPACTS, ISSUES   *Up in Smoke*

Each day 3,000 or so teenagers join the ranks of habitual smokers in the United States. Most are not even fifteen years old. The first time they light up, they cough and choke on irritants in the smoke. They typically become dizzy and nauseated, and develop headaches.

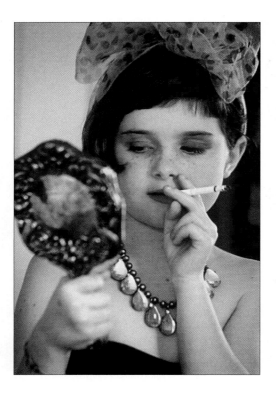

Sound like fun? Hardly. So why do they ignore the signals of the threat to the body and work so hard to be a smoker? Mainly to fit in. To many adolescents, the misguided perception of social benefits overwhelms the seemingly remote threat to health (Figure 35.1).

Yet fulfillment of that threat starts right away. Cilia that line the airways to the lungs normally sweep out airborne pathogens and pollutants. Smoke from just a single cigarette immobilizes them for hours. Smoke also kills the white blood cells that patrol and defend the respiratory system's lining. And so gunk starts to clump up in the airways of young smokers. Pathogens become entrenched in the gunk, as biofilms. More colds, more asthma attacks, and bronchitis are on the way.

Each smoke-filled inhalation delivers nicotine that quickly reaches the circulatory system and brain. This highly addictive stimulant constricts blood vessels, which makes the heart beat harder and raises blood pressure. Smoking also raises the blood level of "bad" cholesterol (LDL) and lowers that of "good" cholesterol

Figure 35.1 Learning to smoke is easy— compared with trying to quit. In one survey, two-thirds of female smokers between ages sixteen and twenty-four wanted to stop entirely. Among those who tried, only about 3 percent remained nonsmokers for an entire year.

the big picture

Overview of Respiration   Animals engage in aerobic respiration, which requires the rapid uptake of oxygen from outside the body and carbon dioxide removal from the inside. Diverse mechanisms help maintain the steep pressure gradients required for the gas exchanges.

Examples From Invertebrates   Tiny invertebrates of aquatic or continually moist habitats use their body covering as the respiratory surface. In other invertebrates, gills, tracheal tubes, and other internal structures increase the surface area available for gas exchange.

(HDL). It makes blood stickier and more sluggish, and it promotes the formation of clots. Clogged arteries, heart attacks, and strokes are among the physiological costs of a social pressure that leads to addiction.

Which smokers have not heard that they are inviting painful, deadly lung cancers? Maybe they haven't heard that carcinogens in cigarette smoke can induce cancers in organs all through their body. For instance, we now know that females who start smoking when they are teenagers are about 70 percent more likely to develop breast cancer than those who never took up the habit.

The damaging effects of cigarette smoke are not confined to smokers. Their families, coworkers, and friends often get an unfiltered dose of carcinogens in tobacco smoke. Urine samples from nonsmokers who live with smokers show that their body contains high levels of carcinogens. Each year in the United States, secondhand smoke causes an estimated 3,000 deaths from lung cancer. Children exposed to secondhand smoke are more prone to develop respiratory ailments such as asthma, as well as middle ear infections.

This chapter focuses on how respiration contributes to homeostasis in the internal environment. If you or someone you know has joined the culture of smoking, you might use the chapter as a preview of what it does to internal operating conditions. For a more graphic preview, find out what goes on every day with smokers in hospital emergency rooms and intensive care units. No glamour there. It is not cool, and it is not pretty.

 How Would You Vote?

Most smokers pick up the habit when they are in their teens. Many get cigarettes from vending machines. Should cigarette vending machines be banned? See the Media Menu for details, then vote online.

Vertebrate Respiration Most vertebrates have a thin, moist respiratory surface inside a pair of lungs, where blood flowing through a circulatory system picks up oxygen and gives up carbon dioxide. Controls govern the active and passive mechanisms involved in breathing.

Effects of Smoking Airborne pollutants or disease agents can damage the lining of the respiratory tract. Inhaling tobacco smoke invites bronchitis, emphysema, and other respiratory disorders, as well as heart disease and a surprising variety of cancers.

35.1 The Nature of Respiration

All living cells in the animal body depend on the high ATP yield of an oxygen-requiring metabolic pathway called aerobic respiration. So how does oxygen get to them? How are the pathway's metabolic wastes swept away?

THE BASIS OF GAS EXCHANGE

Compared to other organisms, animals actively move about far more during at least some stage of their life cycle. Most can move the body, or parts of it, fast and forcefully. It takes a great amount of energy from ATP to support this activity. Aerobic respiration is the only metabolic pathway that can supply it. In Chapter 7, you saw how this pathway requires free oxygen and releases carbon dioxide as a by-product. In all large-bodied animals, organ systems interact in ways that get free oxygen to individual living cells and remove wastes. Figure 35.2 recaps the most important ones.

It all starts with an organ system that specializes in **respiration**. This vital process happens in all animals, no matter what their size. It provides an ongoing flow of oxygen from outside the body right up to a moist boundary with the internal environment, and a flow of carbon dioxide the other way—out from the body.

Respiration starts with concentration gradients—or, as we say for gases, with *pressure* gradients—between an animal and its surroundings. In this case, such a gradient is the difference in the pressure exerted by the number of molecules of a gas between two regions. The steeper the gradient between two areas, the faster the gas will diffuse between them.

At sea level, air contains about 78 percent nitrogen, 21 percent oxygen, 0.04 percent carbon dioxide, and 0.04 percent other gases. The atmospheric pressure at sea level is 760 mm Hg, as measured with a mercury barometer. Figure 35.3 is a diagram of such a device. Oxygen's **partial pressure**—its contribution to the total atmospheric pressure—is 21 percent of that total. It is about 160 mm Hg at sea level.

Gases enter and leave an animal body by crossing a **respiratory surface**. This is a thin layer of epithelium or some other tissue that must be kept moist at all times. Why? Gases can diffuse across a surface only if they are dissolved in fluid. How many molecules can cross a respiratory surface in a given interval? As you will see, the rate depends on several factors, such as surface area, partial pressure gradients, ventilation, and the role of respiratory pigments.

FACTORS INFLUENCING GAS EXCHANGE

SURFACE-TO-VOLUME RATIO The surface-to-volume ratio explained in Section 4.1 puts constraints on gas exchange. Unless an animal has respiratory organs, it must be tiny, tubelike, or flattened. Gases can diffuse directly across the surface of such bodies.

Think back on that flatworm in Figure 24.5. Could it grow in all directions, like a balloon inflating? No. The surface area would not increase at the same rate as its volume. Once its diameter exceeded even a few millimeters, the diffusion distance between the body surface and its internal cells would be too great. Gas exchange would be too slow; the worm would die.

VENTILATION Large-bodied, active animals require faster gas exchange than diffusion alone can provide. Diverse adaptations enhance the exchange rate. For instance, a flap covers many fish gills. The fish forces

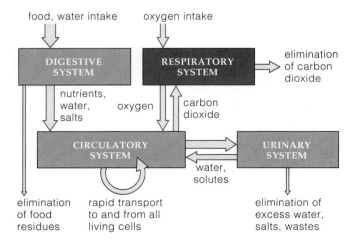

Figure 35.2 Interactions between the respiratory system and other organ systems, the foundation for homeostasis in complex animals.

Figure 35.3 How a mercury barometer can measure the atmospheric pressure. A narrow column of mercury (Hg) rises or falls in a tube, depending on air pressure outside. At sea level, it rises to 760 millimeters (29.91 inches) from the tube's base. Altitude affects the level of atmospheric pressure. On top of Mount Everest, the air pressure is about one-third of what it is at sea level.

Figure 35.4 What a gas! (**a**) An operculum is a stiffened covering over each fish gill. It stirs the surrounding water by opening and closing, which assists in the diffusion of gases dissolved in the water. (**b**,**c**) In mammals, the chest cavity expands and contracts, which helps move gases in and out from their mouth and nose.

operculum

water past these respiratory surfaces by moving the mouth and the flaps. Vertebrate circulatory systems swiftly transport gases to and from the gills or lungs. When dogs and humans breathe in and out, they are ventilating their lungs (Figure 35.4).

RESPIRATORY PIGMENTS Most animals have some type of **respiratory pigment**. Such proteins contain one or more metal ions that reversibly bind oxygen. As you know from Sections 3.6 and 33.2, vertebrate red blood cells are packed with the iron-containing pigment called **hemoglobin**. It is similar to annelid, mollusk, and crustacean hemoglobin, although the one in annelids is six times bigger than yours. It may consist of 200 globin chains that each bind an oxygen molecule. Other invertebrates have hemerythrin (with iron groups) or hemocyanin (copper groups) that reversibly bind oxygen.

Respiratory pigments boost gas exchange rates by helping to maintain steep pressure gradients across a respiratory surface. In human lungs, for instance, the concentration of oxygen is high at the respiratory surface. Blood transports red blood cells right past this surface, and each hemoglobin molecule inside these cells binds as many as four oxygen molecules. In the capillary beds of oxygen-poor tissues, oxygen follows its concentration gradient. It diffuses away from hemoglobin, into interstitial fluid, then into cells.

Myoglobin is another iron-containing respiratory pigment. It acts in oxygen storage and is abundant in vertebrate cardiac muscle and some skeletal muscle. Its folded structure resembles a single globin chain in hemoglobin (Figure 35.5). But myoglobin has a higher oxygen affinity. When hemoglobin releases oxygen near a cardiac muscle cell, the gas diffuses into the cell and binds to myoglobin. If metabolic levels in the

heme

heme

a

b

Figure 35.5 (**a**) Model for myoglobin. This molecule resembles the globin chains of hemoglobin (**b**). By one hypothesis, genes for the four polypeptide chains of hemoglobin genes arose through duplications and mutations of ancestral genes for a myoglobin-like protein. Some bacteria and archaea have such a protein, which helps them detect oxygen levels in the environment.

muscle cell rise and oxygen levels decline, the heme group of myoglobin releases oxygen. In this way, it helps stabilize intracellular oxygen levels.

The process of respiration supplies animal cells with oxygen for aerobic respiration and removes this metabolic pathway's carbon dioxide wastes from the body.

The process moves oxygen from outside the body to a respiratory surface—a thin, moist boundary to the internal environment. It also carries away the carbon dioxide wastes brought to the surface from inside the body.

Gas exchange requires steep partial pressure gradients between the external and internal environments. Ventilation helps keep the gradients steep. So do hemoglobin and other respiratory pigments that reversibly bind oxygen.

35.2 Invertebrate Respiration

Animals exchange gases across moist respiratory surfaces. In the smallest and thinnest species, the respiratory surface is the body wall. Larger species have organs with highly folded surfaces or multiple branchings that increase the area available for the exchanges.

We find the simplest forms of respiration among the small-bodied invertebrates of aquatic or continually moist habitats. Gases diffuse directly across the body surface covering—the integument (Figure 35.6a). This respiration mode is called **integumentary exchange**.

Many aquatic invertebrates have thin-walled, moist respiratory organs called **gills**, although these are not the same in structure as fish gills. Extensively folded gill walls increase the respiratory surface area and gas exchange rates between body fluids and the outside. Figure 35.6b shows the much-folded gills of a sea hare (*Aplysia*) that supplement integumentary exchange.

Crustaceans have feathery gills that evolved from walking legs. In crabs and lobsters, a hard body cover (exoskeleton) protects the gills and other soft parts.

Spiders and other arthropods on dry land have a small, thick, or hardened integument that is not well supplied with blood vessels. Their integument helps conserve water but is not much good at gas exchange. Such animals have an *internal* respiratory surface. For instance, most spiders have internal respiratory organs with thin, folded sheets that vaguely resemble book pages; hence the name, book lungs (Figure 35.7).

Insects, millipedes, centipedes, and some spiders have a **tracheal system** of branching tubes inside the body. Figure 35.8 shows an example from one type of insect. Many tiny openings, called spiracles, perforate the integument on both sides of some body segments. They are the start of tracheal tubes that branch, then branch again. Tips of the finest branchings end right next to cells in all body tissues.

In the tracheal system of small, relatively inactive insects, gases simply diffuse across the tube walls. In a mosquito larva, an aquatic stage, a "breathing tube" connected to the tracheal system extends above the water surface. Even in gilled aquatic insects, gases

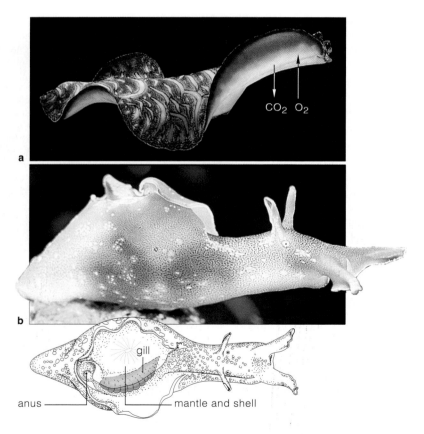

Figure 35.6 Two aquatic invertebrates. (**a**) A flatworm, small enough to get along without an oxygen-transporting circulatory system. Dissolved oxygen in its habitat reaches individual cells simply by diffusing across the body surface. (**b**) Gill of a sea hare (*Aplysia*), one of the gastropod mollusks.

Figure 35.7 Book lungs. In many spiders, including this jumping spider (*Phidippus*), the respiratory surface consists of highly folded, moistened sheets of tissue inside the body. Jumping spiders also have a pair of tracheal tubes.

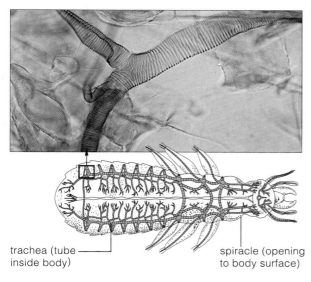

trachea (tube inside body)

spiracle (opening to body surface)

Figure 35.8 Generalized insect tracheal system. Chitin rings reinforce many of the system's branching tubes.

diffuse through a tracheal system. What about large, active insects on land? Like plants, they cannot afford to lose much body water. How do they hold on to it?

Consider the fruit fly. Its spiracles have muscled valves that are open during oxygen-demanding flight but close a bit (and conserve water) in times of rest. What about active animals larger than flies? Consider the grasshopper. It forcefully ventilates its tracheal tubes. When its abdominal muscles contract, internal organs press into the thin, pliable tracheal tubes and force air from them. When the muscles relax, organs spring back, the tracheal lumen expands, and air is sucked in. The largest tracheal tubes also have air sacs attached to them. The sacs permit a greater volume of air to be pulled inside and expelled during each round of muscle contraction and relaxation.

Flatworms and some other invertebrates of aquatic or moist habitats do not have a massive body. By integumentary exchange, gases diffuse directly across their body surface.

Most marine and freshwater invertebrates have gills with moist, thin, and often highly folded walls.

Most insects, millipedes, centipedes, and some spiders use tracheal respiration. Gases flow through open-ended tubes that start at spiracles and end directly next to living cells in all tissues. Most spiders have book lungs.

On land, small, highly active insects control water loss by partially closing spiracles when they are resting. Some large types actively ventilate the tubes with the help of abdominal muscle contractions and air sacs attached to the tubes.

35.3 Gasping for Oxygen

FOCUS ON THE ENVIRONMENT

Pity the aquatic species in lakes, streams, and estuaries polluted with manure, sewage, and fertilizers. The inputs fan the growth of aquatic populations and thereby deplete dissolved oxygen. Species also smother to death when sediments clog or damage their respiratory surface.

All animals tolerate a limited range of environmental conditions. For aquatic species, the dissolved oxygen content of water (DO) is one of the most important factors affecting their survival. Also, how hot is the water? How fast is it flowing? Warmer water holds less oxygen than cooler water. And when water moves fast, more oxygen from the air can be dissolved in it. In warm, still waters, you probably will find that the oxygen content is dangerously low.

As oxygen levels in water fall, so does biodiversity. Pollution can cause DO to plummet. A lake enriched with inflows that contain manure or sewage offers a nutrition boost to aerobic bacteria on the lake bottom. Such bacteria are decomposers. As their populations soar, they use up oxygen, so the amount available to other species plummets. The same thing can happen after phosphate- or nitrogen-rich fertilizers cause an *algal bloom*. This is a population explosion of protists, such as dinoflagellates (Section 20.4). The algal cells multiply, then die. Their metabolism and their later decomposition deplete the water of oxygen.

In freshwater lakes and streams, the aquatic larvae of mayflies and stoneflies are the first invertebrates to disappear when oxygen levels fall. These insect larvae are active predators that demand considerable oxygen. Gilled snails disappear, too. Such declines can have cascading effects on fishes that feed on them. Some fish are more directly affected. Trout and salmon are especially intolerant of low oxygen. Carp (including koi and goldfish) are among the most tolerant of oxygen declines; they can survive even in warm algae-rich ponds or tiny goldfish bowls.

When oxygen levels fall below 4 parts per million, no fish can survive. Leeches thrive as most competing invertebrates disappear. Water with the lowest oxygen level harbors annelids called sludge worms (*Tubifex tubifex*) and bloodworms (*Chironomus*), the aquatic larvae of tiny flies or midges. Both are colored bright red by the hemoglobin that allows them to live in oxygen-poor waters as long as food is plentiful and competitors and predators are few.

Even sludgeworms and bloodworms are vulnerable to suffocation. Deposits of fine silt particles can coat them and prevent gas flow across their integument. Mining, logging, and stripping land for agriculture or development can put silt in runoff from the land.

no oxygen, no fishes

35.4 Vertebrate Respiration

Depending on the species, vertebrates use gills, their skin, paired lungs, or a combination of skin and lungs as their respiratory organs.

GILLS OF FISHES AND AMPHIBIANS

Gills are organs that have a thin, moist, vascularized respiratory surface, and they vary among vertebrates. Larvae of a few fishes have *external* gills that project into water. So do a few amphibians. Adult fishes have a pair of *internal* gills. Figure 35.9 shows overlapping rows of gill filaments, which define slits that extend from the back of the mouth to the body surface.

In fishes, water flowing in from the mouth flows through the gill slits. It passes arrays of gill filaments that house blood vessels. The first vessel is moving oxygenated blood *from* the gill back to the main body mass. This blood has less oxygen than the water does, so oxygen diffuses into it. Next, the water flows over a different vessel moving oxygen-poor blood *into* the gill. The water already gave up some oxygen, but it still has more than the blood inside this vessel. So a second helping of oxygen diffuses in.

Movement of two fluids in opposing directions is called **countercurrent flow**. It lets a fish extract about 80–90 percent of the oxygen dissolved in the water flowing past. That is more oxygen than it would get from a one-way flow mechanism, at less energy cost.

EVOLUTION OF PAIRED LUNGS

Some fishes and all amphibians, birds, and mammals have a pair of **lungs**. A lung is a saclike respiratory organ that is inside some body cavity but connected by airways to the surface. More than 450 million years ago, lungs evolved in some fishes as outpouchings of the gut wall. Lungs increased the surface area for gas exchange in oxygen-poor habitats. Also, vertebrates that moved onto dry land could not depend on gills. Gill filaments get stuck together when flowing water is not available to keep them moist.

Most amphibian larvae have gills, and adults have lungs. Some adult salamanders have neither lungs nor gills. Their skin functions in integumentary exchange. Frogs and toads use small lungs for oxygen uptake, but carbon dioxide diffuses outward across their skin. A frog breathes, but not like you do. Air flows in its nostrils, then it is gulped down into the lungs. Stale air is forced out from the lungs when muscles in the body wall contract (Figure 35.10).

Figure 35.9 (**a–c**) Fish gill ventilation. Water is forced past a gill when the mouth opens and the operculum, a lid, closes (Figure 35.4). It flows out when the mouth closes and the lid opens.

(**d**) Gas exchange in fish gills. Each gill filament is a vascularized respiratory surface divided into a series of smaller surfaces that expose more surface area to water flowing past them.

(**e**) A blood vessel moves oxygen-poor blood from the body into each filament. Another vessel carries oxygenated blood into the body. Capillaries move blood from the second vessel into the first, counter to the direction of water flow. The countercurrent flow favors movement of oxygen (down its partial pressure gradient) from water into the blood.

Figure 35.10 (**a**) Frog respiration. (**b**) The frog lowers the floor of its mouth and inhales air through its nostrils. (**c**) It closes its nostrils, opens the glottis, and elevates the mouth's floor. This *forces* air into the lungs. (**d**) Rhythmic ventilation assists in gas exchange. (**e**) Air is forced out when muscles in the body wall above the lungs contract and lungs elastically recoil.

amphibian salamander; like fishes and early amphibians

amphibian frog; only adults are adapted to dry habitats

reptile lizard; adapted to dry habitats

mammal human; adapted to dry habitats

Figure 35.11 Lung structure of three kinds of vertebrates, suggestive of an evolutionary trend from sacs for simple gas exchange to larger, more complex respiratory surfaces.

Frogs use air flowing from lungs to make sounds. So do all mammals except whales. Sounds start at the entrance of an airway to lungs, as explained shortly.

Paired lungs are the dominant respiratory organs in living reptiles, birds, and mammals (Figure 35.11). Breathing moves air by bulk flow into and out of the lungs, where blood capillaries thread lacily around extensive respiratory surfaces. In response to steep gradients, oxygen and carbon dioxide rapidly diffuse across the surfaces. Oxygen entering the capillaries is circulated through the body. In tissues where its level is low, this gas diffuses into interstitial fluid, then into cells. Carbon dioxide diffuses from cells, to interstitial fluid, then blood, then is expelled from the lungs.

This mode of gas exchange is notably embellished only in birds (Figure 35.12). Birds have a system of air sacs that help force air continuously *through* a pair of lungs. With this exception in mind, we turn next to the human respiratory system. Its operating principles apply to most vertebrates.

A countercurrent flow mechanism in fish gills boosts oxygen uptake from the water. Gills cannot function in dry land habitats, where internal air sacs— lungs—are more efficient.

Amphibians force oxygen into small lungs, but carbon dioxide diffuses out across the skin. Reptiles, birds, and mammals use paired lungs as their respiratory organs. Birds also have air sacs that help keep air flowing continuously through air tubes in the respiratory surface of each lung.

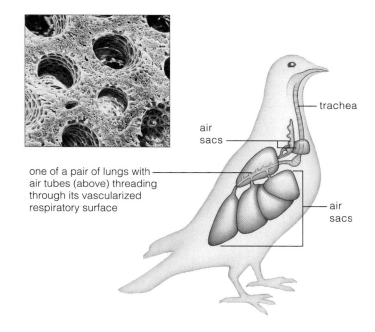

one of a pair of lungs with air tubes (above) threading through its vascularized respiratory surface

trachea

air sacs

air sacs

Figure 35.12 Respiratory system of a representative bird (pigeon). Many air sacs are attached to two small, inelastic lungs. The bird inhales and draws air into the sacs through tubes, open at both ends, that thread through vascularized lung tissue. This tissue is the respiratory surface.

The bird exhales, forcing air from the sacs, through the tubes, and out of the trachea. Air is not just drawn into bird lungs. It is drawn continuously through them and across the respiratory surface. This unique ventilating system supports the high metabolic rates birds require for flight and other energy-intensive activities.

35.5 Human Respiratory System

It will take at least 300 million breaths to get you to age seventy-five. The system that brings about each breath also functions in speech, in the sense of smell, and in homeostatic control of the internal environment.

THE SYSTEM'S MANY FUNCTIONS

Figure 35.13 shows the human respiratory system and lists its functions. Its two paired lungs have about 600 million air sacs, the **alveoli** (singular, alveolus). Rates of air inflow and outflow to the alveoli are adjusted to match metabolic demands for gas exchange.

The respiratory system functions in more than gas exchange. It assists in sense of smell and in speech. It helps blood in veins return to the heart and helps the body dispose of excess heat and water. Controls over breathing also contribute to maintaining the internal environment's acid–base balance. Parts of the system intercept many potentially dangerous airborne agents and substances before they can reach the lungs. Many of the bloodborne substances are removed, modified, or inactivated before they circulate through the body.

The respiratory system ends at alveoli, and there it exchanges gases with the pulmonary capillaries. (The

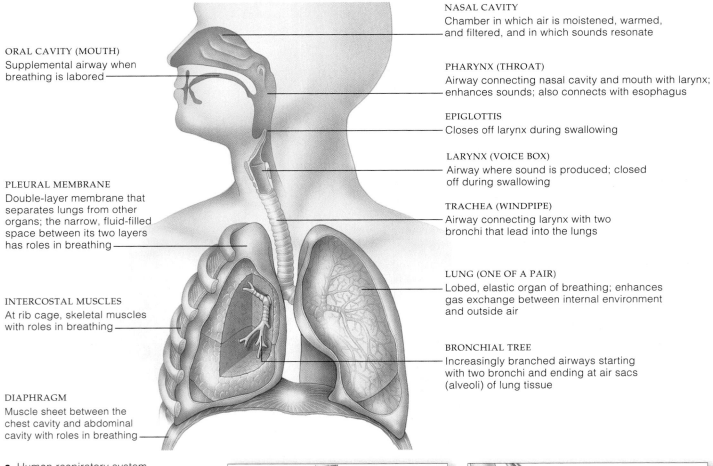

NASAL CAVITY
Chamber in which air is moistened, warmed, and filtered, and in which sounds resonate

ORAL CAVITY (MOUTH)
Supplemental airway when breathing is labored

PHARYNX (THROAT)
Airway connecting nasal cavity and mouth with larynx; enhances sounds; also connects with esophagus

EPIGLOTTIS
Closes off larynx during swallowing

LARYNX (VOICE BOX)
Airway where sound is produced; closed off during swallowing

PLEURAL MEMBRANE
Double-layer membrane that separates lungs from other organs; the narrow, fluid-filled space between its two layers has roles in breathing

TRACHEA (WINDPIPE)
Airway connecting larynx with two bronchi that lead into the lungs

LUNG (ONE OF A PAIR)
Lobed, elastic organ of breathing; enhances gas exchange between internal environment and outside air

INTERCOSTAL MUSCLES
At rib cage, skeletal muscles with roles in breathing

BRONCHIAL TREE
Increasingly branched airways starting with two bronchi and ending at air sacs (alveoli) of lung tissue

DIAPHRAGM
Muscle sheet between the chest cavity and abdominal cavity with roles in breathing

a Human respiratory system

Figure 35.13 (**a**) Human respiratory system: its components and their functions. Muscles, including the diaphragm, and parts of the axial skeleton have secondary roles in respiration. (**b,c**) Location of alveoli relative to bronchioles and to lung (pulmonary) capillaries.

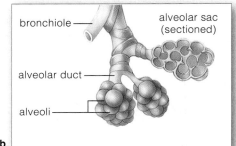

bronchiole

alveolar sac (sectioned)

alveolar duct

alveoli

b

alveolar sac

pulmonary capillary

c

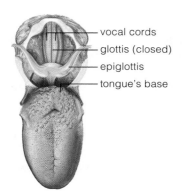

vocal cords
glottis (closed)
epiglottis
tongue's base

glottis closed   glottis open

Figure 35.14 Human vocal cords, where the sounds used in speech originate. Commands from the nervous system coordinate contractions in skeletal muscles that can change the glottis, the gap between them. This series of photographs shows how the glottis changes when we take a normal breath.

Latin *pulmo* means lung.) This is where the circulatory system takes over the task of gas exchange.

FROM AIRWAYS INTO THE LUNGS

Take a deep breath. Now look at Figure 35.13 to get an idea of where the air will travel in your respiratory system. Unless you are out of breath and panting, air has entered two nasal cavities, not your mouth. In the nose, mucus warms and moistens the air. Hairs and cilia of the nasal lining filter dust and particles from it, and sensory receptors detect odors. Air flows into the **pharynx**, or throat. This is the entrance to the **larynx**, a large airway with two paired folds of mucus-covered membrane projecting into it. The lower pair are vocal cords (Figure 35.14). When you breathe, you force air in and out past the glottis, the gap between the cords. The air flow causes these cords to vibrate, and we make different sounds by controlling the vibrations.

While the larynx muscles contract and relax, elastic ligaments in the vocal cords tighten or slacken to alter how much the folds are stretched. The nervous system coordinates the narrowing and widening of the glottis. When its signals cause increased tension in the larynx muscles, for example, the gap between the vocal cords narrows and so high-pitched sounds emerge. The lips, teeth, tongue, and the soft roof above the tongue can be enlisted to modify the basic patterns of sounds.

When infected or irritated, vocal cords swell, which interferes with their capacity to vibrate. Severe swelling can cause *laryngitis*, or a hoarse throat.

At the entrance to the larynx is an **epiglottis**. When this tissue flap points up, air moves into the **trachea**, or windpipe. When you swallow, the epiglottis flops over and points downward. In this position, it covers the entrance to the larynx, so food and fluids can enter a different tube. That tube, the esophagus, is not one of the airways. It connects the pharynx to the stomach.

The trachea branches into two airways: one to each lung. Each airway is a **bronchus** (plural, bronchi). Its epithelial lining has many cilia and mucus-secreting cells that work against infection. Bacteria and airborne particles stick to the mucus; cilia sweep the mucus to the mouth, where it can be expelled.

Human lungs are elastic, cone-shaped organs in the thoracic cavity, one on each side of the heart. The rib cage protects them. Below them is the **diaphragm**, a broad muscle that partitions the coelom into thoracic and abdominal cavities (Figure 28.11). A moist, thin, saclike pleural membrane lines the outer lung surface and the inner surface of the thoracic cavity wall. The wall and lungs press the sac's two surfaces together.

A film of lubricating fluid cuts friction between the pleural membrane surfaces. In *pleurisy*, a respiratory ailment, the membrane gets inflamed and swollen. Its two surfaces rub each other, so breathing is painful.

Inside each lung, air moves through finer and finer branchings of the "bronchial tree." The tree's airways are **bronchioles**. Its finest branchings, the *respiratory* bronchioles, end in cup-shaped alveoli. Most alveoli are clustered as pouches, or alveolar sacs. Collectively, the great numbers of sacs afford a tremendous surface area for gas exchange with blood. If all alveolar sacs in your body could be stretched out in one layer, they would cover the surface of a racquetball court!

The human respiratory system gets oxygen from air to a pair of lungs and gets carbon dioxide out of them, and out of the body. In the lungs, the circulatory system takes over the task of gas exchange in the internal environment.

The respiratory system also has roles in moving venous blood to the heart, vocalizing, adjusting the body's acid–base balance, defending against harmful airborne agents or substances, and removing or modifying certain blood-borne substances.

35.6 The Respiratory Cycle

Breathing ventilates the lungs in a cyclic pattern that enhances the flow of gases between the atmosphere and alveolar sacs.

CYCLIC REVERSALS IN AIR PRESSURE GRADIENTS

Every **respiratory cycle** is one breath in (inhalation) and one breath out (exhalation). *Inhalation is always active; muscle contractions drive it.* Changes in lung and thoracic cavity volume during each respiratory cycle shift the pressure gradients between air inside and outside the respiratory tract. After each breath taken, pressure inside the alveoli is the same as the pressure in airways and the outside environment. Atmospheric pressure, again, is about 760 mm Hg at sea level.

Two mechanisms expand the lungs as the thoracic cavity's volume increases. First, pressure inside the pleural sac helps keep a lung close to the cavity wall.

In a person who is resting, pressure in the pleural sac averages 756 mm Hg, just a bit less than atmospheric pressure. It pushes in on the wall of both lungs even as the pressure inside all the alveoli pushes outward. The difference between the two (4 mm Hg) is enough to make the lungs stretch and fill the thoracic cavity.

Second, fluid cohesion helps each lung stick to the thoracic cavity wall. Press together two wet panes of glass. They slide back and forth easily, but they resist being pulled apart. Similarly, fluid within the pleural sac "glues" the lungs to the wall.

Figure 35.15*b* shows what happens when you start to inhale. The diaphragm flattens and moves down. External *intercostal* muscles (between the ribs) contract and lift the rib cage up and outward. As the thoracic cavity expands, so do the lungs. When pressure in the alveoli decreases below atmospheric pressure, air flows down the pressure gradient, into the airways.

Read Me First!
and watch the narrated animation on the respiratory cycle

INWARD
BULK FLOW
OF AIR

OUTWARD
BULK FLOW
OF AIR

(a)

(b) Inhalation. The diaphragm contracts, moves down. External intercostal muscles contract and lift rib cage upward and outward. The lung volume expands.

(c) Exhalation. Diaphragm, external intercostal muscles return to resting positions. Rib cage moves down. Lungs recoil passively.

Figure 35.15 **(a)** Location of human lungs relative to the diaphragm. **(b,c)** Changes in the thoracic cavity's size in one respiratory cycle. The x-ray images show how maximum inhalation changes the lung volume.

Exhalation is passive when you breathe quietly. The muscles that brought about inhalation relax and the lungs passively recoil. As the lung volume decreases, air in the alveolar sacs is compressed. This compression increases the pressure in the alveoli, which now becomes greater than atmospheric pressure. The new pressure gradient draws air out from the lungs, as Figure 35.15c shows.

Exhalation becomes active and energy-demanding only when you are exercising vigorously and so must expel more air. The abdominal wall muscles contract during active exhalation. Abdominal pressure increases and exerts an upward-directed force on the diaphragm, which moves upward. The internal intercostal muscles contract, also. As they do, they pull the thoracic wall inward and down. This action puts the chest wall into a flattened position and decreases the volume of the thoracic cavity. The lung volume also decreases when the lung's elastic tissue passively recoils.

LIVING THE HIGH LIFE

At high altitudes, atmospheric pressure is so low that oxygen cannot diffuse as easily into the blood in lung capillaries. Recall, from Section 24.3, that people who travel from low elevations to mountains 2,400 meters (about 8,000 feet) or more above sea level risk *hypoxia*, or cellular oxygen deficiency. Sensing the deficiency, the brain calls for *hyperventilation*, or breathing faster and deeper than usual. Hyperventilating can disrupt ion balances in cerebrospinal fluid. The imbalances typically cause shortness of breath, headaches, nausea, vomiting, and arrhythmias. It can even lead to death.

By the mechanisms of **acclimatization**, lowlanders can make physiological and behavioral adaptations to conditions at higher altitudes. They gradually alter their pattern and magnitude of breathing, and their cardiac output. The initial, acute compensations for hypoxia subside in response.

Within a few days, the reduced oxygen delivery stimulates kidney cells to secrete more erythropoietin, a hormone that induces stem cells in bone marrow to divide repeatedly. Their descendants mature into red blood cells. Stepped-up erythropoietin secretion can increase the pace by as much as six times. With more red blood cells, blood's oxygen deliveries increase. And when enough oxygen steadily reaches cells, the kidneys slow down erythropoietin secretion.

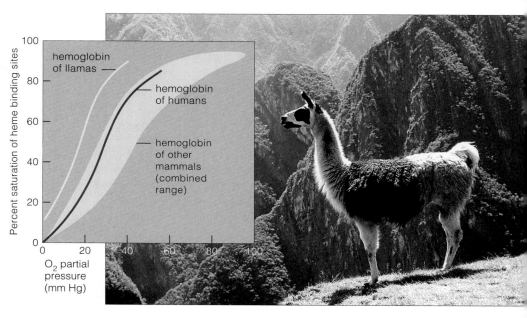

Figure 35.16 Comparison of hemoglobin's binding and releasing capacity for humans, llamas, and other animals.

Human populations that evolved at high altitudes are adapted to low oxygen levels. Compared to those who evolved at lower altitudes, for instance, Tibetan mountain dwellers take more breaths per minute, and their cells have fewer mitochondria. People born in the Andes have higher levels of hemoglobin and more red blood cells than those living in the lowlands.

What about animals that live at high altitudes? As Section 16.4 mentions, llama hemoglobin differs from ours. It binds oxygen more efficiently than ours when oxygen concentration is low (Figure 35.16).

In most mammals, a metabolic product called DPG (2,3-diphosphoglycerate) binds to hemoglobin, which cuts its oxygen-binding capacity. Llama hemoglobin has a reduced affinity for DPG. High-altitude deer mice adapted differently. They keep the DPG level low, which increases the oxygen-binding capacity of hemoglobin without modifications to the molecule.

The respiratory cycle consists of inhalation (one breath in) and exhalation (one breath out). Breathing reverses pressure gradients between the lungs and air outside the body.

Inhalation is always active; it requires muscle contractions. During quiet breathing, exhalation is a passive process. Forceful exhalation requires muscle contractions.

Oxygen levels decrease as altitude increases. Short-term adjustments alter breathing patterns and red blood cell production. Over generations, a high altitude selects for changes in physiology and biochemistry.

35.7 Gas Exchange and Transport

A partial pressure gradient drives oxygen from alveolar sacs into interstitial fluid in the lungs, then into blood, then into tissues throughout the body. Carbon dioxide's pressure gradient drives it in the opposite direction.

EXCHANGES AT THE RESPIRATORY MEMBRANE

An alveolus is a cupped sheet of epithelial cells with a basement membrane. A pulmonary capillary consists of endothelial cells. Its basement membrane is fused with that of the alveolus. The alveolar and capillary endothelia, together with their basement membranes, form the **respiratory membrane** (Figure 35.17). Gases diffuse quickly across this thin membrane.

Both oxygen and carbon dioxide passively diffuse across the respiratory membrane. An inward-directed gradient of oxygen is maintained because inhalations replenish oxygen and cells take it up continually. An outward-directed gradient of carbon dioxide prevails, because cells continually produce carbon dioxide and exhalation removes it.

OXYGEN TRANSPORT

Oxygen does not dissolve well in blood, so it cannot be transported efficiently on its own. In vertebrates, as in other animals, oxygen transport is enhanced by hemoglobin molecules. Red blood cells are packed with this respiratory pigment. A hemoglobin molecule, recall, has four polypeptide chains and four **heme groups**. An iron atom in each heme can reversibly bind oxygen. *Compared to plasma, hemoglobin carries about sixty-seven times more oxygen.* Normally, inhaled air that reaches alveoli has a lot of oxygen, but blood in pulmonary capillaries does not. In lungs, oxygen tends to diffuse into blood's plasma. From there, it

diffuses into red blood cells and binds reversibly with hemoglobin. The resulting **oxyhemoglobin**, or HbO_2, is the form in which oxygen is transported.

The amount of HbO_2 that forms in a given interval depends on oxygen's partial pressure. The higher the pressure, the greater the oxygen concentration, and the more frequently oxygen molecules will randomly collide with heme binding sites.

HbO_2 molecules bind oxygen weakly, and they give it up where oxygen's partial pressure is less than it is in the lungs. The oxygen-binding capacity is low in regions where the blood is warmer, the pH is lower, and carbon dioxide's partial pressure is high. These conditions prevail in metabolically active tissues, and that is where blood gives up oxygen the fastest.

CARBON DIOXIDE TRANSPORT

Carbon dioxide diffuses into blood capillaries in any tissue where its partial pressure is higher than it is in the blood. From there, three mechanisms transport it to lungs. About 10 percent stays dissolved in blood. Another 30 percent binds with some amino acids in hemoglobin to form carbaminohemoglobin ($HbCO_2$). However, most of it—60 percent—is transported as bicarbonate (HCO_3^-). How does bicarbonate form? Carbon dioxide combines with water to form carbonic acid, but this separates into bicarbonate and H^+:

$$CO_2 + H_2O \rightleftharpoons \underset{\text{CARBONIC ACID}}{H_2CO_3} \rightleftharpoons \underset{\text{BICARBONATE}}{HCO_3^-} + H^+$$

A few HCO_3^- molecules form spontaneously, but the vast majority form in red blood cells. Here, action of the enzyme carbonic anhydrase makes carbon dioxide levels fall rapidly. This helps maintain the gradient

a Surface view of capillaries associated with alveoli

pore for air flow between adjoining alveoli

red blood cell inside pulmonary capillary

air space inside alveolus

b Cutaway view of one of the alveoli and adjacent pulmonary capillaries

alveolar epithelium

capillary endothelium

fused basement membranes of both epithelial tissues

c Three components of the respiratory membrane

Figure 35.17 Zooming in on the respiratory membrane in human lungs.

Figure 35.18 The partial pressure gradients for oxygen (*blue* boxes) and carbon dioxide (*pink* boxes) in the respiratory tract and tissues.

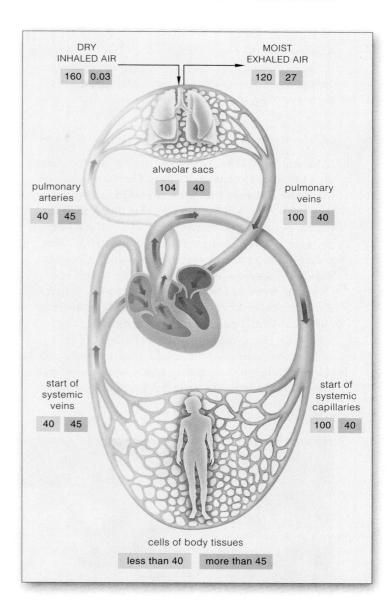

that causes carbon dioxide to diffuse from interstitial fluid into the bloodstream.

The HCO_3^- that forms in red blood cells tends to diffuse into plasma. Hemoglobin acts as a buffer for H^+ ions that form, and so keeps blood from becoming too acidic. A *buffer*, recall, is a molecule that combines with or releases H^+ whenever cellular pH shifts.

The reactions are reversed in alveoli, where carbon dioxide's partial pressure is lower than it is in lung capillaries. Along with water, carbon dioxide forms in the alveolar sacs and is exhaled.

At this point, reflect on Figure 35.18. It summarizes the partial pressure gradients for oxygen and carbon dioxide through the human respiratory system.

BALANCING AIR AND BLOOD FLOW RATES

The nervous system controls both the *rhythm* and the *magnitude* (rate and depth) of breathing. A group of neurons in the brain's medulla oblongata sets the pace for signals that cause the muscles involved in inhalation to contract. When exhalation must be active and strong, other neurons in this brain center issue signals to the muscles involved in exhalation.

Fast, deep breaths expel more carbon dioxide. But carbon dioxide builds up in the body when breaths are shallow and slow. Chemoreceptors in the brain monitor CO_2. A high CO_2 level triggers active, strong contractions by the muscles of exhalation. O_2 and H^+ concentrations also are cues for the mechanisms that adjust breathing. For instance, receptors at the start of carotid arteries notify the brain when oxygen's partial pressure falls below 60 mm Hg. Below this level, the delivery of oxygen to tissues is inadequate for normal function. In response to the chemoreceptor signal, the brain calls for an increase in the rate of breathing.

In some situations, a person "forgets" to breathe. With *apnea*, breathing stops and then spontaneously restarts. Some individuals are affected by sleep apnea. Their breathing stops for a few seconds or minutes, or in rare cases, as much as 500 times a night.

With *sudden infant death syndrome* (SIDS), a sleeping infant cannot wake up from an apneic episode and dies. SIDS has been linked to mutations that result in an irregular heartbeat. The risk is three times higher among infants of women who smoked cigarettes while pregnant. Infants who sleep on their back or sides are less vulnerable than those who sleep stomach-down.

Driven by its partial pressure gradient, oxygen diffuses from alveolar air spaces, through interstitial fluid, and into lung capillaries. Carbon dioxide, driven by its partial pressure gradient, diffuses in the opposite direction.

Hemoglobin in red blood cells enormously enhances the oxygen-carrying capacity of blood. Most carbon dioxide is transported in blood in the form of bicarbonate, nearly all of which forms by enzyme action in red blood cells.

Respiratory centers in the brain stem control the rhythmic pattern and magnitude of breathing, in response to carbon dioxide, oxygen, and hydrogen ion levels in arterial blood.

35.8 When the Lungs Break Down

In large cities, in certain workplaces, and even in the cloud around a cigarette smoker, airborne particles and gases are present in abnormally high concentrations. They put extra workloads on the respiratory system.

BRONCHITIS AND EMPHYSEMA

Ciliated, mucus-producing epithelium lines the inner wall of your bronchioles (Figure 35.19). It is one of the many defenses that protect you from respiratory infections. Chronic irritation of this lining may lead to *bronchitis*. In this disease, epithelial cells become irritated and they secrete too much mucus. Mucus accumulates and bacteria may grow in it. Coughing expels the secretions.

Initial attacks of bronchitis are treatable. When the aggravation continues, bronchioles can become chronically inflamed. Bacteria, chemical agents, or both may attack the wall of bronchioles. Ciliated cells of the walls are killed, and mucus-secreting cells multiply. Fibrous scar tissue forms and in time may narrow or obstruct the airways. Breathing becomes labored and difficult.

Chronic bronchitis can pave the way for *emphysema*. In this condition, tissue-destroying bacterial enzymes go to work on the thin, stretchable walls of alveoli. The walls crumble, and inelastic fibrous tissue builds up around them. Alveoli become enlarged, and not as many exchange gases. In time, the lungs are distended and inelastic, so the balance between air flow and blood flow is permanently compromised. It becomes hard to catch a breath. In any given year, emphysema affects about 3 million people in the United States.

A few people are genetically predisposed to develop emphysema. They do not have a workable gene for the enzyme antitrypsin, which can inhibit bacterial attack on alveoli. Poor diet and persistent or recurring colds and other respiratory infections also invite emphysema later in life. Air pollution and chemicals in the workplace may contribute to the problem. But smoking is by far the major cause of the disease. Emphysema may develop slowly, over twenty or thirty years. When severe damage is not detected in time, lung tissues cannot be repaired (Figure 35.20). Emphysema kills more than 16,000 people each year, and debilitates millions more.

free surface of a mucus-secreting cell

free surface of a cluster of ciliated cells

a b

Figure 35.19 (**a**) Cigarette smoke in the airways irritates the ciliated and mucus-secreting cells that line the two bronchi that lead into the paired lungs (**b**). Repeated irritation can cause chronic bronchitis.

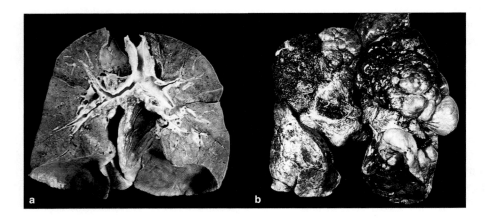

a b

Figure 35.20 (**a**) Appearance of normal lung tissue in humans. (**b**) Appearance of lung tissues from someone affected by emphysema.

Risks Associated With Smoking	Reduction in Risks by Quitting
SHORTENED LIFE EXPECTANCY: Nonsmokers live 8.3 years longer on average than those who smoke two packs daily from their midtwenties on.	Cumulative risk reduction; after 10 to 15 years, life expectancy of ex-smokers approaches that of nonsmokers.
CHRONIC BRONCHITIS, EMPHYSEMA: Smokers have 4–25 times higher risk of dying from these diseases than do nonsmokers.	Greater chance of improving lung function and slowing down rate of deterioration.
CANCER OF LUNGS: Cigarette smoking is the major cause.	After 10 to 15 years, risk approaches that of nonsmokers.
CANCER OF MOUTH: 3–10 times greater risk among smokers.	After 10 to 15 years, risk is reduced to that of nonsmokers.
CANCER OF LARYNX: 2.9–17.7 times more frequent among smokers.	After 10 years, risk is reduced to that of nonsmokers.
CANCER OF ESOPHAGUS: 2–9 times greater risk of dying from this.	Risk proportional to amount smoked; quitting should reduce it.
CANCER OF PANCREAS: 2–5 times greater risk of dying from this.	Risk proportional to amount smoked; quitting should reduce it.
CANCER OF BLADDER: 7–10 times greater risk for smokers.	Risk decreases gradually over 7 years to that of nonsmokers.
CORONARY HEART DISEASE: Cigarette smoking is a major contributing factor.	Risk drops sharply after a year; after 10 years, risk reduced to that of nonsmokers.
EFFECTS ON OFFSPRING: Women who smoke during pregnancy have more stillbirths, and weight of liveborns averages less (which makes babies more vulnerable to disease, death).	When smoking stops before fourth month of pregnancy, risk of stillbirth and lower birth weight eliminated.
IMPAIRED IMMUNE SYSTEM FUNCTION: Increase in allergic responses, destruction of defensive cells (macrophages) in respiratory tract.	Avoidable by not smoking.
BONE HEALING: Evidence suggests that surgically cut or broken bones require up to 30 percent longer to heal in smokers, possibly because smoking depletes the body of vitamin C and reduces the amount of oxygen reaching body tissues. Reduced vitamin C and reduced oxygen interfere with production of collagen fibers, a key component of bone. Research in this area is continuing.	Avoidable by not smoking.

a

b

Figure 35.21 (**a**) From the American Cancer Society, a list of the major risks incurred by smoking and benefits of quitting. (**b**) A child in Mexico City already proficient at smoking cigarettes, a behavior that ultimately will endanger her capacity to breathe.

SMOKING'S IMPACT

Each year, tobacco use kills 4 million people around the world. By 2030 the number may rise to 10 million, 70 percent of whom will be in developing countries.

Each year in the United States, the direct medical costs of treating smoke-induced respiratory disorders drain 22 billion dollars from the economy. In the view of G. H. Brundtland, the head of the World Health Organization and a medical doctor, tobacco is now the only legal consumer product that kills half of its regular users. If you are a smoker, you may wish to reflect on the information in Figure 35.21.

Cigarettes do more than sicken and kill smokers. Nonsmokers die of ailments brought on by breathing *secondhand smoke*—by prolonged exposure to tobacco smoke in the surrounding air. Children who breathe secondhand smoke at home are far more vulnerable to allergies and lung problems than other children.

In the United States, smoking is being banned from enclosed spaces, such as airline cabins and airports, restaurants, and theaters. Cigarette sales to minors are prohibited. There are ongoing attempts to restrict advertising near elementary schools.

Are tobacco companies now viewing children and women of developing countries as untapped markets? Mark Palmer, former ambassador to Hungary, thinks so. He argues that selling tobacco is the worst single thing we do to the rest of the world.

In 2000 in the United States, 10.7 million individuals twelve years and older were smoking *pot*—marijuana (*Cannabis*)—as a way to induce light-headed euphoria. Many are chronic users. The collection of toxic particles being inhaled is called tar, and marijuana cigarettes contain far more tar than tobacco cigarettes. Smokers of marijuana also tend to inhale deeply and to smoke the cigarettes down to stumps, where tar accumulates. They put large doses of irritants and carcinogens into their throat and lungs. In addition to the risk of developing psychological dependency, long-term marijuana users may have chronic throat problems and coughing, bronchitis, and emphysema.

Summary

Section 35.1 Aerobic respiration requires oxygen and produces carbon dioxide wastes. These gases enter and exit an animal body by crossing a respiratory surface. In very small animals, gases simply diffuse across the body surface. Large animals typically exchange gases across the moist membranes of respiratory organs, such as gills and lungs. Oxygen-binding pigments help move gases and maintain gradients favorable to gas exchange.

Section 35.2 Invertebrates of aquatic or continually moist habitats use integumentary exchange; the body wall is their respiratory surface. Gills and tracheal respiration are common among invertebrates. Most spiders also have book lungs, with leaflike folds.

Section 35.3 Water's dissolved oxygen content can be a limiting factor for aquatic animals. A decline in the oxygen content is usually accompanied by a decline in biodiversity. Pollution can cause such a decline.

Section 35.4 Countercurrent flow at a pair of gills allows fish to take up more oxygen from water. Paired lungs, derived from outpouchings of the gut wall of some early vertebrates, are the dominant means of respiration in reptiles, birds, and mammals. Amphibians have lungs but expel carbon dioxide across their skin.

Section 35.5 In human lungs, millions of alveoli help form the respiratory membrane. Air flows through nasal cavities, the pharynx, through the larynx, trachea, bronchi, and bronchioles, which end at alveolar sacs.

Section 35.6 Each respiratory cycle consists of one inhalation and one exhalation. Inhalation is an active, energy-requiring process. Muscle contractions expand the chest cavity, lung pressure decreases below atmospheric pressure, and air flows in. The events are reversed during exhalation, which is usually passive. In acclimatization, people adapt to high altitudes by adjusting breathing patterns, which is accompanied by physiological changes.

Section 35.7 In the lungs, oxygen follows its partial pressure gradient from alveolar air spaces into pulmonary capillaries, then into red blood cells, where it binds with hemoglobin. In capillary beds, especially in metabolically active tissues, the hemoglobin releases oxygen, which diffuses across interstitial fluid into cells.

Carbon dioxide diffuses from cells into the interstitial fluid, then into the blood. Most reacts with water in red blood cells to form bicarbonate. Carbonic anhydrase catalyzes the reactions, which are reversed in the lungs. Here, carbon dioxide diffuses from the blood into air in alveoli, then is expelled. Respiratory centers in the brain control the normal rate and depth of breathing.

Section 35.8 Respiratory disorders, such as bronchitis and emphysema, can be caused or worsened by smoking. Worldwide, smoking is a leading cause of debilitating diseases and deaths.

Self-Quiz   *Answers in Appendix III*

1. The most abundant gas in the atmosphere is _____ .
   a. nitrogen   c. oxygen
   b. carbon dioxide   d. hydrogen

2. Respiratory pigments _____ .
   a. contain metal ions
   b. occur only in vertebrates
   c. increase the efficiency of oxygen transport
   d. both a and c

3. In crustaceans, most gas exchange occurs at _____ .
   a. the tips of tracheal tubes   c. gills
   b. the body surface   d. paired lungs

4. Countercurrent flow of water and blood increases the efficiency of gas exchange in _____ .
   a. fish   c. birds
   b. amphibians   d. all of the above

5. In human lungs, gas exchange occurs at the _____ .
   a. two bronchi   c. alveolar sacs
   b. pleural sacs   d. both b and c

6. When you breathe quietly, inhalation is _____ and exhalation is _____ .
   a. passive; passive   c. passive; active
   b. active; active   d. active; passive

7. During inhalation _____ .
   a. the thoracic cavity expands
   b. the diaphragm relaxes
   c. atmospheric pressure declines
   d. both a and c

8. Most oxygen in blood is transported _____ .
   a. bound to hemoglobin
   b. as carbon dioxide
   c. as bicarbonate
   d. dissolved in the plasma

9. Match the words with their descriptions.
   ____ trachea   a. muscle of respiration
   ____ pharynx   b. gap between vocal cords
   ____ alveolus   c. fine branch of bronchiole tree
   ____ hemoglobin   d. windpipe
   ____ bronchus   e. respiratory pigment
   ____ bronchiole   f. site of gas exchange
   ____ glottis   g. airway leading to lung
   ____ diaphragm   h. throat

Critical Thinking

1. When you swallow food, contracting muscles force the epiglottis down, to its closed position. This prevents food from going down the trachea and blocking air flow. Yet each year, several thousand people choke to death after food enters the trachea and blocks air flow for as little as four or five minutes.

The *Heimlich maneuver*—an emergency procedure only— may dislodge food stuck in the trachea. The rescuer stands behind the victim, makes a fist with one hand, and then positions the fist, thumb-side in, against the abdomen of the person who is choking (Figure 35.22). The fist must be just a bit above the navel and well below the rib cage. The fist is pressed into the abdomen with a sudden upward thrust. The thrust can be repeated if necessary.

Figure 35.22 The Heimlich maneuver being performed on an adult. This emergency procedure may dislodge food stuck in the trachea.

a Place one fist just above the person's navel with your thumb against the abdomen.

b Cover the fist with your other hand. Thrust both up and in with sufficient force to lift the person off his or her feet.

When correctly performed, the Heimlich maneuver forcibly elevates the diaphragm. Repeated thrusts are often able to dislodge the obstruction. Once the obstacle is dislodged, a doctor must see the person at once because an inexperienced rescuer can inadvertently cause internal injuries or crack a rib.

Reflect now on our current social climate in which lawsuits abound. Would you risk performing the Heimlich maneuver to save a relative's life? A stranger's life? Why or why not?

2. When human divers use tanks of compressed air and make deep dives, they must take into account the effects of nitrogen gas in their body. Starting at depths of about 45 meters (150 feet), more gaseous nitrogen than usual becomes dissolved in body tissues. Being lipid soluble, it crosses the lipid bilayer of cell membranes and disrupts nerve impulses. This can cause *nitrogen narcosis*. A diver may become euphoric and then drowsy, as if drunk.

Nitrogen that has entered tissues during a deep dive can also cause problems as the diver moves toward the surface. A too rapid ascent can cause nitrogen bubbles to form in the blood and tissues, causing decompression sickness. Bubbles in the joints cause extreme pain that makes it difficult to stand up, thus giving this ailment its common name, "the bends." If nitrogen bubbles disrupt blood flow to the brain, confusion, unconsciousness, and then death may be the outcome.

A diver who takes plenty of time to return to the sea surface will not have problems with nitrogen bubbles forming. As the diver ascends, pressure gradually drops and so does the amount of nitrogen in the tissues. Explain how the nitrogen that comes out of the tissues during this slow ascent could be removed from the body.

3. Carbon monoxide is a colorless, odorless gas in exhaust fumes from burning fossil fuel, wood, and tobacco. Carbon monoxide can bind to hemoglobin's oxygen-binding sites. It does so far more strongly, so it does not dissociate from hemoglobin as easily as oxygen does. Exposure to even a small amount of this gas can make up to 50 percent of the oxygen-binding sites in hemoglobin unavailable for oxygen transport. What effect would this *carbon monoxide poisoning* have on body tissues?

4. A *hiccup* is a sudden involuntary contraction of the diaphragm. Contraction of this muscle draws air inward during normal inhalation. With hiccups, the glottis closes abruptly and stops the air flow. The cutoff makes the sound associated with a hiccup.

Humans start to hiccup long before birth, by about two months into a pregnancy. Contraction of the muscles of inhalation at the same time as glottis closure is of no use to humans. In lungfish and some amphibians that have gills and a glottis, however, a hiccuping reflex pushes water over the gills while preventing it from entering the lungs. Humans do not have functional gills at any point in their development. Speculate on why the hiccuping reflex persists in the human population.

IMPACTS, ISSUES   *Hips and Hunger*

Americans are among the fattest people in the world, with 60 percent of adults overweight or obese. *Obesity* refers to an overabundance of fat in adipose tissue. Excess weight invites heart disease, diabetes, and some forms of cancer. What to do? The seemingly obvious answer is, "Lose the weight." But it is not that simple.

More and more studies support what dieters have always thought. Extra pounds are tough to lose. Our evolutionary past has a lot to do with it. Like all mammals, we have fat-storing cells in our adipose tissue. This energy warehouse evolved among our hominid ancestors, who could not always be certain of where the next meal was coming from. Stores of fat helped them through the lean times. Now, once fat cells form, they are in the body to stay. Eating just a tiny bit or a lot can only change how empty or full each adipose cell gets. Take in more calories than you burn, and cells of adipose tissue plump up with fat droplets (Section 28.2).

Figure 36.1 Ballooned mouse body or not? It depends partly on leptin.

Adipose cells synthesize leptin. This hormone acts on a part of the brain that helps control appetite. Some mutant mice cannot synthesize leptin. They eat and eat, until they end up looking like inflated balloons. Check out the normal mouse and the leptin-deficient mouse in Figure 36.1. What would happen if you were to inject leptin into the obese mutant? Its appetite would subside and its body would slim down.

Weight watchers can't get too excited about this. Overweight people do not seem to be lacking leptin. If anything, they have more of it. Possibly their leptin receptors are not working or their body cells do not have enough functional leptin receptors.

The stomach and brain also get in on the act. Some of their cells secrete ghrelin. This hormone makes you feel hungry. Ghrelin secretion rises when the stomach is empty and it slows after a big meal.

Consider what happened to the ghrelin level in the blood of obese people who volunteered to stay on a low-fat, low-calorie diet. After six months, they lost an average of 17 percent of their weight. Unfortunately, the amount of ghrelin in the blood climbed 24 percent. Dieters were slimmer but hungrier than ever!

Gastric bypass surgery is a drastic solution for the extremely obese. This "stomach-stapling" surgery closes off part of the stomach and most of the small intestine. It reduces the amount of food a person can eat before feeling full, and therefore cuts down the amount of nutrients that can be absorbed.

the big picture

Digestive Systems   A digestive system is a sac or tube that breaks down food into smaller nutritious bits for absorption into the internal environment, and that expels the unused residues. Most are complete systems, with a mouth and anus.

Human Digestive System   Humans have a complete digestive system with specialized regions and accessory glands and organs. Digestion begins in the mouth and is completed in the small intestine, where most nutrients are absorbed.

In contrast to standard methods of weight loss, gastric bypass reduces ghrelin levels. In one study, ghrelin levels dropped 77 percent in patients who had the procedure. Compared with the volunteers in other weight loss studies, gastric bypass patients are far less likely to regain weight. Lowered ghrelin may be part of the reason. If a drug that slows ghrelin synthesis or release can be developed, it might provide a similar effect, without the surgical risks.

Not everyone is struggling to keep weight off. Some folks have the opposite problem. Appetites generally decline with age. Among some of the elderly, the loss of appetite is so great that it endangers health. The hormone called cholecystokinin (CCK) seems to play a role. When released by the small intestine, CCK stimulates pancreatic secretions, the release of bile, and appetite suppression. A new drug that blocks CCK may help prevent malnutrition in seniors. Anti-obesity drugs that stimulate CCK are also being researched.

Questions about food intake and body weight lead us into the world of **nutrition**. The word encompasses all those processes by which an animal ingests and digests food, then absorbs the released nutrients to provide energy and building blocks for its individual cells. A digestive system typically performs these tasks and also helps maintain the internal environment, regardless of whether food is scarce or nutrients are flooding in. When all works well, the body's caloric inputs balance the outputs.

 How Would You Vote?

Many nutritionists suspect that increasing consumption of "fast foods" is contributing to rising levels of obesity. Should fast-food labels carry consumer warnings, as alcohol and cigarette labels do? See the Media Menu for details, then vote online.

Human Nutrition A balanced diet includes recommended amounts of carbohydrates, proteins, fats, vitamins, minerals, and fiber. Excessive intake of simple sugars or saturated fats may increase the risk of health problems.

About Body Weight To maintain a particular body weight, energy intake must equal energy output. Increasing numbers of people are losing a battle against obesity. Genetic variation in hormones that affect appetite and metabolism may be a factor.

36.1 The Nature of Digestive Systems

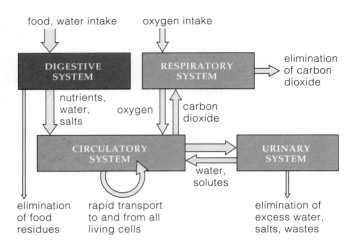

Figure 36.2 Functional links between the digestive, respiratory, circulatory, and urinary systems. These organ systems interact in supplying cells with raw materials and eliminating wastes.

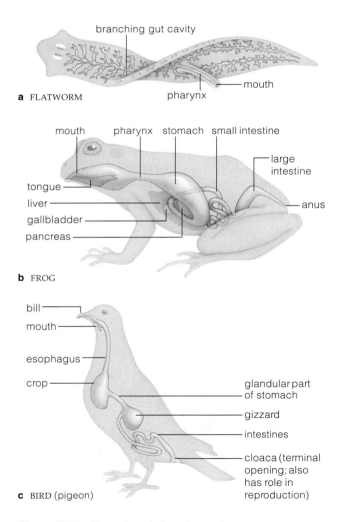

a FLATWORM

b FROG

c BIRD (pigeon)

Figure 36.3 Examples of digestive systems.

Every animal ingests and digests food, then absorbs the released nutrients for later conversion to the body's own carbohydrates, lipids, proteins, and nucleic acids. Many have specializations that adapt them to a particular diet.

Nutritional processes proceed in a **digestive system**. This body cavity or tube mechanically and chemically reduces food to small particles, and then to molecules that are small enough for absorption into the internal environment. The system also eliminates unabsorbed residues. Other organ systems directly or indirectly contribute to nutritional processes (Figure 36.2).

INCOMPLETE AND COMPLETE SYSTEMS

Recall, from Chapter 22, that flatworms are among the simplest animals with organ systems. Their digestive system is *incomplete*, as it is in a few other invertebrates. This means that food enters and wastes leave through a single opening at the body surface. In a flatworm, the saclike, branching gut cavity opens at the start of a pharynx, a muscular tube (Figure 36.3*a*). Food enters the sac, is partly digested, and circulates to cells even as wastes are sent out. This two-way traffic does not favor regional specialization.

Most species of animals have a *complete* digestive system. Basically, they have a tube with an opening at one end for taking in food (a mouth) and an opening at the other end for eliminating unabsorbed residues (an anus). The tube is divided into specialized regions for food breakdown, absorption, and storage.

Figure 36.3*b* shows the complete digestive system of a frog. Between the mouth and anus are a stomach, large and small intestines, and other organs. The liver, gallbladder, and pancreas are organs with accessory roles. They secrete enzymes or other products into the small intestine. Birds, too, have a complete digestive system, with specialized regions (Figure 36.3*c*).

Regardless of its complexity, a complete digestive system carries out five overall tasks:

1. *Mechanical processing and motility.* Movements that break up, mix, and propel food material.

2. *Secretion.* Release of digestive enzymes and other substances into the *lumen* (the space inside the tube).

3. *Digestion.* Breakdown of food into particles and then into nutrient molecules small enough to be absorbed.

4. *Absorption.* Passage of digested nutrients and water across the tube wall and into body fluids.

5. *Elimination.* Expulsion of undigested, unabsorbed residues from the end of the gut.

Figure 36.4 (**a**) A pronghorn antelope (*Antilocapra americana*), its molar, and a human molar. (**b**) Multiple chambered antelope stomach. In the first two chambers, food is mixed with fluid and exposed to fermentation by microbial symbionts. Some symbionts degrade cellulose; others synthesize organic compounds, fatty acids, and vitamins. The host uses part of these substances. Partially digested food is regurgitated into the mouth, chewed, and swallowed. It enters the third chamber and is digested again before entering the last stomach chamber.

CORRELATIONS WITH FEEDING BEHAVIOR

It is possible to correlate the specialized regions of a digestive system with its owner's feeding behavior. Consider the pigeon in Figure 36.3*c*. Its *food-gathering* region is a tough, thickened bill that can peck seeds off the ground. Tubular *food-processing* regions of the pigeon's digestive system are compactly centered in the body mass, which helps balance the bird during flight. As in other seed eaters, a *crop* bulges out from a part of the tube, between the mouth and the stomach. This storage organ is stretchable and allows the bird to gulp a lot of food fast and make quick getaways from predators.

Part of the bird's stomach, a *gizzard*, is a muscular organ that grinds and cycles food. Pigeon intestines are not lengthy; seeds do not require much time for processing. Ducks, ostriches, and other birds with longer intestines eat plants that contain lots of tough cellulose fibers, which require longer processing time.

Or what about the digestive system of pronghorn antelope (*Antilocapra americana*)? Fall through winter, on the mountain ridges from central Canada down into northern Mexico, they browse on wild sage. In spring, they move to open grasslands and deserts, there to browse on new growth. Consider their cheek teeth, or molars. The flattened crown of an antelope molar serves as a grinding platform. Proportionately, it is much larger than the crown on the molars in your mouth (Figure 36.4*a*). Why the difference? Most likely, you do not brush your mouth against dirt as you eat. The antelope does. Abrasive soil particles enter its mouth along with tough plant parts, and they wear down the crown quickly. Natural selection has favored more crown to wear down.

Antelopes are **ruminants**, hoofed mammals having multiple stomach chambers in which cellulose slowly undergoes digestion, as in Figure 36.4*b*. The chambers steadily accept food during extended feeding periods and slowly release nutrients when the animal rests.

Predatory and scavenging mammals gorge on food when they can get it, and may not eat again for some time. One part of their digestive system specializes in storing a lot of food, which is digested and absorbed at a leisurely pace. As you will see, other organs with accessory roles in digestion help ensure an adequate distribution of nutrients between meals.

Digestive systems mechanically and chemically break up food into small molecules that can be absorbed, along with ingested water, into the internal environment. They also expel undigested residues from the body.

Incomplete digestive systems are a saclike cavity with only one opening. Complete digestive systems are a tube with two openings and regional specializations in between.

Digestive systems are adapted in ways that suit different species to their particular diet and style of feeding.

36.2 The Human Digestive System

If it were fully stretched out, the complete digestive system of an adult human would extend 6.5 to 9 meters (21 to 30 feet). That is one big tube! Accessory glands and organs secrete enzymes and other substances that are required for digestion into the tube's lumen.

Humans have a complete digestive system, a tubular tract that has many specialized organs between two openings (Figure 36.5). Mucus-coated epithelium lines the entire surface facing the tract's lumen. Thick, moist mucus helps protect the wall of the tract and

Read Me First!

and watch the narrated animation on the human digestive system

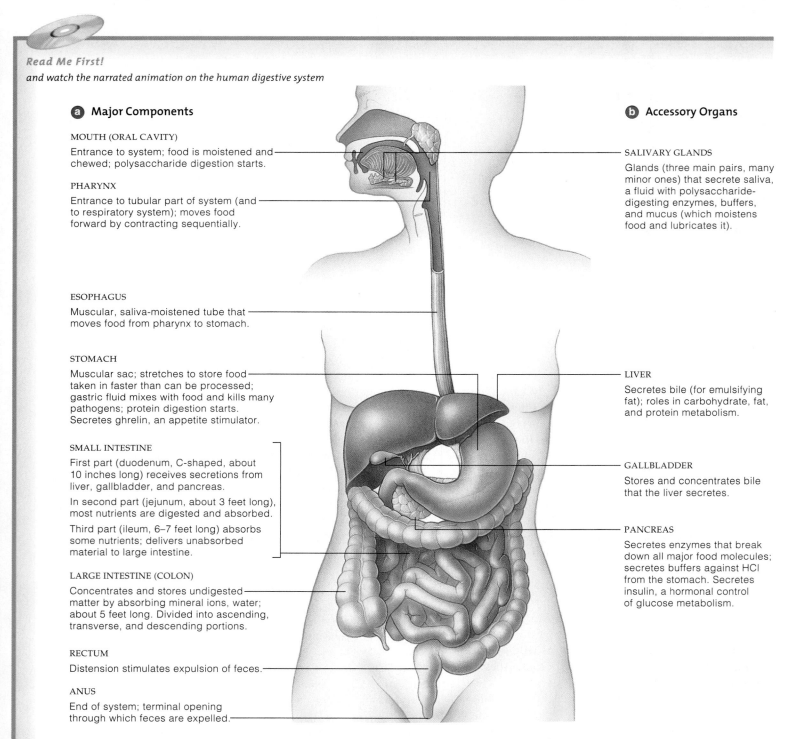

a Major Components

MOUTH (ORAL CAVITY)
Entrance to system; food is moistened and chewed; polysaccharide digestion starts.

PHARYNX
Entrance to tubular part of system (and to respiratory system); moves food forward by contracting sequentially.

ESOPHAGUS
Muscular, saliva-moistened tube that moves food from pharynx to stomach.

STOMACH
Muscular sac; stretches to store food taken in faster than can be processed; gastric fluid mixes with food and kills many pathogens; protein digestion starts. Secretes ghrelin, an appetite stimulator.

SMALL INTESTINE
First part (duodenum, C-shaped, about 10 inches long) receives secretions from liver, gallbladder, and pancreas.

In second part (jejunum, about 3 feet long), most nutrients are digested and absorbed.

Third part (ileum, 6–7 feet long) absorbs some nutrients; delivers unabsorbed material to large intestine.

LARGE INTESTINE (COLON)
Concentrates and stores undigested matter by absorbing mineral ions, water; about 5 feet long. Divided into ascending, transverse, and descending portions.

RECTUM
Distension stimulates expulsion of feces.

ANUS
End of system; terminal opening through which feces are expelled.

b Accessory Organs

SALIVARY GLANDS
Glands (three main pairs, many minor ones) that secrete saliva, a fluid with polysaccharide-digesting enzymes, buffers, and mucus (which moistens food and lubricates it).

LIVER
Secretes bile (for emulsifying fat); roles in carbohydrate, fat, and protein metabolism.

GALLBLADDER
Stores and concentrates bile that the liver secretes.

PANCREAS
Secretes enzymes that break down all major food molecules; secretes buffers against HCl from the stomach. Secretes insulin, a hormonal control of glucose metabolism.

Figure 36.5 (**a**) Major components and (**b**) accessory organs of the human digestive system.

promotes the diffusion of materials across its inner lining, into the internal environment.

Ingested substances are propelled in one direction, from the mouth, on through the pharynx, esophagus, and gastrointestinal tract, or the gut. The human **gut** starts at the stomach and extends through the small intestine, the large intestine (or colon) and rectum, to the anus. Salivary glands and a gallbladder, liver, and pancreas function as accessory organs. These organs secrete enzymes and other substances into particular regions of the human gut.

INTO THE MOUTH, DOWN THE TUBE

Figure 36.5 lists major organs of the human digestive system and briefly describes their functions.

Food gets chewed in the mouth, which in adults is equipped with thirty-two teeth (Figure 36.6). Each tooth has a coat of enamel (calcium deposits), dentin (a thick, bonelike layer), and nerves and blood vessels inside a pulpy matrix. Chisel-shaped incisors shear chunks of food. Cone-shaped canines tear food. The premolars and molars, with their broad crowns and rounded cusps, grind and crush it.

Also in the mouth is a tongue. This organ consists of membrane-covered skeletal muscles that function in positioning of food, swallowing, and speech. It has many chemoreceptors on its surface (Section 30.3).

Chewing mixes food with **saliva**, mostly water in which salivary amylase, bicarbonate (HCO_3^-), and mucins are dissolved. Salivary amylase digests starch. The buffer HCO_3^- maintains the pH in the mouth while you are eating acidic foods. Mucins are proteins. They help make up the mucus that binds food into a softened ball. Pairs of salivary glands in back of and beneath the tongue secrete saliva through ducts that open at the mouth's free surface.

When you swallow, contractions of tongue muscles force food into the **pharynx**, a tubular entrance to the esophagus *and* to the trachea, an airway to the lungs. As food leaves the pharynx, the epiglottis (a flaplike valve) and the vocal cords close the trachea to prevent breathing. That is why you normally do not choke as you swallow food.

Contractions of a muscular tube, the **esophagus**, propel food from the pharynx and past the sphincter at the start of the stomach. A **sphincter** is a ring of smooth muscles along the length of a tube or at one of its openings. It blocks movement through the lumen when its muscles contract.

The stomach empties into the small intestine. The contractions of smooth muscle layers in both organs

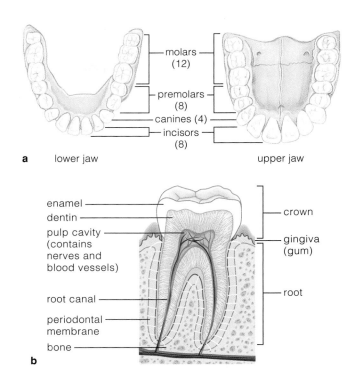

a lower jaw   upper jaw

b

Figure 36.6 (**a**) Number and arrangement of human teeth in the upper and lower jaw. (**b**) A molar's main regions: the crown and the root. Enamel capping the crown is made of calcium deposits; it is the hardest substance in the body.

break up, mix, and move food. Their lumen receives digestive enzymes and other secretions that function in breaking down nutrients. Remember, carbohydrate digestion starts inside the mouth. Protein breakdown starts in the stomach. However, digestion of nearly all carbohydrates, lipids, proteins, and nucleic acids in food is completed in the small intestine. Also, most of the nutrients are absorbed in the small intestine.

Undigested and unabsorbed material now enters the large intestine. Here, waste is compacted as water and ions are absorbed. The last part of the human gut is the rectum, where wastes are briefly stored before being expelled through the terminal opening (anus).

The human digestive system is complete. It begins with the mouth, where food is mechanically broken into particles by the teeth and tongue and mixed with saliva. Enzymes in the saliva start the digestion of carbohydrates.

Swallowing forces food into the pharynx and through the esophagus to the stomach. Protein digestion begins here.

Most digestion is completed in the small intestine and breakdown products are absorbed here. Water and ions are absorbed in the large intestine (colon). Compacted waste is stored in the rectum, then expelled via the anus.

36.3 Digestion in the Stomach and Small Intestine

Chemical digestion of food goes into high gear in the stomach and small intestine. In these highly muscular organs, food mixes with acid and enzymes that break it down into small, absorbable subunits.

We arrive now at the stomach and small intestine. The lumen of each organ receives digestive enzymes and other secretions that help break down nutrients into fragments, then into molecules small enough to be absorbed. Table 36.1 summarizes the major digestive enzymes. Again, carbohydrate breakdown *starts* inside the mouth. Protein breakdown *starts* in the stomach. Breakdown of most carbohydrates, lipids, proteins, and nucleic acids is *completed* in the small intestine.

THE STOMACH

The **stomach** is a muscular, stretchable sac (Figure 36.7). It mechanically mixes and stores ingested food. It secretes substances that help dissolve and degrade food, especially proteins. It helps control the passage of food into the small intestine.

Glandular epithelium lines the inner stomach wall. Each day, its cells secrete about two liters of mucus, hydrochloric acid (HCl), pepsinogens, and some of the other substances that make up **gastric fluid**. This liquid's high acidity (about pH 2), along with strong stomach contractions, partly digests food, forming a semiliquid mass called *chyme*. The acidity kills some food-borne pathogens. It also makes proteins unfold, which exposes peptide bonds. It converts pepsinogens to pepsins—enzymes that cleave the peptide bonds. That is how smaller polypeptides accumulate in the lumen of the stomach. Some cells in the stomach wall release the hormone gastrin, which calls for secretion of HCl and pepsinogen by other cells.

THE SMALL INTESTINE

The small intestine has three regions: the duodenum, jejunum, and ileum (Figure 36.5). Each day, about nine liters of fluid enter the duodenum from the stomach

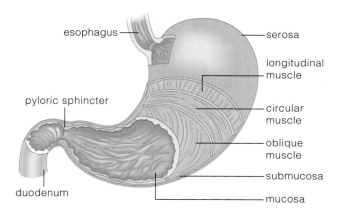

Figure 36.7 Structure of the stomach wall. Like most of the gut wall, it has an inner mucosa resting on a connective tissue with a mesh of nerves inside that locally controls digestion. Next to this tissue are smooth muscle layers that differ in orientation, hence in direction of contraction. The wall's outer layer of connective tissue is the serosa.

Table 36.1	Major Digestive Enzymes and Their Breakdown Products			
Enzyme	Source	Where Active	Substrate	Main Breakdown Products
Carbohydrate Digestion				
Salivary amylase	Salivary glands	Mouth, stomach	Polysaccharides	Disaccharides
Pancreatic amylase	Pancreas	Small intestine	Polysaccharides	Disaccharides
Disaccharidases	Intestinal lining	Small intestine	Disaccharides	MONOSACCHARIDES* (such as glucose)
Protein Digestion				
Pepsins	Stomach lining	Stomach	Proteins	Protein fragments
Trypsin and chymotrypsin	Pancreas	Small intestine	Proteins	Protein fragments
Carboxypeptidase	Pancreas	Small intestine	Protein fragments	AMINO ACIDS*
Aminopeptidase	Intestinal lining	Small intestine	Protein fragments	AMINO ACIDS*
Fat Digestion				
Lipase	Pancreas	Small intestine	Triglycerides	FREE FATTY ACIDS, MONOGLYCERIDES*
Nucleic Acid Digestion				
Pancreatic nucleases	Pancreas	Small intestine	DNA, RNA	NUCLEOTIDES*
Intestinal nucleases	Intestinal lining	Small intestine	Nucleotides	NUCLEOTIDE BASES, MONOSACCHARIDES*

* Breakdown products small enough to be absorbed into the internal environment.

and three glandular organs: the liver, gallbladder, and pancreas. At least 95 percent of the fluid is absorbed across the intestinal lining, shown in Figure 36.8a.

Intestinal and pancreatic enzymes break down the large organic compounds in food to monosaccharides, monoglycerides, fatty acids, amino acids, nucleotides, and nucleotide bases. For instance, pancreatic enzymes called trypsin and chymotrypsin digest proteins into peptides, as pepsin does. Others break down peptides to free amino acids. In addition, the pancreas secretes bicarbonate, which helps to neutralize HCl. The small intestine is less acidic than the stomach because of it.

Fat digestion requires lipases (pancreatic enzymes) and **bile**, which the liver secretes continuously. This fluid consists of water, bile salts and bile pigments, cholesterol, and the phospholipid lecithin. After the stomach has emptied, a sphincter closes the main bile duct from the liver. Bile backs up into the gallbladder, which stores and concentrates it.

Bile salts enhance fat digestion by **emulsification** of fat droplets. Most fats in food are triglycerides. The triglycerides, which are insoluble in water, tend to cluster together as fat globules. But rings of circular muscles in the wall of the small intestine contract in an oscillating pattern known as **segmentation** (Figure 36.8b). While the contractions mix chyme, fat globules break up into small droplets that become coated with bile salts. Bile salts carry negative charges, so these coated droplets repel each other and stay separated. Tiny fat droplets suspended in fluid is the emulsion.

Compared to large fat globules, emulsion droplets offer fat-digesting enzyme molecules a greater surface area. Enzymes break down triglycerides faster to fatty acids and monoglycerides. The next section explains how these digestion products are absorbed.

CONTROLS OVER DIGESTION

The nervous system, endocrine system, and meshes of nerves in the gut wall control the pace of digestion. When food is distending the stomach wall, it activates many mechanoreceptors, which send signals along short reflex pathways to smooth muscles and glands in local tissues. (Longer reflex pathways also signal the brain.) Gut wall muscles contract and glandular cells secrete enzyme-rich fluid into the intestinal tract or hormones into blood. The responses depend partly on the chyme volume and composition. When a large meal activates many mechanoreceptors, contractions become more forceful and the stomach empties faster. High acidity or a high fat content in chyme triggers hormonal secretions that can slow stomach emptying.

Figure 36.8 (**a**) Structure of the small intestine. Its wall has a highly folded inner lining, the mucosa. (**b**) Rings of circular muscle in the wall contract and relax repeatedly in a pattern called segmentation. The pattern of back and forth movement propels, mixes, and forces chyme against the wall, which enhances both digestion and absorption.

What are some of the gastrointestinal hormones? Gastrin secreted from the stomach lining prods other cells to release acid into the stomach. Cholecystokinin (CCK) calls for release of pancreatic enzymes. It also triggers gallbladder contractions that squirt bile into the small intestine. Secretin prompts the pancreas to release bicarbonate. Gastric inhibitory peptide (GIP) slows contraction of intestinal muscles and increases glucose uptake by calling for insulin secretion.

Recall, from Section 29.6, that sympathetic signals, which predominate in times of high activity or stress, can slow the contractions of the smooth muscle of the digestive tract. This slows digestion and the passage of material from one segment of the tube to another. Chronic stress can aggravate digestive disorders.

Carbohydrate breakdown starts in the mouth and protein breakdown starts in the stomach. In the small intestine, most large organic compounds are digested to molecules small enough to be absorbed.

Signals from the nervous system, nerves in the gut wall, and hormones control aspects of digestion.

36.4 Absorption From the Small Intestine

Folds and projections from cell surfaces of the small intestine enormously increase the surface area for interactions with chyme and for nutrient absorption.

STRUCTURE OF THE INTESTINAL LINING

Figure 36.9*a* zooms in on the profusely folded lining of the small intestine, the mucosa. Millions of **villi** (singular, villus) cover the free surface of the folds and function as absorptive structures. Each villus is about one millimeter long. The villi give the lining a somewhat velvety appearance. Inside each villus is one arteriole, one venule, and one lymph vessel. Each is functionally linked with the circulatory system.

Many cells on each villus secrete mucus, hormones, bicarbonate, or the bacteria-killing lysozyme (Figure 36.9*b*). Many others, the "brush border cells," have a dense array of fine, threadlike absorptive structures on the free surface facing the lumen. These ultrafine structures are **microvilli** (singular, microvillus). There can be 1,700 or so microvilli on such "brush border" cells. Without that immense surface area, absorption would proceed hundreds of times more slowly, and that rate would not be enough to sustain human life.

WHAT ARE THE ABSORPTION MECHANISMS?

By the time a meal travels halfway through the small intestine, most is pummeled, digested, and ready for absorption. Absorption, remember, refers to the uptake of nutrients, water, vitamins, and minerals from the gut lumen, into the internal environment. Molecules of water cross the intestinal lining by way of osmosis. Mineral ions are transported into cells of the lining.

Read Me First!

and watch the narrated animation on the structure of the small intestine

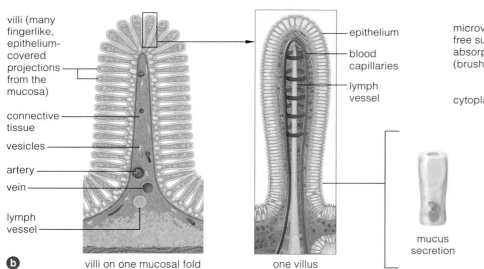

Figure 36.9 (**a**) Structure of the lining of the small intestine, the mucosa. The repeated circular folds of this inner lining are permanent.

(**b**) At the free surface of each mucosal fold are many absorptive structures called villi, fingerlike structures that contain blood capillaries and lymph vessels. Several types of epithelial cells with specialized functions cover the free surface of each villus. At the free surface of one type, the absorptive cells, a crown of microvilli extends into the intestinal lumen.

mucosa
(inner lining)

villi (many fingerlike, epithelium-covered projections from the mucosa)

connective tissue

vesicles

artery

vein

lymph vessel

villi on one mucosal fold

epithelium

blood capillaries

lymph vessel

one villus

microvilli at the free surface of an absorptive cell (brush border cell)

cytoplasm

mucus secretion

hormone secretion

lysozyme secretion

nutrient absorption

Transport proteins in the plasma membrane of brush border cells actively shunt some breakdown products, including monosaccharides and amino acids, across the lining (Figure 36.10a,b). Products of fat digestion —lipid-soluble, fatty acids and monoglycerides—can diffuse across the plasma membrane's lipid bilayer. Bile salts help them do this (Figure 36.10c–f).

By a process called **micelle formation**, bile salts combine with fatty acids and monoglycerides to form tiny droplets (micelles). Molecules inside the micelles continuously exchange places with those suspended in chyme. However, the micelles *concentrate* them in the fluid right next to the intestinal lining. When they become concentrated enough, gradients promote their diffusion out of the micelles and into epithelial cells. The fatty acids and monoglycerides recombine inside cells to form triglycerides. Then triglycerides combine with proteins into particles (chylomicrons), which exocytosis delivers to the internal environment.

Once absorbed, glucose and amino acids directly enter blood vessels. Triglycerides enter lymph vessels that eventually drain into the general circulation. The bloodstream then distributes the absorbed products of digestion throughout the body, along the circuits described in Sections 33.1 and 33.5.

With its richly folded mucosa, millions of villi, and hundreds of millions of microvilli, the small intestine has a vast surface for absorbing nutrients.

Substances pass through the brush border cells that line the free surface of each villus by active transport, osmosis, and diffusion across the lipid bilayer of plasma membranes.

Read Me First!
and watch the narrated animation on digestion and absorption

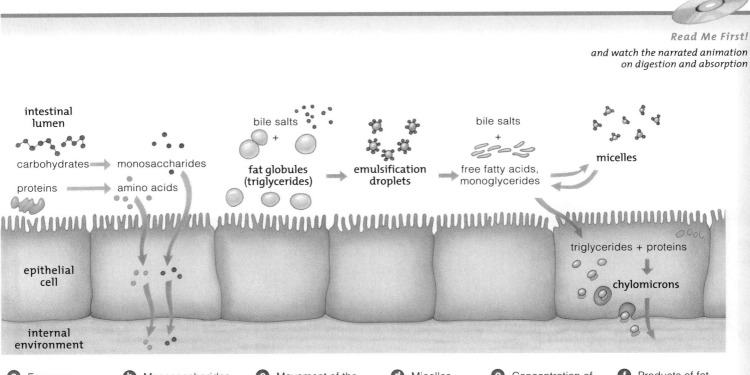

a Enzymes secreted by the pancreas and cells of the epithelial lining complete the digestion of carbohydrates to monosaccharides, and proteins to amino acids.

b Monosaccharides and amino acids are actively transported across the plasma membrane of the epithelial cells, then out of the same cells and into the internal environment.

c Movement of the intestinal wall breaks up fat globules into small droplets. Bile salts prevent the globules from re-forming. Pancreatic enzymes digest the droplets to fatty acids and monoglycerides.

d Micelles form as bile salts combine with digestion products and phospholipids. Products can readily slip into and out of the micelles.

e Concentration of monoglycerides and fatty acids in micelles enhances gradients. This leads to diffusion of both substances across the lipid bilayer of the plasma membrane of cells making up the lining.

f Products of fat digestion reassemble into triglycerides in cells of the intestinal lining. These become coated with proteins, then are expelled (by exocytosis from the cells) into the internal environment.

Figure 36.10 Digestion and absorption in the small intestine.

36.5 What Happens to Absorbed Organic Compounds?

Small organic compounds—sugars, amino acids, and triglycerides—are absorbed in the small intestine. They are distributed to body cells and burned as fuel, stored, or used in synthesis of larger organic compounds.

In Section 7.6, you considered some mechanisms that govern organic metabolism—specifically, disposition of glucose and other organic compounds in the body as a whole. You saw examples of pathways by which carbohydrates, fats, and proteins are degraded into intermediates in an ATP-producing pathway, aerobic respiration. Here, Figure 36.11 rounds out the picture by showing all of the major routes by which organic compounds obtained from food can be shuffled and reshuffled in the body as a whole.

All cells continually recycle some carbohydrates, lipids, and proteins by breaking them apart. They use breakdown products as energy sources and building blocks. This massive molecular turnover is integrated by the nervous and endocrine systems.

Reflect for a moment on a few key points about organic metabolism. When you eat, your body adds to its pool of raw materials. Excess carbohydrates and other organic compounds absorbed from the gut are transformed mostly into fats, which become stored in adipose tissue. Some are converted to glycogen in the liver and in all muscles. While the raw materials are being absorbed and stored, most cells use glucose as the main energy source. There is no net breakdown of protein in muscle tissue or any other tissue during this period, nor is there any net fat breakdown.

Between meals, the brain takes up two-thirds of the circulating glucose. Now most body cells tap their fat and glycogen stores. Adipose tissue cells break down fats to glycerol and fatty acids, and release these to the blood. Liver cells break down glycogen to release glucose, which also enters blood. Most body cells take up the fatty acids as well as the glucose and use them for ATP production.

The liver stores and converts organic compounds to different forms and helps maintain concentrations in the blood at required levels. The liver inactivates most hormones, which it sends on to the kidneys for excretion in the urine. It also removes worn-out blood cells and deactivates certain toxins, including alcohol (Chapter 5). Ammonia (NH_3) is a potentially toxic product of amino acid breakdown. The liver prevents its accumulation by converting it to a somewhat less toxic form, urea, which can be excreted in the urine.

Absorbed sugars and glycogen are the body's most accessible energy source. Excess carbohydrates are converted to fats, which are stored in adipose tissue.

Figure 36.11 Summary of the major pathways of organic metabolism. Cells continually synthesize and tear down carbohydrates, fats, and proteins. Most urea forms in the liver.

36.6 The Large Intestine

Not all components of food can be broken down and absorbed. Undigestible residues and other solid wastes are concentrated and briefly stored before they are expelled from the body.

Material not absorbed in the small intestine is moved into the large intestine, or **colon**. This part of the gut starts at a cup-shaped pouch (cecum), ascends on the right side of the abdominal cavity, extends across to the other side, and descends and connects with the rectum (Figures 36.5 and 36.12). The colon concentrates and stores feces, which are a mixture of water, bacteria, and undigested and unabsorbed material.

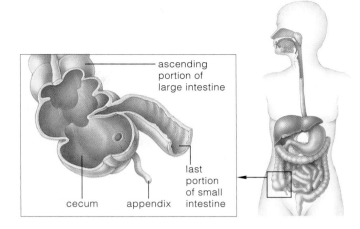

Figure 36.12 Cecum and appendix of the large intestine (colon).

COLON FUNCTION

As waste material moves through the colon, it loses water and becomes more and more concentrated. The cells of the colon's lining actively transport sodium ions out of the lumen, and water follows by osmosis. The same cells secrete mucus that lubricates feces and helps keep them from scratching the colon wall. The cells also secrete bicarbonate, which buffers the acidic products of enteric (gut-dwelling) bacteria. Generally, these bacteria are harmless unless they breach the wall of the colon and enter the abdominal cavity.

As bands of smooth muscle contract and relax, the lumen's contents are moved back and forth against the absorptive surface of the colon wall. Networks of nerves in the wall control the motion, which is similar to the process of segmentation in the small intestine but has a much slower pattern of oscillation. The environment in the small intestine restricts bacterial growth, but 500 or so bacterial species are adapted to the leisurely conditions in the colon. Some species, including *Escherichia coli*, make essential vitamins.

The volume of cellulose and all other undigested material that cannot be decreased by absorption in the colon is known as **bulk**. It adds to the volume of material and slows transit time through the colon.

Following each new meal, secretion of gastrin and signals from the nervous system's autonomic nerves stimulate large parts of the ascending and transverse colon to contract in synchrony. Within seconds, the lumen's contents can be propelled through much of the colon's length.

Bulk builds up in the rectum, the last part of the colon. Eventually, it distends the rectum wall enough to stimulate reflex action and expulsion. The nervous system controls defecation by signals that stimulate or inhibit the contraction of a sphincter at the anus, the last opening of the gut.

COLON MALFUNCTION

The frequency of defecation ranges from three times a day to once a week, on average. Aging, emotional stress, a low-bulk diet, injury, or disease can delay it, a condition called *constipation*. The longer the delay, the more water leaves feces, which harden and dry out. The result is abdominal discomfort, along with loss of appetite, headaches, and often nausea.

Hard feces may become lodged in the **appendix**, a narrow projection from the cecum (Figure 36.12). The appendix has no known digestive function. However, like the ileum, the appendix is packed with bacteria-fighting white blood cells. When feces obstruct normal blood flow and mucus secretion to this projection, it can cause *appendicitis*—an inflamed appendix. Unless the inflamed appendix is surgically removed, it can rupture. Bacteria can then enter the abdominal cavity and cause infections that may be life-threatening.

Also, the colon is vulnerable to cancer. The world's wealthiest countries have the highest rates of *colon cancer*. Diets here tend to be rich in sugar, cholesterol, and salt, and low in bulk. Too little bulk extends the time feces spend in the colon. Food contains natural and artificial carcinogens, which cause more damage the longer they remain in contact with the colon wall. Symptoms of colon cancer include a change in bowel function, rectal bleeding, and blood in feces. Some people are genetically predisposed to develop colon cancer, but a diet that does not have adequate fiber is also known to increase the risk.

The colon functions in the absorption of water and mineral ions. It also compacts undigested residues into feces, which are stored in the rectum, the final part of the digestive tract.

A reflex action triggers expulsion of feces from the body.

36.7 Human Nutritional Requirements

There is an old saying, "You are what you eat," meaning that diet profoundly affects the structure and function of your body. So what do you eat?

Since the 1950s, the United States government has been working to establish nutritional guidelines for good health. The **food pyramid** in Figure 36.13 was created a decade ago as the result of these efforts. It is one of many charts recommending daily amounts of food in certain categories. For an average-sized adult male, the pyramid divides the daily food intake this way: 55–60 percent complex carbohydrates, 15–20 percent proteins (less for females), and 20–25 percent fats.

This particular pyramid is criticized for lumping together all fats as "bad" and all carbohydrates as "good." Increasingly, nutritionists are concluding that it is not the best reflection of a healthy diet. However, the food pyramid is updated periodically and a new one is expected to be presented by the United States government in 2005.

A CARBOHYDRATE–INSULIN CONNECTION

Complex carbohydrates are readily broken down to glucose, the body's main source of energy. Starch is abundant in fleshy fruits, cereal grains, and legumes, such as peas and beans. Foods that are rich in complex carbohydrates are typically also high in fiber, which helps to prevent constipation and colon problems. As epidemiologist Denis Burkitt has put it, when people pass small volumes of feces, you have large hospitals.

Unlike complex carbohydrates, simple sugars do not provide any fiber, vitamins, or minerals. Also, highly refined carbohydrates have a *high glycemic index*. This means that, within minutes of being absorbed, they cause a surge in blood levels of sugar and insulin.

An insulin surge makes cells take up sugar quickly. A few hours later, blood sugar is lower than before, so we get hungry. But circulating insulin keeps cells from dipping into fat stores. Cells metabolize fat only when insulin is low. So we eat more, secrete more insulin,

MILK, CHEESE, YOGURT

Two servings/day (three/day for teenagers or pregnant, lactating, post-menopausal women)

1 cup nonfat or lowfat milk, fortified soy milk
1 cup yogurt or cottage cheese
1–1/2 ounces cheese

FRUITS

Two to four servings/day

1 medium apple, orange, banana, mango
1 wedge melon, pineapple
1 cup fresh berries
3/4 cup unsweetened fruit juice
1/2 cup unsweetened fruit
1/4 cup dried fruit

ADDED FATS AND SIMPLE SUGARS
Other foods have plenty. Restrict intake of butter, salad dressings, sweets, soft drinks, snacks

LEAN MEAT, POULTRY, FISH, LEGUMES, EGGS

Two to three servings/day

2–3 ounces trimmed, cooked, lean meat
2–3 ounces cooked poultry, no skin
2–3 ounces cooked fish
2 egg whites or 1 whole egg (limit yolks per week)
1/2 cup legumes (counts as 1/3 serving)
2 tablespoons peanut butter (counts as 1/3 serving)

VEGETABLES

Three to five servings/day

1/2 cup cooked or raw vegetables
1 cup dark green, leafy raw vegetables
3/4 cup vegetable juice

GRAIN PRODUCTS (BREADS, CEREALS, PASTA)

Six to eleven servings/day

1 biscuit, small roll, or muffin
1 slice bread
1/2 cup cooked cereal, pasta
1 ounce dry unsweetened cereal
1/2 cup rice (brown rice has more nutrients)
1/2 bagel, English muffin

Figure 36.13 A low-fat, high-carbohydrate food pyramid, based on prevailing USDA guidelines.

and thus keep on storing fat—mainly in the form of triglycerides. Over time, high triglyceride levels can increase the risk of heart disease and type 2 diabetes.

Excessive intake of high fructose corn syrup in soft drinks and other foods and drinks is also a concern. Eating or drinking a product sweetened with glucose stimulates the release of leptin from fat cells and also decreases the stomach's ghrelin output. The combined result is a decrease in feelings of hunger. In contrast, fructose delivers just as many calories, but does not affect these hormones, so a person still feels hungry.

GOOD FAT, BAD FAT

You cannot stay alive without fats. Cell membranes incorporate the phospholipid lecithin. Also, other fats act as energy reserves, cushion many internal organs, and provide insulation beneath the skin. Dietary fats help store fat-soluble vitamins. However, the body can make most of its own fats by converting proteins and carbohydrates. The ones that it cannot make, such as linoleic acid, are **essential fatty acids**. A teaspoonful or two each day of some polyunsaturated fat, such as olive oil or corn oil, may turn out to be enough of a dietary fat supplement.

Like other animal fats, butter is a saturated fat and contains cholesterol. This sterol is a precursor for the synthesis of bile acids and steroid hormones. It occurs in all animal cell membranes. But consuming a lot of saturated fat can elevate levels of LDL cholesterol. As described in Section 33.9, this increases the risk for heart disease and stroke.

BODY-BUILDING PROTEINS

Your body also cannot function without proteins. It requires amino acid components of dietary proteins for its own protein-building programs. Of the twenty common types, eight are **essential amino acids**. Your cells cannot synthesize them; you must get them from food. They are methionine (or cysteine, its metabolic equivalent), isoleucine, leucine, lysine, phenylalanine (or tyrosine), threonine, tryptophan, and valine. Most proteins in animal tissues are *complete*; their amino acid ratios match human nutritional needs. Nearly all plant proteins are *incomplete*, in that they lack one or more amino acids that are essential for humans.

ALTERNATIVE DIETS

Alternative diets abound. One, the *Mediterranean diet* pyramid, emphasizes grain products first, and then

fruits and vegetables, legumes and nuts, then *olive oil* as the fat group, and then cheese and yogurt. It limits weekly intakes of fish, poultry, eggs, simple sugars, and—at its tiny top—red meat. Olive oil is as much as 40 percent of the total energy intake. Because this oil is monounsaturated, it is less likely than saturated fats to raise cholesterol levels. It also is an antioxidant that helps eliminate free radicals. You might call this an "olive-oil-is-good" diet.

"Low-carb" diets are now wildly popular. About 10 million Americans report that they are following such a program. These diets call for a reduced intake of carbohydrates and an increased intake of proteins and fats. There is no question that they can promote rapid weight loss. However, the long-term effects of low-carb diets are still being debated, in some cases contentiously, by health professionals.

As described in Section 36.5, digestion of proteins produces toxic ammonia. Why? Enzymes in the liver convert ammonia to urea, which is then filtered out by kidneys and excreted in urine. When the body uses fats instead of carbohydrates as its primary source of energy, waste products called ketones also enter the blood. Ketones must be filtered out of the blood and excreted. The National Kidney Foundation warns that acidic ketones and all that extra filtering can increase risk of kidney stones and may otherwise harm these organs. Anyone who has impaired kidney function should avoid a high-protein diet.

Another possible concern with some low-carb diets is the effect of increased consumption of saturated fats on cholesterol levels. But numerous studies have shown that following a low-carb diet for six months does not increase levels of LDLs ("bad" cholesterol). In some studies, dieters' cholesterol levels actually declined over the course of the study. Still, given the amount of evidence that a diet high in saturated fat increases the risks of heart disease, low-carb dieters are advised to choose lean meats and fish over fattier high-protein foods.

We still do not have all the answers. Here is what we do know. Americans are getting fatter. We will return to this hefty subject in Section 36.9.

For three decades, nutritionists have been advocating a low-fat, high-carbohydrate diet. Too many individuals have mistakenly taken "carbohydrate" to mean it is acceptable to load up on refined sugars and starches. Some carbohydrates may be contributing to rises in obesity and diabetes.

New studies may result in major shifts in nutritional guidelines.

36.8 Vitamins and Minerals

Too little of a vitamin or mineral can disrupt normal development and metabolism. Recall, from Chapter 15, that a deficiency in vitamin A can cause blindness. Vitamin deficiencies impair growth and development of about one-third of the world's children.

Vitamins are *organic* substances that are essential for growth and survival; no other substance can carry out their metabolic functions. Unlike most plant species, which synthesize all the vitamins they require, most animals lost the ability to do so and must obtain all vitamins from their food. At a minimum, human cells require the thirteen vitamins listed in Table 36.2. Each type has specific metabolic roles. For example, the coenzyme NAD is derived from the B vitamin niacin.

Minerals are *inorganic* substances that are essential for growth and survival because no other substances can serve their metabolic functions. For instance, all of

Table 36.2 Major Vitamins: Sources, Functions, and Effects of Deficiencies or Excesses*

Vitamin	Common Sources	Main Functions	Effects of Chronic Deficiency	Effects of Extreme Excess
Fat-Soluble Vitamins				
A	Its precursor comes from beta-carotene in yellow fruits, yellow or green leafy vegetables; also in fortified milk, egg yolk, fish, liver	Used in synthesis of visual pigments, bone, teeth; maintains epithelia	Dry, scaly skin; lowered resistance to infections; night blindness; permanent blindness	Malformed fetuses; hair loss; changes in skin; liver and bone damage; bone pain
D	Inactive form made in skin, activated in liver, kidneys; in fatty fish, egg yolk, fortified milk products	Promotes bone growth and mineralization; enhances calcium absorption	Bone deformities (rickets) in children; bone softening in adults	Retarded growth; kidney damage; calcium deposits in soft tissues
E	Whole grains, dark green vegetables, vegetable oils	Counters effects of free radicals; helps maintain cell membranes; blocks breakdown of vitamins A and C in gut	Lysis of red blood cells; nerve damage	Muscle weakness, fatigue, headaches, nausea
K	Enterobacteria form most of it; also in green leafy vegetables, cabbage	Blood clotting; ATP formation via electron transport	Abnormal blood clotting; severe bleeding (hemorrhaging)	Anemia; liver damage and jaundice
Water-Soluble Vitamins				
B$_1$ (thiamin)	Whole grains, green leafy vegetables, legumes, lean meats, eggs	Connective tissue formation; folate utilization; coenzyme action	Water retention in tissues; tingling sensations; heart changes; poor coordination	None reported from food; possible shock reaction from repeated injections
B$_2$ (riboflavin)	Whole grains, poultry, fish, egg white, milk	Coenzyme action	Skin lesions	None reported
B$_3$ (niacin)	Green leafy vegetables, potatoes, peanuts, poultry, fish, pork, beef	Coenzyme action	Contributes to pellagra (damage to skin, gut, nervous system, etc.)	Skin flushing; possible liver damage
B$_6$	Spinach, tomatoes, potatoes, meats	Coenzyme in amino acid metabolism	Skin, muscle, and nerve damage; anemia	Impaired coordination; numbness in feet
Pantothenic acid	In many foods (meats, yeast, egg yolk especially)	Coenzyme in glucose metabolism, fatty acid and steroid synthesis	Fatigue, tingling in hands, headaches, nausea	None reported; may cause diarrhea occasionally
Folate (folic acid)	Dark green vegetables, whole grains, yeast, lean meats; enterobacteria produce some folate	Coenzyme in nucleic acid and amino acid metabolism	A type of anemia; inflamed tongue; diarrhea; impaired growth; mental disorders	Masks vitamin B$_{12}$ deficiency
B$_{12}$	Poultry, fish, red meat, dairy foods (not butter)	Coenzyme in nucleic acid metabolism	A type of anemia; impaired nerve function	None reported
Biotin	Legumes, egg yolk; colon bacteria produce some	Coenzyme in fat, glycogen formation and in amino acid metabolism	Scaly skin (dermatitis); sore tongue; depression; anemia	None reported
C (ascorbic acid)	Fruits and vegetables, especially citrus, berries, cantaloupe, cabbage, broccoli, green pepper	Collagen synthesis; possibly inhibits effects of free radicals; structural role in bone, cartilage, and teeth; used in carbohydrate metabolism	Scurvy; poor wound healing; impaired immunity	Diarrhea, other digestive upsets; may alter results of some diagnostic tests

* Guidelines for appropriate daily intakes are being worked out by the Food and Drug Administration.

your cells require iron as a component of electron transfer chains. Red blood cells do not function at all without the oxygen-transporting iron in hemoglobin. Neurons stop functioning in the absence of sodium and potassium (Table 36.3). People who are in good health get all the vitamins and minerals they require from a balanced diet of whole foods.

Generally, vitamin and mineral supplements are essential only for strict vegetarians, the elderly, and people who are chronically ill or taking medicine that interferes with how the body processes nutrients. For example, supplemental vitamin K promotes calcium retention and lessens osteoporosis in elderly women.

Some studies suggest that supplementing vitamins A, C, and E may counter some of the effects of aging and protect the immune system by inactivating free radicals. A *free radical*, recall, is a molecular fragment —an atom or a group of atoms having an unpaired electron. A free radical is so reactive that it can disrupt how molecules function.

Excessive amounts of many vitamins and minerals can harm individuals. Large doses of vitamins A and D are examples. Like all fat-soluble vitamins, they can accumulate in body tissues and interfere with normal metabolic activity if taken in excessive amounts.

Vitamins are organic substances. Minerals are inorganic. Both have vital metabolic roles.

A balanced diet provides the required amounts of vitamins and minerals for most people. Deficiencies or excesses can cause health problems.

Table 36.3 Major Minerals: Sources, Functions, and Effects of Deficiencies or Excesses*

Mineral	Common Sources	Main Functions	Effects of Chronic Deficiency	Effects of Extreme Excess
Calcium	Dairy products, dark green vegetables, dried legumes	Bone, tooth formation; blood clotting; neural and muscle action	Stunted growth; possibly diminished bone mass (osteoporosis)	Impaired absorption of other minerals; kidney stones in susceptible people
Chloride	Table salt (usually too much in diet)	HCl formation in stomach; contributes to body's acid–base balance; neural action	Muscle cramps; impaired growth; poor appetite	Contributes to high blood pressure in susceptible people
Copper	Nuts, legumes, seafood, drinking water	Used in synthesis of melanin, hemoglobin, and some transport chain components	Anemia, changes in bone and blood vessels	Nausea, liver damage
Fluorine	Fluoridated water, tea, seafood	Bone, tooth maintenance	Tooth decay	Digestive upsets; mottled teeth and deformed skeleton in chronic cases
Iodine	Marine fish, shellfish, iodized salt, dairy products	Thyroid hormone formation	Enlarged thyroid (goiter), with metabolic disorders	Toxic goiter
Iron	Whole grains, green leafy vegetables, legumes, nuts, eggs, lean meat, molasses, dried fruit, shellfish	Formation of hemoglobin and cytochrome (transport chain component)	Iron-deficiency anemia, impaired immune function	Liver damage, shock, heart failure
Magnesium	Whole grains, legumes, nuts, dairy products	Coenzyme role in ATP–ADP cycle; roles in muscle, nerve function	Weak, sore muscles; impaired neural function	Impaired neural function
Phosphorus	Whole grains, poultry, red meat	Component of bone, teeth, nucleic acids, ATP, phospholipids	Muscular weakness; loss of minerals from bone	Impaired absorption of minerals into bone
Potassium	Diet alone provides ample amounts	Muscle and neural function; roles in protein synthesis and body's acid–base balance	Muscular weakness	Muscular weakness, paralysis, heart failure
Sodium	Table salt; diet provides ample to excessive amounts	Key role in body's salt–water balance; roles in muscle and neural function	Muscle cramps	High blood pressure in susceptible people
Sulfur	Proteins in diet	Component of body proteins	None reported	None likely
Zinc	Whole grains, legumes, nuts, meats, seafood	Component of digestive enzymes; roles in normal growth, wound healing, sperm formation, and taste and smell	Impaired growth, scaly skin, impaired immune function	Nausea, vomiting, diarrhea; impaired immune function and anemia

* Guidelines for appropriate daily intakes are being worked out by the Food and Drug Administration.

36.9 Weighty Questions, Tantalizing Answers

Weight is a touchy subject. Part of the problem is that the standard of what constitutes an "ideal weight" varies from culture to culture. Here we consider weight as it relates to health. Thinner people really do live longer, on average, than heavier ones.

In the United States, 300,000 or so people die annually from weight-related conditions that are preventable. As weight rises, so does the risk of type 2 diabetes, heart disease, hypertension, breast cancer, colon cancer, gallstones, and numerous other ailments.

WHAT IS A GOOD BODY WEIGHT?

Figure 36.14 has weight guidelines for women and men. A *body mass index* (BMI) is another measurement designed to help determine the health risk of weight. It is calculated by this formula:

$$BMI = \frac{weight\ (pounds) \times 703}{height\ (inches)^{2}}$$

Generally, individuals with a BMI of 25 to 29.9 are considered to be overweight. A scored of 30 or more indicates **obesity**, an overabundance of fat in adipose tissue that may lead to severe health problems. How fat is distributed also helps predict risks. Fat that gets stored above the belt, as in "beer bellies," is associated with an increased likelihood of heart problems.

Dieting alone cannot lower a person's BMI value. Eat less, and the body will slow its metabolic rate to conserve energy. So how do you function normally over the long term while maintaining acceptable weight? *You must balance caloric intake with energy output.* For most people, this means eating only recommended portions of low-calorie, nutritious foods *and* exercising regularly.

Bear in mind, energy stored in food is expressed as kilocalories or Calories (with a big C). A **kilocalorie** is 1,000 calories, which are units of heat energy.

To figure out how many kilocalories you should take in daily to maintain a preferred weight, multiply that weight (in pounds) by 10 if you are not active physically, 15 if moderately active, and by 20 if highly active. Next, subtract one of the following amounts from the multiplication result:

Age:	Subtract:
25–34	0
35–44	100
45–54	200
55–64	300
Over 65	400

Suppose you are 25 years old, highly active, and weigh 120 pounds. This means you will require $120 \times 20 = 2,400$ kilocalories daily to maintain your weight. If you want to gain weight you will require more; to lose, you will require less. But this is only a rough estimate. Other factors, including height, must be considered. A person who is 5 feet, 2 inches tall and is active doesn't require as much energy as an active 6-footer who weighs exactly the same amount.

Figure 36.14 How to estimate "ideal" weights for adults. Values shown are consistent with a long-term Harvard study into the link between excess weight and risk of cardiovascular disorders. The "ideal" varies. It is influenced by specific factors such as having a small, medium, or large skeletal frame; bones are heavy.

Weight Guidelines for Women

Starting with an ideal weight of 100 pounds for a woman who is 5 feet tall, add five additional pounds for each additional inch of height. Examples:

Height (feet)	Weight (pounds)
5' 2"	110
5' 3"	115
5' 4"	120
5' 5"	125
5' 6"	130
5' 7"	135
5' 8"	140
5' 9"	145
5' 10"	150
5' 11"	155
6'	160

Weight Guidelines for Men

Starting with an ideal weight of 106 pounds for a man who is 5 feet tall, add six additional pounds for each additional inch of height. Examples:

Height (feet)	Weight (pounds)
5' 2"	118
5' 3"	124
5' 4"	130
5' 5"	136
5' 6"	142
5' 7"	148
5' 8"	154
5' 9"	160
5' 10"	166
5' 11"	172
6'	178

GENES, HORMONES, AND OBESITY

Some people can eat almost anything without ever adding weight. Others would swear that they put on pounds just by looking at food. Genes have a lot to do with it. In 1990, Claude Bouchard reported results of a study of experimental overeating by twelve pairs of male twins. At the start of the study, all were lean young men in their early twenties. For 100 days, they lived in a dormitory, they were told not to exercise, and they followed a diet that delivered 6,000 Calories a week more than they normally consumed.

Not surprisingly, all the men gained weight. But some gained three times as much as others. What accounted for the difference? Those who were twins tended to gain similar amounts of weight. It seemed that genetic differences affected how the individual body responded to the overfeeding.

As another test, Bouchard started with sets of obese twins and put them on a limited calorie diet. Once again he found variation in how the men responded, with twins losing similar amounts.

We are learning more and more about genes that contribute to obesity. This story started in 1950 with the discovery of an unusually fat mouse (Figure 36.15). A lineage of these mice was established and, in 1995, molecular geneticists identified one mouse gene that affected the tendency to put on body fat. They named it the *ob* gene. The DNA sequence in the gene of obese mice differs from that in normal mice.

In 1994, Jeffrey Friedman showed that the mouse *ob* gene encoded the hormone *leptin*. As described in the opening section, humans have a nearly identical hormone. We now know it is one of a number of factors that affect the appetite center in the brain.

Cholecystokinin (CCK) is another. It is secreted by cells in the first part of the small intestine. As noted in the chapter opener, *ghrelin* is a peptide hormone secreted by cells of the stomach lining and the hypothalamus. Unlike leptin and cholecystokinin, which make you feel full, ghrelin makes you hungry and prevents you from burning fat. Your body is adapted to secrete more after you lose weight.

These are only some of the players in complex hormonal pathways that involve the appetite and metabolism centers in the brain. Researchers hope to identify all steps of the ancient pathways that control weight. They also plan to identify the contributions of all those components to other aspects of the body's physiology. Maybe then, someone will be able to design drugs to help those of us who are genetically predisposed to putting on weight and keeping it.

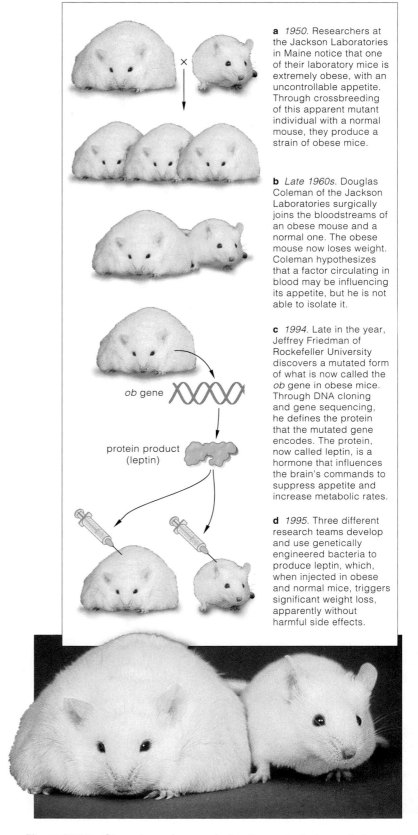

ob gene

protein product
(leptin)

a *1950*. Researchers at the Jackson Laboratories in Maine notice that one of their laboratory mice is extremely obese, with an uncontrollable appetite. Through crossbreeding of this apparent mutant individual with a normal mouse, they produce a strain of obese mice.

b *Late 1960s*. Douglas Coleman of the Jackson Laboratories surgically joins the bloodstreams of an obese mouse and a normal one. The obese mouse now loses weight. Coleman hypothesizes that a factor circulating in blood may be influencing its appetite, but he is not able to isolate it.

c *1994*. Late in the year, Jeffrey Friedman of Rockefeller University discovers a mutated form of what is now called the *ob* gene in obese mice. Through DNA cloning and gene sequencing, he defines the protein that the mutated gene encodes. The protein, now called leptin, is a hormone that influences the brain's commands to suppress appetite and increase metabolic rates.

d *1995*. Three different research teams develop and use genetically engineered bacteria to produce leptin, which, when injected in obese and normal mice, triggers significant weight loss, apparently without harmful side effects.

Figure 36.15 Chronology of research developments that revealed the identity of leptin, a key heritable factor that affects body weight.

Summary

Section 36.1 A digestive system breaks down food to molecules small enough to be absorbed into the internal environment, and it eliminates unabsorbed residues. It works with other organ systems in ways that promote homeostasis. Most animals have a complete digestive system. This is a tube with two openings (mouth and anus) and specialized areas between them.

Section 36.2 Table 36.4 lists the components and accessory organs of the human digestive system, along with their functions.

Section 36.3 Starch digestion starts in the mouth. Chewing mixes food with enzyme-containing saliva. Protein digestion starts in the stomach, a muscular sac with a glandular lining that secretes gastric fluid. Digestion is finished in the small intestine.

Ducts leading from the pancreas and gallbladder empty into the small intestine. The pancreas secretes digestive enzymes. Bile, which assists in fat digestion, is made in the liver and stored in the gallbladder.

Local controls as well as the nervous and endocrine systems respond to the volume and composition of food in the gut. They cause changes in muscle activity and in secretion rates for hormones and enzymes.

Section 36.4 Most nutrients are absorbed in the small intestine. Glucose and amino acids are actively transported across the intestinal lining, then enter the blood. Monoglycerides and fatty acids diffuse into cells of the lining. They recombine to form triglycerides, which are secreted into interstitial fluid. From there, they enter lymph vessels that deliver them to blood.

Section 36.5 Small organic compounds absorbed from the small intestine are stored, used to build larger compounds, burned for energy, or excreted.

Section 36.6 The large intestine absorbs minerals and water, and concentrates undigested residues for expulsion.

Section 36.7 Dietary guidelines are in flux. There is general agreement that high intake of saturated fats and refined carbohydrates should be avoided.

Section 36.8 Vitamins are organic compounds and minerals are inorganic. Both are necessary for normal metabolism and are usually obtained from eating a balanced diet.

Section 36.9 Obesity raises the risk of health problems and shortens life expectancy. To maintain body weight, energy (caloric) intake must balance energy (caloric) output. Heritable factors make it more difficult for some people to maintain a good body weight that promotes overall health. Leptin, ghrelin, and other hormones are among the gene products that are now known to affect appetite centers and metabolism.

Table 36.4	Summary of the Human Digestive System
Mouth (oral cavity)	Start of digestive system, where food is chewed, and moistened; polysaccharide digestion begins
Pharynx	Entrance to digestive and respiratory tract tubes
Esophagus	Muscular tube, moistened by saliva, that moves food from pharynx to stomach
Stomach	Stretchable sac where food mixes with gastric fluid and protein digestion starts; stores food taken in faster than can be processed; its fluid kills many microbes
Small intestine	Receives secretions from liver, gallbladder, pancreas; digests most nutrients; delivers unabsorbed material to colon
Colon (large intestine)	Concentrates and stores undigested matter (by absorbing mineral ions and water)
Rectum	Distension triggers expulsion of feces
Anus	Terminal opening of digestive system
Accessory Organs:	
Salivary glands	Glands that secrete saliva, a fluid with polysaccharide-digesting enzymes, buffers, and mucus
Pancreas	Secretes enzymes that digest all major food molecules; buffers against HCl secretions from stomach lining
Liver	Secretes bile; roles in carbohydrate, fat, and protein metabolism
Gallbladder	Stores and concentrates bile from the liver

Self-Quiz
Answers in Appendix III

1. A digestive system functions in _____ .
   a. secreting enzymes   c. eliminating wastes
   b. absorbing compounds   d. all of the above

2. Protein digestion begins in the _____ .
   a. mouth   c. small intestine
   b. stomach   d. colon

3. Most nutrients are absorbed in the _____ .
   a. mouth   c. small intestine
   b. stomach   d. colon

4. Bile has roles in _____ digestion and absorption.
   a. carbohydrate   c. protein
   b. fat   d. amino acid

5. Monosaccharides and amino acids are both absorbed _____ .
   a. by active transport   c. at lymph vessels
   b. by diffusion   d. as fat droplets

6. The largest number of bacteria thrive in the _____ .
   a. stomach   c. large intestine
   b. small intestine

7. _____ are inorganic substances with metabolic roles that no other substance can fulfill.
   a. Fats   d. Vitamins
   b. Minerals   e. Simple sugars
   c. Proteins   f. both b and d

Figure 36.16 One success story: In 2000, after recovering from anorexia, Dutch cyclist Leontien Zijlaard won three Olympic gold medals. Four years earlier, she had been too malnourished and weak to compete.

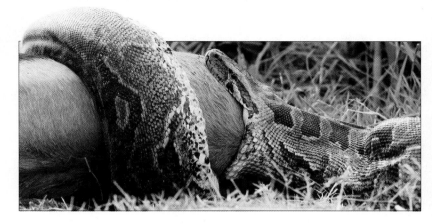

Figure 36.17 Exceptionally Big Mac of the snake world.

8. Match each organ with a digestive function.
   _____ gallbladder   a. makes bile
   _____ colon   b. compacts undigested residues
   _____ liver   c. secretes most digestive enzymes
   _____ small   d. absorbs most nutrients
          intestine   e. secretes gastric fluid
   _____ stomach   f. stores, secretes bile
   _____ pancreas

Critical Thinking

1. *Anorexia nervosa* is an eating disorder in which people, most often young women, starve themselves. The name means "nervous loss of appetite," but affected people are usually obsessed with food and continually hungry.

The causes of anorexia nervosa are complicated, but genes certainly play a role. Australian researchers looked at a gene for a protein that transports norepinephrine, one of the brain's signaling molecules, across cell membranes. Individuals carrying one mutant allele were twice as likely to be anorexic. This discovery may lead to new treatments. Even now, many recovered anorexics enjoy normal lives (Figure 36.16).

Like people who undergo stomach stapling to lose weight, anorexics also are vulnerable to bone loss and brittle bones. Speculate on the underlying causes of the problems in these individuals.

2. How can a python swallow something that is wider than it is? Consider Figure 36.17, then do some research and write up a brief description of your findings.

3. A glassful of whole milk contains lactose (a sugar), proteins, butterfat, vitamins, and minerals. Explain what will happen to each component in your digestive tract.

4. A *peptic ulcer* is a hole in the lining of the stomach or the first part of the small intestine (Figure 36.5*a*). Acid leaking from an ulcer can cause great pain and may even be life threatening if it erodes a blood vessel. Most ulcers arise in people infected by *Helicobacter pylori* bacteria (Figure 19.3). Ulcers are now treated with a course of antibiotics that can extend for as long as two weeks. Side effects are common and include cramps and diarrhea. Women who are being treated have double the normal risk of vaginal infections. Explain why a lengthy course of antibiotics would cause such symptoms.

IMPACTS, ISSUES   *Truth in a Test Tube*

Light or dark? Clear or cloudy? A lot or a little? Asking about and examining urine is an ancient art (Figure 37.1). As early as 1500 B.C., physicians in India recorded that some people were producing copious amounts of sweet-tasting urine that attracted insects. In time, the disorder was named *diabetes mellitus*, which loosely translates as "passing honey-sweet water." Doctors still diagnose it by testing for sugar in urine, although they have replaced the taste test with chemical analysis.

Modern physicians routinely check the sugar level, pH, protein content, and concentration of urine. Acidic urine can signal metabolic problems. Alkaline urine can signal a bacterial infection. Too much protein dissolved in it may mean the kidneys are faltering. Too many solutes may be a sign of dehydration or problems with hormones that control kidney function. Specialized urine tests also detect chemicals produced by cancers of the kidney, bladder, and prostate gland.

Do-it-yourself urine tests have become popular. If a woman wishes to become pregnant, she can use one such test to keep track of the amount of LH, or luteinizing hormone, in her urine. About midway through a menstrual cycle, LH triggers ovulation, a release of an egg from an ovary. Still another over-the-counter urine test can tell her that she is or is not pregnant. Older women can

Figure 37.1 Seventeenth-century physician and nurse examining a urine specimen. The color, consistency, odor, and—at least in the past—the taste of urine afford clues to health problems. By cleansing the blood, kidneys help maintain the internal environment. Urine that forms inside them can reflect abnormal changes in the volume and composition of that crucial environment.

the big picture

The Balancing Act   All animals gain and lose water and solutes. When the two are balanced, the body is maintaining the volume and composition of its extracellular fluid. Mammals control the gains and losses mainly by urinary excretion.

Human Urinary System   Vertebrates have a pair of kidneys that continually filter the blood. The body reclaims conserved and adjusted amounts of the filtrate. The kidneys excrete excess amounts of water and solutes as a fluid called urine.

test for declining hormone levels in urine, a sign that they are entering menopause.

Not everyone is in a hurry to have their urine tested. Olympic athletes have been stripped of medals after mandatory urine tests revealed they had used prohibited drugs. The National Collegiate Athletic Association (NCAA) tests urine samples from about 3,300 student athletes per year for any performance enhancing substances as well as "street drugs."

Use marijuana, cocaine, Ecstasy, or other drugs, and your urine tells the tale. THC, the active ingredient of marijuana, enters the blood, then the liver converts it to another compound. When kidneys filter blood, the THC is dumped into the newly forming urine. It can take up to ten days for all of the THC to be metabolized and removed. Until then, urine tests can detect it.

That urine is such a remarkable indicator of health, hormonal status, and drug use is a tribute to the urinary system. Every day, the body's volume of blood is filtered more than thirty times through its pair of fist-sized kidneys. When all goes well, the two kidneys get rid of excess water and excess or harmful solutes, including many metabolites, toxins, hormones, and drugs.

So far in this unit, you have considered several organ systems that work to keep cells supplied with oxygen, nutrients, water, and other substances. Turn now to the ones that interact specifically to maintain the volume, composition, and even the temperature of the internal environment itself.

 How Would You Vote?

Certain companies use urine testing to screen for drug and alcohol use among prospective employees. Some people say this is an invasion of privacy. Do you think employers should be allowed to require a person to undergo urine testing before being hired? See the Media Menu for details, then vote online.

How Kidneys Work Urine forms by the filtration of blood into kidney tubules, the reabsorption of most water and solutes, and the secretion of some ions and wastes into tubules. Hormones adjust the volume and composition of urine.

Thermal Homeostasis The internal environment gains heat from metabolism. It also gains heat from the external environment and loses heat to it. Certain aspects of the body's form, physiology, and behavior help keep the internal temperature within tolerable limits.

37.1 Maintaining the Extracellular Fluid

The cells of a fish or some other marine animal are adapted to life in the seas. And yet, about 375 million years ago, some lineages of animals that evolved in salty water moved onto dry land. They were able to do so partly because they brought salty fluid along with them, as an internal environment for their cells.

As a starting point, remember that most animals have **interstitial fluid** filling tissue spaces between all their living cells. And remember, most have a circulatory system that moves **blood** to and from tissues. Together, intercellular fluid and blood are **extracellular fluid**—an internal environment for the body's cells. It does not matter if an animal lives on land or in the sea. It must keep the volume and composition of its internal environment within tolerable ranges. Figure 37.2 is an overview of the organ systems that do this.

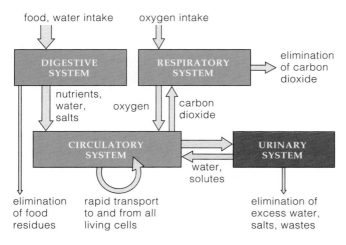

Figure 37.2 Overview of the functional links between the digestive, respiratory, circulatory, and urinary systems. Guided by the nervous and endocrine system, interactions among these systems contribute to maintaining homeostasis in the internal environment.

CHALLENGES IN WATER

Osmosis, recall, is a movement of water across some semipermeable membrane in response to differences in solute concentrations (Chapter 5). In most marine invertebrates, such as crabs, interstitial fluid is like seawater in its solute concentrations, so there is no net inward or outward movement of water by osmosis.

In vertebrates, body fluids are about one-third as salty as seawater. Vertebrates have a **urinary system**, a collection of interacting organs that counter shifts in the composition and volume of extracellular fluid. All have paired **kidneys** that filter blood, form urine, and help maintain the body's water–solute balance. **Urine** is a fluid that rids the body of water and solutes that exceed amounts required for the extracellular fluid.

Freshwater fishes and amphibians constantly gain water and lose solutes (Figure 37.3a). But they do not gain water by drinking it. Rather, water moves into the internal environment by way of osmosis, diffusing across thin respiratory surfaces and skin. Excess water leaves the body as dilute urine. In both groups, solute losses are balanced when more solutes come in with food, and gill and skin cells pump in sodium.

Marine bony fishes have less salt inside than their surroundings. They constantly lose water by osmosis. They make it up by drinking seawater and pumping out excess solutes through gills (Figure 37.3b). Highly concentrated wastes leave in a tiny volume of urine.

CHALLENGES ON LAND

Pioneers on land and their descendants faced intense sunlight, drying winds, more pronounced swings in temperature, water of dubious salt content, and, on occasion, no water at all. How did they conserve or replace the water and salts lost in everyday life? They

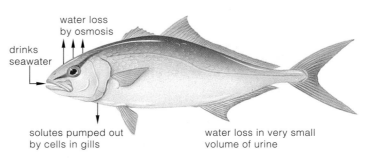

a Freshwater bony fish (body fluids far saltier than surroundings)

b Marine bony fish (body fluids less salty than surroundings)

Figure 37.3 Water–solute balance in freshwater fish (**a**) and saltwater fish (**b**).

Figure 37.4 Differences in how humans and kangaroo rats gain and lose water. Notice how the gains and losses still balance out. They do so in all healthy animals. The next few sections focus on the balancing act in the human body. For comparative purposes, focus here on the aspects of body form, physiology, and behavior that make a kangaroo rat a master of water conservation in this New Mexico desert.

After a brief rainy season, the sun bakes the desert sand for months and free water is scarce. The kangaroo rat waits out the heat of the day inside a burrow, then forages in the cool night. It hops rapidly and far, searching for dry seeds, maybe a succulent, and fleeing from coyotes or snakes. All that hopping requires ATP energy and water. Seeds, chockful of energy-rich carbohydrates, offer both. Metabolic reactions that release energy from carbohydrates and other organic compounds also yield water. "Metabolic water" makes up a whopping 90 percent of a kangaroo rat's daily water intake. It is only about 12 percent of the total water intake for the human body.

A kangaroo rat resting in a cool burrow conserves and recycles water. When it inhales, air is moistened and warmed. When it exhales, water condenses in its cooler nose, and some diffuses back into the body. When seeds are emptied from the kangaroo rat's cheek pouches, they soak up water dripping from the nose. The body reclaims water when dripped-upon seeds are eaten.

A kangaroo rat has no sweat glands. Its feces hold only half the water that human feces do. The kangaroo rat loses water when urinating, but its two kidneys don't let it piddle away much. Each kidney is packed with tiny tubes—nephrons. The parts of the tubes called loops of Henle are amazingly long. They pump a lot of salt back into the body, and nearly all the water follows, by osmosis. The urine ends up three to five times more concentrated than yours.

	KANGAROO RAT	HUMAN
WATER GAIN (milliliters)		
by ingesting solids	6.0	850
by ingesting liquids	0.0	1,400
by metabolism	54.0	350
	60.0	2,600
WATER LOSS (milliliters)		
in urine	13.5	1,500
in feces	2.6	200
by evaporation	43.9	900
	60.0	2,600

must have done so, otherwise the composition and volume of their internal environment would have spun out of control. *How did land-dwelling descendants of marine animals maintain internal operating conditions and so prevent cellular anarchy?*

Consider how humans and other mammals *gain water*. Eating food and drinking fluids put water into the gut, from which it is absorbed. Water also forms as a product of many metabolic reactions. The volume of water that enters the mammalian gut in the first place is influenced by a thirst mechanism, which you will be reading about later in the chapter.

Unlike fishes, land vertebrates do not *lose water* by osmosis. They lose water in controlled ways, largely by **urinary excretion**—the elimination of excess water and solutes in the urine. Additional water evaporates from respiratory surfaces such as those in the lungs (Chapter 35). Also, mammals are the only animals that lose some water by sweating and evaporative water loss. Very little water is lost from food that passes through the mammalian gut; most is absorbed into the internal environment, not eliminated in feces.

Mammals *gain solutes* by absorption from the gut, cell secretions, respiration, and metabolism. Nutrients from food and secretions from cells enter interstitial fluid, then the blood. The respiratory system delivers oxygen to blood. All mammals lose carbon dioxide, the most abundant waste, by respiration.

Mammals *lose solutes* by urinary excretion, respiration, and, in some species, by sweating. Urine includes metabolic wastes. For instance, **urea**, a major waste, forms in the liver when carbon dioxide combines with two ammonia molecules, which formed when amino groups split off of amino acids. Urine also includes hemoglobin breakdown products, food additives, and, as you read earlier, assorted drugs.

Figure 37.4 shows how two different species gain and lose water and solutes in different ways. Despite the differences, the daily gains and losses balance out. And this happens by way of a urinary system.

In all animals, daily gains in water and solutes must balance the daily losses.

In all vertebrates, a urinary system counters unwanted shifts in the volume and composition of extracellular fluid. Paired kidneys filter blood, form urine, and help maintain solute levels within tolerable limits.

37.2 The Human Urinary System

Different solid foods and fluids enter a mammal's gut intermittently. Then variable amounts of absorbed water, nutrients, and other substances move into blood, then into interstitial fluid. All through the day and night, the body makes compensatory adjustments for large and small shifts in that internal environment.

COMPONENTS OF THE SYSTEM

The human urinary system maintains the body's salty extracellular fluid. Drink too much or too little water, wolf down salty potato chips, lose too much sodium in sweat—the responses are quick. The urinary system precisely adjusts how much water and sodium ions the body excretes or saves. It does so by way of two kidneys, two ureters, a urinary bladder, and a urethra (Figure 37.5). The kidneys are a pair of bean-shaped organs, each about as large as an average adult's fist. They are located just outside the peritoneal lining of the abominal cavity, one on each side of the backbone.

Kidneys have an outer capsule of dense irregular connective tissue (Figure 37.6*a*). When you are at rest, about 20 percent of the blood pumped out of your heart's left side reaches the kidneys by way of the renal artery. Blood capillaries thread through the outer cortex and inner medulla of each kidney.

Kidneys filter water, mineral ions, organic wastes, and other substances from the blood flowing through them. Then they adjust the composition of the filtrate and return all but about 1 percent or so to the blood. The remaining small volume of unreclaimed water, solutes, and wastes is urine.

Urine flows from each kidney into a ureter, a tube leading to a muscular sac, the urinary bladder. Urine is briefly stored here before flowing into the urethra, a muscular tube that opens at the body surface.

Flow from the bladder (urination) is a reflex action. When the bladder is full, a sphincter around its neck opens. Smooth muscle in the bladder wall contracts, forcing urine out into the urethra. Another sphincter encircles the urethra where it leaves the pelvic cavity. This sphincter is composed of skeletal muscle, and it is under voluntary control.

NEPHRONS—FUNCTIONAL UNITS OF KIDNEYS

Each kidney contains more than a million **nephrons**. These small, slender tubules are packed in lobes that extend from the kidney cortex down into the central medulla (Figure 37.6*b*). *It is at the nephrons that water and solutes are filtered from blood, and the amounts to be reclaimed by the blood are adjusted.*

A nephron starts where its wall cups around a set of *glomerular* capillaries. The cup, Bowman's capsule, interacts with the capillaries as a blood-filtering unit called a **renal corpuscle** (Figures 37.6*d* and 37.7).

These glomerular capillaries are highly permeable, and blood pressure inside them is very high. The

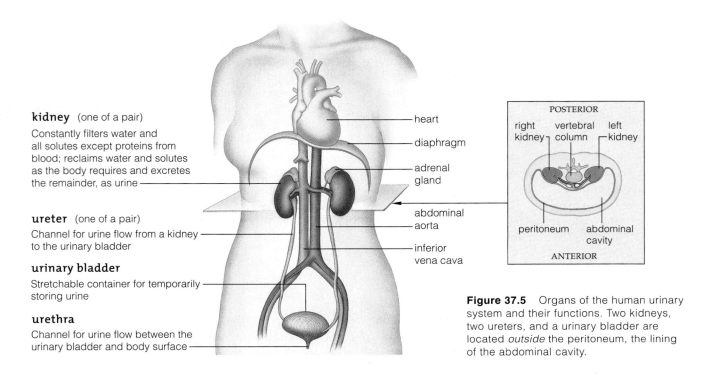

kidney (one of a pair)
Constantly filters water and all solutes except proteins from blood; reclaims water and solutes as the body requires and excretes the remainder, as urine

ureter (one of a pair)
Channel for urine flow from a kidney to the urinary bladder

urinary bladder
Stretchable container for temporarily storing urine

urethra
Channel for urine flow between the urinary bladder and body surface

heart

diaphragm

adrenal gland

abdominal aorta

inferior vena cava

POSTERIOR
right kidney   vertebral column   left kidney

peritoneum   abdominal cavity

ANTERIOR

Figure 37.5 Organs of the human urinary system and their functions. Two kidneys, two ureters, and a urinary bladder are located *outside* the peritoneum, the lining of the abdominal cavity.

Figure 37.6 (a) Human kidney. (b) Its functional units are nephrons. For clarity, the nephron size is exaggerated. (c) A nephron's functional zones. (d) Blood capillaries associated with nephrons. The start of a nephron, Bowman's capsule, balloons around a set of glomerular capillaries. Another set of blood vessels, the peritubular capillaries, threads around tubular parts of the nephron.

small arterioles that branch from the renal artery and deliver blood to nephrons have a wider diameter than the vessels that carry blood away. As a result, blood dams up inside glomerular capillaries. The resulting elevated blood pressure forces fluid out through the leaky capillary walls, into Bowman's capsule.

Each minute, about 1.5 liters of blood flow through the renal artery and into glomerular capillaries. About 120 milliliters of water and small solutes are filtered out and they collect inside Bowman's capsule. That is about 180 liters of filtrate every day!

From Bowman's capsule, fluid enters the **proximal tubule**, a tubular region closest to the cup. Next is the hairpin-shaped **loop of Henle**, then the **distal tubule** (most distant from the capsule). The nephron's last region, a **collecting duct**, is part of a duct system that leads to the kidney's central cavity, then into a ureter. Again, only some of the fluid in the blood that enters

the glomerular capillaries is forced out of them and into Bowman's capsule. The unfiltered portion of the blood continues flowing from them, into peritubular capillaries that thread lacily around the nephron's tubular parts.

During urine formation, blood inside peritubular capillaries reclaims most of the water and solutes that left the distal and proximal tubules. Then veins carry this blood back to the general circulation.

The human urinary system consists of two kidneys, two ureters, and a urinary bladder. The kidneys filter water and solutes from blood. The body reclaims most of this. The rest flows, as urine, through ureters to the bladder, which stores it. The urethra carries urine to the body surface for excretion.

Millions of nephrons are the functional units of the kidney.

37.3 Urine Formation

Urine formation depends on differences in permeability and membrane transport mechanisms in different parts of the nephron wall, which are under hormone control.

THREE RELATED PROCESSES

Urine forms by processes called glomerular filtration, tubular reabsorption, and tubular secretion. The blood pressure that originates with heart contractions drives **glomerular filtration**, which is a movement of water and small solutes from the glomerular capillaries into Bowman's capsule. The filtrate that forms moves next into the proximal tubule of nephrons (Figure 37.7b).

Filtration rates are high, mainly because glomerular capillaries are 10 to 100 times more permeable than other capillaries to water and small solutes. The rates are altered by changes in blood pressure. Arterioles that deliver blood to nephrons are sensitive to short-

term changes in blood pressure and tend to smooth out the fluctuations. They vasodilate if the local blood pressure falls a bit and vasoconstrict if it rises. Even so, the nervous system overrides vasodilation when other organs demand more of the blood volume.

By **tubular reabsorption**, peritubular capillaries reclaim most of the water and solutes in the proximal tubule (Figure 37.7c,d). Blood inside them still has too many ions (mainly H^+ and K^+) and wastes, such as urea. By the process of **tubular secretion**, membrane transport proteins in capillary cells move these wastes into interstitial fluid. Then transport proteins of cells in the nephron wall shuttle them into the urine.

REABSORPTION OF WATER AND SODIUM

As filtrate enters the proximal tubules, cells actively transport some sodium out. Other ions follow. So

Read Me First!
and watch the narrated animation on urine formation

a Blood pumped from the heart travels to the renal artery, then into kidneys, where water and some solutes will be filtered from it. Most of the filtrate will return to the general circulation.

b **Filtration**. At the start of the nephron, blood enters glomerular capillaries. Water and small solutes are filtered into Bowman's capsule.

NEPHRON

c **Tubular Reabsorption**. Water and many solutes cross the proximal tubule's wall and enter interstitial fluid of the kidney cortex. Membrane transport proteins pump most solutes across the wall. Then the solutes enter the peritubular capillaries.

blood vessel entering

blood vessel leaving

BOWMAN'S CAPSULE + GLOMERULAR CAPILLARIES = RENAL CORPUSCLE

d **Tubular Secretion**. Cells of the nephron's tubular wall regions secrete excess H^+ and a few other solutes into the fluid in the nephron's lumen.

e After the hairpin turn, the loop of Henle's wall is impermeable to water. Its cells pump chloride ions out of the loop; sodium passively follows. Pumping makes interstitial fluid saltier, so more water is drawn out of the collecting ducts in the medulla.

f Hormonal action adjusts the concentration of urine:

ADH promotes water reabsorption, so the urine is concentrated. When controls inhibit ADH secretion, urine that forms is dilute.

Aldosterone activates sodium pumps of cells in walls of distal tubules and collecting ducts. More sodium is reabsorbed, so urine has less of it. When controls block aldosterone secretion, more sodium is excreted in urine.

COLLECTING DUCT

g **Urinary Excretion**. What happens to water and solutes not reabsorbed or secreted into the tubule? They flow on through a collecting duct to the renal pelvis, then are eliminated from the body by way of the urinary tract.

Figure 37.7 Urine formation.

a Stimulus
Losing a lot of water lowers the blood volume. Sensory receptors in the hypothalamus detect the deviation from the set point.

b The hypothalamus signals pituitary gland to secrete more ADH.

c ADH circulates to kidneys, makes cells of nephron distal tubules and collecting ducts more permeable to water.

HYPOTHALAMUS

PITUITARY GLAND

f Response
Sensory receptors in hypothalamus detect the increase in blood volume. Signals calling for ADH secretion stop.

d Kidneys reabsorb more water, so less water leaves in urine.

e The blood volume rises.

Figure 37.8 Feedback control of ADH secretion, one of the negative feedback loops from kidneys to the brain that helps adjust the volume of extracellular fluid. Nephrons in the kidneys reabsorb more water when we don't take in enough or lose too much, as by profuse sweating.

does water, by osmosis. The body reabsorbs about two-thirds of the filtrate's water here (Figure 37.7c,d).

Interstitial fluid is saltiest around the hairpin turn of the loop of Henle. Water moves out of the filtrate by osmosis *before* the turn. The fluid left behind gets saltier until it matches interstitial fluid. The loop *after* the turn is impermeable to water, but here transport proteins pump out chloride. As the interstitial fluid gets saltier, it attracts more water out of the filtrate that is just entering the loop (Figure 37.7e).

Adjusting the permeability of the tubules to water and sodium makes the urine more or less dilute. Two hormones are central to the adjustments.

Near the hypothalamus a rise in sodium levels is detected by osmoreceptors, a type of sensory neuron. They signal a *thirst center*, a hypothalamic region that responds by calling for water-seeking behavior. The hypothalamus also stimulates the pituitary gland to secrete **ADH**, or antidiuretic hormone. ADH acts on distal tubules and collecting ducts, making them more permeable to water. The body reabsorbs more water, so less departs in urine (Figure 37.8). ADH secretion slows when the volume of extracellular fluid rises and the sodium concentration declines. A big drop in blood pressure, sensed by baroreceptors in the wall of some arteries, also can lead to ADH secretion (Section 33.7). So can nausea and vomiting.

Cells in arterioles that carry blood into the kidney also respond to a drop in extracellular fluid volume. They release an enzyme, renin, which sets in motion a chain of reactions. The end result is that the adrenal glands above each kidney are stimulated to secrete **aldosterone**. This hormone acts on cells of the distal tubules and increases sodium reabsorption. The result? Less sodium is excreted, and as water follows sodium back into blood, urine becomes more concentrated.

The thirst center can also be stimulated by a dry mouth. In this case, nerve endings send signals to the hypothalamus, which promotes thirst. As water intake is being encouraged, urinary output is reduced.

Diabetes insipidus arises when the pituitary secretes too little ADH or when receptors cannot respond to it. Urine formation is excessive, sometimes as much as sixteen liters a day. The condition is life-threatening only when not enough water replaces the losses.

Filtration, reabsorption, and secretion in the kidneys help maintain the composition and volume of extracellular fluid.

Blood filtration rates depend largely on blood pressure.

The proximal tubule reabsorbs most of the filtered water and solutes. Some adjustments occur at the distal tubule.

Certain excess solutes diffuse out of peritubular capillaries and into interstitial fluid. Then, by tubular secretion, they move into the nephron for subsequent excretion.

ADH promotes water conservation by increasing the amount of water reabsorbed by distal tubules and the collecting duct. Aldosterone then promotes the reabsorption of sodium. This indirectly increases water retention.

37.4 When Kidneys Break Down

Good health depends on kidney function. End-stage renal disease occurs when kidney function declines to less than 10 percent of normal and life is threatened. In the United States, the incidence of renal disease is on the rise.

Kidneys can fail through diabetes, chronic high blood pressure, bacterial or viral infection, or exposure to certain drugs or poisons, such as arsenic. As kidney function falters, waste metabolic products—especially from proteins—build up in the blood. This *uremic toxicity* causes nausea, vomiting, and diarrhea. It also affects mental state and impairs blood clotting.

Kidney stones can lead to renal failure. These hard deposits form as uric acid, calcium, and other wastes settle out of urine and collect in the renal pelvis. They usually leave in urine, but they can become lodged in the ureter or the urethra. In that case, they must be dissolved, broken up, or surgically removed.

About 300,000 people in the United States have end-stage renal failure. A *kidney dialysis machine* helps restore proper solute balances. Like kidneys, it adjusts solutes in blood selectively. "Dialysis" is an exchange of solutes across any artificial membrane positioned between two different solutions.

In *hemodialysis*, a machine is connected to a vein or an artery and pumps blood into membranous tubes submerged in a saline bath. As blood flows through the tubes, wastes diffuse out and salts, glucose, and other vital substances diffuse in. The solute-balanced, cleansed blood then flows back inside the body.

In *peritoneal dialysis*, fluid of suitable composition is pumped into a patient's abdominal cavity, then later drained out. The cavity's lining functions as a permeable membrane for dialysis.

Kidney dialysis is a temporary fix. In chronic kidney failure, it must be used for the rest of the patient's life or until a transplant provides a working kidney. Even so, with treatment and special diet, many patients resume fairly normal activity (Figure 37.9).

Figure 37.9 At age thirteen, Karole Hurtley was the national champion for her age group and karate belt level. Karole lives with renal failure. Eight or nine times a night, while she sleeps, a peritoneal dialysis machine beside her bed pumps a special solution into her torso and filters out waste products that her damaged kidneys are no longer able to remove.

37.5 Acid–Base Balance

Besides maintaining the volume and composition of extracellular fluid, kidneys keep it from getting too acidic or too basic (alkaline). Only the kidneys can rid the body of the acidic products of normal metabolism.

Each meal and each day's worth of metabolism put H^+ and other ions in extracellular fluid even as other events withdraw them. The body depends on controls over the **acid–base balance** of extracellular fluid. The importance of the balancing act becomes evident with *metabolic acidosis*. This condition arises when kidneys cannot secrete enough H^+ to keep pace with all that forms during metabolism. It can be life threatening.

Buffer systems, respiration, and urinary excretion do the balancing. A buffer system, remember, consists of some weak acids or bases that reversibly bind and release ions, thus minimizing shifts in pH.

Normally, extracellular pH is maintained between 7.35 and 7.45. Any excess acids dissociate into H^+ and other fragments, and pH falls. The effect is minimized when excess hydrogen ions react with buffers, as in the bicarbonate–carbonic acid buffer system:

$$H^+ + HCO_3^- \rightleftharpoons H_2CO_3 \rightleftharpoons CO_2 + H_2O$$

BICARBONATE   CARBONIC ACID

Buffers can neutralize excess H^+. Carbon dioxide that forms in the reactions is exhaled from the lungs. Like other buffer systems it has temporary effects only; it doesn't *eliminate* H^+. A great excess of H^+ can overload the buffer system. Only the urinary system can eliminate H^+ and thereby restore the buffers.

The same reactions occur in reverse at a nephron's tubular wall. HCO_3^- formed by the reverse reactions moves into peritubular capillaries and buffers excess acid. The H^+ that forms in the reactions is secreted into the nephron and may join with HCO_3^-. The CO_2 that formed may be returned to the blood, and then exhaled. The H^+ also may combine with phosphate ions or ammonia, then leave the body in urine.

The kidneys, in concert with buffering systems and with the respiratory system, neutralize acids. The action of these organs and mechanisms helps keep the extracellular fluid from becoming too acidic or too basic (alkaline).

A bicarbonate–carbon dioxide buffer system temporarily neutralizes excess hydrogen ions. The urinary system alone eliminates the excess ions and thus restores these buffers.

The bicarbonate–carbon dioxide buffer system is one of the key mechanisms that help maintain the acid–base balance.

37.6 Maintaining the Core Temperature

We turn now to another major aspect of homeostasis. How does the body maintain the temperature of the internal environment at its core within a comfortable range when the surroundings become too hot or cold?

HEAT GAINS AND LOSSES

The core temperature of an animal body rises when heat from the surroundings or metabolism builds up. When warm, a body loses heat to cool surroundings. The core temperature stabilizes when the rate of heat loss balances the rate of heat gain and production.

Heat is gained and lost through exchanges at the body surfaces. Radiation, conduction, convection, and evaporation are processes that drive these exchanges.

Thermal radiation is the emission of heat from any object in the form of radiant energy. Radiant energy from the sun can heat animals. Also, a metabolically active animal itself produces and radiates heat.

By **conduction**, heat is transferred between objects in direct contact with each other. An animal loses heat when it rests on objects cooler than it is. If it contacts objects that are warmer, the animal will gain heat.

By **convection**, moving air or water transfers heat. Conduction plays a part in this process; heat moves down a thermal gradient between the body and air or water next to it. Mass transfer also plays a part, with currents carrying heat away from or toward the body. When it is heated, air becomes less dense and moves away from the body.

By **evaporation**, a liquid converts to gaseous form and heat is lost in the process. Evaporation from the body surface has a cooling effect, because the water molecules that are escaping carry away some energy with them. Dry air and breezes increase evaporative heat loss. High humidity and still air slow it.

ENDOTHERM? ECTOTHERM? HETEROTHERM?

Fishes, amphibians, and reptiles warm their insides mostly with heat gained from the environment, not from metabolic activities. We classify these animals as **ectotherms**, meaning "heated from outside." When the outside temperature changes, ectotherms are only able to make behavioral adjustments. They generally have low metabolic rates and very little insulation. A rattlesnake (Figure 37.10*a*) is like this. When cold, it basks in the sun. When hot, it moves into shade.

Most birds and mammals are **endotherms** ("heat from within"). They have relatively high metabolic rates. Compared to a foraging lizard of the same weight, a foraging mouse requires about thirty times

Figure 37.10 (**a**) Sidewinder, an ectotherm. (**b**) Pine grosbeak, an endotherm, using fluffed feathers as insulation against winter cold.

more energy. Metabolic heat lets endotherms remain active across a wider range of temperatures. Fur, fat, or feathers can function in insulation and minimize heat loss (Figure 37.10*b*).

Some birds and mammals are **heterotherms**. They can maintain a fairly constant core temperature some of the time but let it shift at other times. For example, tiny hummingbirds have very high metabolic rates. They fly fast when foraging for nectar during the day. At night, hummingbird metabolic activity decreases so much that the hummingbird's body may become almost as cool as the surroundings. They flirt with the limit on the amount of heat that can be lost.

Warm, humid regions favor ectotherms, which do not have to spend as much energy as all endotherms do on maintaining the core temperature. In the tropics, reptiles have the advantage, because they can spend more energy on reproduction and other tasks. There they exceed mammals in numbers and diversity. In all cool or cold regions, though, the competitive edge tilts to endotherms. For example, you will not find reptiles running around in the arctic.

Core temperature is maintained when heat gains and losses are in balance.

Animals and their environment gain and lose heat by way of thermal radiation, conduction, convection, and evaporation.

Animals have morphological, physiological, and behavioral adaptations to environmental temperatures. Ectotherms gain heat mainly from the outside, endotherms generate most from within. Heterotherms do both at different times.

37.7 Temperature Regulation in Mammals

Chapter 24 introduced the negative feedback controls that govern increases in human body temperature. Here, let's compare the controls that come into play when mammals in general are subjected to heat stress and cold stress.

The hypothalamus, again, contains centers that help control the core temperature of the mammalian body. These brain centers receive continual input from many thermoreceptors in skin and deeper inside the body (Figure 24.8). When the temperature deviates from a set point, the centers integrate complex responses that involve skeletal muscles, smooth muscle in arterioles that service skin, and, often, sweat glands. Negative feedback loops back to the hypothalamus shut off responses when a favorable temperature is reinstated. Dromedary camels actually reset the set point *higher* during the hottest time of day (Figure 37.11).

RESPONSES TO HEAT STRESS

When a mammal gets too hot, the temperature control centers in the hypothalamus send out signals that call for **peripheral vasodilation**. This causes the blood vessels in the skin to dilate. As blood flow to the skin increases, so does the amount of heat that is lost from the skin to the surroundings (Table 37.1).

Evaporative heat loss is a response to heat stress that takes place at moist respiratory surfaces and across skin. Animals that sweat lose some water this way. For instance, humans and some other mammals have sweat glands that release water and solutes through pores at the skin's surface. An average-sized adult human has 2–1/2 million or more sweat glands. For every liter of sweat that is produced, about 600 kilocalories of heat energy depart the body through evaporative heat loss.

Bear in mind, sweat dripping from skin does not dissipate heat. Only high external temperatures can cause water from sweat to *evaporate*. On hot, humid days, evaporation rates cannot match sweat secretion; air's high water content slows it down. In strenuous exercise, sweating may help balance heat production in skeletal muscle. Extreme sweating, as in a marathon race, forces the body to lose an essential salt—sodium chloride—and water. Large solute losses can disrupt extracellular fluid composition and volume so much that runners may collapse and faint.

With the exception of marine species, just about all mammals (and only mammals) can sweat. Many have additional behavioral responses, such as licking fur or panting. "Panting" is shallow, rapid breathing. It increases the rate of evaporative water loss from the respiratory tract. The body cools when water evaporates from the nasal cavity, mouth, and tongue.

Sometimes peripheral blood flow and evaporative heat loss cannot counter heat stress, and *hyperthermia* follows. With this condition, core temperature rises above normal. An elevation of even a few degrees can be harmful to humans and other endotherms.

RESPONSES TO COLD STRESS

Selectively distributing blood flow, fluffing hair, and shivering are some mammalian responses to cold. Birds make similar responses but maintain a higher body temperature. When peripheral thermoreceptors detect a decrease in outside temperature, they signal the hypothalamus, which commands smooth muscle in arterioles in the skin to contract. The response is **peripheral vasoconstriction**. The arterioles constrict, which decreases the blood's delivery of heat to the body surfaces. When your fingers or toes are chilled,

Figure 37.11 Short-term adaptation to desert heat stress. Dromedary camels are heterotherms; they let the core temperature rise by as much as 14 degrees in the hottest hours of the day, then let it decline at night. A hypothalamic mechanism raises the internal thermostat, so to speak, and doing so has no adverse effects.

Table 37.1	Mammalian Responses to Core Temperature Shifts	
Stimulus	Main Responses	Outcome
Heat stress	Widespread vasodilation in skin; behavioral adjustments; in some species, sweating, panting	Dissipation of heat from body
	Decreased muscle action	Heat production decreases
Cold stress	Widespread vasoconstriction in skin; behavioral adjustments (e.g., minimizing surface parts exposed)	Conservation of body heat
	Increased muscle action; shivering; nonshivering heat production	Heat production increases

Figure 37.12 Two cases of responses to cold stress.

Above, polar bears (*Ursus maritimus*, "bear of the sea"). Unlike other bears, which hibernate, the polar bear remains active during the severe Arctic winter. It does not get too chilled after swimming because the coarse, hollow guard hairs of its coat shed water quickly. Thick, soft underhair traps heat. Brown adipose tissue, about 11.5 centimeters (4–1/2 inches) thick, insulates and helps generate metabolic heat.

Below: In 1912, the *Titanic* left Europe on her maiden voyage to America—and a chunk of the leading edge of a glacier in Greenland broke away and floated out to sea. Late at night, off the Newfoundland coast, the iceberg and ocean liner made an ill-fated rendezvous. The *Titanic* was said to be unsinkable. Survival drills were neglected. There were not enough lifeboats to hold even half the 2,200 passengers. It took about 2–1/2 hours for the *Titanic* to sink, and rescue ships arrived in less than two hours. Even so, 1,517 bodies were recovered from a calm sea. All of the dead had on life jackets; none had drowned. Hypothermia killed them.

all but 1 percent of the blood that would usually flow to skin is diverted to other regions of the body.

Also, muscle contractions make hairs (or feathers) "stand up." This **pilomotor response** creates a layer of still air next to the skin and reduces convective and radiative heat loss. Behavior can minimize exposed surface areas and reduce heat loss, as when polar bear cubs cuddle against their mother (Figure 37.12*a*).

Prolonged exposure to cold induces a **shivering response**. Signals from the hypothalamus that call for rhythmic tremors cause skeletal muscles to contract ten to twenty times per second. Although shivering increases heat production severalfold, it does have a high energy cost and is not effective for long.

Long-term or severe cold exposure also leads to a hormonal response—an increase in thyroid activity that raises the rate of metabolism. This **nonshivering heat production** targets *brown* adipose tissue most often. This connective tissue is richly endowed with blood vessels and mitochondria. In the mitochondria of its cells, the flow of hydrogen ions is used mostly to generate heat, not to make ATP.

Polar bears and other animals that hibernate or are active in cold environments have brown adipose tissue (Figure 37.12*a*). Human infants have this tissue. Adults have very little unless they are well adapted to cold habitats. For instance, some adult Japanese and Korean commercial divers who spend a long time in frigid water collecting shellfish retain this tissue.

Failure to protect against cold leads to *hypothermia*, a condition in which core temperature plummets. In humans, a drop by even a few degrees changes brain function and can cause confusion. Severe cooling can lead to coma and eventually death (Figure 37.12*b*).

Many mammals can recover from hypothermia. However, any frozen cells may die unless the tissues thaw under close medical supervision. Local freezing and destruction of tissues—usually of the hands, feet, nose, and ears—is known as *frostbite*.

Temperature shifts are detected by thermoreceptors that send signals to an integrating center in the hypothalamus.

This center serves as the body's thermostat and calls for adjustments that maintain core temperature.

Mammals counter cold stress by vasoconstriction in skin, behavioral adjustments, increased muscle activity, and shivering and nonshivering heat production.

Mammals counter heat stress by widespread peripheral vasodilation in skin and evaporative water loss.

Summary

Section 37.1 Blood and interstitial fluid make up the extracellular fluid. Maintaining the volume and solute concentration of this fluid is a vital aspect of homeostasis. All organisms must balance solute and fluid gains with solute and fluid losses. In vertebrates, this is the main function of the kidneys and other components of the urinary system.

Mammals gain water by absorption from the gut and metabolism. They lose it by urinary excretion, evaporation from body surfaces, and in feces. They gain solutes by absorption from the gut, secretion, respiration, and metabolism. Solutes are lost in urine and feces, and by respiration and sweating.

Section 37.2 The human urinary system consists of a pair of kidneys, a pair of ureters, a urinary bladder, and the urethra. The renal artery delivers blood to kidneys.

Kidneys contain an enormous number of nephrons, or blood-filtering units. The start of a nephron is the cup-shaped Bowman's capsule. It continues as three tubular regions (proximal tubule, loop of Henle, and distal tubule) and empties into a collecting duct.

Bowman's capsule and the set of highly permeable glomerular capillaries within it are a blood-filtering unit: a renal corpuscle. Most of the filtrate that enters Bowman's capsule gets reabsorbed at the nephron's tubular regions and is returned to the blood. The portion that is not returned is excreted as urine.

Section 37.3 Urine forms in nephrons by three processes: filtration, reabsorption, and secretion.

Filtration. Blood pressure drives water and small solutes (not proteins) out of the capillaries at the start of a nephron. These enter the nephron's tubular parts.

Reabsorption. Water and solutes to be conserved move out of the tubular parts, then into the capillaries that thread around the nephron. A small volume of water and solutes remains in the nephron.

Two hormones regulate the composition and volume of the urine. ADH secreted by the posterior pituitary promotes water reabsorption across the walls of the proximal tubule and collecting duct. As a result, urine is more concentrated. Aldosterone secreted by the adrenal cortex promotes sodium reabsorption into the distal tubule. Because water follows sodium gradients, aldosterone indirectly increases reabsorption of water.

Secretion. A few substances move from peritubular capillaries into the nephron, for excretion.

Section 37.4 When kidneys fail, frequent dialysis or a kidney transplant are the only alternatives.

Section 37.5 The urinary system, buffer systems, and the respiratory system all contribute to maintenance of the body's acid–base balance. Only the urinary system can eliminate H^+ from the body.

Section 37.6 Animal metabolism produces heat. To maintain core temperature, animals must balance this metabolic heat with heat gained from and lost to the surrounding environment.

For ectotherms, core temperature depends more on heat exchange with the environment than on metabolic production of heat. Core temperature is regulated mainly by changes in behavior.

For endotherms (most birds and mammals), a high metabolism is the main source of heat. In these animals, core temperature is governed largely by controls over metabolic heat production and loss.

For heterotherms, core temperature is sometimes tightly controlled and other times allowed to fluctuate with the environment.

Section 37.7 In mammals, the hypothalamus is the main integrating center for control of temperature. It receives signals from thermoreceptors and stimulates response in effector organs such as smooth muscle in arterioles, sweat glands, and muscles.

Self-Quiz   Answers in Appendix III

1. A freshwater fish gains most water by _____ .
   a. drinking   c. osmosis
   b. eating food   d. transport across the gills

2. Fluid filtered into Bowman's capsule flows directly into the _____ .
   a. renal artery   c. distal tubule
   b. proximal tubule   d. loop of Henle

3. Water and small solutes enter nephrons during _____ .
   a. filtration   c. tubular secretion
   b. tubular reabsorption   d. both a and c

4. Kidneys send water and small solutes back to blood by way of _____ .
   a. filtration   c. tubular secretion
   b. tubular reabsorption   d. both a and b

5. ADH binds to receptors on distal tubules and collecting ducts making them _____ permeable to _____ .
   a. more; water   c. more; sodium
   b. less; water   d. less; sodium

6. Match each structure with a function.
   ____ ureter   a. start of nephron
   ____ Bowman's capsule   b. delivers urine to
   ____ urethra   body surface
   ____ distal tubule   c. carries urine from
   ____ pituitary   kidney to bladder
        gland   d. secretes ADH
                                   e. target of aldosterone

7. The main control center for maintaining the temperature of the mammalian body is in the _____ .
   a. anterior pituitary   c. adrenal cortex
   b. posterior pituitary   d. hypothalamus

8. Rising mammalian core temperature causes _____ .
   a. peripheral vasodilation   c. a pilomotor response
   b. shivering   d. both a and c

interstitial fluid
filtrate in nephron

Na⁺

Cl⁻

H₂O

peritubular capillary

cells of the nephron wall

Figure 37.13 Transport proteins in the nephron wall, where sodium and water are reabsorbed in response to, say, gulping too much water or eating far too much salt.

9. Match each term with the most suitable description.
   _____ endotherm
   _____ ectotherm
   _____ convection
   _____ conduction
   _____ thermal radiation

   a. environment dictates core temperature
   b. metabolism dictates core temperature
   c. heat transfer between objects that are in direct contact
   d. water, air current transfers heat
   e. emission of radiant energy

Critical Thinking

1. Label the components of the diagrams below.

2. The breakdown of proteins produces toxic ammonia as a waste product. In mammals, ammonia is converted to urea, which can only be eliminated when it is dissolved in water in the urine. In birds and reptiles, it is converted to uric acid. This is a solid that can be excreted when mixed with just a small amount of water. It takes about twenty to thirty times more water to excrete 1 gram of urea than it does to excrete 1 gram of uric acid. Formulate a hypothesis on how this difference might relate to the species diversity of birds, mammals, and reptiles in different environments. Devise a way to test your hypothesis.

3. Recall, from Section 37.1, that the kangaroo rat has a highly efficient kidney that excretes only a tiny volume of urine. Compared to a human nephron, the kangaroo rat nephron has a loop of Henle that is proportionally far longer. Explain how a longer loop of Henle can help the kangaroo rat conserve water.

4. Proximal tubules are the main site for reabsorption of essential ions and water. As Figure 37.13 shows, each proximal tubule is made up of a single layer of epithelial cells joined together by tight junctions. The cell surface that faces the lumen is lined with tiny projections that

increase its surface area. Sodium ions diffuse from the filtrate into tubule cells, then active transport proteins pump them out into the interstitial fluid. This pumping action sets up an osmotic gradient, so water moves in the same direction. Proximal tubule cells have an abundance of a certain type of organelle. Which organelle is it?

IMPACTS, ISSUES   *Mind-Boggling Births*

In December 1998, Nkem Chukwu of Texas gave birth to octuplets ahead of schedule. The six girls and two boys survived their premature birth, and they were the first octuplets to do so (Figure 38.1).

The *combined* weight of those eight newborns was just a bit more than 10 pounds, or 4.5 kilograms. Odera, the smallest, weighed less than 1 pound (20 grams). She died of heart and lung failure six days later. The other seven had to remain in the hospital for three months before going home. Two required abdominal surgery.

Chukwu had requested hormone injections, which caused many eggs in her ovaries to mature and be released at the same time. She was offered the option to reduce the number of embryos but wanted to carry them all. The first, delivered naturally, was thirteen weeks premature. Chukwu's doctor ordered drugs to stop labor, then surgically delivered the rest.

Over the past two decades, the incidence of multiple births increased by almost 60 percent. The incidence of higher order multiple births—triplets or more— quadrupled. What is going on?

A woman's fertility peaks during her mid-twenties. By age thirty-nine, her chances of natural conception have been about halved. Yet the number of first-time

Figure 38.1   (**a**) Human embryo at an early stage of development. (**b**) Testimony to the potency of fertility drugs—seven survivors of a set of octuplets. Besides manipulating so many other aspects of nature, humans are now manipulating their own reproduction.

the big picture

Reproductive Modes   Different animal species reproduce sexually, asexually, or both. Sexual reproduction, the most prevalent mode, has biological costs as well as selective benefits.

The Nature of Development   Animal life cycles typically unfold through stages of embryonic development. Each stage builds on the successful completion of the stage that preceded it.

mothers who are more than forty years old doubled during the past decade. Many probably would not have given birth at all without the assistance of fertility drugs and other reproductive interventions, such as *in vitro* fertilization.

Fertility drugs are driving the increase in higher order multiple births, and they worry many doctors. Carrying more than one embryo increases the risk of miscarriage, premature delivery, and surgical delivery by cesarean section. Compared to single births, the weight of such newborns is lower and the mortality rates are higher. A woman who carries quintuplets faces a 50 percent risk of miscarriage, and there is a 90 percent chance that her newborns will develop complications resulting from an abnormally low birth weight. Higher order multiple births also present far more physical, emotional, and financial challenges to the parents.

With this chapter, we turn first to basic principles that govern animal reproduction and development, from gamete formation on through the formation of specialized cells, tissues, and organs of the adult. We then focus on humans as a case study.

There are many sections to choose from, some core material, some optional. We offer this much detail because it can directly and indirectly impact your own future in profound ways. What you learn here may help you work through health-related problems and ethical issues concerning how we individually and collectively deal with human reproduction and development.

 How Would You Vote?

Fertility drugs induce multiple ovulations at the same time and increase the likelihood of high-risk multiple pregnancies. Should we restrict the use of such drugs to conditions that limit the number of embryos formed? See the Media Menu for details, then vote online.

Human Reproduction Human males and females have a pair of primary reproductive organs, and accessory ducts and glands. Hormones control reproductive functions and the development of secondary sexual traits.

Human Development A human life cycle extends from gamete formation and fertilization, through birth and adulthood, to reproduction and gradual aging until death. Master genes control the development of the basic body plan.

38.1 Reflections on Sexual Reproduction

Sexual reproduction dominates the life cycle of most animals, even those that also can reproduce asexually. We therefore can expect that the benefits of sexual reproduction outweigh the costs. What are they?

SEXUAL VERSUS ASEXUAL REPRODUCTION

In earlier chapters, we considered the genetic basis of **sexual reproduction**. Again, meiosis and the formation of gametes typically occur in two prospective parents. At fertilization, a gamete from one parent fuses with a gamete from the other and forms the zygote, the first cell of the new individual. We also looked at **asexual reproduction**, whereby a single organism—one parent only—produces offspring. We turn now to examples of the structural, behavioral, and ecological aspects of these two modes of animal reproduction.

Think about a fragment torn away from a sponge body. It can grow, by mitotic cell divisions, into a new sponge. Think of a flatworm that spontaneously splits

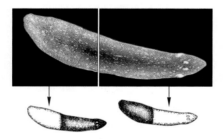

Figure 38.2 One of the flatworms (*Dugesia*) that can reproduce asexually by spontaneous fission. If it divides into two pieces, each piece will replace what is missing and will be genetically identical to the original flatworm.

in two. If its body constricts at its midsection, the part below grips a substrate, starts a tug-of-war with the front, then splits off a few hours later. Both parts go their separate ways and each grows what is missing, thus becoming a whole flatworm (Figure 38.2).

Mutation aside, in cases of *asexual* reproduction, all offspring are genetically the same as their individual parent. Phenotypically, they also are much the same. Phenotypic uniformity helps when the gene-encoded traits are finely adapted to fairly consistent conditions in the environment. Under the circumstances, drastic variations introduced into finely tuned gene packages would not do much good and could do a lot of harm.

Most animals live where opportunities, resources, and danger are variable. They reproduce sexually, and offspring inherit different mixes of alleles from female and male parents. The resulting variation in traits can improve the likelihood that at least some offspring will survive and reproduce if prevailing conditions change.

COSTS OF SEXUAL REPRODUCTION

Separation into sexes is costly. Energy and resources must be allocated to forming and nurturing gametes. Often, reproductive structures that can help deliver or accept sperm must be built. A potential mate might have to be courted. The timing of gamete formation and mating must be synchronized between the sexes.

Reflect on *reproductive timing*. How do sperm in one individual mature at the same time that eggs are maturing in a different individual? Timing requires energy outlays to construct, maintain, and use neural and hormonal control mechanisms in both parents. It requires responsiveness to environmental cues, such as daylength, that signal the best time to start making gametes and to produce offspring.

For example, moose become sexually active only during the late summer and early fall. This timing is adaptive; it ensures that the offspring will be born the following spring, when the weather becomes milder and food will be plentiful for many months.

Think about what it takes to find and recognize a likely mate. Many animals invest energy to produce sex attractants called pheromones or make receptors for them. They invest in visual signals such as richly colored and patterned feathers. Many males attract mates and fend off rivals, as with bonding rituals or claws, horns, or a larger body mass that may make a difference in territorial defense (Figure 38.3).

Producing enough offspring so that at least some survive is also costly (Figure 38.4). Many species of invertebrates, bony fishes, and frogs release sperm,

Figure 38.3 Biological costs associated with sexual reproduction. (**a**) Body mass of male northern elephant seals. The bulls are fighting for access to the far smaller female, lower right. Reflect on the energy and raw materials directed into making (**b**) sable antelope horns.

Figure 38.4 A look at where some invertebrate and vertebrate embryos develop, how they are nourished, and how (if at all) parents protect them.

Snails (**a**) and spiders (**b**) release eggs from which the young later hatch. Most snails abandon their laid eggs. Spider eggs develop in a silk egg sac that the female anchors or carries around with her. Females often die soon after they make the sac. Some species guard the sac, then cart spiderlings about for a few days while they feed them.

(**c**) Ruby-throated hummingbirds and all other birds lay fertilized eggs with big yolk reserves. The eggs develop and hatch outside the mother. Unlike snails, one or both parent birds expend energy feeding and caring for the young.

(**d**) Embryos of most sharks, most lizards, and some snakes develop in their mother, receive nourishment continuously from yolk reserves, and are born live. Shown here, live birth of a lemon shark.

Embryos of most mammals draw nutrients from maternal tissues and are born live. (**e**) In kangaroos and other marsupials, embryos are born "unfinished." They finish their embryonic development inside a pouch on the mother's ventral surface. (**f**) Juvenile stages (joeys) continue to be nourished from mammary glands inside the pouch. A human female (**g**) retains a fertilized egg in her uterus. Her own tissues nourish the developing individual until birth.

eggs, or both into their environment. If each adult were to make only one sperm or one egg each season, the chances for fertilization would not be good. These animals invest energy in making many gametes, often thousands of them.

As another example, nearly all animals on land use internal fertilization, or the union of sperm and egg *within* the female body. They invest metabolic energy to construct elaborate reproductive organs, such as a penis and a uterus. A penis deposits sperm inside the female, and a uterus is a chamber in which an embryo develops inside certain mammalian females.

Finally, animals set aside energy in forms that can *nourish the developing individual* until it has developed enough to feed itself. Nearly all animal eggs contain **yolk**. This thick fluid has an abundance of proteins and lipids that nourish embryonic stages.

The eggs of some species have much more yolk than others. Sea urchins make enormous numbers of tiny eggs with little yolk. Each fertilized egg develops into a freely moving, self-feeding larva in less than a day. Very few escape predators. For sea urchins, then,

reproductive success means allocating small amounts of energy and resources to making each egg.

Birds put a lot of energy and resources into making eggs with a lot of yolk, which has to nourish the bird embryo through an extended time inside an eggshell that forms after fertilization. *Your* mother placed huge demands on her body to nourish a nearly yolkless, fertilized egg through nine months of development. Physical exchanges with her own tissues sustained your embryonic development (Figure 38.4*g*).

As these examples suggest, animals show great diversity in reproduction and development. However, as you will see in the sections to follow, some patterns are widespread throughout the animal kingdom.

Separation into male and female sexes requires special reproductive cells and structures, neural and hormonal control mechanisms, and forms of behavior.

A selective advantage—variation in traits among offspring —offsets biological costs related to separation into sexes.

38.2 Stages of Reproduction and Development

For all animals more complex than sponges, the life cycle has six developmental stages, from gamete formation through growth and tissue specialization.

Figure 38.5 is an overview of the six stages of animal reproduction and development. In *gamete formation*, the first stage, eggs or sperm develop inside parental reproductive tissues or organs. *Fertilization* starts as a sperm penetrates a mature egg. This second stage ends when sperm and egg nuclei fuse, forming a zygote.

The third stage is *cleavage*. Cell divisions increase the number of cells but not the egg's original volume. The cuts produce a **blastula**. In this early embryonic form, the cells are called **blastomeres**, and they enclose a fluid-filled cavity called the blastocoel.

As cleavage draws to a close, cell division slackens. The blastula undergoes *gastrulation*. This fourth stage is a time of structural reorganization. Cells become arranged into a **gastrula**, an early embryonic form with two or three primary tissue layers (germ layers). Cellular descendants of these primary tissues give rise to all tissues and organs of the adult.

Ectoderm, the outermost primary tissue, emerges first. It is the source of the nervous system and the outer part of the integument. Innermost is **endoderm**, source of the gut's inner lining and organs derived from it. In most animal embryos, **mesoderm** forms in between the outer and inner primary tissue layers. It gives rise to muscles, to most of the skeleton, to the circulatory, reproductive, and excretory systems, and to connective tissues of the gut and integument. As you read in Section 22.1, mesoderm evolved hundreds of millions of years ago. It was a key innovation in the evolution of nearly all large, complex animals.

After the primary tissue layers appear, their cells become distinct subpopulations. This marks the onset of *organ formation*. By this process, newly forming cells become increasingly specialized in their structure and function. Their descendants will be the ancestors of all tissues and organs of the adult.

Growth and tissue specialization is the sixth stage of animal development. Over time, tissues and organs mature in size, shape, proportion, and function. This final stage extends into adulthood.

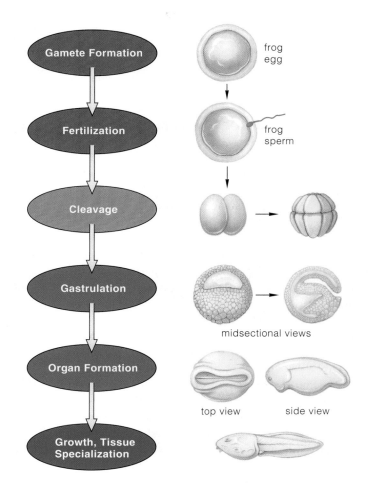

a Eggs form and mature in female reproductive organs, and sperm form and mature in male reproductive organs.

b A sperm and an egg fuse at their plasma membrane, then the nucleus of one fuses with the nucleus of the other to form the zygote.

c By a series of mitotic cell divisions, different daughter cells receive different regions of the egg cytoplasm.

d Cell divisions, migrations, and rearrangements produce two or three primary tissues, the forerunners of specialized tissues and organs.

e Subpopulations of cells are sculpted into specialized organs and tissues in prescribed spatial patterns at prescribed times.

f Organs increase in size and gradually assume specialized functions.

Figure 38.5 Overview of the stages of animal reproduction and development.

Read Me First!
and watch the narrated animation
on the leopard frog life cycle

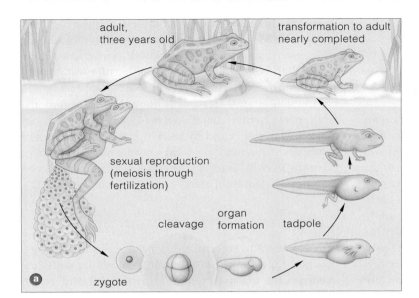

adult, three years old

transformation to adult nearly completed

sexual reproduction (meiosis through fertilization)

cleavage

organ formation

tadpole

zygote

Figure 38.6 Reproduction and development during the life cycle of the leopard frog, *Rana pipiens.*

(**a**) We zoom in on the life cycle as a female releases her eggs into the surrounding water, and a male releases sperm over the eggs. A zygote forms at fertilization.

Appearance of a frog embryo over time. (**b**) About one hour after fertilization, a surface feature—the gray crescent—appears on this type of embryo; it establishes the frog's head-to-tail axis. Gastrulation will start here.

(**c–f**) Cleavage produces a blastula, a ball of cells that contains a fluid-filled cavity or blastocoel.

(**g–j**) During gastrulation, three primary tissue layers form through cell migrations and rearrangements. A primitive gut cavity opens up. Cell differentiation gets under way. A notochord and other organs form; the embryo is on its way to becoming a larval stage called a tadpole.

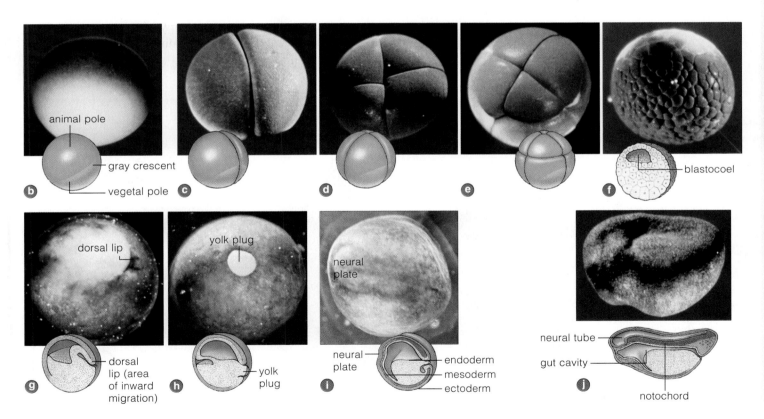

animal pole

gray crescent

vegetal pole

blastocoel

dorsal lip

yolk plug

neural plate

neural tube

dorsal lip (area of inward migration)

yolk plug

neural plate

endoderm

mesoderm

ectoderm

gut cavity

notochord

Figure 38.6 shows examples of these stages for one kind of vertebrate. They offer visual clues to a major principle: *Structures that form during one developmental stage are an essential foundation for the stage that comes after it.* In the sections to follow, you will come across evidence that reinforces this principle.

For most kinds of animals, the life cycle includes six stages of development—gamete formation, fertilization, cleavage, gastrulation, organ formation, and finally growth and tissue specialization. Normally, each stage is successfully completed before the next begins.

38.3 Early Marching Orders

Why don't you have an arm attached to your nose or toes growing from your navel? The patterning of body parts starts with messages in immature, unfertilized eggs.

INFORMATION IN THE EGG

A **sperm**, recall, consists only of paternal DNA and a bit of equipment that helps it reach and penetrate an egg. An **oocyte**, or immature egg, is much larger and more complex than sperm (Section 9.5). As it grows in volume, enzymes, mRNA transcripts, and other factors get stockpiled in its cytoplasm. These "maternal messages" are typically used after fertilization in early rounds of DNA replication and cell division. The angle and timing of microtubule assembly into early bipolar spindles is preordained by tubulins in the cytoplasm. How the cytoplasm gets divided up during early cell divisions depends on such factors, as well as on the amount of yolk and its distribution in the egg.

Consider a frog egg, which shows polarity. Yolk is concentrated near its vegetal pole; pigment granules are concentrated near its animal pole. The egg cortex,

a cytoskeletal mesh just under the plasma membrane, structurally reorganizes itself after a sperm penetrates and fertilizes the egg. Part of the mesh shifts toward the penetration site, creating a crescent-shaped gray area (Figure 38.7a). The crescent's location establishes the anterior–posterior axis for the frog embryo.

In itself, a gray crescent is not evidence of regional differences in maternal messages. Such evidence comes from experiments and by tracking the development of embryos with obvious local cytoplasmic differences (Figure 38.7b,c). Track a frog egg, for instance, and you see that gastrulation starts at the gray crescent.

CLEAVAGE—THE START OF MULTICELLULARITY

Once an oocyte is fertilized, a zygote enters cleavage. By this process, recall, a band of microfilaments just beneath the plasma membrane at the cell midsection contracts and pulls the plasma membrane inward. The band contracts until the cell splits entirely, into two blastomeres, each with a copy of the nucleus (Section 8.4). Remember, these cuts only divide the volume of

Read Me First!

and watch the narrated animation on cytoplasmic localization

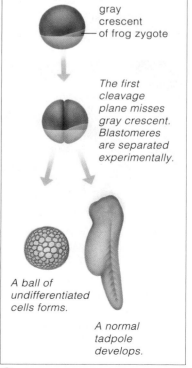

Figure 38.7 (a) A frog egg's cortex contains granules of dark pigment concentrated near one pole. At fertilization, part of the granule-containing cortex shifts toward the point of sperm entry. The shift exposes lighter colored, yolky cytoplasm, as a crescent-shaped gray area.

Normally, the first cleavage puts part of the gray crescent in both of the first two blastomeres. Two experiments, explained next, offer clues to the impact of cytoplasmic localization on the fate of the embryo's cells.

(**b**) For one experiment, the first two blastomeres that formed were physically separated from each other. Each blastomere still gave rise to a whole tadpole.

(**c**) For another experiment, a fertilized egg was manipulated so the cut through the first cleavage plane missed the gray crescent. Only one of the first two blastomeres received the gray crescent. It alone developed into a normal tadpole. Deprived of the maternal messages in the cytoplasm beneath the gray crescent, the other cell could not develop normally.

Figure 38.8 Early cleavage planes for fertilized eggs of (**a**) sea urchins and (**b**) mammals. Cuts are radial in sea urchins and rotational in mammals. (**c**) To sense cleavage's impact, consider Sabra and Nina. These genetically identical twins started life as the same zygote. The first two blastomeres formed at cleavage, the inner cell mass, or another early stage split and started two look-alike individuals.

cytoplasm into ever smaller cells. The blastomeres do not grow in size during cleavage.

Simply by virtue of where they form, the different blastomeres receive different maternal messages. This outcome of cleavage, called **cytoplasmic localization**, helps seal the developmental fate of each cell lineage. Cells of one lineage alone might have a protein that activates, say, a gene coding for a certain hormone.

In most animal zygotes, genes remain silent during early cleavage; the stockpiled maternal proteins and mRNAs help control the rate and angle of the cuts. In mammals, however, certain genes must be activated first. Cleavage cannot even be completed without the protein products of those genes.

CLEAVAGE PATTERNS

Observable differences in how animals develop start showing up during cleavage. The early cuts follow species-specific patterns. The simplest pattern, *radial* cleavage, starts with cuts perpendicular to the mitotic spindle. The resulting blastomeres are similar in size, but they get different parts of the cytoplasm (Figure 38.8*a*). The cuts go completely through some eggs, as in sea urchins. The fluid-filled cavity forms near the center of a ball-shaped array of cells.

In other eggs, cuts do not go all the way through. For instance, yolk impedes the cuts near the vegetal pole of a frog egg. Cuts are faster near the less yolky animal pole, where more blastomeres form. As one outcome, the fluid-filled cavity inside the resulting ball of cells is offset (Figure 38.5*e,f*). Early cuts are also

incomplete in the fertilized eggs of reptiles, birds, and most fishes. Yolk confines the cuts to a small, disk-shaped region. Two flattened layers of cells form, one atop the other, with only a narrow cavity in between.

The nearly yolkless eggs of mammals are cut all the way through, in a *rotational* pattern. The first cut is vertical. One of the two resulting blastomeres is cut vertically, and the other horizontally (Figure 38.8*b*). The next cuts make a loose arrangement of eight cells with spaces in between. Tight junctions start forming among these cells and keep them huddled together. More cuts produce a hollow ball of cells. The outer cells secrete fluid that fills the inner cavity. Other cells cluster against one part of the cavity wall. This type of blastula is called a **blastocyst**. In all mammals, its inner cell mass gives rise to the embryo proper.

Once in a while, the first two blastomeres, the inner cell mass, or even the blastocyst splits in two. *Identical twins*, which have identical genes and are the same sex, may be the result (Figure 38.8*c*). *Fraternal twins* form when two oocytes released during the same menstrual cycle are fertilized by two sperm. Fraternal twins may be the same sex or different sexes.

The egg cytoplasm contains maternal messages in the form of regionally located enzymes and other proteins, mRNAs, cytoskeletal elements, yolk, and other components.

Cleavage divides a zygote into blastomeres. Simply by virtue of the cuts, each ends up with different maternal messages. Cleavage patterns differ among the major animal groups.

38.4 How Do Specialized Tissues and Organs Form?

Nearly all animals have a gut that digests nutrients for absorption. They have surface parts that protect organs inside and detect what is going on outside. Most have organs in between that function in structural support, motion, and circulation. This three-layered body plan emerges after cleavage, as gastrulation gets under way.

During gastrulation, the embryonic cells migrate and rearrange themselves into the three primary tissues of the gastrula—ectoderm, endoderm, and mesoderm. Figure 38.9 shows an example. The first cavity to form in protostome gastrulas becomes a mouth, but it will become an anus in deuterostomes (Section 22.1). Also, the anterior-to-posterior body axis forms in gastrulas. In vertebrates, this axis precedes the formation of a **neural tube**, the forerunner of the brain and spinal cord (Figures 38.10 and 38.11). Such specializations arise through cell differentiation and morphogenesis.

CELL DIFFERENTIATION

All cells of a normal embryo have the same number and kinds of genes, having descended from the same zygote. They all activate the genes for products that

Figure 38.9 Gastrulation in a fruit fly (*Drosophila*), cross-section. After cleavage, the blastula is transformed into a gastrula. Notice how some cells (stained *gold*) migrate inward through an opening. Fruit flies are protostomes and this opening will later become the mouth.

Figure 38.10 Two examples of the types of events involved in morphogenesis.

(**a**) Cell migration. We show the same embryonic neuron (*orange*) at successive times during its "climb" through developing brain tissue. The neuron is responding to chemical cues on the surface of a glial cell (*yellow*), which guide it to its destination in the embryo.

(**b**) Neural tube formation. By the end of gastrulation, ectoderm is a uniform sheet of cells. Along the axis of the future tube, microtubules in ectodermal cells lengthen. Rings of microfilaments constrict in some of these cells, which become wedge-shaped. The part of the ectodermal sheet where they are located folds back on itself, over the wedge-shaped cells, which then disengage from it as a separate tube.

a

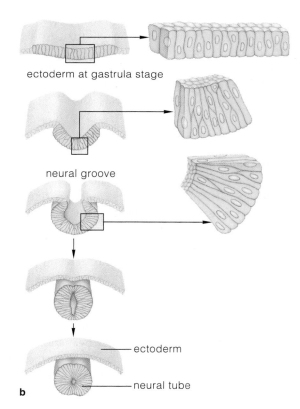

ectoderm at gastrula stage

neural groove

ectoderm

neural tube

b

Figure 38.11 Onset of organ formation in a chick embryo during the first seven days of development. The heart begins to beat between thirty and thirty-six hours. You may have observed such embryos at the yolk surface of raw, fertilized eggs.

22 HOURS   29 HOURS   45 HOURS   55 HOURS   72 HOURS

neural groove

eye
brain
heart
wing bud
neural tube
leg bud
tail

168 HOURS (SEVEN DAYS OLD)

assure their survival, such as histones and glucose-metabolizing enzymes. But from gastrulation onward, cell lineages also engage in **selective gene expression**: they express some groups of genes and not others. This is the start of **cell differentiation**. By this process, cell lineages become specialized in their composition, structure, and function. Remember how one cell type activates genes for crystallins, proteins that form a lens in each eye? It is only one of about 200 differentiated cell types in the human body (Section 14.3).

MORPHOGENESIS

Specialized tissues and organs form in a programmed, orderly sequence by a process called **morphogenesis**. During this process, cells of different lineages divide, grow, migrate, and change in size. Tissues lengthen or widen and fold over. And some cells die in controlled ways at specific locations.

Think about active cell migration. *Cells send out and use pseudopods that move them along prescribed routes.* When they reach their destination, they connect with cells already there. Embryonic neurons migrate this way in a developing nervous system (Figure 38.10a).

How do these cells know where to move and when to stop? They respond to adhesive cues and chemical gradients. Their migrations are coordinated by the synthesis, release, deposition, and removal of specific chemicals in the extracellular matrix.

For example, the neuron shown in Figure 38.10a is responding to the "stickiness" of its surroundings. Adhesion proteins on a patch of its plasma membrane stuck to proteins on the surface of a glial cell. Now, cytoskeletal elements in the neuron lengthen from the attachment site and shove the cytoplasm forward. The neuron migrates until its adhesion proteins reach the spot in the embryo that is "stickiest" to them.

Also, *whole sheets of cells expand and fold inward and outward as their cells change in shape.* Within these cells, microtubules grow longer, and rings of microfilament constrict. The controlled assembly and disassembly of these cytoskeletal components cause the changes.

Figures 38.10b and 38.11 hint at what happens after three primary tissues form in vertebrate embryos. At the embryo's midline, some ectodermal cells elongate and form the neural plate, the start of nervous tissue. The plate sinks inward as cells lengthen and become wedge shaped. At the edges of the resulting groove in the surface, flaps of tissue fold over and meet at the midline, forming the neural tube.

Finally, *programmed cell death (apoptosis) helps sculpt body parts.* Section 24.5 introduced the molecular basis of this process. Basically, signaling molecules from some cells activate tools of self-destruction stockpiled in other, target cells. Remember how apoptosis shapes a human hand from a paddle-shaped body part?

In cell differentiation, some cells selectively use certain genes that other cells do not use. Selective gene expression is the basis of cell differentiation. It results in cell lineages with distinctive structures, products, and functions.

Morphogenesis is a program of orderly changes in body size, shape, and proportion. It produces specialized tissues and organs at prescribed times in prescribed locations.

Morphogenesis involves cell division, active cell migration, tissue growth and foldings, changes in cell size and shape, and programmed cell death or apoptosis.

38.5 Pattern Formation

Maternal messages in the egg cytoplasm guide the earliest stages of development. Later on, communication signals among embryonic cells cause tissues and organs to form according to a mapped-out body plan.

EMBRYONIC INDUCTION

An embryonic cell locked in a tissue selectively reads its genes. It may make and secrete signaling molecules that affect its neighbors. Signals diffusing through its neighborhood may induce changes in its composition, structure, or both. Its descendants inherit a chemical memory of the changes, and they go on to form tissues and organs in places where we expect them to be.

This is a time of **embryonic induction**, of changes in the developmental fate of embryonic cell lineages exposed to gene products from cells of different tissues. Through these changes, specialized tissues and organs emerge from clumps of embryonic cells in predictable places, in the proper order. Their emergence is called **pattern formation**.

We have experimental evidence of cell memory. In one experiment, researchers excised the dorsal lip of a normal axolotl embryo, then grafted it into a novel location in a different axolotl embryo. Gastrulation proceeded at the recipient's own dorsal lip *and at the graft*. Conjoined twins—a double embryo with two sets of body parts—developed (Figure 38.12). The dorsal lip organizes this amphibian's main body axis. It was the first embryonic signaling center discovered. We also have evidence of cell memory from experiments involving chick wing development (Figure 38.13).

Many signals are short-range, involving cell-to-cell contacts. For instance, signals that activate or inhibit genes for adhesion proteins and recognition proteins directly affect how cells interact in tissues and organs. Other signals cause cytoskeletal elements in the target cell to lengthen in some direction. *Such signals can cause a cell to stick to its neighbors, or break free and migrate to a different location, or become segregated from the cells of an adjoining tissue.*

Other signals are long-range. They act on control elements in the DNA of embryonic cells that are some distance away. The **morphogens** are among the long-range signals. These degradable molecules diffuse out of a signaling center, and their concentration weakens with distance. They create a gradient that helps a cell chemically assess its position in the embryo and thus influence how it will differentiate. *Such differences in a signaling molecule's concentration induce lineages of cells positioned at different locations along the gradient to read different parts of the same genome.*

A THEORY OF PATTERN FORMATION

The same kinds of genes are the mapmakers for all major groups of animals. According to one theory of pattern formation, the products of these genes interact in ways that map out the basic body plan:

1. The formation of tissues and organs in ordered, spatial patterns starts with cytoplasmic localization. It continues through short-range, cell-to-cell contacts and through long-range signals, or chemical gradients

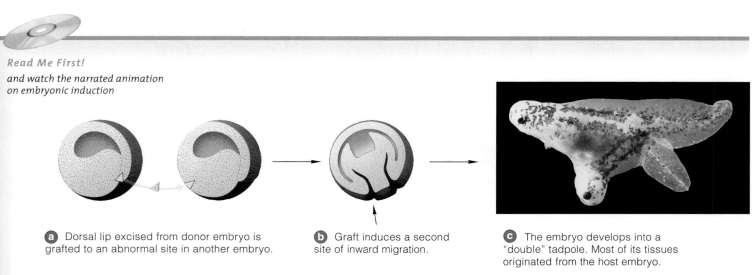

Read Me First!

and watch the narrated animation on embryonic induction

a Dorsal lip excised from donor embryo is grafted to an abnormal site in another embryo.

b Graft induces a second site of inward migration.

c The embryo develops into a "double" tadpole. Most of its tissues originated from the host embryo.

Figure 38.12 Experimental evidence that a dorsal lip controls amphibian gastrulation. A dorsal lip region of an axolotl embryo was transplanted to a different site in another axolotl embryo. It organized the formation of another set of body parts.

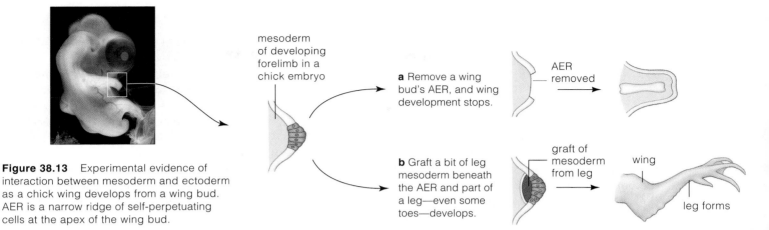

Figure 38.13 Experimental evidence of interaction between mesoderm and ectoderm as a chick wing develops from a wing bud. AER is a narrow ridge of self-perpetuating cells at the apex of the wing bud.

mesoderm of developing forelimb in a chick embryo

a Remove a wing bud's AER, and wing development stops.

AER removed

b Graft a bit of leg mesoderm beneath the AER and part of a leg—even some toes—develops.

graft of mesoderm from leg

wing

leg forms

that weaken with distance from the source. Cells at the start, middle, and end of a gradient are exposed to different chemical information.

2. Morphogens and other inducer molecules diffuse through embryonic tissues, and they activate classes of **master genes** in sequence. The products of these genes interact with control elements to lay the foundation for the basic body plan.

3. Products of **homeotic genes** and other master genes interact with control elements to map out the overall body plan (Sections 14.4 and 14.5). They form when blocks of genes are activated and suppressed in cells along the anterior–posterior axis and dorsal–ventral axis of the embryo. Other gene products interact to fill in details of specific body parts.

Master genes and gene products act in similar ways in all major animal groups. If they fail in mapping out the overall body plan, disaster follows, as when a heart forms in the wrong place. Once the plan is locked in, inductions that follow have only localized effects. If a lens fails to form, only the eye will be affected.

EVOLUTIONARY CONSTRAINTS ON DEVELOPMENT

How long have master genes been around? The basic body plan for sponges, worms, flies, vertebrates, and all other major animal groups has not changed much for about 500 million years. All of the many millions of species are variations on a few dozen plans!

Why is this so? There are few new master genes, *so variations in form might be more of an outcome of how the control genes control each other.* Consider that insect, squid, and vertebrate eyes differ structurally, and yet genes governing eye formation are nearly identical in these groups. An *eyeless* gene controls eye formation in fruit flies (*Drosophila*). In humans, mutation of a nearly identical gene causes eyeless babies. The same gene controls the fate of cells that give rise to legs of fruit flies, crabs, and beetles, butterfly wings, sea star arms, fish fins, and mouse feet. All of these structures start out as buds from the main body axis.

For a long time, we have known that body plans cannot change much because of *physical* constraints (such as the surface-to-volume ratio) and *architectural* constraints (as imposed by body axes). We now know there are *phyletic* constraints on change. These are the constraints imposed on each lineage by interactions of master organizer genes, which operate when organs form and control induction of the basic body plan.

That may be why we have so many species and so few body plans. Once the master genes evolved and started interacting in intricate ways, it would have been hard to change the basic parts without killing the embryo. Mutations have indeed added marvelous variations to animal lineages. But the basic body plans have prevailed through great spans of time, so maybe it simply proved unworkable to start all over again.

Pattern formation is the ordered sculpting of embryonic cells into specialized tissues and organs. Cells differentiate as a result of cytoplasmic localization of materials and chemical cues regarding their relative position in the embryo.

Cytoplasmic localization and then inductive interactions among classes of master genes map the basic body plan. Gene products specify where and how body parts develop. They are long-range and short-range beacons that help cells assess their position and how they will differentiate.

Physical, architectural, and phyletic constraints limit the evolution of animal body plans. Among all major groups, master genes that guide development are similar and are sometimes identical.

38.6 Reproductive System of Human Males

So far, you have considered some principles of animal reproduction and development. The remainder of this chapter applies the principles to humans, starting with the male reproductive system.

WHEN GONADS FORM AND BECOME ACTIVE

In nature, the function of sex is not recreational, it is the perpetuation of one's genes. A human male's genes become packaged in sperm. These male gametes form in a pair of gonads, or primary reproductive organs, that also secrete sex hormones. Male gonads are called **testes** (singular, testis). As Table 38.1 indicates, they are part of a system of reproductive organs and glands.

Testes start to form on the wall of an XY embryo's abdominal cavity. Before birth, they descend into the scrotum, a skin pouch below the pelvic girdle (Figures 11.3 and 38.14). Muscle contractions draw the pouch closer to the main body mass, and muscle relaxation lowers it. Such adjustments help maintain an internal temperature that is suitable for sperm formation.

Packed within each testis are many small, highly coiled tubes called seminiferous tubules. Section 38.7 explains how sperm cells form and start maturing in these tubules in hormone-guided ways.

Sperm production and the emergence of secondary sexual traits start at **puberty**. This stage of postnatal development usually begins in boys between ages twelve and sixteen. Signs that it is under way include enlarging testes, growth spurts, a deepening voice, and more hair growth on the face, chest, and elsewhere.

STRUCTURE AND FUNCTION OF THE REPRODUCTIVE SYSTEM

Mammalian sperm travel from the testes through a series of ducts that lead to the urethra (Figure 38.15). They are not quite mature when they enter one of a pair of long, coiled ducts, the epididymes. Secretions from glandular cells in the duct wall trigger events that put finishing touches on maturing sperm cells. The last part of each epididymis stores mature sperm. During a male's reproductive years, about 100 million sperm mature every day. Unused ones are resorbed or excreted in the urine.

In a sexually aroused male, muscles in the wall of reproductive organs contract, which propels mature sperm into and through a pair of thick-walled tubes called vasa deferentia (singular, vas deferens). More contractions propel the sperm farther, through a pair of ejaculatory ducts, then through the urethra. This last tubular duct runs through the male sex organ— the penis—and opens at its tip. The urethra, remember, also functions in urinary excretion.

Sperm traveling to the urethra mix with glandular secretions and form semen, a thickened fluid that is expelled from the penis during sexual activity. Paired seminal vesicles secrete fructose, a sugar that sperm use as an energy source. The prostate gland and, to a lesser extent, seminal vesicles secrete prostaglandins into the mix. These signaling molecules might induce contractions in the female reproductive tract and so help sperm reach an egg. They may also function in helping sperm slip past the female's immune system.

Figure 38.14 Position of the human male reproductive system relative to the pelvic girdle and urinary bladder.

Labels: pelvic girdle · urinary bladder · scrotum

Table 38.1	Organs and Accessory Components of the Human Male Reproductive System	

Reproductive Organs	
Testis (2)	Sperm, sex hormone production
Epididymis (2)	Sperm maturation site and subsequent storage
Vas deferens (2)	Rapid transport of sperm
Ejaculatory duct (2)	Conduction of sperm to penis
Penis	Organ of sexual intercourse
Accessory Glands	
Seminal vesicle (2)	Secretion of large part of semen
Prostate gland	Secretion of part of semen
Bulbourethral gland (2)	Production of mucus that functions in lubrication

Read Me First!
and watch the narrated animation on the human male reproductive system

PROSTATE GLAND
Secretion of substances that become part of semen

urinary bladder

URETHRA
Dual-purpose duct; serves as channel for ejaculation of sperm during sexual arousal, also for urine excretion at other times

urethra

anterior — posterior

erectile tissue

PENIS
Organ of sexual intercourse

TESTIS
One of a pair of primary reproductive organs; packed with sperm-producing tubules and cells that secrete testosterone and other hormones

anus

EJACULATORY DUCT
One of a pair of sperm-conducting ducts

SEMINAL VESICLE
One of a pair of glands that secrete fructose and prostaglandins, which become part of semen

BULBOURETHRAL GLAND
One of a pair of glands that secrete a lubricating mucus

VAS DEFERENS
One of a pair of ducts for rapid transport of sperm

EPIDIDYMIS
One of a pair of ducts in which sperm complete maturation; the portion farthest from testis stores mature sperm

Figure 38.15 Components of the human male reproductive system and their functions.

Secretions from the prostate gland may also help buffer the acidic conditions that prevail in the female reproductive tract. The pH of vaginal fluid is about 3.5–4.0, but sperm swim more efficiently at pH 6.0. A pair of bulbourethral glands secrete mucus-rich fluid into the urethra during sexual arousal.

CANCERS OF THE PROSTATE AND TESTES

Prostate cancer is a leading cause of death among men, surpassed only by lung cancers. In 2003, more than 220,000 males were diagnosed in the United States; 29,000 died. In the same year, there were about 7,600 cases of *testicular cancer*. In early stages, both cancers are painless and they may spread silently into lymph nodes of the abdomen, chest, neck, then lungs. If the cancer metastasizes, prospects are not good.

Doctors may detect prostate cancer by blood tests for increases in prostate-specific antigen (PSA) and by physical examinations. Adult males should examine their testes monthly—after a warm shower or bath—to check for enlargement, hardening, or new lumps. Treatment of testicular cancer has a high success rate, provided the cancer is caught before it spreads to other parts of the body.

Human males have a pair of testes, or primary reproductive organs that produce sperm and sex hormones. The male reproductive system also has specialized accessory glands and ducts.

The ducts carry sperm from the testes to the urethra, from where they leave the body. The accessory glands secrete the assorted noncellular components of the semen.

38.7 Sperm Formation

Signaling pathways that involve hormonal secretions by the hypothalamus, pituitary gland, and testes control sperm formation. Negative feedback loops are part of these pathways.

Each testis is smaller than a golfball, and contains hundreds of wedge-shaped lobes partitioning the interior. Each lobe contains 1 to 3 coiled tubules and, if all were placed end to end, would stretch out 125 meters (Figure 38.16). Inside each tubule wall are spermatogonia (singular, spermatogonium)—undifferentiated, diploid cells that divide continuously. Descendant cells force older ones away from the wall, toward the interior. During their forced departure, the daughter cells enter meiosis. These cells are primary spermatocytes.

Cytoplasmic divisions that accompany the nuclear divisions are incomplete, and thin cytoplasmic bridges connect the cells (Figure 38.16c). Ions and molecules concerned with development diffuse freely across the bridges that connect the cells of each generation and cause them to mature at the same time. All the while, the primary spermatocytes receive nourishment and signals from Sertoli cells, a type of supporting cell in the seminiferous tubules.

Secondary spermatocytes form after each primary spermatocyte finishes meiosis I (Figure 38.16c). These are haploid cells with chromosomes in the duplicated state. (Here you may wish to review the overview of spermatogenesis in Section 9.4.) Sister chromatids of each chromosome move away from each other during meiosis II. After this, haploid daughter cells called spermatids form. Each cell now starts to mature into a sperm. The entire process—from spermatogonium to mature sperm—takes about 100 days.

An adult male produces sperm on a continuous basis, so many millions of cells are in different stages of development on any given day.

Each mature sperm is a flagellated cell. Its head is packed with DNA, and its tail has a microtubular core (Figure 38.16d). Mitochondria in a midpiece supply energy for the tail's whiplike motions. Extending over most of the head is an acrosome, a cap with enzymes inside. The enzymes can digest extracellular material around an oocyte prior to fertilization.

Four hormones—testosterone, LH, FSH, and GnRH —are essential for sperm formation. **Testosterone**, a steroid hormone, governs the growth, structure, and function of the male reproductive tract as well as the

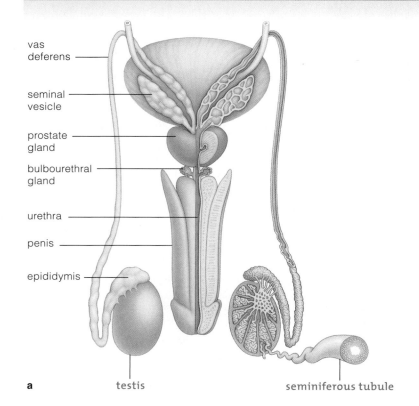

vas deferens

seminal vesicle

prostate gland

bulbourethral gland

urethra

penis

epididymis

a testis seminiferous tubule

wall of seminiferous tubule   Leydig cells between tubules

b

Figure 38.16 (**a**) Male reproductive tract, posterior view. Arrows show the route that sperm take before ejaculation.

Sperm formation. (**b**) Light micrograph of cells in three adjacent seminiferous tubules, cross-section. Leydig cells occupy tissue spaces between tubules. (**c**) How sperm form, beginning with a diploid germ cell. (**d**) Structure of a mature sperm, the male gamete.

Figure 38.17 Negative feedback loops from the testes to the hypothalamus and to the anterior lobe of the pituitary.

formation of sperm. It triggers development of male secondary sexual traits, and it stimulates sexual and aggressive behavior. Leydig cells in the testes secrete testosterone. The anterior lobe of the pituitary gland makes and secretes **LH** and **FSH** (Section 31.4).

The hypothalamus governs secretion of all three hormones (Figure 38.17). It secretes **GnRH** when the blood levels of testosterone and other factors are low. This releasing hormone makes the anterior lobe step up the LH and FSH secretions. In turn, LH stimulates Leydig cells to secrete testosterone, which helps sperm form and mature. At puberty, the FSH binds to Sertoli cells and jump-starts sperm formation.

Sperm formation depends on the hormones LH, FSH, and testosterone. Negative feedback loops from the testes to the hypothalamus and pituitary gland control their secretion.

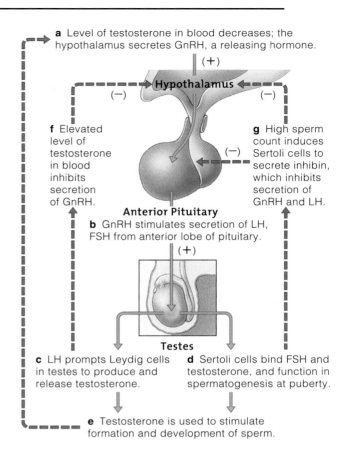

a Level of testosterone in blood decreases; the hypothalamus secretes GnRH, a releasing hormone.

(+)

Hypothalamus

(−)   (−)

f Elevated level of testosterone in blood inhibits secretion of GnRH.

g High sperm count induces Sertoli cells to secrete inhibin, which inhibits secretion of GnRH and LH.

(−)

Anterior Pituitary
b GnRH stimulates secretion of LH, FSH from anterior lobe of pituitary.

(+)

Testes

c LH prompts Leydig cells in testes to produce and release testosterone.

d Sertoli cells bind FSH and testosterone, and function in spermatogenesis at puberty.

e Testosterone is used to stimulate formation and development of sperm.

Read Me First!
and watch the narrated animation on sperm formation

mitosis   meiosis I   meiosis II   lumen

Sertoli cell

spermatogonium (diploid)

c Part of the lumen of one seminiferous tubule

primary spermatocyte

secondary spermatocyte

early spermatids

late spermatid

immature sperm (haploid)

d Sketch of a mature human sperm. The head consists of DNA inside an acrosome, an enzyme-rich cap.

head   midpiece with mitochondria   tail, with its core of microtubules

38.8 Reproductive System of Human Females

The reproductive system of human females functions in the production of gametes and sex hormones. It also has a chamber in which a new, developing individual is protected and nourished until birth.

COMPONENTS OF THE SYSTEM

Figures 38.18 and 38.19 show the reproductive system of a human female, and Table 38.2 lists their functions. The primary reproductive organs are a pair of **ovaries** that produce oocytes (immature eggs) and secrete sex hormones. An oocyte released from an ovary enters one of a pair of oviducts, or Fallopian tubes. Oviducts open onto the **uterus**, a hollow organ shaped a bit like a pear. Fertilization usually takes place in an oviduct, but the embryo develops mainly inside the uterus.

A thick layer of smooth muscle, the myometrium, makes up most of the uterine wall. The uterine lining, or **endometrium**, consists of connective tissues, blood vessels, and glands. The cervix, the narrowed-down part of the uterus, connects with a muscular tube, the vagina. Mucus-secreting epithelium lines the vagina, which extends from the cervix to the body's surface. The vagina receives sperm from the male and serves as part of the birth canal.

At the body's surface are genital organs of sexual stimulation. Outermost are the labia majora, a pair of skin folds padded with adipose tissue. They enclose the labia minora, a pair of smaller folds having many blood vessels but no adipose tissue. These folds partly enclose the clitoris, a female sex organ derived from the same embryonic tissue as the male penis. Like the penis, the clitoris has a great abundance of sensory receptors and is very sensitive to sexual stimulation. The urethra opens at the body surface about midway between the vaginal opening and the clitoris.

Figure 38.18 Location of the human female reproductive system relative to the pelvic girdle and urinary bladder.

OVERVIEW OF THE MENSTRUAL CYCLE

Females of most mammalian species follow an *estrous* cycle meaning they are fertile and "in heat" (sexually receptive to males) at only certain times in the cycle. Females of humans and some other primates follow a **menstrual cycle**. They are fertile intermittently, on a cyclic basis. Their fertile periods are not synchronized with sexual receptivity. Even though they can become pregnant at only certain times in the cycle, they may be receptive to sex at any time.

The next section explains the cycle, but here is a brief overview (Table 38.3). First, an oocyte matures and is released from an ovary, and the endometrium is primed for pregnancy. When fertilization does not occur, blood and bits of the uterine lining flow from the vagina. The flow means "there's no embryo at this time," and marks the start of a new cycle.

Table 38.2	Organs of the Human Female Reproductive Tract
Ovaries	Oocyte production and maturation, sex hormone production
Oviducts	Ducts for conducting oocyte from ovary to uterus; fertilization normally occurs here
Uterus	Chamber in which new individual develops
Cervix	Secretion of mucus that enhances sperm movement into uterus and (after fertilization) reduces embryo's risk of bacterial infection
Vagina	Organ of sexual intercourse; birth canal

Table 38.3	Events of a Menstrual Cycle Lasting Twenty-Eight Days	
Phase	Events	Days of Cycle
Follicular phase	Menstruation; endometrium breaks down	1–5
	Follicle matures in ovary; endometrium rebuilds	6–13
Ovulation	Oocyte released from ovary	14
Luteal phase	Corpus luteum forms, secretes progesterone; the endometrium thickens and develops	15–28

Read Me First!
and watch the narrated animation on the human female reproductive system

OVARY
One of a pair of primary reproductive organs in which oocytes (immature eggs) form and mature; produces hormones (estrogens and progesterone), which stimulate maturation of oocytes, formation of corpus luteum (a glandular structure), and preparation of the uterine lining for pregnancy

OVIDUCT
One of a pair of ciliated channels through which oocytes are conducted from an ovary to the uterus; usual site of fertilization

UTERUS
Chamber in which embryo develops; its narrowed-down portion (the cervix) secretes mucus that helps sperm move into uterus and that bars many bacteria

MYOMETRIUM
Thick muscle layers of uterus that stretch enormously during pregnancy

urinary bladder

urethra

CLITORIS
Small organ responsive to sexual stimulation

LABIUM MINOR
One of a pair of inner skin folds of external genitals

LABIUM MAJOR
One of a pair of outermost, fat-padded skin folds of external genitals

opening of cervix

anus

ENDOMETRIUM
Inner lining of uterus; site of implantation of blastocyst (early embryonic stage); becomes thickened, nutrient-packed, highly vascularized tissue during a pregnancy; gives rise to maternal portion of placenta, an organ that metabolically supports embryonic and fetal development

VAGINA
Organ of sexual intercourse; also serves as birth canal

Figure 38.19 Components of the human female reproductive system and their functions.

The cycle has three phases. In the *follicular* phase, menstruation occurs, the uterine lining breaks down and starts rebuilding, and an oocyte starts maturing. In the next phase, **ovulation**, an oocyte escapes from the ovary. In the *luteal* phase, a glandular structure called the **corpus luteum** forms in an ovary. It starts to secrete hormones that cause the endometrium to thicken in preparation for pregnancy.

Hormones control these phases by way of feedback loops from the ovaries to both the hypothalamus and pituitary gland. FSH and LH promote changes in the ovaries and stimulate them to secrete **estrogens** and **progesterone**. These sex hormones bring about cyclic changes in the endometrium.

A female enters puberty between the ages of ten and sixteen. Pubic hair forms, fat is deposited inside her breasts and around the hips, and menstrual cycles start. Each cycle lasts for about twenty-eight days, on average, but it is longer or shorter for some females. The cycles usually continue until the late forties or early fifties, when sex hormone secretions start to dwindle. The decline in hormone secretions correlates with the onset of *menopause*, the twilight of a female's reproductive capacity.

Ovaries, the primary reproductive organ of human females, produce immature eggs and sex hormones. Endometrium lines the uterus, a chamber in which embryos develop.

Estrogens and progesterone guide the cyclic growth and release of oocytes from the ovary as well as the breakdown and rebuilding of the endometrium, which depends on whether pregnancy occurs. These are the key events of menstrual cycles, which start at puberty.

38.9 Preparations for Pregnancy

Before starting this section, take a moment to review Figure 9.5, the overview of meiosis in an oocyte.

CYCLIC CHANGES IN THE OVARY

During her reproductive years, a female will release about 400 to 500 primary oocytes. A **primary oocyte** is an immature egg that was arrested in prophase I of meiosis. The only ones in a female's ovaries formed while she herself was an embryo—or so it was thought.

Female mice are now known to produce some new oocytes even when they are adults. We do not know whether human females do so.

We do know that oocytes released late in life are at greater risk for abnormal changes in the structure or number of chromosomes as meiosis resumes in them. One serious risk is Down syndrome (Section 11.8).

A follicle consists of a primary oocyte and a layer of cells that surround and nourish it. An ovary holds many follicles. Figure 38.20 shows what happens to

Read Me First!

and watch the narrated animation on ovarian function

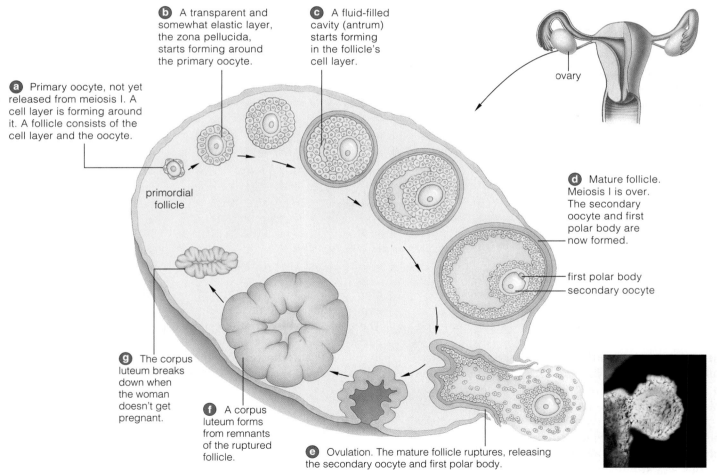

b A transparent and somewhat elastic layer, the zona pellucida, starts forming around the primary oocyte.

c A fluid-filled cavity (antrum) starts forming in the follicle's cell layer.

ovary

a Primary oocyte, not yet released from meiosis I. A cell layer is forming around it. A follicle consists of the cell layer and the oocyte.

primordial follicle

d Mature follicle. Meiosis I is over. The secondary oocyte and first polar body are now formed.

first polar body
secondary oocyte

g The corpus luteum breaks down when the woman doesn't get pregnant.

f A corpus luteum forms from remnants of the ruptured follicle.

e Ovulation. The mature follicle ruptures, releasing the secondary oocyte and first polar body.

Micrograph of a secondary oocyte escaping from the surface of an ovary

Figure 38.20 Cyclic events in a human ovary, cross-section. The follicle does not "move around" as in this diagram, which simply shows the *sequence* of events. In the cycle's first phase, a follicle grows and matures. At ovulation, the second phase, the mature follicle ruptures and releases a secondary oocyte. In the third phase, a corpus luteum forms from the follicle's remnants.

one of them during a menstrual cycle. As the cycle starts, the hypothalamus is secreting enough GnRH to make cells in the anterior lobe of the pituitary step up FSH and LH secretions. These hormones trigger the growth of the oocyte and the formation of more cells around it. Glycoprotein molecules accumulate under the surrounding cells and form the zona pellucida, a noncellular layer (Figure 38.20*b*).

Eight to ten hours before being released, the oocyte completes meiosis I. Its cytoplasm divides unevenly. One of the two haploid daughter cells, a **secondary oocyte**, receives most of the cytoplasm. The other cell is the first of three **polar bodies** that form by way of meiosis. It will degenerate later in the cycle.

FSH and LH collect in fluid in the follicle and prod its cells to secrete estrogens, so the levels of estrogens in blood increase. About halfway through the cycle, the pituitary responds to the increases. It secretes LH in a brief pulse. The follicle swells in response, and its wall weakens and ruptures. Fluid—and the secondary oocyte—are released. *The midcycle surge of LH triggers ovulation, the release of a secondary oocyte from the ovary.*

CYCLIC CHANGES IN THE UTERUS

The estrogens released early in the cycle also call for growth of the endometrium and its glands, which sets the stage for pregnancy. Before the midcycle surge of LH, follicle cells were busy secreting progesterone and estrogens. Blood vessels grew fast in the thickened endometrium. At ovulation, estrogens prompted cells of the cervical canal to release a thin, clear mucus—an ideal medium for sperm to swim through.

The midcycle LH surge that brings about ovulation also stimulates cells in the ruptured follicle to form a corpus luteum. Progesterone and estrogens secreted by this structure cause the endometrium to thicken, in preparation for pregnancy.

All the while, the hypothalamus has been stopping other follicles in the ovary from maturing by slowing down FSH secretion. If a blastocyst does not burrow into the endometrium, the corpus luteum will last for no more than twelve days or so. In the last days of the cycle, it will secrete prostaglandin—local signaling molecules that trigger this gland's self-destruction.

Without the corpus luteum, levels of progesterone and estrogen decline fast, and the endometrium starts to break down. Deprived of oxygen and nutrients, its blood vessels constrict and tissues die. Blood escapes as weakened capillary walls start to rupture. Blood and sloughed endometrial tissues form a menstrual flow, which lasts for three to six days. After this, the

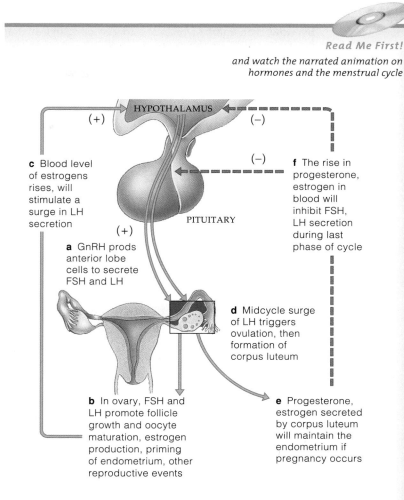

HYPOTHALAMUS

(+)   (−)

(−)

c Blood level of estrogens rises, will stimulate a surge in LH secretion

f The rise in progesterone, estrogen in blood will inhibit FSH, LH secretion during last phase of cycle

PITUITARY

(+)

a GnRH prods anterior lobe cells to secrete FSH and LH

d Midcycle surge of LH triggers ovulation, then formation of corpus luteum

b In ovary, FSH and LH promote follicle growth and oocyte maturation, estrogen production, priming of endometrium, other reproductive events

e Progesterone, estrogen secreted by corpus luteum will maintain the endometrium if pregnancy occurs

Figure 38.21 Hormone secretions during a menstrual cycle. A positive feedback loop from an ovary to the hypothalamus (*green*) triggers ovulation. After a secondary oocyte escapes, a negative feedback loop from the ovary to the hypothalamus and pituitary gland (*red*) inhibit hormone secretions and keep another follicle from maturing until the cycle ends.

rising estrogen levels invite repair and growth of the endometrium. Figure 38.21 summarizes the effects of changing hormone levels for one menstrual cycle.

In a menstrual cycle, FSH and LH stimulate growth of an ovarian follicle. The first meiotic cell division in the oocyte results in a secondary oocyte and the first polar body.

A midcycle surge of LH triggers ovulation—the release of the secondary oocyte and the polar body from the ovary.

Feedback loops to the hypothalamus and pituitary from the ovaries and, later, the corpus luteum control the cyclic changes in the ovary and uterine lining.

38.10 Visual Summary of the Menstrual Cycle

By now, you may have come to the conclusion that the human menstrual cycle is not a simple tune on a biological banjo. It's more like a full-blown hormonal symphony! Before continuing with your reading, take a moment to review Figure 38.22. It correlates cyclic changes in the ovary and the uterus with changing concentrations of hormones that trigger the events of each menstrual cycle.

Read Me First!
and watch the narrated animation
on the menstrual cycle summary

Figure 38.22 Summary of changing hormone levels and their effects during a menstrual cycle that lasts twenty-eight days.

(**a**) The hypothalamus secretes a releasing hormone, GnRH, which stimulates the anterior lobe of the pituitary to secrete FSH and LH.

(**b**) FSH and LH stimulate a follicle to grow, an oocyte to mature, and ovaries to secrete the progesterone and estrogens that induce rebuilding of the endometrium.

(**c**) A midcycle surge of LH triggers ovulation and formation of a corpus luteum. The corpus luteum maintains the endometrium through its secreting progesterone and some estrogens, but it stops doing so if pregnancy does not occur (**d**).

38.11 Pregnancy Happens

When a female and male engage in sexual intercourse, or coitus, the hormonal fog of the moment may obscure what can happen if a secondary oocyte is in an oviduct.

SEXUAL INTERCOURSE

The male sex act begins with an erection (a stiffened and lengthened penis). It culminates in ejaculation, or the forceful expulsion of semen from the penis. As Figure 38.15 shows, the penis contains long cylinders of spongy tissue. The penis of unaroused males stays limp because the large blood vessels that supply the spongy tissue remain vasoconstricted. When a male becomes aroused, the vessels vasodilate; blood flow into the penis exceeds blood flow out. As the spongy tissue inside becomes engorged with blood, the penis stiffens and lengthens. This facilitates insertion into a female's vaginal canal.

Repetitive pelvic thrusting mechanically stimulates friction-activated receptors that abound on the tip of the penis. It also stimulates the female's clitoris and vaginal wall. In the male, the response is involuntary muscle contractions that force sperm-laden semen into the urethra, from which it is ejaculated.

Emotional intensity, hard breathing, strong heart pounding, and the contraction of skeletal muscles in general accompany a rhythmic throbbing of the pelvic muscles. During *orgasm*, the end of the sex act, strong sensations of physical release, warmth, and relaxation dominate. Similar sensations typify female orgasm.

You may have heard that a female cannot become pregnant if she doesn't reach orgasm. Don't believe it.

FERTILIZATION

On average, an ejaculation can put 150 million to 350 million sperm in the vagina. Fertilization occurs if they arrive a few days before or after ovulation or any time in between. Less than thirty minutes after sperm arrive in the vagina, contractions move them deep into the female's reproductive tract. A few hundred actually reach the upper part of the oviduct, where eggs usually are fertilized (Figure 38.23).

Many sperm bind to the oocyte's zona pellucida. Binding triggers the release of enzymes from the cap over each sperm's head. Collectively, these digestive enzymes make a passage through the zona pellucida. Usually only one sperm enters the secondary oocyte. Only its nucleus and centrioles do not degenerate.

Upon sperm penetration, the secondary oocyte and the first polar body complete meiosis II. Now there are three polar bodies and one mature egg, or **ovum**

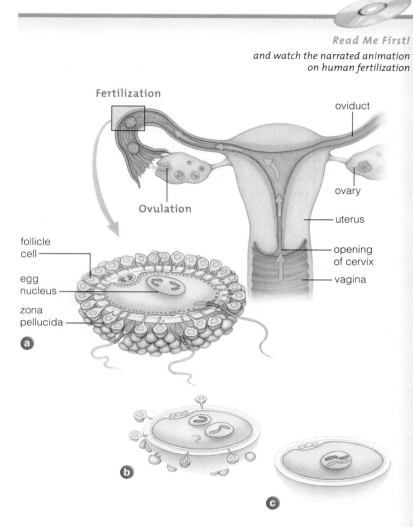

Read Me First!
and watch the narrated animation on human fertilization

Figure 38.23 Fertilization. (**a**) Many human sperm travel rapidly from the vagina to an oviduct (*blue* arrows), where they surround a secondary oocyte. Digestive enzymes released from each sperm's cap clear a path through the zona pellucida.

(**b**) A single sperm penetrates the secondary oocyte, which releases substances that make the zona pellucida impenetrable to the other sperm. Penetration also stimulates meiosis II of the oocyte's nucleus. (**c**) The sperm's tail degenerates; its nucleus enlarges and fuses with the nucleus of its target. Fertilization is over; a zygote has formed.

(plural, ova). The egg nucleus fuses with the sperm nucleus. Together, chromosomes of both nuclei restore the diploid number for a brand new zygote.

The intense physiological events that accompany coitus put sperm on a collision course with an egg.

Fertilization is over when a sperm nucleus and egg nucleus fuse. The diploid zygote that forms is the start of a new individual.

38.12 Control of Human Fertility

The motivation to engage in sex has been evolving for hundreds of millions of years. A few centuries of moral arguments for self-control have not suppressed it. How will we reconcile our biological past with the need for a stabilized cultural present?

THE ISSUE

The transformation of a zygote into an adult raises profound questions. *When does development begin?* As you have seen, major developmental events unfold even before fertilization. *When does life begin?* During her lifetime, a woman produces as many as 500 eggs, all of which are alive. With one ejaculation, a man releases a quarter of a billion sperm, which are alive. Before sperm and egg merge by chance and establish a new individual's genetic makeup, they are as much alive as any other form of life. It is scarcely tenable, then, to say that life begins at fertilization. *Life began more than 3.8 billion years ago—and each gamete, each zygote, and each sexually mature individual is but a fleeting stage in the continuation of that beginning.*

This greater perspective on life cannot diminish the meaning of conception. It is no small thing to entrust a new individual with the gift of life, wrapped in the unique evolutionary threads of our species and handed down through an immense sweep of time.

Yet human population growth is outstripping resources. Many millions already face the horrors of starvation. Living where we do, few of us know what it means to give birth to a child, to give it the gift of life, and have no food to keep it alive. Whether and how fertility should be controlled is a volatile issue. We return to this issue in Chapter 39. Here we simply present the options available to those who decide to postpone or forgo pregnancy (Figure 38.24).

SOME OPTIONS

Abstinence, or no sex at all, is the most reliable way to avoid pregnancy, but it takes great self-discipline to override the neural and hormonal sense of urgency associated with sex. Different versions of the *rhythm method* are forms of abstinence; a woman avoids sex in her fertile period. She calculates when she is fertile by recording how long her menstrual cycles last, by taking her core temperature each morning, or both. Miscalculations are frequent. Sperm deposited in the vagina a few days before ovulation may survive long enough to meet up with an egg.

Withdrawal, or removing the penis from the vagina before ejaculation, dates at least to biblical times. It requires a lot of willpower and still may fail, because fluid released from the penis before ejaculation can contain some sperm.

Douching (chemically rinsing the vagina right after intercourse) is too chancy. Sperm move out of

The Most Effective

Total abstinence	100%
Tubal ligation or vasectomy	99.6%
Hormonal implant (Norplant)	99%

Highly Effective

IUD + slow-release hormones	98%
IUD + spermicide	98%
Depo-Provera injection	96%
IUD alone	95%
High-quality latex condom + spermicide with nonoxynol–9	95%
Oral contraceptive (the Pill)	94%

Effective

Cervical cap	89%
Latex condom alone	86%
Diaphragm + spermicide	84%
Billings or Sympto-Thermal Rhythm Method	84%
Vaginal sponge + spermicide	83%
Foam spermicide	82%

Moderately effective

Spermicide cream, jelly, suppository	75%
Rhythm method (daily temperature)	74%
Withdrawal	74%
Condom (cheap brand)	70%

Unreliable

Douching	40%
Chance (no method)	10%

Figure 38.24 Comparison of the effectiveness of some methods of contraception. These percentages also indicate the number of unplanned pregnancies per 100 couples who use only that method of birth control for a year. For example, "94% effectiveness" for oral contraceptives means that 6 of every 100 women will still become pregnant, on average.

reach of a douche ninety seconds after ejaculation. Frequent douching irritates the reproductive tract.

Controlling fertility by surgical intervention is less chancy. Men who do not want children may opt for a *vasectomy*, a procedure that requires only a local anesthetic. A doctor makes a small incision in the scrotum, then cuts and ties off each vas deferens. Sperm no longer can leave the testes and become part of semen. Surgical reversal of the procedure is not always successful. Only about 60 percent of those who have reversed a vasectomy have been able to father a child.

Tubal ligation nearly always guarantees permanent infertility. The oviducts are cauterized or cut and tied. This procedure is now more common than vasectomy. Surgical reversal is about 70 percent successful.

Less drastic fertility control methods are based on physical and chemical barriers that stop sperm from reaching an egg. *Spermicidal foam* and *spermicidal jelly* poison sperm. An applicator is used to insert either one into the vagina before sex. These products are not always reliable, but using them with a condom or a diaphragm makes them more effective.

A *diaphragm* is a flexible, dome-shaped device. It is inserted into the vagina and positioned over the cervix before intercourse. It is relatively effective if fitted initially by a doctor, used with a spermicidal foam or jelly, inserted correctly each time, and left in place for a prescribed length of time.

Condoms are thin, tight-fitting sheaths worn over the penis during intercourse. Good brands may be as much as 95 percent effective when used with a spermicide. Only latex condoms protect against sexually transmitted diseases. However, condoms can tear or leak, at which time they become useless.

A *birth control pill* delivers synthetic estrogens and progesterone-like hormones that block maturation of oocytes and ovulation. "The Pill" reduces menstrual cramps but can cause nausea, headaches, and weight gain. With at least 50 million users, it is the most common fertility control method in the United States. Used correctly, it is at least 94 percent effective. Use lowers the risk of ovarian cancer but increases the risk of breast, cervical, and liver cancers.

A *birth control patch* is an adhesive patch applied to skin. It delivers the same hormones as an oral contraceptive, and it blocks ovulation the same way.

Progestin injections or implants block ovulation. A *Depo-Provera* injection is effective for three months. *Norplant* works for five years. Both methods are very effective. However, they may cause sporadic, heavy bleeding, and removing Norplant rods can be tricky.

Figure 38.25 Doctor using a microscope and monitor to inject a human sperm into an egg during *in vitro* fertilization (IVF).

Morning-after pills are sometimes taken after a condom tears, or after unprotected consensual sex or rape. One brand, Previn, is a set of two pills with hormones that suppress ovulation and block the secretions from the corpus luteum. Morning-after pills work best when taken early, but are somewhat effective for as long as five days after intercourse.

SEEKING OR ENDING PREGNANCY

Some couples cannot conceive naturally. *In vitro fertilization* refers to conception outside the body ("in glass," as in a glass petri dish). Hormone injections prepare the ovaries for ovulation. Before an oocyte is released, it is withdrawn and a sperm is injected into it (Figure 38.25). Cleavage may follow fertilization. Two to four days later, the resulting tiny ball of cells may be transferred to the woman's uterus.

The fertilization attempts each cost about 8,000 dollars—and most fail. With our population size and health care costs soaring, is this a cost society should bear? Court battles have been waged over this issue.

At the other extreme is *induced abortion*: dislodging and removing an embryo or a fetus from the uterus. Each year in the United States alone, about 50 percent of more than 3.1 million unwanted pregnancies end in abortion. From a clinical standpoint, abortion usually is a rapid, relatively painless procedure and free of complications if performed in the first trimester. The drug cmifepristone (RU 486) and a prostaglandin can induce it during the first nine weeks of pregnancy. The two chemicals block progesterone receptors in the uterus. The uterine lining cannot be maintained, and neither can pregnancy.

Aborting a fetus in the third trimester is highly controversial unless the mother's life is threatened. This textbook cannot offer the "right" answer to a question about the morality of abortion or some other option. It can only offer a serious explanation of how a new individual develops. Your choice of how to answer the question will be just that—your choice.

38.13 Sexually Transmitted Diseases

Having unprotected sex exposes you to potential infection by any pathogens that your partner unknowingly may have picked up from a previous sexual partner.

CONSEQUENCES OF INFECTION

Each year, pathogens that cause **sexually transmitted diseases**, or STDs, infect about 15 million Americans (Table 38.4). Two-thirds of those infected are under age twenty-five; one-quarter are teenagers. Over 65 million Americans now live with an incurable STD. Treating STDs and secondary complications costs a staggering 8.4 billion dollars in an average year.

The social consequences are enormous. Women are more easily infected than men, and they develop more complications. Example: Pelvic inflammatory disease (PID), a secondary outcome of some bacterial STDs, affects about 1 million women annually. It scars the reproductive tract and can cause infertility, tubal pregnancies, and chronic pain (Figure 38.26*a*). Some fetuses that acquire STDs before or during birth abort on their own or develop abnormally. Women with a chlamydial infection commonly transmit it to their newborn (Figure 38.26*b*). Besides causing skin sores, type II *Herpes* virus kills half of the infected fetuses and causes neural defects in one-fourth of the rest.

MAJOR AGENTS OF STDS

HPV Infection by human papillomaviruses (HPV) is the most widespread and fastest growing STD in the United States. At least 20 million are already infected. Of about 100 HPV strains, a few cause *genital warts*. These bumpy growths form on the vagina, cervix, and external genitals, and around the anus (Figure 38.26*c*). In men, they form on the penis and scrotum. A few strains cause cervical cancer. Sexually active women should have an annual pap smear to check for any cervical changes.

TRICHOMONIASIS Infection by *Trichomonas vaginalis*, a flagellated protozoan, causes a yellowish discharge, soreness, and itching of the vagina. Infected men are usually symptom-free. Untreated infections damage the urinary tract, result in infertility, and invite HIV infection. A single dose of an antiprotozoal drug can quickly cure the infection. To prevent reinfection, both sexual partners must be treated.

CHLAMYDIA *Chlamydial infection* is primarily a young person's disease. Forty percent of those infected are between ages fifteen and nineteen; 1 in 10 sexually active teenage girls is infected. *Chlamydia trachomatis* causes the disease; antibiotics quickly kill it. Most infected women have no symptoms and are not even diagnosed. Between 10 and 40 percent of those who are untreated will develop PID. Abnormal discharges from the penis and painful urination are symptoms in about 50 percent of infected men. Untreated men risk inflammation of the epididymes and infertility.

GENITAL HERPES About 45 million Americans have *genital herpes*, caused by type II *Herpes simplex* virus. Transmission to new hosts requires direct contact with active *Herpes* viruses or with sores that contain them. Mucous membranes of the mouth and genitals are vulnerable. Symptoms are often mild or absent. Painful, small blisters may form on the vulva, cervix, urethra, or anal tissues of infected women. Blisters form on the penis and anal tissues of infected men. Within three weeks, the virus enters latency. Sores crust over and heal, but the virus is still in the body.

The virus is reactivated sporadically, which causes painful sores at or near the original site of infection. Sexual intercourse, menstruation, emotional stress, or other infections trigger flare-ups. One antiviral drug, Acyclovir, decreases healing time and often the pain.

Figure 38.26 A few downsides of unsafe sex. (**a**) STDs that scar the oviduct are one cause of tubal pregnancy. Scarring makes the embryo implant itself in an oviduct, not the uterus. Untreated tubal pregnancies can rupture an oviduct and cause bleeding, infection, and death. (**b**) One sign of a chlamydial infection transferred from a mother to her infant. (**c**) Genital warts. (**d**) Chancres typical of secondary syphilis.

Table 38.4	Estimated New STD Cases Per Year *	
STD	U.S. Cases	Global Cases
HPV infection	5,500,000	30,000,000
Trichomoniasis	5,000,000	174,000,000
Chlamydia	3,000,000	92,000,000
Genital herpes	1,000,000	20,000,000
Gonorrhea	650,000	62,000,000
Syphilis	70,000	12,000,000
AIDS	40,000	5,000,000

*Global data on HPV and genital herpes were last compiled in 1990.

GONORRHEA   *Gonorrhea* is caused by the bacterium *Neisseria gonorrhoeae*, which typically crosses mucous membranes of the urethra, cervix, or anal canal during sexual intercourse. An infected female may notice a slight vaginal discharge or burning sensation while urinating. If the pathogen enters her oviducts, it may cause cramps, fever, vomiting, and even scarring that can cause sterility. Less than a week after a male is infected, yellow pus oozes from the penis. Urination becomes more frequent and may also be painful.

Prompt treatment with antibiotics quickly cures this bad disease, yet it still is rampant. Why? Many women ignore early symptoms. Also, people wrongly believe infection confers immunity. A person can get gonorrhea over and over again, probably because there are at least sixteen strains of *N. gonorrhoeae*. Also, use of oral contraceptives invites infection by altering vaginal pH. Populations of resident bacteria decline, so *N. gonorrhoeae* is free to move in.

SYPHILIS   The spirochete *Treponema pallidum* causes *syphilis*, a dangerous STD. Having sex with an infected partner puts this bacterium on the surface of genitals or into the cervix, vagina, or oral cavity. *T. pallidum* slips into the body even through tiny cuts. One to eight weeks later, treponemes are twisting about in a flattened, painless chancre (a localized ulcer).

This first chancre is a sign of the primary stage of syphilis. It usually heals, but treponemes multiply in the spinal cord, brain, eyes, bones, joints, and mucous membranes. In the infectious secondary stage, a skin rash develops and more chancres form (Figure 38.26d). In about 25 percent of the cases, immune responses succeed and symptoms subside. Another 25 percent are symptom-free. In the remainder, lesions and scars appear in the skin and liver, bones, and other internal organs. Few treponemes form in this tertiary stage, but a host's immune system is hypersensitive to them. Chronic immune reactions may damage the brain and spinal cord and cause paralysis.

Possibly because the symptoms are so alarming, more people seek early treatment for syphilis than gonorrhea. Later stages require prolonged treatment.

AIDS   Infection by HIV (human immunodeficiency virus) leads to *AIDS* (*Acquired Immune Deficiency Syndrome*). Someone can become infected and not know it. Infection marks the start of a titanic, long-term battle that the immune system almost always loses; AIDS is an incurable STD (Section 34.10).

At first there may be no outward symptoms. Five to ten years later, a set of chronic disorders develops.

Figure 38.27   NBA legend Magic Johnson, one of the torch bearers of the 2002 Winter Olympics. He was diagnosed as HIV positive in 1991. He contracted the virus through heterosexual sex, and credits his survival to AIDS drugs and informed medical care. He continues to campaign to educate others about AIDS.

The immune system weakens, which opens the door to opportunistic infectious agents. Normally harmless bacteria that are already living in the body are the first to take advantage of the lowered resistance. Then dangerous pathogens take their toll. Eventually they overwhelm the compromised immune system.

Most often, HIV spreads by anal, vaginal, and oral intercourse and intravenous drug use. Virus particles in blood, semen, urine, or vaginal secretions enter a new host through cuts and abrasions in the lining of the penis, vagina, rectum, or oral cavity.

Free or low-cost, confidential testing for exposure to HIV is available through public health facilities. It takes a few weeks to six months or more after first exposure before detectable amounts of antibodies can form in response to the infection. Anyone who tests positive for HIV can spread the virus.

Public education may help slow the spread of HIV (Figure 38.27). Most health care workers advocate safe sex, although there is confusion over what "safe" means. The use of high-quality latex condoms *together with* a spermicide that contains nonoxynol–9 helps prevent viral transmission but carries a slight risk. Open-mouth kissing with an HIV positive individual carries a risk. Caressing is not risky *if* there are no lesions or cuts where HIV-laden body fluids can enter the body. Skin lesions caused by any of the other sexually transmitted diseases are vulnerable points of entry for the virus.

New, costly drug therapies are prolonging some lives. In the late 1990s, the infection rate started to climb again, possibly because of a misperception that AIDS is no longer a threat. But AIDS kills, and the virus keeps on mutating. How long today's drugs can keep a lid on deaths is anybody's guess.

38.14 Formation of the Early Embryo

Nine months or so after the time of fertilization, a woman gives birth. Besides taking longer to develop inside its mother, the new individual will require more intense care for a much longer time compared to other primates.

Pregnancy lasts an average of thirty-eight weeks from the time of fertilization. It takes about one week for a blastocyst to form. All major organs form during the *embryonic* period—the third to the end of the eighth week of pregnancy. When this period ends, the new individual is called a **fetus**. It has distinctly human features. In the *fetal* period, from the start of the ninth week until birth, organs grow and become specialized. We often refer to the first three months of pregnancy as the first trimester. The second trimester extends from the start of the fourth month to the end of the sixth, and the third trimester ends at birth.

CLEAVAGE AND IMPLANTATION

Three to four days after fertilization, cleavage already has started on the zygote. Genes are being expressed; the early cuts require their products. At the eight-cell stage, the cells huddle into a ball. A blastocyst forms by the fifth day. It consists of a trophoblast (a surface layer of cells), a blastocoel filled with their secretions, and an inner cell mass (Figure 38.28*d*). One or two days later, **implantation** is under way. The blastocyst adheres to the uterine lining and invades the mother's tissues, forming connections that will metabolically support the pregnancy. By this time, the inner cell mass has become two flattened layers of cells in the shape of a disk. This embryonic disk will give rise to the embryo proper.

EXTRAEMBRYONIC MEMBRANES

As implantation continues, membranes start to form outside the embryo. First, a fluid-filled *amniotic* cavity opens up between the embryonic disk and part of the blastocyst surface (Figure 38.28*f*). Many cells migrate around the wall of the cavity and form the **amnion**, a membrane that will enclose the embryo. Fluid in the cavity will function as a buoyant cradle in which an embryo can grow, move freely, and be protected from abrupt temperature changes and mechanical impacts.

a **DAYS 1–2.** The first cleavage furrow extends between the two polar bodies. Later cuts are angled, so cells become asymmetrically arranged. Until the eight-cell stage forms, they are loosely organized, with space between them.

b **DAY 3.** After the third cleavage, cells abruptly huddle into a compacted ball, which tight junctions among the outer cells stabilize. Gap junctions formed along the interior cells enhance intercellular communication.

c **DAY 4.** By 96 hours there is a ball of sixteen to thirty-two cells shaped like a mulberry. It is a morula (after *morum*, Latin for mulberry). Cells of the surface layer will function in implantation and will give rise to a membrane, the chorion.

d **DAY 5.** A blastocoel (fluid-filled cavity) forms in the morula as a result of surface cell secretions. By the thirty-two-cell stage, differentiation is occurring in an inner cell mass that will give rise to the embryo proper. This embryonic stage is the blastocyst.

e **DAYS 6–7.** Some of the blastocyst's surface cells attach themselves to the endometrium and start to burrow into it. Implantation has started.

actual size

As the amnion forms, other cells migrate around the inner wall of the blastocyst, forming a lining that becomes a **yolk sac**. This extraembryonic membrane speaks of the evolution of land vertebrates. For most animals that produce shelled eggs, the yolk sac holds nutritive yolk. In humans, one portion of the yolk sac becomes a site of blood cell formation. Another will give rise to germ cells, the forerunners of gametes.

Before a blastocyst is fully implanted, spaces open in maternal tissues and become filled with blood that seeps in from ruptured capillaries. In the blastocyst, a new cavity opens up around the amnion and yolk sac. Fingerlike projections form on the lining of the cavity. It becomes the third membrane, called a **chorion**. This membrane will become part of a placenta, a spongy, blood-engorged tissue to be described shortly.

After the blastocyst is implanted, an outpouching of the yolk sac will become the fourth extraembryonic membrane—the **allantois**. The allantois has different roles in different groups. Among reptiles, birds, and some mammals, it serves in respiration and in storing metabolic wastes. In humans, the urinary bladder and blood vessels for a placenta develop from it.

The blastocyst itself stops menstruation. Cells of the blastocyst secrete a hormone called human chorionic gonadotropin, or HCG. This hormone stimulates the corpus luteum to continue secreting progesterone and estrogens. It does so until the placenta takes over the secretion of HCG, about eleven weeks later.

By the start of the third week, HCG can be detected in samplings of the mother's blood or urine. At-home *pregnancy tests* have a treated "dip-stick" that changes color when urine contains this hormone.

Cleavage of the human zygote produces a cluster of cells that develops into the blastocyst.

Six or seven days after fertilization, the blastocyst implants itself in the endometrium.

Projections from the blastocyst's surface invade maternal tissues, and connections start to form that in time will metabolically support the developing embryo.

Some parts of the blastocyst give rise to an amnion, yolk sac, chorion, and allantois. These extraembryonic membranes serve different functions. Together they are vital for the structural and functional development of the embryo.

Read Me First!
and watch the narrated animation on cleavage and implantation

f **DAYS 10–11.** The yolk sac, embryonic disk, and amniotic cavity have started to form from parts of the blastocyst. *actual size*

g **DAY 12.** Blood-filled spaces form in maternal tissue. The chorionic cavity starts to form. *actual size*

h **DAY 14.** A connecting stalk has formed between the embryonic disk and chorion. Chorionic villi, which will be features of a placenta, start to form. *actual size*

Figure 38.28 From fertilization through implantation. A blastocyst forms, and its inner cell mass will give rise to a disk-shaped early embryo. Three extraembryonic membranes (the amnion, chorion, and yolk sac) start forming. A fourth membrane (allantois) forms after the blastocyst is implanted.

38.15 Emergence of the Vertebrate Body Plan

By the time a woman misses a first menstrual period after fertilization, cleavage is over. Gastrulation is under way.

Gastrulation, recall, is the stage when cell divisions, migrations, and rearrangements give rise to primary tissue layers. Until now, the inner cell mass has been developing much as it did in reptilian ancestors of mammals. The cells have rearranged themselves into a flattened, two-layer embryonic disk, as they still do in yolky eggs of living reptiles and birds.

By now, the embryonic disk is surrounded by the amnion and chorion except where a stalk joins it to the chorion wall. The yolk sac lining has formed from one of the two layers. The other layer now starts to become the embryo proper. A depression appears on the disk, and its sides thicken. This is the primitive streak, which lengthens and thickens the next day. Its appearance marks the onset of gastrulation (Figure 38.29a). It defines an embryo's anterior–posterior axis and, in time, its bilateral symmetry.

Endoderm and mesoderm form from cells that are migrating inward along this axis. Pattern formation starts. Embryonic inductions and interactions among classes of master genes map out the basic body plan of all vertebrates. Tissues and organs begin to form in orderly steps, according to predictable patterns.

For example, by the eighteenth day, the embryonic disk has two folds that will merge to form the neural tube (Figure 38.29b). Neural tube defects arise when this tube does not form as it should. *Spina bifida*, the most common problem, results when the spinal cord protrudes out of an abnormally formed covering.

Some mesoderm folds into a tube that develops into a notochord. In vertebrates, the notochord is only a structural model; bony tissue of a vertebral column is assembled on it. Toward the end of the third week, multiple paired segments called **somites** form from part of the mesoderm. Somites are embryonic sources of most bones, of skeletal muscles of the head and trunk, and of the dermis overlying these body parts.

Pharyngeal arches start to form that will contribute to the face, neck, mouth, nose, larynx, and pharynx. Tiny spaces open up in parts of the mesoderm. In time the spaces will interconnect as a coelomic cavity.

> *The basic vertebrate body plan emerges early in the development of the new individual.*
>
> *A primitive streak, neural tube, somites, and pharyngeal arches form during the embryonic period of all vertebrates. Formation of the primitive streak establishes the body's anterior–posterior axis and its bilateral symmetry.*

yolk sac
embryonic disk
amniotic cavity
chorionic cavity

primitive streak

paired neural folds

neural groove (below, notochord is forming)

future brain

somites

pharyngeal arches

a **DAY 15.** A faint band appears around a depression along the axis of the embryonic disk. This is the primitive streak, and it marks the onset of gastrulation in vertebrate embryos.

b **DAYS 18–23.** Organs start to form through cell divisions, cell migrations, tissue folding, and other events of morphogenesis. Neural folds will merge to form the neural tube. Somites (bumps of mesoderm) appear near the embryo's dorsal surface. They will give rise to most of the skeleton's axial portion, skeletal muscles, and much of the dermis.

c **DAYS 24–25.** By now, some embryonic cells have given rise to pharyngeal arches. These will contribute to the formation of the face, neck, mouth, nasal cavities, larynx, and pharynx.

Figure 38.29 Hallmarks of the embryonic period of humans and other vertebrates. A primitive streak and then a notochord form. Neural folds, somites, and pharyngeal arches form later. (**a**,**b**) Dorsal views of the embryo's back. (**c**) Side view.

38.16 Why Is the Placenta So Important?

Even before the embryonic period starts, the uterus has been interacting with extraembryonic membranes in ways that will sustain the embryo's rapid growth.

By the third week, tiny fingerlike projections from the chorion have grown into the maternal blood that has pooled in endometrial spaces. These projections are the chorionic villi. They enhance the rate of exchange of substances between the mother and the embryo. The villi are functional components of the placenta.

The **placenta** is a blood-engorged organ composed of the uterine lining and extraembryonic membranes. At full term, it will make up about one-fourth of the inner surface of the uterus (Figure 38.30).

A placenta is the body's way of sustaining the new individual while allowing its blood vessels to develop separately from the mother's blood vessels. Oxygen and vital nutrients diffuse out of the maternal blood vessels, across the placenta's blood-filled spaces, then into embryonic blood vessels. The vessels converge in an umbilical cord, the lifeline between the placenta and the new individual. Carbon dioxide and other wastes diffuse in the other direction. The mother's lungs and kidneys dispose of the wastes.

After the third month of pregnancy, the placenta starts secreting progesterone and estrogens. The action of these sex hormones maintains the uterine lining.

The placenta is a blood-engorged organ of endometrial and extraembryonic membranes. It allows the new individual to take up oxygen and nutrients from the mother and give up wastes to her. It does so while allowing embryonic blood vessels to develop separately from the mother's.

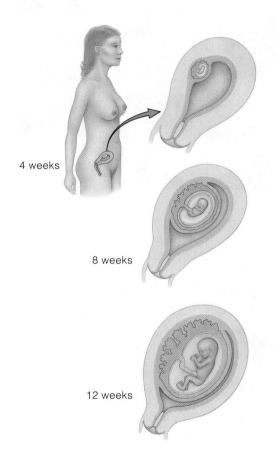

4 weeks

8 weeks

12 weeks

appearance of the placenta at full term

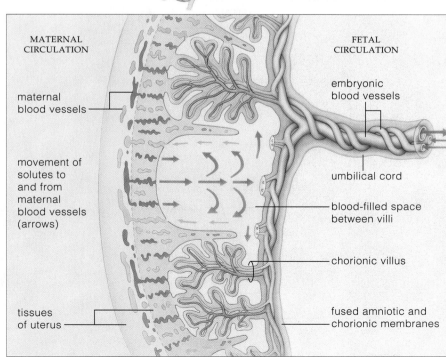

MATERNAL CIRCULATION

FETAL CIRCULATION

maternal blood vessels

movement of solutes to and from maternal blood vessels (arrows)

tissues of uterus

embryonic blood vessels

umbilical cord

blood-filled space between villi

chorionic villus

fused amniotic and chorionic membranes

Figure 38.30 Relationship between fetal and maternal blood circulation in a full-term placenta. Blood vessels extend from the fetus, through the umbilical cord, and into chorionic villi. Maternal blood spurts into spaces between villi, but the two bloodstreams do not intermingle. Oxygen, carbon dioxide, and other small solutes diffuse across the placental membrane surface.

38.17 Emergence of Distinctly Human Features

Early on, a human embryo—with its gill arches and long tail—has a distinctly vertebrate appearance. The tail soon disappears, and by the beginning of the fetal period the developing individual has distinctly human features.

When the fourth week ends, the embryo is 500 times its starting size. Weeks five and six are the boundary between the embryonic and fetal periods. Now growth slows as details of organs fill in. Limbs form; toes and fingers are sculpted from paddles. The umbilical cord forms, and so does an intricate circulatory system. Growth of the all-important head now surpasses that of all other regions (Figure 38.31). Reproductive organs start forming, as explained in Section 11.1. At the end of the eighth week, the individual is no longer just "a vertebrate." Its features define it as a human fetus.

In the second trimester, as developing nerves and muscles connect up, reflexive movements begin. Legs

WEEK 4

yolk sac
connecting stalk
embryo

WEEKS 5–6

forebrain
future lens
pharyngeal arches
developing heart
upper limb bud
somites
neural tube forming
lower limb bud
tail

actual length

a

head growth exceeds growth of other regions
retinal pigment
future external ear
upper limb differentiation (hand plates develop, then digital rays of future fingers; wrist, elbow start forming)
umbilical cord formation between weeks 4 and 8 (amnion expands, forms tube that encloses the connecting stalk and a duct for blood vessels)
foot plate

actual length

b

Figure 38.31 (a) Human embryo at successive stages of development.

kick, arms wave about, and fingers grasp. The fetus frowns, squints, puckers its lips, sucks, and hiccups. When the fetus is five months old, its heartbeat can be heard clearly through a stethoscope positioned on the mother's abdomen. The mother can sense movements of fetal arms and legs.

By now the skin of the fetus is covered with soft, fuzzy hair, the lanugo. A thick, cheeselike coating protects the wrinkled, reddish skin from abrasion. In the sixth month, delicate eyelids and eyelashes form. The eyes open during the seventh month, the start of the final trimester. By this time, too, all portions of the brain have formed and have begun to function.

In the fetal period, the primary tissues that formed in the early embryo become sculpted in ways that transform this vertebrate embryo into one with distinctly human features.

placenta

WEEK 8

final week of embryonic period; embryo looks distinctly human compared to other vertebrate embryos

upper and lower limbs well formed; fingers and then toes have separated

primordial tissues of all internal, external structures now developed

tail has become stubby

c | actual length |

WEEK 16
Length: 16 centimeters
          (6.4 inches)
Weight: 200 grams
          (7 ounces)

WEEK 29
Length: 27.5 centimeters
          (11 inches)
Weight: 1,300 grams
          (46 ounces)

WEEK 38 (full term)
Length: 50 centimeters
          (20 inches)
Weight: 3,400 grams
          (7.5 pounds)

During fetal period, length measurement extends from crown to heel (for embryos, it is the longest measurable dimension, as from crown to rump).

d

38.18 Mother as Provider, Protector, Potential Threat

Each pregnant woman is committing much of her body's resources to the growth and development of a brand-new individual. From fertilization until birth, her future child is at the mercy of her diet, health habits, and life-style.

NUTRITIONAL CONSIDERATIONS

A human embryo gets all the proteins, carbohydrates, and lipids needed for growth and development when the mother-to-be eats a well-balanced diet, as Section 36.7 outlines. However, the demands for vitamins and minerals increase as the placenta preferentially absorbs them for the fetus from her blood. Medically supervised increases in B-complex vitamins before conception and in early pregnancy reduce the risk of severe neural tube defects in the embryo. Folate (folic acid) is especially important in this regard.

Nutritional deficiencies adversely affect many developing organs. For example, the brain undergoes its greatest expansion in the weeks just before and after birth. Poor nutrition during this span may impair intelligence and other brain functions later in life.

A pregnant woman must eat enough to gain 20 to 25 pounds, on average. If she does not, her newborn may be seriously underweight, at risk of postdelivery complications and, in time, impaired brain function.

INFECTIOUS DISEASES

IgG antibodies in a pregnant woman's blood cross the placenta. They protect her developing child from all but the most serious bacterial infections. Some viral diseases can be dangerous in the first six weeks after fertilization, a crucial time of organ formation.

Suppose she contracts *rubella* (German measles) in this critical period. There is a 50 percent chance that some organs will not form properly. For instance, if she is infected as embryonic ears are forming, her

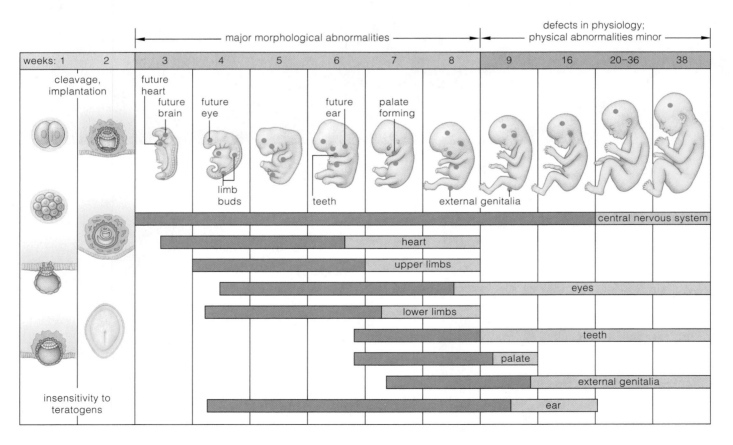

Figure 38.32 Teratogen sensitivity. Teratogens are drugs, infectious agents, and environmental factors that invite embryonic or fetal deformities, usually after organs form. They adversely affect growth, tissue remodeling, and tissue resorption. Dark blue shows the highly sensitive period; light blue shows periods of less severe sensitivity to teratogens. For example, the upper limbs are most sensitive to damage during weeks 4 through 6, and somewhat sensitive during weeks 7 and 8.

newborn may be deaf (Figure 38.32). If she is infected at any time from the fourth month of pregnancy onward, this particular disease will have no notable effect. A woman may avoid the risk entirely by getting vaccinated against the virus before pregnancy.

ALCOHOL, TOBACCO, AND OTHER DRUGS

ALCOHOL   Alcohol passes freely across membranes of cells. It passes freely across the placenta. When a pregnant woman drinks, her developing embryo or fetus quickly absorbs alcohol. Excessive intake invites *fetal alcohol syndrome* (FAS). Symptoms of this disorder include reduced brain size, mental impairment, facial deformities, a small head, slow growth, possible heart problems, and poor coordination (Figure 38.33). In some parts of the United States, the incidence of FAS is as high as 1.5 cases per 1,000 live births. The symptoms are permanent; children affected by FAS never do catch up, physically or mentally.

Even moderate drinking during pregnancy may have negative effects. Increasingly, doctors are urging total abstinence from alcohol during pregnancy.

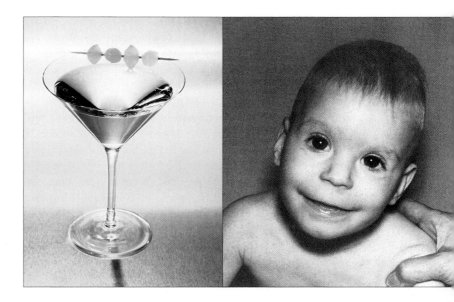

Figure 38.33   An infant with fetal alcohol syndrome—FAS. The obvious symptoms are low and prominently positioned ears, improperly formed cheekbones, and an abnormally wide, smooth upper lip. Growth-related complications and abnormalities of the nervous system can be expected.

TOBACCO   Smoking increases the risk of miscarriage and adversely affects fetal growth and development. Carbon monoxide competes with oxygen for binding sites on hemoglobin, so the embryo or fetus cannot get enough oxygen. Nicotine levels in the amniotic fluid can actually be higher than those in the mother's blood. Toxic compounds accumulate even inside the fetuses of pregnant nonsmokers who are exposed to *secondhand smoke* at home or at work.

Smoking any tobacco regularly during pregnancy results in underweight newborns. This happens even when the woman's weight, nutrition, and other key variables match those of pregnant nonsmokers.

Tobacco smoke adversely affects nutrition. In one study, pregnant women who smoked lowered the blood concentration of vitamin C for themselves *and* for their fetuses, even when intake of that vitamin matched the intake of a control group.

The effects may be long-term. Researchers tracked a group of children born in the same week. Children of smokers were smaller, had twice as many heart defects, and died of more postdelivery complications. By age seven, they were nearly half a year behind children of nonsmokers in their "reading age."

COCAINE   A pregnant woman who uses any form of cocaine increases her likelihood of miscarriage and premature delivery. The drug disrupts development of the nervous system. A child of a cocaine addict is likely to be abnormally small and irritable during early life. Some, but not all, studies suggest that prenatal exposure to cocaine has long-term negative effects on intelligence and behavior.

PRESCRIPTION DRUGS   Pregnant women should not take any drugs except under medical supervision. To underscore this point, the tranquilizer *thalidomide* was routinely prescribed in Europe. Infants of some of the women who used it during the first trimester had severely deformed arms and legs, or none at all. This drug has been withdrawn from the market. But other tranquilizers, sedatives, and barbiturates are still being prescribed, and they may cause some less severe damage. Some *anti-acne drugs* increase the risk of facial and cranial deformities.

Depression during pregnancy is not uncommon and can itself have negative health effects on a fetus. A mother who doesn't feel like eating or otherwise taking care of herself can impair fetal development. To avoid these problems, some pregnant women are treated with antidepressants. So far, these drugs have not been found to increase the rate of miscarriage, slow fetal growth, or increase the incidence of birth defects. However, infants of women who used these drugs right up to the time of delivery may show some withdrawal symptoms.

38.19 From Birth Onward

Human growth and development do not end with birth. Both processes continue as the newborn embarks on a course of extended dependency and learning. As with all mammals, its early survival depends on nutritious milk, typically provided by the mother.

A fetus born too prematurely (before 22 weeks) will not survive. The risk also is great for births before 28 weeks, mainly because the lungs have not developed enough. The risk starts to drop after this. By 36 weeks, the survival rate is 95 percent. A fetus born between 36 and 38 weeks still has some trouble breathing and maintaining a core temperature even with the best of medical care. The most favorable birthing time, on average, is 38 weeks after fertilization.

GIVING BIRTH

The birth process is known as **labor**. Typically, the amnion ruptures just before birth, so amniotic fluid drains from the vagina. The cervical canal dilates, the fetus moves out of the uterus, then out of the vagina and into the outside world (Figure 38.34).

Mild uterine contractions start in the last trimester. **Relaxin** is a hormone secreted by the ovaries and the placenta. It softens cervical connective tissues and loosens ligaments between the pelvic bones. The fetus "drops," or shifts down; its head usually touches the cervix. Rhythmic contractions at the onset of labor get stronger and closer together during the next two to eighteen hours. Normally they cause expulsion of the fetus within an hour after the cervix is fully dilated.

Strong contractions also detach the placenta from the uterus and expel it, as the afterbirth. Contractions help stop the bleeding at the site where the placenta attached to the wall of the uterus. They constrict the

blood vessels at the ruptured attachment site. The umbilical cord is cut and tied off, and a few days after shriveling up, the cord's stump has become the navel.

Corticotropin-releasing hormone (CRH) affects the timing of labor, and it may contribute to *post-partum depression*. The hypothalamus makes this hormone in all humans, but the placenta also makes it during a pregnancy, which can increase its level in the blood by threefold. CRH stimulates the adrenal cortex to secrete **cortisol** (Section 31.4). This hormone may help a mother cope with many strains that pregnancy and labor place on her.

During pregnancy, a high blood level of cortisol can suppress CRH production by the hypothalamus. After birth, when the placenta has been expelled, the CRH level briefly plummets. In many women, this decline triggers a short-term depression that typically continues until the hypothalamus resumes its normal production of CRH.

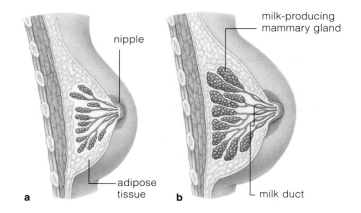

Figure 38.35 (**a**) Breast of a woman who is not pregnant. (**b**) Breast of a lactating woman.

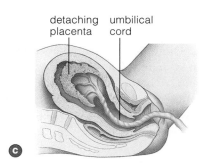

Figure 38.34 Expulsion of (**a**,**b**) a human fetus and (**c**) afterbirth during labor. The afterbirth consists of the placenta, tissue fluid, and blood.

NOURISHING THE NEWBORN

Once the lifeline to the mother is severed, a newborn enters the extended time of dependency and learning that is typical of all primates. Early survival requires an ongoing supply of milk or a nutritional equivalent. **Lactation**, or milk production, occurs in mammary glands inside a mother's breasts (Figure 38.35). Before pregnancy, breast tissue is largely adipose tissue and a system of undeveloped ducts. Their size depends on how much fat they hold, not on their milk-producing ability. During pregnancy, estrogens and progesterone stimulate the development of a glandular system for milk production.

For the first few days after birth, mammary glands produce a fluid rich in proteins and lactose. **Prolactin**, a hormone that calls for synthesis of enzymes used in milk production, is secreted by the anterior lobe of the mother's pituitary gland. As a newborn suckles, the pituitary releases **oxytocin**. This hormone triggers contractions that force fluid into milk ducts. It also causes contractions of uterine muscles that help make the uterus shrink to its pre-pregnancy size.

In addition to nutrients, human breast milk has immunoglobulins that enhance resistance to infection. Other components of milk stimulate the growth of symbiotic bacteria in the infant's gut. Alcohol, drugs, mercury, and other toxins in a mother's body can also be secreted in milk. As during pregnancy, a nursing mother should tailor her life-style and diet.

POSTNATAL DEVELOPMENT

As is the case for many species, humans change in size and proportion until reaching sexual maturity. Figure 38.36 shows a few of the proportional changes that occur as the life cycle unfolds. Table 38.5 defines the prenatal ("before birth") stages and the postnatal ("after birth") stages of a human life.

Postnatal growth is most rapid between the years thirteen and nineteen. Sex hormone secretions step up and bring about the development of secondary sexual traits as well as sexual maturity. Not until adulthood are the bones fully mature.

Hormones control the timing of labor and the onset of lactation, which nourishes the developing newborn.

Human growth and development do not end with birth. Sexual maturity occurs after puberty, when the production of sex hormones is stepped up dramatically.

8-week embryo | 12-week embryo | newborn | 2 years | 5 years | 13 years (puberty) | 22 years

Figure 38.36 Observable, proportional changes in the human body during prenatal and postnatal periods of development. Changes in overall physical appearance are slow but noticeable until the teenage years. For example, compared to an embryo, the legs of teenagers are longer and the trunk shorter, so the head is proportionally smaller. Correlate these drawings with the stages in Table 38.5.

Table 38.5 Stages of Human Development

Prenatal period

Zygote	Single cell resulting from fusion of sperm nucleus and egg nucleus at fertilization.
Morula	Solid ball of cells produced by cleavages.
Blastocyst (blastula)	Ball of cells with surface layer, fluid-filled cavity, and inner cell mass.
Embryo	All developmental stages from two weeks after fertilization until end of eighth week.
Fetus	All developmental stages from ninth week to birth (about 38 weeks after fertilization).

Postnatal period

Newborn	Individual during the first two weeks after birth.
Infant	Individual from two weeks to about fifteen months after birth.
Child	Individual from infancy to about ten or twelve years.
Pubescent	Individual at puberty; secondary sexual traits develop; girls between 10 and 15 years, boys between 12 and 16 years.
Adolescent	Individual from puberty until about 3 or 4 years later; physical, mental, emotional maturation.
Adult	Early adulthood (between 18 and 25 years); bone formation and growth finished. Changes proceed slowly after this.
Old age	Aging processes result in expected tissue deterioration.

38.20 Why Do We Age and Die?

*As the years pass, all multicelled species undergo **aging**; tissues become harder to maintain and repair. Each species has a maximum life span—122 years for humans, 20 years for dogs, 12 weeks for butterflies, 35 days for fruit flies, and so on. The verifiably oldest human lived 122 years. We can expect that genes influence aging processes.*

PROGRAMMED LIFE SPAN HYPOTHESIS

Do biological clocks influence aging? If so, an animal body might be analogous to a clock shop, with each type of cell, tissue, and organ ticking away at its own genetically set pace. Many years ago, Paul Moorhead and Leonard Hayflick tested this hypothesis. They cultured human embryonic cells—which divided about fifty times before dying out.

Hayflick also took cultured cells that were part of the way through the series of divisions and froze them for a few years. After he thawed the cells and placed them in a culture medium, they completed an *in vitro* cycle of fifty doublings and died on schedule.

No cell in a human body divides more than eighty or ninety times. You may well wonder: If an internal clock ticks off their life span, then how can *cancer* cells go on dividing? The answer provides us with a clue to why *normal* cells can't beat the clock.

Cells, remember, duplicate their chromosomes before dividing. Capping the chromosome ends are **telomeres** made of DNA and proteins. Telomeres keep the chromosome ends from unraveling. A small piece of each one is lost with each nuclear division. When only a nub is left, cells stop dividing and die.

Cancer cells and germ cells are exceptions; both make telomerase, an enzyme that makes telomeres lengthen. Expose cultured cells to telomerase, and they go on dividing well beyond normal life span.

CUMULATIVE ASSAULTS HYPOTHESIS

Another hypothesis: Over the long term, aging is the outcome of cumulative damage at the molecular and cellular levels. Environmental assaults as well as spontaneous mistakes by DNA repair mechanisms are at work here.

For example, the rogue molecular fragments called free radicals attack all biological molecules, including DNA. This includes the DNA of mitochondria, the power plants of eukaryotic cells. Structural changes in DNA compromise the synthesis of enzymes and other proteins necessary for normal life processes.

Free radicals are implicated in many age-related problems, including cataracts, atherosclerosis, and

Alzheimer's disease. In one experiment, researchers increased the average life span of roundworms by 50 percent simply by providing them with synthetic antioxidant enzymes.

Other studies suggest that extending the life span may come at a cost. Researchers were able to double the roundworm life span by knocking out a single gene. However, the mutant worms were sterile.

DNA replication and repair problems also have been implicated in aging. *Werner's syndrome*, an aging disorder, has been correlated with a mutation in the gene that specifies a helicase. Such enzymes unwind nucleotide strands. The mutated helicase probably does not compromise DNA replication, because affected people do not die right away. They start aging fast in their thirties and die before age fifty.

However, the nonmutated gene may be crucial for repairs. People with Werner's syndrome accumulate gene mutations at high rates. Sooner or later, the damage interferes with cell division. Like skin cells of the elderly, skin cells of Werner's patients just do not divide many times.

It may be that both hypotheses have merit. Aging may be an outcome of many interconnected processes in which genes, hormones, environmental assaults, and a decline in DNA repair mechanisms come into play. Consider how living cells of all tissues depend upon exchanges of materials with extracellular fluid. Also consider how collagen is a structural component of many connective tissues. If something shuts down or mutates collagen-encoding genes, then missing or altered gene products may disrupt flow of oxygen, nutrients, hormones, and so forth to and from living cells through every connective tissue. Repercussions from such a mutation would ripple through the body.

Similarly, if mutations cause altered self markers on the body's cells, do T cells of the immune system perceive them as foreign and attack? If autoimmune responses were to become more frequent over time, they would promote greater vulnerability to disease and stress associated with old age.

In evolutionary terms, reproductive success means living long enough to produce and raise offspring. Humans can reach sexual maturity in fifteen years *and* help their children reach adulthood. We don't *need* to live longer than we do. But we among all animals have the capacity to think about it, and most of us have decided we like life better than the alternative. Eventually, however, even the most stubborn among us must confront their own mortality. And perhaps, given time, we may all learn to accept the inevitable with wisdom and grace.

Summary

Section 38.1 Asexual reproduction yields offspring that are all genetically identical to their parent. Sexual reproduction requires energetically costly structures, controls, and behavior but creates variation that may assure reproductive success of at least some offspring.

Section 38.2 Most animal life cycles have six stages of embryonic development. Gametes form, fertilization occurs, cleavage results in a blastula, gastrulation results in an early embryo with two or three primary tissue layers (ectoderm, mesoderm, and endoderm), organs form, and tissues and organs become specialized.

Section 38.3 Maternal messages are localized in the egg cytoplasm. After fertilization, cleavage distributes different messages to different daughter cells. Cleavage patterns differ among animal groups.

Section 38.4 Cytoplasmic localization followed by short-range and long-range signals cause embryonic cells to selectively activate genes and differentiate: the cells become specialized in structure, composition, and behavior. Organs form by morphogenesis, as when cells migrate, tissues expand and fold over, and patches of cells undergo programmed cell death.

Section 38.5 Pattern formation is the sculpting of embryonic cells in a certain order, in expected positions in the embryo. Similar master genes govern pattern formation in all animals.

Section 38.6 Testes, the male primary reproductive organs in humans, produce sperm and testosterone.

Section 38.7 Feedback loops from the testes to the hypothalamus and pituitary govern testosterone, LH, and FSH secretion, all of which affect sperm formation.

Section 38.8 Ovaries, the primary reproductive organs in human females, produce oocytes and secrete the sex hormones estrogen and progesterone.

Sections 38.9, 38.10 Feedback loops from the ovaries to the hypothalamus and the pituitary control hormonal secretions of the menstrual cycle.

FSH causes an ovarian follicle to mature. The follicle secretes estrogens, which cause the endometrium to thicken in preparation for pregnancy. A midcycle surge of LH triggers ovulation.

A corpus luteum forms from follicle remnants and secretes progesterone and estrogens, which maintain the endometrium. If fertilization does not occur, it degenerates, the endometrium breaks down and is sloughed off with blood, and the cycle starts again.

Section 38.11 Fertilization occurs after a secondary oocyte meets up with a sperm, usually in the oviduct. The oocyte completes meiosis II and becomes an ovum (mature egg). Its nucleus fuses with the sperm nucleus.

Section 38.12 Fertility is controllable by behavior, surgery, physical or chemical barriers, or manipulations of female sex hormones.

Section 38.13 Sexually transmitted diseases (STDs) are caused by protozoan, bacterial, and viral pathogens and are spread by unsafe sex and other behaviors.

Section 38.14 After fertilization, a blastocyst forms by cleavage and implants itself in the endometrium. Extraembryonic membranes—the amnion, chorion, yolk sac, and allantois—form. They assist in the structural and functional development of the embryo.

Section 38.15 Gastrulation lays down the vertebrate body axis. A neural disk gives rise to the neural tube, the forerunner of the brain and spinal cord. Somites are embryonic precursors of muscles and the skeleton.

Section 38.16 A placenta made of maternal tissues and extraembryonic membranes allows exchanges between the mother and embryo while keeping their blood vessels separated.

Section 38.17 The embryo has distinctly human features by the end of the eighth week of pregnancy.

Section 38.18 A mother's health, nutrition, and life-style affect fetal growth and development.

Section 38.19 During labor, uterine contractions expel the fetus and afterbirth. Hormones trigger labor, the maturation of mammary glands, and milk flow. Development and growth continue until adulthood.

Section 38.20 Aging may be partly programmed in the genes and partly an outcome of cumulative assaults on DNA and other biological molecules.

Self-Quiz   *Answers in Appendix III*

1. Sexual reproduction among animals is _____ .
   a. biologically costly   c. evolutionarily beneficial
   b. diverse in its details   d. all of the above

2. A cell formed during cleavage is a _____ .
   a. blastula   b. morula   c. blastomere   d. gastrula

3. _____ produces three primary tissue layers.
   a. Gametogenesis   c. Gastrulation
   b. Implantation   d. Pattern formation

4. Homeotic genes map out the _____ .
   a. cleavage planes   c. basic body plan
   b. primary tissue layers   d. all of the above

5. Meiotic divisions of _____ produce mature sperm.
   a. Leydig cells   c. both a and b
   b. Sertoli cells   d. neither a nor b

6. During a menstrual cycle, a midcycle surge of _____ triggers ovulation.
   a. estrogens   b. progesterone   c. LH   d. FSH

7. The corpus luteum secretes _____ .
   a. LH   b. FSH   c. progesterone   d. prolactin

8. A _____ implants in the lining of the uterus.
   a. zygote   b. gastrula   c. blastocyst   d. fetus

Figure 38.37 (**a**) Adult zebrafish. (**b**) Normal embryo and (**c**) embryo bearing a mutation that interferes with formation of somites, the bumps of mesoderm that give rise to bones and most skeletal muscle.

Media Menu

Student CD-ROM

Impacts, Issues Video
   Mind-Boggling Births
Big Picture Animation
   Basic principles of development and human examples
Read-Me-First Animation
   Leopard frog life cycle
   Cytoplasmic localization
   Embryonic induction
   Human male reproductive system
   Sperm formation
   Human female reproductive system
   Ovarian function
   Hormones and the menstrual cycle
   Menstrual cycle summary
   Human fertilization
   Cleavage and implantation

InfoTrac

- Could This Be the End of the Monthly Period? *Science News*, August 2001.
- Embryos Stick to Mothers With Sugar. *United Press International*, January 2003.
- The Serious Search for an Anti-Aging Pill. *Scientific American*, August 2002.

Web Sites

- The Virtual Embryo: www.acs.ucalgary.ca/~browder
- American College of Obstetricians and Gynecologists: www.acog.org
- The Visible Embryo: www.visembryo.com
- CDC National Prevention Information Network: www.cdcnpin.org

How Would You Vote?

Fertility drugs can cause multiple-order births. Women who are carrying many fetuses are given the option of reducing the number, but some decline. Some women have chosen to carry nonuplets. Birth is always premature and some or all infants die. Some people have suggested restricting fertility drugs to use with IVF, in which the number of fertilized eggs put in the uterus is controlled. Would you support such a restriction?

9. Which of the following puts human developmental stages in the correct order?
  a. zygote, blastocyst, embryo, fetus
  b. zygote, embryo, blastocyst, fetus
  c. zygote, embryo, fetus, blastocyst
  d. blastocyst, zygote, embryo, fetus

10. Which of the following is (are) caused by bacteria?
  a. chlamydia   d. trichomoniasis
  b. gonorrhea   e. a and b
  c. genital warts   f. c and d

11. Match each term with the most suitable description.
  ____ gamete formation   a. mitotic cell divisions
  ____ fertilization   b. cellular rearrangements form primary tissues
  ____ cleavage
  ____ gastrulation   c. eggs and sperm form
  ____ cell differentiation   d. sperm nucleus and egg nucleus fuse
  ____ morphogenesis
      e. orderly changes in body size and shape
      f. in most species, genes are now selectively activated

12. Match each term with the most suitable description.
  ____ testis   a. maternal and fetal tissues
  ____ cervix   b. stores mature sperm
  ____ placenta   c. produces testosterone
  ____ vagina   d. produces estrogen and progesterone
  ____ ovary
  ____ oviduct   e. usual site of fertilization
  ____ epididymis   f. lining of uterus
  ____ endometrium   g. birth canal
      h. entrance to uterus

Critical Thinking

1. The zebrafish (*Danio rerio*) is special to developmental biologists. This small freshwater fish is easily maintained in tanks. A female produces hundreds of eggs, which can develop and hatch in three days. The transparent embryos let researchers directly observe developmental events (Figure 38.37). Cells can be injected with dye to see how they change position, or they can be killed or injected with genes and the effects observed.

Evolutionarily speaking, *D. rerio* is only remotely related to humans. Explain why researchers might expect the early development of this fish to yield useful information about human development.

2. Fraternal twins, which are nonidentical genetically, arise when two oocytes mature, are released, and are fertilized at the same time. Such twins run in families. The incidence varies among ethnic groups and is highest among blacks and lowest in Asians. Variation may be an outcome of differences in gene products that affect the FSH level in blood. Explain how a high FSH level would increase the likelihood of fraternal twins.

3. By UNICEF estimates, each year 110,000 people are born with abnormalities as a result of rubella infections. Major symptoms of *congenital rubella syndrome,* or CRS, are deafness, blindness, mental impairment, and heart problems. A nonvaccinated woman who is infected in the first trimester of pregnancy is at risk, but not later. Review the developmental events that unfold during pregnancy and explain why this is the case.

VII Principles of Ecology

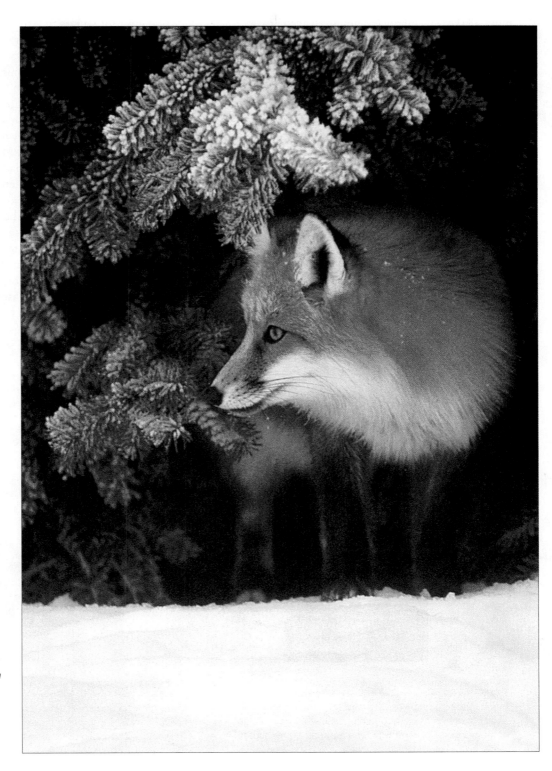

Two organisms—a fox in the shadows cast by a snow-dusted spruce tree. What are the consequences of their interactions with each other, with other kinds of organisms, and with their environment? By the end of this last unit, you might find worlds within worlds in such photographs.

IMPACTS, ISSUES   *The Human Touch*

In 1722, on Easter morning, an explorer landed on a small, remote volcanic island in the Pacific and found a few hundred skittish, hungry Polynesians living in caves. He saw grasses and scorched shrubs, but no trees. He came across hundreds of massive stone statues near the coast and unfinished and abandoned ones in inland quarries. Some weighed fifty tons (Figure 39.1). What did they represent?

When James Cook visited the island two years later, he counted only four canoes. Nearly all of the statues had been tipped over, often onto face-shattering spikes.

Later, researchers solved the mystery of the statues. Easter Island, as it came to be called, is only 165 square kilometers (64 square miles) in size. Voyagers from the Marquesas discovered this eastern outpost of Polynesia around 1,600 years ago. The place was a paradise. Its fertile soil supported dense forests and lush grasses. New arrivals built canoes from long, straight palms, strengthened with rope made of fibers from hauhau trees. They used wood as fuel to cook fish; they cleared forests to plant crops, and they had many children. By 1400, between 10,000 and 15,000 were living on the island. Crop yields had declined. Erosion and harvesting had depleted the nutrients in soil. Fish had vanished from nearshore waters. All native birds had been eaten, and people were starting to raise rats as food.

Survival was at stake and those in power appealed to the gods. They directed people to carve stone images of unprecedented size and use a system of greased logs to move them over miles of rough terrain to the coast.

Figure 39.1 Enormous statues carved from volcanic rock line the shore of Easter Island. Ancestors of the people who built them were skilled navigators who must have traveled more than a thousand miles across the open ocean to reach this small, remote island.

the big picture

Describing a Population   Each population has a characteristic size, density, and age structure. It shows patterns of distribution and growth. The defining characteristics change with time and can be monitored through field studies.

How Populations Grow   When resources are plentiful and unrestricted, population size increases exponentially. Populations eventually return to the carrying capacity, the maximum number of individuals that can be sustained indefinitely in a given environment.

By about 1550, no one fished offshore. No one could build new canoes because all the palms had been cut down and all the hauhau trees had been burned as firewood. Easter Islanders turned to their last source of protein. They started to hunt and eat one another.

Central authority crumbled and gang wars raged. Those on the rampage burned the remaining grasses to destroy hideouts. The dwindling population retreated to caves and launched raids against perceived enemies. Winners ate losers and tipped over the statues. There was nowhere else to go. What could they have been thinking when they chopped down the last palm?

And so, from archaeological and historical records, we know a large Easter Island society flourished amid abundant resources, then abruptly fell apart. Its loss is a footnote in life's greater evolutionary story, compared to the millions of species that disappeared in the past.

Yet there is a lesson here. Ultimately, there are principles that govern the growth and sustainability of all populations. These principles are the bedrock of **ecology**, the systematic study of how organisms interact with the physical and chemical environment.

Interactions start within and between populations and extend through communities, ecosystems, and the biosphere. They are the focus of this last unit of the book. In this chapter we first consider the general characteristics of populations. Later, we will apply the basic principles of population growth to the past, present, and future of the human species.

 How Would You Vote?

In the United States, immigration—legal and illegal—is a significant part of the rapid increase in population size. Undocumented immigrants are still entering the country in large numbers. Should the 7 million who have made a home here for some time be granted legal status? See the Media Menu for details, then vote online.

Life Histories All populations or species show a life history pattern, a set of adaptations that influence each individual's survival, fertility, and time of reproduction. Life history patterns may evolve by way of natural selection.

Human Populations Through agricultural, medical, and technological advances, the human population bypassed the carrying capacity and has been postponing natural limits on population growth. Its numbers have soared, but the operative word is *postpone*.

39.1 Characteristics of Populations

*By this point in the book, you know that a population is a group of individuals of the same species. Ecological interactions begin with characteristics of populations. We call these vital statistics **demographics**.*

Each population has a gene pool and an evolutionary history, as explained in Chapters 16 and 17. It also has a characteristic size, density, distribution, and number of individuals in its various age categories.

Population size is the number of individuals that potentially or actually contribute to the gene pool. The **age structure** is the number of individuals in each of several age categories. For instance, individuals are often grouped into *pre-reproductive*, *reproductive*, and *post-reproductive* ages. Those in the first category have the capacity to produce offspring when they mature. Together with the individuals in the second category, they make up the population's **reproductive base**.

Population density is the number of individuals in some specified area or volume of a habitat. A *habitat*, remember, is the type of place where a species lives. We characterize a habitat by its physical and chemical features and its particular array of species. **Population distribution** is the pattern in which the individuals are dispersed in a specified area.

Crude density is a measured number of individuals in some specified area. It does not reveal how much of a habitat is actually being used as living space. Even areas that appear uniform, such as a long, sandy shoreline, are more like tapestries of light, moisture, temperature, mineral composition, and many other variables. One portion of a habitat might be far more suitable for a population than other portions all of the time or some of the time, as in changing seasons.

Different species occupying the same area typically compete for energy, nutrients, living space, and other resources. Such *interspecific* interactions help shape a population's density and dispersion through a habitat.

Theoretically, a population has a clumped, nearly uniform, or random distribution pattern (Figure 39.2). Clumping is the most common, for several reasons. First, each species is responsive to certain conditions and resources that often are patchy through a habitat. Think of how some animals gather near a water hole or seeds sprout only where there is adequate soil and water. Second, animals may form social groups that offer protection and mating opportunities. A school of fish is like this. Third, the young of many plants and a lot of animals cannot disperse far from the parents.

With nearly uniform distribution, individuals are more evenly spaced than expected based on chance alone. Uniform distribution is relatively rare in nature. We do come across it when competition for resources or territory is fierce, as in a nesting colony of seabirds. Figures 39.2 and 43.15 have splendid examples.

We observe random dispersion only when habitat conditions are nearly uniform, resource availability is fairly steady, and individuals of a population or pairs of them neither attract nor avoid one another. Wolf spiders are ground dwellers that hunt at night, not far from their randomly located burrow (Figure 39.2).

Each population has characteristic demographics: size, density, distribution pattern, and age structure.

Environmental conditions and species interactions shape these characteristics, which may change over time.

clumped

nearly uniform

random

Figure 39.2 Three patterns of population distribution: clumped, as in squirrelfish schools; more or less uniform, as in a royal penguin nesting colony; and random, as when wolf spiders prowl on a forest floor.

39.2 Elusive Heads to Count

Ecologists go into the field to test theories about species interactions and population dynamics, and to monitor the health of threatened or endangered populations.

Once upon a time someone counted a million or so deer living in the United States. Now, a hundred years later, 18 million are living in forests, grasslands, golf courses, and suburban gardens. How would you go about counting the ones living near you?

A full count would measure absolute population density. Census takers supposedly make such a count of human populations every ten years, although not everyone answers the door. Ecologists make counts of large species in small areas, such as birds in a forest, northern fur seals at their breeding grounds, and sea stars in a tidepool. More often, however, a full count is impractical, and they must sample just part of a population and then estimate its total density.

For example, you could get a map of your county and divide it into small plots, or quadrats. **Quadrats** are sampling areas of the same size and shape, such as rectangles, squares, and hexagons. You could then count individual deer in several plots and, from that, extrapolate the average for the specified population. Ecologists often conduct such counts for plants and other species that stay put (Figure 39.3). Some of their counts in small areas help them estimate the population sizes of migrating animals.

Deer are among the animals that do not stay put. So how can you be sure that the individuals you are counting in a given plot are not the same ones you counted earlier in a different plot?

For mobile animals, ecologists sample population density with **capture–recapture methods**. The idea is to capture mobile individuals and mark them in some way. Deer get collars, squirrels get tattoos, salmon get tags, birds get leg rings, butterflies get wing markers, and so on (Figure 39.4). Marked animals are released at time 1. At time 2, traps are reset. The proportion of marked animals in this second sample is taken to be representative of the proportion marked in the whole population:

$$\frac{\text{Marked individuals in sampling at time 2}}{\text{Total captured in sampling 2}} = \frac{\text{Marked individuals in sampling at time 1}}{\text{Total population size}}$$

Ideally, marked and unmarked individuals of the population are captured at random, none of the marked animals dies during the study interval, and none leaves the population or gets overlooked.

In the real world, recapturing marked individuals might *not* be random. Squirrels marked after being attracted to bait in boxes might now be trap-happy or trap-shy. Such individuals may overrepresent or underrepresent their population. Another example: Instead of mailing back the tags of marked fish to ecologists, a few fishermen keep them as good-luck charms. And birds lose their leg rings.

Your estimate also depends on the time of year when you make a sampling. Population distribution varies over time, as during migratory responses to environmental rhythms. Few places yield abundant resources all year long, so many populations move between habitats as seasons change. Canada geese are like this. So are deer. In such cases, then, the capture–recapture methods might be used more than once a year, for several years.

Figure 39.3 Near the Sierra Nevada, a population of creosote bushes showing nearly uniform distribution. The plants compete for scarce water.

Figure 39.4 Two individuals marked for population studies. (**a**) Florida Key deer and (**b**) Costa Rican owl butterfly (*Caligo*).

39.3 Population Size and Exponential Growth

Populations are dynamic units of nature. Depending on the species, they may add or lose individuals every minute of every day, season, or year. Sometimes they glut portions of their habitat with individuals. Other times, individuals are scarce. Populations even drive themselves or are driven to extinction.

GAINS AND LOSSES IN POPULATION SIZE

We can measure change in population size in terms of birth rates, death rates, and how many individuals are entering and leaving during a specified interval.

Population size increases as a result of births and **immigration**, the arrival of new residents from other populations of the same species. Its size decreases as a result of deaths and **emigration**, the departure of individuals that take up permanent residence in some other place. As an example, Arnold Schwarzenegger emigrated from Austria to the United States, where he became a celebrated immigrant. His permanent move decreased the Austrian population by 1 and increased the United States population by 1.

For many species, population size changes during seasonal or daily migrations. However, **migration** is a recurring round trip between two distinct regions, so we need not consider its transient effects in this initial look at population size.

FROM ZERO TO EXPONENTIAL GROWTH

To keep things simple, let's assume that immigration and emigration balance each other over time. This way we can ignore the effects of both on population size. Doing so allows us to define **zero population growth** as an interval during which the number of births is balanced by the number of deaths. Population size is stable during such an interval, with no net increase or decrease in the number of individuals.

Births, deaths, and the other variables that might change population size can be measured in terms of **per capita** rates, or rates per individual. *Capita* means heads, as in head counts.

Visualize 2,000 mice living in a cornfield. Twenty or so days after their eggs are fertilized, the females give birth to a litter, then they nurse the offspring for a while, and then they get pregnant again. Say 1,000 mice are born in one month. The birth rate is 0.5 per mouse per month (1,000 births/2,000 mice). If 200 of the 2,000 die during that interval, the death rate will be 200/2,000 = 0.1 per mouse per month.

Assume further that the birth rate and death rate remain constant. By doing so, we can combine both variables into a single variable—the **net reproduction per individual per unit time**, or r for short. For our mice, r is $0.5 - 0.1 = 0.4$ per mouse per month.

Read Me First!
and watch the narrated animation on exponential growth

Figure 39.5 (**a**) Net monthly increases in a population of field mice living in a cornfield. Start to finish, the list shows a pattern typical of exponential growth. (**b**) Graph the numerical data and you end up with a J-shaped growth curve.

		Net Monthly Increase:	New Population Size:
$G = r \times$	3,920 =	1,568 =	5,488
$r \times$	5,488 =	2,195 =	7,683
$r \times$	7,683 =	3,073 =	10,756
$r \times$	10,756 =	4,302 =	15,058
$r \times$	15,058 =	6,023 =	21,081
$r \times$	21,081 =	8,432 =	29,513
$r \times$	29,513 =	11,805 =	41,318
$r \times$	41,318 =	16,527 =	57,845
$r \times$	57,845 =	23,138 =	80,983
$r \times$	80,983 =	32,393 =	113,376
$r \times$	113,376 =	45,350 =	158,726
$r \times$	158,726 =	63,490 =	222,216
$r \times$	222,216 =	88,887 =	311,103
$r \times$	311,103 =	124,441 =	435,544
$r \times$	435,544 =	174,218 =	609,762
$r \times$	609,762 =	243,905 =	853,667
$r \times$	853,677 =	341,467 =	1,195,134

a

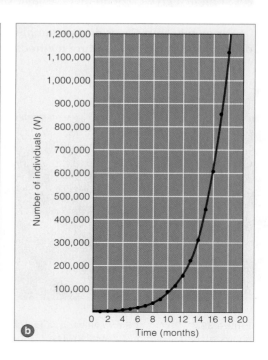

b

Figure 39.6 Effect of deaths on the rate of increase in two hypothetical populations of bacteria. Plot population growth for bacterial cells that reproduce every half hour and you get growth curve *1*. Plot the growth of a population of cells that divide every half hour, with 25 percent dying between divisions, and you get growth curve *2*. Deaths do slow the rate of increase, but as long as birth rate exceeds death rate, exponential growth will continue.

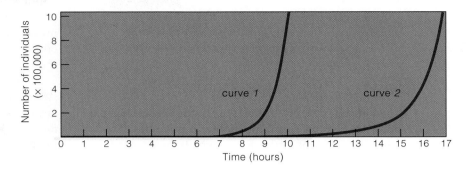

And so our mice in a cornfield give us a simple way to represent population growth as:

| population growth per unit time | = | net population growth rate per individual per unit time | × | number of individuals |

or, more simply, $G = rN$.

The next month begins, and 2,800 mice are in this field. With that net increase of 800 fertile mice, the reproductive base is larger. Population size expands again during the month, for a net increase of 0.4 × 2,800 = 1,120 mice. Population is now 3,920. The mice keep on reproducing month after month. Figure 39.5*a* shows, *in less than two years, the number of mice in the cornfield will increase from 2,000 to more than a million!*

Plot the monthly increases against time and you end up with a graph line in the shape of a "J," as in Figure 39.5*b*. When the growth of any population over time plots out as a J-shaped curve, you know that you are tracking exponential growth.

Exponential growth can be defined as a quantity increasing by a *fixed* percentage of the total in each specified interval. The numbers of births and deaths correlate with population size, so they increase in each interval as long as *r* remains constant. If it does, the increase in population size during any interval depends on the size of the existing reproductive base. *The larger the reproductive base, the greater will be the expansion in population size during a specified interval.*

Now look at other aspects of exponential growth. Start by supplying one bacterium in a culture flask with all the nutrients required for growth. After thirty minutes, the cell divides in two. Its two daughter cells divide, and so on every thirty minutes. Assume none of the cells dies between divisions. The population size doubles in each interval—from 1 to 2, then 4, 8, 16, 32, and so on. The time it takes for a population to double in size is its **doubling time**.

After 9–1/2 hours (nineteen doublings), there are more than 500,000 cells. Ten hours (twenty doublings) later, there are more than a million. Curve *1* in Figure 39.6 is a plot of this outcome.

Will death put the brakes on exponential growth? Suppose 25 percent of the descendant cells die every thirty minutes. Now, it takes about seventeen hours (not ten) for population size to reach 1 million. *Deaths slowed but did not stop population growth* (curve *2* in Figure 39.6). Exponential population growth will continue as long as birth rates exceed death rates.

WHAT IS THE BIOTIC POTENTIAL?

Now, visualize a population occupying a place where conditions are ideal. Every one of its individuals has adequate shelter, food, and other vital resources. No predators, pathogens, or pollutants lurk anywhere in the habitat. That population might well display its **biotic potential**—the maximum rate of increase per individual under ideal conditions.

Every species has a characteristic maximum rate of increase. For many bacteria, it is 100 percent every half hour or so. For humans and other large mammals, the estimated biotic potential is about 2 to 5 percent per year. The *actual* rate depends on the age at onset of reproduction, how frequently the individuals are reproducing, and how many offspring are produced over a lifetime. Athough the human population is not now displaying its full biotic potential, it is growing exponentially, for reasons we'll discuss shortly.

During a specified interval, population size is generally an outcome of births, deaths, immigration, and emigration.

With exponential growth, the growth rate increases over time as the reproductive base becomes ever larger.

As long as the per capita birth rate remains above the per capita death rate, a population will grow exponentially.

39.4 Limits on the Growth of Populations

Complex interactions occur within and between populations in nature, where it's not easy to identify all the factors working to limit population growth.

WHAT ARE THE LIMITING FACTORS?

Most of the time, environmental circumstances keep a population from fulfilling its biotic potential. That's why sea stars—the females of which could produce 2,500,000 eggs each year—do not overflow the oceans.

To get a sense of what some of the constraints may be, start again with a bacterial cell in a culture flask, where you can control the variables. First you enrich the culture medium with glucose and other nutrients necessary for bacterial growth. Then you sit back and let bacterial cells reproduce for many generations.

At first, growth seems to be exponential. Then it slows, and population size remains relatively stable. After the stable period, the size plummets until all the bacterial cells are dead. *What happened?* As the cell population grew ever larger, it used up more and more of the nutrients. Nutrient scarcity became an environmental signal for cells to stop dividing. But even with growth stopped, the population eventually used up all the nutrients and died out.

Any essential resource that is in short supply is a **limiting factor** on population growth. Food, mineral ions, refuge from predators, living space, and even a pollution-free habitat are examples (Figure 39.7). The number of limiting factors can be extensive, and their effects can vary. Even so, one factor alone often puts the brakes on population growth.

Suppose you kept freshening the nutrient supply. After growing exponentially, the population collapsed anyway. Like all other organisms, bacteria produce metabolic wastes. That huge population produced so much, it drastically compromised living conditions in the culture. Collectively its individuals polluted the experimentally designed habitat and put a stop to any further exponential growth.

CARRYING CAPACITY AND LOGISTIC GROWTH

Visualize a small population, with individuals spread through the habitat. As its size increases, more and more individuals must compete for nutrients, living quarters, and other resources. The share available to each diminishes, fewer offspring are born, and more die from starvation or nutrient deficiencies. Now the population's growth rate declines until the births are balanced or even outnumbered by deaths.

Ultimately, the *sustainable* supply of resources will determine the population's size. **Carrying capacity** is the maximum number of individuals of a population that a given environment can sustain indefinitely.

The pattern of **logistic growth** shows how carrying capacity can affect population size. By this pattern, a small population starts growing slowly in size, then it grows rapidly, and finally its size levels off once the carrying capacity is reached. Figure 39.8 shows how this pattern plots out as an S-shaped curve. Here is a way to represent the pattern as an equation:

| population growth per unit time | = | maximum net population growth rate per individual per unit time | × | number of individuals | × | proportion of resources not yet used |

An S-shaped curve is an approximation of what actually goes on in nature. For instance, a population that grows too fast usually overshoots the carrying capacity. The death rate skyrockets and the birth rate plummets. These two outcomes drive numbers down to the carrying capacity or lower (Figure 39.9).

The logistic growth equation just described deals with **density-dependent controls**. Such controls are any factors that operate when increases in population density lower the survival odds for individuals.

When a small, rapidly growing population reaches its carrying capacity, dwindling resources work as a

Figure 39.7 Response to a scarcity of nesting sites, a limiting factor for weaver bird populations in Africa.

(**a**) African weavers construct densely woven, cup-shaped nests that are only wide and deep enough for a hen and her nestlings. Many nests hang from the same spindly limbs. (**b**) This nest in Namibia is like an apartment house in a place where few trees are available. Between 100 and 300 pairs of sparrow weavers occupy their own flask-shaped nests, each with its own tubular entrance.

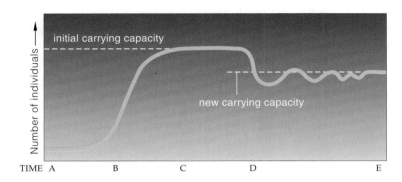

Read Me First!
and watch the narrated animation on logistic population growth

Figure 39.8 The idealized S-shaped curve that is characteristic of logistic growth. Growth slows after a phase of rapid increase (time A to C). The curve flattens as the carrying capacity is reached (time C to D). S-shaped growth curves can show variations, as when changes in the environment lower the carrying capacity (time D to E).

This happened to the human population when bubonic plague swept through Europe's crowded cities in the fourteenth century. One epidemic claimed 25 million lives.

control to stabilize or decrease its numbers. Besides environmental factors, species interactions also may drive the number of individuals below the maximum sustainable level.

Overcrowding invites interactions that increase the likelihood of individual deaths. Predators, parasites, and pathogens interact more intensely when prey or host populations are dense. They usually reduce the numbers of prey or hosts. By thinning a population, they remove conditions that invited their controlling effect, and population size may increase again.

DENSITY-INDEPENDENT FACTORS

A **density-independent factor** causes more deaths or fewer births regardless of a population's density. For instance, each year, millions of monarch butterflies fly down from Canada to Mexico's forested mountains, where they spend the winter. In 2002 a sudden freeze, exacerbated by deforestation, killed millions of the butterflies. The freeze had nothing to do with butterfly density; it was a density-independent event.

Resources in short supply are limiting factors that restrict population growth. Carrying capacity is the maximum number of individuals of a population that can be sustained indefinitely by the resources in a given environment.

With logistic growth, population growth is rapid at low density, then slows as the population approaches carrying capacity, where numbers will level off.

Both density-dependent controls and density-independent factors can bring about decreases in population size.

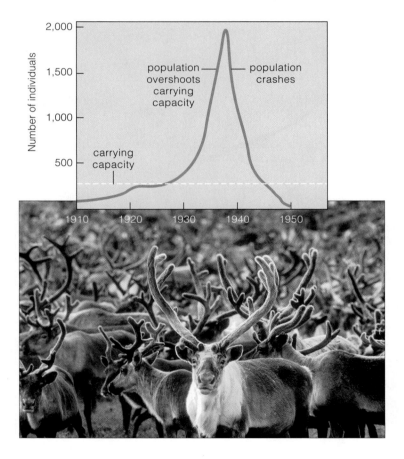

Figure 39.9 Carrying capacity and a reindeer herd. In 1910, four male and twenty-two female reindeer were introduced on St. Matthew Island in the Bering Sea. In less than thirty years, the size of the herd increased to 2,000. Individual reindeer had to compete for dwindling vegetation, and overgrazing destroyed most of it. In 1950, herd size plummeted to eight. The growth curve reflects how the reindeer population size overshot the carrying capacity for the island, then crashed.

39.5 Life History Patterns

Researchers have identified age-specific adaptations that affect the survival, fertility, and reproduction of individuals for many kinds of species.

So far, we have looked at populations as if all of their members are identical during any given interval. For most species, however, individuals of a population are at different stages of development. They interact in different ways with other organisms and with their environment. At different times in the life cycle, they may be adapted to different resources, as when larvae eat leaves and butterflies sip nectar. They may also be more or less vulnerable to danger at different stages.

In short, each species has a **life history pattern**, or a set of adaptations that influence survival, fertility, and age at first reproduction. Each pattern reflects the individual's schedule of reproduction. In this section and the next, we look at a few of the environmental variables that underlie age-specific patterns.

LIFE TABLES

Each species has a characteristic life span, but few of its individuals survive to the maximum age possible. Death looms larger at certain ages. Individuals tend to reproduce during an expected age interval, and in some species they emigrate at an expected time.

Age-specific patterns in populations intrigue life insurance and health insurance companies, as well as ecologists. Such investigators typically track a **cohort**, or a group of individuals, recorded from the time of birth until the last one dies (Table 39.1). They also record the number of offspring born to individuals in each age interval. Life tables list the data for an age-specific death schedule, which are typically converted to much cheerier "survivorship" schedules that show the number of individuals actually reaching specified ages. Table 39.2 is a typical example. It lists data for the 2001 human population of the United States.

Dividing a population into age classes and noting age-specific birth rates and mortality risks can yield useful information. Unlike a crude head count, for example, such data can shape policy decisions about pest management, protection of endangered species, or social planning for specified human populations. Birth and death schedules for the northern spotted owl were cited in federal court rulings that halted mechanized logging in the owl's habitat—old-growth forests of the Pacific Northwest.

PATTERNS OF SURVIVAL AND REPRODUCTION

Evolution proceeds by differences in survival and reproductive success. We measure the reproductive success of individuals in terms of the number of their surviving offspring. But that number varies among species, which differ in how much energy and time are allocated to production of gametes, locating and securing mates, and parenting, and in the size of the

Table 39.1	**Life Table for a Cohort of Annual Plants (*Phlox drummondii*)***			
Age Interval (days)	Survivorship (number surviving at start of interval)	Number Dying During Interval	Death Rate (number dying/ number surviving)	"Birth" Rate during interval (number of seeds from each plant)
0–63	996	328	0.329	0
63–124	668	373	0.558	0
124–184	295	105	0.356	0
184–215	190	14	0.074	0
215–264	176	4	0.023	0
264–278	172	5	0.029	0
278–292	167	8	0.048	0
292–306	159	5	0.031	0.33
306–320	154	7	0.045	3.13
320–334	147	42	0.286	5.42
334–348	105	83	0.790	9.26
348–362	22	22	1.000	4.31
362–	0	0	0	0
		996		

* Data from W. J. Leverich and D. A. Levin, 1979.

Table 39.2	**Life Table for the United States Human Population in 2001**			
Age Interval	Number at Start of Interval	Number Dying During Age Interval	Life Expectancy at Start of Interval	Reported Live Births
0–1	100,000	684	77.2	
1–5	99,316	132	76.7	
5–10	99,184	76	72.8	
10–15	99,108	96	67.9	7,315
15–20	99,012	330	62.9	525,493
20–25	98,682	468	58.1	1,022,106
25–30	98,214	471	53.4	1,060,391
30–35	97,743	554	48.6	951,219
35–40	97,189	801	43.9	453,927
40–45	96,388	1,154	39.2	95,788
45–50	95,234	1,682	34.7	5,244
50–55	93,552	2,373	30.3	263
55–60	91,179	3,474	26.0	
60–65	87,705	5,186	21.9	
65–70	82,519	7,397	18.1	
70–75	75,122	10,018	14.6	
75–80	65,104	13,284	11.5	
80–85	51,820	15,877	8.8	
85–90	35,943	16,147	6.5	
90–95	19,796	11,906	4.8	
95–100	7,890	5,845	3.6	
100+	2,045	2,045	2.7	

Figure 39.10 Three generalized survivorship curves. (**a**) Elephants are Type I populations. They have high survivorship until some age, then high mortality. (**b**) Snowy egrets are Type II populations. They have a fairly constant death rate. (**c**) Sea star larvae represent Type III populations, which show low survivorship early in life.

offspring. It seems that trade-offs have been made in response to selection pressures, such as the conditions prevailing in the habitat and species interactions.

A **survivorship curve** is a graph line that emerges when ecologists plot a cohort's age-specific survival in a habitat. Each species has a characteristic curve, and three types of curves are common in nature.

Type I curves reflect high survivorship until fairly late in life, then a large increase in deaths. Like many annual plants, the phlox in Table 39.1 show this type of pattern. So do large mammals that bear only one or a few large offspring at one time, then engage in extended parental care (Figure 39.10*a*). For example, a female elephant gives birth to four or five calves and devotes several years to parenting each.

Type I curves are typical of human populations in which individuals have access to good health-care services. However, today as in the past, infant deaths cause a sharp drop at the start of the curve in regions where health care is poor. After the drop, the curve levels off from childhood to early adulthood.

Type II curves reflect a fairly constant death rate at all ages. They are typical of organisms just as likely to be killed or die of disease at any age, such as lizards, small mammals, and large birds (Figure 39.10*b*).

Type III curves signify a death rate that is highest early on. We see this for species that produce many small offspring and do little, if any, parenting.

Figure 39.10*c* shows how the curve plummets for sea stars. Sea stars release mind-boggling numbers of eggs. The tiny larvae must eat fast, grow, and finish developing on their own without support, protection, or guidance from parents. Corals and other animals quickly eat almost all of them. Such a plummeting survivorship curve is common among many marine invertebrates, insects, fishes, plants, and fungi.

At one time, ecologists thought selection processes favored *either* early, rapid production of many small offspring *or* late production of a few large offspring. Researchers now comprehend that these patterns are extremes at opposite ends of a range of possible life histories. Also, both life history patterns—as well as intermediate ones—sometimes occur among different populations at different times.

Tracking a cohort (a group of individuals) from birth until the last one dies reveals patterns of reproduction, death, and migration that typify the populations of a species.

Survivorship curves can reveal differences in age-specific survival among species. In some cases, such differences exist even between populations of the same species.

39.6 Natural Selection and Life Histories

Earlier you read that jaws evolved among certain fishes during the Cambrian. This key innovation led to adaptive radiations among predators and diverse defenses among prey. No one witnessed that coevolutionary arms race. But experimental studies show that predators are still acting as selective agents and prey are still evolving.

Several years ago, two evolutionary biologists, with fishnets in hand and drenched in sweat, were doing fieldwork in the mountains of Trinidad, an island in the southern Caribbean Sea. They wanted to capture small fishes that live in shallow freshwater streams of this habitat (Figure 39.11a). The fish were guppies (*Poecilia reticulata*). David Reznick and John Endler were starting their eleven-year study of the variables that affect guppy life history patterns.

Male guppies are generally smaller than females of the same age and have bright-colored scales. The colors function as visual signals for mating during the guppy's complex courtship rituals. Females are drab colored and, unlike the males, they continue to grow even after they reach sexual maturity.

Reznick and Endler were interested in the effects of predation on guppy evolution. They chose their study site because in the streams that run through the mountains of Trinidad, different populations of guppies deal with different predators. Even within a stream, waterfalls can keep predators from moving from one area to another.

Two major predators of guppies are killifishes (*Rivulus hartii*) and pike-cichlids (*Crenicichla*), shown in Figure 39.11b,c. A killifish is not a very big fish. It preys efficiently on small, immature guppies but not on the larger adults. Pike-cichlids live in other streams. They prey on larger and sexually mature guppies, and tend to ignore small guppies.

Reznick and Endler hypothesized that predation is a selective agent that acted to shape guppy life history patterns. They knew that in pike-cichlid streams, guppies grow faster and are smaller at maturity, compared to guppies in killifish streams (Figure 39.12). Also, guppies hunted by pike-cichlids reproduce earlier in life, reproduce more often, and have more young per brood (Figure 39.13).

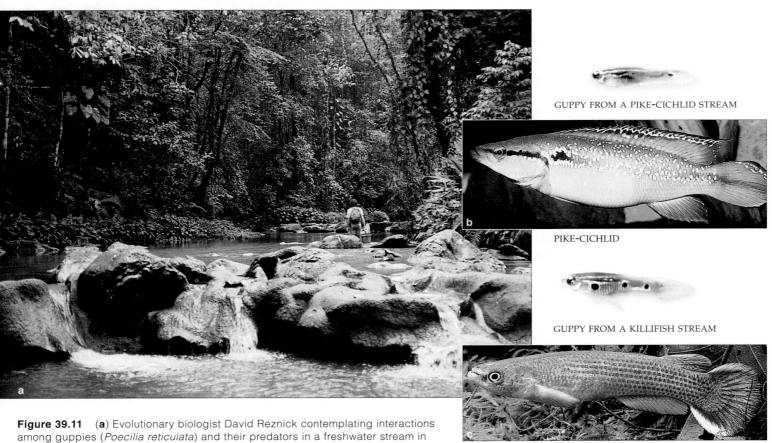

GUPPY FROM A PIKE-CICHLID STREAM

PIKE-CICHLID

GUPPY FROM A KILLIFISH STREAM

KILLIFISH

Figure 39.11 (**a**) Evolutionary biologist David Reznick contemplating interactions among guppies (*Poecilia reticulata*) and their predators in a freshwater stream in Trinidad. (**b**,**c**) Two guppies, and two of the guppy eaters.

Figure 39.12 Some representative guppies that inhabit streams with killifishes (**a**) and with pike-cichlids (**b**).

Figure 39.13 Graphs of some of the experimental evidence of natural selection among guppy populations subjected to different predation pressures. Compared to guppies raised with killifish (*red* bars), guppies raised with pike-cichlids (*green* bars) differed in body size and length of time between broods. Killifish are small and prey on smaller guppies, pike-cichlids are large fish and prey on larger guppies, and so the two predators select for the differences.

Were these differences genetic or did some other variables influence life history patterns in killifish and pike-cichlid streams? To find out, the researchers shipped live guppies from each stream back to their laboratory in the United States. They let the guppies reproduce in separate, predator-free aquariums for two generations. All other physical and chemical conditions in the artificial habitats were identical for the different experimental groups.

As it turned out, the offspring of the experimental guppy populations displayed the same differences as the natural populations. The conclusion? Differences between guppies preyed upon by different predators have a genetic foundation.

So what would happen if the selective pressure on a guppy population changed? Reznick and Endler answered the question with a set of field experiments. In one set, they introduced guppies upstream from a small waterfall. Before the experiment, the waterfall had been a barrier to dispersal. It prevented guppies

and large pike-cichlids from emigrating upstream, where killifish were present. Guppies introduced to the upstream experimental site were taken from a population that had evolved downstream from the waterfall, with the larger pike-cichlids.

Eleven years (thirty to sixty guppy generations) later, researchers revisited the stream. They found the experimental population had evolved. Guppies now had traits that resembled those of guppies that had been living with killifishes for a longer time. The change in predominant predator had influenced body size, frequency of reproduction, and other aspects of guppy life history patterns. Laboratory experiments involving two generations of guppies confirmed that the differences had a genetic basis.

Reznick and Endler showed that life history traits, like other characteristics, can be inherited. They also demonstrated that these traits can evolve. Traits that affect life history can be altered over a surprisingly short time in response to the right selection pressure.

39.7 Human Population Growth

Human population size surpassed 6 billion in 1999. In the five years since then, 400 million more individuals were added to the total.

In 2003, the average rate of increase for the human population was 1.3 percent. Now think about this: As long as the birth rate continues to exceed the death rate, annual additions will result in a *larger* absolute increase each year into the foreseeable future.

Our staggering population growth continues even though more than a billion are malnourished, without clean drinking water or adequate shelter. It continues even though more than a billion of us still do not have access to health-care delivery systems or sewage treatment facilities. It continues primarily in already overcrowded regions, on 10 percent of the land. Figure 39.14 is a graphic clue to what our expansion means with respect to the carrying capacity.

Even if it were possible to double food supplies to keep pace with growth, living conditions would still be marginal for most people. At least 10 million of us would continue to die each year from starvation.

For a time, it will be like the Red Queen's garden in Lewis Carroll's *Through the Looking Glass*, where one is forced to run as fast as one can to stay in the same place. What happens when our population doubles again? Can you brush the doubling aside as being too far in the future? *It is no further removed from you than the sons and daughters of the next generation.*

How did we get into this predicament? For most of its history, the human population grew slowly. Things started to pick up about 10,000 years ago, but in the past two centuries, growth rates skyrocketed. Three trends contributed to the expansion. First, humans steadily developed the capacity to expand into new habitats and climate zones. Second, humans increased the carrying capacity of their existing habitats. Finally, human populations have sidestepped limiting factors that restrain the growth of other species.

Reflect on the first point. Early humans evolved in woodlands, then in savannas. They were vegetarians, mostly, but also scavenged for bits of meat. Bands of hunter–gatherers moved out of Africa about 2 million years ago. By 40,000 years ago, the descendants were established in much of the world (Section 23.12).

Few species can expand into such a broad range of habitats. Having a truly complex brain, early humans drew on learning and memory to figure out how to build fires, make shelters, make clothing, make tools, and cooperate in hunts. With the advent of language, knowledge did not die with the individual. It spread quickly among groups. *The human population expanded into diverse environments far more rapidly than the long-term geographic dispersals of other species.*

Reflect on the second point. Starting about 11,000 years ago or so, many hunter–gatherer bands shifted to agriculture. Instead of simply following migratory game herds, they settled in fertile valleys and other regions that favored seasonal harvesting of fruits and grains. In this way, they developed a more dependable basis for life. A pivotal factor was the domestication of wild grasses, including species ancestral to modern wheat and rice. People harvested, stored, and planted seeds in one place. They domesticated animals for food and pulling plows. They dug irrigation ditches and diverted water to croplands.

Agricultural productivity was a basis for increases in population growth rates. Towns and cities formed. Later in time, food supplies increased again, and yet again, by the use of chemical fertilizers, herbicides, and pesticides. Transportation improved, as did food distribution. *Thus, even at its simplest, management of food supplies through agriculture increased the carrying capacity for the human population.*

What about sidestepping of limiting factors? Until about 300 years ago, poor hygiene, malnutrition, and infectious diseases kept death rates high enough to more or less balance birth rates. Infectious diseases became density-dependent controls. Epidemics swept through overcrowded settlements and cities that were infested with fleas and rodents. Then came plumbing and new methods of sewage treatment. Over time,

Figure 39.14 Human population growth in a small part of the world since 1900. Red signifies dense populations in and around the San Francisco Bay area and Sacramento, based on historical data and satellite imaging.

1900

1940

1954

1962

1990

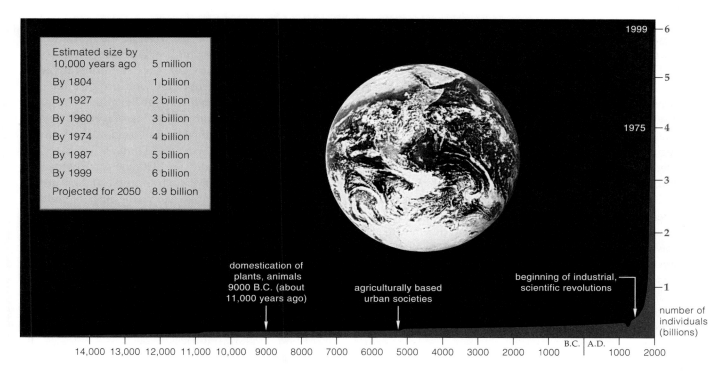

Estimated size by 10,000 years ago	5 million
By 1804	1 billion
By 1927	2 billion
By 1960	3 billion
By 1974	4 billion
By 1987	5 billion
By 1999	6 billion
Projected for 2050	8.9 billion

domestication of plants, animals 9000 B.C. (about 11,000 years ago)

agriculturally based urban societies

beginning of industrial, scientific revolutions

number of individuals (billions)

14,000 13,000 12,000 11,000 10,000 9000 8000 7000 6000 5000 4000 3000 2000 1000   B.C. | A.D.   1000 2000

Figure 39.15 Growth curve (*red*) for the world human population. (The dip between the years 1347 and 1351 is when 60 million people died during a bubonic plague.) The *blue* box lists how long it took for the human population to increase from 5 million to 6 *billion.*

vaccines, antibiotics, and other drugs were developed as weapons against many pathogens. The death rates dropped sharply. Births began to exceed deaths—and rapid population growth was under way.

In the industrial revolution of the mid-eighteenth century, people discovered how to harness the energy stored in fossil fuels, starting with coal. Within a few decades, large industrialized societies began to form in western Europe and in North America. Even more efficient technologies were devised with the start of World War I. Factories began mass-producing cars, tractors, and many other affordable goods. Advances in agriculture meant that fewer farmers were required to support a larger population.

And so, by controlling disease agents and tapping into concentrated, existing forms of energy—fossil fuels—the human population has managed to sidestep major factors that had previously limited its rate of increase.

Where have the far-flung dispersals and stunning advances in agriculture, industrialization, and health care taken us? Starting with *Homo habilis*, it took about 2.5 million years for human population size to reach 1 billion. As Figure 39.15 shows, it took just 123 years to reach 2 billion, another 33 to reach 3 billion, 14 more to reach 4 billion, and 13 more to get to 5 billion. It

took only 12 more years to arrive at 6 billion! Given the principles governing population growth, we may expect the rate of increase to decline as birth rates fall or as death rates rise. Alternatively, we may expect continued increases if technological breakthroughs can expand the carrying capacity. *Even so, continued growth cannot be sustained indefinitely.*

Why? Continuing increases in population size can set the stage for density-dependent controls. Already, pathogens, such as the one that causes SARS, can spread between the continents in a matter of weeks. The problem is compounded by human emigration on a vast scale. Economic hardship and civil strife have put an estimated 50 million people on the move within and between countries. Will relocations of so many individuals be peaceable? Will they find food, clean water, and other basic resources where they end up?

Through expansion into new habitats, cultural interventions, and technological innovations, the human population has temporarily skirted environmental resistance to growth.

As population increases, density-dependent controls, such as disease and competition for resources, may slow growth.

39.8 Fertility Rates and Age Structure

Acknowledgment of the risks posed by rising populations has led to increased family planning in almost every region. Putting the brakes on population growth is not easy and numbers are expected to continue to rise.

Most governments recognize that population growth, resource depletion, pollution, and the quality of life are interconnected. Most are working to lower long-term birth rates, as with family planning programs. Details vary among countries, but most are offering information on available methods of fertility control.

These attempts are having impact; birth rates have been slowing worldwide. Death rates are declining, mainly because improved nutrition and health care are lowering infant mortality rates (the number of infants per 1,000 who die in the first year). However, AIDS has sent death rates soaring in some African countries.

We still expect the world population to reach 8.9 billion by 2050. Think about the resources that will be required. We will have to boost food production and find more sources of energy and fresh water to meet even basic needs, something that still eludes almost half of the population. And large-scale manipulations of resources will intensify pollution.

India, China, Pakistan, Nigeria, Bangladesh, and then Indonesia are expected to show the most growth, in that order. China (with 1.3 billion people) and India (with 1.1 billion) dwarf all other countries; together, they make up 38 percent of the world population. The United States is next in line, with 295 million.

The **total fertility rate** (TFR) is the average number of children born to the women of a population during their reproductive years. TFR estimates are based on current age-specific rates. In 1950, the worldwide TFR averaged 6.5. By 2003, it had declined to 2.8 but was still above replacement levels, the number of children a couple must bear to replace themselves. At present, the replacement level is 2.1 for developed countries and as high as 2.5 in some developing countries. (It is higher in developing countries because more female children die before reaching reproductive age.)

These numbers are averages. TFRs are at or below replacement levels in many developed countries; the developing countries in western Asia and Africa have the highest. Figure 39.16 has some examples of the disparities in demographic indicators.

Comparing the age structure diagrams for different populations is revealing. In Figure 39.17, notice the reproductive age category for the next fifteen years. The average range for childbearing years is 15–49. The broader the base of the diagram, the faster the population can be expected to increase in size. The United States population has a relatively narrow base and is undergoing slow growth.

Even if every couple decides to bear no more than two children, world population growth will not slow for sixty years, because 1.9 billion are about to enter the reproductive age bracket. *More than one-third of the world population is in the broad pre-reproductive base.*

China has the most far-reaching family planning program. Its government discourages premarital sex; it urges people to delay marriage and limit families to one or two children. It offers abortions, contraceptives, and sterilization at no cost to married couples. Even in remote rural areas, paramedics and mobile units offer access to these measures. Couples who follow these guidelines receive more food, free medical care, better housing, and salary bonuses. Their offspring receive free tuition and preferential treatment when they are old enough to enter the job market. Parents who have more than two children lose government benefits and pay more taxes.

Although the policy may sound harsh, it works. Since 1972, China's TFR has fallen sharply, from 5.7 to 1.8. An unintended consequence has been a shift in the country's sex ratio. Traditional cultural preference

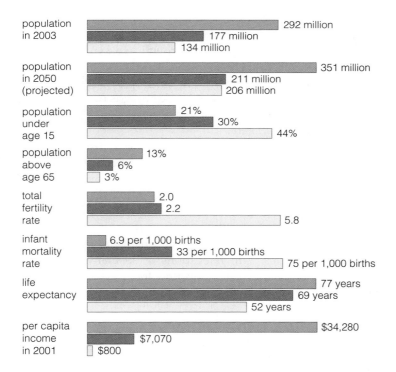

Figure 39.16 Key demographic indicators for three countries, mainly in 2003. The United States (*brown* bar) is highly developed, Brazil (*red* bar) is moderately developed, and Nigeria (*gold* bar) is less developed.

Read Me First!
and watch the narrated animation on age structure diagrams

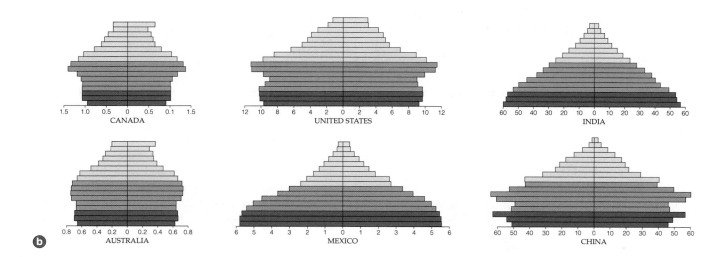

Figure 39.17 (a) General age structure diagrams for countries with rapid, slow, zero, and negative rates of population growth. The pre-reproductive years are *green* bars; reproductive years, *purple;* and the post-reproductive years, *light blue*. A vertical axis divides each graph into males (*left*) and females (*right*). Bar widths correspond to proportions of individuals in each age group. (b) Age structure diagrams for a few representative countries in 1997. Population sizes are measured in millions.

for sons, especially in the rural areas, has led some parents to abort developing females or even commit infanticide. Worldwide, 1.06 boys are born for every girl, but in China the latest census reports 1.19 boys per girl. Also, more than a hundred thousand girls are abandoned each year. In response, the government is now offering additional cash and tax incentives to parents of girls. Meanwhile, China's population time bomb is still ticking. About 150 million of its females are in the pre-reproductive age category.

The worldwide total fertility rate has been dropping, but is still above the replacement level required to arrive at zero population growth.

Most countries support family planning programs of some sort. Even with slowdowns, the human population will continue to increase, because of its pre-reproductive base.

At present, more than one-third of the human population is in a very broad pre-reproductive base.

39.9 Population Growth and Economic Effects

The most highly developed countries have the slowest growth rates and use the most resources.

DEMOGRAPHIC TRANSITIONS

Changes in population growth rates often correlate with four stages of economic development, the heart of the **demographic transition model**. By this model, living conditions are harshest in a *preindustrial* stage, before technology and medical advances spread. Birth and death rates are high, so the growth rate is low. In the *transitional* stage, industrialization begins. Food production and health care improve and death rates slow. Birth rates stay high in the agricultural societies, where big families provide help in the fields. Annual growth rates are between 2.5 and 3 percent. As living conditions improve and birth rates begin to decline, growth generally starts to level off (Figure 39.18).

In the *industrial* stage, industrialization is in full swing and growth slows. People move to cities, and couples often want small families. As they accumulate goods, many decide the time and cost of raising more than a few children conflict with their goals.

In the *postindustrial* stage, population growth rates become negative. The birth rate falls below the death rate, and population size slowly decreases.

The United States, Canada, Australia, and most of western Europe, Japan, and much of the former Soviet Union are in the industrial stage. Most developing countries, such as Mexico, are now in the transitional stage, without enough skilled workers to complete the transition to a fully industrial economy.

If population growth continues to exceed economic growth, then we can expect death rates to rise. Many countries may get stuck in the transitional stage or fall back to the conditions of the preceding stage.

A QUESTION OF IMMIGRATION POLICIES

Figure 39.19 charts the legal immigration to the United States between 1820 and 1997. The greatest increase came after the Immigration Reform and Control Act of 1986 gave legal status to undocumented immigrants who proved they had lived in the country for years. Economic downturns in the 1980s and 1990s fanned resentment against newcomers. Many want to limit legal immigration to 300,000–450,000 per year and to deport undocumented immigrants. Others want the borders open. They also say that such a policy would discriminate against legal immigrants of the same ethnic background. This is a volatile issue, one with social and economic ramifications.

A QUESTION OF RESOURCE CONSUMPTION

Industrialized nations use the most resources. For example, the United States with about 4.6 percent of the world population produces about 21 percent of all goods and services. Yet it requires thirty-five times more goods and services than India. It uses 25 percent or so of the world's processed minerals and much of the energy supplies. It is not alone in this. China and India are now demanding an ever increasing share of the economic pie.

Figure 39.18 The demographic transition model for changes in population growth rates and in sizes, correlated with long-term changes in the economy.

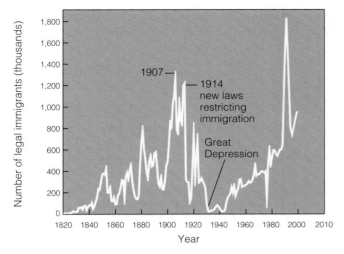

Figure 39.19 Chart of legal immigration to the United States between 1820 and 1997.

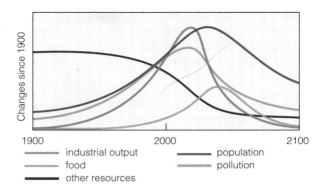

Figure 39.20 Computer-based projection of what may happen if human population size continues to skyrocket without dramatic policy changes and technological innovation. The assumptions were that the population has already overshot the carrying capacity and current trends will continue unchanged.

G. Tyler Miller once estimated that it would take 12.9 billion impoverished individuals living in India to have as much impact on the environment as 284 million people in the United States. He also pointed out that the projected increase in human population growth rates raises serious questions. Will there be enough food, energy, water, and other resources to sustain so many people? Will governments be able to provide adequate education, housing, medical care, and other social services for them? Computer models suggest not (Figure 39.20). Yet some people claim we can adapt politically and socially to a more crowded world if innovative technologies improve harvests, food resources are distributed more equitably, and diets are shifted away from animal products.

There are no easy answers. If you have not been doing so, start following the arguments in the media. It is a good idea to become an informed participant in a debate that will have impact on your future.

IMPACTS OF NO GROWTH

In a growing population, most people are in younger age brackets. When living conditions favor moderate population growth over time, then age distribution should guarantee a future workforce. However, this has social implications. Why? *It takes a large workforce to support individuals in the higher age brackets.*

In the United States, most of the retired seniors expect the government to subsidize their medical care and provide low-cost housing and many other social programs. However, as a result of improved medicine and hygiene, people in these brackets are living far

longer than seniors did when the nationwide social security program was established. Cash benefits now exceed the contributions that individuals made to the program when *they* were younger.

If the population ever does reach and maintain zero growth, a larger proportion of individuals will end up in higher age brackets. Even slower growth poses problems. Will these people continue to receive goods and services as the workforce carries more and more of the economic burden? Put it to yourself. How much economic hardship are you willing to bear for the sake of your parents? Your grandparents? How much will your children bear for you?

We have arrived at a turning point, not only in our biological evolution but in our cultural evolution as well. The decisions awaiting us are among the most pressing and difficult we will ever have to make.

All species face limits to growth. We might think we are different from the rest, for our special ability to undergo rapid cultural evolution has enabled us to postpone the action of most factors that limit growth. The crucial word is *postpone*. No amount of cultural intervention can sidestep the ultimate check imposed by limited resources and a damaged environment.

Differences in population growth among countries correlate with levels of economic development. Growth rates are low in preindustrial, industrial, and post-industrial stages. They are greatest during the transition to industrialization.

Per capita resource consumption is greatest in the most developed countries, which also must sustain an increasing proportion of people in older age brackets.

Summary

Section 39.1   A population is a group of individuals of the same species occupying a given area. Typically we measure its size, density, distribution, and age structure. Most populations in nature show a clumped distribution pattern.

Section 39.2   Counting the number of individuals in representative quadrats and capture–recapture methods are used to estimate population density.

Section 39.3   The growth rate for a population during a specified interval is determined by the rates of birth, death, immigration, and emigration. If we put aside effects of immigration and emigration, we may represent population growth (G) as $G = rN$, where N is the number of individuals in a specified interval and r is the net reproduction per individual per unit time.

In cases of exponential growth, the reproductive base of a population increases and its size expands by a certain percentage during successive intervals. This trend plots out as a J-shaped growth curve. As long as a population's per capita birth rate is above its per capita death rate, it will show exponential growth.

Section 39.4   Any essential resource that is in short supply is a limiting factor on population growth. The logistic growth equation describes how population growth is affected by density-dependent controls, such as competition for resources or disease. The population slowly increases in size, goes through a rapid growth phase, then levels off in size once carrying capacity is reached. Carrying capacity is the maximum number of individuals that can be sustained indefinitely by the resources available in their environment.

Density-independent factors affect populations regardless of how crowded or uncrowded they are.

Section 39.5   Each species has a life history pattern characterized by traits such as age at first reproduction, number of offspring per brood, and life span. There are three general types of survivorship curves.

Section 39.6   Life history traits can have a genetic basis and vary among populations because of differing selective pressures from predation or other factors.

Section 39.7   The human population has surpassed 6.4 billion. Its rapid growth in the past two centuries occurred through expansion into many diverse habitats and through agricultural, medical, and technological developments that raised carrying capacity.

Section 39.8   Family planning is helping to slow population growth. The worldwide total fertility rate is declining. However, the broad pre-reproductive base of the world population means that human population size will continue to increase for at least sixty years.

Section 39.9   The demographic transition model correlates industrial and economic development with changes in population growth rates. Growth rates are lower in developed nations than in developing nations. But, the per capita resource consumption of developed nations is many times higher.

Self-Quiz
Answers in Appendix III

1. The rate at which a population grows or declines depends on the rate of _____ .
   - a. births
   - c. immigration
   - e. a and b
   - b. deaths
   - d. emigration
   - f. all of the above

2. Populations grow exponentially when _____ .
   - a. population size expands by ever increasing increments through successive time intervals
   - b. size of low-density population increases slowly, then quickly, then levels off once carrying capacity is reached
   - c. a and b are characteristics of exponential growth

3. For a given species, the maximum rate of increase per individual under ideal conditions is its _____ .
   - a. biotic potential
   - c. environmental resistance
   - b. carrying capacity
   - d. density control

4. Resource competition, disease, and predation are _____ controls on population growth rates.
   - a. density-independent
   - c. age-specific
   - b. population-sustaining
   - d. density-dependent

5. A life history pattern for a population is a set of adaptations that influence the individual's _____ .
   - a. survival
   - c. age when it starts reproducing
   - b. fertility
   - d. all of the above

6. In 2003, the worldwide average rate of increase for the human population at midyear was _____ percent.
   - a. 0
   - c. 1.3
   - e. 3.8
   - b. 0.5
   - d. 2.7
   - f. 4.6

7. Match each term with its most suitable description.
   - ____ carrying capacity
   - ____ exponential growth
   - ____ biotic potential
   - ____ limiting factor
   - ____ logistic growth
   - a. maximum rate of increase per individual under ideal conditions
   - b. population growth plots out as an S-shaped curve
   - c. maximum number of individuals sustainable by the resources in a given environment
   - d. population growth plots out as a J-shaped curve
   - e. essential resource that restricts population growth when scarce

Critical Thinking

1. If house cats that haven't been neutered or spayed live up to their biotic potential, two can be the start of many kittens—12 the first year, 72 the second year, 429 the third, 2,574 the fourth, 15,416 the fifth, 92,332 the sixth, 553,019 the seventh, 3,312,280 the eighth, and 19,838,741 kittens the ninth year. Is this a case of logistic growth? Exponential growth? Irresponsible cat owners?

Figure 39.21 Saguaros (*Canegiea gigantea*) growing very slowly in Arizona's Sonoran desert.

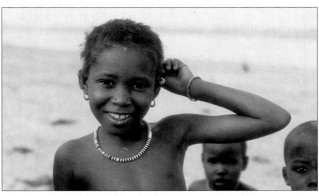

Figure 39.22 A young Malian, with a 10 percent chance of becoming a mother by age fifteen, and a 50 percent chance by age nineteen. Mali's FTR, about 7, is one of the world's highest.

2. As described in Section 39.6, when guppies from pools with pike-cichlids were moved to other pools dominated by smaller killifish, the guppies evolved. Over time, they came to resemble populations that had historically been faced with cichlid predation. Age at first reproduction increased, as did body size. There was also a change in the males' coloration. Compared to their ancestors, the transplanted males were far gaudier. They had bigger and more colorful spots. What could have been the selection pressure driving the change in male coloration? Why might a change in predator pressure affect guppy coloration?

3. Each summer, the giant saguaro cactus produces tens of thousands of tiny black seeds. Most die, but a few land in a sheltered spot and sprout the following spring. The saguaro is a CAM plant (Section 6.6) and growth is slow. After 15 years, the saguaro may be only knee high, and it won't flower until about age 30. The cactus may survive to age 200. Saguaros share their habitat with annuals, such as poppies, that sprout, produce seeds, and die in just a few weeks (Figure 39.21). Explain how these very different life histories could both be adaptive in a desert environment.

4. A third of the world population is younger than fifteen (Figure 39.22). Describe the effect of this age distribution on our population's future growth rate. If you think it will have severe impact, what humane recommendations would you make to encourage individuals of this age group to limit their family size? What are some social, economic, and environmental factors that might prevent them from following the recommendations?

5. Write a short essay about a population having one of the age structures shown below. Speculate on the population's current economic status and the social and economic problems its population may face in the future.

IMPACTS, ISSUES   *Fire Ants in the Pants*

Two species of Argentine ants, *Solenopsis richteri* and *S. invicta*, entered the United States during the 1930s, probably as stowaways on cargo ships. They infiltrated communities throughout the Northern Hemisphere, starting with the southeastern states. *S. invicta*, the red imported fire ant, recently colonized Southern California and New Mexico.

Disturb a fire ant on your skin and it will bite down even as a stinger on its abdomen pumps venom into you. Searing pain precedes the formation of a pus-filled bump on the skin (Figure 40.1). At one time or another, about one-half of the Americans who live where fire ants are common have been stung. More than eighty have died from the attacks.

Imported fire ants menace more than people. These insects will attack just about anything that disturbs them, including livestock, pets, and wildlife. They also are more competitive than native ant species and other animals that feed on insects. They may be contributing to the declines of some native wildlife species.

For example, the Texas horned lizard (*Phrynosoma cornutum*) disappeared from most of its home range when the red fire ant moved in. Native ants have been its food of choice. When the native ant populations dwindled, the lizard ran out of luck, because it cannot eat the invaders. As other examples, we have reports of fire ants directly killing young lizards and hatchlings of quail and other birds that nest on the ground.

Figure 40.1   *From left to right*, fire ant mounds in west Texas, fire ants swarming all over a leather boot, and the type of skin eruptions that follow a concerted attack by these exotic imports.

the big picture

Community Characteristics   Species interactions help to organize the community of any given habitat. In that community, each species has a unique niche, which may shift in small and large ways over time.

Species Interactions   Diverse species interact indirectly or as mutualists, as predators and prey, or as parasites and hosts. Through their interactions, they exert selection pressures on one another, and in many cases they coevolve.

Invicta means invincible in Latin. So far, *S. invicta* is living up to its species name. Pesticides have not slowed its expansion into new habitats and may actually be helping it by knocking off the native populations.

Ecologists are enlisting biological controls. Two phorid fly species attack *S. invicta* in its native habitat. Both species are parasitoids, a rather gruesome type of parasitic predator. The female fly pierces an adult ant's cuticle and lays an egg in its soft tissues. When the fly larva hatches, it eats the tissues as it burrows along toward the head. After growing large enough, the larva secretes an enzyme that makes the ant head fall off. Sheltered inside the detached, cuticle-covered head, the larva undergoes metamorphosis into an adult. The flies are choosy about their host/prey; they have not targeted ants native to the United States.

In 1997, ecologists released one parasitic fly species in Florida. In 2001, they released a second species in other southern states. It is too soon to know whether the biological controls are working. As researchers wait for results, they are exploring other options. One idea is to use imported fungal or protistan pathogens that will infect *S. invicta* but not native ants. Another is to introduce a parasitic ant species that invades *S. invicta* colonies and decapitates the queens.

In any given habitat, species interact as part of a dynamic community. As you will see in this chapter, their interactions shift in small and large ways—some predictable, and others entirely unexpected.

 How Would You Vote?

Currently, only a fraction of the crates being imported into the United States are inspected for exotic species. Would the cost of added inspections be worth it? See the Media Menu for details, then vote online.

Stability and Change   Stable arrays of species develop over time and dominate a habitat as a climax community. Introduction of exotic species and other forces often tip the balance and destabilize communities.

Focus on Biodiversity   Biologists have identified environmental and historical factors that shape global patterns of biodiversity. The current extinction crisis, partly a result of human activities, is threatening global biodiversity. Many conservation efforts are under way.

40.1 Which Factors Shape Community Structure?

*The type of place where each organism normally lives is its **habitat**. Habitats have physical and chemical features, such as temperature, and an array of species. Directly or indirectly, populations of all species in each habitat associate with one another as a **community**.*

A community has a characteristic structure. It arises as an outcome of five factors. *First*, the climate and topography determine temperatures, rainfall, type of soil, and other conditions. *Second*, the kinds and quantities of food and other resources that become available through the year affect which species live there. *Third*, adaptive traits of each species help it to survive and exploit specific resources in the habitat. *Fourth*, species engage in competition, predation, and mutually beneficial actions (Figure 40.2). *Fifth*, the overall pattern of population sizes affects community structure. This pattern reflects the history of changes in population size, the arrivals and disappearances of species, and physical disturbances to the habitat.

Such factors affect the number of species, starting with producers and continuing through consumers. They influence population sizes and help determine the overall number of species. High solar radiation, temperatures, and humidity in tropical habitats favor the growth of many kinds of plants, which support many kinds of animals. Arctic habitats have conditions that do not favor great numbers of species.

THE NICHE

All species of a community share the same habitat, the same "address," but each has a "profession" that sets it apart from the rest. It has a distinct **niche**, the sum of its activities and interactions as it goes about acquiring and using the resources required to survive and reproduce. Its *fundamental* niche is the one that would prevail in the absence of competition and any other factors that could limit how individuals get and use resources. Constraining factors that do come into play tend to bring about a more limited, *realized* niche. The realized niche for a species is dynamic; it may shift over time, in small or large ways.

CATEGORIES OF SPECIES INTERACTIONS

Even in the simplest communities, dozens to hundreds of species typically interact in one of five ways. Some are indirect interactions. For instance, when Canadian lynx eat a lot of snowshoe hares, they indirectly help plants that the hares eat. Plants indirectly help the lynx by fattening hares and providing cover for ambushes.

Commensalism directly helps one species but does not affect the other much, if at all. When a bird uses a tree for a roosting site, the tree gets nothing but is not hurt. In **mutualism**, both species benefit, but not from cozy cooperation. This is a two-way exploitation. In **interspecific competition**, both species can be harmed. **Predation** and **parasitism** directly benefit one species and harm the other. Predators kill prey, and parasites weaken hosts. Commensalism, mutualism, predation, and parasitism are all forms of *symbiosis*, defined as a close association between two or more species during part or all of the life cycle.

> Community structure arises from a habitat's physical and chemical features, resource availability over time, adaptive traits of its species, how its species interact, and the history of the habitat and its occupants.
>
> A niche is the sum of all activities and relationships in which individuals of a species engage as they secure and use the resources necessary to survive and reproduce.
>
> Indirect interactions among species as well as symbiotic ones—commensalism, mutualism, competition, predation, and parasitism—help shape community structure.

Figure 40.2 Three of twelve fruit-eating pigeon species in New Guinea's tropical rain forests. *Left to right*, the tiny pied imperial pigeon, the superb crowned fruit pigeon, and the turkey-sized Victoria crowned pigeon. The forest's trees differ in the size of fruit and fruit-bearing branches. The big pigeons eat big fruit. Smaller ones, with smaller bills, can't peck open big, thick-skinned fruit. They eat small, soft fruit on branches too spindly to hold big pigeons.

The trees feed the birds, which help the trees. Seeds in fruit resist digestion in the pigeon gut. Flying pigeons dispense seed-containing droppings, often some distance from the parent trees. Seedlings don't compete well against mature, well-developed trees for water, mineral ions, and sunlight.

40.2 Mutualism

In a mutualistic interaction, two species take advantage of their partner in ways that benefit both, as when one filches nutrients from the other while sheltering it.

Interactions in which positive benefits flow both ways abound in nature. As rain forest trees provide pigeons with food, the pigeons help disperse the trees' seeds to new sites (Figure 40.2). Many plants and animals enter into such interactions. Most flowering plants are mutualists with insects, birds, bats, and other animals that pollinate them or disperse their seeds. Chapters 21 and 27 provide vivid examples.

Mutualism may be obligatory; each species must have access to the other in order to complete its life cycle and reproduce. This is the case for yucca plants and the yucca moths that pollinate them (Figure 40.3).

A lichen is a case of obligatory mutualism between certain fungi and a photosynthetic protist or bacteria. So is a mycorrhiza, in which fungal hyphae penetrate root cells or form a dense, velvety mat around them. As earlier chapters explain, the fungus is efficient at absorbing mineral ions from soil, and the plant grows better when it pilfers some of the ions. In turn, the fungus pilfers some of the sugar molecules the plant makes by way of photosynthesis. The fungus depends on the plant for reproductive success; it will not make spores if the plant stops photosynthesizing.

An anemone fish cannot live and reproduce away from sea anemones. These cnidarians have nematocyst-laden tentacles that shelter and protect the fish from many reef organisms. A lot of sea anemones survive and reproduce without help, but the species shown in Figure 40.4 gets help from *Amphiprion perideraion*—a tiny partner that chases away butterflyfishes. Like the partner, butterflyfishes are impervious to nematocysts, but they have a penchant for nipping off the tentacles.

And what about the apparent endosymbiotic origin of eukaryotes? Long ago, phagocytes were engulfing aerobic bacterial cells—but some resisted digestion, tapped host nutrients, and then kept on reproducing independently of the host cell body. In time the hosts came to depend on the ATP produced by the guests—which evolved into mitochondria and chloroplasts. If those ancient prokaryotic cells had not coevolved as such intimate mutualists, you and all other eukaryotic species would not be around today (Section 20.4).

In cases of mutualism, both of the participating species benefit from the two-way exploitation. Some mutualistic relationships are obligatory for one or both partners.

Figure 40.3 Mutualism on a rocky slope of Colorado's high desert.

Only one yucca moth species pollinates plants of each *Yucca* species; it cannot complete its life cycle with any other plant. The moth matures when yucca flowers blossom. The female has specialized mouthparts that collect and roll sticky pollen into a ball. She flies to another flower and pierces its ovary, where seeds will form and develop, and lays eggs inside. As she crawls out, she pushes a ball of pollen onto the flower's pollen-receiving platform.

After germination, pollen tubes grow through the ovary tissues and deliver sperm to the plant's eggs. Seeds develop after fertilization. Meanwhile, the moth eggs develop into larvae that eat a few seeds, then gnaw their way out of the ovary. Seeds that larvae do not eat give rise to new yucca plants.

Figure 40.4 Mutualism near the Solomon Islands. This species of sea anemone (*Heteractis magnifica*) interacts with about a dozen species of fishes, including the pink anemone fish (*Amphiprion perideraion*). This tiny, aggressive fish chases away predatory butterflyfishes that can bite off the sea anemone's tentacles. In return, the fish and its eggs get protection and shelter—scarce commodities on coral reefs.

40.3 Competitive Interactions

Where you find limited supplies of energy, nutrients, living space, and other natural resources, you can expect to find organisms competing for a share of them.

Chapters 39 and 43 address *intraspecific* competition between members of the same species, which can be fierce. *Interspecific* competition, between populations of different species, usually isn't as intense. Why not? *The requirements of two species may be similar but are never as close as they are among individuals of the same species.*

In cases of *interference* competition, populations of a species control or prevent access of another species to a resource, regardless of its abundance. Figure 40.5 is one example. Hummingbirds are another. In spring and early summer, a male broadtailed hummingbird evicts birds of his own species from his flower-rich territory in the Rocky Mountains. In August, rufous hummingbirds migrate through the Rockies on the way to wintering grounds in Mexico. Rufous males are more aggressive, stronger competitors; they force male broadtails to give up territory until they move on.

One other example: Certain plants, such as many sages (*Salvia*), exude aromatic compounds from their leaves. The compounds taint the soil around the plant and prevent potential competitors from taking root.

In cases of *exploitative* competition, populations of different species have equal access to a resource, but one is better at using it. Various water fleas (*Daphnia*) eat the same alga. In one experiment, a large-bodied species increased in body mass and numbers, but a smaller species fed the same alga in the same culture actually decreased in body mass and numbers. And this leads us into a theory about competition.

COMPETITIVE EXCLUSION

Any two species differ to a greater or lesser extent in their capacity to secure and use resources. The more they overlap in these respects, the less likely they are to coexist in the same habitat.

Years ago, G. Gause found evidence of this when he grew two species of *Paramecium* separately and then together (Figure 40.6). Both ciliated protozoans require the same food—bacteria—and compete intensely for it. Gause's species, which require identical resources, could not coexist indefinitely. Later experiments with water fleas and many other species yielded the same results, in support of what ecologists now call the theory of **competitive exclusion**.

Gause also studied two other *Paramecium* species that did not overlap much in requirements. When he grew them together, one species tended to feed upon bacteria suspended in culture tube liquid. The other ate yeast cells at the tube bottom. Population growth

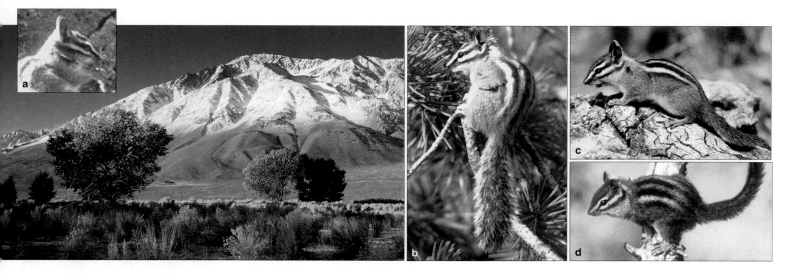

Figure 40.5 Example of interspecific competition in nature. On the slopes of the Sierra Nevada, competition helps keep nine species of chipmunks (*Tamias*) in different habitats.

The alpine chipmunk (**a**) lives in the alpine zone, the highest elevation. Below it are the lodgepole pine, piñon pine, and then sagebrush habitat zones. Lodgepole pine chipmunks (**b**), least chipmunks (**c**), and other species live in the forest zones. Merriam's chipmunk (**d**) lives at the base of the mountains, in sagebrush. Its traits would allow it to move up into the pines, but the aggressively competitive behavior of forest-dwelling chipmunks won't let it. Food preferences keep the pine forest chipmunks out of the sagebrush habitat.

a Time (days)

b Time (days)

c Time (days)

Paramecium caudatum

P. aurelia

Figure 40.6 Results of competitive exclusion between two protozoan species that compete for the same food. (**a**) *Paramecium caudatum* and (**b**) *P. aurelia* were grown in separate culture tubes and established stable populations. The S-shaped graph curves indicate logistic growth and stability.

(**c**) Then the two species were grown together. *P. aurelia* (*brown* curve) drove *P. caudatum* toward extinction (*green* curve in **c**). This experiment and others suggest that two species cannot coexist indefinitely in the same habitat *when they require identical resources*. If their requirements do not overlap much, one might influence the population growth rate of the other, but they may still coexist.

rates slowed for both species—but the overlap in use of resources was not enough for one species to fully exclude the other. The two continued to coexist.

RESOURCE PARTITIONING

Think back on those fruit-eating pigeon species. They all use the same resource: fruit. Yet they overlap only a bit in their use of it, because each prefers fruits of a certain size. They are a case of **resource partitioning** —a *subdividing* of some category of similar resources that lets competing species coexist.

Similarly, three annual plant species live in the same plowed, abandoned field. All require sunlight, water, and minerals. Each exploits a slightly different part of the habitat (Figure 40.7). Bristly foxtail grasses have a shallow, fibrous root system that absorbs water fast during rains. They grow where moisture shifts daily, and are drought-tolerant. Indian mallow has a taproot system in deeper soil that is moist early in spring and drier later. The taproot system of smartweed branches in topsoil and soil below the roots of other species. It grows where soil is always moist.

In some competitive interactions, one species controls or blocks access to a resource, regardless of whether it is scarce or abundant. In other interactions, one is better than another at exploiting a shared resource.

When two species overlap too much in their requirements, they cannot coexist in the same habitat unless they share the required resources in different ways or at different times.

bristly foxtail

Indian mallow   smartweed

Figure 40.7 Resource partitioning among three annual plant species in an abandoned field. The plants differ in how they are adapted to secure soil water and mineral ions. The roots of each species tap into a different region of the soil.

40.4 Predator–Prey Interactions

Predators are consumers that obtain energy and nutrients from living organisms—their prey—which they generally capture and kill. The quantity and types of prey species affect predator diversity and abundance, and the types of predators and their numbers do the same for prey.

COEVOLUTION OF PREDATORS AND PREY

Coevolution influences the characteristics of prey and predator. The term, recall, refers to the joint evolution of two or more species that exert selection pressure on one another through their close ecological interaction over the generations. Suppose a gene mutation arises in a prey population and leads to improved defense against predators. The mutant gene will increase in frequency, because its bearers will tend to survive in greater numbers, as will their offspring.

Suppose a heritable trait also makes individuals of a predator population better at overcoming the novel prey defense. Those individuals will eat better and so survive and leave more descendants. Over time, the predators are selective agents that favor improved defenses in prey; and prey with better defenses are selective agents that favor more effective predators.

MODELS FOR PREDATOR–PREY INTERACTIONS

The extent to which predators limit numbers of prey depends on several factors. A key factor is the response of individual predators to increases or decreases in prey density. Figure 40.8*a* is an overview of the three general patterns of functional responses.

By the type I model, a predator removes a constant proportion of prey over time, regardless of levels of prey abundance. The number of prey killed during any given interval depends only on prey density. This model applies to passive predators, such as spiders. The more flies there are, the more get caught in webs.

By the type II model, the ability of predators to consume and digest prey determines how many prey they capture. As prey density rises, the proportion captured rises steeply at first, then slows as predators are exposed to more prey than they can deal with at one time. Figure 40.8*b* gives an example. A wolf that has just killed a caribou will not hunt another until it has eaten and digested the first one.

By the type III model, predator response is lowest when prey density is low. It is highest at intermediate prey densities, then levels off. This type of response is observed for predators that can switch to other prey when low abundance makes a prey species hard to find. Predators that show type I and type II responses can limit prey at a stable equilibrium point.

Other factors besides individual predator response to prey density are at work. For example, predator and prey reproductive rates affect the interaction. So do hiding places for prey, the presence of other prey or predator species, and carrying capacities.

THE CANADIAN LYNX AND SNOWSHOE HARE

Shifts in environmental conditions can sometimes cause predator and prey densities to oscillate. At the lowest level, predation strongly depresses the prey density. At the highest level, predation is absent and the prey population nears the carrying capacity.

Consider a ten-year oscillation in populations of one predator, the Canadian lynx, and the snowshoe

Figure 40.8 (**a**) Three alternative models for responses of predators to prey density. Type I: Prey consumption rises linearly as prey density rises. Type II: Prey consumption is high at first, then levels off as predator bellies stay full. Type III: When prey density is low, it takes longer to hunt prey, so the predator response is low. (**b**) A type II response in nature. For one winter month in Alaska, B. W. Dale and his coworkers observed four wolf packs (*Canis lupus*) feeding on caribou (*Rangifer tarandus*). The interaction fit the type II model for the functional response of predators to prey density.

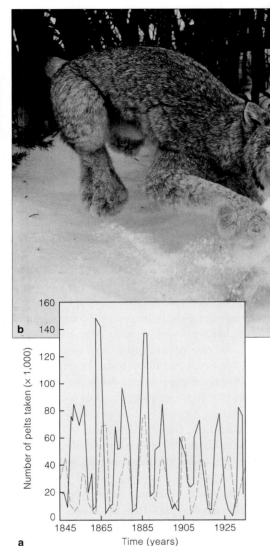

Figure 40.9 (**a**) Correspondence between abundances of Canadian lynx (*dashed* line) and snowshoe hares (*solid* line), based on counts of pelts sold by trappers to Hudson's Bay Company during a ninety-year period. (**b**) Charles Krebs observed that predation causes mass paranoia among snowshoe hares, which continually look over their shoulders during the declining phase of each cycle. (**c**) This photograph supports the Krebs hypothesis that there is a three-level interaction going on, one that involves plants.

The graph might be a good test of whether you tend to accept someone else's conclusions without questioning their basis in science. Remember those sections in Chapter 1 that introduced the nature of scientific methods?

What other factors may have had impact on the cycle? Did the weather vary, with more severe winters imposing greater demand for hares (to keep lynx warmer) and higher death rates? Did the lynx compete with other predators, such as owls? Did the predators turn to alternative prey during low points of the hare cycle? When fur prices rose in Europe, did the trapping increase? When the pelt supply outstripped the demand, did trapping decline?

hare that is its main prey (Figure 40.9). To identify the causes of this pattern, the ecologist Charles Krebs and his coworkers tracked hare population densities for ten years in Alaska, in the Yukon River Valley. They set up 1-square-kilometer control plots and experimental plots. Electric fences kept all predatory mammals out of some plots. Extra food or fertilizers that fanned plant growth were placed in other plots. The team captured and released more than a thousand snowshoe hares, lynx, and other animals, giving each a radio collar.

In predator-free plots, the hare density doubled. In plots with extra food, it tripled. In plots having extra food and fewer predators, it increased elevenfold.

The experimental manipulations delayed the cyclic declines in population density but did not stop them. Why not? Fences could not keep out owls and other raptors. Only 9 percent of the collared hares starved to death; predators devoured most of the rest. Krebs concluded a simple predator–prey or plant–herbivore model cannot explain his results in the Yukon River Valley. For the lynx and snowshoe hare cycle, other variables are at work, during multilevel interactions.

Predator and prey populations tend to exert coevolutionary pressures on one another.

Predators may affect prey density. There are three general patterns of response to changes in prey density. Population levels of prey may also show periodic oscillations.

Predator and prey numbers often vary in complex ways that reflect the multiple levels of interaction in a community.

40.5 An Evolutionary Arms Race

As explained in the preceding section, predators and prey exert selective pressure on one another. One defends itself and the other must overcome defenses. Such interactions started an evolutionary arms race that has resulted in adaptations that often amaze us.

ADAPTATIONS OF PREY

Some prey gain protection from **camouflaging**, or hiding in the open. Their body form, patterning, color, behavior, or some combination of these blend with the surroundings and help avert detection. For example, *Lithops* is a desert plant that looks like a rock (Figure 40.10). It flowers only during a brief rainy season, when other plants grow profusely and distract herbivores from it—and when drinking water (not juicy plant tissues) becomes available.

Many prey species taste bad or contain chemical toxins that sicken their predators. They may have **warning coloration**, conspicuous patterns and colors that predators learn to recognize as *AVOID ME!* signals. A young, inexperienced bird may spear an orange-and-black patterned monarch butterfly once. It learns quickly to associate this coloration and patterning with vomiting foul-tasting toxins.

Dangerous or repugnant species often make little or no attempt to conceal themselves. Skunks, which spray one of the most odious repellents, are like this. So are certain frogs (*Dendrobates*), which are among the most poisonous and vividly colored species.

Many prey species closely resemble a dangerous, unpalatable, or hard-to-catch species. **Mimicry** is the name for the close resemblance in form, behavior, or both between a species that is the model for deception and a different species (the mimic). After predators learn to ignore a model species because of a repellent taste, toxic secretion, or painful bite or sting, they tend also to avoid the mimic. Figure 40.11 shows three tasty but weaponless mimics. All strongly resemble a very aggressive wasp that can sting repeatedly, with painful results.

When luck runs out and a prey animal is cornered or under attack, survival may turn on a last-chance trick. Animals may startle the predator, as by hissing, puffing up, showing teeth, or flashing big eye-shaped spots. A lizard or salamander tail may detach from the body and still wiggle a bit as a distraction. Opossum pretend to be dead. So do hognose snakes, which, like many other prey, also secrete smelly fluid. Many cornered animals secrete or squirt irritating chemical repellents or toxins (Figure 40.12*a*).

Figure 40.10 Prey camouflage. (**a**) What bird? When a predator approaches its nest, the least bittern stretches its neck (which is colored like the surrounding withered reeds), thrusts its beak upward, and sways gently like reeds in the wind. (**b**) An inedible bird dropping? No. This caterpillar's body coloration and the rigid body positions that it assumes help camouflage it from predatory birds. (**c**) Find the plants (*Lithops*) hiding in the open from herbivores with their stonelike form, pattern, and coloration.

a The dangerous model   **b** One of its edible mimics   **c** Another edible mimic   **d** And another edible mimic

Figure 40.11 An example of mimicry. Edible insect species often resemble toxic or unpalatable species that are not at all closely related. (**a**) A yellowjacket can deliver a painful sting. It may be the model for nonstinging wasps (**b**) beetles (**c**), and flies (**d**) that have a similar appearance.

Also, the foliage and seeds of many plants contain bitter, hard-to-digest, or dangerous repellents. Peach, apricot, and rose seeds are loaded with cyanide. And remember that castor bean plant (Chapter 13)? It did not develop its capacity to make the lethal chemical ricin in an evolutionary vacuum. Ricin protects this plant from herbivores that would otherwise eat it.

ADAPTIVE RESPONSES OF PREDATORS

In a coevolutionary arms race, predators counter prey defenses with their own adaptations. Stealth, camouflage, and ingenious ways of avoiding the repellents are some countermeasures. Consider the edible beetles that direct sprays of noxious chemicals at attackers. A grasshopper mouse grabs the beetle and plunges the sprayer end into the ground, and eats the tasty unprotected head (Figure 40.12*b*).

Some prey can outrun predators when they get a head start. But consider the cheetah, the world's fastest land animal. One cheetah was clocked at 114 kilometers per hour (70 mph). Compared to other big cats, the cheetah has longer legs relative to its body size and nonretractable claws that act like cleats to increase traction. Its main prey, Thomson's gazelle, can run longer, but not as fast (80 kilometers per hour); it needs the head start.

Camouflaged predators often lay in wait for their prey. Think of white polar bears stalking seals over ice, striped tigers crouched in tall-stalked, golden grasses, and scorpionfish well hidden on the seafloor (Figure 40.12*c*). Many predatory insects, too, are deception experts (Figure 40.12*d*). Such camouflaged predators select for enhanced sensory systems in prey. By one theory, primate color vision may have evolved in part to enhance detection of predators.

Figure 40.12 Predator responses to prey defenses. (**a**) Some beetles spray noxious chemicals at attackers, which deters them some of the time. (**b**) At other times, grasshopper mice plunge the chemical-spraying tail end of their beetle prey into the ground and feast on the head end. (**c**) Find the scorpionfish, a venomous predator with camouflaging fleshy flaps, multiple colors, and profuse spines. (**d**) Where do the pink flowers end and the pink praying mantis begin?

40.6 Parasite–Host Interactions

Parasites spend all or part of their life cycle in or on other living organisms, from which they draw nutrients. The host is harmed, but usually is not killed outright. Parasites may be adapted to one particular host species, or able to survive in any of a number of hosts.

PARASITES AND PARASITOIDS

Parasites have pervasive impact on populations. By draining nutrients from hosts, they alter the amount of energy and nutrients the host population demands from a habitat. Also, weakened hosts are usually more vulnerable to predation and less attractive to potential mates. Some parasite infections cause sterility. Others shift the ratio of host males to females. In such ways, parasitic infections lower birth rates, raise death rates, and affect intraspecific and interspecific competition.

Sometimes the gradual drain of nutrients during a parasitic infection indirectly leads to death. The host becomes so weakened that it can't fight off secondary infections. Nevertheless, in evolutionary terms, killing a host too quickly is bad for a parasite's reproductive success. A parasitic infection must last long enough to give the parasite time to produce some offspring. The longer it lives in the host, the more offspring. We can therefore expect selective agents to favor parasites that have less-than-fatal effects on their hosts.

Only two types of interactions typically kill a host. First, a parasite may attack a novel host, one with no coevolved defenses against it. Second, the host may be supporting too many individual parasites at the same time. The collective assault may overwhelm the body.

Previous chapters introduced diverse parasites that live in or on a host. Some spend their entire life cycle with a single host. Others are free-living some of the time, or residents of different hosts at different times. Many parasites ride inside insects and other arthropods, the taxis between one host organism and the next.

All viruses and some of the bacteria, protists, and fungi are parasites. Figure 40.13 shows one parasitic fungus that infects amphibians of tropical rain forests.

Even a few plants are parasitic. Nonphotosynthetic types, such as dodders, obtain energy and nutrients from other plants (Figure 40.14). Other types carry out photosynthesis but still tap into the nutrients and water in a host plant's tissues. Mistletoe is like this; its modified roots invade the sapwood of host trees.

Many tapeworms, flukes, and some roundworms are notorious invertebrate parasites (Figure 40.15). So are many insects and crustaceans, as are all ticks.

Parasitoids are insects that develop inside another species of insects, which they devour from the inside out as they mature. Unlike parasites, the parasitoids always kill their hosts. About 15 percent of all insects may be parasitoids.

Social parasites are animals that take advantage of the social behavior of a host species in order to carry out their life cycle. The cuckoos and North American cowbirds, described shortly, are like this.

Figure 40.13 (**a**) Harlequin frog (*Atelopus varius*) from Central America. (**b**) Section through skin of a harlequin frog affected by *chytridiomycosis*, an infectious disease caused by a fungus. *Arrows* point to flask-shaped cells of the fungal parasite. Each cell holds spores that will be released to the skin surface. They may be dispersed to other host frogs.

Figure 40.14 Dodder (*Cuscuta*), also known as strangleweed or devil's hair. Sporophytes of this parasitic flowering plant lack chlorophylls. They wind around a host plant during growth. Modified roots penetrate the host's vascular tissues and draw water and nutrients from them.

Figure 40.15 Adult roundworms (*Ascaris*), an endoparasite, packed inside the small intestine from a host pig. Sections 22.5 and 22.8 give more examples of parasitic worms.

Figure 40.16 A biological control agent—a commercially raised parasitoid wasp about to deposit an egg in an aphid. This wasp reduces aphid populations by crippling the egg laying capacity of its host even before its larva kills it.

USES AS BIOLOGICAL CONTROLS

Parasites and parasitoids are commercially raised and released in target areas as *biological controls*. They are promoted as a workable alternative to pesticides. The chapter introduction and Figure 40.16 give examples.

Effective biological controls display five attributes. The agents are adapted to a specific host species and to its habitat; they are good at locating hosts; their population growth rate is high compared to the host's; their offspring are good at dispersing; and they make a type III functional response to prey, without much lag time after shifts occur in the host population size.

Biological control is not without risks of its own. Releasing more than one kind of biological control agent in an area may trigger competition among them and lessen their effectiveness. Introduced parasites sometimes parasitize nontargeted species as well as, or instead of, species they were expected to control.

In Hawaii, introduction of several parasitoids to control a non-native stink bug resulted in the decline of the koa bug, the state's largest native bug. Few koa bugs have been collected since 1978. Apparently, the parasitoids found the koa bugs, which congregate in large groups, a more tempting target than the pests they were imported to attack. Introduced parasitoids have also been implicated in the ongoing decline of many native Hawaiian butterflies and moths.

Natural selection favors parasitic species that temper their demands in ways that ensure an adequate supply of hosts.

Parasites occur in a vast variety of groups, including bacteria, protists, invertebrates and plants. Parasitoids are insects that feed on and kill other insects. Social parasites use the social behavior of another species to their own benefit.

40.7 Cowbird Chutzpah
FOCUS ON EVOLUTION

The brown-headed cowbird's genus name (Molothrus) means "intruder" in Latin. This bird intrudes, sneakily, into the life cycle of other species. But let us ask: Why?

Brown-headed cowbirds (*Molothrus ater*) evolved in the North American Great Plains. There they lived as commensalists with bison. Great herds of these hefty ungulates stirred up plenty of insects as they migrated through the grasslands, and, being insect-eaters, the cowbirds wandered around with them (Figure 40.17a).

A vagabond way of life did not lend itself to nesting in any one place. However it happened, the cowbirds learned to lay their eggs in nests built by other species, then leave them and move on with the herds. "Host" species did not have the neural wiring to recognize differences between cowbird eggs and their own eggs. Cowbird hatchlings became innately wired for hostile takeovers; even before their eyes open, they shove the owner's eggs out of the nest and demand to be fed as rightful occupants (Figure 40.17b). For thousands of years, the cowbirds successfully perpetuated their genes by way of such parasitic chutzpah.

Then pioneers moved west and cleared swaths of woodlands for pastures. Cowbirds moved in the other direction, easily adapting to the new ungulates—cattle —in the man-made grasslands; hence the name. They started penetrating woodlands and parasitizing novel species. Today, brown-headed cowbirds parasitize at least 15 species of native North American birds, some threatened or endangered. Besides being opportunists, they are big-time reproducers. A female can lay an egg a day for ten days, give her ovaries a rest, do the same again, and then again in one season. As many as thirty eggs in thirty nests—that is a lot of cowbirds.

Figure 40.17 Oh give me a home, where the buffalo roam—brown-headed cowbirds (*Molothrus ater*) originally evolved as commensalists with bison and as social parasites of other bird species of the North American Great Plains. When conditions changed, they expanded their range. They became nest usurpers in woodlands as well as grasslands in much of the United States.

40.8 Forces Contributing to Community Stability

*How does a community come into being? By the classical model for **ecological succession**, a community will slowly develop through a sequence of predictable stages to a final, stable array of species. But fire, treefalls, and other small-scale changes also affect community structure.*

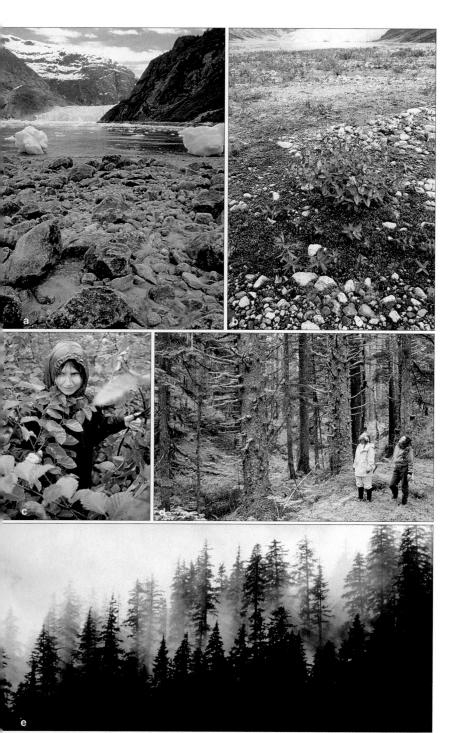

A SUCCESSION MODEL

A community begins with **pioneer species**. These are all opportunistic colonizers of new or newly vacated habitats. They have high dispersal rates; they grow and mature quickly, and they produce many offspring. In time, more competitive species replace them. Then the replacements are replaced until the array stabilizes under prevailing conditions. Such a persistent species composition in a habitat is the **climax community**.

Primary succession is a process that begins when pioneer species colonize a barren habitat, such as a new volcanic island and land exposed when a glacier retreats (Figure 40.18). Pioneers include lichens and plants, such as club mosses, that are small, have short life cycles, and can survive intense sunlight, extreme temperature changes, and nutrient-poor soil. During the early years, hardy annual flowering plants put out many small seeds, which are quickly dispersed.

Once established, pioneers improve soil and other conditions. In doing so, they typically set the stage for their own replacement. Many of the new arrivals are mutualists with nitrogen-fixing bacteria, so they can grow in nitrogen-poor habitats. Seeds of later species find shelter inside mats of the pioneers, which do not grow high enough to shade out the new seedlings.

In time, organic wastes and remains accumulate. By adding volume and nutrients to soil, they favor invasions of new species. Later successional species crowd out the earlier ones, whose spores and seeds travel as fugitives on the wind and water—destined, perhaps, for another new but temporary habitat.

In **secondary succession**, a disturbed area within a community recovers. If improved soil is still there, secondary succession can be quite rapid. It commonly occurs in abandoned fields, burned forests, and tracts of land cleared by volcanic eruptions (Figure 40.19).

Figure 40.18 Primary succession in Alaska's Glacier Bay region. (**a**) A glacier is retreating from the sea. Meltwater leaches nitrogen and other minerals from the newly exposed soil. (**b**) The first invaders of nutrient-poor soil are lichens, horsetails, mosses, and fireweed and mountain avens. Some pioneer species are mutualists with nitrogen-fixing microbes. They grow and spread quickly over the glacial till.

Within twenty years, alder, cottonwood, and willow seedlings have taken hold in drainage channels. Alders, too, benefit from nitrogen-fixing symbionts. (**c**) Within fifty years, alders have formed dense, mature thickets in which cottonwood, hemlock, and a few evergreen spruce grow quickly. (**d**) After eighty years, western hemlock and spruce trees crowd out the mature alders. (**e**) In areas deglaciated for more than a century, forests of Sitka spruce dominate.

Figure 40.19 Secondary succession after Mount Saint Helens erupted in 1980 (**a**). At the base of this Cascade volcano, nothing remained of the climax community. (**b**) In less than a decade, pioneer species were getting established. (**c**) Twelve years later, seedlings of the dominant climax community species, Douglas firs, were taking hold.

THE CLIMAX PATTERN MODEL

At one time, ecologists thought the same general type of community always develops in a region because of climate constraints. However, diverse communities persist in similar climates, as when tallgrass prairie grades eastward into Indiana's deciduous forests.

By the **climax-pattern model**, each community is adapted to many environmental factors—topography, climate, soil, species interactions, periodic fires, and so on—that differ in their effects across a region. Even within a geographic region, climax communities may merge one into another along gradients of particular environmental conditions, such as moisture.

CYCLIC, NONDIRECTIONAL CHANGES

Many small-scale changes recur in small patches of habitats and affect a community's internal dynamics. Observe a disturbed area of habitat, and you might conclude that great shifts in species composition are afoot. Observe the community on a larger scale, and you might find that the overall composition includes all pioneer species and the dominant climax species.

Consider that a tropical forest develops through phases of colonization by pioneers, then increases in species diversity and matures. Whipping sporadically through the forest are strong winds that can knock down trees. Where trees fall, small gaps open in the dense tree canopy, and more light reaches patches on the forest floor. There, conditions favor growth of the previously suppressed small trees and germination of pioneers or shade-intolerant species.

Ice storms, tornadoes, hurricanes, and wildfires are among many other natural disturbances that can alter community structure.

A climax community is a stable, self-perpetuating array of species in equilibrium with one another and their habitat. It develops as a sequence of species, starting with pioneers.

Similar climax stages persist along gradients dictated by environmental factors and by species interactions. Small-scale, recurring changes help shape many communities.

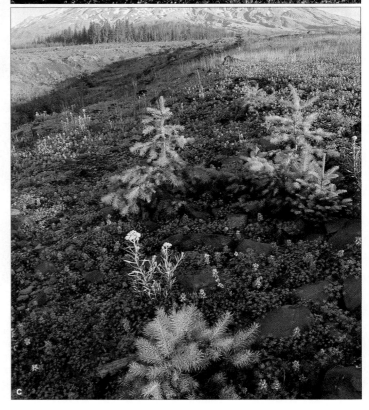

40.9 Forces Contributing to Community Instability

The preceding sections might lead you to believe that all communities become stabilized in predictable ways. This is not always the case.

Community stability is an outcome of forces that have come into uneasy balance. Resources are sustained as long as populations do not reach carrying capacity. Predators and prey coexist as long as neither wins. Competitors continually engage in tugs-of-war over resources. Mutualists are stingy, as when plants make as little nectar as necessary to attract pollinators, and pollinators get as much nectar as they can for the least effort. In the short term, disturbances might slow the growth of some populations. Also, long-term changes in climate typically have destabilizing effects.

If instability becomes great enough, the community might change in ways that will persist even when the disturbance ends or is reversed. If some of its species happen to be rare or do not compete well with others, they might be driven to extinction.

THE ROLE OF KEYSTONE SPECIES

The uneasy balancing of forces in a community comes into focus when we observe the effects of a **keystone species**—a species that has a disproportionately large effect on a community relative to its abundance. Robert Paine was the first to describe the role of a keystone species after his experiments on the rocky shores of California's coast. Species in this rocky intertidal zone survive by clinging to rocks, and access to spaces to cling to is a limiting factor. Paine set up control plots with the sea star *Pisaster ochraceus* and its main prey—chitons, limpets, barnacles, and mussels. He removed all sea stars from his experimental plots.

With the sea stars gone, the mussels (*Mytilus*) took over Paine's experimental plots, and they crowded out seven other invertebrate species. The mussels are the preferred prey of the sea star, and the strongest competitors in their absence. Normally, predation by sea stars keeps the diversity of prey species high,

d Algal diversity in tidepools

e Algal diversity on rocks that become exposed at high tide

Figure 40.20 Effect of competition and predation in an intertidal zone. (**a**) Grazing periwinkles (*Littorina littorea*) affect the number of algal species in different ways in different marine habitats. (**b**) *Chondrus* and (**c**) *Enteromorpha*, two kinds of algae in their natural habitats. (**d**) By grazing on the dominant alga in tidepools (*Enteromorpha*), the periwinkles promote the survival of less competitive algal species that would otherwise be overgrown. (**e**) *Enteromorpha* doesn't grow on rocks. Here, *Chondrus* is dominant. Periwinkles find *Chondrus* tough and dine instead on less competitive algal species. By doing so, periwinkles decrease the algal diversity on the rocks.

Table 40.1   Detrimental Effects of Some Species Introduced into the United States

Species Introduced	Origin	Mode of Introduction	Outcome
Water hyacinth	South America	Intentionally introduced (1884)	Clogged waterways; other plants shaded out
Dutch elm disease:			
Ophiostoma ulmi (fungus)	Asia (by way	Accidental; on infected elm timber (1930)	Millions of mature elms destroyed
Bark beetle (vector)	of Europe)	Accidental; on unbarked elm timber (1909)	
Chestnut blight fungus	Asia	Accidental; on nursery plants (1900)	Nearly all eastern American chestnuts killed
Zebra mussel	Russia	Accidental; in ballast water of ship (1985)	Clog pipes and water intake valves of power plants; displacing native Great Lake bivalves
Japanese beetle	Japan	Accidental; on irises or azaleas (1911)	Over 250 plant species (e.g., citrus) defoliated
Sea lamprey	North Atlantic	Ship hulls, through canals (1860s, 1921)	Trout, whitefish destroyed in Great Lakes
European starling	Europe	Intentional release, New York City (1890)	Outcompete native cavity-nesting birds; crop damage; swine disease vector
Nutria	South America	Accidental release of captive animals being raised for fur (1930)	Crop damage, destruction of levees, overgrazing of marsh habitat

because it restricts competitive exclusion by mussels. Remove all the sea stars, and the community shrinks from fifteen species to eight.

The impact of a keystone species can vary between habitats that differ in their species arrays. Periwinkles (_Littorina littorea_) are alga-eating snails of intertidal zones. Jane Lubchenco showed that their removal can increase _or_ decrease the diversity of algal species in different habitats (Figure 40.20).

In tidepools, the periwinkles prefer to eat the alga _Enteromorpha_, which can outgrow other algal species. By keeping _Enteromorpha_ in check, periwinkles help less competitive algal species survive. On exposed rocks in the lower intertidal zone, _Chondrus_ and other tough, unpalatable red algae that periwinkles avoid are dominant. In this habitat, periwinkles prefer to graze on the competitively weaker algal species. So periwinkles _promote_ algal diversity in tidepools but _reduce_ it on exposed rock surfaces.

SPECIES INTRODUCTIONS TIP THE BALANCE

Instabilities also are set in motion when residents of established communities move out from their home range and successfully take up residence elsewhere. This type of directional movement, called **geographic dispersal**, happens in three ways.

First, over a number of generations, a population might expand its home range by slowly moving into outlying regions that prove hospitable. Second, some individuals might be rapidly transported across great distances, an event called _jump_ dispersal. This often takes individuals across regions where they could not survive on their own, as when insects travel from the

mainland to Maui in a ship's cargo hold. Third, some population might be moved away from a home range by continental drift, at an almost imperceptibly slow pace over long spans of time.

Successful dispersal and colonization of a vacant adaptive zone can be remarkably rapid. Consider one of Amy Schoener's experiments in the Bahamas. She set out plastic sponges on barren sand at the bottom of Bimini Lagoon. How fast did aquatic species take up residence on these artificial habitats? Schoener recorded occupancy by 220 species within thirty days.

When you hear someone bubbling enthusiastically about an exotic species, you can safely bet the speaker isn't an ecologist. An **exotic species** is a resident of an established community that dispersed from its home range and became established elsewhere. Unlike most imports, which never do take hold outside the home range, an exotic species permanently insinuates itself into a new community.

More than 4,500 exotic species have established themselves in the United States after jump dispersal. We put some of the new arrivals, including soybeans, rice, wheat, corn, and potatoes, to use as food crops.

Accidental imports can alter community structure. You learned about imported fire ants in the chapter introduction. Table 40.1 lists others. The next section describes the unintended effects of a few more.

A keystone species is one that has a major effect on the species composition in particular habitats.

Long-term climate shifts, species introductions, and other disturbances can permanently alter community structure.

40.10 Exotic Invaders

Nonnative species are on the loose in communities on every continent. They can alter habitats; they often outcompete and displace native species.

THE PLANTS THAT ATE GEORGIA

In 1876, a vine called kudzu (*Pueraria montana*) from Japan was introduced to the United States. In the temperate regions of Asia where it evolved, it is a well-behaved legume with a well-developed root system. It *seemed* like a good idea to use it for forage and to control erosion. But kudzu grew faster in the Southeast, where herbivores, pathogens, and less competitive plants posed no serious threat to it.

With nothing to stop it, kudzu shoots can grow sixty meters a year. Its vines now blanket streambanks, trees, telephone poles, houses, and almost everything else in their path (Figure 40.21*a*). It withstands burning, and its deep roots resist being dug up. Grazing goats and herbicides help. But goats eat most other plants along with it, and herbicides taint water supplies. Kudzu may spread as far north as the Great Lakes by 2040.

On the bright side, Asians use a starch extracted from kudzu in drinks, herbal medicines, and candy. A kudzu processing plant in Alabama may export this starch to Asia, where demand currently exceeds the supply. Also, kudzu may help save trees; it can be an alternative source for paper. Today, about 90 percent of Asian wallpaper is kudzu-based.

THE ALGA TRIUMPHANT

They looked so perfect in saltwater aquariums, those bright green, long, feathery branches of the green alga *Caulerpa taxifolia*. So researchers from Germany's Stuttgart Aquarium developed a hybrid, sterile strain of *C. taxifolia*, and then they magnanimously shared it with other marine institutions. Was it from Monaco's Oceanographic Museum that the hybrid strain escaped into the wild? Some say yes, Monaco says no.

The aquarium strain of *C. taxifolia* can't reproduce sexually. Yet. It grows asexually by runners, just a few centimeters a day. Bits caught on boat propellers and fishing nets assisted its dispersal. Since 1984, this alga has blanketed more than 30,000 hectares of the seafloor along the Mediterranean coast (Figure 40.21*b*). It thrives along sandy or rocky shores and in mud; it lives ten days after being discarded in meadows. Unlike its tropical parents, this strain does well in cool water, even polluted water. It displaces endemic algae and poisons invertebrates and fishes. Its toxin repels herbivores that might keep it in check.

C. taxifolia also has invaded the coastal waters of the United States. It has the potential to overgrow reefs and destroy marine food webs. In 2000, scuba divers discovered the alga growing in two areas off the coast of Southern California. Someone might have drained the water from a home aquarium into a storm drain or into the lagoon itself. In response to

Figure 40.21 (**a**) Kudzu (*Pueraria montana*) taking over part of Lyman, South Carolina. The plant has now spread from East Texas to Florida and Pennsylvania. Ruth Duncan of Alabama, who makes 200 kudzu vine baskets a year, just can't keep up. (**b**) *Caulerpa taxifolia* suffocating yet another richly diverse marine ecosystem.

Figure 40.22 Rabbit-proof fence? Not quite. This is part of a fence built to hold back the 200 million to 300 million rabbits that are wreaking havoc with Australia's vegetation. It didn't work.

the discovery, a coalition of governmental and private groups sprang into action. They tarped over the area and shut out sunlight, pumped chlorine into the mud to poison the alga, and used welders to boil it. So far, the measures may have worked. As of autumn 2003, there was no sign of the exotic invader.

It is now illegal to import the harmful strain of *C. taxifolia* into the United States. Interstate sale also is prohibited. Some still slip into the country because the aquarium industry successfully lobbied against a ban on all *Caulerpa* species, and it is hard to distinguish the invasive strain without genetic analysis.

THE RABBITS THAT ATE AUSTRALIA

During the 1800s, British settlers in Australia just couldn't bond with koalas and kangaroos, and so they imported familiar animals from home. In 1859, in what would be the start of a major disaster, a landowner in northern Australia imported and then released two dozen European rabbits (*Oryctolagus cuniculus*). Good food and sport hunting—that was the idea. An ideal rabbit habitat with no natural predators—that was the reality.

Six years later, the landowner had killed 20,000 rabbits and was besieged by 20,000 more. The rabbits displaced livestock and caused the decline of native wildlife. Now 200 to 300 million are hippity-hopping through the southern half of the country. They graze on grasses in good times and strip bark from shrubs and trees during droughts. Thumping herds of them can turn shrublands as well as grasslands into eroded

deserts. And their burrows undermine the soil and set the stage for widespread erosion.

Rabbit warrens have been shot at, plowed under, fumigated, and dynamited. The first all-out assaults killed 70 percent of them, but the rabbits rebounded in less than a year. When a fence 2,000 miles long was built to protect western Australia, rabbits made it from one side to the other before workers could finish the job (Figure 40.22).

In 1951, the government introduced a myxoma virus that normally infects South American rabbits. The virus causes *myxomatosis*. This disease has mild effects on its coevolved host but nearly always kills *O. cuniculus*. Mosquitoes and fleas transmit the virus from host to host. Having no coevolved defenses against the import, European rabbits died in droves. But natural selection has since favored a rise in rabbit populations resistant to the imported virus.

In 1991, on an uninhabited island in Spencer Gulf, Australian researchers released rabbits that they had injected with a calicivirus. Rabbits died from blood clots in their lungs, heart, and kidneys. But the test virus escaped from the island in 1995, perhaps on insect vectors. It has been killing 80 to 95 percent of adult rabbits in some parts of Australia. In 1996, the virus was deliberately released in other areas.

Some Australians are now questioning whether the calicivirus should be used on such a large scale. They wonder whether it can jump boundaries and infect animals other than rabbits—one study suggests it might cause illness in pigs. What will be the long-term effects of these viral releases? We don't know.

40.11 Biogeographic Patterns

Biodiversity is the sum of all species occupying a specified area during a specified interval, past as well as present. It differs from one habitat or geographic realm to another, often in predictable patterns.

In large part, the biodiversity of a given area is an outcome of the evolutionary history of each member species and its resource requirements, its physiology, and its capacity for dispersal. Rates of birth, death, immigration, and emigration also shape the number of species. In turn, those rates are influenced by local and regional habitat conditions, and by interactions among species.

Ecologists have identified patterns of biodiversity throughout the world. For example, species richness corresponds to sunlight intensity, moisture levels, and temperature along gradients in latitude, elevation, and depth. Differences between microenvironments give rise to local variations in overall patterns.

MAINLAND AND MARINE PATTERNS

The most striking pattern of biodiversity corresponds to distance from the equator. *For most groups of plants and animals, the number of coexisting species on land and in the seas is greatest in the tropics, and it systematically declines from the equator to the poles.* Figure 40.23 shows examples of this pattern. Let's focus on a few factors that help bring about the pattern and maintain it.

First, for reasons explained in Section 42.1, tropical latitudes intercept more intense sunlight and receive more rainfall, and their growing season is longer. As one outcome, resource availability tends to be greater and more reliable in the tropics than elsewhere. This favors a degree of specialized interrelationships not possible where species are active for shorter periods.

Second, tropical communities have been evolving for a longer time than temperate ones, some of which did not start forming until the end of the last ice age.

Third, species diversity might be self-reinforcing. Diversity of tree species in tropical forests is much greater than in comparable forests at higher latitudes. When more plant species compete and coexist, more kinds of herbivores also compete and coexist, partly because no single kind of herbivore can overcome all chemical defenses of all the different plants. More predatory and parasitic species evolve in response to the bigger buffet of prey and hosts. The same effect applies to diversity on tropical reefs.

Figure 40.23 Two patterns of species diversity corresponding to latitude. The number of ant species (**a**) and breeding birds (**b**) in Central and North America.

Figure 40.24 Surtsey, a volcanic island, at the time of its formation. Newly formed, isolated islands are natural laboratories for ecologists. The chart gives the number of colonizing species between 1965 and 1973.

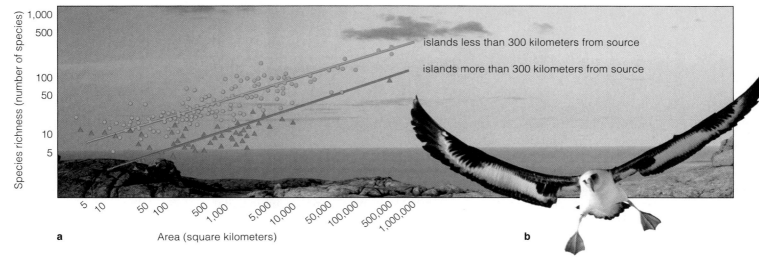

Figure 40.25 (**a**) Two island biodiversity patterns. Distance effect: Biodiversity on islands of a specified size declines with increasing distance from a source of colonizing species. *Green* circles signify islands less than 300 kilometers from the colonizing source. *Orange* triangles signify islands more than 300 kilometers from the source areas. Area effect: Among islands the same distance from a source of colonizing species, the larger ones support more species.

(**b**) Wandering albatross, one travel agent for jump dispersals. Seabirds that island-hop long distances often have seeds stuck to their feathers. Seeds that successfully germinate in a new island community may give rise to a population of new immigrants.

ISLAND PATTERNS

Islands are fine laboratories for biodiversity studies. For instance, a 1965 volcanic eruption created Surtsey, an island southwest of Iceland. Within six months, bacteria, fungi, seeds, flies, and some seabirds became established on the island. A vascular plant appeared two years after the island formed; the first mosses came along two years after that (Figure 40.24). As the island soil became enriched, more and more plant species began to take hold.

As is the case for other islands, the number of new species on Surtsey will not increase indefinitely. Why not? Models based on studies of island communities around the world provide some answers.

First, islands that are far from a source of potential colonists receive few colonizing species, and the few that do arrive are adapted for long-distance dispersal (Figure 40.25*a*). This is known as the **distance effect**.

Second, larger islands tend to support far more species than smaller islands the same distances from a colonizing source. This is known as the **area effect**. Larger islands tend to offer varied habitats and more of them. Most have complex topography and higher elevations. Such variations promote diversity. Also, being bigger targets, larger islands intercept more of the accidental tourists that winds and ocean currents move from the mainland but offer no way back.

Extinctions lower diversity, and they happen more on small islands. Small populations are vulnerable to severe storms, droughts, disease, and genetic drift.

Again, on any island, the number of species will reflect a balance between immigration rates for new species and extinction rates for any established ones. Small islands distant from a source of colonists have low immigration rates and high extinction rates, so they support fewer species once the balance has been struck for their populations.

Biogeographers know a lot more about the patterns of diversity and our disruptions of them. For a lyrical account, read David Quammen's *Song of the Dodo*.

Biodiversity shows global patterns, as when the number of species correlates with gradients in latitude, elevation, and depth. Microenvironments along the gradients often introduce variations in the overall patterns.

Species richness in a given area also is an outcome of the evolutionary history of each species, its requirements for resources, its physiology, its capacity for dispersal, and its rates of birth, death, immigration, and emigration.

Generally, species richness is highest in the tropics and lowest at the poles. The number of species on an island also depends on its size and distance from a colonizing source.

40.12 Threats to Biodiversity

Based on many lines of evidence accumulated over the past few centuries, an estimated 99 percent of all species that have ever lived are extinct. Even so, the full range of biodiversity is greater now than it has ever been at any time in the past.

Reflect on life's evolution, as laid out in Unit Three. Remember the five greatest mass extinctions that we use as boundaries for geologic time? After each event, biodiversity plummeted on land and in the seas. In the aftermath, many surviving species embarked on adaptive radiations. Each time, though, recovery to the same level of species richness was extremely slow, on the order of 20 million to 100 million years.

ON THE NEWLY ENDANGERED SPECIES

No biodiversity-shattering asteroids have struck the Earth for 65 million years. *Yet the sixth major extinction event is under way*. There is evidence that humans are driving many species over the edge. Consider this: Mammals survived the asteroid impact that killed the last of the dinosaurs. But 2 million years ago, humans started hunting them in earnest. About 11,000 years ago, human populations were changing habitats in a big way. Agriculture and domesticated species were favored over wild animals. Only 4,500 or so species of Mammals are with us today. Of these, about 300 are endangered. Among them are most of the whales, otters, wild cats, and primates.

Figure 40.26 Humans as competitor species—an example of dominance in the competition for living space in an arid habitat.

An **endangered species** is any endemic species that is highly vulnerable to extinction. *Endemic* means it originated in one region and is found nowhere else.

We are one species among millions. Yet, primarily as an outcome of our rapid population growth over the past forty years, we are threatening other species with large-scale habitat losses, species introductions, and overharvesting.

HABITAT LOSSES AND FRAGMENTATION

Edward O. Wilson defines **habitat loss** as the physical reduction in suitable places to live, as well as the loss of suitable habitat through pollution. Habitat loss is putting major pressure on more than 90 percent of the endemic species that now face extinction.

Biodiversity is greatest in the tropics, where lands are most threatened by deforestation (Chapter 21). As other examples, 98 percent of the tallgrass prairies, 50 percent of the wetlands, and as much as 95 percent of the United States old-growth forests have disappeared. Even arid lands are not exempt (Figure 40.26).

Habitats often become fragmented, or chopped into patches that each have a separate periphery. Species at any habitat's periphery are more exposed to wind, fire, temperature changes, predators, and pathogens. Also, fragmented patches often are not large enough to support the population sizes required for successful breeding. Patches may be too small to have sufficient food or other resources to sustain a species.

Together with Robert MacArthur, Wilson devised an **equilibrium model of island biogeography**. By this model, a 50 percent loss of a habitat will drive about 10 percent of its endemic species to extinction. A 90 percent loss will drive about 50 percent of the species in a habitat to extinction. The model is being used to estimate the number of current and future extinctions in "islands of shrinking habitat." Such islands need not be true islands; many are patches of natural habitat surrounded by degraded areas or by encroaching suburban developments. Many national parks, tropical forests, reserves, and lakes are like this.

Some species help warn us of changes in habitats and impending loss of diversity. Birds are one of the key **indicator species**. Different types live in all major land regions and climate zones, they react quickly to changes, and they are relatively easy to track.

Species introductions are another major threat to endemic species. Remember, each habitat has only so many resources, and its occupants all compete for a share. Introduced species often have adaptations that make them highly competitive. They have been a key factor in almost 70 percent of cases where endemic species are being driven to extinction.

Also, in a sad commentary on human nature, the more rare a wild animal becomes, the more its value soars in the black market. Figure 40.27 gives you an idea of the money involved in illegal wildlife trading.

As a final point, biodiversity also is threatened by overharvesting of species that have commercial value. Whales are a prime example. Humans kill whales for food, lubricating oil, fuel, cosmetics, fertilizer, even pet food. Substitutes exist for all whale products. Yet some nations ignore a moratorium on hunting large whales. Some whalers poach inside the boundaries of sanctuaries set aside for recoveries.

CONSERVATION BIOLOGY

Awareness of the impending extinction crisis gave rise to **conservation biology**. We define this field of pure and applied research as (1) a systematic survey of the full range of biological diversity, (2) efforts to decipher biodiversity's evolutionary and ecological origins, and (3) attempts to identify methods of maintaining and using biodiversity in ways that can benefit the human population. Its major goal is to conserve and utilize, in sustainable ways, as much biodiversity as possible.

Figure 40.28 depicts the approximate number of named species in some major groups. Our current taxonomic knowledge is fairly complete for many groups, including the flowering plants, conifers, and fishes, birds, mammals, and other vertebrates. We tend to know more about large organisms, especially land dwellers, and about the showy ones, including birds and butterflies. Yet we have a lot of evidence that staggering numbers of fungi, protists, bacteria, and archaea are still out there, waiting to be identified.

For example, biologists recently surveyed a single hot spring in Yellowstone National Park. They turned up more species of archaea than had previously been known for the rest of the Earth—which gives you an idea of how much work remains to be done.

The current range of global biodiversity is an outcome of an overall pattern of extinctions and slow recoveries.

Human activities have raised rates of extinction through habitat losses, species introductions, and overharvesting.

Conservation biology entails a systematic survey of the full range of biodiversity, analysis of its evolutionary and ecological origins, and identification of methods to maintain and use biodiversity for the benefit of humans.

Figure 40.27 Humans as predator species—confiscated tiger skins, leopard skins, rhinoceros horns, and elephant tusks. An illegal wildlife trade threatens the survival of more than 600 species, yet all but 10 percent of the trade escapes detection. A few of the black market prices: Live, mature saguaro cactus ($5,000–15,000), bighorn sheep, head only ($10,000–60,000), polar bear ($6,000), grizzly ($5,000), bald eagle ($2,500), peregrine falcon ($10,000), live chimpanzee ($50,000), live mountain gorilla ($150,000), Bengal tiger hide ($100,000), and rhinoceros horn ($28,600 per kilogram).

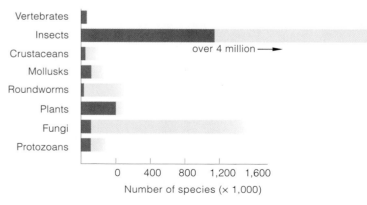

Figure 40.28 Current species diversity for a few major taxa. *Red* signifies the number of named species; *gold* signifies the estimated number of species not yet discovered and named.

40.13 Sustaining Biodiversity

Every species is part of a web of interactions with others of its community. Increasingly, efforts to save endangered species are focusing on identifying and protecting the most threatened communities having the most diversity.

IDENTIFYING AREAS AT RISK

As noted in the preceding section, the vast majority of existing species have not even been described, and we don't know where most of them live. Much effort is now focused on identifying **hot spots**, or habitats of a large number of species that are found nowhere else and are in greatest danger of extinction.

At the first survey level, workers target a limited area, such as an isolated valley. A complete survey is impractical, so they make an inventory of flowering plants, birds, mammals, fishes, butterflies, and other indicator species in the habitat. At the next level up, broader areas that are major or multiple hot spots are systematically explored. The widely separated forests of Mexico are an example of a key hot spot. Research stations have been set up at many different latitudes and at different elevations across the region.

At the highest survey level, hot spot inventories are combined with existing knowledge of biodiversity in ecoregions. An **ecoregion** is a broad land or ocean region defined by climate, geography, and producer species. Figure 40.29, a map of regions identified as essential reservoirs of biodiversity, includes 142 land, 53 freshwater, and 43 marine habitats. Of these, 25 are high-priority targets for conservation efforts.

ECONOMIC FACTORS AND SUSTAINABLE DEVELOPMENT

Every nation enjoys three forms of wealth—material, cultural, and biological wealth. Until recently, not very many countries comprehended the value of biological wealth, which can be the source of food, medicine, and other products. Much of the world's biological wealth has already been lost.

In the next fifty years, human population size may reach 8.9 billion, with most growth in the developing countries. Each person requires raw materials, energy, and living space. How many species will be crowded out? The countries with the greatest monetary wealth also have the least biological wealth and use most of the world's natural resources. Countries with the least monetary wealth and the most biological wealth have the fastest-growing populations.

People who are locked in poverty often must choose between saving themselves or an endangered species. They hunt animals and dig up plants in "protected" parks and reserves. They try to raise crops and herds

▪ Tropical forest	▪ Tropical grassland and savanna	▪ Mediterranean shrub	▪ Mangrove swamp
▪ Temperate forest	▪ Temperate grassland	▪ Desert	▪ Marine ecoregion
▪ Northern coniferous forest	▪ Mountain grassland	▪ Arctic tundra	▪ Freshwater ecoregion

Figure 40.29 Map of the most vulnerable regions of land and seas, compiled by the World Wildlife Fund.

uncut
forest

cut 1
year ago

dirt road

cut 3–5
years ago

cut 6–10
years ago

uncut forest

stream in watershed

Figure 40.30 Strip logging. The practice may protect biodiversity as it permits logging on tropical slopes. A narrow corridor paralleling the land's contours is cleared. A roadbed is made at the top to haul away logs. After a few years, saplings grow in the cleared corridor. Another corridor is cleared above the roadbed. Nutrients leached from exposed soil trickle into the first corridor. There they are taken up by saplings, which benefit from all the nutrient input by growing faster. Later, a third corridor is cut above the second one—and so on in a profitable cycle of logging, which the habitat sustains over time.

Figure 40.31 Riparian zone restoration. This example comes from Arizona's San Pedro River, shown before and after restoration efforts.

on marginal land. Conservation biologists work to identify ways for such people to earn a decent living from biological wealth—by using the biodiversity in threatened habitats in a sustainable way.

It *is* possible to protect a habitat and still withdraw resources in sustainable fashion. For example, besides yielding wood for local economies, tropical rain forest trees yield diverse exotic woods that are prized by developed countries. Gary Hartshorn was the first to propose **strip logging** in portions of forests that are sloped and have a number of streams. As explained in Figure 40.30, logging can be sustained even while the maximum biodiversity is being maintained.

Developed countries also benefit from sustainable methods, as in a **riparian zone**—a narrow corridor of vegetation along a stream or a river (Figure 40.31). The plants afford a major line of defense against flood damage by sponging up water during spring runoffs and summer storms. Shade cast by a canopy of taller shrubs and trees in the riparian zone helps conserve water during droughts.

A riparian zone can provide wildlife with food, shelter, and shade, particularly in arid and semiarid regions. In the western United States, 67 to 75 percent of the endemic species spend all or part of their life cycle in riparian zones. Among them are 136 kinds of songbirds, some of which will nest only in the plants of riparian zones.

Cattle tend to congregate at rivers and streams, where they trample and feed until grasses and shrubs are gone. Even a few head of cattle can ruin a riparian zone. Restricting access and providing water at areas away from streams helps to conserve some riparian zones. Rotating livestock and placing supplemental feed at different grazing areas is also useful.

Preserving biodiversity requires identifying and protecting regions that support the highest levels of biodiversity.

Sustainable development involves utilizing biodiversity in ways that benefit humans, without shredding the intricate web of interactions that sustains a natural community.

Summary

Section 40.1 A habitat is the type of place where individuals of a species normally live. A community is an association of all populations of species that occupy a habitat. Each species in the community has its own niche, the sum of all activities and relationships in which its individuals engage as they secure and use the resources required for their survival and reproduction.

Indirect and direct species interactions influence community structure. They include commensalism, mutualism, competition, predation, and parasitism.

Section 40.2 Mutualism is a species interaction that benefits both participants. Some mutualists can't complete their life cycle without the interaction.

Section 40.3 Two species that require the same resource tend to compete. By the competitive exclusion theory, when two (or more) species require identical resources, they cannot coexist indefinitely. Species are more likely to coexist when they differ in their use of resources. Many coexist by using a shared resource in different ways or at different times.

Section 40.4 Predators and prey exert selection pressure on each other. Densities of predator and prey populations often oscillate. The carrying capacity, density dependencies, refuges, predator efficiency, and often alternative prey sources influence the cycles.

Section 40.5 Threat displays, chemical weapons, camouflage, stealth, and mimicry are outcomes of coevolution between predators and their prey.

Sections 40.6, 40.7 Parasites live in or on other living hosts and withdraw nutrients from host tissues for part of their life cycle. Hosts may or may not die as a result. Parasitoids kill their hosts, and social parasites take over some aspect of a host's life cycle.

Section 40.8 By a model for ecological succession, a community develops in predictable sequence, from its pioneer species to an end array of species that persists over an entire region. A climax community is a stable, self-perpetuating array of species that are in equilibrium with one another and the environment.

Similar climax stages may persist within the same region yet show variation as a result of environmental gradients and species interactions.

Section 40.9 Community structure reflects an uneasy balance of forces, including predation and competition, that have been operating over time.

Section 40.10 Introduction of species that in one way or another expand their geographic range can harm community structure in permanent ways.

Section 40.11 Biodiversity depends on the size of a region, colonization rates, disturbances, and rates of extinction. It depends on the climate and patterns of resource availability. It tends to be highest in the tropics and declines systematically toward the poles.

Sections 40.12, 40.13 Current biodiversity is an outcome of evolutionary forces, mass extinctions, and slow recoveries. Conservation biologists seek ways to conserve and use biodiversity in sustainable fashion.

Self-Quiz
Answers in Appendix III

1. A habitat _____ .
   a. has distinguishing physical and chemical features
   b. is where individuals of a species normally live
   c. is occupied by various species
   d. a through c

2. A niche is _____ .
   a. the sum of activities and relationships by which individuals of a species secure and use resources
   b. unvarying for a given species
   c. something that shifts in large and small ways
   d. both a and c

3. Two species may coexist indefinitely in some habitat when they _____ .
   a. differ in their use of resources
   b. share the same resource in different ways
   c. use the same resource at different times
   d. all of the above

4. A predator population and prey population _____ .
   a. always coexist at relatively stable levels
   b. may undergo cyclic or irregular changes in density
   c. cannot coexist indefinitely in the same habitat
   d. both b and c

5. Parasites _____ .
   a. weaken their hosts   c. feed on host tissues
   b. can kill novel hosts   d. all of the choices

6. In _____ , a disturbed site in a community recovers and moves again toward the climax state.
   a. the area effect   c. primary succession
   b. the distance effect   d. secondary succession

7. The biodiversity of a region is an outcome of _____ .
   a. climate and topography   d. both a and b
   b. possibilities for dispersal   e. a through c
   c. evolutionary history

8. Following mass extinctions, recovery to the same level of biodiversity has taken many _____ of years.
   a. hundreds   b. millions   c. billions

9. Match the terms with the most suitable descriptions.
   ____ geographic dispersal   a. opportunistic colonizer of barren or disturbed places
   ____ area effect   b. greatly affects other species
   ____ pioneer species   c. individuals leave home range, become established elsewhere
   ____ climax community   d. more biodiversity on large islands than small ones at same distance from colonizing source
   ____ keystone species   e. stable, self-perpetuating array of species
   ____ endemic species   f. river or stream's vegetation area
   ____ riparian zone   g. originated in one place and found nowhere else

Figure 40.32 Two phasmids. (**a**) A stick insect from South Africa. (**b**) A leaf insect from Java. (**c**) Phasmid eggs often resemble seeds.

Figure 40.33 One of the realities of the modern world. Would you prefer to be buried in a casket in the ground or mixed with concrete as part of a long-lasting foundation for an artificial reef?

Critical Thinking

1. With antibiotic resistance rising, researchers are looking for ways to reduce use of antibiotics. Some cattle that once would have been fed antibiotic-laced food are now being given feed that contains cultures of bacteria. These *probiotic feeds* are designed to establish or bolster populations of helpful bacteria in the animal's gut. The idea is that if a large population of beneficial bacteria is in place, then the harmful bacteria cannot become established or thrive. Which ecological theory is guiding this research?

2. Insects called phasmids are extraordinary masters of disguise. Most resemble sticks or leaves. Even their eggs may be camouflaged (Figure 40.32). The phasmids are all herbivores. Most spend their days motionless, moving about and feeding only at night. If disturbed, a phasmid will fall to the ground and not move.

   Speculate on the selective pressures that might have shaped phasmid morphology and behavior. Suggest an experiment with one species to help determine whether the appearance and behavior have adaptive value.

3. Figure 40.28 compares the current estimated number of species for some major taxa. Given what you learned from chapters in this unit, which taxa are most vulnerable in the current extinction crisis? Which are likely to pass through it with much of their biodiversity intact? List reasons why. For example, compare the global distribution of species.

4. Mentally transport yourself to a tropical rain forest of South America. Imagine you do not have a job. There are no jobs in sight, not even in overcrowded cities a long distance away. You have no money or contacts to get you anywhere else. And yet you are the sole supporter of a large family. Today a stranger approaches you. He tells you he will pay good money if you can capture and keep alive a certain brilliantly feathered parrot in the forest. You know it is illegal to capture this endangered parrot. You also have an idea of where it lives. What will you do?

5. Here's an alternative to formal burial on land. Ashes of cremated people are mixed with pH-neutral concrete and used to build *artificial reefs*. Each perforated concrete ball weighs 400 to 4,000 pounds and is designed to last 500 years. So far, 100,000 of these reefs are sustaining marine life in 1,500 locations (Figure 40.33). Proponents say that the burial costs less, does not waste land, and is better for the environment. Your thoughts?

IMPACTS, ISSUES *Bye-Bye, Blue Bayou*

Each Labor Day, the coastal Louisiana town of Morgan City celebrates the region's economic mainstays with the Louisiana Shrimp and Petroleum Festival. The state is the nation's third-largest petroleum producer and the leader in shrimp harvesting. But the petroleum industry's success may be contributing indirectly to the possible disappearance of the state's fisheries.

The global air temperature is rising, and fossil fuel burning is a culprit. Warmer air heats water near the sea surface, heated water expands, and so the sea level is rising. Warmer air also is melting ancient glaciers and ice caps, and meltwater is adding to the sea volume.

Since the 1940s, Louisiana has lost an area the size of Rhode Island to the sea. Low elevations along the United States coastline—including *14,720,000 acres* next to the Gulf of Mexico and Atlantic Ocean—may be one to three feet under water within fifty years!

With more than 40 percent of the nation's saltwater marshes, Louisiana has the most to lose (Figure 41.1). Its wetlands are already sinking; dams and levees interfere with the deposition of sediments that could replace those washed out to sea. The rise in sea level will make 70 percent of the wetlands *really* wet, with no land.

Is an ecological and economic disaster unfolding? Consider: More than 3 billion dollars' worth of shellfish and fish are harvested from Louisiana's wetlands each year. About 40 percent of North America's migratory ducks—more than 5 million birds—overwinter here. Also,

Figure 41.1 Glimpse through a cypress swamp in Louisiana. Inland saltwater intrusions are threatening these trees, which are adapted to freshwater habitats.

the big picture

energy input

producers

nutrient cycling

consumers

energy output

Ecosystem Structure Energy flows only one way through an ecosystem—into, through, and out of food webs. Nutrients enter an ecosystem, and some are lost to the environment, but most are cycled among the ecosystem's species.

Food Webs Food chains are linear sequences of the transfer of energy and nutrients, from autotrophs through heterotrophs. They cross-connect in grazing and detrital food webs, which consist of diverse producer and consumer species.

Louisiana's wetlands buffer low elevations inland from storm surges and hurricanes.

More bad news: Warmer water may promote algal blooms and huge fish kills. Also, populations of many pathogenic bacteria increase in warmer water, so more people may get sick after swimming in contaminated water or eating contaminated shellfish.

Inland, heat waves and wildfires will become more intense. Deaths related to heat stroke will climb. Warmer temperatures will permit mosquitoes to extend their inland ranges. Some mosquitoes are vectors for agents of malaria, West Nile virus, and other diseases.

Global warming is expected to increase evaporation and alter weather patterns. We can predict severe floods for some regions and prolonged drought for others, including Louisiana. Worldwide, 3 billion people may run out of fresh water within twelve years.

This chapter can get you thinking about energy flow through ecosystems, starting with energy inputs from the sun. It will show how ecosystems depend on inputs, cycling, and outputs of nutrients—and how nutrients are cycled on a global scale.

The chapter also can get you thinking more about a related concept of equal importance. We have become players in the global flows of energy and nutrients even before we fully comprehend how the game plans work. Decisions we make today about global warming and other environmental issues may affect the quality of human life and the environment far into the future.

 How Would You Vote?

Emissions from motor vehicles are a major source of greenhouse gases. Many people buy large vehicles that use more fuel but are viewed as safer and more useful. Should such vehicles be additionally taxed to discourage sales and offset their environmental costs? Can we expect better fuels as well as more of the fuel-efficient, larger vehicles that are becoming available? See the Media Menu for details, then vote online.

Ecological Pyramids Ecologists construct pyramid diagrams based on measurements of how much energy and nutrients enter an ecosystem, how much is actually captured by the primary producers, and how much of it becomes stored at each trophic level.

Global Cycles Hydrogen, oxygen, carbon, nitrogen, and phosphorus are among the most essential elements for all organisms. Environmental reservoirs hold the greatest amounts of these elements, which move slowly in global cycles and rapidly through ecosystems.

41.1 The Nature of Ecosystems

In the preceding chapter, we focused on how species interact over time and within the space of communities. This chapter continues with those interactions, but at a higher level of organization. Each community and its physical environment function as an integrated system.

OVERVIEW OF THE PARTICIPANTS

Diverse natural systems abound on Earth's surface. In climate, landforms, soil, vegetation, animal life, and other features, deserts differ from hardwood forests, which differ from tundra and the prairies. Reefs differ from the open ocean, which differs from streams and lakes. Despite their differences, all of the systems are alike in many aspects of their structure and function. Ecologists start from these shared similarities to make predictions about prospects for stability and change over the long term.

With few exceptions, the systems run on sunlight that autotrophs, or self-feeders, capture. Plants and phytoplankton are the main autotrophs. Remember, from Chapter 6, that both convert sunlight energy to chemical bond energy and use it to synthesize organic compounds from inorganic raw materials. They are **primary producers** for the system (Figure 41.2).

All other organisms in the system are **consumers**. These heterotrophs feed on the tissues, products, and remains of other organisms. We describe consumers by their diets. *Herbivores* eat plants. *Carnivores* eat flesh. *Parasites* live in or on a host and feed on its tissues. **Detritivores**, such as crabs, eat detritus—particles of decomposing organic matter. Finally, the wastes and remains of all organisms are degraded to inorganic compounds by cells of bacterial, protist, and fungal **decomposers**. Decomposers absorb most breakdown products, but some are released into the environment.

Some consumers do not fall into simple categories. Many are *omnivores*, which dine on animals, plants, fungi, protists, and even bacteria. A red fox is like this (Figure 41.3). If it comes across a dead bird, the fox will scavenge it. A *scavenger* is an animal that ingests dead plants, animals, or both all or some of the time. Vultures, termites, and hyenas are typical scavengers.

How are nutrients cycled in the system? Primary producers get hydrogen, oxygen, and carbon from water and carbon dioxide in the environment. They get phosphorus, nitrogen, and other minerals as well. These are the raw materials for biosynthesis. Later, decomposition of their organic wastes and remains releases nutrients back to the environment. Unless the

Figure 41.2 Model for ecosystems. Energy flows in one direction: into an ecosystem, through living organisms, and out from it. Nutrients are cycled among autotrophs and heterotrophs. In nearly all ecosystems, energy flow starts with autotrophs that capture energy from the sun.

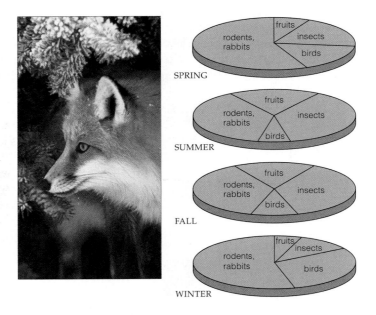

Figure 41.3 Red fox, an omnivore. Its diet shifts with seasonal changes in available food. Rodents, rabbits, and some birds make up the bulk of its diet in spring and winter. It eats more fruits and insects in the summer and fall.

released substances leave from the system, as when mineral ions end up dissolved in a stream that flows out of a meadow, producers typically take them up once again.

What we have just outlined is an **ecosystem**—an array of organisms and their physical environment, interacting through a one-way flow of energy and a cycling of raw materials. It is an open system, unable to sustain itself. *Energy inputs* to most ecosystems are in the form of sunlight. There may be *nutrient inputs*, as from a creek delivering dissolved minerals to a lake. There are also *energy outputs* and *nutrient outputs*.

Energy transfers, remember, cannot be 100 percent efficient (Section 5.1). Over time, the energy originally captured by producers escapes to the environment, mainly as metabolically generated heat.

STRUCTURE OF ECOSYSTEMS

We can classify all organisms of an ecosystem by their functional roles in a hierarchy of feeding relationships called **trophic levels** (*troph*, nourishment). "Who eats whom?" we may ask. If organism **B** eats organism **A**, energy is transferred from **A** to **B**. All organisms at a given trophic level are the same number of transfer steps away from the energy input into an ecosystem.

As one example, think about some organisms of a tallgrass prairie ecosystem. The flowering plants and other producers that tap energy from the sun are at the first trophic level. Plants are eaten by herbivores, such as cutworms, which are at the next trophic level. Cutworms are one of the primary consumers that are eaten by carnivores at the third trophic level, and so on up through tiers of trophic levels.

At each trophic level, organisms interact with the same sets of predators, prey, or both. Omnivores feed at several levels, so we would partition them among different levels or assign them to a level of their own.

A **food chain** is a straight-line sequence of steps by which energy originally stored in autotroph tissues moves to higher trophic levels. In one tallgrass prairie food chain, energy from a plant flows to a cutworm that eats its juicy parts, then to a garter snake that eats the cutworm, to a sandpiper that eats the snake, and finally to a marsh hawk that consumes the sandpiper. Figure 41.4 shows the chain.

Identifying a food chain is a simple way to start thinking about who eats whom in ecosystems. Bear in mind, however, that many different species are usually competing for food in complex ways. Tallgrass prairie producers (mainly flowering plants) do feed grazing mammals and herbivorous insects. Yet many more

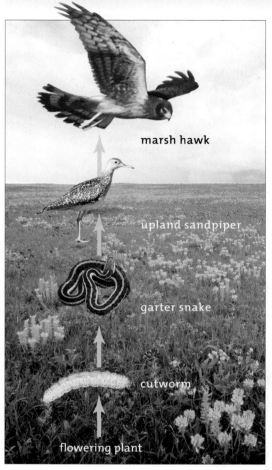

fifth trophic level
top carnivore
(fourth-level consumer)

marsh hawk

fourth trophic level
carnivore
(third-level consumer)

upland sandpiper

third trophic level
carnivore
(second-level consumer)

garter snake

second trophic level
herbivore
(primary consumer)

cutworm

first trophic level
autotroph
(primary producer)

flowering plant

Figure 41.4 Example of a simple food chain and its corresponding trophic levels in a tallgrass prairie.

species interact in nearly all ecosystems, especially at a lower trophic level. A number of food chains *cross-connect* with one another—as **food webs**. And that is the topic of the next section.

Primary producers, or self-feeders, use environmental energy and inorganic raw materials to make organic compounds.

Producers, consumers, and their physical environment make up an ecosystem. A one-way energy flow and a cycling of raw materials interconnect them.

A food chain, a straight-line sequence of who eats whom in an ecosystem, starts with autotrophs (producers) and proceeds through one or more levels of heterotrophs.

In ecosystems, a number of food chains cross-connect, forming food webs.

41.2 The Nature of Food Webs

Food chains cross connect with one another in food webs. By untangling the chains of many food webs, ecologists discovered patterns of organization. Those patterns reflect environmental constraints and the inefficiency of energy transfers from one trophic level to the next.

Plants capture only a fraction of the energy from the sun. They store half of that energy in chemical bonds of organic compounds in new plant tissues. They lose the rest as metabolic heat. Consumers tap into the energy that became stored in plant tissues, remains, and wastes. Then they too, lose metabolic heat. Taken together, *all of these heat losses represent a one-way flow of energy out of the ecosystem.*

When ecologists compared food chains in different kinds of food webs, a pattern emerged. In most cases, energy initially captured by producers passes through no more than four or five trophic levels. Even the rich ecosystems with complex food webs, such as the one in Figure 41.5, do not have lengthy food chains. The inefficiency of energy transfers may limit the sequence.

Field studies and computer simulations of aquatic and land food webs reveal more patterns. Chains in food webs tend to be shortest where environmental conditions can vary widely over time. Food chains tend to be longer in habitats that are more stable, such as ocean depths. The most complex webs tend to have many herbivorous species, as happens in grasslands. By comparison, the food webs with fewer connections tend to have more carnivores.

higher trophic levels

Complex array of carnivores, omnivores, parasites, detritivores, decomposers, and other consumers. Many feed at more than one trophic level all the time, seasonally, or whenever an opportunity presents itself.

second trophic level

Primary consumers (e.g., herbivores, detritivores, and decomposers)

first trophic level

Primary producers

marsh hawk

upland sandpiper

crow

garter snake

frog

spider

weasel

badger

coyote

clay-colored sparrow

prairie vole

pocket gopher

ground squirrel

earthworms, insects (e.g., grasshoppers, cutworms)

grasses, composites

Figure 41.5 Sample of some of the interactions in a tallgrass prairie food web.

Read Me First!
and watch the narrated
animation on food webs

Figure 41.6 The one-way flow of energy through (**a**) a grazing food web and (**b**) a detrital food web.

Energy from producers, the organisms closest to a primary source, flows in one direction through two kinds of webs. In a **grazing food web**, energy flows mostly into herbivores, carnivores, then decomposers. In a **detrital food web**, energy from producers flows mainly into detritivores and decomposers. Figure 41.6 summarizes the flow through these food webs.

In nearly all ecosystems, both kinds of webs cross-connect. For example, in a rocky intertidal ecosystem, energy captured by algae flows to snails, which are eaten by herring gulls as part of a grazing food web. But the gulls also hunt crabs, which are among the primary consumers in the detrital food web.

The amount of energy that moves through the two kinds of food webs differs among ecosystems, and it often varies with the seasons. In most cases, however, most of the energy stored in producer tissues moves through *detrital* food webs. Think of cattle, grazing heavily in a pasture. About half the energy stored in the grass plants enters the grazers. But cattle cannot tap into all of the stored energy. A lot is still present in undigested plant parts and feces, and decomposers and detritivores go to work. Similarly, in marshes, most of the energy initially stored in the marsh grass tissues enters detrital food webs when the plants die.

The inherent inefficiency in energy transfers between trophic levels limits the length of food chains.

Tissues of living photosynthetic organisms are the basis for grazing food webs. Remains and wastes of these organisms are the basis for detrital food webs.

In nearly all ecosystems, both types of food webs prevail and interconnect.

41.3 DDT in Food Webs

We have released thousands of synthetic chemicals and other new organic compounds into the environment. Ecosystem analysis helps us predict their side effects. The danger is that researchers may not have identified all of the variables. The most crucial variable may be the one they do not yet know.

Consider how DDT once entered food webs. This synthetic organic pesticide is nearly insoluble in water, so you might expect it to act only where it is applied. But winds carry DDT vapor, and water can move fine particles of it. Also, DDT is highly soluble in fats, so it can accumulate in animal tissues—which makes it a candidate for **biological magnification**. By this occurrence, a substance that degrades slowly or not at all becomes more concentrated in consumer tissues as it moves through ever higher trophic levels (Figure 41.7). This is not good. DDT and its modified forms interfere with metabolic activities and are toxic to many aquatic and terrestrial animals.

Decades ago, DDT infiltrated food webs and acted on organisms in ways no one had predicted. Where it was sprayed to control Dutch elm disease, songbirds died. In small streams flowing in forests where it was applied to kill budworm larvae, fishes died. In fields sprayed to control one kind of pest, new pests moved in. DDT was also killing the natural predators that help keep pest populations in check!

Figure 41.8 Rachel Carson, who helped awaken public interest in the impact of human activities on nature.

Side effects of biological magnification showed up far away from places where DDT was applied, and much later in time. Most vulnerable were the brown pelicans, bald eagles, ospreys, and other birds that are top carnivores. Some DDT breakdown products disrupted their physiology. As one outcome, bird eggs had brittle shells, many chick embryos did not hatch, and some bird species faced extinction.

Rachel Carson, a respected scientist, sounded the alarm (Figure 41.8). Carson began looking into the effects of pesticides on wildlife after learning that dead birds inside a local wildlife sanctuary showed symptoms of poisoning. The sanctuary had been sprayed with DDT to control mosquitoes.

Independent, critical research on pesticides was almost nonexistent. Carson methodically developed information about the harmful effects of widespread pesticide use. She published her findings in 1962 as a small book, *Silent Spring*. The public embraced it. But the manufacturers of chemicals viewed Carson as a threat to pesticide sales and quickly mounted a campaign to discredit her. At the time, Carson knew she had terminal cancer. Yet she vigorously defended her position. Later, her work became the impetus for the environmental movement in the United States. One line of investigation became the field known as conservation biology.

The United States has banned DDT except when required to protect the public health against insect vectors for disease. Today, there is consensus that DDT was overused and has no place in agriculture. Developing countries still use it mainly to control mosquitoes. Traces are still found in many organisms all over the world, including humans.

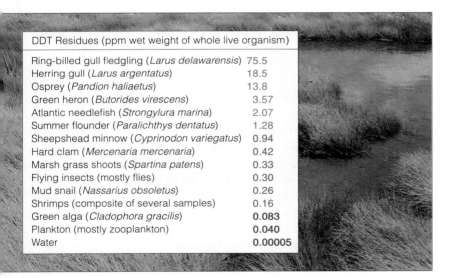

DDT Residues (ppm wet weight of whole live organism)	
Ring-billed gull fledgling (*Larus delawarensis*)	75.5
Herring gull (*Larus argentatus*)	18.5
Osprey (*Pandion haliaetus*)	13.8
Green heron (*Butorides virescens*)	3.57
Atlantic needlefish (*Strongylura marina*)	2.07
Summer flounder (*Paralichthys dentatus*)	1.28
Sheepshead minnow (*Cyprinodon variegatus*)	0.94
Hard clam (*Mercenaria mercenaria*)	0.42
Marsh grass shoots (*Spartina patens*)	0.33
Flying insects (mostly flies)	0.30
Mud snail (*Nassarius obsoletus*)	0.26
Shrimps (composite of several samples)	0.16
Green alga (*Cladophora gracilis*)	0.083
Plankton (mostly zooplankton)	0.040
Water	0.00005

Figure 41.7 Biological magnification in a Long Island, New York, estuary. In 1967, George Woodwell, Charles Wurster, and Peter Isaacson were aware that DDT exposure and wildlife deaths were connected. For instance, residues in birds that died from DDT poisoning were 41–295 parts per million (ppm) and 1–26 ppm in some fishes. In their study, some measured DDT levels were below lethal thresholds but still high enough to interfere with reproduction.

41.4 Studying Energy Flow Through Ecosystems

Ecologists measure the amount of energy and nutrients entering an ecosystem, how much is captured, and the proportion stored in each trophic level.

WHAT IS PRIMARY PRODUCTIVITY?

The rate at which producers capture and store energy in their tissues during a given interval is the **primary productivity** of an ecosystem. How much energy gets stored depends on (1) how many producers there are and (2) the balance between photosynthesis (energy trapped) and aerobic respiration (energy used). *Gross* primary production is all energy initially trapped by the producers. *Net* primary production is the fraction of trapped energy that producers funnel into growth and reproduction. **Net ecosystem production** is the gross primary production *minus* the energy used by producers and by detritivores and decomposers in soil. (In both cases, the subtracted amount of energy would not be available to organisms that eat plants.)

Many factors impact net production, its seasonal patterns, and its distribution in the habitat. This is true both on land and in the seas (Figure 41.10). For example, energy acquisition and storage depend in part on the body size and form of primary producers, mineral availability, the temperature range, and the amount of sunlight and rainfall during each growing season. The harsher the conditions are, the less new plant growth per season, and the lower productivity.

ECOLOGICAL PYRAMIDS

Ecologists often represent the trophic structure of an ecosystem in the form of an ecological pyramid. In such pyramids, the primary producers form a base for successive tiers of consumers above them.

A **biomass pyramid** depicts the dry weight of all of an ecosystem's organisms at each tier. Figure 41.9 shows a biomass pyramid for one aquatic ecosystem. The amounts measured are grams per square meter at some specified time. Most commonly, the primary producers have most of the biomass in pyramids like this, and top carnivores are few. But some biomass pyramids are "upside-down," with the smallest tier on the bottom. This happens in springtime blooms of phytoplankton, which grow and reproduce quickly. The primary producers of these aquatic communities support a larger biomass of zooplankton, which eat them about as fast as they can reproduce.

An **energy pyramid** illustrates how the amount of usable energy diminishes as it is transferred through an ecosystem. Sunlight energy is captured at the base

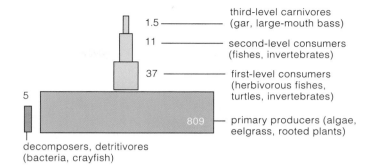

Figure 41.9 Biomass pyramid for Silver Springs, a small aquatic ecosystem. Biomass decreases in successively higher tiers. In some ecosystems, autotrophs are eaten almost as fast as they grow and reproduce. Biomass accumulates faster in consumers, so a biomass pyramid would be upside down.

*Labels on figure:*
1.5 — third-level carnivores (gar, large-mouth bass)
11 — second-level consumers (fishes, invertebrates)
37 — first-level consumers (herbivorous fishes, turtles, invertebrates)
5
809 — primary producers (algae, eelgrass, rooted plants)
decomposers, detritivores (bacteria, crayfish)

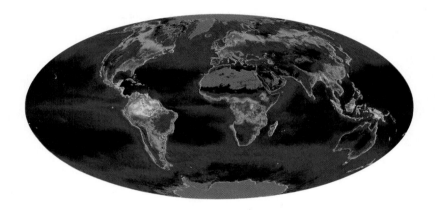

Figure 41.10 Summary of satellite data on net primary productivity during 2002. Productivity is coded as *red* (highest) down through *orange, yellow, green, blue,* and *purple* (lowest). Although average productivity per unit of sea surface is lower than it is on land, total productivity on land and in seas is about equal, because most of Earth's surface is covered by water.

(first trophic level) and declines through successive levels to its tip (the top carnivores). Energy pyramids have a large energy base at the bottom, so they are always "right-side up." As you will see in the next section, they can give a clear picture of energy flow.

Gross primary productivity is an ecosystem's total rate of photosynthesis during a specified interval. The net amount is the rate at which primary producers store energy in tissues in excess of their rate of aerobic respiration. Heterotrophic consumption affects the rate of energy storage.

The trophic structure of an ecosystem can be represented by an ecological pyramid. Biomass pyramids may be top- or bottom-heavy depending on the ecosystem. In contrast, an energy pyramid always has the largest tier on the bottom.

41.5 Energy Flow Through Silver Springs

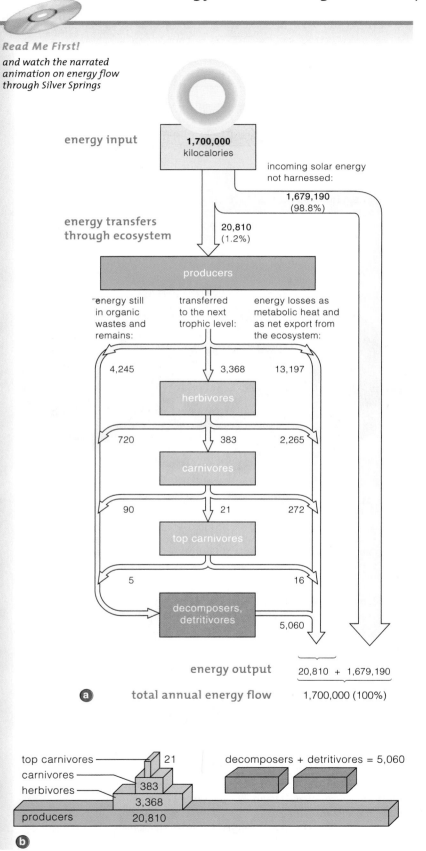

energy input **1,700,000** kilocalories

incoming solar energy not harnessed:

1,679,190 (98.8%)

energy transfers through ecosystem

20,810 (1.2%)

producers

energy still in organic wastes and remains: | transferred to the next trophic level: | energy losses as metabolic heat and as net export from the ecosystem:

4,245 | 3,368 | 13,197

herbivores

720 | 383 | 2,265

carnivores

90 | 21 | 272

top carnivores

5 | | 16

decomposers, detritivores

5,060

energy output   20,810 + 1,679,190

total annual energy flow   1,700,000 (100%)

(a)

top carnivores — 21
carnivores —
herbivores — 383
3,368
producers   20,810

decomposers + detritivores = 5,060

(b)

Energy flows into food webs from an outside source, most often the sun. Energy leaves mainly by losses of metabolic heat, which each organism generates.

Imagine you are with ecologists who want to gather the data to construct an energy pyramid for a small freshwater spring over the course of one year. You observe them as they measure the energy that an individual organism of each type takes in, loses as metabolic heat, stores in its body tissues, and loses in waste products. You see that they multiply the energy per individual by population size, then calculate energy inputs and outputs. In this way, they're able to express the flow of energy per unit of water (or land) per unit of time.

Figure 41.11*a* shows the data from a long-term study of a grazing food web in an aquatic ecosystem. It shows some of the calculations that the ecologists used to depict the energy flow in pyramid form in Figure 41.11*b*.

Given the metabolic demands of organisms and the amount of energy lost in their organic wastes, only 6 to 16 percent of the energy entering one trophic level becomes available for organisms at the next level.

Figure 41.11 (**a**) Breakdown of the annual energy flow through Silver Springs, Florida, as measured in kilocalories/square meter/year.

Most of the primary producers in this small spring are aquatic plants. Most of the carnivores are insects and small fishes; the top carnivores are larger fishes. The original energy source, sunlight, is available all year. The spring's detritivores and decomposers cycle organic compounds from the other trophic levels.

The producers trapped 1.2 percent of the incoming solar energy, and only a little more than a third of that amount became fixed in new plant biomass. The producers used more than 63 percent of the fixed energy for their own metabolism.

About 16 percent of the fixed energy was transferred to the herbivores, and most of this was used for metabolism or transferred to detritivores and decomposers.

Of the energy that transferred to herbivores, only 11.4 percent reached the next trophic level (carnivores). About 5.5 percent of the energy in lower-level carnivores flowed to the top carnivores.

By the end of the specified time interval, all of the 20,810 kilocalories of energy that had been transferred through the system appeared as metabolically generated heat.

(**b**) Pyramid of energy flow through Silver Springs, in kilocalories/square meter/year. This is a simple summary of the data shown in more detail above. Compare Figure 41.9, the biomass pyramid for this same ecosystem.

41.6 Overview of Biogeochemical Cycles

Primary productivity depends on availability of fresh water and on nutrients dissolved in it.

In a **biogeochemical cycle**, an essential element moves from the environment, through ecosystems, then back to the environment. No other element can directly or indirectly fulfill the metabolic role of such elements, or **nutrients**, which is why we call them essential. As you read earlier, oxygen, hydrogen, carbon, nitrogen, and phosphorus are among them.

Figure 41.12 is a model for these cycles. Transfers to and from environmental reservoirs are usually far slower than rates of exchange among organisms of an ecosystem. Water is the main source for hydrogen and oxygen. Gaseous or ionized forms of other elements are dissolved in it. Solid forms of elements are tied up in rocks or sediments.

Nutrients move into and out of ecosystems by way of natural geologic processes. Weathering of rocks is a common source of nutrient inputs into an ecosystem. Erosion and runoff put nutrients into streams that carry them away. Most often, the quantity of a nutrient being cycled through an ecosystem each year is greater than the amount entering and leaving.

Decomposers help cycle nutrients in ecosystems. Various prokaryotic species help transform solids and ions into gases and then back again. By their actions, they convert some of the elements that are nutrients to forms that primary producers can take up.

Three types of biogeochemical cycles are based on parts of the environment that act as reservoirs for specific elements. In the *hydrologic* cycle (global water cycle), oxygen and hydrogen move as molecules of water. In *atmospheric* cycles, some gaseous form of the nutrient is the one available to ecosystems. Carbon and nitrogen cycles are examples. Phosphorus and other solid nutrients that have no gaseous form move in *sedimentary* cycles. They accumulate on the seafloor and eventually return to land by way of geological uplifting, which often takes millions of years. Earth's crust is the largest reservoir for nutrients that have sedimentary cycles.

Nutrient availability influences the primary productivity on which ecosystems depend. In a biogeochemical cycle, a nutrient moves slowly through the environment, then rapidly among organisms, and back to environmental reservoirs.

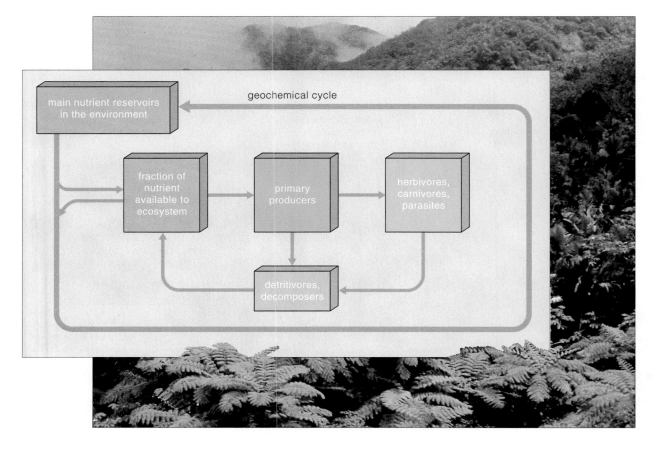

Figure 41.12 One generalized model of nutrient flow through a land ecosystem. The overall movement of nutrients from the physical environment, through organisms, and then back to the environment is a biogeochemical cycle.

41.7 Global Cycling of Water

On land, the availability of water, and nutrients dissolved in it, is not plentiful all of the time in all ecosystems. The variation affects primary productivity.

THE HYDROLOGIC CYCLE

Driven by solar energy, Earth's waters slowly move from the ocean into the atmosphere, to land, and back to the ocean—the main reservoir. Figure 41.13 shows this **hydrologic cycle**. Water evaporating into the lower atmosphere stays aloft as vapor, clouds, and ice crystals, then falls mainly as rain and snow. Ocean circulation and wind patterns influence the cycle.

Water moves nutrients into and out of ecosystems. A **watershed** is any region where precipitation flows into a single stream or river. Watersheds may be as small as the area that drains into a stream or as vast as the Amazon River or Mississippi River basin. Most water entering a watershed seeps into soil or joins surface runoff into streams. Plants take up water from soil and lose it by transpiration.

You might think that surface runoff would quickly leach mineral ions. But studies in New Hampshire's Hubbard Brook watershed show this is not the case. In undisturbed forests in this watershed, each hectare lost only about eight kilograms of calcium per year. Calcium loss was replaced by rainfall and weathering. Also, tree roots mine the soil, so calcium was continually being stored in the biomass of tree tissues. Experimental deforestation caused an imbalance in the nutrient inputs and outputs (Figure 41.14).

THE WATER CRISIS

Our planet has a lot of water, but most of it is too salty for consumption or agriculture. If all of Earth's waters were contained in a bathtub, the freshwater portion from lakes, rivers, reservoirs, groundwater, and other places would barely fill a teaspoon.

Agriculture accounts for two-thirds of our water usages, and irrigation alters the suitability of land for agriculture. The piped-in water commonly has high

Read Me First!
and watch the narrated animation on the hydrologic cycle

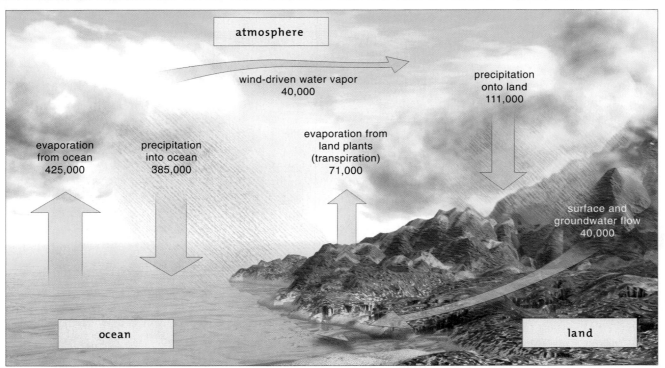

Figure 41.13 The hydrologic cycle. Water moves from the ocean to the atmosphere, land, and back. Arrows identify processes that move water, as measured in cubic kilometers per year. With 1,370,000,000 cubic kilometers, the ocean is the main reservoir. The next largest, polar ice and glaciers, locks up 29,000,000. Groundwater makes up only 4,000,000, lakes and rivers only 241,000, soil 67,000, and the atmosphere 14,000 cubic kilometers.

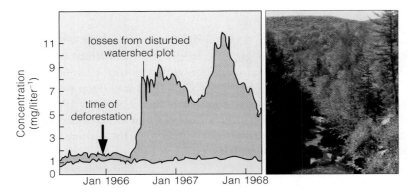

Figure 41.14 Experimental deforestation of New Hampshire's Hubbard Brook watershed. Researchers stripped vegetation from forest plots but did not disturb the soil. They applied herbicides for three years to prevent regrowth. They monitored surface runoff flowing over concrete catchments on its way to a stream below. Concentrations of calcium ions and other minerals were compared against concentrations in water passing over a control catchment, positioned in an undisturbed area.

Calcium losses were *six times* greater in deforested plots. Removing all vegetation from such forests clearly alters nutrient outputs in ways that can disrupt nutrient availability for an entire ecosystem.

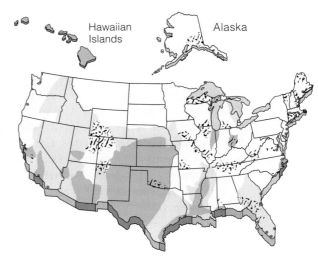

Figure 41.15 Aquifer depletion, seawater intrusion, and groundwater contamination in the United States. *Green* signifies high overdrafts, *gold*, moderate overdrafts, and *pale yellow*, insignificant withdrawals. Shaded areas are sites of major groundwater pollution. *Blue* squares indicate saltwater intrusion from nearby seas.

concentrations of mineral salts. Where the soil drains poorly, evaporation causes **salinization**, a build-up of salt in soil that stunts crop plants and decreases yields.

Soil and aquifers hold **groundwater**. About half of the United States population taps into groundwater as a source of drinking water. Chemicals leached from landfills, hazardous waste dumps, and underground tanks that store gasoline, oil, and some solvents often contaminate it. Unlike flowing streams that recover fast, polluted groundwater is difficult and expensive to clean up.

Groundwater overdrafts, or the amount that nature does not replenish, are high in many areas. Figure 41.15 shows some regions of aquifer depletion in the United States. Overdrafts have now depleted half of the great Ogallala aquifer, which supplies irrigation water for 20 percent of the Midwest's croplands.

Inputs of sewage, animal wastes, and many toxic chemicals from power-generating plants and factories make water unfit to drink. Sediments and pesticides run off from fields into water, along with phosphates and other nutrients that promote algal blooms. The pollutants accumulate in lakes, rivers, and bays before reaching the ocean. Many cities all over the world are still dumping untreated sewage into coastal waters.

If current rates of human population growth and water depletion continue, the amount of fresh water available for everyone will soon be 55 to 66 percent less than it was in 1976. In this past decade, thirty-three nations have already engaged in conflicts over reductions in water flow, pollution, and silt buildup

in aquifers, rivers, and lakes. Among the squabblers are the United States and Mexico, Pakistan and India, and Israel and the Palestinian territories.

Could **desalinization**—the removal of salt from seawater—meet our water needs? Salt can be removed by distillation or pushing water through membranes. The processes require fossil fuels, which makes them more feasible in Saudi Arabia and other countries with small populations and big fuel reserves. Most likely, desalinization won't be cost-effective for large-scale agriculture. It also produces mountains of salts.

We may be in for upheavals and wars over water rights. Does this sound far-fetched? Consider the new dam across the Euphrates River. By building huge dams and irrigation systems at the headwaters of the Tigris and Euphrates rivers, Turkey can, in the view of one of its dam site managers, shut off water flow into Syria and Iraq for as long as eight months "to regulate their political behavior." Regional, national, and global planning for the future is long overdue.

In the hydrologic cycle, water slowly moves on a global scale from the world ocean (the main reservoir), through the atmosphere, onto land, then back to the ocean.

Plants stabilize the soil and absorb dissolved minerals. By doing so, they minimize the loss of soil nutrients in runoff.

Aquifers that supply much of the world's drinking water are becoming polluted and depleted. Regional conflicts over access to clean, drinkable water are likely to increase.

41.8 Carbon Cycle

Most of the world's carbon is locked in ocean sediments and rocks. It moves into and out of ecosystems in gaseous form, so its movement is said to be an atmospheric cycle.

Carbon moves through the lower atmosphere and all food webs on its way to and from the ocean's water, sediments, and rocks. Its global movement is called the **carbon cycle** (Figure 41.16). Carbon dioxide gas is released as cells engage in aerobic respiration, when fossil fuels or forests burn, and when volcanoes erupt. It is removed from air and water when it is taken up by organisms that carry out photosynthesis.

Carbon dioxide gas, or CO_2, is the most abundant form of carbon in the atmosphere. Carbon dissolved in ocean water is mainly combined with oxygen, as carbonate or bicarbonate. For that reason, the cycle is sometimes called the *carbon–oxygen cycle*.

Why doesn't all the CO_2 dissolved in warm ocean surface waters escape into the atmosphere? Driven by the winds and regional differences in water density, water makes a gigantic loop from the surface of the Pacific and Atlantic oceans down to the Atlantic and Antarctic seafloors. The CO_2 moves into deep storage reservoirs before the seawater loops back up (Figure 41.17). This loop affects carbon's global distribution.

When photosynthetic autotrophs fix carbon, they lock billions of metric tons of carbon atoms in organic compounds each year (Chapter 6). The average time that an ecosystem holds a given carbon atom varies greatly. As one example, organic wastes and remains decompose so quickly in tropical rain forests that not much carbon accumulates at the soil surfaces. In bogs, marshes, and similar anaerobic habitats, decomposers cannot degrade organic compounds to smaller bits, so carbon gradually accumulates in peat.

Also, in ancient aquatic ecosystems, carbon became incorporated into the shells and other hard body parts of many marine species. Those organisms died,

Read Me First!
and watch the narrated animation on the carbon cycle

Figure 41.16 Global carbon cycle. Carbon's movement through typical marine ecosystems (**a**) and through ecosystems on land (**b**). *Gold* boxes show the main carbon reservoirs. The vast majority of carbon atoms are locked in sediments and rocks, followed by increasingly lesser amounts in ocean water, soil, and atmosphere, and finally in biomass. The unit of measure for the lists below is 10^{15} grams:

Main carbon reservoirs and holding stations

Sediments and rocks	77,000,000
Ocean (dissolved forms)	39,700
Soil	1,500
Atmosphere	750
Biomass on land	715

Annual fluxes in the global distribution of carbon

From atmosphere to plants (carbon fixation)	120
From atmosphere to ocean	107
To atmosphere from ocean	105
To atmosphere from plants	60
To atmosphere from soil	60
To atmosphere from fossil fuel burning	5
To atmosphere from net destruction of plants	2
To ocean from runoff	0.4
Burial in ocean sediments	0.1

diffusion between atmosphere and ocean

combustion of fossil fuels

bicarbonate and carbonate dissolved in ocean water

photosynthesis   **aerobic respiration**

marine food webs
producers, consumers, decomposers, detritivores

incorporation into sediments   death, sedimentation   uplifting over geologic time

sedimentation

marine sediments, including formations with fossil fuels

a

sank through water, then were buried in sediments. Carbon stayed buried for many millions of years until geologic forces slowly uplifted part of the seafloor.

At present, deforestation, the burning of wood and fossil fuels, and other human activities release more carbon into the atmosphere than can be cycled back to ocean reservoirs by natural processes.

Only about 2 percent of the excess carbon entering the atmosphere will become dissolved in ocean water. Most researchers now suspect that the carbon build-up in the atmosphere is amplifying the greenhouse effect. In other words, the increase may be a contributor to global warming. The next section takes a look at this possibility and its environmental implications.

Most carbon is trapped in the Earth's rocks and sediments, with some dissolved in ocean waters. Carbon moves between the main reservoirs and ecosystems mostly as carbon dioxide.

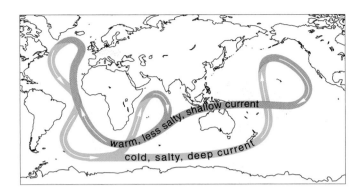

Figure 41.17 One loop of ocean water that delivers carbon dioxide to carbon's deep ocean reservoir. It sinks in the cold, salty North Atlantic and rises in the warmer Pacific.

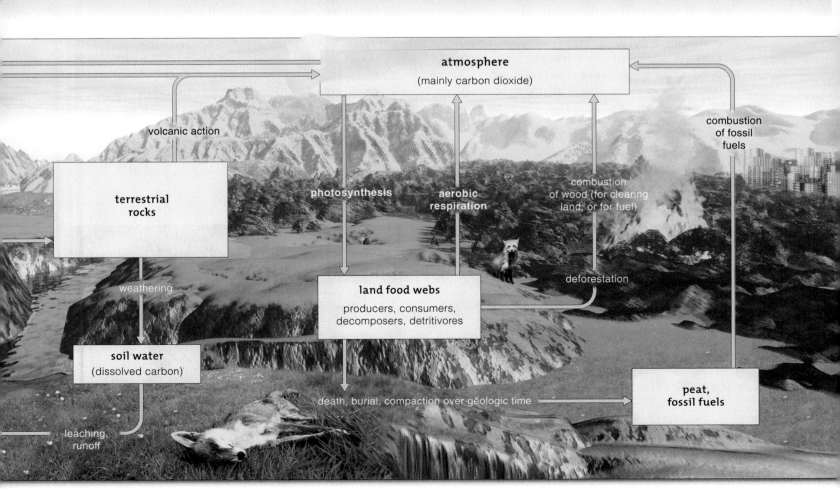

atmosphere (mainly carbon dioxide)

volcanic action

combustion of fossil fuels

terrestrial rocks

photosynthesis

aerobic respiration

combustion of wood (for clearing land; or for fuel)

weathering

deforestation

land food webs
producers, consumers, decomposers, detritivores

soil water (dissolved carbon)

death, burial, compaction over geologic time

peat, fossil fuels

leaching, runoff

b

41.9 Greenhouse Gases, Global Warming

The atmospheric concentrations of gaseous molecules help shape the average temperature near the Earth's surface. Human activities are altering atmospheric gas levels in ways that may cause dramatic climate change.

Concentrations of some gaseous molecules in Earth's atmosphere play a profound role in determining the average temperature near its surface. Temperature, in turn, has far-reaching effects on the global climate.

Atmospheric molecules of carbon dioxide, water, nitrous oxide, methane, and chlorofluorocarbons are among the main players in interactions that affect global temperature. Collectively, these gases act like the panes of glass in a greenhouse—hence the name, "greenhouse gases." Wavelengths of visible light pass through these gases to Earth's surface, which absorbs them and emits longer, infrared wavelengths—heat. Greenhouse gases impede the escape of heat energy from Earth into space. How? The gaseous molecules absorb the longer wavelengths, then radiate much of it back toward Earth (Figure 41.18).

Constant reradiation of heat by greenhouse gases occurs lockstep with the constant bombardment and absorption of wavelengths from the sun. As heat builds up in the lower atmosphere, the temperature near the Earth's surface rises. The warming action is called the **greenhouse effect**. Without it, Earth's surface would be cold and unable to support life.

In the 1950s, laboratory researchers on Hawaii's highest volcano set out to measure the atmospheric concentrations of greenhouse gases. That remote site is almost free of local airborne contamination; it is also representative of overall atmospheric conditions

Figure 41.19 Graphs of recent increases in four categories of atmospheric greenhouse gases. A key factor is the huge numbers of gasoline-burning vehicles in large cities. *Above,* Mexico City on a smoggy morning. With 10 million residents, it is the world's largest city.

for the Northern Hemisphere. What did they find? Briefly, carbon dioxide concentrations follow annual cycles of primary production. They drop during the summer, when photosynthesis rates are highest. They rise in the winter, when photosynthesis rates decline but aerobic respiration is still going on.

Alternating troughs and peaks along the graph line in Figure 41.19*a* are annual lows and highs of global carbon dioxide concentrations. For the first time, we could see the integrated effects of carbon balances for an entire hemisphere. Notice the midline of the troughs and peaks in the cycle. It shows that carbon dioxide concentration is steadily increasing—as are the concentrations of other major greenhouse gases.

Atmospheric levels of greenhouse gases are far higher than they were in most of the past. Carbon

ⓐ Wavelengths in rays from the sun penetrate the lower atmosphere, and they warm the Earth's surface.

ⓑ The surface radiates heat (infrared wavelengths) to the atmosphere. Some heat escapes into space. But greenhouse gases and water vapor absorb some infrared energy and radiate a portion of it back toward Earth.

ⓒ Increased concentrations of greenhouse gases trap more heat near Earth's surface. Sea surface temperatures rise, so more water evaporates into the atmosphere. Earth's surface temperature rises.

Figure 41.18 The greenhouse effect.

a Carbon dioxide (CO_2). Of all human activities, the burning of fossil fuels and deforestation contribute the most to increasing atmospheric levels.

c Methane (CH_4). Production and distribution of natural gas for use as fuel adds to the methane released by some bacteria that live in swamps, rice fields, landfills, and in the digestive tract of cattle and other ruminants.

b CFCs. Until restrictions were in place, CFCs were widely used in plastic foams, refrigerators, air conditioners, and industrial solvents.

d Nitrous oxide (N_2O). Denitrifying bacteria produce N_2O in metabolism. Also, fertilizers and animal wastes release enormous amounts; this is especially so for large-scale livestock feedlots.

Figure 41.20 Recorded changes in global temperature between 1880 and 2000. At this writing, the hottest year on record was 1998.

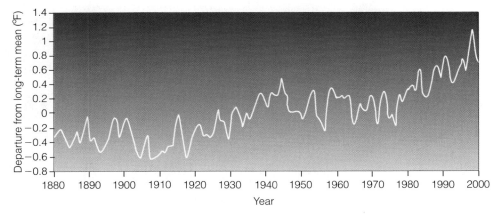

dioxide may be at its highest level since 420,000 years ago, and possibly since 20 million years ago. There is a growing consensus that the rise in greenhouse gases is caused by some human activities, mainly burning of fossil fuels. The big worry is that the increase may have widespread environmental consequences.

The increase in greenhouse gases may be a factor in **global warming**, a long-term rise in temperature near Earth's surface. Since direct atmospheric readings started in 1861, the lower atmosphere's temperature has risen by more than 1°F, mostly since 1946 (Figure 41.20). Also since then, nine of the ten hottest years on record have occurred between 1990 and the present. Data from satellites, weather stations and balloons,

research ships, and supercomputer programs suggest that irreversible climate changes are already under way. Polar ice is melting; glaciers are retreating. This past century, the sea level may have risen as much as twenty centimeters (eight inches). We can expect continued temperature increases to have drastic effects on climate. As evaporation increases, so will global precipitation. Intense rains and flooding are expected to become more frequent in some regions.

It bears repeating: As investigations continue, a key research goal is to investigate all of the variables in play. With respect to the consequences of global warming, the most crucial variable may be the one we do not know.

41.10 Nitrogen Cycle

Gaseous nitrogen makes up about 80 percent of the lower atmosphere. Successively smaller reservoirs are seafloor sediments, ocean water, soil, biomass on land, nitrous oxide in the atmosphere, and marine biomass.

THE CYCLING PROCESSES

Gaseous nitrogen (N_2) travels in an atmospheric cycle called the **nitrogen cycle**. Triple covalent bonds join its two atoms ($N\equiv N$). Volcanic action and lightning convert some gaseous nitrogen into forms that enter food webs. Far more enters by **nitrogen fixation**. By this metabolic process, certain bacteria in soil split all three bonds in gaseous nitrogen, then use the atoms to form ammonia (NH_3). Later, ammonia is converted to ammonium (NH_4^+) and nitrate (NO_3^-), the two forms of nitrogen that most plants can easily take up (Section 26.2 and Figure 41.21).

Nitrogen-fixing bacteria live in the soil, in aquatic habitats, and as photosynthetic mutualists in lichens. Others are mutualists with plants, as when *Rhizobium* forms nodules on the roots of peas or other legumes. Collectively, such bacteria fix about 200 million metric tons of nitrogen each year.

Nitrogen is often a limiting factor for plant growth. Soils that are low in nitrogen put legumes, with their

Read Me First!

and watch the narrated animation on the nitrogen cycle

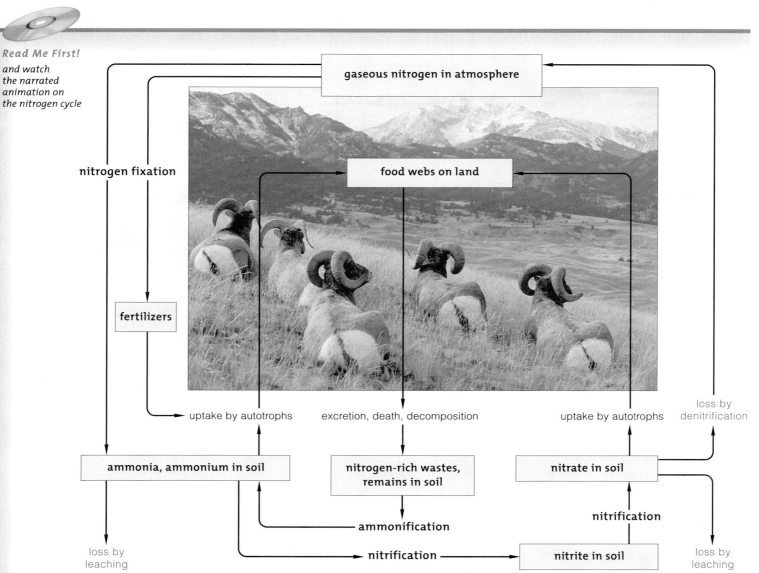

Figure 41.21 The nitrogen cycle in an ecosystem on land. Activities of nitrogen-fixing bacteria make nitrogen available to plants. Other bacterial species cycle nitrogen to plants. They break down organic wastes to ammonium and nitrates.

bacterial symbionts, at a competitive advantage. The legumes have a nitrogen source that other plants do not. In itself, supporting bacteria in root nodules is costly in some cases; legumes lose some metabolic products to the bacteria. In nitrogen-rich soils, plant species that don't have bacterial symbionts and don't pay the metabolic price often outcompete the legumes.

The nitrogen incorporated into plant tissues moves through trophic levels of ecosystems and ends up in nitrogen-rich wastes and remains. By **ammonification**, bacteria and fungi break down nitrogenous materials, forming ammonium. They use some ammonium and release the rest to soil. Plants take it up, as do some nitrifying bacteria. In the first step of **nitrification**; these bacteria produce nitrite (NO_2^-) when they strip electrons from ammonium. Other nitrifying bacteria use nitrite in reactions that form nitrate (NO_3^-).

Ecosystems lose nitrogen through **denitrification**. By this process, certain bacteria convert nitrate or nitrite to gaseous nitrogen or nitrogen oxide (NO_2). Most denitrifying bacteria are anaerobic; they live in waterlogged soils and aquatic sediments.

Ammonium, nitrite, and nitrate are also lost from a land ecosystem through leaching and runoff. Leaching, recall, is the removal of some soil nutrients as water runs through it. Leaching removes nitrogen from land ecosystems and adds it to aquatic ones.

HUMAN IMPACT ON THE NITROGEN CYCLE

Nitrogen losses through deforestation and grassland conversion for agriculture are huge. With each forest clearing and harvest, nitrogen stored in plant biomass is removed. Erosion and leaching from exposed soils take even more away.

Many farmers counter nitrogen losses by rotating crops. For example, a farmer might grow wheat in a field some years, and plant legumes in the same field during alternate years. The nitrogen-rich remains of legumes (and their bacterial symbionts) are plowed under to enrich the soil for wheat production.

Increasingly, farmers are also using nitrogen-rich chemical fertilizers. The fertilizers are manufactured by combining hydrogen gas and nitrogen gas at high pressure and temperature to yield ammonia. Use of chemical fertilizers can greatly increase crop yields per square area, but may also have harmful effects.

For example, fertilizer applications can alter soil chemistry. Use of nitrogen fertilizers adds hydrogen ions to the soil, making it more acid. These ions can displace other positively-charged mineral ions, such as magnesium and calcium, from binding sites on soil

Figure 41.22 Dead and dying trees in Smoky Mountain National Park. Forests are among the casualties of nitrogen oxides and other forms of air pollution.

particles. As a result, the rates at which these vital mineral nutrients are leached from the soil increase.

Deposition of nitrogen in acid rains can have the same effect as overfertilization. Fossil fuel burned in power plants and vehicles releases nitrogen oxides, which contribute to global warming and to acid rain. Winds often carry these air pollutants far from their sources (Figure 41.22). It is estimated some Eastern European forests are now receiving 10 times their normal level of nitrogen inputs from pollutants.

The responses of plant species to high nitrogen levels vary. As a result, any increases in nitrogen can disrupt the balance among plants that are competing in communities, and diversity may decline. The plant species best able to deal with additional nitrogen can outgrow species that are less responsive.

The addition of nitrogen also can disrupt aquatic ecosystems. About half of the nitrogen in fertilizers applied to fields is not taken up by crops. It runs off into rivers, lakes, and estuaries. Sewage from human cities and animal wastes adds more nitrogen to these waters. Enrichment of aquatic ecosystems by nitrogen and phosphorus can encourage harmful algal blooms, as you will see in the next section.

Most nitrogen is in gaseous form in the atmosphere, not in the ionized forms that plants generally can take up. These forms are mainly ammonium and nitrate.

Nitrogen-fixing bacteria convert nitrogen gas to ammonium. Bacteria and fungi break down wastes and remains and release ammonium.

Nitrates are produced by nitrifying bacteria. Denitrifying bacteria convert nitrite and nitrate back to gaseous forms.

Human activities are depleting nitrogen in some ecosystems and adding it to others.

41.11 Phosphorus Cycle

Unlike carbon and nitrogen, phosphorus does not cycle into and out of ecosystems in gaseous form. Like nitrogen, phosphorus can be taken up by plants only in its ionized form and is often a limiting factor for plant growth.

In the **phosphorus cycle**, phosphorus passes quickly through food webs as it moves from land to ocean sediments, then slowly back to the land. The Earth's crust is the largest reservoir of phosphorus.

In rock formations, phosphorus occurs mainly as phosphate ions (PO_4^{3-}). Weathering and erosion put these ions into streams and rivers. Eventually they become part of the ocean sediments (Figure 41.23). Phosphates accumulate slowly. They form insoluble deposits on submerged continental shelves. Millions of years go by. As movements of crustal plates uplift part of the seafloor, phosphates become exposed on land surfaces. Over time, weathering and erosion will release phosphates from exposed rocks once again.

All organisms use phosphates to synthesize ATP, phospholipids, nucleic acids, and other compounds. Plants take up dissolved phosphates from soil water.

Herbivores get them by eating plants; carnivores get them by eating herbivores. Animals lose phosphate in urine and feces. Bacterial and fungal decomposers release phosphate from organic wastes and remains, then plants take them up again.

The hydrologic cycle helps move phosphorus and other minerals through ecosystems. Water evaporates from the ocean and falls on land. As it flows back to the ocean, it transports silt and dissolved phosphates that primary producers require for growth.

Of all minerals, phosphorus is often the limiting factor in natural ecosystems. Only newly weathered, young soils are high in phosphorus, and in aquatic habitats most phosphorus is tied up in sediments. Only a negligible amount of phosphorus occurs in gaseous form, so little is lost to the atmosphere.

Especially in tropical and subtropical parts of the developing countries, phosphorus is being depleted from natural ecosystems that often have phosphorus-poor soils. Phosphorus stored in biomass and slowly released from decomposing organic matter normally sustains the undisturbed forests or grasslands. When

Read Me First!
and watch the narrated animation on the phosphorus cycle

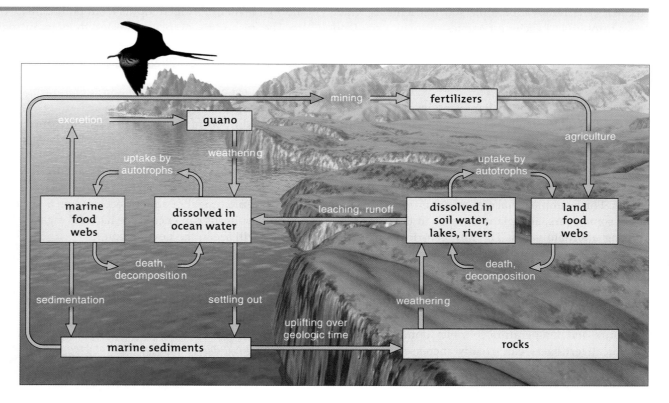

Figure 41.23 Phosphorus cycle. In this sedimentary cycle, most of the phosphorus moves in the form of phosphate ions (PO_4^{3-}) and reaches the oceans. Phytoplankton of marine food webs take up some of it, then fishes eat the plankton. Seabirds eat the fishes, and their droppings (guano) accumulate on islands. Guano is collected and used as a phosphate-rich fertilizer.

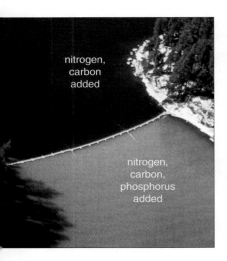

Figure 41.24 One of the eutrophication experiments. Researchers put a plastic curtain across a channel between two basins of a natural lake. They added nitrogen, carbon, and phosphorus on one side of the curtain (here, the *lower* part of the lake) and added nitrogen and carbon on the other side. Within months, the phosphorus-rich basin was eutrophic; with a dense algal bloom covering its surface.

trees are harvested or land is cleared for agriculture, phosphorus is lost. Crop yields start out low and soon become nonexistent. After the fields are abandoned, natural regrowth is sparse. An estimated 1 to 2 billion hectares are already depleted of phosphorus.

What about the developed countries? After years of fertilizer applications, many soils have phosphorus overloads. It is concentrated in eroded sediments and runoff from agricultural fields. Phosphorus also is present in outflows from sewage treatment plants and factories, and in the runoff from fields. Dissolved phosphorus that gets into streams, rivers, lakes, and estuaries can promote destructive algal blooms. Like plants, all photosynthetic algae require phosphorus, nitrogen, and other ions to grow. In many freshwater ecosystems, nitrogen-fixing bacteria keep the nitrogen levels high, so phosphorus becomes the limiting factor. When phosphate-rich pollutants pour in, populations of algae soar. As aerobic decomposers break down the remains of the algae, the water becomes depleted of the oxygen that fishes and other organisms require.

Eutrophication refers to the nutrient enrichment of any ecosystem that is otherwise low in nutrients. It is a natural successional process. Phosphorus inputs, as from agriculture, accelerate it, as the experiment shown in Figure 41.24 demonstrated.

Sedimentary cycles, in combination with the hydrologic cycle, move most mineral elements, such as phosphorus, through terrestrial and aquatic ecosystems.

Agriculture, deforestation, and other human activities upset the nutrient balance of an ecosystem when they accelerate the amount of minerals entering and leaving.

Summary

Section 41.1 An ecosystem consists of an array of organisms together with their physical and chemical environment. There is a one-way flow of energy into and out of an ecosystem, and a cycling of materials among the organisms. All ecosystems have inputs and outputs of energy and nutrients.

Sunlight is the initial energy source for almost all ecosystems. Primary producers convert energy from the sun to ATP and assimilate nutrients that they, and all consumers require. Consumers include herbivores, carnivores, omnivores, decomposers, and detritivores.

Organisms in an ecosystem are classified by trophic levels. Those at the same level are the same number of steps away from the energy input into the ecosystem.

Section 41.2 Linear sequences by which energy and nutrients move through ever higher trophic levels are food chains, which interconnect as food webs. Because the efficiency of energy transfers is always low, most ecosystems support no more than four or five trophic levels away from the original energy source.

In a grazing food web, energy captured by producers flows to herbivores. In a detrital food web, energy from producers flows directly to detritivores and decomposers. In nearly all ecosystems detrital and grazing food webs interconnect.

Section 41.3 By biological magnification, some chemical substance is passed from organisms at one trophic level to those above and becomes increasingly concentrated in body tissues. DDT is an example.

Section 41.4 A system's primary productivity is the rate at which producers capture and store energy in their tissues. It varies with climate, season, nutrient availability and other factors.

Energy pyramids and biomass pyramids are used to show the distribution of energy and materials in an ecosystem. All energy pyramids are largest at the base. The lowest trophic level has the greatest proportion of the energy in an ecosystem.

Section 41.5 Long-term studies of the Silver Springs ecosystem in Florida illustrate the inefficiency of energy transfers. At each trophic level, far more energy was lost to the environment or in wastes and remains than was passed on to the next trophic level.

Section 41.6 In a biogeochemical cycle, water or a nutrient moves through the environment, then through organisms, then back to an environmental reservoir.

Section 41.7 In the hydrologic cycle, water moves from the ocean into the atmosphere, to land, and back to the ocean—the main reservoir. Human actions are disrupting the cycle in ways that result in shortages and pollution of water.

Section 41.8 The carbon cycle moves carbon from its main reservoirs in rocks and seawater, through its gaseous form (carbon dioxide) in the atmosphere, and then through ecosystems. Deforestation and the burning of wood and fossil fuels are adding more carbon dioxide to the atmosphere than the oceans can absorb.

Section 41.9 Greenhouse gases trap heat in the Earth's lower atmosphere and thus make life on Earth possible. But human activities have been increasing the levels of some greenhouse gases, including carbon dioxide, methane, CFCs, and nitrous oxide. The rise in greenhouse gases correlates with a rise in global temperatures and other climate changes.

Section 41.10 The atmosphere is the main reservoir for N_2, a gaseous form of nitrogen that plants cannot use. In nitrogen fixation, some soil bacteria degrade N_2 and assimilate the two nitrogen atoms into ammonia. Other reactions convert ammonia to ammonium and nitrate that plants take up. Some is lost to the atmosphere by the action of denitrifying bacteria. Humans add nitrogen to ecosystems with fertilizers and nitrogen oxides released by the burning of fossil fuels.

Section 41.11 The phosphorus cycle is a sedimentary cycle; gaseous forms are insignificant. Earth's crust is the largest reservoir. Phosphorus can be a limiting factor for producers. Excess inputs of phosphorus to aquatic ecosystems contribute to eutrophication.

Self-Quiz   *Answers in Appendix III*

1. Organisms at the lowest trophic level in a tallgrass prairie are all _____ .
   a. at the first step away from the original energy input
   b. autotrophs   d. both a and b
   c. heterotrophs   e. both a and c

2. Decomposers are commonly _____ .
   a. fungi   b. animals   c. bacteria   d. a and c

3. Primary productivity on land is affected by _____ .
   a. nutrient availability   c. temperature
   b. amount of sunlight   d. all of the above

4. If biological magnification occurs, the _____ will have the highest levels of toxins in their systems.
   a. producers   c. primary carnivores
   b. herbivores   d. top carnivores

5. Disruption of the _____ cycle is depleting aquifers.
   a. hydrologic   c. nitrogen
   b. carbon   d. phosphorus

6. Earth's largest carbon reservoir is _____ .
   a. the atmosphere   c. seawater
   b. sediments and rocks d. living organisms

7. The _____ cycle is a sedimentary cycle.
   a. hydrologic   c. nitrogen
   b. carbon   d. phosphorus

8. _____ is often a limiting factor for plant growth.
   a. Nitrogen   d. both a and c
   b. Carbon   e. all of the above
   c. Phosphorus

Figure 41.25 A few of the participants in a marine food web in the Antarctic seas.

leopard seal

killer whale

skua

emperor penguin

Weddell seal

Adélie penguin

fishes, squid

petrel

blue whale

krill

phytoplankton

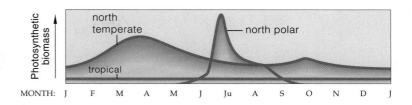

Figure 41.26 *Left:* Antarctica's Larson B ice shelf in (**a**) January and (**b**) March 2002. About 720 billion tons of ice broke from the shelf, forming thousands of icebergs.

MONTH: J F M A M J Ju A S O N D J

Figure 41.27 Annual primary productivity in three ecosystems in the ocean.

9. Nitrogen fixation converts _____ to _____ .
   a. nitrogen gas; ammonia
   b. nitrates; nitrites
   c. ammonia; nitrogen gas
   d. ammonia; nitrates
   e. nitrites; nitrogen oxides

10. Match the terms with suitable descriptions.
   ____ producers
   ____ herbivores
   ____ decomposers
   ____ detritivores

   a. feed on plants
   b. feed on small bits of organic matter
   c. degrade organic wastes and remains to inorganic forms
   d. capture sunlight energy

Critical Thinking

1. Most food chains have at most four or five links. How many links are in the chains in the Antarctic food web shown in Figure 41.25? Decide which of these organisms are producers and which are consumers. What categories of consumers are depicted here? Which are not shown?

2. Polar ice shelves are vast, thickened sheets of ice that float on seawater. In March 2002, 3,200 square kilometers (1,410 square miles) of Antarctica's largest ice shelf broke free from the continent and shattered into thousands of icebergs (Figure 41.26). Scientists knew the ice shelf was shrinking and breaking up, but this was the single largest loss ever observed at one time. Why should this concern people who live in more temperate climates?

3. Fish is a great source of protein and of omega-3 fatty acids, which are necessary for normal nervous system development. This would seem to make fish a good choice for pregnant women. But coal-burning power plants put mercury into the environment and some of it ends up in fish. Eating mercury-tainted fish during pregnancy can adversely affect development of a fetal nervous system.

   Tissues of predatory marine fishes, such as swordfishes, tunas, marlins, and sharks, have especially high levels of mercury. The Environmental Protection Agency has issued health advisories to pregnant women, suggesting that they limit their consumption of fish species most likely to be tainted with mercury. Although sardines are harvested from the same ocean, they have lower mercury levels and are not on the warning list. Explain why two species of fishes that live in the same place can have very different levels of mercury in their tissues.

4. Figure 41.27 graphs the annual primary productivity in three ocean ecosystems. Name some factors that may cause the differences among them. Propose some experiments to test your hypotheses.

Media Menu

Student CD-ROM

Impacts, Issues Video
   Bye-Bye, Blue Bayou
Big Picture Animation
   Energy, nutrients, and global cycles
Read-Me-First Animation
   Trophic levels
   Food webs
   Energy flow through Silver Springs
   Hydrologic cycle
   Carbon cycle
   Nitrogen cycle
   Phosphorus cycle

InfoTrac

- Anthropogenic Disturbance of the Terrestrial Water Cycle. *BioScience*, September 2000.
- Be Fruitful, Multiply, and Lose Nitrogen. *Global Environmental Change Report*, October 2002.
- Too Much of a Good Thing. *National Wildlife*, April–May 2000.
- Spring Forward: Warmer Climates Accelerate Life Cycles of Plants, Animals. *Science News*, March 2003.

Web Sites

- Ecological Society of America: www.esa.org
- Water Resources of the United States: water.usgs.gov
- Environmental Literacy Council: www.enviroliteracy.org
- Energy Efficiency and Renewable Energy: www.eere.energy.gov
- Earth Observatory: earthobservatory.nasa.gov

How Would You Vote?

The United States excels in the production of greenhouse gases. Per person, we produce 19 times as much as India and 5 times as much as Mexico. Given the correlation between rising greenhouse gases and global warming, some argue that the government should discourage the inefficient use of gasoline. Would you support a tax surcharge on the sale of the most energy-inefficient private vehicles, such as large, heavy SUVs? Why or why not?

IMPACTS, ISSUES *Surfers, Seals, and the Sea*

The stormy winter of 1997–1998 was a very good time for surfers on the lookout for the biggest waves, and a very bad time for seals and sea lions. As Ken Bradshaw rode a monster wave (Figure 42.1), half the sea lions on the Galápagos Islands were dying, including nine out of ten pups of the sort shown in the filmstrip at right. In California, the number of liveborn Northern fur seals plummeted. Most of the ones that made it through birth were dead within a few months. Diverse forms of life throughout the world became connected by the full fury of El Niño—a recurring event that ushers in an often spectacular seesaw in the world climate.

That winter, a massive volume of warm water from the southwestern Pacific moved east. It piled into coasts from California down through Peru, displacing currents that otherwise would have churned up tons of nutrients from the deep. Without nutrients, primary producers for marine food webs declined. The scarcity of producers combined with the warm water drove away consumers—anchovies and other migratory fishes. The fishes and

Figure 42.1 Ken Bradshaw surfing a monster wave, more than twelve meters high, during the most powerful El Niño of the past century. In January 1998, a storm formed off the coast of Siberia and generated an ocean swell, which became the biggest wave known to hit Hawaii.

the big picture

Biogeographic Patterns   Sunlight energy drives global circulation patterns in the atmosphere and ocean. Interactions among circulating air, ocean currents, and topography shape regional climates, which influence biodiversity on land and in aquatic habitats.

Major Biomes   Climate influences the formation of soils and other distinct properties of habitats at different latitudes and elevations. Biomes are broad regions of land, each characterized by plant and animal species that dominate its communities.

squids that could not migrate starved to death. So did many seals and sea lions, because fishes and squids happen to be the mainstay of their diet.

Marine mammals did not dominate the headlines, because the 1997–1998 El Niño gave humans plenty of other things to worry about. It battered Pacific coasts with fierce winds and torrential rains, massive flooding, and landslides. As another rippling effect, an ice storm in New York, New England, and central and eastern Canada crippled regional electrical grids. Three weeks later, 700,000 people still had no electricity. Meanwhile, the global seesaw caused drought-driven crop failures and raging wildfires in Australia and Indonesia.

All told, that one El Niño episode killed thousands of people and drained tens of billions of dollars from economies around the world.

El Niño is one of many environmental events that influence life. What causes them? *In this chapter, we connect environmental processes with broad patterns in the distribution of organisms through the biosphere.* We build on Chapters 16 and 17, which introduced these biogeographic patterns in an evolutionary context.

The **biosphere**, again, is the sum of all places where we find life on Earth. Organisms live in the hydrosphere — the ocean, ice caps, and other bodies of water, liquid and frozen. They live on and in sediments and soils of the lithosphere—Earth's outer, rocky layer. Many lift off into the lower region of the atmosphere— the gases and airborne particles that envelop Earth.

 How Would You Vote?

We cannot stop an El Niño event, but we might be able to minimize the social impact and economic damage it causes. Would you support the use of taxpayer dollars to fund research into causes and effects of El Niño? See the Media Menu for details, then vote online.

The Water Provinces   Water covers more than 71 percent of Earth's surface and we find organisms in nearly all of it. Gradients in temperature, light, and dissolved gases and nutrients dictate the composition and organization of freshwater and marine communities.

Cause and Effect: A Case Study   Global weather patterns vary in complex ways over time. Sea surface temperatures and air circulation patterns change during an El Niño event. The changes impact ecosystems, both natural and man-made. They impact human health.

42.1 Global Air Circulation Patterns

The biosphere encompasses ecosystems that range from continent-straddling forests to rainwater pools in cup-shaped clusters of leaves. Except for a few ecosystems at hydrothermal vents, climate influences all of them.

CLIMATE AND TEMPERATURE ZONES

Climate refers to average weather conditions, such as cloud cover, temperature, humidity, and wind speed, over time. It arises from variations in solar radiation, Earth's daily rotation and annual path around the sun, the distribution of land masses and seas, and land elevations. Interactions among these factors are the source of prevailing winds and ocean currents, even the composition of soils.

First, sunlight warms air near Earth's surface (Figure 42.2). The sun's rays are spread out over a greater area near the poles (which intercept them at an angle) than they are at the equator (which intercepts them head-on). The intense light at the equator warms air a lot more. Warm air rises, then spreads north and south toward cooler regions. And so begins a global pattern of air circulation.

Earth's rotation and its curvature alter the initial air circulation pattern. Air masses moving north or south are not attached to Earth—which, like any ball, spins fastest at its equator during each spin around its axis. Air masses moving north or south from a given point *seem* to be deflected east or west relative to the curve of the surface spinning below them. This is the basis of prevailing east and west winds (Figure 42.3c).

Also, land absorbs and gives up heat faster than the ocean, so air parcels above it sink and rise faster. Warm air expands, so its pressure is *lower* where it rises and *greater* where it cools and sinks. The uneven distribution of land and water between regions causes pressure differences. Regional winds arise and disrupt the overall flow of air from the equator to the poles.

In short, latitudinal variations in solar heating lead to a north–south pattern of air circulation, which gets modified by east–west deflections and by air pressure differences over land and water.

Latitudinal differences in rainfall accompany these circulation patterns. Warm air holds more moisture than cool air. At the equator, warm air masses pick up moisture from the seas. As they rise and cool, they

Figure 42.2 *Above*, Earth's atmosphere. The air circulates mainly in the lower atmosphere (troposphere), where temperatures cool rapidly with altitude. An ozone layer between 17 and 27 kilometers above sea level absorbs most of the ultraviolet (UV) wavelengths in the sun's rays. Ozone and oxygen in the atmosphere absorb most of the incoming UV light.

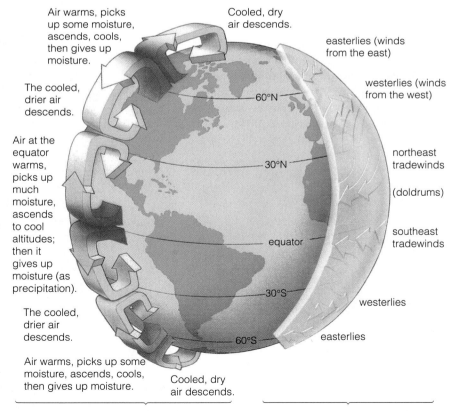

Figure 42.3 (**a**) Concentration of incoming rays of the sun, by latitude. (**b,c**) Gobal air circulation patterns. The latitudinal differences in the solar heating start a north–south air circulation pattern. As Earth's rotation and its curvature deflect the pattern, prevailing east and west winds arise.

Air warms, picks up some moisture, ascends, cools, then gives up moisture.

The cooled, drier air descends.

Air at the equator warms, picks up much moisture, ascends to cool altitudes; then it gives up moisture (as precipitation).

The cooled, drier air descends.

Air warms, picks up some moisture, ascends, cools, then gives up moisture.

Cooled, dry air descends.

Cooled, dry air descends.

easterlies (winds from the east)

westerlies (winds from the west)

northeast tradewinds

(doldrums)

southeast tradewinds

westerlies

easterlies

60°N

30°N

equator

30°S

60°S

b Initial pattern of air circulation as air masses warm and rise, then cool and fall.

c Deflections in the initial pattern near Earth's surface.

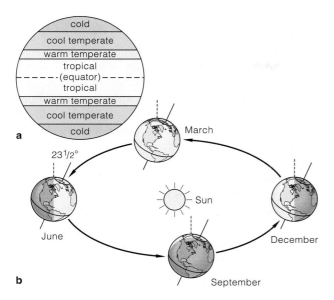

Figure 42.4 (**a**) World temperature zones. (**b**) Incoming solar radiation varies annually. The northern end of Earth's fixed axis tilts toward the sun in June and away from it in December, changing the equator's position relative to the day–night boundary of illumination. Variations in sunlight intensity and daylength cause seasonal temperature shifts.

Figure 42.5 (**a**) Extensive array of electricity-producing photovoltaic cells capturing solar energy. (**b**) A field of turbines harvesting wind energy.

give up some moisture as rain, which supports forest ecosystems. The now-drier air moves north or south. It warms and gets more dry as it descends at latitudes of 30°, where deserts typically form. Farther north and south, air picks up moisture, ascends, and gives up rain at latitudes of about 60°. It descends in polar regions, where cold temperatures and a near-absence of precipitation result in cold, dry polar deserts.

We divide the temperature gradients from Earth's equator to the poles into **temperature zones** (Figure 42.4*a*). Temperatures shift in the zones as Earth's orbit puts it closer and then farther from the sun in each annual cycle (Figure 42.4*b*). Temperatures, the hours of daylight, and winds change with the seasons. The changes are far greater inland (away from the ocean's moderating effects) and farther from the equator. The more pronounced the change, the more that primary productivity rises and falls on land and in the seas.

HARNESSING THE SUN AND WIND

Paralleling the human population's J-shaped growth curve is a steep rise in its total and per capita energy consumption. You may think we have an abundance of energy, but there is a big difference between total and net amounts. *Net* is what is left after subtracting the energy it takes to find, extract, transport, store, and deliver energy to consumers. Some sources, such

as coal, are not renewable. Others, such as direct solar and wind energy, are. The annual amount of incoming solar energy surpasses, by about 10 times, the energy stored in all known fossil fuel reserves. We can harness it in something besides crop plants.

We already know how to get *solar–hydrogen energy*. Photovoltaic cells (Figure 42.5*a*) hold electrodes that, when exposed to the sun's rays, generate an electric current that splits water molecules into oxygen and hydrogen gas, which can be stored efficiently. The gas can directly fuel cars. It can heat and cool buildings. Water is the only waste. It also costs less to distribute hydrogen gas than electricity. Space satellites run on it.

Unlike fossil fuels, the sun's energy and seawater are unlimited. A transition to a solar–hydrogen age might end smog, oil spills, acid rain, and the reliance on nuclear energy, and may reduce global warming.

Solar energy also is converted into the mechanical energy of winds. Arrays of turbines at "wind farms" exploit the wind patterns that arise from latitudinal variations in the intensity of incoming sunlight. One percent of California's electricity already comes from wind power. The winds of North and South Dakota alone could potentially meet 80 percent of the current energy needs of the United States. Winds do not blow constantly, but when they do, wind energy can be fed into utility grids. Wind power also has potential for islands and other areas remote from utility grids.

The world's major temperature zones and climates start with global patterns of circulation. Those patterns arise through interacting factors: latitudinal variations in incoming solar radiation, Earth's rotation and annual path around the sun, and the distribution of land masses and the seas.

42.2 Air Circulation Patterns and Human Affairs

Through activities that pollute the air, human populations interact with global air circulation patterns in unexpected ways, with unintended consequences. **Pollutants** *are any natural or synthetic substances that have accumulated in harmful or disruptive amounts because organisms have had no prior evolutionary experience with them.*

A FENCE OF WIND AND OZONE THINNING

The ozone layer (Figure 42.2) is nearly twice as high above sea level as Mount Everest. From September to October, it thins above both poles. Seasonal **ozone thinning** is so vast that it was once called an "ozone hole" in the atmosphere. Figure 42.6 has an example. With less ozone, more UV radiation reaches Earth and triggers skin cancers, cataracts, and weakened immunity. It also alters the atmosphere's composition by killing phytoplankton, thus causing huge declines in their oxygen-releasing activity.

Chlorofluorocarbons, or CFCs, are major ozone destroyers. These odorless gases have been used as propellants in aerosol cans, coolants in refrigerators and air conditioners, and in solvents and plastic foam. They slowly seep into the air, and they resist breakdown. A free CFC molecule gives up a chlorine atom when it absorbs UV light. Reaction of this atom with ozone yields oxygen and chlorine monoxide. Chloride monoxide in turn reacts with free oxygen and releases another chlorine. Each chlorine atom can break apart more than 10,000 ozone molecules!

Chlorine monoxide levels above polar regions are 100 to 500 times higher than at midlatitudes. Why?

Like a dynamic fence, winds rotate around the poles for most of the winter. Chlorine compounds are split apart on ice crystals in the clouds. After the almost perpetually dark polar winter ends, UV light in the sun's rays invites chlorine to start destroying ozone.

Methyl bromide is even worse for the ozone layer. Each year in the United States, farmers spray about 60 million pounds of this fumigant over croplands to kill insects, nematodes, and other pests.

Developed countries phased out CFC production. The developing countries may phase it out by 2010. Methyl bromide production is expected to end at that time as well. Even so, it will be one or two centuries before the ozone layer fully recovers.

NO WIND, LOTS OF POLLUTANTS, AND SMOG

Certain weather conditions trap a layer of cool, dense air under a warm air layer. Such **thermal inversions** intensify smog, an atmospheric condition in which winds cannot disperse pollutants that accumulate in the trapped air, often to harmful levels (Figure 42.7).

Where winters are cold and wet, *industrial* smog develops as a gray haze over cities that burn a lot of coal and other fossil fuels. Burning releases smoke, soot, ashes, asbestos, oil, particles of lead and other heavy metals, and sulfur oxides. Most industrial smog now forms in China, India, and eastern Europe.

Big cities in warm climate zones form *photochemical* smog as a brown haze, especially where land forms a natural basin. Los Angeles and Mexico City are two classic cases. Nitric oxide, a key pollutant, escapes in

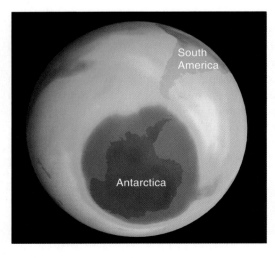

Figure 42.6 Seasonal ozone thinning above Antarctica during 2001. *Darkest blue* indicates the area with the lowest ozone level, at that time the largest recorded.

Figure 42.7 (**a**) Normal pattern of air circulation in smog-forming regions. (**b**) Air pollutants are trapped under a thermal inversion layer.

vehicle exhaust. This combines with oxygen to form nitrogen dioxide. When nitrogen dioxide is exposed to sunlight, it reacts with hydrocarbon gases to form photochemical oxidants. Most hydrocarbon gases are released from spilled or partly burned gasoline.

WINDS AND ACID RAIN

Coal-burning power plants, smelters, and factories emit sulfur dioxides. Vehicles, power plants that burn gas and oil, and nitrogen-rich fertilizers all emit nitrogen oxides. In dry weather, airborne oxides fall as dry acid deposition. In moist air, they form nitric acid vapor, sulfuric acid droplets, and sulfate and nitrate salts. Winds typically disperse them far from their source. They fall to Earth in rain and snow. We call this wet acid deposition, or **acid rain**.

The pH of normal rainwater is 5 or so. Acid rain can be 0 to 100 times more acidic, as potent as lemon juice! It corrodes metals, marble, rubber, plastics, nylon stockings, and other materials. It damages organisms and alters the chemistry of ecosystems.

Depending on the soil type and vegetation cover, some regions are more sensitive to acid rain (Figure 42.8). Highly alkaline soil neutralizes acids before they enter streams and lakes. Also, highly alkaline water can neutralize the acid inputs. But many of the watersheds of northern Europe, southeastern Canada, and regions throughout the United States have thin soil layers on top of solid granite. These soils cannot buffer much of the acidic inputs.

Rain in much of eastern North America is thirty to forty times more acidic than it was several decades ago. Crop yields are declining. Fish populations have already vanished from more than 200 lakes in New York's Adirondack Mountains. Air pollutants from industrial regions are changing the acidity of rainfall. The change is contributing to the decline of forest trees and mycorrhizae that support new growth.

As Harvard and Brigham Young researchers report, living with airborne particles of dust, soot, smoke, or acid droplets shortens the human life span by a year or so. Smaller airborne particles damage lung tissues. High levels of ultrafine particles can increase the risk of lung cancer (Figure 42.9).

At one time the world's tallest smokestack, in the Canadian province of Ontario, produced 1 percent by weight of the world's annual emissions of sulfur dioxide. Today Canada gets more acid deposition from the midwestern United States than it sends across its southern border. *Prevailing winds—hence air pollutants—do not stop at national boundaries.*

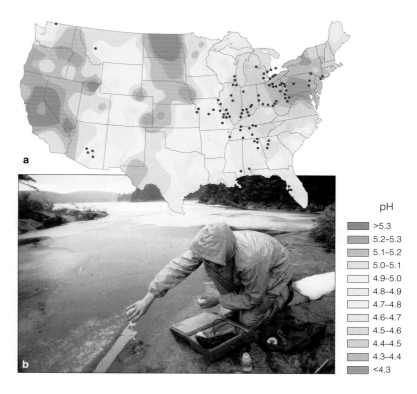

	pH
	>5.3
	5.2–5.3
	5.1–5.2
	5.0–5.1
	4.9–5.0
	4.8–4.9
	4.7–4.8
	4.6–4.7
	4.5–4.6
	4.4–4.5
	4.3–4.4
	<4.3

Figure 42.8 (a) Average 1998 precipitation acidities in the United States. *Red* dots mark large coal-burning power and industrial plants. (b) Biologist measuring the pH of New York's Woods Lake during a spring melt of acidified snow. Acid rain has already altered the lake.

Figure 42.9 Suspended particulate matter: solids and liquid droplets small enough to stay aloft for variable intervals.

Ultrafine particles contribute to respiratory disorders. Carbon black is a powdered form of carbon used in paints, tires, and other goods. About 6 million tons are manufactured annually. About 45 million tons of fly ash are generated annually by coal combustion.

42.3 The Ocean, Landforms, and Climates

*The **ocean** is a continuous body of water that covers more than 71 percent of the Earth's surface. Driven by solar heat and wind friction, its upper 10 percent moves in currents that distribute nutrients through marine ecosystems.*

OCEAN CURRENTS AND THEIR EFFECTS

Latitudinal and seasonal variations in sunlight warm and cool water. At the equator, where vast volumes of water warm and expand, the sea level is about eight centimeters (three inches) higher than at either pole. The volume of water in this "slope" is enough to get sea surface water moving in response to gravity, most often toward the poles. The moving water warms air parcels above it. At midlatitudes it transfers *10 million billion* calories of heat energy per second to the air!

Ocean currents are large volumes of water flowing in response to the tug of trade winds and westerlies. Their direction and properties are outcomes of Earth's rotation and topography. They circulate clockwise in the Northern Hemisphere and counterclockwise in the Southern Hemisphere (Figure 42.10).

Swift, deep, and narrow currents of nutrient-poor water parallel the east coast of continents. Along the eastern coast of North America, warm water moves northward, as the Gulf Stream. Slower, shallow, broad currents of cold water paralleling the western coast of continents flow toward the equator.

Ocean currents influence climate zones. Why are Pacific Northwest coasts cool and foggy in summer? Offshore, the cold California Current is giving up moisture as warm winds near the coast transfer heat to it. Why is Baltimore or Boston muggy in summer? Air masses above the warm Gulf Stream gain heat and moisture that southerly and easterly winds move to those cities. Why are winters milder in London and Edinburgh than in Ontario and central Canada—which are at the same latitude? The North Atlantic Current picks up warm water from the Gulf Stream, then gives up heat to prevailing winds, which warm northwestern Europe.

This all can change. We can expect the long-term rise in atmospheric temperature to disrupt the global system of ocean currents and climate zones.

Figure 42.10 Major climate zones correlated with surface currents and surface drifts of the world ocean. The warm surface currents start moving from the equator toward the poles, but prevailing winds, Earth's rotation, gravity, the shape of ocean basins, and land masses all influence the direction of flow. Water temperatures, which differ with latitude and depth, contribute to differences in air temperature and in rainfall patterns.

warm surface current   cold surface current

dry   warm temperate   subpolar
tropical   cool temperate   polar (ice)
cold

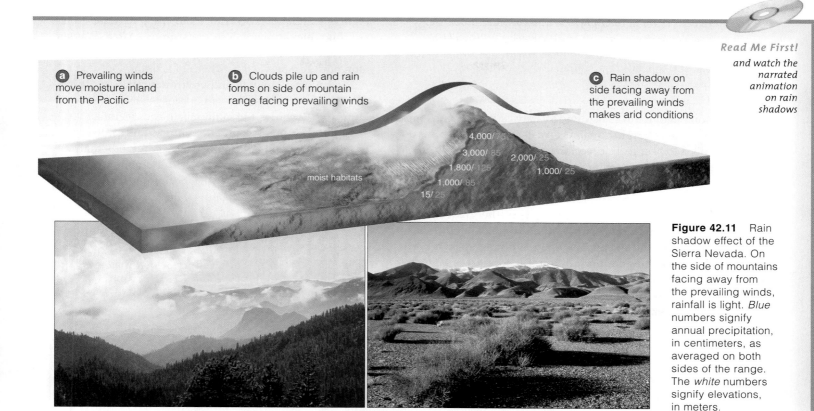

Read Me First!
and watch the narrated animation on rain shadows

a Prevailing winds move moisture inland from the Pacific

b Clouds pile up and rain forms on side of mountain range facing prevailing winds

c Rain shadow on side facing away from the prevailing winds makes arid conditions

4,000/ 75
3,000/ 85
2,000/ 25
1,800/ 125
1,000/ 25
1,000/ 85
moist habitats
15/ 25

Figure 42.11 Rain shadow effect of the Sierra Nevada. On the side of mountains facing away from the prevailing winds, rainfall is light. *Blue* numbers signify annual precipitation, in centimeters, as averaged on both sides of the range. The *white* numbers signify elevations, in meters.

RAIN SHADOWS AND MONSOONS

Mountains, valleys, and other topographical features affect regional climates. *Topography* refers to a region's surface features, such as elevation. Track a warm air mass after it picks up moisture off California's coast. It moves inland, as wind from the west, and piles up against the Sierra Nevada. This high mountain range parallels the distant coast. The air cools as it rises in altitude and loses moisture as rain (Figure 42.11). The result is a **rain shadow**, a semiarid or arid region of sparse rainfall on the leeward side of high mountains. *Leeward* is the side facing away from the wind.

Belts of vegetation at different elevations relate to temperature and moisture differences. Arid grasslands form at the western base of the Sierra Nevada. Higher up, in the cooler air, deciduous and evergreen species reign. Higher still we see a few evergreen species that can withstand a rigorously cold habitat. Above the subalpine belt, hardy, dwarfed plants are all that can withstand the temperature extremes.

There are rain shadows on the leeward side of the Andes, Rockies, Himalayas, and other great mountain ranges. On the side of Hawaii's high volcanic peaks facing into the wind (*windward*), lush tropical forests flourish. Desert conditions prevail on the other side.

Air circulation patterns called **monsoons** influence continents north or south of warm oceans. As land heats intensely in summer, low-pressure air parcels form above it. The low pressure draws moisture-laden air from the ocean, as from the Bay of Bengal on into Bangladesh. The equatorial sun and converging trade winds result in intense heating and heavy rainfall. The air moving northward and southward across the land is the source of alternating dry and wet seasons, of alternating drought conditions and flooding.

Recurring coastal breezes are like mini-monsoons. Water and coastal land have different heat capacities. In the morning, water does not warm as fast as land. When warm air above land rises, cooler marine air moves in. After sunset, the land loses heat faster than water, so land breezes flow in the reverse direction.

Surface ocean currents also influence regional climates and help distribute nutrients in marine ecosystems.

Air circulation patterns, ocean currents, and landforms interact in ways that influence regional temperatures and moisture levels. Thus they also influence the distribution and dominant features of ecosystems.

42.4 Realms of Biodiversity

Regions differ in their physical and chemical properties and evolutionary history. The differences help explain why deserts, grasslands, forests, and tundra form where they do, and why some water provinces are richer than others in biodiversity.

Long ago, naturalists divided Earth's land masses into six **biogeographic realms**—vast expanses where you could expect to find certain plants and animals, such as palm trees and camels in the Ethiopian realm. Which organisms became the poster species was subjective; the realms were not uniform from border to border.

Biomes are finer subdivisions of the great realms. Their distribution is partly an outcome of evolutionary processes and events, as when evolving species were slowly rafted about on different land masses after the splitting of Pangea (Chapter 17). Their distribution is also an outcome of physical and chemical conditions that differ even in the same land region. Start with

Figures 42.12 and 42.13. Tundra forms at high latitudes and high elevations where moisture is scarce. Forests form where there are certain amounts of rain, certain temperatures, and certain types of soils.

Evolutionary events and environmental conditions also underlie the distribution of ecosystems in the seas. Figure 42.12 shows the main marine ecoregions.

Conservationists locate, inventory, and protect the **hot spots**, portions of biomes and ecoregions that are richest in biodiversity and most vulnerable to species losses. More than half of all species on land live in regions identified as hot spots, and nowhere else.

Biomes are vast expanses of land characterized by distinct arrays of species; marine ecoregions are realms of biodiversity in the seas. Their distribution is an outcome of evolutionary history and the prevailing physical and chemical conditions in different parts of the world.

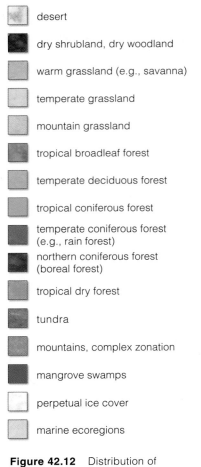

- desert
- dry shrubland, dry woodland
- warm grassland (e.g., savanna)
- temperate grassland
- mountain grassland
- tropical broadleaf forest
- temperate deciduous forest
- tropical coniferous forest
- temperate coniferous forest (e.g., rain forest)
- northern coniferous forest (boreal forest)
- tropical dry forest
- tundra
- mountains, complex zonation
- mangrove swamps
- perpetual ice cover
- marine ecoregions

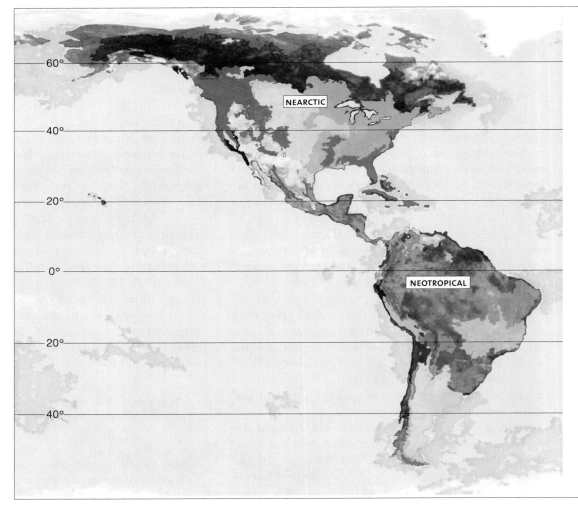

Figure 42.12 Distribution of major biomes within classical biogeographic realms.

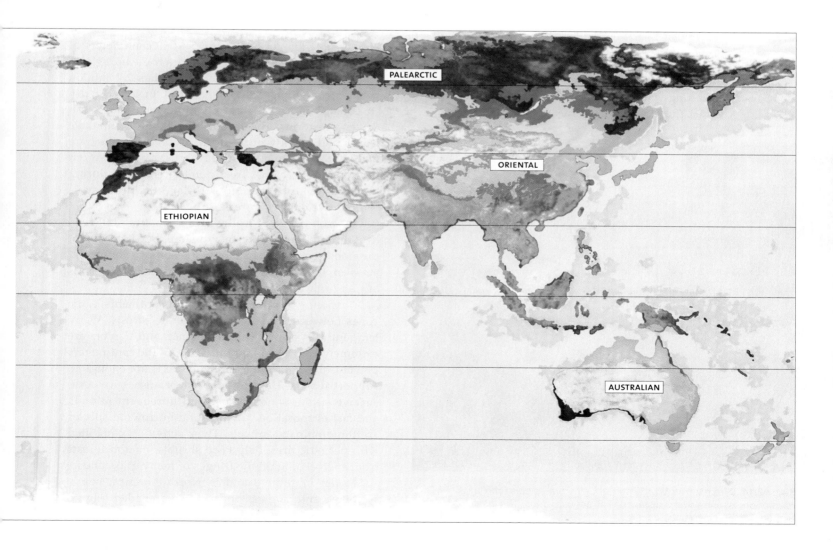

Figure 42.13 Changes in plant form along environmental gradients for North America. Gradients in elevation and water availability affect primary productivity.

42.5 Moisture-Challenged Biomes

At certain latitudes with sparse rainfall and high rates of evaporation, drought-tolerant plants predominate.

DESERTS, NATURAL AND MAN-MADE

Deserts form between latitudes of about 30° north and south. Annual rainfall is less than ten centimeters or so, and evaporation rates are high. Brief, infrequent pulses of rain swiftly erode the topsoil. Humidity is so low that the sun's rays easily penetrate the air. The ground heats fast, then cools fast at night. The arid or semiarid conditions do not support large, leafy plants, but biodiversity is often high. Annuals and perennials flower briefly after the rains (Figure 42.14).

Desertification is the conversion of grasslands to desertlike conditions by long-term shifts in climate or human activities. Extended droughts and overgrazing are factors. Over the past fifty years, 9 million square kilometers were desertified. Each year, at least 200,000 square kilometers are still being converted.

In Africa, for instance, many imported cattle drink far more water than wild herbivores do. Cattle move more often between watering holes and grazing areas, and the movements trample grasses and pack down the soil. Gazelles and other wild herbivores are better at getting water from their food and conserving body water; they lose little in feces, compared to cattle.

Figure 42.15 Satellite image of an immense dust storm that formed in the hot, dry Sahara Desert, then continued past the west coast of Africa, and is well on its way across the Atlantic.

Desertification can impact distant ecosystems. Dust from Africa may be contributing to the decline of coral reefs in the Caribbean Sea. Winds lift dust into the air and transport it across the Atlantic (Figure 42.15). In that dust are fungal, bacterial, and viral pathogens. Since the 1980s, severe dust storms have become more frequent in Africa. So have disease outbreaks among the Caribbean corals.

DRY SHRUBLANDS, DRY WOODLANDS, AND GRASSLANDS

Dry shrublands get less than 25 to 60 centimeters of rain per year. We see them in Mediterranean regions, South Africa, and California, which has 2.4 million hectares (6 million acres) of chaparral. In summer, dry shrublands are vulnerable to lightning-sparked, wind-driven firestorms. The shrubs have highly flammable leaves and quickly burn to the ground. But they are highly adapted to episodes of fire and soon resprout from the root crowns. Trees don't fare as well during firestorms. Shrubs that can withstand the fires have a competitive edge (Figure 42.16).

Dry woodlands prevail where annual rainfall is 40 to 100 centimeters. Trees may be tall, but they do not provide a continuous canopy. Eucalyptus-dominated areas of southwestern Australia and oak woodlands of California and Oregon are examples.

Grasslands form in the interior of continents in the zones between deserts and temperate forests. Warm temperatures prevail in the summer, and winters are extremely cold. Annual rainfall of 25–100 centimeters prevents deserts from forming, but it is not enough to support forests. Drought-tolerant primary producers survive strong winds, sparse and infrequent rainfall, and fast evaporation. Grazing and burrowing species are the dominant animals. Their activities, combined with periodic fires, help keep shrublands and forests from encroaching on the fringes of many grasslands.

On flat or rolling land in North America, native *shortgrass* and *tallgrass* prairie once prevailed (Figure 42.17). Roots of perennial grasses extended deep into

Figure 42.14 From the Mojave Desert of California, cactus (cholla), golden poppies, and—in the distance—tall saguaro cacti. The inset shows flowers of a claret cup cactus.

Figure 42.16 Views of California chaparral. The dominant plants are multibranched, woody, and typically only a few meters tall. They are adapted to periodic fires; some produce seeds that germinate only after a fire. *Far right,* fires race through chaparral in a canyon above Malibu.

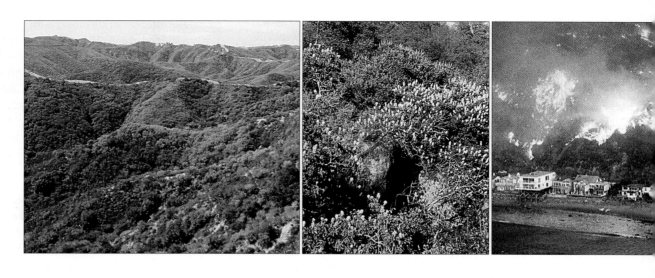

Figure 42.17 *Below,* shortgrass prairie, east of the Rocky Mountains, that once sustained 60 million bison.

Right, African savanna. Herbivores include this herd of wildebeests and giraffes, Cape buffalos, zebras, impalas, and Thomson's gazelles.

the rich topsoil. In the 1930s, shortgrass prairie in the Great Plains was overgrazed. It also was plowed to grow wheat. Prolonged droughts killed this crop, and strong winds stripped away the topsoil. The region became "the Dust Bowl." John Steinbeck's novel *The Grapes of Wrath* is a glimpse into the human tragedies that unfolded because of this environmental disaster.

Grasses two meters tall gave the tallgrass prairie its name. Legumes and composites, including daisies, thrived alongside grasses in the continent's eastern interior. Nearly all original tallgrass prairie has been plowed and converted to cereal croplands. A few parts that escaped the plow are now protected reserves; a few other patches are being restored.

Between the tropical forests and the hot deserts of Africa, South America, and Australia are savannas, broad belts of grasslands with a few shrubs and trees (Figure 42.17). Rainfall averages 90 to 150 centimeters a year, and droughts are seasonal. Where rainfall is low, fast-growing grasses dominate. With more rain, savannas grade into tropical woodlands with low trees, shrubs, and coarse grasses that die back and often burn during the dry season.

Plants of all deserts, dry shrublands, dry woodlands, and grasslands survive sparse rainfall, dry air, strong winds, grazing animals, and recurring episodes of drought and fire.

42.6 More Rain, More Trees

In forest biomes, tall trees grow close together and form a fairly continuous canopy over a broad region. Rainfall and distance from the equator influence which trees dominate.

BROADLEAF FORESTS

Evergreen broadleaf forests form between latitudes 20° north and south, the tropical zones of Africa, the East Indies, Malaysia, Southeast Asia, South America, and Central America. Rainfall is heavy, between 130 and 200 centimeters yearly. Regular rains, an annual mean temperature of 25°C, and at least 80 percent humidity support tropical rain forests (Figure 42.18). In this highly productive forest, evergreens grow new leaves and shed old ones all year. The climate promotes rapid decomposition and mineral cycling; litter can't accumulate. Soils are highly weathered, deficient in organic matter, and poor nutrient reservoirs.

Mild temperatures and predictable seasonal rains support *deciduous broadleaf* forests. In Southeast Asia, India, and other humid tropical regions, some or all leaves drop from trees in a pronounced dry season.

Decomposition is slow; many nutrients are conserved in accumulated litter. Farther from the equator, in the temperate zone, cold, dry winters and less rainfall support temperate deciduous forests (Figure 42.19). In northeastern North America, vast deciduous forests were largely cleared for agriculture. Maple and beech now dominate the forests. To the west, oak–hickory forests flourish, then grade into oak woodlands.

CONIFEROUS FORESTS

Conifers—cone-bearing trees—are primary producers of *coniferous* forests. Most have thick cuticles, needle-shaped leaves, and recessed stomata, adaptations that help conserve water during droughts and icy winters. They dominate the boreal forests, montane coniferous forests, temperate rain forests, and pine barrens.

Boreal forests stretch across northern Europe, Asia, and North America. They are known as *taigas*, meaning "swamp forests." Most form in glaciated regions with cold lakes and streams (Figure 42.20). It rains mostly in summer, and little water evaporates into the cool

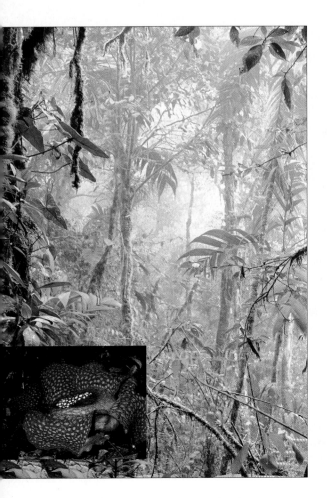

Figure 42.18 Tropical rain forest. Latin America, Southeast Asia, and Africa have the fastest-growing populations but not enough food, fuel, and lumber. Of necessity, they turn to their tropical forests for growth-sustaining resources. Most of the forests may disappear within our lifetime. Yet they have the greatest variety and numbers of insects, and the world's largest ones. They are home to the most species of birds and primates and to plants with the largest flowers. Massive vines twist around trees. Orchids, lichens, and other species grow on tree branches; entire communities live, breed, and die in small pools of water that collect in furled leaves.

The species losses are our losses. As examples, we culture and engineer newly discovered genes of forest species to develop pathogen-resistant strains of food crops and livestock, to develop more effective antibiotics and vaccines. An extract from Madagascar's rosy periwinkle greatly increases the odds of surviving childhood leukemia. Many ornamental plants and spices and foods, including cocoa, cinnamon, and coffee, originated in the tropics. So did latex, gums, resins, waxes, and oils for tires, shoes, toothpaste, ice cream, shampoo, compact discs, condoms, and perfumes.

Conservation biologists rightly decry the mass extinction, the assaults on species diversity, the depletion of much of the world's genetic reservoir. Yet something else is going on here. Too many of us grow uneasy when we pass through obliterated forests in our own country. Is it because we are losing the comfort of our heritage, a connection with our evolutionary past? Many millions of years ago, our earliest primate ancestors moved into the trees of tropical forests. Through countless generations, their nervous and sensory systems evolved and became highly responsive to information-rich, arboreal worlds.

Does our neural wiring still resonate with rustling leaves, with shafts of light and mosaic shadows? Are we innately attuned to the forests of Eden—or have time and change buried recognition of home?

autumn   winter   spring   summer

Figure 42.19 Temperate deciduous forest in North America. The series above shows seasonal changes in a deciduous tree's foliage.

Figure 42.20 Taiga in Alberta, Canada. At right, a firefighter retreating from a wall of flames in a forest of pines, oaks, and palmettos by Daytona Beach, Florida.

summer air. Winters are cold, dry, and more severe in eastern parts of these biomes than in the west, where ocean winds moderate climate. Spruce and balsam fir dominate boreal forests of North America. Pine, birch, and aspen forests form in burned or logged regions. In poorly drained soil, acidic bogs with peat mosses, shrubs, and stunted trees prevail. Farther to the north, boreal forests thin and grade into arctic tundra.

In the Northern Hemisphere, montane coniferous forests extend south through great mountain ranges. In the north and at higher elevations, spruce and firs dominate. They give way to pines in the south and at lower elevations. Coniferous forests flourish in some temperate lowlands and near the Pacific coast from Alaska into northern California. These forests hold the world's tallest trees—Sitka spruce to the far north and redwoods to the south. As recounted in Chapter 21, much of these forests has been logged over.

Forests of pine and scrub oak grow in New Jersey. Southern pine forests dominate the coastal plains of the southern Atlantic and Gulf states. Palmettos grow below loblolly pines in the Deep South. Many pine species are adapted to dry, sandy, nutrient-poor soil and to periodic, lightning-sparked fires. Suppression of all fires can result in a buildup of dry undergrowth that can fuel uncontrollable wildfires (Figure 42.20). Each year, Florida alone carries out controlled burns of about 2,500,000 acres to clear fuel-rich understory. Fire-adapted species survive and often flourish after controlled burns.

Conditions in forest biomes favor dense stands of trees that form a continuous canopy over broad regions. Trees that differ in structure and growth patterns are adapted to rainfall patterns, temperature, and other conditions.

42.7 Brief Summers and Long, Icy Winters

Tundra lies between the polar ice cap and belts of boreal forests in the Northern Hemisphere. It is the youngest biome, having first evolved about 10,000 years ago.

The Northern Hemisphere's treeless plain, the *arctic* tundra, is extremely cold. Annual snowmelt and rain are usually less than 25 centimeters. Plants grow fast in the nearly continuous sunlight of a brief growing season (Figure 42.21*a,b*). Hardy lichens and shallow-rooted, low-growing plants are a base for food webs that include voles, arctic hares, caribou, arctic foxes, wolves, and polar bears. Great numbers of migratory birds nest here in summer, when the air is thick with mosquitoes and other flying insects.

Not much more than the surface soil thaws during summer. Just below is **permafrost**, a frozen layer 500 meters thick in some places. It prevents drainage, so the soil above is perpetually waterlogged. The cool, anaerobic conditions do not favor nutrient cycling. Plant remains accumulate in soggy masses and break down slowly. The accumulation of undecayed organic matter in permafrost makes the arctic tundra one of Earth's greatest stores of carbon. However, the frozen layer is starting to melt. Global warming is melting ice and snow, which reflect sunlight. Newly exposed soil is absorbing heat from the sun, and warming up.

Alpine tundra is a similar biome, but it develops in the high mountains around the world (Figure 42.21*c*). At night, the below-freezing temperatures make it too difficult for trees to grow. Even in summer, shaded patches of snow persist in this biome, but there is no permafrost. Alpine soil is thin and well drained, but it is nutrient poor. As a result, primary productivity is low. Grasses, heaths, and small-leafed shrubs grow in patches where better soil has formed. These low plant species are adapted to withstand strong winds.

Arctic tundra prevails at high latitudes, where short, cold summers alternate with long, cold winters. Alpine tundra prevails in high, cold mountains, regardless of whether the seasons differ with latitude.

Figure 42.21 Arctic tundra in summer (**a**) and winter (**b**). About 4 percent of Earth's land mass is arctic tundra, blanketed with snow for as long as nine months of the year. Most arctic tundra is in northern Russia and Canada, followed by Alaska and Scandinavia.

Small bands of humans have herded reindeer, hunted, and fished in these sparsely populated regions for hundreds of thousands of years. Now more people, and machines, are moving in to extract mineral and fossil fuels. When extraction operations damage soils and plants, it might take decades for a region to recover. Why? Plants of the Arctic tundra grow slowly, and their growth is limited to only a few months of the year.

(**c**) Compact, low-growing, hardy plants typical of alpine tundra in Washington's Cascade Range.

42.8 Don't Forget the Soils

Dig down into the dirt and you will find a soil profile that helps define a particular kind of biome. Desert, woodland, forest, tundra—all have distinctive soils.

Soils are mixtures of mineral particles and varying amounts of the decomposing organic material we call humus. When rocks are weathered and broken down, they form coarse-grained gravel, then sand, silt, and finely grained clay. Water and air fill spaces between soil particles. As explained Section 26.1, the types and proportions of particles, and how tightly the particles are compacted, differ within and between regions.

Soils have a layered structure—a *soil profile*—that reflects how they formed (Figure 42.22). Uppermost is topsoil, which has the most humus and is most prone to erosion. It is less than a centimeter thick on steep slopes and more than a meter thick in grasslands. The deepest layer has rocks that are still breaking down.

Soils too high in one type of particle hamper plant growth. Clay is rich in minerals, but its fine, closely packed particles drain poorly and don't leave much room for essential oxygen. Gravelly or sandy soils are prone to leaching, which depletes them of water and mineral ions.

Loam topsoils have the best mix of sand, silt, and clay for agriculture. Their coarse particles promote drainage, and their fine particles retain water-soluble mineral ions and nutrients required for growth.

Soils influence primary productivity. That is why farms now abound in topsoil-rich former grasslands. Tropical forests are cleared for agriculture too, but they have little topsoil above layers that drain poorly. Clearing exposes topsoil to rains that leach nutrients, so crops perform poorly unless heavily fertilized.

A biome's soil profile and its proportions of sand, silt, clay, gravel, and humus influence its primary productivity.

DESERT SOIL

O horizon: Pebbles, little organic matter

A horizon: Shallow, poor soil

B horizon: Leaching results in salinization (accumulated calcium, sodium)

C horizon: Rock fragments from uplands

GRASSLAND SOIL

A horizon: Alkaline, deep, rich in humus

B horizon: Percolating water enriches layer with calcium carbonates

TROPICAL RAIN FOREST SOIL

O horizon: Sparse litter

A–E horizons: Continually leached; iron, aluminum left behind impart red color to acidic soil

B horizon: Clays with silicates, other residues of chemical weathering

DECIDUOUS FOREST SOIL

O horizon: Scattered litter

A horizon: Rich in organic matter above humus layer unmixed with minerals

B horizon: Accumulated minerals leached from above

C horizon: Poorly weathered rocks

CONIFEROUS FOREST SOIL

O horizon: Well-defined, compacted mat of organic deposits resulting mainly from activity of fungal decomposers

A horizon: Acidic humus; most minerals leached out, silica retained

B horizon: Accumulated clays with oxides of iron and aluminum

Figure 42.22 Soil profiles from a few representative biomes.

42.9 Freshwater Provinces

Freshwater and saltwater provinces cover more of Earth's surface than all biomes combined. They include the world ocean, lakes, ponds, wetlands, and coral reefs. There are no "typical" regions. Ponds are shallow; Siberia's Lake Baikal is 1.7 kilometers deep. All aquatic ecosystems have gradients in light penetration, temperature, and dissolved gases, but values differ greatly. All we can do here is sample the diversity, starting with the freshwater provinces.

LAKE ECOSYSTEMS

A lake is a body of standing freshwater, as in Figure 42.23. Over time, erosion and sedimentation alter its dimensions; it usually ends up filled in or drained. A young lake has littoral, limnetic, and profundal zones (Figure 42.24a). Its littoral zone extends all around the shore to a depth where rooted aquatic plants cannot grow. The well-lit, shallow, warm water supports high biodiversity. The limnetic zone encompasses the sunlit, open water away from shore to depths where light and photosynthesis are limited. Its *phyto*plankton includes cyanobacteria, green algae, and diatoms. *Zoo*plankton includes rotifers and copepods. The profundal zone is all open water at depths that are impenetrable to the most efficient wavelengths for photosynthesis. Below it, sediments house communities of decomposers.

SEASONAL CHANGES IN LAKES In temperate regions, many lakes undergo seasonal changes in density and temperature gradients, from surface to bottom. Often a layer of ice forms in midwinter. Water under the ice is near the freezing point and is the least dense. Water at 4°C is the most dense; it forms deeper layers that are a bit warmer than the surface layer.

In spring, the number of daylight hours increases and the air warms. The lake ice melts, surface water warms to 4°C, and temperature gradients disappear. Winds blowing across the lake surface cause a **spring overturn**, in which strong vertical movements deliver dissolved oxygen from surface waters to the depths. At the same time, nutrients released by decomposers of the lake bottom sediments move to the surface.

The lake becomes thermally stratified in summer. The midlayer cools, becoming a *thermocline* that stops the vertical mixing between the warmer, oxygen-rich surface layer and the cold, oxygen-poor layer below it (Figure 42.24b). Decomposers now deplete the oxygen dissolved near the lake bottom.

In autumn, the upper layer cools and gets denser. It sinks and the thermocline vanishes. During this **fall overturn**, water mixes vertically, and again dissolved oxygen moves down and nutrients move up.

Primary productivity is seasonal. After the spring overturn, longer daylengths and the cycled nutrients favor increased primary productivity. Phytoplankton

Figure 42.23 Lake in Chile's Torres del Paine National Park.

Figure 42.24 (**a**) Lake zonation. The littoral zone extends all around the shore to a depth where aquatic plants stop growing. The profundal zone is all water below the depth of light penetration. Above the profundal are the open, sunlit waters of the limnetic zone. (**b**) Thermal layering occurs in many lakes of temperate zones during the summer.

Figure 42.25 Stream habitats. (**a**) Pool. (**b**) Pool leading into a riffle. (**c**) A run, Sinking Creek. (**d**) Fallen leaves add nutrients to water in a stream. (**e**) A closer look at a riffle.

and aquatic plants take up phosphorus, nitrogen, and other nutrients. During the growing season, vertical thermocline mixing ends. Nutrients do not move up, and photosynthesis slows. By late summer, shortages of nutrients limit growth. A fall overturn does cycle nutrients back to the surface, and there is a sudden burst of primary productivity. But a sustained burst is not possible, given the fewer daylight hours. Primary productivity will not rise again until spring.

TROPHIC NATURE OF LAKES Each lake's topography, climate, and geologic history dictate the numbers and kinds of resident species, how they are dispersed in the lake, and how they cycle nutrients. Soils of the surrounding region and the lake basin contribute to the type and amount of nutrients available.

Interplays among climate, soil, basin shape, and metabolic activities of the lake's residents contribute to a lake's trophic status. *Oligotrophic* lakes typically are newly formed, deep, clear, and nutrient poor, with low primary productivity. *Eutrophic* lakes are older, shallower, and nutrient rich, with a higher primary productivity. As a lake ages, sediments accumulate, water becomes less transparent and more shallow, and phytoplankton come to dominate the community. A filled-in basin is the final successional stage.

Eutrophication, recall, refers to natural or artificial processes that enrich a body of water with nutrients. For example, human activities caused eutrophication of Seattle's Lake Washington. Phosphate-rich sewage drained into the lake from 1941 to 1963. This made nitrogen the limiting factor for photosynthetic species. Cyanobacteria are at an advantage when nitrogen is scarce; they can fix N_2. They became dominant and formed slimy mats in summer, which made the lake useless for recreation. Sewage inputs were stopped. By 1975, the lake neared full recovery.

STREAM ECOSYSTEMS

Flowing-water ecosystems called **streams** start out as freshwater springs or seeps. As they flow downslope, they grow and merge, then often converge as a river. Between a river's headwaters and end, we find riffles, pools, and runs (Figure 42.25). The *riffles* are shallow, turbulent stretches where the water flows swiftly over a rough, sandy, rocky bottom. In *pools*, deep water flows slowly over a smooth, sandy, or muddy bottom. *Runs* are smooth-surfaced, rapidly flowing stretches over bedrock or rock and sand.

Rainfall, snowmelt, geography, altitude, even the shade cast by plants influence a stream's average flow volume and temperature. Streambed composition as well as agricultural, industrial, and urban wastes influence the solute concentrations in the water.

A stream imports organic matter into food webs, especially in forest ecosystems. Trees cast shade and thus hamper photosynthesis, but their litter starts detrital food webs (Section 41.2). Aquatic organisms take up and then release nutrients as the water flows downstream. Nutrients move upstream only in the tissues of migratory fishes and other animals. These nutrients spiral between aquatic organisms and the water as it flows on its one-way course to the sea.

The water provinces are far more extensive than the biomes. Their aquatic ecosystems differ, but they all show gradients in light penetration, temperature, and dissolved gases.

42.10 Life at Land's End

Near the coasts of continents, around islands and reefs, concentrations of nutrients support some of the world's most productive ecosystems.

WETLANDS AND THE INTERTIDAL ZONE

Like freshwater ecosystems, estuaries and mangrove wetlands have distinct physical and chemical features that include depth, water temperature, salinity, and the light penetration. **Estuaries** are partly enclosed coast regions where seawater mixes with nutrient-rich fresh water from rivers, streams, and runoff (Figure 42.26*a*). The confined region, slow mixing of water, and tidal action combine to trap dissolved nutrients. Water flow continually replenishes nutrients, which is one reason estuaries can support highly productive ecosystems.

Primary producers are phytoplankton, plants that tolerate submergence at high tide, and algae. Detrital food webs are common. So many larval and juvenile stages of invertebrates and some fishes develop here that estuaries are sometimes called marine nurseries. Many migratory birds use the estuaries as rest stops.

Estuaries range from broad, shallow Chesapeake Bay, Mobile Bay, and San Francisco Bay to the narrow, deep fjords of Norway and others like them in Alaska and British Columbia. Many estuaries are in decline because fresh water is being diverted upstream, and agricultural runoff and wastes are flowing in.

In tidal flats at tropical latitudes, we find nutrient-rich mangrove wetlands. "Mangrove" refers to forests of salt-tolerant plants in sheltered areas along tropical coasts. These plants have shallow or branching prop roots that extend from the trunk (Figure 42.26*b*). Many have root extensions that take up oxygen.

Net primary productivity of a mangrove wetland depends partly on the tidal volume and flow rate, as well as on salinity and nutrient availability. But these are all rich ecosystems that support notable biomass.

Figure 42.26 (**a**) South Carolina salt marsh. A marsh grass (*Spartina*) is a major producer. (**b**) In the Florida Everglades, a mangrove wetland lined with red mangroves (*Rhizophora*).

Figure 42.27 (**a**) From one of the Hawaiian Islands, sea urchins and a few algae, including sea lettuce (*Ulva*), exposed by a low tide. (**b**) Coral fragment that washed up on the sandy shore of Australia's Heron Island.

ROCKY AND SANDY COASTLINES

Rocky and sandy coastlines support ecosystems of the intertidal zone, which is not renowned for creature comforts. Waves batter its residents; tides alternately submerge and expose them. The higher up they are, the more they dry out, freeze in winter, or bake in summer, and the less food comes their way. The lower they are, the more they compete in limited spaces. At low tides, the birds, rats, and raccoons move in and feed on them. High tides bring the predatory fishes.

Generalizing about coastlines is not easy, for waves and tides continually resculpt them. One feature that rocky and sandy shores share is vertical zonation.

Rocky shores have three zones. The *upper* littoral zone is submerged only during the highest tide of the lunar cycle and is sparsely populated. The *mid*littoral zone is submerged during the highest regular tide and exposed during the lowest. Algae, fishes, hermit crabs, nudibranchs, sea stars, and sea urchins occupy its tidepools (Figure 42.27a). Biodiversity is greatest in the *lower* littoral zone, exposed only in the lunar cycle's lowest tide. In all three zones, erosion is so swift that detritus cannot accumulate. For that reason, you can expect to find only grazing food webs.

Waves and currents continually rearrange loose sediments on the *sandy* and *muddy* shores. Few large plants can grow in unstable places, so you won't find grazing food webs here. Detrital food webs start with inputs from land or offshore.

CORAL REEFS

Coral reefs develop in clear, warm waters near coasts or around volcanic islands, mainly between latitudes 25° north and south (Figure 42.28b). Each is a wave-resistant formation made of the slowly accumulated remains of marine organisms. Hard corals, cemented together, formed the spine. Also, mineral-hardened cell walls of red algae, such as the coralline species in Figure 42.28a, added to the spine. Secretions from many other reef organisms helped hold the collective hard parts together.

Figure 42.28 hints at the wealth of warning colors, tentacles, and stealth of reef species—signs of danger and fierce competition for resources by individuals packed together in a limited space.

Life thrives where land meets sea. Wetlands and coral reefs show high primary productivity. Rocky and sandy shores are not renowned for their creature comforts.

CORALLINE ALGA

SEA FAN

MORAY EEL

CHAMBERED NAUTILUS

CLOWNFISH WITH SEA ANEMONE

CROWN-OF-THORNS SEA STAR ON CORAL

a

b

Figure 42.28 (**a**) A sampling of biodiversity from coral reefs around the world. (**b**) Global distribution of reefs (*orange*), coral banks (*gold*), and solitary corals (*red*). Nearly all reef-building corals form in warm seas (within dark lines).

42.11 Coral Bleaching

FOCUS ON THE ENVIRONMENT

About 6,000 years ago, when Egypt's first dynasty was gaining political power, coral reefs were forming off the Florida coast. Today, countless numbers and kinds of marine invertebrates as well as fishes and algae live there. Will these reefs and others like them around the world endure much longer? Maybe not.

Dinoflagellate symbionts live inside the tissues of most reef-building corals. Dinoflagellates, recall, are single-celled photoautotrophs. In return for shelter, they provide coral polyps with oxygen and recycle nutrients that otherwise would be lost as wastes.

The interaction works well in warm seas, where oxygen and nutrient levels are typically low. When corals are stressed, however, they expel the symbionts. When stressed for more than a few months, corals die. Only their white, hardened chambers remain, an outcome called *coral bleaching* (Figure 42.29).

Coral bleaching is on the rise all over the world, partly because of the higher sea surface temperatures. Other factors also are at work. In 2002, an algal bloom ninety-six kilometers across drifted through the Florida Keys. Toxins released by free-living dinoflagellates of the bloom killed 70 percent of reef-building corals. At about the same time, researchers started transplanting lab-grown pink sea fans (*Gorgonia ventalina*) to a coral reef that a ship had run over. Florida's sea fans are also dying of aspergillosis, a fungal disease. Figure 42.28*a* shows what a healthy sea fan is supposed to look like.

Figure 42.29 Coral bleaching. In 2002, water temperature around Australia's Great Barrier Reef was the highest recorded since 1870. Warmer water caused the most widespread bleaching ever observed on this barrier reef. Nearly 60 percent of the corals bleached out.

42.12 The Open Ocean

The world ocean has two vast provinces (Figure 42.30c). Its benthic (bottom) province starts at continental shelves and extends to deep-sea trenches. Its pelagic province is the full volume of ocean water. The neritic zone is the volume above continental shelves, and the oceanic zone is the volume above the ocean basins.

SURPRISING DIVERSITY

Photosynthesis is seasonal and intense near the ocean surface. Drifting in seawater are phytoplankton, the "pastures" that feed copepods, krill, whales, squids, fishes, and other members of marine food webs. Also near the surface, photosynthetic bacteria collectively called ultraplankton might account for 70 percent of the ocean's primary productivity.

Deeper ocean water is too dark for photosynthesis. There, food webs start with **marine snow**. These tiny bits of organic matter drift down from communities above. They are the base for staggering biodiversity; midoceanic water may be home to 10 million species!

Also, in what might be the greatest of all circadian migrations, a number of species rise thousands of feet to feed in upper waters at night and move down the next morning. Carnivores at the top of the food webs range from familiar types, including sharks and giant squids, to the visually jarring (Figure 42.30*a*).

Hydrothermal vent ecosystems occur in the benthic province. Near-freezing water seeping into fissures in the seafloor becomes superheated. As it spews back out, it leaches mineral ions from the rocks. Dissolved in the outpouring are iron, zinc, copper sulfides, and sulfates of magnesium and calcium. These minerals settle out as rich deposits that function as the energy source for chemoautotrophic species of bacteria and archaea. These prokaryotes are the primary producers for rich food webs that include tube worms (Figure 42.30*b*), as well as crustaceans, clams, and fishes.

Many biologists suspect that life originated in such hot, nutrient-rich places on the seafloor. Section 18.2 sketched out some evidence of this possibility.

UPWELLING AND DOWNWELLING

As you read earlier, prevailing winds that parallel the western coasts of continents tug on the ocean surface. Wind friction gets the surface waters moving. Earth's rotational force deflects masses of slow-moving water away from the coasts. Cold, deep, often nutrient-rich water moves in vertically in its place (Figure 42.31).

Cold, deep water moving up this way is called an **upwelling**. Upwelling occurs in equatorial currents

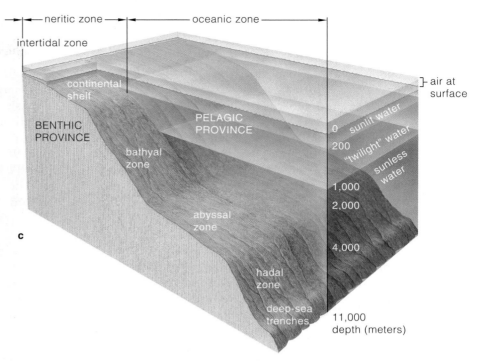

Figure 42.30 The ocean. (**a**) Deep-sea angler fish. (**b**) Tube worms at a hydrothermal vent on the ocean floor. (**c**) Oceanic zones. The seafloor extends from the continental shelves to deep-sea trenches. Dimensions of each zone are not to the same scale.

and along continents, and it cools air masses above it. When thick fogbanks form near California's coast, an upwelling of cold water is interacting with warm air.

In the Southern Hemisphere, commercial fisheries depend on wind-induced upwelling along Peru and Chile. Prevailing coastal winds blow from the south and southeast, tugging surface water away from the shore. Cold, deeper water carried to the continental shelf by the Humboldt Current moves up near the surface. Huge quantities of nitrate and phosphate are pulled up and carried north by the cold Peru Current. The nutrients sustain phytoplankton that are the basis of one of the world's richest fisheries.

Every three to seven years, warm surface waters of the western equatorial Pacific Ocean move eastward. This massive displacement of warm water acts on the prevailing wind direction. The eastward flow speeds up so much that it hampers the vertical movement of water along the coasts of Central and South America.

Water piling into a coastline is forced downward and flows away from it. Near Peru's coast, prolonged **downwelling** of nutrient-poor water displaces cooler waters of the Humboldt Current and puts a stop to upwelling. The warm current typically arrives around Christmas. Fishermen in Peru named it **El Niño** ("the

a North wind starts coastal water moving

b Earth's rotational force deflects the moving water west

c Cold water moves up as replacement

Figure 42.31 Coastal upwelling in the Northern Hemisphere.

little one," in reference to the baby Jesus). The name became incorporated into a more inclusive, scientific explanation. As described next, this recurring event is the El Niño Southern Oscillation, or ENSO.

From the ocean's coral reefs down to hydrothermal vents, throughout the pelagic province, we find staggering degrees of primary productivity and biodiversity.

42.13 Applying Knowledge of the Biosphere

We turn now to an application that reinforces a unifying ecological concept. Events in the atmosphere and ocean, and on land, interconnect in ways that can profoundly influence the world of life.

An El Niño Southern Oscillation, or **ENSO**, is defined by changes in sea surface temperatures and in the air circulation patterns. "Southern oscillation" refers to a seesawing of atmospheric pressure in the western equatorial Pacific, the world's largest reservoir of warm water and warm air. It is the source of heavy rainfall, which releases enough heat energy to drive global air circulation.

Between ENSOs, the warm waters and heavy rains move westward (Figure 42.32a). *During* an ENSO, prevailing surface winds in the western equatorial Pacific pick up speed and "drag" surface waters east (Figure 42.32b). As it does, the westward transport of water slows. The sea surface temperatures increase,

evaporation accelerates, and air pressure falls. These changes have global repercussions.

El Niño episodes typically last for six to eighteen months. Then another oscillation known as **La Niña** starts up, and the weather seesaws again.

As you read in the chapter opening, 1997 ushered in the most powerful ENSO event of the century. The average sea surface temperatures in the eastern Pacific rose 9°F (about 5°C). The warmer water extended 9,660 kilometers west from the coast of Peru.

The 1997–1998 El Niño/La Niña rollercoaster had record-breaking impact on primary productivity in the equatorial Pacific. With the massive eastward flow of nutrient-poor warm water, populations of photosynthesizers shrank to almost undetectable numbers (Figure 42.32c). During the La Niña rebound, cool, nutrient-rich water welled up to the surface and was displaced westward, along the equator. As satellite images clearly revealed, the upwelling had

Read Me First!

and watch the narrated animation on El Niño Southern Oscillations

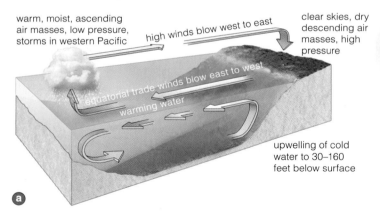

warm, moist, ascending air masses, low pressure, storms in western Pacific

high winds blow west to east

clear skies, dry descending air masses, high pressure

equatorial trade winds blow east to west

warming water

upwelling of cold water to 30–160 feet below surface

(a)

clear skies, descending air masses, high pressure

high winds blow west

warm, moist, ascending air masses, low pressure, storms

rain falls in central Pacific

trade winds weaken; warm water flows east

no upwelling; cold water as deep as 500 feet below surface

(b)

Figure 42.32 (**a**) Westward flow of cold, equatorial surface water in between ENSOs. (**b**) Eastward dislocation of warm ocean water during an ENSO.

Satellite data on primary productivity in the equatorial Pacific Ocean. The concentration of chlorophyll was used as the measure.

(**c**) During the 1997–1998 El Niño event, a massive amount of nutrient-poor water moved to the east, and photosynthetic activity was negligible.

(**d**) During the subsequent La Niña episode, massive upwelling and westward displacement of nutrient-rich water led to a huge algal bloom that stretched all the way to the coast of Peru.

(c) Near-absence of phytoplankton in the equatorial Pacific during an El Niño.

(d) Immense algal bloom in the equatorial Pacific in the La Niña rebound event.

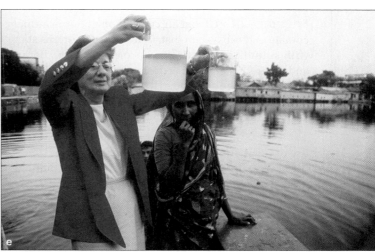

Figure 42.33 (**a**) *Vibrio cholerae*, agent of cholera. (**b**) Copepods host a dormant stage of *V. cholerae* that waits out adverse environmental conditions that don't favor its growth and reproduction. (**c**) Typical Bangladesh waterway from which water samples were drawn for analysis. (**d**) Satellite data on rising sea surface temperatures in the Bay of Bengal correlated with cholera cases in the region's hospitals. *Red* signifies warmest summer temperatures. (**e**) In Bangladesh, Rita Colwell comparing unfiltered and filtered drinking water.

sustained a vast algal bloom, one that stretched across the equatorial Pacific (Figure 42.32*d*).

During the 1997–1998 El Niño event, 30,000 cases of *cholera* were reported in Peru alone, compared to only 60 cases from January to August in 1997. People knew water contaminated by *Vibrio cholerae* causes cholera epidemics (Figure 42.33*a*). The disease agent triggers severe diarrhea; it enters water supplies in feces. People who use the tainted water get infected.

What people did *not* know was where *V. cholerae* lurked between cholera outbreaks. It could not be found in humans or in water supplies. Then it would show up simultaneously in places that could be far apart—usually coastal cities, where the urban poor draw water from rivers flowing to the sea.

Marine biologist Rita Colwell had been thinking about cholera for some time. She suspected that humans were not the host between outbreaks. Was there an environmental reservoir for the pathogen? Maybe. But nobody had detected it in water samples that had been subjected to standard culturing.

Then she had a flash of insight: What if no one could find the pathogen because it changes form and enters a dormant stage between outbreaks?

During a cholera outbreak in Louisiana, Colwell realized that she could use an antibody-based test to detect a protein unique to *V. cholerae*'s surface. Later, similar tests in Bangladesh revealed bacteria in fifty-one of fifty-two samples of water. Standard culture methods had missed it in all but seven samples.

V. cholerae lives in rivers, estuaries, and seas. As Colwell knew, plankton also thrive in these aquatic environments. She decided to restrict her search for the unknown host to warm waters near Bangladesh,

where outbreaks of cholera occur seasonally (Figure 42.33*c*). It was here that she detected the dormant stage of *V. cholerae* in copepods. These tiny marine crustaceans graze on algae and other species that make up phytoplankton (Figure 42.33*b*). The number of copepods—and of *V. cholerae* inside them—rises and falls with shifts in phytoplankton abundances.

Colwell already knew about seasonal variations in sea surface temperatures. Remember the old saying, *Chance favors the prepared mind*? In one sense, she was prepared to recognize a connection between cholera cases and seasonal temperature peaks in the Bay of Bengal. She compared data from the 1990–1991 and 1997–1998 El Niño episodes. Her correlation held. Four to six weeks after the sea surface temperatures go up, so do cases of cholera!

Now Colwell and Anwarul Huq, a Bangladeshi scientist, are studying salinity and other factors that may influence outbreaks. Their goal is to design a model for predicting where cholera will break out next. They have advised women in Bangladesh to use sari cloth as a filter to remove *V. cholerae* cells from the water (Figure 42.33*e*). Copepod hosts are too big to pass through the thin cloths, which can be rinsed in clean water, sun-dried, and used again and again. By using this simple and inexpensive method, cholera outbreaks have been reduced by 50 percent.

Summary

Section 42.1 Global air circulation patterns influence climate and ecosystem distribution. The patterns start with latitudinal variations in incoming solar radiation and are influenced by Earth's daily rotation and annual path around the sun, the distribution of continents and seas, and elevations of land masses. Solar energy, and the winds it drives, are renewable, clean sources of energy.

Section 42.2 Humans influence the atmosphere. The use of CFCs and methyl bromide has depleted ozone in the upper atmosphere. With seasonal ozone thinning, more damaging UV radiation reaches Earth's surface.

Smog, a form of air pollution, arises in areas where large amounts of fossil fuels are burned. Coal-burning power plants are also the main contributors to acid rain, which alters habitats and kills many organisms.

Section 42.3 Latitudinal and seasonal variations in sunlight warm ocean waters and set currents in motion. The currents distribute heat energy around the seas and affect weather patterns. Ocean currents, air currents, and topography interact to shape global climate zones.

Section 42.4 The world's land masses are realms of biodiversity, each with an evolutionary history and a tapestry of physical and chemical conditions. Biomes are vast expanses characterized by specific arrays of species, mainly plants and animals. Regional variations in climate, landforms, and soils influence them. Marine ecoregions are comparable realms of biodiversity.

Section 42.5 Deserts form around latitudes 30° north and south if annual rainfall is sparse. Slightly moister southern or western coastal regions support dry woodlands and shrublands. In the midlatitudinal interiors of continents, vast deserts or grasslands form.

Section 42.6 In equatorial regions of high rainfall, high humidity, and mild temperatures, evergreen tropical forests form. Deciduous broadleaf forests grow in regions with hot summers and cool winters. Where a cold, dry season alternates with a cold, rainy season, coniferous forests dominate.

Section 42.7 Low-growing, hardy plants of the tundra dominate at high latitudes and high altitudes.

Section 42.8 Soil characteristics vary among biomes and help determine their primary productivity.

Section 42.9 Lakes, streams, and other aquatic ecosystems show gradients in penetration of sunlight, water temperature, salinity, and dissolved gases. These factors vary over time and affect primary productivity.

Sections 42.10, 42.11 Coastal zones and tropical reefs support diverse ecosystems. Primary productivity is high in coastal wetlands and on coral reefs. Coral reefs take thousands of years to build. Most are now threatened by global warming and human activities.

Section 42.12 Life persists throughout the ocean. Diversity is highest in sunlit waters. Mineral-rich waters support communities at deep-sea hydrothermal vents.

Upwelling is an upward movement of deep, cool, often nutrient-rich ocean water, typically along coasts of continents. An El Niño event disrupts upwelling and triggers massive, reversible changes in rainfall and other weather patterns around the world.

Section 42.13 Drawing on knowledge of microbial ecology as well as biogeographic patterns, Rita Colwell found a crucial bit of information that led to effective countermeasures against cholera outbreaks.

Self-Quiz
Answers in Appendix III

1. Solar radiation drives the distribution of weather systems and so influences _____ .
   a. temperature zones   c. seasonal variations
   b. rainfall distribution   d. all of the above

2. _____ shields life against the sun's UV wavelengths.
   a. A thermal inversion   c. The ozone layer
   b. Acid precipitation   d. The greenhouse effect

3. Regional variations in the global patterns of rainfall and temperature depend on _____ .
   a. global air circulation   c. topography
   b. ocean currents   d. all of the above

4. A rain shadow is a reduction in rainfall _____ .
   a. on the leeward side of a mountain range
   b. during an El Niño event
   c. that occurs seasonally in the tropics

5. The major cause of acid rain is operation of _____ .
   a. nuclear power plants   c. coal burning
   b. automobiles   d. agricultural machinery

6. Biomes are _____ .
   a. water provinces   d. partly characterized
   b. water and land zones   by dominant plants
   c. vast expanses of land   e. both c and d

7. Biome distribution depends on _____ .
   a. climate   c. soils
   b. topography   d. all of the above

8. Grasslands most often predominate _____ .
   a. near the equator   c. in interior of continents
   b. at high altitudes   d. b and c

9. During _____ , deeper, often nutrient-rich water moves to the surface of a body of water.
   a. spring overturns   c. upwellings
   b. fall overturns   d. all of the above

10. Match the terms with the most suitable description.
    ____ tundra   a. equatorial broadleaf forest
    ____ chaparral   b. partly enclosed land where
    ____ desert   freshwater and seawater mix
    ____ savanna   c. type of grassland with trees
    ____ estuary   d. has low-growing plants at
    ____ boreal forest   high latitudes or elevations
    ____ tropical rain   e. at latitudes 30° north and south
         forest   f. mineral-rich, superheated water
    ____ hydrothermal   supports communities here
         vents   g. conifers dominate
                         h. dry shrubland

Pacific
Ocean

North
America
Arctic
Ocean
Asia

Atlantic
Ocean
Europe

Africa

April 27   April 30   May 2   May 6

Figure 42.34 Global distribution of radioactive fallout after the 1986 meltdown of the Chernobyl nuclear power plant. It put 300 million to 400 million people at risk for leukemia and other radiation-induced disorders. By 1998, the rate of thyroid abnormalities in children living downwind from the site was nearly seven times as high as for those upwind; their thyroid gland concentrated the iodine radioisotopes.

Critical Thinking

1. On April 26, 1986, in Ukraine, a meltdown occurred at the Chernobyl nuclear power plant. Nuclear fuel burned for nearly ten days and released 400 times more radioactive material than the atomic bomb that dropped on Hiroshima. Thirty-one people died right away. An estimated 15,000 to 30,000 have died since of cancers and other effects of radiation. Winds carried radioactive fallout around the globe (Figure 42.34).

The Chernobyl accident stiffened opposition to *nuclear power* in the United States, but recent developments have some people reconsidering. Increasing the use of nuclear energy would diminish the country's dependence on oil from the politically unstable Middle East. Nuclear power does not contribute to global warming, acid rain, or smog. It does produce highly radioactive wastes. Investigate the pros and cons of nuclear power, and decide if you think the environmental benefits outweigh the risks. Would you feel differently if a nuclear power plant were about to be built ten kilometers upwind from your home?

2. Use of off-road recreational vehicles may double over the next twenty years. Many off-road enthusiasts would like increased access to government-owned desert areas. Some argue that it's just the perfect place for off-roaders because "There's nothing there." Do you agree? If not, how would you counter this argument?

3. Write a short description of how global warming might affect spring overturn and thermocline formation in a Minnesota lake. What would be some ecological effects?

4. *Wetlands,* remember, are transitional zones between aquatic and terrestrial ecosystems. Wetlands everywhere are being converted for agriculture, home building, and other human activities. Does the ecological value of wetlands for the nation as a whole outweigh the rights of the private citizens who own many of the remaining wetlands in the United States? Should private owners be forced to transfer title of their land to the government? If so, who should decide the value of a parcel of land *and* pay for it?

IMPACTS, ISSUES *My Pheromones Made Me Do It*

A few years ago, as Toha Bererub walked down a street near her Las Vegas home, she felt a sharp pain above her right eye. Then another. And another. Seconds later, stinging bees had encased the upper half of her body. Firefighters wearing protective gear rescued her, but not before she had been stung more than 500 times.

Bererub's attackers were Africanized honeybees, a hybrid between the European honeybee and a more aggressive African strain (Figure 43.1). The African bees were imported to Brazil in the 1950s for cross-breeding experiments. Breeders were after a gentle but zippier pollinator for orchards. But some of the bees from Africa escaped and started mating with the locals.

Then, in a grand example of geographic dispersal, some descendant bees buzzed all the way from Brazil to Mexico and on into the United States. So far, Africanized bees have established themselves in Texas, New Mexico, Nevada, California to the west, and Alabama, Virginia, and Florida to the east.

Honeybees sting only once, and all make the same kind of venom. Africanized bees make a bit less venom but get riled up faster and mount collective attacks. One squadron reportedly chased a perceived threat for a quarter of a mile.

The Africanized bees became known as "killer bees," although they rarely kill their target. To be sure, their stings are extremely painful, but adults in good health usually can survive a collective attack. Bererub was seventy years old when attacked, and she recovered fully after spending a week in the hospital.

So what makes Africanized bees so testy? Isopentyl acetate. This chemical, which smells like bananas, is a

Figure 43.1 Africanized honeybees standing guard at the entrance to their hive. When a potential intruder appears, they release an alarm pheromone that stimulates other colony members to join them in an attack. In the filmstrip at far right, Toha Bererub is describing how stinging bees swarmed over her face. She recovered fully.

the big picture

The Basis of Behavior Animal behavior starts with gene products that build and operate the nervous and endocrine systems. Animals are prewired to execute innate behaviors. Most are prewired to learn, to modify their behavior through experience with the environment.

Natural Selection and Behavior Communication signals, mating systems, and other forms of behavior have evolved by way of natural selection. From the evolutionary perspective, a behavior is adaptive when it increases the reproductive success of the individual that displays it.

key component of honeybee alarm pheromone.

A *pheromone*, recall, is a chemical signal released by one individual that affects the behavior of another individual of the same species. A honeybee releases an alarm pheromone when threatened and when it stings something. As the signaling molecules diffuse through the air, they form a concentration gradient that guides other bees to the individual sounding the alarm.

Researchers once compared hundreds of colonies of Africanized and European honeybees for their alarm pheromone responses. They positioned a tiny target in front of each colony, then released a small quantity of artificial pheromone. Compared to European bees, the Africanized bees were faster to fly out of the colony and zero in on the perceived threat. They plunged six to eight times as many stingers into it.

The two kinds of honeybees show other behavioral differences. Compared to European bees, Africanized bees are less picky about where they establish a colony. They are more likely to abandon it after being disturbed. Of more concern to beekeepers, they are less interested in stashing large amounts of honey.

Such differences among honeybees lead us into the world of *animal behavior*, to the coordinated responses that animal species make to stimuli. We will look first at the genetic basis of behavior, then at instinctive and learned mechanisms that arise from it. As you will see, natural selection theory helps explain how and why animals behave as they do.

 How Would You Vote?

Africanized bees are slowly expanding their range in North America. Some think the more we know about them, the better we will be able to protect ourselves. Should we fund more research into the genetic basis of their behavior? See the Media Menu for details, then vote online.

Group Living Life in social groups has reproductive costs as well as benefits for the individual. Costs are highest in societies where self-sacrificing behavior is extreme. Selection theory helps explain how such altruistic behavior may still be adaptive.

What About Humans? Microevolutionary forces influence behavioral traits that promote reproductive success. Yet, a human behavior that is adaptive in the sense of perpetuating one's genes may be socially maladaptive; it may be viewed as unacceptable or morally wrong.

43.1 Behavior's Heritable Basis

The nervous and endocrine systems govern behavioral responses to stimuli. Because genes specify the substances required for constructing and operating those systems, they are the heritable foundation for animal behavior.

GENES AND BEHAVIOR

Nervous systems are wired to detect, interpret, and issue commands for response to stimuli. A **stimulus**, remember, is a piece of information about the external or internal environment that a specific type of sensory receptor has detected. It takes gene products to build and operate sensory receptors, nerve pathways and, in most species, a brain. Gene products also influence behavioral responses to stimuli.

Stevan Arnold found evidence of the genetic basis of behavior in the feeding preferences of coastal and inland snake populations in California. Garter snakes living near the coast hunt banana slugs (Figure 43.2*a*). Snakes inland hunt tadpoles and fishes. Offer them a banana slug and they ignore it.

In one set of experiments, Arnold offered captive newborn snakes a bit of slug as a first meal. Newborn coastal snakes usually ate it and flicked their tongue at cotton swabs drenched in essence of slug. (Snakes "smell" by tongue-flicking, which pulls odors into the mouth.) Newborn inland snakes ignored the swabs and rarely ate bits of slug. Here was a clear difference between captive snakes that had no prior experience with slugs. The snakes are programmed before birth to accept or reject slugs; they did not learn feeding preferences by taste trials.

Allelic differences may affect how odor-detecting mechanisms form during garter snake development. To test this hypothesis, Arnold crossed coastal and inland snakes. He predicted that the hybrid offspring

Figure 43.2 (**a**) Banana slug, food for (**b**) an adult garter snake of coastal California. (**c**) Newborn garter snake from a coastal population, tongue-flicking at a cotton swab drenched with tissue fluids from a banana slug.

would make an *intermediate* response to slug chunks and odors. Test results matched his prediction.

Compared with a typical newborn inland snake, many baby snakes of mixed parentage tongue-flicked more often at slug-scented cotton swabs. However, they did so less often than newborn coastal snakes.

HORMONES AND BEHAVIOR

Hormones secreted by the endocrine system also are gene products that influence behavior. Let's focus on oxytocin, a hormone that has roles in birth, lactation, and mate selection, aggression, territoriality, and other social behaviors. All mammals, including voles, make and secrete it.

Prairie voles (*Microtus ochrogaster*) are among the few long-term monogamous mammals. Pair-bonding arises during a single night of repeated matings, with oxytocin secretions providing the ambiance. After the "honeymoon," a male and female have little interest in other potential partners.

In one experiment, researchers kept a male and female together for a few hours but prevented them from mating. Later on, they injected oxytocin into the female. The result was love at first sight; pair-bonding without the requisite sex. In another experiment, pair-bonded female prairie voles were injected with an oxytocin blocker. They promptly dumped their long-term partners and went off to mate with others.

Figure 43.3 Hormones and prairie voles. Distribution of oxytocin receptors (*red*) in the brain of (**a**) a mate-for-life prairie vole and (**b**) a promiscuous mountain vole.

Figure 43.4 A few instinctive behaviors. (**a**) A social parasite, the European cuckoo, lays eggs in the nests of other birds. Even before this cuckoo hatchling opens its eyes, it responds instinctually to a round shape—a host's egg—and shoves it out of the nest. (**b**) The clueless foster parents show instinctual behavior as they respond to the usurper's gaping mouth, not to its size and other traits that characterize their own species. (**c**) A human baby instinctively imitates adult facial expressions.

Not all vole species are monogamous. Differences in the number and distribution of oxytocin receptors are part of the reason. Monogamous prairie voles have more of the receptors than the mountain voles, which are highly promiscuous (Figure 43.3).

Antidiuretic hormone (ADH, or vasopressin) also plays a role. Researchers isolated the gene for an ADH receptor in monogamous prairie voles and transferred copies of it into forebrain cells of male meadow voles (*M. pennsylvanicus*). After the transfer, the males of this far more promiscuous species showed an increased tendency to partner with just one female.

Male meadow voles used as a control group also got copies of the gene, but in a brain region not known to be involved in pair-bonding. Unlike the experimental group, the males retained their playboy ways.

INSTINCTIVE BEHAVIOR

Slug-loving garter snakes and pair-bonding voles are lead-ins to **instinctive behavior**. The term refers to a behavior that is performed without having first been learned through experience in the environment.

Like many other animals, coastal garter snakes are prewired to recognize sign stimuli before being born or hatched. *Sign stimuli* are one or two simple, well-defined cues that trigger an expected response. For these snakes, the cue is a certain slug scent that calls for a *fixed action pattern*—a stereotyped motor program of coordinated muscle activity that runs to completion independently of feedback from the environment. The snake is compelled to strike, capture, and eat a slug.

Similarly, hormone molecules wafting through the air are cues for mating behavior, a fixed action pattern between male and female vole.

Cuckoos, a type of social parasite, offer another case of instinctive behavior. Female cuckoos lay eggs in nests of other species. Cuckoos are blind when they first hatch, but contact with an egg or another round object stimulates a fixed action pattern. The hatchling maneuvers the egg onto its back, then pushes the egg from the nest (Figure 43.4*a*). This behavior ensures that the hatchling receives undivided attention.

The cuckoo's "foster parents" are oblivious to the odd color and size of the usurper. They respond only to one sign stimulus—the gaping mouth of a chick—and continue with their parenting (Figure 43.4*b*).

Humans, too, show instinctive behavior. Three days after birth, a human infant already displays a capacity to mimic facial expressions of an adult who comes close to it (Figure 43.4*c*). The infant cannot see its own face, nor can it feel which facial muscles the adult is using. Somehow it is able to open its mouth, protrude its tongue, or rotate its head the same way as the adult. Infants will also respond to a simplified stimulus—a flat, face-sized mask with two dark spots for eyes. One "eye" won't do the trick.

Genes underlie animal behavior, or coordinated responses to stimuli. Certain gene products have roles in constructing and operating the nervous system, which governs behavior. Other gene products, such as hormones, influence the mechanisms required for particular forms of behavior.

Animals start out life neurally wired to recognize vital cues and to make an instinctively suitable response, one that has not been learned through actual experience.

Many animals execute a fixed action pattern, a stereotyped program of coordinated muscle activity in response to one or two simple, well-defined environmental cues.

43.2 Learned Behavior

*With **learned behavior**, responses vary or change as a result of an individual's experiences. Whether learned or instinctive, behavior develops through interactions between genes and environmental inputs.*

Animals process information about experiences and then use it to change or vary responses to stimuli. **Imprinting** is a classic example of learned behavior. This time-dependent form of learning is triggered by exposure to a sign stimulus. Exposure usually takes place during a sensitive period when the animal is young. Imprinting of baby geese on their mother is a favorite example of animal behaviorists (Figure 43.5).

Learned behavior arises as the environment directly or indirectly alters gene expression. Sensory input and good or bad nutrition are typical factors that lead to alterations in how and what an animal learns.

For example, birdsong is an instinctive behavior. Yet in different habitats, songbirds learn variations, or dialects, of the species song. As Peter Marler showed, many male birds learn their full song ten to fifty days after hatching by listening to other birds sing it. The male nervous system is prewired to recognize a species song; a learning mechanism is primed to select and respond to acoustical input. But what he hears during a sensitive period shapes his *rendition* of the song.

In one study, Marler raised white-crown nestlings in soundproof chambers so they could not hear adult males. The captives sang when mature, but not with the exact structure of a typical adult song. Marler also isolated and exposed captives to recorded white-crown songs *and* song sparrow songs. At maturity, captives sang only the white-crown song and even mimicked the species dialect of the unseen tutor.

In another experiment, Marler did not use taped songs. He let young, hand-reared male white-crowns interact with a "social tutor" of a different species. The males tended to learn the tutor's song.

Many such experiments have been performed. The results support this hypothesis: Birdsong starts with a genetically based capacity to learn from acoustical cues. Think about that when you hear songbirds in spring (Figure 43.6).

Imprinting is a case of learned behavior. It occurs during a genetically determined sensitive period. Like all behavior, it develops by interactions between genes and an animal's social and physical environment.

Figure 43.6 Male marsh wren belting out a territorial song that has been likened to a loud gurgle. Males of this species start to imitate their species song when they are about fifteen days old. Unike many species, marsh wrens continue to learn songs throughout their life.

Figure 43.5 No one can tell these imprinted baby geese that Konrad Lorenz is not Mother Goose!

In response to a moving object and probably to certain sounds, baby geese imprint on the mother goose and follow her during a short, sensitive period right after hatching. They are neurally wired to learn some crucial information —the identity of the one individual that will help protect them in the months ahead. Usually that will be their mother.

Konrad Lorenz, one of the early investigators of animal behavior, presented the baby geese in this figure with sign stimuli that made them form an attachment to him.

43.3 The Adaptive Value of Behavior

If forms of behavior have a genetic basis, then they may evolve in various ways through natural selection. Alleles that encode the most adaptive versions of a trait will tend to increase in frequency in a population, and alternative alleles will not. In time, genetic changes in behavior that yield greater reproductive success will be favored.

Natural selection theory helps us develop and test explanations of why a behavior persists and how it offers reproductive benefits that offset reproductive costs (disadvantages) associated with it. If a behavior is adaptive, it promotes the *individual's* production of offspring. Here are five definitions to keep in mind:

1. **Reproductive success**. An individual produces some surviving offspring.

2. **Adaptive behavior**. Behavior that helps perpetuate an individual's genes; its frequency in a population is maintained or increases over the generations.

3. **Social behavior**. Interdependent interactions among individuals of one or more populations of a species.

4. **Selfish behavior**. A form of behavior that improves an individual's chance to produce or protect its own offspring regardless of the impact on the population.

5. **Altruism**. Self-sacrificing behavior. An individual behaves in a way that helps others in the population but lowers its own chance of producing offspring.

When biologists speak of selfish or altruistic behavior, they don't mean an individual is consciously aware of some behavior or its reproductive goal. A lion doesn't have to know that eating zebras will be good for its reproductive success. Its nervous system just screams HUNTING BEHAVIOR! when the lion is hungry and sees a zebra. Hunting behavior persists in a lion population because genes for neural mechanisms that command hunting behavior are persisting.

To assess the adaptive value of any behavior, look for how it might promote reproductive success. For example, starlings (*Sturnus vulgaris*) nest in cavities of trees and *decorate* the nest bowl with sprigs of fresh leaves of certain strongly scented plants, such as the wild carrot (*Daucus carota*).

Larry Clark and Russell Mason suspected that nest decorating behavior suppresses nest parasites. Some mites parasitize birds and infest their nests. Even a few mites produce thousands of descendants. In large numbers, mites suck enough blood from a nestling to weaken it and affect its growth and survival.

Clark and Mason tested the hypothesis that nest decorating behavior is adaptive because it helps deter mites. They devised a set of experimental nests, some

Figure 43.7 An experiment to test the adaptive value of the nest-decorating behavior of starlings (**a**). Nests designated *A* did not have fresh sprigs of wild carrot (**b**) and other plants that make aromatic compounds. Nests designated *B* had fresh sprigs added every seven days. (**c**) Twenty-one days after the start of the experiment, when chicks had left, researchers made counts of the mites (*Ornithonyssus sylviarum*) infesting each nest. Results showed that aromatic compounds suppress development of juvenile mites into (**d**) adult mites.

with freshly cut wild carrot leaves and some without. They removed natural nests that starlings constructed and were using. Half of the nesting starling pairs got new nests with wild carrot sprigs. Replacement nests for the other pairs of starlings were sprigless.

Figure 43.7c shows the results. The number of mites in sprig-free nests was far greater than the number in sprig-festooned nests. At the end of one experiment, the sprig-free nests teemed with an average of 750,000 mites. Nests with sprigs had only 8,000 mites. Why?

Wild carrot shoots contain a highly aromatic steroid compound that almost certainly repels herbivores and helps plants survive. By coincidence, it prevents mites from maturing sexually and thus prevents population explosions of mites in nests festooned with wild carrot.

Aromatic greenery is a fumigant that decreases the numbers of mites, increases the survival of nestlings, and promotes reproductive success of the decorators.

A genetically determined behavior may persist or increase in frequency in a population when it is adaptive. A behavior is adaptive when it increases the number of descendants that an individual successfully produces.

43.4 Communication Signals

Competing for food, defending territory, alerting others to danger, advertising sexual readiness, forming bonds with a mate, caring for the offspring—such intraspecific behaviors require unambiguous forms of communication.

THE NATURE OF COMMUNICATION SIGNALS

Communication signals are unambiguous cues sent and received among individuals of the same species, and they involve instinctive and learned behaviors. Information-laden cues from a signaler are meant to alter the behavior of receivers. Chemical, acoustical, and visual cues are among the most common.

Pheromones, again, are chemical communication signals. *Signaling* pheromones induce the receiver to respond fast. They include chemical alarms, such as the honeybee call to action against a potential threat. They also include sex attractants. Bombykol is one of them. Bombykol molecules released by a female silk moth can summon males that are kilometers away. *Priming* pheromones bring about physiological (not behavioral) responses. For example, a volatile odor in urine of certain male mice triggers and enhances estrus in female mice of the same species.

Acoustical signals abound in nature. Male birds, frogs, grasshoppers, whales, and other animals make sounds that attract females. Prairie dogs bark alarms. Wolves howl and kangaroo rats drum their feet on the ground when advertising possession of territory.

Some signals never vary. Zebra ears pressed flat to the head convey hostility; ears pointing up convey its absence. Different signals convey the intensity of the message. A zebra with ears laid back is not too riled up as long as its mouth is open only a bit. When the ears are laid back and its mouth gapes, watch out. That combination is a type of *composite* signal. Such signals have information encoded in two or more cues.

Signals often take on different meaning in different contexts. A lion emits a spine-tingling roar to keep in touch with its pride *or* to threaten rivals. Also, a signal can convey information about signals to follow. Dogs and wolves solicit play behavior with their play bow, as in Figure 43.8*a*. Without the bow, a signal receiver may construe the behaviors that follow as aggressive, sexual, or even exploratory—but not playful.

Signals evolve or persist in a population when they promote reproductive success of both the sender *and* receiver. If a signal is harmful, natural selection will favor individuals that don't send it or respond to it.

COMMUNICATION DISPLAYS

The play bow is a *communication* display, a pattern of behavior that is a social signal. The *threat* display is another common pattern. It announces that a signaler is prepared to attack a signal receiver. When a rival for a receptive female confronts him, a dominant male baboon will roll his eyes upward and "yawn," which exposes his canines (Figure 43.8*b*). The signaler often benefits when the rival backs down, for he retains access to the female without having to engage in a fight. The signal receiver benefits because he avoids a serious beating, infected wounds, and possibly death.

Such displays are ritualized, with intended changes in the function of common behavior patterns. Normal movements may be exaggerated or frozen. Feathers,

Figure 43.8 Communication displays. (**a**) Play bow of a young male wolf soliciting a romp. (**b**) Part of a male baboon's threat display: exposed canines. (**c**) Courtship display of Adele penguins.

Read Me First!
and watch the narrated animation on honeybee dances

Figure 43.9 Honeybee dances, a type of tactile display. (**a**) Honeybees that have visited a food source close to their hive perform a *round* dance on the hive's honeycomb. Worker bees that maintain contact with the forager throughout the dance will go out and search for food near the hive.

(**b**) A bee that visits a feeding station more than 100 meters distant from the hive performs a *waggle* dance. During the dance, it makes a straight run and waggles its abdomen. A waggle dancer also varies the dance speed to convey more information about distance to a food source. For example, when food is 150 meters away, a bee dances much faster, and with more waggles per straight run, compared to a dance about a food source that is 500 meters away.

(**c**) As Karl von Frisch discovered, a *straight* run's orientation varies, depending on the direction in which a food source is located. He put one dish of honey on a direct line between a hive and the sun. Foragers that located it returned to the hive and oriented their straight runs right up the honeycomb. He put another dish of honey at right angles to a line between the hive and the sun. Foraging bees made their straight runs 90 degrees to vertical on the honeycomb. Thus, a honeybee "recruited" into foraging orients its flight *with respect to the sun and the hive*. By doing so, it wastes less time and energy during its food-gathering expedition.

When bee moves straight down comb, recruits fly to source directly away from the sun.

When bee moves to right of vertical, recruits fly at 90° angle to right of the sun.

When bee moves straight up comb, recruits fly straight toward the sun.

---

manes, claws, and other body parts are often notably enlarged, patterned, and colored. Ritualization is well developed in *courtship* displays, the steps that must precede pair formation. Many birds have elaborate courtship displays (Figures 16.20, 17.17, and 43.8c).

In *tactile* displays, a signaler touches the receiver in ritualized ways. After locating a source of pollen or nectar, a foraging honeybee returns to its colony, a hive, and performs a complex dance. It moves in a defined pattern, jostling a crowd of workers that stay in close physical contact with it. Signals give others information about the general location, distance, and direction of a source of food (Figure 43.9).

ILLEGITIMATE SIGNALERS AND RECEIVERS

Unintended recipients can intercept communication signals. Male tungara frogs make two kinds of calls—one simple, the other complex. The calls say "come hither" to female frogs, but they also mean "dinner here" to fringe-lipped bats. Complex calls are more attractive to females but make it easier for the bats to locate the caller. When bats are around, male frogs call less and are more likely to make the simpler call.

There are also illegitimate signalers. Some assassin bugs acquire the scent of their prey—termites—by hooking a dead termite on their back. By signaling that they "belong" to a termite colony, they can more easily hunt termites. Another example: If a female of certain predatory firefly species sees a flash from a male of a different species, she will flash back. If she lures him into attack range, she will capture and eat him. Getting eaten is an evolutionary cost of having an otherwise useful response to a come-hither signal.

A communication signal is a transfer of information between two members of the same species. Such signals evolve only when they benefit both the signaler and receiver.

Members of a different species sometimes act illegitimately as a communication signaler or receiver.

43.5 Mates, Offspring, and Reproductive Success

For reasons we need not explore here, many people find mating and parenting behaviors of animals fascinating. How useful is selection theory in helping us interpret such behavior? Let's take a look.

SEXUAL SELECTION AND MATING BEHAVIOR

Competition among members of one sex for access to mates is common. So is choosiness in selecting a mate. Such activities, recall, are forms of **sexual selection**. This microevolutionary process favors traits that give the individual a competitive edge in attracting and often holding on to mates (Section 16.9).

But *whose* reproductive success is it—the male's or female's? Male animals, remember, produce many tiny sperm, and females produce far larger but fewer eggs. For a male, success generally depends on how many eggs he can fertilize. For a female, it depends more on how many eggs she produces or how many offspring she can raise. Usually, the most important factor in a female's sexual preference is the quality of the mate, not the quantity of partners.

Female hangingflies (*Harpobittacus apicalis*) provide an instructive example. They choose males that offer superior food. A male hunts and kills a moth or some other insect. Then he releases a sex pheromone, which attracts females to him and his "nuptial gift" (Figure 43.10*a*). A female tends to select the male that offers a large calorie-rich gift. Only after the female has been eating the gift for five minutes or so does she start to accept sperm from her partner. She lets the male continue inseminating her—but only as long as it takes for her to devour the gift.

Before twenty minutes are up, a female hangingfly can break off the mating at any point. If she does, she may well mate with a different male hangingfly and accept his sperm. Doing so dilutes the reproductive success of her first partner.

Females of other species shop around for the good-looking males. Consider the fiddler crabs that live along muddy shores from Massachusetts to Florida. One of the male's two claws is enlarged. Sometimes it is enlarged enough to make up more than 50 percent of his body weight (Figure 43.10*b*). When spring tides are favorable, the male crabs build elaborate mating burrows near one another. Each male stands beside his burrow, waving his oversized claw. Females stroll by, checking out the males and their burrows. When a female likes what she sees, she follows the male into his burrow and engages in sex.

Many female birds are just as choosy. Male sage grouse (*Centrocercus urophasianus*) come together at a lek, a type of communal display ground. Each male stakes out a few square meters. With tail feathers erect, males use their large, puffed-out neck pouches to emit booming calls (Figure 43.10*d*). As they do this,

Figure 43.10 (**a**) Male hangingfly dangling a moth as a nuptial gift for a potential mate. Females of some hangingfly species choose sexual partners that offer the largest gift to them. By waving his enlarged claw, a male fiddler crab (**b**) may attract the eye of a female fiddler crab (**c**). A male sage grouse (**d**) showing off as he competes for female attention at a communal display ground.

they stamp about on their patch of prairie, a bit like wind-up toys. Females tend to select and mate with just one male sage grouse. Then they go off to nest and raise the young by themselves. Many females often choose the same male, so most of the males never do mate.

In another behavioral pattern, sexually receptive females of some species cluster in defendable groups. Where you come across such a group, you are likely to observe males competing for access to clusters. The competition for ready-made harems favors highly combative male lions, sheep, elk, elephant seals, and bison, to name a few types of animals (Figure 43.11).

PARENTAL CARE

If females fight for males, then we can expect the males to provide more than sperm delivery—say, parental care. Midwife toads are an example. A male wraps strings of fertilized eggs around his legs until the eggs hatch (Figure 43.12a). With her eggs being cared for, a female is free to mate with other males—if she can find some with no eggs. As the breeding season progresses, such males become rare, and females fight for access to them. They even attempt to pry mating pairs apart.

Parental behavior is costly. It drains away time and energy that parents could spend on living long enough to reproduce another time. However, for many species, parenting improves the odds that their offspring will live. The benefit of giving time and energy to offspring (immediate reproductive success) may offset reduced reproductive success at a later time.

For amphibians and reptiles, parenting is rare once the young are hatched. Crocodilians are an exception. As with birds, crocodilian parents build a nest. The young call when they are ready to hatch. Parents dig them up and care for them for some time as they grow.

Most birds are monogamous and both parents often care for the young (Figure 43.12b). In mammals, males typically leave after mating. Females raise the young alone, and the males attempt to mate again or conserve energy for the next breeding season (Figure 43.12c). Mammalian species in which males do help care for the young tend to be monogamous. Only about five percent of all mammals fall into that category.

Selection theory explains some aspects of mating behavior. Male or female preferences for certain behavioral traits can give the individual a competitive edge and promote its reproductive success.

Figure 43.11 (**a**) Sexual competition between male elephant seals, which fight for access to a cluster of females. (**b**) Male bison locked in combat during the breeding season.

Figure 43.12 (**a**) Male midwife toad with developing eggs wrapped around his legs. (**b**) Male and female Caspian terns cooperate in the care of their chick. (**c**) A female grizzly will care for her cub for as long as two years. The male takes no part in its upbringing.

43.6 Costs and Benefits of Social Groups

Survey the animal kingdom and you will find a range of social groups, with evolutionary costs and benefits.

COOPERATIVE PREDATOR AVOIDANCE

Cooperative responses to predators help some groups reduce the net risk to all. Vulnerable individuals, too, scan for predators, join in a counterattack, or engage in more effective defenses (Figure 43.13).

Vervet monkeys, meerkats, prairie dogs, and many other mammals cooperate with their alarm calls, as in Figure 43.13*a*. A prairie dog barks a certain signal when it sights an eagle and a different signal when it sights a coyote. Others dive into burrows to escape an eagle's attack or they stand erect, the better to scan the horizon and zero in on the coyote.

Ecologist Birgitta Sillén-Tullberg found evidence of group benefits for Australian sawfly caterpillars that live in clumps on branches. When something disturbs them, the individuals collectively rear up and writhe about (Figure 43.13*b*). They also vomit partly digested eucalyptus leaves, which are toxic to songbirds and other animals that prey on them.

As Sillén-Tullberg hypothesized, individual sawfly caterpillars benefit from their coordinated repulsion of predatory birds. She used her hypothesis to predict that birds are more likely to eat a solitary caterpillar.

To test her prediction, she offered caterpillars to young hand-reared birds. Birds offered one caterpillar at a time ate an average of 5.6. Birds offered a clump of them ate an average of 4.1. As Sillén-Tullberg had predicted, individuals were safer in the group.

THE SELFISH HERD

Simply by their physical position in the group, some individuals form a living shield against predation on others. They belong to a **selfish herd**, a simple society that benefits their reproductive self-interest. Selfish-herd behavior has been studied in bluegill sunfishes. Male sunfishes build adjacent nests on lake bottoms. Females deposit eggs where the males have used their fins to scoop out depressions in mud.

If a colony of bluegill males is a selfish herd, then we can predict competition for the "safe" sites—at the center of a colony. Compared to eggs at the periphery, eggs in nests at the center are less likely to be eaten by snails and largemouth bass. Competition does indeed occur. The largest, most powerful males tend to claim centermost locations. Other, smaller males assemble around them and bear the brunt of predatory attacks. Even so, they are better off in the group than on their own, fending off a bass single-handedly, so to speak.

COOPERATIVE HUNTING

Many predatory mammals, including wolves, lions, and wild dogs, live in social groups and cooperate in hunts (Figure 43.14). Are group hunts more successful than solitary hunts? Often they are not. Researchers

Figure 43.13 Group defenses. (**a**) Black-tailed prairie dogs bark an alarm call that warns others of predators. Does this put the caller at risk? Not much. Prairie dogs usually act as sentries only if they are done feeding and are standing beside their burrows. (**b**) Australian sawfly caterpillars form clumps and collectively regurgitate a fluid (yellow blobs) that is toxic to most predators. (**c**) Musk oxen adults (*Ovibos moshatus*) form a ring of horns, often around the young.

Figure 43.14 Members of a wolf pack (*Canis lupus*). Wolves cooperate in hunting, caring for the young, and defending a territory. Benefits are not distributed equally. Only the highest ranking individuals, the alpha male and alpha female, breed.

Figure 43.15 Nearly uniform spacing in a crowded cormorant colony.

observed a solitary lion that captured prey about 15 percent of the time. Two lions hunting together did capture prey twice as often, but they had to share it, so the number of successful hunts per lion balanced out. When more lions joined the hunt, the success rate per lion fell. Wolves show a similar pattern. Among many cooperative hunters, hunting success in itself may not explain group living. Individuals do hunt together, but they also may fend off scavengers, care for one another's young, and protect territory.

DOMINANCE HIERARCHIES

Many social groups share resources unequally among some individuals that are subordinate to others. Most wolf packs, for instance, have one dominant male that breeds with just one dominant female. Other wolves in the pack are nonbreeding brothers and sisters, or aunts and uncles. They all hunt and bring food to the individuals that guard the young in their den.

Baboons live in large troops. A female stays with the group into which she was born and inherits social standing from her mother. Dominant females get more food, water, and grooming. Their young grow and mature faster than those of lower-ranking females.

Why do subordinates relinquish resources and, often, breeding privileges? They might do so to avoid being hurt and to belong to a group. Possibly they can't survive out on their own. Or possibly challenging a strong individual can invite injury or death. Besides, a subordinate might get a chance to reproduce if it lives long enough or if a predator or old age removes its dominant peers. Some subordinate wolves and baboons do manage to move up the social ladder when the opportunity arises.

REGARDING THE COSTS

If social behavior is advantageous, *then why are there so few social species?* In most places, the costs outweigh benefits. For instance, when more individuals do live together, they compete more for a share of the food. Cormorants, puffins, and many other seabirds form enormous breeding colonies (Figure 43.15). All must compete for a share of the same ecological pie.

Large social groups also attract more predators. If individuals are crowded together, they invite parasites and contagious diseases that jump from host to host. The individuals may also be at risk of being killed or exploited by others. Given the opportunity, breeding pairs of herring gulls cannibalize a neighbor's eggs and any chicks that wander away from their nest.

Living in a social group can provide benefits, as through cooperative defenses or shielding against predators.

Group living has costs, in terms of increased competition, increased vulnerability to infections, and exploitation by others of the group.

43.7 Why Sacrifice Yourself?

Extreme cases of sterility and self-sacrifice have evolved in only two groups of insects and one group of mammals. How are genes of the nonreproducers perpetuated?

SOCIAL INSECTS

Honeybees and fire ants (Chapter 40), are among the true social (eusocial) insects. Like termites, they stay together for generations in a group that has a division of labor. Many permanently sterile individuals care cooperatively for the offspring of just a few breeding individuals. Often they are highly specialized in form and function (Figure 43.16).

Consider a honeybee colony, or hive. Its queen bee, the only fertile female, secretes a pheromone that other female bees distribute through the hive. This signaling molecule suppresses the development of ovaries in all other females, rendering them sterile. The queen bee is larger than the worker bees, partly because of her enlarged egg-producing ovaries (Figure 43.17a).

About 30,000 to 50,000 female workers feed larvae, clean and maintain the hive, and build honeycomb from waxy secretions. Adult workers live for about six weeks in the spring and summer. When foragers return to the hive after finding a rich source of nectar or pollen, they engage others in a dance. This tactile display recruits more foragers (Figure 43.9). Workers also cooperate by transferring food to one another. They guard the entrance to the hive and will sacrifice themselves to repel intruders.

Males, the stingless drones, develop only in spring and summer. They have no part in the day-to-day work and subsist on food gathered by their worker sisters. Drones live for sex. Each day, they fly out in search of a mate. If one is lucky, he will find a virgin queen on her single flight away from her colony. The sole function of her flight is to meet up with and mate with a drone. A drone dies right after he inseminates a virgin queen, which then starts a new colony. She will store and use his sperm for years, perpetuating his genes and those of his original colony.

Like honeybees, termites live in enormous family groups with a queen specialized for producing eggs (Figure 43.17c). Unlike a honeybee hive, each termite colony has sterile individuals of both sexes. A king supplies the female with sperm. Winged reproductive termites of both sexes develop seasonally.

SOCIAL MOLE RATS

Vertebrates are not known for sterility and extreme self-sacrifice. The only eusocial mammals are African mole-rats. The best studied is *Heterocephalus glaber*, the naked mole-rat. Clans of this nearly hairless rodent occupy burrows in arid parts of East Africa.

One reproducing female dominates the clan and mates with one to three males (Figure 43.17b). Other, nonbreeding members live to protect and care for the "queen" and "king" (or kings) and their offspring. The sterile diggers excavate subterranean tunnels and chambers that serve as living rooms or waste-disposal sites. When a digger locates a tasty root or tuber, it hauls a bit of it back to the main chamber and emits a series of chirps. The chirps recruit others, which help carry the food back to the chamber. In this way, the queen, her retinue of males, and her offspring get fed.

Digger mole-rats also deliver food to other helpers that usually loaf about, shoulder to shoulder and belly to back, with the reproductive royals. These "loafers" actually spring to action when a snake or some other enemy threatens the clan. Collectively, and at great risk, they chase away or attack and kill the predator.

Figure 43.16 Specialized ways of serving and defending the colony. (**a**) An Australian honeypot ant worker. This sterile female is a living container for her colony's food reserves. (**b**) Army ant soldier (*Eciton burchelli*) with formidable mandibles. (**c**) Eyeless soldier termite (*Nasutitermes*). It bombards intruders with a stream of sticky goo from its nozzle-shaped head.

Figure 43.17 Three queens. (**a**) A queen honeybee with her court of sterile worker daughters. (**b**) This queen naked mole-rat has twelve mammary glands, the better to feed her many offspring. In a laboratory colony at Cornell University, one female produced a litter of twenty-eight pups. And she gave birth to more than 900 offspring during her lifetime. (**c**) A termite queen (*Macrotermes*) dwarfs her offspring and her mate. Her body pumps out thousands of eggs a day.

Figure 43.18 Damaraland mole-rats in a burrow. Like their relatives, the naked mole-rats, they live in colonies having nonbreeding workers. But unlike the naked mole-rats, these fuzzy burrowers are not highly inbred.

INDIRECT SELECTION FOR ALTRUISM

None of the altruistic individuals of a honeybee hive, termite colony, or naked mole-rat clan directly passes genes to the next generation. So how are genes that underlie altruistic behavior perpetuated? According to William Hamilton's theory of **inclusive fitness**, genes associated with altruism can be favored by selection if they lead to behavior that will increase the number of offspring produced by an altruist's closest relatives.

A sexually reproducing, diploid parent caring for offspring is not helping exact genetic copies of itself. Each of its gametes, and each of its offspring, inherits one-half of its genes. Other individuals of the social group that have the same ancestors also share genes with their parents. Two siblings (brothers or sisters) are as genetically similar as a parent and its offspring. Nephews and nieces have inherited about one-fourth of their uncle's genes.

Sterile workers may be indirectly promoting genes for "self-sacrifice" through altruistic behavior that will benefit their relatives. All of the individuals in honeybee, termite, and ant colonies are members of an extended family. Nonbreeding family members support siblings, a few of which are future kings and queens. Although a guard bee dies after driving her stinger into a bear, siblings in the hive will perpetuate some of her genes.

Does close relatedness explain why naked mole-rats are the only eusocial mammals? Through DNA fingerprinting studies, we know that all individuals in one naked mole-rat clan are *very* close relatives and differ genetically from individuals of other clans. Each clan is highly inbred through generations of brother–sister, mother–son, and father–daughter matings.

However, inbreeding may not be necessary for mole-rat eusociality. The social organization of the Damaraland mole-rat (*Cryptomys damarensis*) resembles that of *H. glaber* (Figure 43.18). Nonbreeding members of both sexes cooperatively assist one breeding pair. However, breeding pairs of wild Damaraland mole-rat colonies usually are unrelated.

Researchers are now searching for other factors that select for mole-rat eusocial behavior. According to one hypothesis, an arid habitat and patchy food supplies favor the mole-rat genes that underlie cooperation in digging burrows, searching for food, and fending off competitors of other species for resources.

Altruistic behavior may persist when individuals pass on genes indirectly, by helping relatives survive and reproduce.

By the theory of inclusive fitness, genes associated with altruistic behavior that is directed toward relatives may spread through a population in certain situations.

43.8 An Evolutionary View of Human Social Behavior

Evolutionary forces have shaped the behavior of all animals, including humans. However, only humans consistently make moral choices about their behavior.

HUMAN PHEROMONES

When women live together, as in a college dormitory, their menstrual cycles typically become synchronized. Martha McClintock and Kathleen Stern have evidence that pheromones in sweat play a role. The researchers exposed women to the sweat from others who were in different menstrual cycle phases. Sweat secreted by women in the early phase accelerated ovulation in other women exposed to it and shortened their cycles. Sweat from women who were near ovulation had the opposite effect; it delayed ovulation and lengthened the cycles. Sweat had no effect on the menstrual cycle in about 30 percent of the women.

Do humans respond to pheromones? Possibly, but we don't know how. In most mammals, pheromones bind to specific receptors in a vomeronasal organ (VNO). Neurons connect the VNO to brain regions that control behavior. In human adults, VNOs are tiny ductlike structures on a septum that divides the nose into two nostrils. Many researchers think that human VNOs are vestigial structures (no longer functional). Some human DNA sequences are similar to sequences that encode pheromone receptors in rodents, but most of them are not activated.

HORMONES AND BONDING BEHAVIOR

The hormone oxytocin, recall, facilitates pair bonding in prairie voles. It also influences social attachments between humans. Consider *autism*. A person who has this behavioral disorder does not form normal social relationships. Compared to unaffected individuals, an autistic child has less oxytocin circulating in blood.

Today, researchers are studying oxytocin's role in the formation of bonds between a mother and infant and between sexual partners. Nursing stimulates the secretion of oxytocin. So does orgasm, even a friendly massage. Do brief rises in oxytocin levels promote long-term bonding in humans? We don't know yet.

EVOLUTIONARY QUESTIONS

If we are comfortable with studying the evolutionary basis of the behavior of termites, naked mole-rats, and other animals, why do so many people resist the idea of analyzing human behavior in the same way? Often they fear that attempts to identify the adaptive value of some human trait will be used to define its morality. However, there is a clear difference between trying to explain behavior in terms of its evolutionary history and attempting to justify it. To biologists, "adaptive" does not mean "morally right." It only means useful in perpetuating an individual's genes.

An example: Infanticide is morally repugnant. Is it unnatural? No. It happens in many animal groups and human cultures. Male lions often kill the offspring of other males when they take over a pride. Doing so frees up the females to breed with them, increasing the infanticidal male's reproductive success.

Biologists may predict that unrelated human males are a threat to infants and evidence supports this. The absence of a biological father and the presence of an unrelated male increases risk of death for an American child under age two by sixty times.

What about parents who kill their own offspring? In her book on maternal behavior, primatologist Sarah Blaffer Hrdy cites a study of a village in New Guinea in which approximately 40 percent of the newborns were killed by parents. She argues that when resources or social support are hard to come by, a mother might increase her fitness by killing a newborn. She can then allocate child-rearing energy to her other offspring or save it for children she may have in the future.

Do most of us find such behavior appalling? Yes. Does such behavior warrant attention? Think about all you have learned in this book, then decide.

Biologists describe human behavior as it is, not as it ought to be. They take into consideration our cultural roots as a society of hunter–gatherers, and our evolutionary roots as social primates.

Summary

Section 43.1 Animal behavior starts with genes that specify products required for development of the nervous, endocrine, and muscular systems. Hormones are among the gene products that affect behavior.

Instinctive behavior is performed without having been learned by experience in the environment. It is a prewired response to one or two simple, well-defined environmental cues.

Section 43.2 An animal learns when it processes and integrates information from experiences, then uses it to vary or change how it responds to stimuli. Imprinting is learning that occurs only during a sensitive period early in life.

Section 43.3 Behavior with a genetic basis is subject to evolution by natural selection. It evolved as a result of individual differences in reproductive success in past generations. Behavior persists when its reproductive benefits exceed the reproductive costs.

Section 43.4 Communication signals are meant to change the behavior of individuals of the same species. Pheromones are signaling molecules that have roles in social communication.

Visual signals are key components of courtship displays and threat displays. Acoustical signals are sounds that have precise, species-specific information. Tactile signals are specific forms of physical contact between a signaler and a receiver.

Section 43.5 Sexual selection favors traits that give an individual a competitive edge in attracting and often holding on to mates. Females of many species select for males that have traits or engage in behaviors they find attractive. When large numbers of females cluster in defensible areas, males may compete with one another to control the areas.

Parental care has reproductive costs in terms of future reproduction and survival. It is adaptive when benefits to a present set of offspring offset the costs.

Section 43.6 Animals that live in social groups may benefit by cooperating in predator detection, defense, and rearing the young. Benefits of group living are often distributed unequally. Species that live in large groups incur costs, including increased disease and parasitism, and increased competition for resources.

Section 43.7 Ants, termites, and some other insects as well as two species of mole-rats are eusocial. They live in colonies with overlapping generations and have a reproductive division of labor. Most colony members do not reproduce; they assist their relatives and rear their offspring.

According to the theory of inclusive fitness, such extreme altruism is perpetuated because altruistic individuals have some portion of their genes in common with their reproducing relatives. Altruistic individuals in the social group pass on "by proxy" the genes that underlie this behavior.

Section 43.8 Behavioral biologists ask questions about the mechanisms and adaptive significance of human behavior. A behavior that is adaptive in the evolutionary sense may still be judged by society to be morally wrong.

Self-Quiz
Answers in Appendix III

1. "Starlings minimize nest mites by festooning nests with wild carrot," said Clark and Mason. Their statement was _____ .
   a. an untested hypothesis
   b. a prediction
   c. a test of a hypothesis
   d. a proximate conclusion

2. Genes affect the behavior of individuals by _____ .
   a. influencing the development of nervous systems
   b. affecting the kinds of hormones in individuals
   c. governing development of muscles and skeletons
   d. all of the above

3. Generally, living in a social group costs the individual, in terms of _____ .
   a. competition for food, other resources
   b. vulnerability to contagious diseases
   c. competition for mates
   d. all of the above

4. A statement that overcrowding causes lemmings to disperse to areas that are more favorable for reproduction is _____ .
   a. consistent with Darwinian evolutionary theory
   b. based on a theory of evolution by group selection
   c. supported by the finding that most animals behave altruistically during their lives

5. Eusocial insects _____ .
   a. live in extended family groups
   b. are found among almost all insect orders
   c. show a reproductive division of labor
   d. a and c
   e. all of the above

6. Helping other individuals at a reproductive cost to oneself might be adaptive if those helped are _____ .
   a. members of another species
   b. competitors for mates
   c. close relatives
   d. illegitimate signalers

7. Match the terms with their most suitable description.
   _____ fixed action pattern
   _____ altruism
   _____ basis of instinctive and learned behavior
   _____ imprinting
   _____ pheromone

   a. time-dependent form of learning requiring exposure to key stimulus
   b. genes plus actual experience
   c. stereotyped motor program that runs to completion independently of feedback from environment
   d. assisting another individual at one's own expense
   e. one communication signal

Figure 43.19 A Nazca booby attends to its single surviving chick.

Figure 43.20 Behaviorally confused rooster.

Media Menu

Student CD-ROM

Impacts, Issues Video
    My Pheromones Made Me Do It
Big Picture Animation
    Causes and examples of behavior
Read-Me-First Animation
    Honeybee dances

InfoTrac

- The Armpit of the Avian World. *American Scientist*, May 2001.
- Female Song Sparrows Prefer Males That Learn Well. *BioScience*, December 2002.
- Sex and the Single Fly. *Newsweek*, August 14, 2000.
- Chemical Talk. *Natural History*, July 2001.
- Staying Close to Home. *International Wildlife*, May–June 2001.
- Mothers and Others. *Natural History*, May 2001.

Web Sites

- Animal Behavior Society: www.animalbehavior.org
- Inside the Animal Mind: www.pbs.org/wnet/nature/animalmind
- Tales from the Hive: www.pbs.org/wgbh/nova/bees
- Living Links: www.emory.edu/LIVING_LINKS
- Human Behavior and Evolution Society: www.hbes.com

How Would You Vote?

Honeybees are important pollinators for orchards and many crop plants throughout the world, including the United States. As the hybrid Africanized honeybee continues to expand its range, local colonies that are maintained by commercial beekeepers are becoming Africanized. Quarantines on bees from colonized regions might help slow the geographic dispersal of Africanized bees, but wild hives will continue to become established in new habitats. The Africanized bees currently pose a threat to humans, pets, and livestock. Given the threat, would you support increased funding for research on their genetics and behavior?

Critical Thinking

1. Nazca boobies (*Sula granti*) lay two eggs several days apart. No matter how much food is available, only one chick survives to adulthood (Figure 43.19). The first chick to hatch pushes its younger sibling from the nest, and that sibling dies of starvation and neglect. Formulate a hypothesis on how it might be adaptive for parents to lay two eggs if one sibling is expected to kill the other. Design an experiment to test your hypothesis.

2. *Sexual imprinting* is common in birds. During a short sensitive period in early life, a bird learns the features it will look for later in a mate. The rooster in Figure 43.20 was observed wading after ducks with amorous intent. Speculate on what might have caused this behavior.

3. Svante Paabo has correlated a gene mutation with the origin of *language*. All mammals have the *FOXP2* gene, which has a 715 base-pair sequence. The mutant human gene differs from the *FOXP2* gene in mice by 3 base pairs and by only 2 base pairs in chimpanzees. This mutation, and others, altered brain regions that control muscles in the face, throat, and vocal cords. It apparently arose about 200,000 years ago and swiftly replaced the nonmutated form in the populations immediately ancestral to *Homo sapiens*. Speculate on how the rapid rise in its frequency may have affected social behavior and learning.

4. A cheetah scent-marks plants in its territory with certain exocrine gland secretions. What evidence would you require to demonstrate that the cheetah's action is an evolved communication signal?

5. Among primates, differences in sexual behavior tend to be related to the size of a male's gonads. Gorillas have relatively tiny testicles. Testicles of a male weighing 450 pounds may weigh about an ounce. Gorillas live in groups that consist of a male, a few females, and their offspring. When a female is ready to mate, only a single adult male is usually around to inseminate her.

In contrast, a female chimpanzee advertises her fertile period and will mate with many males. A 100-pound chimpanzee male has gonads about four times as weighty as a gorilla's. By producing large amounts of sperm, this male increases the chance that his sperm, rather than sperm from a rival, will fertilize a female's egg.

A human male's testicles weigh about half as much as a chimpanzee's. Speculate on the possible evolutionary link with female promiscuity and male sperm competition.

Epilogue

BIOLOGICAL PRINCIPLES AND THE HUMAN IMPERATIVE

Molecules, single cells, tissues, organs, organ systems, multicelled organisms, populations, communities, ecosystems, and the biosphere. These are architectural systems of life, assembled in increasingly complex ways over the past 3.8 billion years. We are latecomers to this immense biological building program. Yet within the relatively short span of 10,000 years, our activities have been changing the character of the land, ocean, and atmosphere, even the genetic character of species.

It would be presumptuous to think that we alone have had profound impact on the world of life. As long ago as the Proterozoic, photosynthetic organisms were irrevocably changing the course of biological evolution by enriching the atmosphere with oxygen. During the past as well as the present, competitive adaptations led to the rise of some groups, whose dominance assured the decline of others. Change is nothing new. What *is* new is the capacity of one species to comprehend what might be going on.

We now have the population size, technology, and cultural inclination to use up energy and modify the environment at rapid rates. Where will this end? Will feedback controls operate as they do, for instance, when population growth exceeds carrying capacity? In other words, will negative feedback controls come into play and keep things from getting too far out of hand?

Feedback control will not be enough, for it operates after deviation. Our patterns of resource consumption and population growth are founded on an illusion of unlimited resources and a forgiving environment. A prolonged, global shortage of food or the passing of a critical threshold for the global climate can come too fast to be corrected; in which case the impact of the deviation may be too great to be reversed.

What about feedforward mechanisms, which might serve as early warning systems? For example, when sensory receptors near the surface of skin detect a drop in outside air temperature, each sends messages to the nervous system. That system responds by triggering mechanisms that raise the body's core temperature before the body itself becomes dangerously chilled.

Extrapolating from this, if we develop feedforward control mechanisms, would it not be possible to start corrective measures before we do too much harm?

Feedforward controls alone will not work, for they operate after change is under way. Think of the DEW line—the Distant Early Warning system. It is like a vast sensory receptor for detecting missiles launched against North America. By the time it does what it is supposed to, it may be too late to stop widespread destruction.

It would be naive to assume we can ever reverse who we are at this point in evolutionary time, to de-evolve ourselves culturally and biologically into becoming less complex in the hope of averting disaster. Yet there is reason to believe we can avert disaster by using a third kind of control mechanism—a capacity to anticipate events even before they happen. We are not locked into responding only after irreversible change has begun. We have the capacity to anticipate the future—it is the essence of our visions of utopia and hell. *We all have the capacity to adapt to a future that we can partly shape.*

For instance, we can stop trying to "beat nature" and learn to work with it. Individually and collectively, we can work to develop long-term policies that take into account biotic and abiotic limits on population growth. Far from being a surrender, this would be one of the most intelligent behaviors of which we are capable.

Having a capacity to adapt and using it are not the same thing. We have already put the world of life on dangerous ground because we have not yet mobilized ourselves as a species to work toward self-control.

Our survival depends on predicting possible futures. It depends on preserving, restoring, and constructing ecosystems that fit with our definition of basic human values and available biological models. Human values can change; our expectations can and must be adapted to biological reality. *For the principles of energy flow and resource utilization, which govern the survival of all systems of life, do not change.* It is our biological and cultural imperative that we come to terms with these principles, and ask ourselves what our long-term contribution will be to the world of life.

Appendix I. Classification System

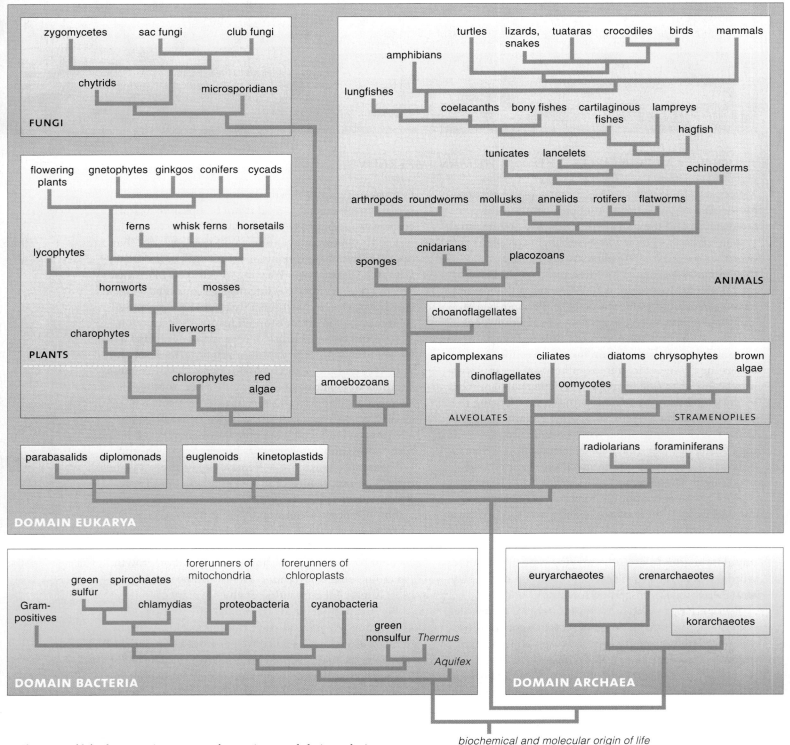

This tree of life shows major groups of organisms and their evolutionary connections. Each set of organisms (taxon) has living representatives. Each branch point represents the last common ancestor of the set above it. Small boxes within domains Archaea and Eukarya highlight taxa that are currently being recognized as the equivalent of kingdoms in earlier classification systems. This tree is based on morphological, genetic, and molecular comparisons of the most important groups. It is a work in progress and subject to refinements as more information comes in.

DOMAIN OF BACTERIA

KINGDOM BACTERIA
Gram-negative and Gram-positive prokaryotic cells. Collectively, great metabolic diversity; photosynthetic autotrophs, chemosynthetic autotrophs, and heterotrophs.

PHYLUM FIRMICUTES   Typically Gram-positive, thick wall. Heterotrophs. *Bacillus, Staphylococcus, Streptococcus, Clostridium, Actinomycetes.*

PHYLUM GRACILICUTES   Typically Gram-negative, thin wall. Autotrophs (photosynthetic and chemosynthetic) and heterotrophs. *Anabaena* and other cyanobacteria. *Escherichia, Pseudomonas, Neisseria, Myxococcus.*

PHYLUM TENERICUTES   Gram-negative, wall absent. Heterotrophs (saprobes, pathogens). *Mycoplasma.*

DOMAIN OF ARCHAEBACTERIA (ARCHAEA)

KINGDOM ARCHAEBACTERIA
Methanogens, extreme halophiles, extreme thermophiles. Evolutionarily closer to eukaryotic cells than to eubacteria. All strict anaerobes living in habitats as harsh as those that probably prevailed on the early Earth. Compared with other prokaryotic cells, all archaebacteria have a distinctive cell wall and unique membrane lipids, ribosomes, and RNA sequences. *Methanobacterium, Halobacterium, Sulfolobus.*

Table A   Representative Bacteria and Archaea Grouped on the Basis of Numerical Taxonomy

Some Major Groups	Main Habitats	Characteristics	Representatives
BACTERIA			
Photoautotrophs:			
Cyanobacteria, green sulfur bacteria, and purple sulfur bacteria	Mostly lakes, ponds; some marine, terrestrial habitats	Photosynthetic; use sunlight energy, carbon dioxide; cyanobacteria use oxygen-producing noncyclic pathway; some also use cyclic route	*Anabaena, Spirulina, Rhodopseudomonas,*
Photoheterotrophs:			
Purple nonsulfur and green nonsulfur bacteria	Anaerobic, organically rich muddy soils, and sediments of aquatic habitats	Use sunlight energy; organic compounds as electron donors; some purple nonsulfur may also grow chemotrophically	*Rhodospirillum, Chlorobium*
Chemoautotrophs:			
Nitrifying, sulfur-oxidizing, and iron-oxidizing bacteria	Soil; freshwater, marine habitats	Use carbon dioxide, inorganic compounds as electron donors; influence crop yields, cycling of nutrients in ecosystems	*Nitrosomonas, Nitrobacter, Thiobacillus*
Chemoheterotrophs:			
Spirochetes	Aquatic habitats; parasites of animals	Helically coiled, motile; free-living and parasitic species; some major pathogens	*Spirochaeta, Treponema*
Gram-negative aerobic rods and cocci	Soil, aquatic habitats; parasites of animals, plants	Some major pathogens; some fix nitrogen (e.g., *Rhizobium*)	*Pseudomonas, Neisseria, Rhizobium, Agrobacterium*
Gram-negative facultative anaerobic rods	Soil, plants, animal gut	Many major pathogens; one bioluminescent (*Photobacterium*)	*Salmonella, Escherichia, Proteus, Photobacterium*
Rickettsias and chlamydias	Host cells of animals	Intracellular parasites; many pathogens	*Rickettsia, Chlamydia*
Myxobacteria	Decaying organic material; bark of living trees	Gliding, rod-shaped; aggregation and collective migration of cells	*Myxococcus*
Gram-positive cocci	Soil; skin and mucous membranes of animals	Some major pathogens	*Staphylococcus, Streptococcus*
Endospore-forming rods and cocci	Soil; animal gut	Some major pathogens	*Bacillus, Clostridium*
Gram-positive nonsporulating rods	Fermenting plant, animal material; gut, vaginal tract	Some important in dairy industry, others major contaminators of milk, cheese	*Lactobacillus, Listeria*
Actinomycetes	Soil; some aquatic habitats	Include anaerobes and strict aerobes; major producers of antibiotics	*Actinomyces, Streptomyces*
ARCHAEA			
Methanogens	Anaerobic sediments of lakes, swamps; animal gut	Chemosynthetic; methane producers; used in sewage treatment facilities	*Methanobacterium*
Extreme halophiles	Brines (extremely salty water)	Heterotrophic; also, unique photosynthetic pigments (bacteriorhodopsin) form in some	*Halobacterium*
Extreme thermophiles	Acidic soil, hot springs, hydrothermal vents	Heterotrophic or chemosynthetic; use inorganic substances as electron donors	*Sulfolobus, Thermoplasma*

DOMAIN OF EUKARYOTES (EUKARYA)

KINGDOM "PROTISTA"

Diverse single-celled, colonial, and multicelled eukaryotic species. Existing types are unlike prokaryotes and most like the earliest forms of eukaryotes. Autotrophs, heterotrophs, or both. Reproduce sexually and asexually (by meiosis, mitosis, or both). Not a monophyletic group. The kingdom may soon be split into multiple kingdoms, and some of its groups (chlorophytes and charophytes) are classified by some botanists as plants.

PHYLUM "MASTIGOPHORA" Flagellated protozoans. Free-living heterotrophs; many are internal parasites. They have one to several flagella. A non-monophyletic grouping of ancient lineages, including the diplomonads, parabasalids, and kinetoplastids. *Trypanosoma, Trichomonas, Giardia.*

PHYLUM EUGLENOPHYTA Euglenoids. Mostly heterotrophs, some photoautotrophs, some both depending on conditions. Most with one short, one long flagellum. Pigmented (red, green) or colorless. Related to kinetoplastids. *Euglena.*

PHYLUM SARCODINA Amoeboid protozoans. Heterotrophs, free-living or endosymbionts, some pathogens. Soft-bodied, with or without shell, pseudopods. Rhizopods (naked amoebas, foraminiferans), actinopods (radiolarians, heliozoans). *Amoeba, Entamoeba.*

ALVEOLATES

PHYLUM CILIOPHORA Ciliated protozoans. Heterotrophs, predators or symbionts, some parasitic. All have cilia. Free-living, sessile, or motile. *Paramecium.*

PHYLUM APICOMPLEXA Heterotrophs, sporozoite-forming parasites. Complex structures at head end. Most familiar types known as sporozoans. *Cryptosporidium, Plasmodium, Toxoplasma.*

PHYLUM PYRRHOPHYTA Dinoflagellates. Photosynthetic, mostly, but some heterotrophs. *Noctiluca, Karenia brevis.*

STRAMENOPILES

PHYLUM OOMYCOTA Water molds. Heterotrophs. Decomposers, some parasites. *Saprolegnia, Phytophthora, Plasmopara.*

PHYLUM CHRYSOPHYTA Golden algae, yellow-green algae, diatoms, coccolithophores. Photosynthetic. Some flagellated. *Mischococcus, Synura, Vaucheria.*

PHYLUM PHAEOPHYTA Brown algae. Photosynthetic, nearly all endemic to temperate or marine waters. *Macrocystis, Laminaria, Sargassum, Postelsia.*

GROUPS CLOSELY RELATED TO PLANTS

PHYLUM RHODOPHYTA Red algae. Mostly photosynthetic, some parasitic. Nearly all marine, some in freshwater habitats. *Porphyra, Antithamion.*

PHYLUM CHLOROPHYTA Green algae. Mostly photosynthetic, some parasitic. Most freshwater, some marine or terrestrial. *Chlamydomonas, Spirogyra, Ulva, Volvox, Codium, Halimeda.*

PHYLUM CHAROPHYTA Closest relatives of plants. Desmids, stoneworts. *Micrasterias, Chara.*

GROUPS OF SLIME MOLDS

PHYLUM MYXOMYCOTA Plasmodial slime molds. Heterotrophs. A multinucleated mass (the plasmodium) arises by mitosis without cell division, feeds and migrates as a unit, then forms spore-bearing structures. *Physarum.*

PHYLUM ACRASIOMYCOTA Cellular slime molds. Heterotrophs; free-living, phagocytic amoeboid cells aggregate into a mass that migrate as a unit, then form spore-bearing structures. *Dictyostelium.*

FUNGI

Nearly all multicelled eukaryotic species with chitin-containing cell walls. Heterotrophs, mostly saprobic decomposers, some parasites. Nutrition based upon extracellular digestion of organic matter and absorption of nutrients by individual cells. Multicelled species form absorptive mycelia within substrates and structures that produce asexual spores (and sometimes sexual spores).

PHYLUM CHYTRIDIOMYCOTA Chytrids. Primarily aquatic; saprobic decomposers or parasites that produce flagellated spores. *Chytridium.*

PHYLUM ZYGOMYCOTA Zygomycetes. Producers of zygospores (zygotes inside thick wall) by way of sexual reproduction. Bread molds, related forms. *Rhizopus, Philobolus.*

PHYLUM ASCOMYCOTA Ascomycetes. Sac fungi. Sac-shaped cells form sexual spores (ascospores). Most yeasts and molds, morels, truffles. *Saccharomycetes, Morchella, Neurospora, Sarcoscypha, Claviceps, Ophiostoma, Candida, Aspergillus, Penicillium.*

PHYLUM BASIDIOMYCOTA Basidiomycetes. Club fungi. Most diverse group. Produce basidiospores inside club-shaped structures. Mushrooms, shelf fungi, stinkhorns. *Agaricus, Amanita, Craterellus, Gymnophilus, Puccinia, Ustilago.*

"IMPERFECT FUNGI" Sexual spores absent or undetected. The group has no formal taxonomic status. If better understood, a given species might be grouped with sac fungi or club fungi. *Arthobotrys, Histoplasma, Microsporum, Verticillium.*

"LICHENS" Mutualistic interactions between fungal species and a cyanobacterium, green alga, or both. *Lobaria, Usnea.*

KINGDOM PLANTAE

Multicelled eukaryotes. Nearly all photosynthetic autotrophs with chlorophylls *a* and *b*. Some parasitic. Nonvascular and vascular species, generally with well-developed root and shoot systems. Nearly all adapted to survive on land; a few in aquatic habitats. Sexual reproduction predominant with spore-forming chambers and embryos in life cycle; also asexual reproduction by vegetative propagation and other mechanisms.

PHYLUM "BRYOPHYTA" Bryophytes; mosses, liverworts, hornworts. Not a monophyletic group. Seedless, nonvascular, haploid dominance, sperm are flagellated; require water for fertilization. *Marchantia, Polytrichum, Sphagnum.*

SEEDLESS VASCULAR PLANTS

Diploid sporophyte dominates, flagellated sperm require water for fertilization.

PHYLUM "RHYNIOPHYTA" Earliest known vascular plants; muddy habitats. A polyphyletic group, some are primitive lycophytes. Extinct. *Cooksonia, Rhynia.*

PHYLUM LYCOPHYTA Lycophytes, club mosses. Small single-veined leaves, branching rhizomes. *Lepidodendron* (extinct), *Lycopodium, Selaginella.*

PHYLUM SPHENOPHYTA Horsetails. Reduced scalelike leaves. Some stems photosynthetic, others nonphotosynthetic, spore-producing. *Calamites* (extinct), *Equisetum.*

PHYLUM PTEROPHYTA Ferns. Large leaves, usually with sori. Largest group of seedless vascular plants (12,000 species), mainly tropical, temperate habitats. *Pteris, Trichomanes, Cyathea* (tree ferns), *Polystichum.*

PHYLUM PSILOPHYTA Whisk ferns. No obvious roots, leaves on sporophyte, very reduced. *Psilotum.*

PHYLUM "PROGYMNOSPERMOPHYTA" The progymnosperms. Ancestral to early seed-bearing plants; extinct. *Archaeopteris.*

SEED-BEARING VASCULAR PLANTS

PHYLUM "PTERIDOSPERMOPHYTA"   Seed ferns. Fernlike gymnosperms; extinct. *Medullosa.*

PHYLUM CYCADOPHYTA   Cycads. Group of gymnosperms (vascular, bear "naked" seeds). Tropical, subtropical. Compound leaves, simple cones on male and female plants. Plants usually palm-like. *Zamia, Cycas.*

PHYLUM GINKGOPHYTA   Ginkgo (maidenhair tree). Type of gymnosperm. Seeds with fleshy outer layer. *Ginkgo.*

PHYLUM GNETOPHYTA   Gnetophytes. Only gymnosperms with vessels in xylem and double fertilization (but endosperm does not form). *Ephedra, Welwitchia, Gnetum.*

PHYLUM CONIFEROPHYTA   Conifers. Most common and familiar gymnosperms. Generally cone-bearing species with needle-like or scale-like leaves.

Family Pinaceae. Pines (*Pinus*), firs (*Abies*), spruces (*Picea*), hemlock (*Tsuga*), larches (*Larix*), true cedars (*Cedrus*).

Family Cupressaceae. Junipers (*Juniperus*), Cypresses (*Cupressus*), Bald cypress (*Taxodium*), redwood (*Sequoia*), bigtree (*Sequoiadendron*), dawn redwood (*Metasequoia*).

Family Taxaceae. Yews. *Taxus.*

PHYLUM ANTHOPHYTA   Angiosperms (the flowering plants). Largest, most diverse group of vascular seed-bearing plants. Only organisms that produce flowers, fruits. Some families from several representative orders are listed:

BASAL FAMILIES

Family Amborellaceae. *Amborella.*
Family Nymphaeaceae. Water lilies.
Family Illiciaceae. Star anise.

MAGNOLIIDS

Family Magnoliaceae. Magnolias.
Family Lauraceae. Cinnamon, sassafras, avocados.
Family Piperaceae. Black pepper, white pepper.

EUDICOTS

Family Papaveraceae. Poppies.
Family Cactaceae. Cacti.
Family Euphorbiaceae. Spurges, poinsettia.
Family Salicaceae. Willows, poplars.
Family Fabaceae. Peas, beans, lupines, mesquite.
Family Rosaceae. Roses, apples, almonds, strawberries.
Family Moraceae. Figs, mulberries.
Family Cucurbitaceae. Squashes, melons, cucumbers.
Family Fagaceae. Oaks, chestnuts, beeches.
Family Brassicaceae. Mustards, cabbages, radishes.
Family Malvaceae. Mallows, okra, cotton, hibiscus, cocoa.
Family Sapindaceae. Soapberry, litchi, maples.
Family Ericaceae. Heaths, blueberries, azaleas.
Family Rubiaceae. Coffee.
Family Lamiaceae. Mints.
Family Solanaceae. Potatoes, eggplant, petunias.
Family Apiaceae. Parsleys, carrots, poison hemlock.
Family Asteraceae. Composites. Chrysanthemums, sunflowers, lettuces, dandelions.

MONOCOTS

Family Araceae. Anthuriums, calla lily, philodendrons.
Family Liliaceae. Lilies, tulips.
Family Alliaceae. Onions, garlic.
Family Iridaceae. Irises, gladioli, crocuses.
Family Orchidaceae. Orchids.
Family Arecaceae. Date palms, coconut palms.
Family Bromeliaceae. Bromeliads, pineapples.
Family Cyperaceae. Sedges.
Family Poaceae. Grasses, bamboos, corn, wheat, sugarcane.
Family Zingiberaceae. Gingers.

KINGDOM ANIMALIA

Multicelled eukaryotes, nearly all with tissues, organs, and organ systems; show motility during at least part of their life cycle; embryos develop through a series of stages. Diverse heterotrophs, predators (herbivores, carnivores, omnivores), parasites, detritivores. Reproduce sexually and, in many species, asexually as well.

PHYLUM PORIFERA   Sponges. No symmetry, tissues. *Euplectella.*

PHYLUM PLACOZOA   Marine. Simplest known animal. Two cell layers, no mouth, no organs. *Trichoplax.*

PHYLUM CNIDARIA   Radial symmetry, tissues, nematocysts.
Class Hydrozoa. Hydrozoans. *Hydra, Obelia, Physalia, Prya.*
Class Scyphozoa. Jellyfishes. *Aurelia.*
Class Anthozoa. Sea anemones, corals. *Telesto.*

PHYLUM MESOZOA   Ciliated, wormlike parasites, about the same level of complexity as *Trichoplax.*

PHYLUM PLATYHELMINTHES   Flatworms. Bilateral, cephalized; simplest animals with organ systems. Saclike gut.
Class Turbellaria. Triclads (planarians), polyclads. *Dugesia.*
Class Trematoda. Flukes. *Clonorchis, Schistosoma.*
Class Cestoda. Tapeworms. *Diphyllobothrium, Taenia.*

PHYLUM ROTIFERA   Rotifers. *Asplancha, Philodina.*

PHYLUM NEMERTEA   Ribbon worms. *Tubulanus.*

PHYLUM MOLLUSCA   Mollusks.
Class Polyplacophora. Chitons. *Cryptochiton, Tonicella.*
Class Gastropoda. Snails (periwinkles, whelks, limpets, abalones, cowries, conches, nudibranchs, tree snails, garden snails), sea slugs, land slugs. *Aplysia, Ariolimax, Cypraea, Haliotis, Helix, Liguus, Limax, Littorina, Patella.*
Class Bivalvia. Clams, mussels, scallops, cockles, oysters, shipworms. *Ensis, Chlamys, Mytelus, Patinopectin.*
Class Cephalopoda. Squids, octopuses, cuttlefish, nautiluses. *Dosidiscus, Loligo, Nautilus, Octopus, Sepia.*

PHYLUM BRYOZOA   Bryozoans (moss animals).

PHYLUM BRACHIOPODA   Lampshells.

PHYLUM ANNELIDA   Segmented worms.
Class Polychaeta. Mostly marine worms. *Eunice, Neanthes.*
Class Oligochaeta. Mostly freshwater and terrestrial worms, many marine. *Lumbricus* (earthworms), *Tubifex.*
Class Hirudinea. Leeches. *Hirudo, Placobdella.*

PHYLUM NEMATODA   Roundworms. *Ascaris, Caenorhabditis elegans, Necator* (hookworms), *Trichinella.*

PHYLUM TARDIGRADA   Water bears.

PHYLUM ONYCHOPHORA   Onychophorans. *Peripatus.*

PHYLUM ARTHROPODA
Subphylum Trilobita. Trilobites; extinct.
Subphylum Chelicerata. Chelicerates. Horseshoe crabs, spiders, scorpions, ticks, mites.
Subphylum Crustacea. Shrimps, crayfishes, lobsters, crabs, barnacles, copepods, isopods (sowbugs).
Subphylum Uniramia.
Superclass Myriapoda. Centipedes, millipedes.
Superclass Insecta.
Order Ephemeroptera. Mayflies.
Order Odonata. Dragonflies, damselflies.
Order Orthoptera. Grasshoppers, crickets, katydids.
Order Dermaptera. Earwigs.
Order Blattodea. Cockroaches.

Order Mantodea. Mantids.
Order Isoptera. Termites.
Order Mallophaga. Biting lice.
Order Anoplura. Sucking lice.
Order Hemiptera. Cicadas, aphids, leafhoppers, spittlebugs, bugs.
Order Coleoptera. Beetles.
Order Diptera. Flies.
Order Mecoptera. Scorpion flies. *Harpobittacus.*
Order Siphonaptera. Fleas.
Order Lepidoptera. Butterflies, moths.
Order Hymenoptera. Wasps, bees, ants.
Order Neuroptera. Lacewings, antlions.

PHYLUM ECHINODERMATA Echinoderms.

Class Asteroidea. Sea stars. *Asterias.*
Class Ophiuroidea. Brittle stars.
Class Echinoidea. Sea urchins, heart urchins, sand dollars.
Class Holothuroidea. Sea cucumbers.
Class Crinoidea. Feather stars, sea lilies.
Class Concentricycloidea. Sea daisies.

PHYLUM HEMICHORDATA Acorn worms.

PHYLUM CHORDATA Chordates.

Subphylum Urochordata. Tunicates, related forms.
Subphylum Cephalochordata. Lancelets.

CRANIATES

Superclass "Agnatha." Jawless fishes, including ostracoderms (extinct).

Class Myxini. Hagfishes.

Class Cephalaspidomorphi. Lampreys.

Subphylum Vertebrata. Jawed vertebrates.

Class "Placodermi." Jawed, heavily armored fishes; extinct.

Class Chondrichthyes. Cartilaginous fishes. (sharks, rays, skates, chimaeras).

Class "Osteichthyes." Bony fishes. Not monophyletic.
Subclass Dipnoi. Lungfishes.
Subclass Crossopterygii. Coelacanths, related forms.
Subclass Actinopterygii. Ray-finned fishes.
Order Acipenseriformes. Sturgeons, paddlefishes.
Order Salmoniformes. Salmon, trout.
Order Atheriniformes. Killifishes, guppies.
Order Gasterosteiformes. Seahorses.
Order Perciformes. Perches, wrasses, barracudas, tunas, freshwater bass, mackerels.
Order Lophiiformes. Angler fishes.

TETRAPODS (A subgroup of craniates)

Class Amphibia. Amphibians.
Order Caudata. Salamanders.
Order Anura. Frogs, toads.
Order Apoda. Apodans (caecilians).

AMNIOTES (A subgroup of tetrapods)

Class "Reptilia." Skin with scales, embryo protected and nutritionally supported by extraembryonic membranes.
Subclass Anapsida. Turtles, tortoises.
Subclass Lepidosaura. *Sphenodon*, lizards, snakes.
Subclass Archosaura. Dinosaurs (extinct), crocodiles, alligators.

Class Aves. Birds. In some of the more recent classification systems, dinosaurs, crocodilians, and birds are grouped in the same category, the archosaurs.
Order Struthioniformes. Ostriches.
Order Sphenisciformes. Penguins.
Order Procellariiformes. Albatrosses, petrels.
Order Ciconiiformes. Herons, bitterns, storks, flamingoes.
Order Anseriformes. Swans, geese, ducks.
Order Falconiformes. Eagles, hawks, vultures, falcons.
Order Galliformes. Ptarmigan, turkeys, domestic fowl.
Order Columbiformes. Pigeons, doves.
Order Strigiformes. Owls.
Order Apodiformes. Swifts, hummingbirds.
Order Passeriformes. Sparrows, jays, finches, crows, robins, starlings, wrens.
Order Piciformes. Woodpeckers, toucans.
Order Psittaciformes. Parrots, cockatoos, macaws.

Class Mammalia. Skin with hair; young nourished by milk-secreting glands of adult.

Subclass Prototheria. Egg-laying mammals (monotremes; duckbilled platypus, spiny anteaters).
Subclass Metatheria. Pouched mammals or marsupials (opossums, kangaroos, wombats, Tasmanian devil).
Subclass Eutheria. Placental mammals.
Order Edentata. Anteaters, tree sloths, armadillos.
Order Insectivora. Tree shrews, moles, hedgehogs.
Order Chiroptera. Bats.
Order Scandentia. Insectivorous tree shrews.
Order Primates.
Suborder Strepsirhini (prosimians). Lemurs, lorises.
Suborder Haplorhini (tarsioids and anthropoids).
Infraorder Tarsiiformes. Tarsiers.
Infraorder Platyrrhini (New World monkeys).
Family Cebidae. Spider monkeys, howler monkeys, capuchin.
Infraorder Catarrhini (Old World monkeys and hominoids).
Superfamily Cercopithecoidea. Baboons, macaques, langurs.
Superfamily Hominoidea. Apes and humans.
Family Hylobatidae. Gibbon.
Family "Pongidae." Chimpanzees, gorillas, orangutans.
Family Hominidae. Existing and extinct human species (*Homo*) and humanlike species, including the australopiths.
Order Lagomorpha. Rabbits, hares, pikas.
Order Rodentia. Most gnawing animals (squirrels, rats, mice, guinea pigs, porcupines, beavers, etc.).
Order Carnivora. Carnivores.
Suborder Feloidea. Cats, mongooses, hyenas.
Suborder Canoidea. Dogs, weasels, skunks, otters, raccoons, pandas, bears.
Order Pinnipedia. Seals, walruses, sea lions.
Order Proboscidea. Elephants; mammoths (extinct).
Order Sirenia. Sea cows (manatees, dugongs).
Order Perissodactyla. Odd-toed ungulates (horses, tapirs, rhinos).
Order Tubulidentata. African aardvarks.
Order Artiodactyla. Even-toed ungulates (camels, deer, bison, sheep, goats, antelopes, giraffes, etc.).
Order Cetacea. Whales, porpoises.

Appendix II. Units of Measure

Metric-English Conversions

Length

English		Metric
inch	=	2.54 centimeters
foot	=	0.30 meter
yard	=	0.91 meter
mile (5,280 feet)	=	1.61 kilometer

To convert	multiply by	to obtain
inches	2.54	centimeters
feet	30.00	centimeters
centimeters	0.39	inches
millimeters	0.039	inches

Weight

English		Metric
grain	=	64.80 milligrams
ounce	=	28.35 grams
pound	=	453.60 grams
ton (short) (2,000 pounds)	=	0.91 metric ton

To convert	multiply by	to obtain
ounces	28.3	grams
pounds	453.6	grams
pounds	0.45	kilograms
grams	0.035	ounces
kilograms	2.2	pounds

Volume

English		Metric
cubic inch	=	16.39 cubic centimeters
cubic foot	=	0.03 cubic meter
cubic yard	=	0.765 cubic meters
ounce	=	0.03 liter
pint	=	0.47 liter
quart	=	0.95 liter
gallon	=	3.79 liters

To convert	multiply by	to obtain
fluid ounces	30.00	milliliters
quart	0.95	liters
milliliters	0.03	fluid ounces
liters	1.06	quarts

Appendix III. Answers to Self-Quizzes

CHAPTER 1

1.	cell	1.1
2.	Metabolism	1.2
3.	Homeostasis	1.2
4.	adaptive	1.4
5.	mutation	1.4
6.	domains	1.3
7.	a	1.2
8.	d	1.2, 1.4
9.	d	1.4
10.	a	1.6
11.	c	1.4
	e	1.4
	d	1.5
	b	1.5
	a	1.5

CHAPTER 2

1.	F; hydrogen atoms have no neutrons	2.1
2.	b	2.1
3.	d	2.2
4.	c	2.4
5.	a	2.4
6.	f	2.5
7.	f	2.6
8.	acid, base	2.6
9.	c	2.6
10.	e	*Impacts, Issues*
	d	2.6
	b	2.4
	c	2.5
	a	2.1

CHAPTER 3

1.	See table 3.1	
2.	d	3.1
3.	c	3.1
4.	f	3.3
5.	b	3.4
6.	d	3.4
7.	e	3.4
8.	d	3.5, 3.7
9.	d	3.5
10.	d	3.7
11.	b	3.7
12.	c	3.5
	e	3.7
	b	3.4
	d	3.7
	a	3.3

CHAPTER 4

1.	c	4.3
2.	See Figs. 4.19, 4.20	
3.	d	4.5
4.	d	4.11
5.	False; many cells have a wall	4.11
6.	c	4.1
7.	e	4.8
	d	4.8
	a	4.1, 4.4
	b	4.7
	c	4.7

CHAPTER 5

1.	c	5.1
2.	b	5.2
3.	d	5.3
4.	a	5.7
5.	d	5.5
6.	b	5.5
7.	d	5.8
8.	d	5.5
9.	c	5.2
	g	5.6
	a	5.2
	d	5.2
	e	5.2
	b	5.2
	f	5.1

CHAPTER 6

1.	carbon dioxide; sunlight	*Impacts, Issues*
2.	a	6.2, 6.3
3.	b	6.2, 6.5
4.	c	6.3
5.	c	6.2, 6.5
6.	b	6.5
7.	c	6.5
	a	6.3, 6.4
	b	6.5

CHAPTER 7

1.	d	7.1
2.	c	7.1, 7.2
3.	b	7.3
4.	c	7.4
5.	See Fig. 7.3	
6.	c	7.5
7.	b	7.5
8.	d	7.6
9.	b	7.2
	c	7.5
	a	7.3
	d	7.4

CHAPTER 8

1.	d	8.1
2.	b	8.1
3.	c	8.1
4.	d	8.2
5.	a	8.2
6.	c	8.2
7.	a	8.3
8.	b	8.2
9.	d	8.3
	b	8.3
	c	8.3
	a	8.3

CHAPTER 9

1.	c	9.2
2.	b	9.2
3.	a	9.2
4.	d	9.1, 9.5
5.	d	9.2, 9.4
6.	b	9.2
7.	d	9.2
8.	c	9.3
9.	d	9.3
10.	d	9.2
	a	9.1
	c	9.3
	b	9.2

CHAPTER 10

1.	a	10.1
2.	b	10.1, 10.2
3.	a	10.1, 10.6
4.	b	10.1, 10.2
5.	c	10.2
6.	a	10.2, 10.4
7.	d	10.3
8.	b	10.3
	d	10.2
	a	10.1
	c	10.1

CHAPTER 11

1.	c	11.1, 11.3
2.	a	11.3
3.	e	11.7
4.	a	11.4
5.	c	11.1
6.	d	11.8
7.	c	11.8
8.	c	11.3
	e	11.7
	d	11.4
	b	11.7
	a	11.2
	f	11.3

CHAPTER 12

1.	c	12.2
2.	d	12.2
3.	c	12.2
4.	a	12.4
5.	d	12.4
6.	c	12.5

CHAPTER 13

1.	c	13.1
2.	b	13.1
3.	c	13.1
4.	c	13.2
5.	a	13.2
6.	e	13.4
	c	13.3
	a	13.1
	f	13.2
	d	13.2
	g	13.1
	b	13.2

CHAPTER 14

1.	d	14.1, 14.3
2.	d	14.1
3.	d	14.3, 14.5
4.	d	14.3
5.	d	14.3
6.	d	14.1, 14.2, 14.3
7.	a	14.2
8.	d	14.2
9.	a	14.1, 14.2, 14.3
10.	h	14.3
11.	d	14.3
12.	b	14.4
13.	c	14.4
14.	e	14.4, 14.5
	a	14.2
	d	*Impacts, Issues*
	b	14.3
	c	14.4

CHAPTER 15

1.	d	15.7
2.	c	15.2
3.	plasmid	15.2
4.	b	15.2
5.	a	15.3
6.	b	15.4
7.	d	15.5
8.	b	15.7, 15.8
9.	d	15.4
	c	15.6
	b	*Impacts, Issues*
	e	15.3
	a	15.7

CHAPTER 16

1.	populations	16.5
2.	d	*Impacts, Issues*
3.	a	16.5
4.	c	16.3
5.	b	16.7
6.	c	16.8
7.	b	16.11
8.	c	16.11
	d	16.3
	a	16.5
	b	16.10

CHAPTER 17

1.	a	17.4
2.	d	17.7
3.	c	17.10
4.	c	17.11
5.	c	17.5
6.	d	17.7
7.	d	17.11
8.	c	17.11
9.	d	17.11
10.	See Fig. 17.31	
11.	e	17.11
	a	17.1
	f	17.1
	c	17.4
	b	17.11
	d	17.4
	g	17.10

CHAPTER 18

1.	c	18.1
2.	c	18.2
3.	c	18.3
4.	b	18.4
5.	f	18.1
	c	18.2
	d	18.2
	a	18.3
	b	18.4
	e	18.4

CHAPTER 19

1.	Left, head (top), sheath (middle), tail fiber (bottom). Right: capsid (left), envelope (right)	19.4
2.	d	19.1
3.	c	19.3
4.	c	19.1
5.	c	19.2
6.	d	19.2
7.	b	19.2
8.	d	19.2
9.	b	19.4
10.	false	19.4
11.	d	19.4
12.	d	19.1
13.	d	19.3
	e	19.2
	b	19.4
	f	19.1
	c	19.3
	g	19.4
	a	19.5

CHAPTER 20

1.	b	20.2
2.	d	20.2
3.	b	20.3
4.	b	20.5
5.	d	20.5
6.	d	20.4
7.	a	20.6
8.	c	20.7
9.	a	20.8
10.	c	20.9
11.	c	20.9
12.	b	20.2
	e	20.2
	f	20.5
	i	20.3
	h	20.5
	j	20.7
	d	20.5
	a	20.10
	c	20.12
	g	20.12

CHAPTER 21

1.	a	21.3
2.	c	21.5
3.	b	21.2
4.	c	21.3
5.	a	21.3
6.	b	21.5
7.	c	21.6
	e	21.1
	g	21.3
	h	21.5
	f	21.2
	a	21.1
	b	21.1
	d	21.7

CHAPTER 22

1.	d	22.1
2.	a	22.1
3.	a	22.4
4.	a	22.5
5.	b	22.12
6.	c	22.5, 22.6
7.	b	22.1, 22.13
8.	b	22.13
9.	i	22.2
	c	22.3
	h	22.4
	b	22.5
	a	22.8
	f	22.6
	d	22.9
	e	22.7
	g	22.13

CHAPTER 23

1.	b	23.1
2.	a	23.2
3.	d	23.3, 23.4
4.	c	23.6
5.	f	23.6, 23.7
6.	a	23.7
7.	c	23.8
8.	f	23.2, 23.6
		23.9, 23.11
9.	c	23.12
10.	g	23.3
	a	23.4
	e	23.7
	b	23.8
	c	23.9
	f	23.9
	d	23.12

CHAPTER 24

1.	a	24.1
2.	c	24.1
3.	d	24.2
4.	d	24.5
5.	b	24.4
	c	24.1
	a	24.5
	d	24.3

CHAPTER 25

1.	Left; right	25.1
2.	a	25.1
3.	d	25.1
4.	Eudicot; monocot; see Fig. 25.4	
5.	c	25.2
6.	c	25.2
7.	b	25.2
8.	b	25.1, 25.2
9.	See Fig. 25.21	
10.	d	25.6
11.	b	25.3
	d	25.6
	e	25.2
	c	25.2
	f	25.4
	a	25.6

CHAPTER 26

1.	e	26.1
2.	c	26.3
3.	b	26.2
4.	a	26.2
5.	c	26.3
6.	d	26.3
7.	a	26.4
8.	d	26.4
9.	c	26.4
	g	26.1
	e	26.5
	b	26.2
	d	26.3
	a	26.3
	f	26.5

CHAPTER 27

1.	a	27.1
2.	c	27.3
3.	b	27.2
4.	c	27.3
5.	c	27.4
6.	e	27.6
7.	c	27.6
8.	c	27.9
9.	e	27.8
	f	27.9
	b	27.2
	c	27.2, 27.3
	d	27.1
	a	27.10

CHAPTER 28

1.	a	28.1
2.	c	28.1
3.	a	28.1
4.	b	28.2
5.	b	28.2
6.	c	28.2
7.	c	28.3
8.	d	28.3
9.	d	28.4
10.	b	28.1
	g	28.1
	a	28.2
	c	28.6
	d	28.3
	f	28.2
	e	28.1

CHAPTER 29

1.	a	29.1
2.	c	29.1
3.	d	29.2
4.	a	29.3
5.	d	29.1
6.	c	29.6
7.	b	29.6
8.	b	29.7
9.	a	29.7
10.	b	29.8
11.	e	29.4
	d	29.3
	f	29.8
	b	29.7
	g	29.8
	a	29.7
	c	29.7

CHAPTER 30

1.	c	30.1
2.	c	30.2
3.	e	30.1, 30.3
4.	a	30.2
5.	b	30.4
6.	b	30.5
7.	d	30.7
	g	30.5
	f	30.6
	a	30.5
	c	30.7
	e	30.3
	b	30.4

CHAPTER 31

1.	f	31.1
2.	b	31.3
3.	See Figs. 31.4, 31.5	31.3
4.	a	31.3
5.	e	31.4, 31.5
6.	b	31.5
7.	b	31.4
8.	b	31.5
9.	d	31.4
	f	31.4
	c	31.5
	e	31.5
	a	31.6
	b	31.5

CHAPTER 32

1.	d	32.1
2.	b	32.4
3.	d	32.5
4.	b	32.3
5.	b	32.4
6.	d	32.4
7.	e	32.2
	f	32.6
	g	32.6
	h	32.2
	a	32.4
	c	32.2
	b	32.1
	i	32.4
	d	32.6

CHAPTER 33

1.	c	33.1
2.	b	33.1
3.	d	33.2
4.	b	33.4
5.	d	33.2
6.	b	33.6
7.	c	33.6
8.	a	33.7
9.	c	33.7
10.	d	33.10
11.	f	33.8
	a	33.10
	e	33.2
	g	33.6
	b	33.6
	c	33.7
	d	33.5
12.	See Fig. 33.11	33.6
13.	See Fig. 33.17	33.7

CHAPTER 34

1.	e	34.1
2.	c	34.3
3.	d	34.4
4.	d	34.4, 34.5
5.	e	34.3
6.	e	34.6
7.	a	34.6
8.	b	34.7
9.	c	34.3
	b	34.6
	a	34.1
	e	34.5
	d	34.9

CHAPTER 35

1.	a	35.1
2.	d	35.1
3.	c	35.2
4.	a	35.4
5.	c	35.5, 35.7
6.	d	35.6
7.	a	35.6
8.	a	35.7
9.	d	35.5
	h	35.5
	f	35.5, 35.7
	e	35.1
	g	35.5
	c	35.5
	b	35.5
	a	35.5

CHAPTER 36

1.	d	36.1
2.	b	36.3
3.	c	36.4
4.	b	36.3, 36.4
5.	a	36.4
6.	c	36.6
7.	b	36.8
8.	f	36.3
	b	36.6
	a	36.3
	d	36.4
	e	36.3
	c	36.3

CHAPTER 37

1.	c	37.1
2.	b	37.2
3.	d	37.3
4.	b	37.3
5.	a	37.3
6.	c	37.2
	a	37.2
	b	37.2
	e	37.2
	d	37.3
7.	d	37.7
8.	a	37.7
9.	b	37.6
	a	37.6
	d	37.6
	c	37.6
	e	37.6

CHAPTER 38

1.	d	38.1
2.	c	38.2, 38.3
3.	c	38.2, 38.4
4.	c	38.5
5.	d	38.7
6.	c	38.9
7.	c	38.9
8.	c	38.14
9.	a	38.14
10.	e	38.13
11.	c	38.2
	d	38.2, 38.11
	a	38.2, 38.3
	b	38.2, 38.15
	f	38.2, 38.4
	e	38.2, 38.4
12.	c	38.6
	h	38.8
	a	38.16
	g	38.8, 38.19
	d	38.9, 38.10
	e	38.8, 38.11
	b	38.6, 38.7
	f	38.8

CHAPTER 39

1.	f	39.1
2.	a	39.3
3.	a	39.3
4.	d	39.4
5.	d	39.5
6.	c	39.7
7.	c	39.4
	d	39.3
	a	39.3
	e	39.4
	b	39.4

CHAPTER 40

1.	d	40.1
2.	d	40.1
3.	d	40.3
4.	b	40.4
5.	d	40.6
6.	d	40.8
7.	e	40.11
8.	b	40.12
9.	c	40.9
	d	40.11
	a	40.8
	e	40.8
	b	40.9
	g	40.12
	f	40.13

CHAPTER 41

1.	d	41.1
2.	d	41.1
3.	d	41.4
4.	d	41.3
5.	a	41.7
6.	b	41.8
7.	f	41.11
8.	c	41.10, 41.11
9.	a	41.10
10.	d	41.1
	a	41.1
	c	41.1
	b	41.1

CHAPTER 42

1.	d	42.1
2.	c	42.2
3.	d	42.3
4.	a	42.3
5.	c	42.2
6.	e	42.4
7.	d	42.4, 42.8
8.	c	42.5
9.	d	42.9, 42.10
10.	d	42.7
	h	42.5
	e	42.5
	c	42.5
	b	42.10
	g	42.6
	a	42.6
	f	42.12

CHAPTER 43

1.	a	43.3
2.	d	43.1
3.	d	43.6
4.	a	43.3
5.	d	43.7
6.	c	43.7
7.	c	43.1
	d	43.3
	b	43.1, 43.2
	a	43.2
	e	43.4

Appendix IV. Answers to Genetics Problems

CHAPTER 10

1. a. Both parents are heterozygotes (*Aa*). Their children may be albino (*aa*) or unaffected (*AA* or *Aa*).

b. All are homozygous recessive (*aa*).

c. Homozygous recessive (*aa*) father, and heterozygous (*Aa*) mother. The albino child is *aa*, the unaffected children *Aa*.

2. a. *AB*   **c.** *Ab, ab*

b. *AB, aB*   **d.** *AB, Ab, aB, ab*

3. a. All offspring will be *AaBB*.

b. 1/4 *AABB* (25% each genotype)
1/4 *AABb*
1/4 *AaBB*
1/4 *AaBb*

c. 1/4 *AaBb* (25% each genotype)
1/4 *Aabb*
1/4 *aaBb*
1/4 *aabb*

d. 1/16 *AABB* (6.25%)
1/8 *AaBB* (12.5%)
1/16 *aaBB* (6.25%)
1/8 *AABb* (12.5%)
1/4 *AaBb* (25%)
1/8 *aaBb* (12.5%)
1/16 *AAbb* (6.25%)
1/8 *Aabb* (12.5%)
1/16 *aabb* (6.25%)

4. A mating of two M^L cats yields 1/4 MM, 1/2 M^LM, and 1/4 $M^L M^L$. Because $M^L M^L$ is lethal, the probability that any one kitten among the survivors will be heterozygous is 2/3.

5. Yellow is recessive. Because F_1 plants have a green phenotype and must be heterozygous, green must be dominant over the recessive yellow.

6. a. *ABC*

b. *ABc, aBc*

c. *ABC, aBc, ABc, aBc*

d. *ABC, aBC, AbC, abC, ABc, aBc, Abc, abc*

7. Because all F_1 plants of this dihybrid cross had to be heterozygous for both genes, then 1/4 (25%) of the F_2 plants will be heterozygous for both genes.

8. a. The mother must be heterozygous $I^A i$. The male with type B blood could have fathered the child if he were heterozygous $I^B i$.

b. Genotype alone cannot prove the accused male is the father. Even if he happens to be heterozygous, *any* male who carries the *i* allele could be the father, including those heterozygous for type A blood ($I^A i$) or type B blood ($I^B i$) and those with type O blood (*ii*).

9. A mating between a mouse from a true-breeding, white-furred strain and a mouse from a true-breeding, brown-furred strain would provide you with the most direct evidence.

Because true-breeding strains of organisms typically are homozygous for a trait being studied, all F_1 offspring from this mating should be heterozygous. Record the phenotype of each F_1 mouse, then let them mate with one another. Assuming only one gene locus is involved, these are possible outcomes for the F_2 offspring:

a. All F_1 mice are brown, and their F_2 offspring segregate:
3 brown : 1 white.

Conclusion: Brown is dominant to white.

b. All F_1 mice are white, and their F_2 offspring segregate:
3 white : 1 brown.

Conclusion: White is dominant to brown.

c. All F_1 mice are tan, and the F_2 offspring segregate:
1 brown : 2 tan : 1 white.

Conclusion: The alleles at this locus show incomplete dominance.

10. You cannot guarantee that the puppies will not develop the disorder without more information about Dandelion's genotype. You could do so only if she is a heterozygous carrier, if the male is free of the alleles, and if the alleles are recessive.

11. Fred could use a testcross to find out if his pet's genotype is *WW* or *Ww*. He can let his black guinea pig mate with a white guinea pig having the genotype *ww*.

If any F_1 offspring are white, then the genotype of his pet is *Ww*. If the two guinea pig parents are allowed to mate repeatedly and all the offspring of the matings are black, then there is a high probability that his pet guinea pig is *WW*.

(If, say, ten offspring are all black, then the probability that the male is *WW* is about 99.9 percent. The greater the number of offspring, the more confident Fred can be of his conclusion.)

12. a. 1/2 red, 1/2 pink

b. All pink

c. 1/4 red, 1/2 pink, 1/4 white

d. 1/2 pink, 1/2 white

13. 9/16 walnut comb
3/16 rose comb
3/16 pea comb
1/16 single comb

14. Because both parents are heterozygotes ($Hb^A Hb^S$), the following are the probabilities for each child:

a. 1/4 $Hb^S Hb^S$

b. 1/4 $Hb^A Hb^A$

c. 1/2 $Hb^A Hb^S$

CHAPTER 11

1. a. Human males (XY) inherit their X chromosome from their mother.

b. A male can produce two kinds of gametes. Half carry an X chromosome and half carry a Y chromosome. All the gametes that carry the X chromosome carry the same X-linked allele.

c. A female homozygous for an X-linked allele produces only one kind of gamete.

d. Half of the gametes of a female who is heterozygous for an X-linked allele carry one of the two alleles at that locus; the other half carry its partner allele for that locus.

2. If only one parent is heterozygous for the autosomal dominant allele, the chance of a child inheriting that allele is 50 percent.

3. a. Nondisjunction can occur at anaphase I or anaphase II of meiosis.

b. As a result of translocation, chromosome 21 may get attached to the end of chromosome 14. The new individual's chromosome number would still be 46, but its somatic cells would have the translocated chromosome 21 in addition to two normal chromosomes 21.

4. Because females (XX) could be white-eyed, the recessive allele had to be on one of their X chromosomes. What if white-eyed males (XY) had the recessive allele on their X chromosome and their Y chromosome had no corresponding eye-color allele? In that case, they would have white eyes. They would have no dominant allele to mask the effect of the recessive one.

5. Because the phenotype appeared in every generation shown in the diagram, this must be a pattern of autosomal dominant inheritance.

However, if a son bears the allele for the disorder on his X chromosome, then it will be expressed. He will develop the disorder, and most likely he will not father children because of his early death.

6. A daughter could develop this muscular dystrophy only if she were to inherit two X-linked recessive alleles—one from each parent. Males who carry the allele are unlikely to father children because they develop the disorder and die an early death.

7. In the mother, a crossover between the two genes at meiosis generates an X chromosome that carries neither mutant allele.

Appendix V. Closer Look at Some Major Metabolic Pathways

ENERGY-REQUIRING STEPS OF GLYCOLYSIS

(two ATP invested)

ENERGY-RELEASING STEPS OF GLYCOLYSIS

(four ATP produced)

Splitting of fructose-1,6-bisphosphate yields two 3-carbon molecules that are interconvertible. For simplicity, we show how only one of those molecules is converted to pyruvate, although both are.

Figure A   Glycolysis, ending with two 3-carbon pyruvate molecules for each 6-carbon glucose molecule entering the reactions. The *net* energy yield is two ATP molecules (two invested, four produced).

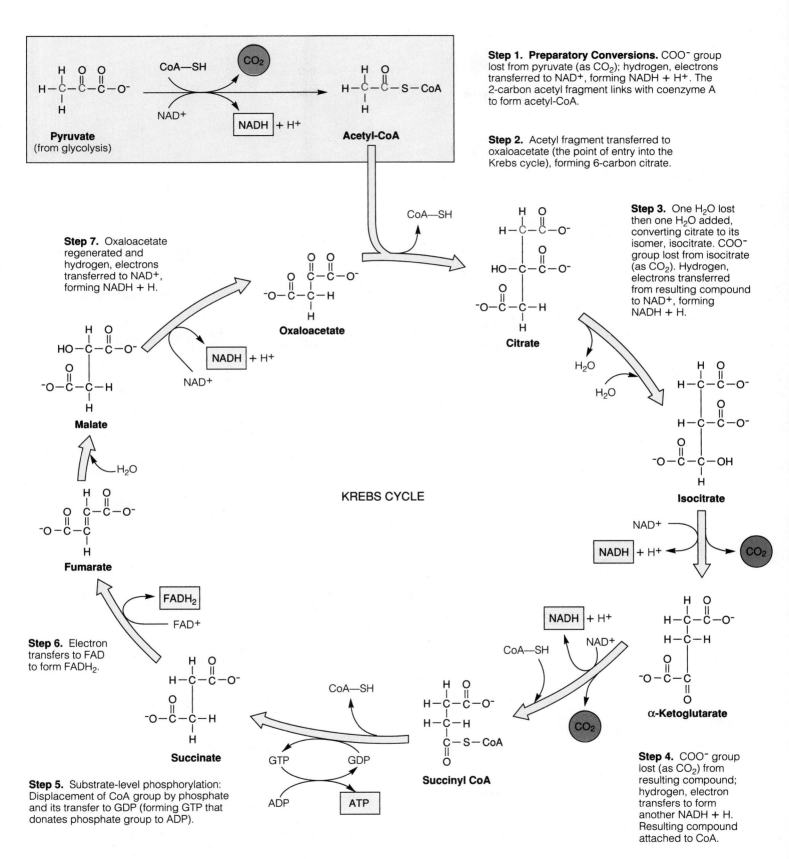

Step 1. Preparatory Conversions. COO^- group lost from pyruvate (as CO_2); hydrogen, electrons transferred to NAD^+, forming $NADH + H^+$. The 2-carbon acetyl fragment links with coenzyme A to form acetyl-CoA.

Step 2. Acetyl fragment transferred to oxaloacetate (the point of entry into the Krebs cycle), forming 6-carbon citrate.

Step 3. One H_2O lost then one H_2O added, converting citrate to its isomer, isocitrate. COO^- group lost from isocitrate (as CO_2). Hydrogen, electrons transferred from resulting compound to NAD^+, forming $NADH + H$.

Step 7. Oxaloacetate regenerated and hydrogen, electrons transferred to NAD^+, forming $NADH + H$.

Step 6. Electron transfers to FAD to form $FADH_2$.

Step 5. Substrate-level phosphorylation: Displacement of CoA group by phosphate and its transfer to GDP (forming GTP that donates phosphate group to ADP).

Step 4. COO^- group lost (as CO_2) from resulting compound; hydrogen, electron transfers to form another $NADH + H$. Resulting compound attached to CoA.

KREBS CYCLE

Pyruvate (from glycolysis)

Acetyl-CoA

Oxaloacetate

Citrate

Isocitrate

Malate

α-Ketoglutarate

Fumarate

Succinyl CoA

Succinate

Figure B Krebs cycle, also known as the citric acid cycle. *Red* identifies carbon atoms entering the cyclic pathway (by way of acetyl-CoA) and leaving (by way of carbon dioxide). These cyclic reactions run twice for each glucose molecule that has been degraded to two pyruvate molecules.

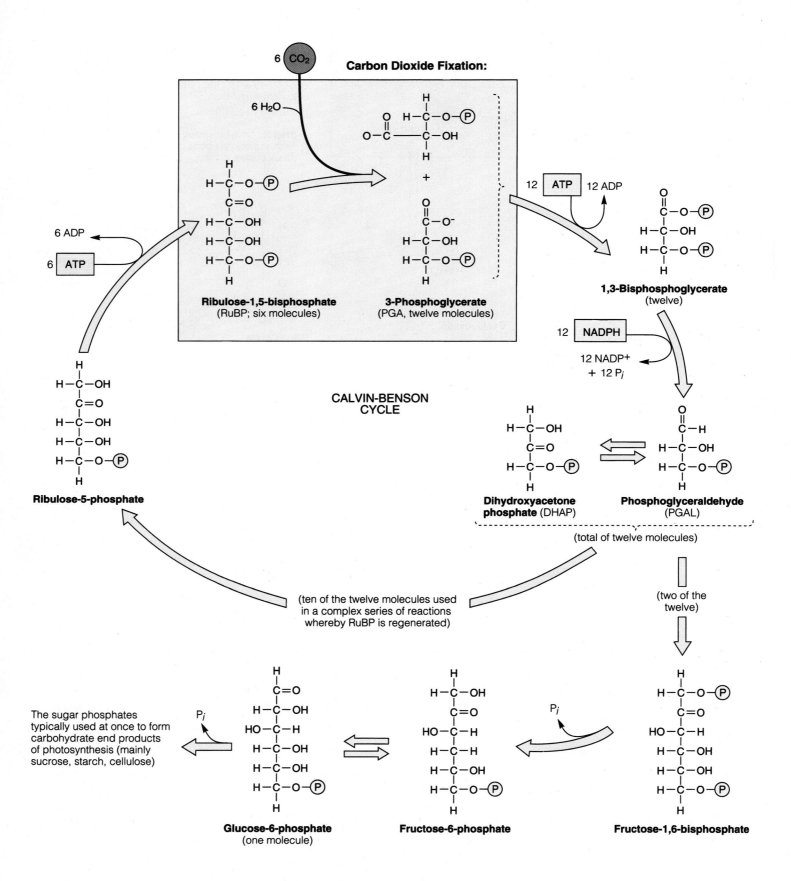

Figure C   Calvin–Benson cycle of the light-independent reactions of photosynthesis.

Appendix VI. The Amino Acids

Neutral, nonpolar side group

glycine (gly)

alanine (ala)

valine (val)

isoleucine (ile)

leucine (leu)

phenylalanine (phe)

proline (pro)

methionine (met)

Neutral, polar side group

serine (ser)

threonine (thr)

tyrosine (tyr)

tryptophan (trp)

asparagine (asn)

glutamine (gln)

cysteine (cys)

Acidic side group

aspartic acid (asp)

glutamic acid (glu)

Basic side group

lysine (lys)

arginine (arg)

histidine (his)

Appendix VII. Periodic Table of the Elements

Group

Noble Gases

IA(1)

(18)

Atomic number →	11
Symbol →	Na
Atomic mass →	22.99

Atomic masses are based on carbon-12. Numbers in parentheses are mass numbers of most stable or best known isotopes of radioactive elements.

Transition Elements

Period

Inner Transition Elements

Lanthanide Series 6

| * | 58 Ce 140.1 | 59 Pr 140.9 | 60 Nd 144.2 | 61 Pm (145) | 62 Sm 150.4 | 63 Eu 152.0 | 64 Gd 157.3 | 65 Tb 158.9 | 66 Dy 162.5 | 67 Ho 164.9 | 68 Er 167.3 | 69 Tm 168.9 | 70 Yb 173.0 | 71 Lu 175.0 |

Actinide Series 7

| ** | 90 Th 232.0 | 91 Pa 231.0 | 92 U 238.0 | 93 Np 237.0 | 94 Pu (244) | 95 Am (243) | 96 Cm (247) | 97 Bk (247) | 98 Cf (251) | 99 Es (252) | 100 Fm (257) | 101 Md (258) | 102 No (259) | 103 Lr (260) |

Appendix VIII. Annotations to A Journal Article

This journal article reports on the movements of a female wolf during the summer of 2002 in northwestern Canada. It also reports on a scientific process of inquiry, observation and interpretation to learn where, how and why the wolf traveled as she did. In some ways, this article reflects the story of "how to do science" told in section 1.5 of this textbook. These notes are intended to help you read and understand how scientists work and how they report on their work.

(1) ARCTIC

(2) VOL. 57, NO. 2 (JUNE 2004) P. 196–203

(3) Long Foraging Movement of a Denning Tundra Wolf

(4) Paul F. Frame,[1,2] David S. Hik,[1] H. Dean Cluff,[3] and Paul C. Paquet[4]

(5) (Received 3 September 2003; accepted in revised form 16 January 2004)

(6) ABSTRACT Wolves (*Canis lupus*) on the Canadian barrens are intimately linked to migrating herds of barren-ground caribou (*Rangifer tarandus*). We deployed a Global Positioning System (GPS) radio collar on an adult female wolf to record her movements in response to changing caribou densities near her den during summer. This wolf and two other females were observed nursing a group of 11 pups. She traveled a minimum of 341 km during a 14-day excursion. The straight-line distance from the den to the farthest location was 103 km, and the overall minimum rate of travel was 3.1 km/h. The distance between the wolf and the radio-collared caribou decreased from 242 km one week before the excursion to 8 km four days into the excursion. We discuss several possible explanations for the long foraging bout.

(7) *Key words:* wolf, GPS tracking, movements, *Canis lupus*, foraging, caribou, Northwest Territories

(8) RÉSUMÉ Les loups (*Canis lupus*) dans la toundra canadienne sont étroitement liés aux hardes de caribous des toundras (*Rangifer tarandus*). On a équipé une louve adulte d'un collier émetteur muni d'un système de positionnement mondial (GPS) afin d'enregistrer ses déplacements en réponse au changement de densité du caribou près de sa tanière durant l'été. On a observé cette louve ainsi que deux autres en train d'allaiter un groupe de 11 louveteaux. Elle a parcouru un minimum de 341 km durant une sortie de 14 jours. La distance en ligne droite de la tanière à l'endroit le plus éloigné était de 103 km, et la vitesse minimum durant tout le voyage était de 3,1 km/h. La distance entre la louve et le caribou muni du collier émetteur a diminué de 242 km une semaine avant la sortie à 8 km quatre jours après la sortie. On commente diverses explications possibles pour ce long épisode de recherche de nourriture.

Mots clés: loup, repérage GPS, déplacements, *Canis lupus*, recherche de nourriture, caribou, Territoires du Nord-Ouest

Traduit pour la revue *Arctic* par Nésida Loyer.

(9) Introduction

Wolves (*Canis lupus*) that den on the central barrens of mainland Canada follow the seasonal movements of their main prey, migratory barren-ground caribou (*Rangifer tarandus*) (Kuyt, 1962; Kelsall, 1968; Walton et al., 2001). However, most wolves do not den near caribou calving grounds, but select sites farther south, closer to the tree line (Heard and Williams, 1992). Most caribou migrate beyond primary wolf denning areas by mid-June and do not return until mid-to-late July (Heard et al., 1996; Gunn et al., 2001). Conse-

quently, caribou density near dens is low for part of the summer.

During this period of spatial separation from the main caribou herds, wolves must either search near **(10)** the homesite for scarce caribou or alternative prey (or both), travel to where prey are abundant, or use a combination of these strategies.

Walton et al. (2001) postulated that the travel of **(11)** tundra wolves outside their normal summer ranges is a response to low caribou availability rather than a pre-dispersal exploration like that observed in territorial wolves (Fritts and Mech, 1981; Messier, 1985). The authors postulated this because most such travel was directed toward caribou calving grounds. We report details of such a long-distance excursion by a breeding female tundra wolf wearing a GPS radio collar. We discuss the relationship of the excursion to movements of satellite-collared caribou (Gunn et al., 2001), supporting the hypothesis that tundra wolves make directional, rapid, long-distance movements in response to seasonal prey availability.

[1] Department of Biological Sciences, University of Alberta, Edmonton, Alberta T6G 2E9, Canada
[2] Corresponding author: pframe@ualberta.ca
[3] Department of Resources, Wildlife, and Economic Development, North Slave Region, Government of the Northwest Territories, P.O. Box 2668, 3803 Bretzlaff Dr., Yellowknife, Northwest Territories X1A 2P9, Canada; Dean_Cluff@gov.nt.ca
[4] Faculty of Environmental Design, University of Calgary, Calgary, Alberta T2N 1N4, Canada; current address: P.O. Box 150, Meacham, Saskatchewan S0K 2V0, Canada

196

Annotations (right margin):

1 Title of the journal, which reports on science taking place in Arctic regions.

2 Volume number, issue number and date of the journal, and page numbers of the article.

3 Title of the article: a concise but specific description of the subject of study—one episode of long-range travel by a wolf hunting for food on the Arctic tundra.

4 Authors of the article: scientists working at the institutions listed in the footnotes below. Note #2 indicates that P. F. Frame is the *corresponding author*—the person to contact with questions or comments. His email address is provided.

5 Date on which a draft of the article was received by the journal editor, followed by date one which a revised draft was accepted for publication. Between these dates, the article was reviewed and critiqued by other scientists, a process called peer review. The authors revised the article to make it clearer, according to those reviews.

6 ABSTRACT: A brief description of the study containing all basic elements of this report. First sentence summarizes the *background* material. Second sentence encapsulates the *methods* used. The rest of the paragraph sums up the *results*. Authors introduce the main *subject* of the study—a female wolf (#388) with pups in a den—and refer to later *discussion* of possible explanations for her behavior.

7 Key words are listed to help researchers using computer databases. Searching the databases using these key words will yield a list of studies related to this one.

8 RÉSUMÉ: The French translation of the abstract and key words. Many researchers in this field are French Canadian. Some journals provide such translations in French or in other languages.

9 INTRODUCTION: Gives the background for this wolf study. This paragraph tells of known or suspected wolf behavior that is important for this study. Note that (a) major species mentioned are always accompanied by scientific names, and (b) statements of fact or *postulations* (claims or assumptions about what is likely to be true) are followed by references to studies that established those facts or supported the postulations.

10 This paragraph focuses directly on the wolf behaviors that were studied here.

11 This paragraph starts with a statement of the *hypothesis* being tested, one that originated in other studies and is supported by this one. The hypothesis is restated more succinctly in the last sentence of this paragraph. This is the *inquiry* part of the scientific process—asking questions and suggesting possible answers.

12 This map shows the study area and depicts wolf and caribou locations and movements during one summer. Some of this information is explained below.

13 STUDY AREA: This section sets the stage for the study, locating it precisely with latitude and longitude coordinates and describing the area (illustrated by the map in Figure 1).

14 Here begins the story of how prey (caribou) and predators (wolves) interact on the tundra. Authors describe movements of these nomadic animals throughout the year.

15 We focus on the denning season (summer) and learn how wolves locate their dens and travel according to the movements of caribou herds.

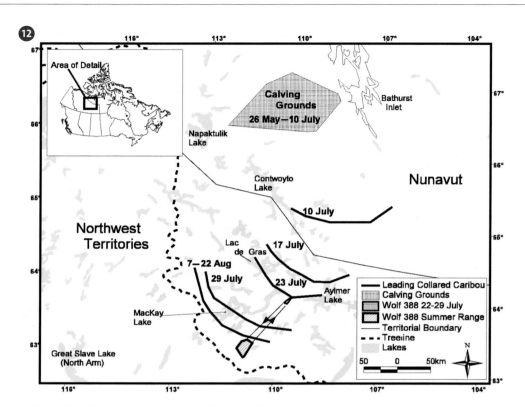

Figure 1. Map showing the movements of satellite radio-collared caribou with respect to female wolf 388's summer range and long foraging movement, in summer 2002.

🟢 Study Area

Our study took place in the northern boreal forest–low Arctic tundra transition zone (63° 30′ N, 110° 00′ W; Figure 1; Timoney et al., 1992). Permafrost in the area changes from discontinuous to continuous (Harris, 1986). Patches of spruce (*Picea mariana, P. glauca*) occur in the southern portion and give way to open tundra to the northeast. Eskers, kames, and other glacial deposits are scattered throughout the study area. Standing water and exposed bedrock are characteristic of the area.

🟢 Details of the Caribou-Wolf System

The Bathurst caribou herd uses this study area. Most caribou cows have begun migrating by late April, reaching calving grounds by June (Gunn et al., 2001;

Figure 1). Calving peaks by 15 June (Gunn et al., 2001), and calves begin to travel with the herd by one week of age (Kelsall, 1968). The movement patterns of bulls are less known, but bulls frequent areas near calving grounds by mid-June (Heard et al., 1996; Gunn et al., 2001). In summer, Bathurst caribou cows generally travel south from their calving grounds and then, parallel to the tree line, to the northwest. The rut usually takes place at the tree line in October (Gunn et al., 2001). The winter range of the Bathurst herd varies among years, ranging through the taiga and along the tree line from south of Great Bear Lake to southeast of Great Slave Lake. Some caribou spend the winter on the tundra (Gunn et al., 2001; Thorpe et al., 2001).

In winter, wolves that prey on Bathurst caribou do **15** not behave territorially. Instead, they follow the herd throughout its winter range (Walton et al., 2001; Musiani, 2003). However, during denning (May–

Table 1. Daily distances from wolf 388 and the den to the nearest radio-collared caribou during a long excursion in summer 2002.

Date (2002)	Mean distance from caribou to wolf (km)	Daily distance from closest caribou to den
12 July	242	241
13 July	210	209
14 July	200	199
15 July	186	180
16 July	163	162
17 July	151	148
18 July[1]	144	137
19 July[1]	126	124
20 July	103	130
21 July	73	130
22 July[2]	40	110
23 July[2]	9	104
29 July[3]	16	43
30 July	32	43
31 July	28	44
1 August	29	46
2 August[4]	54	52
3 August	53	53
4 August	74	74
5 August	75	75
6 August	74	75
7 August	72	75
8 August	76	75
9 August	79	79

[1] Excursion starts.
[2] Wolf closest to collared caribou.
[3] Previous five days' caribou locations not available.
[4] Excursion ends.

August, parturition late May to mid-June), wolf movements are limited by the need to return food to the den. To maximize access to migrating caribou, many wolves select den sites closer to the tree line than to caribou calving grounds (Heard and Williams, 1992). Because of caribou movement patterns, tundra denning wolves are separated from the main caribou herds by several hundred kilometers at some time during summer (Williams, 1990:19; Figure 1; Table 1).

 Muskoxen do not occur in the study area (Fournier and Gunn, 1998), and there are few moose there (H.D. Cluff, pers. obs.). Therefore, alternative prey for wolves includes waterfowl, other ground-nesting birds, their eggs, rodents, and hares (Kuyt, 1972; Williams, 1990:16; H.D. Cluff and P.F. Frame, unpubl. data). During 56 hours of den observations, we saw no ground squirrels or hares, only birds. It appears that the abundance of alternative prey was relatively low in 2002.

⑰ Methods

Wolf Monitoring

⑱ We captured female wolf 388 near her den on 22 June 2002, using a helicopter net-gun (Walton et al., 2001). She was fitted with a releasable GPS radio collar (Merrill et al., 1998) programmed to acquire locations at 30-

minute intervals. The collar was electronically released (e.g., Mech and Gese, 1992) on 20 August 2002. From 27 June to 3 July 2002, we observed 388's den with a 78 mm spotting scope at a distance of 390 m.

Caribou Monitoring

In spring of 2002, ten female caribou were captured by helicopter net-gun and fitted with satellite radio collars, bringing the total number of collared Bathurst cows to 19. Eight of these spent the summer of 2002 south of Queen Maud Gulf, well east of normal Bathurst caribou range. Therefore, we used 11 caribou for this analysis. The collars provided one location per day during our study, except for five days from 24 to 28 July. Locations of satellite collars were obtained from Service Argos, Inc. (Landover, Maryland).

Data Analysis

Location data were analyzed by ArcView GIS soft- ⑲ ware (Environmental Systems Research Institute Inc., Redlands, California). We calculated the average distance from the nearest collared caribou to the wolf and the den for each day of the study.

Wolf foraging bouts were calculated from the time 388 exited a buffer zone (500 m radius around the den) until she re-entered it. We considered her to be traveling when two consecutive locations were spatially separated by more than 100 m. Minimum distance traveled was the sum of distances between each location and the next during the excursion.

We compared pre- and post-excursion data using Analysis of Variance (ANOVA; Zar, 1999). We first tested for homogeneity of variances with Levene's test (Brown and Forsythe, 1974). No transformations of these data were required.

Results ⑳

Wolf Monitoring

Pre-Excursion Period: Wolf 388 was lactating when ㉑ captured on 22 June. We observed her and two other females nursing a group of 11 pups between 27 June and 3 July. During our observations, the pack consisted of at least four adults (3 females and 1 male) and 11 pups. On 30 June, three pups were moved to a location 310 m from the other eight and cared for by an uncollared female. The male was not seen at the den after the evening of 30 June.

Before the excursion, telemetry indicated 18 foraging bouts. The mean distance traveled during these bouts was 25.29 km (± 4.5 SE, range 3.1–82.5 km). Mean greatest distance from the den on foraging

16 Other variables are considered—prey other than caribou and their relative abundance in 2002.

17 METHODS: There is no one scientific method. Procedures for each and every study must be explained carefully.

18 Authors explain when and how they tracked caribou and wolves, including tools used and the exact procedures followed.

19 This important subsection explains what data were calculated (average distance ...) and how, including the software used and where it came from. (The calculations are listed in Table 1.) Note that the behavior measured (traveling) is carefully defined.

20 RESULTS: The heart of the report and the *observation* part of the scientific process. This section is organized parallel to the Methods section.

21 This subsection is broken down by periods of observation. Pre-excursion period covers the time between 388's capture and the start of her long-distance travel. The investigators used visual observations as well as telemetry (measurements taken using the global positioning system (GPS)) to gather data. They looked at how 388 cared for her pups, interacted with other adults, and moved about the den area.

22 The key in the lower right-hand corner of the map shows areas (shaded) within which the wolves and caribou moved, and the dotted trail of 388 during her excursion. From the results depicted on this map, the investigators tried to determine when and where 388 might have encountered caribou and how their locations affected her traveling behavior.

23 The wolf's excursion (her long trip away from the den area) is the focus of this study. These paragraphs present detailed measurements of daily movements during her two-week trip—how far she traveled, how far she was from collared caribou, her time spent traveling and resting, and her rate of speed. Authors use the phrase "minimum distance traveled" to acknowledge they couldn't track every step but were measuring samples of her movements. They knew that she went at least as far as they measured. This shows how scientists try to be exact when reporting results. Results of this study are depicted graphically in the map in Figure 2.

Figure 2. Details of a long foraging movement by female wolf 388 between 19 July and 2 August 2002. Also shown are locations and movements of three satellite radio-collared caribou from 23 July to 21 August 2002. On 23 July, the wolf was 8 km from a collared caribou. The farthest point from the den (103 km distant) was recorded on 27 July. Arrows indicate direction of travel.

bouts was 7.1 km (± 0.9 SE, range 1.7–17.0 km). The average duration of foraging bouts for the period was 20.9 h (± 4.5 SE, range 1–71 h).

The average daily distance between the wolf and the nearest collared caribou decreased from 242 km on 12 July, one week before the excursion period, to 126 km on 19 July, the day the excursion began (Table 1).

23 **Excursion Period:** On 19 July at 2203, after spending 14 h at the den, 388 began moving to the northeast and did not return for 336 h (14 d; Figure 2). Whether she traveled alone or with other wolves is unknown. During the excursion, 476 (71%) of 672 possible locations were recorded. The wolf crossed the southeast end of Lac Capot Blanc on a small land bridge, where she paused for 4.5 h after traveling for 19.5 h (37.5

km). Following this rest, she traveled for 9 h (26.3 km) onto a peninsula in Reid Lake, where she spent 2 h before backtracking and stopping for 8 h just off the peninsula. Her next period of travel lasted 16.5 h (32.7 km), terminating in a pause of 9.5 h just 3.8 km from a concentration of locations at the far end of her excursion, where we presume she encountered caribou. The mean duration of these three movement periods was 15.7 h (± 2.5 SE), and that of the pauses, 7.3 h (± 1.5). The wolf required 72.5 h (3.0 d) to travel a minimum of 95 km from her den to this area near caribou (Figure 2). She remained there (35.5 km2) for 151.5 h (6.3 d) and then moved south to Lake of the Enemy, where she stayed (31.9 km^2) for 74 h (3.1 d) before returning to her den. Her greatest distance from the den, 103 km, was recorded 174.5 h (7.3 d) after the excursion began, at 0433 on 27 July. She was 8 km from a collared caribou on 23 July, four days after the excursion began (Table 1).

The return trip began at 0403 on 2 August, 318 h (13.2 d) after leaving the den. She followed a relatively direct path for 18 h back to the den, a distance of 75 km.

The minimum distance traveled during the excursion was 339 km. The estimated overall minimum travel rate was 3.1 km/h, 2.6 km/h away from the den and 4.2 km/h on the return trip.

24 **Post-Excursion Period:** We saw three pups when recovering the collar on 20 August, but others may have been hiding in vegetation.

Telemetry recorded 13 foraging bouts in the post-excursion period. The mean distance traveled during these bouts was 18.3 km (+ 2.7 SE, range 1.2–47.7 km), and mean greatest distance from the den was 7.1 km (+ 0.7 SE, range 1.1–11.0 km). The mean duration of these post-excursion foraging bouts was 10.9 h (+ 2.4 SE, range 1–33 h).

When 388 reached her den on 2 August, the distance to the nearest collared caribou was 54 km. On 9 August, one week after she returned, the distance was 79 km (Table 1).

Pre- and Post-Excursion Comparison

25 We found no differences in the mean distance of foraging bouts before and after the excursion period (F = 1.5, df = 1, 29, p = 0.24). Likewise, the mean greatest distance from the den was similar pre- and post-excursion (F = 0.004, df = 1, 29, p = 0.95). However, the mean duration of 388's foraging bouts decreased by 10.0 h after her long excursion (F = 3.1, df = 1, 29, p = 0.09).

26 *Caribou Monitoring*

Summer Movements: On 10 July, 5 of 11 collared caribou were dispersed over a distance of 10 km, 140 km south of their calving grounds (Figure 1). On the same day, three caribou were still on the calving grounds, two were between the calving grounds and the leaders, and one was missing. One week later (17 July), the leading radio-collared cows were 100 km farther south (Figure 1). Two were within 5 km of each other in front of the rest, who were more dispersed. All radio-collared cows had left the calving grounds by this time. On 23 July, the leading radio-collared caribou had moved 35 km farther south, and all of them were more widely dispersed. The two cows closest to the leader were 26 km and 33 km away, with 37 km between them. On the next location (29 July), the most southerly caribou were 60 km

200 *P.F. Frame, et al*

farther south. All of the caribou were now in the areas where they remained for the duration of the study (Figure 2).

A Minimum Convex Polygon (Mohr and Stumpf, 1966) around all caribou locations acquired during the study encompassed 85 119 km².

Relative to the Wolf Den: The distance from the nearest collared caribou to the den decreased from 241 km one week before the excursion to 124 km the day it began. The nearest a collared caribou came to the den was 43 km away, on 29 and 30 July. During the study, four collared caribou were located within 100 km of the den. Each of these four was closest to the wolf on at least one day during the period reported. **27**

28 **Discussion**

Prey Abundance

Caribou are the single most important prey of tundra wolves (Clark, 1971; Kuyt, 1972; Stephenson and James, 1982; Williams, 1990). Caribou range over vast areas, and for part of the summer, they are scarce or absent in wolf home ranges (Heard et al., 1996). Both the long distance between radio-collared caribou and the den the week before the excursion and the increased time spent foraging by wolf 388 indicate that caribou availability near the den was low. Observations of the pups' being left alone for up to 18 h, presumably while adults were searching for food, provide additional support for low caribou availability locally. Mean foraging bout duration decreased by 10.0 h after the excursion, when collared caribou were closer to the den, suggesting an increase in caribou availability nearby. **29**

Foraging Excursion

One aspect of central place foraging theory (CPFT) **30** deals with the optimality of returning different-sized food loads from varying distances to dependents at a central place (i.e., the den) (Orians and Pearson, 1979). Carlson (1985) tested CPFT and found that the predator usually consumed prey captured far from the central place, while feeding prey captured nearby to dependants. Wolf 388 spent 7.2 days in one area near caribou before moving to a location 23 km back towards the den, where she spent an additional 3.1 days, likely hunting caribou. She began her return trip from this closer location, traveling directly to the den. While away, she may have made one or more successful kills and spent time meeting her own energetic needs before returning to the den. Alternatively, it may have taken several attempts to make a kill,

24 Post-excursion measurements of 388's movements were made to compare with those of the pre-excursion period. In order to compare, scientists often use *means*, or averages, of a series of measurements—mean distances, mean duration, etc.

25 In the comparison, authors used statistical calculations (F and df) to determine that the differences between pre- and post-excursion measurements were *statistically insignificant*, or close enough to be considered essentially the same or similar.

26 As with wolf 388, the investigators measured the movements of caribou during the study period. The areas within which the caribou moved are shown in Figure 2 by shaded polygons mentioned in the second paragraph of this subsection.

27 This subsection summarizes how distances separating predators and prey varied during the study period.

28 DISCUSSION: This section is the *interpretation* part of the scientific process.

29 This subsection reviews observations from other studies and suggests that this study fits with patterns of those observations.

30 Authors discuss a prevailing *theory* (CBFT) which might explain why a wolf would travel far to meet her own energy needs while taking food caught closer to the den back to her pups. The results of this study seem to fit that pattern.

31 Here our authors note other possible explanations for wolves' excursions presented by other investigators, but this study does not seem to support those ideas.

32 Authors discuss possible reasons for why 388 traveled directly to where caribou were located. They take what they learned from earlier studies and apply it to this case, suggesting that the lay of the land played a role. Note that their description paints a clear picture of the landscape.

33 Authors suggest that 388 may have learned in traveling during previous summers where the caribou were. The last two sentences suggest ideas for future studies.

34 Or maybe 388 followed the scent of the caribou. Authors acknowledge difficulties of proving this, but they suggest another area where future studies might be done.

35 Authors suggest that results of this study support previous studies about how fast wolves travel to and from the den. In the last sentence, they speculate on how these observed patterns would fit into the theory of evolution.

36 Authors also speculate on the fate of 388's pups while she was traveling. This leads to . . .

which she then fed on before beginning her return trip. We do not know if she returned food to the pups, but such behavior would be supported by CPFT.

31 Other workers have reported wolves' making long round trips and referred to them as "extraterritorial" or "pre-dispersal" forays (Fritts and Mech, 1981; Messier, 1985; Ballard et al., 1997; Merrill and Mech, 2000). These movements are most often made by young wolves (1–3 years old), in areas where annual territories are maintained and prey are relatively sedentary (Fritts and Mech, 1981; Messier, 1985). The long excursion of 388 differs in that tundra wolves do not maintain annual territories (Walton et al., 2001), and the main prey migrate over vast areas (Gunn et al., 2001).

Another difference between 388's excursion and those reported earlier is that she is a mature, breeding female. No study of territorial wolves has reported reproductive adults making extraterritorial movements in summer (Fritts and Mech, 1981; Messier, 1985; Ballard et al., 1997; Merrill and Mech, 2001). However, Walton et al. (2001) also report that breeding female tundra wolves made excursions.

Direction of Movement

32 Possible explanations for the relatively direct route 388 took to the caribou include landscape influence and experience. Considering the timing of 388's trip and the locations of caribou, had the wolf moved northwest, she might have missed the caribou entirely, or the encounter might have been delayed.

A reasonable possibility is that the land directed 388's route. The barrens are crisscrossed with trails worn into the tundra over centuries by hundreds of thousands of caribou and other animals (Kelsall, 1968; Thorpe et al., 2001). At river crossings, lakes, or narrow peninsulas, trails converge and funnel towards and away from caribou calving grounds and summer range. Wolves use trails for travel (Paquet et al., 1996; Mech and Boitani, 2003; P. Frame, pers. observation). Thus, the landscape may direct an animal's movements and lead it to where cues, such as the odor of caribou on the wind or scent marks of other wolves, may lead it to caribou.

33 Another possibility is that 388 knew where to find caribou in summer. Sexually immature tundra wolves sometimes follow caribou to calving grounds (D. Heard, unpubl. data). Possibly, 388 had made such journeys in previous years and killed caribou. If this were the case, then in times of local prey scarcity she might travel to areas where she had hunted successfully before. Continued monitoring of tundra wolves may answer questions about how their food needs are met in times of low caribou abundance near dens.

34 Caribou often form large groups while moving south to the tree line (Kelsall, 1968). After a large aggregation of caribou moves through an area, its scent can linger for weeks (Thorpe et al., 2001:104). It is conceivable that 388 detected caribou scent on the wind, which was blowing from the northeast on 19–21 July (Environment Canada, 2003), at the same time her excursion began. Many factors, such as odor strength and wind direction and strength, make systematic study of scent detection in wolves difficult under field conditions (Harrington and Asa, 2003). However, humans are able to smell odors such as forest fires or oil refineries more than 100 km away. The olfactory capabilities of dogs, which are similar to wolves, are thought to be 100 to 1 million times that of humans (Harrington and Asa, 2003). Therefore, it is reasonable to think that under the right wind conditions, the scent of many caribou traveling together could be detected by wolves from great distances, thus triggering a long foraging bout.

Rate of Travel

35 Mech (1994) reported the rate of travel of Arctic wolves on barren ground was 8.7 km/h during regular travel and 10.0 km/h when returning to the den, a difference of 1.3 km/h. These rates are based on direct observation and exclude periods when wolves moved slowly or not at all. Our calculated travel rates are assumed to include periods of slow movement or no movement. However, the pattern we report is similar to that reported by Mech (1994), in that homeward travel was faster than regular travel by 1.6 km/h. The faster rate on return may be explained by the need to return food to the den. Pup survival can increase with the number of adults in a pack available to deliver food to pups (Harrington et al., 1983). Therefore, an increased rate of travel on homeward trips could improve a wolf's reproductive fitness by getting food to pups more quickly.

Fate of 388's Pups

36 Wolf 388 was caring for pups during den observations. The pups were estimated to be six weeks old, and were seen ranging as far as 800 m from the den. They received some regurgitated food from two of the females, but were unattended for long periods. The excursion started 16 days after our observations, and it is improbable that the pups could have traveled the distance that 388 moved. If the pups died, this would have removed parental responsibility, allowing the long movement.

Our observations and the locations of radio-collared caribou indicate that prey became scarce in the area of the den as summer progressed. Wolf 388 may have abandoned her pups to seek food for herself. However, she returned to the den after the excursion, where she was seen near pups. In fact, she foraged in a similar pattern before and after the excursion, suggesting that she again was providing for pups after her return to the den.

(37) A more likely possibility is that one or both of the other lactating females cared for the pups during 388's absence. The three females at this den were not seen with the pups at the same time. However, two weeks earlier, at a different den, we observed three females cooperatively caring for a group of six pups. At that den, the three lactating females were observed providing food for each other and trading places while nursing pups. Such a situation at the den of 388 could have created conditions that allowed one or more of the lactating females to range far from the den for a period, returning to her parental duties afterwards. However, the pups would have been weaned by eight weeks of age (Packard et al., 1992), so nonlactating adults could also have cared for them, as often happens in wolf packs (Packard et al., 1992; Mech et al., 1999).

Cooperative rearing of multiple litters by a pack could create opportunities for long-distance foraging movements by some reproductive wolves during summer periods of local food scarcity. We have recorded multiple lactating females at one or more tundra wolf dens per year since 1997. This reproductive strategy may be an adaptation to temporally and **(38)** spatially unpredictable food resources. All of these possibilities require further study, but emphasize both the adaptability of wolves living on the barrens and their dependence on caribou.

(39) Long-range wolf movement in response to caribou availability has been suggested by other researchers (Kuyt, 1972; Walton et al., 2001) and traditional ecological knowledge (Thorpe et al., 2001). Our report demonstrates the rapid and extreme response of wolves to caribou distribution and movements in summer. Increased human activity on the tundra (mining, road building, pipelines, ecotourism) may influence caribou movement patterns and change the interactions between wolves and caribou in the region. Continued monitoring of both species will help us to assess whether the association is being affected adversely by anthropogenic change.

(40) Acknowledgements

This research was supported by the Department of Resources, Wildlife, and Economic Development, Government of the Northwest Territories; the Department of Biological Sciences at the University of Alberta; the Natural Sciences and Engineering Research Council of Canada; the Department of Indian and Northern Affairs Canada; the Canadian Circumpolar Institute; and DeBeers Canada, Ltd. Lorna Ruechel assisted with den observations. A. Gunn provided caribou location data. We thank Dave Mech for the use of GPS collars. M. Nelson, A. Gunn, and three anonymous reviewers made helpful comments on earlier drafts of the manuscript. This work was done under Wildlife Research Permit – WL002948 issued by the Government of the Northwest Territories, Department of Resources, Wildlife, and Economic Development.

(41) References

BALLARD, W.B., AYRES, L.A., KRAUSMAN, P.R., REED, D.J., and FANCY, S.G. 1997. Ecology of wolves in relation to a migratory caribou herd in northwest Alaska. Wildlife Monographs 135. 47 p.

BROWN, M.B., and FORSYTHE, A.B. 1974. Robust tests for the equality of variances. Journal of the American Statistical Association 69:364–367.

CARLSON, A. 1985. Central place foraging in the red-backed shrike (*Lanius collurio* L.): Allocation of prey between forager and sedentary consumer. Animal Behaviour 33:664–666.

CLARK, K.R.F. 1971. Food habits and behavior of the tundra wolf on central Baffin Island. Ph.D. Thesis, University of Toronto, Ontario, Canada.

ENVIRONMENT CANADA. 2003. National climate data information archive. Available online: http://www.climate.weatheroffice.ec.gc.ca/Welcome_e.html

FOURNIER, B., and GUNN, A. 1998. Musk ox numbers and distribution in the NWT, 1997. File Report No. 121. Yellowknife: Department of Resources, Wildlife, and Economic Development, Government of the Northwest Territories. 55 p.

FRITTS, S.H., and MECH, L.D. 1981. Dynamics, movements, and feeding ecology of a newly protected wolf population in northwestern Minnesota. Wildlife Monographs 80. 79 p.

GUNN, A., DRAGON, J., and BOULANGER, J. 2001. Seasonal movements of satellite-collared caribou from the Bathurst herd. Final Report to the West Kitikmeot Slave Study Society, Yellowknife, NWT. 80 p. Available online: http://www.wkss.nt.ca/HTML/08_ProjectsReports/PDF/Seasonal MovementsFinal.pdf

HARRINGTON, F.H., and ASA, C.S. 2003. Wolf communication. In: Mech, L.D., and Boitani, L., eds. Wolves: Behavior, ecology, and conservation. Chicago: University of Chicago Press. 66–103.

HARRINGTON, F.H., MECH, L.D., and FRITTS, S.H. 1983. Pack size and wolf pup survival: Their relationship under varying ecological conditions. Behavioral Ecology and Sociobiology 13:19–26.

HARRIS, S.A. 1986. Permafrost distribution, zonation and stability along the eastern ranges of the cordillera of North America. Arctic 39(1):29–38.

HEARD, D.C., and WILLIAMS, T.M. 1992. Distribution of wolf dens on migratory caribou ranges in the Northwest

37 Discussion of cooperative rearing of pups and, in turn, to speculation on how this study and what is known about cooperative rearing might fit into the animal's strategies for survival of the species. Again, the authors approach the broader theory of evolution and how it might explain some of their results.

38 And again, they suggest that this study points to several areas where further study will shed some light.

39 In conclusion, the authors suggest that their study supports the hypothesis being tested here. And they touch on the implications of increased human activity on the tundra predicted by their results.

40 ACKNOWLEDGEMENTS: Authors note the support of institutions, companies and individuals. They thank their reviewers ad list permits under which their research was carried on.

41 REFERENCES: List of all studies cited in the report. This may seem tedious, but is a vitally important part of scientific reporting. It is a record of the sources of information on which this study is based. It provides readers with a wealth of resources for further reading on this topic. Much of it will form the foundation of future scientific studies like this one.

Territories, Canada. Canadian Journal of Zoology 70:1504–1510.

HEARD, D.C., WILLIAMS, T.M., and MELTON, D.A. 1996. The relationship between food intake and predation risk in migratory caribou and implication to caribou and wolf population dynamics. Rangifer Special Issue No. 2:37–44.

KELSALL, J.P. 1968. The migratory barren-ground caribou of Canada. Canadian Wildlife Service Monograph Series 3. Ottawa: Queen's Printer. 340 p.

KUYT, E. 1962. Movements of young wolves in the Northwest Territories of Canada. Journal of Mammalogy 43:270–271.

———. 1972. Food habits and ecology of wolves on barren-ground caribou range in the Northwest Territories. Canadian Wildlife Service Report Series 21. Ottawa: Information Canada. 36 p.

MECH, L.D. 1994. Regular and homeward travel speeds of Arctic wolves. Journal of Mammalogy 75:741–742.

MECH, L.D., and BOITANI, L. 2003. Wolf social ecology. In: Mech, L.D., and Boitani, L., eds. Wolves: Behavior, ecology, and conservation. Chicago: University of Chicago Press. 1–34.

MECH, L.D., and GESE, E.M. 1992. Field testing the Wildlink capture collar on wolves. Wildlife Society Bulletin 20:249–256.

MECH, L.D., WOLFE, P., and PACKARD, J.M. 1999. Regurgitative food transfer among wild wolves. Canadian Journal of Zoology 77:1192–1195.

MERRILL, S.B., and MECH, L.D. 2000. Details of extensive movements by Minnesota wolves (Canis lupus). American Midland Naturalist 144:428–433.

MERRILL, S.B., ADAMS, L.G., NELSON, M.E., and MECH, L.D. 1998. Testing releasable GPS radiocollars on wolves and white-tailed deer. Wildlife Society Bulletin 26:830–835.

MESSIER, F. 1985. Solitary living and extraterritorial movements of wolves in relation to social status and prey abundance. Canadian Journal of Zoology 63:239–245.

MOHR, C.O., and STUMPF, W.A. 1966. Comparison of methods for calculating areas of animal activity. Journal of Wildlife Management 30:293–304.

MUSIANI, M. 2003. Conservation biology and management of wolves and wolf-human conflicts in western North America. Ph.D. Thesis, University of Calgary, Calgary, Alberta, Canada.

ORIANS, G.H., and PEARSON, N.E. 1979. On the theory of central place foraging. In: Mitchell, R.D., and Stairs, G.F., eds. Analysis of ecological systems. Columbus: Ohio State University Press. 154–177.

PACKARD, J.M., MECH, L.D., and REAM, R.R. 1992. Weaning in an arctic wolf pack: Behavioral mechanisms. Canadian Journal of Zoology 70:1269–1275.

PAQUET, P.C., WIERZCHOWSKI, J., and CALLAGHAN, C. 1996. Summary report on the effects of human activity on gray wolves in the Bow River Valley, Banff National Park, Alberta. In: Green, J., Pacas, C., Bayley, S., and Cornwell, L., eds. A cumulative effects assessment and futures outlook for the Banff Bow Valley. Prepared for the Banff Bow Valley Study. Ottawa: Department of Canadian Heritage.

STEPHENSON, R.O., and JAMES, D. 1982. Wolf movements and food habits in northwest Alaska. In: Harrington, F.H., and Paquet, P.C., eds. Wolves of the world. New Jersey: Noyes Publications. 223–237.

THORPE, N., EYEGETOK, S., HAKONGAK, N., and QITIRMIUT ELDERS. 2001. The Tuktu and Nogak Project: A caribou chronicle. Final Report to the West Kitikmeot/Slave Study Society, Ikaluktuuttiak, NWT. 160 p.

TIMONEY, K.P., LA ROI, G.H., ZOLTAI, S.C., and ROBINSON, A.L. 1992. The high subarctic forest-tundra of northwestern Canada: Position, width, and vegetation gradients in relation to climate. Arctic 45(1):1–9.

WALTON, L.R., CLUFF, H.D., PAQUET, P.C., and RAMSAY, M.A. 2001. Movement patterns of barren-ground wolves in the central Canadian Arctic. Journal of Mammalogy 82:867–876.

WILLIAMS, T.M. 1990. Summer diet and behavior of wolves denning on barren-ground caribou range in the Northwest Territories, Canada. M.Sc. Thesis, University of Alberta, Edmonton, Alberta, Canada.

ZAR, J.H. 1999. Biostatistical analysis. 4th ed. New Jersey: Prentice Hall. 663 p.

Glossary

ABC model Idea that products of three groups of master genes direct a flower's development.

ABO blood typing Method of characterizing an individual's blood based on proteins (A, B, or their absence) at surface of red blood cells.

abortion Expulsion of a pre-term embryo or fetus from the uterus.

abscisic acid ABA. Plant hormone; induces bud and seed dormancy; makes stomata close.

abscission Dropping of leaves, flowers, fruits, or other parts from plants.

acclimatization Long-lasting physiological and behavioral adaptation to a new habitat.

acid Any dissolved substance that donates H+ to other solutes or to water molecules.

acid rain Falling of rain (or snow) rich in acidic sulfur and nitrogen oxides.

acid–base balance State in which extracellular fluid is not too acidic or too basic, an outcome of controls over the concentrations of dissolved ions.

actin Cytoskeletal protein; the main component of thin filaments in a sarcomere.

action potential Brief reversal in the resting membrane potential of an excitable cell.

activation energy Minimum amount of energy required to start a reaction; enzyme action lowers this energy barrier.

activator Regulatory protein that enhances a cell activity (e.g., gene transcription).

active transport Pumping of a specific solute across a membrane against its concentration gradient, through a transport protein's interior. Requires energy input.

acute inflammation Nonspecific defense response to tissue injury in animals. Signs include localized redness, heat, swelling, pain.

adaptation Any aspect of form, function, behavior, or development that improves an individual's capability of surviving and reproducing in a given environment.

adaptive behavior Behavior that contributes to the individual's reproductive success.

adaptive immunity Capacity of B cell and T cell populations to tailor their defenses against previously unencountered pathogens.

adaptive radiation Burst of diversification from a single lineage; gives rise to new species adapted to specific environmental niches.

adaptive zone Some way of life available to species physically and behaviorally able to live it (e.g., "catching flying insects at night").

adenine Nitrogen-containing base component of nucleic acids; also a nucleic acid containing an adenine base. Base-pairs with thymine in DNA or uracil in RNA.

ADH Antidiuretic hormone. Hypothalamic hormone that promotes water conservation.

adhering junction Structure that forms a site of adhesion between cells. Found in tissues subject to stretching or abrasion.

adipose tissue Connective tissue that specializes in fat storage.

adrenal cortex Outer portion of adrenal gland; secretes cortisol and aldosterone.

adrenal medulla Inner portion of adrenal gland; secretes epinephrine, norepinephrine.

aerobic respiration Oxygen-dependent pathway of ATP formation in which glucose is broken down to carbon dioxide and water in several steps, including glycolysis, the Krebs cycle, and electron transfer phosphorylation. Typical net yield: 36 ATP.

age structure Of a population, distribution of individuals among different age categories.

alcoholic fermentation Anaerobic ATP-forming pathway. NADH transfers electrons to acetaldehyde, forming ethanol. Reactions start with pyruvate from glycolysis and regenerate NAD^+. Net yield: 2 ATP.

aldosterone Adrenal cortex hormone; promotes sodium reabsorption in the kidney.

alkylating agent A substance that replaces a hydrogen with an alkyl (saturated organic) group in a biological molecule.

allantois Extraembryonic membrane that exchanges gases and stores metabolic wastes of embryos of reptiles, birds, some mammals. In humans, it forms urinary bladder, placental blood vessels.

allele One of two or more molecular forms of a gene that arise by mutation and specify slightly different versions of the same trait.

allergen Normally harmless substance that can provoke inflammation, excess mucus secretion, and often immune responses in susceptible people.

allergy Hypersensitivity to an allergen.

allopatric speciation Divergences end in speciation after a physical barrier that arises between populations of a species stops gene flow between them.

altruism Behavior that lowers an individual's chance of reproductive success but also helps others of its species.

alveolus, plural **alveoli** In bronchioles, a cup-shaped sac where gases are exchanged between blood and interstitial fluid.

amino acid Organic compound with an amino group (NH_2), a carboxylic acid group (COOH), and a side group bonded covalently to the same carbon atom. Subunit of proteins.

ammonification Of nitrogen cycle, process by which soil fungi and bacteria break down nitrogenous wastes and remains to ammonia compounds that plants may absorb.

amnion An extraembryonic membrane; encloses a fluid-filled sac in which amniote embryos develop.

amniote A tetrapod that produces amniote eggs. A reptile, bird, or mammal.

amoeboid protozoan Protist that forms pseudopods; some "naked," others "shelled" (e.g., amoeba, foraminiferan, radiolarian).

amphibian Vertebrate with four legs (or four-legged aquatic ancestor); its body plan and reproductive mode are somewhere between fishes and reptiles.

anagenesis Speciation pattern; changes occur within an unbranched line of descent.

analogous structures Similar body parts in distantly related lineages that arise as a result of similar environmental pressures.

anaphase Of mitosis, stage when sister chromatids of each chromosome move to opposite spindle poles. During anaphase I (meiosis), each duplicated chromosome and its homologue move to opposite poles. During anaphase II, sister chromatids of each chromosome move to opposite poles.

anemia Disorder resulting from deformed or insufficient quantity of red blood cells.

aneuploidy In cells, too many or too few chromosomes relative to the parental number.

angiosperm Flowering plant.

animal Multicelled, motile heterotroph that has embryonic stages and usually tissues, organs, and organ systems.

annelid Bilateral, highly segmented worm with well-developed organ systems (e.g., oligochaete, polychaete, leech).

anther Part of a stamen; pollen forms in it and is dispersed from it.

anthocyanin Red to blue photosynthetic accessory pigment.

antibiotic A natural or synthetic chemical agent that kills or inhibits growth of micro-organisms, especially bacteria.

antibody Antigen-binding receptor made and secreted by B cells.

anticodon Series of three nucleotide bases in tRNA; can base-pair with an mRNA codon.

antigen A substance chemically recognized as nonself; triggers an immune response.

antigen-presenting cell Cell that can break down and display antigen attached to self markers on its surface (e.g., macrophage, dendritic cell, or B cell).

antioxidant Enzyme or cofactor that can help neutralize free radicals, which may otherwise damage DNA and other molecules of life.

aorta Main artery of systemic circulation.

apicomplexan One of a group of parasitic protists that has a unique device used in penetrating host cells.

apoptosis Programmed cell death.

appendix Narrow projection from the cecum.

archaea Evolutionarily distinct domain of prokaryotic organisms.

Archean Eon in which life arose (3.9–2.5 billion years ago).

archipelago An island chain some distance away from a continent.

area effect Idea that larger islands support more species than smaller ones at similar distances from sources of colonizer species.

arteriole Type of blood vessel between arteries and capillaries where control mechanisms govern the distribution of blood flow through the body.

artery Thick-walled, muscular, transport vessel; carries blood away from heart.

arthropod Invertebrate with exoskeleton, jointed appendages (e.g., crustacean, insect).

artificial selection Selection of traits among a population under contrived conditions.

asexual reproduction Any reproductive mode by which offspring arise from and inherit genes from just one parent.

atom Fundamental form of matter that has mass and takes up space, and cannot be broken apart by everyday means.

atomic number The number of protons in the nucleus of an atom; identifies an element.

ATP Adenosine triphosphate. Nucleotide made of adenine, ribose, and three phosphate groups; main energy carrier in cells.

ATP/ADP cycle Alternating formation of ATP and ADP through phosphate group transfers.

ATP synthase Membrane-bound active transport protein that acts as an enzyme of ATP formation.

australopith Type of early hominid; extinct.

autoimmune response An immune response of the body against its own normal cells.

autoimmunity Failure of self-recognition in which the immune system attacks normal body cells.

automated DNA sequencing Robotic method of determining the nucleotide sequence of a region of DNA. Uses gel electrophoresis and laser detection of fluorescent tracers.

autonomic nerve One of the nerves that carry signals to and from internal organs.

autonomic nervous system All nerves from central nervous system to smooth muscle, cardiac muscle, and glands of viscera.

autosome Any chromosome of a type that is the same in males and females of a species.

autotroph An organism that makes its own food using an environmental energy source and carbon from carbon dioxide.

auxin Plant hormone of apical meristems; induces stem lengthening and responses to gravity and light.

axon Neuron's signal-conducting zone.

B cell See B lymphocyte.

B lymphocyte White blood cell that secretes antibody molecules. Also called plasma cell.

bacteria The most widespread and diverse group of prokaryotic organisms.

bacterial conjugation Transfer of plasmid DNA from one prokaryotic cell to another.

bacteriophage A virus that infects bacteria.

balanced polymorphism Outcome of selection; two or more alleles for a trait are being maintained in a population over time.

bark All tissues external to the vascular cambium in older woody stems.

basal body An organelle that gives rise to cilia or flagella; resembles a centriole.

base Any substance that accepts hydrogen ions (H^+) when dissolved in water, thus forming hydroxyl ions (OH^-). Also, the nitrogen-containing component of a nucleic acid.

base-pair substitution Mutation in which one nucleotide is incorrectly substituted for another during DNA replication.

basophil Fast-acting white blood cell that secretes histamine during inflammation.

big bang Model for origin of universe in which all matter, energy, and time originated from a single point in a gigantic explosion.

bilateral symmetry Body plan with main axis from anterior to posterior, separated into right and left sides, and dorsal and ventral surfaces.

bile Liver secretion required for fat digestion.

biogeochemical cycle Movement of element from environmental reservoirs, through living things, then back to the environment.

biogeographic realm One of six vast land areas with distinctive species.

biogeography Study of the patterns of distribution of species.

biological clock Internal time-measuring mechanism; adjusts activities seasonally, daily, or both in response to environmental cues.

biological magnification Some substance becomes more concentrated in body tissues as it moves up through food chains.

biological species concept The definition of a species is based on its reproductive isolation.

biology The scientific study of life.

bioluminescence Production of fluorescent light by a living organism.

biomass Combined weight of all organisms at a given trophic level in an ecosystem.

biomass pyramid Diagram showing the dry weight of all organisms at each trophic level of an ecosystem.

biome A large land region characterized by dominant plant species and habitat conditions.

biosphere All regions of the Earth's waters, crust, and atmosphere where organisms live.

biotic potential The maximum rate of increase of a population under ideal conditions.

bipedalism Routinely walking upright.

bipolar mitotic spindle Dynamic array of microtubules that moves chromosomes in precise directions during mitosis or meiosis.

bird Only animal (besides some extinct related dinosaurs) that grows feathers.

blastocyst Early mammal development stage; two layers of cells surround a fluid-filled cavity.

blastomere One of the small, nucleated cells that form during cleavage of animal zygote.

blastula Early outcome of cleavage; one layer of blastomeres encloses a fluid-filled cavity.

blood Liquid connective tissue that has fluid and cellular components.

blood pressure Fluid pressure, generated by heart contractions, that circulates blood.

blood–brain barrier Mechanism that controls which solutes enter cerebrospinal fluid.

bone remodeling Ongoing mineral deposits and withdrawals from bone.

bone tissue Calcium-hardened connective tissue that composes the vertebrate skeleton.

bottleneck A sudden, drastic reduction in population size.

Bowman's capsule Part of a nephron; receives water and solutes being filtered from blood.

brain An integrating center that receives and processes sensory input and issues commands for responses by muscles and glands.

brain stem Most ancient nerve tissue in the vertebrate hindbrain, midbrain, and forebrain.

bronchiole One of the finely branched airways in the lung.

bronchus, plural **bronchi** Airway that branches from trachea and leads into lungs.

brown alga One of the stramenopiles; mostly marine, photosynthetic (e.g., kelp).

bryophyte A nonvascular land plant (e.g., moss, liverwort, hornwort).

bud Undeveloped shoot, consisting mostly of meristematic tissue.

buffer system A weak acid and the base that forms when it dissolves in water. The two work as a pair to counter slight shifts in pH.

bulk Of the vertebrate gut, the volume of undigested material in the small intestine that cannot be decreased by absorption.

bulk flow Mass movement of one or more substances in the same direction, in response to pressure, gravity, or another external force.

C3 plant Plant that makes three-carbon PGA in the first step of carbon fixation.

C4 plant Plant that makes four-carbon oxaloacetate in the first step of carbon fixation.

calcium pump Membrane-bound active transport protein specific for calcium ions.

Calvin–Benson cycle Light-independent cyclic reactions of photosynthesis. Forms sugars from CO_2 using ATP and NADPH.

CAM plant Plant that conserves water by opening stomata only at night, when it fixes carbon by repeated turns of the C4 pathway.

camouflage Coloration, patterning, or other aspects of form or behavior that make an individual blend with its surroundings.

cancer Malignant neoplasm; mass of cells that divide abnormally and can spread in the body.

capillary Smallest of the blood vessels; site of diffusion between blood and interstitial fluid.

capillary reabsorption Movement of fluid from interstitial fluid into capillaries.

capture–recapture method For population counts; collecting, marking, releasing, then recapturing representative animals.

carbon cycle Movement of carbon from the atmosphere, through food webs and ocean's waters and rocks, and back to atmosphere.

carbon fixation Autotrophic cell secures carbon atoms from the air and incorporates them into a stable organic compound.

cardiac cycle Sequence of muscle contraction and relaxation in a single heartbeat.

cardiac muscle Muscle of the heart wall.

cardiac pacemaker Sinoatrial (SA) node; a mass of self-excitatory cardiac muscle cells that set the rate for a normal heartbeat.

carotenoid Red to yellow accessory pigment.

carpel Female reproductive part of a flower.

carrying capacity The maximum number of individuals in a population (or species) that a given environment can sustain indefinitely.

cartilage A specialized connective tissue that is dense, pliable, yet resists compression.

cartilaginous fish Jawed fish with a skeleton of cartilage (e.g., shark, ray).

Casparian strip In roots; waxy, impermeable band in cell walls of endodermis or exodermis where water and mineral uptake is controlled.

catastrophism Idea that abrupt changes in the geologic or fossil record resulted from large scale disasters that were divinely invoked.

CD4 lymphocyte White blood cell having a receptor to which HIV may bind (e.g., T-cell).

cDNA DNA made from an mRNA transcript, using reverse transcriptase.

cell Smallest living unit, with a capacity to survive and reproduce on its own.

cell cortex Three-dimensional mesh of actin filaments and other proteins just under the plasma membrane.

cell count The number of cells of a given type in one microliter of blood.

cell cycle Series of events from one cell division to the next. Interphase, mitosis, and cytoplasmic division constitute one cycle.

cell differentiation Cell lineages become specialized in structure and function by selectively activating genes.

cell junction A site where adjoining cells interact physically, chemically, or both.

cell plate In a dividing plant cell, a disk-like structure that becomes a crosswall with new plasma membrane on both sides.

cell plate formation Mechanism of cytoplasmic division in plant cells.

cell theory Idea that all organisms consist of similar units of organization called cells.

cell wall A semirigid, permeable structure encloses the plasma membrane of many cells; helps cell retain its shape and resist rupturing.

central vacuole Fluid-filled storage organelle of a plant cell.

centriole Organelle that organizes formation and direction of cilia, flagella, and spindles.

centromere Constricted part of a chromosome to which spindle microtubules attach.

centrosome Cell region where microtubules are produced.

cephalization Evolutionary trend toward the concentration of sensory structures and nerve cells in a head.

cerebellum Hindbrain region; maintains balance and coordinates limb movements.

cerebrum Forebrain region that deals with olfactory input and motor responses. In mammals, offers complex level of integration.

charophyte Type of green alga; the closest living relatives of plants.

chemical bond A union between the electron structures of two or more atoms or ions.

chemical equilibrium The state at which the concentrations of reactants and products in a reversible chemical reaction remain constant.

chemical synapse Thin cleft between a presynaptic neuron and a postsynaptic cell. Neurotransmitter molecules diffuse across it.

chemoreceptor A sensory receptor which responds to chemical stimuli.

chemotaxin In animals, a chemical signal that attracts phagocytic white blood cells.

chlamydias A group of bacteria that are obligate intracellular parasites of vertebrates.

chlorophyll Primary photosynthetic pigment.

chlorophyte A member of the most diverse group of green algae.

chloroplast Organelle of photosynthesis in plants and many protists.

chordate Animal having a notochord, dorsal hollow nerve cord, pharynx, and gill slits in pharynx wall during at least part of life cycle.

chorion Extraembryonic membrane. In some mammals, becomes part of placenta.

chromatin All DNA molecules and associated proteins in a nucleus.

chromosome A coherent structure consisting of a DNA molecule and associated proteins.

chromosome number Sum of all chromosomes in a given type of cell.

chrysophyte A category of photosynthetic protists (e.g., golden algae).

ciliate Ciliated protozoan; type of protist that usually has profuse cilia at its surface.

cilium, plural cilia In some eukaryotic cells, a short motile structure or sensory structure.

circadian rhythm Biological activity repeated in cycles, each about twenty-four hours long.

circulatory system Organ system that moves substances to and from cells; can help stabilize body temperature and pH. In many animals, consists of a heart, blood vessels, and blood.

cladogenesis Speciation pattern; a lineage splits, and populations diverge genetically.

cladogram Evolutionary tree diagram; groups taxa on the basis of their shared derived traits.

classification system A way of organizing and retrieving information about species.

cleavage Early stage of animal development. Mitotic cell divisions divide a fertilized egg into many smaller, nucleated cells; original volume of egg cytoplasm does not increase.

climate Prevailing weather conditions in some region.

climax community Array of species that can, if habitat remains stable, persist indefinitely without being replaced by other species.

climax-pattern model Idea that more than one stable community can persist in the same region when environmental factors vary.

cloning vector DNA molecule that can accept foreign DNA and replicate in a host cell.

club fungus Fungus having club-shaped cells that produce and bear basidiospores.

cnidarian Radial invertebrate at the tissue level of organization; makes nematocysts.

coal Nonrenewable energy source formed over 280 million years ago from submerged, compacted plant remains.

codon Sequence of three bases in an mRNA strand that codes for an amino acid, or acts as a start or stop signal for translation.

coelom In most animals, a cavity between the gut and body wall that is lined with a tissue.

coenzyme Small molecule that participates in an enzymatic reaction, and is reversibly modified during the reaction (e.g., a vitamin).

coevolution Joint evolution of two interacting species, brought about by changes in selection pressures operating between the two.

cohesion The capacity to resist rupturing when placed under tension (stretched).

cohesion–tension theory Idea that evaporation from plants exerts tension in xylem, pulling cohesive columns of water upward from roots to leaves.

cohort A group of same-aged individuals, tracked throughout their life spans.

collecting duct Last tubular region of a kidney nephron.

collenchyma Simple plant tissue that imparts flexible support during primary growth.

colon Large intestine; concentrates undigested and unabsorbed material.

commensalism Interaction between species that benefits one and has no effect on other.

communication signal Information-laden cue directed by one member of a species to another; chemical, visual, acoustic, or tactile.

community All species living and interacting in some habitat.

companion cell Specialized parenchyma cell; helps load sugars into sieve tubes of phloem.

comparative morphology Scientific study of the body form and structures of major groups.

compartmentalization Defensive response by a woody plant; resin or toxins are secreted in response to injury or an attack.

competitive exclusion Idea that, when two species require exactly the same resources, competition will drive one or the other to extinction in a shared habitat.

complement system Plasma proteins that circulate in blood in inactive form; has roles in nonspecific defenses and immune responses.

compound Molecule consisting of two or more elements in unvarying proportions.

concentration gradient A difference in the number of molecules (or ions) of a substance between two adjoining regions.

condensation reaction Covalent bonding of two molecules into a larger molecule, often with the formation of water as a by-product.

conduction In temperature studies, exchange of heat between two touching objects owing to a thermal gradient between them.

cone Reproductive structure made of modified leaves (e.g., a cluster of ovule-bearing, woody scales of a pine tree).

cone cell Vertebrate photoreceptor; allows sharp daytime vision, color vision.

conifer Evergreen gymnosperm; tree or shrub with needlelike or scalelike leaves.

connective tissue Most abundant, pervasive animal tissue. Specialized types are cartilage, bone tissue, adipose tissue, and blood.

conservation biology Field devoted to surveying biological diversity, studying its origins, and attempting to maintain it.

consumer Heterotroph that obtains carbon and energy by feeding on other organisms.

continuous variation Of a population, a more or less continuous range of small differences in a given trait among its individuals.

control group A group used as a standard for comparison with an experimental group.

convection Movement of air or water next to an object aids conductive heat loss from it.

cork cambium A lateral meristem; replaces epidermis with cork on woody plant parts.

corpus luteum A glandular structure that forms from cells of a ruptured ovarian follicle; secretes progesterone and estrogen.

cortex A rindlike layer. In vascular plants, a ground tissue; supports parts and stores food.

cortisol Adrenal cortex hormone; roles in glucose regulation and stress responses.

cotyledon Seed leaf. Part of a plant embryo. Two form in eudicot seeds, one forms in monocot seeds.

countercurrent flow Movement of two fluids in opposite directions, as when water and blood flow in opposite directions in a fish gill.

covalent bond Sharing of one or more electrons between two atoms.

craniate Chordate having a brain inside a cranium. Includes living fishes, amphibians, reptiles, birds, and mammals.

creatine phosphate Energy source for muscle cells; transfers phosphate to ADP to form ATP.

crossing over At prophase I of meiosis, the reciprocal exchange of segments between two nonsister chromatids of a pair of homologous chromosomes; results in novel combinations of alleles.

culture Sum of behavior patterns of a social group, passed between generations by way of learning and symbolic behavior.

cuticle Body cover. Of plant surfaces, a layer of waxes and cutin. Of roundworms and annelids, a thin, flexible coat. Of arthropods, a lightweight exoskeleton hardened with protein and chitin.

cyanobacteria Bacteria that carry out noncyclic (oxygen-producing) photosynthesis.

cycad A gymnosperm that forms pollen- and seed-bearing structures on different plants; an ancient lineage that coexisted with dinosaurs.

cyst Resting structure; encloses certain small organisms or spores for part of the life cycle.

cytokine Signaling molecules of the immune system (e.g., interferons, interleukins).

cytokinin Plant hormone; stimulates cell division and leaf expansion, slows leaf aging.

cytoplasm All cell parts, particles, and semifluid substances between the plasma membrane and the nucleus (or nucleoid).

cytoplasmic localization Polar distribution of gene products in a cell. In an egg, determines body axis of the future embryo.

cytoskeleton Interconnected system of protein filaments that structurally supports, organizes, and moves a eukaryotic cell and its internal structures.

cytotoxic T cell T lymphocyte that touch-kills virus infected, cancerous, and other altered body cells with antigen bound to self markers.

day-neutral plant Plant that flowers when mature, rather than in response to daylength.

decomposer Fungal or bacterial heterotroph that obtains carbon and energy from remains, products, or wastes of organisms.

deforestation Removal of all trees from a large area.

deletion Loss of a chromosome segment. Also, a mutation involving the loss of one or more bases of a DNA molecule.

demographic transition model Explanation of the effects of industrialization on human population growth.

demographics A population's vital statistics (e.g., size, distribution, density, age structure).

denaturation The three-dimensional shape of a protein or some other complex molecule unravels as its hydrogen bonds are disrupted.

dendrite Short, slender extension from cell body of a neuron; a signal input zone.

dendritic cell Type of antigen-presenting white blood cell.

denitrification Conversion of nitrate or nitrite by certain soil bacteria to gaseous nitrogen (N_2) and nitrous oxide (N_2O).

dense, irregular connective tissue Animal tissue with fibroblasts, many asymmetrically positioned fibers in a ground substance. In skin, some capsules.

dense, regular connective tissue Animal tissue with rows of fibroblasts between bundles of fibers. In tendons, ligaments.

density-dependent control Factor that operates at high population density to slow birth rate and/or increase death rate (e.g., disease, competition).

density-independent factor Factor that impacts a population's birth rate or death rate regardless of density (e.g., fire or flood).

deoxyribonucleic acid See DNA.

derived trait A novel feature that arose in a species and is shared only by its descendants.

dermis Skin layer beneath the epidermis; consists mainly of dense connective tissue.

desalinization Removal of salt from seawater.

desertification Conversion of grassland or cropland to a desertlike condition.

detritivore Heterotroph that feeds on bits of decaying organic matter (e.g., an earthworm).

deuterostome Bilateral animal in which the first indentation in the embryo becomes the anus (e.g., echinoderms, chordates).

development Series of stages by which structurally and functionally distinct body parts emerge, in orderly patterns, in a new multicelled individual.

diaphragm Muscular partition between thoracic and abdominal cavities. Also, a contraceptive device inserted into the vagina to prevent sperm from entering uterus.

dicot Dicotyledon. Flowering plant having embryos with two cotyledons.

diffusion Net movement of like molecules or ions down their concentration gradient.

digestive system Body sac or tube, often with regions where food is ingested, digested, and absorbed, and where residues are eliminated.

dihybrid cross Intercross between two individuals, each heterozygous for two genes (e.g., *AaBb*).

dinoflagellate Single-celled, flagellated, cellulose-plated protist; most photosynthetic.

dinosaur One of a group of extinct reptiles that originated in the Triassic and dominated land environments for 140 million years.

diploid Presence of two of each type of chromosome (i.e., pairs of homologs) in a cell nucleus at interphase.

diploid number Total chromosome number in cells that have a pair of each type of chromosome characteristic of the species.

directional selection Mode of natural selection by which forms of a trait at one end of a range of phenotypic variation are favored and all others are selected against.

disaccharide A common oligosaccharide; two covalently bonded sugar monomers.

disease Illness caused by an infectious, dietary, or environmental factor.

disruptive selection Mode of natural selection by which forms at both ends of the range of phenotypic variation are favored and intermediate forms are selected against.

distal tubule Tubular part of nephron where water and sodium are selectively reabsorbed.

distance effect Idea that the farther away an island is from a source of potential colonizing species, the lower its species diversity will be.

DNA Deoxyribonucleic acid. Carries the primary hereditary information for all living organisms and many viruses.

DNA chip Microarray of DNA spots on a glass plate; used to study gene expression.

DNA fingerprint Unique cleavage pattern of an individual's DNA; also called restriction fragment length polymorphism (RFLP).

DNA ligase Repair enzyme that joins breaks in DNA molecule; connects new DNA fragments during replication.

DNA polymerase Enzyme that catalyzes replication and repair of DNA.

DNA replication Process by which a cell duplicates its DNA molecules before dividing.

dormancy A time of metabolic inactivity in spores, cysts, seeds, plants, and some animals.

dosage compensation Control mechanism that balances gene expression between the sexes, starting at early stages of development.

double fertilization Of flowering plants only, fusion of sperm and egg nuclei, plus fusion of another sperm nucleus with nuclei of a cell that gives rise to endosperm.

doubling time Length of time it takes a population to double in size.

downwelling Downward movement of water along a coast, from ocean surface to its depths.

duplication DNA sequence repeated several to many hundreds or thousands of times.

ecdysone Hormone of many insect life cycles; roles in metamorphosis, molting.

echinoderm Invertebrate with calcified spines and plates in the body wall. Radial with some bilateral features (e.g., sea stars, sea urchins).

ecology Study of how organisms interact with one another and with their environment.

ecoregion Broad land or ocean region defined by climate, geography, and producer species.

ecosystem An array of species and their physical environment.

ectoderm First-formed, outermost primary tissue layer of animal embryos; gives rise to nervous system tissues and integument's outer layer.

ectotherm Animal that maintains core temperature by absorbing environmental heat.

Ediacarian One of a group of soft-bodied animals that arose in the precambrian.

effector Muscle (or gland); helps bring about movement (or chemical change) in response to neural or endocrine signals.

effector cell Differentiated lymphocyte that, during immune responses, engages and destroys antigen-bearing agents.

egg Female gamete.

El Niño A recurring warm current that displaces cool, nutrient-rich water along South America's coast; has global effects.

electric gradient A difference in electric charge between adjoining regions.

electron Negatively charged unit of matter, with particle-like and wavelike properties; occupies an orbital around atomic nucleus.

electron transfer chain Array of membrane-bound enzymes and other molecules that accept and give up electrons in sequence; allows the release and capture of energy in small, useful increments.

electron transfer phosphorylation Last stage of aerobic respiration; electrons flow through mitochondrial electron transfer chains, to O_2. The flow sets up an electrochemical gradient that drives ATP formation.

element Material consisting of atoms all with the same atomic number.

embryonic induction Signaling molecules released from one embryonic tissue affect the development of an adjacent tissue.

emigration Individuals leave a population.

emulsification In the vertebrate gut, the coating of fat droplets with bile salts, so that the droplets become suspended in chyme.

endangered species An endemic (native) species that is highly vulnerable to extinction.

endocrine gland One of the ductless glands that secrete hormones into some body fluid.

endocrine system Integrative system of cells, tissues, and organs, functionally linked to the nervous system, that exerts control by way of its hormones and other chemical secretions.

endocytosis Cellular uptake of a substance; plasma membrane forms a vesicle around it.

endoderm Inner primary tissue layer of most animal embryos; source of inner gut lining and organs that are derived from it.

endometrium Inner lining of uterus.

endophytic fungi Fungi that live inside the bodies of plants.

endoplasmic reticulum ER. Organelle that starts at the nuclear envelope and extends through cytoplasm. Smooth ER assembles membrane lipids, breaks down fatty acids, and inactivates some toxins; Rough ER (has ribosomes on its cytoplasmic side) modifies new polypeptide chains.

endoskeleton Of chordates, the internal framework of cartilage and bone; works with skeletal muscle to support and move the body.

endosperm Triploid nutritive tissue in flowering plant seeds.

endospore Resistant resting structure that forms in some bacteria. It encloses the bacterial chromosome and some cytoplasm.

endosymbiosis One species spends its entire life inside another, in an interaction that benefits both.

endotherm Animal that can generate and maintain its body temperature using metabolic rate, body form and behavior.

energy Capacity to do work.

energy pyramid Diagram of an ecosystem's trophic structure; shows usable energy at each trophic level.

enhancer A small sequence in DNA that binds transcription regulating molecules.

ENSO El Niño Southern Oscillation. The eastward movement of warm surface waters of the western equatorial Pacific, displaces cold water off South America; has widespread climatic effects.

enzyme A type of protein (or, rarely, RNA) that accelerates a chemical reaction.

eosinophil White blood cell that acts against extracellular parasites, such as worms.

epidermis Outermost tissue layer; occurs in plants and all animals above the sponge level of organization.

epiglottis Flaplike structure between pharynx and larynx; its controlled positional changes direct air into trachea or food into esophagus.

epithelial tissue Animal tissue that covers external surfaces and lines internal cavities and tubes. One surface is free and the other rests on a basement membrane.

equilibrium model of island biogeography Model that predicts the number of species an island will support, based on island size and its distance from a source of colonists.

esophagus Muscular tube; in vertebrates it connects the pharynx and stomach.

essential amino acid Amino acid an animal cannot synthesize and must get from food.

essential fatty acid Fatty acid an animal cannot synthesize and must get from food.

estrogen Female sex hormone secreted by ovaries; helps oocytes mature, primes uterine lining for pregnancy, maintains secondary sexual traits, affects growth and development.

estuary Partly enclosed coast region where seawater mixes with fresh water and runoff from land, as from rivers.

ethylene Gaseous plant hormone; promotes fruit ripening and abscission of leaves, flowers, fruits.

eudicot True dicot; one of the flowering plants generally characterized by embryos with two cotyledons.

euglenoid Single-celled, flagellated protist of freshwater habitats; most are photosynthetic.

Eukarya Domain of eukaryotic cells; all protists, plants, fungi, and animals.

eutherian Placental mammal.

eutrophication Nutrient enrichment of a body of water (e.g., a lake or pond).

evaporation Process of conversion of a liquid to a gas; requires energy input.

evaporative heat loss Thermoregulation mechanism: cooling due to evaporation of water.

evolution, biological Genetic change in a line of descent. Outcome of microevolutionary events: gene mutation, natural selection, genetic drift, and gene flow.

evolutionary tree Diagram of evolutionary relationships; each branch signifies a separate line of descent from a common ancestor; each branch point a time of divergence.

exocrine gland Any gland that secretes products (e.g., milk) to a free epithelial surface, usually through ducts or tubes.

exocytosis Release of a vesicle's contents outside the cell surface when it fuses with and becomes part of the plasma membrane.

exodermis Cylindrical sheet of cells inside the root epidermis of most flowering plants; helps control uptake of water and solutes.

exon One of the base sequences of an mRNA transcript that eventually will be translated.

exoskeleton External skeleton (e.g., hardened arthropod cuticle).

exotic species A species that has left the community in which it evolved and become established elsewhere.

experimental group A group upon which an experiment is performed, and compared with a control group.

exponential growth, population An increase in population size by a fixed percentage of the whole in a given interval.

extinction Irrevocable loss of a species.

extracellular fluid Of most animals, all fluid not in cells; plasma (blood's liquid portion) plus interstitial fluid.

extreme halophile An archaean or other species of a notably salty habitat.

extreme thermophile An archaean or other species of a notably high-temperature habitat.

eye Sensory organ that incorporates a dense array of photoreceptors.

FAD Flavin adenine dinucleotide. Nucleotide coenzyme; transfers electrons and unbound protons (H^+) between reaction sites.

fat Type of lipid with a glycerol head attached to one, two, or three fatty acid tails.

fatty acid Organic compound with a backbone of up to 36 carbon atoms, and a carboxyl group at the end.

feedback inhibition Of cells, an activity causes a change in cellular conditions, and that change in turn causes the activity to slow down or stop.

fern A seedless vascular plant with fronds that often are divided into leaflets.

fertilization Fusion of a sperm nucleus and an egg nucleus, the result being a zygote.

fetus Stage of animal development; in humans, the start of the ninth week to birth.

fibrous root system The lateral branchings of adventitious roots that form on a young stem.

filter feeder Animal that filters food from a current of water directed through a body part (e.g., through a sea squirt's pharynx).

first law of thermodynamics The total amount of energy in the universe is constant; energy can be converted from one form to another, but cannot be created or destroyed.

fitness Degree of adaptation to an environment, as measured by relative genetic contribution to future generations.

fixation Only one kind of allele remains at a given locus in a population; all individuals have become homozygous for it. One outcome of genetic drift in small populations.

fixed action pattern Program of coordinated, stereotyped muscle activity that is completed independently of feedback from environment.

flagellated protozoan One of the most ancient heterotrophic protists; has one or more flagella.

flagellum, plural flagella A whip-like motile structure of many free-living eukaryotic cells.

flower Reproductive shoot of an angiosperm.

fluid mosaic model A cell membrane is fluid because of the motions and interactions of its component lipids and proteins.

food chain A linear flow of energy captured by primary producers (autotrophs) into ever higher trophic levels of an ecosystem.

food pyramid Charts showing proportions of different foods that compose a healthy diet.

food web Cross-connecting food chains.

fossil Recognizable, physical evidence of an organism that lived in the distant past.

fossilization Extremely slow transformation of an organism's remains to stony hardness as a result of pressure and chemical changes.

founder effect A form of bottlenecking. By chance, allele frequencies of a few individuals that establish a new population differ from the frequencies in the original population.

free radical Highly reactive molecule with at least one unpaired electron.

FSH Follicle-stimulating hormone. Produced and secreted by the anterior lobe of pituitary gland; has reproductive roles in both sexes.

functional group An atom or a group of atoms with characteristic properties that is covalently bonded to an organic compound's carbon backbone.

fungus, plural fungi Eukaryotic heterotroph that obtains nutrients by extracellular digestion and fabsorption; notable for prolific spore formation.

gametophyte Multicelled, gamete-producing body (haploid) that forms during life cycles of some algae and all plants.

ganglion plural ganglia Distinct cluster of cell bodies of neurons.

gap junction A channel through a complex of abutting membrane proteins of two cells; permits ions and small molecules to move rapidly between the cells.

gastric fluid Acidic mix of secretions (e.g., HCl, mucus) from the stomach's glandular epithelium.

gastrula Earliest developmental stage in which cells are arranged as two or three primary tissue layers.

gel electrophoresis Molecules that migrate through a gel matrix in response to an electric force become separated by size and charge.

gene Unit of information for a heritable trait in DNA, passed from parents to offspring.

gene control A molecular mechanism that governs if, when, or how a specific gene is used (transcribed or translated).

gene flow Physical flow of alleles into or out of a population by immigration or emigration.

gene library Mixed collection of host cells that contain cloned DNA fragments representing all or most of a genome.

gene locus A gene's chromosomal location.

gene mutation Small-scale change in the nucleotide sequence of a gene that can result in an altered protein product.

gene pool All the genes in a population.

genetic code The correspondence between triplets of nucleotides in DNA (then mRNA) and specific sequences of amino acids in a polypeptide chain; the basic language of protein synthesis in cells.

genetic disorder Heritable defect in one's genetic material; causes mild to severe medical problems.

genetic divergence Accumulation of differences in the gene pools of populations after something stops gene flow between them. Over time, if some of the differences promote reproductive isolation, speciation may follow.

genetic drift Random change in allele frequencies over time brought about by chance alone; its effect is greatest in small populations.

genetic engineering Manipulation of an organism's DNA, usually with the intent of altering at least one aspect of phenotype.

genome All the DNA in a haploid number of chromosomes for a given species.

genomics The study of genes and gene function in humans and other organisms.

genus A group of related species.

geographic dispersal Individuals move away from their home range and successfully establish themselves elsewhere.

geologic time scale Chronology of Earth history; major subdivisions correspond to mass extinctions.

germination Of seeds and spores, resumption of growth after dormancy, dispersal from the parent organism, or both.

gibberellin Plant hormone; stimulates stem lengthening, helps seeds and buds break dormancy, promotes flowering in some plants.

gill slit Opening in a thin-walled pharynx. Serves in food-trapping and respiration; jaws evolved from certain gill slit supports.

gills Type of respiratory organ that occurs in some invertebrates and most fishes.

ginkgo Deciduous gymnosperm of an ancient lineage that produces fleshy plum-sized fruits.

gland Secretory organ derived from epithelial tissue (e.g., sweat gland, thyroid gland).

global broiling hypothesis Idea that asteroid impact caused the K–T mass extinction by creating a colossal fireball, causing the global temperature to rise thousands of degrees.

global warming Long-term rise in the temperature of the Earth's lower atmosphere.

glomerular filtration Movement of water and small solutes out of glomerular capillaries and into the first part of the kidney tubule.

glycolysis Breakdown of glucose to two pyruvate molecules. First stage of aerobic respiration and fermentation.

gnetophyte Tropical tree, vine, or shrub of an ancient gymnosperm group.

GnRH Gonadotropin-releasing hormone; triggers hypothalamus to release LH and FSH.

Golgi body Organelle of endomembrane system; final modification of polypeptide chains into proteins, lipid assembly, and packaging of both in vesicles for secretion or for use inside cell.

Gondwana Paleozoic supercontinent that collided with other large land masses and formed Pangea.

Gram-positive bacteria One of the several bacterial groups having cell walls that are colored purple by Gram-staining procedure.

gravitropism In plants, directional growth of roots and shoots in response to gravity.

grazing food web Food web in which most energy from producers flows to herbivores, then to carnivores.

greenhouse effect Warming of the lower atmosphere as greenhouse gases reradiate heat energy back toward the Earth's surface.

groundwater Water in soil and aquifers.

growth Of multicelled species, increases in the number, size, and volume of cells. Of bacteria, increases in cell number.

growth and tissue specialization Stage of animal development when the body enlarges and organs assume their functions.

growth ring One of the alternating bands of early and late wood in a tree with extensive secondary growth.

gut Of animals, a sac or tube from which food is absorbed into the internal environment.

habitat Type of place where a species lives.

habitat loss Reduction of habitat for a species as a result of environmental destruction.

half-life The time it takes for half of a given quantity of atoms of a radioisotope to decay.

haploid number Total chromosome number in cells that have one of each type of chromosome characteristic of the species.

Hardy–Weinberg rule Allele frequencies will be stable over the generations if there is no mutation, the population is infinitely large and isolated from other populations of the same species, mating is random, and all individuals reproduce equally and randomly.

heart Muscular, pressure-generating pump that keeps blood flowing through the vessels of a circulatory system.

heartwood Dry core tissue of older stems and roots; functions in support and the storage of metabolic wastes.

helper T cell T lymphocyte that stimulates B cells and other T cells to divide in response to antigen recognition; critical component of all immune responses.

heme group Iron-containing functional group that reversibly binds oxygen.

hemoglobin Respiratory protein in red blood cells; consists of four polypeptide chains and four heme groups.

hemostasis Process that stops blood loss after injury by way of coagulation, vessel spasm, platelet plug formation, and other effects.

herbicide Natural or synthetic toxin that can kill plants or inhibit their growth.

heterotherm An endotherm that can lower metabolic rates; activities idle as it cuts energy cost of maintaining its core temperature.

heterotroph Organism unable to make its own organic compounds; feeds on autotrophs, other heterotrophs, or organic wastes.

heterozygous condition For a specified trait, having two different alleles at a locus.

higher taxa One of ever more inclusive groupings of species based on relatedness (e.g., phylum, kingdom, domain).

histamine Chemical released by mast cells during an immune response; contributes to inflammation and other symptoms of allergy.

homeostasis Maintenance of physical and chemical aspects of the internal environment within ranges suitable for cell activities.

homeotic gene One of a class of master genes; helps determine identity of body parts during embryonic development.

hominid A human or an extinct humanlike primate (e.g., an australopith).

hominoid Apes, humans, and their extinct recent ancestors.

homologous chromosome One of a pair of chromosomes, identical in size, shape, and gene sequence, each inherited from a different parent. Nonidentical sex chromosomes are also considered homologs.

homologous structure Similar body part that occurs in different species as a result of descent from a common ancestor.

homozygous dominant condition Having a pair of dominant alleles at a gene locus.

homozygous recessive condition Having a pair of recessive alleles at a gene locus.

hormone Chemical signaling molecule formed by one part of the body, acting upon another. In animals, a product of endocrine glands, endocrine cells, or neurons. In plants, a product of cells in primary shoots or roots.

horsetail Seedless vascular plant with tiny scale-like leaves, branching rhizomes, and silica-reinforced stems.

hot spot Habitat for a large number of species found nowhere else and facing extinction.

human Member of the genus *Homo*.

human gene therapy The transfer of normal or modified genes into a person to correct a genetic defect, or boost resistance to a disease.

humus Decomposing organic matter in soil.

hybrid Individual having a nonidentical pair of alleles for a trait being studied.

hydrogen bond An intermolecular interaction between a covalently bonded hydrogen atom and a different atom bearing a negative charge (e.g., oxygen, fluorine, or nitrogen).

hydrologic cycle Driven by solar energy, water evaporates from the ocean into the atmosphere, moves onto the land, then back to the ocean.

hydrolysis An enzymatic cleavage reaction in which a molecule is split, and the components of water (—OH and —H) become attached to each of the fragments.

hydrophilic substance Polar molecule (e.g., glucose) that easily dissolves in water.

hydrophobic substance Nonpolar molecule (e.g., oil) that resists dissolving in water.

hydrostatic pressure Pressure exerted by a volume of fluid against a wall, membrane, or some other structure that encloses it.

hydrostatic skeleton Of many soft-bodied invertebrates, a fluid-filled cavity or cell mass against which contractile cells act.

hypertonic solution Of two fluids, the one having the higher solute concentration.

hypha, plural **hyphae** Fungal filament with a chitin-reinforced wall; part of a mycelium.

hypothalamus Brain center of homeostatic control over the internal environment, viscera, and emotions; also produces some hormones.

hypothesis In science, a possible explanation of a phenomenon, one that has the potential to be proven false by experimental tests.

hypotonic solution Of two fluids, the one having the lower solute concentration.

immigration New individuals permanently move into an area.

immune system Body system that recognizes antigen and mounts attacks against specific threats. Adaptive in jawed vertebrates.

immunoglobulin One of the five classes of antibody proteins (e.g., IgG).

implantation In placental mammals, the burrowing of a blastocyst into the uterus wall.

imprinting Learning that occurs during a sensitive period for a young animal, triggered by exposure to a simple stimulus.

in vitro fertilization IVF. Combining sperm and eggs outside the body, as in a petri dish.

inbreeding Mating between close relatives.

inclusive fitness Genetic contribution made to the next generation by an individual and its close relatives.

independent assortment In meiosis, each homologous chromosome and its partner are assorted into different gametes independently of other pairs.

indicator species Any species that provides warning of changes in habitat and impending widespread loss of biodiversity.

induced-fit model An enzyme changes shape to fit a bound substrate, and the resulting tension destabilizes the substrate's bonds.

infection Invasion and multiplication of a pathogen in a host. Disease follows if defenses are not mobilized fast enough; the pathogen's activities interfere with normal body function.

inflammation See acute inflammation.

inhibitor A hypothalamic hormone that slows secretion by cells in a target gland.

insertion Mutation involving insertion of one to a few bases into a DNA strand. Also, a movable attachment of muscle to bone.

instinctive behavior Any behavior that an animal performs without having first learned it through experience.

integrator Control center (e.g., brain) for the animal body that receives, processes, and stores sensory input; issues commands for coordinated responses.

integumentary exchange Respiration across a thin, moist, and often vascularized surface layer of animal tissue.

intermediate filament Cytoskeletal element; mechanically strengthens some animal cells.

internal environment In animals, blood and interstitial fluid (extracellular fluid).

interneuron Neuron of brain or spinal cord.

interphase Cell cycle interval between nuclear divisions; a cell grows in mass and roughly doubles the number of cytoplasmic components. DNA replication occurs during the interphase that precedes mitosis.

interspecific competition Competition between members of different species.

interstitial fluid Extracellular fluid in the spaces between animal cells and tissues.

intron Noncoding gene sequence that is removed from a pre-mRNA before translation.

inversion Mutation in which a section of chromosome becomes oriented in reverse.

invertebral disk Cartilaginous flex point and shock absorber between vertebrae.

invertebrate An animal without a backbone.

ionic bond Interaction between ions held together by attraction of opposite charges.

ionizing radiation Radiation with enough energy to eject electrons from atoms.

isotonic solution A fluid having the same solute concentration as another fluid to which it is being compared.

isotopes Two or more forms of an element's atoms differing in the number of neutrons.

jaw Of chordates, a hinged pair of cartilaginous or bony feeding structures.

joint Area of contact between bones.

karyotype Preparation of an individual's metaphase chromosomes sorted by length, centromere location, and shape.

key innovation Change in body form or function that allows a lineage to exploit the environment in more efficient or novel ways.

keystone species A species that has a major role in shaping community structure.

kidney One of a pair of vertebrate organs that filter substances from blood and form urine.

kilocalorie A thousand calories of heat energy, the amount needed to raise the temperature of 1 kilogram of water by 1°C. Standard unit of measure for the energy content of foods.

knockout experiment Experiment in which, to study the function of a gene, an organism is engineered to lack its expression.

Krebs cycle The second stage of aerobic respiration in which pyruvate is broken down to carbon dioxide and water. Two ATP form. Occurs only in mitochondria.

K–T asteroid impact theory Idea that an asteroid impact caused a mass extinction at the Cretaceous–Tertiary boundary (65 million years ago).

La Niña Cool climatic event that occurs between El Niño episodes; has global effects.

labor Process by which a placental mammal gives birth.

lactate fermentation Anaerobic pathway of ATP formation. Pyruvate from glycolysis is converted to three-carbon lactate, and NAD^+ is regenerated. Net energy yield: 2 ATP.

lactation In mammals only, secretion of milk by mammary glands.

Langerhans cell Immune cell in skin; engulfs pathogens, alerts immune system to threats.

larva, plural **larvae** A sexually immature stage of many invertebrates.

larynx Tubular airway from the pharynx to lungs. Contains vocal cords in some animals.

lateral gene transfer Movement of genetic material between existing cells by conjugation or other processes.

leaching Loss of nutrients from soil as water percolates through it.

lethal mutation Mutation with drastic effects on phenotype; usually causes death.

LH Luteinizing hormone. Anterior pituitary hormone that has reproductive roles in males and females.

lichen Mutualistic interaction between a fungus and a photoautotroph.

life-history pattern Adaptations affecting life span, fertility, and age at first reproduction.

ligament Strap of dense connective tissue that bridges a joint and attaches to bones.

light-dependent reactions The first stage of photosynthesis. Sunlight energy is trapped and converted to chemical energy of ATP, NADPH, or both, depending on the pathway.

light-independent reactions Second stage of photosynthesis in which sugars are formed from CO_2 using ATP and NADPH. Also called the Calvin-Benson cycle.

limbic system Brain center that governs emotions and has roles in memory.

limiting factor Any essential resource that can halt population growth when supplies of it dwindle.

lineage An ancestor–descendant sequence of cells, populations, or species.

linkage group All genes on a chromosome.

lipid Nonpolar hydrocarbon; fats, oils, waxes, phospholipids, and sterols are lipids.

lipid bilayer Mainly phospholipids arranged tail-to-tail in two layers; structural basis of all cell membranes.

loam Soil with roughly the same proportions of sand, silt, and clay.

local signaling molecule A cell secretion that alters chemical conditions in nearby tissues.

logistic growth, population The size of a population increases slowly, then rapidly, then levels off as the carrying capacity is reached.

loop of Henle Hairpin-shaped, tubular part of a nephron that reabsorbs water and solutes.

loose connective tissue Animal tissue with fibers, fibroblasts loosely arrayed in semifluid ground substance.

lungs Internally moistened sacs specialized for gas exchange.

lycophyte Seedless vascular plant having leaves with a single vein (e.g., club moss).

lymph Fluid in vessels of lymphatic system.

lymph node Lymphoid organ packed with lymphocytes; filters lymph.

lymph vascular system Parts of lymphatic system; delivers excess tissue fluid, absorbed fats, and reclaimable solutes to blood.

lymphatic system Organ system that returns fluid and solutes from interstitial fluid to blood; also functions in body defense.

lysis Gross damage to a plasma membrane, cell wall, or both that lets the cytoplasm leak out and so causes cell death.

lysosome Organelle of intracellular digestion.

lysozyme Infection-fighting enzyme present in mucous membranes.

macroevolution Large-scale patterns, trends, and rates of change among higher taxa.

macrophage Phagocytic white blood cell with roles in nonspecific defense and immunity.

magnoliid One of the three flowering plant groups (e.g., magnolias, avocado trees).

marine snow Organic matter that drifts to ocean depths and supports food webs there.

marsupial Pouched mammal.

mass extinction Catastrophic event or phase in geologic time when entire families or other major groups are irrevocably lost.

mass number The sum of all protons and neutrons in an atom's nucleus.

mast cell Histamine-secreting white blood cell with roles in inflammation and allergies.

master gene A gene whose product has widespread effects on other genes.

mechanoreceptor Sensory cell or nearby cell that detects mechanical energy (e.g., a change in pressure).

medulla oblongata Hindbrain region with reflex centers that influence functions basic to survival (e.g., sleeping, breathing, coughing).

megaspore Haploid spore formed by meiosis in the ovary of a seed-bearing plant; one of its cellular descendants develops into an egg.

meiosis Only nuclear division process that halves the chromosome number of a parental cell, to the haploid number. Forms gametes in animals and spores in plants.

melanin Brownish-black pigment; protects human skin from ultraviolet radiation.

memory The capacity to store and retrieve information about past sensory experience.

memory cell B or T cell that formed during an immune response; it remains in a resting phase until a secondary immune response.

menstrual cycle Recurring cycle in human females. Includes menstruation, ovulation, repair and priming of uterus for pregnancy.

meristem Region of actively dividing, undifferentiated cells in plants; descendants of meristematic cells give rise to mature plant tissues (e.g., leaves, stems, roots).

mesoderm Primary tissue layer that occurs in most animal embryos; gives rise to internal organs and part of the integument; pivotal in the evolution of large, complex animals.

mesophyll Photosynthetic ground tissue of a leaf; a type of parenchyma.

messenger RNA mRNA. Single-stranded ribonucleotide product of gene transcription; encodes protein-building instructions.

metabolic pathway Sequence of enzyme-mediated reactions by which cells assemble and build or break down organic compounds.

metabolism All the controlled, enzyme-mediated chemical reactions by which cells acquire and use energy to synthesize, store, degrade, and eliminate substances.

metamorphosis Of certain animals (e.g., many insects), drastic transformation of an immature stage to the adult through tissue reorganization and remodeling of parts.

metaphase Of meiosis I, stage when all pairs of homologous chromosomes have become positioned at the spindle equator. Of mitosis or meiosis II, all duplicated chromosomes are positioned at the spindle equator.

methanogen Type of archaebacterium that produces methane gas as a metabolic product.

micelle formation Clustering of bile salts, fatty acids, and monoglycerides into droplets; enhances fat absorption in the small intestine.

microevolution Of a population, any change in allele frequencies resulting from mutation, genetic drift, gene flow, natural selection, or some combination of these.

microfilament Cytoskeletal element; consists of actin subunits. Involved in movement and structural integrity of cells.

microspore Walled, haploid spore that gives rise to a pollen grain in a seed-bearing plant.

microtubule Cytoskeletal element; consists of tubulin subunits. Contributes to cell shape, growth, and motion; constituent of spindles.

microvillus, plural **microvilli** Slender extension from free surface of certain cells; increases surface area (e.g., for absorption).

migration Recurring movement of organisms from one region to another, then back.

mimicry Close resemblance of one species to another; confers a selective advantage upon one or both species by deceiving predators.

mineral An element or inorganic compound formed by natural geologic processes; many are required for normal metabolic function.

mitochondrion, plural **mitochondria** Organelle of ATP formation; site of aerobic respiration's second and third stages.

mitosis Nuclear division mechanism that maintains the parental chromosome number for forthcoming daughter cells. Basis of growth, tissue repair, and often asexual reproduction of eukaryotes.

mixture Two or more elements intermingled in proportions that can and usually do vary.

model Theoretical description of something that has not been directly observed.

molecular clock The time of origin of one lineage or species relative to others may be estimated by comparing the number of neutral mutations; assumes that accumulation of neutral mutations occurs at a fixed rate.

molecule Two or more atoms of the same or different elements joined by chemical bonds.

mollusk Invertebrate having a unique tissue flap (mantle) draped over a soft, fleshy body; most have an external or internal shell (e.g., a gastropod, bivalve, cephalopod).

molt Periodic shedding of worn-out or too-small body structures. Permits some animals to grow in size or renew parts.

monocot Monocotyledon; flowering plant with one embryonic seed leaf (cotyledon), floral parts usually in threes (or multiples of three), and often parallel-veined leaves.

monohybrid cross Intercross between two individuals, each heterozygous for one gene (e.g., *Aa*).

monophyletic group A group descended from a common ancestor in which the derived trait that characterizes the group first evolved.

monosaccharide One of the simple sugars (e.g., glucose) that are unit components of oligosaccharides or polysaccharides.

monotreme Egg-laying mammal.

monsoon Air circulation pattern that moves moisture-laden air arising from warm oceans to continents north or south of them.

morphogen Type of inducer molecule that diffuses through embryonic tissues, activating master genes in sequence; it contributes to mapping out the overall body plan.

morphogenesis Programmed, orderly changes in body size, proportion, and shape of an animal embryo through which all specialized tissues and early organs form.

morphological convergence In response to similar selective pressures, evolutionarily distant lineages evolve in similar ways and end up resembling each other in appearance, function, or both.

morphological divergence In response to differing selective pressures, diverging lineages undergo gradual change from the body form of their common ancestor.

motor neuron Neuron that relays signals from the central nervous system to muscle or gland cells.

motor protein Protein that associates with microtubules or microfilaments and has a role in cell movement.

motor unit A motor neuron and all muscle cells that form junctions with its endings.

multiple allele system Three or more slightly different molecular forms of a gene that occur among the individuals of a population.

muscle fatigue Decline in a muscle's capacity to generate force after prolonged contraction.

muscle fiber Of skeletal muscle, a cylindrical multinucleated fiber that develops through the fusion of many cells during development.

muscle spindle A stretch-detecting sensory organ associated with skeletal muscle.

muscle tension Mechanical force exerted by a contracting muscle; resists opposing forces.

muscle tissue Tissue with arrays of cells able to contract under stimulation, then passively lengthen and return to their resting position.

muscle twitch Single, brief contraction of a muscle in response to a stimulus.

mutation Heritable change in DNA.

mutation rate Occurrence of mutations in a particular gene as a function of time.

mutualism Symbiotic interaction that benefits both partners.

mycorrhiza, plural **mycorrhizae** Mutualistic interaction between a fungus and root.

myofibrils Threadlike, cross-banded structures within a skeletal muscle fiber.

myoglobin A respiratory pigment abundant in cardiac and skeletal muscle cells.

myosin Motor protein that makes up the thick filaments of a sarcomere.

NAD+ Nicotinamide adenine dinucleotide. A nucleotide coenzyme; abbreviated NADH when carrying electrons and H+.

natural killer cell NK cell. Lymphocyte that touch-kills tumor cells and viral-infected cells.

natural selection Microevolutionary process; the outcome of differences in survival and reproduction among individuals that differ in the details of their heritable traits.

negative control In gene expression, regulatory proteins slow or stop transcription or translation.

negative feedback mechanism A homeostatic mechanism by which a condition that changed as a result of some activity triggers a response that reverses the change.

nematocyst Cnidarian capsule that has a dischargeable, tube-shaped thread, sometimes barbed; releases a toxin or sticky substance.

neoplasm Mass of cells (tumor) that have lost control over growth and division.

nephridium plural **nephridia** Unit that controls composition and volume of fluid in some invertebrates (e.g., earthworms).

nephron One of the kidney's urine-forming tubules; filters water and solutes from blood, then selectively reabsorbs portions of both.

nerve Sheathed bundle of the axons of sensory neurons, motor neurons, or both.

nerve cord A longitudinal nerve; most animals have one to three. Chordate nervous systems arise from a dorsal nerve cord.

nerve net Simple nervous system of cnidarians and some other invertebrates; a diffuse mesh of nerve cells in epithelial tissue.

nervous system Organ system with nerve cells interacting in signal-conducting and information-processing pathways. Detects and processes stimuli, and elicits responses from effectors (e.g., muscles and glands).

nervous tissue Tissue of excitable neurons and supporting neuroglia.

net ecosystem production Of ecosystems, total net energy accumulated in producers through their growth and reproduction in a given interval (net primary production).

net population growth rate per individual (r) For population growth equations, a variable combining birth rates and death rates; assumes that both remain constant in specified interval.

neural tube The embryonic and evolutionary forerunner of the brain and spinal cord.

neurotransmitter Signaling molecule secreted by axon endings of a neuron.

neutral mutation Mutation that has little or no effect on phenotype.

neutron Subatomic particle found in an atom's nucleus; has mass but no charge.

neutrophil Most abundant type of white blood cell; engulfs pathogens and has a role in inflammatory responses.

niche Sum of all activities and relationships by which a species obtains and uses resources.

nitrification The chemical conversion of ammonia to nitrate by soil bacteria.

nitrogen cycle Movement of nitrogen from the atmosphere, through the ocean, sediments, soils, and food webs, then back to atmosphere.

nitrogen fixation Conversion of nitrogen gas to forms that plants can take up from soil.

nondisjunction Failure of sister chromatids or homologous chromosomes to separate during meiosis or mitosis. Daughter cells end up with too many or too few chromosomes.

non-ionizing radiation Radiation that carries enough energy to boost electrons to higher energy levels, but not enough to remove them.

nonshivering heat production Hormone-induced increase in metabolic rate in response to prolonged or severe cold exposure.

notochord Of chordates, a rod of stiffened tissue (not cartilage or bone) that is a supporting structure for the body.

nuclear envelope Lipid bilayer membrane enclosing the nucleus of eukaryotes.

nucleic acid Single-stranded or double-stranded molecule composed of nucleotides joined at phosphate groups (e.g., DNA, RNA).

nucleic acid hybridization Any base-pairing between DNA or RNA from different sources.

nucleoid Of bacterial cells, the region in which DNA is physically organized; not separated from the cytoplasm by a membrane.

nucleosome A small stretch of eukaryotic DNA wound around histone proteins.

nucleotide Small organic compound with a five-carbon sugar, a nitrogen-containing base, and a phosphate group.

nucleus Organelle that physically separates DNA from the cytoplasm in a eukaryotic cell.

numerical taxonomy Method of determining the relationship between an unidentified organism and a known group by comparing traits. The greater the number of traits in common, the greater the inferred relatedness.

nutrient Element with a direct or indirect role in metabolism that no other element fulfills.

nutrition Processes of selectively ingesting, digesting, absorbing, and converting food into the body's own organic compounds.

obesity Excess of fat in adipose tissue; caloric intake has exceeded the body's energy output.

ocean A continuous body of saltwater that covers more than 71 percent of the Earth.

olfactory receptor Chemoreceptor for water-soluble or volatile substance.

oocyte Immature egg.

oomycote A nonphotosynthetic stamenopile protist; many are plant pathogens.

operator Part of an operon; a binding site for a regulatory protein.

operon Promoter–operator sequence that controls transcription of more than one bacterial gene.

organ Two or more tissues arrayed in a specific pattern and interacting in some task.

organ system Two or more organs interacting chemically, physically, or both in a task.

organelle Membrane-bound compartment in the eukaryotic cytoplasm; has one or more specialized metabolic functions.

organic compound Molecule containing carbon and hydrogen; may also contain oxygen, nitrogen, and other elements.

osmoreceptor Sensory cell that detects change in solute concentration of surrounding fluid.

osmosis Diffusion of water between two regions separated by a selectively permeable membrane.

osmotic pressure Hydrostatic pressure that counters inward diffusion of water through a selectively permeable membrane inside a cell or enclosed body region.

ovary In flowering plants, enlarged base of one or more carpels. In most animals, a female gonad in which eggs form.

ovulation Release of a secondary oocyte from an ovary, induced by an LH surge.

ovule Tissue mass in which an egg forms in a plant ovary; immature seed.

ovum Mature secondary oocyte; mature egg.

oxidation–reduction reaction Transfer of electrons between reactant molecules.

oxytocin Posterior pituitary hormone; induces lactation and shrinkage of uterus after pregnancy; also affects social behavior of some mammals.

ozone thinning Seasonal thinning of the atmospheric ozone layer; most pronounced above polar regions.

pain Perception of injury to a body region.

pain receptor A nociceptor; a sensory receptor that detects tissue damage.

pancreatic islet Any of the 2 million or so clusters of endocrine cells of the pancreas.

Pangea Paleozoic supercontinent.

parapatric speciation Mode of speciation in which subpopulations of a species that are maintaining contact along a common border evolve into distinct species.

parasitism Interaction in which one organism (the parasite) lives on or in another (the host) and feeds on its tissues.

parasitoid Type of insect larva that grows and develops in a host organism (usually another insect), consumes its soft tissues, and kills it.

parasympathetic nerve An autonomic nerve; its signals cause a slowdown in overall activity and divert energy to basic tasks.

parathyroid gland One of four endocrine glands; its secretions cause a rise in blood calcium levels.

parenchyma Type of simple tissue that makes up the bulk of a plant.

parthenogenesis An unfertilized egg gives rise to an embryo.

partial pressure The contribution of any gas to total atmospheric pressure.

passive transport Diffusion of a solute across a cell membrane, through a transport protein.

pathogen Disease-causing agent that can infect a target species and multiply inside it.

pattern formation During development, sculpting of embryonic cells into specialized animal tissues and organs at expected times, in expected places.

PCR Polymerase chain reaction. A method that rapidly amplifies the number of specific DNA fragments.

peat bog Acidic wetland where peat mosses grow; peat is their compressed remains.

pedigree Chart of genetic connections.

pellicle Of some protists, a flexible body covering of protein-rich material.

peptide hormone Short chain of amino acids that acts as a hormone.

per capita Per individual.

periodic table Tabular arrangement of elements based on their chemical properties.

peripheral vasoconstriction Diameter of arterioles decreases and blood flow to body surfaces is reduced.

peripheral vasodilation Diameter of arterioles increases and blood flow to body surfaces is increased.

permafrost A perpetually frozen layer of soil.

peroxisome Enzyme-filled vesicle that breaks down amino acids and fatty acids to hydrogen peroxide, which is converted to harmless products.

pH scale A measure of the H^+ concentration (acidity) of blood, water, and other solutions. pH 7 is neutral.

pharynx A muscular tube for filter-feeding in invertebrate chordates, and, in some species, gas exchange. In many other animals, part of the digestive tract.

phenotype Observable trait or traits of an individual that arises from gene interactions and gene–environment interactions.

pheromone Hormone-like exocrine gland secretion; diffuses through air and affects a different member of the same species.

phloem A complex tissue that conducts sugars, solutes through a vascular plant.

phospholipid Lipid with a phosphate group. Major constituent of biological membranes.

phosphorus cycle Movement of phosphorus from land, through food webs, to ocean sediments, then back to land.

phosphorylation Enzyme-mediated transfer of a phosphate group between molecules.

photoperiodism Biological response to change in the relative amounts of daylight and darkness.

photoreceptor Light-sensitive sensory cell.

photosynthesis Process by which organisms use sunlight energy to convert carbon dioxide and water to sugars.

photosystem In photosynthetic cells, a cluster of membrane-bound, light-trapping pigments and other molecules.

phototropism Change in the direction of cell movement or growth in response to light.

phycobilin Red to blue photosynthetic accessory pigment.

phylogeny Evolutionary relationships among species, starting with an ancestral form and including branches leading to descendants.

phytochrome A light-sensitive pigment. Its controlled activation and inactivation affect hormones governing many plant activities, including growth, branching, and flowering.

phytoplankton Aquatic community of floating or swimming photoautotrophs.

pigment Any light-absorbing molecule.

pilomotor response Hairs or feathers stand up; creates a layer of still air next to the skin.

pineal gland Light-sensitive endocrine gland; secretes melatonin.

pioneer species Species that can colonize newly formed or newly vacated habitats.

pituitary gland Endocrine gland that interacts with the hypothalamus to control other glands and organs.

placenta Organ that forms from endometrial tissue and extraembryonic membranes. Permits exchange of substances between a pregnant female and her fetus while keeping their bloodstreams separate.

plankton Community of mostly microscopic species that swim or float in lakes or seas.

plant A multicelled photoautotroph with well-developed roots and shoots.

plasma Liquid portion of blood; mainly water in which ions, proteins, sugars, gases, and other substances are dissolved.

plasma membrane Outermost cell membrane; structural and functional boundary between the cytoplasm and fluid surrounding the cell.

plasmid A small, circular molecule of bacterial DNA that carries a few genes and is replicated independently of the chromosome.

plate tectonic theory Idea that great slabs (plates) of the Earth's outer layer float about slowly on the mantle beneath them and have rafted continents to new positions over time.

platelet Cell fragment that circulates in blood; acts in clot formation.

pleiotropy Positive or negative effects that alleles at a single gene locus have on two or more traits.

polar body In vertebrates, one of four cells that forms by meiotic cell division of an oocyte but that does not become the ovum.

pollen grain Sperm-bearing gametophyte of a gymnosperm or angiosperm.

pollination Transfer of pollen to a female part of a flower (stigma).

pollinator Agent that transfers pollen to female floral parts.

polypeptide chain Three or more amino acids linked by peptide bonds.

polyploidy Having three or more of each chromosome type characteristic of a species.

polysaccharide Straight or branched chain of many covalently linked sugar units of the same or different kinds. Most common types are cellulose, starch, and glycogen.

population Group of individuals of the same species in a specified area.

population density Number of individuals per specified area or volume of a habitat.

population distribution Pattern of dispersion of individuals of a population.

population size Number of individuals making up a population.

positive control In gene expression, regulatory proteins enhance transcription or translation.

positive feedback mechanism An event intensifies as a result of its own occurrence.

predation Interaction in which one organism (the predator) eats another (prey), typically killing it.

predator Free-living organism that captures and feeds on other organisms (its prey).

prediction Statement about what you expect to observe in nature.

pressure flow theory Organic compounds flow through phloem in response to pressure and concentration gradients between source regions (e.g., leaves) and sinks (e.g., regions where sugars are being used or stored).

pressure gradient Difference in pressure being exerted in two adjoining regions.

prey Organism that predators can capture and eat.

primary growth Plant growth originating at root tips and shoot tips.

primary immune response Defensive actions by white blood cells elicited by first-time recognition of antigen. Includes antibody- and cell-mediated responses.

primary oocyte An immature egg that is stopped in prophase I of meiosis.

primary producer Type of autotroph that secures energy directly from the environment.

primary productivity Of ecosystems, the rate at which primary producers capture and store energy in their tissues during some interval.

primary succession Sequence of community development from pioneer species to climax stage in a previously barren habitat.

primary wall Of young plant cells, a thin, flexible wall that permits division and changes in shape; consists of cellulose, polysaccharides, glycoproteins.

primate Mammalian lineage; includes prosimians, tarsioids, anthropoids (monkeys, apes, humans).

primer Short nucleotide sequence designed to serve as a site of initiation for DNA synthesis on DNA or RNA.

prion Small infectious protein that causes fatal degenerative diseases of nervous system.

probability The chance that each outcome of an event will occur is proportional to the number of ways in which the outcome can be reached.

probe Short nucleotide sequence, labeled with a tracer, designed to hybridize with part of a specific gene or mRNA.

producer Autotroph (self-feeder); nourishes itself using sources of energy and carbon from the environment. Photoautotrophs and chemoautotrophs are examples.

progesterone Female sex hormone secreted by ovaries and the corpus luteum.

prokaryotic cell Archaean or bacterium; single-celled organism, most often walled; lacks a nucleus and other organelles.

prokaryotic fission Bacterial mode of reproduction. Involves DNA replication, accumulation of new membrane (and usually wall material) at or near the cell midsection, then cytoplasmic division.

prolactin Anterior pituitary hormone that stimulates milk production.

promoter Short stretch of DNA to which RNA polymerase binds and initiates transcription.

prophase Of mitosis, a stage when duplicated chromosomes start to condense, a spindle forms, and the nuclear envelope starts to break up. Duplicated pairs of centrioles move to opposite spindle poles. In prophase I of meiosis, crossing over also occurs.

protein Organic compound consisting of one or more polypeptide chains folded and twisted into a three-dimensional shape.

proteobacteria Most diverse bacterial group.

protist One of the mainly single-celled species of eukaryotes traditionally grouped in the catch-all "kingdom Protista." Currently being classified into groupings that reflect evolutionary relationships.

proto-cell Membrane-bound metabolic machinery; transitional stage that may have preceded the origin of living cells.

proton Positively charged subatomic particle found in an atom's nucleus.

proto-oncogene A gene that, when mutated or overexpressed, helps turn a normal cell into a cancerous one.

protostome Lineage of coelomate, bilateral animals that includes mollusks, annelids, and arthropods. The first indentation to form in protostome embryos becomes the mouth.

proximal tubule Nephron's tubular portion extending from Bowman's capsule.

pseudocoel A body cavity that is not fully lined with tissue derived from mesoderm.

pseudopod Flexible, temporary lobe of membrane-enclosed cytoplasm that amoebas, amoeboid cells, and phagocytic white blood cells use for motility or engulfing food.

puberty Of human development, a post-embryonic stage when gametes start to mature and secondary sexual traits emerge.

pulmonary circuit A blood circulation route that moves blood from the heart's right side, through lungs, then to the heart's left half.

punctuation model of speciation Idea that most morphological changes occur in a brief surge, as populations start to diverge.

Punnett-square method Construction of a simple diagram to predict the probable outcomes of a genetic cross.

pyruvate Three-carbon compound that forms as the end product of glycolysis.

quadrat One of many areas of a given size and shape in which samples are taken or individuals of a population are counted.

radial symmetry Animal body plan having four or more roughly equivalent parts around a central axis, as in a sea anemone.

radioactive decay An atom emits energy as subatomic particles and x-rays as its unstable nucleus disintegrates spontaneously. The process transforms one element into another.

radioisotope Isotope with an unstable nucleus (too many or too few neutrons).

radiometric dating Method of determining the age of a fossil by comparing the relative proportions of parent and daughter radioisotopes in rock samples or fossils.

rain shadow Reduction in rainfall on the leeward side of a high mountain range; results in arid or semiarid conditions.

recombinant DNA A DNA molecule that contains genetic material from more than one organism.

red alga Type of photoautotrophic protist; most are multicelled and aquatic; phycobilins mask their chlorophyll *a*.

red marrow Site of blood cell formation in the spongy tissue of many bones.

reflex Stereotyped, involuntary movement in response to a stimulus.

refractory period Brief interval following an action potential when a small patch of neural membrane is insensitive to stimulation.

regulatory protein Protein that enhances or inhibits protein synthesis.

relaxin Hormone secreted by ovaries and placenta that prepares the cervix and pelvis for giving birth.

releaser Hypothalamic releasing hormone that stimulates secretion by a target gland.

renal corpuscle Bowman's capsule plus the glomerular capillaries it encloses.

reproductive base All individuals of a population that are in the pre-reproductive and reproductive age brackets.

reproductive isolating mechanism Any heritable aspect of body form, function, or behavior that prevents interbreeding between populations of the same species.

reproductive success Production of viable offspring by the individual.

reptile Tetrapod vertebrate that has scaly skin and water-conserving kidneys, and produces amniote eggs (e.g., dinosaur, crocodilian, turtle, snake, lizard, tuatara).

resource partitioning A subdividing of resources in time or space that allows similar species to coexist in a habitat.

respiration Movement of oxygen into an animal's internal environment and carbon dioxide out of it.

respiratory cycle One inhalation and one exhalation.

respiratory membrane The alveolar and capillary endothelia, together with their basement membranes.

respiratory pigment A molecule in body fluids that reversibly binds oxygen (e.g., hemoglobin).

respiratory surface Thin, moist epithelium that functions in gas exchange.

resting membrane potential Of a neuron and other excitable cells, a voltage difference across the plasma membrane that holds steady in the absence of outside stimulation.

restriction enzyme A protein that recognizes and cuts specific sequences of nucleotides in double-stranded DNA.

retina Dense array of photoreceptors at the back of the eye.

reverse transcriptase Enzyme that assembles a single strand of DNA from free nucleotides on an RNA template; found in RNA viruses.

Rh blood typing Method of characterizing red blood cells by a certain membrane surface protein. Rh^+ cells have it; Rh^- cells do not.

rhizoid Simple rootlike absorptive structure of some fungi and nonvascular plants.

ribonucleic acid See RNA.

ribosomal RNA rRNA. Structural and functional RNA component of ribosomes.

ribosome Structure upon which polypeptide chains are built. An intact ribosome consists of two subunits of rRNA and proteins.

riparian zone Narrow corridor of vegetation on either side of a stream or river.

RNA Ribonucleic acid. Any of a class of single-stranded nucleic acids with roles in transcription, translation, and catalysis.

RNA polymerase Enzyme that catalyzes the addition of nucleotides to a growing strand of RNA (transcription).

RNA world A presumed period before the origin of life when RNA may have stored protein-building information.

rod cell Vertebrate photoreceptor sensitive to dim light; contributes to the coarse perception of movement across the visual field.

root Plant part, typically belowground; absorbs water and minerals, anchors aboveground parts, and often stores food.

root hair Thin extension of a root epidermal cell; collectively, root hairs greatly increase the surface area for absorbing water and ions.

root nodule Localized swelling on a root that contains mutualistic nitrogen-fixing bacteria.

roundworm Cuticle-covered, bilateral worm with a false coelom and complete digestive system; a nematode.

rubisco RuBP carboxylase. Enzyme that catalyzes attachment of a carbon atom from carbon dioxide to RuBP and starts the C3 photosynthetic pathway.

ruminant Hoofed, herbivorous mammal with multiple stomach chambers.

sac fungus Fungus that forms its sexual spores (ascospores) in sac-shaped cells.

salinization Salt buildup in soil through poor drainage, evaporation, and heavy irrigation.

saliva Glandular secretion that mixes with food and starts starch breakdown in mouth.

salt Compound that releases ions (other than H^+ and OH^-) in solution.

sampling error An experimental pitfall; arises when the sample or subset of a population, an event, or some other aspect of nature under study is not representative of the whole.

sap Sugary solution that circulates through vascular tissues of plants.

sapwood Functioning secondary xylem of an older woody plant.

sarcomere Basic unit of contraction in skeletal and cardiac muscle. Shortens as a result of ATP-driven interactions between actin and myosin filaments.

scientific theory An explanation of the cause of a range of related phenomena; has been rigorously tested but is still open to revision.

sclerenchyma Simple plant tissue that supports mature plant parts; often protects seeds. Most of its cells have thick, lignin-impregnated walls.

second law of thermodynamics A law of nature stating that the spontaneous direction of energy flow is from organized forms to less organized forms; with each conversion, some energy is randomly dispersed in a form (usually heat) not as useful for doing work.

second messenger Molecule within a cell that mediates a hormonal signal from outside.

secondary oocyte An oocyte that has finished meiosis I just before being released from an ovary at ovulation.

secondary succession Recovery of a community to its climax stage following a habitat disturbance, such as a forest fire.

secondary wall A rigid, permeable wall that forms inside the primary wall of some older plant cells.

seed Mature ovule; an embryo sporophyte and endosperm surrounded by a seed coat.

segmentation Of animal body plans, a series of units that may or may not be similar to one another in appearance. Of tubular organs, an oscillating movement produced by rings of circular muscle in the tube wall.

selective gene expression Controlled activation or suppression of transcription and translation; leads to cell identity.

selective permeability Built-in capacity of a cell membrane to stop some substances from crossing, and to allow others to cross it, at certain times, in certain amounts.

selfish behavior An individual protects or increases its own chance to produce offspring regardless of consequences to its social group.

selfish herd Individuals that cluster together for protection against predators or some other environmental danger.

senescence Processes leading to the natural death of an organism or parts of it (e.g., abscission of leaves from a deciduous tree).

sensation Conscious awareness of a stimulus.

sensory adaptation Decrease in response to a stimulus maintained at constant strength.

sensory neuron Type of neuron that detects a stimulus and relays information about it to an integrating center such as a brain.

sensory receptor Any cell or some part of it that can detect a stimulus.

sex chromosomes Chromosomes that, in certain combinations, determine a new individual's sex.

sexual dimorphism Occurrence of distinctive female and male phenotypes.

sexual reproduction Production of genetically variable offspring by meiosis, gamete formation, and fertilization.

sexual selection Mode of natural selection; favors a trait that gives the individual a competitive edge in attracting or keeping a mate, hence in reproductive success.

sexually-transmitted disease STD. Any one of the diseases spread by sexual intercourse.

shivering response Rhythmic tremors in response to cold; increases heat production.

shoot A stem, leaf, or flower.

sieve tube Conducting tube of phloem.

sister chromatids Two identical DNA molecules (and associated proteins) attached at the centromere until they are separated from each other at mitosis or meiosis; each is then a separate chromosome.

six-kingdom system Classification of all species into the kingdoms Bacteria, Archaea, Protista, Fungi, Plantae, and Animalia.

skeletal muscle tissue Striated contractile tissue; main component of muscles that attach to and move bones.

skin External integument of vertebrates; an outer epidermis and underlying dermis.

sliding-filament model Idea that muscle contraction occurs as a result of ATP-driven interactions between myosin and actin filaments in sarcomeres.

slime mold Predatory protist; amoebalike cells form fruiting bodies during life cycle.

smooth muscle tissue Nonstriated contractile tissue found in soft internal organs.

social behavior Diverse interactions among individuals of a species, which display, send, and respond to forms of communication that have genetic and learned components.

sodium–potassium pump Active transport protein that moves sodium and potassium ions across the cell membrane.

soil Mixture of mineral particles and decomposing organic material, with air and water occupying spaces between the particles.

soil erosion Movement of land under the force of wind, running water, and ice.

solute Any substance dissolved in a solution.

somatic nerve Type of nerve that carries signals from the central nervous system to skeletal muscles, and from sensory receptors into that system.

somites Paired bumps of mesoderm in a vertebrate embryo that give rise to skeletal muscles and bones, and part of the dermis.

species One kind of organism. Of species that reproduce sexually, one or more groups of natural populations in which individuals interbreed and are reproductively isolated from other such groups.

sperm Mature male gamete.

sphincter A ring of smooth muscles that can alternately contract and relax to close off and open a passageway to the body surface.

spinal cord The part of the central nervous system that runs through a canal in the vertebral column.

spirochete Member of a group of spring-shaped bacteria; some are human pathogens.

spleen A lymphoid organ that is a filtering station for blood, a reservoir of red blood cells, and a reservoir of macrophages.

sponge A filter-feeding animal with no body symmetry and no tissues.

spore A reproductive or resting structure of one or a few cells, often walled or coated; protects against harsh conditions, aids in dispersal, or both.

sporophyte A vegetative body that grows by way of mitotic cell divisions from a plant zygote and produces spore-bearing structures.

spring overturn Downward movement of oxygen-rich surface waters and upward movement of nutrient-rich waters from the depths of temperate-zone lakes in spring.

stabilizing selection Mode of natural selection; intermediate phenotypes in the range of variation are favored and extremes are selected against.

stamen A male reproductive part of a flower.

statolith A gravity-sensing mechanism based on clusters of dense particles.

stem cell Undifferentiated animal cell that can divide indefinitely; a portion of daughter cells differentiate into specialized cell types.

steroid hormone Lipid-soluble hormone derived from cholesterol.

sterol Lipid with a rigid backbone of four fused carbon rings.

stimulus A form of energy that can be detected by a sensory receptor.

stoma, plural **stomata** A gap between two guard cells in leaf or stem epidermis; allows the diffusion of water vapor and gases across the epidermis.

stomach Muscular, stretchable sac that mixes and stores ingested food, helps break it up mechanically and chemically, and controls its entry into the small intestine.

stramenopile One of a group of protists having flagella with tinsel-like filaments.

stratification Layering of sedimentary rock; results from deposition of materials over time.

strip logging Logging of forested slopes in a narrow corridor to lessen negative impacts.

strobilus, plural **strobili** A conelike cluster of spore-producing structures (e.g., of cycads).

stromatolite Fossilized mats of shallow-water microbial communities, mainly cyanobacteria.

substrate-level phosphorylation The direct, enzyme-mediated transfer of a phosphate group from a substrate to another molecule.

surface-to-volume ratio Physical relationship in which volume increases with the cube of the diameter, but surface area increases with the square; constrains increases in cell size.

survivorship curve A graph that reflects how many individuals of a cohort survive, on average, at successive ages in their life span.

sympathetic nerve An autonomic nerve; deals mainly with increasing overall body activities at times of excitement or danger.

sympatric speciation In the absence of a physical barrier, a new species arises within the home range of an existing species.

synaptic integration Summation of all excitatory and inhibitory signals arriving at the trigger zone of a neuron or some other excitable cell.

syndrome A set of symptoms that characterize an abnormality or a disorder.

systemic circuit Circulatory route that carries oxygenated blood from the left side of the heart, through all tissues, then carries oxygen-poor blood back to the right half of the heart.

T lymphocyte T cell; a type of white blood cell vital to immune responses (e.g., helper T cells and cytotoxic T cells).

tandem repeat One of many short DNA sequences, occurring one after the other, in a chromosome. Used in DNA fingerprinting.

taproot system A primary root together with its lateral branchings; typical of eudicots.

taste receptor A chemoreceptor for substances dissolved in fluid bathing it.

TCR Receptor on T cell surface; binds to an antigen fragment that has become attached to a self marker on an antigen-presenting cell.

telomere Repetitive DNA cap at chromosome tip; shorter after each nuclear division.

telophase Of meiosis I, a stage when one member of each pair of homologous chromosomes reaches a spindle pole. Of mitosis and of meiosis II, the stage when chromosomes decondense into threadlike structures and daughter nuclei form.

temperature A measure of molecular motion.

temperature zone Globe-spanning latitudinal bands of temperature (e.g., tropical, warm temperate, cool temperate).

tendon Cord or strap of dense connective tissue that attaches muscle to bone.

test, scientific A means to determine the accuracy of a prediction, as by conducting experiments, making observations, or developing models.

testcross Experimental cross to determine whether an individual of unknown genotype that shows dominance for a trait is either homozygous dominant or heterozygous.

testis, plural **testes** A type of gonad (primary male reproductive organ); produces male gametes and sex hormones.

testosterone Male sex hormone produced in testes; functions in sperm formation and development of secondary sexual traits.

tetanus Of a muscle, sustained contraction that results from repeated stimulation of a motor unit. In a disease by the same name, a toxin prevents muscles from being released from contraction.

tetrapod Vertebrate that is a four-legged walker or a descendant of one (e.g., amphibian, reptile, bird, mammal).

thalamus Brain region; a coordinating center for sensory input and a relay station for signals to the cerebrum.

theory of uniformity Theory that Earth's surface changes in slow, uniformly repetitive ways. Helped change Darwin's view of evolution. Has since been replaced by plate tectonics theory.

thermal inversion A layer of dense, cool air trapped beneath a layer of warm air; can keep air pollutants close to the ground.

thermal radiation The surface of a warm body emits heat in the form of radiant energy.

thermoreceptor Sensory cell that detects radiant energy (heat).

thigmotropism Orientation of the direction of growth in response to physical contact with a solid object (e.g., a vine curls around a post).

three-domain system Classification of all species into the domains Bacteria, Archaea, and Eukarya.

threshold Of excitable cells (e.g., a neuron or muscle cell), the minimum amount of change in the resting membrane potential that causes an action potential.

thylakoid membrane In plants, internal portion of a chloroplast's membrane system, often folded into flattened sacs, that forms a single compartment. Pigments and enzymes are embedded in it; site of photosynthesis.

thymus gland Lymphoid organ; its hormones influence T cells, which form in bone marrow but migrate to and differentiate in the thymus.

thyroid gland Endocrine gland; its hormones influence overall growth, development, and metabolic rates of warm-blooded animals.

tight junction Animal cell junction that prevents substances from leaking between adjoining cells.

tissue Of multicelled organisms, a group of cells and intercellular substances that function together in one or more specialized tasks.

tissue culture propagation Inducing a tissue or organism to grow from an isolated cell of a parent tissue, in vitro.

topsoil Uppermost soil layer, the one that is most essential for plant growth.

total fertility rate TFR. Within a population, the average number of children born to a woman during her reproductive years.

tracer Substance with attached radioisotope that researchers can track after delivering it into a cell, a multicelled body, ecosystem, or other system.

trachea A type of air-conducting tube that occurs in respiratory systems. Of land vertebrates, the windpipe that connects the larynx to bronchi.

tracheal system Of insects and some other land-dwelling arthropods, a system of tubes that carry gases from body surfaces to tissues.

tracheid One of two types of water-conducting cells in xylem.

transcription First stage of protein synthesis, in which a strand of RNA is assembled on a DNA template (gene).

transfer RNA tRNA. A class of small RNA molecules that deliver amino acids to a ribosome; each pairs with an mRNA codon during translation.

translation Second stage of protein synthesis. Information encoded in an mRNA transcript guides the synthesis of a new polypeptide chain from amino acids; occurs at ribosomes.

translocation Of cells, a repositioning of a stretch of DNA to a new chromosomal location with no molecular loss. Of vascular plants, distribution of organic compounds through phloem.

transpiration Evaporative water loss from the aboveground parts of a plant.

transposon Transposable element. A stretch of DNA that can jump spontaneously and randomly to a different location in the genome; may cause mutation.

triglyceride A lipid that has three fatty acid tails attached to a glycerol backbone.

trophic level All organisms the same number of transfer steps away from the energy input into an ecosystem.

tubular reabsorption In kidneys, return of water and solutes from a nephron to blood.

tubular secretion In kidneys, excess ions and other substances move from interstitial fluid into the nephron, and are excreted in urine.

tumor Tissue mass with cells dividing at an abnormally high rate. If benign, cells stay in place; if malignant, they metastasize, or move to form tumors in new places in the body.

ultrafiltration Bulk flow of a small amount of protein-free plasma from a blood capillary when the outward-directed effect of blood pressure exceeds the inward-directed osmotic movement of interstitial fluid.

upwelling Upward movement of cold, nutrient-rich water from the ocean depths to the ocean surface.

uracil Nitrogen-containing base of a nucleotide of RNA molecules. Can base-pair with adenine.

urea Waste product formed in the liver from ammonia and CO_2; excreted in urine.

urinary excretion Mechanism by which excess water and solutes are removed from the body through the urinary system.

urinary system Organ system that adjusts the volume and composition of blood, and thereby helps maintain extracellular fluid.

urine Fluid consisting of any excess water, wastes, and solutes; it forms in kidneys by filtration, reabsorption, and secretion.

uterus In pouched and placental female mammals, organ in which embryos develop and are nurtured.

variable A specific aspect of an object or event that may differ over time and among individuals. In an experimental test, a single variable is directly manipulated in an attempt to support or disprove a prediction.

vascular bundle A strand-like array of primary xylem and phloem that threads through a plant's ground tissue system.

vascular cambium A lateral meristem that gives rise to secondary xylem and phloem, which increases stem or root diameter.

vascular cylinder Arrangement of vascular tissues as a central cylinder inside a root.

vasoconstriction Shrinking of blood vessel diameter, especially arterioles.

vasodilation Enlargement of blood vessel diameter, especially arterioles.

vegetative growth A new plant grows from an extension or fragment of its parent.

vein In plants, a vascular bundle inside a leaf. In animals, a vessel that returns blood to the heart and acts as a blood volume reservoir.

venule A small blood vessel that serves as a transitional conducting tube between a small-diameter capillary and a larger diameter vein.

vernalization The induction of flowering by exposure to low temperature.

vertebra, plural **vertebrae** A series of hard bones that function as a skeletal backbone and that protect the spinal cord.

vertebrate Animal having a backbone.

vessel member One cell type in xylem.

vestibular apparatus Organ of equilibrium in the inner ear.

villus, plural **villi** Fingerlike projections from the free surface of an epithelium, as into the lumen of the small intestine.

viroid A bit of RNA that infects plants.

virus A noncellular infectious agent that can be replicated only if its genetic material enters a host cell and subverts metabolic machinery.

viscera All soft organs inside an animal body (e.g., heart, lungs, and stomach).

vision Perception of visual stimuli.

visual accommodation Adjustments in a lens position or shape that focus light on a retina.

vitamin Any of more than a dozen organic substances that an organism requires in small amounts for metabolism but generally cannot synthesize for itself.

warning coloration Pattern or coloration that makes a toxic organism (or its mimics) easy to detect and avoid.

watershed Region from which water drains into a single stream or river.

water–vascular system Of sea stars and sea urchins, a system of many tube feet that are deployed in synchrony for smooth movement.

wavelength A wavelike form of energy in motion. The horizontal distance between the crests of every two successive waves.

wax A type of lipid with long-chain fatty acids attached to long-chain alcohols or carbon rings.

white blood cell Leukocyte. A type of blood cell that functions in basic housekeeping, nonspecific defenses, and adaptive immunity (e.g., eosinophil, neutrophil, macrophage, or lymphocyte).

X chromosome inactivation The programmed condensation of one of the X chromosomes in somatic cells of a mammalian female.

xanthophyll Yellow-orange carotenoid. An accessory pigment.

xenotransplantation Transfer of an organ from one species to another.

x-ray diffraction image Pattern formed when x-rays that have been directed at a molecule are scattered; the resulting pattern of streaks and dots is used to calculate positions of atoms in the molecule.

xylem Of vascular plants, a complex tissue that conducts water and solutes through pipelines of interconnected walls of cells, which are dead at maturity.

yellow marrow A fatty tissue in the cavities of most mature bones.

Y-linked gene Gene on a Y chromosome.

yolk Protein-rich and lipid-rich substance that nourishes embryos in animal eggs.

yolk sac Extraembryonic membrane. In most shelled eggs, it holds nutritive yolk; in humans, part becomes a site of blood cell formation and some cells give rise to forerunners of gametes.

zero population growth No overall increase or decrease in population size during a specified interval.

zygomycete Fungus that forms a thick-walled sexual spore (a diploid zygote) in a thin cover.

zygote First cell of a new individual, formed by fusion of a sperm nucleus with an egg nucleus at fertilization; a fertilized egg.

Art Credits and Acknowledgments

Page i, iii, Frans Lanting/Minden Pictures

TABLE OF CONTENTS **Page iv** bottom left, Art by Raychel Ciemma; **Page v** top, from left, Lisa Starr; Larry West/FPG/Getty Images; © Professors P. Motta and T Naguro/SPL/Photo Researchers, Inc. **Page vi** bottom left, © Jennifer W. Shuler/Science Source/Photo Researchers, Inc. **Page vii** top, from left, © George Lepp/CORBIS; (both) Courtesy of Carl Zeiss MicroImaging, Thornwood, NY; © Mc Leod Murdo/Corbis Sygma/Model, courtesy of Thomas A. Setitz from *Science*; Courtesy of Joseph DeRisa from *Science*, 1997 Oct. 24; 278 (5338) 680–686. **Page viii** left, (above) © Alan Solem; (below) Courtesy of Professor Martin F. Yanofsky, UCSD; **Page ix** top, from left, © P. Hawtin, University of Southampton/SPL/Photo Researchers, Inc.; © Science Photo Library/Photo Researchers, Inc.; © John Clegg/Ardea, London; © Robert C. Simpson/Nature Stock; Gerry Ellis/The Wildlife Collection. **Page x** Top © R. J. Erwin/Photo Researchers, Inc. **Page xi** Top, from left, © Cory Gray; © Illustration by Karen Carr; © Bruce Iverson; © Jim Christensen, Fine Art Digital Photographic Images. **Page xii** bottom left, © Science Photo Library/Photo Researchers, Inc. **Page xiii** Top, from left, © AP/Wide World Photos; © Kevin Fleming/CORBIS; © Michael Neveux; © NSIBC/SPL/Photo Researchers, Inc. **Page xiv** Top, © NIBSC/SPL/Photo Researchers, Inc. **Page xv** top, from left, © Francois Gohier/Photo Researchers, Inc.; © Archivo Iconografico, S.A./CORBIS; © Dow W. Fawcett/Photo Researchers, Inc.; © Amos Nachoum/CORBIS **Page xvi** bottom left, © C. James Webb/Phototake USA. **Page xvii** top, from left, © Pr. Alexande Meinesz, University of Nice-Sophia Antipolis; © Eric and David Hosking/CORBIS; © James Marshall/CORBIS. © Hank Fotos Photography

INTRODUCTION NASA Space Flight Center

CHAPTER 1 **Page 2** Left, © Mark M. Lawrence/CORBIS. **the big picture** (page 2) From left, Lisa Starr with PDB ID:1BNA; H.R. Drew, R.M. Wing, T. Takano, C. Broka, S. Tanaka, K. Itakura, R.E. Dickerson; Structure of a B-DNA Dodecamer. Conformation and Dynamics, PNAS; right, © Peter Scoones; (page 3) From left, © Nick Brent; right, © Raymond Gehman/CORBIS. **Page 3** Top, © Peter Turnley/CORBIS. **1.1** (a) Lisa Starr, rendered with Atom In A Box, © Dauger Research, Inc. 1.1 (b, above left) Lisa Starr with PDB file courtesy of Dr. Christina A. Bailey, Department of Chemistry & Biochemistry, California Polytechnic State University, San Luis Obispo, CA (b, above center) Lisa Starr with PDB ID: 1BBB; Silva, M. M., Rogers, P. H., Arnone, A; A third quaternary structure of human hemoglobin A at 1.7-A resolution; J Biol Chem 267 pp. 17248 (1992); (b, above right) Lisa Starr with PDB file from Klotho Biochemical Compounds Declarative Database; (b, below) Lisa Starr; (c) Lisa Starr; (d) ©Science Photo Library/Photo Researchers; (e) © Bill Varie/CORBIS; (f) Jeffrey L. Rotman/CORBIS; (g) © Jeffrey L. Rotman/CORBIS; (h) © Jeffrey L. Rotman/CORBIS; (i) © Peter Scoones; (j) NASA; (k) NASA. **1.2** Lisa Starr. **1.3** (a-e) Jack de Coningh. **1.4** Gary Head and Lisa Starr **1.5** © Y. Arthrus-Bertrand/Peter Arnold, Inc. **1.7** (page 8) clockwise from above, © Lewis Trusty/Animals Animals; © Emiliania Huxleyi photograph, Vita Pariente, scanning electron micrograph taken on a Jeol T330A instrument at Texas A&M University Electron Microscopy center; © Carolina Biological Supply Company; © R. Robinson/Visuals Unlimited, Inc.; (e) © Oliver Meckes/Photo Researchers; Courtesy of James Evarts; (page 9) clockwise from above left, © John Lotter Gurling/Tom Stack & Associates; (h) © Edward S. Ross; (i) © Robert C. Simpson/Nature Stock; (j) © Edward S. Ross; (k) © Joe McDonald/CORBIS; (l) © CNRI/SPL/Photo Researchers, Inc.; (m) © P. Hawtin, University of Southampton/SPL/Photo Researchers, Inc. **1.8** Top, All Courtesy of Derrell Fowler, Tecumseh, Oklahoma; bottom, © Nick Brent, enhanced by Lisa Starr. **1.9** Lisa Starr. **Page 11** Left, © Raymond Gehman/CORBIS; center, © LWA-Stephen Welstead/CORBIS; right, © Lester Lefkowitz/CORBIS. **Page 12** Right, © Royalty-Free/CORBIS. **1.10** (a) © Chris D.

Jiggins; (b) © www.thais.it; top background, © Wolfgang Kaehler/CORBIS; bottom right, © Martin Reid. Art, Lisa Starr. **Page 15** © Digital Vision/PictureQuest. **Page 16** (a–d) © Gary Head.

Page 17 Unit I © Wim van Egmond/Micropolitan Museum.

CHAPTER 2 **Page 18** Top, (a) © David Arky/CORBIS. **the big picture** (page 18) Lisa Starr; (page 19) From left, Art, Lisa Starr with PDB ID:1BNA; H.R. Drew, R.M. Wing, T. Takano, C. Broka, S. Tanaka, K. Itakura, R. E. Dickerson; Structure of a B-DNA Dodecamer. Conformation and Dynamics; PNAS; center, © Steve Lissau/Rainbow; right, © FoodPix. **Page 19** Top, © Dinodia. **2.2** (a, b) Lisa Starr; (c) Lisa Starr, rendered with Atom In A Box, © Dauger Research, Inc. **2.3** Art, Lisa Starr. **2.4** (a) © John Greim/Medichrome; (b, c) Art by Raychel Ciemma; (d) © Harry T. Chugani, M.D., UCLA School of Medicine **Page 22** © Michael S. Yamashita/CORBIS. **2.5** Lisa Starr, rendered with Atom In A Box, © Dauger Research, Inc. **2.6** Lisa Starr. **2.7** Gary Head and Lisa Starr. **2.8** Art, (a, b) Lisa Starr; photographs, (a) above, Gary Head; below, © Bruce Iverson; (c) Lisa Starr with PDB ID:IBNA; H.R. Drew, R.M. Wing, T. Takano, C. Broka, S. Tanaka, K. Itakura, R. E. Dickerson; Structure of a B-DNA Dodecamer. Conformation and Dynamics, PNAS. **2.9** (a,b,c, left) Lisa Starr with PDB file from NYU Scientific Visualization Lab; (b, right) © Steve Lissau/Rainbow; (c, right) © Kennan Ward/CORBIS. **2.10** Lisa Starr. **2.11** (a) © Lester Lefkowitz/CORBIS; (b) Lisa Starr. **2.12** Lisa Starr. **2.13** © Michael Grecco/Picture Group **Page 31** © National Gallery Collection; By kind permission of the Trustees of the National Gallery, London/CORBIS

CHAPTER 3 **3.1** Left, © 2002 Charlie Waite/Stone/Getty Images; center, John Collier. Great Britain, 1850-1934, *Priestess of Delphi*, 1891, London, oil on canvas, 160.0 x 80.0cm. Gift of the Rt. Honourable, the Earl of Kintore,1893; right, Lisa Starr, with PDB files from NYU Scientific Visualization Lab. **the big picture** (page 32) From left © Wayne Bennett/CORBIS; right, Lisa Starr; (page 33) From left, Lisa Starr with PDB file courtesy of Dr. Christina A. Bailey, Department of Chemistry & Biochemistry, California Polytechnic State University; San Luis Obispo, CA; center, Lisa Starr with PDB ID: 1BBB; Silva, M. M., Rogers, P. H., Arnone, A.; A third quaternary structure of human hemoglobin A at 1.7-A resolution; J Biol Chem 267, pp. 17248; right, Lisa Starr with PDB ID: 1BNA; H. R. Drew, R. M. Wing, T. Takano, C. Broka, S. Tanaka, K. Itakura, R. E. Dickerson; Structure of a B-DNA Dodecamer. Conformation and Dynamics, PNAS. **Page 33** Top right, Lisa Starr with PDB file from NYU Scientific Visualization Lab **3.2** (a) Lisa Starr with PDB file from NYU Scientific Visualization Lab (b) Lisa Starr with PDB file from Klotho Biochemical Compounds Declarative Database **3.3** Gary Head **3.4** © Eric and David Hosking/CORBIS; right, © Wayne Bennett/CORBIS; below, Lisa Starr. **3.5** Lisa Starr. **3.6** Lisa Starr with PDB file from NYU Scientific Visualization Lab. **3.7** (a) © Dr. W. Michaelis/Universitat Hamburg; (b) both, Courtesy of K.O. Stetter & R. Rachel, University of Regensburg © Boetius et all. 2000, Nature 407, 623-626; (b, right) Seth Gold. **3.8** Ian R. MacDonald **3.9** © John Sibbick. **3.10** Lisa Starr **3.11** Lisa Starr **3.12** Art, Lisa Starr; photograph, © Steve Chenn/CORBIS **3.13** Art, Lisa Starr; photograph, © David Scharf/Peter Arnold, Inc. **3.14** (a) Lisa Starr with PDB file courtesy of Dr. Christina A. Bailey, Department of Chemistry & Biochemistry, California Polytechnic State University, San Luis Obispo, CA; (b-d) Art, Precison Graphics. **3.15** (a, b) Art, Precision Graphics; photograph, © Kevin Schafer/CORBIS **3.16** (a) Lisa Starr with PDB file courtesy of Dr. Christina A. Bailey, Department of Chemistry & Art, Lisa Starr; (b) Precision Graphics (c) Lisa Starr **3.17** (a) © Scott Camazine/Photo Researchers, Inc. (b) Lisa Starr (c) Precision Graphics. **3.18** (a) Gary Head; (b-e) Lisa Starr and Chris Keeney with PDB files from NYU Scientific Visualization Lab. **3.19** (a) Lisa Starr; (b, left) Lisa Starr; (b, right) Lisa Starr After: Introduction to Protein Structure 2nd ed., Branden & Tooze, Garland Publishing, Inc.; (c, left) Lisa Starr with PDB ID: 1BBB; Silva, M. M., Rogers, P. H., Arnone, A.; A third quaternary structure of human hemoglobin A at 1.7-A resolution; J Biol Chem 267 pp. 17248 (1992); (c, right) Lisa Starr After: Introduction to Protein Structure 2nd ed., Branden & Tooze, Garland Publishing, Inc. **3.20** (a, b) Lisa Starr with PDB ID: 1BBB; Silva, M. M., Rogers, P. H., Arnone, A.; A third quaternary structure of human hemoglobin A at 1.7-Å resolution; J Biol Chem 267 pp.

17248 (1992). **3.21** (a, b) Gary Head and Lisa Starr with PDB files from New York University Scientific Visualization; (c) Dr. Gopal Murti/SPL/Photo Researchers, Inc. (d) photograph, Courtesy of Melba Moore. **3.22** Gary Head. **3.23** (a) Gary Head; (b) Lisa Starr **3.24** Lisa Starr with PDB ID:1BNA; H.R. Drew, R.M. Wing, T. Takano, C. Broka, S. Tanaka, K. Itakura, R.E. Dickerson; Structure of a B-DNA Dodecamer. Conformation and Dynamics; PNAS V. 78 2179, 1981 Lisa Starr. **Page 49**, left, Art by Lisa Starr with PDB ID: 1AKJ; Gao, G. F., Tormo, J., Gerth, U. C., Wyer, J. R., McMichael, A. J., Stuart, D. I., Bell, J. I., Jones, E. Y., Jakobsen, B. K.; Crystal structure of the complex between human CD8alpha(alpha) and HLA-A2; Nature 387 pp. 630 (1997) **3.25** Art, Lisa Starr; photograph, © Kevin Fleming/CORBIS

CHAPTER 4 **4.1** (a-c) © Tony Brian, David Parker/SPL/Photo Researchers, Inc. **the big picture** (page 50) From left, Art by Raychel Ciemma; right, Lisa Starr; (page 51) Left and right, Art, Lisa Starr. **Page 51** Top, © Tony Brian, David Parker/SPL/Photo Researchers, Inc. **4.2** Lisa Starr. **4.3** Art assembly, Gary Head; Art, Lisa Starr. Photographs: (hummingbird) © Robert A. Tyrrell; (human); © Pete Saloutos/CORBIS; (redwood), © Sally A. Morgan, Ecoscene/CORBIS **4.5** Left, © Armed Forces Institute of Pathology; right, © National Library of Medicine. **4.6** Bottom left, © Leica Microsystems, Inc., Deerfield, IL; right, Gary Head **4.7** Left, Gary Head; right, © Geoff Tompkinson/Science Photo Library/Photo Researchers, Inc. **4.8** (a) © Jeremy Pickett-Heaps, School of Botany, University of Melbourne; (b) © Jeremy Pickett-Heaps, School of Botany, University of Melbourne; (c) © Jeremy Pickett-Heaps, School of Botany, University of Melbourne; (d) © Jeremy Pickett-Heaps, School of Botany, University of Melbourne. **4.9** (a) Precision Graphics (b) Lisa Starr (c) Raychel Ciemma **4.10** (page 56) Left, Lisa Starr with integrin: PDB ID:1JV2; Xiong, J.-P., Stehle, T., Diefenbach, B., Zhang, R., Dunker, R., Scott, D. L., Joachimiak, A., Goodman, S. L., Arnaout, M. A.: Crystal Structure of the Extracellular Segment of Integrin αVβ3 Science 294 pp. 339 (2001); (page 57), Lisa Starr, Chris Keeney, and Leif Buckley with Human growth hormone: PDB ID:1A22; Clackson, T., Ultsch, M. H., Wells, J. A., de Vos, A. M.: Structural and functional analysis of the 1:1 growth hormone:receptor complex reveals the molecular basis for receptor affinity. J Mol Biol 277 pp. 1111 (1998); HLA: PDB ID: 1AKJ; Gao, G. F., Tormo, J., Gerth, U. C., Wyer, J. R., McMichael, A. J., Stuart, D. I., Bell, J. I., Jones, E. Y., Jakobsen, B. K.; Crystal structure of the complex between human CD8alpha and HLA-A2; Nature 387 pp. 630 (1997); glut1: PDB ID:1JA5; Zuniga, F. A., Shi, G., Haller, J. F., Rubashkin, A., Flynn, P., Iserovich, P., Fischbarg, J.: A Three-Dimensional Model of the Human Facilitative Glucose Transporter Glut1 J. Biol. Chem. 276 pp. 44970 (2001); calcium pump: PDB ID:1EUL; Toyoshima, C., Nakasako, M., Nomura, H., Ogawa, H.: Crystal Structure of the Calcium Pump of Sarcoplasmic Reticulum at 2.6 Angstrom Resolution. Nature 405 pp. 647 (2000); ATPase: PDB ID:18HE; Menz, R. I., Walker, J.E., Leslie, A.G.W.: Structure of Bovine Mitochondrial F1-ATPase with Nucleotide Bound to All Three Catalytic Sites: Implications for the Mechanism of Rotary Catalysis. Cell (Cambridge, Mass.)106 pp. 331 (2001). **4.11** (a) © K.G. Murti/Visuals Unlimited; (b) © R. Calentine/Visuals Unlimited; (c) © Gary Gaard and Arthur Kelman; (d) Lisa Starr. **4.12** top, (a) © Russell Kightley/Science Photo Library/Photo Researchers, Inc.; (b) © University of California Museum of Paleontology; (c) © University of California Museum of Paleontology. **4.13** © M.C. Ledbetter, Brookhaven National Laboratory. **4.14** © Micrograph, Gl L. Decker **4.15** Top left, Lisa Starr; top right, © Stephen L. Wolfe; (a) © Stephen L. Wolfe; (b) Lisa Starr. **4.16** Top left, Lisa Starr; (a) © Stephen L. Wolfe; (b) Stephen L. Wolfe; (c) right, Art, computer enhanced by Lisa Starr; (d) right, Art, computer enhanced by Lisa Starr (e) Gary Grimes, computer enhanced by Lisa Starr; photograph (right), © Keith R. Porter. **4.18** Art, Lisa Starr; photograph, © Dr. Jeremy Burgess/SPL/Photo Researchers, Inc. **4.19** Lisa Starr. **4.20** Lisa Starr. **4.21a-c** Lisa Starr; photograph, courtesy Mary Osborn, Max Planck Institute for Biophysical Chemistry, Goettingen, FRG. **4.22** Lisa Starr. **4.23** (a) Precision Graphics after Stephen L. Wolfe, Molecular and Cellular Biology, Wadsworth, 1993. (b) © CNRI/SPL/Photo Researchers, Inc. **4.24** (a) Raychel Ciemma & Lisa Starr; (c-e) Raychel Ciemma **4.25** (a) © George S. Ellmore (b) © Science Photo Library/Photo Researchers, Inc.; right, © Bone Clones, www.boneclones.com. **4.26** Ronald Hoham, Dept. of Biology, Colgate University. **4.26** Lisa Starr and Leif Buckley. **4.27** (a,b) From "Tissue & Cell," Vol. 27, pp.421-427, Courtesy of Bjorn Afzelius, Stockholm University.

CHAPTER 5 5.1 Left, © Chris Keeney; right, Lisa Starr with PDB ID: 1DGF; Putnam, C. D., Arvai, A. S., Bourne, Y., Tainer, J. A.: Active and Inhibited Human Catalase Structures: Ligand and Nadph Binding and Catalytic Mechanism J.Mol.Biol. 296 pp. 295 (2000). **the big picture** (page 72) From left, © William Dow/CORBIS; Lisa Starr; (page 73) From left, © Scott McKiernan/ZUMA Press; Lisa Starr. **Page 73** © Paul Edmondson/CORBIS. 5.2 Top, © Craig Aurness/CORBIS; bottom, © William Dow/CORBIS. 5.3 Lisa Starr. 5.4 Lisa Starr with PDB file from Klotho Biochemical Compounds Declarative Database. 5.5 Lisa Starr. 5.6 Lisa Starr, from B. Alberts, et al., Molecular Biology of the Cell, 1983, Garland Publishing. 5.7 Lisa Starr. 5.8 Lisa Starr. 5.9 Lisa Starr. 5.10 Lisa Starr with PDB ID: 1DGF; Putnam, C. D., Arvai, A. S., Bourne, Y., Tainer, J. A.: Active and Inhibited Human Catalase Structures: Lingand Nadph Binding and Catalytic Mechanism J Mol Biol. 296, p. 295 (2000). 5.11 Lisa Starr. 5.12 Lisa Starr. 5.13 (a) Gary Head; (b) © Scott McKiernan/ZUMA Press. 5.14 (a) Gary Head; (b) © Foodpix/Bill Boch; (c) © Woods Hole Oceanographic Institution. 5.15 Lisa Starr. 5.16 Top, © Andrew Lambert Photography/Science Photo Library/Photo Researchers; Art, Raychel Ciemma. 5.17 Lisa Starr. 5.18 Lisa Starr and Chris Keeney with PDB files from NYU Scientific Visualization Lab. 5.19 Lisa Starr, After: David H. MacLennan, William J. Rice and N. Michael Green, " The Mechanism of Ca2+ Transport by Sarco (Endo)plasmic Reticulum Ca2+-ATPases."JBC Volume 272, Number 46, Issue of Nov. 14, 1997 pp. 28815-28818. **Page 86** © Hubert Stadler/CORBIS. 5.20 Precision Graphics. 5.21 Art, top, Raychel Ciemma; bottom (all), © M. Sheetz, R. Painter, and S. Singer, J of Cell Biol., 70:193 (1976) by permission, The Rockefeller University Press. 5.22 Lisa Starr. 5.23 Both, © R.G.W. Anderson, M.S. Brown and J.L. Goldstein. Cell 10:351 (1977). 5.24 Chris Keeney. 5.25 (a) © Juergen berger/Max Planck Inst./SPL/Photo Researchers, Inc.; (b) Lisa Starr. **Page 90** Lisa Starr with PDB ID: 1CBJ; Hough, M.A. Hasnain, S.S. Crystallographic structures of bovine copper-zinc superoxide dismutase reveal asymmetry in two subunits: functionally important three and five coordinate copper sites captured in the same crystal. J. Mol. Biol. v287 pp. 579, 1999. 5.26 © Frieder Sauer/Bruce Coleman. 5.27 © Prof. Marcel Bessis/SPL/Photo Researchers, Inc.

CHAPTER 6 6.1 Both, NASA. **the big picture** (page 92) From left, © 2002 PhotoDisc; Lisa Starr. **Page 93** © Wendy A. Kozlowski. 6.2 Left, © 2002 PhotoDisc; right, Precision Graphics. 6.3 Lisa Starr with PDB files from NYU Scientific Visualization Lab. 6.4 (a,b) Lisa Starr after Stephen L. Wolfe, Molecular and Ceullar Biology, Wadsworth. 6.5 © Larry West/FPG/Getty Images. 6.6 Art, Raychel Ciemma; micrograph, Carolina Biological Supply Company. 6.7 Left, © Craig Tuttle/CORBIS; (a-c) Art, Lisa Starr with Preface, Inc.; (d) Lisa Starr with Light Harvesting Complex PDB ID: 1RWT; Liu, Z., Yan, H., Wang, K., Kuang, T., Zhang, J., Gui, L., An, X., Chang, W.: Crystal Structure of Spinach Major Light-Harvesting Complex at 2.72 A Resolution Nature 428 pp. 287 (2004). 6.8 Lisa Starr with Preface, Inc. **Page 97** © Darron Luesse, Department of Biology, Indiana University. 6.9 (a, top) Harindar Keer, Thorsten Ritz Laboratory at UC Irvine, Dept. of Physics and Astronomy, using VMD proprietary software; (a, bottom) Lisa Starr; (b-d) Lisa Starr with Light Harvesting Complex PDB ID 1RWT; Liu, Z., Yan, H., Wang, K., Kuang, T., Zhang, J., Gui, L., An, X., Chang, W.: Crystal Structure of Spinach Major Light-Harvesting Complex at 2.72 A Resolution Nature 428 pp. 287 (2004). **Page 100** Top right, Lisa Starr with Light Harvesting Complex PDB ID: 1RWT; Liu, Z., Yan, H., Wang, K., Kuang, T., Zhang, J., Gui, L., An, X., Chang, W.: Crystal Structure of Spinach Major Light-Harvesting Complex at 2.72 A Resolution Nature 428 pp. 287 (2004). 6.10 Lisa Starr. 6.11 Chris Keeney. 6.12 Top left, Courtesy of John S. Russell, Pioneer High School; center left, © Foodpix/Bill Boch; bottom left, © Chris Hellier/CORBIS; top center, micrograph, Bruce Iverson, computer-enhanced by Lisa Starr; center center, micrograph, Ken Wagner/Visuals Unlimited, computer-enhanced by Lisa Starr; bottom center, micrograph, James D. Mauseth, University of Texas; Art, (all) Gary Head. **Page 104** (top left) Lisa Starr. 6.13 Gary Head. 6.14 © E.R. Degginger. 6.15 (a) © Herve Chaumeton/Agence Nature; (b) © Douglas Faulkner/Sally Faulkner Collection.

CHAPTER 7 7.1 © Louise Chalcraft-Frank and FARA. **the big picture** (page 106), right © Professors P. Motta and T Naguro/SPL/Photo Researchers, Inc.; art, Lisa Starr; (page 107) From left, © Randy Faris/CORBIS; © Gary Head; **Page**

107 Top, © Professors P. Motta and T Naguro/SPL/Photo Researchers, Inc. 7.2 Raychel Ciemma and Gary Head. 7.3 (page 108) Top left, © Jim Cummins/CORBIS; top right, © John Lotter Gurling/Tom Stack & Associates; bottom, © Chase Swift/CORBIS; (page 109) Art, Lisa Starr with Gary Head. 7.4 (page 110), Lisa Starr; (page 111) (a-f) Lisa Starr and Gary Head, after Ralph Taggart. 7.5 from left, Raychel Ciemma; Lisa Starr; © Professors P. Motta and T. Naguro/SPL/Photo Researchers, Inc. 7.6 Art, Lisa Starr. 7.7 Gary Head. 7.8 Top, Raychel Ciemma, (below) Lisa Starr. 7.9 Lisa Starr with Preface, Inc. 7.10 (a) © Adrian Warren/Ardea, London; (b) © Foodpix/Ben Fink; (c) ©Foodpix/Ben Fink; (d) Lisa Starr with Gary Head. 7.11 Lisa Starr. 7.12 Left, © Randy Faris/CORBIS; right, Gladden Willis, MD/Visuals Unlimited. 7.13 © Gary Head; art, Lisa Starr. **Page 120** © R. Llewellyn/SuperStock, Inc. **Page 121** Lisa Starr with Gary Head.

Page 123 Unit II © Francis Leroy, Biocosmos/Science Photo Library/Photo Researchers, Inc.

CHAPTER 8 8.1 Left, micrograph, Dr. Pascal Madaule, France; inset, Courtesy of the family of Henrietta Lacks. **the big picture** (page 124) from left, © L. Willatt, East Anglian Regional Genetics Service/SPL/Photo Researchers, Inc.; © Jennifer W. Shuler/Science Source/Photo Researchers, Inc.; (page 125) from left, Gary Head; © Science Photo Library/Photo Researchers, Inc. **Page 125** Micrograph, Dr. Pascal Madaule, France. 8.2 (a,b) Gary Head; bottom © Divital Vision/Getty Images. 8.3 (a) From Allen TD, Jack EM, Harrison CJ (1988). The three dimensional structure of human metaphase chromosomes determined by scanning electron microscopy. In Adolph KW (1988) Chromosomes and Chromatin Volume II, CRC Press. Boca Raton, FL. Chapter 10, Fig.5, p.58; (b) Raychel Ciemma and Lisa Starr; (c) © B. Hamkalo; (d) © O. L. Miller, Jr., Steve L. McKnight. 8.4 Raychel Ciemma and Gary Head. 8.5 Left © L. Willatt, East Anglian Regional Genetics Service/SPL/Photo Researchers, Inc.; (a-c) Lisa Starr. **Page 129** Bottom right, Raychel Ciemma. 8.6 (pages 130-131) Photographs, © Jennifer W. Shuler/Science Source/Photo Researchers, Inc.; Art, Raychel Ciemma. 8.7 (a) Art, Raychel Ciemma; Micrograph, D. M. Phillips/Visuals Unlimited; (b) Art, Lisa Starr; Micrograph, R. Calentine/Visuals Unlimited. 8.8 © Jennifer W. Shuler/Science Source/Photo Researchers, Inc. 8.9 Bottom, © Lennart Nilsson from Behold Man, 1974 by Albert Bonniers Forlog and Little, Brown & Company, Boston. **Page 133** Top, © Lennart Nilsson from A Child Is Born 1966, 1977 Dell Publishing Company, Inc. 8.10 (a, b) © Phillip B. Carpenter, Department of Biochemistry and Molecular Biology. 8.11 © Science Photo Library/Photo Researchers, Inc. 8.12 Betsy Palay. 8.13 Left, © Ken Greer/Visuals Unlimited; center, © Biophoto Associates/Science Source/Photo Researchers, Inc.; right, © James Stevenson/SPL/Photo Researchers, Inc. **Page 136** Raychel Ciemma and Gary Head. 8.14 From Allen TD, Jack EM, Harrison CJ (1988) *The three dimensional structure of human metaphase chromosomes determined by scanning electron microscopy.* In Adolph KW (1988) *Chromosomes and Chromatin Volume II,* CRC Press. Boca Raton FL. Chapter 10, Fig. 5, p. 58.

CHAPTER 9 9.1 (a) © Andrew Syred/Photo Researchers, Inc.; (b) © George D. Lepp/CORBIS; (c) © Dan Kline/Visuals Unlimited. **the big picture** (page 138) From left, Courtesy of Carl Zeiss MicroImaging, Thornwood, NY; Lisa Starr; (page 139) From left, Lisa Starr; © Francis Leroy, Biocosmos/Science Photo Library/Photo Researchers, Inc. **Page 139** © Tomohiro Kono/Tokyo University of Agriculture, enhanced by Lisa Starr. 9.2 Image courtesy of Carl Zeiss MicroImaging, Thornwood, NY. 9.3 Raychel Ciemma. 9.4 © L. Willatt, East Anglian Regional Genetics Service/SPL/Photo Researchers, Inc. **Page 141** Art, Raychel Ciemma. 9.5 (pages 142-143) Photography, Courtesy John Innes Foundation Trustees, computer enhanced by Gary Head; Art, Raychel Ciemma. 9.6 (a,b) Raychel Ciemma; (c–f) Lisa Starr. 9.7 Raychel Ciemma. 9.8 Seth Gold. 9.9 Lisa Starr. 9.10 Art, Lisa Starr; right, © Francis Leroy, Biocosmos/Science Photo Library/Photo Researchers, Inc. 9.11 Lisa Starr. 9.12 © Ron Neumeyer, www.microimaging.ca. 9.13 (a,b) © Lisa O'Connor/ZUMA/Corbis.

CHAPTER 10 10.1 Art, Lisa Starr; left, © Children's Hospital & Medical Center/CORBIS; right © Simon Fraser/RVI, New Castle-Upon-Tyne/SPL/Photo Researchers, Inc. **the big picture** (page 150) From left, © George Lepp/CORBIS; Raychel Ciemma and Precision Graphics; (page 151) From left, © Ted Somes; D. & V. Hennings. **Page 151** Science Photo Library/Photo Researchers, Inc. 10.2 © The Moravian Museum, Brno. 10.3 © Jean M. Labat/Ardea, London; Art, Jennifer Wardrip.

10.4 Lisa Starr. 10.5 Precison Graphics. 10.6 Raychel Ciemma and Precison Graphics. 10.7 Raychel Ciemma and Precison Graphics. 10.8 Raychel Ciemma. 10.9 Raychel Ciemma and Precison Graphics. 10.10 © David Scharf/Peter Arnold, Inc.; Art, Precision Graphics. 10.11 Top, William F. Ferguson; bottom, © Francesc Muntada/CORBIS; Art, Raychel Ciemma. 10.12 All, Ted Somes. 10.13 © Bettman/CORBIS. 10.14 From top, © Frank Cezus/FPG; © Frank Cezus/FPG; © Ted Beaudin/FPG; © Michael Prince/CORBIS. 10.15 (a,b) Gary Head; (c) Courtesy of Ray Carson, University of Florida News and Public Affairs. 10.16 D. & V. Hennings. 10.17 © Pamela Harper/Harper Horticultural Slide Library; Art, Lisa Starr from Prof. Otto Wilhelm Thome, Flora von Deutschland Osterreich und der Schweiz. 1885, Gera, Germany. **Page 163** © Gideon Mendel/CORBIS. 10.18 Left, © Tom and Pat Leeson/Photo Researchers, Inc.; right, © Rick Guidotti, Positive Exposure. 10.19 © Leslie Faltheisek. Clacritter Manx.

CHAPTER 11 11.1 © Reuters/CORBIS. **the big picture** (page 166) From left. © 2001 PhotoDisc, Inc.; © 2001 EyeWire, Inc.; © Stapleton Collection/CORBIS; (page 167) From left, Raychel Ciemma; © Saturn Stills/SPL/Photo Researchers, Inc. **Page 167** © Daniel Weinberger, M.D., E. Fuller Torrey, M.D., Karen Berman, M.D., NIMH Clinical Brain Disorders Branch, Division of Intramural Research Programs, NIMH 1990. From: When Someone Has Schizophrenia, A brief overview of the symptoms, treatments, and research findings. 2001. **Page 168** © Jose Luis Pelaez, Inc./CORBIS. 11.2 Raychel Ciemma and Preface, Inc. 11.3 (a) Precision Graphics and Gary Head; (b) Robert Demarest with permission from M. Cummings, Human Heredity: Principles and Issues, 3rd Edition, p. 126. © 1994 by Brooks/Cole. All rights reserved; (c) Robert Demarest after Patten, Carlson & others; bottom right (girl), © 2001 PhotoDisc, Inc.; (boy), © 2001 EyeWire, Inc. 11.4 Art, Gary Head and Raychel Ciemma; (b) © Charles D. Winters/Photo Researchers, Inc.; (f) © Omikron/Photo Researchers, Inc. 11.5 Raychel Ciemma. 11.6 © Stapleton Collection/CORBIS. 11.7 Art, Precision Graphics; right, © Dr. Victor A. McKusick. 11.8 © Steve Uzzell. 11.9 Top left, © Frank Trapper/Corbis Sygma; (a,b) Lisa Starr. 11.11 Lisa Starr; 11.13 Art, After V. A. McKusick, Human Genetics, 2nd Ed., © 1969. Reprinted by permission of the author; right, © Bettman/CORBIS. 11.14 © Eddie Adams/AP Wide World Photos. **Page 176** Precision Graphics. **Page 177** Precison Graphics. 11.15 (a,b) Courtesy G. H. Valentine. 11.16 From "Multicolor Spectral Karyotyping of Human Chromosomes," by E. Schrock, T. Ried, et al, Science, 26 July 1996, 273:495. Used by permission of E. Schrock, T. Reid and the American Association for the Advancement of Science. 11.17 (a) Raychel Ciemma; (b) © CNRI/Photo Researchers, Inc. 11.18 Preface, Inc. 11.19 © UNC Medical Illustration and Photography. 11.20 © Saturn Stills/SPL/Photo Researchers, Inc. 11.21 From Lennart Nilsson, A Child is Born, (c) 1966, 1977 Dell Publishing Company, Inc.; Art, Lisa Starr. 11.22 © Matthew Alan/CORBIS; inset © Fran Heyl Associates/Jacques Cohen, computer-enhanced by © Pix Elation. 11.23 (a) © Carolina Biological/Visuals Unlimited; (b) © Terry Gleason/Visuals Unlimited; (c) Raychel Ciemma and Preface, Inc. **Page 183** Bottom left, Precison Graphics.

CHAPTER 12 12.1 (a) © C. Barrington Brown, 1968 J. D. Watson; (b) © Mc Leod Murdo/CORBIS Sygma. **the big picture** (page 184) From left, Lisa Starr; Lisa Starr with PDB ID: 1BBB; Silva, M. M., Rogers, P. H., Arnone, A.: A third quaternary structure of human hemoglobin A at 1.7-A resolution. J Biol Chem 267 pp. 17248 (1992); (page 185) From left, Art, Precision Graphics; © James King-Holmes/SPL/Photo Researchers, Inc. **Page 185** © PA News Photo Library. 12.2 Raychel Ciemma; 12.3 Art, Lisa Starr; bottom, © Eye of Science/Photo Researchers, Inc. 12.4 Gary Head 12.5 © SPL/Photo Researchers, Inc. 12.6 Lisa Starr with PDB ID: 1BBB; Silva, M. M., Rogers, P. H., Arnone, A.: A third quaternary structure of human hemoglobin A at 1.7-A resolution. J Biol Chem 267 pp. 17248 (1992). **Page 189** Bottom left, Preface Inc. 12.7 © 1956-2004 The Novartis Foundation (www.novartis-found.org.uk), formerly the Ciba Foundation, reproduced with permission. 12.8 Precision Graphics. 12.9 Precision Graphics. 12.10 Top, (1-3), © James King-Holmes/SPL/Photo Researchers, Inc.; (4) © PA News Photo Library; bottom, from left, © Reuters/Landov; © P.A. NEWS/CORBIS KIPA; © Photo courtesy of the College of Veterinary Medicine, Texas A & M University; © Photo courtesy of the College of Veterinary Medicine, Texas A & M University.

CHAPTER 13 13.1 © Vaughan Fleming/SPL/Photo Researchers, Inc. **the big picture** (page 194), From left (top left) Lisa Starr with PDB ID: 1BBB; Silva, M. M., Rogers, P.H., Arnone, A.: A third quaternary structure of human hemoglobin A at 1.7- Å resolution, J Mol Biol Chem 267 pp. 17248 (1992); Lisa Starr with, Theoretical model; PDB ID: 1K7N; C.Q.LIU, S.X.LIU, SQ.LIU, M.W.JIA, JHE, The Research of MRNA's Conformation and Interaction Between MRNA and TRNA in Translation (To be Published); right/left, © Model by Dr. David B. Goodin, The Scripps Research Institute; right; Model, courtesy of Thomas A. Setitz from Science; (page 195) From left, Lisa Starr; right/left © Nik Kleinberg; right © Steve Terrill/ CORBIS. **Page 195** Lisa Starr with PDB ID: 2AAI; Rutenber, E., Katzin, B.J., Ernst, S., Collins, E.J., Mlsna, D., Ready, M.P., Robertus, J.D.: Crystallographic Refinement of ricin to 2.5 A. Proteins 10pp. 240 (1991). **13.2** (a,b) Precison Graphics; (c) Lisa Starr. **13.3** (a-d) Art, Lisa Starr. **13.4** Gary Head. **13.5** Lisa Starr. **13.6** Precison Graphics and Gary Head. **13.7** Top, Model by Dr. David B. Goodin, The Scripps Research Institute; bottom, Lisa Starr. **13.8** (a) Model, courtesy of Thomas A. Steitz from Science; (b) Lisa Starr. **13.9** (a-k) Lisa Starr. **13.10** Lisa Starr. **13.11** background © Steve Terrill/CORBIS; left inset © Nik Kleinberg; middle inset; © Wayne Armstrong. **13.12** (c) John W. Gofman and Arthur R. Tamplin. From Poisoned Power: The Case Against Nuclear Power Plants Before and After Three Mile Island, Rodale Press, PA, 1979. **13.13** Lisa Starr. **13.14** © Dr. M.A. Ansary/Science Photo Library/Photo Researchers, Inc.

CHAPTER 14 **14.1** (a) From the archives of www.breast-path.com, courtesy of J.B. Askew, Jr., M.D., P.A. Reprinted with permission, © 2004 Breastpath.com; (b) Courtesy of Robin Shoulla and Young Survival Coalition. **the big picture** (page 206) From left, Raychel Ciemma; Lisa Starr; (page 207) From left, From the collection of Jamos Werner and John T. Lis; Courtesy of Edward B. Lewis, California Institute of Technology. **Page 207** Lisa Starr with PDB ID:1N5O; Williams, R. S., Glover, J. N. M.: "Structural Consequences of a Cancer-Causing Brca1-Brct Missense Mutation" J. Biol. Chem. 278 pp. 2630 (2003). **14.2** Raychel Ciemma and Gary Head. **14.3** (left) Lisa Starr with PDB ID: 1CJG; Spronk, C. A. E. M., Bonvin, A. M. J. J., Radha, P. K., Melacini, G., Boelens, R., Kaptein, R.: The Solution Structure of Lac Repressor Headpiece 62 Complexed to a Symmetrical Lac Operator. Structure (London) 7 pp. 1483 (1999). Also PDB ID: 1LBI; Lewis, M., Chang, G., Horton, N. C., Kercher, M. A., Pace, H. C., Schumacher, M. A., Brennan, R. G., Lu, P.: Crystal structure of the lactose operon repressor and its complexes with DNA and inducer. Science 271 pp. 1247 (1996); lactose pdb files from the Hetero-Compound Information Centre - Uppsala (HIC-Up). **Page 209** Lois Ellen Frank/CORBIS. **14.4** Chris Keeney. **14.5** From the collection of Jamos Werner and John T. Lis. **14.6** (a) © Visuals Unlimited; (b) UCSF Computer Graphics Laboratory, National Institutes, NCRR Grant 01081. **14.7** (a) © Dr. Karen Dyer Montgomery; (b) Raychel Ciemma. **14.8** © Jack Carey. **14.9** Top © Lisa Starr; (a-d) © Carolina Biological/Visuals Unlimited. **14.10** Left, © Walter J. Ghering/University of Basel, Switzerland; right, Courtesy of Edward B. Lewis, California Institute of Technology. **14.11** (a) © Palay/Beaubois after Robert F. Weaver and Philip W. Hedrick, Genetics. © 1989 W.C. Brown Publishers; (b,c) © Jim Langeland, Jim Williams, Julie Gates, Kathy Vorwerk, Steve Paddock and Sean Carroll, HHMI, University of Wisconsin-Madison. **14.12** Jim Langeland, Jim Williams, Julie Gates, Kathy Vorwerk, Steve Paddock and Sean Carroll, HHMI, University of Wisconsin-Madison. **14.13** Lawrence Berkeley National Laboratory.

CHAPTER 15 **15.1** Left, © AP/Wide World Photos; center © Charles O'Rear/CORBIS; right © ScienceUV/Visuals Unlimited. **the big picture** (page 218) From left, © Jim Bourg/Reuters/Corbis; right © Professor Stanley Cohen/ SPL/Photo Researchers, Inc; (page 219) From left, © Lowell Georgis/CORBIS; right © Jeans for Gene Appeal. **Page 219** © IRRI Photo Bank Institute (IRRI). **15.2** Left, Lisa Starr; right © Volker Steger/SPL/Photo Researchers, Inc. **15.3** © Jim Bourg/Reuters/Corbis. **Page 222** © Professor Stanley Cohen/SPL/Photo Researchers, Inc. **15.4** (a) Art, Lisa Starr; (b) Lisa Starr. **15.5** Chris Keeney with permission of © QIAGEN, Inc. **15.6** Lisa Starr. **15.7** Lisa Starr. **15.8** Lisa Starr. **15.9** Lisa Starr. **15.10** Left © David Parker/SPL/ Photo Researchers, Inc.; right © Cellmark Diagnostics, Abingdon, UK. **15.11** Top left, © TEK IMAGE/Photo Researchers, Inc.; art, Lisa Starr. **15.12** (a) Courtesy Calgene LLC; (b) © Dr. Vincent Chaing, School of Forestry and Wood Projects, Michigan Technology University. **15.13** (a–c) Lisa Starr; (e) © Keith V. Wood; (d) © Lowell Georgis/CORBIS. **15.14** (a) Transgenic goat produced using nuclear transfer at GTC Biotherapeutics. Photo used with

permission; (b) © Work of Atsushi Miyawaki, Qing Xiong, Varda Lev-Ram, Paul Steinbach, and Roger Y. Tsien at the University of California, San Diego; (c) © Adi Nes, Dvir Gallery Ltd. **15.15** (a) © Jeans for Gene Appeal; (b) Courtesy Dr. Paola Leone, Cell & Gene Therapy Center/ University of Medicine 7 Dentistry of New Jersey. **15.16** © Matt Gentry/Roanoke Times. **15.17** Courtesy of Joseph DeRisa. From Science, 1997 Oct. 24; 278 (5338) 680–686. **15.18** (a) © Mike Stewart/Corbis Sygma; (b) © Simon Kwong/REUTERS/Landov.

Page 235 Unit III © Wolfgang Kaehler/CORBIS.

CHAPTER 16 **16.1** Left, © Bettmann/CORBIS; right, © St Bartholomew's Hospital/Science Photo Library/Photo Researchers, Inc. **the big picture** (page 236) From left, © Christopher Ralling; right; © David Parker/SPL/Photo Researchers, Inc.; (page 237) Left/left, © Peter Bowater/ Photo Researchers, Inc.; left/right, © Owen Franken/ CORBIS; right, Gary Head. **Page 237** Top, © Bettmann/ CORBIS. **16.2** (a) © Wolfgang Kaehler/CORBIS; (b) © Earl & Nazima Kowall/CORBIS; (c) © Wolfgang Kaehler/ CORBIS; (d) © Edward S. Ross; (e) © Edward S. Ross. **16.3** (left) Art, Gary Head; (right) © Bruce J. Mohn; (inset) © Phillip Gingerich, Director, University of Michigan. Museum of Paleontology. **16.4** (a) © Jonathan Blair/CORBIS; (b) © Biophoto Associates/Photo Researchers, Inc.; (c) © John Shaw/Photo Researchers, Inc. **Page 16.5** (a) Courtesy George P. Darwin, Darwin Museum, Down House; (b) © Christopher Ralling; (c) © Heather Angel; (c,d) Art, Precision Graphics (e) © Dieter & Mary Plage/Survival Anglia. **16.6** (a) © Joe McDonald/CORBIS; (b) © Karen Carr Studio www.karencarr.com. **16.7** (a) © Heather Angel; (b) © Kevin Schafer/CORBIS; (c) © Dr. P. Evans/Bruce Coleman, Ltd.; (d) © Alan Root/Bruce Coleman Ltd.; **16.8** © Down House and The Royal College of Surgeons of England; (b) © Gary Head; (b) © John M. Merck, Jr., University of Maryland. **16.9** (a) © Gary Head; (b) © John M. Merck, Jr., University of Maryland. **16.10** Left, © Thomas Mangelsen; right, © Theo Allofs/CORBIS. **16.11** Left, © Francois Photo Researchers, Inc. **16.12** Left, © Peter Bowater/Photo Researchers, Inc.; top center © Owen Franken/CORBIS; top right © Sam Kleinman/CORBIS; center, © Jim Cornfield/ CORBIS; center right, © Christopher Briscoe/Photo Researchers, Inc.; bottom, © Alan Solem. **16.13** Gary Head. **Page 248** © Terry Whittaker, Frank Lane Picture Agency/ CORBIS. **16.14** Gary Head. **16.15** all, Courtesy of Hopi Hoekstra, University of California, San Diego. **16.16** Gary Head. **16.17** Precision Graphics, using NIH data. **16.18** © Peter Chadwick/Science Photo Library/Photo Researchers. **16.19** Left, © Thomas Bates Smith; right, © Thomas Bates Smith. **16.20** © Bruce Beehler, enhanced by Chris Keeney. **16.21** (a, b) Precision Graphics after Ayala and others; bottom, © Michael Freeman/CORBIS. **Page 254** © Steve Bronstein/The Image Bank/Getty Images. **16.22** Precision Graphics, after computer models developed by Jerry Coyne. **16.23** © Frans Lanting/Minden Pictures (computer-modified by Lisa Starr); Art, Raychel Ciemma. **16.24** Left, © David Neal Parks; right, © W. Carter Johnson. **16.25** John Kalusmeyer, University of Michigan Exhibit of Natural History.

CHAPTER 17 **17.1** (a) NASA Galileo Imaging Team; (b) top, Art by Don Davis; (b) bottom NASA Galileo Imaging Team. **the big picture** (page 258) from left, Raychel Ciemma; right (both) © Jack Jeffrey Photography; (page 259), from left © Carnegie Museum of Natural History; right, Gary Head. **Page 259,** Top, © David A. Kring, NASA/Univ. Arizona Space Imagery Center. **17.2** left © H. P. Banks; right, © Jonathan Blair. **17.3** © Jonathan Blair/ CORBIS. **17.4** (a) Gary Head; (b) © 2001 PhotoDisc, Inc; (c,d) Lisa Starr. **Page 263, 17.6** © CORBIS. **17.7** (a) Leif Buckley; (b) Leif Buckley and Lisa Starr. **17.8** (a-d) Lisa Starr, after A. M. Ziegler, C. R. Scotese, and S. F. Barrett, "Mesozoic and Cenozoic Paleogeographic Maps," and J. Krohn and J. Sundermann (Eds.), Tidal Frictions and the Earth's Rotation II, Springer-Verlag, 1983; (e) © Martin Land/Photo Researchers, Inc.; (f) © John Sibbick. **17.9** Raychel Ciemma. **17.10** (a) J. Scott Altenbach, University of New Mexico, computer enhanced by Lisa Starr; (b) © Frans Lanting/Minden Pictures; computer enhanced by Lisa Starr; (c) © Stephen Dalton/Photo Researchers, Inc.; art, Lisa Starr and Raychel Ciemma, art reference, Natural History Collection, Royal BC Museum; **17.11** (a) Courtesy of Professor Richard Amasino, University of Wisconsin-Madison; (b) Courtesy of Professor Richard Amasino, University of Wisconsin-Madison; (c) Courtesy of Professor Martin F. Yanofsky; (d) © Jose Luis Riechmann; **17.12** Raychel Ciemma; **17.13** (a) top, © Courtesy of Prof. Richard Amasino, University of Wisconsin-Madison; (a) bottom, © Jennifer Grenier, Grace Boekhoff-Falk and Sean Carroll, HMI, University of Wisconsin-Madison; (b) top, © Herve

Chaumeton/Agence Nature; (b) bottom, © Jennifer Grenier, Grace Boekhoff-Falk and Sean Carroll, HMI, University of Wisconsin-Madison; (c) top, © Peter Skinner/Photo Researchers, Inc.; (c) bottom, Courtesy of Dr. Giovanni Levi; (d) © Dr. Chip Clark. **Page 270** © TEK IMAGE/Photo Researchers, Inc.; **17.14** Precision Graphics. **17.15** left © Kjell B. Sandved/Visuals Unlimited; center, © Jeffrey Sylvester/FPG/Getty Images; right, © Thomas D. Mangelsen/Images of Nature. **17.16** (a, b) Art, Jennifer Wardrip. **17.17** (a) © John Alcock, Arizona State University; (b) © Tui Roy/Minden Pictures; **17.17** (c) © Alvin E. Staffan/ Photo Researchers, Inc. **17.18** Top left, © Digital Vision/ PictureQuest; top, © Joe McDonald/CORBIS. **17.19** (a) © Graham Neden/CORBIS; (b) © Kevin Schafer/CORBIS; center, © Ron Blakey, Northern Arizona University (c) © Rick Rosen/Corbis SABA. **17.20** (a-c) Preface, Inc. (d) all, © Jack Jeffrey Photography. **17.21** Top, © Steve Gartlan; bottom left and right © Below Water Photography/ www.belowwater.com. **17.22** left © Lance Nelson/CORBIS; center, © Eric Crichton/CORBIS; right, © Maximilian Stock Ltd./Foodpix. **17.23** Lisa Starr after W. Jensen and F. B. Salisbury, Botany: An Ecological Approach, Wadsworth, 1972. **17.24** (a) Courtesy of Dr. Robert Mesibov; (b) Courtesy of Dr. Robert Mesibov; **17.24** (c) Lisa Starr. **17.25** Courtesy of Daniel C. Kelley, Anthony J. Arnold, and William C. Parker, Florida State University Department of Geological Science. **Page 278** (bottom left) Gary Head. **17.26** Left, Lisa Starr; right © Carnegie Museum of Natural History. **17.27** from left © Science Photo Library/Photo Researchers, Inc.; © Galen Rowell/CORBIS; © Kevin Schafer/CORBIS; Courtesy of Department of Entomology, University of Nebraska-Lincoln; Bruce Coleman, Ltd. **17.28** Gary Head. **17.29** Gary Head. **17.30** from left, © Phillip Colla Photography; © Randy Wells/CORBIS © Cousteau Society/Getty Images; © Robert Dowling/CORBIS; Art, Gary Head. **17.31** Lisa Starr. **17.32** Lisa Starr. **17.33** © Gulf News, Dubai, UAE

CHAPTER 18 **18.1** Inset, Courtesy of Agriculture Canada; © Raymond Gehman/CORBIS. **the big picture** (page 286) From left, © Jeff Hester and Paul Scowen, Arizona State University, and NASA; ©Sidney W. Fox; (page 287) From left, © Chase Studios/Photo Researchers, Inc.; Raychel Ciemma. **Page 287** © Philippa Uwins/The University of Queensland. **18.2** © Jeff Hester and Paul Scowen, Arizona State University, and NASA. **18.3** © Chesley Bonestell; inset art, Raychel Ciemma. **18.4** (a) © Tim Thompson/ CORBIS; (b) © Dr. Ken MacDonald/SPL/Photo Researchers, Inc.; (c) © Micheal J. Russell, Scottish Universities Environmental Research Centre. **18.5** (a) © Sidney W. Fox; (b) From Hanczyc, Fujikawa, and Szostak, Experimental Models of Primitive Cellular Compartments: Encapsulation, Growth, and Division;www.sciencemag.org Science 24 October 2003; 302;529, Figure 2, page 619. Reprinted with permission of the authors and AAAS; (c) Preface, Inc. **18.6** (a) © Stanley M. Awramik; (b,c) Bruce Runnegar, NASA Astrobiology Institute; (d) © N.J. Butterfield, University of Cambridge. **18.7** (a) © Chase Studios/Photo Researchers, Inc.; (b) © John Reader/SPL/Photo Researchers, Inc.; (c) © Sinclair Stammers/SPL/Photo Researchers, Inc. **18.8** Raychel Ciemma. **18.9** Left, © Robert Trench, Professor Emeritus, University of British Columbia; (b) right, Courtesy of Isao Inouye, Institute of Biological Sciences, University of Tsukuba. **18.10** Raychel Ciemma and Precision Graphics.

Page 299 Unit IV © Layne Kennedy/CORBIS.

CHAPTER 19 **19.1** © The Bridgeman Art Library/Getty Images. **the big picture** (page 300) From left, Lisa Starr; (page 301) From left, © NIBSC/SPL/Photo Researchers, Inc.; © Camr, Barry Dowsett/Science Photo Library/Photo Researchers, Inc. **Page 301** © Carl Cook. **19.2** (a,b) Lisa Starr. **19.3** (a) © P. Hawtin, University of Southampton/ SPL/Photo Researchers, Inc.; (b) © Dr. Dennis Kunkel/ Visuals Unlimited. **19.4** Raychel Ciemma. **19.5** Lisa Starr. **19.6** (a) © Dr. Jeremy Burgess/SPL/Photo Researchers, Inc.; (b) © P. W. Johnson and J. MeN. Sieburth, Univ. Rhode Island/BPS. **19.7** © Dr. Terry J. Beveridge, Department of Microbiology, University of Guelph, Ontario, Canada. **19.8** © Dr. Manfred Schloesser, Max Planck Institute for Marine Microbiology. **19.9** (a) © Stem Jems/Photo Researchers, Inc.; (b) © California Department of Health Services; (c) © Bernard Cohen, M.D., Dermatlas; http://www.darmatlas .org. **19.10** Lisa Starr. **19.11** (a) © Courtesy Jack Jones, Archives of Microbiology, Vol. 136, 1983, pp. 254-261. Reprinted by permission of Springer-Verlag. (b) © Dr. John Brackenbury/Science Photo Library/Photo Researchers, Inc. **19.12** (a) © Martin Miller/Visuals Unlimited; (b) © Alan L. Detrick, Science Source/Photo Researchers, Inc.; (c) © Dr. Harald Huber, Dr. Michael Hohn, Prof. Dr. K.O. Stetter, University of Regensburg, Germany. **19.13** (a-d)

Wagner/University of Delaware, www.udel.edu/Biology/Wags; right/left © Pat Johnson Studios Photography; right/right © Darrell Gulin/Getty Images; (page 403) From left, Gary Head; Slim Films. **Page 403** © Star Tribune/Minneapolis-St. Paul. **24.2** Left, Raychel Ciemma; top, Courtesy of Charles Lewallen; center & bottom, © Bruce Iverson. **24.3** left © 2000 PhotoDisc, Inc, with art by Lisa Starr; top right © Cnri/Spl/Photo Researchers, Inc.; bottom right © Dr. Robert Wagner/University of Delaware, www.udel.edu/Biology/Wags. **24.4** Left © Pat Johnson Studios Photography; right, © Darrell Gulin/The Image Bank/Getty Images. **24.5** (a) © Cory Gray; (b) © PhotoDisc/Getty Images; (c) © Heather Angel; (d) © Biophoto Associates/Photo Researchers, Inc. **24.6** Gary Head. **24.7** © Galen Rowell/Peter Arnold, Inc. **24.8** Gary Head. **24.9** Left art, Gary Head; right, © Niall Benvie/CORBIS. **24.10** Left © Kennan Ward/CORBIS; right © G. J. McKenzie (MGS). **24.11** © Frank B. Salisbury. **24.12** Slim Films. **24.13** (a,b) Courtesy of Dr. Kathleen K. Sulik, Bowles Center for Alcohol Studies, the University of North Carolina at Chapel Hill; (c) © John DaSiai, MD/Custom Medical Stock. **24.14** (a) Courtesy of Dr. Consuelo M. De Moraes; (b-d) © Andrei Sourakov and Consuelo M. De Moraes.

Page 415 Unit V © Jim Christensen, Fine Art Digital Photographic Images.

CHAPTER 25 **25.1** Left, © Michael Westmoreland/CORBIS; right, © Reuters/CORBIS. **the big picture** (page 416) From left, left/left, © Ernest Manewal/Index Stock Imagery. left/right © Darrell Gulin/CORBIS; right, © Andrew Syred/Photo Researchers, Inc.; (page 417) From left, © Mike Clayton/University of Wisconsin Botany Department, Lisa Starr. **Page 417** © Allstock/Stone/Getty Images. **25.2** Raychel Ciemma. **25.3** Raychel Ciemma. **25.4** (a) From top, © Bruce Iverson; (a) © Ernest Manewal/Index Stock Imagery; (a) © Simon Fraser/Photo Researchers, Inc.; (a) © Andrew Syred/Photo Researchers, Inc.; Art, Gary Head; (b) from top, © Mike Clayton/University of Wisconsin Department of Botany; (b) © Darrell Gulin/CORBIS; (b) © Gary Head; © Andrew Syred/Photo Researchers, Inc.; (b) art, Gary Head. **25.5** Art, Precision Graphics. **25.6** Left, © Donald L. Rubbelke/Lakeland Community College; right © Andrew Syred/Photo Researchers, Inc. **25.7** (a) © Dr. Dale M. Benham, Nebraska Wesleyan University; (b) © D. E. Akin and I. L. Risgby, Richard B. Russel Agricultural Research Center, Agricultural Research Service, U.S. Dept. Agriculture, Athens, GA; (c) © Kingsley R. Stern. **25.8** (a-c) Lisa Starr; bottom © Andrew Syred/Photo Researchers, Inc. **25.9** © George S. Ellmore. **25.10** (a) top © Gary Head; (a) bottom, © Dale M. Benham, Ph.D., Nebraska Wesleyan University; (b-d) Raychel Ciemma. **25.11** All art, D. & V. Hennings; (a) center; © Mike Clayton/University of Wisconsin Botany Department; © James W. Perry; (b) center, © Carolina Biological Supply Company; right © James W. Perry. **25.12** All art, D. & V. Hennings. **Page 424** © 2001 PhotoDisc, Inc./Getty. **25.13** © David Cavagnaro/Peter Arnold, Inc. **25.14** (a) © N. Cattlin/Photo Researchers, Inc.; (b) Raychel Ciemma; (c) © C. E. Jeffree, et al., Planta, 172(1): 20-37, 1987. Reprinted by permission of C. E. Jeffree and Springer-Verlag; (d) © Jeremy Burgess/SPL/Photo Researchers, Inc. **25.15** (a) © Simon Fraser/Photo Researchers, Inc.; (b) © Gary Head. **25.16** (a) Left, © John Limbaugh/Ripon Microslides, Inc.; art, After Salisbury and Ross, Plant Physiology, Fourth Edition, Wadsworth; (b) top, © Mike Clayton/University of Wisconsin Department of Botany; (b) bottom, © Mike Clayton/University of Wisconsin Department of Botany. **25.17** © Carolina Biological Supply Company; **25.18** Left, © Omikron/Photo Researchers, Inc.; center, © Omikron/Photo Researchers, Inc.; right, © Omikron/Photo Researchers, Inc. **25.19** (a) Raychel Ciemma; **25.20** Raychel Ciemma **25.21** (a,c) Lisa Starr; (b) © H. A. Core, W. A. Cote, and A. C. Day, Wood Structure and Identification, 2nd Ed., Syracuse University Press, 1979. **Page 430** Bottom right, D. & V. Hennings. **25.22** (a,b) © Edward S. Ross. **Page 431** Top left, Lisa Starr. **25.23** (a) © Jon Pilcher; (b) © George Bernard/SPL/Photo Researchers, Inc.; (c) © Peter Ryan SPL/Photo Researchers, Inc.

CHAPTER 26 **26.1** Left, © OPSEC Control Number #4 077-A-4; right, © Billy Wrobel, 2004. **the big picture** (page 432) From left, © William Ferguson; © Andrew Syred/Photo Researchers, Inc.; (page 433) From left, left/left, © Jeremy Burgess/SPL/Photo Researchers, Inc.; left/right © Jeremy Burgess/SPL/Photo Researchers, Inc.; right © J.C. Revy/Science Photo Library/Photo Researchers, Inc. **Page 433** © Keith Weller/USDA-ARS. **26.2** (a) © William Ferguson; (b) © Robert Frerch/Stone/Getty Images; (c) Courtesy of NOAA. **26.3** © Andrew Syred/Photo Researchers, Inc.; **26.4** © Prof. DJ Read, University of Sheffield; **26.5** (a) © Adrian P. Davies/Bruce Coleman; (a)

art, Jennifer Wardrip (b) © NifTAL Project, Univ. of Hawaii, Maui. **26.6** (a, c) © Leonard Morgan; (b) © Mike Clayton/University of Wisconsin Department of Botany; (b, c-right) Lisa Starr; **26.7** Left, © Alison W. Roberts, University of Rhode Island; center, © H. A. Core, W. A. Cote and A. C. Day, *Wood Structure and Identification*, 2nd Ed., Syracuse University Press, 1979; right, © H. A. Core, W. A. Cote and A. C. Day, *Wood Structure and Identification*, 2nd Ed., Syracuse University Press, 1979; art, Lisa Starr. **26.8** Left, © The Ohio Historical Society, Natural History Collections; Raychel Ciemma. **26.9** (a,b) © Claude Nuridsany & Marie Perennou/Science Photo Library/Photo Researchers, Inc.; **26.10** © micrograph by Ken Wagner/Visuals Unlimited, computer-enhanced by Lisa Starr. **26.13** (a) © Don Hopey/Pittsburgh Post-Gazette, 2002, all rights reserved. Reprinted with permission; (b) © Jeremy Burgess/SPL/Photo Researchers, Inc.; (c) © Jeremy Burgess/SPL/Photo Researchers, Inc. **26.14** (a) © James D. Mauseth, MCDB; (b) © J.C. Revy/Science Photo Library/Photo Researchers, Inc.; (c) © Martin Zimmerman, Science, 1961, 133:73-79, © AAAS. **26.15** Left, Gary Head; right, Palay/Beaubois; bottom right, Precision Graphics. **26.16** (a) © NOAA; (b) USDA/Forestry Service; (c) © David W. Stahle, Department of Geosciences, University of Arkansas.

CHAPTER 27 **27.1** Left, Courtesy of Caroline Ford, School of Plant Sciences, University of Reading, UK; right © James L. Amos/CORBIS. **the big picture** (page 446) From left, Raychel Ciemma; © Herve Chaumeton/Agence Nature; (page 447) From left, © Cathlyn Melloan/Stone/Getty Images; © Roger Wilmshurst; Frank Lane Picture Agency/CORBIS. **Page 447** © Dr. John Hilty. **27.2** Art, Raychel Ciemma and Precision Graphics; (b) © John McAnulty/CORBIS; (b) right, © Robert Essel NYC/CORBIS. **27.3** (a) © David M. Phillips/Visuals Unlimited; (b) © Dr. Jeremy Burgess/SPL/Photo Researchers, Inc.; (c) © David Scharf/Peter Arnold, Inc. **27.4** Left. © John Alcock, Arizona State University; center, (bat) © Merlin D. Tuttle, Bat Conservation International; Top & bottom right, © Thomas Eisner, Cornell University. **27.5** Raychel Ciemma. **27.6** Left/left, © Michael Clayton, University of Wisconsin; bottom left, Raychel Ciemma; center/left, © Michael Clayton, University of Wisconsin; center, © Dr. Charles Good, Ohio State University, Lima; center/right, © Michael Clayton, University of Wisconsin; right/right, © Michael Clayton, University of Wisconsin. **27.7** (a) © R. Carr; (b) © Gregory K. Scott/Photo Researchers, Inc.; (c) © Robert H. Mohlenbrock USDA-NRCS PLANTS Database/USDA SCS. 1989. "Midwest wetland flora; field office illustrated guide to plant species." Midwest national Technical Center, Lincoln, NE; (c) From left, © Dr. Dan Legard, University of Florida GCREC, 2000; (d) © Andrew Syred/SPL/Photo Researchers, Inc.; (d) © Richard H. Gross. **27.8** © Darrell Gulin/CORBIS; **27.9** (a-e) © Professor Dr. Hans Hanks-Ulrich Koop. **27.10** © Mike Clayton/University of Wisconsin Department of Botany. **27.11** Art, Raychel Ciemma; top right, © Barry L. Runk/Grant Heilman, Inc.; center, © James D. Mauseth, University of Texas; bottom right, © Herve Chaumeton/Agence Nature. **27.12** © Sylvan H. Wittwer/Visuals Unlimited; **27.13** Top left, © Adam Hart-Davis/Photo Researchers, Inc.; (a-e) Art, Gary Head; (f) both, © Eric B. Brennan. **27.14** (a,b) Lisa Starr; © Cathlyn Melloan/Stone/Getty Images. **Page 460** Bottom left, Gary Head. **27.15** (a) © Michael Clayton, University of Wisconsin; (b,c) © Muday, GK and P. Haworth (1994) "Tomato root growth, gravitropism, and lateral development: Correlations with auxin transport." "Plant Physiology and Biochemistry 32, 193-203" with permission from Elsevier Science. **27.16**: (a,b) Micrographs courtesy of Randy Moore from "How Roots Respond to Gravity" M. L. Evans, R. Moore, and K. Hasenstein, Scientific American, December 1986. **27.17** © Gary Head. **27.18** (a-d) Art, Gary Head; (e) © Ray Evert, University of Wisconsin. **27.20** (a) Top, © Juergen Berger, Max Planck Institute for Developmental Biology? Tuebingen, Germany; (a, below and b-all) © Jose Luis Riechmann; (c) art, Lisa Starr. **27.21** Left, © Roger Wilmshurst; Frank Lane Picture Agency/CORBIS; right, © Dr. Jeremy Burgess/Photo Researchers, Inc. **27.22** Left, © Eric Welzel/Fox Hill Nursery, Freeport, Maine; art, Lisa Starr. **27.23** © Grant Heilman Photography, Inc.

Page 467 Unit VI © Kevin Schafer.

CHAPTER 28 **28.1** Left, © Ohlinger Jerry/CORBIS Sygma; right, © Sachs Ron/CORBIS Sygma. **the big picture** (page 468) From left, © Ray Simmons/Photo Researchers, Inc.; © Ed Reschke; ©Ed Reschke; © Triarch/Visuals Unlimited; (page 469) From left, © Mauro Fermariello/Photo Researchers, Inc.; Robert Demarest. **Page 469** © Science Photo Library/Photo Researchers, Inc. **28.2** (a) © Manfred Kage/Bruce Coleman, Ltd.; (b) Top right, © Focus on Sports; (c) left © Ray Simmons/Photo Researchers, Inc.; (c)

center, © Ed Reschke/Peter Arnold, Inc.; (c) right © Don W. Fawcett; all art, Lisa Starr. **28.3** Top, © Gregory Dimijian/Photo Researchers, Inc.; bottom, Raychel Ciemma, adapted from C. P. Hickman, Jr., L. S. Roberts, and A. Larson, *Integrated Principles of Zoology*, Ninth Edition, Wm. C. Brown, 1995. **28.4** Leif Buckley. **28.5** (a) top, © John Cunningham/Visuals Unlimited; (b,c) top © Ed Reschke; (d) top, © Science Photo Library/Photo Researchers, Inc.; (e) top, © Michael Abbey/Photo Researchers, Inc.; (f) top, © University of Cincinnati, Raymond Walters College, Biology (a-f) bottom, Lisa Starr and Leif Buckley. **28.6** © Roger K. Burnard; Art, left, Joel Ito; right, L. Calver. **28.7** © Science Photo Library/Photo Researchers, Inc. **28.8** Top, © Tony Mcconnell/Science Photo Library/Photo Researchers, Inc.; (a,b) © Ed Reschke; (c) © Biophoto Associates/Photo Researchers, Inc. **28.9** Left, © Kim Taylor/Bruce Coleman, Ltd.; right, © Triarch/Visuals Unlimited. **28.10** (a) © Mauro Fermariello/Photo Researchers, Inc.; (b) Courtesy Nature Biotechnology. **28.11** (a-c) Art, Palay/Beaubois; (d) Art, Seth Gold and Lisa Starr. **28.12** (a) Robert Demarest; (b) © John D. Cunningham/Visuals Unlimited. **28.13** Robert Demarest. **28.14** (a) Lisa Starr; (b) © Frank Trapper/CORBIS Sygma; (c) © AFP/CORBIS. **Page 479** Bottom right, © Pascal Goetgheluck/Science Photo Library/Photo Researchers, Inc. **Page 480** © Dr. Preston Maxim and Dr. Stephen Bretz, Department of Emergency Services, San Francisco General Hospital. **28.15** © Sean Sprague/Stock, Boston. **28.16** © David Macdonald.

CHAPTER 29 **29.1** (a) © CORBIS; (b) © PA Photos. **the big picture** (page 482) Left and right, Lisa Starr; (page 483) From left, Lisa Starr; Kevin Somerville and Precision Graphics; © C. Yokochi and J. Rohen, *Photographic Anatomy of the Human Body*, 2nd Ed., Igaku-Shoin, Ltd., 1979. **Page 483** © Manni Mason's Pictures. **29.2** (a) Robert Demarest; (b) Lisa Starr. **29.3** Kevin Somerville. **29.4** Lisa Starr. **29.5** Lisa Starr. **29.6** (a,b,d,e) Art, Precision Graphics; (c) © Jeff Greenberg/Index Stock Imagery. **29.7** (a, c) Lisa Starr; (b) © Dr. Constantino Sotelo from International *Cell Biology*, p. 83, 1977. Used by permission of the Rockefeller University Press. **29.8** Left, Micrograph by Don Fawcett, Bloom and Fawcett, 11th edition, after J. Desaki and Y. Uehara/Photo Researchers, Inc.; right, Kevin Somerville. **29.9** Gary Head. **29.10** (a, b) Robert Demarest. **29.11** Robert Demarest. **29.12** (a–c) Raychel Ciemma and Lisa Starr. **29.13** Kevin Somerville and Precision Graphics. **29.14** Robert Demarest and Precision Graphics. **29.15** Left art, Robert Demarest; right, © Manfred Kage/Peter Arnold, Inc. **29.16** Kevin Somerville. **29.17** Kevin Somerville. **29.18** (a) art, Raychel Ciemma and Lisa Starr; (a) © Colin Chumbley/Science Source/Photo Researchers, Inc.; (b) © C. Yokochi and J. Rohen, *Photographic Anatomy of the Human Body*, 2nd Ed., Igaku-Shoin, Ltd., 1979. **29.19** (a) Palay/Beaubois after Penfield and Rasmussen, *The Cerebral Cortex of Man*, © 1950 Macmillan Library Reference. Renewed 1978 by Theodore Rasmussen. Reprinted by permission of The Gale Group.; (b) © Colin Chumbley/Science Source/Photo Researchers, Inc. **29.20** (a) Raychel Ciemma; (b) © Marcus Raichle, Washington Univ. School of Medicine. **29.21** Lisa Starr. **29.23** (a,b) © E. D. London, et al., *Archives of General Psychiatry*, 47:567-574, 1990. **29.24** © Ed Kashi/CORBIS. **29.25** Art, Robert Demarest. © Herve Chaumeton/Agence Nature. **29.26** © Kathy Plonka, *The Spokesman-Review*.

CHAPTER 30 **30.1** © AP/Wide World Photos. **the big picture** (page 504) From left, © David Turnley/CORBIS; Art, Palay/Beaubois after Penfield and Rasmussen, *The Cerebral Cortex of Man*, © 1950 Macmillan Library Reference. Renewed 1978 by Theodore Rasmussen. Reprinted by permission of The Gale Group; © Colin Chumbley/Science Source/Photo Researchers, Inc; (page 505) From left, Robert Demarest; © Lennart Nilsson © Boehringer Ingelheim International GmbH. **Page 505** © Phillip Colla, Ocean-Light.com. All Rights Reserved Worldwide. **30.2** (a) From Hensel and Bowman, *Journal of Physiology*, 23:564-568, 1960; (b) © David Turnley/CORBIS. **30.3** Art, Palay/Beaubois after Penfield and Rasmussen, *The Cerebral Cortex of Man*, © 1950 Macmillan Library Reference. Renewed 1978 by Theodore Rasmussen. Reprinted by permission of The Gale Group; right, © Colin Chumbley/Science Source/Photo Researchers, Inc. **30.4** Raychel Ciemma. **30.5** Precision Graphics. **30.6** Robert Demarest. **30.7** Top, Kevin Somervill; Lisa Starr. **Page 509** © AFP Photo/Timothy A. Clary/CORBIS. **30.8** Gary Head. **30.9** (a,b,e) Art, Robert Demarest; (a) right, © Fabian/CORBIS Sygma; (c) Precision Graphics; (d) © Medtronic Xomed; (e) top, © Dr. Thomas R. Van De Water, University of Miami Ear Institute. **30.10** Both, © Robert E. Preston, courtesy Joseph E. Hawkins, Kresge Hearing Research Institute, University of Michigan Medical School; **30.11** (a) top, © E. R. Degginger; (a) art, Raychel Ciemma after M. Gardiner, The Biology of Vertebrates, McGraw-Hill, 1972; (b, c) Raychel Ciemma.

30.12 Robert Demarest. 30.13 (a) © Richard Megna/ Fundemental Photographs, NYC; (b,c) Kevin Somerville, enhanced by Lisa Starr. 30.14 Top, © Lennart Nilsson © Boehringer Ingelheim International GmbH; art, Micrograph, Lennart Nilsson © Boehringer Ingelheim International. 30.15 www.2.gasou.edu/psychology/ courses/muchinsky and www.occipita.cfa.cmu.edu. 30.16 Kevin Somerville. 30.17 Courtesy of Dr. Bryan Jones, University of Utah School of Medicine. 30.18 © Eric A. Newman, enhanced by Lisa Starr. 30.19 © Chase Swift. 30.20 © F. Spoor using Voxel-man.

CHAPTER 31 31.1 Left, Courtesy of Michael Lannoo; right; © David Aubrey/CORBIS. **the big picture** (page 518) From left, Lisa Starr with PDB ID: 1MSO; Smith, G. D., Pangborn, W. A., Blessing, R. H.: *The Structure of T6 Human Insulin at 1.0 A Resolution Acta Crystallogr., Sect.D 59 pp. 474 (2003)*; (page 519) From left, Courtesy of Dr. Erica Eugster; © James King-Holmes/Photo Researchers, Inc. **Page 519** Courtesy of Wandering Eye Production. Used by permission. **Page 520** Top left, Lisa Starr with PDB ID: 1MSO; Smith, G. D., Pangborn, W. A., Blessing, R. H.: *The Structure of T6 Human Insulin at 1.0 A Resolution Acta Crystallogr., Sect.D 59 pp. 474 (2003)*. 31.2 Kevin Somerville. 31.3 Lisa Starr 31.4 Robert Demarest. 31.5 Art, Robert Demarest; top right, Courtesy of Dr. Erica Eugster; bottom right, © Lisa Starr. 31.6 Leornard Morgan. 31.7 (a) © Gary Head; (a) Art, Raychel Ciemma; (b) © Bettmann/CORBIS. 31.8 © Biophoto Associates/SPL/Photo Researchers, Inc. 31.9 Leonard Morgan. 31.10 (a) © The Stover Group/D. J. Fort; (b) © Joeseph Kiesecker, Penn State University. 31.11 (a,b) © James King-Holmes/Photo Researchers, Inc. 31.12 Top, © Frans Lanting/Bruce Coleman, Ltd.; art, From R. C. Brusca and G. J. Brusca, *Invertebrates,* © 1990 Sinauer Associates. Used by permission. 31.13 © Dr. Carlos J. Bourdony. Page 534 Robert Demarest 31.14 Courtesy of G. Baumann, MD, Northwestern University. 31.15 © Kevin Fleming/CORBIS.

CHAPTER 32 32.1 © Michael Neveux. **the big picture** (page 536) Feom left, D. & V. Hennings; Joel Ito; (page 537) From left, Robert Demarest; Robert Demarest. **Page 537** Steve Cole/PhotoDisc Green/Getty Images. 32.2 © Linda Pitkin/Planet Earth Pictures. 32.3 Bottom left, © Stephen Dalton/Photo Researchers, Inc.; art, Precision Graphics. 32.4 Bottom left, © Yokochi and J. Rohen, *Photographic Anatomy of the Human Body*, 2nd Ed., Igaku-Shoin, Ltd., 1979; top art, Raychel Ciemma; bottom art, D. & V. Hennings. 32.5 (a,b) Joel Ito; top right © Ed Reschke. 32.6 K. Kasnot. 32.7 (a,b) © Professor P. Motta/Department of Anatomy/ La Sapienza, Rome/SPL/Photo Researchers, Inc. 32.8 Robert Demarest. 32.9 © N.H.P.A./ANT Photolibrary. 32.10 (a,b) Raychel Ciemma. 32.11 © bottom, Dance Theatre of Harlem, by Frank Capri; (b,c) © Don Fawcett/ Visuals Unlimited, from D. W. Fawcett, The Cell, Philadelphia; W. B. Saunders Co., 1966; (a) Art, Robert Demarest; compilation by Gary Head. 32.12 Nadine Sokol and Gary Head. 32.14 Gary Head. 32.15 Kevin Somerville and Gary Head. **Page 547** © Sarto-Lund/Green/Getty Images. 32.16 Painting by Sir Charles Bell, 1809, courtesy of Royal College of Surgeons, Edinburgh. 32.17 (a) © Paul Sponseller, MD/Johns Hopkins Medical Center; (b) Courtesy of the family of Tiffany Manning.

CHAPTER 33 33.1 © Mark Thomas/Science Photo Library. **the big picture** (page 550) From left, Precision Graphics; right © National Cancer Institute/Photo Researchers, Inc.; (page 551) From left, © Sheila Terry/ SPL/Photo Researchers, Inc.; Lisa Starr. **Page 551** Top right, © Faye Norman/Science Photo Library; bottom left, From A. D. Waller *Physiology: The Servant of Medicine*, Hitchcock Lectures, University of London Press, 1910. 33.2 Top left, © Darlyne A Murawski/Getty Images; bottom left, © Getty Images; (a,c) art, Raychel Ciemma; (b,d) Precision Graphics. 33.3 (a-c) Precision Graphics; (d) After M. Labarbera and S. Vogel, American Scientist, 1982, 70:54-60. 33.4; Top, © National Cancer Institute/Photo Researchers, Inc.; 33.4 Art, Lisa Starr. 33.5 Left, © 2001 EyeWire; Art, Lisa Starr with art references from *Bloodline Image Atlas*, University of Nebraska-Omaha/Sherri Wicks, Human Physiology and Anatomy, University of Wisconsin Biology Web Education System, and others. 33.6 (a) © Lester V. Bergman & Associates, Inc.; (b) © Lester V. Bergman & Associates, Inc.; (c) Gary Head after A. Ayala and J. Kiger, Modern Genetics, © 1980 Benjamin-Cummings. 33.7 Nadine Sokol after G.J. Tortora and N. Anagnostakos, *Principles of Anatomy and Physiology*, 6th Edition. © 1990 by Biological Sciences Textbooks, Inc., A & P Textbooks, Inc., and Ellia-Sparta, Inc. Reprinted by permission of John Wiley & Sons, Inc. 33.8 Kevin Somerville. 33.9 Precision Graphics. 33.10 Gary Head. 33.11 (a) © C. Yokochi and J. Rohen, *Photographic Anatomy of the Human Body, 2nd Ed.,* Igaku-Shoin, Ltd., 1979; (b,c) Raychel Ciemma. 33.12 Precision Graphics. 33.13 (a) left © Don W. Fawcett; (a) Art, Lisa Starr; (b) Raychel Ciemma. 33.14 Robert Demarest. 33.15 Precision Graphics. 33.16 Art, Precision Graphics; right, © Jose Pelaez, Inc./CORBIS. 33.17 Left, © Sheila Terry/SPL/Photo Researchers, Inc.; inset, Courtesy of Oregon Scientific, Inc. 33.18 Left, © Biophoto Associates/ Photo Researchers, Inc.; right, Lisa Starr. 33.19 (a) Left art, Lisa Starr, using © 2001 PhotoDisc, Inc.; Dr. John D. Cummingham/Visuals Limited; (b,c) Kevin Somerville. 33.20 Photograph, Professor P. Motta/Department of Anatomy/University La Sapienca, Rome/SPL/Photo Researchers, Inc. 33.21 (a) © Ed Reschke; (b) © Biophoto Associates/Photo Researchers, Inc. 33.22 © Lester V. Bergman/CORBIS. 33.23 Precision Graphics. 33.24 (a,b) Raychel Ciemma; (c) Lisa Starr. **Page 570** Below right, Raychel Ciemma and Robert Demarest. 33.25 © Lennart Nilsson from Behold Man, (c) 1974 by Albert Bonniers Forlag and Little, Brown and Company, Boston.

CHAPTER 34 34.1 © Lowell Tindell. **the big picture** (page 572) From left, © Biology Media/Photo Researchers, Inc.; © Mark L Stephenson/CORBIS; (page 573) Both, Lisa Starr. **Page 573** © NIBSC/Photo Researchers, Inc. 34.2 Art, Lisa Starr, with art references from *Bloodline Image Atlas*, University of Nebraska-Omaha/Sherri Wicks, Human Physiology and Anatomy, University of Wisconsin Biology Web Education System, and others.; bottom right, © Biology Media/Photo Researchers, Inc. **Page 575** © Mark L Stephenson/CORBIS. 34.3 Art, Raychel Ciemma; bottom right © NSIBC/SPL/Photo Researchers, Inc. 34.4 Art, Lisa Starr; 34.4: right © Robert R. Dourmashkin, courtesy of Clinical Research Centre, Harrow, England. 34.5 © David Scharf/Peter Arnold, Inc.; Art, Lisa Starr. 34.6 Art, Lisa Starr; top center, © Ken Cavanagh/Photo Researchers, Inc. 34.7 Precision Graphics. 34.8 Lisa Starr. 34.9 Preface, Inc. 34.10 Raychel Ciemma. 34.11 Top, Lisa Starr, from Harris, L. J.; Larson, S. B.; Hasel, K. W.; McPherson, A.; *Biochemistry 36*, p. 1581 (1997). of rendered with RIBBONS. (a–c) Lisa Starr. **Page 582** Lisa Starr. 34.12 Lisa Starr. 34.13 Lisa Starr. 34.14 © Dr. A. Liepins/SPL/Photo Researchers, Inc. 34.15 © Lowell Georgia/Science Source/Photo Researchers, Inc. 34.16 Top, © David Scharf/Peter Arnold, Inc.; bottom, © Mednet/Phototake, Inc. 34.17 © Greg Ruffing. 34.18 © Zeva Oelbaum/Peter Arnold, Inc. 34.19 Top left, © NIBSC/Photo Researchers, Inc.; art, Raychel Ciemma after Stephen Wolfe, *Molecular Biology of the Cell*, Wadsworth, 1993. 34.20 Photo courtesy of MU Extension and Agricultural Information.

CHAPTER 35 35.1 © Ariel Skelley/CORBIS. **the big picture** (page 592) From left, © John Lund/Getty Images; right; © Peter Parks/Oxford Scientific Films; (page 593) From left, Kevin Somerville; © CNRI/SPL/Photo Researchers, Inc. **Page 593** © James Stevenson/Photo Researchers, Inc. 35.2 Gary Head and Precision Graphics. 35.3 Precision Graphics. 35.4 (a) © David Nardini/Getty Images; (b) © John Lund/Getty Images; (c) © Joe McBride/Getty Images. 35.5 Lisa Starr with PDB ID 1A6M:Vojtechovsky, J., Berendzen, J., Chu, K., Schlichting, I., Sweet, R. M.: Implications for the Mechanism of Ligand Discrimination and Identification of Substates Derived from Crystal Structures of Myoglobin-Ligand Complexes at Atomic Resolution To be Published. 35.6 (a) © Peter Parks/Oxford Scientific Films; (b) © Herve Chaumeton/ Agence Nature; art, Precision Graphics. 35.7 Top, © Joe Warfel/Eight-Eye Photography; center; © D. E. Hill; art, Redrawn from *Living Invertebrates*, V & J. Pearse/M. & R. Buchsbaum, The Boxwood Press, 1987. 35.8 © Ed Reschke; art, Precision Graphics. **Page 597** © C. C. Lockwood. 35.9 Raychel Ciemma and Precision Graphics. 35.10 Raychel Ciemma. 35.11 Lisa Starr. 35.12 Top left, © H. R. Duncker, Justus-Liebig University, Giessen, Germany; art, Raychel Ciemma. 35.13 Kevin Somerville. 35.14 Left art, modified from A. Spence and E. Mason, *Human Anatomy and Physiology*, Fourth Edition, 1992, West Publishing Company; Courtesy of Kay Elemetrics Corporation. 35.15 (a) © 2000 PhotoDisc, Inc.; (b) © SIU/Visuals Unlimited; (c) © SIU/ Visuals Unlimited; (a,b,c) art, Lisa Starr. 35.16 Art, Precision Graphics; right, © Francois Gohier/Photo Researchers, Inc. 35.17 (a) © R. Kessel/Visuals Unlimited; (b, c) Art, Lisa Starr. 35.18 Leonard Morgan. 35.19 (a) © Lennart Nilsson from Behold Man, © 1974 by Albert Bonniers Forlag and Little, Brown and Company, Boston; (b) © CNRI/SPL/ Photo Researchers, Inc. 35.20 (a) © O. Auerbach/Visuals Unlimited; (b) © O. Auerbach/Visuals Unlimited. 35.21 Courtesy of Dr. Joe Losos. 35.22 Leif Buckley.

CHAPTER 36 36.1 Courtesy of Kevin Wickenheiser, University of Michigan. **the big picture** (page 610) From left, © W. Perry Conway/CORBIS; Kevin Somerville; (page 611) From left, © Elizabeth Hathon/CORBIS; © Gary Head.

Page 611 © Gusto/Photo Researchers, Inc. 36.2 Gary Head and Precision Graphics. 36.3 Raychel Ciemma. 36.4 (a) © W. Perry Conway/CORBIS; (a,b) Art, adapted by Lisa Starr from A. Romer and T. Parsons, *The Vertebrate Body*, Sixth Edition, Saunders Publishing Company, 1986. 36.5 Kevin Somerville. 36.6 Nadine Sokol. 36.7 After A. Vander, et al., Human Physiology: Mechanisms of Body Function, Fifth Edition, McGraw-Hill, 1990. Used by permission. 36.8 (a) © Microslide courtesy Mark Nielsen, University of Utah; (b) After A. Vander, et al., *Human Physiology: Mechanisms of Body Function, Fifth Edition*, McGraw-Hill, 1990. Used by permission. 36.9 Art, Lisa Starr after Sherwood and others; top center, © Microslide courtesy Mark Nielsen, University of Utah; center right, © D. W. Fawcett/Photo Researchers, Inc. 36.10 Raychel Ciemma and Lisa Starr. 36.11 Left, © Ralph Pleasant/FPG/Getty Images, right, Precision Graphics. 36.12 Kevin Somerville and Precision Graphics. 36.13 Top, (pyramid) © 2001 PhotoDisc, Inc.; bottom, © Elizabeth Hathon/CORBIS. **Page 623** © Gary Head. 36.14 © Gary Head. 36.15 © Dr. Douglas Coleman, The Jackson Laboratory; art, Precision Graphics and Gary Head. 36.16 © Reuters NewsMedia/CORBIS. 36.17 © Gunter Ziesler/ Bruce Coleman, Inc.

CHAPTER 37 37.1 © Archivo Iconografico, S.A./CORBIS. **the big picture** (page 630) From left, Robert Demarest; From T. Garrison, *Oceanography: An Invitation to Marine Science*, Brooks/Cole, 1993. All rights reserved; (page 631) From left, Robert Demarest; © Dan Guravich/CORBIS. **Page 631** @ Lawrence Lawry/Science Photo Library/Photo Researchers, Inc. 37.2 Gary Head and Precision Graphics. 37.3 (a,b) From T. Garrison, *Oceanography: An Invitation to Marine Science*, Brooks/Cole, 1993. All rights reserved. 37.4 Top, © David Noble/FPG/Getty Images; center, © Claude Steelman/Tom Stack & Associates; bottom, © Gary Head. 37.5 Robert Demarest. 37.6 Robert Demarest. 37.7 Top, Robert Demarest; bottom, Precision Graphics. 37.8 © Evan Cerasoli; top art, Kevin Somerville; bottom art, Robert Demarest. 37.9 © Air Force News/Photo by Tech. Sgt. Timothy Hoffman. 37.10 (a) © Bob McKeever/Tom Stack & Associates; (b) © S. J. Krasemann/Photo Researchers, Inc. 37.11 © David Parker/SPL/Photo Researchers, Inc. 37.12 Top, © Dan Guravich/CORBIS; bottom, © CORBIS-Bettmann. 37.13 Art, Precision Graphics; © Gary Head. **Page 643** Bottom left, Robert Demarest.

CHAPTER 38 38.1 (a) © Lennart Nilsson from *A Child is Born*, © 1966, 1977 Dell Publishing Company, Inc.; (b) © 1999 Dana Fineman/CORBIS Sygma. **the big picture** (page 644) From left, © Ron Austing; Frank Lane Picture Agency/CORBIS; © Carolina Biological Supply Company; (page 645) From left, Raychel Ciemma; © Lennart Nilsson, *A Child is Born*, © 1966, 1977 Dell Publishing Company, Inc. **Page 645** © Dow W. Fawcett/Photo Researchers, Inc. 38.2 © Fred SaintOurs/University of Massachusetts-Boston; art, Lisa Starr. 38.3 (a) © Marc Moritsch; (b) © Photo Researchers, Inc. 38.4 (a) © Frieder Sauer/Bruce Coleman, Ltd; (b) © Matjaz Kuntner; (c) © Ron Austing; Frank Lane Picture Agency/CORBIS; (d) © Doug Perrine/seapics.com; (e) © Carolina Biological Supply Company; (f) © Fred McKinney/FPG/Getty Images; (g) © Gary Head. 38.5 Art, Palay/Beaubois and Precision Graphics. 38.6 (a) Art, Raychel Ciemma; (b–j) © Carolina Biological Supply Company; (b–j) art, L. Calver. 38.7 Art, Robert Demarest. 38.8 (a, b) Art, Precision Graphics; (c) © Gary Head. 38.9 (a–d) © Dr. Maria Leptin, Institute of Genetics, University of Koln, Germany. 38.10 (a) Lisa Starr; (b) Raychel Ciemma after B. Burnside, *Developmental Biology*, 1971, 26:416-441. Used by permission of Academic Press. 38.11 Art, Precision Graphics; (Photographic series) © Carolina Biological Supply Company; far right © Peter Parks/Oxford Scientific Films/Animals, Animals. 38.12 (a,b) Art, Lisa Starr; (c) © Professor Jonathon Slack. 38.13 Left, © Peter Parks/Oxford Scientific Films/Animals, Animals; (a,b) Art, Raychel Ciemma after S. Gilbert, Developmental Biology, Fourth Edition. 38.14 Art, Raychel Ciemma with Lisa Starr; right, © Laura Dwight/ CORBIS. 38.15 Raychel Ciemma with Lisa Starr. 38.16 (a,b,c) Art, Raychel Ciemma; (b) © Ed Reschke. 38.19 Raychel Ciemma. 662: 38.20 Left art, Raychel Ciemma; top right art, Robert Demarest; bottom right, Photograph Lennart Nilsson from *A Child is Born*, © 1966, 1977 Dell Publishing Company, Inc. 38.21 Raychel Ciemma with Precision Graphics. 38.22 K. Sommerville, Robert Demarest, and Preface, Inc. 38.23 Raychel Ciemma; 38.24 Preface, Inc. 38.25 © Lester Lefkowitz/CORBIS. 38.26 (a) © Dr. E. Walker/Photo Researchers, Inc.; (b) © Western Ophthalmic Hospital/Photo Researchers, Inc.; (c) © Kenneth Greer/Visuals Unlimited; (d) © CNRI/Photo Researchers, Inc. 38.27 © Todd Warshaw/Getty Images. 38.28 Raychel Ciemma. 38.29 Raychel Ciemma. 38.30 Raychel Ciemma. 38.31 (a-d) Art, Raychel Ciemma; (a–d) top, From Lennart Nilsson, *A Child is Born,* © 1966, 1977 Dell Publishing Company, Inc. **38.32** Raychel Ciemma, modified from K. L. Moore, *The Developing Human: Clinically Oriented Embryology,* Fourth Edition, Philadelphia: W. B. Saunders Co., 1988. **38.33** Left, © Zeva Oelbaum/CORBIS right, © James W. Hanson, M.D. **38.34** Robert Demarest. **38.35** Raychel Ciemma. **38.36** Top art, Raychel Ciemma adapted from L. B. Arey, Developmental Anatomy, Philadelphia, W. B. Saunders Co., 1965; bottom, Lisa Starr. **Page 680** © Denis Scott/CORBIS. **38.37** (a) © David M. Parichy; (b) © Dr. Sharon Amacher; (c) © Dr. Sharon Amacher.

Page 683 Unit VII © Alan and Sandy Carey.

CHAPTER 39 **39.1** © David Nunuk/Photo Researchers, Inc. **the big picture** (page 684) From left, © Amos Nachoum/CORBIS; Gary Head; (page 685) From left, © Joe McDonald/CORBIS; © Don Mason/CORBIS. **Page 685** © Tom Till/Stone/Getty Images. **39.2** Art, Precision Graphics and Gary Head; bottom left, © Amos Nachoum/CORBIS; bottom center, © A. E. Zuckerman/Tom Stack & Associates; right inset, © CORBIS. **39.3** © E. R. Degginger; inset, © Jeff Fott Productions/Bruce Coleman, Ltd. **39.4** (a) © Cynthia Bateman, Bateman Photography; (b) © Tom Davis. **39.5** left © Jeff Lepore/Photo Researchers, Inc.; art, Precision Graphics. **39.6** Precision Graphics. **39.7** (a) © Lilliam Lampas/Omni-Photo; (b) © Christine Pemberton/Omni-Photo. **39.8** Gary Head. **39.9** Top, Gary Head; bottom, © Jacques Langevin/CORBIS Sygma. **39.10** (a) © Joe McDonald/CORBIS; (b) © Wayne Bennett/CORBIS; (c) © Douglas P. Wilson/CORBIS; art, Preface, Inc. **39.11** (a) © Helen Rodd; (b, above, c, above) © David Reznick/University of California - Riverside; computer enhanced by Lisa Starr; (b, below, c, below) © Hippocampus Bildarchiv. **39.12** (b) © John A. Endler. **39.13** Precision Graphics. **39.14** © U.S. Geological Survey. **39.15** NASA; art, Precision Graphics. **39.16** Gary Head. **39.17** Preface Inc. **39.18** Left, © Adrian Arbib/CORBIS; center, Precision Graphics; right, © Don Mason/CORBIS. **39.19, 39.20** Ater T. Miller, Jr., *Environmental Science,* Sixth Edition, Brooks/Cole, 1997. All rights reserved. **39.21** © John Alcock/Arizona State University. **39.22** © Wolfgang Kaehler/CORBIS. **Page 703** Precision Graphics.

CHAPTER 40 **40.1** Left & center, Photography by B. M. Drees, Texas A&M University. Http://fireant.tamu.edu; right, © Daniel Wojak/USDA. **the big picture** (page 704) from left, © Martin Harvey, Gallo Images/CORBIS; right, Thomas W. Doeppner; (page 705) from left, © Pat O'Hara/CORBIS; right © James Marshall/CORBIS. **Page 705:** Top, © John Kabashima. **40.2** Top left, © Donna Hutchins; center left, © B. G. Thomson/Photo Researchers, Inc.; bottom left, © Len Robinson, Frank Lane Picture Agency/CORBIS; bottom right, © Martin Harvey, Gallo Images/CORBIS. **40.3** Left, © Bob and Miriam Francis/Tom Stack & Associates; right © Harlo H. Hadow. **40.4** Thomas W. Doeppner. **40.5** (mountain peak) © Richard Cummins/CORBIS; (a,d) © Don Roberson; (b) © Kennan Ward/CORBIS; © D. Robert Franz/CORBIS. **40.6** Top, right © Michael Abbey/Photo Researchers, Inc.; bottom right © Eric V. Grave/Photo Researchers, Inc.; (a–c) art, Gary Head. **40.07** Top, © Joe McDonald/CORBIS; bottom left, © Hal Horwitz/CORBIS; bottom right, © Tony Wharton, Frank Lane Picture Agency/CORBIS; top art, Precision Graphics after N. Weldan and F. Bazazz, Ecology, 56:681-688, © 1975 Ecological Society of America. **40.8** (a, b) Art, Gary Head, after Rickleffs & Miller, Ecology, Fourth Edition, page 459 (Fig 23.13a) and page 461 (Fig 23.14); right, © W. Perry Conway/CORBIS. **40.9** (a) Precision Graphics; (b) © Ed Cesar/Photo Researchers, Inc.; (c) © Kennan Ward. **40.10** (a) © JH Pete Carmichael; (b) © Edward S. Ross; (c) © W. M. Laetsch. **40.11** (a-c) © Edward S. Ross; (d) © Nigel Jones. **40.12** (a,b) © Thomas Eisner, Cornell University; (c) © Jeffrey Rotman Photography; (d) © Bob Jensen Photography. **40.13** (a) Courtesy of Ken Nemuras; (b) © CDC. **40.14** Left, © The Samuel Roberts Noble Foundation, Inc.; right, Courtesy of Colin Purrington, Swarthmore College. **40.15** © C. James Webb/Phototake, U.S.A. **40.16** © Peter J. Bryant/Biological Photo Service. **40.17** Left, © Richard Price/Getty Images; right ©E.R. Degginger/Photo Researchers, Inc. **40.18** (a) © Doug Peebles/CORBIS; (b) © Pat O'Hara/CORBIS; (c) © Tom Bean/CORBIS; (d) © Tom Bean/CORBIS; (e) © Duncan Murrell/Taxi/Getty Images. **40.19** (a) © R. Barrick/USGS; (b,c) © 1980 Gary Braasch. **40.20** (a,c) © Jane Burton/Bruce Coleman, Ltd.; (b) © Heather Angel; (d,e) Art, Precision Graphics based on Jane Lubchenco, *American Naturalist,* 112:23-29, © 1978 by The University of Chicago Press. Used with permission. **40.21** Left, © The University of Alabama Center for Public TV; (a) © Angelina Lax/Photo Researchers, Inc.; (b) © Pr. Alexande Meinesz, University of Nice-Sophia Antipolis.

40.22 © John Carnemolla/Australian Picture Library. **40.23** Art, Gary Head after W. Dansgaard, et al., Nature, 364:218-220, 15 July 1993; D. Raymond, et al., Science, 259:926-933, February 1993; W. Post, American Scientist, 78:310-326, July-August 1990. **40.24** Left, both, © Pierre Vauthey/CORBIS Sygma; right art, Gary Head after S. Fridriksson, Evolution of Life on a Volcanic Island, Butterworth, London, 1975. **40.25** (a) © Susan G. Drinker/CORBIS; (a) art, computer-modified by Lisa Starr (b) © Frans Lanting/Minden Pictures. **40.26** © James Marshall/CORBIS. **40.27** Left, © Bagla Pallava/CORBIS Sygma; right inset, © A. Bannister/Photo Researchers, Inc. **40.28** Lisa Starr. **40.29** Lisa Starr. **40.30** Art, Lisa Starr with photographs © 2000 PhotoDisc, Inc. **40.31** Both, © Bureau of Land Management. **40.32** (a) © Anthony Bannister, Gallo Images/CORBIS; (b) © Bob Jensen Photography; (c) © Cedric Vaucher. **40.33** Both, Courtesy of Eternal Reefs, Inc., www.eternalreefs.com

CHAPTER 41 **41.1** Courtesy of Louisiana Department of Wildlife. **the big picture** (page 730) Left, Gary Head; right/left © D. Robert Franz/Planet Earth Pictures; right/right © Frans Lanting/Bruce Coleman, Ltd; (page 731) Left, Preface, Inc.; right, Gary Head. **Page 731** Top, © Harold Naideau/CORBIS. **41.2** Art, Gary Head and Lisa Starr; photographs, above, © PhotoDisc, Inc.; below, © David Neal Parks. **41.3** © Photograph Alan and Sandy Carey; Art, Gary Head after R. L. Smith, *Ecology and Field Biology,* Fifth Edition. **41.4, 41.5** Art, Lisa Starr with photographs, flowering plants, © Frank Oberle/Stone/Getty Images; marsh hawk, © J. Lichter/Photo Researchers, Inc.; crow, © Ed Reschke; upland sandpiper, © O. S. Pettingill Jr./Photo Researchers, Inc.; garter snake, © Michael Jeffords; frog, © John H. Gerard; yellow spider, © Michael Jeffords; weasel, Courtesy of Biology Department, Loyola Marymount University; sparrow, © Rod Planck/Photo Researchers, Inc.; cutworm, © Nigel Cattlin/Holt Studios International/Photo Researchers, Inc.; gopher and prairie vole, © Tom McHugh/Photo Researchers, Inc.; all others, © PhotoDisc, Inc. **41.6** Art, Precison Graphics and Gary Head; photographs, left © D. Robert Franz/Planet Earth Pictures; right © Frans Lanting/Bruce Coleman, Ltd. **41.7** © Gary Head. **41.8:** © Eric Hartmann/Magnum Photos. **41.9** Preface, Inc. **41.10:** NASA/GSFC. **41.11** Precision Graphics. **41.12** © Gerry Ellis/The Wildlife Collection; Art, Gary Head. **41.13** Lisa Starr. **41.14** (left) Gary Head after G. E. Likens and F. H. Bormann, "An Experimental Approach to New England Landscapes," in A. D. Hasler (ed.), Coupling of Land and Water Systems, Champman & Hall, 1975; right, © Dr. Elizabeth Hane. **41.15** Water Resources Council; **41.16** Lisa Starr after Paul Hertz; photograph © 2000 PhotoDisc, Inc. **41.17** Lisa Starr and Gary Head, based on NASA photographs from JSC Digital Image Collection. **41.18** Lisa Starr and Gary Head, based on NASA photographs from JSC Digital Image Collection. **41.19** © Yann Arthus-Bertrand/CORBIS. **41.20** (a) Lisa Starr and Gary Head, compilation of data from Mauna Loa Observatory, Keeling and Whorf, Scripps Institute of Oceanography; (b) Lisa Starr and Gary Head, compilation of data from Prinn, et al., CDIAC, Oak Ridge National Laboratory;world Resources Institute; Khalil and Rasmussen, Oregon Graduate Institute of Science and Technology, CDIAC DB-1010; Leifer and Chan, CDIAC DB-1019; (c,d) Lisa Starr and Gary Head, compilation of data from World Resources Institute; Law Dome ice core samples, Etheridge, Pearman, and Fraser, Commonwealth Scientific and Industrial Research Organisation; Prinn, et al., CDIAC, Oak Ridge National Laboratory; Leifer and Chan, CDIAC DB-1019. **41.21** © Jeff Vanuga/CORBIS; Art, Gary Head. **41.22** © Frederica Georgia/Photo Researchers, Inc. **41.23** Art, Gary Head and Lisa Starr; photograph, © 2000 PhotoDisc, Inc. **41.24** Fisheries & Oceans Canada, Experimental Lakes Area; **41.2** Art, Gary Head; photographs © Bruce Coleman. **41.26** Courtesy of NASA's Terra satellite, supplied by Ted Scambos, National Snow and Ice Data Center, University of Colorado, Boulder. **41.27** Gary Head, after T. Garrison.

CHAPTER 42 **42.1** © Hank Fotos Photography. **the big picture** (page 752) from left, L. Calver; © Douglas Peebles/CORBIS; (page 753) from left © Pat O'Hara/CORBIS; right, Lisa Starr. **Page 753,** Top, © Wolfgang Kaehler/CORBIS. **42.2** Preface, Inc. **42.3** L. Calver. **42.4** Precision Graphics; **42.5:** (a) © Alex MacLean/Landslides; (b) © Alex MacLean/Landslides. **42.6** NASA. **42.7** Lisa Starr. **42.8** (a) Gary Head, adapted from *Living in the Environment* by G. Tyler Miller, Jr., p. 428. © 2002 by BrooksCole, a division of Thomson Learning; (b) © Ted Spiegel/CORBIS. **42.9** Gary Head. **42.10** NASA. **42.11** Left, © Sally A. Morgan/Ecoscene/CORBIS; right, © Bob Rowan, Progressive Image/CORBIS; above, art, Lisa Starr. **42.12** NASA. **42.13** Lisa Starr using photographs © 2000 PhotoDisc, Inc. **42.14**

Bottom, © George H. Huey/CORBIS. **42.14** Left inset, © John M. Roberts/CORBIS. **42.15** © Orbimage Imagery. **42.16** Left, © John C. Cunningham/Visuals Unlimited; center, © Jack Wilburn/Animals Animals; right, © AP/Wide World Photos. **42.17** Bottom left, © Tom Bean Photography; top right © Jonathan Scott/Planet Earth Pictures. **42.18** © 1991 Gary Braasch Photography; bottom inse,t © Karl Lehmann/LostWorldArts.com. **42.19** Left, © James Randklev/CORBIS; inset, All, © Randy Wells/CORBIS. **42.20** © Raymond Gehman/CORBIS; inset right, © Nigel Cook/Dayton Beach News Journal/Corbis Sygma. **42.21** (a) © Darrell Gulin/CORBIS; (b) © Paul A. Souders/CORBIS; (c) © Pat O'Hara/CORBIS. **42.22** Art, D & V. Hennings after Whittaker, Bland, and Tilman. **42.23** © Onne van der Wal/CORBIS. **42.24** (a) Precision Graphics; (b) Precision Graphics after E. S. Deevy, Jr., Scientific American, October 1951. **42.25** (a) © E. F. Benfield, Virginia Tech; (b) © E. F. Benfield, Virginia Tech; (c) © E. F. Benfield, Virginia Tech; (d) © E. F. Benfield, Virginia Tech; (e) © E. F. Benfield, Virginia Tech. **42.26** (a) © Annie Griffiths Belt/CORBIS; (b) © Douglas Peebles/CORBIS. **42.27** (a) © Mike Zens/CORBIS. **42.27** (b) © Paul A. Souders/CORBIS. **42.28 (a)** Top left, © Douglas Faulkner/Sally Faulkner Collection; top right, © Royalty-Free/CORBIS; center left, © Jeff Rotman; center right, © Alex Kirstitch; bottom left, © Amos Nachoum/CORBIS; bottom right, © Douglas Faulkner/Sally Faulkner Collection; (b) T. Garrison, *Oceanography: An Invitation to Marine Science,* Third Edition, Brooks/Cole, 2000. All rights reserved. **42.29** © Greenpeace/Grace. **42.30** (a) © Peter David/FPG/Getty Images. **42.30** (b) © Robert Vrijenhoek, MBARI; (c) Art, L. K. Townsend. **42.31** Lisa Starr. **42.32** (a,b) Lisa Starr; (c) NASA/Goddard Space Flight Center Scientific Visualization Studio; (d) NASA/Goddard Space Flight Center Scientific Visualization Studio. **42.33** (a) © Eye of Science/Photo Researchers, Inc.; (b) © CHAART, at NASA Ames Research Center; (c) © Douglas P. Wilson; Frank Lane Picture Agency/CORBIS; (d) © Raghu Rai/Magnum Photos. **42.34** Precision Graphics after M. H. Dickerson, "ARAC: Modeling an Ill Wind," in Energy and Technology Review, August 1987. Used by permission of University of California Lawrence Livermore National Laboratory and U. S. Dept. of Energy.

CHAPTER 43 **43.1** Top, © Scott Camazine; **the big picture** (page 778), from left, © James Zipp/Photo Researchers, Inc.; © Monty Sloan, www.wolfphotography.com; (page 779) from left, © Australian Picture Library/CORBIS; © Matthew Alan/CORBIS. **Page 779** Top, © Ralph Fountain. **43.2** (a) © Eugene Kozloff; (b) © Stevan Arnold; (c) © Stevan Arnold. **43.3** Left, © Robert M. Timm & Barbara L. Clauson, University of Kansas (a) Reprinted from *Trends in Neuroscience,* Vol. 21, Issue 2, 1998, L.J. Young, W. Zuoxin, T.R. Insel, "Neuroendocrine bases of monogamy," Pages 71-75, ©1998, with permission from Elsevier Science; (b) Reprinted from *Trends in Neuroscience,* Vol. 21, Issue 2, 1998, L.J. Young, W. Zuoxin, T.R. Insel, "Neuroendocrine Bases of Monogamy," Pages 71-75, ©1998, with permission from Elsevier Science. **43.4** (a) © Eric Hosking; (b) © Stephen Dalton/Photo Researchers, Inc.; (c) © Jennie Woodcock; Reflections Photolibrary/CORBIS. **43.5** © Nina Leen/TimePix. **43.6** © James Zipp/Photo Researchers, Inc. **43.7** (a) © Robert Maier/Animals Animals; (b) © John Bova/Photo Researchers, Inc.; (c) From L. Clark, Parasitology Today, 6(11), Elsevier Trends Journals, 1990, Cambridge, UK. (d) © Jack Clark/Comstock, Inc. **Page 43.8** (a) © Monty Sloan, www.wolfphotography.com; (b) © Tom and Pat Leeson, leesonphoto.com; (c) © Kevin Schafer/CORBIS. **43.09** © Stephen Dalton/Photo Researchers, Inc.; art, D. & V. Hennings. **43.10** (a) © John Alcock, Arizona State University; (b) © Pam Gardner; Frank Lane Picture Agency/CORBIS; (c) © Pam Gardner; Frank Lane Picture Agency/CORBIS; (d) © D. Robert Franz/CORBIS. **43.11** (a) © Ingo Arndt/Nature Picture Library; (b) © Michael Francis/The Wildlife Collection. **43.12** (a) © B. Borrell Casals; Frank Lane Picture Agency/CORBIS; (b) © Steve Kaufman/CORBIS; (c) © John Conrad/CORBIS. **43.13** (a) © Tom and Pat Leeson, leesonphoto.com; (b) © John Alcock, Arizona State University; (c) © Paul Nicklen/National Geographic/Getty Images. **43.14** © Jeff Vanuga/CORBIS. **43.15** © Eric and David Hosking/CORBIS. **43.16** (a) © Australian Picture Library/CORBIS; (b) © Alexander Wild; (c) © Professor Louis De Vos. **43.17** (a) © Kenneth Lorenzen; (b) © Nicola Kountoupes/Cornell University; (c) © Peter Johnson/CORBIS. **43.18** © Dr. Tim Jackson, University of Pretoria. **Page 792** © Matthew Alan/CORBIS. **43.19** © Brad Bergstrom. **43.20** © F. Schutz.

EPILOGUE © Joseph Sohm, Visions of America/CORBIS.

Index

The letter i *designates illustration;* t *designates table;* **bold** *designates defined term;*
▪ *highlights the location of applications contained in text.*

A

Aardvark, 393, 393i
ABA, 440, 440i, 458t, 459, 464
Abalone, 512, 512i
ABC model, **463**, 463i
Abdomen, insect, 370
Abdominal aorta, 558i
Abdominal cavity, 476i
▪ ABO blood typing, 158, 158i, 164, **557**
▪ Abortion, 151, 165, 179, **180**, 181, 667
Abscisic acid, **440**, 440i, 458t, **459**, 464
Abscission, **458**, 458t, **464**, 464i
Absorption, 612
of ethanol, in stomach, 72
in small intestine, 618–619, 618i, 619i
Absorption spectra, 95i
▪ Abstinence, sexual, 137, 666, 666i
Abyssal zone, 773i
Acacia tortilis, 414
Acanthostega, 383i
Acceptor molecule, 90
Accessory fruits, **452**, 453i
Accessory heart, 365, 365i
Accessory organs, 628t
Accessory pigment, **94**, 95i, 105
Accessory protein, 66
Acclimatization, **603**
Accommodation, visual, **513**
Acer (maple), 424i, 452, 453, 453i
Acetaldehyde, 72, 73
Acetate, 72
Acetic acid, 72
Acetyl-CoA, 112, 113i, 118
Acetyl group, 208, 211
Acetylcholine (ACh), 488–489, 501, 548
Achillea millefolium, 162–163, 162i
▪ Achondroplasia, 173t, 174, 174i
Acid, **28**–29, **30t**
Acid-base balance, **638**
Acid deposition, 757
▪ Acid rain, 29, 331, 747, 747i, **757**, 757i
▪ Acid stomach, 28
Acidic solution, **28**
Acidophilus, 304
▪ Acidosis, 29, 638
▪ Acne drugs, 677
Acoelomate animal, 355i
Acorn, 421, 452, 452t
Acoustical cues, 504–505
Acoustical signals, 784
Acquired characteristics, inheritance of, 240
▪ Acquired Immunodeficiency Syndrome. *See* AIDS
Acromegaly, 525
ACTH, 521i, 524, 526, 532t
Actin, 66, 66i, **544**, 544i, 545i, 547
Action potential, **485**, 506, 506i, 509
Action potential propagation, 486–487, 486i–487i, 490
Activated helper T cells, 578
Activation energy, 78–**79**, 78i, 81
Activator protein, 80, 80i, **208**–209
Active binding site, enzyme, 78–80, 80i, 84
Active immunization, 586
Active transport, **83**, 83i, 85, 85i, **407**, 443i
Active transporter, 57, 57i
Acute inflammation, **576**, 576i
▪ Acute respiratory infection, 310t
▪ Acyclovir, 668

▪ ADA deficiency, 587, 587i
Adaptation, **236**, **244**
evolutionary, 244–245, 336–337, 367
of fishes, 382
invertebrate, 362, 364–365, 367
long- *vs.* short-term, 244–245
of prey, 712–713, 712i, 713i
sensory adaptation, **506**
Adaptive behavior, **783**, 792
Adaptive immunity, **574**, 575t, 578–579
evolution of, 574–575
Adaptive radiation, 263i, **278**–279
animal, 279i, 293, 356, 365
mammals, 392, 393, 393i
plant, 334, 336, 343, 346, 347
Adaptive trait, **10**, 252
Adaptive zone, **278**–279
Addiction, 500
Addictive drugs, 500–501
Adductor longus, 543i
Adenine (A), 46, 46i, 75, 184i, 188–189, 188i, 189i, 196–197, 199, 202, 227i
▪ Adenosine deaminase deficiency, 587, 587i
Adenosine diphosphate, 75, 85i, 97i, 104i
Adenosine phosphate, 48t
Adenosine triphosphate. *See* ATP
Adenovirus, 308i
Adenylate cyclase, 523i
ADH, 521i, 522, 524, 524i, 525, 532t, 636i, **637**, 637i, 781
Adhering junction, 69, 69i, **471**, 471i
Adhesion protein, 56, 56i
Adipose tissue, 40–41, 118, 122, **473**, 473i, 610
brown, 122, 641, 641i
Adolescent, 679t
ADP, 75, 85i, 97i, 104i
Adrenal gland, 495, 521i, 524, 637
cortex, **526**–527, 526i, 533t
medulla, **527**, 533t
Adrenaline. *See* Epinephrine
Adult, 679t
Adventitious root, **427**, 457i
AEDs (automated external defibrillators), 551
Aerobe, 305
Aerobic bacteria, 295, 320, 597
▪ Aerobic exercise, 547
Aerobic metabolism, 90–91, 263i
Aerobic reaction, 307
Aerobic respiration, 51, 64, 75, 77, 77i, 93, 97, 102, **108**–109, 108i, 109i, 293, 296, 296i–297i, 546, 546i, 594, 737
animal, 367
in cells, 294, 329
plant, 427, 434, 456
second stage of, 112–113, 112i, 113i
third stage of, 114–115, 114i
Aesculus, 424i
A. hippocastanum, 464i
Africa, 398–399, 572–573, 572i
African clawed frog, 518–519, 530, 530i, 571
African emergence model, 399
▪ African sleeping sickness, 316, 317i
African violet, 454, 454t
▪ Africanized bees, 778, 778i
▪ Afterbirth, 678, 678i
▪ Agar, 322
Agaris bisporus, 327
▪ Age-related macular degeneration (AMD), 515
Age-specific life history patterns, 692–695

Age structure, **686**, 698, 699i, 703, 703i
▪ Agent Orange, 460
Agglutination, **556**–557, 556i
Aggregate fruit, **452**, 452t, 453i
▪ Aging, 680
appetite and, 611
▪ cardiovascular disorders, 567
hormones and, 459, 464
▪ muscles and exercise, 547
process, 80
▪ skin and, 479
▪ visual disorders, 515
Agouti gene, 249
▪ Agriculture
cereal croplands, 763
crop failures, 416–417, 445
▪ deforestation and, 349, 446
domestication of plants, 343
endangered species and, 724
erosion and, 435, 435i
▪ farming methods, 228
▪ fertilizer and, 29, 333, 597, 747, 749
genetic engineering in, 460, 466
grassland conversion for, 747
herbicides and pesticides, 518–519
hormones and, 458–459, 458i, 458t, 459i
human population growth and, 696
insects/pests and, 320, 333, 366, 370, 460
ozone layer and, 756
▪ pesticide use, 447, 460, 466
phosphorus and, 748–749
▪ probiotic feeds, 729
runoff, 597, 747, 749
soil and, 767
water and, 91, 416–417, 740–741
Agrobacterium tumefaciens, 228i, 229, 305
Aguilegia, 348i
▪ AIDS, 2, 308, 310, 310t, 481, 501i, 572–573, 587–589, 668t, 669, 669i
drug therapies, treatments, 352
drugs and vaccines for, 589
▪ AIDSVAX, 589
Air, at sea level, 594, 594i
Air circulation patterns
global, 754–755, 754i
humans and, 756–757
▪ Air pollution, 747i
▪ deforestation and, 334–335, 334i, 335i
▪ fossil fuels and, 2, 29i, 93, 441i
▪ fungi and, 330, 331, 339
Air space, in plant tissue, 103i, 420, 423, 423i, 427
Aix sponsa, 35i
Ajellomyces capsulatus, 329
Akkadian civilization, 416
Alanine (ala), 42i, 198i
Albatross, 255i, 391i, 723i
▪ Albinism, 159, 164, 164i, 173t
Albumin, denaturation of, 44
Alcohol, 72, 91, 501
in blood, 72
▪ consumption of, 63, 72, 541, 541i
ethyl alcohol, 501
▪ fetal alcohol syndrome, 677, 677i
pregnancy and, 677, 677i
structure, 34
toxicity and, 72
▪ Alcoholic beverages, making of, 329
▪ Alcoholic cirrhosis, 72
▪ Alcoholic fermentation, **116**, 116i
▪ Alcoholic hepatitis, 72
Aldehyde, 34i, 38
Aldosterone, 521i, 533t, **637**

Age structure, **686**, 698, 699i, 703, 703i
Alexander the Great, 255, 300, 300i
Alfalfa leaf-cutting bee, 446
Alfalfa plant, 423i, 446, 466
Alga, 17i, 53, 138, 286, 287, 362, 708, 718i, 719, 720–721, 720i. *See also* Green alga; *specific types*
brown, 316i, 320, 321, 321i
▪ commercial uses, 321, 322
evolution, 336, 336i
golden and yellow-green, 321
life cycle, 336
as photosynthesizer, 93, 96, 101
red, 105, 105i, 287, 292i, 293, 297i, 307i, 316i, 319, 322, 322i
snow alga, 286, 333, 333i
▪ Algal bloom, 92, 92i, 320, 333, 597, 731, **749**, 772, 775
Alginic acid, 321
Alkaline solution, **28**
Alkaloid, 327, 329, 330
▪ Alkalosis, 29
▪ Alkylating agent, **203**
All-or-nothing response, 486
Allantois, **671**
Allele, **140**, **153**, **164**, **168**
codominance of, **158**, 158i
dominant, 151, **153**, 155–156, 155i, 156i, 164
dominant/recessive interaction, 151, 153–159, 154i, 155i, 156i, 158i, 159i, 164, 168, 183, 213, 237
incomplete dominance of, **158**, 158i
lethal, **164**, 174
multiple allele system, **158**
mutant, 158, 168, 183
recessive, 151, 153–155, 154i, 155i, 156i, 158, 159, 183, 213
trait variation and, 144–145, 144i, 153, 153i, 154–157, 159, 211, 250
wild-type, 168, 183, 254
Allele frequency, **246**–248, 247i, 252, 254, 254i, 255, 272, 278
Allergen, 586
▪ Allergy, 369, **586**–587, 586i
Alligator, 388i, 389, 497i, 519
Allolactose, 208–209, 209i
Allopatric speciation, **274**–275, 274i, 275i
Allosteric binding site, **80**, 80i
Allosteric control, 211
Almond, 452t
Alpha-1,3-galactose, 231
Alpha cell, pancreatic, 528, 529i
Alpha globin, 44, 44i
Alpine pennycress, 433
Alpine tundra, 761i, 766, 766i
Alternative diets, 623
Alternative splicing, of genes, **197**
▪ Altitude sickness, 122
Altruism, **783**
indirect selection for, 791
Alu element, 269
Aluminosilicate crystal, 434
Alveolate, 316i, 318, 319
Alveolus, **600**, 600i, 601
▪ Alzheimer's disease, 107, 230, 499, 680
Amacrine cell, retina, 514, 514i
Amanita muscaria, 327i
Amber, insects in, 261
Amborella, 348i
Amino acid, **42**–44, 42i, 63, 79i, 118, 180, 460, 526, 616t, 617, 623
origin of cell and, 290
protein synthesis and, 195i, 196, 198–199, 198i, 200, 200i
residue, 42i

sequence, 196, 196i, 198, 201, 202, 202i, 270, 271
in space, 289
Amino group, 34i, 42i
Aminopeptidase, 616t
■ Amish, Older Order, 255
Ammonia, 25i, 28i, 29, 81i, 118, 288, 289, 304, 436, 620, 623, 643, 746
Ammonification, **747**
Ammonite, 239i
Ammonium, 746–747
■ Amniocentesis, 180i, 181
Amnion, 181, **670**–671
Amniote, **386**–387, 386i
Amniotic cavity, 670, 671i
Amoeba, 66, 67, 89i, 128
Amoeba discoides, 295
A. proteus, 317i
■ Amoebic dysentery, 317
Amoeboid cell, 325
Amoeboid protozoan, 316i, **317**
AMP, 325i
Amphetamine, 500
Amphibian, 381i, **384**, 384i. *See also specific types*
brain, 497i
characteristics, 211, 354, 354t
circulation, 553, 553i
deformities of, 530, 530i
evolution, 263i, 266, 383i, 384–385
gills, **598**
■ habitat contamination, 385, 385i
reproduction, 385
respiration, 599i
spinal reflexes, 495
Amphiprion perideraion, 707, 707i
Amplification, DNA, 220, 224–227, 225i, 227i, 286, 307
Amplitude, sound, 510, 510i
Amygdala, 499, 499i
Amylose, 38i, 39i
■ Amyotrophic lateral sclerosis, 107
Anabaena, 304, 304i
Anabolic pathway, **77**
Anaerobe, 263i, 305, **306**, 307
Anaerobic atmosphere, 296i–297i
Anaerobic energy-releasing pathways, 108, 108i
Anaerobic habitat, 293
Anaerobic reaction, **50**, 51
Anagenesis, **278**
■ Analgesic, 501
Analogous structures, **267**, 267i
Anaphase, 124, 128i, 129, **130**, 131, 131i, 132, 141i, 142i–143i, 148, 178i
■ Anaphylactic shock, 587
Anatomy, **403**
Ancient human remains, 339
■ ANDi, 234, 234i
Androgen, 533t
■ Androgen insensitivity syndrome, 173i, 522
Androstenedione, 536
■ Anemia, **556**
■ Aneuploidy, **178**–179, 179i
Angina pectoris, 567
Angioplasty, 567
Angiosperm, 334, 336i, **346**, 346i. *See also* Flowering plant
Angiotensin, 533t
■ Anhidrotic ectodermal dysplasia, 213, 213i
Animal, 6, **9**, 39, 282, 332t
behavior, 779
characteristics, 92, 105, 146, **354**–355, 354i, 355i
commonalities with plants, 402–414
evolution, 297i, 356–357
inheritance patterns, 168, 178

Animal cell, 52i, 53i, 60i, 66, 68, 70t, 132–133, 132i, 133i
Animal cuticle, 39, 39i
Animal kingdom, 354–355, 354t
Animal survival experiment, 12–13, 251, 251i
■ Animal testing, 481
Aniridia gene, 214–215
Ankle bone, 239i, 257i, 539i
Annelid, 254t, 355, 356i, **362**–363, 362i
Annual plant, 428
Anopheles, 318, 319i
■ Anorexia nervosa, 629, 629i
Ant, 370, 447, 704–705, 722i, 790i, 791
■ Antacid, 28
Antarctica, 338, 751, 751i, 756i
Anteater, 393, 393i
Antelope, 414, 646i
Antenna, 368, 370, 370i
Antennapedia gene, 212
Anterior, 476i
Anterior-posterior axis, 212, **354**, 354i
Anther, 140i, 268, 448, 448i, 450, 450i
Anthocyanin, **95**, 95i, 449
■ Anthrax, 304, 573
Anthropoid, 394, 395
■ Anti-acne drugs, 677
■ Anti-anxiety drugs, 489
■ Antibiotic resistance, 59, 222, 237, 248, 310, 318, 573, 729
■ Antibiotics, **248**, 304, **310**, 575, 629
■ Antibody, 48t, 231, **578**, 580i
antibody-mediated response, 582–583
immunoglobulins, 582–583, 582i
■ monoclonal antibodies, 583
Antibody-mediated response, 578, 579i, 582–583, 582i, 583i
antibody function, 582
classes of immunoglobulins, 582–583
Antibody synthesis, 580
Anticipatory behavior, 795
Anticodon, **199**, 199i, 201
■ Antidepressant, 677
Antidiuretic hormone, 521i, 522, 524, 524i, 525, 532t, 636i, **637**, 637i, 781
Antigen, **578**, 578i, 579–581
Antigen-MHC complex, 580, 585
Antigen-presenting cell, **578**, 578i
Antigen receptors, 580, 580i
■ Antihistamine, 587
Antilocapra americana, 613, 613i
Antimicrobial protein, 577
■ Antioxidant, **80**
■ Antisense drug, **205**
■ Antiseptic, natural, 339
Antithamnion plumula, 322i
Antithrombin, 229
Antitrypsin, 606
Anus, 355, 614i, 628t
evolution, 356i
invertebrate, 364i, 365i, 366i, 373
Anvil, 510, 510i
■ Anxiety, 163
Aorta, 159, **558**, 558i
Apatosaurus, 387i
Ape, 394, 396
Aphid, 138–140, 138i, 140i, 715i
Apical meristem, 417, 418i, **419**–422, 422i, 426i, 427, 428, 456, 456i, 458
Apicomplexan, 316i, **318**, 322
Apis mellifera. See Bee
Aplysia californica, 503, 503i, 596i
■ Apnea, 605
Apoptosis, **412**–413, **585**, 653

Appendage, invertebrate, 367–369, 370i
■ Appendicitis, 366i, 621
Appendix, **621**, 621i
Appetite, 610–611, 627
Apple, 329, 452–453
Aquatic ecosystem, 738, 738i
Aquatic snail, 105i, 361i, 364i
Aqueous humor, 512, 513i, 515
Aquifer depletion, 741, 741i
Arabian oryx, 414
Arabidopsis thaliana, 268i, 463, 463i
Arachnid, 369, 369i
Arceuthobium, 348i, 714
Archaea, **8**
Archaean eon, 263, 263i, **292**
Archaeanthus linnenbergeri, 346i
Archaebacteria (Archaea), 725
characteristics, 8, 8i, 70t, 81i, 306–307, 306i, 332t
as domain of life, **8**, 8i, 58, 280, 280i, 282i, 286, 304i, 306
evolution, 287, 296i–297i
methanogenic, 36–37, 36i, 306, 306i
structural adaptations, 59, 59i
Archaeopteryx, 376–377, 376i, 387i, 390
Archafructus sinensis, 346, 346i
Archipelago, **274**–275, 275i
Architectural constraints, 655
Arctic ice cap, 26i
Arctic tern, 391
Arctic tundra, 761i, 766, 766i
Area effect, **723**, 723i
Arginine, 198i, 202i
Arid habitat, 724
Arid lands, 724
Arm (ray), invertebrate, 365i, 372, 372i
Arm bones, 539i
Armadillo, 242, 242i
Armillaria ostoyae, 327
Army ant, 790i
Arnold, S., 780
Aromatic compound, 783, 783i
Aromatic oil, 429, 449
■ Arrhythmia, 567, 567i
■ Arsenic poisoning, 432, 638
Artemisia, 318
Artemisinin, 318
Arteriole, **562**, 562i, 563
■ Arteriosclerosis, 566
Artery, 558i, **562**, 562i, 634, 635i
■ Arthritis, 541
Arthropod, 354t, 355, 356i, **367**–371, 367i–371i
Artidactyl, 257
Artificial reef, 729, 729i
Artificial selection, **10**, 10i, 218
Artificial twinning, **192**
Ascaris. See Roundworm
A. lumbricoides, 366, 366i
Ascending aorta, 558i
Ascorbic acid, 624t
Ascus, 328, 329i
Asexual reproduction, **140**, **646**
animals, 126, 357, 360
cloning, 454–455, 454t
flowering plants, 126, 357, 446, 454–455, 454t
fungal, 326, 327i, 328, 328i, 329, 329i
by mitotic cell division, 324i, 325i, 327i, 328i
protistan, 318, 319, 319i, 321, 322i, 323, 324i, 325i
sexual *vs.*, 138–140, 138i, 140i, 646
Asparagine, 198i
Asparagus, 449
■ Aspartame, 180
Aspartate, 198i

Aspen tree, 228, 228i, 454
Aspergillosis, 772
■ Aspirin, 459, 577, 764i
Assassin bug, 785
Association areas, 498, 499
Asteroid, 258–259, 258i, 263i, 264, 265, 279, 285, 292, 347, 384, 386–387
Asthma, 329, 528, 586, 592, 593
Atala, A., 475
■ Ataxia, 106–107, 106i
Atelopus varius, 714i
■ Atherosclerosis, 49, 107, 566–567, 566i, 680
Atherosclerotic plaque, 567
■ Athletes, and drug testing, 631
Athlete's foot, 329i
Atlantic Ocean, 92i
Atmosphere, 740i, 754i
early, 286, 288–289
Atmospheric cycle, 739
Atmospheric pressure, 594, 594i
Atom, 4, 4i, **18**, 20, 20i, 22i, 24, **30t**
radioactive, **21**, 262
Atomic bonding, 22–23, 74
Atomic number, **20**
ATP, **75**, 522
cyclic pathway, 99, 100i
formation, 60, 64, 70t, 74, 77, 77i, 83, 92–93, 98–99, 98i, 99i, 100i, 209, 295, 306, 307, 427
function, 48t, 59, 66, 67, 72i, 75, 76, 83i, 85, 85t, 89t, 93, 96, 97i, 101, 102, 130, 142i, 295, 307
muscle contraction and, 546, 546i
net yield of, 115
noncyclic pathway, 99, 100i
release of chemical energy from cells and, 106–120
structure, 46, 46i
ATP/ADP cycle, **75**, 75i
ATP synthase, 57i, **98**, 99i, 100i
ATPase, 83i
Atrazine, 460, 460i, 518–519
Atrial contraction, 560
■ Atrial fibrillation, 567
Atrial natriuretic hormone, 533t
Atrioventricular valve, 560
Atrium, heart, 560, 560i
Attalea, 424
Auditory canal, 510, 510i
Auditory nerve, 510i
Australia, 699i, 721, 721i
Australian, 761i
Australian sawfly caterpillar, 788, 788i
Australopith, **396**, 397
Australopithecus, 396
A. afarensis, 396, 396i
A. garhi, 397
■ Autism, 166, 792
■ Autoimmune disorder, 587
Autoimmune response, **587**, 680
Automated DNH sequencing, **227**
Automated external defibrillators, 551
Autonomic nerve, **494**
Autonomic subdivision, 493i
■ Autosomal dominant disorder, 206, 206i
■ Autosomal dominant inheritance, 159, 166, 173, 173t, 174, 174i
Autosomal recessive inheritance, 535
Autosome, **168**, 171
Autotroph, **92**, 93, 294, 732, 732i, 733i
Autumn crocus, 66
Auxin, **458**–459, 458t, 459i, 460, 461, 464
AV node, 561, 561i
AV valve, 560i, 561i

Avery, O., 186
Avocado, 348
Axillary bud, 418i, 424i, 454t
Axolotl, 385
Axon, **485**, 490i
  endings, 484i, 485
■ AZT (azidothymidine), 589

B

B cell (B lymphocyte), 555, 555i,
  574, 574i, **578**, 580–581, 581i, 587
  antibody-mediated response and,
    582, 582i
B vitamins, 624t, 676
Baboon, 784, 784i, 789
Bacillus anthracis, 304
Bacillus shape, 302i
Backbone, 538. *See also* Vertebral
    column
  carbon, 34, 38, 40i, 43, 94i, 101, 188
    sterol, 41, 41i
    sugar-phosphate, 189i, 191, 191i
Bacterial cell, 50i, 52i, 58i, 70t, 222
Bacterial chromosome, 303i, 304
Bacterial colony, 224, 224i
Bacterial conjugation, **302**
Bacterial flagellum, 58i, 59, 70t,
    302, 302i
Bacteriophage, **186**, 186i, 308, 308i
  multiplication cycle, 309, 309i
  relative size, 53i
Bacteriorhodopsin, 307
Bacterium, 332t. *See also* Prokaryotic
    cell; *specific names*
  aerobic, 295, 320, 597
  bacterial spore, 304i, 305
  at body surfaces and in gut
    lumen, 575
  characteristics, **8**, 8i, 9i, 58, 92,
    303, 303i, 305, 332t
  DNA in, 58, 58i, 59, 184i, 222, 294
  as domain of life, **8**, 8i, 9i, 58, 280,
    280i, 282i
  functions, 7, 8i, 209, 295, 302, 304,
    304i, 305, 330, 339, 436
  nitrogen-fixing, 746, 746i
  origin, 54, 64, 282i, 287, 293i,
    296i–297i, 304–305
  pathogenic, 59, 150, 301, 302i, 310
  size, 50i, 53i, 58i
  structure, 52, 52i, 58i, 70t, 223i,
    294i, 302, 302i, 305i, 332t
  types, 36i, 37, 59, 88, 89i, 94,
    97, 105, 186, 294–295, 301, 304i,
    305, 306
Bahamas, 719
Balance, 509, 517, 517i
Balanced polymorphism, **252**
Balancing selection, 252
Bald cypress, 445
Bald eagle, 725i
Baleen whale, 368
Ball-and-stick model, 34i, 36i
■ Balloon angioplasty, 567
Banana, 452t
Banana slug, 780, 780i, 781
Banding pattern, DNA, 226, 226i,
    227, 227i
Bangiomorpha pubescens, 292i, 293
Bangladesh, 775, 775i
Barbed thread, cnidarian, 358i
Bark, **429**, 429i
Barnacle, 368, 368i
Baroreceptor, 563, 637
Barr body, 213, 213i
Basal body, **67**, 67i
■ Basal cell carcinoma, 135i
Basal group, flowering plants,
    348i, 350t
Base, **28**–29, 30t, 75, 75i
  acid-base balance, **638**
  nucleotide, 616i, 617

Base-pair substitution, **202**, 202i
Base-pairing rule, 190, 196, 199
Base pairings, 46i, 184i, 189, 189i,
    196i, 198, 199, 202, 203, 224, 225
Basement membrane, 69i
Basic solution, **28**
Basilar membrane, 510–511, 511i
Basilisk lizard, 9i
Basilosaurus, 239i
Basophil, 555, 555i, **574**, 574i, 590t
Basswood, 102i, 103i, 440i
Bat, 239, 266i, 279, 346, 785
  as pollinator, 447, 449, 449i
  wing, 267, 267i
Bathyal zone, 773i
Bay of Bengal, 775, 775i
Bayliss, W., 520
Bcl-2 gene, 413
Bdelloid rotifer, 149, 149i
Beagle voyage, 240–241, 240i, 241i
Bean plant, 102, 419, 425i, 436, 457i
  castor bean, 713
  soybean, 231, 436i, 455, 460
Bear
  brown, 271i
  grizzly, 725i, 787i
  polar, 26i, 244–245, 244i, 271i, 641,
    641i, 725i
Beavertail cactus, 103i
Becquerel, H., 21
Bedrock, 435i
Bedstraw plant, 453
Bee, 41, 371i, 446, 447, 449, 466,
    778–779, 785, 785i, 790, 790i,
    791, 791i
  Africanized, 778, 778i
Bee hummingbird, 390
■ Beef, 49, 361, 361i
Beeswax, 41, 41i
Beetle, 138i, 269, 371i, 446, 449,
    713, 713i
Behavior
  adaptive, 783, 783i, 792
  animal, 779
  anticipatory, 795
  in bacteria, 305
  birds, 273
  communication signals, 784–785
  courtship, 252, 252i, 272i, 273
  culture and, 394
  genes and, 780
  heritable basis of, 780–781
  hormones and, 780–781
  humans, 166–167, 792
  instinctive, 781, 781i
  learned, 782
  mammals, 392
  melatonin and, 530
  parental, 787, 787i, 792
  primates, 395
  self-sacrificing, 790–791
  sexual, 252, 252i, 365, 786–787
  social behavior, 508, 520, 783
Behavioral ecology, 778–795
Behavioral flexibility, 392
Behavioral isolation, 272i, **273**, 273i
Behavioral trait, 246
Bell, cnidarian, 358i, 359
Bell curve, 160i, 161, 161i
■ Bends, 609
Bengal tiger, 725i
■ Benign tumor, 135, 135i
Benthic province, 773, 773i
Bererub, T., 778, 778i, 779i
Berg, P., 220
Bermuda grass, 454t
Berry, **452t**, 453
Beta-carotene, 94–95, 95i, 105, 218
Beta cell, pancreatic, 528, 529i
Beta globin, 44, 44i
Bias in report results, 13, 16
Bicarbonate, 29, 604–605, 615, 617

Biceps, 491, 491i, 542i
Biceps brachii, 543i
Biceps femoris, 543i
Biennial plant, **428**, 464
Big bang, **288**
Big laughing mushroom, 326i
Bighorn sheep, 725i
Bilateral animal, 355, 356i, 360,
    362, 364
  invertebrate, 353i, 354, 366, 367
  vertebrate, 354
Bilateral symmetry, **354**, 354i
Bile, **617**
Bile salt, 41
Bill, bird, 377
Binary fission, **318**
Binding energy, 79
■ Binge drinking, 73
■ Bioartificial bladder, 475, 475i
■ Biochemical weapon, 194–195, 195i
Biochemistry, comparative, 303
Biodiversity, **722**. *See also* Diversity
  conservation biology and,
    349, 725
  disruptions in, 259, 261, 279
  island patterns, 722i, 723, 723i
  mainland and marine patterns,
    722, 722i
  oxygen levels in water and, 597
  patterns of, 722–723
  range of, 261
  realms of, 760–761, 760i–761i
  sustaining, 726–727
  threats to, 724–725
Bioethics
■ abortion, 151, 165
■ allocating scarce resources, 73, 91,
    137, 417, 431
■ bioprospecting, 287, 298
■ cloning, 184–185, 192, 192i, 193
■ cost of conservation, 353, 375
■ gene therapy, 230–231
■ genetic engineering, 93, 105, 139,
    139i, 149, 229, 443, 445, 447
■ genetic screening, 151, 165,
    180–183
■ human impact on biosphere, 301,
    313, 315, 350, 466
■ patenting genomes, 218–219, 218i,
    219i, 221
■ public health, 195, 206, 207, 313
■ tax breaks for recycling, 335, 350
■ test tube organisms, 51, 125, 137,
    220, 234, 234i, 298
■ tracking asteroids, 259, 285
■ waste disposal, 333
  wildlife trading, 725, 725i
Biogeochemical cycles, **739**, 739i
  carbon cycle, 742–743, 742i–743i
  hydrologic cycle, 740–741
  nitrogen cycle, 746–747
  phosphorus cycle, 748–749
Biogeographic patterns, 722–723
Biogeographic realms, **760**–761,
    760i–761i
Biogeography, **238**
  evolutionary evidence from,
    264–265
  island patterns, 722i, 723, 723i
  mainland and marine patterns,
    722, 722i
Biological clock, 391, **530**, 680
Biological control, 705, 715, 715i
Biological inquiry, nature of, 11
Biological magnification, **736**, 736i
Biological molecule, 24–25, 29,
    38, 289
Biological principles and human
  imperative, 795
Biological species concept, **272**
Biological wealth, 726–727
■ Biological weapon, 573

Bioluminescence, 234, 319, 319i
Biomass pyramid, **737**, 737i
Biome, **760**–766 760i–761i, 762i,
    763i, 766i, 767i
Bioprospecting, 286, 287, 298
Biosphere, **5**, 5i, 77, 96–97, 334,
    752–777, **753**
■ air circulation patterns, 754–757
■ applying knowledge of, 774–775
    climate and temperature zones,
      754–755
    freshwater provinces, 768–769
    human impact on, 301, 313, 315,
      350, 466, 795
    ocean currents, 758
    open ocean, 772–773
    realms of biodiversity, 760–761
Biosynthesis pathway, **77**, 80,
    84, 128
■ Bioterrorism, 2–3, 236
Biotic potential, **689**, 702
Biotin, 624t
Bipedalism, 383, 392–**394**, 396, 396i
Bipolar cell, retina, 514, 514i
■ Bipolar disorder, 166–167
Bipolar spindle, **129**, 129i, 130, 130i,
    132i, 143i
Birch tree, 424
Bird, 279, 281, 281i, 381i, 387, **390**,
    390i, 497i, 553i, 639, 722i, 724
  behavior, 273, 786–787, 787i,
    794, 794i
  characteristics, 354, 354t
  cowbirds, 715, 715i
  digestive system, 612, 612i, 613
  dinosaurs and, 376–377, 376i, 390
  evolution, 390–391, 390i
  flight, 390–391, 391i
  flightless, 238, 238i, 239
  migration, **391**
  as pollinator, 346, 449
  reproduction, 139, 647, 647i
  reptiles and, 376–377, 376i
  respiration, 599, 599i
Bird of paradise, 252i
Birds of prey, 517, 517i
Birdsong, 782, 782i
Birth, 120, 678, 678i
  multiple, 644–645, 644i
  premature, 678
■ Birth control, 666–667, 666i
Birth rate, 698, 700
Birth weight, 250, 251i
Bison, 715, 715i, 763i, 787i
Bivalve, 364
Black-bellied seedcracker, 251, 251i
Black bread mold, 328, 328i
■ Black Death, 3, 236i
Black locust, 424i
Black market, 725i
Black widow spider, 369, 369i
Bladder, 321, 321i, 361
  invertebrate, 363, 363i
■ lab-grown, 475, 475i
  swim bladder, 382
  urinary, 634, 634i
Blade, 321, 321i, 424, 424i, 425
Blastocoel, 670, 670i
Blastocyst, **651**, 670–671,
    670i–671i, 679t
Blastomere, **648**
Blastopore, 355, 356i
Blastula, **648**
Bleaching, coral, 772, 772i
Blind spot, 513i
■ Blindness, 515
Blood, 28, 86, 367, 405i, **473**,
    552, **632**
  alcohol in, 72
  cellular components, 473i, 555i
  characteristics, 554–555
  coagulation, 566

composition and volume, 554–555, 554i
oxygen in, 44, 318
pH of, 28i, 29
pressure, transport, and flow distribution, 562–563
Blood-brain barrier, **496–497**, 501, 564
Blood circulation. *See* Circulatory system
■ Blood clotting, 73, 353, 566–567, 566i, 571
■ Blood clotting disorder, 169, 173t, 175, 229
■ Blood disorder, 556
■ Blood doping, 556
Blood flow distribution, 562–563, 563i
Blood fluke, 361i
Blood glucose level, 7, 19, 39, 118, 522–523, 526, 528–529, 529i
■ Blood-letting, 363i
Blood pressure, **562**, 562i
■ controlling, 563, 563i
hypertension, 250, 528, 566, 567, 626, 638
kidneys and, 563
measuring, 563i
Blood typing, 556–557, 556i
Blood vessel, 562i
human, 135, 135i, 159
invertebrate, 362, 363, 363i, 367, 558i
Bloodworm, 586
Blue crab, 535, 535i
Blue-footed booby, 241i, 272i, 273
Blue jay, 255, 255i
Blue light, 441, 460
■ Blue offspring, 173t, 257
Blue whale, 93, 392i
Blueberry, 95, 447
Bluegill sunfish, 788
BMI (body mass index), 626
Body-building proteins, 623
Body cavities, 354, 355, 355i, 476i
Body fat distribution, 169
Body fluid, 29, 405
Body hair distribution, 169
Body height, 161, 161i
Body mass, 251
Body mass index (BMI), 626
■ Body piercing, 481, 481i
Body plan, 212, 215, 239, 240
animal, 354–355, 354i, 355i
annelid, 362
bird, 390i
Caenorhabditis elegans, 366, 366i
cuttlefish, 365i
earthworm, 363i
emergence during pregnancy, 672, 672i
euglenoid, 317i
evolution of, 356, 362, 364–365, 372
evolutionary constraints on development, 655
flatworm, 360
giant kelp, 321i
invertebrate, 358, 358i, 359, 359i, 360, 360i, 362, 363i, 364i, 365i, 366, 366i
lichen, 330i
pattern formation and, 654–655
plant, 330i, 416, 418, 418i
prokaryotic cell, 302, 302i
protist, 317i, 321i
reptile, 388, 388i
roundworm, 366
tomato plant, 418i
vertebrate, 672, 672i
virus, 308i
Body symmetry, 352i, 354

Body temperature, 402–403, 409i, 577
Body wall, 355, 355i, 357i, 366i
invertebrate, 363, 363i, 372
Body weight, 610–611, 626–627
Bombykol, 784
Bonding, chemical, 18, 22
atomic bonding, 22–23, 74
carbon, 34
covalent, 24–25, 25i, 34, 46, 47i, 56, 75, 95, 436
hydrogen, 25–27, 25i, 26i, 34, 43, 47i, 56, 79, 81, 189–191, 270, 438–439, 439i
ionic, 24, 24i
molecular, 24–25
peptide, 42i, 43, 200, 200i, 352
Bonding behavior
hormones and, 792
pair-bonding, 780–781
Bone
formation and remodeling, 540–541, 541i
human skeletal system, 539i
movement and, 540–541
structure and function, 540, 540t, 541
Bone density, 541, 541i
Bone marrow, 540, 540i, 555i, 568i, 574, 585
Bone remodeling, **540–541**
Bone tissue, 4i, 69, 69i, **473**, 473i, 540, 540i
Bonobo, 394
Bony fish, 381i, 382i, 383
Book lung, 596, 596i
Boreal forest, 307, 760i–761i, 764–765
Boron, 434t
Borrelia burgdorferi, 305, 305i, 369
■ Botox injections, 549
Bottleneck, **254**, 255, 278
Bottlenose dolphin, 320
Bottom dweller, 372
■ Botulism, 116, 194, 304, 313, 548
Bouchard, C., 627
■ Bovine spongiform encephalopathy, 312, 312i
Bowman's capsule, 634–635, 635i
Brachial artery, 558i
Bradshaw, K., 752, 752i
■ Bradycardia, 567, 567i
Brain, 354, 493i
drugs and, 500–501
emotional-visceral, 499
energy sources for, 118
evolution, 493
human, 128, 166, 174, 180, 396–397, 497, 497i, 498–499
invertebrate, 360, 360i, 363i, 365, 367
mammal, 393, 497i
primate, 395
subdivisions, 496, 496i
vertebrate, 493, 496–497, 497i
Brain case volume, 401
Brain scan, 21i
Brain stem, **496**
Branch, plant, 422
Branch point, 278, 278i
Branch root, 457i
Branched evolution, 278, 281, 281i
Brassica oleraceo, 268, 268i
Brassinolides, 459
Brazil, 698i
BRCA gene, 207
Bread mold, 328, 328i
Breast, 678i, 679
■ precocious development, 534, 534i
tissue, 206i
■ Breast cancer, 206–207, 206i, 208, 217, 585, 593, 626

Breastbone, 539i
Breastfeeding, 679
Breathing, 408, 601–603, 601i, 605. *See also* Respiration; Respiratory system
Breeding ground, fish, 321
Brevetoxin, 320
Bristle, animal, 362, 375i
Bristlecone pine, 244i, 344i, 410
Bristly foxtail, 709, 709i
Brittle star, 372, 372i
Broadleaf forest, 760i–761i, 764, 765i
Broca's area, 498, 498i
Brock, T., 286
Bronchial airway, 150
Bronchial tree, 600i, 601
Bronchiole, 600i, **601**
■ Bronchitis, 592, 606, 606i
Bronchus, **601**
Brown, R., 54, 54i
Brown adipose tissue, 122, 641, 641i
Brown alga, 316i, 320, **321**, 321i
Brown bear, 271i
Brown-headed cowbird, 715, 715i
Brown recluse spider, 369, 369i
Brundtland, G. H., 607
Brush border cell, 618, 618i
Bryophyte, 336, **338–339**, 350t. *See also* Nonvascular plant
■ BSE, 312, 312i
■ Bubonic plague, 2–3, 236, 691i, 697i
Bud, animal appendage, 269
Bud, plant, 418i, **422**, 428, 454, 456
axillary, 418i, 424i, 454t
lateral, 418i, 422, 422i, 458, 458i, 459
terminal, 418i
Bud scale, 422
Budding, 329, 332t, 357, 361
Buffer, 29, 605, 638
Buffer system, **29**, 638
Bulb, 454t
Bulb of Krause, 507, 507i
Bulbourethral gland, 656i, 657, 657i, 658i
Bulk, **621**
Bulk flow, **86**, 564i, **565**
Bulk-phase endocytosis, **88**
Bull's-eye rash, 305i
Bundle-sheath cell, 102i, 103i
Bur clover, 453
■ Burial, 729, 729i
Burkitt, D., 622
■ Burn, skin, 29, 309, 479
Bursa, 543i
Bushbaby, 517i
Buttercup plant, 419, 420i, 426i
Butterfly, 12–13, 269, 370i, 371i, 447, 687i, 691
as pollinator, 446, 447, 449
wing, 246, 247i, 249i, 250, 267
Butterfly fish, 707, 707i
Byrd, R., 16

C
C3 plant, **102**, 102i
C4 plant, **102**, 102i
Cabbage plant, 348, 458i
Cacao tree, 446, 446i, 453
Cactus, 102, 103i, 238i, 407, 426, 703, 703i, 725i, 762i
Cadmium ion, 433
Caecilian, 385, 385i
Caenorhabditis elegans, 366, 366i
Caffeine, 500
Caiman, 389, 389i
Calamites, 342, 342i
Calcitonin, 533t, 541
Calcitriol, 533t
Calcium, 22, 29, 159, 368, 434t, 528, 540–541, 625, 625t, 740, 741i

Calcium carbonate, 314, 321, 359, 367, 372, 373
Calcium ion, 82i, 85, 85i, 434, 440, 440i, 488, 540–541
Calcium pump, 57i, **85**, 85i
Calcivirus, 721
California Current, 758
California poppy, 427i
California red-legged frog, 385
Calisher, C., 300
Callinectes sapidus, 535, 535i
Callus, **454**, 455i
■ Calorie, 626
Calvin, M., 21, 323
Calvin-Benson cycle, 93, **101**, 101i, 102, 102i, 104i
Calyx, **448i**
CAM plant, **102**, 440, 703, 703i
Camarhynchus pallidus, 243i
Cambrian period, 263i, 269, 269i, 356, 380
Camel, 245, 245i, 274i, 285, 640, 640i
Camelid, 274, 274i
Camellia plant, 424
Camellia sinensis, 343i
Camera eye, 512, 512i
Camouflaging, **712**, 712i, 713
cAMP, 46, 48t, 209, 325i, 522, 523, 523i
■ Camptodactyly, 160, 173t
Camptosaurus, 387i
Campylobacter, 312
Canada goose, 108, 122
Canadian lynx, 706, 710–711, 711i
■ Canavan's disease, 230i
■ Cancer, 88, **135**, 777, 777i
■ basal cell carcinoma, 135i
■ breast, 206–208, 206i, 217, 585, 593, 626
■ carcinogens, 206
■ carcinoma, 135i, 206i
■ causes, 2, 19, 177, 203, 232, 308, 460
■ cell cycle and, 125, 134–135, 134i, 135i
cell death and, 413
cell-mediated response, 584–585
■ cervix, 134i, 137, 668
■ chemotherapy, 585
■ colon, 517, 621, 626
■ HeLa cells, 124i, 135, 214–125
HPV and, 668
■ immunotherapy and, 585
■ Kaposi's sarcoma, 588, 588i
■ leukemias, 173t, 177i, 230, 556
■ lung, 593, 757, 757i
■ malignant tumor, 135, 135i
■ prostate, 657
■ skin, 134–135, 135i
■ squamous cell carcinoma, 135i
■ survival rates, 135, 137
■ testes, 657
treatment, 66, 135, 137, 206, 217, 352
■ vaccines, 585, 591
Cancer cell, 135, 578, 680
Canegiea gigantea, 703, 703i, 725i, 762i
Canine, 784
Canis lupus (Canidae), 710i, 784, 784i, 789, 789i
Cannabis, 501, 607, 631
Cannibalism, 685, 789
Canola plant, 460
CAP, 208–209
CAP, **102**, 102i
Capillary, 402i, 405i, 407i, **562**, 562i, 564–565
function, 564–565
glomerular, 634–635, 635i, 636
lymph, 568i, 569

pulmonary, 600i
reabsorption, 564i, **565**
Capillary bed, 558, 559i, 564–565, 564i, 569
inflammatory response and, 576–577
Capsella, 452, 452i
Capsule
Bowman's, 634–635, 635i
renal, 634, 635i
Capture-recapture method, **687**, 687i
Carapace, 369
Carbohydrate, 32i, 48t, 96, 128
complex, 38, 48t
digestive enzymes, 616t
early Earth and, 289
function, 39, 73, 101, 101i, 442
organic metabolism, 620, 620i
short-chain, 38, 48t
structure, 38–39
Carbohydrate-insulin connection, 622–623
Carbon, 20, 22, 23i, 188, 342, 434t, 766
annual fluxes in global distribution, 742i
bonding behavior, 34
global cycling of, 306, 349
isotopes, 21
reservoirs and holding stations, 742i
Carbon 12 (12^C), 21, 262, 262i
Carbon 14 dating, 262, 262i
Carbon black, 757i
Carbon cycle, **742–743**, 742i–743i
Carbon dioxide, 29, 37, 75, 82i, 288, 290, 302
animals and, 365
as gas, 742
global concentrations, 744, 745i
in photosynthesis, 92, 93, 101, 104i
plants and, 421, 424, 425i, 440, 441
transport, 604–605
Carbon fixation, **101**, 102, 102i–103i
Carbon monoxide, 288
■ Carbon monoxide poisoning, 609
Carbon-oxygen cycle. See Carbon cycle
Carbonic acid, 28, 29
Carbonic anhydrase, 604–605
Carboniferous period, 263i, 334, 337i, 340–343, 342i, 381, 386, 392
Carbonyl group, 34i
Carboxyl group, 34i, 42i
Carboxypeptidase, 616t
■ Carcinogen, 206
■ Carcinoma, 135i, 206i
Cardiac cycle, **560**, 561i
Cardiac muscle, 43, 56i, 69, **474**, 474i, **561**, 561i
■ Cardiac pacemaker, **561**
■ Cardiopulmonary resuscitation, 550–551
Cardiovascular circuit, 558, 559i
■ Cardiovascular disorder, 566–567
Cardiovascular system, 558–559. *See also* Circulatory system
major blood vessels, 558i
Caribou, 710i
Carnivore, 9, 732, 733i
Carotene, 478
Carotenoid, **94**, 95i, 105, 441, 449
Carotid artery, 558i
Carpal, 539i
Carpel, 152i, 346i, **448**–450, 448i, 450i, 463, 463i
Carrageenan, 322
Carroll, L., 696
Carrot, 94, 427, 452i, 454, 454t, 783, 783i
Carrying capacity, **690**, 691, 691i, 696

Carson, R., 736, 736i
Cartenoid pigment, 333
Cartilage, 43, 69, 174, **472**, 472i, 473i
Cartilaginous fish, 381i, **382**, 382i
Cascade Range, 766i
Casparian strip, **437**, 437i
Caspian tern, 787i
Castor bean plant, 713
Castor oil plant, 194–195, 194i, 195i
Cat, 81i, 164, 213, 213i, 702
Bengal tiger, 725i
civet, 313, 313i
cloned, 192, 192i
lion, 7i, 281i, 784, 788–789, 792
Manx, 164, 165i
Siamese, 81i
wild, 724
Catabolic pathway, **77**
Catabolite activator protein, 208–209
Catalase, 72, 72i, 79i, 80, 90–91
■ Cataracts, 515, 680, 756
Catastrophism, **240**
Catecholamine, 523
Caterpillar, 6i, 367, 414, 414i, 712i, 788, 788i
Cattle, 49, 306, 311, 361, 361i, 715, 727, 729
Caulerpa taxifolia, 720–721, 720i
Cause and effect, 12, 392
Cave paintings, 399i
CCK, 611, 617, 627
CD4 lymphocyte, **588**
cDNA, **223**, 223i, 224
Cecum, 621, 621i
Cedar waxwing, 453i
Celera Genomics, 220i, 221
Celery plant, 420i, 466
Cell, **4i**, **52**, 210
animal, 86, 125, 130, 132–133, 132i, 133i
bacterial, 287, 294i
components, 42–43, 52, 54–55, 65i, 66, 68–69, 70t, 295i
cortex, 66
energy flow, **74–78**, 74i, 75i, 77i, 78i, 83i, 85i
eukaryotic. *See* Eukaryotic cell
evolution, 294, 295
excitable, 485
flowering plant, 420i, 421i, 423i, 425i, 426i, 427i, 429i
mechanisms, 67, 70t, 125, 130, 132–133, 132i, 133i, 427, 456, 458–459
membrane. *See* Cell membrane
origin, 50–51, 72i, 77i, 286–293, 291i, 306
prokaryote. *See* Prokaryotic cell
relative size, 52–53, 53i, 58
release of chemical energy, 106–120
shape, 53, 66, 83
size, 52–53, 53i, 58, 332t
as unit of organization, 4–5, 4i
Cell communication, 407, 412–413
Cell count, **555**
Cell cycle, 125, **128**–129, 128i, 134–135, 134i, 206
cancer and, 125, 134–135, 134i, 135i
Cell death, 134, 191, 412, 412i, 413, 653
cancer and, 413
Cell differentiation, 192, **210**, **652–653**
Cell division, 124–128, 126t, 332t, 428i. *See also* Meiosis; Mitosis
Cell junctions, **69**, 69i, 471, 471i
Cell-mediated immune response, 578, 579i, 584–585, 584i, 585i
cancer and, 584–585

Cell membrane, 26, 50i, 56–57, 56i, 57i, 58i, 289
components, 41, 41i, 82, 82i
crossing, 73i, 80, 82–86, 82i, 83i, 84i, 85i, 89t
Cell memory, 654, 655i
Cell migration, 652i, 653
Cell plate formation, 132i, **133**
Cell suicide, 412–413, 586, 653
Cell surface specialization, 68–69
Cell theory, **54**
Cell-to-cell communication, 407
■ Cell typing, 66
Cell wall, 58, 59, 65i, **68**, 70t, 332t, 420
Cellular slime mold, 325, 325i, 333
Cellulose, 38i, 39, 39i, 48t, 68, 68i, 132i, 133, 323, 420, 440, 464
Cenozoic era, 263, 263i, 279i, 393
Central nervous system, 493, 493i, 494i, 502t
Central vacuole, **63**, 65i, 70t, 87, 96i, 433, 440i
Centrifugation, **170i**
Centriole, 65i, 67, 130, 130i, 142i
Centrocercus urophasianus, 786–787
Centromere, **127**, 130, 141, 141i
Centrosome, **130**, 142i–143i
Cephalization, **354**, 360, **492–493**
Cephalochordate, 378
Cephalopod, 364, 365, 365i, 512i
Cereal croplands, 763
Cerebellum, 393, **496**, 496i, 497i, 502t
Cerebral cortex, 497–499, 498i, 507, 507i, 508
Cerebral hemispheres, 497
Cerebrospinal fluid, 496–497, 496i
Cerebrum, **496**, 497, 502t
Certhidea olivacea, 243i
■ Cervical cancer, 134i, 137, 668
Cervical nerve, 493i
Cervix, 137, 660, 660t, 661i
Cetacean, 257
CFCs, 745i, 756
CFTR protein, 229
Chaetae, 362
Chaetodipus intermedius, 249, 249i
■ Chagas disease, 316
Chain fern, 341, 341i
Chain of Being, 238, 240
Chalk deposit, 314, 314i, 321
Chambered nautilus, 239i, 365i
Chameleon, 388, 475i
Channel protein, 352
Chaparral, 762, 763i
Chara, 323i
Chargaff, E., 188
Charge, of proton/electron, **20**, 24
Charophyte, 316i, **323**, 323i, 336, 337i
Chase, M., 187i
Checkpoint gene, 134–135, 134i
Cheetah, 713, 794
Chelicerate, 367, 369
Chemical basis of life, 30t
Chemical bond, 18, **22**
■ Chemical burn, 29
Chemical energy, 7, 59, 74, 92–93, 96, 98
Chemical equation, 23i
Chemical equilibrium, **76**, 76i
Chemical formula, **23**, 23i
Chemical message, 46, 120, 409
Chemical synapses, **488–489**, 488i
Chemical work, 74, 75i
Chemoautotroph, 292, **302**, 304, 305
Chemoreceptor, **506**, 508
■ Chemotherapy, 585
■ Chernobyl nuclear power plant, 777, 777i

Cherry tree, 95, 448i, 450, 451i, 452, 452t
Chesapeake Bay, 432
Chestnut, 329
Chestnut blight fungus, 719t
Chicken, 158, 159i, 229i, 266i, 270, 571, 653i, 655i
comb, 158, 159i, 165
■ Chicken pox, 591
Chickweed, 449i
Chigger mite, 369
Child, 679t
Childbirth. *See* Birth
Chiloglottis trapeziformis, 252
Chimpanzee, 129, 177, 268i, 269, 394, 396i, 725i, 794
Chin, 398
China, population and family planning, 698–699, 699i
Chipmunk, 708, 708i
Chironex, 358i
Chironomus, 597
Chitin, 39, 39i, 267, 362, 367
Chiton, 364, 364i
Chlamydia trachomatis, 305, 668
■ Chlamydial infection, 305, 668, 668i, 668t
Chlamydias, **305**
Chlamydomonas, 323i, 324, 324i
C. nivalis, 286, 333, 333i
Chlorella, 17i, 323
Chloride, 625t
Chlorine, 23i, 24, 24i, 434t, 756
Chlorine ion, 24, 27, 28, 82i
Chlorine monoxide, 756
Chlorofluorocarbons, 745i, 756
Chlorophyll, 64, 92i, **94**, 94i, 95i, 98–99, 290, 304, 317, 320, 322, 323
Chlorophyte, **323**, 323i
Chloroplast, 39i, 53i, 66, 67, 70t, 92, 96, 96i, 97i, 98, 98i, 101, 317i, 319, 322
function, 60, **64**, 64i, 65i, 318, 424
origin, 59, 287, 294, 295, 297i, 304
Choanoflagellate, 356
Chocolate, 446–447
Choking, 608–609, 609i
Cholecalciferol, 479
Cholecystokinin, 611, 617, 627
■ Cholera, 2, 248, 775, 775i
Cholesterol, 35, 41, 41i, 48t, 49, 522, 566, 567, 623
■ hypercholesterolemia, 173t
Cholla cactus, 762i
Chondrus, 718i, 719
Chordate, 282, 353, 354t, 355, 356i, **378–379**, 381i
Chorion, 670i, **671**, 671i
Chorionic villi, 671i, 673, 673i
■ Chorionic villi sampling, 181
Choroid, 512, 513i
Chromatid, 126, 126i, 130–131, 130i, 131i, 134, 141, 141i, 143i, 144–145, 144i, 148i, 168, 168i
Chromatin, **61**, 61i, 62i
Chromosomal DNA, 222, 223
Chromosome, **61**, 127, 127i, 129, 129i, 141, 141i, 147. *See also* Sex chromosome; X chromosome
abnormality, 168, 173t
bacterial, 303i, 304
deletion, **176**, 176i
duplication, **176**, 176i
inversion, **176**, 176i
meiosis and change in, 168, 168i, 171, 171i, 178i
metaphase chromosome, 170, 170i
Philadelphia chromosome, 177, 177i
polytene chromosome, 211, 211i
primate, 129, 177

structural change, 173t, 176–177, 176i, 177i
translocation, **177**, 177i
Y chromosome, 141i, 144i, 168–169, 169i, 179
Chromosome, eukaryotic, 126–127, 126i, 127i, 140
homologous, **140–141**, 141i, 142i, 144, 150, 153i, 154, 156–157, 168, 168i, 171, 247i
human, 127, 127i, 137i, 141i, 147, 202i
meiosis and, 140–145, 140i–143i
mitosis and, 124, 130–131, 130i–131i
structure, 126–127
Chromosome number, **129**, **140**, 146, 147, 152, 153i, 167, 168
changes in, 173t, 178–179, 178i, 179i
diploid, **129**, 130i, **140**, **153**, 153i, 168, 169i
haploid, **141**, 142i, 143i, 146, 221
■ Chronic anemia, 556
■ Chronic myelogenous leukemia (CML), 173t, 177i
■ Chronic viral hepatitis, 481
Chrysanthemum, 462, 462i
Chrysophyte, 316i, 320, **321**, 321i
Chukwu, N., 644, 644i
Chyme, 616
Chymotrypsin, 616t, 617
■ Chytridiomycosis, 714i
Cicada, 272i, 273
Cichlid, 276, 276i
■ Cigarette smoking. *See* Smoking
Ciliary body, 512, 513i
Ciliary muscle, 513, 513i
Ciliated cell, 575
Ciliated protozoan (ciliate), 67, 316i, **318**
Cilium (cilia), 17i, **67**, 67i, 70t, 130, 360, 360i, 375, 375i, 511
Cingulate gyrus, 499, 499i
Ciona savignyi, 401, 401i
Circadian migration, 772
Circadian rhythm, **411**
Circuit, 490
cardiovascular, 558, 559i
pulmonary, 553, 558, 559i
systemic, 553, 558, 559i
Circular papillae, 508i
Circulatory system, 550–571, **552**
amphibian, 553, 553i
bird, 553i
blood disorders, 556
blood typing, 556–557
capillaries, 402i, 405i, 407i, 562, 562i, 564–565
cardiovascular disorders, 566–567
characteristics of blood, 554–555
closed, **363**, 363i, 365, 552, 552i, 553i
digestive system and, 558–559, 559i, 612i
earthworm, 552, 552i
evolution of vertebrate circulation, 552–553
between fetus and mother, 673, 673i
fish, 552–553, 553i
heart and, 560–561
human, 477i
human cardiovascular system, 558–559
invertebrate, 363, 363i, 365, 367
lymphatic system and, 553, 568–569
mammal, 553i
nature of, 552–553
open, **367**, 552, 552i

pressure, transport, and flow distribution, 562–563
respiratory system and, 558–559, 559i
urinary system and, 558–559, 559i, 632i
■ Cirrhosis, 72
cis Fatty acid, 49, 49i
Citric acid cycle, 112–113
Citrus fruit, 371i
Civet cat, 313, 313i
Clade, **282**
Cladistics, **281**
Cladogenesis, **278**
Cladogram, **281**, 281i, 304i
Cladonia rangiferina, 330i
Claret cup cactus, 762i
Clark, L., 783
Classification system, 259i, **280–281**, 280i, 282i
current species diversity, 725, 725i
of prokaryotes, 303
six-kingdom, 280, 280i
three-domain, 8, 280, 280i, 282i
Clavaria, 326i
Claviceps purpurea, 329, 329i
Clavicle, 539i
Clay soil, 434, 767
Cleavage, 648, 648i, 650–651, 651i, 670
animal reproduction, **132–133**, 132i, 133i
patterns of, 651, 651i
reaction, 35, 35i, **79**, 354
Cliffs of Dover, 314i
Climate, **754**
carbon fixation and, 102, 102i
change in, 745
change in, global, 93, 261, 263i, 279, 334, 337, 347, 356
change in, regional, 349, 416–417, 759
El Niño, 752–753, 752i, 773–775, 774i
ice age, 261, 263i
ocean currents and, 758
temperature zones and, 754–755
zones, 758, 758i
Climax community, **716**, 717i
Climax-pattern model, **717**
■ Clinical trials, 591
Clitoris, 660, 661i, 665
Cloaca, 388, 390
Clonal selection, 580, 581i
Clone, **140**
animal, 184–185, 184i, 192, 192i, 193, 234, 234i
DNA, 184, 185i, 192, 222, 222i, 223, 223i, 224
plant, 140, 432–433, 432i, 454–455, 454t, 455i
Cloning, 140, 222–225, 225i
bioethics of, 184–185, 192, 192i, 193
Cloning site, 272
Cloning vector, **222**, 222i
Closed circulatory system, **363**, 363i, 365, 552, 552i, 553i
Clostridium botulinum, 304, 548
C. tetani, 304, 305i, 548, 548i
Clot, 73, 353, 566–567, 566i, 571, 576–577
■ Clotting disorder, 169, 173t, 175, 229
Clotting factor, 576–577
Clover plant, 436
Club drug, 500
Club fungus, **326–327**, 326i, 327i, 329, 329i
Club moss, 334, 340, 340i, 342, 342i, 350t
■ CML, 173t, 177i

Cnidarian, 352i, 354t, 355, 356i, **358–359**, 358i, 359i, 365, 492, 492i
Coagulation, 566
Coal, 29i, **342**, 342i, 757i
Coastal redwood forest, 334
Cobra lily, 242i, 424i
Coca plant, 333
Cocaine, 333, 489, 500, 500i
pregnancy and, 677
Cocci, 305
Coccolithophore, 321, 321i, 341
Coccus shape, 302i
Coccygeal nerve, 493i
Coccyx, 239, 239i, 285
Cochlea, 510, 510i
Cochlear duct, 510, 511i
Cocklebur plant, 453, 453i, 462, 462i
Coconut palm tree, 421, 453, 455
■ Codeine, 501
Codium, 105i
Codominance, allelic, **158**, 158i
Codon, **198–200**, 199i, 200i, 201i
Coelacanth, 383, 383i
Coelom, 353, **355**, 355i, 363i, 364, 367
Coelomate animal, 355i
Coelomic chamber, 362, 363
Coenzyme, 46, 48t, 77, **80**, 88, 96
Coevolution, **346**, 446, 447, 449, **710**
Cofactor, **76**, 80, **89t**
Cohesion, 27, 27i, **438**, 439i
Cohesion-tension theory, **438–439**, 438i, 439i
Cohort, **692**
Colchicine, 66, 170, 170i
Colchicum autumnale, 66
Cold, 507
■ Cold sore, 309, 478
Cold stress, 640–641, 641i
Coleman, D., 627i
Coleoptera order, 371i
Coleoptile, 456i, 457i, **458**, 458t, 460, 461
Coleus, 422i
Collagen, 48t, 187, 189, 229, 248, 472, 680
Collarbone, 539i
Collecting duct, **635**, 635i
Collenchyma, 416, 419, **420**, 420i, 423, 423i, 430t
Collins, F., 221
Colon, 209, 575, 614i, **621**, 621i, 628t
■ Colon cancer, 517, 621, 626
■ Colon polyp, 517
Colonizing organism, 358i, 359
Colonizing species, 722i, 723, 723i
Colony, 358i
bacterial, 224, 224i
predatory, 305
■ Color blindness, 173t, 175, 175i, 515
Columbine, 348i
Columnar cell, 470, 470i
Colwell, R., 775, 775i
■ Coma, 29
Comb type, chicken, 158, 159i, 165
Comet, 292, 386
Commensalism, **706**, 715, 715i
■ Commercial shipping, 505
Committed cells, 555i
■ Common cold, 308, 592
Common wall cress, 463, 463i
Communication display, 784–785, 785i
Communication line, neuron, 484i
Communication protein, 56, 56i
Communication signal, **784–785**, 784i
communication displays, 784–785
Community, 5, 5i, **706**
biogeographic patterns, 722–723
biological controls, 715

climax pattern model, 717
competitive exclusion, 708–709
competitive interactions, 708–709
conservation biology, 725
cowbirds, 715
cyclic, nondirectional changes, 717
economic factors and sustainable development, 726–727
endangered species, 724
exotic invaders, 720–721
factors shaping structure, 706
habitat losses and fragmentation, 724–725
identifying risk areas, 726
instability, 718–719
island patterns, 723
keystone species, 718–719
mainland and marine patterns, 722
mutualism, 707
niche, 706
parasite-host interactions, 714–715
parasites and parasitoids, 714
population size and, 706
predator-prey coevolution, 710
predator-prey interaction models, 710
predator-prey interactions, 710–711
predators, adaptive responses, 713
prey adaptations, 712–713
resource partitioning, 709
species interactions, 706
species introductions, 719
stability, 716–717
succession model, 716
sustaining biodiversity, 726–727
threats to biodiversity, 724–725
tropical, 722
Compact bone tissue, 540, 540i
Companion cell, 421, 421i, 423, 423i, **442**, 442i, 443
Comparative biochemistry, 303
Comparative genomics, 232
Comparative morphology, 239, 242, 245, 246, 266, **266**
Compartmentalization, **410**, 410i
Competition
dominant competitor, 321
exploitative, 708
humans as competitor species, 724, 724i
interference competition, 708, 708i
interspecific, 706, 708, 708i
intraspecific, 708
keystone species and, 718–719, 718i
molecular, 296i–297i
plant competition for sunlight, 417, 424, 429
for resources, 417, 429
sexual, 787, 787i
Competitive exclusion, **708**, 708–709, 709i
Competitive interaction, 708–709
Complement proteins, 577, 577i, 579
Complementary DNA, 223, 223i, 224
Complete digestive system, **355**, 355i, 362, 364i, 370, 375, 375i, 612
Complete protein, 623
Complex carbohydrate, **38**, 48t
Complex plant tissue, 419, **420**, 420i, 421, 421i, 430t
Composite flower, **449**
Composite signal, 784
Compound, **23**, **30t**
Compound eye, 370i, 512, 512i

Compound leaf, **424**, 424i
Compound light microscope, 54, 54i
Concentration gradient, **82**–86, 83i, 85i, 98, 100i, 212, 407, 433, 440i, 443i
Conception, 666–667
Condensation reaction, **35**, 35i, **79**, 401
■ Condom, 666i, 667
Conducting tube, plant, 421
Conducting zone, 484i, 485
Conduction, **639**
Cone, **344**, 344i, 345i
Cone cell, **514**, 514i, 515
Cone snail, 352–353, 352i, 353i
Confuciusornis sanctus, 376i, 377
■ Congenital rubella syndrome, 682
Conifer, 282, 334, 337i, 343, **344**–346, 344i, 345i, 350i, 350t, 429
Coniferous forest, 760i–761i, 764–765
   soil, 767i
Conjugation, 302, 303
Connective tissue, **472**–473
   characteristics, 472i–473i
   vertebrate, 159
Conotoxin, 352, 353
Conservation
■   cost of, 353, 375
   of genes, 190, 190i, 191i, 199i, 353
   water conservation in plants, 410–411, 421, 433, 440–441, 440i, 441i
Conservation biology, 349, **725**, 736
Conserved gene, 270, 353
■ Constipation, 621
Consumer, **6**–8, 7i, 74i, 342, 349, 354, 370, **732**
Continental drift, 264, 265, 265i, 719
Continental shelf, 773i
Continuous assembly, in DNA, 191i
Continuous variation, **160**–161, 160i, 161i
■ Contraception, 666–667, 666i
Contractile cell, 359
Contractile protein, 211
Contractile vacuole, 91i, 317i, 318
Contraction, 62, 66, 67, 85, 544–545, 544i, 545i, 546
   ATP and, 546, 546i
   atrial, 560
   cardiac muscle, 561
   energy for, 546, 546i
   isometric/isotonic, 546, 547i
   skeletal muscle, 544–545
   tetanic, 546, 546i
Control gene, 413
Control group, **12**
Control pathway, 206
Control protein, 211
Controlled burn, 765
Conus, 352–353, 352i, 353i
C. geographicus, 352i
C. magnus, 352
Convection, **639**
Convergence, morphological, **267**, 267i
Convergent evolution, 393, 512i
Converging circuit, 490
Cook, J., 684
Cooksonia, 260i, 336, 336i, 340i
Cooperative hunting, 788–789, 789i
Cooperative predator avoidance, 788, 788i
Copepod, 368, 368i, 775, 775i
Copernican model of solar system, 14
Copernicus, N., 14
Copper, 434t, 625t
Copper ion, 90, 90i
Copper mine, 81i
Coprolite, 260
Coral, 138, 319, 358, 358i, 359, 374i

Coral bleaching, 772, 772i
Coral fungus, 326i
Coral grouper, 382i
Coral reef, 2, 3, 5i, 322, 359, 762, 770i, 771, 771i
Core temperature, 639
   mammalian response to shifts in, 640t
   temperature regulation in mammals, 640–641
Cork, 54i, **429**, 429i, 430t
Cork cambium, 418i, **428**, 428i, 429, 429i, 430t
Corm, 454t
Cormorant, 789, 789i
Corn, 102i, 103i, 212, 218i, 229, 231, 329, 348, 416i, 423i, 424, 427i, 432i
   characteristics, 171, 436, 449, 452t, 454t
   genetically modified, 453, 455, 460
   grain (kernel), 203i, 456, 456i, 457i, 461, 461i
Cornea, 160, 512–513, 512i, 513i
Cornell University, 791i
Corolla, 448i
Coronary artery, 558i
■ Coronary bypass surgery, 567, 567i
Coronavirus, 310, 313, 573
Corpus callosum, 497, 497i
Corpus luteum, **661**, 663, 664i, 671
Cortex, 422i, **423**, 423i, 427i, 429, 437, 437i, 439i. *See also* Cerebral cortex
   adrenal, 526–527, 526i, 533t
   cell, 66
   kidney, 634, 635i
Corticotropin, 521i, 524, 526, 532t
Corticotropin-releasing hormone, 526, 678
Cortisol, 250, 521i, 522, 524, 526–527, 533t, **678**
■ Cosmetics, 481
Cosmic cloud, 288–290
Costa Rican owl butterfly, 687i
Cotton plant, 228, 228i, 231, 276i, 460
Cotyledon, **418i**, **419**, 419i, 452, 452i, 456, 456i, 457i
Cotylosaur, 266, 266i
Countercurrent flow, **598**
Courtship behavior, 252, 252i, 272i, 273
Courtship display, 252i, 272i, 273, 785
Covalent bond, **24**–25, 25i, 34, 46, 47i, 75, 95, 436
Covas, R., 251
Cowbird, 714, 715, 715i
■ CPR, 550–551
Crab, 269, 368, 531, 531i
   blue, 535, 535i
   fiddler, 786, 786i
   horseshoe, 369, 369i
   life cycle, 368i
Crack cocaine, 500
Cranberry, 81i, 339
Cranial bone, 539i
Cranial cavity, 476i
Cranial nerve, 493i, 494
Craniate, **380**, 380i
Cranium, 380
Crank, 500
Craterellus, 326i
Crayfish, 146
■ Creatine, 537
Creatine phosphate, **546**, 546i
Crenarchaeota, 306, 306i, 307
Crenicichla, 694–695, 694i, 695i, 703
Creosote bush, 454, 687i
Cretaceous period, 258i, 263i, 337i, 343, 386, 388

■ Creutzfeldt-Jakob disease, 312, 312i
CRH, 526, 678
■ Cri-du-chat syndrome, 173t, 176–177, 176i, 177i
Crick, F., 184i, 187, 189, 220
Crinoid, 365i, 372
Crocodilian (crocodile), 281, 281i, 387, 388i, 389, 389i, 787
Crop, 455, 459, 613, 719, 763. *See also* Agriculture
Crop, invertebrate, 363i
Crop failures, 416–417, 445
Crop plant, 466. *See also specific names*
Crop rotation, 747
Cross-fertilization, **153**, 156i–157i
Crossing over, **144**–145, 144i, 147, 148i, 166, 168, 168i, 171, 171i
Crow, 300
CRTR gene, 150
Crude density, 686
Crustacean, 93, 354t, 367, 368, 368i, 512, 531, 531i
   current species diversity, 725i
   life cycle, 368
   respiration, 596, 596i
Crustal movement, 261, 263i, 264–265, 264i, 274
Cryphonectria parasitica, 329
Cryptomys damarensis, 791
Crystal, arrangement of, 24, 24i
Crystallin, 210
Crystalline cone cell, 512i
Cuboidal cell, 470, 470i
Cuckoo, 781, 781i
Cucumber plant, 449, 452t
Culture, **394**
   primate, 395
Cumulative assaults hypothesis, 680
Cuscuta, 714i
Cuticle
   animal, 39, 39i
   invertebrate, 362, 363i, 366–368
   plant, 69, 69i, 102, 102i, 336, 343, 350i, **421**, 421i, 424, 425i, 440, 440i
Cutin, 421, 440
Cutshwall, C., 587i
Cuvier, G., 240
Cyandium caldarium, 287
Cyanide, 90
■ Cyanide poisoning, 90, 122
Cyanobacteria, 59, 282i, 292, 293i, 295, **304**–305, 306i, 322, 330, 769
   origin, 291
Cyanophora paradoxa, 295i
Cyathea, 340i
Cycad, 334, 337i, 343, **344**, 344i, 346i, 350t
Cycas, 344i
Cyclic adenosine monophosphate, 46, 48t, 209, 325i, 522, 523, 523i
Cyclic pathway
   ATP formation, 99, 100i
   photosynthesis, 292, 296i–297i, 349
Cypress swamp, 730i
Cypress tree, 344
Cyst, **316**
Cysteine, 198i, 623
■ Cystic fibrosis, 150–151, 150i, 159, 165, 173t, 181, 229
Cytochrome *c*, 270
Cytokine, **574**, 585
Cytokinesis, **132**
Cytokinin, 458t, **459**
Cytoplasm, 50, **52**, 53, 126, 304, 317, 437, 440, 440i
   eukaryotic, 52, 52i, 60, 62i, 69, 424, 427
   function, 47, 62i, 80, 83, 85, 87, 88, 89i, 195, 199, 201, 204i, 211
   prokaryotic, 52, 52i, 58i, 59, 294

Cytoplasmic division, 125, 128, 129, 129i, 132–133, 132i, 133i, 138i, 140–143, 146, 147i
Cytoplasmic division/fusion, 324i, 326, 327i, 450, 450i, 451i
Cytoplasmic localization, **651**
Cytosine (C), 46, 46i, 184i, 188, 188i, 189, 189i, 196, 199, 203, 227, 227i
Cytoskeleton, 60, 65i, **66**, 67, 70t, 316, 317
Cytotoxic cell, 575t, **578**, 584–585, 585i

D

Da Vinci, L., 260, 261
D'Agostino, S., 249
Daisy, 348, 449
Dalai Lama, 221i
Dale, B. W., 710i
Damselfly, 371i
Dandelion, 427, 453
Danio rerio, 571, 682, 682i
Daphnia, 708
Darlingtonia californica, 242i, 424i
Darwin, C., 10, 11, 14, 152, 240–243, 240i, 246, 264, 376, 460
Date rape, 73
Daucus carota, 94, 427, 452t, 454, 454t, 783, 783i
Daughter cell, 125, 126, 128–129, 131, 131i, 133, 141, 143i
   plant, 456
Day-neutral plant, **462**
Daylength
   pineal gland and, 530
   reproduction and, 646
DDT, 531, 736, 736i
De Moraes, C., 414, 414i
Dead Sea, 307
Death, 120, 680
   cell, 134
   exponential growth and, 689, 689i
   plant, 428, 464
Death rate, 698, 700i
Deciduous, 344, 424
Deciduous broadleaf forest, 764, 765i
Deciduous forest, 760i–761i, 764, 765i
Deciduous forest soil, 767i
Declining species, 373, 375, 446
Decomposer, 7, 7i, 342, 349, **732**
   bacterial, 339
   fungal, 314, 315, 326, 327, 339
   invertebrate, 366, 371
   protistan, 314, 316, 320
■ Decompression sickness, 609
Deep-sea angler fish, 773i
Deep-sea food web, 36
Deep-sea trench, 773i
Deer, 687, 687i
Deer mice, 603
Deer tick, 305i
Deerfly, 512i
Defecation, 621
■ Defibrillator, 550–551, 561, 567
Definitive host, **360**
Deforestation, 334–335, 334i, 335i, **349**, 435i, 446, 724, 740, 741i, 743, 743i, 745i, 747
Degradative pathway, **77**
Dehiscent, **452t**
Deletion
   chromosomal, **176**, 176i
   in DNA, 202, 202i
Delphi Oracle, 32–33, 32i
Delta cell, pancreatic, 528
Deltoid, 543i
Demographic indicator, 698, 698i
Demographic transition model, **700**, 700i
Demographics, **686**
Denaturation, **44**–45
Denatured DNA, **224**–225, 224i, 225i

Dendrite, 484i, **485**
Dendritic cell, 555, 555i, **574**, 574i, 578i, 580–581, 581i, 590t
Dendrobates, 471i, 712
Dendroclimatology, 445
Denitrification, **747**
Dense connective tissue, **472**, 472i
Density-dependent control, **690**–691, 697
Density-independent factor, **691**
Deoxyribonucleic acid. *See* DNA
Deoxyribonucleotide, component bases, 46, 46i
Deoxyribose, 38
■ Depo-Provera injection, 666i, 667
■ Depression, 163, 166, 547
  post-partum, 678
  during pregnancy, 677
Derived trait, **281**
Dermal tissue, plant, 416, 418i, 419, 421, 422, 427
Dermis, **478**, 478i
Desalinization, **741**
Desert, 338, 345, 481, 481i, 760i–761i, 762, 762i
Desert soil, 767i
Desertification, **762**
■ Designer organ, 475
■ Designer plant, 228
Desmid, 323, 323i
Detrital food web, **735**, 735i, 769
Detritivore, **732**
Detritus, **362**
Deuterostome, **353i**, 355, 356i
Development, **6**, **404**. *See also* Embryonic development
  animal, 367, 370i
  body plan, 655
  eudicot, 420, 422–423, 422i, 423i, 428–429, 428i, 429i, 452, 452i, 457i
  hands, 412–413, 413i
  heart, 571
  human, 647i, 679i, 679t
  leopard frog, 649i
  monocot, 422, 426, 428, 452, 457i
  patterns, 268–269
  patterns, speciation, 238, 278–279, 283i
  plant, 338i, 454, 456–459, 456i, 457i, 458i, 459i, 463
  postnatal, 679, 679i, 679t
  prenatal, 679i, 679t
Devil's hair, 714i
Devonian period, 263i, 337i, 340, 340i, 343, 369, 383, 383i, 384
DEW (Distant Early Warning) system, 795
■ Diabetes, 7, 107, 231, 250, 638
  blood glucose level, 7, 19, 39, 118, 522–523, 526, 528–529, 529i
  insipidus, 525, 637
  mellitus, 528–529, 535, 567, 630
  Type 1, 528–529, 535
  Type 2, 529, 623, 626
■ Diagnostic tool, 166, 170, 170i
■ Dialysis, 638
■ Diaphragm (birth control), 666i, 667
Diaphragm (muscle), 600i, **601**, 602i
■ Diarrhea, 209, 304, 310t, 316, 329
Diastolic pressure, 563
Diatom, 287, 321, 321i
Dickensonia, 356i
Dicot, 419i, 422, 424i
Dictyostelium, 412
Dictyostellum discoideum, 325, 325i, 333
Diet. *See also* Nutrition
  alternative, 623
■ fat-rich, 41i
■ iodine-deficient, 527, 527i
■ low-carb, 49

■ Mediterranean, 623
■ vegan, 571
Dietary fat, 623
■ Dietary supplement, 536–537, 625
Differentiated cell, **192**
Diffusion, **82**–83, 82i, 83i, **406**, 564–565
  facilitated, 83, 84
  metabolism and, 82–83
  plants, 102, 421, 424, 425i, 427, 434, 434t, 440
  plasma membrane, 83, 84i, 86, 88, 88i, 437, 440, 440i, 442
  water molecule, in cells, 52, 68, 82, 82i, 86–87, 86i, 87i
  water molecule, in plants, 421, 425, 425i, 428–429, 433, 437, 437i, 439i, 440, 443i
Diffusion rate, 83
Digestion, 612. *See also* Nutrition
  absorption in small intestine, 618–619, 618i, 619i
  colon function and malfunction, 621
  controls over, 617
  extracellular, 326
  fat, 616t
  intracellular, 88
  large intestine, 621
  organic metabolism, 620
  stomach and small intestine, 616–617
■ Digestive disorder, 617
Digestive enzyme, 60, 62, 63, 72, 135i, 174, 186, 424i, 453, 464, 616, 616t, 617
Digestive gland, invertebrate, 364i
Digestive system, **355**, 370, **612**–613
  bird, 612, 612i, 613
  circulatory system and, 558–559, 559i
  complete, **355**, 355i, 362, 364i, 370, 375, 375i, 612
  flatworm, 612, 612i
  frog, 612, 612i
  human, 477i, 614–615, 614i, 628t
  incomplete, **355**, 355i, 359, 360, 366i, 612
  invertebrate, 360i, 362, 363i, 375, 375i
  organ systems and, 612, 612i
  urinary system and, 632i
Dioon, 344i
Diploid cell, 130i, 211
Diploid chromosome number (2*n*), **129**, 130i, **140**, **153**, 153i, 168, 169i
Diploid dominance, 350t
Diploid organism, 150, 153, 155
Diploid stage, 146i, 156, 322i, 324i, 327i, 328i, 338i, 341i, 345i, 347i, 350t, 451i
Diplomonad, 316i
Directional change, 278
Directional selection, **248**, 249i, 278, 310
Disaccharidases, 616t
Disaccharide, **38**, 616t

Discontinuous assembly, in DNA, 191i
Discoverer XVII, 125
■ Disease, **172**, 310–311, 315
  endemic, 310
  foodborne, 310–311
  infectious, 236, 310–311, 310t, 676–677, 696–697
  sporadic, 310
Dispersing agent, 453, 466
Disruptive selection, **251**, 251i
Dissolved oxygen (DO), 597
Dissolved substance, **27**
Distal, 476i
Distal tubule, **635**, 635i
Distance effect, **723**, 723i
Distant Early Warning (DEW) system, 795
Disulfide bridge, 479i
Diverging circuit, 490
Diversity, **407**. *See also* Biodiversity
  current species diversity, 725, 725i
  eukaryotes, 316, 326–329
  evolutionary view, **10**, 14t, 243, 254
  genetic, 149, 248, 249, 254, 283t
  insect, 370–371, 370i
  invertebrate, 362, 364, 366, 367
  of life, 2, 6, 8, 8i, 9, 10, 139, 184i, 187, 189
  ocean, 772
  plants, 337, 338, 341, 343, 344i, 346, 346i, 348, 348i, 424
  prokaryotes, 302–305
■ Diving, 609
Division of labor, 356, 367, 371
Dixon, H., 438
■ Dizziness, 509
DNA, **6**, 680
  amplification/sequencing, 220, 224–227, 225i, 227i, 286, 307
  animal sequences, 356
  bacterium in, 58, 58i, 59, 184i, 222, 294
  banding pattern, 226, 226i, 227, 227i
  cDNA, 223, 223i, 224
  chromosomal, 222, 223
  cloned, 184, 185i, 192, 222–224, 222i, 223i
  complementary, 223, 223i, 224
  conservation of genes, 190, 190i, 191i, 199i, 353
  continuous assembly, 191i
  deletion, 202, 202i
  denatured, **224**–225, 224i, 225i
  discovery of, 186–187
  disruption of, 80, 94, 134, 191, 202–203, 202i
  enzyme function in replication, 126–128, 134, 190–191, 191i, 197, 211, 222, 222i, 309
  eukaryotic, 58, 58i, 126–127, 197, 202, 270, 316
  evolution and, 64, 177, 189, 200, 287, 291, 294, 294i, 295
  forensics science and, 226
  function, 6, 10, 43, 70t, 126–127, 186–187, 195, 197, 200, 210i, 211
  gene expression and, 208, 209i, 210i
  in human body cell, 61
  hydrogen bond in, 25i, 189, 190, 191
  insertion, 202
  mtDNA, 270, 276
  primate, 177, 269, 270
  prokaryotic, 58, 59, 64, 222, 222i, 223i, 286, 302, 303, 303i, 307, 308, 308i
  recombinant, 220–222, 222i, 223i, 309

repair mechanisms, 191, 203, 220, 223i
replication, 61, 128, 128i, 134, 142–143, 185, **190**–191, 190i, 191i, 195, 196, 203, 208i, 222, 223, 303i, 309
research guidelines, 220
sticky end of fragment, 222, 223i
supercoiling, 127i
technologies, 218, 220–222, 224–232
template, 185, 191, 191i, **195**, 196i–197i, 197, 202i
transcription, 208–212, 209i, 210i
viral, 223, 308, 308i, 309, 309i
DNA chip, **232**, 232i
DNA database, 221
■ DNA fingerprinting, 164, **226**, 226i, 270
DNA ligase, **191**, 222, 222i
DNA polymerase, 22, **191**, 203, 223, 225, 226, 307
DNA structure, 25, 25i, 46, 47i, 58, 58i, 127i, 184i, 188–189, 188i, 189i, 208, 223
  discovery of, 184, 186–190, 220–221
  double helix, 184, 185i, 189–191, 189i, 196i–197i, 220, 223i
  hydrogen bonds in, 25i, 189, 190, 191
Dodder, 714i
Dog, 784
■ Dolly, 184–185, 184i, 192, 192i
Dolphin, 320, 517, 517i
Domain
  of life, 8, 8i, 58, 296i–297i
  in polypeptide chains, **43**
Domestication
  of animals, 724
  of seed plants, 343
Dominance hierarchy, 789
Dominant allele, 151, **153**, 155, 155i, 156, 156i, 164
Dominant competitor, 321
■ Donor organ, 151
Dopamine, 489, 500, 503
Dormancy, 146, 304, 343, 447, 458t, **464**, 464i
Dorsal nerve cord, 360
Dorsal surface, **354**, 354i, 476i
Dosage compensation, **213**
Dosidicus, 214, 354t, 365, 365i, 487i, 512, 512i
Double-blind study, **16**
Double covalent bond, **25**, 56
Double fertilization, 347i, **450**, 451i
Double helix, 25i, 46, 47i, 127i, 220, 291
Doubling time, **689**, 689i
■ Douching, 666–667, 666i
Douglas fir tree, 320
■ Down syndrome, 173t, 178–179, 179i, 181, 662
Downwelling, **773**
Downy mildew, 320
DPG, 603
Dragonfly, 370
Drew-Baker, K., 322
Drinking culture, 73
Drinking water, **741**
Dromaeosaurus, 376i
Dromedary camel, 245, 245i, 640, 640i
Drosophila, 211i, 212, 212i, 270, 275, 353, 370i, 371i, 571, 597, 652i
D. melanogaster, 171, 182–183, 183i, 214–215, 214i, 215i, 217i, 571
■ Drought, 331, **416**–417, 435i, 438i, 445, 445i, 464, 572–573, 727, 763
■ Drug abuse, 72–73, 482–483, 489, 500–501

Drug addiction, 500, 503, 503i
  warning signs of, 500t
Drug production, and yeast, 329
■ Drug research, 352, 357, 361
■ Drug testing, 631
Drupe, **452t**
Dry acid deposition, 757
Dry fruit, 452, 452t, 456, 456i
Dry mouth, 637
Dry shrubland, 760i–761i, 762, 763i
Dry weight, plant, **434**
Dry woodland, 760i–761i, 762
■ Duchenne muscular dystrophy, 217, 547
Duck, 730
Duck-billed platypus, 392i, 401, 401i
Duck louse, 371i
Duckweed, 348, 424
Dugesia. See Flatworm
Duke University, 468
Duncan, R., 720i
Dunce gene, 214
Dunkleosteus, 380i
Duodenum, 614i, 616–617
Duplication, chromosomal, **176**, 176i
Dust Bowl, 435i, 763
Dust mite, 369, 369i
Dust storm, 762, 762i
Dutch elm disease, 719t
■ Dwarfism, 525, 534–535, 535i
Dynamic equilibrium, 509, 509i
Dynastes, 371i
Dynein, 66, 67, 67i, 71, 71i, 130, 131
■ Dysentery, amoebic, 317

E

Eagle nebula, 288i
Ear, human, 509–511, 509i, 510i–511i
■ Ear infection, 593
Eardrum, 510, 510i, 511i
Early atmosphere, 288–289
Early wood, **429**, 429i
Earth
  age, 241, 262
  asteroids and, 258
  crust, 19, 19i, 263i, 264, 264i, 265, 274, 288, 289
  early, 286, 288–289, 293
  history, 241, 261–263, 288–291, 289i
  magnetic poles, 264
Earthworm, 354t, 362–363, 362i, 363i, 492i, 493, 512
  body plan, 363i
  circulatory system, 552, 552i
Earwig, 371i
Easter Island, 684–685, 684i, 685i
Eastern hognose snake, 386i
Ecdysone, 211, **531**, 531i
ECG, 550, 551i, 567, 567i
Echinoderm, 282, 353, 354t, 355, 356i, **372**–373, 372i, 373i
Echolocation, 504
Eciton burchelli, 790i
Ecological access, and lineage, 279
Ecological isolation, **273**, 273i
Ecological pyramid, 737
Ecological separation, 276
Ecological success, **716**
  climax-pattern model, 717
  cyclic, nondirectional changes, 717
  succession model, 716
Ecology, 683i, **685**
Economic factors
  population growth, 700–701, 700i
  sustainable development, 726–727
■ Economy-class syndrome, 571
Ecoregion, **726**, 726i
Ecosystem, **5**, 5i, 228, 286, 313, 730–731, **733**
  aquatic, 738, 738i
  biogeochemical cycles, 739

carbon cycle, 742–743
DDT in food webs, 736
disruption, 28–29
ecological pyramids, 737
energy flow, 74i, 77i, 732i, 737
energy flow, Silver Springs, 738
food webs, 734–735
forest, 331, 435
global cycling of water, 740–741
global warming, greenhouse gases, 744–745
lake, 768–769, 768i
model for, 732i
nature of, 732–733
net production, 737, 737i
nitrogen cycle, 746–747
nutrient cycle, 366, 732–733
ocean, 751
participants, 732–733
phosphorus cycle, 748–749
primary productivity, 737
recovery, 335
stream, 769
structure of, 733
tallgrass prairie, 733, 733i, 734i
water crisis, 740–741
■ Ecstasy, 482–483, 500
Ectoderm, **355**, 360, **476**, **648**, 652, 652i, 655i
Ectotherm, **639**, 639i
■ Edema, 366, 565, 576, 576i
Ediacaran, **356**
Eel, 321
Effector, **408**, 408i, 409i, 578, 579i
Effector B cell, 582, 582i
Egg, **146**
  animal, 53i, 146, 147i, 152, 201, 211, 650
  bird, 390i
  human, 123i, 124, 145, 146, 169
  invertebrate, 361, 361i, 364, 366i, 368i, 373
  plant (gamete), 153–155, 337, 338i, 341i, 345, 345i, 347i, 448, 448i, 450, 451i, 454
  protist, 320
  yolk, 647
Egg-laying mammal, 393
Eincorn, 277i
Ejaculation, 665
Ejaculatory duct, 656, 656i, 657i
El Niño Southern Oscillation, 752–753, 752i, **773**, **774**–775, 774i
Elastin, 159, 472
Electric gradient, **83**, 84, 98, 100i
Electrocardiogram, 550, 551i, 567, 567i
Electrochemical work, 74
Electromagnetic energy, 94
Electromagnetic spectrum, 94, 94i
Electron, 18i, 20, 22, 24, **30t**, 55, 77, 77i, 80, 98
Electron density cloud, 20i
Electron microscope, 55, 55i
Electron orbital, 22, 22i, 23, 23i
Electron transfer chain, **77**, 77i, 90, 97i, 98, 99, 100i
Electron transfer phosphorylation, **109**, 114, 114i
Electron transfer reaction, **35**, **79**
Element, **18**, **30t**, 288
Elephant, 266, 693i
Elephant seal, 254, 787i
■ Elephantiasis, 366, 366i, 565
Elimination, 612
■ Ellis-van Creveld syndrome, 172i, 173t, 255
Elodea, 105i
Elongation, RNA translation, 200, 200i–201i
Embolus, 567

Embryo, **181**, 476, 644i, 679t
  early embryo formation, 670–671, 670i–671i
  emergence of human features, 674–675, 674i–675i
  somatic, 455i
  teratogen sensitivity, 676i
  tissue layers, 354–355, 358, 360
Embryo sac, **450**, 451i
Embryonic development
  animal, 128, 267
  animal, and gene controls, 207, 210–212, 214–215, 215i
  animal, human, 133i, 137, 169, 169i, 178
  animal, invertebrate, 353–355, 360
  animal, patterns and processes, 239, 268–269, 273, 273i
  animal, vertebrate, 35
  plant, 452, 452i, 456, 456i, 458i, 459
  plant, patterns in, 268, 343, 345, 345i, 347i, 419, 422, 446, 452–454, 454t, 456–457, 456i, 457i
  stages of, 674i–675i
Embryonic disk, 670, 670i–671i, 672, 672i
Embryonic induction, **654**
Embryonic period, 670, 672, 672i, 674i–675i
■ Embryonic stem cell, 468
Emigration, **255**, **688**
Emotion, 499
Emotional stress, 309
Emotional-visceral brain, 499
Emperor penguin, 40i, 41
■ Emphysema, 606, 606i
Emu, 238i
Emulsification, **617**
Encapsulated receptor, 507
■ Encephalitis, 369
Encrusting sponge, 357i
■ End-stage renal failure, 638
Endangered species, 185, 335, 344, 373, 385, **724**, 726–727
Endemic disease, **310**
Endemic species, 724–725
Endergonic reaction, **74**, 75i
Endler, J., 694–695
Endocrine cell, 528
■ Endocrine disrupter, 518–519, 530, 530i
Endocrine glands, **471**
Endocrine system, 477i, **520**, 521i
  direct responses to chemical change, 528–529
  feedback loops in hormone secretions, 526–527
  hormones and the environment, 530–531
  hypothalamus and pituitary gland, 524–525
  nature of hormone action, 522–523
  vertebrate, 520–521, 521i
Endocytic pathway, 60, 63i, 88
Endocytic vesicle, 63i
Endocytosis, **83**, 83i, 88–89, 88i, 309
Endoderm, **355**, 360, **476**, **648**, 652
Endodermis, 426i, **427**, 427i, 437i, 439i
Endogenous pyrogen, 577
Endomembrane system, 62–63, 62i, 158, 201, 201i, 296i–297i
Endometrium, **660**, 661i, 663, 664i, 670i
Endophytic fungus, **330**
Endoplasmic reticulum, **62**, 62i, 63, 63i, 70t, 88i, 201, 201i, 287, 294, 294i, 316, 317i
Endorphin, 501
Endoskeleton, **538**

Endosperm, 347, 347i, 450–**451**, 451i, 452, 452i, 456, 456i
Endospore, **304**, 304i
Endosymbiont, 318, 322
Endosymbiosis, 198, **294**–295, 304, 317, 319, 322
Endothelium, 560, 560i
Endotherm, **639**, 639i
Energy, **6**, 72, 74
  alternative sources for humans, 118–119, 119i
  basis of metabolism, 6–7, 7i
  from fats, 118
  for human body, 74
  for muscle contraction, 546, 546i
  from proteins, 118
  source of life, 286–287
  sources, 6, 7, 29, 342
Energy carrier, 76, **89t**
Energy consumption, 755
Energy flow, 732i
  in cell function, **74**–78, 74i, 75i, 77i, 78i, 83i, 85i
  food web, 735i
  life's organization and, 6–7, 7i
  through ecosystems, 737
  through Silver Springs, 738, 738i
Energy hill, 75, 78i
Energy inputs and outputs, 733
Energy levels of electron, 22
Energy pyramid, **737**, 738, 738i
Energy-releasing pathway, 108–109
  comparison of main types, 108
  evolution of life and, 120
Englemann, T., 95i
Enhancer, **208**, 212
Enkephalin, 501
ENSO, 752–753, 752i, **773**, **774**–775, 774i
Entamoeba histolytica, 317
Enterobius vermicularis, 366
Enteromorpha, 718i, 719
Entropy, **74**, 75i
Enveloped virus, 308, 309
Environment. *See also* Internal environment
  carrying capacity, 690–691
  effects on gene expression, 160, 162–163, 166, 167, 176–177, 176i, 177i, 192
  extreme, 307i
■ farming methods and, 228
  fluid environment, 405
■ hormones and, 530–531
  remediation, 432–433, 445
■ thyroid function and, 530, 530i
Environmental cue, for plants, 446, 464
Environmental estrogens, 530–531
Environmental movement, 736
Environmental Protection Agency, 432, 519, 751
Enzyme, **76**, **89t**
  controls in, 80–81, 80i, 81i, 208, 210, 458
  evolution, 290
  structure, 35, 43, 48t, 78, 79, 79i
  types, 72, 186, 222, 286, 305, 309, 316, 324, 326
Enzyme function, 6, 35i, 73, 78–79, 295
  blood clotting, 73, 353
  in cancer, 135
  digestive, 60, 62, 63, 72, 135i, 174, 186, 424i, 453, 464, 616t
  in DNA replication, 126–128, 134, 190–191, 191i, 197, 211, 222, 222i, 309
  impaired, 180
  melanin production, 159, 162, 162i
  in metabolism, 76–77, 90–91, 160
  in nitrogen fixation, 436

in photosynthesis, 77, 101
protein synthesis, 196, 200
in reproduction, 324
sugar attachment, 158
in toxins, 194
waste removal, 206, 370
Eocene epoch, 245, 274i, 299, 395
Eomaia scansoria, 279i
Eosinophil, 555i, **574**, 574i, 584, 590t
Ephedra viridins, 344, 344i
Epidemic, **310**, **329**
Epidermal growth factor, 134
Epidermal hair, 424
Epidermis, **478**, 478i
animal, 355i, 358i, 359
■ lab-grown, 475, 475i
plant, 96i, 103i, 416, 418i, 419, 420i, **421**, 421i, 423–427, 423i, 425i, 426i, 429, 430i, 433, 437, 439i, 440, 440i
Epidermophyton floccosum, 329i
Epididymes, 656
Epididymis, 656i, 657i, 658i
Epiglottis, 600i, **601**, 601i
■ Epilepsy, 352
Epinephrine, 495, 521i, 523, 527, 533t
Epiphyte, **341**
Epithelial tissue, 69, 69i, 134, 159, 405i, 470–471
Epithelium, **470**, 470i
Epstein-Barr virus, 556
Equator, 722
Equilibrium, 509
chemical, 76, 76i
dynamic, 509, 509i
genetic, 246–247
static, 509, 509i
Equilibrium model of island biogeography, **724**
Equisetum, 334, 337, 337i, 340–342, 340i, 342i, 350t
Equus, 266, 285
ER, **62**, 62i, 63, 63i, 70t, 88i, 201, 201i, 287, 294, 294i, 316, 317i
Erectile tissue, 657i
Erection, 665
Ergotism, 329
Erosion, 349
■ Erythroblastosis fetalis, 557
Erythrocytes. *See* Red blood cell
Erythropoietin, 533t, 603
Escherichia coli, 58i, 187i, 193, 208–209, 224, 302, 302i, 305, 312, 621
Esophagus, 601, 614i, **615**, 628t
invertebrate, 363i
Essential amino acid, **623**
Essential fatty acid, **623**
Estrogen, 35, 35i, 169, 179, 269, 521i, 533i, 536, 575, **661**, 663, 664i
environmental, 530–531
Estrogen receptor, 587
Estrous cycle, 660
Estuary, **91**, 290, 445, **770**
Ethane, 72, 228
Ethanol, 72, 116, 116i, 228
Ethiopian, 761i
Ethyl alcohol, 501
Ethyl group, 203
Ethylene, 432, 458t, **459**, 461, 464
oracle of Delphi and, 32–33, 32i
Eubacterium. *See* Bacterium
Eucalyptus tree, 348
Euchlanis, 375i
Eudicot (true dicot), 229, 282, 460
characteristics, **348**, 348i, 350t, 416, 419, 419i, 425, 425i, 427, 427i, 429, 431
development, 420, 422–423, 422i, 423i, 428–429, 428i, 429i, 452, 452i, 457i

■ Eugenic engineering, **230**
Euglena, 317i
Euglenoid, **316**–317, 316i, 317i
Eukarya, as domain of life, **8**, 8i, 140, 280, 280i, 282i
Eukaryote
characteristics, 8, 8i, 9i, 86, 332t
DNA. *See* DNA
endosymbiotic origin of, 707
gene control in, 171, 207, 210–211, 210i
origin, 198, 199i, 263i, 293, 296i–297i
types, 314–315, 315i
Eukaryotic cell, 41, 51, 51i, 52i, 53i, 60, 64, 204i, 263i, 291–295, 292i, 295i
components, 41, 47, 58, 64i, 65i, 66, 68, 70t
cycle, 124, 126, 128i
photosynthetic, 64, 64i
Euphrates River, 741
European earwig, 371i
European rabbit, 721, 721i
European starling, 719t
Euryarchaeota, 306, 306i
Eusocial insect, 790, 791i
Eusocial mammal, 790, 791, 791i
Eutherian, **393**
Eutrophic lake, 769
Eutrophication, **749**, **769**
Evans, R., 230i
Evaporation, **27**, 438, 439i, 440, **639**
Evaporative heat loss, **640**
Everglades, 770i
Evergreen, **344**, **424**
Evergreen broadleaf forest, 764
Evolution, **10**, 45, 278–279, 310–311, 373
adaptation and, 244–245, 336–367
AIDS and, 573
asteroids and, 258
of body plans, 356, 362, 364–365, 372, 655
branched *vs.* unbranched, 278, 281, 281i
coevolution, 346, 446, 447, 449, 710
convergent, 393, 512i
diversity and, 10, 14t, 243, 254
DNA and, 64, 177, 189, 200, 287, 291, 294, 294i, 295
of early cells, 292–293, 292i, 293i
emergence of early humans, 396–397
emergence of modern humans, 398–399
evidence of, 212, 215, 257i, 264–265, 270–271
exceptions to trend, 373
immunity, 574–575
influencing factors, 140, 144–145, 203, 265, 343, 365
interpreting and misinterpreting, 376–377
of land vertebrates, 671
macroevolution, 263, 266, 283t
microevolution, 236, 237, 244, 247, 258, 272, 283t, 353
"missing links," 376–377, 376i, 379
natural selection and life histories, 694–695, 694i, 695i, 703
of parasites and pathogens, 236–237
primate trends, 394–395
radiometric dating, 377
social behavior, 792
of structural organization, 404–405
success in, 362, 364, **367**
theories of, 11, 14, 238–240, 242
weight and, 610

Evolutionary tree, 270–271, 271i, **278**, 278i, 282, 282i
animal, 356i
Archaea, 306i
how to construct, 281, 281i
invertebrate, 357i, 358i, 360i, 362i, 364i, 366i, 367, 372i
plant, 334i, 337, 337i, 348i
primate, 399i
protist, 316, 316i
Excitable cell, 485
Excitation energy, 512
Excitatory effects, 488–489
Excretion, 471
Excretory organ, 364i
Exergonic reaction, **75**, 75i, 77
Exhalation, 602–603, 602i
Exocrine glands, **471**, 471i, 478
Exocytic vesicle, 63i
Exocytosis, **83**, 83i, **88–89**, 88i
Exodermis, root, **437**
Exon, **197**, 197i, 211, 221, 223
Exoskeleton, **367**, 367i, 368, **538**, 538i
Exotic imports, 704–705, 704i, 705i
Exotic invaders, 720–721
Exotic species, **719**, 719t
Expansion mutation, **174**
Experiment, **12**, 12i
Experimental design, examples
animal research, 214–215, 249, 254, 254i
genetics, 171, 183, 193
human health, 12, 16
plant research, 442, 459i, 462, 464, 464i
search for DNA, 186, 186i, 187i
seawater viruses, 313
Experimental evidence, 152, 243
Experimental group, **12**
■ Experimental organism, 162, 172, 182–183, 186, 186i, 366, 442
Experimental test, 11, 12
Exploitative competition, 708
Exponential growth, **688–689**, 688i, 689i, 702
External oblique, 543i
External skeleton, and chitin, 39
■ Extinction, **279**, 388, 398, 723–726. *See also* Mass extinction
Extinction crisis, 252, 263i, 265, 278–**279**, 335
Extracellular digestion, 326
Extracellular fluid, 405, 408, **632–633**, 632i
acid-base balance, 638
challenges in water, 632, 632i
challenges on land, 632–633
Extraembryonic membrane, 670–671
Extreme halophile, 297i, **306**, 307, 307i
Extreme phenotype, and selection, 250
Extreme thermophile, **306**, 307, 307i
Eye, 159, 210, **512**
camera eye, 512, 512i
■ color, 160–161, 160i
compound, 370i, 512, 512i
as first line of defense, 575
formation of, 655
human, 160–161, 160i, 512–513, 513i
invertebrate, 364, 367, 370i

retina, 514
simple, 512
third, 388
vision, 512–513, 513i
Eyeless gene, 214–215
Eyespot, 316, 317i, 360, 364i, 375, 375i

F

Facial bones, 539i
Facilitated diffusion, 83, 84
FAD, 46, 46i, 48t, **109**
Fall overturn, **768–769**
Fallopian tube, 660
False coelom, 355, 366, 366i
Family pedigree, 166, 171
Family planning, 698–699
■ Fanconi anemia, 173t
■ Farming methods, and environment, 228
Farsightedness, 515, 515i
■ FAS, 677, 677i
■ Fast-food, 49
Fast-twitch muscle fiber, 117
Fat, **40**, 48t
body fat distribution, 169
deposits, 473
■ diet high in, 41i
dietary, 623
digestion, 616t, 617
digestive enzymes, 616t
energy from, 118
■ low-fat diet, 49
neural, 40
nutrition and, 622–623
organic metabolism, 620, 620i
saturated, 623
■ synthetic replacement for, 12
Fat-soluble vitamin, 624t, 625
Fate map, 215i
Fatty acid, **40**, 48t, 49, 49i, 62, 118, 122, 526, 546, 617
essential, 623
as precursor of life, 289
saturated, 40, 40i, 56
structure, 34, 40, 56, 56i
trans fat, 49, 49i
unsaturated, 40, 40i, 56
Fatty acid chain, 209
Fatty acid tail, 34, 40, 40i, 49
Feather, 41, 105, 267, 281, 281i, 390
Feather star, 372
Feces, 361, 370, 453
Feedback inhibition, **80**, 80i
Feedforward mechanism, 795
Feeding behavior, 613
Femoral artery, 558i
Femoral vein, 558i
Femur, 539i, 540, 540i
Fermentation, 329
Fermentation pathway, 116–117
Fern, 265, 265i, 282, 336i, 337, 337i, **340**, 340i, 341, 341i, 342i, 343, 346i, 350t, 458
Fertility, 644–645
control of human fertility, 666–667
■ Fertility drug, 644–645
Fertility rate, 698–699
Fertilization, 138i, 144, 146–147, 146i, 154, 157, 476, 647, 648, 648i, 665, 665i, 670i
cross-fertilization, 153, 156i–157i
double, 347i, 450, 451i
flowering plant, 153, 155, 448, 448i, **450–451**, 450i–451i
human, 169i, 178, 665, 665i
protist, 324i, 338, 341i
■ in vitro, 181, 181i, 192, 645, 667, 667i
■ Fertilizer, 29, 333, 597, 747, 749
Fescue, 330

Festuca arundinacea, 330
- Fetal alcohol syndrome (FAS), 677, 677i
Fetal period, 670, 675i
- Fetoscopy, 180i, 181
Fetus, **181, 670**, 674–675, 674i–675i, 679t
circulation between fetus and mother, 673, 673i
Rh blood typing, 557, 557i
teratogen sensitivity, 676i
- Fever, 81, 577
- Fever blister, 309
Fiber, plant, **421**, 425, 429
Fibrillin, 159
Fibrinogen, 566
Fibroblast, 472
Fibrous protein, 48t
Fibrous root system, **427**, 427i
Fibrous tunic, 513i
Fibula, 539i
Fiddler crab, 786, 786i
Field experiment, examples
animal survival, 12–13, 251, 251i
environmental factors, 162–163, 162i
Mendel's peas, 152–157
Field mustard, 268, 268i
Fig tree, 452t, 466
Fight-flight response, 495, 527
Filament, 448, 448i, 450i
Filter feeder, 357, 358i, **378**
Fin, **381**, 382i
Finch, 242, 243i, 251, 251i, 275
Fir tree, 320, 344
Fire, 398, 762, 763i, 765, 765i
Fire ant, 704–705, 704i, 705i, 790
Firefly, 228i, 319, 785
First law of thermodynamics, **74**
Fish, 27, 93, 263i, 269, 509. *See also* Gill
adaptation of, 382
body plan, 365i
bony, 381i, 382i, 383
brain, 497i
breeding ground, 321
cartilaginous, 381i, 382, 382i
characteristics, 354, 354t, 361
circulation, 552–553, 553i
gas exchange, 594–595, 595i
jawed, 380–383, 380i
jawless, 380, 380i
lobe-finned, 383, 383i
- mercury-tainted, 751
oxygen level in water and, 597
ray-finned, 383
reproduction, 139
scales, 382
school of, 5, 5i, 686i
- shellfish poisoning, 320
water-solute balance, 632, 632i
Fish kill, 319, 320, 320i, 375, 731
Fisheries, 730, 773
Fission, transverse (invertebrate), 360
Fixation, **254**
Fixed action pattern, **781**
Flabellina iodinea, 364i, 365
Flagellated protozoans, **316**
Flagellum, **67**
bacterial, 58i, 59, 70t, 302, 302i
choanoflagellate, 356
dinoflagellate, 316i, 319, 319i, 320, 320i, 359, 597, 772
eukaryotic, 59i, 67, 70t
invertebrate, 356, 357i
protist, 67, 316, 317i, 320, 320i, 323i
sperm, 67, 130, 146, 147i
Flame cell, 360, 360i
Flamingo, 105

Flat structural model, 34, 34i, 35i
Flatworm, 406, 407i, 492, 492i, 594, 596i, 612
body plan, 360
characteristics, 138, 354t, 355, 360–361, 360i
digestive system, 612, 612i
evolution, 356i
reproduction, 646, 646i
Flavivirus, 300, 301
Flavoprotein, 460
Flax, 420i
Flea, 236, 371i
Flight, 390–391, 391i
Flightless bird, 238, 238i, 239
Float (cnidarian), 358i, 359
Flooding, 349
Florida Everglades, 770i
Florida Key deer, 687i
Florida Keys, 772
Florigen, 459
Flower, 348i, **418**, **448**
composite, 449
formation, 268, 268i, 462–463, 462i, 463i
perfect vs imperfect, 449
structures, **346**, 346i, 418–430, 418i–429i, 430t, 448, 448i
Flowering plant
basal group, 348i, 350t
cell, 420i, 421i, 423i, 425i, 426i, 427i, 429i
characteristics, 276–277, 346–347, 350t
cloning, 454–455, 454t, 455i
eudicot (true dicot) *vs.* monocot, 419, 419i
evolution, 263i, 282, 334, 337i, 343, 346, 350t, 446
fertilization, 153, 155, 448, 448i, 450–451, 450i–451i
fruit, 452–453, 452i
growth patterns, 347, 440, 456–457, 456i, 457i
life cycle, 428, 448, 448i, 450–451, 450i–451i, 464
nonphotosynthetic, 348, 348i
ovaries, 346i, 347, 347i, 448, 448i, 450, 451i
pollinization, 272i, 273, 343, 347i, 350t, 449, 449i, 450
primary growth, 418i, **419**–423, 420i–423i, 428
reproduction, 126, 140i, 347, 357, 446, 448–449, 448i, 449i, 454–455, 454t
root systems, 426–427, 426i, 427i
secondary growth, **428**–429, 428i, 429i
seed, 418i, 419, 419i, 428, 448
venation, 425i
Flowing-water ecosystems, 769, 769i
- Flu shot, 308
Fluid environment, 405
Fluid mosaic model, **56**, 56i, 57i
Fluid pressure, 86–87, 440
Fluke, 354t, 360
Fluorescent light, 98
- Fluoride, 19
Fluorine, 625t
Fly, 280i, 317, 370i, 446, 538i, 713i
Fly agaric mushroom, 327i
Fly ash, 757i
Flying fish, 382i
Focusing mechanism, 513i
- Focusing problem, 515, 515i
Folate (folic acid), 479, 624t, 676
Follicle, 662–663, 662i, 664i
Follicle-stimulating hormone, 521i, 524, 532t, 659, 663, 664i, 682
Follicular phase, 660t, 661

Food
- as energy source, 118–119
- Frankenfood, 218–219, 231
genetically modified, 218–219, 248
- glucose and, 118
processing, 613
world production, 417
Food and Drug Administration, 536
Food chain, **733**, 733i, 734
- Food crop, 455, 459, 719
Food gathering, 613
Food irradiation, 313
- Food poisoning, 304, 312, 548
- Food production, and algal extracts, 321, 322
Food pyramid, **622**, 622i
Food-spoiling mold, 329, 329i
Food web, 105, **733**–735
DDT in, 736, 736i
detrital, 735, 735i, 769
energy flow in, 735i
fungi and, 326
grazing, 735, 735i, 738, 738i
ocean, 36, 262, 307, 321
tallgrass prairie, 734i
Foodborne disease, 310–311
Foot, invertebrate, 372
Foot bones, 539i
Foot muscle, mollusk, 364, 364i
- Football, and heat stroke, 402–403
Foraminiferan, 314, 315, 315i, 317, 317i
Forebrain, 496, 496i, 497, 497i, 502t
Forelimb, vertebrate, 266–267, 266i
Forensics science, and DNA, 226
Forest, 760, 760i–761i
boreal, 307, 760i–761i, 764–765
broadleaf, 760i–761i, 764, 765i
coast redwood, 334
coniferous, 760i–761i, 764–765, 767i
deciduous, 760i–761i, 764, 765i, 767i
deforestation, 334–335, 334i, 335i, 349, 435i, 446, 724, 740, 741i, 743, 745i, 747
ecosystems, 331, 435
evergreen, 764
evolution, 334
logging, 334–335, 334i, 335i
old-growth, 334, 692, 724
products, 334, 335, 349, 350, 350i, 431
rain forest, 446, 727, 760i–761i, 764, 764i, 767i
swamp, 251, 263i, 334, 337, 340–342, 340i, 434–435, 445i
temperate, 349, 760i–761i, 765i
tree fern, 340i
tropical, 349, 349i, 760i–761i
watershed, 349
Formaldehyde, 228, 290
FormyImethionine, 58
Fossil
formation, **260**–261
megafossil, 292i
microfossil, 294
- radiometric dating and, 21, 262, 262i
trace fossil, 260
Fossil, examples
ammonite, 239i
Archaefructus sinensis, 346, 346i
bacteria, 59i
Bangiomorpha pubescens, 292i, 293
Basilosaurus, 239i
bird, 299
Cooksonia, 260i, 336, 336i, 340i
cyanobacteria, 293i
Dickensonia, 356i
early prokaryotic cells, 292i

edicarans, 356i
Eomaia scansoria, 279i
foraminiferan, 278, 278i, 315i
Ginkgo biloba, 344i
Glossopteris, 265, 265i
Grypania spiralis, 292i
ichtyosaur, 260i
Lystrosaurus, 265, 265i
Myxococcoides minor, 59i
Palaeolyngbya, 59i
Psilophyton, 336, 336i, 340i
Spriggina, 356i
Tawuia, 292i
trilobite, 356i
Fossil fuel, 29, 93, 730, 743, 745, 745i, 747
formation, 342
human population growth and, 697
Fossil record, 292, 293, 303, 376–377
interpreting, 239, 239i, 242, 257, 258, 258i, 260, 265, 265i, 266, 269, 271, 274, 278, 281, 365
predatory species and, 356
transitional forms, 262–263, 278i, 293
Fossilization, **260**
Founder effect, **255**, 255i, 278
Fovea, 513i, 515, 515i
Fox, 683i, 709, 709i, 732, 732i
FOXP2 gene, 794
Fragaria, 453, 453i, 454, 454t
- Fragile X syndrome, 173t
Fragmentation, 357
Fragmented habitats, 724–725
Frameshift mutation, **202**, 202i
Francke, W., 252
- Frankenfood, 218–219, 231
Franklin, R., 188, 190, 190i
Fraser, C., 234, 298
Fraternal twins, 651, 682
Free fatty acid, 616t
Free nerve ending, 507, 507i
Free radical, **80**, 107, 134, 203, 625, 680
Fregata minor, 235i
Frequency, 510, 510i
Frequency distribution, 160i, 161
Freshwater habitat, 50
Freshwater ecosystem, 768–769
Friedman, J., 627, 627i
- Friedreich's ataxia, 106–107, 106i
Frigate bird, 235i
Frisch, K. von, 785i
Frog, 27, 53i, 61, 340, 384i, 385, 385i, 471i, 495, 497i, 518–519, 530, 530i, 542i, 598–599, 712
development, 649i
digestive system, 612, 612i
reproduction, 648i, 649i, 650, 650i, 651
Frond, 341, 341i
Frontal lobe, 498, 498i
Frontal plane, 476i
Frostbite, 641i
Fructose, 38i, 623, 656
Fruit, 343i, 347, **452**–453, 452i
abscission, 458–459, 458i, 459i
accessory, 452, 453i
aggregate, 452, 452t, 453i
citrus, 371i
dry, 452, 452t, 456, 456i
multiple, 452, 452t
ripening, 459
simple, 452, 452t, 453i
Fruit-eating pigeon, 706i, 709
Fruit fly. *See Drosophila*
Fruiting body, 305, 325, 325i, 326, 326i, 328, 329i
FSH, 521i, 524, 659, 663, 664i, 682
Fucoxanthin, 320
Fumigant, 756, 783

Functional group, **34**–35, 34i, 42i, 85, 89t
Functional-group transfer, **35**, **78–79**
Fundamental niche, 706
Fungus
  characteristics, **9**, 68, 70t, 92, 138, 314, 315, 315i, 326–329, 326i, 327i, 332t
  current species diversity, 725i
  decomposers, 6, 9, 9i, 326, 327
  endophytic, 330
  evolution, 59, 263, 263i, 282i, 297i
  life cycle, 326–327, 326i, 327i
  mutualism and, 707
  parasites, 326, 327
  ▪ pathogens, 314, 315, 316i, 329, 329i, 333, 420, 458
  symbionts, 304, 326, 328, 330–331, 331i, 348, 436
Fur, 281, 281i
Fur color, 81i, 159, 162, 162i
Furrow, 132, 419i
Fusarium oxysporum, 333
Fusion, in plant production, 138, 450

G

GABA, 489, 548
Galactose, 174
▪ Galactosemia, 173t, 174, 174i
Galápagos Islands, 241i, 242, 244, 274–275, 752
Galápagos tortoise, 389i
Galilei, G., 14, 54
▪ Gall tumor, 305
Gallbladder, 614i, 628t
▪ Gallstone, 626
Gametangia, 328i
Gamete, 141, 146, 246, 248, 322i, 324i, 337, 350t, 448i
Gamete formation, 648, 648i
  animal, 138i, 139, 140, 142i, 146, 171, 171i
  genetic variation and, 144–147, 147i, 203
  human, 145
  meiosis and, 322i, 324i
  plant, 146, 156i–157i, 338, 338i, 339, 339i, 448i
  protistan, 318, 319i, 322i, 324i
Gamete mortality, 273, 273i
Gametophyte, **321**, **336**, 446, 448, 448i
  plant, 146, 153, 334i, 336i, 337–339, 341, 341i, 343, 345, 345i, 347i
  protist, 318, 319i, 322i, 324i
Gamma aminobutyric acid, 489, 548
Gamma-glutamyl carboxylase, 353
Gamma ray, 94i, 202i
Ganglion, 494i, 495i
  bilateral nervous system, **492**
  invertebrate, **363**
Ganglion cell, 514, 514i
Gap gene, 215i
Gap junction, 56i, 69, 69i, 412, **471**, 471i, 561i
Garter snake, 780, 780i, 781
Gartshorn, G., 727
Gas exchange, 594–595, 594i, 595i
  fish, 594–595, 595i
  large bodies, 406
  mammals, 595, 595i
  plants, 417, 420, 421, 424, 425i, 427, 440
  at respiratory membrane, 604
  respiratory pigments and, 595
  surface-to-volume ratio and, 594
  transport and, 604–605
  ventilation and, 594–595
Gaseous nitrogen, 746, 747
▪ Gastric bypass surgery, 610–611

Gastric fluid, 28, 28i, 69, 81, **616**
Gastric inhibitory peptide, 617
Gastrin, 533t, 616, 617
Gastrocnemius, 543i
Gastrodermis, 358i
Gastrointestinal hormone, 617
Gastropod, 364
Gastrula, **648**
Gastrulation, 648, 652, 652i, 654, 654i, 672, 672i
Gause, G., 708–709
GCC, 353
Gel electrophoresis, **226**, 227
Gemmule, 357
Gene, **44**, **140**, **150**, 153, 168, 195, **212**
  behavior and, 780
  checkpoint, 134–135, 134i
  components, 332t
  conservation, 190, 190i, 191i, 199i, 353
  conserved, 270, 353
  control gene, 413
  cumulative interactions in, 160
  gap gene, 215i
  globin, 177
  highly conserved, 270, 353
  homeotic, 212, 268, 269, 269i, 463, 655
  lateral gene transfer, 303, 310
  locus gene, 153, 153i, 158, 159, 171
  master, 169, 212, 463
  of nonreproducers, 790–791
  oncogene, 134
  pair-rule, 215i
  positive control, 208
  proto-oncogene, 206
Gene control, **207**, 212–213
  behavior and, 166–167, 173t
  eukaryotic, 210–215, 210i
  loss of, 234–235
  mechanisms, 206–209, 234–235
  in plant development, 456, 458, 463
  prokaryotic, 208–209, 271
Gene expression, 192, 208, 210–212, 215, 232, 246, 269, 277, 456, 458, 463
  environmental factors and, 151, 160, 162–163, 166, 167, 176–177, 176i, 177i
Gene flow, 247, **255**, 255i, 256t, 266, 272, 274, 276, 278, 283t
Gene library, **224**
Gene linkage, human, 171
Gene mutation, **202**, 202i, 203
Gene pair, **153i**, 156–159
Gene pool, 140, 237i, **246**
Gene product, 159, 160, 163, 203, 780
Gene region, **194i**, 195i, 196i, 197i
Gene sequencing, 211, 303
Gene swapping, 144–145, 144i, 271, 307, 310
▪ Gene therapy, 230–232, 230i, 231, 587i
Gene transcription, 208, 210–211, 210i, 234
Gene translation, 211
Genetic abnormality, **172**
▪ Genetic analysis, 172–173, 172i
Genetic code, **194i**, 195i, **198**, 198i
▪ Genetic counseling, 167, 181
▪ Genetic disorder, 167, **172**–173, 172i, 173t, 191
  ▪ phenotypic treatment, 180–181, 180i, 181i
  ▪ porphyria, 480–481, 480i
  screening for, 150–151, 167, 180
  ▪ types, 44, 71, 133, 150–151, 150i, 159, 165, 173t, 174–179, 183, 202, 202i, 206, 213, 217, 230, 257
Genetic distance, 399

Genetic divergence, **272**, 283t, 353, 356
Genetic drift, 247, **254**–255, 254i, 255, 255i, 256t, 278, 283t
▪ Genetic engineering, 93, 105, 139, 139i, 149, 218–219, **228**–230, 305, 308, 432–433, 432i, 443, 445, 447, 460, 466
Genetic equilibrium, 246–247
Genetic persistence, 283t
Genetic recombination, **168**, 168i
▪ Genetic screening, 150–151, 165, 167, 180–181, 183
▪ Genetic testing, 181
Genetic variation, 157
Genetically modified food, 218–219, 248
Genetics experiment, 171, 183, 193
▪ Genital herpes, 585, 668, 668t
Genital pore, invertebrate, 360i
▪ Genital warts, 137, 668, 668i
Genome, **221**, 303, 307
  animal, 356i, 366
  human, 202, 218, 220i, **221**, 269, 270
  ▪ patenting, 218–219, 218i, 219i, 221
Genomic library, **224**
Genomics, 232
Genotype, **153**, 155, 157, 163, 171i, 179, 181, 192, 247i
Genus (taxon), **8**, **280**
Geographic dispersal, **719**, 778–779
Geographic distribution patterns, 238, 238i, 274
Geographic isolation, 274
Geologic change, 264–265
Geologic time, 258–259, 262–263
Geologic time scale, **262**, 263i
Geological uplifting, 739, 743
Geology, 239
Geospiza conirostris, 243i
G. scandens, 243i
Geringer, F., 176
Germ cell, 126, 141, 144–146, 169i, 203, 374, 446, 680
Germ layer, animal, 354–355, 356i
▪ German measles, 676–677
Germinating spore, 322i
Germination, 146, 324i, 428, 446, 450–452, 450i–451i, **456**, 464
Gey, G., 124–125
Gey, M., 124–125
Ggta1 gene, 231
GH, 524, 532t
Ghengis Khan, and gene flow, 172i, 255
Ghrelin, 610–611, 623, 627
Giant anteater, 393, 393i
Giant horsetail, 342, 342i
Giant kelp, 315i, 321, 321i
Giant squid, 365
Giardia lamblia, 316, 316i, 317i
▪ Giardiasis, 316
Gibberella fujikuroi, 458
Gibberellin, **458**, 458i, 458t, 464
Gibbon, 394
Gilbert, W., 220, 221
▪ Gigantism, 525, 525i
Gill, **381**, 594–**596**, 595i, 596i, **598**, 598i
  invertebrate, 364, 364i, 365, 365i, 367
  mushroom, 327i
Gill slit, **378**, 378i, 380i, 382i
Gilled snail, 597
Gingerich, P., 257
Ginkgo, 344, 350t
Ginkgo biloba, 344, 344i
Ginkgo tree, 337i, 344, 344i
GIP, 617
Giraffe, 240, 473i
Gizzard, 281, 281i, 363i, 613

Genetic distance, 399

Glacier, 261, 274, 286, 333, 716i, 730, 740i, 745
Glacier Bay, 716i
Gladiolus plant, 454t
Gland, 73, **470**–471
Gland cell, 211, 364
Glandular epithelium, 470–471, 520
▪ Glaucoma, 515
Glaucophyte, 295, 295i
Gleditsia, 424i
Global broiling hypothesis, **386–387**
Global climate change, 93, 261, 263i, 279, 334, 337, 347, 356
Global precipitation, 745
Global temperature, 730–731, 745, 745i
  asteroids and, 264, 265
▪ Global warming, 2, 26, 33, 731, 743, **745**, 766
Globin, 44, 44i
Globin gene, 177
Globular protein, 48t
Glomerular capillary, 634–636, 635i
Glomerular filtration, **636**, 636i
Glossopteris, 265, 265i
Glottis, 601, 601i
Glucagon, 521i, 522, 523i, 528, 533t
Glucocorticoid, 533t
Glucose, 34, 34i, 38, 38i, 48t, 64, 82i, 84i, 92, 174, 208–209, 526, 528, 546, 546i, 620, 620i, 623
  aerobic respiration and, 108–109
  breakdown, 77
  cell uptake of, 76, 84, 84i
  conversion, 101, 101i
  ▪ food intake and, 118
  glycolysis and, 110i–111i
  synthesis, 74, 75, 75i, 77, 96, 97i, 104i
Glucose metabolism, 528–529, 529i
Glucose transporter, 84, 84i
Glutamate, 44, 45i, 198, 198i, 202i
Glutamine, 198i
Gluteus maximus, 543i
Glycemic index, 622
Glyceride, 48t
Glycerol, 38, 40i, 118
Glycine, 198i, 202i, 548
Glycogen, 39, 39i, 48t, 118, 526, 529i, 546, 546i
Glycolipid, 158
Glycolysis, 76, 106i, **108**, 110–111, 110i–111i, 135, 174, 174i, 209, 210
  fermentation and, 116, 116i
Glycoprotein, 43, 308, 308i
Glyphosate, 460
Glyptodont, 242, 242i
Gnetophyte, 337, 337i, **344**, 344i, 350t
GnRH, **659**, 663, 664i
Goat, 229i
▪ Goiter, 527, 527i
▪ Golden rice, 218
Golgi body, 62–63, 62i, 63i, 65i, 70t, 88i, 316, 317i
Golgi complex, 522
Gonad, **359**, 360, 530, 533t
  human male, 656, 656i, 656t
  invertebrate, 366i
  males, size of, 794
Gondwana, **265**, 265i
▪ Gonorrhea, 668t, 669
Goose, 497i, 782i
Goose barnacle, 368i
Gorgonia ventalina, 772
Gorilla, 129, 177, 394, 394i, 725i, 794
Gowero, C., 572–573, 572i
Gradual model of speciation, **278**
Gram bacteria, 305

Gram-positive bacteria, **304**, 304i, 305, 306i
Gram staining, 304
Grape, 452t
Grapevine, 320, 454
Grass plant, 348, 427, 427i, 449i
Grasshopper, 370i, 552i, 597
Grasshopper mouse, 713, 713i
Grassland, 307, 435, 747, 760i–761i, 762–763
Grassland soil, 767i
Gravelly soil, 767
▪ Graves' disease, 587
▪ Graves' disorder, 527
Gravitropism, **458**, 461, 461i
Gravity-sensing mechanism, 461
Gray matter, 493, 495i, 498
Grazing food web, **735**, 735i, 738, 738i
Great Barrier Reef, 772i
Great Dying, Permian, 37
▪ Great Lakes, 380i
Great Plains, 763
Great Salt Lake, 307, 307i
Green alga, 55i, 94, 324, 324i, 330, 333i, 336i
    evolution, 297i, 316i, 317, 319, 334
    functions/uses, 323, 325, 333
    life cycle, 324, 324i
    photosynthesis in, 95i, 105, 105i
Greenhouse effect, 349, 743, **744–745**, 744i–745i
Greenhouse gas, 33, 744–745, 744i–745i
Griffith, F., 186, 186i
Grip, 395, 395i
Grizzly bear, 725i, 787i
Ground pine, 340
Ground substance, 472
Ground tissue, 416, 418i, 419, 419i, 422–424, 427, 427i, 449
Groundwater, 740i, **741**
▪   contamination, 432–433, 432i, 460, 741i
    overdrafts, 741, 741i
Group defense, 788, 788i
Growth, **404**, **420**
Growth and tissue specialization, 648, 648i
Growth curves, 688–689, 688i, 689i, 690, 691i, 697i
Growth hormone, 524, 532t
Growth-inhibiting hormone, 459, 461
Growth pattern, plants, 420–429, 420i–429i, 464
Growth regulator, 459
Growth ring, **429**, 429i, 431, 431i, 445, 445i
Growth-stimulating hormone, 456
Grypania spiralis, 292i
Guanine (G), 46, 46i, 184i, 185, 188, 188i, 189, 189i, 196, 197, 199, 227, 227i,
Guano, 748i
Guar, 185
Guard cell, 96i, 421, 430t, 440, 440i
Guillemin, R., 525
Guinea pig, 217, 511i
Gular sac, 235i
Gulf of Aqaba, 5i
Gulf Stream, 758
Guppy, 694–695, 694i, 695i, 703
Gut, **615**
    animal, 354, **355**, 355i
    human, 137
    invertebrate, 362, 363, 367, 370
Gut cavity, 355i, 359
Gymnophilus, 326i
Gymnosperm, 263i, 334, 336i, 340, 343–345, 346i, 350t, 428, 449, 458

H
Habitat, **407**, 686, **706**
    aquatic, 316, 317, 319–321, 323, 336, 362, 368
    land, 318, 323, 326, 336, 343, 362
▪ Habitat contamination, and amphibians, 385, 385i
▪ Habitat destruction, and sea turtles, 388
Habitat loss, 446, **724–725**
Habropoda laboriosa, 447
Hadal zone, 773i
Haemophilus influenzae, 222
Hagar Qim temple, 314i, 315
Hagfish, 379, 379i
Hair, 43, 392, 478
    body hair distribution, 169
    epidermal, 424
    mammal, 41
    structure, 479i
    texture, 479i
Hair cell, 509, 511, 511i
Hair follicle, 478i
▪ Hair thinning, 478
Half-life, **262**, 262i
Hallucinogen, 327, 327i, 501
Halophile, extreme, 297i, **306**, 307, 307i
Hamilton, W., 791
Hammer, 510, 510i
Hand, 133i, 266
    apoptosis and development of, 412–413, 413i
    bones, 539i
    grip, 395, 395i
Hangingfly, 786, 786i
Haploid chromosome number (*n*), **141**, 142i, 143i, 146, 221
Haploid dominance, 350t
Haploid resting cell, 146
Haploid spore, 336, 448, 448i, 450, 450i
Haploid stage, 138i, 146, 146i, 322i, 324i, 327i, 328i, 338i, 341i, 345i, 347i, 451i
Hardwood tree, **429**, 431
Hardy-Weinberg rule, 247i
Harlequin frog, 714i
Harpobittacus apicalis, 786
Hawaii, 715
Hawaiian Archipelago, 264i, 274–275
Hawk, 517, 517i
▪ Hay fever, 586
Hayes, M., 176
Hayes, T., 518–519, 519i
Hayflick, L., 680
HbS molecule, 44, 45i
HCG, 671
HDL, 49, 566, 592–593
Head
    invertebrate, 354, 360, 362, 362i, 363i, 364, 368, 370, 375, 375i
    nervous system and, 492–493
    vertebrate, 354
Hearing, 510–511, 510i
▪ Hearing loss, 511, 511i
Heart, 19, 64, 69, 159, 245, 281, 281i, **552**, 560–561, 560i
▪   abnormalities, 159, 159i
    accessory heart, 365, 365i
    blood pressure, transport, and flow distribution, 86, 562–563
    cardiac muscle contraction, 561
    development, 571
    hormones from, 533t
    human, 558
    invertebrate, 363i–365i, 365, 367
    muscle, 43, 56i, 69, 474, 474i, 561, 561i
    structure, 560

▪ Heart attack, 49, 229, 468, 528, 567, 593
▪ Heart disease, 49, 250, 571, 623, 626
Heartwood, **429**, 429i
Heat energy, **26**, 27, 74, 82, 83, 105
Heat gains and losses, 639
Heat stress, 640
▪ Heat stroke, 402–403, 414
▪ Heimlich maneuver, 608–609, 609i
HeLa cell, 124–125, 124i, 135
Helianthus, 102, 452t, 464, 466i
Helical coiling, advantages, 291
Helicobacter felis, 280i
H. pylori, 302i, 629
Heliconius cydno, 12–13, 13i, 22, 23i
H. eleuchia, 13, 13i
Helper T cell, **578**, 579, 584–585
▪ Hematopoietic growth factors, 585
Heme group, **44**, 44i, 79i, 80, **604**
Hemerythrin, 595
Hemlock tree, 331i
Hemocyanin, 595
▪ Hemodialysis, 638
Hemoglobin, 43–45, 44i, 45i, 48t, 165, 177, 202, 210, 245, 253, 257, 318, 554–555, **595**, 603, 603i, 604
Hemolymph, 571
▪ Hemolytic anemia, 556
Hemorrhage, 329
▪ Hemorrhagic anemia, 556
▪ Hemorrhoid, 565
Hemostasis, **566**
Henslow, J., 241
Hepatic portal vein, 558i
▪ Hepatitis, 585
    alcoholic, 72
    chronic viral, 481
▪ Hepatitis B, 501i, 586
▪ Hepatitis D, 309, 310t
HER2 protein, 585
Her2 receptor, 206–207
Herbaceous plant, **428**
▪ Herbicide, 335, 349, **460**, 518–519
Herbivore, 9, 330, 732, 733i
▪ Herceptin, 585
Heredity. *See* Inheritance
Herman-Giddens, M., 534
Hermaphrodite, **360**, 518
    invertebrate, 361, 364i, 365, 366, 368
▪ Herniated disk, 538
Heroin, 501, 501i
Heron Island, 770i
▪ Herpes, 585, 668, 668t
Herpes simplex virus, 308i, 309, 668
Herpes virus, 478, 588
Herring gull, 789
Hershey, A., 187i
Hesperidium, 452t
Heteractis magnifica, 358, 358i, 359, 492, 492i, 538i, 707, 707i
Heterocephalus glaber, 790, 791, 791i
Heterocyst, 304, 304i
Heterosporous, 337
Heterotherm, **639**, 640i
Heterotroph, **92**, 93, 294, 315, 315i, **316**, 317, 320, 324, 326, 354, 732, 732i
Heterozygous condition, **153**, 155–158, 157i, 164, 173, 174, 252–253
▪ Hiccups, 609
Hickory tree, 429
Higgins, T., 550–551, 567
▪ High altitude, 603, 603i
High-density lipoprotein, 49, 566, 592–593
High fructose corn syrup, 623
High glycemic index, 622
Himalayan rabbit, 162, 162i
Hindbrain, 496, 496i, 497, 497i, 502t

Hippocampus, 499, 499i
Hirudo medicinalis, 363i
Histamine, **576**, 576i
Histidine, 45i, 79i, 198i
Histone, 58, 126–127, 127i, 208
▪ Histoplasmosis, 329
▪ HIV, 308–310, 308i, 572–573, 585, 588–589, 669, 669i
    drugs and vaccines for, 589
    replication cycle, 588, 589i
    transmission, 588
HLA, 49, 49i
Hoekstra, H., 249
Hognose snake, 386i
Hognosed bat, 392i
Holdfast, **293**, 321, 321i
Holly plant, 441i
Homeodomain, **212**, 212i
Homeostasis, **7**, **405**, 559
    animals, 408–409
    plants, 410–411
Homeotic gene, **212**, 268, 269, 269i, 463, **655**
Hominid, **394–397**, 396i, 538
Hominoid, 394, 395
Homo erectus, 343, 398, 398i, 399
H. habilis, 397, 397i, 697
H. neanderthalensis, 398i
H. sapiens, 398–399, 398i. *See also* Human
Homologous chromosome, **140–141**, 141i, 142i, 144, 150, 153i, 154, 156–157, 168, 168i, 171, 247i
Homologue, 141, 141i, 142i, 144–145, 148i, 157, 168, **266**, 266i, 267
Homosporous, 337
Homozygous condition, **153**, 156, 162, 162i, 164, 174, 254, 255, 257
    dominant/recessive, **153–154**, 153i, 154i
Honey locust tree, 424i
Honey mushroom, 327
Honeycomb, 41, 41i
Honeycreeper, lineage, 275, 275i, 278
Hooke, R., 54
Hookworm, 354t
Horizon, soil, 435, 435i
Horizontal cell, 514, 514i
Hormone, 7, 19, 88, 208, 250, 406, 471, **520**. *See also specific types*
    behavior and, 780i
    bonding behavior and, 792
    categories of, 522t
    discovery of, 520
    endocrine disrupters, 518–519
▪   environment and, 530–531
    feedback loops in secretions, 526–527
    nature of hormone action, 522–523
    peptide hormones, 522–523
    released from mammalian pituitary gland, 532t
    responses to, 522
    signaling mechanisms, 522–523
    sources of, 533t
    steroid hormones, 522
Hormone disrupter, 530, 530i
Hormone function
    in animals, 367
    in plants, 440, 447, 454, 456, 458t, 460, 460i, 462, 464
    in sexual traits, 169
Hornwort, 338, 339, 339i
Horse, 266, 285
Horse chestnut tree, 464i
Horseshoe crab, 369, 369i
Horsetail, 334, 337, 337i, **340–342**, 340i, 350t

Host, **294**, 295, 310, 314, **360**
  human, 361, 361i, 366i
Host cell, 198, 308, 309
Hot pepper, 466
Hot spot, **726**, 726i, **760**
Hot spring, 286, 307, 725
Housefly, 280i
Household mold, 329
Houseplant, 454
■ HPV, 668, 668t
Hrdy, S. B., 792
Hubbard Brook watershed,
  740, 741i
■ Hudson River, 380i
Hudson's Bay Company, 711i
Hughes, S., 509, 509i
Human, 53i, 214, 246, 246i, 250, 255,
  280i, **396**
  behavior, 166–167
  chromosome, 127, 127i, 129, 129i,
    141, 141i, 147
  as competitor species, 724, 724i
  development, 647i
  development stages, 679i, 679t
  early forms, 258–259, 261, 343,
    396–397
  effects on biosphere, 301, 313, 315,
    350, 466, 795
  embryo, 644i
■ emotional health, 163
  evolution, 214–215, 239,
    258–259, 263i
  gamete, 246
  glands, 73
  immune system, 230, 231
  inheritance patterns, 164, 168–169,
    168i, 169i, 171–179
  karyotype, 170i
  modern forms, 398–399
  origins of modern humans,
    398–399
  as predator species, 725, 725i
  primate evolution, 394–395
  reproduction, 140, 140i
  sex determination, 168–169, 169i
  skeletal structure, 239, 239i, 266,
    266i, 268i, 269, 394i
  traits, 394–395
Human body, chemical elements in,
  18, 18i, 19i
Human body cell, DNA in, 61
Human chorionic
  gonadotropin, 671
Human collagen, 229
■ Human gene therapy, **232**
Human Genome Project, 221
Human health experiment, 12, 16
■ Human papilloma virus, 137,
  668, 668t
Human population growth,
  696–697, 696i, 697i, 795
  biological wealth and, 726–727
  computer models, 701, 701i
  effects of, 334, 349
  energy consumption and, 755
  water depletion and, 741
Human social behavior, 792
Humboldt Current, 773
Humerus, 539i
Hummingbird, 53i, 348i, 447, 639,
  647i, 708
Humphead parrotfish, 8
Humus, **434**, 767
Hunger, 610, 623, 627
■ Huntington's disease, 107, 173t, 174
Huq, A., 775
Hurtley, K., 638i
■ Hutchinson-Gilford progeria
  syndrome, 176, 176i
Hybrid, **153**
Hybrid molecule, **223**, 270, 272, 285
Hybridization, 153, 271

Hydra, 354t
Hydrangea, 163
Hydration, sphere of, 27, 27i
Hydrocarbon, 34
Hydrochloric acid, 28, 29
Hydrogen, 20, 20i, 22, 23i, 24, 25,
  25i, 288, 289, 306, 434t
Hydrogen atom, relative size, 4,
  4i, 76
Hydrogen bond, **25–27**, 26i, 34, **43**,
  47i, 56, 79, 81, 270, 438–439, 439i
  in DNA, 25i, 189, 190, 191
Hydrogen cyanide, 288
Hydrogen ion (H⁺), 28, 28i, 30t, 34i,
  64, 76, 77i, 82i, 98, 99, 100i
Hydrogen peroxide, 63, 79i, 80,
  90–91
Hydrogen sulfide, 307
Hydroid, 358, 359
Hydrologic cycle, 739, **740**, 740i
  phosphorus cycle and, 748
Hydrolysis, **35**, 35i, 39
Hydrophilic head, 41i, 56, 56i
Hydrophilic substance, **26**, **30t**
Hydrophobic substance, **26**, **30t**,
  34, 79
Hydrophobic tail, 41i, 56i, 82
Hydrostatic pressure, **87**
Hydrostatic skeleton, 359, **538**, 538i
Hydrothermal vent, 59i, 105, 286,
  289, 290i, 291, 292, 306, 307,
  772, 773i
Hydroxide ion (OH⁻), 28, 30t
Hydroxyl group, 39i
Hydrozoan planula, 492
Hygrophorus, 326i
Hyman, F., 159i
■ Hypercholesterolemia, 173t
Hypericum, 430i
■ Hypertension, 250, 528, 566, 567,
  626, 638
Hyperthermia, 640
■ Hyperthyroidism, 527
Hypertonic solution, **86–87**, 87i
Hypertonicity, 287
■ Hypertrophic cardiomyopathy, 106i
■ Hyperventilation, 603
Hypha, **326**, 327i, 331, 436
Hypnotics, 500–501
Hypocotyl, **457i**
Hypodermis, 478, 478i
Hypothalamus, **496**, 496i, 497, 497i,
  499, 499i, 502t, 521i, **524–525**,
  524i, 525i, 637, 637i, 640, 659,
  659i, 663
  menstrual cycle and, 664i
Hypothermia, 641, 641i
Hypotheses, examples of, 239–243,
  252, 264
  Alexander the Great's death, 300
  cell evolution, 294, 295
  inheritance, 152–153
  origin of cells, 288, 290, 291, 291i
  origin of life, 289
  photosynthesis, 95i, 105
  prions and degenerative
    disease, 311
Hypothesis, **11**
■ Hypothyroidism, 527
Hypotonic solution, **86–87**, 87i, **440i**
■ Hypoxia, 603
■ Hysterectomy, 137

I
IAA, 459
Ice, 26i, 27
Ice age, 261, 263i
Ice cap, 730
Ice cover, perpetual, 760i–761i
Ice shelf, 751, 751i
Iceberg, 751, 751i
Ichthyolestes, 517i

Ichthyosaur, 386, 387i
Ichthyostega, 383i
Ideal weight, 626i
Identical twins, 651, 651i
IgA immunoglobulin, 582i, 583
IgD immunoglobulin, 582, 582i
IgE immunoglobulin, 582i, 583, 586
IgG antibody, 676
IgG immunoglobulin, 582–583, 582i
IgM immunoglobulin, 582, 582i
Ignicoccus, 307i
Iguana, 388
Ileum, 614i, 616–617
Iliac artery, 558i
Iliac vein, 558i
IL2RG gene, 230
Immigration, **255**, 685, **688**, 700,
  701i, 723
Immigration Reform and Control
  Act of 1986, 700
Immune response
  control of, 579
■ deficient responses, 587
  enhanced or compromised,
    586–587
Immune system, 49, 73, 230,
  574–575, **578**
  human, 230, 231
  memory of, 578
  vertebrate, 308, 310
Immunity, 572–591, 756
  adaptive immunity, 574–575,
    578–579
  AIDS and, 588–589
  allergies, 586–587
  antibody-mediated response,
    582–583
  antimicrobial proteins, 577
  autoimmune disorders, 587
  cell-mediated response, 584–585
  defenses, 574–575
  defenses enhanced or
    compromised, 586–587
  deficient immune responses, 587
  evolutionary background,
    574–575
  fever, 577
  first line of defense, 575
  immunization, 586
  immunotherapy, 585
  inflammatory response, 576–577
  intact skin and mucous
    membranes, 574
  lines of defense, 575t
  lymphocytes, 580–581
  nonspecific internal defenses, 574
  nonspecific internal responses,
    576–577
Immunization, **586**, 586i
Immunoglobulins (Igs), **582**
  classes of, 582–583, 582i
Immunological memory, 580, 581i
■ Immunotherapy, 585
■ Immunotoxin, 585
Imperfect flower, **449**
Imperfect fungus, 326
Implantation, **670**, 670i
Imported pest, 447
Imprinting, **782**, 782i
■ In vitro fertilization, **181**, 181i, 192,
  645, 667, 667i
Inbreeding, **255**, 791
Inclusive fitness, **791**
Incomplete digestive system, **355**,
  355i, 359, 360, 366i, 612
Incomplete dominance, of alleles,
  158, 158i
Incomplete protein, 623
Incremental energy release, 114
Indehiscent, **452t**
Independent assortment theory,
  156–**157**, 156i, 157i

India, 698, 699i
Indian mallow, 709, 709i
Indian pipe, 348i
Indicator species, **724**
Indirect interaction, 706
Indocetus, 517i
Indoleacetic acid, 459
■ Induced abortion, 667
Induced-fit model, **79**
Inductive logic, 15
Industrial smog, 756–757
Industrial stage, 700, 700i
Inert element, **20**
Infant, 679i, 679t
Infant death, 230
Infanticide, 792
Infection, **310**, 434t
Infectious disease, 236, 310t
  evolution and, 310–311
  human population growth and,
    696–697
■ Infectious mononucleosis, 556
Inferior vena cava, 558i, 560i
Inflammation, 575t
Inflammatory response, 576–577
■ Influenza, 2, 308, 308i
Infolding, 294, 294i
Infrared radiation, 94i, 105, 292
Inhalation, 602, 602i
Inheritance, **6**, 125, 152–153, 166
  chromosal basis, 152, 166,
    168–169, 168i, 169i
Inheritance patterns, 150–153,
  171, 171i
  human, 172–175, 172i, 173i
Inhibitor, 80, 80i, **524**, 525i
Inhibitory effects, 488, 489
Initiation, in RNA translation,
  200, 200i
Ink sac, 365i
Inner ear, 509, 510, 510i, 517i
Input zone, 484i, 485
Insect
  agriculture and, 320, 333, 366,
    370, 460
  in amber, 261
  characteristics, 354t, 367, 367i,
    369–371, 369i, 370i
  current species diversity, 725i
  disease vectors, 301, 305, 316–318,
    319i, 366, 369, 371, 375
  diversity, 370–371, 371i
  eusocial, 790, 791i
  evolution, 263i, 346, 353
  flight among, 367, 370, 371i
  larvae, 211
  life cycle, 370i
  nymph, 370
  parasite, 446
  parasitoids, 714
  pesticides and, 248, 371, 375
  plants and, 446, 459, 466
  pollinators, 344, 346, 371, 424i, 449
  pupa, 370
  respiration, 596–597, 597i
  social, 790, 790i
  survivorship curve, 424i
  vision, 512
Insertion, in DNA, **202**
Instinctive behavior, **781**, 781i
Insulin, 7, 19, 48t, 118, 520, 521i,
  522–523, 528–529, 529i, 533t
Integrator, **408**, 408i, 409i
Integument, 450, 452, 512
Integumentary exchange, **596**, 596i
Integumentary system, 477i
Intelligence, as evolutionary factor,
  365, 401
Interaction. *See* Species interaction
Intercostal muscle, 600i, 602, 602i
Interference competition, 708, 708i
Interferon, 574, 577, 584, 585

Interleukin, 574, 584, 585
Interleukin-2, 229, 585
Intermediate, **76**, 77, **89t**
Intermediate filament, **66**
Intermediate host, **360**
Internal body temperature, 409
Internal environment, 7, 29, 73, 81, 81i, 210, **405**, 630–643
  acid-base balance, 638
  human urinary system, 634–635
  kidney disorders, 638
  maintaining core temperature, 639
  maintaining extracellular fluid, 632–633
  temperature regulation in mammals, 640–641
  urine formation, 636–637
Internal temperature, 402–403
Internal transport, large bodies, 406
Interneuron, 475, **485**, 490
Internode, 418i, 422, 457i
Interphase, **128**, 128i, 130–131, 130i–131i, 138i, 141, 141i, 142i, 148i
Interspecific competition, **706**, 708, 708i
Interspecific interaction, 686
Interstitial fluid, **408**, **552**, **632**
Intertidal zone, 718–719, 718i, 770, 773i
Intervertebral disk, 495i, 538, 539i
Intestinal enzyme, 617
Intestinal nucleus, 616t
Intestine, 41, 209, 248, 304, 317, 360, 366i
  vertebrate, 361i
Intracellular digestion, 88
Intraspecific competition, 708
Intron, **197**, 197i, 221, 223, 309
Inversion, chromosomal, **176**, 176i
Invertebrate, 138, 352–372, **354**
  hormones and environment, 531, 531i
  respiration, 596–597, 596i, 597i
  vision, 512i
Involuntary muscle, 474
Iodide, 527
Iodine, 625t
■ Iodine-deficient diet, 527, 527i
Ion, **24**, **30t**
Ionic bond, **24**, 24i
Ionizing radiation, **203**
Iraq war, 316
Iridium, 258, 259i
Iris, 430i, 440i, 462
Iris, eye, 160–161, 160i, 512, 513i
Iron, 434t, 625t
Iron atom, 80
■ Iron deficiency, 571
■ Iron deficiency anemia, 556
Iron-sulfide rock, 290i, 291
Irrigation, 740–741
Isaacson, P., 736i
Island, 722i, 723, 723i
Isolation, 272–274, 272i, 273i
Isoleucine, 198i, 199, 623
Isometric contraction, 546, 547i
Isopentyl acetate, 778–779
Isotonic contraction, 546, 547i
Isotonic solution, **86**
Isotope, **21**, **30t**, 193, 262
Isthmus of Panama, 274
Ixodes (deer tick), 305i

J

J-field, 432
J-shaped growth curve, 688i, 689
Jackson Laboratories, 627i
Jade plant, 454t
Jamestown colony, 445
Japanese beetle, 719t
Jasmonate, 459

Jaw, **380**, 381, 392
  invertebrate, 365i
  mammal, 392
Jawed fish, 380–383, 380i
Jawless fish, 380, 380i
Jawlike structure, invertebrate, 375
Jay, 246, 255, 255i
Jejunum, 614i, 616–617
Jellyfish, 234, 352, 354t, 355, 358, 358i, 509
Jellyfish bell, 359
Jet propulsion, cephalopod, 365
Johns Hopkins University, 12, 124
Johnson, M., 669i
Joint, **367**, **541**
Jugular vein, 558i
Jump dispersal, 719, 723i
Jumping spider, 596i
Juniper seedling, 331i
Juniperus occidentalis, 280
Jupiter, and asteroids, 258
Jurassic period, 263i, 337i, 343, 386, 387i, 393
Juvenile, **367**
  invertebrate, 366, 368i, 370

K

K-T asteroid impact theory, **386–387**
■ Kaguya, 139, 139i
Kalahari Desert, 481, 481i
Kangaroo, 647i
Kangaroo rat, 633i, 643
Kangaroo rat cell, 66i
Kapan, D., 12–13
■ Kaposi's sarcoma, 588, 588i
Karenia brevis, 320, 320i
■ Kartagener syndrome, 71, 71i
Karyotype, **170**
■ Karyotype, spectral, 170, 170i, 177i, 178i
Kelp, 315i, 321, 321i
Kepone, 531
Keratin, 48t, 478
Keratinocyte, 413, 478
Kereru, 466
Ketoacidosis, 528
Ketone, 34i, 38, 528, 623
Key innovation, **279**, 293, 337, 355, 367
Keystone species, **718–719**
Kidman, N., 479
Kidney, 63, 361, 623, **632**, 634, 634i, 635i
  blood pressure and, 563
  cortex, 634, 635i
■ disorders of, 638
  hormones from, 533t
  maintaining acid-base balance, 638
  medulla, 634, 635i
  parathyroid glands and, 528
■ Kidney dialysis machine, 638
■ Kidney failure, 638, 638i
■ Kidney stones, 638
■ Killer bees, 778, 778i
Killifish, 694–695, 694i, 695i, 703
Kilocalorie, **626**
Kinase, 134
Kinesin, 66–67
Kinetochore, **127**, 130
Kinetoplast, **316**, 316i
Kitti's hognosed bat, 392i
Kiwi plant, 449
■ Klinefelter syndrome, 173t, 179
Knee joint, 539i, 541
■ Knockout experiment, **214**
Knockout mice, 229
Koa bug, 715
Koala, 392i
Korana, G., 198
Korarchaeota, 306, 306i

Krebs, C., 711i, 711i
Krebs cycle, **109**, 112–113, 113i
Krill, 93
Kubicek, M., 124
Kudzu, 720, 720i
Kwang, J., 295

L

La Niña, **774**, 774i
Labia majora, 660, 661i
Labia minora, 660, 661i
Labium, 370i
Labor, 409, **678**, 678i
Labrador retriever, 159
Lacks, H., 124–125, 124i, 137
Lactate, 546
■ Lactate fermentation, **117**, 117i
Lactation, **679**
Lactobacillus, 117, 575
  L. acidophilus, 304
Lactose, 38, 174, 208–209, 209i
■ Lactose intolerance, 209
LacZ gene, 222i
Ladybird beetle, 371i
Lake ecosystem, 740i, 768–769, 768i
  seasonal changes in, 768–769
  thermal layering in, 768i
  trophic nature of, 769
  zonation of, 768, 768i
Lake Washington, 769
Lamarck, J., 240
Lamin, 66
Laminaria, 321
Lamprey, 380, 380i, 381i
Lancelet, 354t, 378, 378i, 591
Land plant, origin, 343
■ Land rights, 777
Land snail, 364i
Langerhans cell, **478**, 578i
■ Language, 395, 794
Lanugo, 675
Larch tree, 344
Large intestine. *See* Colon
Largest organism, 327
Larrea divaricata, 454, 687i
Larva, invertebrate, **357**, 359, 361, 361i, 368i, 370, 372
■ Laryngitis, 601
Larynx, 600i, **601**
Lascaux, France, 399i
■ Laser, uses, 137, 227, 227i
■ Laser angioplasty, 567
■ LASIK, 515
■ Late blight, 315
Late wood, **429**, 429i
Latency, 309
Lateral bud, 418i, 422, 422i, 458, 458t, 459
Lateral gene transfer, **303**, 310
Lateral meristem, 417, 418i, **419**, 428, 428i
Lateral root, 418i, 426, 427, 427i
Lateral shoot, 422, 423
Latimeria, 383, 383i
Latissimus dorsi, 543i
Latrodectus, 369, 369i
Lattice structure, 26i, 59i
Lava basalt flow, 249i
Lawn grass, 425
Laysan albatross, 391i
LDL, 49, 566, 592
LDL mutation, 49
Leaching, **435**, 747
Lead, 262
■ Lead poisoning, 432
Leaf, 96i, 102, 102i
  C3 plant *vs.* C4 plant, 102, 102i
  characteristics, 27, 103i, 272i, 418i, 424–425, 424i, 425i, 459
  epidermis, 69i, 421, 421i, 424, 424i, 425i
  formation, 422, 422i, 464

function, 424–425, 425i, 440
  monocot *vs.* eudicot, 424–425
Leaf blade, 425i
Leaf color, 95, 95i
Leaf decay, 407i
Leaf fine structure, 424–425, 425i
Leaf folding, 411, 411i
Leaf hair, 424i
Leaflet, 424i
Leafy gene, 463
Leakey, M., 396i
Learned behavior, **782**
■ Learning disability, 179
Least bittern, 712i
Lecithin, 623
Leder, P., 198
Leech, 354t, 362, 363i, 597
Leg
  bones, 69, 539i
  formation, 269
  invertebrate, 368–370
Legume, 410, 436, 452t, 746–747
  bacteria and, 305
Leishmani donovani, 316
Leishmania mexicana, 89i
■ Leishmaniasis, 89i, 316
Lek, 786–787, 786i
Lens, 512–513, 512i, 513i, 515
Leopard frog, 518–519, 649i
Lepidodendron, 342i
Leptin, 610, 610i, 623, 627, 627i
Lesser long-nosed bat, 447
Lethal allele, **164**, 174
Leucine (leu), 42i, 45i, 198i, 623
■ Leukemia, 173t, 177, 230, 468, 556, 777i
Leukocyte. *See* White blood cell
Leydig cell, 658i, 659
LH, 521i, 524, 532t, 630, 659, 663, 664i
Lichen, 304, 323, **330**, 330i, 707
Lid, cnidarian, 358i
Life, **6**
  beginning of, 666–667
  chemical basis, 18–30, 30t
  diversity of, 2, 6, 8, 8i, 9, 10, 139, 184i, 187, 189
  energy-releasing pathways and, 120
  in extreme environments, 307i
  history, 296i–297i
  organizational levels, 4–5
  origin, 287–293, 307
  synthesis of, 291
  unity of, 2, 6, 8, 9
Life cycle
  alga, 336
  animal, 139, 146, 146i
  arthropod, 367
  black bread mold, 328, 328i
  blood fluke, 361, 361i
  brown alga, **321**
  chain fern, 341i
  club fungus, 326i
  cnidarian, 359, 359i
  conifer, 344–345
  crab, 368i
  crustacean, 368
  flowering plant, 428, 448, 448i, 450–451, 450i–451i, 464
  fungus, 326–327, 327i
  green alga, 324, 324i
  insect, 370i
  invertebrate, 359, 359i, 367, 368
  lily, 347i
  moss, 338i
  parasite, 319i, 361i
  plant, 139, 146, 146i, 336, 336i
  Plasmodium, 319i
  red alga, 322i
  seedless vascular plant, 340, 341i

slime mold, 324–325, 325i
tapeworm, 361i
Life history patterns, **692**–693
  natural selection and, 694–695, 694i, 695i, 703
Life table, 692, 692t
Ligament, 539i, **541**
Light, properties, 94–95, 94i
Light-activated pigment, 307
Light-dependent reaction, **96**, 97i, 98–99, 100i, 104i
Light energy, 74, 96, 98, 460
Light-energy conversion, 70t
Light-harvesting complex, 97i, 98–99, 98i–99i
Light-independent reaction, **96**, 97i, 100i, 104i
Lightning, simulation of, 289
Lignin, 68, 228, 336, 420–421, 420i, 438
Lilac, 464i
Lily, 242i, 347i, 348, 372, 424i, 454t, 455
Lily pad, 406, 407i
Limb, evolution, 239, 266i, 269, 269i, 279, 383, 383i, 392
Limbic system, 496i, **499**, 502t, 508
  components of, 499i
Limestone, 314–315, 314i, 321
Limiting factor, **690**, 690i, 696–697
Limnetic zone, 768, 768i
Limulus, 369, 369i
Lineage, 124, 203i, 210, **261**, 304i, 306, 332t. *See also* Speciation
  animal, 352–356
  divergence in, 232, 266–267, 266i, 271, 280, 282i
  eukaryote, 316, 320, 332t
  fungi, 315, 326
  patterns and processes, 259, 268–271, 278–279
  plant, 102, 334i, 336, 348i
  prokaryote, 293, 332t
  protist, 8, 282i, 316i, 320, 356
  true-breeding, 153, 155i, 183
  vertebrate, 266–269
Linkage group, **171**, 171i
Linnaeus, C., 8
Linoleic acid, 623
Linolenic acid, 40i
Lion, 7i, 281i, 784, 788–789, 792
Lipase, 616t, 617
Lipid
  characteristics, 35, **40**–41, 48t, 53i, 440
  early Earth and, 289, 293
  function, 33i, 48t, 73, 158
  synthesis, 63, 128
Lipid bilayer, **56**, 56i, 57, 59, 62i, 82–84, 82i, 83i, 94i
Lipid envelope, 308i
Lipoprotein, 43
Lisosome, 60
Listeria, 312
Lithops, 712, 712i
Littoral zone, 768, 768i
Liver, 41, 63, 64, 361, 614i, 628t
  function, 72–73
  hormones from, 533t
  organic metabolism, 620, 620i
Liver cell, 39, 60i, 118, 210, 318, 319i
■ Liver disease, 72
Liver donor, 73
Liverwort, 282, 337i, 338, 339, 339i
Livestock, 727
Lizard, 387, 388, 389i, 467i, 647i
Llama, 245, 245i, 274, 274i, 285, 603, 603i
Loam, **434**
Loam topsoil, 767
Lobaria oregana, 330i

Lobe-finned fish, 383, 383i
Lobster, 368, 368i, 375
Local feedback, 527
Local signaling molecule, **520**, 528
■ Lockjaw, 548, 548i
Locomotion, 354, 362, 612
  invertebrate, 369, 372–373, 373i
Locus gene, **153**, 153i, 158, 159, 171
Lodgepole pine, 436i
Loggerhead turtle, 321
Logging runoff, 597
Logistic growth, **690**, 691i, 702
Loligo, 487i
Long-day plant, **462**
Long-term memory, 499
Loop of Henle, **635**, 635i, 636i, 637, 643
Loose connective tissue, **472**, 472i
Looy, C., 261i
Lorenz, K., 782i
Lottorina littorea, 718i, 719
Lotus, 348i, 415i
■ Lou Gehrig's disease, 107
Louisiana, 730–731, 730i, 775
■ Low-carb diet, 623
Low-density lipoprotein, 49, 566, 592
■ Low-fat diet, 49
Lower limb bones, 539i
Loxosceles reclusa, 369, 369i
LSD, 501
Lubchenco, J., 719
Luciferase, 228, 319
"Lucy," 396i
Luft, R., 106
■ Luft's syndrome, 106
Lumbar nerve, 493i
Lumen, 496
Luna moth, 371i
Lung, 67, 67i, 150–151, 159, 245, 281, 281i, 381, 405i, **598**, 600i, 601, 602i
  evolution of paired lungs, 598–599, 599i
■ Lung cancer, 593, 757, 757i
■ Lung disease, 2
Lungfish, 383, 383i
Lupinus arboreus, 410–411, 411i
Luteal phase, 660, 661
Luteinizing hormone (LH), 521i, 524, 532t, 630, 659, 663, 664i
Lycophyte, 337, 337i, **340**, 340i, 342, 342i
Lycopodium, 340, 340i
Lyell, C., 241, 243
■ Lyme disease, 305, 369
Lymph, **568**
Lymph capillary, 568i, 569
Lymph node, **366**, 555, 568, 568i, **569**, 581
Lymph vascular system, **568**–569, 568i
Lymph vessel, 135, 568i, 569
Lymphatic duct, 568i
Lymphatic system, 477i, **553**
  circulation and, 553, 568–569
  components, 568i
  immunity and, 581, 581i
  lymphoid organs and tissues, 569
Lymphocyte, 555, 580–581, 590t
  B cells, 590t
  cytotoxic T cells, 590t
  helper T cells, 590t
Lymphoid organs and tissues, 569
Lysine, 198i, 202i, 623
Lysis, **309**, 309i, 577, 577i
Lysogenic pathway, 309, 309i
Lysosome, **63**, 65i, 70t, 88, 88i, 412
Lysozyme, **574**, 575, 618, 618i
Lystrosaurus, 37, 37i
Lytic pathway, 309, 309i

Ⓜ
MacArthur, R., 724
Macrocystis, 321, 321i
Macroevolution, **263**, 266, 283t
Macronutrient, soil, **434**
Macrophage, 67, 88, 89i, 555, 555i, **574**, 574i, 576, 584, 590t
Macrotermes, 791i
Macula, 515
■ Macular degeneration, 515
■ Mad cow disease, 229, 312
Madrone tree, 315
Magicicada septendecim, 272i, 273
Magnesium, 434t, 625t
Magnetic field, 517
Magnetotactic bacteria, 305
Magnolia, 348
Magnoliid, 423
Maiasaura, 386i
Maidenhair tree, 344, 344i
■ Malaria, 3, 253, 253i, 310t, 318, 319i, 731
Mali, 703i
■ Malignant tumor, 135, 135i
■ Malnutrition, 218–219, 572–573
Malpighian tubule, 370
Malthus, T., 242
Mammal, 381i, 392–393
  adaptive radiation, 392, 393, 393i
  behavior, 392
  brain, 393, 497i
  characteristics, 354
  circulation, 553i
  cleavage, 651, 651i
  core temperature, 640–641, 640t
  dinosaurs and, 392, 393
  egg-laying, 393
  eusocial, 790, 791, 791i
  evolution, 392–393
  gas exchange, 595, 595i
  hair, 41
  hormones released from pituitary gland, 532t
  jaw, 392
  origin, 279, 279i
  parental care, 787, 787i
  placental, 212, 279i, 392i, 393, 393i
  polyploid, 277
  pouched, 393
  reproduction, 139, 647i
  respiration, 599i
  teeth, 392, 392i
  temperature regulation, 640–641
  traits, 392i
Mammary glands, 478, 678i, 679
Manatee, 320
Mandible, 370i
Manganese, 434t
Mangrove swamp, 760–761i
Mangrove wetland, 770, 770i
■ Manic depression, 166–167
Mantle, **364**, 364i, 365i
Mantle cavity, 364i, 365
Manx cat, 164, 165i
Manzanita, 273
Maple tree, 419, 424i, 452t, 453, 453i
Marchantia, 339i
■ Marfan syndrome, 159, 159i, 173t
Margulis, L., 294
Marigold, 276i, 449i
Marijuana, 333, 501, 607, 631
Marine ecoregion, 760, 760i–761i
Marine nursery, 770
Marine snow, **772**
Marine worm, 362, 375i
Marjoram plant, 436i
Markov, G., 194
Marler, J., 782i
Marr, J., 300
Mars, 258, 289
Marsh grass, 770i
Marsh marigold, 449i

Marsh wren, 782i
Marsupial, 212, **393**, 393i, 647i
Mason, R., 783
Mass extinction, 37, 252, 258, 263i, **279**, 283t, 343, 384, 386, 392, 724, 764i
Mass number, **20**
Mast cell, 555i, **574**, 574i, 576, 576i, 590t
■ Mastectomy, 207
Master gene, 212, **655**
Maternal behavior, 792
Maxam, A., 220
Maxilla, 370i
Mayan civilization, 416, 416i
Mayfly, 597
Mayr, E., 272
McCain, G., 15
McClintock, B., 202, 203i
McClintock, M., 792
McGwire, M., 536
McIr gene, 249
MDMA, 482–483, 500
Mealybug, 138i
■ Measles, 308, 310t, 676–677
Measles virus, 588
Mechanical energy, 7
Mechanical isolation, 272i, **273**, 273i
Mechanical work, 74
Mechanoreceptor, **506**, 506i, 507, 507i, 509
Medicago, 423i, 446, 466
Medici, M. de, 527i
■ Medicine, radioisotopes in, 21
■ Mediterranean diet, 623
Mediterranean fruit fly, 371i
Medulla oblongata, **496**, 496i, 497i, 502t
Medullosa, 342i, 343
Medusa, 358, 358i, 359, 359i
Meerkat, 481, 481i
Megachile rotundata, 446
Megafossil, 292i
Megakarocyte, 555, 555i
Megaspore, 337, 342–345, 346i, 347i, 448i, **450**, 451i
Meiosis
  chromosomal change and, 168, 168i, 171, 171i, 178i
  eukaryotic, 277, 316, 320, 322i, 322t, 324i, 327i, 328i
  gene swapping and, 144–145
  plant, 276, 338i, 341i, 345i, 347i, 448i, 450, 451i
  process, **126**, 156–157, 156i, 246, 247i
  sexual reproduction and, 138i, 139, 140–143, 146, 147i, 169, 169i
  vs. mitosis, 148i
Meissner's corpuscle, 507, 507i
Melanin, 81i, 105, 159–162, 160i, 161i, 162i, 164, 164i, 213i, **478**, 479
Melanocyte, 478
Melanocyte-stimulating hormone, 532t
■ Melanoma, 135, 585
Melatonin, 521i, 530, 533t
Membrane crossing, 485
Membrane crossing mechanisms, 83–85, 83i, 84i, 85i
Membrane protein, 34, 98
Membrane receptor, 206
Memory, **499**, 499i
  cell, 654, 655i
■ drug abuse and, 483
  of immune system, 578
  immunological, 580, 581i
Memory cell, 578, 579i, 580, 581i
Mendel, G., 150–157, 152i, 158
Mendeleev, D., 20, 20i

Mendelian inheritance pattern, 154–157, 154i, 155i, 156i, 157i, 172
Mendelian pattern, 226
Meninges, 495, 497i
■ Meningitis, 495
Menopause, 541, 541i, 575, 630–631, 661
Menstrual cycle, **660**, 660t, 663, 671, 792
  hormone secretions during, 663i
  overview, 660–661
  visual summary, 664i
Mercury barometer, 594, 594i
■ Mercury poisoning, 432, 586
■ Mercury-tainted fish, 751
Meristem, **417**, **419**, 423, 423i, 426i, 427–429, 428i, 452, **454**, 458, 459, 463
Merozoite, 318, 319i
Meselson, M., 193
Mesoderm, **355**, 360, **476**, **648**, 652, 655i, 672, 672i
Mesoglea, 358i, 359, 492i
Mesophere, 754i
Mesophyll, 102, 102i, 297i, 420, **424**–425, 425i, 439i, 442
Mesozoic, 263, 263i, 279i, 343, 346i
Mesquite shrub, 426
Messenger RNA, 47, 194i, 196, **197**, 197i, 198, 200, 200i, 206, 209i, 210i, 223, 522, 523i, 580i
Metabolic acidosis, 638
Metabolic pathway, **77**, 77i, 80, 80i, 89t, 174, 208, 290
Metabolic reaction, 60, 263i, 291
Metabolic water, 97i
Metabolism, **73**, **290**, 424, 464
  diffusion and, 82–83
  nature of, 6, 63, 72, 74–76, 295, 332t
  origin of, 290
Metacarpal, 539i
Metal ion, 80
Metamorphosis, **367**, 370, 370i
Metaphase, 124, 128i–131i, 129, **130**, 131, 137i, 139, 141i, 142i–143i, 145, 145i, 147, 148i, 156–157, 156i, 178i
Metaphase chromosome, 170, 170i
Metastasis, **135**
Metatarsal, 539i
Meteorite, 289
■ Methamphetamine, 503, 503i
Methane, 26i, 32i, 33, 34, 36, 36i, 306, 745i
  early Earth and, 288
  as precursor of life, 289
Methane hydrate deposit, 36, 37, 37i, 307
Methane seeps, 36–37, 36i
Methanocaldococcus jannaschii, 306
Methanogen, 8i, 36, 36i, 37, 297i, **306**, 306i
Methanogenic Archaea, 36–37, 36i, 306, 306i
Methionine (met), 42i, 58, 198, 198i, 623
Methyl bromide, 756
Methylation, 208, 211, 213
Mexico, 699i, 726
Mexico City, 744i
MHC markers, 578, 578i, 585
Micelle formation, **619**
Micrasterias, 323i
Microevolution, 236, 237, **244**, **247**, 258, 272, 283t, 353
Microfilament, 65i, **66**, 66i, 67, 88, 132, 132i, 133i
Microfossil, 294
Micronutrient, plant, **434**
Microphyll, **340**, 341

Micropropagation, 455i
Microscope, 54–55, 54i, 55i
Microspore, 337, 342, 343, 345, 346i, 347i, 448i, **450**–451, 450i–451i
Microtubule, 59, 65i, **66**–67, 67i–68i, 124, 127, 129–131, 129i–131i, 142i–143i
Microtus ochrogaster, 780, 780i
M. pennsylvanicus, 780i, 781
Microvilli, **356**, 357i, **618**, 618i
Microwave, 94i
Midbrain, 496, 496i, 497, 497i, 502t
Middle Ages, epidemics in, 329
Middle ear, 510, 510i
Middle lamella, 68i, 132i
Midge, 446
Midrib, leaf, 425i
Midsagittal plane, 476i
Midwife toad, 787, 787i
Miescher, J., 186
■ Migraine, 329
Migration, **391**, 447, 517, **688**
Milk, 208, 392, 679
Miller, G. T., 701
Miller, S., 289, 289i
Millipede, 367
Mimicry, **12**, **712**, 713i
Mineral, **624**–625, 625t
  in cell function, 88, 434
Mineralocorticoid, 533t
Minimal organism, **234**
Mining runoff, 597
Miocene, 263i, 274i, 314, 395
Missing link, 376–377, 376i, 379
Mississippi River, 274
Mistletoe, 348i, 714
Mite, 369, 369i, 783, 783i
Mitochondria, 51, 53i, 60, **64**, 64i, 65i, 70t, 310, 317i
  aerobic respiration and, 112i
  characteristics, 295, 295i
  origin, 198, 287, 294–295, 295i
■ Mitochondrial disorder, 106–107
Mitochondrial DNA, 270, 276
Mitosis, 124, **126**, 128–131, 130i–131i, 134
■ artificial, 170, 170i
  eight-cell stage, 181, 181i, 192
  eukaryotic, 210, 276, 316, 324, 324i, 325, 325i, 328i, 332t, 354
  plant, 420, 448i, 450, 450i, 454, 455
  vs. meiosis, 148i
Mixture, **23**, **30t**
Model, **11**
Model for creation of living cell, 291i
Model organism, 333
Modification enzyme, **222**
Mojave Desert, 762i
Mold, 315, 316i, 320, 324–325, 325i, 328, 328i, 329, 329i, 333
Mole
  mammal, 266
  skin, 134
Mole-rat, 790, 791, 791i
Molecular biology, beginnings, 220–221
Molecular bond, 24–25
Molecular clock, **271**
Molecular competition, 296i–297i
Molecular model, 34i, 36i
Molecular tool, 222–223
Molecule, **4i**, 22, 23, **30t**, 53i
  of life, **32**, 34–35, 34i
  as unit of organization, 4–5, 4i, 74
Mollusk, 262i, 352, 352i, 353, 354t, 355, 356i, **364**–365, 364i, 365i
  current species diversity, 725i
Molothrus ater, 715, 715i
Molting, 356i, 367, 367i, 368, 368i, 370, 370i, **531**, 531i
Molybdenum, 434t
Monarch butterfly, 447, 691

Mongoose, 481, 481i
Monitor lizard, 388
Monkey, 394, 394i
■ Monoclonal antibody, 583, 585
Monocot
  characteristics, 42, 282, 348, 348i, **350t**, 416, 419, 419i, 423, 423i, 425, 425i, 427i
  development, 422, 426, 428, 452, 457i
  life cycle, 347, 347i, 428
Monocyte, 555i, 576
Monoglyceride, 616t, 617
Monohybrid cross, **154**–155, 154i, 155i, 157
Monomer, **35**, 38, 46
Monophyletic group, **281**, 282, 316, 356
Monosaccharide, 38, 48t, 616t, 617
Monosodium glutamate, 508
Monosomy, 178
Monotreme, **393**, 393i
Monotropa uniflora, 348i
Monsoon, **759**
Montane coniferous forest, 761i, 765
Moore, M., 45i
Moorhead, P., 680
Morgan, T., 182–183
■ Morning-after pill, 666i, 667
Morph, **246**
Morphine, 352
Morphogen, **654**
Morphogenesis, 652i, **653**, 653i, 672i
Morphological convergence, **267**, 267i
Morphological divergence, **266**–267, 266i, 272, 272i, 278, 281
Morphological trait, 246
Morphology, comparative, 239, 242, 245, 246, **266**
Mortensen, V., 149, 149t
Morula, 670i, 679t
Mosaic tissue effect, 213, 213i
Mosquito, 253, 301, 318, 319i, 366, 370i, 375, 375i, 731, 736
Moss, 282, 338, 338i, 339, 339i, 458
Moth larva, **215i**, 446, 449
Mother cell, 450–451
Motile structure, 67, 67i, 317, 318
  invertebrate, 359, 362, 364, 365
Motility. *See* Locomotion
■ Motion sickness, 509
Motor area, 498, 498i
Motor neuron, 475, 475i, 484i, **485**
Motor protein, **66**–67, 66i, 130, 142i
Motor skill, 498, 499
Motor unit, **546**
Moufet, T., 238
Mount Saint Helens, 717i
Mountain, complex zonation, 760i–761i
Mountain ash, 453i
Mountain gorilla, 725i
Mountain grassland, 760i–761i
Mouse, 130i, 139, 139i, 214, 229, 229i, 249, 249i, 269, 269i, 571, 603, 627, 627i, 688–689, 688i, 713, 713i
  cloned, 185, 192, 192i
Mouse-ear cress, 268i
Mouth, 600i, 614i, 615, 615i, 628t
  as first line of defense, 575
  invertebrate, 355, 356i, 363i, 370, 370i
Mouthpart, 375, 375i, 442i
Movement, 536–549
  bones and joints, 540–541
  clostridium botulinum, 548
  contraction of skeletal muscle, 544–545
  energy for contraction, 546

  properties of whole muscles, 546–547
  skeletal-muscular systems, 542–543
  skeletons, 538–539
mRNA, 47, 194i, 196, **197**, 197i, 198, 200, 200i, 206, 209i, 210i, 223, 522, 523i, 580i
MSG, 508
MSH, 532t
mtDNA, 270, 276
Mucin, 615
Mucosa, 618, 618i
Mucous membrane, 574, 575
Mucus, 150–151, 150i
Mudpuppy, 385
Mule, 273
Müller, F., 12
Multicelled organism, **5**, 5i, 356
Multiple allele system, **158**
Multiple births, 644–645, 644i
Multiple fruit, **452**, 452t
■ Multiple sclerosis (MS), 490, 587
Multiplication cycle, virus, 308–309, 309i
Multiregional model, 399
Musca domestica, 280i
Muscle, 355
  aging and, 547
■ exercise and, 547
  intercostal muscles, 600i
  invertebrate, 363, 363i, 366i, 367
  properties of whole muscles, 546–547
  skeletal-muscular systems, 542–543
  types of contractions, 546
Muscle cell, 29, 39, 43, 53, 66, 67, 211, 522–523
Muscle contraction, 62, 66, 67, 85, 544–545, 544i, 545i
  energy for, 546, 546i
■ Muscle cramp, 547
Muscle fatigue, **547**
Muscle fiber, 474, **542**
Muscle spindle, **491**, 491i
Muscle tension, **546**
Muscle tissue, 474
Muscle twitch, **546**, 546i
■ Muscular dystrophy, 173t, 183, 217, 547
Muscular system, 477i
Mushroom, 326, 326i, 327, 327i
Musk ox, 788i
Muskweed, 323i
Mussel, 364, 718–719
Mutant allele, 158, 168, 183
Mutation, **10**, 140, **153**, 248, **256t**, 283t
  accumulation of, 248, 270–271, 292
■ cancer and, 206–207
  causes, 44–45, 160, 246–247
  evolution and, 140, 203
  outcomes, 44, 134, 134i, 135i, 163, 176, 177, 177i, 180, 183, 195i, 212–213, 230, 253, 254, 257, 268–269, 353, 355, 356, 455, 458, 463
  rate, **248**
  spontaneous, 167, 168, 174, 176, 177i, 203
  types, 159, 172, 174, 176, 202, 248, 270, 271, 308
Mutualism, 330, **436**, **706**, 707, 707i
Mutualist species, 316, 436, 446–447
Mycelium, **326**, 327i, 328i
Mycoherbicide, 333
Mycoplasma genitalium, 234, 298
Mycorrhiza, **330**, 331i, 436, 707
Mycorrizal fungi, 348
Myelin sheath, 490, 490i
Myocardium, 560, 560i

Myofibril, 474, **544**, 544i, 545i
Myoglobin, 117, **595**, 595i
Myometrium, 660, 661i
Myosin, 66, 67, **544**, 544i, 545, 545i
Myriapod, 367
Mytilus, 718–719
Myxobacteria, 305
Myxococcoides minor, 59i
Myxoma virus, 721
■ Myxomatosis, 721

N

Nachman, M., 249
NAD^+, 46, 46i, 48t, 76, 80, 89i, **109**
$NADP^+$, 48t, 80, 96, 97i, 99i, 104i
NADPH, 64, 92, 93, 96, 97i, 99i, 100i, 101, 104i
Naive T cell, 580
Naked amoeboid, 317, 317i
Naked seed, 344
Nanoarchaeum equitans, 307, 307i
Nanobe, 287, 287i
Nasal cavity, 600i
Nash, J. F., 166, 166i
Nasutitermes, 790i
■ Natural antiseptic, 339
Natural gas, 37, 745i
Natural killer cell, 555, 555i, **574**, 574i, 584, 585, 590t
Natural selection, **10**, **243**, 256t, 283t, 380
  life histories and, 694–695, 694i, 695i, 703
  theory of, 11, 152, 243, 376–377
  types, 247–253
Natural selection theory, 783
■ Nausea, 509
Nautiloid, 365
Navel orange tree, 454, 454t
Nazca booby, 794, 794i
■ NBD, 166–167
Neandertal, 398
Nearctic, 760i
■ Nearsightedness, 515, 515i
Needle sharing, 501i
Negative feedback, 526
  adrenal cortex and, 526–527, 526i
  bone formation and, 541
  controls, 795
  mechanisms, **408–409**, 408i
  sperm formation, 659i
Negative gene control, **208**, 209i
Neisseria gonorrhoeae, 669
Nelumbo nucifera, 348i, 415i
Nematocyst, **358**, 358i, 365
Nematode, 366
Neoplasm, **134–135**, 134i
Neotropical, 760i
Nephridium, **363**, 363i
Nephron, **634–635**, 635i
Neritic zone, 773i
Nerve, **490**, 493i, 495i
  structure of, 490i
Nerve cell, 66, 85, 299, 354, 359, 360, 363
Nerve cord, invertebrate, 360, 360i, **363**, 363i, 378, 378i
Nerve fascicle, 490i
Nerve net, **492**, 492i
Nervous system, 83ǐ, 174, 477i, **484**
■ disorder, 206
  functional divisions, 493i
  head and, 492–493
  human, 493i
  invertebrate, 359, 360i, 362, 363, 363i, 372
  peripheral system, 493, 493i, 494–495
  types of, 492–493
Nervous tissue, **475**
Nest decorating, 783, 783i
Net ecosystem production, **737**, 737i

Net reproduction per individual per unit time, **688–689**, 688i
Neural control, 482–503
  action potential propagation, 486–487
  chemical synapses, 488–489
  drugging the brain, 500–501
  human cerebral cortex, 498–499
  major expressways, 494–495
  neurons, 484–485
  paths of information flow, 490–491
  synaptic integration, 489
  types of nervous systems, 492–493
  vertebrate brain, 496–497
Neural fold, 672, 672i
Neural groove, 652i
Neural plate, 653
Neural tube, **652**, 652i, 653, 672, 672i, 676
  formation of, 652i
■ Neurobiological disorder, 166–167
■ Neurofibromatosis, 173t, 206, 206i
Neuroglia, 475, 485
■ Neurological disorder, 21i
Neuromuscular junction, 489i, 491i
Neuron, 89i, 128, **475**, 484–485
  circuits and nerves, 490
  communication line, 484i
  direction of information flow, 484i
  functional zones, 484–485
  plasma membrane, 485, 485i
■ Neurotoxic shellfish poisoning, 320
Neurotoxin, 369
Neurotransmitter, 89i, **488**, **520**
Neurotransmitter junction, 488–489
Neutral fat, 40
Neutral mutation, **248**, 271
Neutrality, 28
Neutron, 20, **30t**
Neutrophil, 555, 555i, **574**, 574i, 576, 584, 590t
New Guinea, 792
Newborn, 679t
NF1 gene, 206
Niacin, 624t
Niagara Falls, 86, 86i
Niche, **706**
Nicotine, 500, 592
Nigeria, 698i
Night length, and plants, 462, 464
9+2 array, 67, 67i, 70t, 302
Nirenberg, M., 198
Nitrate, 746, 747
Nitrate reductase, 81i
Nitric oxide, 756–757
Nitrification, **747**
Nitrite, 747
Nitrobacter (bacterial cell), 294i
Nitrogen, 22, 25, 81i, 138, 188, 193, 206, 288, 324, 434t
Nitrogen cycle, **746–747**, 746i
  human impact on, 747
  in seas, 305
Nitrogen dioxide, 757
Nitrogen fixation, 305, 330, 410, **436**, 436i, **746**
Nitrogen-fixing bacteria, 746, 746i
Nitrogen gas, 304
Nitrogen loss, 747
■ Nitrogen narcosis, 609
Nitrogen oxide, 747, 747i, 757
Nitrous oxide, 745i
NK cell, 555, 555i, **574**, 574i, 584, 585, 590t
Noctiluca scintillans, 319i
Node, plant, 418i, 422, 424i, 428
Noncyclic pathway, ATP formation, 99, 100i
Nondirectional change, 717
Nondisjunction, **178–179**, 178i
Nonionizing radiation, **203**

Nonphotosynthetic plant, flowering, 348, 348i
Nonpolar covalent bond, **25**, 25i, 34
Nonpolar hydrocarbon, **40**
Nonshivering heat production, **641**
Nonspecific response, 575t
Nonvascular plant, 334i, 335
  characteristics, 338–339, 350t
  evolution, 336i, 337, 337i
  metabolic products, 339, 339i
Norepinephrine, 489, 500, 501, 521i, 527, 533t, 629
■ Norplant, 666i, 667
North Atlantic Current, 758
Northern coniferous forest, 760i–761i
Northern spotted owl, 692
Northwestern University, 534–535
Nose, as first line of defense, 575
Nostoc, 58i
Notl enzyme, 222
Notochord, 378, 378i, 672, 672i
Notophthalamus, 384i
NSP, 320
Nucelotide, 46–47, 46i
Nuclear division, 125, 127, 138i, 140, 332t
Nuclear envelope, **61**, 61i, 62i, 65i, 130, 131, 131i, 142i, 211, 294, 294i
Nuclear fusion, cell, 324i, 327i, 328i
Nuclear fusion reaction, 288
Nuclear pore, **61i**, 62i, 211
■ Nuclear power, 777
Nucleic acid, 33i, **46–47**, 46i, 48t, 270, 289
Nucleic acid digestion, 616t
Nucleic acid hybridization, **224**, 270
Nucleoid, **52**, 59, 302i
Nucleolus, **61**, 61i, 62i, 65i, 70t
Nucleoplasm, **61**, 61i, 65i
Nucleosome, **127**, 127i
Nucleotide, **188–191i**, 188i, 189i, 191i, 195, 198, 203, 616t, 617
Nucleotide base, 616i, 617
Nucleus
  atomic, 20i, 22, 23, 23i, 24i
  cell, **8**, **52**, 52i, 61, 61i, 142i, 148i, 295i, 317i, 360i, 450, 450i, 451
  function, 20, 51, 62i, 65i, 70t
  origin, 287, 296i–297i
Nudibranch, 364–365
Numerical taxonomy, **303**
Nursing, 792
Nutmeg, 348
Nutria, 719t
Nutrient, **739**
  in cell cycle, 128
  cycling, 74i, 366
  inputs and outputs, 733
  plant, 425, **434**, 434t, 464
  recycling in ecosystems, 732–733
Nutrient flow, model for, 739i
Nutrition, **611**. *See also* Diet; Digestion
  alternative diets, 623
  body-building proteins, 623
  carbohydrate-insulin connection, 622–623
  fats, 623
  human requirements, 622–623
  importance of, 160
  vitamins and minerals, 624–625
  weight and, 626–627
Nutritional labeling, 219
Nymph, insect, 370

O

Oak tree, 255, 315, 328, 424i, 427, 429, 429i, 431
Oat seed, 459i
Ob gene, 627, 627i
Obelia, 359

Obese mice, 627, 627i
■ Obesity, 567, 610–611, **626–627**. *See also* Weight
  genes and, 627
  hormones and, 627
Occipital lobe, 498, 498i
Ocean, 740i, 751, **758–759**, 772–773
  currents, 758, 758i
  diversity of, 772
  food webs, 36, 262, 307, 321
  monsoons, 759
  open ocean, 772
■ pollution, 2, 320, 432
  primary productivity in, 751
  proportion of elements in, 19i
  rain shadows, 759
  upwelling and downwelling, 772–773
Oceanic zones, 773i
Ochoa, S., 198
Octopus, 354t, 365, 512, 512i
Octuplets, 644, 644i
Odor, 780
  in plants, 424i, 429, 449
Offspring (F_1, F_2), 152–155, 154i, 157, 158, 158i, 171, 171i, 183
Ogallala aquifer, 741
Oil gland, 478, 478i
Okazaki, R., 191i
Old age, 679t. *See also* Aging
Old-growth forest, 334, 692, 724
Old man's beard, 330i
■ Older Order Amish, 255
Olduvai Gorge, 397, 397i
Oleic aid, 40i
Olestra®, 12
Olfaction, 508
Olfactory bulb, 508, 508i
Olfactory cue, 508
Olfactory lobe, 496, 496i, 497i, 499, 502t
Olfactory receptor, 508, 508i
Olfactory tract, 499i, 508, 508i
Oligochaete, 362
Oligomer, **224**, 225
Oligosaccharide, 38, 48t, 158
Oligotrophic lake, 769
Olive, 452
■ Olive oil, 623
Ommatidium, 512i
Omnivore, 732, 732i
Oncogene, 134
Onion plant, 454t
Oocyte, **146**, 147i, **650**, 660, 662–663, 662i, 663i, 664i
Oomycote, 316i, **320**, 330
Open circulatory system, **367**, 552, 552i
Operator, **208**, 209i
Operon, **208–209**, 209i
Opossum, 146, 393
Opposable thumb, 395
Optic disk, 513i
Optic nerve, 497i, 513i, 514
Opuntia basilaris, 103i
Oracle of Delphi, 32–33, 32i
Oral cavity, 318, 318i, 600i, 614i
Orange (fruit), 452t, 453
Orangutan, 394
Orbital, **22**, 22i, 23, 23i
■ Orchard tree, 454, 459
Orchid, 140, 252, 280i, 348, 453, 454t, 455
Ordovician period, 263i, 337i, 365i
Organ, **4i**, 140i, 354, 355, **360**, **404**, 469
  designer organs, 475
■ Organ donor, 591, 591i
Organ formation, 648, 648i, 652–653, 653i, 654–655
Organ of Corti, 511, 511i

Organ systems, **4i**, 60, 354, **360**, 360i, 363, **404**, 469, 476–477, 476i–477i
functions, 477i
human skin, 478–479
■ Organ transplant, 151, 231, 578, 585, 591, 591i
Organelle, 51, **60**, 66i, 201, 295, 318i
function, 60, 64, 332t
origin, 294–295, 295i
Organic acid, in humus, 434
Organic compound, **34**, 34i, 48t, 64, 80, 93, 126, 160
early Earth and, 288–289, 293
in plants, 432, 442–443, 443i
Organic litter, **434**
Organic metabolism, 620, 620i
Organization, animal, 357, 358
Orgasm, 665, 792
Oriental, 761i
Ornithonyssus sylviarum, 783, 783i
Ornithorhynchus anatinus, 401i
Oryctolagus cuniculus, 721, 721i
Oryx leucoryx, 414
Oscilloscope, 487i
Osmoreceptor, **506**, 637
Osmosis, **86**, **86i**, **87**, 440i, 443, 632
Osmotic pressure, **87**
■ Osteoarthritis, 541
Osteoblast, 540, 541
Osteoclast, 540, 541
Osteocyte, 540
Osteogenesis imperfecta, 549, 549i
■ Osteoporosis, 541, 541i, 625
Ostracoderm, 380
Ostrich, 146, 238i, 390
Otter, 724
Outer ear, 510, 510i
Output zone, 484i, 485
Oval window, 510, 510i, 511i
Ovary, 521i, 533t, **660**, 660t, 661i
animal, 360i
flowering plant, 346i, 347, 347i, **448**, 448i, 450, 451i
human, 140i, 169, 169i
Ovary wall, 451i, 452, 452i
Ovum, **665**
invertebrate, 359i
Owl, 249, 517, 517i, 692
Ox, wild, 185
Oxalis, 460i
Oxaloacetate, 102i, 103i
Oxidation-reduction reaction, **76**–77, 77i
Oxygen, 64, 82i, 205, 408
aerobic respiration and, 114
in atmosphere, 37, 50–51, 97, 263i, 287–289, 292–294, 304, 349
in blood, 44, 318
diffusion in plants, 102, 421, 424, 425i, 427, 434, 434t, 440
environmental problems, 597
free, 50, 96, 97, 100i, 292–294, 306, 321, 335
in photosynthesis, 95i, 98
structure, 22, 23i, 25, 25i, 31, 83
uptake in animals, 365
Oxygen transport, 604
Oxyhemoglobin (HbO₂), 604
Oxytocin, 409, 521i, 524, 524i, 532t, **679**, 780–781, 780i, 792
Oyster, 364
Ozone, 31
Ozone hole, 756, 756i

Ozone layer, 336, 754i, 756
Ozone thinning, **756**, 756i

P

Paabo, S., 794
Pacific yew, 137
Pacinian corpuscle, 507, 507i
Pain, **507**
Pain receptor, **506**, 507
■ Pain treatment, 352
Paine, R., 718–719
Pair-bonding, 780–781
Pair-rule gene, 215i
Palaeolyngbya, 59i
Palearctic, 761i
Paleocene, 395
Paleomap, 274i
Paleozoic era, 263, 263i, 369
Palisade mesophyll, 103i, **424**, 425i
Palm, 348, 424
Palmer, M., 607
Palp, 370i
Pancreas, 7, 61i, 62, 118, 520, 522–523, 614i, 628t
endocrine cells, 528
exocrine cells, 528
Pancreatic amylase, 616t
Pancreatic cell, 61i, 520
Pancreatic enzyme, 617
Pancreatic hormone, 528–529, 529i
Pancreatic islet, 521i, **528**–529, 533t
Pancreatic nuclease, 616t
Panda, 271i
Pandemic, **310**, 313
Pangea, 263i, **264**, 265i, 393, 393i, 401, 760
Pansy, 348i
Panting, 640
Pantothenic aid, 624t
■ Pap smear, 137, 668
■ Papillomavirus, 137, 668, 668t
Parabasalid, 316i
Paradisaea raggiana, 252i
■ Paralysis, 468
Paramecium, 91i, 318, 318i, 708–709, 709i
■ Paranoid schizophrenia, 166
Paranthropus, 396
Parapatric speciation, **277**
Parapod, 362
Parasite, 89i, 302, 307, 314–316, 360, 368i, 369, **714**–715, 732
in animals, 9, 253, 305, 318–320, 360–361, 361i, 366, 366i, 368
defenses against, 357
good, 371
in humans, 89i, 236, 316–318, 366, 366i
insect, 362
in plants, **326**, 327, 329, 329i, 366
Parasite-host interaction, 714–715, 714i, 715i
Parasitism, **706**
Parasitoid, 705, **714**–715
Parasitoid wasp, 414, 414i, 715i
Parasympathetic nerve, 494–**495**, 494i
Parathyroid gland, 521i, 527i, **528**, 533t, 541
secretions from, 528
Parathyroid hormone, 521i, 528, 533t, 541
Parenchyma, 416, 419, **420**, 420i, 421, 424, 428, 429, 429i, 430t, 440
Parent, true-breeding, 156i, 158, 158i, 183
Parent cell, 125, 129, 131, 133
Parent generation, 153, 154, 171, 171i
Parental behavior, 787, 787i
Parental care, 787, 787i
Parietal lobe, 498, 498i
■ Parkinson's disease, 107, 231, 468

Parrotfish, 4i, 5i, 8
Parthenogenesis, 138–139, 138i, **454**, 454t
Partial pressure, **594**
Parvovirus, 308
Passiflora (passion flower), 461i
Passive immunization, 586
Passive transport, **83**, 83i, 84, 84i, 407
Passive transporter, 57, 57i
Patella, 539i
■ Patenting genomes, 218–219, 218i, 219i, 221
Pathogen, **301**, 303, 306, 308–310
bacterial, 59, 150, 301, 302i, 310
defenses against, 88, 310, 357, 421, 459
fungal, 314, 315, 316i, 329, 329i, 333, 420, 458
human, 151, 232, 236, 302i, 304, 305, 308–311, 316, 371
plant, 305, 309, 315, 320, 330, 333
Pattern formation, **215**, 215i, 217i, **654**–655
Pauling, L., 187, 189, 190
Pax-6 gene, 214–215
PCR, 221, **225**, 225i, 286, 307
Pea plant, 129, 151–154, 152i, 154i, 436, 452
Peach fruit/pit, 421
Peach tree, 329
Pear fruit, 420i, 421
Peat bog, 81i, **339**, 339i, 342, 434–435, 742
Peat moss, 339, 339i
Pectin, 68, 420, 438i, 440, 464
Pectoral girdle, 539i
Pectoralis major, 543i
■ Pedigree, family, **172**, 172i, 173i, 175, 175i, 181
Pelagic province, 773i
Pellide, 316, 317i, 318
Pelvic cavity, 476i
■ Pelvic inflammatory disease, 668
Penguin, 40i, 41, 93, 248, 266, 266i, 310, 686i, 784i
Penis, 169i, 519, 656, 656i, 657i, 658i, 665
invertebrate, **360**, **360i**, 368
Pepo, 452t
Pepper plant, 348
Pepsin, 81, 616, 616t
Pepsinogen, 616
■ Peptic ulcer, 81, 302i, 471, 629
Peptide, 459, 617
Peptide bond, 42i, 43, 200, 200i, 352
Peptide hormone, **522**–523, 522i, 523i
Per capita rate, **688**
Peregrine falcon, 10, 10i, 725i
Perennial plant, **428**, 464
Perez, I., 482i
Perfect flower, **449**
Perforation plate, plant, 438i
Pericardium, 560, 560i
Pericycle, 426i, 427, 427i, 437i
Periderm, 421, 429, 429i, 430t
Periodic table, **20**, 20i
Peripheral nervous system, 493, 493i, 494–495
Peripheral vasoconstriction, **640**
Peripheral vasodilation, **640**
■ Peritoneal dialysis, 638, 638i
Peritoneum, 355, 355i
Periwinkle, 718i, 719
Permafrost, **766**
Permian mass extinction, 37
Permian period, 263i, 337, 337i, 343, 392
Permian-Triassic boundary layer, 261i
Peroxide, 134

Peroxisome, **63**
Perpetual ice cover, 760i–761i
Peru Current, 773
Pest resurgence, 248
■ Pesticide resistance, 248
■ Pesticides, 231, 371, 375, 446, 519, 530–531, 530i, 736, 736i
Pests, defenses against, 459
PET scan, 21, 21i
Petal, 346i, 448, 448i, 449, 463, 463i
Peter the Great, 329
Petiole, 424, 424i, 425i
PGA, 101, 101i, 102, 102i
PGAL, 101, 101i, 104i, 110, 118
pH, 81, 81i, 82, 83, 304, 339, 359
of blood, 28i, 29
extracellular, 638
of rainwater, 29, 757
of seawater, 28, 28i
of vaginal fluid, 657
pH scale, **28**, 28i, 29
Phagocyte, 302, 317, 357i, 555, 575t, 576
Phagocytic cell, 412
Phagocytosis, **88**, 89i
Phagocytotic vesicle, 89i
Phalange, 539i
Pharyngeal arch, 672, 672i
Pharynx, **360**, 360i, 362, 363i, 366i, **378**, 378i, **601**, 614i, **615**, 628t
Phaseolus vulgaris, 425i, 457i
Phasmid, 729, 729i
Phenotype, **153**, 155i, 158–161, 163, 171, 171i, **246**, 248, 250, 252, 255i
Phenotypic ratio, 154i, 155, 155i, 156i, 157i, 159i, 160, 171, 183
Phenylalanine, 198i, 623
■ Phenylketonuria, 173t, 180–181
Pheromone, 508, **520**, 646, 778–779, 778i, **784**
human, 792
priming, 784
sex, 252
signaling, 784
Phidippus, 596i
Philadelphia chromosome, 177, 177i
Philetairus socius, 251, 251i
Phloem, 336, 341, 406, 416, 419, **421**, 430t, 458t
functions, 433, 437, 437i, 439i, 442–443, 442i, 443i
primary, 420–427, 420i–427i, 428i
secondary, 428, 428i, 429, 429i
Phlox, 692t, 693
Phorid fly, 705
Phosphate group, 34i, 46, 75, 75i, 76, 85, 85i, 101, 188–191, 211, 291, 319
Phosphate-group transfer, 85, 85i
Phosphate ion, 748, 748i
Phosphatidylcholine, 48t
Phosphoglycerate, 101, 101i, 102, 102i
Phospholipid, **41**, 41i, 48t, 50i, 56, 56i
Phosphorus, 625t
Phosphorus cycle, **748**–749, 748i
Phosphorylase kinase, 523i
Phosphorylated glucose, 118
Phosphorylation, **75**, 85, 85i, 134, 211
Photoautotroph, 105, 292, **302**, 304, 304i, 316, 317, 321, 323, 330, 335, 359
Photochemical smog, 756–757
Photolysis, 98, 100i
Photon, **94**, 98, 98i–99i, 100i, 514
Photon energy, 98, 98i, 512
Photoperiodism, **462**, 462i
Photoreceptor, **506**, 512, 512i, 514, 514i

Photosynthesis, 92, 95i, 105, 737
 chemistry of, 23i, 77, 77i, 92i,
  96–97, 96i, 97i, 104i
 cyclic pathyway, 292,
  296i–297i, 349
 discovery of, 21, 262, 323
 mechanisms, 441, 441i, 457i, 460,
  466
 nature of, 74, 92, 93, 424, 425, 425i
 noncyclic pathway, 287, 292, 294,
  296i–297i
 origin, 263i
 oxygen-releasing pathway, 50,
  302, 335
Photosynthesizer, 58i, 92, 93, 314,
  316, 317i
Photosynthetic cell, 64, 64i, 69i, 74,
  96i, 287, 295, 321, 421i, 424, 439i,
  441, 443
Photosynthetic pigment, 70t, 92, 94,
  95i, 460
Photosynthetic product, 433, 436,
  440, 442, 458t, 459
Photosynthetic species, 70t, 263i,
  320, 330i, 333, 333i
Photosystem, 96, 97i, 98–99,
  98i–100i
Phototropism, 460, 460i
Photovoltaic cell, 755, 755i
Phrynosoma cornutum, 704
Phthalate, 534
Phycobilin, 95, 105, 322
Phyletic constraint, 655
Phylogeny, 280, 281, 282, 303, 316
Phylum, 282
Physalia utriculus, 358i, 359
Physarum, 325i
Physical access, and lineage, 279
Physical constraint, 655
Physiological trait, 246
Physiology, 403
Phytochrome, 411, 462, 462i
Phytophthora, 330
 P. infestans, 315, 320
 P. ramorum, 315, 320
Phytoplankton, 368, 737, 756,
  768–769, 772, 773, 775
Phytoremediation, 432–433, 432i
Picloram, 460
■ PID, 668
Pig, 231, 231i, 366, 591, 591i
 cloned, 185
Pig virus, 231
Pigeon, 10i, 246, 390i, 706i
■ Pigeon breeding, 10
Pigment, 514
 accessory pigments, 94, 95i, 105
 cartenoid, 333
 light-activated, 307
 photosynthetic, 70t, 92, 94, 95i, 460
 respiratory, 595, 604
 screening pigment, 512i
 visual, 105
Pigment molecule, 94–95, 95i, 97i,
  98, 98i, 105, 471i, 512
Pike-cichlid, 694–695, 694i, 695i, 703
Pillbug, 368
Pillotina, 305
Pilomotor response, 641, 641i
Pilus (pili), 58i, 59, 302, 302i
Pine tree, 146, 343i, 344–345, 344i,
  345i, 429, 431
Pineal gland, 496i, 497, 497i, 502t,
  521i, 530, 533t
 daylength and, 530
Pineapple (fruit), 452, 452t
Pinna, 510, 510i
Pinus contorta, 436i
 P. longaeva, 244i, 344i, 410
Pinworm, 366
Pioneer species, 716, 716i, 717i
Pisaster ochraceus, 718–719

Pisum sativum, 129, 151–154, 152i,
  154i, 436, 452
Pith, 422i, 423, 423i, 427i
■ Pituitary dwarfism, 525,
  534–535, 535i
■ Pituitary gigantism, 525, 525i
Pituitary gland, 496i, 502t, 521i,
  524–525, 524i, 637, 637i, 659i
■ abnormal outputs, 525
 anterior lobe secretions,
  524, 532t
 hormones released from, 532t
 intermediate lobe secretions, 532t
 posterior lobe secretions, 524,
  524i, 532t
■ PKU, 173t, 180–181
Placenta, 181, 671, 671i, 673, 673i,
  678, 678i
Placental mammal, 212, 279i, 392i,
  393, 393i
Placoderm, 380, 380i, 381
Placozoan, 356, 356i
Planarian, 360, 360i
Plankton, 317
Plant. See also specific names
 characteristics, 9, 9i, 39, 92, 332t
 commonalities with animals,
  402–414
 current species diversity, 725i
 death, 428, 464
 evolution, 263i, 282i, 297i,
  336–337, 336i
 life cycle, 139, 146, 146i,
  336, 336i
 as photosynthesizer, 93, 96, 101
 products, 421, 455, 459, 719
 reproduction, 138, 140i, 146, 228i
 stress responses, 414, 414i
 structure, 41, 69i, 94, 416–429
 water uptake in, 27, 436, 438–440,
  438i, 439i
Plant cell, 29, 52i, 53i, 54, 60i, 67, 69,
  70t, 87, 132i, 456
 components, 39, 39i, 63, 68, 68i, 86
Plant growth
 fungi and, 330, 330i, 331, 331i
 nutrients for, 434–435, 434t
 primary, 418i, 419–423, 420i–423i
 secondary, 428–429, 428i, 429i
 soil composition for, 434–435, 434t
Plant research experiment, 442,
  459i, 462, 464, 464i
Planula, 358i, 359, 360, 492
Plasma, 408, 473, 554, 554i
Plasma cell, 582, 582i
Plasma membrane, 287, 294, 294i,
  302, 307
 in cytoplasmic division, 132–133,
  132i, 133i, 142i
 diffusion, 83, 84i, 86, 88, 88i, 437,
  440, 440i, 442
 function, 50i, 52, 52i, 53, 56i,
  57–59, 57i, 63i, 65i, 68i, 70t, 303,
  303i, 307
 origin, 290–291, 291i
Plasma protein, 73
Plasmid, 59, 222, 222i, 223i, 228i,
  229, 302
Plasmodesma, 65i, 68i, 69, 424
Plasmodesmata, 412
Plasmodial slime mold, 324–325, 325i
Plasmodium, 253, 318, 319i
Plasmolysis, 440i
Plasmopara viticola, 320
Plastid, 64, 318, 322, 461
Platanthera, 447, 447i
Plate tectonic theory, 264–265
Platelet, 554i, 555, 555i, 566
Platypus, 392i, 401, 401i
Play bow, 784, 784i
Pleiotropy, 159
Pleural membrane, 600i, 601

Pleurisy, 601
Pliocene period, 393
Plutonium, 194
Pneumocystis carinii, 588, 588i
■ Pnuemonia, 186, 329
Podocarp, 344
Podocoryne, 492
Poecilia reticulata, 694–695, 694i,
  695i, 703
Poison, 66
Polar bear, 26i, 244–245, 244i, 271i,
  641, 641i, 725i
Polar body, 147i, 662i, 663, 665
Polar covalent bond, 25, 25i
Polar ice, 740i, 745
Polar ice shelf, 751, 751i
Polar molecule, transport of, 83, 84
Polarity, 26, 26i, 83
Polio, 3, 125
Pollen grain, 153, 337, 343–345, 343i,
  347i, 419i, 448–450, 449i, 454
Pollen sac, 345i, 347i, 448, 450, 450i
Pollen tube, 345i, 347i, 448, 450,
  450i, 454
Pollination, 272i, 273, 343, 347i,
  350t, 450
 plant strategies, 449, 449i
Pollinator, 337, 343, 344, 346, 348,
  348i, 371, 446, 448, 449, 449i
Pollutant, 756–757
■ Pollution. See also Air pollution
 cleanup, 432–433, 432i
 in estuaries, 432
 in groundwater, 432–433,
  432i, 460
 in oceans, 2, 320, 432
 soils, by toxic metals, 432
 in waterways, 320, 432
Poly-A tail, 197, 211
Polychaete, 354t, 362, 362i
■ Polycythemias, 556
PolydactylyI, 172i, 173t
Polyhedral virus, 308i
Polymer, 35
Polymerase chain reaction (PCR),
  221, 225, 225i, 286, 307
Polymorphism, 246
Polyp, 358, 358i, 359, 359i
Polypeptide, 42i, 43, 43i, 44i, 58, 62,
  66, 66i, 79i, 90, 209i, 302, 616
Polypeptide chain, 194, 195, 198,
  200, 201, 204i
Polyploidy, 178–179, 179i, 276–277,
  276i, 277i
Polysaccharide, 38, 39, 48t, 59, 69,
  302, 420, 472
Polysome, 201
Polytene chromosome, 211, 211i
Polytrichum, 338i
Pome, 452
Ponderosa pine, 344–345, 345i
Pons, 496, 496i, 497i, 502t
Pool, 769, 769i
Poplar tree, 424i, 432–433, 432i
Poppy plant, 138i, 427, 452t,
  703, 762i
Population, 5, 5i, 246, 272, 684–703
 characteristics of, 686
 characteristics of populations, 686
 counting, 687
 fertility rates and age structure,
  698–699
 gene pool, 237i
 life history patterns, 692–693
 life tables, 692
 natural selection and life
  histories, 694–695
 patterns of survival and
  reproduction, 692–693
 variation in, 252–255
Population density, 686, 687
Population distribution, 686, 686i

Population growth, 666, 686,
  696–697
 biotic potential, 689
 carrying capacity and logistic
  growth, 690–691
 community structure and, 706
 demographic transitions, 700
 density-independent factors, 691
 economic effects, 700–701, 700i
 effects of, 242, 303, 334, 349
 exponential growth, 688–689
 gains and losses in, 688
 immigration policies, 700
 impacts of no growth, 701
 limits on, 690–691
 resource consumption, 700–701
 from zero to exponential growth,
  688–689
Population size
 community and, 706
 gains and losses, 688
Populus, 424i, 432–433, 432i
 P. tremuloides, 454, 455i
 P. trichocarpa x deltoides,
  432–433, 432i
Porcupine, 407
Pore, plant, 419i
Poriferan, 354t
Pork, 361
Porphyra tenera, 322, 322i
■ Porphyria, 480–481, 480i
Porphyrins, 480
Porpoise, 266, 266i
Portuguese man-of-war, 358i, 359
Positive feedback, 486, 526
Positive feedback mechanism, 409
Positive gene control, 208
■ Post-partum depression, 678
Post-reproductive age, 686
Postelsia, 321i
Posterior, 354, 354i, 476i
Postindustrial stage, 700, 700i
Postnatal development, 679,
  679i, 679t
Postsynaptic cell, 500
Postzygote isolatiing mechanism,
  272, 273i
Potassium, 29, 434t, 625t
Potassium ion, 85, 434, 440,
  440i, 485
Potato plant, 315, 320, 454t, 466
Pouched mammal, 393
Poverty, 726–727
Power grip, 395, 395i
Power plant, fuel for, 29i, 339
Poxvirus, 308
Prairie
 shortgrass, 762–763, 763i, 773, 773i
 tallgrass, 724, 733, 733i, 734i,
  762–763
Prairie dog, 788, 788i
Prairie white-fringed orchid,
  447, 447i
Praying mantis, 713i
pre-mRNA, 211
Pre-reproductive age, 686
Precision grip, 395, 395i
■ Precocious breast development,
  534, 534i
Precursor, 38
Predation, 706, 718–719, 718i
Predator, 314, 316, 318, 318i, 319,
  352, 356–358, 360, 363, 363i, 365,
  368i, 369, 371, 373, 424i, 710
 adaptive responses of, 713, 713i
 defenses against, 356, 364–365,
  367, 368, 372, 373, 388
 natural selection and, 694–695,
  694i, 695i, 703
Predator-prey interaction, 710–711
Predatory colony, 305

Prediction, **11**
Prefrontal cortex, 498i, 499
Pregnancy, 179, 665, 665i
  alcohol, tobacco, and other drugs, 677
  cleavage and implantation, 670
  ■ contraception, 666–667, 666i
  cyclic changes in ovary, 662–663, 662i
  cyclic changes in uterus, 663
  emergence of distinctly human features, 674–675, 674i–675i
  emergence of vertebrate body plan, 672, 672i
  extraembryonic membranes, 670–671
  fertilization and, 665, 665i
  formation of early embryo, 670–671, 670i–671i
  giving birth, 678, 678i
  infectious diseases and, 676–677
  ■ mercury-tainted fish and, 751
  nourishing the newborn, 679, 679i
  nutritional considerations, 676
  placenta, 673, 673i
  preparations for, 662–663
  Rh blood typing, 557, 557i
  seeking or ending, 667
  sexual intercourse and, 665
  unplanned, 666i
  ■ Pregnancy test, 630, 671
Prehensile movement, 395
■ Preimplantation diagnosis, 181
Preindustrial stage, 700, 700i
Premature birth, 678
Premotor cortex, 498
Prenatal development, 679i, 679t
■ Prenatal diagnosis, 167, 179, 180i, 181
Prescription drugs, and pregnancy, 677
Pressure, 507
Pressure flow theory, **442**
Pressure gradient, **83**, 87, 433, 442, 594
  respiration and, 605, 605i
Prevailing winds, 754, 754i
■ Previn, 667
Prey, adaptations of, 712–713, 712i, 713i
Prey density, 710, 710i
Prezygotic mechanism, 272–273, 273i
Primary growth, plant, **417**, **418i**, 419, 420, 422i
■ Primary immune deficiency, 587
Primary leaf, 457i
Primary motor cortex, 498, 498i
Primary oocyte, **662**
Primary producer, **732**
Primary productivity, **737**, 737i, 751
  gradients and, 761i
  soil and, 767
Primary root, 418i, 422, 426, 456, 456i, 457i
Primary shoot, 422, 423, 456i, 458, 461
Primary somatosensory cortex, 498, 498i, 507i
Primary structure, plant, 422–427, 422i, 423i
Primary succession, **716**, 716i
Primary tissue layer, 354–355
Primary wall, in plant cell, **68**, 68i, 69
Primate, **394**, 394i, 724
  chromosome structure, 129, 177
  evolution, 268i, 269, 394–395
  origins and early divergences, 395
  skeleton, 394i
  traits, 394–395
Primer, **225**, 226, 227

Priming pheromone, 784
Primitive streak, 672, 672i
Prion, 311, **312**, 312i
PRL, 521i, 524, 532t, **679**
Probability, **154–155**
Probe, **224**, 232
Probe hybridization technique, 224, 224i
■ Probiotic feed, 729
Procambium, 418i
Procamelus, **275i**
Producer, **6–8**, 7i, 74i, 92
Product, **76**, 77, 78i, **89t**
Proerythroblast, 555i
Profundal zone, 768, 768i
■ Progeria, 173t
Progesterone, 521i, 533t, **661**, 663, 664i
Progestin, 667
Proglottid, 361, 361i
Programmed life span hypothesis, 680
Prokarya, 280
Prokaryote, **8**, 292
  characteristics, 8i, 9i, 58i, 286, 287, 332t
  origin, 268i, 296i–297i
Prokaryotic cell, **50**, 51, 51i, 52i, 86, 124, 263i, 291, 292i, 294, 300, 302–303, 302i, 302t, 357i
  components, 58–59, 58i, 70t, 204i
Prokaryotic fission, **303**, 303i, 309i, 332t
Prokaryotic growth rate, 303
Prolactin, 521i, 524, 532t, **679**
Proline, 45i, 198i, 202i
Promoter, **196**, 196i, **208**, 209i, 212, 214
Pronghorn antelope, 613, 613i
Prop root, 457i
Propagation, vegetative, 454, 454t
■ Property rights, 777
Prophase, 124, 128–**130**, 130i, 141i, 142i–143i, 144–145, 144i, 147
■ Prophylactic mastectomy, 207
Prosimian, 395, 395i
Prostaglandin, 528, 577, 656, 663
■ Prostate cancer, 657
Prostate gland, 179, 656–657, 656i, 657i, 658i
Prostate-specific antigen, 657
Protease, 81
■ Protease inhibitor, 589
Protein
  energy from, 118
  function, 6, 42–43, 61, 73, 78, 126, 159
  gene expression and, 186, 210, 270
  organic metabolism, 620, 620i
  origin, 289, 290
  structure, 33i, 42–45, 43i, 44i, 53i, 290, 290i
  synthesis, 42i, 43, 77, 80, 128, 291, 459
  types, 34, 48t, 49, 49i, 56, 56i, 57, 57i, 59, 66, 194
Protein cap, and amoebas, 66
Protein coat, 308
Protein digestion, 616t
Protein kinase, 522, 523i
Protein lattice, 59i
Protein synthesis, 248
  mutation during, 202–203, 202i, 203i
  transcription stage, 196–197, 196i–197i, 204i
  translation stage, 198–201, 198i–199i, 200i–201i
Proteobacteria, **305**
Proterozoic eon, 263, 263i, 292, 293, 293i

Protist
  characteristics, 8, 8i, 68, 70t, 91, 92, 138, 280, 280i, 314–323, 332t, 356, 356i
  evolution, 263i, 282, 282i, 295, 295i, 297i, 316i
  functions, 316, 319–322
  major groups, 9i, 17i, 314–323, 316i, 317i, 318i, 319i, 321i, 322i, 323i
Protista, kingdom, 91, 316, 316i
Proto-cell, 263i, 286, 286i, 287, **290**, 291, 291i
Proto-oncogene, **206**
Protoderm, 418i
Proton, 20, 22, **30t**, 307
Protonephridium, 360, 360i, 375, 375i
Protoplast, **455i**
Protosome, 353i, 355, 356i, 375
Protozoan, **89i**, **316**, 317, 317i, 318
  current species diversity, 725i
Proximal, 476i
Proximal tubule, **635**, 635i
Prunus (cherry tree), 95, 448i, 450, 451i, 452, 452t
Pryor, G., 161i
PSA, 657
Pseudocoel, **355**, 355i
Pseudocoelomate animal, 355i
Pseudomonas marginalis, 58i
Pseudopod, 67, 88, 89i, 317, 317i
Psilophyton, 336, 336i, 340i
Psychedelics, 501
Psychoactive alkaloid, 327
Psychoactive drug, 500–501
Pterosaur, 266i
PTH, 521i, 528, 533t, 541
Puberty, **530**, **656**, 661
  ■ early onset, 519
Pubescent, 679t
Public health, 195, 206, 207, 313
■ Public land, 287, 298
Pueraria montana, 720, 720i
Puffball mushroom, 327
Pulmonary artery, 558i, 560i
Pulmonary capillary, 600i
Pulmonary circuit, **553**, 558, 559i
Pulmonary vein, 558i, 560i
Punctuation model of speciation, **278**
Punnett-square diagram, **155**, 155i, 156i, 169i, 247i
Pupa, insect, 370
Pupal stage of moth, 6i
Pupil, 512, 513i
Purine, 188
Puriri tree, 266
Purple coral fungus, 326i
Pyrenestes ostrinus, 251, 251i
Pyrimidine, 188
Pyrococcus horikoshii, 313
Pyruvate, 76, **108**
Python, 516, 516i, 629, 629i

Q

Quadrat, **687**
Quadriceps femoris, 543i
Quaking aspen, 454, 455i
Quamman, D., 243i, 723
Quantitative reporting, 13
Quaternary structure, 90
■ Queen Victoria, 175i
Quercus, 255, 315, 328, 424i, 427, 429, 429i, 431

R

Rabbit, 162, 162i, 721, 721i
■ Rabies, 586
Racoon, 271i
Radial animal, 352i, 354, 372
Radial cleavage, 651, 651i

Radial section of plant, **420i**
Radial symmetry, **354**, 354i, 358, 492
Radiant energy, 94
■ Radiation, 585. *See also* Adaptive radiation
  cosmic, 134, 288, 292
  infrared, 94i, 105, 292
  ■ ionizing, 203
  nonionizing, 203
  solar, 754i, 755i
  thermal, 639
  UV, 94i, 95i, 105, 293, 336
■ Radiation poisoning, 137, 202i
Radio wave, 94i
Radioactive decay, **21**
Radioactive fallout, 777, 777i
Radioisotope, **21**, **30t**, 187i, 262, 262i
Radiolarian, 317, 317i
Radiometric dating, 376–377
Radius, 539i
Radula, **364**, 364i, 365i
Radura, 313i
Ragweed, 449i
Rain forest, 446, 727, 760i–761i, 764, 764i, 767i
Rain shadow, **759**, 759i
Rainfall, 754–755
Rainwater, pH of, 29, 757
Rana pipiens, 518–519, 649i
Randall, M., 230
Rangifer tarandus, 710i
Ranunculus, 419, 420i, 426i
Raspberry (fruit), 452, 452t
Rat, 236–237, 237i
Rat liver cell, 60i
Rat somatotropin, 229
Rattlesnake, 388, 389i, 639, 639i
Rattus, 236–237, 237i
Raven, 300
Ray, **372i**
Ray-finned fish, 383
Reactant, **76**, 77, 78i, **89t**
Reaction chamber, 288i
Reaction rate, 81
Realized niche, 706
Rearrangement reaction, **35**, **79**
Receptade, 448, 448i, **452**, 453i
Receptor, **7**, 29, 408i, 409i, 507, 520
Receptor-mediated endocytosis, 88
Receptor protein, 57, 57i, 59
Recessive allele, 151, 153–155, 154i, 155i, 156i, 158, 159, 183, 213
Reciprocal cross, **183**
Recognition protein, 49, 49i, 57, 57i, 135
Recombinant DNA, **220**–222, 222i, 223i, 309i
Recombination, genetic, **168**, 168i, 171
Rectum, 366, 614i, 621, 628t
Rectus abdominis, 543i
■ Recycling, tax breaks for, 335, 350
Red alga, 105, 105i, 287, 292i, 293, 297i, 307i, 316i, 319, **322**, 322i
Red blood cell, 4, 44, 87i, 128, 158, 202i, 210, 317i, 319i, 473, 554–555, 554i, 555i, 564
  ■ malaria parasite and, 318, 319i
■ Red blood cell disorder, 556
Red buckeye tree, 424i
Red fox, 732, 732i
Red grape, 95
■ Red-green color blindness, 515
Red marrow, **540**
Red oak, 429i
Red-spotted salamander, 384i
Red tide, 320
Redox reaction, **76**–77, 77i
Redwood forest, 334
Redwood tree, 9i, 53i, 108i, 315, 320, 334, 344, 429, 438
Reeve, C., 468, 468i

Reflex, 490–491
Regional climate change, 349, 416–417, 759
Regulatory protein, 206i, **208**, 211, 212
Reindeer, 691i
Reiskind, J., 161i
■ Rejection, 578, 585
Relaxin, **678**
Release factor (protein), 201
Releaser, **524**, 525i
Renal artery, 558i, 634, 635i
Renal capsule, 634, 635i
Renal corpuscle, **634**
■ Renal failure, 638
Renal pelvis, 634, 635i
Renal vein, 558i, 634, 635i
Renin, 637
Replacement level, 698
Replication enzyme, 222, 223
Repressor, **208**, 209i
Reproduction, **6**, 272, 291. *See also* Asexual reproduction; Sexual reproduction
   flowering plant, 126, 140i, 346, 346i, 347, 357, 446, 448–449, 448i, 449i, 454–455, 454t
   fungi, 327i, 328i
   human, 647i. *See also* Reproduction, animal
   patterns, 692–693
   plant, 338i, 341i, 345i
   prokaryotic, 303
   protist, 319i, 322i, 324i, 325i
Reproduction, animal, 644–682, 647i. *See also* Pregnancy
   aging and death, 680
   birth, 678–679
   cell differentiation, 652–653
   cleavage, 650–651, 670
   cleavage patterns, 651
   control of human fertility, 666–667
   cyclic changes in ovary and uterus, 662–663
   egg, 650
   embryonic induction, 654
   emergence of human features, 674–675
   emergence of vertebrate body plan, 672
   evolutionary constraints on development, 655
   extraembryonic membranes, 670–671
   fertilization, 665
   formation of early embryo, 670–671
   formation of specialized tissues and organs, 652–653
   implantation, 670
   menstrual cycle, 664
   morphogenesis, 653
   mother as provider, protector, potential threat, 676–677
   nourishing newborns, 647, 647i, 679
   pattern formation, 654–655
   placenta, 673
   postnatal development, 679
   sexual intercourse, 665
■ sexually transmitted diseases, 668–669
   sperm formation, 658–659
   stages of, 648–649, 648i, 649i
Reproductive age, 686
Reproductive base, **686**
Reproductive isolating mechanism, **272**–273, 272i, 273i, 274
Reproductive isolation, 272–273
Reproductive organ, 140, 140i, 360i, 365i, 366, 422
   human, 169i

Reproductive success, 680, 692–693, **783**
   plant, 429
Reproductive system, 477i
   human females, 660–661, 660i, 660t
   human females, components, 660, 661i
   human females, menstrual cycle, 660–661
   human males, 656–657, 656i, 656t
■ human males, cancers of prostate and testes, 657
   human males, components, 657i
   human males, gonad development, 656
   human males, structure and function, 656–657
Reproductive timing, 646
Reptile, 381i, **386**, 386i, 388–389
   birds and, 376–377, 376i
   body plan, 388, 388i
   brain, 497i
   characteristics, 354, 354t
   origin, 263i
   reproduction, 139
   respiration, 599i
Resin, 429
Resource consumption, 700–701, 795
Resource partitioning, **709**, 709i
Resources
   allocating when scarce, 73, 91, 137, 417, 431
   competition for, 417, 429
Respiration, 405, 405i, 592–609, **594**
   balancing air and blood flow rates, 605
   carbon dioxide transport, 604–605
■ environmental issues, 597
   evolution of paired lungs, 598–599, 599i
   exchanges at respiratory membrane, 604
   gas exchange, 594–595
   gas exchange and transport, 604–605
   human, 600–601
   invertebrate, 596–597
■ lung disorders, 606–607
   nature of, 594–595
   oxygen transport, 604
   respiratory cycle, 602–603
   vertebrate, 598–599, 598i, 599i
■ Respiratory acidosis, 29
Respiratory cycle, **602**–603
   cyclic reversals in air pressure gradients, 602–603
■ high altitude and, 603, 603i
■ Respiratory disorder, 329, 757i
Respiratory membrane, **604**, 604i
Respiratory organ, invertebrate, 364, 365, 367
Respiratory pigment, **595**, 604
Respiratory surface, **594**
Respiratory system, 477i
   circulatory system and, 558–559, 559i
   digestive system and, 612i
   organ systems and, 594, 594i
   urinary system and, 632i
Respiratory system, human, 600–601
   breathing, 601, 601i
   components, 600i
   functions, 600–601, 600i
Responsiveness to change, **7**
Resting membrane potential, **485**
Resting spore, 304i, 324, 336
Resting structure, 304
Restriction enzyme, **222**, 222i, 223

Restriction fragment length polymorphism, 226
Resurrection plant, 340
Retina, **512**, 512i, 513i, 514, 514i, 515i
Retinal stimulation, 513i
Retinol, 105
Retrovirus, 309
Reverberating circuit, 490
Reverse transcriptase, **223**
Reverse transcription, 309
Reznick, D., 694–695, 694i
RFLP, 226
Rh blood typing, **557**, 557i
Rhea, 238i
Rhesus monkey, 234, 234i, 270
■ Rheumatic heart disease, 571
■ Rheumatoid arthritis, 541, 587
Rhinoceros, 725i
Rhizobium, 305, 436i
Rhizoid, 328i, 338, **338**
Rhizome, 340, 341, 341i, 342, 454t
Rhizophora, 770i
Rhizopus stolonifer, 328, 328i
Rhodopsin, 514
Rhubarb, 95, 466
Rhyniophyte, 336, 337i
■ Rhythm method, 666, 666i
Rhythmic leaf folding, 411
Rib, 539i
Rib cage, 539i
Ribeiroia, 385i
Riboflavin, 624t
Ribonucleic acid. *See* RNA
Ribonucleotide, 196, 199
Ribonucleotide monomer, 47
Ribose, 38, 75, 75i, 196
Ribosomal RNA, **196**, 200, 207, 270, 291, 306, 307
Ribosome, **52**, 59, 60, 61, 291
   function, 62i, 70t, 194, 195i, 196, 198, 200–201
   structure, 199, 199i
Ribulose biphosphate, 101, 101i
Rice plant, 229, 348, 454t, 455, 458
Ricin, 194–195, 195i, 585, 713
Ricinus communis, 194–195, 194i, 195i
■ Rickets, 528, 528i
Riffle, 769, 769i
Riparian zone, **727**, 727i
Ritualization, 784–785
River, 740i
River rock, and bacteria, 59
Rivulus hartii, 694–695, 694i, 695i, 703
RNA, 47, 48t, 62i, 64, 70t, 78, 89t, 208–209, 211
   synthesis, 195–199, 196i–199i
RNA polymerase, **196**–197, 196i–197i, 208, 209, 211
RNA virus, 309
RNA world, **291**, 309
Roanoke Island colony, 445
Robinia, 424i
Rock pocket mouse, 249, 249i
Rockefeller University, 627i
Rocky coastline, 770i, 771
■ Rocky Mountain spotted fever, 369
Rod cell, **514**, 514i
Rodenticide, 236–237
Rodhocetus, 257, 257i
Rodinia, 293
Roe, 373
Rooster, 794, 794i
Root, **418**, 420
   function, 336, 341, 417, 418, 426–427, 426i, 427i, 432, 437, 443i, 454
   primary structure, 426–427, 426i, 427i

structure, 418i, 426–427, 426i, 428
   water/nutrient uptake in, 432, 437, 437i
Root cap, 418i, 426i, 427, 461, 461i
Root hair, 418i, 426i, **427**, **436**, 436i, 439i
Root nodule, **436**, 436i, 457i
Root system, **418i**, 454
Root tip, 417, 418i, 419, 420, 426i, 427, 452i, 458i
Rose plant, 329, 429, 454t
Rotational cleavage, 651, 651i
Rotifer, 375, 375i
Rough ER, 62, 62i, 65i
Round window, 510, 510i, 511i
Roundup, 460
■ Roundup Ready plant, 460
Roundworm, 146, 270, 354t, 355, 356i, **366**, 366i, 680, 714i
   current species diversity, 725i
■ Roundworm infection, 565
rRNA, **196**, 200, 207, 270, 291, 306, 307
■ RU 486, 667
Rubber cup fungus, 326i
■ Rubella, 676–677, 682
Rubisco (RuBP), **101**, 101i, 102, 102i, 104i, 105
Ruffini ending, 507, 507i
Rufous hummingbird, 447, 708
Ruminant, **613**, 613i
Run, 769, 769i
Runner, plant, 454, 454t
Runoff, **434**, 435
Russian imperial family, 226
Rust, fungal, 327
Rye plant, 329, 329i, 348, 426
Ryegrass, 424

S
SA node, 561, 561i
Sac, 60
Sac fungus, 326, 326i, **328**–329, 329i
Saccharide, **38**
Saccharomyces cerevisiae, 116, 116i, 329
Saccharum officinarum, 38, 343i, 348
Saccule, 509, 509i
Sacral nerve, 493i
Sacred lotus, 348i, 415i
■ SAD, 530, 535
■ Safe sex, 137, 669
Sage grouse, 786–787, 786i
Sage plant, 273, 708
Sagittaria sagittifolia, 272i
Saguaro cactus, 449i, 703, 703i, 725i, 762i
Sahelanthropus tchadensis, 396
St. John's wort, 430i
St. Matthew Island, 691i
Salamander, 384i, 385, 517, 519, 598
Salicylic acid, 459
Salinity, and enzyme activity, 81
Salinization, **741**
Saliva, 575, **615**
Salivary amylase, 615, 616t
Salivary gland, 211i, 614i, 628t
Salmonella, 312
Salp, 287
Salt, **29**, **30t**
Salt marsh, 770i
Saltwater intrusion, 741i
Saltwater marsh, 730, 730i
Salvia, 273, 708
Sampling error, **16**, 16i, 154, 163, 254
San Pedro River, 727i
Sand dollar, 372
Sand flea, 316
Sandfly, 89i
Sandstone, 261
Sandy coastline, 770i, 771
Sandy soil, 767

Sanger, F., 220
Sap, **429**
Sap flow, 86
Saprobe, 302, **326**
Saprolegnia, 320
Sapwood, **429**, 429i
Sarcomere, **544**, 544i, 545i
Sarcoplasmic reticulum, 62
Sarcosoma, 326i
Sargasso Sea, 321
Sargassum, 321
■ SARS, 2, 206, 310, 310i, 313, 573
Sartorius, 543i
Saturated fat, 623
Saturated fatty acid, **40**, 40i, 56
Saurophaganax, 387i
Sauropsid, 386
Savanna, 760i–761i, 763, 763i
Sawfly caterpillar, 788, 788i
Scala tympani, 510, 511i
Scala vestibuli, 510, 511i
Scallop, 364, 364i
Scalp, 478
Scanning electron microscope, 55
Scapula, 539i
Scarab beetle, 371i
Scarcoscypha coccinea, 329i
Scarlet cup fungus, 329i
Scarlet hood, 326i
Scarus coelestinus, 8
Scarus gibbus, 8
Scavenger, 9, 732
    invertebrate, 360, 362, 363i
Scenedesmus, 55i
Schally, A., 525
Schiestl, F., 252
Schistosoma japonicum, 361i
■ Schistosomiasis, 361
■ Schizophrenia, 166
Schoener, A., 719
School of fish, 5, 5i, 686i
Schwann, T., 54
Schwann cell, 490, 490i
Schwarzenegger, A., 688
Sciatic nerve, 493i
■ SCID, 587
■ SCID-X1 patient, 230, 230i
Scientific approach, 11, 151
Scientific inquiry, steps in, 11
Scientific method, 711i
Scientific testing, **11**
Scientific theory, **11**, 15, 377
Sclera, 512, 513i
Sclereid, **421**
Sclerenchyma, 416, 419, **420–421**,
    420i, 420t
Scolex, 360, 361, 361i
Scorpion, 369, 369i
Scorpionfish, 713, 713i
■ Scrapie, 312
Screening pigment, 512i
Scrotum, 656, 656i
■ Scrub typhus, 369
Sea anemone, 358, 358i, 359, 492,
    492i, 538i, 707, 707i
Sea biscuit, 372
Sea cucumber, 354t, 372, 372i, 373
Sea fan, 772, 772i
Sea hare, 503, 503i, 596i
Sea lamprey, 719t
Sea lettuce, 323, 323i, 770i
Sea level, 730, 745, 758
Sea lily, 372
Sea lion, 752–753
Sea slug, 357
Sea squirt, 378, 379i, 401, 401i
Sea star, 138, 242, 269, 269i, 354t,
    364, 372, 372i, 373, 373i, 690, 693,
    693i, 718–719
Sea temperature, 772, 772i, 774,
    775, 775i
Sea turtle, 320, 388, 517

Sea urchin, 128, 354t, 372, 372i, 373,
    647, 651, 651i, 770i
Sea volume, 730
Sea wasp, 358i
Seabird, 93
Seafloor methane cycle, 36i
Seal, 646i, 752–753, 753i
■ Seasonal affective disorder,
    530, 535
Seasonal change, effects of, 146,
    447, 456, 462, 464
Seawater, pH of, 28, 28i
Seawater intrusion, 741i
Second law of thermodynamics, **74**
Second messenger, **522**
Secondary growth, plant, **417**, **418i**,
    428–429, 428i, 429i
Secondary immune deficiency, 587
Secondary immune response,
    581, 581i
Secondary oocyte, 662i, **663**, 665
Secondary sexual trait, 534, 534i,
    656, 659
Secondary succession, **716**, 717i
Secondary wall
    plant, 420i, 421
    plant cell, **68**, 68i
■ Secondhand smoke, 593, 607, 677
Secretin, 520, 533t, 617
Secretion, 471, 612
Secretory pathway, 60, 63i
Sedge, 251
Sedimentary cycle, 739, 748, 748i
Sedimentary rock, 239, 241, 258,
    258i, 260, 261, 261i, 262, 278
Sedimentation, 261, 342, 349
Seed, **337**, **343**, **452**, 457i
    dispersal, 453
    flowering plant, 418i, 419, 419i,
        428, 448
    germination, 451, 452, 456, 457i
    gymnosperm, 337, 343–345,
        344i, 345i
Seed coat, 42, 347i, 426, 451i, 452,
    456, 456i, 457i, 461
Seed fern, 342i, 343
Seed leaf, 418i, 419, 452, 452i
Seedling, 345i, 347i, 451, 451i, 457i
Segal, E., 15
Segmentation, 354, **355**, 362, 362i,
    368, 375i, **617**, 617i
    fused, 367, 370
Segregation theory, 154–155, 154i
Selaginella lepidophylla, 340
Selection pressure, 236–237, 240,
    252, 293, 356, 466
    oxygen as, 51, 287
Selective advantage, 139, 144,
    189, 429
Selective gene expression, 212, **653**
Selective permeability, **82**, 82i, 86
Selective transport, 86
Self-sacrificing behavior, 783,
    790–791
Self *vs.* nonself, 574, 578
Selfish behavior, **783**, 790–791
Selfish herd, **788**
Semen, 656
Semicircular canal, 509, 509i
Semiconservative replication, DNA,
    190, 190i, 191i, 193
Semilunar valve, 560, 560i, 561i
Seminal vesicle, 656, 656i, 657i, 658i
Seminiferous tubule, 656,
    658i, 659i
Senescence, 458t, **464**
Sensory adaptation, **506**
Sensory area, 498
Sensory cell, 354
Sensory neuron, 475, 478i, **485**, 511i,
    512, 512i
Sensory pathway, 506

Sensory perception, 504–517
    balance, 509
    chemoreceptors and special
        senses, 508
    hearing, 510–511
    retina, 514
    sensing touch, pressure,
        temperature, and pain, 507
    sensory pathways, 506
    vision, 512–513
■ visual disorders, 515
Sensory receptor, **408**
Sensory structure, invertebrate, 367
Sensory tunic, 513i
Sepal, 346i, 448, 448i, 463, 463i
Sephton, M., 261i
Septum, 560i, 561, 792
Sequencing, DNA, 220, 224–227,
    225i, 227i, 286, 307
Serine, 198i
Serotonin, 163, 482–483, 489, 501
Serratus anterior, 543i
Sertoli cell, 658, 659, 659i
Seta, 362, 363i
■ Severe Acute Respiratory
    Syndrome, 2, 206, 310, 310i,
    313, 573
■ Severe combined
    immunodeficiency, 587
Sex chromosome, 129, 129i, 166,
    168, 171
    gene recombination, 168, 179
    human, 169, 173t, 175, 179, 179i
Sex determination, human,
    169, 169i
Sex hormone, 35, 41, 169, 661
Sex-linked inheritance, 166
Sex pheromone, 252
Sex pilus, 302, 302i
Sexual behavior, 252, 252i
Sexual competition, 787, 787i
Sexual dimorphism, **252**
Sexual imprinting, 794, 794i
Sexual intercourse, 665
Sexual reproduction, 138–139, 138i,
    140, **646**–647
    animal, 354
    asexual *vs.*, 138–140, 138i,
        140i, 646
    costs/benefits, 346, 646–647, 646i
    eukaryotic, 126
    flowering plants, 448–449,
        448i, 449i
    fungal, 326, 328
    invertebrate, 357, 359, 367, 368
    plant, 338–340, 338i, 341i, 343,
        345i, 347i
    protist, 293, 318, 321, 324i
    reflections on, 645–646
Sexual selection, 235, 235i, **252**, 401,
    786, 786i, 787i
    mating behavior and, 786–787
■ Sexually transmitted disease (STD),
    305, 316, 588, **668**–669, 668i, 668t
Shale, 261
Shamrock, 460i
Shape
    in enzyme function, 79, 80, 81
    jawed fish, 382
    in membrane crossing, 84, 84i,
        85, 85i
Shark, 281, 281i, 382, 382i, 497i
    reproduction, 647i
Sheath, 424i
Sheep, 312
Shelf fungus, 327
Shell, 314, 317, 317i, 321, 321i, 364,
    364i, 365, 365i, 388, 389i
Shell model of atom, 20i, 22, 23i
■ Shellfish poisoning, 320
Shepherd's purse, 452, 452i

Shifting cultivation, **349**
"Shipping label," of amino acid, 201
Shivering response, **641**
Shoot, **418**, 422, 423, 454, 456i,
    458, 461
Shoot system, 336, 404i, 418i, 436i,
    454–455, 455i
Shoot tip, 417, **418i**, 419, 420,
    422–423, 422i, 452i, 455, 458
Short-chain carbohydrate, 38, 48t
Short-day plant, **462**
Short-term memory, 499
Shortgrass prairie, 762–763, 763i,
    773, 773i
Shoulder blade, 539i
Shoulla, R., 206–207
Showshoe hare, 706
Shrimp, 368
Shrubland, 760i–761i, 762–763, 763i
Siamang, 394
Siamese cat, 81i
Siamese twins, 654, 654i
Sibling species, **273**
■ Sickle-cell anemia, 44–45, 45i, 159,
    165, 173t, 181, 202, 252–253,
    253i, 556
■ SIDS, 605
Sierra Nevada, 759, 759i
Sieve plate, 421i, 442, 442i
Sieve tube, 421i, 423, 423i, 426i,
    437i, **442**, 442i, 443i
Sieve-tube member, 421, 430t, 443
Sight. *See* Vision
Sign stimuli, 781
Signal reception, 412–413, 522
Signal response, 412–413, 522
Signal transduction, 412–413,
    458, 522
Signaling mechanism, 407
Signaling molecule, 89i, 208, 325,
    352, 440, 458, 508, 520
Signaling pheromone, 784
Silent Spring (Carson), 736
Silkworm moth, 6i, 508, 784
Sillén-Tullberg, B., 788
Silt, 434, 445, 597
Silurian, 337i
Silver Springs, Florida, 737i,
    738, 738i
Silverfish, 370i
Simple epithelium, 470, 470i
Simple eye, 512
Simple fruit, **452**, **452t**, 453i
■ Simple goiter, 527, 527i
Simple leaf, **424**, 424i
Simple tissue, plant, 419, **420–421**,
    420i, 430t
Singh, C., 312, 312i
Sink, **442**, 443i
Sink region, 442
Sinoatrial node, 561, 561i
■ Sinus infection, 150
Siphon, 352i, 365, 365i
Sister chromatid, **126**–127, 126i
Skeletal joint, 541
Skeletal muscle, 64, 117, 117i, 498,
    542–543, 542i, 543i
Skeletal muscle tissue, **474**, 474i
Skeletal-muscular system,
    542–543, 543i
Skeletal system, 477i, 672, 672i
Skeleton, 60, 538–539
    bones of human skeletal
        system, 539i
    cytoskeleton, 60, 65i, **66**, 67, 70t,
        316, 317
    hydrostatic, 362
    reptile, 539i
Skin, **478**, 478i
■ aging and, 479
■ body piercing, 481, 481i
    human, 478–479

lab-grown, 475, 475i
mechanoreceptors in, 507i
structure of, 478i
■ sun exposure and, 479
as surface barrier, 574, 575
tattooing, 481, 481i
vitamins and, 479
■ Skin cancer, 134–135, 135i, 479, 585, 756
Skin cell, 413
Skin color, 478, 479
■ Skin tumor, 206
■ Skin ulcer, 475, 475i
Skull bones, 539i
Skunk, 712
■ Sleep apnea, 605
Sliding-filament model, **545**, 545i
Slime mold, 316i, **324**–325, 325i
Slime stalk, 305
Slime streamer, 81i
Slow-twitch muscle fiber, 117
Sludge worm, 597
Slug, 354t, 364
"Slug," of slime mold, 325, 325i
Small intestine, 72, 81, 575, 614i, 617i, 628t
absorption in, 618–619, 618i, 619i
digestion, 616–617
hormones from, 533t
organic metabolism, 620, 620i
Small organic compounds, families, 35
Smartweed, 709, 709i
Smell, 508
Smith, H., 222
Smog, 756–757, 756i
photosynthesis and, 441, 441i
■ Smoking, 206, 479, 500, 541, 541i, 567, 592–593, 592i, 606–607, 606i, 607i
pregnancy and, 677
reduction in risks by quitting, 607t
risks associated with, 607t
secondhand smoke, 593, 607, 677
Smoky Mountain National Park, 747i
Smooth ER, 62, 62i, 63i, 65i
Smooth muscle, 159
Smooth muscle tissue, **474**, 474i
Smut, fungal, 327
Snail, 246i, 354t, 364. *See also individual types*
reproduction, 647i
Snake, 387, 388, 647i, 780. *See also individual types*
■ Snake bite, 388
Snapdragon plant, 158, 158i
Snow alga, 286, 333, 333i
Snowshoe hare, 710–711, 711i
Snowy egret, 693i
Sociable weaver, 251, 251i
Social attachments, between humans, 792
Social behavior, 508, 520, **783**
Social groups
cooperative hunting, 788–789
cooperative predator avoidance, 788
costs and benefits, 788–789
dominance hierarchies, 789
selfish herd, 788
Social insect, 790, 790i
Social parasite, 714, 781, 781i
Sodium, 23i, 24, 24i, 29, 625t
urine formation and, 636–637, 636i, 643, 643i
Sodium chloride, 24, 24i, 27, 29
Sodium hydroxide, 29
Sodium ion, 24, 27, 82i, 83, 85, 409, 485
Sodium-potassium pump, **85**, **485**

Soft connective tissue, 472
Softwood tree, **429**, 431
Soil, **434**, 740i, **767**
acid/alkaline, 434–435
plant growth and, 432
■ pollution by toxic metals, 432
Soil acidity, 163
Soil bacteria, 436
Soil erosion, **435**
Soil formation, and lichens, 330
Soil fungi, 330
Soil profile, 435, 435i, 767, 767i
Soil water, 434, 439i, 440i
Solanum cheesmanii, 244, 244i, 418i
S. lycopersicon, 404i
S. lycopersicum, 244, 244i
Solar energy, 755, 755i
Solar-hydrogen energy, 755
Solar radiation, 754i, 755i
Solar system, 14, 258, 288
Solar tracking, 460, 460i, **466**, 466i
Soldier termite, 520
Solenopsis invicta, 704–705
S. richteri, 704
Solute, **27**, **30t**, 86–87
gaining and losing, 633, 633i
Solute concentration, **86**, 86i, 87, 433, 442, 443i
Solvent, **27**
Somatic cell, **126**, 128, 203, 366
Somatic embryo, 455i
Somatic nerve, **494**
Somatic pain, 507
Somatic sensation, 506, 507
Somatic subdivision, 493i
Somatomedin, 533t
Somatostatin, 528, 533t
Somatotropin, 521i, 532t, 535
Somite, **672**, 672i
■ Sonar system testing, 504–505
Song of the Dodo (Quammen), 723
Songbird, 727, 782, 782i
Sorus, 341, 341i
Sound
intensity of, 511i
lungs and, 599
properties of, 510, 510i
respiratory system and, 601, 601i
Source region, in translocation, 443i
South Carolina, 770i
Soybean plant, 231, 436i, 455, 460
Space-filling model, 36i
Spaceguard Survey, 285
■ Spanish flu epidemic, 3, 231
Spanish shawl nudibranch, 364i, 365
Spartina, 770i
Special senses, 506, 508
Speciation, **272**
allopatric, 274–275, 274i, 275i
models for, 274–279, 275i
parapatric, 277
patterns, 278–279
sympatric, 276–277
Species, **8**, 238, **272**, 282i
Species interaction, 704–705, 706i
categories of, 706
competitive interactions, 708–709
mutualism, 707, 707i
parasite-host interactions, 714–715
predatory-prey interactions, 710–711
Species introduction, 719, 719t, 724–725
Species loss, 764i
Spectacled caiman, 389i
Spectral karyotype, 170, 170i, 177i, 178i
Speech, 179

Speed, as evolutionary factor, 365
Sperm, 530–531, **650**, 656–657, 657i, 659i, 665, 665i
animal, **146**, 147, 147i
formation of, 658–659, 658i–659i
function, 531
human, 54, 123i, 145, 146, 168–169, 169i
inheritance and, 152
invertebrate, 357, 359i, 361, 364
low counts, 179
plant, 153–155, 338i, 339, 341i, 343, 345, 345i, 347i, 448, 448i, 450, 450i
structure, 67, 71i
Sperm count, 179, 519, 530–531
Sperm production, 656
Spermatid, 146, 658, 659i
Spermatocyte, 658, 659i
Spermatogenesis, 658
Spermatogonia, 658
■ Spermicidal foam, 666i, 667
■ Spermicidal jelly, 666i, 667
Sphaerodactylus ariasae, 388, 388i
Sphagnum, 339, 339i
Sphenodon, 389i
Sphenophyte, 341
Sphere of hydration, 27, 27i
Sphincter, **615**
Spicule, 373
Spider, 354t, 369, 369i, 512
reproduction, 674i
respiration, 596, 596i
Spider monkey, 394i
Spider silk protein, 229
■ Spina bifida, 672
Spinach, 462
Spinal canal, 496i
Spinal cavity, 476i
Spinal cord, 491, 493i, **495**, 495i, 502t
evolution, 493
parasympathetic outflow, 494i
sympathetic outflow, 494i
vertebrate, 493
■ Spinal cord injury, 468
Spinal nerve, 494
Spindle, microtubular, 124, 135, 143i, 145, 148i, 170
Spindle equator, 130i, 132i, 133i, 142i–143i
Spine, echinoderm, 372i, 373
Spinks, L., 482, 482i, 483i
Spiny anteater, 393, 393i
Spiny cactus, 238i
Spiny lobster, 517
Spiny spurge, 238i
Spirillum shape, 302i
Spirochete, 305, 305i
Spirogyra, 95i
Spirulina, 304
Spleen, 361, 555, 568i, **569**, 581
Sponge, 352i, 354, 354t, 356, 356i, **357**, 357i, 646
Spongy bone tissue, 540, 540i
Spongy mesophyll, 103i, **424**, 425i
Sporadic disease, **310**
Sporangia, 336
Spore, **304**, **326**, **336**
bacterial, 304i, 305
fungal, 315i, 326–329, 327i, 328i, 329i
meiosis and, 138i, 336, 338i, 341i, 347i
plant, **146**, 336, 337, 338i, 339, 341, 341i, 343, 348i
protist, 293, 315i, 321, 322i, 324, 324i, 325, 325i, 450
types, **304**, **343**
Spore-bearing structure, 305, 324, 325i, 326, 326i, 327, 328
Spore sac, 328i

Sporophyte, 146, 146i, 446, **448**, 448i, 451–453, 451i, 456, 456i, 461
plant, **334i**, **336**, 336i, 338–341, 338i, 339i, 341i, 343, 344, 345, 345i, 347, 347i, 422
protist, **321**, 322i
Sporozoite, 318, 319i
■ Sprain, 541
Spriggina, 356i
Spring overturn, **768**
Spruce tree, 344, 683i
Squamous cell, 470, 470i
■ Squamous cell carcinoma, 135i
Squid, 214, 354t, 365, 365i, 487i, 512, 512i
SRY gene, 169
Stabilizing selection, **250**, 251
Stahl, F., 193
Stalk, 424, 424i
Stamen, 152i, 346i, **448**, 448i, 449, 463, 463i
Staphylococcus aureus, 312
Star, creation of, 288
Star anise, 348i
Starch, 35, 38–39, 38i, 48t, 64, 77, 101, 442
Starch grain, 461
Starling, 719t, 783, 783i
Starling, E., 520
Static equilibrium, 509, 509i
Statolith, **461**, 461i
STD. *See* Sexually transmitted disease
Stearic acid, 40i
Stegosaurus, 387i
Steinbeck, John, 763
Stem. *See* Shoot
■ Stem cell, **230**, 468, 473, **554**, 555i, 574
Sterility, 790–791, 790i
Sterility/self-sacrifice, genes of nonreproducers, 790–791
indirect selection for altruism, 791
social insects, 790
social mole rats, 790
Stern, Kathleen, 792
Sternum, 539i
Steroid, cholesterol in, 41
Steroid hormone, 463, **522**, 522t, 523i
■ Steroid use, 541, 541i
Sterol, **41**, 41i, 48t
Sterol backbone, 41, 41i
Steward, F., 454
STH, 524, 525
Sticky end, of DNA fragment, 222, 223i
Stigma, 448, 448i, 450, 450i, 454
Stimulant, 500
Stimulus, 506, **780**
energy as, 7
neurons and, 485
strength of, 506i
Stingray, 382i
Stink bug, 371i, 715
Stinkhorn, 327
Stipe, 321, 321i
Stirrup, 510, 510i
Stolon, 328i
Stoma, 96i, **102**, 103i, 336, 350t, 404, 421, 424, 425i, 430t, 433, 438–441, 439i, 441i
Stomach, 81, 575, 613i, 614i, **616**, 616i, 628t
absorption of ethanol, 72
acid in, 28
digestion, 616–671
hormones from, 533t
invertebrate, 364i, 373, 375i
■ Stomach acid, 471
Stomach lining, 69

Stone cell, 420i
Stonefly, 597
Stonewort, 323, 323i
Straight-chain structure, 38i
■ Strain, 541
Stramenopile, 315, 316i, 320
Strangleweed, 714i
Stratification, **261**
Stratified epithelium, 470
Strawberry plant, 453, 453i, 454, 454t
Stream ecosystem, **769**, 769i
■ Strength training, 547
Streptoccus pneumoniae, 186, 186i
S. pyrogenes, 571
■ Streptomycin, 248
■ Stress, 163, 526–527, 617
Stress hormone, 250
Stretch receptor, 507
Stretch reflex, 491, 491i
■ Strict vegan diet, 571
■ Stringer, K., 402–403, 409i, 414
Strip logging, **727**, 727i
Strobilus, **340**–345, 340i, 344i, 345i
■ Stroke, 49, 509, 567, 593, 623
Stroma, 64, 96, 97i, 98, 99i, 101
Stromatolite, **292**, 293i
Strong acid, 28
Strong base, 28, 29
Structural formula, 24–25
Structural genomics, 232
Structural organization
  evolutionary history of, 404–405
  levels of, 404–405
Sturnus vulgaris, 783
Style, 448i, 454
Suberin, 429, 464
Subjective *vs.* objective
  viewpoint, 14
Substrate, 77, **78**–81, 79i, 80i, 89t, 91
Substrate-level
  phosphorylation, **110**
Successful organism, 370
Succession model, 716
Succulent, 102
Sucker, annelid, 362
■ Sudden cardiac arrest, 550–551
■ Sudden infant death syndrome, 605
■ Sudden oak death, 315, 320, 320i
Sugar
  structure, 34, 75, 75i, 92, 93, 101, 188, 289
  uptake in plants, 421, 425, 429
Sugar beet, 38
Sugarcane, 38, 343i, 348
Suicidal cell, 412–413, 412i, 413i
Sula granti, 794, 794i
Sulfate-eating bacteria, 36i
Sulfolobus, 307
Sulfur, 187i, 434t, 625t
  in ATP-forming reaction, 307
Sulfur cycle, in seas, 305
Sulfur dioxide, 29, 757
Sulfuric acid, 29, 364
Sun, 754, 754i
  age, 288
■ exposure to, 203
  origin of life and, 288, 289, 290
  plant competition for, 417, 424, 429
  plant development and, 454, 457i
  skin color and, 479
■ Sunburn, 309, 479
Sunfish, 788
Sunflower plant, 102, 452t, 464, 466i
Sunlight energy, 6, 7, 74, 74i, 77i, 92–95, 92i, 97i, 98, 104i, 302, 307
Supercoiling, DNA, 127i
Supercontinent, 263i, 264, 265, 293
Superior vena cava, 558i, 560i
Supernatural *vs.* science, 12, 14, 32–33

Superoxide dismutase, 90, 90i
Superplume, 264i
■ Supertaster, 517
Surface barrier, 574, 575t
Surface runoff, 740, 741i
Surface-to-volume ratio, 52–53, 53i, 326, **406**, 424, 427, 594
■ Surfing, 752, 752i
Surtsey, 722i, 723
Surveillance protein, 134
Survival pattern, 692–693
Survivorship curve, **693**, 693i
■ Sushi, 322, 373
Suspended particulate matter, 757i
Sustainable development, 726–727
Suttle, C., 313
Swallowtail butterfly, 371i
Swamp forest, 251, 263i, 334, 337, 340–342, 340i, 434–435, 445
Sweat, 27, 150, 637i, 640, 792
Sweat gland, 213, 213i, 402i, 478, 478i
Swim bladder, 382
Swine flue virus, 231
Symbiont, **294**, 295, 319, 772
  animal, 306, 357, 359
  fungal, 304, 314, 323, 326, 330–331, 336, 436
  plant, 305, 328, 432, 433
Symbiosis, **330**, 706
Sympathetic nerve, 494–**495**, 494i
Sympatric speciation, **276**–277
Synapse, **168**
Synapsid, 386, 392
Synaptic integration, **489**, 489i
■ Syndrome, **172**
Synovial cavity, 543i
■ Synthetic fat replacement, 12
■ Syphilis, 668i, 668t, 669
Syringa, 464i
Systemic circuit, **553**, 558, 559i
Systemin, 459
Systolic pressure, 563

T
T cell, 555, 580–581, 581i, 587, 680
  antibody-mediated response, 582, 582i
  cell-mediated response, 584i
T-cell receptor, **580**
T lymphocyte, 229, 230, 555i, 569, **574**, 574i, 578
Table salt, 24, 24i, 27, 27i, 29
Table sugar, 38, 343i
■ Tachycardia, 567, 567i
Tactile display, 785, 785i, 790, 790i
Tadpole, 385, 385i, 518, 530, 530i, 649i
Taenia saginata, 361i
Taigas, 764–765, 765i
Tail, chordates, 378, 378i
Tall fescue, 330
Tallgrass prairie ecosystem, 724, 733, 733i, 734i, 762–763
Tamias, 708, 708i
Tandem repeat, **226**
Tangential section, of plant, **420i**
Tannin, 429
■ Tanning, 479
Tanzania, 396, 397
Tapeworm, 354t, 360
Taproot, 344
Taproot system, **427**, 427i
Taq polymerase, 225i
Tarantula, 369i
Target cell, 520
Tarsal, 539i
Tarsier, 394i
Tarsioid, 394
Tasmanipatus anophthalmus, 277
T. barreti, 277i
Taste, 508

Taste bud, 508, 508i
Taste receptor, **508**, 508i, 517
■ Tattooing, 481, 481i
Tawuia, 292i
Taxa
  derived traits and, 281
  higher, 280, 280i
Taxodium distichum, 445
Taxol, 66, 135
Taxonomy, numerical, **303**
Taxus brevifolia, 66
T. brevifolius, 137
TCR, **580**
Tea plant, 343i
Tears, 575
Tectorial membrane, 511, 511i
Tectum, 496i, 502t
Teeth, 613i, 615, 615i
  bacteria and, 59
  in mammals, 392, 392i
  primates, 395
Telesto, 374i
Telomerase, 680
Telomere, **185**, 680
Telophase, 124, 128i, 129, **131**, 131i, 132, 141i, 142i–143i, 148i
Temnodontosaurus, 387i
Temperate coniferous forest, 760i–761i
Temperate deciduous forest, 760i–761i, 765i
Temperate forest, 349
Temperate grassland, 760i–761i
Temperature, 507. *See also* Core temperature
  enzyme activity and, **26**, **81**, 81i, 83
  plants and, 441, 456
  responses to cold stress, 640–641, 641i
  responses to heat stress, 640
Temperature regulation, mammals, 640–641
Temperature zones, 754, **755**, 755i
Template (DNA), 223, 223i, 225, 227
Temple of Apollo, 32i
Temporal isolation, 272i, **273**, 273i
Temporal lobe, 498, 498i
Tendon, **542**, 543i
Tendril, **461**
Tension, **438**, 439i
Tentacle, 358i, 359, 364, 365i
Teratogen sensitivity, 676i
Termination, RNA translation, 200, 201i
Termite, 305, 316, 370, 785, 790, 791, 791i
Test, **11**
■ Test-tube baby, 181
Testcross, **155**
Testes, 521i, 533t, **656**, 656i, 657i, 658i, 659i, 794
  human, 140i, 169, 169i, 179
  invertebrate, 360i
  sperm formation, 658
■ Testicular cancer, 657
Testosterone, 35, 35i, 169, 179, 519, 521i, 522, 533t, 536, **658**–**659**
Tetanic contraction, 546, 546i
■ Tetanus, 304, 310t, **546**, 548, 548i, 586
Tetany, 29
Tetrapod, 382–**383**, 383i
  amphibians, 384–385
Texas horned lizard, 704
Textularia, 315i
TH, 525
Thalamus, 484, **496**, 496i, 497i, 499, 499i, 502t, 508
■ Thalassemias, 202i, 556
■ Thalidomide, 677

THC, 631
Theobroma cacao, 446, 446i
■ Therapeutic vaccine, 591
Therapsid, 392–393
Thermal energy, 82
Thermal inversion, **756**, 756i
Thermal radiation, **639**
Thermocline, 768–769, 768i
Thermodynamics laws, **74**, 74i
Thermogenesis, 122
Thermophile, **286**, 297i, 313
  extreme, **306**, 307, 307i
Thermoreceptor, **506**, 507, 516, 516i
Thermotoga maritima, 313
Thermus aquaticus, 225, 225i, 286, 286i, 307
Thiamin, 624t
Thighbone, 539i, 540, 540i
Thigmotropism, **461**, 461i
Thiomargarita namibiensis, 302, 305, 305i
Third eye, 388
Thirst center, 637
Thlaspi caerulescens, 433
Thomson's gazelle, 713
Thoracic cavity, 476i, 602, 602i
Thoracic cavity/segment, 368
Thoracic duct, 568i
Thoracic nerve, 493i
Thorax, 369, 370
Thorium, 262
Threat display, 784–785
Threonine, 45i, 198i, 202i, 623
Throat, 600i
Thrombus, 567
THS, 525
Thumb, 395
Thylakoid, 64, 64i, **96**
Thymine (T), 46, 46i, 47, 184i, 188, 188i, 189, 189i, 196, 196i, 202, 203, 227, 227i, 291
Thymine dimer, 203
Thymopoietin, 533t
Thymosin, 521i
Thymus gland, 521i, 533t, 568i, **569**, 580
■ Thyroid abnormalities, 527, 777i
Thyroid gland, 19, 269, 521i, **527**, 527i, 533t, 541, 587
■ environmental effects on function, 530, 530i
  skewed feedback from, 527
■ Thyrotoxicosis, 587
Thyrotropin, 532t
Thyroxine, 521i, 527, 533t
Ti plasmid, 228i, 229
Tibia, 539i
Tibialis anterior, 543i
Tick, 39i, 305, 305i, 369
Tidal flat, 50
  early life and, 289–291, 290i
Tight junction, 69, 69i, 471, 471i
Tigris River, 741
Tilia americana, 102i, 103i
Timothy grass cell, 60i
Tinman gene, 571
Tissue, **4i**, **404**, 468–469
  connective tissues, 472–473
  epithelial tissue, 470–471
  muscle tissues, 474
  nervous tissue, 475
■ Tissue culture propagation, **454**–455, 454t
■ Tissue engineering, 475
Tissue layer, in embryos, 354–355, 358, 360
Tissue specialization, 648, 648i
■ Tissue transplant, 585
Titanic, 641i
Toad, 384i, 385
Tobacco. *See* Smoking
Tobacco mosaic virus, 308i

Tolerance, 500
Tomato plant, 228, 244, 244i, 418i, 452t
morphology of, 404i
Tongue, 508i, 601i, 615
Tonicity, **86–87**, 87i
Tonsils, 568i
Tools, 397–399, 397i
Topography, 759
Topsoil, 349, **435**, 767
Torres del Paine National Park, 768i
Total fertility rate (TFR), **698**
Touch, 507
Toxic compound, 90
■ Toxic goiter, 527
Toxic species, eukaryote, 319, 320
Toxic waste, 432–433, 432i
Toxicity
alcohol and, 72
predation and, 364
Toxin, 194–195, 195i, 358, 424, **466**
■ Toxin attack, on body, 72
Toxin diversity/specificity, 352, 466
TPA protein, 229
Trace element, 18
Trace fossil, 260
Tracer, **21**, **30t**
Trachea, 367, 600i, **601**
Tracheal system, 596–597, 597i
Tracheid, 194–195, 195i, 358, 424, **466**
■ Trachoma, 305
Tract, 493
Trade war, 231
Trait, **10**, 171, 244, 250, 272
heritable, 151, 152, 174–175, 174i, 175i, 251
types, 166–167, 169, 171, 246, 281, 353
variation in, 45, 139, 140, 140i, 144–145, 147, 151, 160, 202, 242, 353
Trans Fatty acid, 49, 49i
Transcription
DNA, 208–212, 209i, 210i,
gene, 194i, 196–197, 196i–197i, 204i, 309
Transcription control site, 212i
Transfer RNA, **196**, 199, 199i, 200, 200i
Transgenic animal, 229, 229i, 231, 234, 285, 285i
Transgenic organism, **228**
■ Transgenic pig, 591, 591i
Transgenic plant, 228, 228i
Transition state, **79**
Transitional stage, 700, 700i
Translation of RNA, 195i, 204i, 208, 209i, 210i, 211
Translocation, **442–443**, 442i, 443i
chromosomal, **177**, 177i
Transmission electron microscope (TEM), 55, 55i
Transpiration, **438–439**, 439i
Transport protein, 59, **76**, 82i, 83, 84, 84i, **89t**, 150, 163, 211, 437, 461, 485
Transport tube, 421
Transposon, **202**, 203i, 206, 234, 269
Transverse fission, **360**
Transverse plane, 476i
Transverse section of plant, **420i**
Trapezius, 543i
■ Trastuzumab, 585
Tree
defenses, 410, 410i
fungi and, 331
Tree fern forest, 340i
Tree ring, 429, 429i, 431, 431i
Tree shrew, 395, 395i
Treponema pallidum, 669
Triassic period, 263i, 337i, 392

Triceps, 542i
Triceps brachii, 543i
Trichinella spiralis, 366
T. spiralis, 366i
Trichomonad, 316
Trichomonas, 316i
T. vaginalis, 316, 668
■ Trichomoniasis, 668, 668t
■ Trichonosis, 366
Trichoplax adhaerens, 356, 356i
Triclopyr, 460
Trigger zone, 484i, 485
Triglyceride, 118, 617, 622–623
Triiodothyronine, 521i, 527, 533t
Trilobite, 365i, 367
Trinidad, 694–695
Triple covalent bond, **25**, 25i
Triploid nucleus, 451
Trisomy 21, 178–179, 179i
Triticum, 343i
T. aestivum, 277i
T. monococcum, 277i
T. tauschii, 277i
T. turgidum, 277i
tRNA, **196**, 199, 199i, 200, 200i
Trophic levels, **733**
Trophoblast, 670, 670i
Tropical broadleaf forest, 760i–761i
Tropical community, 722
Tropical coniferous forest, 760i–761i
Tropical dry forest, 760i–761i
Tropical forest, 349, 349i, 761i
Tropical rain forest, 446, 727, 764, 764i
Tropical rain forest soil, 767i
Tropics, 724
Troponin-1, 211
Troposphere, 754i
Trout, 382
Troyer, V., 174i
True bug, 370i
True tissue, 356i
Truffle, 328–330, 329i
Trumpet chanterelle, 326i
Trypanosoma, 316i
T. brucei, 316, 317i
T. cruzi, 316
Trypanosome, 316
Trypsin, 81, 616t, 617
Tryptophan, 42i, 80–81, 80i, 198i, 211, 623
Tsetse fly, 317
TSH, 521i, 524, 527
Tuataras, 387, 388, 389i
■ Tubal ligation, 666i, 667
■ Tubal pregnancy, 668i
Tube foot, 372–373, 372i
Tube worm, 772, 773i
Tuber, 454i
Tuber melanosporum, 329i
■ Tuberculosis, 248, 310, 310i, 310t
Tubifex tubifex, 597
Tubular reabsorption, **636**, 636i
Tubular secretion, **636**, 636i
Tubulin subunit, 129
■ Tularemia, 369
Tulip plant, 454t
Tumor cell, 134, 585i
Tumor formation, 125
Tumor necrosis factor, 574
Tumor supressor, 134, 207
Tundra, 307, 346, 760, 760i–761i, 766, 766i
Tungara frog, 785
Tunicate, 354t, 378, 379i
Turbellarian, 354t, 360
Turgor pressure, **87**, **440**, 440i, 442, 443, 443i, 458
■ Turner syndrome, 173t, 179, 179i
Turtle, 270, 387, 388, 389i
Twain, M., 385
Twig, 428, 428i, 454

Twins, 651, 651i
2,4-D, 460
Type I curve, 693
Type II curve, 693, 693i
Type III curve, 693, 693i
Typhus, 236
Tyrosinase, 81i
Tyrosine, 198i, 623

U

Udotea cyathiformis, 323, 323i
Ulnar nerve, 493i, 539i
Ultrafiltration, 564i, **565**
Ultraplankton, 772
Ultraviolet light, 449, 449i, 479, 754i
Ultraviolet radiation, 94i, 95i, 105, 293, 336
Ulva, 323, 323i, 770i
Umami, 508
Umbilical cord, 673, 673i, 674i, 678, 678i
Umbrella thorn tree, 414
Unbranched evolution, 278
Undulating membrane, 317i
Uniformity theory, 241, **264**
United Nations, 573
U.S.
immigration, 700, 701i
population, 698i, 699i
population and workforce, 701
resource consumption, 700–701
United States Navy, 504–505
Unity of life, 2i, 6, 8, 9, 14i, 187, 189, 283t
University of Tokyo, 139
University of Utah, 353
■ Unsafe sex, 73, 137
Unsaturated fatty acid, **40**, 40i, 56
Upper limb bones, 539i
Upright walking. *See* Bipedalism
Upwelling, **772–773**, 773i
Uracil (U), 4, 196, 196i, 199, 291
Uranium, 21, 262
Urea, 620, 620i, 623, **633**, 643
■ Uremic toxicity, 638
Ureter, 634, 634i, 635i
Urethra, 634, 634i, 656, 657i, 658i, 660
Uric acid, 370, 643
Urinary bladder, 634, 634i
Urinary excretion, **633**
Urinary system, 477i, **632**
circulatory system and, 558–559, 559i
components, 634
digestive system and, 612i
human, 634–635, 634i
nephrons, 634–635
organ systems and, 632i
organs and functions, 634i
Urine, 174, 575, 630–631, 630i, **632**
formation of, 636–637, 636i, 637i
■ Urine testing, 630–631, 630i
Urochordate, 378
Ursa maritimus, 26i, 244–245, 244i, 271i, 641, 641i
Useless body parts, 239, 239i
Usnea, 330i
Uterus, 409, **660**, 660t, 661i, 679
human, 137, 169i
invertebrate, 366i
Utricle, 509, 509i
UV light, 449, 449i, 479, 754i

V

■ Vaccine, 125, 186, 318, **586**, 591
for AIDS, 589
recommended vaccines, 586t
Vacuole, 305i, 316
Vagina, 59, 169i, 248, 302, 304, 575, 660, 660t, 661i
Vaginal fluid, 657

■ Vaginal infection, 575
■ Vaginitus, 304
Valine (val), 44, 45i, 198i, 202i, 623
■ Valium, 489
Valve, 364
■ Vampire, 480–481, 480i
van Helmont, J. B., 105
van Leeuwenhoek, A., 54
Vanilla planifolia, 280i
Varanus komodoensis, 388
Variable, **12**
Variation
conditions for, 246–247, 247i, 252–255
continuous, **160–161**, 160i, 161i
in traits, 45, 139, 140, 140i, 144–145, 147, 151, 157, 158, 160–161, 160i, 161i, 192, 202, 242
■ Varicose vein, 565
Vas deferens, 656, 656i, 657i, 658i
Vascular bundle, 419i, **422–423**, 423i, 425, 425i, 428, 428i
Vascular cambium, 418i, **428**, 428i, 429, 429i, 439i, 458
Vascular cylinder, 426i, **427**, 427i, **437**, 437i, 439i
Vascular plant, **337**
diversity in, 337, 338, 341, 344i, 348, 348i
evolution, 263i, 282, 337i, 343
growth/development, 338i, 341i, 345i, 347i
seed-bearing, 337, 337i, 349, 350t, 416–429
seedless, 337, 337i, 340–341, 350t
water transport in, 432, 438–439, 438i, 439i
Vascular tissue, 27, 42, 336, 406, 418i, 419–422, 422i, 438
Vascular tunic, 513i
■ Vasectomy, 666i, 667
Vasoconstriction, **563**, 566
Vasodilation, **563**, 576, 576i
Vasopressin. *See* ADH
vCJD, 312, 312i
Vector
in genetic research, 232, 253i
insect, 27, 42, 336, 418i, 419–422, 422i, 438
■ Vegan diet, 571
■ Vegetable oil, 40, 49
Vegetative growth, 454–455, 454t
■ Vehicle exhaust, 756–757
Vein, **562**, 562i, 565, 565i
human, 407i
leaf, 96i, 103i, 419i, 425, 425i, 439i, 443
Velvet walking worm, 269i, 277i
Venation, flowering plants, 425i
Venom, 369
Venous pressure, 565, 565i
Venous valve, 565, 565i
Venter, C., 221, 234, 298
Ventilation, 594–595
Ventral surface, **354**, 354i, 476i
Ventricle, 496i, 560, 560i
■ Ventricular fibrillation, 567, 567i
Venule, **562**, 565
Vernalization, **464**
Vertebrae, **381**, 495i, 538, 539i
Vertebral column, 495, 495i, 538, 539i
Vertebrate, **354**, 376–399
amniotes, 386–387
amphibians, 384–385
birds, 390–391
characteristics, 40
chordates and, 378–379, 381
current species diversity, 725i
early primates to hominids, 394–395
embryo development, 35

emergence of body plan, 672, 672i
emergence of early humans, 396–397
emergence of modern humans, 398–399
endocrine system, 520–521, 521i
evolution, 263i, 266–267, 266i, 353, 354t, 380–381
evolution of circulation, 552–553
evolution of hearing, 510–511
existing reptilian groups, 388–389
jawed fishes and tetrapods, 382–383
key innovations, 381
major groups, 381, 381i
mammals, 392–393
respiration, 598–599, 598i, 599i
skeleton, 69i
vision, 512, 512i
■ Vertigo, 509
Vesicle, 60, 62, 62i, 63, 65i, 83i, 87, 88, 88i, 130, 132i, 133, 318
Vessel, 423i, 429, 429i
Vessel member, 421, 421i, 426i, 430t, **438**, 438i
Vestibular apparatus, **509**, 509i, 510i, 517, 517i
Vestibular nerve, 509
Vestigial structure, 285
Vibrio cholerae, 775, 775i
Vicuna, 274, 274i
Vietnam Veterans, 460
Villi, **618**, 618i
Vine, **461**
Viola, 348i
■ Viral disease, treatment, 206
Viral DNA, 184i, 186, 222, 309i, 313
Viral enzyme, 223, 309
■ Viral infection, 308–309
Viral RNA, 309
Virchow, R., 54
Virginia Tech Swine Research, 231i
Viroid, **309**
■ Virus, 125. *See also specific types*
characteristics, 34, 301, 301i, **308**, 308i
defenses against, 88
in gene therapy, 232
multiplication cycle, 303, 303i, 308–309, 309i
■ Virus particle, synthetic, 187i
Visceral pain, 507
Visible light, 95i
Vision, **512–513**
daytime vision, 394, 395
human eye, 512–513, 513i
requirements for, 512
retina, 514
■ Vision impairment, 515
Visual accommodation, **513**
Visual acuity, 515
Visual cortex activity, 498i
■ Visual disorder, 515
Visual field, **512**
Visual pigment, 105
Vitamin A, 105, 624t, 625
■ Vitamin A deficiency, 218
Vitamin C, 38, 624t, 625
Vitamin D, 479, 541, 624t, 625
cholesterol in, 41, 48t
Vitamin D₃, 528
■ Vitamin D deficiency, 528, 528i
Vitamin E, 80, 624t, 625
Vitamin K, 305, 624t, 625

Vitamins, **624–625**, 624t
in coenzyme, 80, 88
fat-soluble, 624t, 625
skin and, 479
Vitex lucens, 466
Vitreous body, 512, 512i, 513i
Vocal cord, 601, 601i
Voice box, 600i, 601
Volcanic ash, 260, 261
Volcanic island, 722i, 723, 723i
Volcano, 717i
Vole, 780–781, 780i
Voluntary motor activity, 498
Voluntary muscle, 474
Volvex, 323, 323i
Volvox, 323, 323i
Vomeronasal organ (VNO), 508, 792

W
Wade, S., 503, 503i
Walking. *See* Bipedalism
Wall cress, 463, 463i
Wall pore, 449i
Wallace, A., 238, 243i, 246
Waller, A., 550, 551i
Walnut tree, 428i
Wandering albatross, 723i
■ War on drugs, 333
Warfarin, 236–237
Warm grassland, 760i–761i
Warmth, 507
Warning coloration, **712**
Wasp, 252, 272i, 358i, 712, 713i
■ Waste disposal, 333
Waste removal, 359
invertebrate, 370, 375, 375i
plant, 429
Wastewater, 81i
Water. *See also* Water molecule
cell metabolism and, 63, 64
early Earth and, 288–290
gaining and losing, 633, 633i
hydrologic cycle, 740–741
oxygen in, 597
uptake in plants, 69, 436, 438–441, 438i, 439i, 440i, 441i
Water-conducting tube, 421, 421i, 433, 438–439, 438i, 439i
Water conservation, plants, 410–411, 421, 433, 440–441, 440i, 441i
Water crisis, 740–741
Water flea, 708
Water hyacinth, 719t
Water lily, 348
Water loss, plants, 102
Water mold, 315, 320
Water molecule
diffusion in cells, 52, 68, 82, 82i, 86–87, 86i, 87i
diffusion in plants, 421, 425, 425i, 428–429, 433, 437, 437i, 439i, 440, 443i
in photosynthesis, 75, 92, 96, 97i, 98, 100i
polarity, 25, 25i, 26, 26i, 56, 82, 82i
properties, 25–27, 27i, 56, 434, 437
structure, 23, 25, 25i, 26i
■ Water pollution
estuaries, 432
groundwater, 432–433, 432i, 460
Water rights, 741

Water-soluble vitamin, 624t
Water-solute balance, 406–407, 632, 632i
Water-vascular system, 372, 373i
Waterlily, 406, 407i
Watermelon, 452t
Watershed, 349, **740**
Watson, J., 184i, 187, 189, 189i, 190, 220, 221
Wavelength, **54**, 55, **94**, 94i
Wax, **41**, 41i, 48t, 69, 367, 421, 424i, 437i, 440, 453
Weak acid, 28, 29
Weaver bird, 690i
Weed, **460**
■ Weight, 610–611, 626–627. *See also* Obesity
■ gastric bypass surgery, 610–611
guidelines for, 626
Welwitschia mirabilis, 344, 344i
Went, F., 460
■ Werner's syndrome, 680
■ West Nile encephalitis, 300–301
■ West Nile virus, 375, 731
Western juniper, 280i
Western yew, 66
Wet acid deposition, 757
Wetland, 724, 730–731, 770, 777
Wexler, N., 172i
Whale, 239, 257, 504–505, 504i, 517, 517i, 724, 725
Wheat plant, 228, 229, 277i, 329, 343i, 348, 417, 454t, 455, 763
White blood cell, 88, 91i, 177i, 230, 361, 406, 473, 554i, 555, 555i, 574, 574i, 590t
■ White blood cell disorder, 556
White matter, 493, 495i
White rust, 320
■ Whooping cough, 310
Whorl, 448, 463, 463i
Wild cat, 724
Wild rock dove, 10i
Wild-type allele, 168, 183, 254
Wildlife trading, 725, 725i
Wilkins, M., 188, 189, 190
Willow, 459
Wilmut, I., 184i
Wilson, E. O., 724
Wilting, plant, **440i**
Wind, and acid rain, 757
Wind energy, 755, 755i
Wind farm, 755
Wind fence, 756
Wind power, 755
Windpipe. *See* Trachea
Wing, 279, 538i
bat, 267, 267i
bird, 267, 267i
butterfly, 246, 247i, 249i, 250, 267
insect, 267, 267i, 370
Witchhunt, and ergotism, 329
■ Withdrawal, 666, 666i
Wobble effect, in RNA translation, 199
Woese, C., 306, 307
Wolf, 710i, 784, 784i, 789, 789i
Wolf spider, 686, 686i
Wood, formation, 417
Wood duck, 35i
Woodland, 760i–761i, 762–763
Woodwardia, 341, 341i
Woodwell, G., 736i

Woody plant, 68, 421, **428–429**, 428i, 429i
Woolly mammoth, 261
Workforce, and population, 701
World food production, 417
World Health Organization, 218, 607
World Heritage site, 314i, 315
World population growth, 698–699
World Trade Center, 226
Wrist, 539i
Wuchereria bancrofti, 366
Wurster, C., 736i

X
X chromosome, 141i, 144i, 168–169, 169i, 175, 175i, 179, 277
inactivation, **212**–213, 213i
■ X-linked anhidrotic dysplasia, 173t
■ X-linked disorder, 547
■ X-linked inheritance, 173t, 175, 175i, 183
x-Ray, 21, 55, 94i, 137, 188, 203
x-Ray diffraction, **188**–190, 188i, 189i
Xanthophyll, **95**, 95i
Xenopus laevis, 518–519, 530, 530i
Xenotransplantation, **231**
XIST gene, 21
■ XTC (Ecstasy), 482–483, 500
■ XXX syndrome, 173t
Xylem, 406
components, 336, 341, 416, 419–**421**, 420i, 421i, 426i, 427i, 430t
function in water uptake, 437, 437i, 438, 439i, 458t, 460
primary, 420–427, 420i–427i, 428i
secondary, 428–429, 429i
■ XYY condition, 173t, 179

Y
Y-box protein, 211
Y chromosome, 141i, 144i, 168–169, 169i, 179
Yarrow plant, 162–163, 162i
Yeast, 116, 116i, 232, 232i, 270, 294, 329
Yellow bush lupine, 410–411, 411i
Yellow marrow, **540**, 540i
Yellowjacket, 713i
Yellowstone National Park, 286, 286i, 307i, 725
Yolk, **647**
Yolk sac, **671**, 671i
Yucca moth, 707, 707i
Yucca plant, 707, 707i

Z
Z band, 544, 544i
Zalmout, I., 257
Zea mays. See Corn
Zebra, 784
Zebra mussel, 719t
Zebrafish, 571, 682, 682i
Zero population growth, **688**, 701
Zijlaard, L., 629i
Zimbabwe, 572–573, 572i
Zinc, 625t
Zona pellucida, 662i, 663, 665, 665i
Zooplankton, 737, 768
Zygomycete, **328**
Zygote, 476, 679t

Warning labels on fast foods?
Higher vehicle taxes on SUVs?
Pesticide spraying to kill mosquitoes?

HOW WOULD *YOU* VOTE?

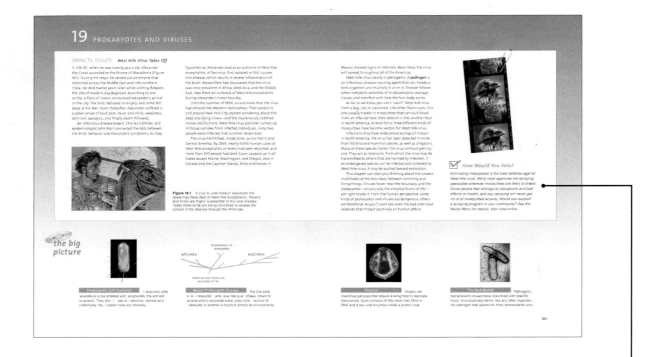

Wake-up calls for today's hot-button issues

There is nothing abstract about the *Impacts, Issues* essays that open each chapter in the book. Directly or indirectly, the questions they raise about biology-related issues are likely to affect you, the people you know, and billions of people you will never meet. Is there a "right" or "wrong" answer to them? That is up to you to decide; we do not presume to tell you what to think.

Each chapter's *How Would You Vote?* question raises a relevant, sometimes controversial issue that you may have read or heard about in the news. As explained on the facing page, it invites you to weigh information on both sides of the issue, broaden your thoughts about it, and cast *your* vote along with many thousands of other students all across the country. As you sift through the literature, you probably will come to understand how knowledge of basic biology can help you formulate your answers.

 How Would You Vote?

Eliminating mosquitoes is the best defense against West Nile virus. Many local agencies are spraying pesticides wherever mosquitoes are likely to breed. Some people fear ecological disruptions and bad effects on health, and say spraying will never get rid of all mosquitoes anyway. Would you support a spraying program in your community? See the Media Menu for details, then vote online.